THE SOLID WASTE HANDBOOK

THE SOLID WASTE HANDBOOK

A PRACTICAL GUIDE

Edited by

WILLIAM D. ROBINSON, P.E.
Consulting Engineer
Trumbull, Connecticut

A Wiley-Interscience Publication

JOHN WILEY & SONS

New York • Chichester • Brisbane • Toronto • Singapore

628.44
S 686

Copyright © 1986 by John Wiley & Sons, Inc.

All rights reserved. Published simultaneously in Canada.

Library of Congress Cataloging in Publication Data:

Main entry under title:

The Solid waste handbook.

"A Wiley-Interscience publication."
Includes index.
1. Factory and trade waste—Handbooks, manuals,
etc. 2. Refuse and refuse disposal—Handbooks,
manuals, etc. I. Robinson, William D.

TD897.S565 1986   628.4'4   85-12454
ISBN 0-471-87711-5

Printed in the United States of America

10 9 8 7 6 5 4 3 2 1

CONTRIBUTORS

Bernays Thomas Barclay
Chadbourne & Parke
New York, New York

George E. Boyhan
Waste Technologies International
Palm Beach, Florida

Rigdon H. Boykin
Chadbourne & Parke
New York, New York

Robert Brickner
Gershman, Brickner & Bratton
Washington, D.C.

Diane Buxbaum
U.S. EPA Solid Waste Management
    Programs
New York, New York

Jay A. Campbell
Henningson, Durham & Richardson
Alexandria, Virginia

Larry Canter
University of Oklahoma
Norman, Oklahoma

Lawrence Chertoff
Pennsylvania Resource Systems
Pittsburgh, Pennsylvania

E. Joseph Duckett
Schneider Consulting Engineers
Bridgeville, Pennsylvania

Miro Dvirka
William F. Cosulich Associates,
    Engineers
Woodbury, New York

James L. Easterly
Meridian Corporation
Falls Church, Virginia

Robert H. Greeley
Aenco, Inc., Division Cargill Corp.
Albany, New York

Warren T. Gregory
Smith Barney, Harris Upham & Co.
New York, New York

H. Lanier Hickman, Jr.
Government Refuse Collection &
    Disposal Association
Silver Spring, Maryland

William L. Kovacs
Jaeckle, Fleischmann & Mugel
Washington, D.C.

Robert Cowles Letcher
U.S. Department of Energy
Morgantown, West Virginia

Calvin Lieberman
Institute of Scrap Iron and Steel
Washington, D.C. and Toledo, Ohio

Michael R. Lissack
Smith Barney, Harris Upham & Co.
New York, New York

R. S. Madenburg
Morrison-Knudsen Co. Inc.
Boise, Idaho

Sergio E. Martínez
Consulting Engineer
Miami, Florida

Heinrich Matthee
Dyckerhoff Engineering
Wiesbaden, West Germany

Walter R. Niessen
Camp, Dresser & McKee, Inc.
Boston, Massachusetts

Anthony R. Nollet
Aenco, Inc., Division Cargill Corp.
Albany, New York

Philip R. O'Leary
University of Wisconsin
Madison, Wisconsin

Charles Peterson
Gershman, Brickner & Bratton
Washington, D.C.

James Reynolds
HDR Techserve, Inc.
Omaha, Nebraska

David G. Robinson
Urban Planning/Waste Survey
   Consultant
Trumbull, Connecticut

William D. Robinson
Consulting Engineer
Trumbull, Connecticut

Fred Rohr
Consulting Engineer
Oak Brook, Illinois

Stuart H. Russell
Thermo-Electron, Inc.
Waltham, Massachusetts

Elizabeth C. Saris
Science Applications, Inc.
McLean, Virginia

Lawrence T. Schaper
Black & Veatch,
   Engineers/Architects
Kansas City, Missouri

David J. Schlotthauer
Henningson, Durham & Richardson,
   Inc.
Omaha, Nebraska

Barry S. Shanoff
Law Offices of Barry S. Shanoff
Silver Spring, Maryland

Mary T. Sheil
New Jersey Office of Recycling
Newark, New Jersey

Roger G. Slutter
Fritz Engineering Laboratory
Lehigh University
Bethlehem, Pennsylvania

Jane C. Souzon
Law Offices of Barry S. Shanoff
Silver Spring, Maryland

Gordon L. Sutin
Gordon Sutin Consultants, Ltd.
Dundas, Ontario, Canada

Donald K. Walter
Office of Energy from Municipal
   Waste
U.S. Department of Energy
Washington, D.C.

David Watson
Engineering/Process Divisions
Blue Circle Industries
Kent, England

Kenneth L. Woodruff
Resource Recovery Consultant
Morrisville, Pennsylvania

Jonathan M. Wooten
Smith Barney, Harris Upham & Co.
New York, New York

FOREWORD

Reflecting on the advancements in solid waste processing and resource recovery since the early 1970s yields both encouraging and disturbing trends. Until a period of environmental awareness began in the late 1960s, landfilling (dumping) and incineration of solid waste were the two principal means of disposal in North America. Many incinerators were closed because of air pollution, dumps were closed or upgraded to sanitary landfills, and landfilling became a more widely utilized method of disposal. Because of incinerator air pollution and groundwater contamination caused by improper landfilling, work began on waste volume reduction and materials recovery in an attempt to conserve landfill space and reclaim nonrenewable resources. This was the primary focus of waste processing and resource recovery until the Arab oil embargo in the fall of 1973, when energy prices skyrocketed. The focus then turned toward the energy content of solid waste, as prices for conventional energy soared for the next eight years. During this period of energy price escalation, several waste processing and resource recovery demonstration facilities with materials recovery and fuel preparation met with difficulties caused by a combination of poor planning, poor system and equipment selection, and improper scale-up decisions. Several plant failures by major firms made industry headlines. In fact, many facilities constructed during the 1970s experienced operational and maintenance difficulties (both mass burn and prepared fuel) that appear to continue in some cases.

We were aware of the European experience with resource recovery. For a number of years incinerators had been fitted with waterwalls and waste heat boilers had been used to recover energy, usually in the form of low-pressure steam. This by-product energy-recovery technology has been demonstrated at many facilities throughout the world and many firms in the United States have adopted the use of this older technology rather than continue developing more sophisticated techniques for more efficient energy recovery. Although the older technology appears to be satisfactory, it has offered little state-of-the-art progress except for necessary North American improvements vis-à-vis higher pressures and temperatures and more difficult raw material. Initially, however, it has appeared to be politically safer.

Theoretically, incineration of waste by mass burning is not the most efficient means of thermal reduction and resource recovery. Mass burning does not require preprocessing of the waste; but a tradeoff is that combustion, heat release, and overall energy recovery are less efficient than for a prepared waste. Thus, it is not deemed acceptable practice to burn run-of-mine coal in a furnace. The coal is basically prepared by sizing, and sometimes to remove tramp, noncombustible contaminants.

Acceptable waste disposal is the primary consideration in resource recovery, and the most efficient, least cost, environmentally acceptable system with maximum on-line availability is required. Maximum waste reduction requires maximum burnout with minimum residue. This is seldom obtained in mass burning (especially small-scale) because of highly heterogeneous raw feed. Residue that must be disposed of is 25–40% by weight of the waste input. The only way to achieve efficient combustion with maximum burnout and minimum ash is by sizing and preparing the waste as a fuel. Further processing of the refuse reduces costs by providing for recovery of materials as well as smaller combustion and air pollution equipment. Lower excess air is required in the combustion process, reducing total volumes of gas to be cleaned and emitted into the atmosphere in most cases. As a result of more uniform material with fewer inerts, there is less downtime due to slagging, abrasion, and wear. Although processing systems have experienced difficulties, the use of new and improved shredder designs and more simplified process flows has greatly improved this area. As a result, the processing approach has become reliable and practical for resource recovery. Burning prepared waste is likely to become obsolete in the future as even more efficient energy conversion processes are developed. This industry must not revert to the past but continue to move forward.

Resource recovery from municipal solid waste is a young industry that has experienced difficulties, and any municipality or agency planning for resource recovery must prepare for a 20 year (or longer) commitment. It is hoped that this book will play a role in decision making and that it will prove useful as it was intended to be—A Practical Guide!

K. L. Woodruff

Morrisville, Pennsylvania

PREFACE

Solid waste disposal has historically been a mundane out-of-sight—out-of-mind necessity, given little heed by the general public until recently when advances in industrial technologies and changing lifestyles have contributed to serious and widespread perception of danger to public health. It has caught the attention of much of the public because of:

Well-publicized land, water, and air contamination from a broader spectrum of increasing quantities of the more complicated wastes from rapid industrial growth and sophistication since World War II.

The increasing ability to identify and measure deleterious compounds and their environmental effects.

Imminent termination of existing land disposal sites and unavailability of new ones.

Public perceptions that resist efforts to establish facilities for solving disposal problems, creating delay and even termination of sorely needed projects.

Therefore, recognition of the unrelenting priority of solid waste disposal has established a permanent and growing solid waste management industry as evidenced by the entrance of multibillion-dollar corporations as full-service turnkey vendor–operators of resource recovery disposal facilities, along with consulting engineering professionals, equipment and systems manufacturers, and academia. This follows, by at least 20 years, a major trend toward privatization in the solid waste collection and landfill industry by independent haulers which expanded rapidly, including consolidations by merger and acquisition. It led to the formation of several multibillion-dollar private collection-landfill corporations that operate throughout North America, with an active and progressive trade association, the N.S.W.M.A. (National Solid Waste Management Association). Concurrently, the G.R.C.D.A. (Government Refuse Collection and Disposal Association) expanded throughout the United States in the interests of the municipal, county, and state solid waste collection, transport, and landfill sector.

The "death and taxes" inevitability of solid waste collection and disposal, with its attractive and somewhat recession-proof economic character in a generally expanding economy, contributed significantly to this agglomeration, and the formation of medium- and large-size profitable corporations with modern business management orientation soon followed.

Thus the stage was set for the entrance of "Fortune 500" corporations, including the collection and landfill operator agglomerates, into the waste-to-energy business. When Congress enacted P.U.R.P.A. (Public Utilities Regulatory Policy Act) in 1978, revenue from refuse-fired steam-electric generation was virtually guaranteed, enhancing the dedicated boiler concept and the adaptability of incineration to electricity generation when the load can adjust to variations in boiler output. In addition, the often futile task of electrical energy market development was practically eliminated, although the tenuous rate negotiating, siting, and permitting procedures remain.

Business projections in the hundreds of millions of dollars are confidently predicted for the solid waste industry during the rest of the century.

As of mid-1985, however, the proposed comprehensive Federal Tax Reform could impede urgent implementation of vital waste disposal services. Revisions in the code could eliminate tax advantages such as investment tax credits and accelerated depreciation for private equity investor full-service entrepreneurs.

No-risk demands of investors and taxpayers appear to be creating a financing Darwinism with survival of the deepest pockets compatible with the private entrepreneurial trend that offers single overall project responsibility as opposed to the A & E (Architect & Engineer) approach. Without the more lucrative A & E core contracts for design and construction, solid waste consulting services are mostly limited to performing environmental and financial contract studies, second opinion, and third party consultation by cadres of systems analysts and economists, with fewer personnel having solid waste system design and operating experience.

If tax reform legislation discourages full service equity partnership that offers single responsibility for overall performance, resurrection of the traditional A & E approach might encourage a tendency to assign unrealistic (and seldom redeemable) project performance guarantees to subsystem suppliers, repelling those with integrity and discouraging competition and innovation.

It must be remembered that the need for acceptable alternatives to land disposal of solid waste became serious in Europe long ago through scarce land and high energy costs. Thus the alternatives of composting and principally thermal reduction with energy recovery were critical, leading to full participation by governments at all levels with minimal private participation and public resistance.

The history is quite different in the United States. Seemingly unlimited land availability, low energy costs, and traditional political practice postponed the inevitable use of alternative disposal methods until economic and environmental obstacles have made it a more difficult and urgent transition. Although conventional refractory furnace incineration flourished since the beginning of the century, it withered and nearly vanished under the onslaught of environmental and public nuisance crusades along with aggressive pursuit by the collection and landfill industries in a successful campaign to abandon or convert incinerators to transfer stations, delaying the day of reckoning.

Closing approximately 101 incinerators since 1974 (and many more previously) certainly represented a valid concern for air quality, but it also included an overkill that prematurely closed many incinerators that might have bought precious time by rehabilitation and retrofit.*

Advocates of source separation and recycling programs to reduce the quantities of solid waste for disposal are beginning to see significant progress. A growing number of such programs are planned for solid waste management projects that will require interfacing with engineered, integrated systems for energy and materials recovery. In this category are Fresno County, California; the state of New Jersey; and the cities of Berkeley, California, Philadelphia, Pennsylvania, and New York, New York. This is in addition to notable expansion of voluntary local and regional source separation-citizen recycling programs.

There has been a common misperception that source separation and recycling are synonymous with the institutionalized resource recovery movement involving centralized energy and materials recovery facilities. They are not synonymous, and despite a somewhat competitive relationship, a trend toward cooperative interfacing is most welcome and solid waste disposal administration will benefit. Chapter 10 in this handbook presents a comprehensive discussion of the subject and the organizations involved.

Now under debate is the question of a necessity for source separation if current investigation indicates that incineration of certain materials, such as chlorinated plastics and sources of heavy metals, contributes significantly to stack emission and ash residue levels of worrisome contaminents now under scrutiny (NO_x, CO, SO_2, dioxins, heavy metals, and so on). This approach could supplement the current research for eliminating or minimizing incinerator pollution through furnace operating techniques for optimal combustion, by treatment devices such as flue gas scrubbers and chemical treatment of raw feed and ash residue, and for improved analytical techniques and test protocols. The following organizations are actively conducting these investigations:

A.S.M.E. (American Society of Mechanical Engineers)
  Solid Waste Processing Division
  Air Pollution Control Division
  Research Committee on Municipal, Commercial and Industrial Waste
A.S.T.M. (American Society for Testing and Materials), E-38 Committee on Resource Recovery
C.A.R.B. (California Air Resources Board)
Environment Canada (An agency of the Canadian Government)

The outlook for improved and environmentally acceptable solid waste disposal has never been better. Not only has the industry attracted exceptional individual talent, it has been enhanced by the long, arduous, and effective dedication of its professional societies, trade associations, publications, ven-

---

*Alvarez, Ronald J. "Status of Incineration and Generation Of Energy from Thermal Processing of M.S.W.," Proceedings, A.S.M.E. Ninth National Waste Processing Conference, Washington, D.C., May 1980, pp. 5–26.

dors, and by its government agencies—federal, state, and local, and, let us hope, by a perceptive congress.

William D. Robinson, P.E.
Editor-in-Chief

Trumbull, Connecticut
January 1986

CONTENTS

PART 1 THE PUBLIC ISSUES

1 Using the Handbook, Who and How   **3**
William D. Robinson

    1.1   Who and How, 4
    1.2   Issues That Are Controversial, Neglected, or Ignored, 4
    1.3   Chapter Abstracts, 4

2 Legislation and Involved Agencies   **9**
William L. Kovacs

    2.1   History of Solid Waste Management Laws, 9
    2.2   Hazardous Waste Aspects of RCRA, 10
    2.3   Solid Waste Guidelines and Planning Process, 13
    2.4   Procurement of Products Containing Recovered Materials, 15
    2.5   Other Federal Statutes Impacting on Solid Waste Management, 16

        2.5.1   Department of Energy, 16
        2.5.2   National Energy Conservation Policy Act, Pub. L. 95-619, 18
        2.5.3   Federal Energy Regulatory Commission and PURPA, 18
        2.5.4   Internal Revenue Service (IRS), 19

    2.6   Implementation of RCRA, 19

        2.6.1   EPA's Implementation of RCRA, 19
        2.6.2   The EPA Solid Waste Program, 20
        2.6.3   The EPA, 20
        2.6.4   The Procurement of Recovered Materials, 21

    2.7   DOE's Impact on Solid Waste Management, 21
    2.8   Impact of New Federalism, 21
    2.9   Current and Future Issues in Solid Waste, 23

        2.9.1   The Liability Issue, 23
        2.9.2   Cost Competitiveness of New Technology, 23
        2.9.3   The Need for Solid Waste Flow Control and the Municipal Action Exemption, 24
        2.9.4   The Absent Parties in EPA Litigation—States and Municipalities, 25
        2.9.5   Procurement and Recycling, 26

    2.10 Summary, 26

3 Public Perceptions and Community Relations   **31**
Lawrence Chertoff and Diane Buxbaum

3.1   Introduction, 31
3.2   Resource Recovery Project Case Study, 31

    3.2.1   Facilities Investigated, 31
    3.2.2   Data-Gathering Technique, 32
    3.2.3   Summary of Interview Questions Asked, 32
    3.2.4   Communities Studied, 32

3.3   Implications of Case Study, 33

    3.3.1   Motivating Forces, 33
    3.3.2   Militating Factors, 33
    3.3.3   Economic Considerations, 33
    3.3.4   Community Education, 34

3.4   Case Analysis, 34

    3.4.1   Abandoned Projects, 34
    3.4.2   Abandoned Sites, 36
    3.4.3   Successful Projects, 38
    3.4.4   Projects in Doubt, 39

3.5   Summary, 40

**4 The Feasibility Study, Procurement, and Construction
Management**   **43**
Stuart H. Russell, Robert Brickner, and Charles Peterson

4.1   The Feasibility Study, 43

    4.1.1   Introduction, 43
    4.1.2   Gathering Basic Data, 44
    4.1.3   Identifying Markets, 52
    4.1.4   Selecting Alternatives, 56
    4.1.5   Net System Cost Modeling, 58
    4.1.6   Comparing Alternatives, 63

4.2   Procurement (and Construction Management), 68

    4.2.1   Introduction, 68
    4.2.2   Approaches: A/E, Turnkey, Full Service, 69
    4.2.3   Procurement Methods, 73
    4.2.4   Construction Management, 77

5 Waste Disposal/Resource Recovery Plant Costs   **93**
W. D. Robinson and Sergio E. Martínez

5.1   Capital Cost, 93

    5.1.1   Spectrum of Facilities Costs, 94
    5.1.2   Preproject Expenses, 95
    5.1.3   Financing a New Plant, 95
    5.1.4   Foreign Financing, 96
    5.1.5   Preproject Engineering, 96
    5.1.6   Selecting a Consulting Engineer, 96
    5.1.7   Concept and System Choices: Reliability and Redundancy, 97
    5.1.8   Cost Control, 97
    5.1.9   Purchasing Procedures, 99

5.2   Construction Cost, 99

    5.2.1   Schedules, 100
    5.2.2   Labor Cost, 104
    5.2.3   Cost Containment, 105
    5.2.4   Cost Controls, 106

5.3   Operating Costs, 107

    5.3.1   Plant Ownership and Operation by Local Government, 109
    5.3.2   Private Operation and Publicly Owned Plants, 109
    5.3.3   Facilities Owned and Operated by Private Industry, 109
    5.3.4   Revenues, 109
    5.3.5   Expense, 113
    5.3.6   Profit, 114
    5.3.7   Cost Control, 115

6   Economics and Financing of Resource Recovery Projects   121
Warren T. Gregory, Jonathan M. Wooten, Michael R. Lissack,
and R. S. Madenburg

6.1   Resource Recovery Financing Structures, 121

    6.1.1   Public Ownership, 121
    6.1.2   Private Ownership Financing, 122
    6.1.3   Leveraged Lease Financing Structures, 124
    6.1.4   Builder/Operator Ownership, 125
    6.1.5   Accounting Considerations, 125

6.2   A Case Analysis: Various Financing Alternatives for a Cogeneration
    Resource Recovery Facility over 20 years, 127

    6.2.1   Landfill Only, No Resource Recovery, 127
    6.2.2   Publicly Owned Resource Recovery Plant Versus Land
         Disposal: Bond Debt Service Level Each Year, 127
    6.2.3   Publicly Owned Resource Recovery Plant Versus Land
         Disposal: Bond Debt Service Lower in Early Years (Beginning
         at Interest Only) and Higher in Later Years, 127
    6.2.4   Leveraged Lease Financings, 131
    6.2.5   Leveraged Lease Financings with Stabilization Fund in Early
         Years, 131
    6.2.6   Vendor Ownership Financings, 131

6.3   Case Analysis Summary, 132

Appendix 6.1   Assessing Waste-To-Energy Project Risks, 133
Appendix 6.2   Resource Recovery Ratings (Bonds) Approach, 139

7   Legal Issues   151
Barry S. Shanoff and Jane C. Souzon

7.1   Waste Flow Control, 151

    7.1.1   Competitive Tipping Fees, 151
    7.1.2   Private Agreements and Contracts, 151
    7.1.3   Legislative Controls, 152

7.2   Interstate Commerce, 153
7.3   Finished Landfill Site Continuing Liabilities, 155

    7.3.1   Insurance, 156
    7.3.2   Surety Bonds, 156
    7.3.3   Trust Funds, 156

Appendix 7.1   Sample Franchise Administration and Rate-Averaging
               Procedure, 156
Appendix 7.2   Sample Intermunicipal Agreement Re: Solid Waste, 162
Appendix 7.3   Sample Indemnity Bond, 171
Appendix 7.4   Sample Provisions—Landfill Environmental Trust Fund, 172

PART 2   IMPLEMENTATION ISSUES: SYSTEMS, HARDWARE, OPERATIONS

8   Collection of Residential Solid Waste   177
H. Lanier Hickman, Jr.

8.1   Introduction and Policy Overview, 177
8.2   Managing Change in a Solid Waste Collection System, 178

      8.2.1   Introduction, 178
      8.2.2   Game Plan for Change, 178

8.3   Cost Accounting Procedures for Solid Waste Collection Systems, 179

      8.3.1   Introduction, 179
      8.3.2   Enterprise Fund Accounting, 179
      8.3.3   System Deficiencies, 179
      8.3.4   Summary, 180

8.4   Unions and Solid Waste Collection, 180

      8.4.1   Introduction, 180
      8.4.2   Collective Bargaining in Residential Solid Waste Collection,
              180
      8.4.3   Managing Change, 182
      8.4.4   Summary, 182

8.5   Contracting for Residential Solid Waste Collection, 183

      8.5.1   Introduction, 183
      8.5.2   Determining Type and Level of Service, 183
      8.5.3   Technical Specifications, 183
      8.5.4   Summary, 185

8.6   Collection Equipment Maintenance Programs, 185

      8.6.1   Introduction, 185
      8.6.2   Planned Maintenance, 185
      8.6.3   Components in a Maintenance Program, 185
      8.6.4   Other Factors to Consider, 186
      8.6.5   Summary, 187

8.7   Optimizing the Performance of Collection Services, 187

      8.7.1   Introduction, 187
      8.7.2   Factors Affecting Productivity and Costs, 187
      8.7.3   Measuring Productivity in Residential Solid Waste Collection
              Systems, 189
      8.7.4   The Five-Stage Process to Improve Residential Solid Waste
              Collection Systems, 190

9   Transfer of Municipal Solid Waste   **195**
Laurence T. Schaper

   9.1   The Transfer Station, 195

         9.1.1   Potential Advantages, 195
         9.1.2   Types of Users, 196

   9.2   Location, 196
   9.3   Design Choices, 197

         9.3.1   Station Concepts, 197
         9.3.2   Sizing Transfer Facilities, 197
         9.3.3   Site Development and Ancillary Facilities, 202

   9.4   Process Options, 203

         9.4.1   Baling, 203
         9.4.2   Shredding, 204

   9.5   Transfer Vehicles, 204

         9.5.1   Compaction Trailers, 204
         9.5.2   Noncompaction Trailers, 204
         9.5.3   Number of Vehicles Required, 205

   9.6   Materials-Handling Equipment, 206
   9.7   Maintenance, 207
   9.8   Cost Analysis and Case Studies, 208

         9.8.1   Cost Analysis, 208
         9.8.2   Case Studies, 211

10   Source Separation and Citizen Recycling   **215**
Robert Cowles Letcher and Mary T. Sheil

   10.1   Perceptions, Analysis, and Status, 215

         10.1.1   Recycling Defined, 216
         10.1.2   Source Separation Programs Defined, 216
         10.1.3   Recycling and the Waste Disposal Industry, 216
         10.1.4   Implications for Both Concepts, 216
         10.1.5   Waste: Perceptions and Perspectives, 217
         10.1.6   The Institutionalization of Waste Disposal, 217
         10.1.7   Benefits of Source Separation, 219
         10.1.8   Benefits of the Recycling System, 220
         10.1.9   Source Separation Versus Centralized Resource-Recovery
                  Process Systems, 221
         10.1.10 Summary of Source Separation Program Incentives and
                  Benefits, 223
         10.1.11 Summary of Materials, Markets, and Programs, 227
         10.1.12 Case Studies, 229

   10.2   Recycling: A Statewide Program for New Jersey, 238

         10.2.1   Background, 240
         10.2.2   Implementing the Recycling Plan, 246
         10.2.3   Meeting the Challenge, 246
         10.2.4   Collection of Recyclables, 247
         10.2.5   Recycling—A Cost Avoidance Mechanism, 247
         10.2.6   Market Expansion and Development, 247
         10.2.7   Education—The Key to Success, 248
         10.2.8   A Total Effort, 248

Appendix 10.1   Sample Contract to Sell Used Papers, 250
Appendix 10.2   State Recycling Associations, 251
Appendix 10.3   Trade Associations of Industries Which Process or Use
                Recycled Materials, 252
Appendix 10.4   State Resource Recovery Agencies, 253
Appendix 10.5   New Jersey Programs, 256
                Program A: Municipal Curbside Collection with a Drop-Off
                             Center, 256
                Program B: Municipal Curbside Collection, 257
                Program C: Drop-Off Centers in Urbanized/Suburban
                             Regional Area Program, 258

11   Land Disposal   259

Philip R. O'Leary, Larry Canter, William D. Robinson

11.1   Landfill Disposal: Theory and Practice, 259

        11.1.1   Definition and Background, 259
        11.1.2   Principles of Operation, 260
        11.1.3   Biological and Chemical Processes, 263
        11.1.4   Environmental Protection Considerations, 266
        11.1.5   Guidelines: Federal and State, 267
        11.1.6   Landfill Development, 267
        11.1.7   Service Area, Waste Quantities, and Land Requirements, 268
        11.1.8   Siting Procedures, 269
        11.1.9   Techniques for Comparing Candidate Sites by Specific Issues,
                  272
        11.1.10   Public Involvement, 274
        11.1.11   Plan Preparation and Regulatory Approval, 280
        11.1.12   Leachate Formation and Control, 286
        11.1.13   Methane Gas Formation and Control, 313
        11.1.14   Landfill Operations, 321
        11.1.15   Landfill Equipment Selection and Utilization, 323
        11.1.16   On-Site Processing, 326
        11.1.17   Operator Safety, 327
        11.1.18   Site Closure and Long-term Care, 329

11.2   Landfill With Bales, 338

        11.2.1   Background, 338
        11.2.2   The Baling Process, 338
        11.2.3   High-Density Balers, 338
        11.2.4   Medium-Density Balers, 341
        11.2.5   Transportation of Bales, 343
        11.2.6   The Balefill, 345
        11.2.7   Approximate Capital and Operating Costs, 345
        11.2.8   Summaries of Balefill Test Results and Testing of Bales as
                  Foundation Material, 346

Appendix 11.1   Key Elements of the Criteria for Classification of Solid Waste
                Disposal Facilities and Practices, 347
Appendix 11.2   Maximum Contaminant Levels for Determining Whether Solid
                Waste Disposal Activities Comply with Groundwater
                Protection Criteria, 349
Appendix 11.3   Sanitary Landfill Inspection Report, 351
Appendix 11.4   Sanitary Landfill Design and Operational Guidelines, 354
Appendix 11.5   Sample of Technical Site Criteria for Chemical Waste
                Disposal, 360

Appendix 11.6   Items to Be Included in the Engineering Report for a Sanitary Landfill, 364

Appendix 11.7   Landfill Site Rating Method, 365

Appendix 11.8   Decision Factors in Sanitary Landfill Site Selection, 369

Appendix 11.9   Evaluation of Solid Waste Baling and Landfilling, 370
Ralph Stone and Richard Kahle

Appendix 11.10   Engineering Study of Baled Solid Waste As Foundation Material, 373
Roger G. Slutter

12   Resource Recovery: Prepared Fuels Energy and Materials   377

David J. Schlotthauer, George E. Boyhan, William D. Robinson, Kenneth L. Woodruff, Jay A. Campbell, Gordon L. Sutin, David G. Robinson, E. Joseph Duckett, Anthony R. Nollet, and Robert H. Greeley

12.1   Energy Recovery Overview, Processed Fuels, 377

      12.1.1   Dedicated Units, 380
      12.1.2   Modification of Existing Units, 380
      12.1.3   Energy Recovery Methods and Products, 382
      12.1.4   Cofiring, 383
      12.1.5   Codisposal, 386
      12.1.6   Economics and Case Histories, 388

12.2   Processed Refuse Fuel Types, 398

12.3   Methods of Combustion or Energy Recovery of Processed Fuels, 400

      12.3.1   Spreader Stoker Firing, 401
      12.3.2   Suspension-Fired Units, 404
      12.3.3   Fluidized Bed Units, 405
      12.3.4   Cyclone Furnace Firing, 409
      12.3.5   Pyrolysis, 410
      12.3.6   Cement Kilns, 411
      12.3.7   Bioconversion, 412

12.4   Fuel Process Systems, 415

      12.4.1   Dry Process, 416
      12.4.2   Wet Process, 417
      12.4.3   Combined Dry/Wet System, 417
      12.4.4   Energy Output Comparison, 419
      12.4.5   Characteristics of Dry/Wet Systems, 419
      12.4.6   Market for RDF Fuel, 419
      12.4.7   RDF Storage, 422
      12.4.8   By-Product Recovery, 423

12.5   Process and Materials-Handling Systems and Equipment: Shredding and Receiving Systems, 423

      12.5.1   Background, 423
      12.5.2   Typical RDF Dry Process Components and Systems, 424
      12.5.3   Shredding and the Air-Classifier Anomalies, 428
      12.5.4   Size Reduction: Key Factors, 429
      12.5.5   Shredders, 430
      12.5.6   Shredder Operating Characteristics, 432
      12.5.7   Design/Operating Factors Common to Topfeed Shredders, 436
      12.5.8   Recent Improvements in Shredder Design, 437
      12.5.9   Flail Mills, 441

12.5.10 Rotary Shear, 442

12.5.11 Front-End Raw Material Receiving Systems, 444

12.5.12 Front-End Receiving Conveyers and Burden Depth Control, 446

12.5.13 Shredder Discharge Conveyers, 449

12.5.14 Summary, 449

12.6 Process and Materials Handling Equipment: Rotary Shear Shredders, Design and Operation, 452

12.6.1 Background and Description, 452

12.6.2 Operating Experience, 453

12.6.3 Operating and Maintenance Costs, 454

12.6.4 Applications, 454

12.6.5 Shear Shredder Manufacturers, 455

12.7 Process and Materials Handling Equipment: Screens for Solid Waste Processing, 455

12.7.1 Background, 455

12.7.2 Vibrating Screens, 455

12.7.3 Trommel Screens, 456

12.7.4 Disc Screens, 458

12.7.5 Summary, 459

12.7.6 Representative Installations, 459

12.7.7 Solid Waste Processing Screen Manufacturers, 460

12.8 Densified Refuse-Derived Fuel (dRDF), 461

12.8.1 Background, 461

12.8.2 Production Technology Status, 462

12.8.3 Densification Equipment Performance and Problems, 462

12.8.4 dRDF Properties and Characteristics, 467

12.8.5 Storage and Handling, 469

12.8.6 Densification Costs, 469

12.8.7 dRDF Combustion Experience, 471

12.9 Refuse Derived Fuel Storage, Retrieval, and Transport, 473

12.9.1 RDF Storage, Retrieval, and Transport, 473

12.9.2 Remote Steam Plant and RDF Transport, 474

12.9.3 Processing Facility and Steam Plant, Same Site, 474

12.9.4 Atlas Storage and Retrieval System, 475

12.9.5 Miller Hofft Bin and Retrieval System, 477

12.9.6 Concrete Bunker Bulk Storage, 477

12.9.7 Floor Bulk Storage, 479

12.9.8 Surge Storage, 479

12.9.9 Miller Hofft Surge Bins, 479

12.9.10 Sprout Waldron Surge Bins, 479

12.9.11 Moving By-Pass Surge Storage Systems, 479

12.9.12 Hooper Live-Bottom Bin, 480

12.9.13 RDF Distribution and Feed, 481

12.10 Recovered Materials Specifications and Markets, 483

12.10.1 Introduction, 483

12.10.2 Ferrous Metals, 483

12.10.3 Glass, 484

12.10.4 Aluminum, 486

12.10.5 Paper and Corrugated, 487

12.10.6 Other Miscellaneous Material, 491

12.10.7 Conclusion, 496

12.11   Recovered Materials-Equipment and Systems, 497

    12.11.1   Introduction, 497
    12.11.2   Air Classifiers, 497
    12.11.3   Ferrous Metal Recovery, 498
    12.11.4   Nonferrous Metals Recovery, 503
    12.11.5   Paper Recovery, 504
    12.11.6   Glass Recovery, 505
    12.11.7   Plastics Recovery, 505
    12.11.8   Ash Processing for Metals and Aggregate Recovery, 506

12.12   Raw Material Quantity and Composition: A Final Check, 507

    12.12.1   Quantification Survey, 507
    12.12.2   Presurvey Planning, 508
    12.12.3   Survey Scope, 509
    12.12.4   Quantification Survey Work Tasks, 513
    12.12.5   Quantification Survey Summary Report, 514
    12.12.6   Waste Composition Survey, 515
    12.12.7   The Sorting Program, 522
    12.12.8   Laboratory Analysis, 527

12.13   Health and Safety: Health Aspects, 530

    12.13.1   Explosion Protection, 532
    12.13.2   Dusts, 536
    12.13.3   Microbiological Aspects, 537
    12.13.4   Noise Control, 538
    12.13.5   Conclusion, 539

12.14   Health and Safety: Implementation, 541

    12.14.1   Background and Scope, 541
    12.14.2   Safety Rules and Practice, 542
    12.14.3   Personnel Safety, 544
    12.14.4   Raw Material Presort, 546
    12.14.5   Raw Material Surveillance, 548
    12.14.6   Explosion Protection, 550
    12.14.7   Remedial Measures: Explosions in Resource-Recovery Plants, 552
    12.14.8   Postexplosion Procedures, 554

13   Resource Recovery: Mass Burn Energy and Materials   557
Miro Dvirka

13.1   Mass Burn Energy Recovery Overview, 557

    13.1.1   Dedicated Unit: Boiler Types, 557

13.2   Existing Units and Retrofits, 560
13.3   Mass Burn Energy Products, 561

    13.3.1   Constraints, 561
    13.3.2   Steam Generation, 562
    13.3.3   Power Generation, 564
    13.3.4   Cogeneration, 565

13.4   Codisposal, Sewage, 567

    13.4.1   Coburning (in suspension) of Predried Sludge Above Grate-Fired Refuse, 567
    13.4.2   Coburning Dewatered Sludge, Layered with Refuse in Furnace Feed, 569

13.5   Field-Erected Units: Systems and Sizing, 571
13.6   Raw Material Receiving and Storage, 572

    13.6.1   Pit/Bunker Sizing, 572
    13.6.2   Oversized Material, 572
    13.6.3   Fire and Ventilation, 573

13.7   Retrieval and Furnace Feed, 573

    13.7.1   Crane Design Criteria, 573
    13.7.2   Crane Feed Cycle Design Criteria, 574

13.8   Stoker and Furnace Design, 575

    13.8.1   Combustion Process Equations, 575
    13.8.2   Stoker Design, 577
    13.8.3   Furnace Design, 581

13.9   Water-Cooled Rotary Combustor, 587
13.10   Small-Scale "Modular" Units, 590

    13.10.1   Combustion Concepts, 590
    13.10.2   Raw Material Receiving and Storage, Modular Units, 590
    13.10.3   Raw Material Retrieval and Feed Systems, 590
    13.10.4   Combustion Systems, 591
    13.10.5   Emissions Control, Modular Units, 591
    13.10.6   Application Constraints, 593

14   Resource Recovery: Air Pollutant Emissions and Control   **595**
Walter R. Niessen

14.1   Regulatory Context—Federal, 595

    14.1.1   National Environmental Policy Act (1969), 596
    14.1.2   Clean Air Act of 1970 and Amendments, 596

14.2   Regulatory Context—State and Local, 597
14.3   Air Pollutant Uncontrolled Emissions, 597

    14.3.1   Inorganic Particulate and Comparison of Firing Methods, 597
    14.3.2   Combustible Particulate, 605
    14.3.3   Total Particulate, 606
    14.3.4   Carbon Monoxide (CO), 607
    14.3.5   Nitrogen Oxides (NO_X), 608
    14.3.6   Sulfur Oxides, 608
    14.3.7   Hydrochloric Acid, 609
    14.3.8   Micropollutants, 609

14.4   Control Technology, 613

    14.4.1   Particulate Matter, 614
    14.4.2   Carbon Monoxide and Hydrocarbons, 617
    14.4.3   Oxides of Nitrogen (NO_X), 617
    14.4.4   Acid Gases, 618
    14.4.5   Micropollutants, 618

15   Marketing Resource Recovery Products   **621**
*Rigdon Boykin, Bernays Thomas Barclay and Calvin
Lieberman*

15.1   Energy, 621

    15.1.1   Energy Marketing Principles, 621
    15.1.2   Federal Energy Law Affecting Marketing Considerations, 625

15.1.3   Energy Values, 628
15.1.4   Negotiating a Power Sales Contract, 636

15.2   Marketing Recovered Materials: A Viewpoint of the Private Scrap
        Processor, 643
        Calvin Lieberman

15.2.1   Choices in Strategic Planning, 643
15.2.2   Identifying and Evaluating Markets, 644
15.2.3   Evaluating Raw Material Supply and Recovery Technologies,
          645
15.2.4   Evaluating Risks, 648
15.2.5   Recovered Materials Quality/Salability, 648
15.2.6   Disincentives in Resource Recovery, 648
15.2.7   Engineering With Unpredictable Raw Material, 649
15.2.8   Raw Material Flow Control: A Word of Caution, 649
15.2.9   Markets For Recovered Materials: The Hard Facts, 650

16   Energy from Refuse in Industrial Plants   653
William D. Robinson and Fred Rohr

16.1   Background, 653
16.2   Industrial Wastes as Boiler Fuel, 653
16.3   Industrial Incinerators, 654

16.3.1   Background, 654
16.3.2   The Early Los Angeles Excess Air Refractory Furnace, 654
16.3.3   Controlled Air Designs, 655

16.4   Energy Recovery Methods, 660

16.4.1   Background, 660
16.4.2   Utilization Choices: Steam, Hot Water, Hot Air, KW, 662
16.4.3   Boiler Types, 663

16.5   Operating and Maintenance Factors, 664

16.5.1   Waterside Tube Failure, 664
16.5.2   Fireside Tube Wastage, 664
16.5.3   Refractory Linings, 664
16.5.4   Stokers, 665
16.5.5   Ram Feed, 665
16.5.6   Ash Removal, 665
16.5.7   Feedwater Treatment, 666

16.6   Industrial Solid Waste Incineration, 666

16.6.1   Concept Choices, 666

16.7   Industry as the Purchaser of Refuse Energy, 668
16.8   Industrial Cogeneration, 668

16.8.1   Background, 668
16.8.2   Technology and Systems, 671
16.8.3   Regulatory Factors, 672
16.8.4   Economic Factors, 672
16.8.5   Operation and Maintenance Cost Factors, 673
16.8.6   Operating Cost Summary, 673

16.9   Conclusions, 675

Appendix 16.1   Two 200 TPD Composite Plant Designs for a Starved Air
                 System and for an Excess Air System, 677

Case Histories, 680

17 **Residential, Commercial and Industrial Bulky Wastes**   **697**
    William D. Robinson

    17.1   Introduction, 697
    17.2   Nature of the Waste, 697

            17.2.1   Residential Bulky Waste, 697
            17.2.2   Commercial Bulky Waste, 698
            17.2.3   Industrial Bulky Waste, 698

    17.3   Present Disposal Status, 698

            17.3.1   Background, 698

    17.4   Bulky Waste Process Experience, 700

            17.4.1   Background, 700

    17.5   Bulky Waste Processing Case Histories, 703

            17.5.1   City of Harrisburg, Pennsylvania, 703
            17.5.2   City of Chicago, Illinois, Goose Island, 706
            17.5.3   Resources Recovery (Dade County) Inc., Miami, Florida, 707
            17.5.4   City of East Chicago, Indiana, 714
            17.5.5   City of Omaha, Nebraska, Solid Waste Recycling Center, 719
            17.5.6   City of Glen Cove, New York, Codisposal/Energy Recovery
                      Facility, 721
            17.5.7   City of Montreal, Quebec, Canada, 725
            17.5.8   City of Kyoto, Japan, 725
            17.5.9   City of Ansonia, Connecticut, 728
            17.5.10   City of Tacoma, Washington, 728

    17.6   Aborted Bulky Waste Process Projects, 734

            17.6.1   Background, 734
            17.6.2   Summary of Aborted Projects, 734
            17.6.3   Analysis of Aborted Bulky Waste Process Projects, 734

    Appendix 17.1   Omaha Shredder Product Screen Analysis and Noise Level
                     Survey, 735

18 **Refuse Fuels in the Portland Cement Industry (Including Tires
    and Shredder Residue)**   **737**
    David Watson, Heinrich Matthee, and William D. Robinson

    18.1   Experience in England, 737

            18.1.1   Refuse versus Other Fuels—Technical Factors, 737
            18.1.2   Development of Blue Circle's Interest, 738
            18.1.3   Resumé of Blue Circle's Experience, 741
            18.1.4   Current Developments, 741
            18.1.5   Questions and Answers, 742

    18.2   Experience in West Germany, 743

            18.2.1   Background: Tires, 743
            18.2.2   Miscellaneous Shredder Wastes, 744
            18.2.3   Auto Shredder Wastes, 745
            18.2.4   Asphaltic Sludge, 746

    18.3   Experience in North America, 746

            18.3.1   Background, 746
            18.3.2   Factors in a Discouraging Outlook, 747

18.3.3   Scrapped Auto Shredding Residues, 747
18.3.4   Conclusion, 748

19 Biological Processes   **749**
Donald K. Walter, James L. Easterly and Elizabeth C. Saris

19.1   Background, 749
19.2   Anaerobic Digestion, 750

    19.2.1   Introduction, 750
    19.2.2   Basic Processes, 750
    19.2.3   Feedstocks, 750
    19.2.4   Products, 751
    19.2.5   Reactor Types, 751
    19.2.6   Design Parameters, 753

19.3   Fermentation Processes, 753

    19.3.1   Background, 753
    19.3.2   Basic Processes, 754
    19.3.3   Feedstocks, 754
    19.3.4   Products, 754
    19.3.5   Design Parameters, 755

19.4   Compost, 755

    19.4.1   Background, 755
    19.4.2   Basic Process, 755
    19.4.3   Process Description, 755
    19.4.4   Feedstocks, 756
    19.4.5   Products, 756
    19.4.6   Design Parameters, 756
    19.4.7   Reactor Types, 757

19.5   Applications and Economics, 757

    19.5.1   Anaerobic Digestion, 757
    19.5.2   Fermentation, 758
    19.5.3   Composting, 758

19.6   Case Histories, 759

    19.6.1   Anaerobic Digestion, 759
    19.6.2   Compost, 761

Appendix 19.1   Biomass as Fuel for Electric Generation: Planned and Existing
Projects in the United States, 763

PART 3   HAZARDOUS SOLID WASTES

20 Federal Regulatory Issues   **773**
William L. Kovacs

20.1   Introduction, 773
20.2   History of the Federal Hazardous Waste Regulatory Program, 773

    20.2.1   Past Practices, 773
    20.2.2   Intent and Development of RCRA, Congressional Debate, 774

20.3 The Act—Its Organization, Scope, and Contents, 775

    20.3.1 Identification and Listing of Hazardous Wastes, 775
    20.3.2 Requirements Imposed On Generators of Hazardous Waste, 775
    20.3.3 Requirements Imposed on Transporters of Hazardous Waste, 775
    20.3.4 Requirements Regulating Those Who Treat, Store, or Dispose of Hazardous Waste, 776
    20.3.5 Permit Authority, 776
    20.3.6 Authorized State Programs, 777
    20.3.7 Enforcement of RCRA, 777
    20.3.8 The Hazardous and Solid Waste Amendments of 1984, 778

20.4 Hazardous Waste Management Regulations under RCRA, 780

    20.4.1 40 C.F.R. Part 260, General Regulations for Hazardous Waste Management, 780
    20.4.2 40 C.F.R. Part 261, Regulations Identifying Hazardous Waste, 780
    20.4.3 40 C.F.R. Part 262, Requirements upon Generators of Hazardous Waste, 782
    20.4.4 40 C.F.R. Part 263, Requirements upon Transporters of Hazardous Waste, 783
    20.4.5 40 C.F.R. Part 264, Requirements upon Owners and Operators of Permitted Hazardous Waste Facilities, 783
    20.4.6 40 C.F.R. Part 265, Interim Status Standards, 789
    20.4.7 40 C.F.R. Part 267, Interim Standards for Owners and Operators of New Hazardous Waste Land Disposal Facilities, 789
    20.4.8 Interface of RCRA Regulations with State Programs (Part 271 Regulations), 790

20.5 EPA, Its Organization, and Regional Offices, 791
20.6 EPA's Permitting Procedures, 791

    20.6.1 The Permit Application, 791

20.7 EPA's Inspection Authority, Reporting Requirements, and Enforcement, 792

    20.7.1 Inspections, 792
    20.7.2 Reporting Requirements, 792
    20.7.3 Enforcement, 793

20.8 The Superfund Program, 793

    20.8.1 Key Superfund Provisions and the Agencies that Implement It, 793
    20.8.2 The Relationship of Superfund to RCRA, 794

20.9 Current Changes and Future Federal Role, 795

    20.9.1 Changes by the Reagan Administration, 795
    20.9.2 Future RCRA Regulatory Program, 796
    20.9.3 Future Superfund Program, 796

20.10 Summary, 796

21 State and Local Regulatory Issues   **799**
James Reynolds and H. Lanier Hickman, Jr.

21.1 Introduction, 799
21.2 State Program Development, 799

21.2.1   Life before the Resource Conservation and Recovery Act, 799
21.2.2   Standardization, 800
21.2.3   Effects of RCRA, 800

21.3   Policy Issues of Concern to Local Government, 801

21.3.1   Introduction, 801
21.3.2   Facility Siting, 801
21.3.3   Economic Impact on Industry, 801
21.3.4   The Exempted (Small) Generator, 802
21.3.5   Closed and Abandoned Hazardous Waste Disposal Sites and
         Orphaned Hazardous Wastes, 802
21.3.6   Emergency Response and Contingency Plans, 802
21.3.7   Summary, 803

Index   **805**

THE SOLID WASTE
HANDBOOK

PART 1
THE PUBLIC ISSUES

CHAPTER 1
USING THE HANDBOOK: WHO AND HOW

W. D. ROBINSON, P.E.

Consulting Engineer
Trumbull, Connecticut

1.1 WHO AND HOW

The Solid Waste Handbook is a practical guide that presents a comprehensive single source reference for current solid waste management public issues, implementation issues, and hazardous waste administration guidelines. It examines public perceptions, legislation, regulation and agencies, planning, finance, technologies, operations, economics, and administration, with methods for choosing alternatives and an assessment of trends for the future.

It is not *a design engineering manual.*

As a practical guide, it is useful at all levels of government: federal, state, county, and local. It is for legislators, administrators, and planners; technical, legal, and finance professionals; public and private facilities operators; environmentalists, concerned citizens groups, and students.

In a single volume, it reviews changes in waste management practice mandated by the Resource Conservation and Recovery Act (RCRA) of 1976 and amended in 1978, 1980, and especially November, 1984, that pervasively revises waste classification with additional land and air pollution identification and liabilities for intensified restriction.

The Handbook examines common and divergent interests at public and private interfaces along with past, current, and state-of-the-art practice for land disposal, resource recovery, and their alternative techniques for enlightened decision making in all facets.

1.2 ISSUES THAT ARE CONTROVERSIAL, NEGLECTED, OR IGNORED

Assessment of technologies in resource recovery: What works, what is marginal, and what has failed (Chapters 12, 13).

Disagreement with certain nostrums and theoretical approaches for materials handling and processing as spawned by government study grants along with private studies, reports and books, that are characterized by ponderous utilization of modeling and statistical analysis in comparing systems and equipment, and proposing guidelines. As scholarly as these exercises appear, they sometimes present an analytical overkill, which is distracting to the reader and may not be very useful (Chapter 12, Section 12.5).

The inaccuracies of most waste quantity and composition estimates, with techniques for adequate evaluation, are discussed in detail (Chapter 12, Section 12.12).

Waste flow control effects on administration, facilities design, operation, and economics are examined as are political and legal aspects, with current and pending court decisions and appeals (Chapter 2).

In-depth review of public perceptions and their effect on planning and implementation, with representative case histories (Chapter 3).

The misunderstanding and controversy over source separation/citizen recycling, including a state-wide plan (Chapter 10).

The vagaries of municipal resource recovery, from the viewpoint of the private secondary materials (scrap) process industry with recommendations for market evaluation, processing implementation, and operations (Chapter 15.2).

Examination of the usually neglected problem of bulky (oversize) waste logistics and its effect on current disposal practice. Included are viable process techniques and case histories largely unknown or ignored by almost everyone in the industry (Chapter 17).

Energy recovery from industrial waste in industry is seldom discussed elsewhere (except for the Purdue University Conferences and Bio-mass industry activity). This chapter reviews the current status and technologies and includes case histories (Chapter 16).

Incinerator stack emissions, particularly the more recent concerns for particulates and exhaust gas containing carbon monoxide (CO), sulfur oxides (SO_x), oxides of nitrogen (NO_x), formation of acid gases, dioxins, and discussion of control media (Chapter 14).

1.3   CHAPTER ABSTRACTS

Chapter 2: Legislation and Involved Agencies

Recent (November 1984) major amendments to the Resource Conservation and Recovery Act (RCRA) and their pervasive effect on all sectors of the industry, including the regulatory agencies, are examined. Focus is on more restrictive waste classification and manifesting, pollution control, and reporting. A review of the evolution of RCRA, with past and present disposal practice relates the impact on effective solid waste management. Included are hazardous solid waste aspects of RCRA, solid waste guidelines, and the planning for standards compliance and enforcement, the roles of the Environmental Protection Agency, Department of Energy, Public Utility Regulatory Policies Act, the Department of Commerce and recovered material, flow control, and the Internal Revenue Service. EPA's recalcitrance, impeding RCRA implementation, is succinctly reviewed in the chapter summary.

Chapter 3: Public Perceptions and Community Relations

Case histories of public resistance, acceptance of facilities, recommendations for coping with the phenomena, and cultivating community relations are discussed in depth.

Chapter 4: The Feasibility Study, Procurement, and Construction Management

The anatomy of these functions with key elements are presented including modeling for competent decisions and implementation.

Chapter 5: Waste Disposal/Resource Recovery Plant Costs

Capital, construction, and operating cost needs and sources with effective management and control are presented for various concepts.

Chapter 6: Financing Land Disposal and Resource Recovery Projects

Various public, private, and combination options of conventional and intricate financing structure are discussed. Graphic and tabular references are displayed, with appendices for risk assessment and bond-rating criteria for various project concepts. The proposed federal tax reform effects on solid waste management are also examined.

Chapter 7: Legal Issues

The legal issues of waste flow control are discussed along with recent and pending court decisions, appeals, interstate commerce issues, finished landfill continuing liabilities, taxation, and insurance. Appendices review franchise administration and rate setting, intermunicipal agreements, landfill environment trust funds, and indemnity bond issue samples. Case references and abstracts are included.

Chapter 8: Collection of Residential Solid Waste

Managing change in a residential collection system is discussed along with cost-accounting procedures, labor unions and collective bargaining, contracting for collection services, equipment maintenance programs, optimizing collection services, routes, and productivity, including formula, matrix, and tabular techniques.

Chapter 9: Transfer of Municipal Waste

Transfer station advantages, location, design parameters, ancillary facilities, and process (shred, bale) options are developed, with criteria and simple formulae included. Transfer vehicle and materials handling equipment selection with maintenance factors are presented along with cost analysis and case studies, including sizing and cost computations.

Chapter 10: Source Separation and Citizen Recycling

Distinct differences between source separation/citizen recycling and resource recovery systems in philosophy and approach are thoroughly explained contrary to a common misperception of their being identical. Motivation and incentives, methodologies, yields, markets and benefits are explored in-depth. Case studies and New Jersey's statewide program is presented in detail. Appendices list address and phone contact for 50 state resource recovery agencies, recycling and trade associations. Potential benefits of integration with centralized resource recovery systems are reviewed.

Chapter 11: Landfill Disposal: Theory and Practice

Detailed discussion, graphic, tabular, and formulae presentation of landfill methods, in situ and migratory biological and chemical processes, environmental protection, government guidelines, siting procedures and public involvement, leachate formation and control, methane gas formation and control, equipment selection and utilization, site closure and long-term care are features of this comprehensive chapter. Appendices for regulated contaminant levels, facilities classification, site-rating methods, and decision matrices for site selection factors are included. High- and medium-density "balefill" logistics, methods, equipment, and benefits are examined along with the "shredfill" concept.

Chapter 12: Resource Recovery: Prepared Fuels, Energy, and Materials

Case histories, systems concepts, operations, and performance are examined including energy recovery methods, prepared refuse fuel types, process systems, and equipment applications for materials handling, storage, and retrieval. Size reduction, screening, magnetics, and air separation are reviewed along with the wisdom of raw material reexamination for quantity and composition. Materials recovery systems are analyzed as well as American Society for Testing and Materials specifications and markets. Workplace and site protection measures for health and safety are evaluated for fire, explosion, and materials handling/processing along with investigation of ambient pathogen/microbiological factors.

Chapter 13: Resource Recovery: Mass-Burn Energy and Materials

The historic background precedes comprehensive investigation of its present status beginning with dedicated boiler units of waterwall furnace and refractory furnace/convection boiler designs, existing unit retrofits, mass-burn operational constraints, energy products of steam, kilowatts, and cogeneration, and codisposal of sewerage sludge and solid waste. Systems-sizing techniques/selection for field-erected units are presented and include raw material receiving, storage, retrieval, furnace feed, stoker, and furnace designs. The water-cooled rotary combustor system design is reviewed along with the application, advantages, and vagaries of small-scale and modular units. Emissions control and application restraints are also included.

Chapter 14: Resource Recovery: Air Pollutant Emission and Control

Three main categories are examined in depth:

1. The regulatory framework: What must be done to analyze an emissions problem, manage it, and secure a permit to build and/or operate.
2. The emissions estimate: How much of each pollutant will be generated in land disposal (equipment, dust, etc.), processing, and/or combustion.
3. The control strategy: What air pollution control methods (a combination of installed hardware and operating technique) will adequately control emissions to acceptable levels.

Discussed in a federal context is the National Environmental Policy Act (1969) and the Clean Air Act of 1970 with amendments, along with state and local impacts of this legislation. Definitions, present status, and the outlook for emissions of uncontrolled pollutants includes inorganic particulate, total particulate, carbon monoxide (CO), nitrogen oxide (NO_x), sulfur oxides (SO_x), hydrochloric acid (HCl), and micropollutant toxicity, including dioxins and acid gases. Control technology and methodologies for preventing formation and/or neutralizing are examined along with alternative control systems.

Chapter 15: Marketing Resource Recovery Products

Authoritative advice for marketing the resource recovery products of energy and materials is presented in depth. Energy marketing principles and the relevant federal laws, utility industry practice and energy values are scrutinized. Long- and short-run avoided-cost calculation methods are presented along with escalators, wheeling, other valuation issues, and electricity to end users. Power sales contract negotiating techniques, information sources on utilities, strategy and tactics in dealing with utilities, and elements of the long-term sales contract are examined. Planning and practice in recovering and marketing materials have been neglected and mismanaged by municipal solid waste entrepreneurs, systems designers, and government. This chapter presents a view by the traditional and experienced collectors and processors of recyclable materials that reviews the hard facts. Choices in strategic planning are described as well as identification and evaluation of markets, evaluating raw material supply and recovery techniques, assessing risks, recovered materials quality and saleability, disincentives, engineering for unpredictable raw material, and the motivations for waste flow control and its effects.

Chapter 16: Energy from Refuse in Industrial Plants

Several case histories of large- and small-scale installations are included along with typical capital and operating costs. The salient backgrounds of industrial wastes as boiler fuel and early industrial incinerator design criteria leading to present-day designs and operations are presented. Included are small-scale controlled air designs for starved air, excess air, and rotary kiln techniques as well as larger units for prepared waste fuel and cofiring with fossil fuel. Energy choices are reviewed for steam, hot air, and water, electricity, and cogeneration. Firetube and watertube waste heat units, and waterwall furnace designs and applications are described. Operating and maintenance factors are presented for each along with stoker types, water treatment, and ash removal. Mass-burn and prepared fuel concept choices are included as well as cogeneration factors of technology, regulation, operating and maintenance costs, and economics.

Chapter 17: Residential, Commercial, and Industrial Bulky Wastes

The predominant practices in bulky waste disposal have remained mostly crude and haphazard despite a number of technologically successful process plants in operation in the 1970s and at present. Although bulky waste has always had a somewhat lower priority, more sophisticated process and disposal methods can no longer be ignored. The solid waste industry (planners, managers, designers, operators, legal and financial professionals, and the public) is mostly unaware of this experience, which can form the basis for improved and necessary programs. This chapter presents 10 case histories along with the scope and nature of the waste, present disposal status and background, process systems, equipment, and experience, including a typical shredding plant noise level survey. Recent cessation and crises in bulky trash collections are listed in Chapter 17 references.

Chapter 18: Refuse Fuels in the Portland Cement Industry (including Tires and Auto Shredder Residue)

Trials and operating experience are reported from England, West Germany, and North America that show mixed results. Experience in England includes at least one successful operational production plant firing prepared municipal solid waste (excluding tires) supplementing conventional fuels. Continuing testing with tires and waste rubber and fluidized-bed pyrolysis of solid waste for generation of combustible gas for kiln firing are described. Tables and illustrations supplement descriptions of equipment, systems, and operating results that appear promising. West Germany shows encouraging results when firing a cement kiln with tires and municipal waste, and limited success with auto shredder residue. Illustrations and data are included. The outlook in North America is more benign. Unlike counterparts in Europe and Japan with higher energy cost, there is little enthusiasm for firing supplemental refuse fuel, that is, municipal solid waste, tires, and auto shredder residue, although test programs are continuing.

Chapter 19: Biological Processes and Biofuels

Comprehensive examination of the several active processes describes anaerobic digestion (gaseous fuels), fermentation processes (liquid fuels), and composting (humus and solids), including design parameters, systems and equipment, feedstocks, products, and operations. Also included is a study of biomass as a fuel to produce electricity in the United States by region, biomass fuel type (wood, agricultural landfill, gas, and municipal solid waste), and total kilowatts produced from each type.

Chapter 20: Hazardous Waste: The Federal Regulator

The focus of this chapter is on the federal government as a regulator of hazardous waste from its point of generation through its transportation, treatment, storage, or disposal. To place the role of federal regulation in perspective, this chapter reviews hazardous waste management prior to the Resource Conservation and Recovery Act (RCRA), the reasons for federal involvement in hazardous waste management, the legislative history of RCRA, the structure and substance of RCRA as it regulates hazardous waste, and the present status of hazardous waste regulations. Also, the organization of the Environmental Protection Agency (EPA) for the management and enforcement of its hazardous waste regulations, the Comprehensive Environmental Response, Compensation and Liability Act of 1980 (hereinafter referred to as "The Superfund Act"), and the Superfund Act as it relates to RCRA, with recent and anticipated changes in federal regulation of hazardous waste.

Chapter 21: Hazardous Waste: State and Local Regulatory Issues

This chapter addresses issues involving state and local hazardous waste practices. It describes general trends in changing state regulatory programs. Also identified are programmatic issues that give the hazardous waste generator, transporter, and owner/operator of hazardous waste treatment, storage, and disposal facilities an insight into the major changes that have been made in state hazardous waste management programs. It discusses issues of vital interest to local government and officials such as facility siting, economic impact on industry, the exempted (small) generator, closed or abandoned sites and orphaned wastes, and emergency response with contingency plans.

CHAPTER 2

LEGISLATION AND INVOLVED AGENCIES

WILLIAM L. KOVACS

Jaeckle, Fleischmann & Mugel
Washington, D.C.

2.1 HISTORY OF SOLID WASTE MANAGEMENT LAWS

2.1.1 Prior to Resource Conservation and Recovery Act

Prior to the enactment of the Resource Conservation and Recovery Act (RCRA) in September of 1976, solid waste management was primarily a local problem and few states had enacted any type of solid waste law.[1] Although prior to RCRA there were several federal statutes concerning solid waste, none of the statutes mandated or offered any guidance as to how solid waste problems could be managed. In particular, the Solid Waste Disposal Act, Pub. L. 89-272, 89th Cong. 2d. Sess. (1965) as amended by the Resource Recovery Act of 1970, Pub. L. 91-512 (1970) provided technical and financial assistance to state and local governments in the planning and development of resource recovery and other solid waste disposal technologies. Such legislation affirmed reliance on local and state action for coping with the solid waste problem.[2] Furthermore, the Solid Waste Disposal Act did not concern itself with hazardous waste of any type.

In addition, Section 208 of the Federal Water Pollution Control Act, Pub. L. 92-500, 92d Cong. 1st Sess. (1972), affected solid waste planning activities by authorizing, as part of water pollution control activities, consideration of the disposal on land of pollutants that affected water quality.

By 1975 the incentives provided under the Solid Waste Disposal Act and the Water Pollution Control Act encouraged 48 states to adopt some form of solid waste management program.[3] However, such state programs exhibited little uniformity. They were staffed by as few as one person or as many as 62 persons, and operated with budgets ranging from zero dollars to 1.2 million dollars. The typical program was staffed and budgeted closer to the lower end of the range.[4]

Notwithstanding the lack of attention being paid to solid waste management at all levels of government, the 94th Congress was approached by municipalities, as well as environmental, public interest, and industry groups, about the lack of landfill capacity and its impact on municipal government, environmental protection, and industrial expansion. The problem was not viewed as an immediate threat to public health and safety, but rather as a future threat to the health of the nation. Congressman Fred B. Rooney of Pennsylvania, Chairman of the Subcommittee on Transportation and Commerce, the panel with jurisdiction over solid and hazardous waste management, began several weeks of hearings on the problem and developed a compilation of facts that formed the underlying justification for enacting RCRA.[5]

The initial findings were startling to a nation that viewed solid waste as a local problem. For the first time, Congress and the nation became aware that approximately 3–4 billion tons of material

The author was Chief Counsel of U.S. House of Representatives Subcommittee on Transportation and Commerce during the enactment of the Resource Conservation and Recovery Act. This Subcommittee has primary jurisdiction over solid and hazardous waste. The author presently practices law in Washington, D.C. with Jaeckle, Fleischmann & Mugel and is Chairman of the Virginia Hazardous Waste Facilities Siting Board and past Chairman of the American Bar Association Energy Resources Law Committee, Section Torts and Insurance Practice.

were discarded each year, of which approximately 135 million tons were municipal waste. Furthermore, it was learned that the cost of disposing of the municipal waste ranged between four to six billion dollars annually and that the amount of solid waste being disposed was so great that 48 of the larger cities in the United States would run out of landfill capacity by 1982.[6]

Also, for the first time, numerous reports were coming to the attention of the Congress and the nation concerning the adverse health impacts of the improper disposal of hazardous waste.[7] Drawing upon experience with other environmental acts—for example, the Water Pollution Control Act, where money was being expended after the nation's waters were already polluted—the Transportation and Commerce Subcommittee decided to act to avoid the potential problems posed by shrinking landfill capacity and the improper disposal of waste rather than attempting to remedy the problems after the harm had been done.[8]

Perhaps the most unique aspect of RCRA was the manner in which it was enacted. Because the Senate and the House had entirely different versions of the legislation and there was only one week left in the 94th Congress, a conference between the House and Senate was impossible. Therefore, Congressman Fred Rooney, Chairman of the House Subcommittee that reported the bill, and Senator Jennings Randolph, Chairman of the Senate Committee with jurisdiction over RCRA, requested that their staffs meet to determine whether or not a compromise bill could be agreed upon before the Congress adjourned. The staffs of the two committees were able to agree upon a compromise version of the House bill. The primary provisions of the substitute bill, those relating to solid and hazardous waste management and the procurement of recovered materials, were almost identical to the provisions contained in H.R. 14496, the original Rooney bill.[9]

When the House of Representatives brought H.R. 14496 to the floor, Congressman Rooney moved to insert the agreed upon substitute in lieu of the provisions contained in H.R. 14496. By inserting and passing the agreed upon substitute in lieu of H.R. 14496, the House was able to send to the Senate legislation that it could agree to without a conference committee. The substitute was approved by both the House and Senate with only hours remaining in the 94th Congress, and it was transmitted to President Ford for his signature.[10] President Ford signed RCRA into law October 21, 1976.

Because RCRA was enacted without ever going to conference, its legislative history is incomplete and somewhat difficult to understand. Therefore, if one is looking for the intent of Congress concerning a particular provision of RCRA, one must first determine if the provision in question originated in the House or the Senate and then look to the respective House and Senate bills and committee reports to determine its meaning. For some of the compromise provisions of RCRA there is no legislative history.[11]

2.1.2   Focus of RCRA

The House version of RCRA, which provided the basis for the final compromise, viewed the solid waste problem as multifaceted whereas the Senate took a strictly regulatory approach. The House reasoned that the free enterprise system should play a large role in solid waste management through the production and procurement of items composed of recovered materials. In addition, as the real costs of land disposal began to rise, it projected there would be incentives on the part of industry either to reduce solid waste generated or to find ways to recycle it or use it for energy generation.

Simply put, the premise of RCRA is that the solid waste problem is a result of our industrial society. As such, the true costs of disposal should be borne by those benefitting from the products that generate the waste. Therefore, once the subsidy of the cheap landfill is removed, other more advanced technologies will develop.[12]

The three main objectives that RCRA addresses are: (1) hazardous waste management; (2) solid waste management; and (3) the procurement of materials made from recovered wastes. RCRA was Congress' first attempt in an environmental statute to have the free market mechanism work for environmental protection. Such a mechanism would work by mandating certain standards for disposal of solid and hazardous waste that would protect public health and safety, thereby requiring those benefitting from the functions that create the waste to pay the cost of its disposal. In effect the new RCRA standards, when implemented, would incorporate all costs (health and safety) into the cost of disposal and not just the cost of the land which was the only cost associated with land disposal at that time. Further, as the cost of land disposal increased there would be an incentive to provide other more environmentally protective technologies.

On the other hand, Congress mandated procurement of items containing recovered materials as a means of encouraging industry to utilize recyclable material in the products sold to the government. By so doing, certain industries would have an advantage over others that do not utilize recyclable materials in their products manufactured for sale to the government.

2.2   HAZARDOUS WASTE ASPECTS OF RCRA

The Congress stated that its overriding concern was the effect of the disposal of discarded hazardous waste on the population and the environment.[13] In addressing this problem, RCRA required the

Environmental Protection Agency (EPA) to promulgate standards for persons who generate and transport hazardous waste, as well as those who own and operate hazardous waste treatment, storage, or disposal facilities.[14]

Although EPA is required to promulgate these regulations concerning all aspects of hazardous waste management, the states retain the authority to establish a state program to administer and enforce those regulations in lieu of a federal program, provided the state program meets federal minimum standards.[15] The state standards can be more stringent than the federal standards, however.

If a state does not receive EPA approval for its hazardous waste program, it is EPA's responsibility to implement and enforce the federal regulations in that state. Once a state's hazardous waste program is approved, however, EPA cannot take legal action against a violator of the federal regulations without giving notice to the state in which the violation occurs.[16] EPA can disapprove a state program only after an opportunity for public hearing and a finding: *"that (1) such State program is not equivalent to the Federal program; (2) such program is not consistent with the Federal and State programs applicable in other States; or (3) such program does not provide adequate enforcement."*[17]

Prior to any state receiving authority to administer its hazardous waste plan, EPA is required to:

2.2.1 Identify What Wastes Are Hazardous

The primary function that makes all of the hazardous waste provisions operative is the development by EPA of criteria for identifying hazardous waste. In making such identification, EPA is required to take into account the toxicity, persistence in nature and degradability, and potential for accumulation of each waste in living tissue. After developing the criteria for identifying such hazardous characteristics, EPA is required to promulgate a list of those wastes that: *"(A) cause, or significantly contribute to an increase in mortality or an increase in serious irreversible, or incapacitating reversible, illness; or (B) pose a substantial present or potential hazard to human health or the environment when improperly treated, stored, transported, disposed of, or otherwise managed."*[18]

In determining which wastes are hazardous and therefore subject to regulation, the only standard to be applied is the protection of human health and the environment. Cost/benefit is not a factor to be considered in the regulation of hazardous waste.[19]

2.2.2 Requirements upon Those Who Generate, Treat, Store, or Dispose of Hazardous Waste

RCRA does not in any way limit the generation of hazardous waste that the House committee termed to be a direct result of national industrial production and a product of the American lifestyle. However, with regard to those who generate, treat, store, or dispose of hazardous waste, RCRA requires cradle-to-grave regulation. For those who generate hazardous waste, RCRA requires that EPA establish requirements respecting recordkeeping practices, labeling practices, use of appropriate containers, information on the general chemical composition, use of a manifest system, submission of reports to EPA or the appropriate state agency setting out the quantities of hazardous waste generated, and the ultimate disposition of all hazardous waste reported.

With regard to those who transport hazardous waste, EPA is required to promulgate regulations that include, but need not be limited to, requirements respecting recordkeeping, source and delivery points, proper labeling, compliance with the manifest system, and transportation of all such hazardous waste only to shipper-designated facilities. Simply put, only labeled wastes may be transported under the act pursuant to a manifest system which requires compliance by the generator, the transporter, and the storage, treatment, or disposal facility.

In effect, the manifest system is a trip-ticket system. The shipper (generator) must initiate the trip ticket when the waste is entrusted to the transporter. The transporter, in turn, must note any transferance of the waste to another carrier or to the disposal facility designated as the recipient by the shipper. Under the act, it is unlawful for the transporter of hazardous waste to unload the waste anywhere but at the facility designated by the shipper. The disposal site must be one authorized to handle hazardous waste.

EPA's most comprehensive regulatory authority, however, concerns the owners and operators of hazardous waste treatment, storage, and disposal facilities. RCRA requires EPA to promulgate regulations establishing performance standards that shall include, but need not be limited to, requirements respecting (1) maintaining records of all hazardous wastes treated, stored, or disposed, and the manner in which such wastes are treated, stored, or disposed; (2) reporting, monitoring, and inspection and compliance with the manifest system; (3) operating methods, techniques, and practices; (4) location, design, and construction; (5) contingency plans to minimize anticipated damage; and (6) qualifications as to ownership, continuity of operation, training for personnel, and financial responsibility. In addition, the act requires that owners of hazardous waste management facilities "provide assurances of financial responsibility and continuity of operation consistent with the degree and duration of risks."[20]

2.2.3 Additional RCRA Standards, Landfill Prohibition

In addition to the standards imposed by the initial enactment of RCRA, Congress enacted the Hazardous and Solid Waste Amendments of 1984, Pub.L. 98-616 (November 8, 1984) which substantially strengthened the EPA's ability to protect health and the environment. Because of the scope of such amendments, certain key provisions of Pub.L. 98-616 should be noted.

Section 3001 of RCRA is amended to require EPA to issue by March 31, 1986 standards for handling wastes from generators of between 100 kilograms and 1000 kilograms per month. Prior to this amendment, generators of less than 1000 kilograms per month were exempt from regulation. EPA however, has the discretion to impose on such small quantity generators standards that are less stringent than the standards imposed on those who generate over 1000 kilograms of hazardous waste per month.

Section 3004 of RCRA is amended to prohibit within six months after the enactment of Pub.L. 98-616 the placement of bulk or non-containerized liquid hazardous waste in any landfill. Further, within 15 months after the enactment of Pub.L. 98-616 the Administrator is required to promulgate regulations: (1) which minimize the disposal of containerized liquid hazardous waste in landfills; and (2) minimize the presence of free liquids in containerized hazardous waste to be disposed of in landfills.

In addition, the new amendments to Section 3004 prohibit within 12 months after the enactment of Pub.L. 98-616 the disposal of non-hazardous liquid waste in any hazardous waste landfill unless such landfill demonstrates it is the only reasonably available alternative and its placement does not pass a present risk of contamination of any underground source of drinking water.

The new amendments to Section 3004 also ban the land disposal of certain enumerated wastes unless EPA, within specific timeframes, issues regulations that find that prohibiting certain methods of land disposal is not required to protect health and the environment. If EPA does not issue its regulations within the required timeframes, then the land disposal of such enumerated wastes is prohibited.

Section 3004 was also amended to provide that each new landfill or surface impoundment or a new landfill or surface impoundment unit at an existing facility and each replacement, lateral expansion of an existing landfill or surface impoundment is required to have at least two liners and a leachate collection system and ground water monitoring system.

Section 3004 was also amended to require the Administrator to regulate the owners and operators of facilities that produce a fuel from hazardous waste; facilities that burn hazardous waste as a fuel and persons who market any fuel produced from hazardous waste. Further, the amendments require a warning lable on such fuel indicating it contains hazardous waste.

Section 3005 of RCRA was amended by Pub.L. 98-616 to require the Administrator to process all land disposal permits within four years, all permits for incinerators within five years and all other permits within eight years. Further, the length of a permit is limited to ten years and the Administrator must review all land disposal permits no later than five years after issuance.

With regard to land disposal facilities operating under interim status, their interim status terminates within 12 months after the enactment of the amendments contained in Pub.L. 98-616 unless the facility applies for a final permit and certifies that such facility is in compliance with all applicable groundwater monitoring and financial responsibility requirements.

Section 3005 is also amended to prohibit within four years existing surface impoundments from receiving, storing or treating any listed hazardous waste unless such impoundment meets the same technical standards (double lines, groundwater monitoring and leak detection) as applied to new surface impoundments.

Additional requirements imposed by Pub.L. 98-616: (1) require the Administrator to issue regulations concerning leakage from underground storage tanks and for the owners of such tanks to notify state agencies of such tanks; (2) prohibit the export of hazardous waste unless the exporter files a notice of what is to be exported and the receiving nation consents to accept such waste; (3) require that after September 1, 1985 all manifests must contain a generator certification that the volume and toxicity of the waste have been reduced to the maximum degree economically practicable and the management of the waste minimizes risk to the extent practicable; (4) ban the underground injection of hazardous waste into or above any formation within a quarter mile of a well that is a source of drinking water; (5) authorize the Administrator to issue permits for experimental facilities for one year without first issuing standards for permits for such experimental facility.

2.2.4 Enforcement of RCRA Standards

Enforcement of the hazardous waste regulations rests with the states, if they obtain EPA approval. EPA's enforcement authority in a state only occurs if the state does not seek such approval or fails to administer its program adequately. In states with their own hazardous waste programs, those state requirements may be more stringent than the federal regulations.

EPA's primary enforcement tools are the compliance order and the civil suit. Civil penalties may be as high as $25,000 per day for each day of noncompliance with an EPA order. RCRA also provides for criminal penalties for specific acts. Under the act, any person who knowingly transports any hazardous waste to a facility that does not have a permit or disposes of any hazardous waste without having obtained a permit or in violation of any material condition or requirement of a permit shall be subject upon conviction to a fine of $50,000 for each day of violation, imprisonment for not more than 2 years, or both. Furthermore, if an individual or other entity makes any false statement or representation in any application, label, manifest, record, report, permit, or other document, that person shall upon conviction be subject to a fine of not more than $25,000 for each day of violation, or to imprisonment, not to exceed 1 year, or both. For subsequent offenders, the penalties double to $50,000 and 2 years imprisonment.[21]

In situations where the violator is convicted of having knowingly violated RCRA provisions in circumstances that manifest an unjustified and inexcusable disregard for human life or an extreme indifference to human life, the punishment may be as high as $250,000 and not more than 2 years in jail, or both. A defendant that is an organization can be fined up to $1,000,000.[22]

In addition to federal or state enforcement of RCRA, Section 7002 of RCRA provides that any citizen may bring a lawsuit against: (1) any person or government agency for failure to comply with any effective regulation, standard, permit, condition, or requirement; (2) the Administrator of EPA for failure to perform any act or duty under the act that is not discretionary; and (3) any past or present generator, transporter, or owner or operator of a treatment, storage or disposal facility, who has contributed, or who is contributing, to the past or present handling, storage, treatment, transportation, or disposal of any solid or hazardous waste that may present an imminent and substantial endangerment to health or the environment. Prior to initiating such lawsuit, the citizen must give notice to EPA, to the state in which the violation occurs, and to the alleged violator.[23]

2.2.5 Financial Assistance

Congresssional support for the hazardous waste program developed and implemented by the States has increased from a $25,000,000 authorization in fiscal year 1978 to a $60,000,000 authorization for each of the fiscal years 1985 through 1988. In allocating appropriated funds the Administrator, after consultation with the states, is to take into account the extent to which hazardous waste is generated, transported, treated, stored, or disposed of within each state and its exposure to human beings.

2.3 SOLID WASTE GUIDELINES AND PLANNING PROCESS

2.3.1 Goals of Solid Waste Plans

The RCRA provisions relating to municipal nonhazardous waste were designed to assist and encourage environmentally sound methods of waste disposal, to maximize utilization of resources recovered from waste, and to encourage resource conservation. These goals were to be achieved through federal financial and technical assistance to state and local jurisdictions for the development of solid waste management plans that fostered cooperation among federal, state, and local governments and private industry.[24]

As an incentive to state, local, and interstate governments to develop solid waste management plans, RCRA provided technical assistance through federal personnel with expertise in the technical, marketing, financial, and legal aspects of solid waste management. Financial assistance was available in the form of grants for the development and implementation of solid waste management plans.[25]

2.3.2 Technical Assistance

In addition to EPA making available technical expertise, it was required to develop a central library for the collection of solid waste management information and to publish specific guidelines describing available solid waste management practices. Congressional intent behind requiring the publication of guidelines and not regulations for the management of solid waste was to provide assistance to state and local authorities while avoiding any attempt to preempt state or local initiatives.

In fulfilling this responsibility, RCRA requires that the guidelines be published after consultation with appropriate federal, state, and local authorities. The guidelines are to assist state and local government in identifying and establishing regions for solving common solid waste problems, develop state solid waste management plans, close open dumps and prohibit the establishment of new open dumps, and prevent surface and groundwater contamination.[26]

In addition to the issuance of guidelines, EPA is also required to establish Resource Conservation and Recovery Panels, each consisting of four persons skilled in the technical, marketing, financial, and legal aspects of solid waste management.[27] Such panels are to be available upon request to state and local governments. The purpose of these panels is to make available to governmental units ". . . expertise unbiased by the profit motive," and to advise such governmental units on the technical and institutional difficulties with various resource recovery systems. Also, such panels are to apprise governmental units of alternative methods of resource recovery and conservation and to assist the local government at all stages of the planning process, including providing information on whether such governmental unit should even construct a particular resource recovery facility or whether it would be better served by alternative processes.[28]

In addition to the EPA assistance panels, EPA is to publish guidelines concerning alternative types of resource recovery and conservation systems, EPA's evaluation of such systems, and available new and additional markets for recovered materials.[29]

Congress considered the importance of such panels to the planning process so great that it mandated that 20% of EPA's administrative funds be expended for their operation.[30]

2.3.3 Financial Assistance

Federal financial assistance of $45,000,000, $55,000,000, $30,000,000, $25,000,000, and $30,000,000, respectively, was authorized in the form of grants for fiscal years 1978, 1979, 1980, 1981, and 1982 to state and local jurisdictions for the development of solid waste plans. Although no funds were authorized for fiscal years 1983 and 1984, Congress authorized $20,000,000 for each of the fiscal years 1985 through 1988 for the implementation of approved solid waste management plans. In addition, $15,000,000 was authorized for fiscal year 1985 and $20,000,000 was authorized for each of the fiscal years 1986 through 1988 for grants to States, and where appropriate state, local or interstate authorities, to implement permit programs for solid waste facilities to assure that if such facilities receive hazardous household waste as a result of the small generators provision, (Section 3001(d) of RCRA) that such facilities shall comply with EPA's future criteria for sanitary landfills published under sections 1008 and 4004 of RCRA. The funds were authorized to the states for most aspects of development and implementation of state plans. However, such funds could not be used for construction of a solid waste facility, the acquisition of land, or any subsidy for the price of the recovered materials. These planning and implementation funds were allotted to the states according to the population of each state. Upon receipt of federal funds, the state was required to reallocate appropriate amounts to the local or regional entities within the state in proportion to their responsibilities under the solid waste management plan.[31]

Another form of financial assistance authorized under RCRA was Rural Communities Assistance. This authorization was limited to municipalities with a population of 5000 or less, or counties with a population of 10,000 or less, or less than 20 persons per square mile and not within a metropolitan area. Allotment of these funds is "on the basis of the average of the ratio which the population of rural areas of each State bears to the total population of rural areas of all the States".[32]

It was the congressional intent that the development and implementation grants authorized by RCRA were to be supplemental to state solid waste budgets and that state solid waste expenditures could not be reduced below the state solid waste expenditures for 1975. A state could be denied federal financial assistance if its expenditures dropped below the 1975 level.[33]

To obtain federal technical or financial assistance under the act, a state is required to: identify regions within such state or between states, determine the state or local agencies that are to develop and implement the solid waste plan for the identified region, and develop and implement a solid waste management plan approved by the administrator.[34] The information that EPA collects and makes available and the published guidelines relating to solid waste management are the sources that can be utilized by state and local officials to meet the requirements of the act. Although each of the requirements offers the state and local participants wide flexibility, each requirement must be met if the state is to receive federal financial assistance.[35]

2.3.4 Minimum Standards for State Solid Waste Plans

For its solid waste management plan to be approved, the state seeking approval must:

Identify the responsibilities for all state and local entities under the plan.

Prohibit the open dumping of solid waste.

Permit all local governmental entities to enter into long-term contracts for the supply of solid waste to resource recovery facilities.

Provide that all solid waste be disposed of by resource conservation, resource recovery, or in sanitary landfills.

Provide that in determining the size of a waste-to-energy facility, adequate provision shall be given to the present and reasonably anticipated future needs of the recycling and resource recovery interests within the area encompassed by the planning process.[36]

One of the effects of RCRA's prohibition on open dumps is to raise the cost of landfills to a point where resource recovery and recycling facilities would be competitive.[37] Each state is required to develop its own plan using resource conservation or recovery methods best applicable to its situation. Therefore, each state can utilize a different approach to solid waste planning, or a combination of approaches.

Additionally, each state must adopt and implement a permit program to insure that solid waste management facilities receiving hazardous waste due to the provisions of section 3001(d) of RCRA for small generators will comply with the Administrators revised criteria for sanitary landfills. If a state does not develop and implement a permit program, then the Administrator may use the enforcement authorities provided him under the hazardous waste provisions of RCRA.[38]

During the implementation of state solid waste management plans, each state must permit public participation in the development of the plan. Also, EPA must provide for and encourage public participation prior to the approval of any state solid waste management plan.[39]

2.3.5 Compliance of Solid Waste Plans

Compliance of the state plan with federal guidelines can be enforced in four ways. First, EPA can withhold federal financial or technical assistance to a state if the state is not implementing or enforcing its plan.[40]

Second, any citizen can bring suit against any person or government instrumentality—including a state, local, or regional authority alleged to be in violation of any requirements of the act, or the plan—or against the administrator for failure to perform any duty that is not discretionary under the act.[41]

A third method of enforcement permits EPA to seek injunctive relief whenever the disposal of solid waste presents an imminent and substantial endangerment to health or the environment.[42] EPA is limited in its enforcement of the municipal solid waste aspects of RCRA because the congressional intent is that municipal solid waste can be best solved by local effort encouraged with federal incentives. However, EPA was recently authorized a fourth method of enforcement whereby it can use its enforcement authorities provided under the hazardous waste sections of RCRA in states that fail to implement permit programs for solid waste facilities receiving hazardous waste as a result of receiving hazardous household waste and the provisions relating to small generators.[43]

2.4 PROCUREMENT OF PRODUCTS CONTAINING RECOVERED MATERIALS

One of the prime goals of RCRA is to require each federal procuring agency with respect to any procurement in excess of $10,000 to procure items composed of the highest percentage of recovered materials practicable, consistent with maintaining a satisfactory level of competition. If a procuring agency decides not to procure the item consisting of the highest percentage of recovered goods practicable, its decision must be based on the determination that such procurement item is not reasonably available within a reasonable period of time, it failed to meet performance standards set forth in the applicable specifications, or that it was only available at unreasonable prices. Additionally, each procuring agency is required to develop an affirmative procurement program which will assure that items composed of recovered materials will be purchased to the maximum extent practicable. The affirmative procurement program shall contain a promotional and preference program for recovered materials; a program requiring estimates of the recovered materials utilized in the performance of a contract and reasonable verification for such estimates and monitoring of such affirmative procurement program.[44]

To achieve this goal of procuring items containing the highest percentage of recovered materials, the Congress made the provision mandatory upon each procuring agency as well as EPA, the Office of Procurement Policy (OPP), and the Department of Commerce (DOC).[45]

2.4.1 EPA's Role in Procurement of Recovered Materials

EPA is required to provide each federal agency with information on the availability, sources of supply, and potential uses of materials recovered from solid waste.[46] Each procuring agency is then required to procure materials composed of the highest percentage of recovered materials practicable.[47] It should be noted that the definition of recovered material only includes material recovered from solid waste and does not include energy recovered from solid waste.[48]

2.4.2 OPP's Role in Procuring Recovered Materials

RCRA mandates that OPP coordinate the various federal agencies to ensure that items composed of the highest percentage of recovered goods practicable are procured. Furthermore, the OPP is to coordinate all other policies for federal procurement in such a way as to maximize the use of recovered resources.[49]

2.4.3   Department of Commerce's Role in Procuring Recovered Materials

The responsibilities of the DOC relating to municipal solid waste are to:

Develop accurate specifications for recovered materials.

Stimulate and develop markets for recovered materials.

Evaluate and promote proven resource recovery technology, including waste-to-energy facilities as well as recycled facilities.

Establish a forum for the exchange of technical and economic data relating to resource recovery facilities.[50]

Accurate Specifications

DOC, through the National Bureau of Standards, is to develop specifications for the classification of materials recovered from waste. The specifications ". . . shall pertain to the physical and chemical properties and characteristics of such materials with regard to their use in replacing virgin materials in various industrial, commercial and governmental uses."[51] The purpose of this section is to establish a system whereby recovered materials can be substituted for virgin materials.[52]

Development of Markets

The DOC is required under Section 5003 of RCRA to:

Identify the geographical location of existing or potential markets for recovered materials.

Identify the economic and technical barriers to the use of recovered materials.

Encourage new uses for recovered materials.[53]

The Congress, in enacting section 5003 of RCRA, recognized that available and stable markets are the key to successful materials recovery. DOC was required to identify both the existing as well as potential markets for such materials.

Technology Promotion

Section 5004 of RCRA authorizes the DOC ". . . to evaluate the commercial feasibility of resource recovery facilities and to publish the results of such evaluation, and to develop a data base for purposes of assisting persons in choosing such a system."[54] The congressional intent of having DOC evaluate the technologies was to enable prospective purchasers, whether public or private, to purchase the respective systems on the basis of its merits rather than on ". . . the hope of obtaining support from a government agency perceived to have a special interest in the proliferation of the specific technology."[55]

Forums to Exchange Data

Another provision permits DOC to sponsor forums, conferences, and other educational arrangements whereby the technical and economic aspects of resource recovery can be discussed by government, industry, and the public. Participation in such forums does not relieve any participant of the restrictions of other laws, e.g., antitrust.[56]

2.5   OTHER FEDERAL STATUTES IMPACTING ON SOLID WASTE MANAGEMENT

In addition to RCRA, the Congress has enacted numerous other statutes that affect solid waste management, in particular the planning and construction of waste-to-energy and recycling facilities used for recovering materials and energy from solid waste. Listed below by agency is a brief description of some other pertinent statutes affecting solid waste management.

2.5.1   Department of Energy

Energy Security Act., Pub.L. 96-294

The primary goals of Title II of the Energy Security Act are to reduce the dependence of the United States on imported oil by:

1. All economically and environmentally sound means, including use of biomass energy.
2. The implementation of a national program for the increased production and use of biomass

that does not impair the nation's ability to produce food and fiber on a sustainable basis for domestic and export use.[57]

To achieve the objectives of the act, the Secretary of Energy, in consultation with EPA and DOC, was to develop, within 90 days of enactment, a comprehensive plan to carry out the goals of the municipal waste-to-energy portions of the act.[58]

To assist in carrying out the comprehensive plan, the Secretary of Energy was given the authority to implement the waste-to-energy provisions of the Energy Security Act by providing to waste-to-energy facilities:

1. Construction loans.[59]
2. Guaranteed construction loans.[60]
3. Price support loans and price guarantees.[61]
4. The establishment of an accelerated research, development, and demonstration program for promoting the commercial viability of the processes for the recovery of energy from municipal waste.[62]

In awarding the financial assistance in a manner necessary to implement the comprehensive waste-to-energy program, the Secretary of Energy is required to establish procedures concerning the review and evaluation of applications for such assistance.

Section 235 of the Energy Security Act limits the Secretary's discretion in the awarding of financial assistance under the act. The primary limitations, in addition to the development of a waste-to-energy program that does not impair the nation's ability to produce food and fiber on a sustainable basis for domestic and export use, are that:

1. The Secretary must assure that the necessary municipal waste feedstocks for the facility are available and will continue to be available for the expected economic life of the project.
2. The Secretary gives due consideration to promoting competition.
3. The Secretary is prohibited from giving financial assistance to any waste-to-energy facility unless he first determines that the project will be technologically and economically viable; that assurances are provided that the project will not use in any substantial quantities waste paper that would otherwise be recycled for use other than as a fuel; and that the facility will not substantially compete with facilities in existence on the date of the financial assistance that are engaged in the separation or recovery of reusable materials from municipal waste.[63]

Department of Energy Organization Act, Pub. L. 95-91

The Department of Energy Organization Act is the organic statute that created the department. In so doing, the Congress declared ". . . that the establishment of the Department of Energy is in the public interest and will promote the general welfare by assuring coordinated and effective administration of federal energy policy and programs."[64] In addition to the general goal of assuring a coordinated federal energy policy, Section 102 of the Department of Energy Organization Act enumerates the goals of the Department of Energy (DOE). Such goals are diverse and include, in addition to promoting the additional production of energy, the requirements that DOE:

1. Promote the maximum possible energy conservation measures.
2. Create a comprehensive energy conservation policy strategy that will receive highest priority.
3. Carry out a comprehensive and balanced research and development program.
4. Place major emphasis on the development and commercialization of recycling.
5. Ensure protection of environmental goals.
6. Ensure that the productive capacity of private enterprise is utilized in the development and achievements of the policies and purposes of the act.[65]

As a mechanism for achieving its goals, the DOE is required to develop a National Energy Plan that summarizes all research and development efforts to forestall energy shortages, reduce waste, foster recycling, encourage conservation, and protect the environment.[66] Furthermore, the plan is to review and appraise the adequacy and appropriateness of available technologies for the treatment of solid waste and develop strategies that will maximize private production and investment in significant supply sectors.

Department of Energy Act of 1978—Civilian Applications, Pub. L. 95-238

The Department of Energy Act of 1978 was the first enactment to give the DOE general authority to award loan guarantees, grants, contracts, and price supports to municipal waste reprocessing demonstration facilities.[67] This act amended the Federal Non-Nuclear Energy Research and Development Act of 1974.

Section 19 of the act authorized up to $300,000,000 in loan guarantees for the construction and start-up of demonstration facilities for the conversion of various domestic resources (coal, shale, and biomass) into fuel.[68]

Section 20 of the Act is directly related to municipal waste reprocessing demonstration facilities. Section 20 authorized $20,000,000 to the DOE to make grants, contracts, and price support payments, or any combination thereof, to such demonstration facilities.[69]

Furthermore, the Non-Nuclear Energy Research and Development Act of 1974 was amended by the Energy Security Act which requires the Secretary of Energy to establish an accelerated research, development, and demonstration program for promoting the commercial viability of processes for the recovery of energy from municipal waste.[70] As part of such an accelerated program, the Secretary is to assess, evaluate, demonstrate, and improve the performance of existing municipal waste-to-energy technologies with respect to the capital costs, operating costs, and maintenance costs. However, the purpose of the program under the Federal Non-Nuclear Research and Development Act of 1974 is solely limited to facilities for demonstration purposes.

2.5.2 National Energy Conservation Policy Act, Pub. L. 95-619

The goal of the DOE under the National Energy Conservation Policy Act is to increase the use of recovered or recycled materials in industrial operations rather than the production of energy from municipal solid waste.[71] In enacting the National Energy Policy Conservation Act, the Congress found that:

1. Significant amounts of energy can be conserved if recovered or recycled materials are utilized as part of an industry's manufacturing operations.

2. Substantial additional volumes of industrial energy and other scarce natural resources will be conserved in future years if such major energy consuming industries increase to the maximum feasible extent utilization of recovered materials in their manufacturing operations.

3. The recovery and utilization of such recovered materials can substantially reduce the dependence of the United States on foreign natural resources and reduce the growing deficit in its balance of payments.

To achieve the above objectives, the Congress required the Secretary of Energy to estabish targets for the use of recovered materials for the metals, paper, textile, and rubber industries.[72] The targets to be achieved by the respective industries are to be set in terms of the percentage of recovered materials compared with the total amount of materials utilized by the respective industry for its manufacturing operation. In achieving the targets for increased utilization of recovered materials, the DOE is to establish procedures and create incentives whereby such industries may cooperate with the federal government in the establishment and achievement of such targets. In establishing such a cooperative arrangement, the DOE is to provide incentives for increased industrial utilization of energy-saving recovered material in its industrial operation.

2.5.3 Federal Energy Regulatory Commission and PURPA

Public Utility Regulatory Policies Act of 1978 (PURPA) provides incentives for cogeneration facilities and small power producers by requiring the Federal Energy Regulatory Commission (FERC) to issue rules requiring electric utilities to purchase electricity from qualified cogenerators and small power facilities.[73] To qualify, a facility must: (1) produce 80 megawatts or less of power; (2) at least one-half of the annual fuel used in the facility must be biomass, waste, renewable resources, or any combination thereof; (3) the total annual input of oil, natural gas, and coal may not exceed 25% in BTU value; and (4) the equity ownership of the small power producer by the utility may not exceed 50%.[74]

For these facilities that qualify as small energy producers, PURPA provides exemption from some federal and state laws and regulations including those relating to financial arrangements and the Federal Power Act. More importantly, however, PURPA requires that electric utilities interconnect with a qualified small energy producer and that the electric utility purchase the electricity produced by the small power producer. Essentially, PURPA requires FERC, through regulation, to guarantee a market for the electricity generated by cogenerators or small power producers.[75]

2.5.4 Internal Revenue Service (IRS)

Statutory Mandate

Federal income tax laws may be used to bring about, at least ostensibly, certain desired behavior through tax incentives or disincentives. However, every time Congress confers a tax incentive (or for that matter fails to give an incentive or enacts a disincentive) in the pursuit of one objective, the accomplishment of another objective is often made more difficult.

The present tax laws and implementing tax regulations offer some benefits to recyclers as well as to waste-to-energy facilities. However, the intricacy of the Internal Revenue Service (IRS) regulations may make it difficult for recyclers to chart a separate path from those involved in waste-to-energy without risking some of their own tax benefits.

Three sections of the Internal Revenue Code concern access to, and use of, the municipal solid waste stream. Those sections are the IRS definition of solid waste, Section 103 of the IRS Code concerning industrial development bonds, and the recycling tax credit established by the Energy Tax Act, Pub. L. 95-618.

Definition of Solid Waste

The IRS regulations adopt the outdated definition of solid waste contained in the Solid Waste Disposal Act of 1965, and its only modification is that the IRS further requires that the discarded material be useless, unwanted, and have no market or other value at the place where it is located. Therefore, if a person is willing to purchase such property at any price, such material is not solid waste within the IRS definition.[76]

Section 103 of Internal Revenue Code, Industrial Development Bonds

Section 103 of the Internal Revenue Code permits private corporations to receive the benefits of tax-exempt industrial development bonds issued by political subdivisions if such bonds are issued to finance the construction of solid waste disposal facilities, even if such facilities are for the use of a private corporation. The proceeds from the bonds can be applied to all aspects of the waste disposal function of the facility, including the processing of the waste in order to put it into the form in which it is salable.

However, revenues from the bonds cannot be applied to further processing or transportation of the waste after it is put in salable form. Therefore, once the solid waste facility converts the waste into a form that has value, it is no longer considered solid waste.[77] Another limitation in the use of industrial development financing is that "substantially all" of the proceeds from the bonds are to be used to provide for solid waste disposal processing capacity, including resource recovery. The IRS interprets "substantially all" to mean 90% of the proceeds must be applied to the solid waste capacity of the facility. However, the IRS only requires that 65% of the materials processed by the facility qualify as "solid waste" under the IRS definition of the term. Implicit in the IRS requirement is that 35% of the materials processed by a solid waste facility financed through industrial development bonds can be materials that have value.

Recycling Credit, Established by Energy Tax Act, Pub. L. 95-618

The Energy Tax Act provides a credit for "recycling equipment," which is defined as equipment used exclusively to sort and prepare solid waste for recycling, or is used in the recycling of solid waste. The recycling credit is applicable to any equipment used in the recycling of material, including some virgin material, provided the amount of such virgin material is 10% or less of the material processed. The recycling credit enacted in Pub. L. 95-618 expired on December 31, 1982.

2.6 IMPLEMENTATION OF RCRA

2.6.1 EPA's Implementation of RCRA

EPA's implementation of RCRA can only be described as tardy, fragmented, at times nonexistent, and consistently inconsistent. Furthermore, EPA has isolated the statutory structure of RCRA by administering the Subtitle C hazardous waste program as though it is not related in any manner to the Subtitle D solid waste program. Such a strict separation, however, is artificial at least with regard to landfills and the small generators exception, which permits generators until 1986 to dispose up to one ton a month of hazardous waste into a state approved municipal or industrial landfill rather than requiring its transportation to a hazardous waste landfill. By March 31, 1986, EPA must promulgate standards for waste generated in quantities greater than 100 and less than 1000 kilograms per month. Such standards may vary from the other hazardous waste regulations but must still protect health and the environment. In addition, the artificial separation fails to consider the mixtures of waste that occur at a municipal solid waste landfill.

The artificial separation between the two programs is important because presently the EPA is devoting little attention to the management of solid waste and landfills receiving municipal solid waste. It has been placing all personnel into the management of its hazardous waste program.

2.6.2 The EPA Solid Waste Program

With regard to the implementation of its solid waste program, EPA, within the first year of the enactment of RCRA, issued its guidelines for the "Identification of Regions and Agencies for Solid Waste Management."[78] However, its guidelines for the development of "Solid Waste Management Plans" and the "Criteria for the Classification of Solid Waste Disposal Facilities and Practices" did not come out until almost 3 years after the enactment of RCRA and almost a year and a half after the mandated congressional timetable.[79] As soon as the above guidelines were issued, litigation ensued between industry and EPA.[80]

The industry groups alleged that the guidelines failed to give owners and operators of facilities classified as open dumps notice and opportunity for comment before their facility was classified as an open dump under the state planning program and that EPA failed to recognize that facilities classified as open dumps are entitled to receive the protection of state-issued compliance schedules for the closure of open dumps.[81] EPA and the industry groups obtained court approval to defer briefing indefinitely in an attempt to obtain a negotiated settlement of their differences on the content of the regulations.[82]

After 1 year of negotiations, EPA and the industry groups arrived at a negotiated settlement that received court approval. The basis of the settlement was that EPA would amend its regulations to require that a state provide opportunity for public participation in determination of whether a facility is classified as an open dump. This participation included written notice to the owner/operator of the facility classified as an open dump 30 days prior to the submission of such classification to the federal government. In addition, EPA permitted states to seek partial EPA approval of the compliance schedules of their state plans to provide for facilities classified as open dumps the protections contained in the compliance schedules concerning the upgrade or closures of such facilities.

EPA published the proposed settlement in the *Federal Register* for comment and issued final regulations that were substantially the same as those agreed upon with the industry groups.[83]

In addition, the states, using federal funds allotted to them pursuant to Section 4007 of RCRA, began developing subtitle D solid waste plans. Presently, 22 state solid waste plans have received EPA approval; an additional 14 state solid waste plans have been adopted and submitted to EPA for approval; and another 12 draft plans have been submitted to EPA.[84]

2.6.3 The EPA

EPA did not issue its first set of regulations concerning the management of hazardous waste until February of 1980.[85] At that time, EPA promulgated its general regulations for hazardous waste management, regulations identifying hazardous waste, and regulations concerning hazardous waste generators and transporters as well as interim status standards for owners and operators of hazardous waste facilities and regulations for owners and operators of permitted hazardous waste facilities.[86] In actuality, EPA's 1980 rules accounted for all of its regulations except for the technical aspects of hazardous waste management, which would include surface impoundments, land treatment, waste piles, landfills, incinerators, and standards for new hazardous waste land disposal facilities. The later regulations were not promulgated until 1982,[87] with the regulations becoming effective in January 1983.[88]

Similar to the development of the solid waste regulations, numerous industry groups have brought suit against EPA challenging the validity of many of their hazardous waste regulations. In particular, in *In Re: Landfill Disposal Regulation Litigation v. EPA*, U.S. Court of Appeals for the District of Columbia Circuit, Docket No. 82-2205, and consolidated cases, industry is challenging EPA's most recent regulation concerning the disposal of hazardous waste on land.

Industry and EPA are attempting the negotiate a compromise of their differences on EPA's regulations concerning hazardous waste management. Such negotiations are similar to those undertaken subsequent to the industry challenge to EPA regulations concerning dumps (see Section III.A.1).

Notwithstanding EPA's delay in issuing its hazardous waste regulations, 37 states have already received authorization to implement phase I of the hazardous waste regulations and 10 states have received authority to implement phase II of the hazardous waste regulations.[89]

2.6.4 The Procurement of Recovered Materials

The RCRA provisions concerning the procurement of items containing recovered materials were complex, requiring EPA, DOC, OPP, and all other federal and state agencies utilizing federal funds to participate. The DOC would help develop standards permitting that items containing recovered materials could be substituted for items containing virgin materials. The EPA was to prepare guidelines for the use of procuring agencies in complying with the procurement requirements of Section 6002 of RCRA. The administrator of the OPP was to coordinate the procurement policy among all federal procuring agencies so as to ensure implementation of the policy.

All three federal agencies failed to issue any guidelines concerning the procurement of items containing the maximum amount of recovered materials practicable. The net effect is the same as if the procurement provisions of RCRA were never enacted. Furthermore, the private sector has not attempted to utilize the procurement provisions of RCRA that require the procuring agency to purchase items containing the maximum amount of recovered materials. Any person having a product comprised of such recovered materials can offer it as a substitute for a material composed of virgin material, and if it meets the statutory criteria, then the agency must procure it or make a specific determination as to its failure to meet such criteria. Oddly enough, the Section 6002 provisions have not been utilized.

In addition, the DOC only exercised its discretion to evaluate and promote proven solid waste, resource recovery or recycling technology on one occasion.[90] However, DOC was very active in the study and development of markets for recovered paper.[91] Presently, the DOC program concerning the development of accurate specifications for recovered materials is terminated.

2.7 DOE'S IMPACT ON SOLID WASTE MANAGEMENT

While EPA and the states were struggling with the implementation of the solid and hazardous waste programs, the Congress enacted the Energy Security Act, which authorized over $200,000,000 in the form of price supports and loan guarantees for the development of waste-to-energy systems to produce energy from municipal waste. To achieve this goal, the DOE was required to develop a comprehensive waste-to-energy plan detailing the energy benefit of new waste-to-energy facilities as well as the structure that would assist in achieving its goal of developing additional energy from solid waste.

The DOE internally issued several versions of its Comprehensive Waste-To-Energy Plan; however, it never published such a plan or submitted it to Congress.[92]

In addition, although DOE issued a Notice of Proposed Rulemaking pursuant to the Energy Security Act concerning its Price Support regulations for municipal waste energy facilities, such regulations were short lived. They were issued on October 29, 1980, while Jimmy Carter was President, and withdrawn on June 14, 1982. Furthermore, President Reagan's first budget reduced the funds available for loan guarantees and price supports for waste-to-energy facilities to almost zero.[93]

In short, notwithstanding the promise of federal funds to build an entire waste-to-energy industry that would create additional available energy for the nation while taking care of society's solid waste problem, it did little, if anything, to assist solid waste management. In fact, the comprehensive waste-to-energy program may have even hurt solid waste management, because management stopped while local authorities hoped to have the federal government pay for their solid waste problems.

2.8 IMPACT OF NEW FEDERALISM

When Congress enacted RCRA in 1976, it established separate authorities for Subtitle D Solid Waste Management Planning Programs and the Subtitle C Hazardous Waste Management Programs.

In the Subtitle D programs, each subsequent year's authorization for the program diminished because Congress intended the management of solid waste to remain a state and local function. The federal role was to give the states the seed money for solid waste planning, and a time period to establish its own mechanism to implement its approved solid waste plan. However, from the date of enactment of RCRA, state and local governments were aware that eventually they were responsible for implementing their solid waste plans. Additionally, the funds appropriated for the Subtitle D program also diminished along with the authorizations. In the Fiscal Year 1983 (FY 83) budget there were zero funds appropriated for solid waste management.[94]

Conversely, the authorizations for the Hazardous Waste Management aspects of the RCRA have increased from $25,000,000 per year in 1978 to $60,000,000 per year in 1985. Such funds were used for the development and implementation of state hazardous waste programs. Appropriations for the state hazardous waste programs also increased from $29.3 million in FY 1981 to over $50 million in FY 1985. This, coupled with the $620,000,000 appropriated under The Comprehensive Environmental Response, Compensation and Liability Act of 1980 (Superfund) in FY 1985 for the clean-up of existing hazardous waste sites, clearly established the focus of federal funding priorities in the management of solid and hazardous waste. This policy is almost the exact opposite of the funding priorities of the federal government in 1980, when the bulk of its funding sources were for DOE price supports to waste-to-energy facilities.

One result of this new policy of not funding the Subtitle D solid waste program will be reduced incentive for states to actually implement the solid waste plans that they developed between 1976 and 1980. Without occasional EPA review, planning authorities for the various regions will be free to implement whatever projects they wish with little or no concern for the Solid Waste Plan submitted to EPA. The federal government will not have the manpower to assure compliance with the plans approved by EPA. Also, without the financial incentive to implement solid waste plans, it is doubtful if the states will continue to follow their solid waste plans or update their state plans when new amendments to RCRA are enacted.

Therefore, states will be left to their own discretion as to the implementation of their solid waste plans. Such a lack of direction from the federal government in some instances will mean no activity. However, some states will be very imaginative in their experimentation. For example, the state of New Jersey, to preserve landfill space before getting its waste-to-energy facilities on line, has established a recycling allowance in which local municipalities will receive a rebate from the state based upon the amount of materials recovered.[95] Not only does New Jersey hope to extend its landfills, but it believes its plan makes economic sense. New Jersey values the recovered materials at $700,000,000 and hopes that it will attract industries that utilize such materials.

On the other hand, other local authorities, for example Akron, Ohio, have enacted flow control ordinances that require all solid waste to be burned for energy recovery. The local private haulers and the recyclers in the case of *Hybud v. City of Akron* challenged this ordinance in court alleging that it violated federal antitrust law, the Commerce Clause of the Constitution, and that it constituted a taking of property.[96] Although the municipality persuaded the lower court and the court of appeals of its position, the U.S. Supreme Court remanded the appeals court decision in light of a case known as *Community Communications v. City of Boulder,* which held that municipalities can be subject to federal antitrust laws.[97]

The federal district court, and the court of appeals upon reconsideration of the *City of Akron* decision in light of the U.S. Supreme Court's decision in *City of Boulder,* again found in favor of the city of Akron. The basis for the court's decision was that the enactment of the flow control ordinance was within the clear intention of the Ohio legislature when it authorized the Ohio Water Development Authority to assist municipalities with the development of solid waste and resource recovery facilities. Therefore, the courts held that the actions of the City of Akron were pursuant to state action and as such the antitrust laws are not applicable.[98] Furthermore, since the Supreme Court only remanded the *City of Akron* case for reconsideration concerning the applicability of the antitrust law, all other issues raised by the private haulers were not to be reconsidered and the disposition of such matters by the Sixth Circuit Court of Appeals stands.

The benefits of states being free to develop solid waste plans that differ from federal guidelines is that there will be considerable experimentation. Conversely, with EPA, DOC, and DOE out of the picture, there is no guidance at the federal level as to the workability of technology or technical assistance in the planning aspects of solid waste management. The only federal incentive left for involvement in the solid waste planning process is the issuance of industrial development bonds.

However, because most lobbying groups involved in waste-to-energy have dissolved there is little effort to move the federal government back into offering financial and technical assistance to the states for solid waste management. Only the National Resource Recovery Association, an affiliate of U.S. Conference of Mayors, is presently concerned with waste-to-energy, and their primary goal is to exempt themselves from the antitrust laws when enacting flow control measures.

With little incentive to get the federal government back into the solid waste business, it is up to the states and industry to manage their own solid waste problems, and to insure the proper sizing of waste-to-energy facilities through the implementation of the standard requiring adequate provision be made for present and reasonably anticipated recycling activities.

On the hazardous waste side and with the enactment of Superfund legislation, and countless federal regulations and the imposition of new congressional standards, the federal government for the long term will be in the business of regulating hazardous waste management and the clean-up of unsafe disposal facilities. The controversy in this area centers around the degree of enforcement needed to protect the environment, the amount of security needed to assure payment for any of the postclosure liability incurred by hazardous waste operators, the use of Superfund monies, and the ability of hazardous waste generators, as well as those who transport, treat, store, and dispose of hazardous waste, to obtain pollution liability insurance.

EPA, as well as the public through citizen suits, will continue to implement RCRA. But because of state rights, states will attempt under RCRA to develop and implement the federal hazardous waste standards in their states. This is evident by the fact that 37 states have already received phase-I approval for implementing RCRA standards and 10 states have phase-II authority for implementing RCRA standards.

2.9 CURRENT AND FUTURE ISSUES IN SOLID WASTE

Although some writers consider the spiraling costs of landfill, as well as the complexities of siting new landfills, to be the key RCRA problems of the future, such views simply ignore the legislative intent of RCRA. One of the primary goals of RCRA was to drive up the costs of landfilling so as to reflect its true economic costs to society as well as help make newer technologies for the treatment, storage, and disposal of solid waste more cost competitive. Therefore, the real future issues will not focus as much on landfills as on who will really pay for the cost of improper disposal methods and will EPA have the courage to impose the true costs of land disposal on the owners and operators of landfills so as to make the new hazardous waste treatment technologies more competitive.

In addition, state and local governments will become increasingly concerned about the development of EPA regulations through the judicial process, in which industry or environmental groups sue the EPA, alleging a final regulation is arbitrary, capricious, and beyond the scope of the agency's authority. EPA and those involved in the lawsuit, absent state or local government, will negotiate a resolution of their dispute. The state and local governments will become more active in this judicial process if their concerns are to be reflected in EPA's regulations.

The final issue of the future will be the need for recycled materials. Each of the above issues will be discussed separately.

2.9.1 The Liability Issue

The overriding issue in the management of hazardous waste will be the liability issue. Both RCRA and Superfund have their limitations. RCRA established a set of standards that imposed liability if there is a present breach. Although RCRA does not apply retroactively, it can be applied to any situation currently causing environmental damage.

Superfund, on the other hand, may apply joint and several liability to any person causing damage to natural resources. However, Superfund's application is limited solely to harm to natural resources and does not apply to harm to individuals.

The missing link in federal legislation is who is responsible for personal injury caused to individuals by the improper disposal of hazardous waste. As citizens become more aware of the problems associated with hazardous waste and more specific occurrences that may cause them harm, they will be initiating law suits. Presently however, the federal courts will be interpreting the applicability of state common law doctrines rather than federal legislation.

The concerned industries may want to consider preventive legislation establishing a limit on liability for past improper disposal practices in return for a mechanism that quickly compensates victims of the improper disposal of hazardous waste. Also, since hazardous waste management has changed tremendously over the years, some consideration should be given to a state-of-the-art defense so that past actions, legal when they occurred, are not judged by 1985 standards.

The most troubling aspects of hazardous waste management concerns the liabilities imposed upon an industry for past actions that complied with all hazardous waste laws and regulations at the time disposal occurred. Superfund and various common law theories make it clear that industry may be subjected to liability if the waste was disposed at any time at a facility that caused harm to natural resources.

The solution to these complex liability issues impacting industry and individuals harmed by the exposure to hazardous waste is to enact a comprehensive federal law which would permit speedy recovery by those harmed by improper disposal of hazardous waste without the tremendous burden of proof that must now be met while limiting the damage award so as to avoid bankrupting industries for past actions that may have conformed to standards applicable at the time disposal occurred but are judged according to a more advanced technology when examined at a later period. Workmans compensation laws would be an appropriate analogy.

2.9.2 Cost Competitiveness of New Technology

The ability of new treatment disposal technologies to compete with or displace land disposal is primarily in hands of EPA. To date, however, EPA's actions have made such new technologies non-competitive with land disposal because EPA failed to regulate land disposal of hazardous wastes in a manner that would increase its costs to include all the costs (injuries to humans from improper disposal, contamination of water supplies, cleanup, monitoring) related to its safe disposal. Since 1976, Congress intended EPA to vigorously regulate land disposal in order that its cost reflect the true costs to society as well as to make new treatment and disposal technologies more competitive. Between 1976–1984, EPA did little to increase the cost of land disposal. Therefore, in 1984 Congress, mistrusting of EPA's ability to regulate land disposal to protect health and the environment, amended RCRA to prohibit many wastes from being disposed of on land unless EPA established that such disposal did not present harm to human health and the environment. As these new amendments become effective, the cheapest disposal methods will be eliminated and new treatment and disposal technologies will hopefully flourish as they become either more competitive or the only means of managing hazardous waste.

2.9.3   The Need for Solid Waste, Flow Control, and the Municipal Action Exemption

As the cost of landfilling becomes greater and new sites become harder to find due to environmental conditions or citizen opposition, municipalities are turning toward the development of waste-to-energy facilities to manage their solid waste problems. In doing so, the municipality quickly learns that a steady stream of solid waste is essential for the economic well being of the facility. This point is impressed upon them by the financial community that will sell the bonds to underwrite the cost of the facility.

The steady flow of solid waste is important for two reasons. First, the solid waste is the raw material necessary to produce energy. Second, the facility is paid a tipping fee for every ton of solid waste delivered to the facility. Therefore, the more tons of solid waste delivered to the facility, the greater the tipping fee as well as the greater the amount of energy that can be produced.[99]

Certain investment bankers have required, as a prerequisite to financing a facility, an ordinance that requires all solid waste, as well as recyclables, to be delivered to the waste-to-energy facility. Obviously, the greater the quantities of solid waste and recyclables delivered, the greater the security for the bondholders in the form of tipping fees and energy produced.

The recycling community opposes such flow control ordinances because they bar their access to recyclable materials, which are the recyclers' raw material for their industrial processes. For example, old newspaper is the raw material in the production of new newsprint.

Furthermore, the recyclers argue that such comprehensive flow control measures violate the federal antitrust laws by imposing an unreasonable restraint on trade as well as erecting barriers to interstate commerce. If all recyclables must flow to one facility, they say, the municipality is prohibiting all commerce in recyclables for its own economic benefit.

Finally, the recyclers cite numerous reports that conclude that waste-to-energy facilities and recycling are compatible. In fact, one recent study even concludes that the energy content of the solid waste is greater when recycling occurs because many of the noncombustible materials (glass) are also removed along with the recyclable paper.[100]

Notwithstanding the compatibility of recycling with waste-to-energy technologies, numerous locations are presently considering the enactment of flow control legislation. In such instances, the communities appear to be more concerned about the economics of the waste-to-energy facility rather than solid waste management.

Both waste-to-energy and recycling reduce the need for landfill. Therefore, if both are compatible, both should be permitted in developing a solution to the solid waste problem, especially where, by permitting recycling, the local taxpayer is saved the tipping fee for disposing of the materials that are being recycled. In many instances, recycling contributes to the local economy since some recyclers actually pay for the recyclable materials.

There are still localities, however, that insist on flow control legislation requiring that all material be delivered to a waste-to-energy facility.[101] The legality of such a broad flow control measure is presently in doubt since the U.S. Supreme Court in January 1982 decided *Community Communications Company, Inc. v. City of Boulder*, 102 S. Ct. 835 (1982), which held that a municipality is subject to the antitrust laws when it restrains trade unless it was acting pursuant to a clear and affirmatively expressed state policy permitting such a restraint of trade.

However, to negate the Supreme Court decision, the U.S. Conference of Mayors and the National League of Cities have lobbied for an exemption from the antitrust laws, which can be titled the "Municipal Action Exemption." The exemption, if enacted, would give municipalities the same exemption from the antitrust laws that the states presently have under *Parker v. Brown*, 317 U.S. 341 (1943). In *Parker*, the U.S. Supreme Court held that states acting in their sovereign capacity are not subject to the federal antitrust laws. As proposed, such an exemption would place municipalities in the same immune status from antitrust liability as states, when such municipality was acting in its sovereign capacity.[102] It would literally make municipalities free of federal and state control in the making of all decisions affecting commerce within their geographic area. Although most industries are sympathetic to the plight of the municipality, they believe that the municipality should not be able to restrain trade solely for its own economic and nongovernmental benefits. Otherwise, it is possible that municipalities could take over profitable enterprises to make up for municipal deficits.

Opponents of the "Municipal Action Exemption" offer an example of the flow control ordinance to illustrate their point. Municipalities will utilize their power to displace competition for recyclables by directing their flow to the waste-to-energy facility, they say. The city would be using its power for its economic benefit while being unconcerned about the adverse impact that is caused to viable recycling operations when competition for recyclables is displaced. The opponents argue that the free market system should prevail and that a municipality should not be able to legislate competitive advantages for itself or others at the expense of viable industry.

In the 98th Congress, both the Municipal Action Exemption legislation and the flow control issues were addressed. First, Congress enacted the Local Government Antitrust of 1984, Pub.L. 98-544 (Oct. 24, 1984) which barred treble damage suits against municipalities for antitrust violations but permits individuals and corporations to seek injunctive relief against municipalities for violations of the antitrust laws. Second, Congressman James Florio (D–N.J.) enacted a compromise solution to the flow control issue as it affects solid waste management. In Pub.L.98-616 (Nov. 8, 1984), Congressman Florio amended Sections 4001 and 4003 of RCRA by adding the following language:

> *In developing such comprehensive plans, it is the intention of this Act that in determining the size of the waste-to-energy facility, adequate provision shall be given to the present and reasonably anticipated future needs of the recycling and resource recovery interests within the area encompassed by the planning process.*

This compromise language was supported by the municipalities and the recycling and waste-to-energy industries. The thrust of this compromise language is to focus the municipality's attention on the most serious problem encountered in the development of a waste-to-energy facility—its size. By addressing the sizing issue prior to construction, when there is ample time to ensure the continued viability of both the waste-to-energy facility as well as the recycling industries located in the planning area, it can reasonably be expected that the needs of the two industries can be reconciled.

Furthermore, the Florio Amendment indirectly addresses the result of an oversized facility—its need to receive recyclables in order to receive the additional tipping fee from the municipality. This legislative compromise will focus the municipality's attention on the size of its waste-to-energy facility prior to construction.

Attention to this single facet of the solid waste planning process could avoid the need to include recyclables as part of solid waste because it eliminates the guesswork in determining the size of the available waste stream. The municipality and the taxpayers then become the real beneficiaries, because they will pay less capital costs for the construction of the facility, as well as less in tipping fees for the life of the facility, while enjoying the ability to market the recyclable materials within their planning area.

2.9.4 The Absent Parties in EPA Litigation—States and Municipalities

Throughout EPA's development of its RCRA regulations, the controversial regulations have been formulated by settlements between interested parties and EPA. Such settlements, which have been reached in the open dump regulations and groundwater standards, and which may occur on the recently promulgated land disposal regulations, take place when industry or environmental groups challenge EPA's final regulations as being arbitrary, capricious and contrary to law. Consistently absent from these settlements are the state and local governments, the principal entities that must administer any settlement agreed upon by EPA and others. Although once a settlement is reached EPA publishes the proposed agreement in the Federal Register, thus permitting the states and local governments to comment. Providing input on an agreement reached between parties over months of negotiations is many times not a realistic method for impacting regulations.

Therefore, because states and municipalities are the primary organizations responsible for administering RCRA regulations, they must begin, through their respective associations, intervening in all judicial challenges to final RCRA regulations. Only through such intervention will states and municipalities be able to participate in the negotiations concerning the substance of EPA's final RCRA regulations.

Another option open to states and municipalities that would permit additional input into final RCRA regulations subject to judicial review, and therefore, open to the possibility of a negotiated settlement, is the use of Section 7002 of RCRA. By utilizing this section, the respective associations for the states and municipalities could petition the administrator of EPA to promulgate a regulation requiring EPA to give notice to those associations representing states and municipalities of each lawsuit that challenges a final RCRA regulation. The states and localities would then be on notice that negotiations might occur between interest groups and EPA concerning the final content of a regulation. It would be up to the state or locality to intervene to protect its interest.

In this way, states and localities could be directly involved with the enforcement of RCRA and would be situated so as to participate in the content of the final regulations. Obviously those parties with a vested interest would oppose intervention and the active participation of the states and municipalities in the negotiations and judicial proceedings. By permitting participation of the states and localities in such negotiations, the courts and EPA will hear the views of those most likely to deal with the problem on the practical level.

The 98th Congress enacted a partial remedy of this issue in Pub.L. 98-616 (Nov. 8, 1984) by amending section 7003 of RCRA to require the Administrator to provide notice, a reasonable opportunity for comment, and an opportunity for a public meeting on the proposed settlement prior to a final order being entered on the proposed settlement.

2.9.5 Procurement and Recycling

Although the procurement regulations have never been promulgated, nor has any person utilized the Section 6002 provision of RCRA that requires the procuring agency to make a determination in those instances it does not procure an item containing recovered and recyclable materials, recyclable materials will be in more demand as the supply of primary materials decreases.[103]

A recent report entitled "Wastepaper, the Future of a Resource: 1980–2000," studies the worldwide supply of available fiber for making paper products. Its conclusion is that because of exports to Third World nations, as well as the increasing use of recovered fiber in the virgin paper process, there could well be a worldwide shortage of fiber.[104] These findings are similar to those contained in an earlier report entitled "The Global 2000 Report to the President" prepared by the Council of Environmental Quality and Department of State. The report predicted that the need for fuel wood will exceed supply by 25% by 1994, and that the growing wood stock per capita will decrease by 47% by 2000 thus causing a decrease in available fiber supply.

It is ironic that with worldwide fiber shortages on the horizon the EPA has not even attempted to promulgate guidelines for the federal government's use in procuring recovered materials.[105] To the contrary, EPA filed an amicus curiae brief in support of the City of Akron's broad flow control ordinance which requires the burning of recyclable paper.[106] Again, because of EPA's failure to implement the congressional intent of RCRA, the 98th Congress amended section 6002 of RCRA to require that all federal agencies enact an affirmative procurement program to ensure that items composed of recovered materials will be purchased to the maximum extent practicable.

Back in 1975, the U.S. House of Representatives Subcommittee on Transportation and Commerce used a fact book on the economic aspects of materials recovery. The same House Committee joined the efforts of EPA, DOC, OPP, and all federal procuring agencies and required them to procure recovered materials whenever practicable. Such mandates were ignored and are now forgotten by the involved government agencies. Unfortunately, the world economic system needs additional materials. Federal, state, or local government, through careful implementation of the procurement provisions of RCRA, can help address world demand, or else shortages will make fiber scarce and drive up its price to lower demand. Hopefully, with their new mandates, the federal agencies will carry out congressional intent and amend their procurement regulations so as to purchase items composed of recovered materials.

2.10 SUMMARY

EPA's implementation of RCRA is tardy, inconsistent, and neglectful of the parties such as the states, whose cooperation it needs for proper program development. In particular, EPA never focused upon the primary foundation of RCRA, which is that the law was based on the free market system with no subsidies being permitted except for the procurement of recovered materials. Simply by making landfills pay the costs of their potential pollution, the cost of disposing at a landfill would rise, thereby making other treatment technologies more competitive.

Also, at each stage of its regulatory process, EPA ignored the economic underpinnings of RCRA. EPA abandoned Subtitle D solid waste planning, thereby ignoring the impact on groundwater by municipal waste landfills. Then, by exempting small generators from the hazardous waste regulations, EPA permitted each such generator to dispose of up to one ton of hazardous waste a month at the municipal waste landfill. Although EPA is attempting to regulate the treatment, storage, and disposal of hazardous waste, it has by both abandonment and incomplete regulations permitted the random disposal of all types of solid and hazardous waste without any concern for a national groundwater policy.

If RCRA is to be implemented, the states, whose primary concern is the health and welfare of its citizens, must integrate the solid and hazardous waste provisions of RCRA so as to protect their own land, water, and air resources. They can use federal common law to protect themselves from pollutants from other states.[107]

Perhaps because the states have a more immediate relationship with their citizens than does the federal government, they can be successful. In the final analysis, if the states do not act, there will be costly regulation without consistent implementation of the nation's solid and hazardous waste laws.

REFERENCES

1. W. L. Kovacs and Klucsik, The New Federal Role in Solid Waste Management: (The Resource Conservation and Recovery Act of 1976). *Columbia Journal of Environmental Law,* **3,** 212–213 (1977).

2. Section 202(b) of the Solid Waste Disposal Act, 42 U.S.C. 3251.

3. E. S. Savas, *Evaluating the Organization of Service Delivery: Solid Waste Collection and Disposal,* Ch. 14, p. 33.

4. H. R. Rep. No. 94-1491, 94th Cong., 2d Sess. 24 (1976).

5. Subcommittee on Transportation and Commerce, Committee on Interstate and Foreign Commerce, U. S. House of Representatives, Materials relating to the Resource Conservation and Recovery Act of 1976, 94th Cong. 2d Sess. (Comm. Print 20); Waste Control Act of 1975: Hearings on H. R. 5487 and H. R. 406 Before Subcommittee on Transportation and Commerce, 94th Cong. 1st Sess. (1975).

6. Subcommittee on Transportation and Commerce, Committee on Interstate and Foreign Commerce, U. S. House of Representatives, *supra* at 1–3.

7. H. R. Report No. 94-1491, *supra,* pp. 17–24.

8. Resource Recovery Implementation, Engineering and Economics sponsored by Engineering Foundation, July 26, 1976, Ringe, N.H.

9. H. R. 14496, 94th Cong. 2d Sess. (1976).

10. Congressional Record: H. R. 14496, 94th Cong. 2d Sess., 122 Cong. Rec. H 11181 (daily ed. Sep. 27, 1976).

11. See W. L. Kovacs and Klucsik, *supra,* pp. 216–220 for a detailed discussion of the legislative history.

12. H. Rep. No. 94-1491, *supra,* p. 11.

13. H. Rep. No. 94-1491, *supra,* p. 3.

14. 42 U.S.C. § 6922, 6923, 6924.

15. 42 U.S.C. § 6925.

16. 42 U.S.C. § 6926.

17. 42 U.S.C. § 6926.

18. 42 U.S.C. § 6921.

19. W. L. Kovacs and Klucsik, *supra,* p. 225; W. L. Kovacs, *Federal Controls on the Disposal of Hazardous Wastes on Land, in Resource Conservation and Recovery Act: A Compliance Analysis,* 23 (1979).

20. 42 U.S.C. § 6924.

21. 42 U.S.C. § 6928.

22. 42 U.S.C. § 6928(a).

23. 42 U.S.C. § 6972(b).

24. W. L. Kovacs and Klucsik, *supra,* p. 230; 42 U.S.C. § 6941.

25. 42 U.S.C. §§ 6913, 6948.

26. 42 U.S.C. §§ 6907, 6942.

27. 42 U.S.C. § 6913.

28. H. Rep. No. 94-1491, *supra,* pp. 8, 14.

29. 42 U.S.C. § 6942(10)(11).

30. Section 2006(b) of the Resource Conservation and Recovery Act, 42 U.S.C. § 6916(b) (1976).

31. 42 U.S.C. § 6948.

32. 42 U.S.C. § 6949.

33. 42 U.S.C. § 6948(b).

34. 42 U.S.C. § 6947.

35. H. Rep. No. 94-1491, *supra,* pp. 7, 33.

36. 42 U.S.C. §§ 6943, 6947.

37. H. Rep. No. 94-1491, *supra,* pp. 11, 32–33, 39.

38. Hazardous and Solid Waste Amendments of 1984, Pub.L. 98-616, § 302, (November 8, 1984).

39. 42 U.S.C. § 7004(b).

40. 42 U.S.C. § 6947(b)(3).

41.   42 U.S.C. § 6972.

42.   42 U.S.C. § 6973.

43.   H. Rep. No. 94-1491, *supra*, p. 33; Pub.L. 98-616, § 302.

44.   42 U.S.C. § 6962; added by Pub.L. 98-616, § 501.

45.   id.

46.   42 U.S.C. § 6962(e).

47.   42 U.S.C. § 6962(c).

48.   42 U.S.C. § 6903(19).

49.   42 U.S.C. § 6962(g).

50.   42 U.S.C. § 6951.

51.   42 U.S.C. § 6952.

52.   H. Rep. No. 94-1491, *supra*, p. 42.

53.   42 U.S.C. § 6953.

54.   42 U.S.C. § 6954.

55.   H. Rep. No. 94-1491, *supra*, p. 44.

56.   Section 505 of H. R. 14996, the Resource Conservation and Recovery Act, 94th Cong. 2d Sess. (1976) which provided a defense to an antitrust claims that such forums, conferences, and other educational arrangements resulted in antitrust violations. Section 505 was dropped at the meeting between the respective House and Senate Committee staffs. *See:* H. Rep. No. 94-1491, *supra*, pp. 44–45. *See also:* W. L. Kovacs and Klucsik, *supra*, pp. 216–220.

57.   42 U.S.C. § 8801.

58.   42 U.S.C. § 8811.

59.   42 U.S.C. § 8832.

60.   42 U.S.C. § 8833.

61.   42 U.S.C. § 8834.

62.   42 U.S.C. § 8837.

63.   42 U.S.C. § 8835 (a)(6)(c).

64.   42 U.S.C. § 7112.

65.   42 U.S.C. § 7112(2)(4)(5)(6)(13)(14).

66.   42 U.S.C. § 7321(c)(4).

67.   42 U.S.C. §§ 5919, 5920.

68.   42 U.S.C. § 5919.

69.   42 U.S.C. § 5920.

70.   42 U.S.C. § 8837.

71.   42 U.S.C. §§ 6344–6344(a).

72.   42 U.S.C. § 6344.

73.   Pub. L. No. 95-617, 92 Stat. 3117.

74.   16 U.S.C. § 824a-3.

75.   16 U.S.C. § 824a-3. The incentives were upheld in *Ferc v. Mississippi,* 50 U. S. L. W. 4566 (June 1, 1982).

76.   Treas. Reg. § 1.103-8(2)(ii)(b); Rev. Rul. 72-190; Rev. Rul. 75-184; Rev. Rul. 76-222.

77.   Rev. Rul. 76-222.

78.   40 C. F. R. 255, (May 16, 1977).

79.   40 C. F. R. 256, (July 31, 1979); 40 C. F. R. 257, (September 13, 1979).

80.   *Chemical Manufacturers Assn. v. EPA,* No. 79-2, 2298 (D. C. Cir.).

81.   45 *Fed. Reg.* 73440-93443 (Nov. 4, 1980).

82.   *Chemical Manufacturers Assn. v. EPA,* No. 79-2298 (D. C. Cir.) Order of Oct. 24, 1980.

83.   46 *Fed. Reg.* 47048–47051 (Sep. 23, 1981).

84.   EPA working document entitled "V. State Activities" and "States Granted Interim Authorization for Phase I and II".

85.   40 C. F. R. Parts 260, 261, 262 concerning EPA General Regulations for Hazardous Waste Management; "EPA Regulations for Identifying Hazardous Waste; and EPA Regulations for Hazardous Waste Generators" were published on February 26, 1980.

86.   40 C. F. R. Parts 260, 261, 262, 263, 264, 265.

87.   47 *Fed. Reg.* 32274–32388 (July 26, 1983).

88. 47 *Fed. Reg.* 32274 (July 26, 1983).

89. Same as 84.

90. DOC reviewed a composting process to determine if its' end product contained any hazardous materials.

91. Office of Recycled Materials, National Bureau of Standards, U.S. Department of Commerce, "Procurement of Products containing recovered material: A Summary of Activities in 7 States," (July 1981).

92. 45 *Fed. Reg.* 52900 (August 8, 1980) requested comments concerning the contents of the plan.

93. Supplemental Appropriations and Rescission Act 1981, Pub. L. 97-12; Interior and Related Agencies Act, 1982, Pub. L. 97-100.

94. Executive Office of Management and Budget, Budget of the United States, Fiscal Year. 1983 Appendix, I-55.

95. Id. at I-56.

96. *Glenwillow Landfill, Inc. v. City of Akron,* 485 F. Supp. 671(N.D. Ohio 1979); *Hybud Equipment Corp. v. City of Akron,* 654 F. 2d 1187 (6th Cir. 1981).

97. *Community Communications Co., Inc. v. City of Boulder, Colorado,* 455 U.S. 40, 102 S. Ct. 835 (1982); *Hybud Equipment Corp. v. City of Akron,* 455 U.S. 931, 102 S. Ct. 1416 (1982).

98. *Hybud Equipment Corp. v. City of Akron,* Case Nos. C78-1733A, C78-65A (N.D. Ohio April 6, 1983); 742 F.2d 949 (6th Cir. 1984).

99. W. L. Kovacs, Flow Control: An Unnecessary Constitutional Conflict in Managing Solid Waste. *Environmental Analyst* 3 (June 1982).

100. Essex County, New Jersey, Analysis of an Integrated Material and Energy Recovery System for Essex County, New Jersey (Aug. 13, 1982).

101. Akron, Ohio; Islip, New York; Chattanooga, Tennessee (proposed).

102. Draft legislation circulated by Senator Thurmond Jan.–June 1983; H. R. 2981, 98th Cong. 1st Sess.

103. Vol. II Council of Environmental Quality and the Department of State, The Global 2000 Report to the President, pp. 318–333.

104. Franklin Associates Ltd., Wastepaper, The Future of a Resource 1980–2000.

105. H. R. 2867, 98th Cong. 1st Sess., §23. The purpose of this proposed amendment is to require the affirmative procurement of items containing recovered materials. This amendment was subsequently enacted into law by Pub.L. 98-616, § 501 (Nov. 8, 1984).

106. Amicus Curiae brief of Environmental Protection Agency in *Hybud Equipment v. City of Akron,* No. 80-3121 (6th Cir. 1981).

107. *City of Milwaukee v. Illinois,* 15 ERC 1908 (1981).

CHAPTER 3

PUBLIC PERCEPTIONS
AND COMMUNITY RELATIONS

LAWRENCE CHERTOFF

Pennsylvania Engineering Corporation
Consultant, Citizens Advisory Committee, New York City
Brooklyn Navy Yard Resource Recovery Facility

DIANE BUXBAUM

EPA Solid Waste Management Program

3.1 INTRODUCTION

During the 1970s, the public's perception of resource recovery rose to a high point boosted by a sense of American ingenuity in overcoming foreign oil cartels. But by decade's end, skepticism had replaced optimism. Some of this doubt may be traced to faulty design or engineering scale-up of earlier projects or to an imperfect understanding of the market for the recovered resources. Most of the aborted attempts, however, can be attributed to public apprehension concerning the impact of a resource recovery facility in "my backyard." This doubt can be seen in extraordinary construction and systems performance insurance, in municipal requirements for long operating histories of the proposed equipment and especially in difficult siting requirements. These restraints on new facility construction, to some degree, may account for the relatively small tonnage of U.S. waste processed through heat recovery incinerators. According to Hagler, Bailly & Company,[1] by 1979, Switzerland, Sweden, and West Germany were each processing over 50% of their municipal solid waste (MSW) in energy-recovery incinerators compared with less than 2% in the United States.

To determine what the principal factors were that led to public acceptance or rejection of proposed plants and to deduce what must be done to overcome public antagonism to an energy recovery waste disposal system, the authors analyzed recent siting cases.

3.2 RESOURCE RECOVERY PROJECT CASE STUDY

3.2.1 Facilities Investigated

The authors reviewed 120 projects and investigated 30 projects recently presented to the public for approval. Twenty were selected for their representative value of the current national situation. Sample projects that were chosen have been announced to the public on a site specific basis and are in the planning, engineering, construction, or operational phase but have not yet completed 1 year of postacceptance operation. The samples conform generally to the national distribution of similarly sized projects of equal maturity. The authors further categorized half of the sample projects as large scale, 1000 TPD (tons/day) capacity or greater, and half as medium scale, 200 TPD capacity or greater. Small-scale projects (less than 200 TPD) are not represented in this analysis. After preliminary investigation, the author believes that issues pertinent to small-scale facilities are significantly different from medium- and large-scale projects and should be analyzed in a separate study.

3.2.2 Data-Gathering Technique

For each of the proposed or operating facilities the project manager or his surrogate was inter-viewed by an author or his research assistant by telephone or at prior meetings. Due to the subjective nature of a study of public perception, the reader should bear in mind that the raw data has been filtered through the perceptions of the project managers.

3.2.3 Summary of Interview Questions Asked

1. Why did your community choose an alternative to existing waste disposal procedures?
2. What was the accuracy of the public's understanding of resource recovery systems prior to the procurement process?
3. How was the project and its site announced?
4. What was the initial public reaction?
   a. What was the basis of a favorable reaction?
   b. What was the basis of a negative reaction?
5. What public educational techniques were used?
6. What is the current public position and why?
7. How is the public position evidenced?
8. What is the current status of the project?

3.2.4 Communities Studied

Large-Scale Facilities (Larger than 1000 TPD)

Community	Major Motivating Factor	Major Militating Factor	Accuracy of Initial Public Understanding	Level of Public Education Program	Project Status
Brooklyn, NY	Landfill	Site	High	High	In doubt
Dade, FL	Landfill	Size	Low	Moderate	Operational
Hartford, CT	Landfill	Past failure	High	Moderate	Under construction
Haverhill, MA	Landfill	None	High	High	Operational
Multitown, NY	Water quality	Size	High	Little	Abandoned
Pinellas, FL	Landfill	None	Low	Moderate	Operational
Portland, OR	Future landfill	Cost	Low	High	Abandoned
San Diego, CA	Future landfill	Site	Low	High	Site abandoned
Springfield, MA	Future landfill	Site	Low	High	Site abandoned
Westchester, NY	Landfill	Site	High	Moderate	Early opera-tional

Medium-Scale Facilities (200–1000 TPD)

Community	Major Motivating Factor	Major Militating Factor	Accuracy of Initial Public Understanding	Level of Public Education Program	Project Status
Berkeley, CA	Future landfill	Cost	High	Moderate	Abandoned
Boca Raton, FL	Landfill	None	Low	Moderate	On schedule
Daytona, FL	Sludge disposal	None	Low	Moderate	On schedule
Dutchess, NY	Future landfill	Site	High	Moderate	On schedule
Fitchburg, MA	Future landfill	Site	Low	High	Site abandoned
Glen Cove, NY	Water quality	Site	Low	Low	Operational

Community	Major Motivating Factor	Major Militating Factor	Accuracy of Initial Public Understanding	Level of Public Education Program	Project Status
Islip, NY	Water quality	Site	High	Moderate	On schedule
Onandaga, NY	Future landfill	Site	Low	High	Site abandoned
Oswego, NY	LF/Water quality	Site	Low	High	On schedule
West Contra Costa, CA	Landfill	None	Low	High	In doubt

3.3 IMPLICATIONS OF CASE STUDY

Of the communities studied, 35% have completely abandoned plans for the project or have abandoned the planned facility site. Thus, 15% are in serious difficulty preceding site or project failure. Twenty-five percent are progressing toward contract signing on a schedule that has not been delayed by more than twice the time planned, and 25% are either under construction or operational. These ratios appear to fairly represent the mid- to large-scale waste-to-energy recovery industry in 1983.

In each case of project or site abandonment, the technology was not at issue and, finally, neither was possible degradation of air quality. Facility size and cost and intense local site opposition may have galvanized resistance to a project, but it was insufficient public appreciation of the need to find a waste disposal alternative that permitted relatively few people to redirect public policy.

3.3.1 Motivating Forces

Proposed projects that have had the greatest public support were in communities that have either (1) reached absolute physical limits on further landfill use and/or (2) have reached a level of groundwater (drinking water) deterioration through saltwater or leachate intrusion that is perceived and acknowledged by the public.

3.3.2 Militating Factors

Projects that have met the broadest resistance were in communities where the public could see no clear need for the proposed facility. Proposed projects that have generated the most intense and most focused public resistance are those to be sited near (within 1 mile of) communities categorized as middle to lower-middle income with a substantial number of tenant-owned dwellings. It is in this type of community where homeowner equity represents the major personal asset of residents and where a fear of reduced rate of land value appreciation is a grave threat. It is in this type of community where variations of the phrase "not in my backyard" appear most frequently in transcripts of public meetings.

To summarize, the four major forces acting for or against proposed energy recovery waste processing facilities are:

1. Absolute physical limitation on further landfilling.
2. Agreement by the public that current disposal practice poses a clear and present danger to community health or welfare.
3. Facility size.
4. Facility site.

3.3.3 Economic Considerations

There are variations of each of these major motivating or militating forces, the principal one being economics and, to a lesser degree, community education. In most communities investigated, where landfill capacity has reached absolute physical limitation, refuse could be remote hauled to outlying landfills at costs sometimes exceeding 500% of the current disposal practice. Where the cost (current or projected) for remote hauling is unquestionably far greater than for resource recovery, the economic argument is so overpowering that this alternative to resource recovery is dropped from consideration. Similarly, if the cost of retrofitting existing landfills to meet RCRA requirements or to improve drinking water quality is so in excess of costs projected for resource recovery systems, landfill up-grading loses public sympathy. Since further landfilling can no longer be considered a viable alternative, for the purpose of this study, landfill operations can be treated as if it had reached absolute physical limitation.

3.3.4 Community Education

In communities where the forces favoring resource recovery are nearly balanced with those forces acting in opposition, timely community education programs may have some impact. In communities where the proposed project is to be sited near an intensely hostile neighborhood, the larger community that would be served by the project may out-vote the opposing neighborhood. For this to occur, the community must be made sufficiently aware of the future costs of continued landfilling, or of the danger to drinking water. The threat of sharply higher landfill disposal costs or the threat of water quality deterioration must be perceived by the community as a "clear and present" danger.

The above conditions notwithstanding it would be naive of community leaders to expect welcome acceptance of the proposed project by those citizens who will live closest to the facility and who anticipate the greatest personal impact.

3.4 CASE ANALYSIS

3.4.1 Abandoned Projects

Of the 20 projects analyzed, three (15%) were completely abandoned: Multitown, New York, Portland, Oregon, and Berkeley, California.

Multitown, New York

This project was planned to serve four communities on Long Island, NY. The proposed project was to have had a daily throughput capacity of approximately 3400 TPD. Ground water degradation was the principal force motivating the four communities to choose an alternative to continued landfilling.

The geology of the southern section of Long Island, where the plant was to be located, is sandy with a relatively high water table. Rapid population growth in this area has resulted in overpumping of groundwater with some resultant salt water intrusion. Leaching from landfill operations has caused an excessively high coliform level in the drinking water. Community residents and government health officials have been aware and concerned about this problem for many years. Therefore, the concept of waste disposal, other than landfilling, is highly regarded in the community.

In 1968, to achieve presumed economies of scale, four townships agreed to support an energy recovery facility that could process all of their MSW. The initial public response to this plan was favorable as evidenced by continued financial support for further feasibility studies, by positive editorials in respected local and regional newspapers, and by the ratio of favorable letters to newspapers.

As a result of jurisdictional disputes among the four towns concerning allocation of costs, availability of residue disposal, and political bickering, the project moved very slowly toward procurement.[2] In 1975, one of the four towns voted to secede from the project to pursue building its own resource recovery facility. The Multitown proposed plant was scaled down to 2200 TPD.

In 1978, the State of New York established a public authority for the project. At that time, public approval was high and a site was chosen that bordered on all three townships. By 1981 intense local opposition had developed among residents who lived nearest the site. Almost all of the residents living within 2 miles of the planned site were homeowners. They opposed the plant site, citing excessive truck traffic, noise, air pollution, odor, and questionable economics. Some local politicians were enlisted in their cause. The issues were narrowed to plant size and cost. Arguments of economies of scale were countered with operating histories showing no appreciable economic advantage of one 2200 TPD plant compared with several smaller ones.

In 1982, 14 years after the project had begun and after approximately $7 million had been spent for development, a second town withdrew to develop its own medium-sized plant, citing excessive project costs, delays, and siting opposition.[3] The plant size was again scaled down from 2200 TPD for three towns to 1800 TPD to conform to the needs of the two remaining towns. Environmental impact statements were issued, hearings held, and approval to commence was sought.

Continued publicity about the proposed plant and many public hearings did not reduce either the strength or the number of residents opposed to the plant. The two remaining towns in the project dissolved their partnership. One of the towns now plans to build a facility to serve only its own needs on a far smaller scale. The other town has contracted to have its waste hauled to a resource recovery facility in another town.

In the case of Multitown, there was never any doubt about the need for a resource recovery facility. The participating towns maintained a high level of awareness about the need for alternatives to landfill and the benefits of a waste-to-energy facility. Although public education programs were at a low level, media coverage of continuing developments was extensive. Nevertheless, as time progressed from 1968, the public attitude turned increasingly negative and increasingly entrenched.

Conclusion. Three of the four original participating towns are now planning to build their own smaller facilities and are progressing on schedule. In the opinion of several town managers, Multitown failed because of its size and jurisdictional disputes.

In Multitown, the motivating force for a resource recovery facility—the need to improve drinking water quality—was matched against jurisdictional and facility size militating factors. The site was located near a neighborhood of middle-class homeowners and the proposed facility was clearly larger than needed to serve the residents of any one town. Adding to these problems, the project planning dragged on for 15 years. In 15 years, dissidents become an institution.

Portland, Oregon

The Metropolitan Service District proposed to build the Metro East Resource Facility, a $150 million, 1400 TPD, mass-burn, energy recovery plant. The principal force motivating this decision was the need to plan an alternative to disposing in a landfill scheduled for completion in 1988. The principal forces acting against the proposal were cost and a lack of public consensus that the need was real or immediate. Although the level of public awareness was considered low, a concerted effort was mounted by the Service District to educate its residents to the need to plan for resource recovery.

After the site was announced, residents who lived nearest the proposed site organized against the proposal. According to local public officials, the perception of the public was that a regional government was forcing an unwarranted and expensive project on local citizens. No convincing case could be made for a clear and present danger to community health or welfare in continuing the present mode of waste disposal. A well-organized program of public education, including newsletters, public meetings, television programs, and informational meetings at civic associations, initiated after site announcement, did change public opinion from highly negative to moderately negative.

In a binding referendum held in November, 1982, Portland voted to add to its City Charter a prohibition against building a mass-burn unit in that city. According to local civic groups organized to combat the resource recovery project, the need to process MSW when the landfill was exhausted, would be met through source separation.[4]

Public education had shown some good effect, but in a city of 1,000,000 people only 5000 voted in the referendum. Resource recovery lost by 150 votes.

Conclusion. Portland could not persuade its residents that a need for resource recovery was clear and present. The attitude of "not in my backyard" was coupled with "not in my life time."

Berkeley, California

The history of this project is much the same as Portland, Oregon, except that Berkeley proposed to build a medium-sized facility, 300 TPD throughput. Landfill was quickly being exhausted; a transfer station was available for remote hauling but at a cost of at least 50% greater than that anticipated for a resource recovery plant.

Interest in the proposed facility was great in this highly educated and environmentally aware community. But the community viewed the facility as environmentally less attractive than a combination of source separation and remote disposal. The public education program is considered by Berkeley officials to have been moderate and was not sufficient to dispel the perception, by the voting public, that source separation can substitute for mass processing. In a public referendum, the voters chose, by a margin of 5 to 4, to ban refuse combustion for a period of not less than 5 years.[5]

Conclusion. In Berkeley, the public did not perceive a clear and present danger to their health or welfare in continuing the practice of remote disposal and source separation. Even in a community so knowledgeable about environmental issues, it is difficult to educate an entire voting populace sufficiently well for them to appreciate fully the economic and technical distinctions between source separation and mass processing. The need for resource recovery must be perceived as an alternative to a genuinely threatening health or financial situation. It appears that this was not the case in Berkeley.

3.4.2 Abandoned Sites

Of the 20 projects analyzed, four (20%) were required to abandon their initial sites: San Diego, California, Springfield, Massachusetts, Fitchburg, Massachusetts, and Onandaga County, New York. In each case, the need for an alternative to existing MSW disposal practices was not perceived by the general population to be "clear and present." Local pressure groups were able to force abandonment of the planned sites primarily because the larger population to be served by these facilities was generally uninterested in or unaware of the need for these facilities. In each case, the public education program was considered by government officials to have been extensive, although they were initiated after site selection. Whether the project sites were abandoned in spite of the public education programs or, in part, because these programs raised more doubt and alerted more people to an already unpopular project can only be the subject of conjecture. But, in each case, public officials believe that the level of public awareness of resource recovery benefits was low prior to announcement of the project site and in each case, after considerable public education efforts, public resistance to the projects increased.

San Diego, California

The city developed the SANDER project to mass-burn and recovery energy from a daily through-put of 1200 TPD of MSW. The facility is to replace the current practice of remote hauling to a rapidly diminishing landfill. Planning ahead to avoid a financial crisis of extremely remote disposal, the city announced the project and site at a press conference in 1981. The facility was to be located on waterfront property zoned for commercial use.

The initial public reaction is now termed as having been very favorable, although the level of public awareness about such facilities is believed to have been low. Extensive public education programs, including press conferences, mailers, brochures, and frequent public meetings increased public awareness of the proposed project but also aroused public antipathy against the proposed site. Anticipated truck traffic and despoilage of a potential recreational area were the most cited objections. No strong constituency developed in favor of the project, although conceptually, resource recovery was considered an improvement over landfilling.[6]

Continued pressure from a vocal minority to reserve the site for use as a marina and with a lack of focused public pressure to build the resource recovery plant caused the host town of Chula Vista to withdraw its offer of the land. The SANDER project is now seeking a new site.

Conclusion. The project did not meet the test of having been perceived as a solution to a "clear and present" threat to the financial welfare of San Diego residents. Public education commenced after public opinion had formed. In the belief that there was no great or immediate need for the project, public opinion sided with opponents of the site.

Springfield, Massachusetts

In 1979, the city of Springfield requested proposals for the construction and operation of a 2000 TPD energy recovery facility. The project was to reduce reliance on a landfill that was both situated on a flood plain and that had a history of spontaneous combustion. The proposed steam customer was to be Monsanto Corporation.

At the time the project was announced, the public awareness of a need for an alternative to landfilling was considered low. Throughout the contractor selection process, a public education program was maintained through media coverage, newsletters, and public meetings. In retrospect, public officials of Springfield concede that the education effort was started late and that results were disappointing.[7]

The neighborhood nearest the plant site can be characterized as homeowners whose major assets are their homes. Intense local site opposition developed a few months after project announcement. The general public, who would have been served by the facility and whose taxes may have been kept lower by the facility's replacement of the nonconforming landfill, showed little interest in the project. Without strong project support and without a "clear and present" need for it, local opposition had a disproportionately strong influence.

Monsanto Corporation could not come to agreeable terms with the city for the purchase of steam. The project was dropped at that site. Springfield is now seeking a new site and a new steam customer.

Conclusion. The need for the project was not viewed by the general public as pressing. Although local sentiment was not against the concept of resource recovery or against the chosen technology, it was against the plant being sited so near their homes. Monsanto, rather than gaining credit for helping the city solve a serious problem was, instead, the recipient of unfavorable public relations with those living nearest its plant. The city of Springfield may have been wiser to have tied its steam sales to a customer less sensitive to public pressure and to have had an alternate energy customer to relieve Monsanto from being the leverage point of the project opponents.

Fitchburg, Massachusetts

At less than one-tenth the size of the Springfield, Massachusetts, project, the Fitchburg project was almost a duplicate of community opposition. The facility, as proposed, was designed to process less than 200 TPD in a mass-burn, energy recovery plant. The city's landfill capacity was due to be exhausted in 3 years; remote disposal or a new conforming landfill would cost more than twice the existing landfill cost. Prior to site announcement, public knowledge of waste processing or of the city's MSW disposal problems was low but generally favorable to resource recovery. The plant site was to be adjacent to James River Paper Company, the sole steam customer. The neighborhood nearest the proposed plant site can be characterized as middle-income homeowners.

In February, 1983, the proposed site was announced. The reaction of residents adjacent to the site was immediate and intensely negative. The project manager's life was threatened, fires were reported at his home, and hate mail and telephone calls were received.

The city responded with an educational program, public meetings, bus tours of successful resource recovery plants, and an offer to finance a citizen advisory committee with outside consultants. The public broadened its vehemence to include the steam customer, accusing it of being a polluter and stating that it should not receive discounted steam rates.[8] The paper mill, assuming it would receive favorable publicity by aiding the resource recovery project, was disheartened to find itself the focal point of severe militance. The city abandoned the initial site and is attempting to find an alternative one.

Conclusion. As in Springfield, the residents closest to the proposed site are middle-class homeowners. The bulk of their life savings is represented by homeowner equity. The steam customer agreed to participate in the project partly to receive discounted energy but, equally important, to gain public favor by assisting the city with a civic project. When this anticipated favorable publicity turned into a public relations liability, the company withdrew from the project. The paper company, as the sole energy customer, was essential to the project, therefore a vulnerable target of project opponents. By maintaining an alternative customer, the pressure would not have been so great on James River Paper Company and the opposition might not have been so successful.

Public education and the offer of public funds for an advisory board came after site selection and after community leaders formed their opinion and organized local opposition. The offer was viewed as disingenuous by skeptical residents. The need for the project was not perceived by the general population as crucial to their continued well being. Without a convinced constituency in favor of a project, it can be stopped by an organized and highly motivated minority.

Onandaga County, New York

Onandaga County encountered much of the same opposition as Fitchburg and Springfield. As in the cases of Fitchburg and Springfield, Syracuse, New York, the principal city of Onandaga County, was running out of landfill capacity. To avoid costly remote disposal, the city influenced its county to build a 200 TPD facility to mass-burn and recover energy from MSW. Feasibility studies started in 1968, and the project size and site were announced in 1974. The facility was to be in an economically depressed area characterized as lower middle-class. The county would combine the facility with an economic redevelopment program for the whole area.

Initial public awareness of the city's waste disposal problems and of resource recovery was low. The immediate reaction of homeowners in the area was hostile. A black neighborhood near the proposed site called the project racist. Demonstrations against the project were organized and media coverage of the opposition was widespread. Some public education was attempted after site selection but no effective effort was mounted. The little public education that was offered opened up the possibility to many residents that "painless" options were available but not chosen by the county. Citizen groups insisted on having the county review many unsolicited and often irresponsible offers to process the city and county MSW.[9]

In 1980, after 6 years, the county abandoned the original site and chose a different one 3 miles away. Prior to site announcement, the county launched an effective public relations campaign. It included weekly newsletters, an information-gathering committee composed of representatives of each opposing citizen group, and 21 days of public hearings.

Using the same mass-burn technology as originally proposed, the project, at its new site, has been able to complete contract negotiations and is preparing for a General Obligation bond issue.

Conclusion. The project was able to move forward when sited away from homeowners and preceded by an intensive public relations campaign.

3.4.3 Successful Projects

In our study, the projects considered successful are those that are delayed no more than 50% in their schedule of project development, those that are under construction and those that are operational. Ten of the 20 projects studied (50%) can thus be categorized as "successful." Four of the successful projects (40%) are larger than 1000 TPD and six (60%) are smaller. Six of the 10 successful projects (60%) have or had a clearly perceived absolute limitation on further landfilling, two (20%) have had a clearly perceived groundwater crisis, one (10%) has a clearly perceived sludge disposal crisis, and one has a future landfill limitation. In nine of the 10 cases (90%), the need for an alternative to landfilling is perceived as real by the public. Only one project, Dutchess County, New York, is to be sited where there is still intense local opposition; it is also the one project where there is no immediate threat to health or welfare in existing waste disposal practice. All but one project (90%) claim to have had a moderate to high level of public education preceding or accompanying the project development. Therefore, in 90% of the successful projects studied, the public perceives a clear need for an alternative to landfilling and in 90% of the successful projects there has been a moderate to high level of public education. Of the 10 "successful" projects, five will be examined for their instructional value: Islip, New York, Dutchess County, New York, Daytona Beach, Florida, Boca Raton, Florida, and Westchester County, New York.

Islip, New York

The town of Islip, on Long Island, was one of the original participants in the original Multitown project and it was the second town to withdraw from it. As a result of its participation in the Multitown project, the residents of Islip are very familiar with the issues requiring an alternative to landfilling and are familiar with resource recovery systems. The proposed Islip project is nominally 500 TPD in throughput capacity, about one-seventh the size of the original Multitown facility. The project is sited adjacent to a local airport, the land is owned by the town. Electricity produced by the facility will be sold to the local utility. Although there is some resistance to the site from the nearest residents, the opposition cannot be termed intense. The nearest community to the site can be categorized as economically middle to upper-middle class. Real estate devaluation is an issue, but the number of residents affected are few and the general public who will be served by the facility appear to recognize its need and support its construction. The project is progressing quite close to schedule.

    Conclusion. Islip is similar to its predecessor facility, Multitown, in its genuine need resulting from groundwater contamination. It differs from the Multitown facility by serving only the community where it will be located and in being small enough, by comparison to the original Multitown plant, to be perceived as an affordable facility. In Islip, the motivating force of improved drinking water was not diffused by jurisdictional disputes.

Dutchess County, New York

The county is planning to build a resource recovery facility to process 400 TPD of MSW. The facility will be an alternative to remote disposal when the present landfill is completed. The community nearest the proposed site does not appear to perceive a clear and immediate need for this alternative. Although community awareness of resource recovery technologies is high, intense local site opposition continues. IBM, the energy customer, believed its offer to accept steam would be viewed positively in terms of public relations. Ironically, the intense local siting opposition is from many of its own employees. As a result, IBM became the leverage point for project opponents.[10] Fortunately for the project, the local utility remains a viable energy customer with minor modifications to plant design. The project is progressing on schedule.

    Conclusion. The project's success suffered from an insufficient perception of "clear and present" danger to the community's health or welfare. It was also jeopardized by a vulnerable energy customer. These two factors alone were enough to cause site changes for Springfield, Massachusetts, and for Fitchburg, Massachusetts. The back-up energy customer, Central Hudson Gas and Electric, a regulated public utility required by federal and state law to purchase electricity, is mitigating excessive public and political pressure on IBM and permitting the project to advance.

Daytona Beach, Florida

The city of Daytona Beach is planning to build a codisposal facility capable of mass-burning 250 TPD of combined MSW and dewatered sludge. Prior to completion of feasibility studies, the city had publicized the need to find an alternative to landfilling of sewage sludge; no environmentally acceptable land is available within reasonable transportation distance. The city also had announced its proposed waste processing site near the existing waste water treatment plant and further than 1 mile from the nearest homes. In explaining the need for the new facility, the city has used newsletters, a mobile video theater, extensive media releases, and public hearings. The project is progressing on schedule.

   Conclusion.   Although the need for an alternative to sludge disposal will soon be pressing, it cannot be termed a "clear and present" threat to the health or welfare of the residents of Daytona Beach. Yet, because of early public relations explaining the techniques of sludge and MSW disposal and the benefits of combining the two to produce energy for municipal use, the community's first impression of the proposed project was favorable.

   Daytona Beach has been fortunate in having a good site. But its early public education and the manner of releasing project data further aided in the acceptance of the proposed facility site.

Boca Raton, Florida

This community anticipates a reduced availability of landfill for its MSW. Although the cost of disposal, in the early years of operation, will be higher than if the available landfill were used to completion, the community is enthusiastically accepting the construction of a 240 TPD codisposal, mass-burn facility.[11] The proposed site is in the geographic center of the city, near a wastewater treatment plant. Although the site is good, some credit for public acceptance must be given to the public education program. The program started early with long-term cost projections showing high future costs for continued landfill operation, carefully explained the proposed technology in press releases, public hearings, and town meetings, and requested and received interim resolutions for continued preprocurement development. The project is on schedule.

   Conclusion.   Good choice of site, early and effective community relations, and high public awareness of the sensitivity of South Florida geology to groundwater contamination combined to make project development in this community smooth and rapid.

Westchester County, New York

For over a decade, the county had tried and failed to site a large-scale waste-to-energy facility. In its most recent attempt, the county offered the city of Peekskill a reduced disposal charge to host the plant. The site is three-quarters of a mile from the nearest residence.

   Anticipating intense local opposition resulting from years of negative publicity, the county conducted an aggressive public relations campaign enlisting the aid of the County Executive to negotiate the facility site. Each of the many environmental groups was given an opportunity to participate in the draft environmental impact statement, all concerned citizens were made fully aware of the health and legal threat of continued operation of the noncomplying local landfill, and of the benefits available through resource recovery. The facility is now operational.

   Conclusion.   The community was aware of the "clear and present" danger of continued landfill disposal. The county mounted an aggressive public information program that stressed understanding of the issues but did not allow latitude in choosing the project site. The energy customer, a local utility, is required by the Public Utilities Regulatory Policies Act of 1978 (PURPA) to accept energy; site opponents would have little leverage in pressuring the utility. The County Executive selected the project site; site opponents would have limited political leverage. As a result, localized siting opposition did not spread to wider opposition.

3.4.4   Projects in Doubt

For every project and for every site that had to be abandoned and for some that were successfully completed, there was a period of doubt when the forces acting in opposition to a project, or its site, were in approximate balance with the forces that favored its continued support. The resolution of these projects is often a function of the degree to which the need for a facility is genuine and the degree to which that need is perceived by the general public.

   Even in such cases, however, where the need is perceived by the general public as genuine and the project parameters are sound (economics, environmental quality, waste flow control, energy customer, technology, and financing), the project, or its site, may be jeopardized by local opponents who succeed in influencing public policy. A 3000 TPD, mass-burn facility, proposed for New York City's Brooklyn Navy Yard, is in such jeopardy, as of mid-1985.

Brooklyn, New York

The city of New York disposes of approximately 24,000 TPD of MSW in three 1000 TPD incinerators and in two landfills. The larger of the two, Fresh Kills in the Borough of Staten Island, accepts wastes only from barges fed through the city's extensive marine transfer network. Through agreements with the New York State Department of Environmental Conservation, the smaller of these landfills, which accepts up to 10,000 TPD of MSW, must close by 1985. The remaining landfill will then have to receive over 21,000 TPD, a physical impossibility given the limitations of marine transfer capabilities and the site's inaccessibility by truck. In the preceding 12 years, various plans were formed to build as many as 10 waste-to-energy incinerators. The first of this group is to be located in a former Navy shipbuilding and repair yard located on the city's heavily commercial East River. All MSW deliveries and residue removal would be by barge.

As a result of much press coverage, the general population is highly aware of the refuse disposal crisis facing the city. Through intensive efforts of the city's Sanitation Department, aided by voluntary environmental groups, resource recovery is the favored means of waste processing. The proposed site is in an industrial complex that has been only partially active in the last decade. There are rental and some cooperatively owned apartments within one-quarter mile of the proposed site. The neighborhood can be characterized as economically middle class.

In 1981, the site choice was announced. Immediate and severely adverse reaction developed in the community nearest the site. Objections focused on the potential health impact of air emissions and the sentiment was overwhelmingly "not in my backyard."

The city then began an intensive public education program which included newsletters, public hearings, tours of the landfill, and a publicly financed citizen advisory committee. Some local politicians, whose constituency included the affected neighborhoods, requested that the city choose another site.

Conclusion. The larger community affected by the waste disposal issue was aware of the clear and present danger of absolute physical limitations of future landfilling. The neighborhoods closest to the proposed site strongly objected to a plant of this scale "in my backyard." The principal difference between this project and the Westchester project appears to be in the timing of the public information program and in the stated commitment to the site by the highest ranking community official, in this case the mayor. By late 1984 it was unclear what the resolution of the issue would be.

3.5 SUMMARY

Of the projects studied, the four major forces acting for or against community acceptance of a resource recovery facility are:

1. Absolute physical limitation on further landfilling.
2. Agreement of public that current disposal practice poses a clear and present threat to community health or welfare.
3. Facility size.
4. Facility site.

The first two conditions act in favor of public acceptance, the last two act against it. Overlayed on these conditions are the personal economic considerations of those who live nearest the facility and political considerations. In communities where homeowner equity represents the major personal asset, localized opposition has been strongest. In communities where site choice is left to citizen advisory committees or where the major political force is not steadfastly in support of the site choice, acceptance of the site has been most difficult.

It is important to the project's success that the energy customer is not placed in a public policy-making position. The energy customer should either be legally obligated to accept energy or no single energy customer should be essential to the success of the project.

Apparently, public education programs that have been most successful are ones that have begun prior to site announcement and have factually detailed the liabilities of not building the proposed facility. Indecision on the part of political or public policy leaders encouraged opponents of project sites.

In planning a resource recovery project, policy makers should accept that homeowners living nearest the site will never be happy about the site choice. Benefits to the larger community may have to justify the imposition of an unwanted public facility in a hostile neighborhood.

REFERENCES

1. H.-C. Bailly, Hagler, Bailly & Company, European and American Experiences—A Point of View. U. S. Environmental Protection Agency, Resource Recovery and Waste Reduction Activities: A Nationwide Survey (SW-432 November, 1979).
2. J. J. King, District Leader, 11th Assembly District, New York, interview.

3. C. Weidner, Islip, New York, Town Engineer, interview.

4. D. Durig, Solid Waste Director, Metropolitan Service District, Portland, Oregon, interview.

5. L. Arnold, Refuse Collection Superintendent, Berkeley, California, interview.

6. L. Lopez, Fiscal Analyst, San Diego County, California, interview.

7. A. Cousins, Assistant Director, Department of Environmental Management, Boston, Massachusetts, interview.

8. J. Joyner, Resource Recovery Project Director, Fitchburg Energy Recovery Facility, Fitchburg, Massachusetts, interview.

9. D. Lawless, Public Information Coordinator, Onandaga County Resource Recovery Management, Syracuse, New York, interview.

10. R. Vrana, Commissioner, Department of Solid Waste Disposal, Dutchess County, Poughkeepsie, New York, interview.

11. J. Yelverton, Sanitation Administrator, Boca Raton, Florida, interview.

CHAPTER 4

THE FEASIBILITY STUDY, PROCUREMENT, AND CONSTRUCTION MANAGEMENT

STUART H. RUSSELL

Thermo Electron, Inc.
Waltham, Massachusetts

ROBERT BRICKNER

CHARLES PETERSON

Gershman, Brickner & Bratton, Inc., Consultants
Washington, D. C.

4.1 THE FEASIBILITY STUDY
Stuart H. Russell

4.1.1 Introduction

In the solid waste field, the question of "what to do" must always be answered before the question of "how to do it." Public officials are often asked to compare various alternatives such as transfer stations versus direct haul, one potential landfill site versus another, resource recovery versus landfilling, one technology for resource recovery versus another, and many other combinations. County or state solid waste management plans are helpful in setting overall goals for waste management but are often times too general to allow public officials at the county or city level to choose among the various alternatives that meet these goals. Furthermore, many times these plans are not updated frequently (see Chapter 12, Section 12.12).

This situation leaves the city or county Public Works Director or Solid Waste Director to select the best course of action. Also, private developers of landfills, transfer stations, and resource recovery facilities must perform an analysis of some of the same alternatives as part of business planning.

The purpose of this chapter is to set out an analysis framework within which alternatives for the handling and disposal of municipal solid waste can be compared. This chapter is not intended to be a discussion of solid waste management planning methodology, although many of the elements of this analysis framework are used in preparing solid waste management plans.

In some situations, only a portion of the analysis framework presented here is needed, whereas in others the entire process may be appropriate. For example, resource recovery may not be included in the analysis, or the investigator may only be interested in comparing two locations for a transfer station. Both of these conditions limit the scope of the analysis considerably. This chapter, however, will present a comprehensive methodology that will accommodate all the elements of a solid waste management system (transfer stations, landfills, resource recovery facilities, etc.) from which the reader can select the elements that best fit his or her particular situation.

There are many considerations that go into "deciding what to do." These considerations are technical, environmental, political, and economic. All of these areas of consideration must be dealt with in the decision-making process, however an economic analysis is often the first step. Indeed, many times a comparison of the economic costs of competing alternatives is the single most important determinant of feasibility. For these reasons, this chapter will present an *economic analysis* methodology rather than attempting to discuss all elements of the decision-making process. This economic analysis methodology, which is many times called a "feasibility study," is a step-by-step procedure that can be used to account for all of the costs and benefits of each alternative using comparable basic data so that valid comparisons can be made.

The discussion in this chapter is limited in two primary ways. First, the analysis methodology deals only with alternative ways to handle and dispose of "municipal solid waste." This category of solid waste refers to the residential and commercial wastes that normally arrive at local or regional landfills. Definitions of these waste categories can be found in the reference literature[1,2] and in other chapters in this handbook. This chapter will not deal with analyzing alternatives for handling and disposing of industrial wastes (except for the packaging, office, and cafeteria wastes that normally are disposed in municipal waste landfills along with residential and commercial wastes). Neither will this chapter deal with the handling and disposal of bulky or hazardous solid wastes. These topics are covered in Chapter 17 and, in Part III, Chapters 20 and 21 of this handbook.

Second, this chapter focuses on alternative waste management practices that occur after collection of the waste. That is, this chapter deals with the analysis of all or a portion of the system of waste transport following collection (including transfer stations), waste processing and/or combustion, and final land disposal of raw or processed waste. This chapter does not include discussions of waste collection or methods to recover materials from wastes before collection (source separation). These two subject areas are presented in other chapters of this handbook.

The sequence of this chapter follows closely the logic of the feasibility study methodology: (1) gathering basic data, (2) identifying markets, (3) selecting alternatives, (4) net system cost modeling, and (5) comparing alternatives. Because of space restrictions, details about many of the techniques presented in this chapter is necessarily limited. However, references are given where appropriate for additional reading on certain subjects. Also, where appropriate, numerical examples from actual studies or hypothetical cases are given to help illustrate some of the concepts presented.

4.1.2 Gathering Basic Data

Gathering a complete set of reliable basic data is the critical first step in any economic feasibility analysis, and is perhaps the step that is most often ignored or put off until after other elements of the study have been started. It is important that the person performing the analysis (the investigator) avoid the temptation to "put the cart before the horse" by starting the other parts of the analysis, such as identifying markets for recovered energy or selecting alternatives, before completing the basic data-gathering step. The accuracy and viability of all calculations performed in other sections of this analysis depend on the data gathered in this step. For example, waste quantity data are the basis for facility sizing which determines capital costs, operating costs, and revenues. The investigator may find it necessary to recalculate almost every number in the analysis if waste quantity or other basic data must be changed late in the study.

The data that should be gathered in this step include: (1) existing waste quantity and composition, (2) projected waste quantity, composition, and geographic distribution, (3) unit costs for hauling wastes, and (4) potential sites for new landfills, transfer stations, and resource recovery facilities. Before beginning to gather these data, the investigator should establish the level of accuracy required for the analysis. For example, if the results of the analysis are to be the basis of a final decision to commit to the financing of a new landfill or resource recovery facility, the level of accuracy must be greater than if the purpose of the analysis is simply to select one or two alternatives for further consideration. The level of accuracy selected will significantly influence the cost of data gathering, so the investigator must balance the costs of data collection against the intended use of the results.

It is also important, before beginning, to define the study area. The study area boundaries may seem to be obvious, but many studies have taken much more time and money to complete because not enough thought was given to the definition of the study area. For example, in the case of a city study, it may be advisable to include certain surrounding communities or unincorporated areas that use the same landfills. In county or multijurisdictional studies in which it is unclear which portions of an area would be most efficiently served by a new facility, it is advisable to include the broadest potential area of participation into the study area so that the data are available at the beginning of the study rather than adding the data from new areas after part of the analysis is complete.

Existing Waste Quantity and Composition

The existing quantity and composition of the solid waste in the study area is arguably the most important set of basic data for any feasibility study. It is also many times the most difficult to define accurately because of unreliable or nonexistent data. The following paragraphs briefly explain how to go about obtaining the best information.

If the investigator is not already familiar with the existing collection/haul/disposal system in the study area, an examination of the existing system should be made first. The investigator should identify the major municipal and private collectors of commercial, residential, and industrial waste in the study area as well as the locations of all landfills in which waste collected in the study area is disposed. Other information that should be gathered for each landfill (if available) includes: (1) incoming waste quantity rates in tons (or metric tons) per day or week, and the portions of the study area from which the waste comes, (2) expected remaining operating life and any anticipated expansion plans, (3) existing environmental problems, (4) types of waste accepted, and (5) existing drop charge rates and any anticipated future increases. Locations and capacities of any other existing solid waste facilities such as transfer stations or incinerators should also be identified. This information will be used in other parts of the feasibility study.

There are three primary sources of data that can be used to establish existing solid waste quantities:

1. Landfill or transfer station scale data.
2. Volume or truck count records.
3. Landfill survey results.

Regardless of the source of data used, it is necessary to work with units of weight (tons or pounds; metric tons or kg) rather than volume (cubic yards or cubic meters). Although landfill operators are most concerned with the volume of the waste, the analysis of waste transport costs, transfer stations, and resource recovery facilities requires the use of units of weight because such units give a much more precise measure of quantity than units of volume.

Landfill or transfer station scale data will usually yield the highest quality estimates of existing quantity. It is useful to observe the actual operation of the scales to determine if there are any operational problems that might make the data inaccurate. Scale data should be collected for at least the most recent 12 months at all of the landfills or transfer stations serving the study area. If a certain landfill or transfer station also accepts waste from outside the study area, the data must be adjusted accordingly.

If scales are not present at a landfill, records of the number of loads and/or volume (cubic yards or cubic meters) are usually kept. Although not as accurate, these records can be used along with appropriate density factors to calculate tonnage. The investigator can use information from local municipal or private haulers or published information to establish density factors (pounds per cubic yard or kg per cubic meter in the collection vehicle). In general, density factors for compacted collection vehicles have ranged from 400 to 600 lb/yd^3 in the vehicle (237–356 kg/m^3). For noncompacted vehicles (pickup trucks and larger, open trucks), density factors have ranged from 100 to 200 lb/yd^3 (59 to 118 kg/m^3).[3] Packer manufacturer specifications for achievable densities should not be used for this calculation because these specifications usually represent the maximums for a particular type of packer body rather than the lower fleet average density factor that occurs in normal operation. It may be advisable to perform a local truck weighing program to confirm density factors for different truck types. Portable axle scales can be used with an acceptable degree of accuracy. A sample size of 10% of the average weekly number of loads for each truck type should result in acceptable precision.

If landfill or transfer station records are either unreliable or nonexistent, a landfill survey may be required (see Chapter 12, Section 12.12). Landfill survey methods involving weekly or seasonal truck counts and weighing programs have been demonstrated.[4,5] The investigator must carefully select the survey period, the number of landfills surveyed and the type of data to be taken depending on the amount of time and money that are available for this task and the level of data precision required. The investigator should consult Refs. 4 and 5 for further explanation of survey methods.

The data from any of these sources (scale data, volume or truck count data, or landfill survey) should result in an estimate of the number of tons collected and disposed by the homes and businesses in the study area in the most recent year. To provide information for sizing resource recovery facilities, the total quantity should be broken down into the residential/commercial waste category (sometimes called the "processible" portion) and other wastes such as white goods, construction/demolition debris, street sweepings, etc. (the "nonprocessibles"). If possible, a breakdown of quantities disposed by month should also be gathered. Monthly data are important because these data can be used to establish the seasonal variation in waste quantity which can be a factor in facility sizing and in matching the energy output of a resource recovery facility with the energy demand of potential energy markets. It is also helpful to compile quantity data for the past 3–5 years so that growth trends can be observed.

Table 4.1 and Figure 4.1 illustrate the kind of data that should result from the gathering of existing waste quantities. Table 4.1 displays annual waste quantity data for the northern portion of San Diego County, California (two landfills). Figure 4.1 illustrates a composite of 4 years of seasonal variation in waste quantity received at the same two landfills expressed as percentages of the total annual quantity.[6]

Table 4.1 Annual Quantities of Solid Waste, Two Landfills, Northern San Diego County, California (Tons per Year)[a]

Calendar Year	San Marcos Landfill[b]	Bonsall Landfill[b]	Total Waste[c]	Processible Waste[d]	Nonprocessible Waste[e]
1980	259,722	108,875	368,597	350,167	18,430
1981	296,692	108,602	405,294	385,029	20,265
1982	283,767	130,191	413,958	393,260	20,698
1983	311,822	164,765	476,587	452,757	23,829

Source: Ref. 6.
[a] Tons × 0.907184 = metric tons (t).
[b] Tonnages derived by truck count and the application of K factors (tons per vehicle) to different vehicle types. Quantities are reduced by 8% to account for K factor prediction error based on County test weighings.
[c] Total of San Marcos and Bonsall data.
[d] Processible waste, estimated at 95% of total.
[e] Nonprocessible waste, 5% of total.

When resource recovery is to be part of a feasibility study, it is necessary to know the composition of the waste in addition to waste quantity. Waste composition by material category and waste combustion characteristics should be estimated for the processible portion of the total waste. Since nonprocessible wastes will normally go directly to a landfill, composition information for these materials is not as important. Processible waste composition is the basis for calculating the quantities of recoverable materials and the potential energy yield.

The investigator should establish a material category breakdown which includes the percent (by weight) of all combustibles, ferrous metals, aluminum, and all other inorganics. A further breakdown of the combustibles into subcategories such as paper, cardboard, yard waste, plastic, rubber, etc., and the noncombustibles into subcategories such as glass, nonferrous metals, etc., is generally not required and is only necessary if the recovery of these fractions is contemplated. It is important to avoid overdetailing the waste composition. Breakdowns into minute subcategories will not add important information to the feasibility study and will require considerably more time and expense to obtain. Also, at a minimum, the following combustion characteristics should be established for *as-received* raw solid waste; higher heating value in BTU/lb (or kJ/kg), moisture content in percent by weight, and ash content in percent by weight. A full chemical analysis of the waste (ultimate analysis) will be necessary before design of the boiler and air pollution control equipment, but is not usually necessary for a feasibility study.

The most accurate way to gather composition data is to collect representative samples from local landfills or transfer stations for composition analysis. Several references discuss the methods for obtaining samples and conducting laboratory analysis.[5,7,8,9] These methods involve taking grab

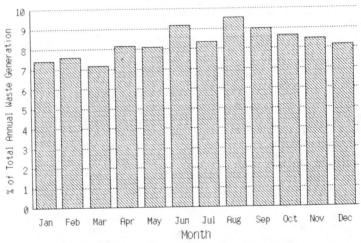

Fig. 4.1 Seasonal variation in municipal solid waste quantity. Composite of data from 1980 through 1983 for two landfills in northern San Diego County, California.

Table 4.2   Solid Waste Composition, San Diego, California, Combined Residential and Commercial Waste

Component	Percent by Weight
Newspaper	6.36
Corrugated	9.03
Other Paper	25.07
Plastic	7.37
Yard Waste	19.61
Other Combustibles	18.10
Ferrous	5.01
Aluminum	1.00
Glass	6.51
Other Noncombustibles	1.94
Total	100.00

Combustion Characteristics:

Higher heating value = 4,932 BTU/lb (11,472 kJ/kg)[a]

Ash content = 6.69%[b]

Moisture content = 26.03%[b]

Source: Ref. 9.

[a] As-received waste.

[b] Based on analysis of combustible fraction only.

samples of 500–1000 lb (227–454 kg) each, hand-sorting by material category and laboratory analysis of the combustible fraction for higher heating value, moisture content, and ash content. These sampling programs require a significant expenditure of time and money and should be undertaken only after careful research and experimental design. Table 4.2 shows the results of such a sampling program in San Diego County, California.[9]

An alternative to a sampling program is to examine the results of landfill sampling analyses in other areas along with other published information to form a judgment of the likely composition of the waste in the study area. Although not as accurate as a landfill sampling program, a reasonably accurate estimate of waste composition can be selected from these data sources because of the surprising uniformity in the composition of the processible portion of municipal solid waste in most parts of the United States. National averages for waste composition have been estimated by the U. S. Environmental Protection Agency (EPA)[10] and sampling program results have been reported for several places in the country.[9,11,12,13,14] Other references can be consulted for combustion characteristics.[15,16] The investigator should make any appropriate adjustments to the results of sampling programs in other areas to reflect more accurately composition in the study area. Factors to consider are as follows:

1. Differences in weather can affect moisture content and heating value.
2. Differences in the amount of source separation may cause differences in composition.
3. The waste composition in an area in which a mandatory beverage container deposit law (bottle bill) is in effect may have a different composition when compared with an area in which such a law is not in effect.

Projections of Waste Quantity and Composition

Making projections of waste quantity and composition is not an exact science, but it is a necessary step in any feasibility study. The investigator will often be confronted with conflicting or inadequate data upon which to make judgments about future waste quantities and composition. After all of the data are collected, the investigator must select the most reasonable data and must make certain simplifying assumptions to complete the analysis. The following general guidelines should be kept in mind while performing this analysis:

1. The investigator should be interested in long-term, average trends over 10–20 years rather than attempting to account for all forseeable short-term trends.
2. The investigator should make projections that reflect the average trend for the study area rather than attempting to account for changes that may occur in specific portions of the study area.

3. It is generally preferable to err on the low side of waste quantity projections rather than overestimating quantities, especially when resource recovery is part of the analysis.

With these things in mind, there are several steps involved in calculating a reasonable and usable set of waste quantity and composition projections: (1) establish unit waste factors, (2) establish waste generation districts, (3) obtain transportation data, employment projections, and/or population projections, (4) calculate future waste quantities, and (5) estimate future composition shifts. The following paragraphs briefly describe activities in each of these steps.

Population- and employment-based unit waste factors (pounds per capita or per employee per day) are widely used and accepted for projecting solid waste quantities. There have been attempts to relate the generation of solid waste to other factors such as household income level, retail sales volume, and others.[17] However, population and employment have been shown to be the best predictors of solid waste generation. These data are also commonly used as the basis for city and regional planning, so they are easily obtainable in projected form and are usually broken down by subareas within the study area. Sources for population and employment data include city or county planning departments, transportation departments, or the regional planning agency with jurisdiction in the study area.

A composite (residential, commercial, and processible industrial) unit waste factor is the most commonly used and the simplest to establish for waste quantity projection. It is calculated by simply dividing the total amount of solid waste in a particular study area by the population in the study area. In making this calculation, the investigator should make certain that the waste quantity represents only quantities generated in the study area and that the population and waste quantity data are both representative of the same year. Depending on the intended use of the unit waste factor, the investigator may wish to exclude all nonprocessible wastes from the waste quantity. Also, to avoid a seasonal bias in the unit waste factor, an annual waste generation quantity should be used in the calculation rather than a monthly or daily quantity. The resulting units of tons (or metric tons) per capita per year are easily converted to pounds (or kg) per capita per day by multiplying by a factor of 5.479 (2000 lb/ton divided by 365 days/year) or 2.740 (1000 kg/metric ton divided by 365 days/year). The magnitude of the composite unit waste factor varies from place to place depending on a number of factors including the amount of commercial and industrial activity in the study area as well as individual residential waste generation habits. Table 4.3 gives composite unit waste factors calculated from scale data in several areas in the United States along with an estimate of the national average by the U. S. EPA.[10,18−22]

The calculation of separate population- and employment-based unit waste factors for residential, commercial, industrial, and commercial/industrial type wastes has been discussed in the literature.[23] This approach to projecting solid waste quantities may provide for a more detailed accounting of waste generation in a study area, but is probably not necessary for most feasibility studies. Furthermore, many times it is difficult to obtain the necessary data to calculate these separate factors. Waste quantity data must be available in separate categories (residential, commercial, industrial), and both population and employment data must be available in projected form for the planning period desired and must be divided into the desired geographic districts within the study area.

In feasibility analyses that include optimizing the locations of solid waste facilities, it is necessary to divide the study area into a number of subdivisions or districts and make quantity projections for each. It is also necessary to obtain the distance and travel time between each district for the waste transport analysis (see Section 4.1.5). The first step in gathering this information is to divide the study area into districts. Census tracts are generally too small for the purposes of this analysis. It is common to find that city and county planning departments or regional planning

Table 4.3   Residential and Commercial (Composite) Unit Waste Factors

Location	Year	Unit Waste Factor lb/capita/day	(kg/capita/day)	Reference
San Diego Co., CA	1980	5.8	(2.6)	(18)
King Co. (Seattle), WA	1982	5.5	(2.5)	(17)
Phoenix, AZ	1978	6.25	(2.83)	(16)
Charlotte, NC	1977	6.7	(3.0)	(20)
Pinellas Co. (St. Petersburg), FL	1983	5.0	(2.3)	(19)
Omaha-Council Bluffs Region, NE and IA	1978	3.27	(1.48)	(16)
EPA national average	1975	3.2	(1.45)	(8)

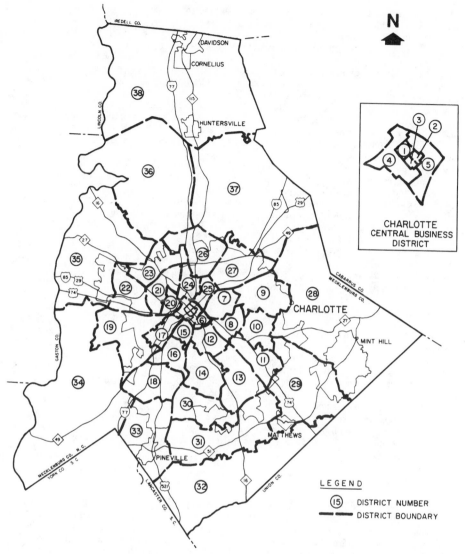

Fig. 4.2 Thirty-eight Waste Generation Districts selected for a study in Charlotte, North Carolina.

agencies will aggregate census tracts into "planning zones" or "transportation analysis zones." The use of these larger zones for "waste generation districts" (WGD) will normally provide more than enough detail for this analysis. The investigator may wish further to aggregate these zones to simplify the analysis. The size of the WGDs should be such that no more than 3–4% of the total waste generation is contained in each. Figure 4.2 illustrates the waste generation districts selected for a study of Charlotte, North Carolina.[22] These WGDs are aggregations of the 328 transportation zones comprising Mecklinburg County used by the North Carolina Department of Transportation for transportation planning purposes.

After establishing the WGDs for the analysis, the investigator should obtain the following data for each WGD: (1) travel time between each WGD and all the other WGDs, (2) travel distance between each WGD and all other WGDs, and (3) employment and/or population projections for the study period desired. The transportation data are usually available from city, county, or state transportation planning departments either in tabular form or on computer tape (or other data storage media). If not available, the time and distance data can be generated by drawing the WGD boundaries on a road map of the study area and using a map wheel to measure the distance between an assumed centroid for each WGD and the centroid for all other WGDs along the major roads.

Distances can be converted into travel times by assuming an average travel speed representative of a solid waste vehicle, or by selecting different travel speeds for the different sections of road that are assumed to be used by the vehicles (i.e., freeway, main street, side street). If a waste transport analysis is not to be part of the feasibility study, these data need not be obtained.

The employment and population projections should be available from city, county, or regional planning agencies in tabular form or on computer storage media. It may be necessary to aggregate these data into the WGDs chosen for the analysis. The investigator should select a projection period for the study that matches the financing period for any potential new facilities (landfills, transfer stations, for resource recovery) that are part of the study. A period of 20–30 years is common practice, but the period might be as short as 10 years. These projections are normally made in 5-year increments, but some are on a 10-year basis. If 5-year increment data are not available, the interstitial years can be calculated by linear interpolation without loss of significant accuracy.

Waste quantity projections are calculated by multiplying the selected unit waste factor(s) by the population or employment (as appropriate) for each WGD and then converting the result into tons (or metric tons) per year. Separate projections of residential, commercial, industrial, or commercial/industrial wastes (if used) can then be summed within each district to derive WGD totals. Summing the WGD totals will result in the total waste generated in the study area for each projected year. Table 4.4 illustrates the results of this type of calculation for the WGDs in Charlotte, North Carolina shown in Figure 4.2.[22]

It is probable that the waste generation habits of households and businesses will cause the solid waste stream to change in both quantity and composition in the future. These changes may render the previously discussed unit waste factor and waste composition assumptions invalid in the future—especially given that a projection of 20–30 years for waste quantity and composition is many times necessary for the kind of feasibility analyses discussed here. There is evidence that unit waste factors have had both up and down trends within the last 20–30 years.[10,11,23] It is usually best, however, to assume that the unit waste factors calculated with existing solid waste quantities and population or employment will remain constant over the study period. There have also been composition shifts over the last 20–30 years. For example, changes in food and other consumer product packaging has increased the amount of plastics and aluminum in the waste stream. Predicting the direction and magnitude of future solid waste composition changes is very difficult because the factors that influence these changes are not well understood. In the absence of any specific information about future changes that may affect the composition of the solid waste in the study area, it is advisable to assume that the existing composition will remain the same over the study period. One specific change that can directly influence both the unit waste factor and the composition is the implementation of a comprehensive source separation program in the study area. Another chapter in this handbook deals with the impacts these programs can have on waste quantity and composition.

Transport Costs

A feasibility study that includes the analysis of transfer stations or that compares different locations for solid waste facilities requires certain basic data for the analysis. Time and distance information between WGDs has been discussed previously. Along with time and distance data, the waste transport analysis requires estimates of the costs per mile and per hour to move the waste after collection from the end of the collection route to the various facility locations in the study.

Depending on the type of analysis, unit transport costs (dollars/mile or km and dollars/hr) may be required for the following vehicle types: (1) collection vehicles (packer trucks), (2) transfer vehicles (vehicles used to transfer larger loads from transfer stations), and (3) residue and/or ash vehicles (usually a dump truck or roll-off truck). If transfer stations are part of the analysis, unit costs for both the collection vehicle and transfer vehicle will be required, and if a resource recovery facility is involved, the residue/ash vehicle unit costs are required.

Methods for calculating these costs have been discussed in the literature[23] and should be consulted for more detail. The following are some general guidelines:

1. Calculate unit costs for collection vehicles only for transport of the waste after collection. That is, maintenance costs for the packer body should not be included, and fuel efficiency during travel after the collection route should be used. Labor costs for the driver and crew (if the crew normally accompanies the truck to the disposal site) should be included as a dollar/hr unit cost.

2. Costs for the type of transfer vehicle that matches the type of transfer station selected should be developed (i.e., open-top or compacted). Labor costs for the driver should not be included as a dollar/hr unit cost. Driver labor should instead be included as part of the annual labor cost of the transfer station.

3. Driver labor should be included as a dollar/hr cost for the residue/ash vehicle.

Table 4.4 Population, Employment, and Processible Waste Projections by WGD for Charlotte, North Carolina

WGD	1980[a]			2000[a]		
	pop.	emp.	waste	pop.	emp.	waste
1	300	12,767	16,768	1,300	22,710	30,161
2	100	9,877	12,914	100	12,200	15,941
3	0	7,384	9,621	0	20,645	26,901
4	1,800	10,965	15,076	2,400	22,684	30,609
5	4,900	10,929	16,388	6,000	17,669	25,652
6	4,100	10,364	15,301	4,100	10,457	15,421
7	15,300	4,057	11,986	15,300	3,917	11,806
8	14,100	6,390	14,502	14,100	6,205	14,261
9	23,100	3,383	14,527	24,200	7,082	19,827
10	13,200	4,570	11,737	13,800	8,018	16,492
11	15,700	4,625	12,904	18,900	5,386	15,297
12	14,500	5,475	13,484	14,700	5,703	13,870
13	19,500	1,965	11,101	22,000	3,652	14,394
14	15,500	4,668	12,870	17,600	10,932	21,954
15	7,400	10,332	16,704	7,400	9,690	15,867
16	15,000	7,625	16,506	15,200	13,395	24,112
17	10,900	8,236	15,506	10,900	7,234	14,198
18	23,000	5,079	16,692	25,900	6,010	19,174
19	13,500	3,430	10,382	13,700	14,867	25,374
20	3,700	9,410	13,882	3,700	8,804	13,093
21	17,600	7,156	17,034	17,600	8,156	18,352
22	8,400	5,293	10,575	12,900	10,466	19,288
23	11,000	6,862	13,760	13,000	16,905	27,721
24	12,900	3,212	9,834	13,600	3,876	11,008
25	6,100	9,932	15,614	6,500	9,480	15,200
26	8,000	3,029	7,452	11,900	5,005	11,734
27	18,400	6,901	17,051	23,600	7,556	20,183
28	18,700	1,569	10,236	49,400	10,782	35,686
29	14,900	4,591	12,510	35,200	6,581	23,992
30	14,200	2,972	10,093	17,800	2,991	11,693
31	16,000	1,065	8,395	42,200	2,967	22,349
32	4,000	0	1,752	10,000	0	4,380
33	6,200	5,396	9,747	11,300	10,164	18,194
34	6,500	8,197	13,527	26,000	29,983	50,454
35	17,300	3,893	12,651	33,500	12,486	30,944
36	7,700	1,676	5,557	29,800	14,480	31,921
37	11,900	3,515	9,792	38,000	14,651	35,734
38	13,500	0	5,913	30,000	0	13,140
Totals	428,900	216,790	470,344	653,600	383,801	786,377

Note: tons × 0.907184 = metric tons (t).
[a] Tons/year.

The result of this analysis will be unit transport costs that will be used in combination with the time and distance data to determine the relative transport costs for different facility locations. Section 4.1.5, Net System Cost Modeling, shows how the transport analysis is integrated with other parts of the economic analysis methodology.

Potential Facility Sites

An essential part of this basic data-gathering step is to establish potential sites for landfills, transfer stations, or resource recovery facilities depending on the scope of the feasibility study. Many times the potential facility sites will be known or established as part of the study in advance. For those studies, in which potential sites are not known, the following paragraphs will provide some guidance.

For a landfill site, it is first necessary to know the size of the area required. Methods for calculating acreage requirements can be found in the literature.[24,25] After calculating the acreage requirement, the investigator should look for parcels of land that meet the size restriction and also have the following general characteristics:

1. Vacant industrial or agricultural land.
2. Close proximity to the study area.
3. Favorable soil type and topography.
4. Good access to major roads.
5. Maximum separation from residential areas.
6. Separation from surface or ground water.
7. Not in a flood zone.
8. Not within 10,000 ft (3048 m) of an airport with jet traffic.

More detail about landfill siting can be found in Chapter 11 of this handbook.

For potential transfer station sites, the investigator first should consider sites adjacent to existing landfills that will be closed in the future if these sites are in relatively close proximity to the study area. In addition to existing landfill sites, the investigator should look for vacant industrial land with good access to major transportation routes. Generally, 2–5 acres are required, depending on the capacity of the facility. A sloping site is preferable, especially if there is access from both the low and high elevations.

Potential sites for resource recovery facilities should have many of the same characteristics as potential sites for transfer stations. Many times, however, the site for a resource recovery facility is constrained by the location of the energy buyer (a facility generating steam for sale to an industrial buyer, for example). For resource recovery facilities not so constrained (facilities that generate electric power or produce a fuel for sale to a distant user), the site of an existing transfer station or adjacent to an existing landfill has many advantages from a public acceptance standpoint. Any potential site should have 10–20 acres available, have good transportation access, and have adequate availability of utilities (gas, water, electric power, sewer) and fire protection. Also, the availability of cooling water is critical to the location of an electric power generation facility.

4.1.3 Identifying Markets

This section only applies to feasibility studies that include resource recovery alternatives. It is necessary to identify and collect certain data for potential buyers of the outputs of a resource recovery facility (materials and energy) for two major reasons. First, the potential buyers, or "markets" determine which forms of energy and materials should be recovered. Second, the available markets influence the choice of a location for the resource recovery facility.

The following paragraphs briefly describe how to calculate the quantities of recoverable resources using the basic data and how to survey both energy and materials markets in the study area (also see Chapters 10 and 15).

Quantities of Recoverable Resources

Before beginning the energy and materials market surveys, it is helpful to calculate, on a preliminary basis, the amounts of the different forms and types of energy and materials that might be recovered from the available waste. These preliminary quantities can help limit the survey scope and can assist in discussions about the proposed resource recovery facility with potential markets.

The amount and form of energy available in the study area depends on the characteristics of the waste and the particular type and size of energy conversion technology utilized. Since alternative energy conversion technologies have not yet been selected in this analysis, typical factors can be defined to calculate the amounts of the common output energy forms. These amounts can then be used to screen potential energy markets. Table 4.5 gives some typical energy conversion factors that may be used to make such estimates.

The amounts of the various materials recoverable from the waste are a function of the waste composition and the recovery efficiency of the separation method used. Table 4.6 displays some example calculations. The investigator should use the waste composition data obtained in the basic data-gathering step for this calculation. Note that these factors apply only to materials separated from mixed solid waste by mechanical means or by a combination of mechanical and manual means

Table 4.5 Typical Energy Conversion Factors

Yield	Solid Fuel[a]	Steam[b]	Electric Power[c]
Quantity per ton of waste	0.5 to 0.75 tons	6,300 lb	500 kWh
Quantity per metric ton of waste	0.5 to 0.75 tons	2,835 kg	551 kWh
BTU per ton of waste	7.5–10.5 million	6.5 million	—
kJ per metric ton of waste	7.2–10.0 million	6.0 million	—

[a]Range based on data from the Ames, Iowa, facility[26] and the Madison, Wisconsin, facility.[27]
[b]Based on 4500 BTU/lb (10,467 kJ/kg) raw waste, 70% combustion efficiency, 10% in-plant steam usage, and 1000 BTU/lb (2326 kJ/kg) steam.
[c]Based on 4500 BTU/lb (10,467 kJ/kg) raw waste, and 18,000 BTU/kWh (18,991 kJ/kWh) total plant heat rate.

in a resource recovery facility. Potential recovery rates for source-separated materials are discussed elsewhere in this handbook (see Chapter 12, Table 12.4.7).

Energy Market Survey

The list of potential energy markets should include the electric power utilities serving the study area and large users of fuel for steam generation in the study area. With information from this list, the full range of potential energy buyers can be established. The state or local Chamber of Commerce generally has a list of the largest industries in the study area. The headquarters of the local electric utility can also be consulted for locations and technical data for all area power plants.

Other sources of information are also available. Much of the information required for the energy market survey can be obtained by gathering data for the existing boilers in or near the study area. The most comprehensive source of information about boilers in a given area is usually the list of all permitted point sources of air pollution kept by the state agency responsible for air quality, or in some states, county health departments or air pollution control districts. These lists are usually computerized and contain the location, rated capacity, owner, and many times the amounts and types of fuels burned in each year. Reference 33 gives a more detailed discussion of the data available and how it can be used. To limit the survey time, the investigator should only consider establishments within a 10- to 20-mile (16–32 km) radius of the study area that have a combined heat input to all boilers on one site of greater than about 50 million Btu/hr (53 million kJ/hr), the energy equivalent of about 100 tons (90.7 metric tons) per day of processible waste.

An initial telephone contact can be helpful in further eliminating certain potential markets from the survey list. The telephone contact should be used to confirm capacity information, types and quantities of fuels used, steam conditions, and to discuss operating schedules (daily, monthly). Another critical piece of information that can be obtained in a telephone call is seasonal variations in steam generation. If steam generation typically drops to zero (or an idling amount) during certain parts of the year for a given potential market, this prospect should probably be dropped from

Table 4.6 Example Material Recovery Factors

Material	Typical % Composition	Recovery Efficiency (%)	% of Total Proc. Waste
Ferrous metals[a]	5.0	74	3.7
Aluminum[b]	1.0	80	0.8
Glass[c]	6.5	50	3.25
Film plastic[d]	2.0	50	1.0
Corrugated[e]	9.0	50	4.5
Compost[f]	—	—	37.5

[a]See Refs. 28 and 29, Chapter 4. Based on tests at Recovery I facility in New Orleans, Louisiana.
[b]Based on raw waste pretrommeling and eddy current separation. See Refs. 28 and 31.
[c]See Ref. 30.
[d]Polyethylene film plastics only, recovery efficiency based on Sorain-Cecchini system. See Ref. 31.
[e]Recovery efficiency for hand picking assumed.
[f]Assumes 25% of waste is noncompostable, and 50% reduction ratio in the digestion process (see Ref. 32).

consideration. This seasonal drop in steam demand is typical of certain industries which have seasonal production schedules or use steam mainly for space heating or cooling.

A personal visit to those potential energy markets remaining under consideration should be made. The information that should be gathered during the visits includes:

1. Site information including available land for a resource recovery facility, and steam piping layout.
2. Fuel usage on an annual, quarterly, and monthly basis.
3. Current and projected fuel prices.
4. Steam generation on an annual, quarterly, and monthly basis for the past 2–3 years.
5. Boiler data including type, rating, age, ash handling system, steam conditions, and air pollution control systems.

Questionnaire forms that can be filled out by the investigator during the plant visit or left with the plant contact person to fill out have been developed.[34]

An analysis of the data gathered in the survey can identify which potential markets are possible users of fuels derived from solid waste which can replace the conventional fuels used to produce steam, or of steam at specified temperature and pressures to supplement or replace steam now generated. Fuels that can be derived from solid waste include the solid variety, which is the separated combustible fraction either in a loose, shredded form, or in pellets, and the liquid and gaseous variety, which are the products of various pyrolysis processes. Solid refuse-derived fuel (RDF) has been successfully co-fired with coal in travelling grate, spreader-stoker-type boilers designed for coal only at a nominal rate of 50% of the input energy in both shredded and pelletized form.[35,36] Solid fuel in shredded form has also been successfully co-fired with coal at rates of up to 10% of input energy in suspension-fired, pulverized coal boilers.[35] Pyrolysis gas and oil have been test-fired in existing natural gas- and fuel oil-fired boilers, but these solid waste pyrolysis facilities have failed to prove feasible economically and technically. None of these pyrolysis systems is currently in operation. Markets for replacement fuel, therefore, should be limited to prospects that have existing spreader stoker, cyclone, or suspension-fired coal boilers. The investigator should be careful to use the co-firing rates previously given for calculating the amount of RDF that may be utilized by each market.

As an alternative to purchasing replacement fuels, many large energy users may prefer purchasing replacement steam. Steam can be generated from the combustion of solid waste in various types of equipment, which is discussed in Chapters 12, 13, and 16 of this handbook. A single user with an annual and seasonal steam demand sufficient to purchase all of the steam available from the solid waste in the study area is preferred. However, facilities such as the Akron, Ohio, and Haverhill/ Lawrence, Massachusetts, plants have been built to serve several steam users and to generate electricity via cogeneration. The investigator should consider industrial steam users, hospitals, university campuses, or other large building complexes served by a central heating/cooling system. Electric power utilities are limited markets for steam because most utility boilers require steam at temperatures above the recommended limit of 750°F (399°C) for solid waste boilers.

Electric power can be produced from solid waste combustion with the addition of a turbine generator to the steam generation equipment. Generally, higher steam temperatures and pressures are required for generating electric power. Chapters 12 and 13 in this handbook describe the available technology in more detail. The most viable markets for electric power are public electric power utilities, investor-owned electric power utilities, or municipal electric power systems. Selling electric power directly to an industrial user is generally not feasible because of the on-line availability usually required by these users. The passage of the Public Utility Regulatory Policies Act of 1978 and the adoption of the associated Federal Energy Regulatory Commission (FERC) regulations in 1980 have made utility markets more accessible to resource recovery facilities.[37] Some utilities have developed "Standard Offers" for the purchase of electric power from solid waste and other qualifying power-producing facilities. The investigator should meet with representatives of the utility to discuss the terms of purchase such as price (energy and capacity), required availability, tie-in locations, voltage, and any other technical requirements. It is important to know all of the buyer's requirements in advance of the conceptual design of the facility.

The investigator should also consider markets for co-generated steam and electric power. Chapters 12, 13, and 16 of this handbook describe co-generation techniques. The markets for the steam and electric power that can be produced in a solid waste-fired co-generation facility are the same as previously discussed for these energy forms separately. The best candidates for a co-generation facility are industrial or institutional steam users with steam demands below the amount of steam available from the solid waste in the study area or those with a highly seasonal steam demand that periodically drops below the seasonal steam availability. With a co-generation facility, the excess steam can be used to generate electric power.[37]

Materials Markets

There are markets for a variety of materials that can be separated or produced from mixed municipal solid waste. Some of these same markets purchase materials separated from solid waste prior to disposal (source separation), but this discussion is limited to materials that are separated mechanically or by a combination of mechanical and manual means after disposal. Another chapter of this handbook discusses markets for source-separated materials (see Chapter 10, Section 10.1.10; Chapter 15, Section 15.2).

Materials for which separation technology exists fall into the following general groups:

1. Ferrous metals.
2. Aluminum.
3. Mixed nonferrous metals.
4. Glass.
5. Polyethylene plastic.
6. Corrugated cardboard.
7. Compost.

Methods and equipment for recovering these materials are discussed in Chapter 12, Section 12.7.

There are two general types of markets for the ferrous metals: tin or copper recovery and remelt. Tin and copper recovery operations chemically process the recovered metal to recover the tin plating from the steel cans which make up the majority of the material or use the cans in the recovery of copper from low-grade copper ore. Steel-making industries remelt the material along with virgin iron in the manufacture of new steel products. Remelt buyers are by far the strongest market, but the demand for scrap fluctuates with the demand for steel. More detail about ferrous markets and material specifications can be found in Chapter 15, Section 15.2.

Markets for aluminum include major aluminum sheet and bar producers and aluminum smelters producing specification ingots. Currently, the market for recovered aluminum is strong and prices are relatively high for material that meets market specifications, primarily with respect to alloy mix.[39] The best markets for mixed nonferrous metals are scrap dealers—especially those operating auto shredders. These dealers may hand pick some of the higher-value materials such as copper and sell the rest or incorporate the product into other loads of nonferrous scrap (see Refs. 29 and 38 for more detail).

Markets for recovered glass are primarily glass container manufacturers who can use recovered glass along with virgin materials in the manufacturing process. The separation technology for recovery of glass has been demonstrated,[40] but sorting the glass by color has not proved feasible as yet.[41] Noncolor-sorted glass can be used in mixtures of up to 20% by manufacturers of green or amber glass containers (see Refs. 29 and 38 for a more detailed discussion).

The markets for polyethylene plastic are manufacturers of plastic bags and other film plastic products. These manufacturers now purchase or use their own off-specification rejects or trim-mings in the manufacturing process along with virgin polymers. These buyers require a clean, pelletized product that meets certain strength and purity requirements. There are mechanical processes for removing film polyethylene plastics (less than 15 mils or 0.038 cm, primarily plastic bags) from mixed municipal solid waste; however, none has been demonstrated in the United States.[31]

The markets for corrugated cardboard include secondary materials brokers and corrugated, kraft, or linerboard manufacturers. Materials brokers will arrange for the sale of the corrugated to either domestic or foreign paper manufacturers. Hand picking, sometimes in conjunction with mechanical concentration, has been used to recover corrugated from mixed solid waste. Market specifications for grade-11 "corrugated containers"[42] must generally be met to market this material successfully.

The technology for the production of compost from solid waste is well-developed and has proved technically feasible.[43,44] Most municipal solid waste composting systems have failed, how-ever, because of the difficulty in marketing the product at a price that can cover the costs of production. Although new areas for market development have been suggested,[45] the current mar-kets for compost continue to be primarily individual homeowners who use the product for lawns and gardens (also see Chapter 19, Biological Processes).

A good source of information on the current market value of recovered materials is a local scrap dealer or a national broker of secondary materials. Local scrap dealers can be found in the tele-phone yellow pages and national brokers can be found in some of the trade publications such as *American Metals Market*, *Iron Age*, and *Official Board Markets*. Because revenues from the sale of recovered materials usually represent only a minor portion of the total revenues to a resource recovery facility, it is not usually necessary to have long-term contacts for the sale of these materials in advance of project financing. However, it is prudent to get written indication of interest in the purchase of recovered materials along with price and delivery terms from potential buyers during the feasibility study stage. Methods for obtaining such written indications of interest have been discussed in the literature.[46]

4.1.4 Selecting Alternatives

This part of the feasibility analysis methodology brings together the basic data and market information gathered previously to select a set of alternatives for economic analysis. The alternatives must be carefully crafted to suit the needs of each particular feasibility study. The purpose of this section is to provide some general guidelines for selecting alternatives that, when analyzed, will give the desired results.

Alternatives that are poorly thought out or incomplete are a common problem in feasibility studies for solid waste management. Consider that the best economic analysis possible will not result in valid conclusions if the selected alternatives lack the proper elements, test the wrong factors, or are not comparable. In selecting alternatives, the investigator must be certain that all of the elements of the solid waste management system that may change as a result of the proposed actions are included. For example, if landfilling and resource recovery are being compared, the differences between the costs of raw waste transport, landfilling costs for residues and nonprocessibles, and revenues must be compared as well as the differences between the costs of building and operating the landfill and the resource recovery facility.

The selected alternatives must also be comparable both with respect to the quantity and types of wastes handled and with respect to time. All selected alternatives must handle all (or the same portion) of the waste generated in the study area over the entire study period. For example, the cost of an existing landfill with 5 years of remaining capacity that disposes of both processible and nonprocessible wastes is not comparable to the cost of a resource recovery plant that will handle only processible wastes over a 20-year period.

In selecting alternatives, the investigator must consider the following elements for each:

1. *The Primary Facility.* The facility to which the majority of the waste will be directed for processing or disposal. This facility can be either a landfill or a resource recovery plant or a combination of more than one facility.

2. *The Final Disposal Landfill.* The landfill used for disposal of all processing residue, ash, nonprocessibles, and by-pass waste when a resource recovery facility is part of the alternative.

3. *The Transport System.* The system of transporting the raw waste from the end of the collection route to the primary facility(ies) and/or to the final disposal landfill (may include transfer stations).

Depending on the investigator's study scope and objectives, some of these elements may not be needed in each alternative. For example, if different primary facilities are being compared which will use the same site, the transport system element can be excluded from the analysis. If there is only one primary facility in the analysis, and the interest is mainly in comparing different transport systems, the primary facility can be removed from the alternatives. Other examples of situations requiring a more limited analysis are numerous. The investigator must design the alternatives for each particular situation.

The following paragraphs give certain general guidelines for establishing the elements of each alternative.

The Primary Facility

For each primary facility, consideration must be given to technology, configuration, and location. Solid waste management technology is discussed in other chapters of this handbook and will not be repeated here; however, the investigator should have a working knowledge of both landfilling and resource recovery technology to select appropriate primary facilities. The investigator may already have some idea about the technologies to be compared. It is essential, however, that the technology selected for any resource recovery alternative first meet the requirements of the potential energy markets. Energy sales comprise 95% of the revenue in a typical resource recovery facility; therefore, it is essential that the selected technology meet the requirements of the energy buyer as a primary goal. Materials markets, while important, should not govern the selection of the basic technology, because most marketable forms of materials can be separated in combination with a variety of different energy recovery technologies. The investigator must know the energy form(s) required (fuel, steam, electric power), delivery rates desired, and the required physical specifications (in the case of RDF production). With this information, the required equipment can be selected and sized.

It is also important that the investigator choose technologies for resource recovery that have been demonstrated successfully in full-scale operation over a significant amount of time. A proven technology is important not only from a reliability and cost standpoint, but also because financing an unproven technology can be difficult.

In addition to selecting general types of technologies for the primary facilities, the configuration of the facility must be determined in preliminary terms. As part of the configuration, the capacity of the facility or facilities must be established. For resource recovery alternatives, the type of receiving arrangement (floor dump or pit) and raw waste surge storage sizing must be determined. In addition, the number and capacities of processing lines, boilers, and other equipment must be determined. Also, the type of energy delivery system (steam line, truck transport, electric power lines, etc.) must be established. This information is necessary to account properly for all facility costs. If the investigator is not knowledgeable in these areas, it may be prudent to engage the services of a qualified engineer.

A particularly important decision at this point in the analysis is the sizing of the primary facility. For a landfill, the investigator should first determine the remaining life of the existing landfills used by the study area. If the remaining capacity will allow disposal beyond the end of the projection period, then the present landfill(s) can be used as the landfill alternative. If multiple landfills are currently used, the investigator should be certain to consider the increased fill rates at some landfills that will result from the closure over time of other landfills when calculating remaining capacity. If the present landfilling system does not have sufficient capacity to dispose of the waste from the study area for the entire projection period, then a new landfill size and location must be postulated. Methods for calculating the required landfill area for a given waste generation rate are discussed in other chapters of this handbook.

For a resource recovery facility, sizing is a critical task. In the situation in which the energy market is a utility purchasing electric power, or a steam market with a demand that is always greater than the amount of steam that can be generated with the available waste, the facility should be sized for the amount of waste available in the start-up year, rather than selecting a higher capacity to allow for future growth in the waste stream. This sizing criterion is advisable because quantity projections are speculative and the economic consequences of over-sizing can be severe. Furthermore, the facility can be designed to accommodate expansion easily at a later date, and any excess waste that is generated can be by-passed to the landfill. If the energy market demand is highly seasonal or always below the amount of energy available from the solid waste, the facility can be sized to match the available waste quantity and either dump the excess energy, or use the cogeneration concept (if technically feasible). Alternatively, the facility might be sized to meet only the energy market demand with the excess waste disposed by landfilling. The choice depends on the revenues which can be derived from cogenerated electric power versus the cost of installing and operating the required equipment. It is important to understand, therefore, that the selection of a resource recovery facility size depends on a number of technical and economic issues such as the preceding, and should be considered carefully when establishing the configuration of the alternatives.

Primary facility location is another decision that must be made in establishing alternatives. Methods for selecting potential sites for landfills have been discussed previously. A new landfill must be part of the primary facilities in the landfill alternative if existing landfill capacity is not sufficient to dispose of all study area waste over the selected projection period. The location of the resource recovery facility depends on site availability and the type of energy being produced. For facilities producing solid fuels (RDF), the location can be as close to the generation of the raw waste as possible because raw waste transport costs to incinerators are much higher than fuel transport costs. If the facility produces steam, it is necessary to select a site close to the energy buyer because of the high cost of building and operating long steam lines. For facilities generating electric power, the facility should be sited close to the waste generation because there are usually many potential locations for utility grid tie-in. In certain cases it may be best to locate a processing plant in one location and the associated combustion facility in another.

The Final Disposal Landfill

A landfill for the final disposal of nonprocessibles, processing residue, and ash is required for any resource recovery alternative. In addition, a landfill is needed to dispose of processible waste which must bypass a resource recovery facility during scheduled and unscheduled facility downtime, or because the facility is sized with a capacity less than the available waste quantity. With a resource recovery facility, the volume of landfill required will be considerably less than a waste management system that relies totally on landfilling, but the need for a landfill is by no means eliminated.

To determine the landfill volume requirements, the investigator should first calculate the tonnage and average density (in-place, compacted in the landfill) for all nonprocessibles over the life of the resource recovery facility. The tonnage of nonprocessibles should have been determined in the basic data-gathering step. Average in-place densities may range from 1500 to 2000 lb/yd³ (890–1187 kg/m³) depending on the material. Second, based on an estimate of the facility operating availability, the investigator should calculate the amount of bypass processible waste that will be disposed of in each operating year of the resource recovery facility. The investigator should consider the effects of seasonal variations in waste quantity because the amount of waste available may exceed the facility throughput capacity during certain parts of the year and may be below this capacity in others. The effects of a waste stream growing with population versus a constant resource recovery facility throughput capacity also will influence the amount of by-passed waste over time. In-place density factors of 1000–1200 lb/yd³ (593–712 kg/m³) are achievable in a well-run landfill for raw waste. Also, a typical cover-to-waste ratio of 4:1 can be used to calculate required landfill volume for both nonprocessible and by-passed waste processible wastes.

Finally, the volume of combustion ash* and/or process residue from the amount of waste processed by the resource recovery facility must be calculated. Between 30% and 35% of the processed tonnage will remain to be landfilled depending on the technology used. This material will have an in-place density of between 1200 and 1800 lb/yd³ (712–1068 kg/m³) and the cover-to-waste ratio will be generally higher than for raw waste. A ratio of about 6:1 can be used because the material may not require covering as frequently as raw waste and because this material has fewer surface voids than raw waste. The sum of all of the required landfill volumes over the life of the resource recovery facility will enable the investigator to determine if the present system of landfills is sufficient, or if a new final disposal landfill must be part of the resource recovery alternatives.

The Transport System

The final configuration of the most economical waste transport system will be determined in the mathematical modelling section of this chapter (see Section 4.1.5). At this point in the analysis, the investigator must make a preliminary judgment as to the need to include transfer stations in the analysis. If the longest distance between a waste generation zone and a potential primary facility location is 10–15 mi (16–24 km) or greater, transfer stations should be included in the analysis. If the study area is geographically smaller than this, it is generally unlikely that transfer stations can be justified economically.

If transfer stations are to be a part of the analysis, the investigator must select the type of transfer station to be used in the analysis (compacted or open-top, noncompacted) and the range of potential sizes for the stations. The selection of transfer station type is a function of the load limitations on the public roads in the study area and other factors. Chapter 9 describes the different transfer station types and selection criteria (see Chapter 9: Transfer Municipal Solid Waste).

The correct capacity for a transfer station or group of transfer stations cannot be determined at this point in the analysis. The proper capacity depends on how many WGDs will show an economic advantage in transporting waste to the transfer station rather than directly to the primary facility. Section 4.1.5 describes a mathematical model for determining transfer station sizing. It is, however, necessary to select two or three alternative transfer station sizes for use in the mathematical modeling. By examining the potential locations for the transfer stations in relation to the locations chosen for the primary facilities, the investigator can estimate visually which WGDs might be closer to the transfer stations than the primary facilities. The sum of the waste generation projections for these groups of WGDs will give an estimate of the upper limit of the transfer station size. One or two smaller sizes should also be selected for cost analysis so that a smaller size can be tested in the mathematical modeling of the transport system.

4.1.5 Net System Cost Modeling

The net system cost (NSC) model is a mathematical modeling technique that can be used to select the most economical facility configuration of each alternative under consideration and compare the initial operating year system costs. The NSC model can be used, for example, to determine the most economical location for a new landfill or resource recovery facility given a number of possible choices. The model could also be used to select the most economical locations, sizes, and service areas for transfer stations given a number of possible sites. Indeed, the model is structured so that all of the elements of a solid waste management system (following collection) can be simulated so that each alternative under consideration can be configured to minimize the net costs for transportation, processing/combustion, and disposal.

Figure 4.3 illustrates all of the possible elements of the net system cost model. Each transportation link and each facility has an associated cost, and certain facilities may have incoming revenues. Not all of the selected alternatives will include all of these elements. For example, a landfill-

---

*Incinerator ash toxic leachate potentials must be thoroughly evaluated for possible secure landfill requirements.

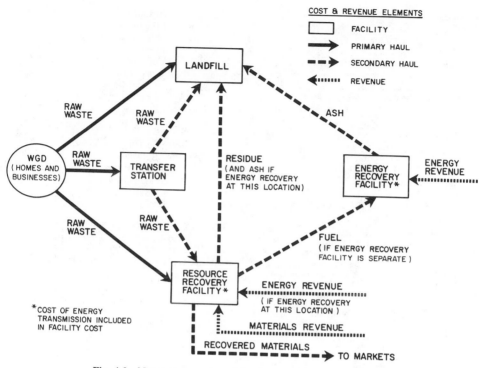

Fig. 4.3 Net system cost model—cost and revenue elements.

only alternative will not include the resource facility and may, or may not include the transfer station. The purpose of the model is to allow the investigator to calculate the system costs of all possible combinations of transportation links, facility locations, and facility sizes within each alternative to select the most economical configuration. The optimized configuration of each alternative can then be compared. Methods for economic comparison are presented in Section 4.1.6.

Figure 4.3 illustrates the possible combinations associated with a single WGD, one transfer station location and size, one landfill location and size, and a single resource recovery location. Given a study area with many WGDs, several potential locations for landfills, transfer stations, and resource recovery facilities, and several potential sizes for transfer stations, the possible combinations of WGD waste assignments, facility sizes and facility locations can become quite numerous. For this reason, system modeling problems of this nature have been reduced to a set of equations for solution by computers. These models, which use linear programming techniques, have been described in the literature[47–52] and will not be repeated here. However, to give the investigator a general understanding of how the model works, the following paragraphs will describe the calculations in simplified terms.

Certain elements of this model require estimates of the capital and annual costs of the various facilities which may be included in the analysis (landfills, transfer stations, and resource recovery facilities). Cost-estimating techniques for these kinds of facilities have been discussed in the literature[53] and will not be described here. It is prudent to utilize qualified professionals for facility cost-estimating so that all of the proper elements of capital and annual costs are included. Cost estimates using capital cost factors in dollars per ton of daily (24 hr) throughput and annual cost factors based on a dollar per ton processed figure have been presented.[54] It is unwise, however, even for an initial feasibility study to use such factors. There are so many local variables that can influence costs, that the investigator risks calculating costs that have no relation to reality by using such factors. With a minimal amount of time and expense, an experienced solid waste professional can perform a preliminary capital and annual cost estimate of much higher quality.

The net system cost model has the following general mathematical structure:

$$NSC = \sum_{j=1}^{J} (P_j + S_j + OM_j + A_j - R_j) \tag{4.1}$$

where

NSC = Net system cost in dollars per unit time.
j = Individual facility j.
J = Total number of facilities in the configuration.
P_j = Primary haul cost to facility j in dollars per unit time (week, year).
S_j = Secondary haul cost for facility j in dollars per unit time.
A_j = Debt service for facility j in dollars per unit time.
OM_j = Operation and maintenance cost for facility j in dollars per unit time.
R_j = Total revenue income to facility j in dollars per unit time.

Equation 4.1 can be interpreted as follows: For any particular configuration of facilities and waste assignments, the NSC is the sum of the transport and facility costs minus the revenues for all facilities. The computer can calculate NSC for all possible configurations of each alternative so that the configuration with the lowest NSC can be selected. The following paragraphs describe each element of the NSC equation in more detail.

Primary Haul Cost (also see Chapter 8)

Primary haul costs are the costs incurred in transporting raw waste from the end of the collection route to a landfill, transfer station, or resource recovery facility (see Figure 4.3). For any particular landfill, resource recovery facility, or transfer station, the primary haul cost is the sum of the primary haul costs from each WGD contributing waste to the facility. Or, for any facility j:

$$P_j = \sum_{i=1}^{I} \frac{W_{ij}}{C_i} [(PMC_i\, D_{ij}) + (PHC_i\, T_{ij})] \tag{4.2}$$

where

i = Individual WGD i.
I = Total number of WGDs in the analysis.
PMC_i = Unit haul cost in dollars/mile (or dollars/km) for WGD i.
D_{ij} = Round-trip distance in miles (or km) between WGD i and facility j.
W_{ij} = Amount of waste transported from WGD i to facility j in tons (or metric tons) per unit time.
C_i = Capacity of the collection vehicle for WGD i in tons (or metric tons).
PHC_i = Unit haul cost in dollars/hour for WGD i.
T_{ij} = Round-trip time in hours between WGD i and facility j plus unloading time at facility j.

Equation 4.2 is constrained in the following ways. First, the sum of all waste transported to any facility must not exceed that facility's throughput capacity. Second, all waste generated in any WGD must be transported to one or more facilities. Note that the equation can be simplified somewhat if a composite per-mile (or per-km) unit haul cost is used instead of both time- and distance-related unit haul costs. Note also that this equation represents the use of a single set of unit haul costs. If different unit haul costs for different waste types are to be used, the equation becomes more complicated. This added complexity is not usually warranted for a feasibility study.

Secondary Haul Cost (also see Chapter 9)

Secondary haul costs are incurred in hauling raw waste from a transfer station to a landfill or resource recovery facility. Also included in this group of costs are the costs to transport processing residue and/or ash from a resource recovery facility. Costs for transporting energy products (RDF) and recovered materials to the markets are sometimes also considered secondary haul costs, but many times either are included as facility costs or subtracted from revenues to produce a revenue net of transport costs (F.O.B. facility location). For a particular facility, the secondary haul cost is the sum of the costs of transporting any raw waste and/or other materials to all other facilities in the configuration or to markets. Mathematically,

$$S_j = \sum_{m=1}^{M} \sum_{k=1}^{K} \frac{W_{mjk}}{C_{jm}} [(SMC_{jm}\, D_{jk}) + (SHC_{jm}\, T_{jk})] \tag{4.3}$$

where

m = Type of material (raw waste, residue, ash, RDF, or recovered material).
M = Total number of materials for which secondary haul is required.
j = Origin facility.
k = Destination facility.

K = Total number of destination facilities in the analysis.

SMC_{jm} = Secondary unit haul cost in dollars/mile (or dollars/km) for facility j and material type m.

D_{jk} = Round-trip distance in miles (or km) between facility j and facility k.

W_{mjk} = Amount of material type m transported between facility j and facility k per unit time.

C_{jm} = Capacity of vehicle hauling material type m from facility j in tons (or metric tons).

SHC_{jm} = Secondary unit haul cost in dollars/hour for facility j and material type m.

T_{jk} = Round-trip travel time in hours from facility j to facility k plus unloading time at facility k.

Note that when facility j is a transfer station, $SHC_{jm} = 0$ because driver labor should be included in the annual operating cost of the transfer station.

Operation and Maintenance Cost

Facility operation and maintenance (O & M) costs include equipment maintenance, operating labor, utilities, and other costs. Some costs are fixed with respect to the amount of waste handled by the facility for a given facility size (site and building maintenance, insurance costs, and others). The other O & M costs vary directly with the amount of waste handled by the facility (within limits). The total O & M costs for a given facility can be calculated by adding the fixed and variable costs for the amount of waste received by the facility

$$OM_j = FC_j + VC_j \left(\sum_{i=1}^{I} W_{ij} + \sum_{t=1}^{I} W_{tj} \right) \tag{4.4}$$

where

FC_j = Fixed costs for facility j in dollars per unit time.

VC_j = Variable costs for facility j in dollars per ton (or metric ton).

W_{tj} = Amount of waste transported from transfer station t to facility j in tons (or metric tons) per unit time.

T = Total number of transfer stations contributing waste to facility j.

Equation 4.4 is a much simplified representation of facility O & M costs. As stated previously, estimating O & M costs for transfer stations, landfills, and resource recovery facilities requires specialized knowledge and should be done by qualified professionals.

Facility Capital Cost and Debt Service

The annual debt service for each facility is a function of the total capital cost, the financing method, and the interest rate. Capital costs include all expenditures related to the facility in question during the design, construction, and start-up of the facility. The following elements must be included:

1. Site development.
2. Buildings.
3. Equipment.
4. Rolling stock.
5. Energy distribution.
6. Contingency.
7. Engineering and legal services.
8. Owner's administration.
9. Construction insurance.
10. Capitalized financing costs.

The investigator must be certain that all facilities in the analysis include all of these elements. A common fault of many feasibility studies is to leave out some of these elements resulting in unrealistically low estimates of capital requirements. For example, certain resource recovery "system" suppliers may quote capital costs that appear attractively low until all the proper elements are added, such as engineering, legal, administration, rolling stock, energy distribution, and capitalized financing costs.

One capital cost element that can have a particularly strong influence on total capital requirements is capitalized financing costs. These costs can add 50% and more to the capital cost of a facility depending on the financing method selected. With the wide variety of financing methods (especially for resource recovery facilities) that have developed over the past few years, it is particularly important to obtain the advice of experienced professionals in the solid waste field during the feasibility study phase. Some of the available financing methods and their impact on capital costs have been discussed in the literature.[53-55]

The annual debt service corresponding to the total capital cost and the financing method selected can be calculated on a "levelized" annual basis by multiplying the total capital cost requirement by the capital recovery factor for the financing interest rate and the financing period:

$$A_j = TCC \frac{f(1 + f)^n}{(1 + f)^n - 1} \tag{4.5}$$

where

A_j = Annual debt service in dollars per year.
TCC = Total capital cost in dollars.
f = Financing interest rate in percent per year.
n = Financing period in years.

Equation 4.5 is a simplified way of calculating annual debt service. The actual amount may be a "stepped" debt service that is lower in the first few years and higher in later years, but this equation will have sufficient accuracy for the feasibility study. Tables for capital recovery factors are available in numerous references.[56]

Revenues

The revenue component of net system cost acts to offset the other components. Revenues are mostly derived from the sale of recovered energy and materials, although the investment of certain bond reserve funds (if used) may also result in revenues that are "fixed" with respect to the amount of waste handled by the facility. If the feasibility study is being done from the perspective of the public, drop charges or tipping fees that may be collected at existing or future landfills, transfer stations, or resource recovery facilities should not be considered "revenue" in the calculation of net system cost. Since such fees are ultimately the burden of the taxpayer or ratepayer in the study area, these fees are not revenues from the standpoint of net system cost to the community. However, in feasibility studies being done for private facilities from the perspective of the owner or operator, the inclusion of tipping fees may be appropriate. The formulation given here assumes that tipping fees are not included in the analysis. The total revenues for a particular facility can be calculated as follows:

$$R_j = FR_j + \sum_{z=1}^{Z} \sum_{x=1}^{X} (RR_{xjz} \, V_{xjz}) \tag{4.6}$$

where

FR_j = Fixed revenue for facility j.
RR_{xjz} = Amount of recovered resource of type x sold by facility j to market z in units appropriate to the type of recovered resource (i.e., tons/day, lb/hr, Kwh/month, etc.).
V_{xjz} = Unit market value for recovered resource x sold to facility j to market z in appropriate units (i.e., dollars/ton, dollars/1000 lb, dollars per Kwh).

Note that RR_{xjz} in Equation 4.6 is variable with respect to the amount of waste processed or handled by the facility. For example, the amount of recovered ferrous metal or the amount of steam generated is a direct function of the amount of waste processed and burned.

Summary

When using Equations 4.1–4.6, the investigator should be certain that all of the components of the NSC for a particular facility j are calculated on a consistent time basis. For example, if facility costs (A_j and OM_j) are calculated on the basis of some future start-up year, transport costs (P_j and S_j) and revenues (R_j) should be escalated using appropriate factors to the same start-up year. The investigator should also be careful to use consistent time variables (dollars/wk or dollars/yr) for each cost element. The use of units of dollars/day is not recommended because different facilities in a particular alternative may operate for a different number of days per week or per year, causing problems in calculation consistency.

The use of the results of this and other similar solid waste system analysis models should be tempered with an understanding of the model's limitations. First, the model will not "optimize" facility locations and sizes. It will only allow the investigator to select the lowest cost configuration from the given choices. Second, the accuracy or applicability of the results is sensitive to the accuracy of the input data (waste quantity projections, WGD time and distance data, facility costs, etc.). The investigator's confidence in the accuracy of these input data should be reflected in his confidence in the results of the net system cost modeling analysis.

4.1.6 Comparing Alternatives

There are many different ways to compare the alternatives selected for analysis. These methods of comparison depend on the investigator's perspective, the amount of time and money available for the analysis, and the intended use of the results. A comparison of only capital costs, or a comparison of the single-year net system cost as described previously have obvious limitations. To provide more information for the decision-making process and to start to investigate financing options, cost, and revenue projections must be made over the selected planning period. By so doing, the investigator can discover if the NSC of alternative systems are likely to change relative to one another over the study period. Furthermore, such projections provide needed information to financial planners.

The following paragraphs describe two methods for analyzing and comparing solid waste management alternatives using cost and revenue projection techniques—the pro forma analysis and the life cycle cost analysis. In addition, a discussion of microcomputer applications and sensitivity analysis is included.

The Pro Forma Analysis

The pro forma type of economic analysis is a common technique in business planning that can be used successfully to compare solid waste management alternatives. Pro forma is Latin for "future form" or something provided in advance to prescribe form or describe items. A pro forma economic analysis is simply a projection of the economics of a specified system or facility over its expected operating life.

For a solid waste system or facility, the pro forma analysis starts with revenues projected over a selected period, such as 20 years, and subtracts all system costs projected over the same period. A pro forma analysis is usually done on an annual basis, but theoretically any time division can be used. The results of such an analysis can be used by a public official to determine the net system costs of both a landfilling and resource recovery alternative over a specified operating lifetime, for example. This type of analysis can also be used by a private entrepreneur to compare the economics of two different facility sizes or technology types. The analysis can also be used by a financial planner to determine if a proposed resource recovery facility would bring in enough revenues to cover costs and debt service over its operating life, assuming different rates of operating cost inflation and energy value escalation.

To perform a pro forma analysis, the investigator must determine which elements of system costs and revenues are subject to escalation due to inflation, and which elements are not subject to inflation. Debt service is not subject to inflation, but may vary from year to year if calculated on a "stepped" basis. Items such as land lease or taxes may also be a fixed amount each year. In addition, certain revenues such as investment income from certain bond reserve funds may also be unaffected by inflation.

For those elements of system cost that are subject to inflation, an inflation rate must be selected. The inflation rate represents a change in the general level of prices or monetary purchasing power over time. An average inflation rate can be assumed for all cost components, or different rates can be applied to individual components to account for differences in expected inflation of labor and equipment maintenance costs for example. In general, an average inflation rate for all components of cost will provide sufficient accuracy for a feasibility study.

Escalation rates for revenues derived from the sale of recovered materials and energy must also be selected for those alternatives that include resource recovery facilities. Energy escalation rates should be chosen carefully. A common mistake in feasibility studies is to be overly optimistic about energy price escalation. There is a temptation to predict the continuation of the rapid rises in energy prices that have occurred in the past several years. Where available, the investigator should use the projections used by the potential energy buyer for projecting energy costs as the basis of energy revenue projections. For example, most electric power utilities make projections of power prices over a 10- to 20-year period for planning purposes. Also, many natural gas and fuel oil suppliers make projections of future price increases. There are also published projections of fuel price increases that can provide guidance.[57] Regardless of the source of data, the investigator should use moderate projections of energy revenue increases, because the feasibility calculations are quite sensitive to this factor.

Escalation rates for recovered material revenues should also be chosen carefully. Prices for secondary materials have varied widely over the past few decades, with only minimal escalation when adjusted for inflation for some materials. For example, corrugated containers (grade 11) had

Table 4.7   Pro Forma Analysis, Landfill Alternative (Dollars per Year × 1000)

Item	1985	1995	2004
Operating revenues	0	0	0
Facility costs			
Landfill O & M[c]	2303	3751	5820
Landfill debt service[a]	930	930	930
Transfer station O & M[c]	928	1512	2345
Transfer station debt service[b]	562	562	562
Haul costs			
Primary[c]	3975	6475	10045
Secondary[c]	1750	2851	4422
Total costs	10448	16080	24123
Net system costs[d]	10448	16080	24123

[a] At 640 ac (2.6 million m^2) landfill, $9.1 million, financed at 8%, 20 years.
[b] Two, 3750 tons/week (3402 t/week) stations, $5.5 million, financed at 8%, 20 years.
[c] Inflation rate: 5%/year.
[d] Total costs minus all revenues.

approximately the same market value in 1983 as in 1973 according to a compilation of published prices.[58] It may be wise, therefore, to use current prices and assume no escalation (or only a few percent per year) for these materials.

To illustrate the pro forma analysis technique numerically, Tables 4.7 and 4.8 were prepared. Table 4.7 is a pro forma analysis of a hypothetical new landfilling system for a study area of about 600,000 population that handles about 400,000 tons (363,000 t) per year of processible waste in the start-up year of 1985, growing to about 500,000 tons (454,000 t) per year in 2004. The proposed system, which replaces current landfills scheduled to close by 1985, also includes two transfer stations located at strategic points in the study area. It is assumed for this analysis that alternative locations for the transfer stations and new landfill have been analyzed to select the configuration with the lowest NSC (see Section 4.1.5). Table 4.8 is a comparable pro forma analysis of a resource recovery system for the same study area. In this hypothetical case, a mass-burn steam plant designed to burn 7200 tons (6532 t) per week of processible solid waste is to be constructed to provide saturated steam to a local industrial buyer. It is assumed that ferrous metal is recovered from the ash and sold to a scrap metal dealer. Also, the transport system for this alternative utilizes two transfer stations for waste delivery to the steam plant. Figure 4.4 compares the results of the pro forma analysis for these two hypothetical alternatives. All NSC values are in nondiscounted ("then-year") dollars.

Table 4.8   Pro Forma Analysis, Steam Plant Alternative (Dollars per Year × 1000)

Item	1985	1995	2004
Operating revenues			
Steam[a]	10502	18807	31775
Ferrous metals[b]	56	56	56
Facility costs			
Steam plant O & M[c]	5315	8658	13431
Steam plant debt service[d]	10372	10372	10372
Transfer station O & M[c]	545	888	1377
Transfer station debt service[e]	373	373	373
Landfill cost (bypass and ash)[c]	1490	2427	3765
Haul costs			
Primary[c]	3459	5634	8741
Secondary[c]	472	769	1193
Total costs	22026	29120	39251
Net system costs[f]	11468	10257	7421

[a] Unit value $5.50/1000 lb in 1985, escalated @ 6%/year.
[b] Unit value $5.00/ton, no escalation.
[c] Inflation rate: 5%/year.
[d] 7200 tons/week (6532 t/week), $94.7 million, 9%, 20 years.
[e] Two stations, $495,000, 8%, 20 years.
[f] Total costs minus all revenues.

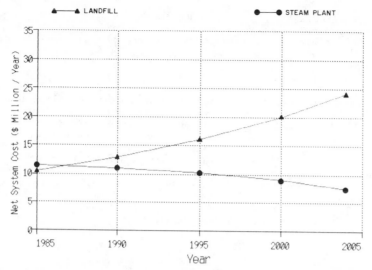

Fig. 4.4 Pro forma analysis. Comparison of two hypothetical alternatives—landfill alternative and solid waste-fired steam plant alternative (most likely case).

Note that the net system cost of the landfilling alternative is lower than the steam plant alternative in the first few years of the analysis, but is higher later in the analysis period. These two analyses were done from a public perspective. That is, the analysis compares the projected net system cost of each alternative to the community over a 20-year period. Tipping fees are not included as revenues because the purpose was to examine and compare the overall cost to the community of each alternative.

A different kind of pro forma analysis would be used to examine the steam plant alternative for possible financing. Table 4.9 illustrates the type of pro forma analysis that a financial advisor might use to determine the financial feasibility of the project (steam plant and associated transfer stations). Note that the haul cost components have been removed and that tipping fees have been added as a revenue source. In this analysis, the investigator is examining project costs and revenues rather than community costs and revenues. The "debt service coverage" factor is of particular interest in this analysis because it is an indication of the risk of the project to potential sources of

Table 4.9 Financial Pro Forma Analysis, Steam Plant Alternative (Dollars per Year × 1000)

Item	1985	1995	2004
Operating revenues			
Tipping fees[a]	9547	10025	24125
Steam[b]	10502	18807	31775
Ferrous metal[b]	56	56	56
Facility costs			
Steam plant O & M[c]	5315	8658	13431
Transfer station O & M[c]	545	888	1377
Landfill cost (bypass and ash)[c]	1490	2427	3765
Net revenues available for debt service	12755	22442	37383
Debt service			
Steam plant[d]	10372	10372	10372
Transfer stations[e]	373	373	373
Debt service coverage[f]	1.19	2.09	3.48

[a] In 1985 $30/ton escalated at 5%/year.
[b] See notes *a* and *b*, Table 4.8.
[c] Inflation rate: 5%/year.
[d] See note *d*, Table 4.8.
[e] See note *e*, Table 4.8.
[f] Net revenues divided by total debt service.

debt financing. A similar kind of pro forma analysis would be used by a private entrepreneur to examine project feasibility for private ownership. The private owner would add a calculation of projected before-tax profit to the pro forma analysis along with other factors for a business analysis.

The Life-Cycle Cost Analysis

The life-cycle cost (LCC) analysis technique is an economic analysis methodology that accounts for both the initial purchase price of a system or facility and the operating costs of the system or facility over its anticipated operating life. This analysis technique goes one step beyond the pro forma analysis by calculating a net present value, or life-cycle cost, for the facility or system in question. The life-cycle cost of a project is the lump sum of money that would have to be set aside at one time (usually at the beginning) to pay for construction and operation over the project's operating life. The LCC analysis starts by projecting a series of costs, revenues, or net costs (such as NSC) according to a selected inflation rate, adjusts these projections for the time value of money to a common year, and then sums the resulting adjusted costs, revenues, or net costs to derive the life-cycle cost. Assuming that $C_1 \ldots, C_n$ is a series of costs (or revenues) in present dollars, the general form of the equation for calculating the life-cycle cost of this series (evaluated in present dollars) is:

$$\text{LCC} = \sum_{n=1}^{N} C_n \frac{(1 + c)^n}{(1 + d)^n} \tag{4.7}$$

where
LCC	=	Life-cycle cost.
n	=	Period of analysis.
N	=	Total number of periods in one analysis.
C_n	=	Cost (or revenue) in period n (present dollars).
c	=	Inflation rate.
d	=	Discount rate.

LCC analysis techniques have been discussed in the literature for resource recovery systems and other facilities[59-63] and will not be discussed here in detail.

To perform an LCC analysis, an inflation rate and a discount rate must be chosen. The selection of inflation (escalation) rates have been discussed. The discount rate and inflation rate are two distinct factors. The discount rate is the interest rate that reflects the time value of money and is used to convert expenditures and revenues occurring at different times to a common time frame. Different approaches to setting the discount rate based on prevailing interest rates and the "lost opportunity" concept have been discussed in the literature.[60,63-66] Others[67] have suggested a technique in which the inflation rate and discount rate are combined into a single factor for adjusting costs in an LCC analysis technique. Historical data on interest rates and inflation rates were used to determine that inflation and interest rates have largely cancelled each other out over the last 25 years, with the interest rate (discount rate) being about 2% per year higher on average.

The LCC analysis technique can be useful in comparing solid waste system alternatives because it allows the investigator to determine, for example, whether the lower costs for a landfilling system versus a resource recovery system in the early years of the analysis will be offset by higher costs for the landfill in later years. To illustrate this point, an LCC analysis was performed for the two hypothetical alternatives presented in Tables 4.7 and 4.8. The life-cycle cost for the landfill alternative is about $155 million (1985 dollars) versus about $105 million for the steam plant alternative. Although the landfill alternative was lower in cost in the first few years (see Figure 4.4), the life-cycle cost is higher than the steam plant alternative. Note that the discount rate selected for this analysis is equal to the assumed bond interest rate for each alternative (8% for the landfill, 9% for the steam plant).

Microcomputer Applications and Sensitivity Analysis

The relatively low cost and wide availability of microcomputers has brought rather sophisticated economic analysis techniques within the reach of almost anyone who can follow simple instructions. The economic analysis methods discussed in the previous paragraphs can be performed quite easily on many microcomputers on the market today with the appropriate software.

Most of the electronic spreadsheet types of software packages will perform all of the functions necessary to do both a pro forma and an LCC analysis for solid waste system alternatives. There are other specialized financial analysis programs that can also be used, and the investigator can consult various reference books for information about spreadsheet and financial analysis software available for a variety of microcomputer systems.[68,69]

Most spreadsheet programs operate with rows and columns (just like a paper spreadsheet). The pro forma analyses presented in Tables 4.7, 4.8, and 4.9 were generated on a microcomputer with one of these spreadsheet programs. The cost and revenue elements were placed in a series of rows and each analysis year was placed in the columns. The initial year numbers were entered in the first column and equations for escalation were programmed for each component. Column sums and differences were programmed to derive the annual totals for costs and revenues.

The LCC calculations were performed with the same economic models developed for the pro forma analysis. The net system cost row was adjusted with a discounting equation in each year and accumulated. Many spreadsheet-type software packages have built-in functions for performing a "Net Present Value" or life-cycle cost calculation. A simple command programs the calculation so that the user need not know the mathematical formulation. The result of this type of calculation was used as the life-cycle cost for both the landfill and steam plant alternatives.

The use of microcomputers (or any other digital computer) for economic analysis provides easy access to information about the economic impacts of changes in assumptions. The investigator can, for example, change assumptions about inflation rates, energy escalation rates and operating costs; make a complete recalculation of the pro forma or LCC analyses; and then receive the results in a matter of seconds. A "sensitivity analysis" of this type is an essential part of analyzing these kinds of projects.

Variables should be chosen that have the greatest economic impact for the sensitivity analysis, such as the energy escalation rate. The investigator might also select a variable for which the projections are the most uncertain, such as facility operation and maintenance cost. It is generally preferable to start with a "base" or "most likely" set of variables, then postulate a "high" or "optimistic" set and a "low" or "pessimistic" set. These cases will give the likely range of the number of interest (NSC, life-cycle cost, debt service coverage, profit, etc.).

Figures 4.5 and 4.6 illustrate how the projected NSC for the landfill and steam plant alternatives presented previously will change under different sets of assumptions. The assumptions used for each case are given in Table 4.10 (note that the "most likely" case assumptions were used for the pro forma analyses presented in Tables 4.7, 4.9, and 4.10). Under the "pessimistic" assumptions, the NSC projections for the landfill and steam plant alternatives become nearly the same, with the steam plant alternative being slightly higher. Under "optimistic" assumptions, the two curves diverge to a greater degree than under "most likely" assumptions.

Figure 4.7 was prepared to show the effect of the different sets of assumptions on the life cycle cost of each alternative. Note that even under the "pessimistic" assumptions, the steam plant alternative has a slightly lower life cycle cost.

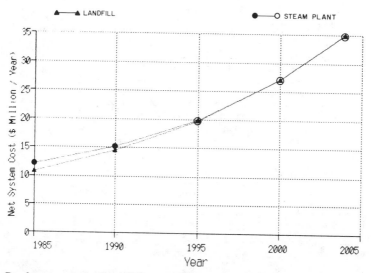

Fig. 4.5 Pro forma analysis. Comparison of landfill and steam plant alternatives: "pessimistic" sensitivity case.

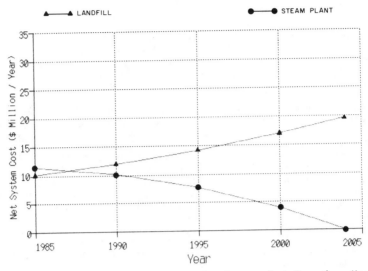

Fig. 4.6 Pro forma analysis. Comparison of landfill and steam plant alternatives: "optimistic" sensitivity case.

4.2 PROCUREMENT AND CONSTRUCTION MANAGEMENT

Robert Brickner

Charles Peterson

4.2.1 Introduction

A land disposal or resource recovery project enters the procurement, or third phase, after the concept has been found to be viable (phase 1) and the project building blocks, such as waste supply, energy market agreements, and so on, are in place (phase 2). Once the procurement process has been completed, and other key components have been or are close to being accomplished, such as financing, construction on the project will begin. Construction management is an important activity during this phase of the project. The key considerations with procurement and construction management are covered in the following subsections.

Procurement refers to the buying, purchasing, renting, leasing, or otherwise acquiring supplies, services, or construction for public use. During this phase of a project, community officials must determine the approach and method to be used in acquiring a land disposal or a resource recovery system. The assignment of responsibility for engineering, design, construction, start-up, and operation between the community, a contractor, and an engineering company will be determined by the approach selected. The method determines the process that will be used to select the sources of equipment and services available from vendors. A community's decision on the procurement approach and method to be used will be based on various factors. Four important considerations are: (1) facility ownership, (2) financing approach, (3) risk allocation, and (4) assumption of responsibility. The relationship between community considerations and procurement approaches and methods is illustrated in Table 4.11.

Table 4.10 Sensitivity Analysis Assumptions

Item	Optimistic Case	Most Likely Case	Pessimistic Case
Inflation Rate (%/year)[a]	4.0	5.0	7.0
Energy Escalation Rate (%/year)	6.5	6.0	3.0
O & M Costs[b]	10% lower than Most Likely Case	as stated[c]	10% higher than Most Likely Case

[a] Applies to all cost items except debt service.
[b] Applies to all O & M items (Landfill, Transfer Stations, and Steam Plant).
[c] See Tables 4.7, 4.8, and 4.9.

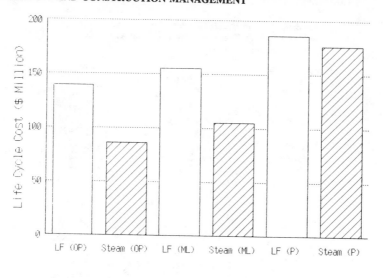

LF = LANDFILL ALTERNATIVE
STEAM = STEAM PLANT ALTERNATIVE
OP = OPTIMISTIC CASE
ML = MOST LIKELY CASE
P = PESSIMISTIC CASE

Fig. 4.7 Life cycle cost analysis. Comparison of landfill and steam plant alternatives: three cases.

4.2.2 Approaches: A/E, Turnkey, Full Service

The three basic approaches that may be used by a public agency to procure a facility and related services are:

Architect/engineering (A/E)
Turnkey
Full service

Architect/Engineering

Traditionally, public agencies have used the A/E approach to procure public works projects such as sewer systems, roads, and schools. A number of communities have also used the A/E approach to procure a waste-to-energy system, see Table 4.12.

The A/E approach involves two separate procurements: (1) engineering services and (2) facility construction. The selected engineering company is responsible for the design of the facility as well as the equipment and materials to be used, see Table 4.13. In addition, this company will prepare the specifications bid document, or invitation for bids (IFB), which is used for the second procurement. After receipt of bids, the public agency will select the same, or a different, engineering company to supervise construction of the recovery facility. This supervision, or construction management, is done to ensure that the specified materials and equipment are used. (Section 4.2.4 contains a discussion on construction management.) Additional activities for the A/E firm include

Table 4.11 Resource Recovery Facility Procurement

Community Considerations	Procurement Approach	Procurement Method
Facility ownership	A/E	Competitive/sealed bid
Financing approach	Turnkey	Competitive negotiation
Risk allocation	Full service	Two-step form advertising
Assumption of responsibility		Sole source negotiation

Table 4.12   Procurement Approach for Selected Energy Recovery Facilities That Have Been Built or Are Under Construction, August 1984

Procurement Approach	Location	Start-up Date
A/E	Chicago, IL (NW)	1970
	Harrisburg, PA	1972
	Nashville, TN	1974
	Ames, IA	1975
	Oceanside, NY	1976
	Chicago, IL (SW)	1977
	N. Little Rock, AR	1977
	Akron, Ohio	1979
	Duluth, MN	1979
	Madison, WI	1979
	Salem, VA	1979
	Albany, NY	1981
	Gallatin, TN	1982
	Glen Cove, NY	1982
	Harrisonburg, VA	1982
	Lakeland, FL	1983
	Columbus, OH	1983
Turnkey	Lane County, OR	1978
	Auburn, ME	1980
	Durham, NH	1980
	Hampton, VA	1980
	Collegeville, MN	1981
	Windham, CT	1982
Full service	Saugus, MA	1975
	Milwaukee, WI	1977
	Hempstead, NY	1978
	Bridgeport, CT	1979
	Monroe County, NY	1980
	Pittsfield, MA	1981
	Niagara Falls, NY	1981
	Dade County, FL	1982
	Portsmouth, NH	1982
	Wilmington, DE	1983
	Pinellas County, FL	1983
	Haverhill/Lawrence, MA	1985
	Baltimore, MD	1985
	Westchester County, NY	1985
	North Andover, MA	1985
	Tampa, FL	1985

facility start-up and operation through acceptance testing. Another assignment could be the preparation of operating manuals.

After the facility has passed the acceptance tests, the public agency assumes responsibility. Generally, the public agency will then operate the facility with its staff. An alternative is for the agency to contract with a private company. This option has been used at only two of the A/E-procured waste-to-energy facilities—Glen Cove, New York, and North Little Rock, Arkansas. However, this may become more common, if the trend toward private operation of publicly owned sewage treatment plants is an indicator.

The public agency that procures an energy recovery plant under the A/E approach is also the owner of the facility. Consequently, the agency assumes all the risk commonly associated with ownership. The only risks actually taken by the A/E firm is that the facility will be built on schedule and will pass the acceptance tests (Table 4.14). The owner (public agency) bears essentially all the risks under the A/E approach. This points out the importance of using proven recovery methods and an experienced A/E firm, if this approach is used.

A summary of the advantages and disadvantages with an A/E procurement are given below:

Table 4.13 Responsibility Assignments under Alternative Procurement Approaches

Responsibility Assignments	Procurement Approaches		
	A/E	Turnkey	Full Service
Planning	PA/E[a]	PA/C[a]	PA/C
System Design	E	C	C
Preparation and Issuance of System Specifications	E	C	C
Construction Supervision	E	C	C
Construction	PA	C	C
Operation	PA	PA	C
Ownership	PA	PA	PA or C

Source: Blyth Eastman Paine Webber, Inc. and Pryor, Cashman, Serhman & Flynn "Resource Recovery Procurement and Financing: A Guidebook for Community Action" New York State Department of Environmental Conservation, Albany, 1981.
[a]PA, Procuring agency; C, contractor; E, engineer as agent for procuring agency.

Advantages

Procurement is relatively "easy" and familiar to the public jurisdictions who are planning the project; and

It may be theoretically the lowest cost procedure available to the procuring agency.

Disadvantages

A public development agency assumes the substantial portion of the risk relative to the facility function since it (through its engineer) is responsible for the major portion of the total facility design and intended long-term function;

This approach will include only "standard" equipment vendor warranties for 30 days to one year. No long-term equipment performance guarantees will be provided;

A public agency becomes financially responsible for any change order or modifications to the preliminary design not reflected in the original contractor specifications package;

This method cannot apply the "public competitive bid scrutiny" to those systems/technologies that are patented by system contractors and that can only be procured by turnkey or full service methods; and

This approach could exclude well-qualified "system vendors" who feel uncomfortable with municipal operation of their equipment package.

Turnkey

Turnkey procurements have been used by industry to obtain process systems. This approach has also been used by communities to procure energy recovery plant, (Table 4.12). As shown in the table, the turnkey approach became an accepted method later in the history of waste-to-energy facility procurement than the A/E approach.

Under the turnkey approach, a single contractor has sole responsibility for the design, construction, and start-up of a facility (Table 4.12). The procuring agency selects a turnkey contractor on the basis of the responses to a request for proposals (RFP). An RFP states, in general terms, the type of system desired by the procuring agency. This differs from the IFB (invitation for bids) used with the A/E approach where the system, equipment, and materials are specified.

The equipment and materials to be used in a facility are selected by the contractor. In addition, the contractor may design and construct the facility or subcontract part of the work. The facility is turned over to the procuring agency after construction is completed and the energy recovery plant passes the acceptance tests. The procuring agency becomes responsible for the facility at this point and may either operate the plant with public employees or contract with a private company for plant operation.

A modification to the turnkey approach is to include a requirement in the RFP for the contractor to operate the facility for an initial period, typically 1 to 3 years after start-up. With this method, the contractor will still have responsibility for the plant in the early years of operation when design failures are most likely to occur. The result of a modified turnkey procurement is to transfer some of the risk of facility problems from the procuring agency to the contractor (see Table 4.14).

The advantages and disadvantages with turnkey procurements are summarized below:

Table 4.14 Risk Assignment under Alternative Procurement Approaches

Risk Elements	Risk Assumed By:[a]		
	A/E Procurement	Turnkey Procurement	Full-Service Procurement
Capital costs risks			
Capital costs overruns	O	C	C
Additional capital investment to achieve required operating performance	O	C	C
Additional facility requirements due to new state or federal legislation	O	O	O
Delays in project completion which leads to delays in revenue flow and adverse effect of inflation	O	C	C
Operating and maintenance costs risks			
Facility technical failure	O	C	C
Excessive facility downtime	O	O*	C
Underestimation of facility O & M requirements (labor, materials, etc.)	O	O*	C
Insufficient solid waste stream	O	O/M	M
Significant changes in the solid waste composition	O	O	O/C
Changes in state and federal legislation which affect family operations	O	O	O
Inadequate facility management	O	O*	C
Underestimation of residue disposal costs	O	O*	C
Tipping fee income risks			
Diversion of waste to other competing facilities	M	M	M
Overestimation of the solid waste stream	O/M	O/M	O/M
Adverse changes in participating community's fiscal condition	O	O	O
Recovered energy income risks			
Overestimation of technology energy recovery efficiency	O	C	C
Significant change in the solid waste composition	O	O	O/C
Changes in legislation that affect energy production and/or use	O	O	O
Overestimation of solid waste quantities	O	O	M
Significant adverse changes in the energy market financial condition or local commitment	O	O	O/EM
Downward fluctuation in the price of energy	O	O	O
Inability to meet energy market specifications	O	O*	C

[a] Participants:
O = Owner (assumes owner's ability to assume risk).
C = Contractor.
EM = Energy market.
M = Participating municipalities.
*Modified turnkey procurements may provide for intermediate or long-term private contractor facility operations which could lead to further risk assumption by the private contractor.

Advantages

Centralization of responsibility for design and construction by one party, the facility contractor.

Owner acceptance of the facility only after determining that it is functioning as intended.

Direct utilization of the system contractor's experience in designing and constructing the facility.

Shift of the initial performance risk related to facility design from the owner to the facility contractor.

Disadvantages

Higher costs, since a system contractor will want to be paid for his risk assumption.

Loss of control of certain aspects of the project that might affect total project economics.

Long-term performance risk remaining with the owner after the facility is tested and accepted.

Full Service

The full-service option involves complete design, construction, operation, and potential ownership by a private firm (Table 4.13). This procurement approach is increasing in importance (Table 4.12). As in the turnkey approach, the project development agency issues an RFP, but the contract is for a solid waste energy recovery service instead of a plant. The selected vendor is responsible for design, construction, equipment supply, start-up, testing, and subsequent operation (and costs) of the facility. The full-service contractor may also own the facility and be solely responsible to the bondholders for repayment of the financing plan. The owner in this case would seek to back up his responsibility to the bondholders with long-term contracts for assured waste supply and tipping fees from participating municipalities, and for sale of energy and/or materials. The potential for transferring risk from a public agency to the contractor is highest under a full service approach (Table 4.14).

Under a modified full-service procurement, a public agency may own the facility rather than the private vendor. In this case, the full-service contractor could lease the plant and repay the bondholders via the public agency for the term of the lease. Under this arrangement, the plant operator may be entitled to all profits or may be required to share revenues in excess of a predetermined return on investment with the public agency (or the participating municipalities where they are represented by the development authority). In such an arrangement, the lessee (plant operator) may or may not be considered as the owner of the plant for tax purposes, since title still resides with the public agency. If the ownership structure conveys tax ownership to the operator as lessee, the lessee of the plant may be entitled to the full investment tax credit, the accelerated building and equipment depreciation, and other tax-related benefits. Alternatively, the contractor may act solely as the plant operator, without leasing, under contract to the public agency.

The advantages and disadvantages associated with the full service approach are listed below:

Advantages

Complete responsibility for the facility development may rest with one party.

The facility function is guaranteed by the full service contractor on a day-to-day basis.

There is direct use of the system developer's experience in the design, construction, and operation of the facility.

Long-term performance risk is shifted from the procuring agency to the full-service contractor.

Disadvantages

Higher costs may be associated with the project development because of the full-service contractor's assumption of long-term operational risk.

The procuring agency may feel that there is insufficient control over certain aspects of the project.

Ownership (i.e., residual value) remains with the full-service contractor at the end of the contract term.

4.2.3 Procurement Methods

Four methods have been developed for a community to procure a resource recovery system. These methods are:

Competitive sealed bidding.

Competitive negotiation.

Two-step formal advertising.

Sole source negotiation.

A comparative summary of the procurement methods is given in Table 4.15. The individual steps in each of these procurement methods are illustrated in Table 4.16.

Competitive Sealed Bidding

Competitive sealed bidding, also known as formal advertising, is the most commonly used method for acquiring equipment, services, and construction for public use. Use of this method requires the

Table 4.15 Comparison of Alternative Procurement Methods[a]

	Competitive Sealed Bidding	Competitive Negotiation	Two-Step Formal Advertising	Sole-Source Negotiation
Nature of requirement	Well-defined system or service	Item or service insufficiently defined to permit detailed specifications	Item or service reasonably defined, but not firm enough for sealed bidding	Item or service ill defined or of a proprietary nature.
Sources of supply or service	Many	More than one	More than one	One
Form of solicitation document	Invitation for bids	Request for proposals (request for qualifications)	Step one: Request for proposals (request for qualifications) Step two: Invitation for bids	Request for proposals (request for qualifications)
Requirement for public notice	Yes	Yes	Yes	Yes
Opening of offers	Bids publicly read aloud	No public disclosure of proposal contents	Step one: No public disclosure of proposal contents Step two: Bids publicly read aloud	No public disclosure of proposal
Permitted changes to offers	None other than correction of inadvertent errors	Substantial changes permitted	Step one: Substantial changes permitted Step two: None other than correction of inadvertent errors	Substantial changes permitted
Evaluation criteria	As defined in IFB; price is determining factor	As defined in RFP; price and other factors considered	Step one: As defined in RFP; no price consideration Step two: Price is determining factor	As defined in RFP
Permitted discussions after submittal of offer	None	Substantial	Step one: Substantial Step two: None	Substantial
Selection and award	Responsive and responsible bidder whose bid is most advantageous to procuring body	Selection from responding offerors; award after negotiation to the offeror whose proposal is most advantageous to the procuring body	Step one: Determines acceptable proposals Step two: Limited to acceptable proposals from step one. Award in step two to responsible bidder whose bid is most advantageous to procuring body	Selection predetermined; award after negotiation

Table 4.16 Procurement Steps by Method

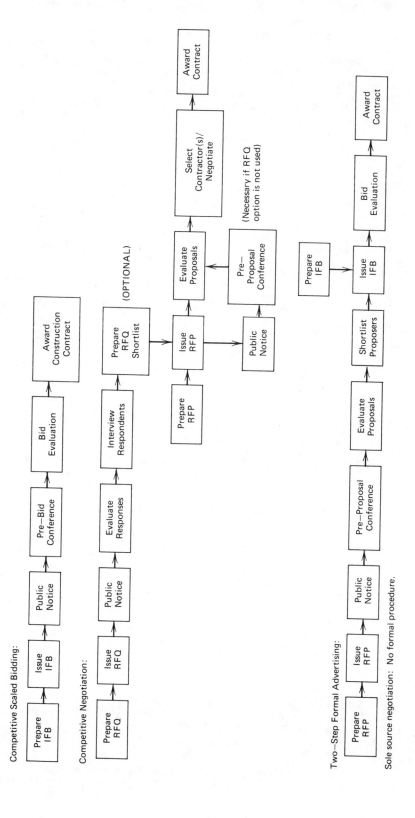

Procurement
Method

Competitive Scaled Bidding:

Prepare IFB → Issue IFB → Public Notice → Pre-Bid Conference → Bid Evaluation → Award Construction Contract

Competitive Negotiation:

Prepare RFQ → Issue RFQ → Public Notice → Evaluate Responses → Interview Respondents → Prepare RFQ Shortlist (OPTIONAL)

Prepare RFP → Issue RFP → Public Notice → Pre-Proposal Conference → Evaluate Proposals → Select Contractor(s)/Negotiate → Award Contract

(Necessary if RFQ option is not used)

Two-Step Formal Advertising:

Prepare RFP → Issue RFP → Public Notice → Pre-Proposal Conference → Evaluate Proposals → Shortlist Proposers → Issue IFB → Bid Evaluation → Award Contract

Prepare IFB → Issue IFB

Sole source negotiation: No formal procedure.

preparation and issuance of an IFB containing detailed specifications and a purchase description of the desired item or service. Upon receipt of bids, the development agency determines if the item or service being offered satisfies and is responsive to the requirements of the IFB and if the bidder is responsible. No change in bids is permitted once they have been opened (with limited exceptions). Once bid evaluation is completed, award is made solely on an objective basis to the lowest responsive bidder. Since it is the traditional method in the A/E approach, competitive procurement is a widely used procedure in public works implementation and is well understood by public procurers. Prior to preparation of an IFB, an A/E firm must be retained through a professional services contract.

Competitive Negotiation

The use of the competitive negotiation method of procurement is generally reserved for those situations where the item or service desired requires extensive discussions with offerers to determine the fairness and reasonableness of offers. This process is used generally to create a marketplace where price is ultimately established by bargaining between the procuring agency and a number of qualified offerers. Under this procedure, an RFP, as opposed to an IFB, is used containing general system and performance specifications and indicating the evaluation criteria to be used (price being only one) and the relative importance of each. The contents of proposals received are not disclosed publicly and information contained in any proposal is not divulged to competing offerers. This procedure differs from competitive sealed bidding in two major respects. First, judgmental factors are used to determine both compliance with the requirements of the RFP as well as to evaluate competitive proposals. The effect of this is that the quality of competing proposals may be compared and trade-offs made between the price and quality of offers. The second difference is that discussions are permitted after the submittal of proposals, and changes in proposals may be made to arrive at final offers that are most responsible to the procuring agency's needs. Award is made, after negotiations, to the offerer whose proposal is most advantageous to the development agency based upon price and other evaluation factors.

An alternative to the issuance of an RFP as the first document is the release of a request for qualifications (RFQ). The RFQ will request specific information on the experience and background of the respondents with similar types of projects. In addition, the RFQ will provide descriptive information on the proposed resource recovery project. An evaluation committee composed of community representatives and their consultants will rank the respondents to the RFQ. A "shortlist" of qualified vendors is prepared. The shortlisted vendors will be sent an RFP from which to prepare a proposal. An advantage with the RFQ method is that vendors are more likely to prepare a rigorous proposal than if only an RFP is issued. The more closely defined competition can attract more attention from a vendor. This is particularly important in the small-scale segment (500 tons/day and less) of the market, where there are many more communities than in the large-scale area. Consequently, it is important that smaller communities are able to attract qualified vendors to their procurement.

Negotiated procurements apply primarily to the turnkey and full-service approaches. Negotiated procurement has not been as widely used in the acquisition of public works, and occasionally it runs into restrictions from laws requiring competitive sealed bidding on the basis of price. Although competitive sealed bid procurement is well suited to systems that can be clearly specified in advance, a growing number of municipalities and development agencies are coming to realize that negotiated procurement is the most effective method for the acquisition of systems and services for solid waste energy-recovery projects (i.e., involving technology, markets, and/or operations contracts).

As noted, the heart of the negotiated procurement method is the RFP, that generally solicits bids on the basis of specifications more broadly drawn than those in an IFB. An RFP for a turnkey approach will usually contain more technical detail than one for a full-service option, since project guarantees and risks are different. Nevertheless, the use of an RFP shifts much of the design responsibility to the contractor and, hopefully, provides the procuring agency with a wide range of proposed technical solutions. Following proposal evaluation, a contractor (or set of finalists) is selected and the procuring agency then enters into contract negotiations with this firm. Any necessary deviations from the RFP must be considered at this point. Furthermore, the contract should contain sufficient flexibility to allow the contractor to adjust to unanticipated technical and financial changes in the systems. Without this flexibility, the increased risks borne by the contractor will be reflected in higher contract costs to the procuring agency.

Two-Step Formal Advertising

The two-step formal advertising procedure was developed by the Federal government and is used in situations where the complexity of the system or service desired prevents the preparation of detailed specifications by the procuring agency. This procedure incorporates features of both the competitive sealed bidding and competitive negotiation methods and involves the issuance of two solicitation documents. In step one, an RFP is issued requesting the submittal of unpriced technical proposals. Discussions are conducted separately with all offerers to ensure complete understanding by the procuring agency of what is being offered and to enable offerers to change proposals so that they are responsive to the RFP. All offerers whose proposals are accepted either as submitted or following modifications based on these discussions are then issued an IFB in step two. Thereafter, the procedure is identical to that in competitive sealed bidding, with award of the contract to the responsible bidder whose bid is most advantageous to the procuring agency. As with the competitive sealed bidding method, an A/E firm must be retained prior to preparation of the RFP. The A/E firm typically will be responsible for preparing the procurement documents.

Sole-Source Negotiation

Sole-source negotiation involves no competition and is used when the procuring agency determines that there is only one source for the desired service and/or equipment. A proposal is made in response to an RFP, the terms and conditions of the contract are negotiated, and award is made. This method is used only on occasion because of the number of competing vendors in this field. A modification of this method is when only one qualified respondent is shortlisted in the RFQ/RFP process used with competitive negotiation. A community could begin negotiation with the lone shortlisted vendor using the RFP as a basis for negotiation. This process was used in Claremont, New Hampshire.

4.2.4   Construction Management

Background

The term "construction management" may have several different meanings and implications depending upon the context within which it is used. The inexperienced public sector may view the phrase as merely the management activities before the actual field construction program begins. Vendors may see construction management activities revolving around their order and delivery dates, which affect a project schedule and other critical path planning activities. Still others have seen construction management firms eliminating the general contractor and in turn acting as the agent of the owner in placing individual contracts and coordinating several contractors involved in a project.

A definition of construction management (CM) is given to clarify this misunderstanding: CM is a process by which a potential project owner engages an agent, referred to as the CM, or construction manager, to coordinate and communicate the entire project process, including project feasibility, design, planning, letting, construction, and project implementation, with the objective of minimizing the project time and cost, and maintaining the project quality.

The construction management activity typically will involve the following tasks:

1. Organize, schedule and coordinate all the work programs and tasks related to the overall construction of the project.
2. Establish close liaison between manufacturers' representatives, installation and erection supervisors, construction engineers and subcontractors' representatives, and construction forces in the field.
3. Review all engineering/design drawings, specifications, and schedules proposed by individual subcontractors and suppliers.
4. Review all plant construction criteria.
5. Organize and direct the overall responsibilities of the various field engineers, specialists and supervisors assigned to the jobsite.
6. Direct all phases of work related to quality assurance and quality control procedures to assure compliance.
7. Conduct meetings and conferences as the need arises to ensure the integration and cooperation of all in meeting installation milestone dates.
8. Provide advice and counsel to the client, team members, and other project participants concerning all phases of work progress.
9. Hold weekly meetings to develop an integrated biweekly schedule of all subcontractors for this duration. This schedule should include all activities and manpower for the work effort of individual subcontractors.
10. Direct and provide liaison between the job site and the design engineer to obtain the necessary information during the course of the project.

11. Conduct regular and periodic job site inspections to assure compliance of work progress with design criteria and to assure conformance with regulatory codes, specifications, and licenses.

12. Establish a high level of cooperative relations among all project participants to achieve the utmost job interest and maximum work performance relative to the overall construction effort.

13. Organize and mobilize the necessary construction management personnel as required.

14. Coordinate construction activities with operations during construction.

The role of the construction management firm will vary with the three general procurement approaches for resource recovery systems which are: (1) architect/engineer, (2) turnkey, and (3) full service. Typically, construction management is similar under the latter two procurement approaches, since one vendor is responsible for construction of the recovery facility.

With an Architect/Engineer

The following diagram depicts position of a construction manager in a typical A/E project where separate waste-to-energy equipment contracts are bid and a separate facility contract is bid for the general construction (concrete, site, civil, etc.) and building installation.

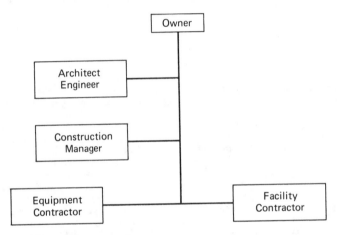

In this case, resource recovery projects have typically had the project A/E firm provide the construction management services.

Depending upon the nature of the procurement specifications, the A/E may completely design the entire facility and require general contractor bids. Furthermore, the equipment suppliers typically are subcontractors to the general contractor and the lines of contractual responsibility are as follows:

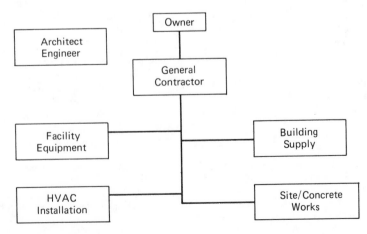

The general contractor would have a specific contract with the owner specifying duties and responsibilities in the example shown above. In turn, the general contractor has separate contracts for equipment supply and facility services to accomplish the construction of a plant following the specifications developed by the architect/engineer. The general contractor will most likely coordinate construction activities through an in-house construction management team.

Another approach to the traditional A/E procurement is for the owner to hire a separate specialized construction management firm independent of the architect/engineer and the general contractor. This firm acts as an "agent" of the owner to expedite the project in a cost-conscious manner while maintaining project quality. In this view, the construction manager can hire and/or purchase materials/services for the owner. A diagram of the separate construction management approach to an A/E procurement is presented below:

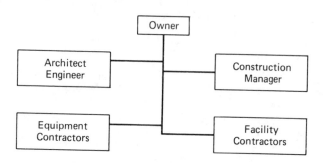

Changing times have placed new demands on construction projects. As a result, there has been significant re-evaluation and experimentation with the role and scope of service provided by various organizations involved in the construction phase of projects. The typical public works project, usually developed through an A/E procurement, has seen the traditional roles of the owner, A/E, and contractor sometimes failing to achieve the desired project results. The procurements variations in this traditional A/E role have been tried under varied circumstances with varying results.

Advantages

The following items highlight the interface and potential benefits of having a separate construction manager using an A/E procurement method.

1. With the conventional A/E procurement, the owner can be assured of professional independence because the construction management firm can be added to the project as independent of and with no direct financial interest in the construction materials, labor, or equipment contracts. Hence, the construction management firm's prime objective is to serve the owner's needs. In the implementation of an A/E project, questions and differences of opinion inevitably arise that involve the owner's interests. His interests may be most effectively protected by having this competent and independent management and construction expertise available to him.

2. In an A/E procurement, owner personnel, designer, and the construction manager can be merged into an integrated project team, specifically adapted to project needs and owner requirements.

3. The owner in an A/E procurement can obtain stronger management of the project. The problems of project implementation are increasingly becoming those of involving the managment of the many functions and activities involved in the project rather than those of a technical nature. This requires sound integrated management from start to finish, with heavy emphasis on items such as scheduling, cost control, purchasing, and expediting. Construction management services can provide this additional effort and emphasis.

4. The owner has greater assurance of quality control. The construction orientation of the construction manager and his staff, the increased level of effort devoted to quality assurance and the professional independence of this effort encourage quality.

5. Project costs may be reduced as a result of several factors. First, the construction manager can develop the various contracts for purchase of equipment and materials and construction, and these can be very definitive as to amount of work included. Thus a minimum amount of contractor guesswork should enter into the project bidding. Second, each contract can be awarded by competitive bidding to meet legal requirements. Third, numerous smaller contracts rather than a few larger contracts can be developed. This means that more

contractors are able to bid each of the contracts because the contract size falls within their capability. The multifaceted A/E bidding process avoids bidding restrictions on extremely large contracts because of limited contractor financial or management capability. In addition, each contractor assumes only the risk of his own work rather than having to assume risk involved in numerous subcontractors. Finally, the cost of the professional service is more than offset by the elimination of the prime contractor mark-up on the various subcontractors. With the A/E procurement and construction management approach, the service fee is significantly less than the normal prime contractor mark-up for overhead and profit.

Activity by Construction Phase

The focus of the construction management firm will vary by the phase of construction. The three phases addressed in this subsection are: (1) design, (2) preconstruction, and (3) construction.

Design Phase. The specific activities of the construction management firm can vary widely from job to job. The following discussion highlights the type of project involvement that could be undertaken.

Design-phase services by the construction management firm can be performed primarily during the design development and working drawings phases of the designer of record. Included in the total design fees for the project are the following potential construction management services.

1. *Consultation during Project Development.* Schedule and attend meetings with the A/E during the development of conceptual and preliminary designs to advise on the selection of materials, building systems, and equipment. The construction management firm can provide recommendations on construction feasibility, availability of materials and labor, time requirements for installation and construction, and factors related to alternative methods of construction.

2. *Project Construction Budget.* Prepare a project budget as soon as the major project requirements have been identified. Additionally, the construction management firm can prepare an estimate based on a quantity survey of drawings and specifications at the end of the designer's schematic design phase for approval by the owner. The firm can update this estimate of the design development phase and advise the owner and the A/E if it appears that the project construction budget will not be met. If the owner's budget will be exceeded, he can make recommendations for corrective action.

3. *Coordination of Contract Documents.* Review the drawings and specifications of the A/E as they are being prepared, recommending alternative solutions, without assuming any of the design professional's responsibilities for design.

4. *Review of Drawings and Specifications.* Review with the A/E, the project drawings and specifications to eliminate areas of conflict and overlapping in the work to be performed by the component contractors.

Preconstruction Phase. Preconstruction phase services by the construction management firm are primarily performed at the completion of the contract documents in preparation for competitive bidding. The following basic services may be provided.

1. *Scheduling.* Develop a project time schedule that coordinates and integrates the A/E's design efforts with construction schedules. Additional scheduling activities involve updating the project time schedule for the construction operations of the project, including activity sequences and durations, and processing of shop drawings and samples.

2. *Construction Planning.* Recommend to the owner the purchase of long-lead items to ensure delivery by the required dates established on the schedule.

3. *Bid Packages.* Recommend the division of work in the drawings and specifications to facilitate the bidding and awarding of component contracts.

4. *Bidding.* The construction management firm receives competitive bids on the work of the various component contractors and assists in bid opening and review process. After analyzing the bids, recommendations can be made to the owner on which contracts and contractors should be awarded the bids.

5. *Insurance and Bonding.* Coordinate the insurance and bonding needs of the various contractors within the guidelines and requirements of the owner.

Construction Phase. Construction phase services will vary depending upon the owner's in-house capabilities. These services are vital to the completion of the project in a quality/schedule/cost-conscious manner. The construction manager may perform the basic services described below.

1. *Control.* Monitor the work of the component contractors and coordinate the work with the activities and responsibilities of the owner and A/E to complete the project in accordance with the owner's objectives of quality, time, and cost.

2. *Full-Time Staff.* Maintain at the project site a project coordinator to provide general direction of the work and progress of the component contractors and act as the owner's representative.

3. *On-Site Organization.* Establish lines of authority in order to carry out the overall plans of the construction team.

4. *Progress Meetings.* Conduct progress meetings with the component contractors, owners, and the A/E to discuss jointly matters of procedures, progress, problems, and scheduling.

5. *Monitoring of Schedule.* Identify potential variances between the completion date and the work progress. Recommend to the owner and component contractors adjustments in the schedule to meet the scheduled completion date.

6. *Personnel.* Determine the adequacy of the component contractor's personnel and equipment, materials, and supplies to meet the scheduled completion date. Recommend courses of action to the owner when requirements of the component contractor are not being met.

7. *Construction.* The construction manager does not provide labor and materials to the actual construction process. Being a full-time representative of the owner, the construction manager provides no conflict of interest by deriving a profit from the construction process.

8. *Cost Control.* Develop and monitor a project cost-control system. The construction management firm can revise and refine the initially approved projection construction cost budget, incorporating changes as they occur. Other activities would be to identify variances between budget and actual expenditures and advise the owner whenever projected cost exceeds budgets or estimates.

9. *Cost Accounting Records.* Maintain cost records performed under actual cost, unit costs, or other bases requiring accounting records. Afford the owner access to these records and preserve them for a period (usually 3 years) after final payment.

10. *Change Orders.* Develop and implement, with the cooperation of the A/E, any and all change orders. Review requests, assist the owner in negotiating the recommended changes.

11. *Payment of Component Contractors.* Implement a procedure for the review and processing of requisitions and make recommendations to the owner for payment.

12. *Permits and Fees.* Assist the owner and A/E in obtaining building permits and special permits for permanent improvements, excluding permits required to be obtained by the component contractors. Assist in obtaining approvals from the authorities having jurisdiction.

13. *Owner's Consultant.* The construction manager may assist the owner in selecting and retaining professional services of special consultants and coordinate their services.

14. *Inspection.* The construction manager will inspect the work of the component contractors for defects and deficiencies in the work without assuming any of the A/E's responsibility to monitor the construction process.

15. *Safety.* It is the responsibility of each component contractor to maintain appropriate safety measures in compliance with all federal, state, and local statutes, rules, regulations, and orders.

16. *Document Interpretations.* Refer all questions for interpretation of documents to the A/E Professional.

17. *Shop Drawings.* Establish a procedure with the A/E for expediting the processing and review of shop drawings.

18. *Reports.* Submit written progress reports to the owner and A/E on the work process and percentage of completion. The construction manager should keep a daily log available to the owner and the A/E if the need arises.

19. *Records.* Maintain, at the project site, records of all contracts, drawings, specifications, samples, maintenance and operating manuals, and other construction-related documents, including all revisions. Maintain current set of record drawings, specifications, and operating manuals. At the completion of the project, the construction manager should deliver all such records to the owner.

20. *Substantial Completion.* Assist the A/E with the determination of substantial completion of the work and prepare a list of incomplete or unsatisfactory items and a schedule for their completion.

21. *Start-Up.* With the owner's maintenance personnel, direct the checkout of utilities, operations systems, and equipment for readiness and assist in their initial start-up and testing by the component contractors.

22. *Final Completion.* Assist the A/E in the determination of final project completion and provide the owner with a written notice of such. The construction manager should secure and transmit to the owner the required guarantees, affidavits, releases, bonds, and waivers. Finally, the construction manager should turn over to the owner all keys, manuals, and record drawings.

Turnkey/Full-Service Procurement

Because of the major acceptance of project risks, the prime contractor under a turnkey or full-service procurement usually assembles an experienced construction management team. Management of the engineering design, construction, construction management, start-up, and testing of a resource recovery facility is organized and staffed to provide for orderly and effective project execution. With some full-service contracts in this industry now being awarded for in excess of $100 million, significant detail and expertise from various project team members must be drawn upon.

The project teams typically are structured as shown in the organization charts in Table 4.17. The prime contractor provides overall project management and support through its corporate home office staff. A project manager is usually assigned prior to the start of negotiations with the client. This person has complete project control and usually reports directly to a corporate officer of the corporation. For implementation of engineering, design, construction, construction management,

Table 4.17 Representative Project Organization

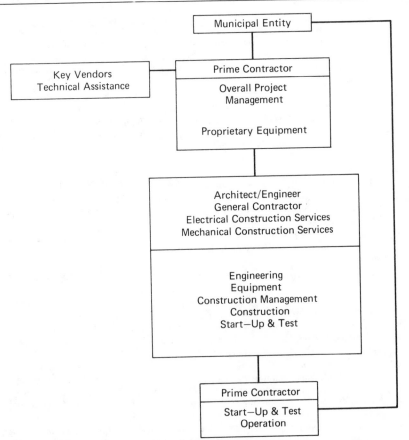

start-up, and performance testing, the contractor will usually have subcontract agreements with the team member. The areas of expertise of each subcontractor are typically as follows:

1. Project management, engineering design, and construction management.
2. Civil/structural construction services.
3. Electrical construction services.
4. Mechanical construction services.

Other construction erection such as the boilers, stack, and some miscellaneous equipment may be accomplished under subcontracts, typically with the vendors of the equipment.

The project team approach is organized to provide the project with an experienced and highly qualified team of companies working in unison to perform the skilled tasks required to design, build, start-up, and test the facility. The participants directly or indirectly share in the costs, risks, and benefits of the project and therefore have a great incentive for assuring timely and efficient performance throughout the completion of all project activities.

The team concept allows a much broader base from which experienced and specialized personnel can be selected to manage and supervise the project. Because of the incentives inherent in this concept, each participant has an incentive to assign the most capable, productive employees to the project.

All project engineering, design, construction, and construction management activities of the project will be managed by the team's project manager. The project manager provides direction to the engineering and design team, support services, and the project construction manager. He coordinates the activities of team members and is the official representative of the prime contractor. The project manager is responsible for the successful completion of the project.

Within the project structure, functional groups may report to the project manager. These are:

1. Engineering and design, which are responsible for the performance of all technical design activities.
2. Construction management, which manages construction and special subcontractors, planning, scheduling, constructibility review, and all site activities, and plant start-up and testing support.
3. Support services, which includes estimating, computer, vendor files, procurement, expediting, and traffic.

The project manager is the principal point of contact for all project entities and activities. In addition, the manager is the contact between the prime contractor and the client and is responsible for the day-to-day management and administration of the project. He, assisted by the assistant project manager, guides the project through the preliminary planning stages, coordination of engineering and design, technical and commercial evaluation of equipment, licensing assistance, procurement, expediting, construction management, construction, start-up, and testing activities as required by the contract. He monitors costs, cost forecasts, and progress to control the project. He maintains scheduled progress by directing and coordinating the efforts of all project participants. It is his responsibility to keep all management parties informed at all times of project developments on both a formal and an informal basis. Formal written reports are issued on a monthly basis. Movement of all personnel on or off the project is subject to the manager's approval.

Engineering/Design Organization

The project engineering personnel structure, as shown in Table 4.18, is usually established with the job engineer from each of the functional disciplines reporting to the project engineer-technical. The functional and discipline groups are responsible for assigned tasks, and the interrelations are on a person-to-person basis, properly documented and controlled by the project engineer-technical, and the engineering manager. The job engineers are responsible to the project engineer-technical, who ensures their availability in meeting the technical, schedule, and cost requirements. Project personnel are usually selected on the basis of specific expertise and experience.

Construction Management Organization

Construction management is performed in a manner that uses the experience of all the team members. A typical construction management organization is shown on the site construction organization chart, see Table 4.19.

The construction manager reports to the project manager. Home office engineering and con-

Table 4.18 Typical engineering organization

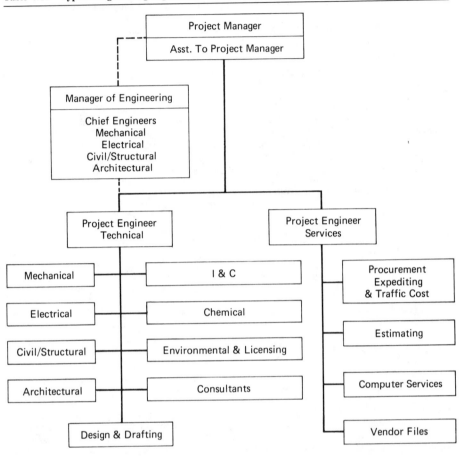

struction management personnel are directly involved in field activities on an as-needed basis to ensure continuity as discussed in the following construction management program.

It is extremely important to have construction and operations expertise involved during early phases of a project to enhance the development of plant layouts that are cost efficient and feasible to construct, and will operate reliably. Consequently, key personnel with construction and operations expertise are integrated with the project team at the outset of the project. These personnel remain with the project for its duration and are transferred to the field at the appropriate time. This procedure enhances the development of a well-conceived, constructible design and encourages good personal communications among the project team members. The single-team concept, from the outset of the project, is an important aspect of the approach to project management and control for this project.

The primary line of communication between the field and the project engineering team is normally between the project construction manager and the project engineer-technical. However, individuals on their respective staffs are contact points for specific areas of the work. Contacts between these technical personnel are recorded by means of letters or telephone confirmation memoranda, and the respective team members are kept informed of the substance of all contacts.

1. *Home Office Support.* Home office support by engineering is provided throughout the entire project. All required reports, schedules, revisions, and cost estimates are sent from the project's construction manager to the project manager, where all such items are processed as required. Home office support includes: participation in conferences and reviews; development of operation and maintenance procedures and manuals, spare parts and services, personnel training, and identification and tagging programs; generation of a list of expendable supplies required during start-up and testing; preparation and issuance of as-built drawings; and administrative support, as required.

Table 4.19 Site Construction Organization

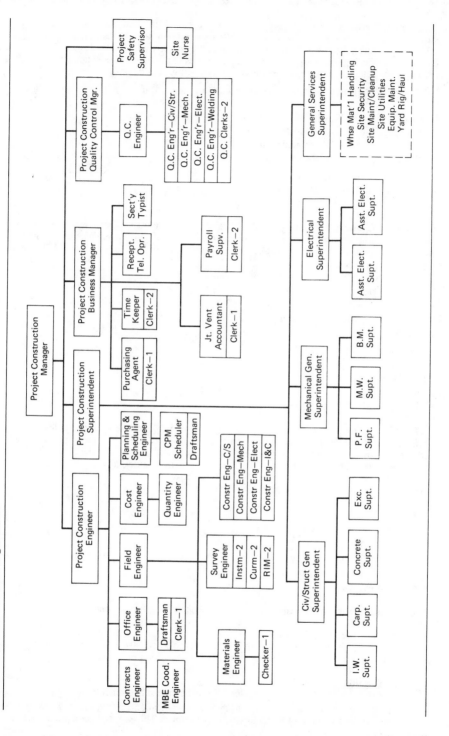

2. *Field Services.* Field services are provided during the construction period by the construction management team as discussed in the construction management plan. These engineers assist in the interpretation of drawings and specifications, review field design changes to the contractor's drawings, prepare as-built design drawings in the field or assist home office staff, and review other field-prepared drawings for conformity with the final design drawings.

Construction management services are usually categorized in the following service areas:

A. Field construction management.
B. Construction engineering.
C. Contract/subcontract administration.
D. Construction labor management.
E. Construction services.
F. Construction quality control.
G. Construction management plan.

A. Field Construction Management. The field staff, under the direction of the construction manager, must be structured to provide the day-to-day supervision and management of the project. Activities of the field staff will typically include, but not be limited to the following:

1. Coordinate with project and vendor representatives
2. Design liaison/coordination
3. Construction engineering support
4. Manage and direct all construction
5. Integrated scheduling and cost program
6. Quality control aspects
7. Contract administration
8. Project safety
9. Plant test
10. Shakedown operation activities

B. Construction Engineering. This service is responsible for the various studies and reviews necessary to ensure that the project is constructed in the shortest feasible time frame using methods of construction to minimize the cost of the project while still maintaining the quality of the work. Typical activities of the group should include constructibility design reviews with a significant effort employed to assure that ease of construction and value engineering concepts are an integral part of the project. Key aspects considered are as follows:

1. Site layout
2. Ease of construction
3. Materials
4. Methods and techniques
5. Interference resolution
6. Cost evaluation

C. Contract/Subcontract Administration. A formal program for construction subcontract administration should be established which defines policies, procedures, and guidelines that are consistent with the requirements of complex contracts. The program basis may start with the concept of project integrity and ultimate cost savings. The construction management team could provide the following construction subcontract administration functions:

1. Preparation of bidders' lists should include shop inspection, financial analysis, performance analysis, and current backlog review of prospective subcontractors and suppliers.
2. Preparation of bid packages should include specification review, instructions to bidders, general conditions, cost breakdowns, and bid form. This preparation is required to assure that:
    a. The bid packages are prepared in sufficient detail to allow a bidder to submit a responsive bid minimizing future changes and claims.
    b. The bidder understands the package.
    c. The desired bid is received.
    d. The construction project manager can properly monitor and control the work.

3. Presiding over meetings, including prebid and preaward meetings, to finalize all aspects of the construction subcontract terms of agreement.

4. Preparation of bid analysis and recommendations for award.

5. Conducting of postaward conferences.

D. Construction Labor Management. All of the onsite labor for construction, erection, and installation of all permanent materials should be provided under the construction manager's direction. The project construction manager will be the focal point for local labor negotiations. Union labor negotiations at the national level should be most likely addressed through the contractor's home office.

The site construction staff should be the focal point of the project control program having the responsibility for the work coordination and the tasks attendant to provide assurance that all work is done in conformance with the drawings and specifications.

The site construction staff, composed of groups of individuals, provide the following important functional services:

1. Construction management
2. Field construction engineering
3. Construction services
4. Construction contract/subcontract administration
5. Construction business services
6. Construction quality control

These groups are under the direction of the project construction manager and should be closely coordinated with one another.

E. Construction Services. The construction services group typically functions at the direction of the project construction engineer and provides the services of cost control, planning, and scheduling. The continued involvement of the construction services group during construction would be to act as the catalyst in establishing and maintaining continuity between the contractors, engineers, and construction management. For the most part, this effort is a continuation and elaboration upon the home office preconstruction activities. Emphasis is in forecasting, monitoring progress, and maintaining records. The construction services group will:

1. Maintain close liaison with construction engineers, subcontractors and home office engineering to develop and update the project schedule.

2. Receive and analyze equipment and material delivery schedules to initiate action to expedite the critical items affecting the schedule (flagging items for more attention).

3. Coordinate the planning of the sequence and target dates for completion of the total project.

4. Monitor the physical progress and report on items critical to schedule.

5. Record progress on a continuing basis and issue an updated schedule at regular and periodic intervals.

6. Define future problems and conflicts.

7. Review and analyze subcontractor's billings.

8. Develop and coordinate programs for continuous monitoring/recording of job costs and progress.

9. Coordinate impacts of job costs due to approved/pending contract change orders.

10. Prepare and update short/long-range cash flow projections.

11. Coordinate cost distribution activities by the code of accounts.

12. Accumulate and report the quarterly forecast.

13. Prepare final cost and property records report.

F. Construction Quality Control. The construction quality control group will implement the procedures in accordance with the quality control manual prepared specifically for each resource recovery facility. The quality control program will include overview procedures for monitoring inspection and documentation of work performed; materials and equipment reviewed and installed, auditing, and monitoring subcontractors and their quality control programs; control of nonconforming material; compilation and liaison with cognizant inspection and regulatory agencies; and include observation of suppliers and shop operations.

The construction quality control group should be supported by a staff of quality control specialists, surveillance engineers, and a technician experienced in each of the construction disciplines and contracted testing laboratories. Quality control personnel have the authority and the responsibility to identify quality control problems, initiate corrective action, and verify implementation of the corrective action.

Prior to construction, the construction management team should prepare detailed schedules, procedures, and construction administration programs tailored to the specific needs of the entire project so that job organization is structured properly to facilitate work progression in an orderly sequence. As construction begins, work progress and expenditures should be systematically monitored so that potential problems can be readily exposed and quickly acted upon.

Quality control surveillance techniques should be applied to the materials and equipment used in the work as well as the construction methods prior to and during their incorporation into construction. The construction activity should also be monitored for code and specification compliances. Materials, equipment, and work that do not meet requirements should be identified and handled according to the prescribed procedures. This should reduce the occurrence of built-in problems and enhance successful project completion.

Strict quality control of operations should also be maintained throughout the entire construction period generally to reduce the amount of rework required and to avoid delays.

To meet successfully the desired construction schedule requires that competent and qualified personnel be assigned to the job. The proposed project arrangement must have the capacity to provide a full range of services and personnel in planning, scheduling, managing, and supervising of construction areas. Plans should be developed for the temporary facilities. The construction plan must include a detailed construction procedure and should exercise the flexibility required in developing procedures and schedules so as to allow for the incorporation of new action into the construction plan without undue delay.

G. Construction Management Plan. To accomplish the objectives of the project, it is mandatory that certain procedures and design criteria be used by all involved within the project. One of the first documents developed is the project manual. It provides all personnel assigned to the project with an overall understanding of the total project, the client's requirements, and the design criteria to be followed. Some of the major procedures are as follows.

1. *Methodology of Work Breakdown.* The engineering schedule identifies each study, procurement specification, and design drawing required for the project and assigns start dates, durations, and finish rates. The interdependence between activities is delineated and each step in the completion of each activity is identified.

2. *Measurement Techniques Used.* Schedule accomplishment and resource expenditures are both individually monitored and compared with the plan. This provides continuous information to the project manager and the client regarding adherence to the project schedule and budget.

3. *Warning Systems to Possible Trouble Spots.* The capabilities of the scheduling and reporting system employed by the construction management organization should include the generation of look-ahead schedules for magnification of short-term future activities, late-item reports which print out items behind schedule, and borderline reports which report items in danger of falling behind schedule.

The project manager directs the planning process and functions as the overall leader in coordinating the activities of the various disciplines. Planning and scheduling specialists should be assigned to the project to assist the project manager. A key project personnel should participate in the initial schedule development process.

Progress is monitored periodically at the working level, and the schedule is updated. Each of the project team members should review the updates for accuracy and to determine the impact of changes on their specific function and responsibilities.

On large projects, computer-assisted systems will process progress information and generate a target schedule that compares the current status with the original target and produces exception reports for the entire project. For the individual working groups, separate exception reports are prepared indicating the area in which the work is or may be deviating from established goals. These exception reports are then analyzed to evaluate the status of the project in critical areas and to determine what corrective action must be taken to keep the project on schedule.

Near-term (30- to 60-day) look-ahead schedules may be generated and distributed to project team members for their use in planning their day-to-day activities. The look-ahead reports are abbreviated extractions of the larger schedule data base.

1. *Document Control.* At the beginning of the project, a standard distribution system may be established indicating the proper distribution of all the different types of project documentation, the action expected from each recipient, and the time permitted for such action. All meetings and trips should be recorded with appropriate minutes stating decisions and on-going assignments and commitments. All telephone conversations should also be recorded and distributed.

2. *Monitoring*. The overall project schedule and subnetwork schedules should be updated periodically, and all activities relating to procurement need to be highlighted to identify potential delays and changes in equipment or material lead times. After orders are placed, expediting reports are used to monitor the vendors' progress. Corrective action by the project manager is initiated where required.

3. *Engineering and Construction Interface*. The periodic updating of the overall schedule on the long-lead list reveals any potential impact of procurement activities on engineering and construction schedules. Modifications are made to any activities required to maintain project schedule.

4. *Quantity Items*. Early in the engineering/design effort, estimates compiled from bills of material are established based on the preliminary design, previous projects of similar design, and development of material specifications. As the design progresses, the bulk-order estimates and the bills of material quantities are revised. A periodic report from the field shows usage of materials and, in combination with updated bills of materials, provides the information necessary to supplement initial bulk orders and maintain adequate field inventory.

REFERENCES

1. D. J. Hagerty, J. L. Pavoni, and J. E. Heer, Jr., *Solid Waste Management,* Van Nostrand Reinholt, New York, 1973, pp. 3–22.

2. W. R. Niessen, Properties of waste materials, in *Handbook of Solid Waste Management.* (D. J. Wilson, Ed.), Van Nostrand Reinholt, New York, 1977, pp. 10–61.

3. S. H. Russell, *Resource Recovery Economics: Methods for Feasibility Analysis,* Marcel Dekker, New York, 1982, p. 23.

4. U. S. Environmental Protection Agency, *Omaha-Council Bluffs Solid Waste Management Plan—Status Report 1969,* sw-3tsg, U. S. Government Printing Office, Stock No. 5502-0012, Washington, D. C., 1971.

5. J. P. Woodyard and A. Klee, Solid waste characterization for resource recovery design, in *Proc. Sixth Mineral Waste Utilization Symposium* (E. Aleshin, Ed.), U. S. Bureau of Mines and IIT Research Institute, Chicago, May 2–3, 1978.

6. Unpublished data from the County of San Diego, Department of Public Works, 1983.

7. N. Chin and P. Franconeri, Composition and heating value of municipal solid waste in the Spring Creek area of New York City, in *Proc. 1980 National Waste Processing Conf.* (K. Woodruff, Ed.), Am. Soc. of Mech. Engineers, Washington D. C., May 11–14, 1980, p. 239.

8. H. I. Hollander et al., A comprehensive municipal refuse characterization program, in *Proc. 1980 National Waste Processing Conf.* Am. Soc. of Mech. Engineers, Washington, D. C., May 11–14, 1980, p. 221.

9. B. J. Trinklein, An applied statistical approach to refuse composition sampling, in *Proc. 1982 National Waste Processing Conf* (L. M. Grillo, Ed.), Am. Soc. of Mech. Engineers, New York, May 2–5, 1982.

10. U. S. Environmental Protection Agency, OSW, *Fourth Report to Congress—Resource Recovery and Waste Reduction,* SW-600, Washington, D. C., 1977.

11. D. Rimberg, *Municipal Solid Waste Management,* Noyes Data Corp., Park Ridge, N. J., 1975.

12. D. J. Hagerty et al., *Solid Waste Management,* Van Nostrand Reinholt, New York, 1973.

13. J. L. Pavoni et al., *Handbook of Solid Waste Disposal,* Van Nostrand Reinholt, New York, 1975.

14. S. H. Russell, *Resource Recovery Economics: Methods for Feasibility Analysis,* Marcel Dekker, New York, 1982, Table 2.2, pp. 28–29.

15. U. S. Environmental Protection Agency, *St. Louis Refuse Processing Plant: Equipment, Facility, and Environmental Evaluations,* EPA-650/2-75-044, May, 1975.

16. U. S. Environmental Protection Agency, *Evaluation of the Ames Solid Waste Recovery System, Part 1: Summary of Environmental Emissions; Equipment, Facilities and Economic Evaluations,* EPA-MERL, EPA-600/2-77-205, Nov., 1977.

17. W. L. Bider and W. E. Franklin, A method for determining processible waste for a resource recovery facility, in *Proc. 1980 National Waste Processing Conf.* (K. Woodruff, Ed.), Am. Soc. of Mech. Engineers, Washington, D. C., May 11–14, 1980, p. 211.

18. S. H. Russell, *Resource Recovery Economics: Methods for Feasibility Analysis,* Marcel Dekker, New York, 1982, Table 2.3, p. 31.

19. Unpublished records from King County, Washington, Department of Public Works, 1982.

20. County of San Diego, *San Diego Regional Solid Waste Management Plan 1982–2000,* 1982.

21. HDR Techserv, Inc., *Pinellas County, Florida Refuse to Energy Addition—Consulting Engineer's Feasibility Report,* in Official Statement by Pinellas County, Florida for Solid Waste and Electric Revenue Bonds, Series 1983, December 20, 1983.

22. Henningson, Durham & Richardson, Inc., *City of Charlotte, NC Solid Waste Disposal and Resource Recovery Study,* Vol. 1, Nov., 1978.

23. S. H. Russell, *Resource Recovery Economics: Methods for Feasibility Analysis,* Marcel Dekker, New York, 1982, Chap. 2, pp. 17–72.

24. U. S. Environmental Protection Agency, *Sanitary Landfill Design and Operation,* Brunner and Keller, SW-287, NTIS No. PB-227-565.

25. S. Wiess, *Landfill Disposal of Solid Waste,* Noyes Data, Park Ridge, N.J., 1975.

26. A. W. Joensen et al., *Seminar Proceedings: Municipal Solid Waste as a Utility Fuel,* Electric Power Research Institute, Nov., 1982.

27. R. J. Vetter, Madison's energy recovery program, the second generation, *J. Resource Management Technol.,* **12** (1), pp 19–23, 1983.

28. K. Runyun and J. F. Bernheisel, *Ferrous Metals Recovery at Recovery I, New Orleans,* Initial Operating Report TR-80-1, National Center for Resource Recovery, Inc., Washington, D. C., 1980.

29. H. Alter, *Materials Recovery from Municipal Waste, Unit Operations in Practice,* Marcel Dekker, New York, 1983.

30. John Arnold, *Initial Mass Balance for Froth Flotation Recovery of Glass from Municipal Solid Waste,* NCRR Technical Report RR77-1, National Center for Resource Recovery, Inc., Washington, D. C., Dec., 1977.

31. C. Noto La Diega and G. Variali, Recovery and utilization of plastic present in solid urban waste, in *Proc. International Conference: Solid Waste, Sludges and Residual Materials: Monitoring, Technology and Management,* Rome, June 17–20, 1980.

32. J. L. Pavoni, et al., *Handbook of Solid Waste Disposal,* Van Nostrand Reinholt, New York, 1975.

33. J. Kohl et al., *A Manual for Locating Large Energy Users for Co-generation and Other Energy Actions,* North Carolina State University and the Research Triangle Institute, March, 1980.

34. S. H. Russell, *Resource Recovery Economics: Methods for Feasibility Analysis,* Marcel Dekker, New York, 1982, pp. 109–111.

35. United States Environmental Protection Agency, *Evaluation of the Ames Solid Waste Recovery System, Part 2: Summary of Environmental Emissions; Equipment, Facilities and Economic Evaluations,* EPA-MERL, EPA-600/2-77/205, Nov., 1977.

36. S. H. Russell, Refuse derived fuel (RDF), in *Proc. Engineering Foundation Conference on Present Status and Research Needs in Energy Recovery from Solid Wastes* (R. Matula, Ed.), ASME, Sept. 22, 1976.

37. G. B. Liss and L. Larochelle, PURPA, cogeneration and resource recovery, in *Proc. 1982 National Waste Processing Conf.* (L. M. Grillo, Ed.), Am. Soc. of Mech. Engineers, New York, May 2–5, 1982.

38. S. H. Russell, *Resource Recovery Economics: Methods for Feasibility Analysis,* Marcel Dekker, New York, 1982, pp. 87–98.

39. American Society for Testing and Materials, *Standard Specification for Municipal Aluminum Scrap (MAS),* ASTM Method E753, ASTM, Philadelphia, 1980.

40. J. H. Heginbitham, Recovery of glass from urban refuse by froth flotation, in *Proc. Sixth Mineral Waste Utilization Symposium* (E. Aleshin, Ed.), U. S. Bureau of Mines and IIT Research Institute, May 2–3, 1978, p. 231.

41. J. P. Cummings, Glass and non-ferrous metal recovery subsystem at Franklin, Ohio—Final report, in *Proc. Fifth Mineral Waste Utilization Symposium* (E. Aleshin, Ed.), U. S. Bureau of Mines and IIT Research Institute, Chicago, April 13–14, 1976, p. 176.

42. Paper Stock Institute of America, *Circular PS-83, Paper Stock Standards and Practices,* Natl. Assoc. of Recycling Industries, New York, 1983.

43. C. G. Golveke, Biological processing: Composting and hydrolysis, in *Handbook of Solid Waste Management* (D. G. Wilson, Ed.), Van Nostrand Reinholt, New York, 1977, p. 197.

44. D. S. Airan and J. H. Bell, Resource recovery through composting—A sleeping giant, in *Proc. 1980 National Waste Processing Conf.,* Am. Soc. of Mech. Engineers, Washington, D. C., May 11–14, 1980, p. 121.

45. R. H. Greely and A. R. Nollet, New horizons in composting, in *Proc. 1980 National Waste Processing Conf.* (K. Woodruff, Ed.), Am. Soc. of Mech. Engineers, Washington, D. C., May 11–14, 1980.

46. H. Alter and J. J. Dunn, *Solid Waste Conversion to Energy, Current European and U.S. Practice,* Marcel Dekker, New York, 1980.

47. G. F. Haddix, Regional solid waste problems, two cases, in *Comput. & Urban Soc.,* Vol. 1, Pergamon Press, 1975, pp. 179–193.

48. G. F. Haddix and M. K. Wees, *Solid Waste Management Planning Models with Resource Recovery,* Proc. ORSA-TIMS Meeting, Las Vegas, 1975.

49. W. J. Baumol and P. Wolfe, A warehouse location problem, *Operations Research,* Vol. 6, 1958.

50. J. F. Hudson, D. S. Grossman, and D. H. Marks, *Analysis Models for Solid Waste Management,* Dept. of Civil Engineering, Massachusetts Institute of Technology, 1973.

51. D. H. Marks and J. C. Liebman, Location models: Solid waste collection example, *Journal of the Urban Planning Division, Proc. Amer. Soc. Civil Engineers,* 97(VPI), 1971.

52. K. Spielburg, Algorithmis for the simple plant-location problem with side conditions, *Operations Research,* Vol. 17, 1969.

53. S. H. Russell, *Resource Recovery Economics: Methods for Feasibility Analysis,* Marcel Dekker, New York, 1982, Chap. 6, pp. 233–267.

54. D. A. Scott and G. Brayton, Economic analysis and financing of resource recovery projects: A guide for public officials, in *Proc. of the Technical Program, GRCDA 21st Annual International Seminar and Equipment Show,* Governmental Refuse Collection and Disposal Assoc. and Environment Canada, Winnepeg, Aug. 29–Sep. 1, 1983, pp. 273–286.

55. U. S. Conference of Mayors, *Proceedings, Resource Recovery Financing Conference,* U. S. Conference of Mayors, Washington, D. C., 1982.

56. E. L. Grand and W. G. Ireson, *Principles of Engineering Economy,* Ronald Press, New York, 1970.

57. Wharton Economics Forecasting Associates, *Nelson Refinery Index,* 1983.

58. *Official Board Markets Newsletter,* Harcourt Brace Jovanovich Pub., 1958 through May, 1983.

59. S. H. Russell, *Resource Recovery Economics: Methods for Feasibility Analysis,* Marcel Dekker, New York, 1982, Chap. 6, pp. 233–267.

60. S. H. Russell and M. K. Wees, Life cycle costing for resource recovery facilities, in *Proc. 1980 National Waste Processing Conf.* (K. Woodruff, Ed.), Am. Soc. of Mech. Engineers, May 11–14, 1980, pp. 259–267.

61. R. T. Ruegg, *Life Cycle Costing Manual for the Federal Energy Management Programs,* Natl. Bureau of Stds., NBS Handbook-135 (Rev.), DE82017356, May, 1982.

62. R. T. Felago et. al., Life cycle costs and public decisions, in *Proc. 1982 National Waste Processing Conference* (L. M. Grillo, Ed.), Am. Soc. of Mech. Engineers, May 2–5, 1982.

63. R. S. Brown et al., *Economic Analysis Handbook,* prepared for Naval Facilities Engineering Command HQ, June, NTIS No. AD-A020859.

64. K. J. Arrow, Discounting and public investment criteria, in *Water Research* (A. V. Kneese and S. C. Smith, Eds.), Baltimore, 1966.

65. M. J. Bailey and M. C. Jensen, Risk and the discount rate for public investments, in *Studies in the Theory of Capital Markets* (M. C. Jensen, Ed.), Praeger, New York, 1972.

66. J. A. Stockfish, *Measuring the Opportunity Cost of Government Investment,* Research Paper P-490, Institute for Defense Analysis, March, 1969.

67. I. Isenberger, D. S. Remer, and G. Lorden, *The Role of Interest and Inflation Rates in Life-Cycle Cost Analysis,* a report prepared for the NASA Jet Propulsion Laboratory, National Aeronautics and Space Administration, NPO-15228, 1983.

68. R. P. Wells, S. Rochowansky, and M. Mellin, *IBM Software 1984,* The Book Company, Los Angeles, 1984.

69. J. Stanton, R. P. Wells, S. Rochowansky, and M. Mellin, *Apple Software 1984,* The Book Company, Los Angeles, 1984.

70. J. J. Adrian, *CM: The Construction Management Process,* Reston Publishing Co., Inc., Reston, Virginia, 1981, p. 2.

CHAPTER 5

WASTE DISPOSAL/RESOURCE RECOVERY PLANT COSTS

W. D. ROBINSON

Consulting Engineer
Trumbull, Connecticut

SERGIO E. MARTÍNEZ

Consulting Engineer
Miami, Florida

5.1 CAPITAL COST

Background

Solid waste disposal and resource recovery plants are expensive. Many projects have cut corners to reduce costs, sometimes with fatal results. Some examples include:

Underestimation of environmental qualifications and public resistance.
Grossly inadequate raw material quality and quantity evaluation.
Inadequate raw material surveillance at recovery/disposal sites.
Receiving pit/crane design insufficiencies.
Marginal stoker sizing.
Deficient ash-handling systems.
Insufficient furnace volume/surface.
Undersized and overpackaged modular units.
Process plant close coupled overcrowding.
Inferior operations telecommunication systems.
Compromised space/building sizes.
Underpowered shredders.
Ineffective dust collection/control.
Insufficient spare capacity.

Smaller communities are less vulnerable to these problems because they require facilities mainly for disposal of up to 100 tons/day with capital costs in a range up to $30,000/ton-day. These are small, compact plants serving populations under 100,000, and having little or no resource recovery. They usually involve size reduction and/or baling at a landfill site or at transfer stations.

North America

Today, large urban areas may require facilities having rated capacities of at least 1000 tons/day with a capital cost, complete and dedicated to energy and materials recovery, from $65,000 to over

The significant contribution of Sergio E. Martinez, P.E., Miami, Florida, to the preparation of this chapter is acknowledged herewith.

$100,000/ton-day. Projects of this scope can require raising over 200 million dollars—enormous sums to be raised by any entity in any money market.

Europe

In contrast, requirements in Europe are quite different for equivalent capability and scope. European facilities built during the late 1970s had price tags of $70,000/ton-day for traditional mass-burn, waterwall energy recovery. This is double an average cost of about $35,000/ton-day in the late 1960s.

Land in Europe is scarce and usually is more expensive compared with North America. Likewise, large sums for aesthetics—architecture and landscaping for more strict community acceptance—have been required, but such differences are diminishing. European waste-to-energy plants are usually owned by cities, nonprofit authorities, or government utilities and there is little private ownership. Tipping fees are comparatively low because total owning and operating costs do not include increments of depreciation or a profit factor. Also, exchange rates have increased comparative cost advantages for European equipment and systems if they are exported to the United States. In addition, source separation is widely practiced that virtually guarantees less difficult raw material free of troublesome items, with easier materials handling and processing (thermal reduction, etc.).

5.1.1 Spectrum of Facilities Costs

Many engineers in the United States, including boiler manufacturers and federal agencies, were disenchanted with the limitations of mass burning techniques historically practiced in Europe and endorsed the innovation and versatility of various prepared refuse-derived fuels (RDF) concepts. Many of these efforts in preparing and utilizing various forms of RDF were not entirely successful, and include a few total failures. The history of projects to produce RDF for sale have found that markets are elusive and unpredictable.

Presently, therefore, most energy recovery concepts must include dedicated (their own) facilities for producing a saleable end product, be it combustible fluids or steam and electricity. This favors plant capacities upwards of 500 tons/day and a utility-type power-generation facility with its inherent high requirements for availability, labor skills, and capital formation.

In today's energy recovery markets, the capital cost of the steam-electric generation complex can be more than double that of any front-end process required. In comparison, the boiler island for prepared refuse fuels for semisuspension firing is essentially standard equipment vis-à-vis the customized European-style mass-burn units, which are somewhat more expensive when total requirements such as stokers vs. RDF preparation, etc., are compared.

1. *Process for Disposal Only.* The least expensive plants are dry-process shred and/or bale for landfill having materials recovery options but no energy generation. An average cost of such facilities is about $30,000/ton-day. There are limited locations where they can compete, i.e., daily cover requirements are waived and/or conventional compaction techniques are inefficient.

2. *RDF Production Only.* Plants for production of RDF for sale to independent customers cost an average of about $35,000/ton-day. Such installations usually have been upwards of 400 tons/day where cooperative integrated ventures exist (city of Albany/State of New York; Baltimore County, Cockeysville/Crane Station, Baltimore Gas & Electric). Not included is the cost of modification at the fuel burning facility, a new boiler unit or plant. Modifications to an existing boiler range from 30% to 50% of the RDF preparation plant.

3. *RDF with Energy Recovery.* RDF dry-process plants which include steam generation only (no KW) can range up to $60,000/ton-day for 500 tons/day and larger. Under 500 tons/day, costs will range from $30,000 to $50,000/ton-day.[1] If electric generation is included, the cost for all sizes increases upwards of 40%, Table 5.3 and Reference 1.

4. *Mass Burn with Energy Recovery.* Mass-burn plants over 500 tons/day for steam generation only can range from $60,000 to $70,000/ton-day. Addition of electric generation increases cost to about $80,000–$100,000. Small scale plants under 500 tons/day for steam generation only range from $25,000 to $40,000/ton-day. The extra for electric generation is 40% to 50%.[1] See Tables 5.1, 5.2, and 5.3.

During the next 5 years, capital costs will probably rise faster than the common escalation indices due to underestimation of construction and operating costs for large plants (> 1000 tons/day). It is expected that future large scale refuse-to-energy facilities will cost upwards of $100,000/ton-day. In addition to geographic and disposal requirement logistics, economies of scale have contributed to the significant number of facilities upwards of 1000 tons/day, with maximum size limits probably established by certain regressions for "superplants" (over 3000 tons/day) such as acceptable locations, space requirements, transfer stations, truck traffic, etc. For example: A 2000 tons/day cogeneration installation will serve a population of 1 million. A population of two million, however, would require a 4000 tons/day "superplant," and all logistics such as raw material delivery, transfer stations, storage, and processing, if involved, might become unmanageable, as would a saleable financing package, construction management, etc.

5.1.2 Preproject Expenses

The time and cost for developing a solid-waste, resource-recovery facility of any persuasion is high and climbing, with a time frame between 3 and 5 years of preliminary work before construction can start, and up to 9 years from initial studies according to an active solid waste plant developer.[2] There are many reasons for long delays, with bureaucracy and regulations heading the list. Planning must be in conjunction with government—a city, county, township, an authority, etc. Inevitably, approval is required by a number of public servants, most with a limited term of office and subject to election within a few years. *Individuals in authority have become very cautious in view of the failures of the last few years, and have overreacted, by and large, by somewhat blindly embracing purportedly proved technologies with their shortcomings and ignoring the advantages of the newer techniques, which have not all failed.*

An apathetic public is often unaware of criticism by knowledgeable dissenters of the choices of officials and their consultants despite revelations in the media, civic organizations, etc. Often, a projected cost is over the limit that the local authority can approve, or the length of the operating contract is beyond present legal limits, whereupon special approvals are required. All these problems tend to delay target dates. The same is true of environmental impact controversy which is also leading to additional and expensive pollution control such as scrubbers and special treatment of ash disposal.

Upon recommendation by their consulting engineer, local authorities must select a process and obtain firm long-term commitments for sufficient raw material to assure operation at design capacity. This procedure has become most critical with regard to financing. The process can extend beyond a reasonable period (several years), and the project may expire in its original form. It then faces new personnel who, if sympathetic, must become familiar with copious details.

During this preproject period, the new enterprise, whether by local government or private investors, incurs expenses. These include office facilities and personnel along with high-cost technical assistance. Concurrently, effective public relations are very important. The new facility must be "sold" to the public through presentations to civic and social organizations, public hearings, etc., as well as by publicity through all available media.

Altogether, preproject expenses will easily exceed $1,000,000 for a large, modern, solid waste plant at a time when the economics and acceptability of it have not been established and with the risk that it may never proceed. *Preproject expense can vary between 1% and 3% of the project total cost regardless of size.*

5.1.3 Financing a New Plant*

Until recently, general obligation bonds have been a popular financing method. Local authorities, township, city or county, will issue general obligation bonds for the specific purpose of financing it. The bonds are tax free and usually must be backed by the state; they are sold in the open market and eventually must be repaid. Depending upon the financial record of the city or local government involved, selling the bonds can be easy or difficult. In many cases, a referendum or special approval is required. Bonding increases debt and is usually a political negative that public officials and voters try to avoid. Once bond money is available, it is invested in short-term obligations at higher interest until the money is actually spent. If there is a long interval between bond sale and project expenditures, this interest income can amount to millions of dollars. Considering the costs and risks, the trend today is toward revenue bond financing, often with fuel service system vendors. General obligation debt secured by the taxpayers has been least favored.

Private enterprise usually borrows from banks or insurance companies; it must be financially sound, provide guarantees to investors, and include very substantial amounts for presently high interest cost during construction. The only way to avoid this high interest cost is for the corporation to receive progress payments. This is a workable solution, and the local authority will withhold a percentage retention from each progress payment in escrow until the plant meets its contractual guarantees.

Local authorities seldom borrow from lending institutions because these institutions are reluctant to lend large amounts to local governments with restricted financing guarantees. Government grants can be of great help in decreasing the original capital cost. For example, the New York State Environmental Quality Bond Act of 1972 provided substantial grants for new solid waste plants for the city of Albany and Westchester County. These grants reduce appreciably the sums to be raised

---

*Also see Chapter 6, Financing Land Disposal and Resource Recovery Projects.

by local authorities and also reduce appreciably the net owning and operating cost. However, such grants are not readily available in most states.

5.1.4 Foreign Financing

Foreign financing might be considered as a possibility for new facilities and several have been partially financed in Europe and Japan. There are various approaches:

1. Most European countries and Japan now have agencies that develop export markets aggressively. These agencies provide low-interest loans to borrowers who pay 10–20% of the total cost upon delivery and the balance over several years. The agency pays suppliers at the time of shipment, with a requirement that much of the equipment must originate in the lending country. Japan has recently become an aggressive source of foreign financing.
2. Equipment suppliers in Europe and Japan looking for new markets can usually offer delayed payment terms up to several years after delivery. In most cases, the supplier borrows from his local bank at relatively low interest and adds a small mark-up to cover financing. As of late 1984, foreign equipment can be 50% to 150% less expensive than that manufactured in the United States.
3. Where licensed equipment or processes are involved, the licensing company (usually financially sound) will offer as part of their "package" a delayed payment for services and/ or equipment. In most cases, these delayed payments occur after the plant is in operation.

5.1.5 Preproject Engineering*

During the preproject period, studies are required to select the process and to determine the size of the plant, its location, systems and equipment, and environmental considerations, etc. This engineering cost is at risk since there is always a possibility that the plant will not be built. A private turnkey vendor corporation usually has a qualified contract from the local authority to cover some of the cost but still must invest a sizable sum on the assumption that the project will proceed and that this preliminary cost will be recovered in the subsequent project profit.

At this time, decisions involving projected operating efficiencies and cost should have top priority. A low original capital expenditure saddled with excessively high operating costs is as fatal as the reverse. Usually, however, compromise can keep capital and operating costs within acceptable limits. Concurrently, a decision must be made regarding architectural aesthetics. Local officials and residents may insist that plant architecture be relevant to the character of the area, with no exception for municipal public works. Although solid waste facilities are functional industrial structures designed for minimal adequate building costs, thoughtful cooperative design can usually satisfy serious concern.

5.1.6 Selecting a Consulting Engineer

Selection of the engineering firm is extremely important for all stages of the project be it A&E (Architect/Engineer) or turnkey vendor. A track record of successful similar facilities is paramount, along with a cadre of experienced professionals who are available as required. With careful attention, dividing the work between two or more firms for A&E projects encourages competition. If one firm does not perform as expected, the other is familiar enough with the project to take over. The same criteria apply to consultants selected for review opinion, environmental impact statements, and construction management or monitoring, when turnkey system vendors are involved.

Contracts for engineering work should be on a lump sum basis with partial payments for work completed and heavy penalties for delays. The scope of work must be stipulated clearly, and as a general rule, medium-size engineering firms are usually most adaptable. Small firms (less than 50 people) may not assign the necessary manpower when really needed, and large firms (over 400) often have tedious chains of command, inflexible ideas, high overhead with overall higher costs, and no particular advantage except perhaps the irrelevancy (engineering-wise) of politically desirable name recognition (which can be helpful for financing, etc.). On the other hand, communications and organization may be loose resulting in a "headless horseman." Likewise, on-the-job-training involving key decisions is not uncommon with many consulting firms and has resulted in less than satisfactory plant performance in many cases.

---

*See Chapter 4, Section 4.1 (The Feasibility Study), and Section 4.2 (Procurement and Construction Management).

5.1.7 Concept and System Choices: Reliability and Redundancy

When a basic disposal/recovery concept has been selected, systems and equipment vendors (turnkey entrepreneurs or combinations thereof) must be chosen, with track record and experience as the vital criteria. *Too many unsuccessful ventures have been the result of inadequate experience and on-the-job training by engineer/contractors who succumbed to vendor euphoria.*

Solid waste energy recovery plants must be designed with utility levels of availability, with adequate redundancy (spare capacity) for continuous 24-hr, 7-day operation in most cases. Raw storage capacity of 2 or 3 days to accommodate nonextraordinary production interruptions is common and seems to be compatible with reasonable redundancy, which varies inversely with raw storage. Overkill in either case can impose prohibitive capital costs not only for equipment and receiving capability, but also for building size and land requirements as well.

Redundancy can be an expensive cushion if extended beyond reasonable requirements.[3] Unless there is no acceptable by-pass disposal alternative, each system should be optimized in this respect, considering that most plants operate at or close to the economic margin.

In this connection, the capital cost of conservative equipment and systems designs can lower maintenance costs and increase availability. For example, boiler fireside corrosion and deposits are perhaps the most vexing and expensive maintenance items aside from lower furnace grate and refractory downtime and costs. More exotic metallurgy is appearing especially in mass-fired designs.

Mass-burn plants in Europe have found redundancy to be a high priority. For example, the largest (Rotterdam) has six furnace trains each with a daily design capacity of 440 tons to provide a total plant daily production requirement of but 1750 tons. After less than one year of operation, a large new waste/energy steam-electric plant in the United States finds it necessary to add another full size boiler unit.

5.1.8 Cost Control

Cost control begins with conception of the plant and must provide the first estimates based mainly on historical data. As the project progresses, realistic cost control, including preproject expenses, permits early development of an accurate budget.

The most important document created by the project manager is the equipment list. It is this list that becomes the basis for the overall equipment budget and develops into the basic control document throughout the life of the project. Figure 5.1a shows an equipment list used on a recent successful project. It is printed on 11½ in. × 17 in. (29 cm × 43 cm) sheets which can be cut in width as required for the various groups that will use the list. Engineering firms and subcontractors require only the left half of the form because they are not interested in deliveries or in cost data. For the general contractor the next two columns are included with the delivery data. Project engineers, accounting, cost control, and owner's personnel receive the whole form. Separate sheets are used for each department and the columns are self-explanatory. A typical entry is shown. A summary sheet is prepared as per Figure 5.1b.

ALPHA RECOVERY COMPANY				EQUIPMENT LIST DEPARTMENT:									PREPARED BY: ORIGINAL ISSUE DATE: REVISION: DATE:
							DATE		COST IN M $				
ITEM	QTY.	CODE NO.	DESCRIPTION	SUPPLIER	P.O.	REQ'D AT SITE	PROMISED DELIVERY	F.O.B. SUPPLIER	FREIGHT	DUTY	C.I.F. SITE		REMARKS
16	3	6025 – 02/04	B'L'R FEED PUMPS	DELPHIN CORP.	S–012	MAY'83	JAN.'83	25.4	1.1	0.4	26.9		ONE IS TURBINE DRIVEN, ITEM 26

Fig. 5.1a Equipment list.

ALPHA RECOVERY COMPANY OMEGA PLANT	EQUIPMENT COST SUMMARY				PREPARED BY: DATE: _____
DEPARTMENT	COST IN M $				REMARKS
	F.O.B. SUPPLIER	FREIGHT	DUTY	C.I.F. SITE	
TOTALS					

Fig. 5.1b   Summary sheet.

Cost Breakdown

There are several ways to present a breakdown of capital costs for an A&E project (turnkey vendors usually do not readily reveal such data). One of the simplest is shown by Table 5.1 for a new 2000 tons/day solid waste plant. Items included are:

1. *Equipment.* This is the cost of all equipment, including piping, instrumentation, and electrical. It is the cost F.O.B. U.S.A. manufacturer; for foreign equipment, it includes overseas packing. As far as possible, all items should be purchased at firm prices without escalation. This usually requires a percentage surcharge that is always worthwhile because it permits an accurate budget without any surprises when the project is well advanced.

2. *Spare Parts.* Total amount usually varies from 6 to 10% depending mainly on suppliers location and whether or not required spares are stock items. Foreign equipment requires more spares due to longer delivery time.

3. *Duty and Freight.* Imported material requires U.S. duty that varies according to the equipment. It is usually in the range of 6–10% of the equipment value. Freight can be inland for domestic equipment plus overseas for foreign equipment. Needless to say, a railroad spur at the site represents an appreciable saving in inland freight.

4. *Engineering.* This includes preproject engineering, conceptual engineering, and detailed engineering. Usually several firms are involved, with costs in the range of 4 to 8% of the total.

5. *Equipment Erection.* This is the actual labor cost of installing equipment, including piping, electrical and instrumentation. As far as possible, subcontractors should be used and contracts should be "lump sum." When this is not possible, contracts should be on "unit cost." By all means avoid "cost plus" contracts that usually result in heavy overruns. Depending upon the complexity of the system components, installation will vary from 35 to 50% of equipment cost.

6. *Construction.* Include all civil and structural work, both labor and materials. Usual materials are structural steel and reinforced concrete with noncombustible siding and roofing. Again "lump sum" or "unit cost" subcontracts are recommended.

7. *Overhead.* This includes a variety of items such as payroll taxes, fringe benefits, supervisory personnel, temporary facilities, consumable stores, construction equipment rent-

Table 5.1   Capital Cost, 2000 Tons/Day RDF Steam Electric Plant (in 1000s)

1. Equipment		$
process area	14,000	
power complex	30,000	
auxiliary services	7,000	
		51,000
2. Spare parts		5,000
3. Duty and freight		4,000
4. Engineering		8,000
5. Equipment erection		
direct labor	14,000	
subcontracts	10,000	
		24,000
6. Construction		
materials	10,000	
labor	10,000	
		20,000
7. Overhead		
general	5,000	
fringe benefits	4,000	
construction equipment	3,000	
		12,000
Subtotal		$124,000
8. Insurance and legal		5,000
9. Interest during construction		28,000
10. Contingency		10,000
Grand Total		$167,000

als, expendable tools, etc. It can vary from 5 to 15% of the total cost and from one contractor to another.

8. *Insurance and Legal.*   Very little can be done to control these costs. They are inherent in any project and usually seem much higher than necessary.

9. *Interest during Construction.*   Since the dollar amount may be very high at today's interest rates, it should be carefully computed. There must be a cash flow for the duration of the project with the best estimate of when expenses will be incurred and when monies will be received. This will indicate cash required and actual interest expenses.

10. *Contingency.*   This allowance depends upon company policy, and a minimum of 5% of total project cost is recommended.

5.1.9   Purchasing Procedures

A common procedure of three proposals with evaluated selection of the lowest is inadequate, especially for large purchases. Unless an extremely aggressive purchasing agent is available, negotiations for large orders (say over $100,000) should be carried out by upper management.

For any large purchase, the list of possible suppliers should be screened to two or three based strictly on technical merits. At that point, engineering should advise if the remaining suppliers are equal technically. If not, they must provide a dollar value for discrepancies between the possible suppliers. Henceforth, the negotiation is mostly commercial. Final negotiations by upper management can bring price reductions of over 10% and delayed payment terms can represent additional savings in interest expenses. Spare parts and field services must be negotiated at the same time as the main equipment, otherwise such items will be priced higher; the same is true for packing, freight, insurance.

5.2   CONSTRUCTION COST

The budget for construction cost should be as accurate as possible but always should be conservative. During construction, usually a period of 2 years or more for large ($\geq$ 1000 tons/day) facilities, unexpected setbacks are not uncommon, but judicious planning and attention to detail can minimize the effects of any "Murphy's Law" episodes.

5.2.1 Schedules

A realistic construction schedule must be prepared with man-hours by trade and by month. This is usually based on historical records of similar plants with necessary corrections for the local labor market. A typical chart is shown by Figure 5.2*a*. The area under each curve is the total number of man-months for each trade. Civil includes all construction including structural work, roads, rail siding, and so on, site improvements and landscaping. Mechanical covers ironworkers, boilermakers, welders, steam and pipe fitters, and millwrights. Electrical includes all instrumentation (electronic or pneumatic) and miscellaneous includes office personnel, first aid, transportation, stores, security, administration, and so on.

When planning the work of various trades, it is important to avoid sharp peaks, keeping construction and erection curves as flat as possible. This permits the maintenance of crews of quality personnel without many replacements. Overtime should not be considered at all in the planning stage. It is not only expensive but usually ineffective and should be scheduled only when absolutely necessary and closely controlled. A common mistake is to schedule a certain number of hours overtime per week to "catch up." Within a couple of weeks, workers are invariably producing, with overtime, just about the same output as during regular hours. If a worker knows that he is going to work overtime, he will slow his pace (sometimes subconsciously). "Spot" overtime without advance notice is an effective remedy. The total man-hours per month are extremely important and by comparing this total with the original estimate it is possible to have a clear picture of the status of the project. A typical distribution of man-hours by month is shown by Table 5.2.

There are always a few items in any project with long deliveries and these are usually boilers, turbogenerators, and certain switchgear. This equipment most often determines the overall construction time and become so called "critical path items." Since delivery time can be in excess of 1 year and with erection time (a boiler for example) also over a year, the overall schedule is over 2 years unless orders can be placed well before construction begins. This approach can be risky and few will take that risk without a fair and reasonable escape proviso. This is accomplished with orders "subject to cancellation," which permits placement in the manufacturers construction schedule and receipt of preliminary drawings, permitting early start of engineering. Such orders are subject to a "cancellation charge," however, which is a risk that must be taken once the probabilities of project completion are acceptable. Only by following this procedure is an overall schedule of less than 3 years possible.

A comprehensive progress chart or "S-curve" is shown by Figure 5.2*b*. This graph shows the actual progress month by month as originally projected. As construction progresses, a record is kept that compares actual performance against the projected curve and a record curve is plotted showing "actual" progress. Most often, the projection is overly optimistic and the actual values fall below the curve with a corresponding extension of the completion date. This is the principal reason for projects finishing over budget. Data from completed projects shows that seldom does monthly completion reach 9% of the total.

When construction starts, there must be constant effort to shorten construction time, and the experience of the general contractor is critical. The record of solid waste plants constructed during the last five years quickly reveals that the majority of these plants have taken much longer than planned, and in a few cases, as much as double the original projection. In very few cases, projects have been completed as scheduled with but two plants operating on or ahead of schedule. The schedules for construction and erection should include necessary allowances for bad weather. In the northern states there are blizzards and snowstorms; in the south it can be hurricanes or torrential rains. Working hours must be adjusted as required by the changing seasons and local customs, and expediting completion is the easiest way to save money.

The original budget must include increments for overhead during construction and for interest on capital taken down. Both items are directly proportional to construction time, with present interest rates, the latter is a very high value, usually millions of dollars.

Many methods have been tried to keep track of construction progress and to decrease elapsed time. One of these is critical path method (CPM), with its many sophisticated forms, long computer printouts, etc. Although CPM is effective when properly utilized, it requires reliable personnel to provide accurate data input, otherwise the system is useless. It is better to have high-caliber supervisory personnel concentrate their efforts on construction and do away with CPM or other sophisticated control methods. Simple bar charts such as shown by Figure 5.2*c* are easily understood by everyone and provide the necessary means of control. Bar charts are prepared for various functions and basically relate to material delivery information.

In addition to an overall schedule for the complete plant, another should be prepared for work required during the following 2 months on the basis of equipment deliveries. These revised schedules can differ appreciably from the original due to variations in equipment deliveries, weather conditions, delays in engineering, etc., with an additional work schedule prepared covering the next 2 weeks of work to anticipate changes in manpower requirements.

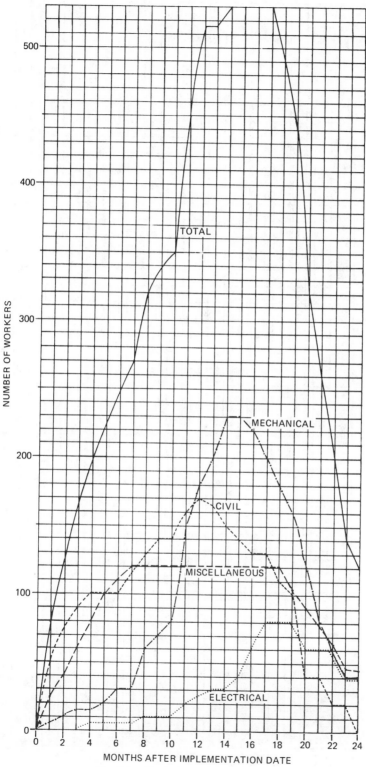

Fig. 5.2a Typical manpower chart.

PROJECT SCHEDULE

DATE: _____

LINE	DESCRIPTION	WEIGHTED VALUE	ACTIVITY % COMPL.	REMARKS
1	ENGINEERING			
2	PROCESS DESIGN	2.0		
3	DETAIL DESIGN	4.0		
4	PROCUREMENT			
5	PURCHASING	3.0		
6	EQUIPMENT DELIVERY	3.0		
7	WAREHOUSING	1.0		
8	MATERIALS OF CONST.	2.0		
9	CONSTRUCTION			
10	SITEWORK	2.5		
11	CONCRETE	7.0		
12	STRUCTURAL	4.0		
13	ARCHITECTURAL	4.5		
14	EQUIPMENT ERECTION	24.0		
15	PIPING	12.0		
16	ELECTRICAL	16.0		
17	INSTRUMENTS	5.5		
18	MISCELLANEOUS	4.5		
19	TEST AND CHECK-OUT	3.0		
20	START UP	2.0		
21	MONTHLY PROJECT COMPLETION %			
22	CUMULATIVE %			
23	ACTUAL %			

PROJECT STATUS: 25 50 75

% COMP. scale: 100, 95, 90, 85, 80, 75, 70, 65, 60, 55, 50, 45, 40, 35, 30, 25, 20, 15, 10, 5, 0

Timeline columns — 1979 (APR–DEC), 1980 (JAN–DEC), 1981 (JAN–JUN), numbered 1 through 27.

Line 21 (MONTHLY): 0.5 0.5 0.5 1.0 1.5 2.0 2.5 3 4 3.5 4.5 5.5 7.0 7.0 8.5 8.5 7.5 7.5 5.0 5.0 3.5 1.5 1.0 1.0 1.0 0.5

Line 22 (CUMULATIVE %): 0.5 1.0 1.5 2.5 4.0 6.0 8.5 11.5 15.5 19.0 23.3 29.0 36.0 43.0 51.5 60.0 67.5 72.0 82.5 87.5 92.5 96.0 97.5 93.3 99.5 100.0

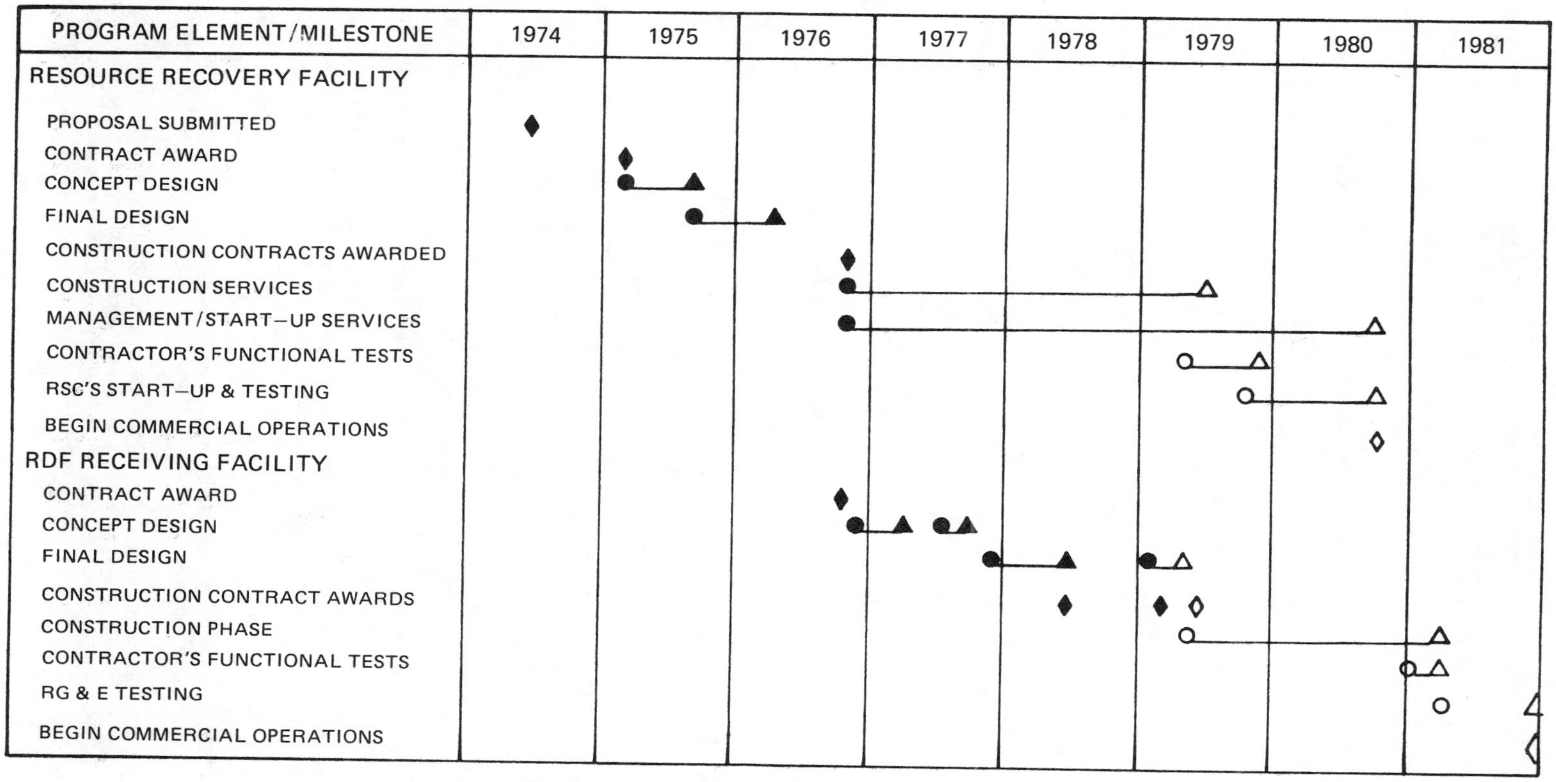

Fig. 5.2c "Bar chart"-type overall progress chart.

Table 5.2   Direct Labor Manhours

Month	Man-Hours	Cumulative
1	16,000	16,000
2	25,000	41,000
3	33,000	74,000
4	40,000	114,000
5	45,000	159,000
6	49,000	208,000
7	54,000	262,000
8	64,000	326,000
9	68,000	394,000
10	70,000	464,000
11	90,000	554,000
12	103,000	657,000
13	103,000	760,000
14	106,000	866,000
15	106,000	972,000
16	106,000	1,078,000
17	106,000	1,184,000
18	98,000	1,282,000
19	88,000	1,370,000
20	64,000	1,434,000
21	52,000	1,486,000
22	40,000	1,526,000
23	28,000	1,554,000
24	24,000	1,578,000

5.2.2   Labor Cost

Basic labor rates depend on the region. Prior to commencing work, a survey should be made in the surrounding areas of wage rates at other construction jobs in the immediate vicinity, and above all, of the availability of skilled workers. In some communities, there is no choice and union labor must be hired. If a choice is available, open shop is definitely preferred because it represents a savings of 25 to 35% over union labor. The difference in final cost is not so much basic hourly rates, which are fairly close between union and nonunion (the difference usually being just a few cents), but rather a number of other items that make union labor much more expensive. Some of these are cost of overtime, transportation cost, coffee breaks, vacation pay, crew composition, ratio of apprentices, trade responsibilities, etc. All of these make a significant difference, but the most important by far is productivity.

Under an open shop, labor must produce. If a worker performs, he is promoted; if he does not, he is fired. Labor is hired and promoted exclusively on the basis of merit. On union jobs, workers are sent by the union hall without any regard for capability. It is a question of seniority, and overall per capita productivity is quite low compared with open shop workers.

On union jobs, the number of helpers versus skilled workers is fixed whether they are needed or not. Equipment received must be unloaded by the trade involved even though it may be ordinary steel that can be unloaded by ordinary labor. The question of strikes is of course of great importance. These are always possible on a union job and can be costly or even fatal on construction jobs.

If a job must be run union, a "labor project agreement" should be negotiated with the local unions prior to starting any construction work even if it means delaying the start. Such an agreement must clearly specify crew compositions, trades responsible for various tasks, working conditions, etc. Such a contract or agreement must be signed with all local unions and should include a "no strike clause."

5.2.3 Cost Containment

Some of the procedures that will help to keep construction cost down are:

1. Avoid intricate layouts and ornate architecture. Solid waste plants are industrial plants and as such should be designed from the industrial point of view and not as an architectural monument. Although architects sometimes tend to produce expensive works of art, solid waste plants need not be ugly and unattractive. Building lines should be simple and functional. Classic exteriors should be reserved for college campuses and public buildings. Solid waste plant construction should generally be a combination of structural steel and concrete, with roofing and siding as required by local climatic conditions, but always within the industrial concept. Ornate finishes should be avoided in offices and sanitary facilities, layouts should allow space for future revision, corridors should be ample, and, above all, provision for equipment maintenance must be incorporated from the beginning.

2. Continuous communication between the construction crew and the architect/engineer is a necessity because not every detail can be decided in advance by the project engineers. As construction progresses, the field construction crew will suggest any miscellaneous small changes that should be incorporated. If the design has been frozen by the architects, this is not possible. It is at this time that coordination between both groups is of extreme importance so that change can be implemented as required. The architect/engineer should be a local firm, easily reached and familiar with local codes and regulations.

3. Cost plus or open end subcontracts should be completely avoided. When the scope of the work can not be exactly defined, subcontracts should be issued on a cost per unit. In general, local subcontractors can usually perform much better because they are familiar with local construction codes and they know the local labor market.

4. Keep the organization and chain of command simple, lean, and trim during construction and have quality as the first priority of supervision.

5. Keep overtime to a minimum. If possible, plan 10-hr workdays, 4 days per week with 3 days off. This schedule is extremely effective on large construction jobs. It provides more productive working time per week and permits making up for a day lost due to inclement weather. Overtime should be sporadic, used only as required, and never as routine. Many jobs have tried 5 work days of 10 hr and have found that after a few weeks, production in 50 hr is nearly the same as a 40-hr week. This does not mean that overtime is not useful; it only means that overtime should be selective. If on Monday morning, a worker knows that he is going to work 50 or 60 hrs that week, he will subconsciously slow down with reduced productivity. If on the other hand, the worker does not know that overtime is imminent, he will work at his normal pace and, when at the end of the week he is told that there will be overtime, the extra hours will be more productive.

6. Keep meetings to the absolute minimum and, when possible, to a minimum number of people. In general, the maximum attendance should be six. Meetings involving large numbers of people represent a tremendous loss of time. Most items discussed involve only two or three of those present and all others are simply wasting time and money. It is better to have several meetings with few people than one meeting with many. It is the responsibility of the person conducting the meeting to keep it short and to the point.

7. Always pay slightly higher wages than other local construction jobs. This helps keep the necessary and best of the skilled local workers.

8. Try to subcontract earth moving, security, etc. Usually, local contractors providing these services can do it cheaper than the general contractor.

9. Do not buy expensive construction equipment. Instead, lease such equipment with a "buy-back provision"; then it may be purchased later with attractive terms.

10. Keep design changes to a minimum. After a system has been engineered, changes are expensive not only for additional engineering but also for the delay incurred and in some cases the additional cost of changes to equipment already ordered. On any job, an engineering "point of no return" should be established. It simply means that no further changes are allowed just to improve the design. Of course, if an error in design is discovered, it must be corrected and changes *must* be made. However, after the "point of no return," changes to produce "a better layout" should not be considered at all.

11. Security at the site is a must and the first requirement is a slide fence with a single gate for all personnel and vehicles. This gate should be monitored continuously by a guard who passes in only those actually working in the project. Any visitors must sign in and out after their visit has been authorized. All vehicles leaving the site should be checked for tools and materials. To remove any material from the site, a pass must be given by the stores supervisor. If the site is isolated from the community, at least two guards should be on duty after working hours and they should be equipped with portable radios.

12. Allow enough free space around the buildings for lay-down area of materials and equipment. Usually not less than five times the building area should be allowed for this purpose.

13.   Keep the number of temporary buildings to an absolute minimum. Use leased trailers for offices. Concentrate early construction on a simple permanent building that can be used for storage as equipment starts arriving. If necessary, use leased trailers and tarpaulins until a permanent building becomes available.

14.   Try to purchase all equipment CIF site. In many cases, if properly negotiated at the time an order is placed, the supplier will absorb the freight. Whenever possible avoid purchasing FOB supplier's plant.

15.   Prefabricate as much as possible, both at the supplier's plant and at the site. Suppliers of certain major components—boilers, etc.—can prefabricate at the factory at much less cost than the contractor on-site for structural steel and ducts; prefabrication at ground level is preferred and should be planned in advance.

16.   Enforce all safety precautions. Most accidents are preventable and a good safety man can save many times his salary. Accidents disrupt the work pattern and lower morale. A good safety record is paramount in a successful project!

5.2.4   Cost Controls

Once the initial budget is established, it should not change during the life of the project. Periodic evaluations will produce an updated "estimated final cost" (which is really a revised budget), but it is important not to stray very far from the original budget.

Figure 5.3a illustrates a typical cost report summary produced each month. This report should be available to management immediately at month's end and certainly not later than the 15th of the following month. The cost report breaks the job into its basic components and shows the amounts committed for each item, the estimated final cost, and any variation from the original budget. It includes a number of supporting sheets with breakdown of various components as well as a "variance analysis" explaining the "overs" and "unders" re, the original budget.

This cost report is analyzed by management when it is issued to take necessary corrective measures, and it may be sufficiently complicated for computerization, in which case there must be every effort to keep it simple. Most computer programs are sophisticated, long, detailed, and tiresome, and they receive less attention than a simple report prepared by hand.

The "estimated final cost" for each item in the report is the sum of actual expenditures to date plus what has been committed but not yet spent, and the estimated spendable remainder. It is this

PAGE _____ OF _____

REPORT NO. _____

COST REPORT – SUMMARY

AS AT _____

BACK UP PAGE	DESCRIPTION	BUDGET	ACTUAL COST		EST. TO COMPL.	EST. FINAL COST	VARIANCE	
			CURRENT	TO DATE			OVER	UNDER
	Equipment – Process Area							
	Equipment – Power Complex							
	Equipment – Auxiliary Servs.							
	Spare Parts							
	Duty & Freight							
	Engineering							
	Sub–Total							
	Erection – Direct Labor							
	Erection – Sub Contracts							
	Civil Material							
	Civil Labor							
	Overhead							
	Fringe Benefits							
	Construction Equipment							
	Sub–Total							
	Legal & Insurance							
	Interest During Construction							
	Sub–Total							
	Contingency							
	Grand Total Costs							

Fig. 5.3a   Typical monthly cost report summary.

last column that must be carefully watched since estimates could be way off, with unexpected surprises at the end of the job. This is particularly true for piping and electrical work.

Equipment is usually shown CIF site with the number of departments as required. Piping, electrical, and instrumentation are seldom broken by departments because it is difficult to have an accurate budget early in the project.

Spare parts needed for operation of the plant do not necessarily have to be in the capital cost and can vary between 5 and 12% of the total project cost. It is common practice to order most of them with the equipment, although they are not required until after the plant starts; a much better price can be obtained from the supplier if they are included in the original order. Also, they can be bought at prices not affected by escalation.

Duty and freight apply only to foreign equipment. The former must be paid in U. S. dollars but overseas freight can usually be paid in foreign currency if the equipment is being financed outside the United States. Engineering includes basic conceptual as well as detailed engineering, often by a local firm as well as specialty consultants, i.e., noise, hazards, etc. Erection covers all manhours required to install the equipment, both normal hours and overtime. Civil refers not only to actual civil work but also to all building construction, inside finishes, and landscaping. Overhead includes office supplies, supervision, administration, computer services, security, etc.

The amounts actually spent for labor should be carefully checked against protected costs of labor agreed upon and actual manhours for various trades. Variances must be explained by the supervisors, including possible remedial steps.

Part of the financial package prepared at the beginning of the project is a "cash flow projection" showing month-by-month receipts and expenses with the consequent projected cash on hand at the end of each month. When the results for the month are known, they must be compared with original projections to determine if a correction is required on the receipts (progress payments, draw down, etc.). Otherwise an adjustment must be made on the rate of expenditures.

As part of overall cost control, the following must be monitored continuously:

1. Rentals paid for construction and transportation equipment. It is important to check if they are reasonable, going rates paid by others in the area. Even more important is whether or not the equipment is still required. Often, equipment is leased, completes its task, and is forgotten, but the rental continues. A monthly report as shown by Fig. 5.3b is recommended with a tabulation of all leased equipment. When this report is distributed, it is usually found that one or more pieces of equipment can be returned.

2. Prices paid by the purchasing department for day-to-day items. Often quotations are obtained at the beginning of the project, but with time prices are increased by the supplier and orders continue without any analysis of alternate suppliers. The cost-control officer must periodically audit orders being placed to assure that competitive prices are paid.

3. Expense accounts of individuals. As a rule, supporting receipts must be provided for any expense over $25.00 and accounts must be submitted within 5 days of completion of any trip. It is surprising how many companies have individuals who are over 6 months late in submitting their expense accounts.

4. Actual quantities being received against quantities ordered. It is common practice for suppliers to deliver slightly more than ordered and as a rule, the extras are received and paid for. In some cases, standard packages or standard lengths are involved. For example, if 18 ft of pipe are ordered, probably 20 or 22 ft will be supplied because it is the standard length. However, if 20 bolts are required and ordered, 144 should not be received and paid for just because 144 is the standard package.

5. Accurate time in and out of all workers is a necessity. Just 4 mins early departure on a 500-man job means 40 man-hours lost/day. Time clocks are a necessity to check daily time sheets submitted by the foremen.

5.3 OPERATING COST

Background

A popular belief during the 1970s was the "gold in garbage" nostrum. At that time there was limited knowledge of the capital cost of solid waste plants and even less of actual operating costs. Private investors and government officials were under the impression that solid waste plants would produce a high return on investment and attractive cash flow. Unfortunately, nearly every project estimate at that time was grossly miscalculated. The finished costs of plants that began construction in the late 1970s were vastly underestimated, with revenues well below projections.

A contemporary summary[2] stated that: "No known resource recovery plant lives up to the technical or financial projections made by its promoters" and also, "we estimate that the private sector has lost more than $200 million in attempting to develop systems to recover resources from America's solid wastes." The American Can Co., Monsanto, and Allis-Chalmers have withdrawn

RENTAL EQUIPMENT REPORT
for month of

P.O. NO.		26A		
Equipment Serial No. Location		Lift truck A4678 Shop		
Supplier Contract length (mo.)		Denwar 12		
Rental/month Rent to apply Purchase value		500 50% 8,500		
Date in Accumulated weeks Estimated date out		1/12/80 16 6/1/81		
Remarks		In use 14 hrs/day		

Fig. 5.3b Typical rental equipment report.

from the market following marginal or unsuccessful ventures and several others have marginal balance sheets. This includes raw incineration as well as prepared fuels. Windham, Connecticut, has lost its steam market and is adding a turbo generator (a possible precarious move for small 200 tons/day plants despite the Public Utility Regulatory Policies Act of 1978 (PURPA), whereas Chicago Northwest and Harrisburg had little or no steam markets until fairly recently which may now have elevated them to marginality or slightly plus.

The city of Madison, Wisconsin, is looking for a steady market for RDF, and Monroe County, New York, has ceased operations pending rehabilitation, RDF market development, or replacement. The CEA/OXY Eco-Fuel Plant at Bridgeport, Connecticut, has been closed, and a decision to replace it with an expensive dedicated waterwall steam electric plant has evolved. A much less expensive alternative would be to adapt the same cyclone-fired utility boiler at United Illuminating Company to an RDF Coarser than the difficult-to-produce (powdery) Eco-fuel and at 20 to 30% of the capital cost of the proposed replacement mass-burn plant costing upwards of $200,000,000. Crane station of Baltimore Gas and Electric Company is doing it successfully in a cyclone unit fired by a nominal 1-inch (25m) RDF produced by National Ecology for Baltimore County at its Cockeysville, Maryland, process plant. Another casualty has been the City of Chicago Southwest Supplemental Fuel Production Facility which was closed after excessive start-up difficulty. However, plans for corrective rehabilitation may be underway.

The Saugus, Massachusetts, Refuse Steam Plant is now receiving sufficient raw material but, like similar plants, found major and expensive modifications to the boiler/stoker design necessary since start-up.[4] The steam customer has withdrawn and will be replaced by a dedicated turbo/generator. Few of the large plants completed in North America during the 1970s can be considered financially successful by private industry standards. With few exceptions, they cannot sustain an average output more than 60% to 70% of design capacity.

The experience in Europe is similar but in a different time frame, and having started earlier, there is more (but qualified) success. In most cases, Europe has continued its traditional incineration of raw municipal solid waste (MSW) in waterwall furnaces. Typical is Munich, Germany, which operates five boilers at 2668 psi at 1000°F (184bar, 538°C) and burns over 500,000 tons/year. All boilers have auxiliary fuel (pulverized coal or gas) and MSW is normally less than 50% of the total BTU input. Due mainly to high temperatures, there has been considerable tube wastage. The five units were installed in late 1960 and early 1970, and on-line availability has been reported at about 52%, resulting in per ton disposal costs about double the original projections.

5.3.1 Plant Ownership and Operation by Local Government

Operation is usually by a quasi-independent authority or a department under strong political influence, technically deficient, and often with low managerial capability. Record keeping is usually poor and accounting is often insufficient to permit determination of accurate operating costs. In many cases, an operating loss can be concealed in an overall cost that includes garbage collection, landfill operation, etc. In general, local governments have difficulty hiring qualified technical personnel to manage and operate such facilities and have turned to a private operator. Small plants, 50–100 tons/day, are less critical in this regard and can be run by the local sanitation department.

5.3.2 Private Operation and Publically Owned Plants

This can be at a fixed annual fee plus expenses or on a cost per ton basis. There are many such contracts and they vary considerably, depending on the scope of services, escalation, and penalty provisions, that is, Pinellas County, Florida, Glen Cove, New York, and so on. Some are beneficial for the private operators who simply perform a service without financial risk but with an incentive for efficient operation via contract renewal and perhaps pride. In any event, local government absorbs overall excess costs (losses), usually reflected in higher tipping fees.

5.3.3 Facilities Owned and Operated by Private Industry

There are several such facilities and, in general, the financial return is marginal, with a few breaking even or with moderate profit. Reasons for marginality are usually underestimation of capital and operating costs, overestimation of revenue, indicating a lack of experience going in, or unforeseen and costly postcontract disputes. Tax advantages for private equity can be a major incentive regardless of the operational and financial success of the venture.

For any of the categories described in 5.3.1, 2, 3 above, the internal organization should be basically the same. It can vary depending upon the government group involved or a parent company, but in general, it should resemble the organization shown in Figure 5.4 (if the plant is to produce RDF only or if the output is going to landfill, there would be no power group).

According to an early private operator,[1] "It costs $35.00/ton to process a ton of garbage, including debt service, taxes, utilities and personnel." With tipping fees of less than $20.00/ton, it is difficult for revenues to cover the difference. For private industry with a large investment, a reasonable return is expected; but for local government performing a necessary service for the community, zero profit is acceptable and revenues need only equal expenses. Even with generous tax concessions for private investors in resource recovery projects (that assist overburdened local government in obtaining necessary infrastructure services) *profit can be elusive!*[5]

The following are critical in the struggle for the "gold" in garbage.

5.3.4 Revenues

The Tipping Fee

Raw material delivery accounting systems vary from handwritten weigh scale delivery tickets to sophisticated computer identification/accounting. In most cases, the net weight delivered to the plant by a vehicle is charged at a fixed rate per ton. This tipping fee varies from low values less than $10.00/ton to over $30.00/ton. Tipping fees in North America are expected to increase to a minimum of $30.00/ton at most resource recovery plants over 1000 tons/day during 1984.

Tipping fees can be collected directly by the operating company, but in most cases, these fees are collected by the local authority and usually do not vary according to material mix, moisture, etc., which can be important operating cost factors. The raw material can be high in paper and organics for energy production, but it can also contain large amounts of noncombustibles including grit, glass, and metals which contribute to furnace fireside corrosion and deposits, increased ash, lower BTU values, and higher excess air requirements. Likewise, high moisture contributes

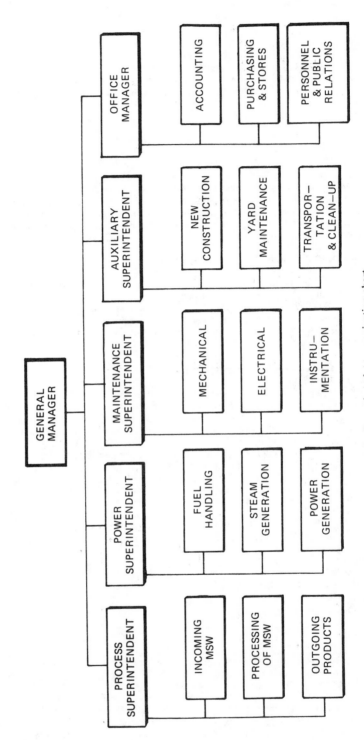

Fig. 5.4 Operating plant organization chart.

significantly not only to these undesirables but to the suspected contributions to precursors of dioxin formation.

For these reasons, plant operators, designers, specification writers, and others are struggling to establish average profiles for raw MSW, RDF, ash, and so on, which are reasonable, fair, and acceptable to equipment and system vendors, and above all, insistence on contractual accountability. It requires periodic random sampling and a sometimes complicated adjusting formula. A compromise is attainable which establishes minimum percentages of combustible and maximum percentage of moisture with adjustment of the tipping fee if the values are exceeded. It is the operators responsibility to monitor, evaluate, and maintain records for corroboration of tipping fee adjustment.

Altogether, continuing experience assures that tipping fees will increase faster than inflation during the next 5 years and that tipping fees of at least $30.00/ton will prevail at resource recovery plants of *any* type over 500 tons/day. With diminishing availability and increasingly stringent and costly land disposal regulations combined with stricter stack emissions and residue control requirements, this increase in fees is an unfortunate fact of life.

Steam/Electricity

By far the highest revenue to be expected from incineration of solid waste is the sale of steam. Unlike Europe, where historically there has been a ready and government-influenced district heating market, the outlook in North America is much more precarious and logistically less attainable. There are a few apparently successful installations including Nashville, Tennessee, Chicago Northwest, Harrisburg, Pennsylvania, Pittsfield, Massachusetts, and Albany, New York, with steam selling from $3.00 to $10.00/1000 lb (454 kg). As stated previously, Saugus, Massachusetts, Windham, Connecticut, and Akron, Ohio, (partially) have lost their steam markets but all are adding turbogenerators.

The stability of electric generation revenues is somewhat different. Although there is an implied certainty of the market via PURPA, the vagaries of the "least avoided cost" provision can be precarious also. Likewise, continuing challenge of the PURPA regulation can present additional uncertainties. In any event, electric generation revenues vary between 3 and 6 cents/kwh, with negotiations in the northeast approaching 11 cents/kwh (Bridgeport). This variation must be constantly monitored with perhaps modulating automatic control of operations to optimize revenue, that is, fire fewest boilers for maximum kilowatt output, using RDF entirely from storage, and cease RDF production, holding process equipment in "spinning reserve," or shutdown during peak "lambda" (least avoided cost) rates to minimize in-plant kilowatt consumption. This requires ample RDF and raw MSW storage.

For mass-burn plants, it would be the same approach, that is, peak generation with fewest boilers, with remaining boilers kept on "hot reserve." The thermal inertia/firing rate controlability of mass-burn units could render such techniques marginal, however.

Present contracts with utilities vary considerably and are sometimes unfair to the operating plants. There are instances where the utility purchases all energy produced (gross generation) at the low price and sells it back for plant use at regular market rates.

Utility contracts are usually complicated (see Chapter 15, Section 15.1) and favor the utility. Before PURPA, contracts were negotiated mostly at the discretion of the utility, which could impose stringent conditions and limitations. In future contract solicitation, solid waste electric energy entrepreneurs will contact several utilities to obtain the best terms. If the contract is with a utility not serving the area, a "wheeling charge" must be paid to the local utility; however, these charges are relatively low and the "foreign" utility offer should be high enough to offset the additional charge (see Chapter 15, Section 15.1).

The price per kilowatt-hour the utility will pay is the "least avoided cost" determined by the lowest of their system averages. Again, this is precarious for the seller since it includes generation at modern and very efficient fossil fuel units as well as hydro and nuclear plants, which may have low production costs. Moreover, higher peak rate adjustments are not passed back to the PURPA generator. Note, however, that PURPA rates of 11 cents/kwh are being discussed in the northeast.

Sale of RDF

The sale price is usually determined by the calorific value of the RDF and/or the cost of fuel being replaced by the customer.

RDF usually has a heating value of 4000 to 6000 BTU/lb (1008 to 1512 kcal/kg), or 5000 BTU/lb (1260 kcal/kg) as an average, or 10^7 BTU/ton (2.52×10^6 kcal). The steam produced can vary depending on the pressure and temperature, but as an average we can assume it to be 1050 BTU/lb

(588 kcal/kg). Boiler efficiency is usually 60 to 65%, say 62% as an average with 10.5 lb (4.7 kg) of steam/kwh as an acceptable value. With these values, we obtain.

$$\text{(English)} \ \frac{10^7 \ \text{BTU/ton RDF} \times .62 \ \text{Boiler Efficiency}}{1050 \ \text{btu/lb stm} \times 10.5 \ \text{lb stm/kwh}} = \frac{562 \ \text{kwh/ton RDF}}{(.62 \ \text{kwh/kg RDF})} \qquad (5.1)$$

$$\text{(SI)} \ \frac{2800 \ \text{kcal/kg RDF} \times .62}{588 \ \text{kcal/kg stm} \times 4.7 \ \text{kg stm/kwh}} = .62 \ \text{kwh/kg RDF}$$

Since a well-run power plant using #6 oil might have an overall heat rate of 10,500 BTU/kwh (2646 kcal/kwh), and using $33.00/bbl. [42 gal (158l)] oil, we have:

$$\text{(English)} \ \frac{\$33.00/\text{bbl} \times 10,500 \ \text{BTU/kwh}}{6 \times 10^6 \ \text{BTU/bbl}} = \$.058/\text{kwh} \qquad (5.2)$$

$$\text{(SI)} \ \frac{\$33.00/\text{bbl} \times 2646 \ \text{kcal/kwh}}{1.5 \times 10^6 \ \text{kcal/bbl}} = \$.058/\text{kwh}$$

The oil equivalent price for RDF is then:

$$562 \ \text{kwh/ton RDF} \times \$.058/\text{kwh} = \$32.60 \ \text{ton} \qquad (5.3)$$

However, a customer, including utilities, would offer considerably less and the reasons usually given are purported:

Slag problems
Quality and quantity of ash
Cost to adapt existing boilers
Added transportation cost
Emissions problems

Utilities in particular are reluctant to use RDF for these reasons which can lead to prohibitively prolonged start-up of RDF process plants and perhaps "bleeding to death." This can be a compelling reason for any waste-to-energy project to include its own dedicated steam electric plant despite the burden of doubling or tripling the capital cost.

Although RDF is a "waste" fuel, its preparation is not without cost. Nevertheless, it is an important adjunct in alleviating the deteriorating landfill outlook and reducing dependency on nonrenewable fuel sources.

Recovered Materials

Ferrous metal averages 6 to 8% by weight in most raw municipal solid waste. It is recovered magnetically and sold in a range from $5.00 to $40.00/ton (1984) depending upon cleanliness and bulk density.

Aluminum averages from 0.7% to 1.0% by weight in most MSW and is recovered in but a few plants. Depending upon beneficiation and markets, the price (1984) ranges from $400.00 to $900.00/ton.

Other nonferrous metal such as copper, brass, and zinc can be recovered by minerals processing techniques, with prices still soft in 1984. In addition to revenue, an important and increasing benefit will be the necessity to remove heavy metals from residues to landfill due to growing concern for ground water contamination from that source.

Glass recovery is technically feasible (except for automatic color sorting). There are limited markets but glass recovery for revenue is seldom practiced. Removal for improving maintenance and operations is not uncommon in RDF preparation, however.

Ash aggregates are produced by many incinerators and include ceramics, stones, and miscellaneous metals. Such residue has been found suitable as aggregate for the bituminous mixture in street paving. Test sections evaluated recently consisting of incinerator residue mixed with asphalt instead of a lime binder to improve adhesion compare favorably with standard mixes and may enhance the economics of future solid waste facilities. Likewise, fly ash and bottom ash fines have found a local market with Portland cement manufacturers in at least one location. These are the smallest increments of overall plant revenues, but they should increase as front-end and back-end recovery systems for all concepts improve and the national urgencies for infrastructure rehabilitation proceed, pending classification (hazardous waste) of incinerator residues.

Geographic location is an important factor in the feasability of materials recovery because shipping costs may discourage markets and preclude profitable revenue. For example, glass removal for revenue is not at all justifiable unless there is a nearby user with quality matching requirements. A railroad siding is a distinct advantage but trucking is always available, perhaps at higher cost. Other industries enjoy special rail rates for virgin raw materials such as steel and paper manufacturing; however, recovered/salvage materials have not enjoyed such government largesse and have suffered accordingly. Also, see Chapter 12, Section 12.11, and Chapter 15, Section 15.2.

5.3.5 Expense

Labor

Small plants (up to 500 tons/day) can operate with payrolls up to 15 whereas large installations over (2000 tons/day) require up to and over 75, including support services. Labor has become a major increment of operating cost, thus the trend toward maximum automation:

1. 1000 tons/day: One full operating shift + one skeletal second shift maintenance crew.
2. 1000 to 2000 tons/day: Two full shifts + one full second shift and maintenance crew.
3. 2000 to 3000 tons/day: Three or four operating shifts with a continuous maintenance cadre (required regardless of plant size if there is steam electric generation).

The degree of automation is rapidly increasing with sophisticated automatic process control systems for combustion, steam and water cycles, and so on, to minimize labor, which is extremely expensive in most areas and carries up to 30% in fringe benefits. Likewise, process monitoring of less critical functions by remote readout (TV, process sensors, etc.) is a trend, but caution is urged in replacing a good process monitor/worker in his zone of critical process functions such as incoming hazardous material, critical transfer/storage points, ash removal/handling (especially in mass-burn plants), or stack/effluent conditions. In some cases, labor cannot be reduced at all for raw material handling, surveillance, and traffic control. Hourly labor rates vary from minimum wage for casual unskilled rovers to over $20.00/hr for skilled crane and experienced control room operators in steam electric plants.

Cost of Materials

These are relatively small increments of total operating costs. Most plants require a supply of fluid fuel* for start-up (light off), water treatment chemicals, spare parts, and miscellaneous operating/maintenance supplies such as lubricants, welding rod, cleaning supplies, etc. At present, these costs are relatively low, but they must be controlled to avoid significant escalation, especially any increasing auxiliary fuel requirement. A combined effort of the process superintendent and plant engineer can produce a tabulation of yearly requirements from which the purchasing department can negotiate yearly bulk orders with partial deliveries as required.

Spare Parts

Spare parts are a major cost item and can represent several million dollars sitting in the warehouse; at present, interest rates can amount to over $1.00/ton processed. Solid waste plants require an ample supply of spares, and maintaining a complete inventory is not a prudent practice, that is, it requires working capital. A few suggestions follow:

For bearings, belts, packing, and so on, with a local supplier who will guarantee a necessary supply and invoice only as used.

For crane spares and conveyer parts, negotiate a similar arrangement.

For items not used frequently and high in cost, require the successful bidder (prior to original equipment purchase) to keep a spare(s) at his plant on consignment for payment when and if shipped. Such an arrangement may involve more than one plant using equipment of the same type and size, such as rotors for fans, pumps, turbines, shredders, and spares for cranes, vehicles, and so on.

Editors Note: An auxiliary burner near secondary air ports (overfire air in small mass burners) is not uncommon now in Europe to maintain temperatures at or above minimums to not only enhance combustion efficiency but also to inhibit formation of the precursors of certain toxic emissions and/or destroy them if they are formed, including dioxins. The State of New Jersey has proposed a requirement for such auxiliary burners in a new incinerator emissions code.

Keep an accurate record of spares usage and reorders based upon most "economical quantity" from an analysis of minimum cost/functional relationships. This method is found in most any economics textbook and includes the cost of placing purchase orders as well as the cost of holding units in inventory. It is easy to use and gives realistic results.

Indirect Expense

Taxes, insurance, supervision, administration, inventory control, and so on, have become major items in most operations, large and small. They are often overlooked and also include security, public relations, housekeeping, etc. Not much can be done about certain of these expenses, such as taxes and insurance, that have no alternatives. Likewise, supervision is a necessity and austerity at that level often produces unacceptable results. Public relations is a must but is easily overdone. It can be prudent and perhaps necessary to conduct plant tours for civic organizations, citizens groups and prospects for new plants, including the skeptics.

An area where crucial control is possible is "administration" (or so-called "general expenses"), which is often a hiding place for many "miscellaneous" expenses. Today, there is a malignant tendency to complicate office procedures via abuse of computerized information systems and copiers, and the like, with unnecessary and irrelevant information beyond what is really needed—a gross manifestation of Parkinson's Law! A good administrator should simplify all paperwork, use printed forms wherever possible, and minimize copies for distribution. Necessary reports must be produced as required but should be as short as possible; computers should be used when *really needed* but additional printouts are often just ego trips.

An area usually ripe for significant saving is inventory control, but it is unwise to invest in a sophisticated system to control nuts and bolts. A popular system is the so called ABC inventory control, which classifies all store items into A, B, or C categories. The A category includes items of high unit cost which may not be used for years, do not require any real control, and constitute a large portion of the total inventory dollar value. The C category involves bolts, nails, sandpaper, rags, and so on—items that are used every day but represent a very low dollar value; again, these items do not require "red control" since the cost is quite low. The B category includes medium-price items with relatively high frequency of use and usually account for more than half of the total inventory value. This is the group that requires accurate control and proper reordering techniques.

Financing Charges*

This item depends, of course, on how the facility is financed. Grants can reduce costs significantly, while borrowed money can be extremely expensive. There are usually two separate money costs. The first involves the initial capital cost which must be repaid with annual interest. It matters not whether the money is raised by a bond issue, by borrowing, by local government, by private enterprise, or for short or long term: *The obligation is crucial and the financing cost for a large, modern facility over 1000 tons/day can be over $15.00/ton.*

The second component, the cost of working capital, is sometimes overlooked or at least underrated. It can easily reach $3 million or more (or upwards of $1.00/ton processed). Working capital is often obtained through a revolving bank loan.

5.3.6 Profit

Question: For such a vital public service as solid waste disposal, should there be any profit at all? The answer is: "Profit" is not a MUST but "zero loss" certainly is.

Processing of MSW is an obligation of the local government and should be performed without a loss. This has not been the case with most plants that have become operational to date (1984). To repeat, revenues were overestimated and expenses were underestimated, so that with few exceptions, they are operating at a loss if all costs are exposed. Not all facilities that expected to reach full capacity in 3 or 4 months are operating at expected rates after several years of operation. This presents a completely different financial picture from initial projections.

Most of the recent installations involve a private system vendor or entrepreneur usually under a long-term design-build-manage contract with local government, and such private operators expect attractive profit for their "technical and managerial know-how." However, several such system vendors have withdrawn from the MSW market following unsatisfactory ventures and some survivors are struggling. *Unless a radical change occurs within the next 2 years, local governments will find few private entities willing and able to accept the risks, thus leading to a Darwinian survival of the deepest pockets and the danger of a token competitive division of the market not*

---

*Also see Chapter 6.

unlike the "conscious parallelism" accusation leveled at the "Big Three" auto makers before the "Foreign Auto Invasion."

5.3.7   Cost Control

Cost control of a MSW plant is very simple. It is mainly dependent on *PRODUCTIVITY*, which is defined as follows:

$$\frac{\text{Output}}{\text{Capital + labor + materials + energy}} = \qquad (5.4)$$

$$\frac{\text{Energy produced + materials recovered}}{\text{Operating cost}} \qquad (5.5)$$

in other words, EFFICIENCY OF OPERATION.

The following analysis includes variables and possible controls:

1. *Quantity of MSW Processed.*   This is limited either by the size of the plant or the availability within an authorized region. In most cases, these factors are inflexible. Unless a plant operates at or close to rated capacity, the cost/ton will be excessive, and many are in this category (at or near 70% of expected rates) that have started up during the past 3 years (most with sufficient raw material).

Several plants are in operation that can not obtain the necessary quantity of MSW from their public sources and have tried to attract private haulers with but limited success. Others have gone to nearby towns for their MSW supply with the corresponding increase in hauling distance.

2. *Tipping Fee.*   This fee is usually fixed in the contract between the operating company and the local government, but is subject to escalation adjustment. In most plants it is low, having been developed from previous landfill disposal fees. It will be much higher (well over $20/ton) for contracts negotiated in the next few years.

3. *Revenue per Unit of Energy Produced.*   It is fixed by contract with the utility or energy user and is usually tied to the cost of fuel replacement but without any other adjustment. For most operations selling energy from wastes, revenue rates very much favor the purchaser.

4. *Revenue per Unit of Recovered Material.*   This income is dependent upon available markets for ferrous, aluminum, glass, and so on, which are limited logistically and competitively (see Chapter 15). Revenue depends largely upon freight rates. Recovered materials have high rates for both truck and rail haul, and it has not been uncommon to the point that in many cases, freight absorbs all or most of the revenue.

5. *Labor Expense.*   This is usually fixed by the design of the plant and local rates. Some minor variations are permissible but not much.

Table 5.3   Capital Cost, Large Scale Refuse Fired Steam-Electric Plants

	A	B	C	D	E
Tons/Year	650000	657000	587000	465000	936000
Tons/Day	2083	2105	1881	1490	3000
Size (T/D)	2 × 1000	3 × 750	3 × 670	2 × 750	
Base Year	1979	1982	1982	1982	1979
Plant Cost M $	103.153	178.950	185.	121.531	131.5
Project Cost M $	160.	236.970	253.875	196.905	183.8
1983 Plant Cost	129.8	178.95	185.	121.53	165.4
1983 Project Cost	201.4	236.97	253.88	196.91	231.2
1000 $/T Plant Cost	62.4	84.8	98.4	81.6	55.1
1000 $/T Project Cost	96.8	112.3	135.0	132.1	61.3

A = Southeastern mass burn, steam-electric plant, operational.
B = Northeastern mass burn, steam-electric plant, operational.
C = Mid-Atlantic mass burn, steam-electric plant, start-up late 1984.
D = Northeastern mass burn, steam-electric plant, start-up early 1985.
E = Southeastern RDF/materials recovery, partial suspension burning, operational.

Table 5.4 Selected Data on Small-Scale U.S. Systems Using the Excess-Air Design, as of July 1983, Reference 1

Location	No. of Modules	Capacity (ton/day) of Each Module	Date of Startup (Past or Projected)	Capital Cost (10^6)	System Vendor[a]	Energy Market
Norfolk, Va.	2	180	5/67	2.2	A&E	Naval Station
Braintree, Mass.	2	200	3/71	2.85	Riley	Weymouth Art & Leather Co.
Portsmouth, Va.	2	80	1976	4.5	A&E	Naval shipyard
Waukesha, Wis.[b]	2	60	1971	1.7	A&E	Unknown
Hampton, Va.[c]	2	100	1980	10.3	J.M. Kenith Co.	NASA-Langley
Pittsfield, Mass.[c]	2	120	7/81	1.0	Vicon/Enercon	Crane Paper
Gallatin, Tenn.[c]	2	100	3/82	9.8	O'Connor	R.R. Donnelly, TVA, others
Prudhoe Bay, Ark.	1	120	1982	Unknown	Basic Env. Eng.	Unknown
Collegeville, Minn.[c]	1	65	1982	2.4	Basic Env. Eng.	St. John's University
Harrisonburg, Va.	2	50	1982	Unknown	A&E	City of Glen Cove
Glen Cove, N.Y. (codisposal)	1	200(MSW) 10 (sludge)	1982	22 + 12	A&E	
Bannock Co., Idaho	2	100	1983–1984	9	Clark-Kenith Co.	FMC, Idaho Power
Lassen County, Calif.	1	96	1984	7.15	Bruun & Sorrenson	Lassen Comm. College

Location					Vendor	Owner
Fayettesville, Ark.	1	150	1984–1985	Unknown	Bruun & Sorrenson	Univ. of Ark.
Savannah, Ga.	2	220	1984–1985	28	Katy-Seghers	American Cynamide
Pasagoula, Miss.	1	150	1984–1985	5.9	Sigoure Freres	Thiokol
Islip, N.Y.	2	255	1984–1985	30–40	O'Connor-Penn. Eng.	Long Island Lighting
Davis County, Utah	2	250	1984–1985	33	Katy-Seghers	Hill AFB
Rutland, Vt.	2	120	1984–1985	11	Vicon	Central Vt. Public Service
Dutchess Co., N.Y.	2	200	1984–1985	Unknown	O'Connor-Penn. Eng.	IBM, utility
Burlington, Vt.	2	60	1984–1985	8	A&E	Univ. of Vermont
Claremont, N.H.	1	200	1984–1985	18	Clark-Kenith	Central Vt. Public Service
New Hanover Co., N.C.	2	100	1984–1985	13.84	Clark-Kenith	W.R. Grace Plus Elec
Washington Co., N.Y.	2	200	1984–1985	Unknown	Vicon	CIBA—GEIGY
Delaware Co., Penn.	1	50	1984–1985	2	N/A	County Geriatric Center
Tri Cities, Calif.	4	120	1984–1985	25	Vicon	

[a] A&E indicates that the system was designed by an architectural and engineering firm and the equipment was provided by various vendors.
[b] Retrofitted in 1979 (for $10^6\cdot3.9$).
[c] Case studies of these installations appear in Reference 1.

Table 5.5 Selected Data on Small-Scale U.S. Systems Using the Starved-Air Design, as of July 1983, Reference 1

Location	No. of Modules	Capacity (ton/day) of Each Module	Date of Startup (Past or Projected)	Capital Cost (10^6)	System Vendor	Energy Market
Siloam Springs, Ark.[a]	2	10.5	6/75	Unknown	Consumat	Allen Canning Co.
Blytheville, Ark.[a]	4	12.5	8/75	Unknown	Consumat	Chrome Plating Co.
Groveton, N.H.	2	12	Unknown	Unknown	ECP	Groveton Paper Mill
North Little Rock, Ark.[b]	4	25	8/77	1.45	Consumat	Koppers
Salem, Va.	4	25	9/78	1.9	Consumat	Mohawk Rubber
Jacksonville, Fla.[c]	1	48	1978	Unknown	SEE	Unknown
Osceola, Ark.	2	25	1/80	1.1	Consumat	Crompton Mills[a]
Genesee, Mich.	2	50	2/80	2	Consumat	Unknown
Durham, N.H.	3	36	9/80	3.3	Consumat	University of New Hampshire
Auburn, Maine	4	50	4/81	3.97	Consumat	Pioneer Plastics
Dyersburg, Tenn.	2	50	8/81	2	Consumat	Colonial Rubber
Windham, Conn.	4	25	8/81	4.125	Consumat	Kendall Co.
Crossville, Tenn.	2	30	12/81	1.11	Env. Services Corp.	Crossville Rubber
Cassia County, Idaho	2	25	1982	1.5	Consumat	J.R. Simplot

Location						
Batesville, Ark.	1	50	1982	1.2	Consumat	General Tire & Rubber
Park County, Mont.	2	36	1982	2.321	Consumat	Yellowstone Park
Waxahachie, Texas[d]	2	25	1982	2.1	Synergy/Clear Air[d]	International Aluminum Co.
Miami Airport, Fla.	2	30	1982	Unknown	Synergy/Clear Air[d]	Miami Airport
Portsmouth, N.H.	4	50	1982	6.25	Consumat	Pease Air Force Base
Red Wing, Minn.	2	36	1983	Unknown	Consumat	S.B. Foote Tanning
Cattaraugus County, N.Y.[d]	3	37.5	1983	5.6	Synergy/Clear Air[d]	Cuba Cheese
Miami, Okla.	3	36	1983	3.14	Consumat	B.F. Goodrich
Oswego County, N.Y.	4	50	1983	Unknown	Consumat	Armstrong Cork
Pasagoula, Miss.	3	50	1983–1984	6+	Consumat	Unknown
Oneida County, N.Y.	4	50	1983–1984	Unknown	Unknown	Griffis AFB
Tuscaloosa, Ala.	1	240	1984	13	Consumat	B.F. Goodrich
Fitchburg, Mass.	1	200	1984	Unknown	Consumat	Paper companies
Fergus Falls, Minn.	3	29	1984	3	Control	Mid-America Dairymen, Inc.

[a] System now shut down and equipment removed.
[b] System now shut down and equipment removed, but Consumat is to supply new equipment.
[c] System now shut down.
[d] Synergy and Clear-Air are now separate companies, each marketing its own system.

6. *Materials Expense.* Again, this is dependent upon the design of the plant. Slight improvements possible by good management.

7. *Indirect and Financing Charges.* Practically no significant improvement is possible.

We are then left with two large variables: (1) Amount of energy recovered and (2) amount of materials recovered. Both of these are a direct consequence of the "efficiency of operation."

1. Operating at maximum permissible rates for optimal combustion and maintenance efficiencies. This will be reflected in less internal energy consumption with a consequent increase in energy available for export, and also lower labor costs.

2. Operating to obtain maximum energy per unit of incoming waste. This is where various processes differ widely and where large operational savings are possible. Steam boilers should be supplied with homogeneous fuel. When moisture and BTU/lb (kcal/kg) are fairly constant, it is possible to operate at high efficiency with optimal excess air.

3. Maintaining an aggressive maintenance program. Solid waste plants are extensive materials handling operations, and lost time can be extremely high due to unscheduled equipment down time. An alert and experienced plant engineer can avoid equipment failures which translate into overtime for operating personnel and increased in-plant energy consumption.

Preventive maintenance is vital for solid waste plants. Equipment overhauls must be scheduled on a regular basis, and wear parts should be replaced at that time even if they are operating satisfactorily. It is wiser to spend $120 replacing a pump bearing when it is open for periodic overhauling than to shut down on an emergency or overtime basis when the cost can easily exceed $500.

An effective way to minimize down time is to establish a bonus based on plant output; this provides an incentive to all workers to keep the plant operating and to expedite repair work. A well-established bonus system will save many times the amount distributed.

4. Aggressive cost accounting will reflect true operating costs and will provide necessary financial information to management. It is surprising how many MSW plants do not have accurate records of operating costs.

In some cases this is due to poor management but in others it is a way of hiding the fact that the operation is overly expensive and subject to heavy criticism if the real values are known. Effective cost accounting requires continuous monitoring of all income and expenses with periodic reports showing trends and future projections.

Table 5.3 gives the capital cost of some U.S. large scale, refuse-fixed, steam-electric plants. Tables 5.4 and 5.5 give selected data on small-scale U.S. systems using, respectively, the Excess-Air Design, and the Starved-Air Design.

REFERENCES

1. Argonne National Laboratory, Thermal systems for conversion of municipal solid waste, Volume 3. Small scale systems: A technology status report, D.O.E. No. ANL/CNSV-TM 120, vol. 3. July, 1983.

2. K. J. Rogers, Vice President Environmental Systems Division, Combustion Equipment Associates, *The New York Times*, Nov. 11, 1979.

3. H. G. Rigo, Simulation as a resource recovery plant design tool, Proceedings, ASME 10th Biennial Conference, May, 1982, New York.

4. L. R. Galese, Wheelabrator finds turning garbage into energy one heap of a problem. *Wall Street Journal*, Nov. 16, 1978, p. 14.

5. A. R. Nollet and R. H. Greeley, The basic economics of energy recovery from solid waste, ASCE Environmental Engineering Conference, Boulder, Colorado, July, 1983.

CHAPTER 6
ECONOMICS AND FINANCING OF RESOURCE RECOVERY PROJECTS

WARREN T. GREGORY
JONATHAN M. WOOTEN
MICHAEL R. LISSACK

Smith, Barney, Harris, Upham & Co., Inc.
New York, New York

R. S. MADENBURG (Appendix 6.1)

Morrison-Knudsen Company, Inc.
Boise, Idaho

With the recent passage of the Deficit Reduction Act of 1984 by the Committee of Conference, the sale of industrial development bonds to finance resource recovery facilities will undergo a dramatic change. Although most of the major tax issues are not affected by this legislation, the ability to set aside a large portion of a state's volume cap for these projects is now in question. Although it is too soon to predict exactly how institutional arrangements will develop in each state, we think it is clear that projects requiring a large portion of the state's allocation will be subject to further scrutiny and competition with other priorities. We can only assume that this will further retard the growth of resource recovery financings.

This chapter is an attempt to present the project economics and financing formats that could be used for a hypothetical resource recovery project. The graphs enclosed represent the ranges of tipping fees that generally result from various project financing scenarios. Clearly many other questions of risk and public policy must be addressed before a decision can be made. However, we think it is useful for a public official to have this type of general comparison so that other risks and public policy questions can be considered within the broader framework of price and tipping fee.

In general, an increase in inflation will tend to steepen the tipping fee curves illustrated in this example and conversely, a decrease in inflation will tend to flatten them. This assumes a linear relationship between inflation, the cost of financing the project, and the cost of alternative energy sources.

6.1 RESOURCE RECOVERY FINANCING STRUCTURES[1]

6.1.1 Public Ownership

General Obligation Bonds

A political jurisdiction may issue bonds secured by the full faith and credit and the taxing power of the jurisdiction. Some bonds may pledge specific project revenues and thus be "double-barreled" with specified taxes and revenue as security. General obligation bonds are considered to be the most secure type of tax-free bonds and consequently may be sold at a lower net interest cost than revenue bonds. A referendum, if required in order to issue the bonds, can be quite time consuming and might delay the project if not properly planned.

Project Revenue Bonds

Revenue bonds are secured by contracts such as put-or-pay agreements and documents such as a trust indenture. This is to control the flow and use of revenues while ensuring that certain covenants are adhered to. Additional security is provided by a debt service reserve fund in case the regular debt service payment is interrupted.

Project Revenue Bond Structure

A typical revenue bond would be issued in one series. The proceeds of the bonds, together with investment income, would be sufficient to:

Complete construction of the project without reliance on additional debt or equity.

Fund a reserve fund equal to maximum annual debt service on the bonds.

Fund interest through completion of construction.

Cover all costs incident to the financing.

The trust indenture will establish funds into which the bond proceeds will be deposited and invested until required. It is anticipated that net bond proceeds (net of anticipated interest earnings on the funds) will be deposited as follows:

Capitalized Interest Fund. The portion of the bond proceeds to be used for the payment of interest on the bonds through the end of the construction period.

Debt Reserve Fund. The portion of the bond proceeds to be used to meet the reserve fund requirement.

Construction Fund. The portion of the bond proceeds to be used for the construction costs. Disbursements will be made in accordance with the schedule provided in the construction contract.

All project revenues are deposited with the indenture trustee for the following funds in the order presented:

Debt Service Fund. An amount sufficient to pay interest, maturing principal, and sinking fund installments on the bonds.

Operating Fund. An amount sufficient to pay O&M expenses.

General Reserve Fund. All amounts remaining.

This flow of funds is called a gross pledge. It has been recommended for this structure primarily to ease comparison with the private ownership structure. A net pledge in which operation and maintenance costs are paid ahead of debt service would also be acceptable in a revenue bond structure.

Security

A. Security During the Construction Phase. The bonds are secured by: (1) a commitment from the builder for liability up to the construction cost of the project and (2) a first lien on the reserve funds. In addition, a jurisdiction would be responsible for such risks as changes in law. For instance, if there is a change of law, the issuing entity will issue additional debt to rectify the problem and tipping fees will be increased to pay the additional debt service incurred as a result of the delay.

B. Security during the Operation Phase. The bonds will be secured by a first lien on the project revenues. Revenues include the base user fees, the revenues from energy sales and recovered materials' sales, liquidated damages payments, proceeds from insurance claims, investment income on all funds held by the trustee under the trust indenture, and the bond reserve fund principal. Under the trust indenture, the jurisdiction will covenant to deliver or cause to be delivered to the project an agreed upon amount of municipal solid waste (excluding hazardous waste), as provided in the disposal agreement.

6.1.2 Private Ownership Financing

The focus of Smith Barney's resource recovery structure effort has been private ownership financing. The majority of our projects are not economically feasible, when judged by the standard of comparable landfilling costs in the early years of the project, without the use of private ownership financing.

Private ownership financing can significantly reduce the financing cost of an energy resource facility by either passing on a major portion of the tax benefits of the project through lower financing charges or enhancing the profitability of the builder, thus allowing lower O&M fees. The financing cost is reduced by:

In the case of leasing, replacing a portion of municipal debt or vendor equity with third-party capital at a nominal cash cost. The equity is compensated for the most part by tax benefits associated with the project. The equity capital reduces the bonds otherwise outstanding, and therefore the debt service costs which are covered by the rental payments can be reduced.

In the case of private ownership financing not involving leases, the owner can either provide equity at a nominal cash cost as in the leasing example, or pass back a rebate which offsets debt servicing costs, such as a reduction of management fees or operating costs.

The magnitude of private ownership financing benefits is such that, depending upon the precise structure chosen and the tax risks involved, an assumed revenue bond interest rate can be reduced roughly 300 to 500 basis points. The private ownership savings vary somewhat as interest rates and other tax assumptions change.

The residual value of the facilities must rest with the private owner at the termination of the lease term or service agreement expiration in any private ownership financing. The residual value loss should be treated as a cost of private ownership financing which must be incorporated before any comparison with direct public ownership. Therefore, tax benefits related to solid waste facilities are an added benefit when coupled with the tax-exempt financing advantages, versus merely representing an offset to the financing of most public facilities where private owners are restricted to taxable debt financing.

Any private owner may obtain tax-exempt financing for solid waste facilities, since they qualify as exempt activities under Section 103 of the Internal Revenue Code.

Private Ownership. The tax benefits available to an owner of an asset will provide an adequate rate of return to some level of equity investment without any significant cash flow forthcoming from the communities. This equity investment replaces debt that would otherwise be required to finance a municipally owned asset. The resulting payment is substantially less than debt service on a 100% debt financing. Since the interest rate on the debt of a private ownership structure is likely to be identical with the interest rate obtainable directly, the payments are reduced approximately in proportion to the equity investment.

The reduction in the payment is dependent upon the tax benefits available in a project. The greater the present value of the tax deductions and credits, the greater the quantity of equity that can be provided with the required rate of return for equity investors, and the lower the payment. In the case of solid waste disposal facilities, the tax benefits are particularly valuable as a result of the additional investment credit on a major portion of the facilities. The present value of these tax benefits will support more private equity, thereby reducing the debt financing requirement and the total financing cost.

Assuming the builder is unwilling or unable to take all of the tax benefits in the project on a current basis and is thus not interested in ownership, leveraged leasing is an efficient mechanism to obtain a lower effective financing cost. These benefits could accrue to the project, depending on the competitive negotiating situation. Under the Tax Equity and Fiscal Responsibility Act (TEFRA) of 1982, certain items such as safe-harbor leasing and the use of ACRS (Accelerated Cost Recovery Systems) depreciation by tax-exempt borrowers have been eliminated. However, solid waste facilities have thus far been protected from significant adverse changes.

The following is a general appraisal of the impact of current tax law upon the availability of third-party equity to the energy recovery facility. Although certain elements of the tax law remain to be clarified, particularly as they may be affected by the Tax Reform Act of 1984, HR4170 now pending before Congress, this represents our best judgment at this time.

Capital Cost Recovery

Under the tax law prior to December 1, 1981, most resource recovery property was depreciable over 8 years or 22.5 years on an accelerated basis. The 1981 law uses a simplified cost recovery approach and provides 3-year, 5-year, 10-year, and 15-year write-offs for most categories of property. Tax advantages may be affected if the property is interpreted to be utility property.

The 1981 cost recovery schedule for solid waste property, essentially straight line over 5 years, provides negligible benefits over the 8 years double-declining balance method used previously. However, the turbine-generator depreciation over 5 years provides much larger benefits due to the extremely long previous write-off of 22.5 years. The depreciation of building and land improvements over 15 years is also significantly more valuable, since most owners and lessors previously would have chosen depreciation lives significantly longer than 15 years. These benefits are under challenge in the current session of Congress; the challenges are directed to the use of "accelerated" depreciation by tax-exempt entities.

The present ability of resource recovery facilities to utilize both ACRS depreciation schedules and tax-exempt debt financing is not available to most other types of tax-exempt activity projects due to TEFRA prohibitions that became effective this year.

Service Agreements

The primary tax issue in most resource recovery projects is whether the disposal agreement between the operator and the jurisdiction is a "service agreement" or a "lease" as construed in Section 48(a)(5) in the Internal Revenue Code. If the agreement is deemed to be a lease, the facilities do not qualify as Section 38 property and do not qualify for tax credits.

The service agreement issue is being directly challenged by the Tax Reform Act of 1984. Attempts to ascribe greater indices of ownership to public entities through a service agreement may cause unacceptable tax risks. Since the tax benefits are typically several times the estimated residual value and financing cost of the plant over the term of the lease, the trade-off of ownership for tax benefits is usually very attractive. Some concessions on service agreement renewals may be given to prevent an excessive economic consequence to system users.

Impact on Equity Marketing

The cost of equity in most leveraged leases has increased significantly over the last year. Fortunately, the structure of a leveraged lease transaction reduces the impact of an increase in equity rate of return on the effective financing cost. The savings of leasing during the financing term are somewhat reduced, but, in our judgment, not sufficiently to make leasing unattractive compared with the employment of 100% revenue bonds without tax benefits, as in public ownership.

6.1.3 Leveraged Lease Financing Structures

The sophisticated leveraged lease structure has been developed over many years and represents very nearly the most efficient device known to obtain the value of tax benefits from third parties. The efficiency of the leveraged lease comes about through the participation of large financial institutions, which have a very predictable tax appetite and enter into transactions expecting only a lender's rate of return. Operating companies, such as resource recovery builder/operators, generally have an unpredictable tax appetite as a result of being in a cyclical business, and thus have a higher cost of equity than the large financial institutions.

The leveraged lease structure does not address the risk-taking function of equity often perceived in the marketplace. Leveraged lease equity investors view their investment as a loan, the primary deviation being that the major form of repayment is through tax advantages rather than through cash. In the leveraged lease structure, risk-taking must be compensated. In this respect, the leveraged lease structure is little different from a municipally owned project, where tax benefits are not captured and risk-taking by parties other than the municipality must be compensated.

Tax Questions

Despite the cost efficiency of the leveraged lease mechanism, a number of tax issues exist for which there is insufficient authority for tax counsels to be in a position to give a "clean" opinion regarding the availability of tax benefits. With most resource recovery projects, the primary tax issue is whether the facilities are "used by" a tax-exempt entity. Section 48 of the Internal Revenue Code provides that facilities that are used by public entities are not "Section 38" property and, therefore, ineligible for investment tax credit.

Tax Perspective

From a practical point of view, viability of the leveraged leasing approach depends on using one of two approaches. The most time consuming, but safest, is the approach of picking a builder/operator and having either the builder/operator or the equity participant of the facilities ask for investment tax credit. The ruling request would ask the Internal Revenue Service to examine the waste disposal agreement and other documents in the transaction to ascertain that the facilities are not used by the jurisdiction. This determination depends on a number of factors, the most important being the possession and control of the builder/operator over the facilities and the amount of risk taken. If a favorable tax ruling is obtained, the Internal Revenue Service is precluded from challenging the tax status of the tax benefits on any grounds addressed in the ruling request. This would mean that the jurisdiction or the builder/operator would not have to indemnify the equity investor for any of these circumstances. However, for certain other circumstances, basically "acts and omissions," indemnifications would still be required.

The more difficult but potentially faster approach is to obtain a clean opinion from a tax counsel, relied upon by the lessor, that the facility qualifies for tax benefits. Having gone through several experiences in this area, we are of the opinion that except for the most radically structured possession and extraordinary amounts of risk, "clean opinions" will not be available. Even where a clean opinion is to be received, it is likely that the jurisdiction would have to indemnify against the loss of tax credits should the Internal Revenue Service ever challenge the transaction.

Revenue Stabilization Fund

A variation on the leveraged lease structure is available to "levelize" the tipping fees ordinarily associated with resource recovery projects. In the leveraged leasing context, rather than repay bonds used to fund construction with the proceeds of the equity contribution, the funds are placed in a Revenue Stabilization Fund. The leveraged lease equity investor is relieved of the obligation to repay bonds used during construction once the funds are deposited into the account with the trustee.

Early in the project, the tipping fee required from the communities is subsidized through the takedown of funds available in the Revenue Stabilization Fund. Later in the project, the tipping fee must rise proportionally to pay interest and principal. Subject to the amount of equity available in the leveraged lease, considerable latitude exists either to make a flat tipping fee schedule or to create an even lower beginning tipping fee, escalating subsequent tipping fees according to an index.

6.1.4 Builder/Operator Ownership

The tax benefits available to the builder/operator are the same as those available to an equity investor in a leveraged lease. The builder/operator also has the ability to put equity into the project pretax, if it desires, since no sale of the facilities need take place. An evaluation of builder/operator ownership structures producing the lowest financing cost depends on a number of factors, including (1) the availability of and predictability of tax-shelterable income, (2) the builder/operator's willingness to price tax benefits aggressively, (3) the need of temporary usage of cash resulting from depreciation benefits, and (4) the builder/operator's marginal cost of funds.

Conceptually, the builder/operator would be in a position to offer an even lower financing cost than the leveraged lease structure, since the builder/operator need not follow certain lease criteria that address cash flow and maximum leverage in leasing arrangements. While a leveraged lease investor may be limited to 80% leverage using tax-exempt debt, for example, the builder/operator can finance the asset entirely with tax-exempt debt and "rebate" a lower cost by subsidizing the debt service on the tax-exempt bonds. This alternative can be very beneficial to the jurisdiction since, unlike the equity investor in a leveraged lease, the jurisdiction can fund the equity portion of construction out of low-cost, tax-exempt debt.

Several factors make aggressive pricing of tax benefits by a builder/operator uncertain. The builder/operator may not have sufficient predictable tax appetite for solid waste projects to be completely confident of using all of the tax benefits under all reasonable economic conditions. Second, although tax benefits may be usable, the builder/operator may require a high rate of return on the investment in the project. Finally, the builder/operator may not want to bear the legal costs and management time to document and become comfortable in the complex area of tax-oriented financing.

6.1.5 Accounting Considerations

Most vendors are unwilling to participate in resource recovery projects unless the project obligations are off the balance sheet. The most common method used to achieve this result is a 50:50 joint venture with another participant. The participants then record the project by the equity method, netting off debt obligations.

The disadvantage of the 50:50 joint venture is that a large portion of the return must be shared with another participant, who often contributes less support to the project. Several methods exist to give a builder/operator a larger share of the project return without bringing the project on its balance sheet.

Leveraged Lease Structure

Under the leveraged lease structure described above, the leveraged lease may be recorded as an operating lease if the lease payments are contingent on the receipt of tipping fees from the communities. Obviously, the tipping contract obligations must be strong in order to make the leveraged lease salable. Any further credit support from the builder/operator may be recorded as a normal-course-of-business obligation, if its impact is not to guarantee a project indebtedness.

Making the leveraged lease payments purely contingent on tipping contract revenues may raise the question of whether the lease is made to the communities directly, rather than to the builder/ operator. This would potentially disqualify the assets for investment tax credit and energy credit. Elements of risk-taking by the builder/operator will mitigate this risk considerably.

Direct Builder/Operator Ownership

If a contract can be negotiated with the jurisdiction whereby the financing element may be distinguished from the operating compensation to the builder/operator, the vendor's accountants may be persuaded that, in substance, the vendor has entered into a "sales-type" lease. In this structure, for accounting purposes only, the builder/operator will be considered to have sold a sales-type lease from the communities to the authority issuing the bonds. It is critical to this accounting treatment that there be no important uncertainties as to the payment obligation of the jurisdiction once the facilities are completed.

If the transaction is considered as a sales-type lease, the project obligations would be off the builder/operator's balance sheet; however, the jurisdiction would continue to be the tax owner. This is an example of where the tax treatment of a transaction is distinctly different from the accounting treatment.

Joint-Venture Leasing Structure

An operating company joint venture typically employed by certain builder/operators is cumbersome to negotiate and implement. A far simpler arrangement may be for a builder/operator to enter into a 50:50 joint-venture partnership with an equity investor to own the project facilities and to lease them to the builder/operator. In this manner, the builder/operator would receive one-half of all tax benefits and all project cash flows in excess of the cost of capital of the project.

The lease would be structured to be a noncapitalizable lease under FASB 13 (Financial Accounting Standards Board). The joint venture would be recorded by the equity method, since its business would be leasing. The builder/operator could also own a lesser amount of the joint venture and still record the project by the equity method, thereby tailoring its project ownership percentage to its tax forecasts.

Deferred Equity Vendor Ownership Structure

We believe that the high tipping fees of vendor ownership transactions in the past have been due to both the high cost of vendor equity and lack of imagination. To reduce the cost of resource recovery financings, we have devised the "deferred equity" financing technique, which simultaneously returns the value of tax benefits to the participating communities and provides for either a level or rising tipping fee structure over the life of the project. Although the project is financed entirely with bonds during the construction period, the vendor is required to contribute substantially more funds over the operating period in order to bring financing costs down.

Using the deferred equity approach, it is possible to structure a financing that may be competitive with leveraged leasing if the structure of the financing eliminates the high cost of equity previously mentioned.

The cash flow available to the investor can be more flexible than that of the equity participant in a leveraged lease. If the high yield required from an investor for equity can be reduced by reducing the need for implicit equity returns during the construction period when the project is not producing any revenues, a deferred equity structure will provide tipping fees comparable to those required using the leveraged lease structure. This is accomplished by having the equity contributed at varying times in the operational stage of the project when such contributions may offset other project revenues.

Project Participants/Legal Relationships

The project will be designed, constructed, and operated by a full-service vendor or designated subsidiaries (the "contractor"). The cost of constructing the project will be guaranteed by the contractor, subject only to cost increases due to (1) agreed- to technical changes in the project, (2) governmental acts or changes in law constituting a *force majeure*, and (3) agreed to escalation formulas pursuant to established cost of construction industry indices.

The beneficial owner will be one or more equity investors who will, in turn, lease the facilities to the contractor.

The project will be constructed pursuant to the "construction contract" between the contractor and the owner. The construction contract includes the technical specifications for the project and the acceptance criteria.

Once the project has been completed and the contractor has met the acceptance criteria, the operation of the project will be governed by a waste disposal service agreement between the contractor and the jurisdiction.

The contractor is responsible for all costs and expenses of operating the project, subject only to cost escalation provisions and cost increases of the contractor due to *force majeure* events which are delineated in the service agreement.

The "issuer" of the "bonds" will be the jurisdiction or an alternate, qualified bond issuer. The bonds will be issued pursuant to a "trust indenture" between the issuer and the "indenture trustee."

The bond proceeds will be loaned to the owner pursuant to a "loan agreement" between the issuer and the owner.

The issuer's rights and duties under the loan agreement will be assigned to the indenture trustee under the trust indenture.

6.2 A CASE ANALYSIS: VARIOUS FINANCING ALTERNATIVES FOR A COGENERATION RESOURCE RECOVERY FACILITY OVER 20 YEARS

The various assumptions selected for this analysis are shown in Table 6.1.

6.2.1 Landfill Only, No Resource Recovery

Figure 6.1 shows the tipping fees for a hypothetical community if no resource recovery plant were built, and garbage was simply landfilled. The fees start at $17 per ton and escalate rapidly over the 20-year period.

6.2.2 Publicly Owned Resource Recovery Plant Versus Land Disposal: Bond Debt Service Level Each Year

Figure 6.2 shows the tipping fees necessary to finance a publicly owned resource recovery plant compared with that of a landfill. In this case, the debt service for the bonds used to finance the plant is level each year. This has the effect of fixing a large portion of the yearly cost of the plant. Since the net revenues generated by the plant (before debt service) escalate over time, the necessary tipping fees decline until they actually become negative in the last year.

6.2.3 Publicly Owned Resource Recovery Plant Versus Land Disposal: Bond Debt Service Lower in Early Years (Beginning at Interest Only) and Higher in Later Years

Figure 6.3 adds another tipping fee curve for a publicly owned resource recovery plant. Debt service on the bonds is lower in the early years (beginning at interest only) and higher in the later years. This has the effect of keeping cost and revenue escalations parallel to each other and allows for a level tip fee schedule throughout the life of the project. By appearance, the lower costs in the beginning do not seem equal to the higher costs at the end. This is because the tip fees on the graphs are shown in future (then year) dollars. The present value of the two tip fee streams is virtually identical.

Table 6.1 Assumptions for 1500 TPD Resource Recovery Plant

Construction cost	$75,000/Ton
Construction period	3 Years
Interest rate	10%
Investment rate	10%
Tons of throughput	1500/Day
Facility availability	85%
O&M cost (first year)	$20/Ton
Steam production (lb steam/lb refuse)	2 lbs/lb (2 kg/kg)
Electricity production	120 kWh/Ton
Steam price (1984)	$6.00/M lb ($13.20/mkg)
Electric price (1984)	$0.05/kWh

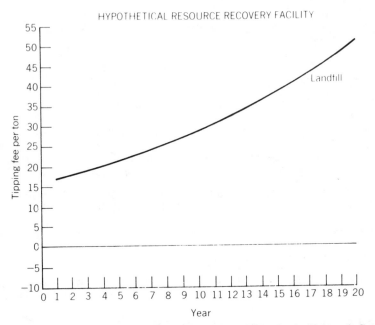

Fig. 6.1 Tipping fees, 1500 TPD, landfill only. (Smith, Barney, Harris, Upham & Co., Inc.)

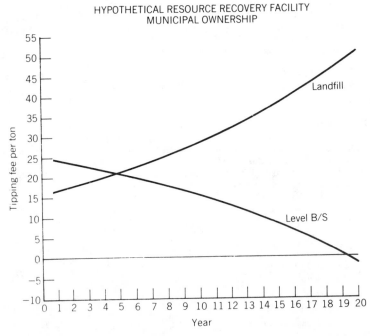

Fig. 6.2 Tipping fees, 1500 TPD publicly owned resource recovery plant, level debt service each year. (Smith, Barney, Harris, Upham & Co., Inc.)

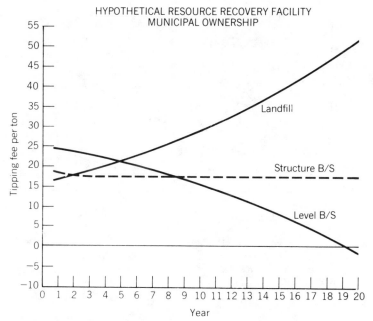

Fig. 6.3 Tipping fees, 1500 TPD publicly owned resource recovery plant, debt service lower early years (beginning at interest only), and higher in later years. (Smith, Barney, Harris, Upham & Co., Inc.)

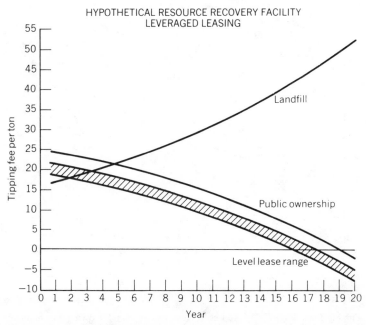

Fig. 6.4 Tipping fees, 1500 TPD resource recovery plant, leveraged lease financing. (Smith, Barney, Harris, Upham & Co., Inc.)

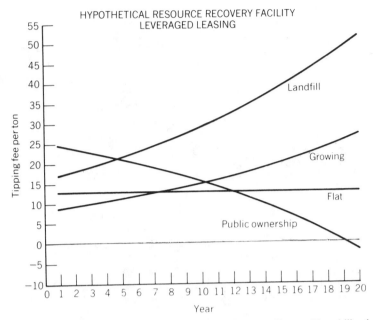

Fig. 6.5 Tipping fees, 1500 TPD resource recovery plant, leveraged lease with stabilization fund in early years. (Smith, Barney, Harris, Upham & Co., Inc.)

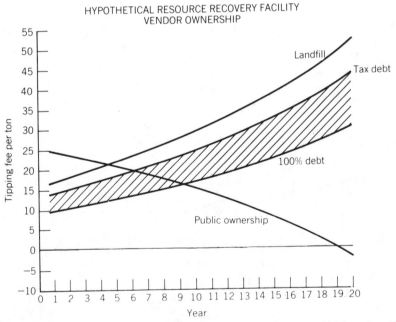

Fig. 6.6 Tipping fees, 1500 TPD resource recovery plant, vendor ownership financing. (Smith, Barney, Harris, Upham & Co., Inc.)

Table 6.2   Tipping Fees: Landfill versus Various Financing Options for a Resource Recovery Plant ($/Ton)

Year	Landfill	Public Ownership		Leveraged Lease (Third Party Owner)				Vendor Ownership	
		Lev. D/S	Str. D/S	No ITC	ITC	Flat	Growing	100% Debt	75% Debt
1	17.00	24.66	17.78	21.52	18.80	12.89	8.89	10.00	14.31
2	18.02	23.89	17.78	20.75	18.03	12.89	9.42	10.60	15.17
3	19.10	23.08	17.78	19.94	17.22	12.89	9.99	11.24	16.08
4	20.25	22.21	17.78	19.07	16.35	12.89	10.59	11.91	17.04
5	21.46	21.29	17.78	18.15	15.43	12.89	11.22	12.62	18.07
6	22.75	20.32	17.78	17.18	14.46	12.89	11.90	13.38	19.15
7	24.11	19.29	17.78	16.15	13.43	12.89	12.61	14.19	20.30
8	25.56	18.20	17.78	15.06	12.34	12.89	13.37	15.04	21.52
9	27.10	17.04	17.78	13.90	11.18	12.89	14.17	15.94	22.81
10	28.72	15.82	17.78	12.68	9.96	12.89	15.02	16.89	24.18
11	30.44	14.52	17.78	11.38	8.66	12.89	15.92	17.91	25.63
12	32.27	13.14	17.78	10.00	7.28	12.89	16.88	18.98	27.16
13	34.21	11.68	17.78	8.54	5.82	12.89	17.89	20.12	28.79
14	36.26	10.13	17.78	6.99	4.27	12.89	18.96	21.33	30.52
15	38.44	8.49	17.78	5.35	2.63	12.89	20.10	22.61	32.35
16	40.74	6.75	17.78	3.61	0.89	12.89	21.31	23.97	34.29
17	43.19	4.91	17.78	1.77	−0.95	12.89	22.58	25.40	36.35
18	45.78	2.95	17.78	−0.19	−2.91	12.89	23.94	26.93	38.53
19	48.52	0.88	17.78	−2.26	−4.98	12.89	25.38	28.54	40.85
20	51.44	−1.37	17.78	−4.51	−7.23	12.89	26.90	30.26	43.30

6.2.4   Leveraged Lease Financings

Figure 6.4 shows two curves representing the cost of leveraged lease financings with the area between them shaded. In one case (the upper end of range), no investment tax credit (ITC) has been taken. In the second case (the lower end of the range), the maximum projected amount of ITC has been taken. Pending legislation may lower the amount of ITC available to resource recovery facilities. Depending on the outcome, the actual tipping fee schedule should fall somewhere within the shaded area. As you can see, the tax advantages utilized by the private owner have kept the tipping fees lower than that of a publicly owned plant. This is due to the equity contribution by the third-party owner. Equity is used to retire bonds and, thus, lower debt service throughout the life of the project.

6.2.5   Leveraged Lease Financings with Stabilization Fund in Early Years

Figure 6.5 also shows leveraged lease financings. However, in these cases, a stabilization fund has been used to lower tipping fees in the early years. The stabilization fund is funded with monies derived from the equity contribution of the private owner. It is drawn down in the early years of the project to lower tip fees. Normally in a leveraged lease, the equity contribution is used to retire bonds on the first day of operations. This lowers the debt service on the bonds and therefore lowers the tipping fee. However, if the equity contribution is used for a stabilization fund, the corresponding principal amount of bonds will not be retired, and therefore the debt service, and consequently the tip fees, will be higher in the later years.

6.2.6   Vendor Ownership Financings

Figure 6.6 shows the range of tipping fee schedules generated by vendor ownership financings. In the first case (the upper end of the range), 75% of the project is financed by debt and the remaining 25% by equity. The equity in this case is contributed during the construction phase of the project and therefore directly lowers construction costs. In the second case, the construction cost of the plant is initially paid for entirely with debt. The equity contribution of the private owner is deferred

Table 6.3 Savings per Ton Compared with Landfill Cost

Year	Public Ownership		Leveraged Lease (Third-Party Owner)				Vendor Ownership	
	Lev. D/S	Str. D/S	No ITC	ITC	Flat	Growing	100% Debt	75% Debt
1	−7.66	−0.78	−4.52	−1.80	4.11	8.11	7.00	2.69
2	−5.87	0.24	−2.73	−0.01	5.13	8.60	7.42	2.85
3	−3.98	1.32	−0.84	1.88	6.21	9.11	7.87	3.02
4	−1.96	2.47	1.18	3.90	7.36	9.66	8.34	3.20
5	0.17	3.68	3.31	6.03	8.57	10.24	8.84	3.40
6	2.43	4.97	5.57	8.29	9.86	10.85	9.37	3.60
7	4.82	6.33	7.96	10.68	11.22	11.50	9.93	3.82
8	7.36	7.78	10.50	13.22	12.67	12.19	10.53	4.04
9	10.06	9.32	13.20	15.92	14.21	12.93	11.16	4.29
10	12.90	10.94	16.04	18.76	15.83	13.70	11.83	4.54
11	15.92	12.66	19.06	21.78	17.55	14.52	12.54	4.82
12	19.13	14.49	22.27	24.99	19.38	15.40	13.29	5.11
13	22.53	16.43	25.67	28.39	21.32	16.32	14.09	5.41
14	26.13	18.48	29.27	31.99	23.37	17.30	14.93	5.74
15	29.95	20.66	33.09	35.81	25.55	18.34	15.83	6.08
16	33.99	22.96	37.13	39.85	27.85	19.44	16.78	6.45
17	38.28	25.41	41.42	44.14	30.30	20.60	17.78	6.83
18	42.83	28.00	45.97	48.69	32.89	21.84	18.85	7.24
19	47.64	30.74	50.78	53.50	35.63	23.15	19.98	7.68
20	52.81	33.66	55.95	58.67	38.55	24.54	21.18	8.14

until the operation phase of the project and is then used to lower tipping fees. The range represents various permutations between these two cases.

6.3 CASE ANALYSIS SUMMARY

Table 6.2 compares landfill (only) tipping fees versus the resource recovery plant financing options. Table 6.3 compares landfill cost per ton versus savings for resource recovery with various financing options.

REFERENCES

1. W. T. Gregory, "Resource Recovery Project Economics," New York State Legislature Commission on Solid Waste Management Conference for Government and Citizen Decisionmakers, Hofstra University, Hempstead, New York, September 12, 1984.
2. R. S. Madenburg, "Assessing Waste-To-Energy Project Risks," Proceedings, ASME 11th National Waste Processing Conference, Orlando, Florida, June 4–6, 1984., p. 28–31.
3. Presented at Standard & Poor's Corporation Symposium, "An Approach to Municipal Utility Ratings," New York, October 15, 1984.

APPENDIX 6.1. ASSESSING WASTE-TO-ENERGY PROJECT RISKS[2]

R. S. Madenburg

ABSTRACT

The decision to proceed with a waste-to-energy project is directly dependent upon the proper identification and mitigation of project risks. Because this technology is new and unfamiliar, project risks should be carefully analyzed.

Disposal of wastes is a community responsibility and the community is ultimately at risk should a waste-to-energy project fail to meet its economic and/or technical expectations. The understanding and fair allocation of risks is essential to the success of the project.

This paper discusses some of the more significant technical, construction, operational, and financial risks associated with the design, construction, operation, and ownership of waste-to-energy facilities. It also discusses how to manage, assign and mitigate these risks.

ASSESSING WASTE-TO-ENERGY PROJECT RISKS

RISK EXPOSURE

To be successful, a waste-to-energy project must be well-conceived, it must have a solid financial structure, and the sponsor must allocate the speculative risks and the related benefits equitably among the participants.

In addition, the roles of the participants must be clearly defined. This is especially true in situations where one party may assume multiple roles, as in the case of the owner-engineer/builder-operator, or in situations where financial agreements dictate who assumes the various responsibilities, as in a leveraged lease.

In the broadest sense, the roles and responsibilities of the key participants are as follows:

Facility Owner

Owns the physical plant; provides financing; provides site and access; acquires permits and licenses; is a key party in the energy sales agreement; utilizes project tax benefits and cash flow.

Engineer/Constructor

Provides preliminary engineering and detailed design services; specifies and procures engineered items; provides construction and construction management.

Process Vendor

Provides the process and combustion technology; supervises plant start-up and performance testing; guarantees process performance; supervises the training of operations personnel.

Operator

Operates and maintains the plant; coordinates acquisition of fuel supply; disposes of nonprocessible wastes; provides operational review of facility design.

Fuel Supplier

Furnishes fuel as specified for quantity and quality. (Normally, this means he enters into a "put or pay" contract which unconditionally guarantees the supply of fuel to the project for a specified term.)

This appendix is taken from *Proceedings of ASME 11th National Waste Processing Conference,* Orlando, Florida, June 1984.

Energy Purchaser

Purchases the steam and/or electric energy produced by the project. Normally, this means he enters into a "take or pay" or "take, if tendered" agreement with the project owner. He is also responsible for the interties between the plant and the energy user.

Financier

Provides for the funding of the plant; arranges for construction and take-out financing; packages the project for financial rating; represents the project in the solicitation of bond offerings.

Insurance Underwriter

Insures the static and/or nonspeculative risks, i.e., those risks where there is a chance of financial loss but no chance of gain.

PROJECT RISKS

Because waste-to-energy technology is new and relatively unfamiliar, extra attention must be given to the analysis of any potential risks that may be associated with it. The decision to proceed—or not to proceed—with a project must be based in large part upon the proper identification, mitigation and allocation of these risks. It is essential that the participant who can best control the outcome of an event or task be assigned responsibility for any risks associated with it. However, we have identified a trend among municipal governments to shift risk to other participants.

If the project is to succeed, all parties must see a potential financial gain that outweighs their risk exposure.

Project risks are those potential situations or events which, if they occur during the execution of a project, negatively affect its financial outcome. Risks must not only be identifiable and understandable by all participants, but they must be within control of the party at risk. Participants cannot be expected to guarantee factors outside their control.

Risks can be broadly categorized as contractual risks, engineering and construction risks, operational risks, financial risks and technological risks.

Contractual Risks

Such risks may be introduced into the project through ambiguous contract wording, poor communication among project participants, failure on the part of some participants to meet contractual obligations, or poor contract administration. Contractual risk is also introduced through the use of onerous language which may require a participant to accept an unassignable risk. Lack of clarity and precision in contractual language can also expose participants to increased costs, because cost contingencies will be applied against the unknown. Contractual risk may also arise because too many or incompatible entities are party to the contract.

Engineering and Construction Risks

The engineer/constructor is expected to design and build the facility on schedule and within contractually-specified parameters and budget. The timely completion of the project, hence its financial success, is directly dependent upon the contractor's ability to forecast, assess and manage those elements of exposure that are directly under his control.

Cost

Cost overruns are normally directly accountable to inaccurate estimation of engineering and construction requirements. They generally result from inadequate or premature scoping of the project — e.g., failure to obtain firm fixed-priced commitments from vendors and subcontractors—or inadequate contingency allowances for engineering changes or unforeseen site conditions. Project cost overruns are frequently the result of delays in the work schedule.

Schedule

Delays in the completion of a project expose the engineer/constructor to increased interest and overhead expenses, and a general escalation of project costs. Delays can result from such conditions as poor construction management, inadequate scheduling estimates, labor strikes or slowdowns, engineering errors in the design or the estimation of site conditions, delays in delivery of equipment, or defective equipment. However, completion of project milestones by specific dates is essential. This is especially true for waste-to-energy projects because debt retirement for these projects is frequently dependent upon the generation of revenues on a predictable and timely basis.

Errors or Omissions

Engineers are responsible for exercising the skill and diligence that is normally rendered by a professional engineering firm. They are responsible and accountable for the design of a fully-functional and workable product. Errors or omissions frequently result in change orders, replacement of equipment, and an increase in construction costs. Such errors place the engineer at risk.

Unknown Site Conditions

The risk of unknown site conditions can be minimized by making a thorough geotechnical cost evaluation of the site during the project development phase. Engineers and constructors can protect themselves against exposure to risk resulting from subsidence, dewatering problems, compaction, archaeological finds, nonrecorded obstructions, etc., by prequalifying the site with assumed site conditions.

Codes and Regulations

The engineer/constructor is liable for the application and interpretation of the proper technical codes and safety and environmental regulations. To identify applicable regulatory requirements, the engineer should consult with an experienced industrial underwriter during the development phase. Consultation with environmental, health and safety, and building officials at the local, state and federal levels is a must. However, the engineer/constructor must also rely heavily on experience with similar projects and own sound judgement to define the applicability of codes to a specific project. Over-kill through the application of extraneous codes and regulations increases project costs without adding value.

Labor Disputes

Work stoppages and labor disputes are frequently attributable to the actions of the constructor or construction manager; however, work stoppages which are not within the control of the constructor normally result in an excusable delay. Because the financial impact of a labor dispute can be so serious, it is essential that the constructor or construction manager have a thorough understanding of site labor conditions, local work rules, and craft jurisdictional policies. Labor risks can be minimized by negotiating a project work agreement with local craft and by coordinating labor jurisdictional areas among the various crafts *prior* to the start of work.

Operational Risks

After construction and acceptance testing, operational risks—the possibility that the facility will fail to provide expected levels of revenue—become the area of major concern. Events which can affect revenue flow during operation include:

Failure of the facility to perform according to design specifications.

Technical operation of a waste-to-energy project is the most pertinent concern affecting the financial success of the project. Performance failures may result from poor design or workmanship, poor maintenance, disruption or changes in the character of the fuel supply; employee work stoppages; natural disasters; condemnation by a public authority; or changes in health, safety and environmental regulations.

Higher than expected operations and maintenance costs.

Higher costs can result from unplanned expenses in the areas of maintenance, labor, services and materials, tax assessments, and insurance premiums.

Regulatory Requirements

The project must be structured to meet the administrative requirements of the Public Utility Regulatory and Policies Act (PURPA), the Federal Energy Regulatory Commission (FERC), public utility commissions, and applicable court rulings. The recent upholding of the principal of avoided cost by the United States Supreme Court is an example of the latter.

Financial Risks

The financing entity in a waste-to-energy project, as in any project, must be assured of the timely repayment of project debt, both interest and principal. Areas of major financial risk include:

(1) Availability and cost of money for project financing.

(2) Changes in interest rates.

(3) Adverse IRS determination of tax benefits.

(4) Revenue shortfalls resulting from the general condition of the economy, i.e., double-digit inflation or reduced avoided cost for power generation due to reduced demand or to over capacity. In addition, the economic viability of the project may be compromised by technological changes or changing local business conditions. For example, the steam purchaser may go out of business.

(5) Unenforceability of power purchase agreements with governmental units. For example, "take or pay" contracts have been invalidated on Washington Public Power Supply System units four and five.

(6) Voidability of contracts due to misrepresentations, mistakes, and fraud.

(7) Lack of creditworthiness or experience of project participants.

(8) Unavailability of insurance and negative impact of deductibles.

Technological Risks

It is of paramount importance that the financial community recognize that the project's technology is well demonstrated and proven. Whether or not a project receives a favorable bond rating depends upon how good a case can be made for the quality and reliability of the technology to be employed. For example, it is essential to demonstrate that similar or identical technology has been used successfully on other projects. The U.S. Environmental Protection Agency has identified the following key technical issues related to waste-to-energy projects:

Waste Preprocessing, Receiving and Handling

The facility must be explosion resistant and not susceptible to "log-jamming" in the material flow. A two-to three-day waste holding capacity should be provided. The material handling system must be of a proven design for the intended service. The grating system should have a demonstrated life of at least two years.

Combustor/Steam Generator

The use of demonstrated technology in the combustor area, as well as the coupling of the combustor and steam generator, is essential. Specific attention must be placed on the furnace volume and injection of over-fire air to ensure proper combustion kinetics. Superheater tube banks should be so located as to maximize tube life. Because the flue gas is highly erosive, the thickness of tube walls should exceed ASME standards. Convective tube spacing should be adequate to prevent plugging and to facilitate tube cleaning.

Ash Removal

The ash drag-out system generally is both maintenance and labor intensive. Utilization of ash removal systems with good operational histories is essential.

Equipment Redundancy

Major equipment redundancy is necessary to enable implementation of a well-planned preventative maintenance program. Consideration should also be given to using both steam and electric drives on the major prime movers. Revenue projections should assume no more than 80 percent plant availability. Provisions should be allowed for a fully equipped on-site maintenance facility and an adequate stock of replacement parts.

Basically, the financial community and insurors must be convinced that the technology is demonstrated and that there is virtually no chance that the system will not perform as expected.

MITIGATION OF RISKS

The key to a successful project lies in the sponsor's ability to identify and equitably distribute risks during project development. It is essential that all areas of concern be identified and discussed, no matter how unlikely their occurrence may be. It is also necessary to establish an efficient system for managing these risks should they become eminent. This is the basis for the contractual agreements which formalize risk allocation and management.

The first step in mitigation of risk is to segregate the speculative risks from the nonspeculative risks. Generally, nonspeculative risks are those which threaten loss and offer no potential for gain (e.g., fire, earthquake, flood, and force majeure). Such risks are generally insurable. Speculative risks offer the risk taker economic gain in return for effective performance and proper risk management. Good management and prudence are the risk taker's insurance against loss.

To minimize the impact of uncertainties, project participants must recognize risk elements, understand risk accountability, know how to manage risk effectively, and be able to share risk equitably through the contractual process.

Naturally, each participant wants to protect his own interests. The governmental unit does not want to obligate its tax base and wants the project to be self-sufficient; the engineer/constructor wants to receive reasonable compensation for his services; the bond holders want to be assured of timely payment of principal and interest. However, risk mitigation is a give and take situation. If any risks are assigned without regard to a participant's ability to control them, the project has little hope of success.

Managing Risks

Managing risks means aggressively approaching each risk situation in such a manner as to reduce the probability of its occurrence and to minimize its impact should it occur. Techniques for risk management include:

(1) Obtaining firm price quotations from vendors and subcontractors for all major engineered items in order to assign these risks outside the project.

(2) Performing sufficient preliminary engineering during project development to clearly define scope, determine actual site conditions and to understand and examine all risks.

(3) Assigning risks prudently by means of appropriate contractual formats. Various formats ranging from "fixed price" through "fully cost reimbursable" to "incentive" type contracts can be used to assign risks.

TABLE 1 RISK ALLOCATION GUIDELINE

RISK ELEMENT	OWNER	ENGINEER	CONSTRUCTOR	PROCESS VENDOR	OPERATOR	MITIGATION METHOD
Engineering & Construction Related:						
Technology Selection	P	S		S		Thorough pre-project analysis
Design of System Involving new Technology		S		P		Contingency allowances
Design Errors and Omissions		P				Insurance
Design Changes	P	S		P		Contingency allowances
Ambiguous Specifications and Plans		P	S	S		Contingency allowances
Mechanical Completion		S	P	S		Insurance and contingency
Completion by Date Certain	S	S	P	S		Contingency allowances
Unknown Site Conditions	P	S	S			Thorough pre-project analysis
Unilateral Owner Directed Changes	P	S	S	S		Contingency allowances
Acts of God	P	S	S	S	S	Insurance
Effect of Delays	P	S	S	S		Insurance and contingency allowances
Court Injunction, War or Civil Disturbances	P					Contingency allowances
Patent Infringement				P		Contingency allowances
Coordination of Subcontractors & Suppliers			P	S		Contingency and avoid through related project experience
Labor Disputes & Strikes & Productivity			P			Contingency and avoid through related project experience
Changes of Law	P	S	S	S	S	Pre-project analysis and contingency
Operations Related:						
Selection of O&M Contractor	P				S	Thorough pre-project analysis
Higher Than Expected O&M Cost	S			S	P	Contingency and contractual incentives
Changes in Supply & Character of Waste	P				S	Thorough pre-project analysis
Process Performance		S	S	P	S	Insurance
Change in Price of Energy	P				S	Contractual incentives
Waste Fuel Supply Shortfall	P				S	Contractual incentives
Changes in Environmental, Health & Safety Regulations	S				P	Contingency allowance and pre-project analysis
Availability of Plant	S				P	Insurance, contractual incentives
Natural Disasters	S				P	Insurance
Increases in Taxes	S				P	Contingency allowances
Increases in Insurance Premiums	S				P	Contingency allowances
Poor Management					P	Contractual incentives
Inability to Meet Specifications of Energy Market	S	S	S	P	S	Insurance and contingency allowances
Financial Related:						
Financing Availability, Cost, Arbitrage	P			S		Thorough pre-project analysis
Inaccurate Cash Flow Projections	S	P				Contingency allowances
Changes in Interest Rates	P					Contingency allowances
Adverse IRS Determination	P					Thorough pre-project analysis
Revenue Shortfalls Due to Off Spec. Performance		S	S	P	S	Insurance
Revenue Short Falls Due to General Economic Factors	P					Contractual incentives

LEGEND: P - Primary Responsibility
S - Secondary Responsibility

(4) Dedicating specific contingency budgets and reserve margin to each risk area.

(5) Using outside consultants to evaluate risks, identify potential cost impacts, and develop contingency plans.

(6) Conceptualizing the project thoroughly so that all objectives and restraints are understood.

(7) Making certain that no risk, unless purposely shared, is included in more than one participant's scope of responsibility.

(8) Re-evaluating potential risk situations as the project progresses. Uncertainty will decrease as the project nears completion.

Insuring Risks

Commerical insurance is available to protect project participants from economic consequences associated with nonspeculative risks.

The following table outlines the types of insurance that are available for various kinds of exposure.

Exposure	Insurance Coverage
Repair or replace damaged equipment	Property Damage/Builder Risk /Transit
Pay for continuing expenses, including profit, while repairs are bing made	Business Interruption
Pay for project expenses if procured equipment is delayed by an insured peril	Contingent Business Interruption/Extra Expense
Protect project from legal costs and judgements to third parties	Comprehensive General Liability
Provide funds if a contractor, engineer or vendor fails to perform within the scope of this responsibility	Performance and Payment Bonds
Pay for project expenses resulting from uncontrollable circumstances	Cost Overrun/Delayed Opening
Pay project equity participants if tax credits are recaptured by IRS because facility is destroyed	Investment Tax Credit Recapture
Modify the plant to meet specified performance guarantees	Efficacy*
Pay project exposure while plant operates below specified performance levels	Efficacy*
Pay consequential damages due to engineering errors or omissions	Professional Liability
Injuries to employees during course of employment	Workers Compensation

*Efficacy insurance market is limited with large deductibles and high premiums relating to State of the Art Projects.

CAVEAT. DO NOT OVER INSURE. Insurance should be viewed as a risk management technique intended to protect project participants from *catastrophic* financial losses. Deductibles and limited cost sharing with the underwriters will keep premiums low while providing needed protection. Contingency allowances should include insurance deductibles. Overuse of insurance as a mechanism to reduce risk increases costs, hence reduces profits.

Project insurance requirements must be coordinated with experienced financial advisors, and must be consistent with the overall exposure of the project participants.

Allocating Risks

(1) Allocate sufficient risk to each participant to motivate him to perform in a professional manner.

(2) Consider the degree of control over the risk to be allocated when assigning risk responsibility.

(3) Consider each participant's ability to protect himself against risks allocated to him.

(4) In general, allocate risks of a national or international character (i.e., a currency devaluation, or an oil embargo) primarily to the project owner.

(5) Share mutually dependent risks. This will prevent conflict and inadvertent assumptions of loss because of inability to determine fault.

BIBLIOGRAPHY

The Business Roundtable, "A Construction Industry Cost Effectiveness Project Report," Report A-7; October, 1982

Anonymous, "Financial Considerations Affecting Implementation of a Large Multiparty Cogeneration Project," Kidder Peabody & Company, New York, New York; April, 1979

Tirello, Jr., E. K., et al., "Cogeneration – Industry Comment," Lehman Brothers Kuhn Loeb Research; August, 1982

Thayer, D., Private Correspondence, Marsh & McLennan, Incorporated, Seattle, Washington

Erikson, C. A., "Risk Sharing in Construction Contracts," thesis presented to the University of Illinois at Urbana, Illinois, in 1979, in partial fulfillment of the requirements for the degree of Doctor of Philosophy

Stukhart, D., "Sharing the Risks of the Cost of Inflation," *Transactions of the American Society of Cost Engineers;* June, 198

Sauvage, I. G., "Issues in Arranging Project Financing for Waste-to-Energy Facilities," Lehman Brothers Kuhn Loeb Incorporated, New York, New York; April, 1983

Deitchman, J. V., "Packaging the Full Insurance Program," Johnson & Higgins, Los Angeles, California; April, 1983

Chrisman, R., "An Overview of U.S. & Canadian Experience with European Mass Burning Waterwall Incinerator Systems," U.S. EPA, Office of Solid Waste Report No. 68-10-6071; December 1981

Key Words: Economics • Management • Planning •
Power Generation • Revenue Pricing • Risks

APPENDIX 6.2. RESOURCE RECOVERY RATINGS (BONDS) APPROACH

STANDARD & POOR'S **CreditWeek** THE AUTHORITY ON CREDIT QUALITY

CreditComment

S&P'S DEBT RATING DEFINITIONS

A Standard & Poor's corporate or municipal debt rating is a current assessment of the creditworthiness of an obligor with respect to a specific obligation. This assessment may take into consideration obligors such as guarantors, insurers, or lessees.

The debt rating is not a recommendation to purchase, sell, or hold a security, inasmuch as it does not comment as to market price or suitability for a particular investor.

The ratings are based on current information furnished by the issuer or obtained by S&P from other sources it considers reliable. S&P does not perform an audit in connection with any rating and may, on occasion, rely on unaudited financial information. The ratings may be changed, suspended, or withdrawn as a result of changes in, or unavailability of, such information, or for other circumstances.

The ratings are based, in varying degrees, on the following considerations:

1. Likelihood of default—capacity and willingness of the obligor as to the timely payment of interest and repayment of principal in accordance with the terms of the obligation;
2. Nature of and provisions of the obligation;
3. Protection afforded by, and relative position of, the obligation in the event of bankruptcy, reorganization, or other arrangement under the laws of bankruptcy and other laws affecting creditors' rights.

AAA Debt rated 'AAA' has the highest rating assigned by Standard & Poor's. Capacity to pay interest and repay principal is extremely strong.

AA Debt rated 'AA' has a very strong capacity to pay interest and repay principal and differs from the highest rated issues only in small degree.

A Debt rated 'A' has a strong capacity to pay interest and repay principal although it is somewhat more susceptible to the adverse effects of changes in circumstances and economic conditions than debt in higher rated categories.

BBB Debt rated 'BBB' is regarded as having an adequate capacity to pay interest and repay principal. Whereas it normally exhibits adequate protection parameters, adverse economic conditions, or changing circumstances are more likely to lead to a weakened capacity to pay interest and repay principal for debt in this category than in higher rated categories.

BB, B, CCC, CC Debt rated 'BB', 'B', 'CCC', or 'CC' is regarded, on balance, as predominantly speculative with respect to capacity to pay interest and repay principal in accordance with the terms of the obligation. 'BB' indicates the lowest degree of speculation and 'CC' the highest degree of speculation. While such debt will likely have some quality and protective characteristics, these are outweighed by large uncertainties or major risk exposures to adverse conditions.

C This rating is reserved for income bonds on which no interest is being paid.

D Debt rated 'D' is in default, and payment of interest and/or repayment of principal is in arrears.

Plus (+) or minus (−): The ratings from 'AA' to 'B' may be modified by the addition of a plus or minus sign to show relative standing within the major rating categories.

Provisional ratings: The letter 'p' indicates that the rating is provisional. A provisional rating assumes the successful completion of the project being financed by the debt being rated and indicates that payment of debt service requirements is largely or entirely dependent upon the successful and timely completion of the project. This rating, however, while addressing credit quality subsequent to completion of the project, makes no comment on the likelihood of, or the risk of default upon failure of, such completion. The investor should exercise his own judgment with respect to such likelihood and risk.

L The letter 'L' indicates that the rating pertains to the principal amount of those bonds where the underlying deposit collateral is fully insured by the Federal Savings & Loan Insurance Corp. or the Federal Deposit Insurance Corp.

* Continuance of the rating is contingent upon S&P's receipt of an executed copy of the escrow agreement or closing documentation confirming investments and cash flows.

N.R. Indicates no rating has been requested, that there is insufficient information on which to base a rating, or that S&P does not rate a particular type of obligation as a matter of policy.

Debt Obligations of Issuers outside the United States and its territories are rated on the same basis as domestic corporate and municipal issues. The ratings measure the creditworthiness of the obligor but do not take into account currency exchange and related uncertainties.

Bond Investment Quality Standards: Under present commercial bank regulations issued by the Comptroller of the Currency, bonds rated in the top four categories ('AAA', 'AA', 'A', 'BBB', commonly known as "investment grade" ratings) are generally regarded as eligible for bank investment. In addition, the laws of various states governing legal investments impose certain rating or other standards for obligations eligible for investment by savings banks, trust companies, insurance companies and fiduciaries generally.

This appendix is taken from *Standard & Poor's Creditweek*, October 15, 1984, pp. 44–54.[3]

Resource recovery ratings approach

Disposal of the estimated 450,000 tons of municipal solid waste, discarded daily in the United States, poses serious challenges to both municipal and state governments. Landfills and incineration options, the traditional methods of waste removal, are becoming impractical due to environment, space or cost constraints.

Resource recovery, defined as burning waste for conversion to energy, represents an alternative for meeting long-term waste disposal needs which is gaining in use. Over 51 plants are in the planning stages, including facilities to be located in most major metropolitan areas. Currently, S&P rates 15 municipal financings representing almost $2 billion in debt for resource recovery projects.

Rating criteria

Credit assessment of resource recovery financings focus on four interrelated factors:
- —Legal provisions
- —Technological exposure
  - Construction risk
  - Operating risk
- —Waste stream and demand for service
  - Waste stream composition
  - Economic base
  - Waste stream control
- —Economic and financial feasibility

Resource recovery revenue bonds involve risks. Vulnerabilities during project start-up and operation are especially acute because technologies used in these projects have limited experience in the U.S. In addition, to support an economically feasible facility, competitive pressures must be mitigated. Both the availability of waste to maintain a fully operational project and the continued demand for project output may affect the facility's competitive position. Finally, the capacity of the participants to support debt under both best and worst case operating scenarios forms an important part of the credit assessment.

A new development in public finance, the public-private partnership often seen in resource recovery, requires a different analytical focus. In the determination of credit quality, analysis centers on the likelihood of a strong, economically viable project. Revenues derived from operations must not only comfortably amortize debt and maintain a fee structure competitive with alternative disposal and energy sources, but also provide economic incentives for both private and public sector participants.

Legal provisions

Analysis of each project begins with a review of the legal documents. Regardless of the ownership option, common legal protections are present. The bond indenture should require that the project's revenues be at least sufficient to cover operating expenses and debt obligations. Stronger rate covenants, typically used when public ownership is retained, provide high coverage levels. A fully funded debt service reserve in addition to working capital and replacement and renewal funds are among the essential credit elements. Each of these security features provide flexibility in the event of revenue stream interruption. Timely replenishment of these funds in the event of draw down is important. Furthermore, when proceeds from guarantee payments, liquidated damages, or insurance claims are specifically allocated for debt repayment, adequate time must be allowed to make payment to the trustee before a default under the indenture is triggered.

Legal limits imposed for additional debt are also a credit concern. Historical revenues available for debt service should cover outstanding and new fixed costs. Weaker tests may not impose safeguards against additional indebtedness and allow the issuance of senior rather than parity or subordinate obligations. However, provisions must also be allowed to facilitate the sale of completion bonds.

Where private ownership occurs, and in cases with a third-party equity contributor, pertinent legal documents may also include the lease, loan, mortgage, and assignment agreements. For such structures, unconditional assignment of these documents, including lease rental and other payments derived from project participants to the trustee, are important for a strong legal structure. Also, assurances must be obtained that bankruptcy of the equity participant will not affect the bondholder's claim to lease payments. In addition, any return on the equity contributor's investment must be made after debt service payment.

The two major contracts securing the revenue stream are the waste disposal and energy agreements. Under the *disposal contracts*, the municipalities are required to deliver waste and pay disposal or "tip fees" to support project operations. S&P evaluates the capacity of each participant to meet this obligation which continues in many cases even if the project ceases operations. Stronger agreements are structured to allow tip fee adjustments to reflect operating expense escalation and sharing of project profits. In effect for the life of the bonds, commit-

Resource recovery flow diagram

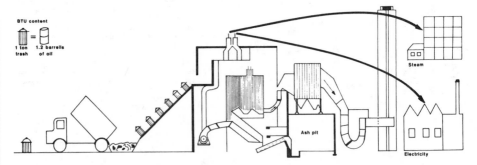

BTU content

1 ton trash = 1.2 barrells of oil

Steam

Ash pit

Electricity

Waste is delivered to boilers as fuel, where it creates energy which is converted to steam or electricity.

Revenue bond financing–private ownership

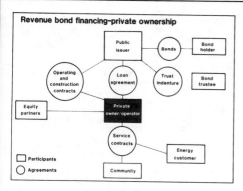

To date, most resource recovery projects have been financed with at least a partial municipal or corporate guarantee for debt repayment. The limited operating history of these facilities often necessitates a "deep pocket" credit support. When there is a full commitment for debt repayment in place, the rating reflects the credit strength of the guarantors. With partial commitments, S&P focuses on the 'weak link' project participant with the ultimate responsibility for servicing the debt. For example, in several recent financings, the corporate partner insured facility performance but committed to repay debt only in certain isolated instances. It was the responsibility of the municipal partners to assume the full obligation in the event of prolonged facility outages.

ments under these contracts enhance credit quality if guaranteed waste delivery insures a fully operational facility. However, agreements are reviewed in tandem with existing waste control legislation.

Energy contracts, which govern the payment for project output, are generally conditioned upon the facility's performance. The credit strength of the energy customer in meeting this long-term obligation to purchase steam and/or electriciay is an important analytical factor. In assessing the likelihood that project output will remain in demand and pricing formulas for both steam and electric output are reviewed for sensitivity to competing energy sources and inflationary or regulatory changes. Since electric prices are often related to the purchaser's incremental or avoided fuel cost or in some cases system-wide average costs, this formula should be able to adjust to compete with the availability of higher or lower priced alternatives. Any major declines in electric or steam revenues may place upward pressure on tip fees and reduce profitability. Regulation of electric rates would also be seen as an additional exposure.

Revenue bond financing–public ownership

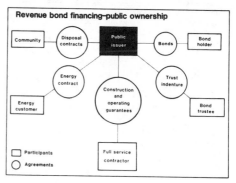

Resource recovery ratings

	Current rating	Debt outstanding ($000)
Nashville Thermal Transfer Corp., Tenn. series 1972	BBB–	14.5
Saugus, Mass. solid waste disposal rev. bonds 1975 series	A	42.0
Connecticut Resource Recovery Auth. Greater Bridgeport system bonds series A†	AA–	47.5
Ohio Water Development Auth. recycle energy rev. bonds Akron project	CC	46.0
Pittsfield, Mass. solid waste disposal rev. bonds 1979 series (Vicon Recovery Associates Resource Recovery Project)	BBB+	6.2
Delaware Solid Waste Auth. resource recovery rev. bonds series 1979 (Delaware Reclamation Project)	A–p	56.4
Pinellas County, Fla. solid waste and electric rev. bonds series 1980, 1984 resource recovery system	A*	243.1
Westchester County, N.Y. Industrial Development Agy. resource recovery rev. bonds series A, 1982 (Westchester Resco Project)	Ap*	237.5

	Current rating	Debt outstanding ($000)
North East Maryland Waste Disposal Auth. resource recovery rev. bonds series 1983	Ap	190.7
North Andover, Mass. resource recovery rev. bonds, 1983 (Mass. Refuse Technical Project)	A–p	167.7
Tampa, Fla. solid waste system rev. bonds series 1983 (McKay Bay Refuse-to-Energy Project)	A*	115.6
Summer County Resource Auth., Tenn. solid waste disposal facility rev. bonds series 1983	A	12.0
Tulsa Public Facilities Auth., Okla. solid waste steam and electric rev. bonds series 1984	AAA*	60.0
Mass. Industrial Finance Agy. (Lawrence/Haverill) solid waste disposal rev. bonds (Refuse Fuels Associates Project) series 1982	AAA*	58.2
Marion County, Oregon fltg./fixed rate solid waste and electric rev. bonds (Ogden Martin Systems of Marion Inc.), series 1984	AAA*/A-1+	56.6

*Fully or partially insured by 'AAA' rated guarantors.
†See CreditWatch Section of June 4 CreditWeek.

(continued on next page)

Resource recovery definition

Resource recovery is the conversion of solid waste into saleable energy. Garbage and other refuse is burned to produce either steam or electricity. Facilities constructed to perform this function use one of two methods. "Mass-burn" incinerates almost totally unprocessed refuse. The "refuse derived fuel (RDF)" technique shreds and compacts the refuse to achieve greater energy intensity. It is generally estimated that one ton of garbage contains slightly more energy (in BTUs) than a barrel of oil.

Technological exposure

Another key to the assessment of project feasibility is an analysis of the technological risk characteristic to resource recovery facilities. S&P considers construction operating risk and the extent to which contractual guarantees diminish this exposure. The ability of the project participants to meet these obligations from the initial planning stages throughout the life of the bonds is a key analytical concern.

Construction risk

Construction and related risk is present in all phases of the project from design through final acceptance. Vulnerabilities exist in the design state particularly if a facility is "scaled up" or enlarged from a demonstration model. The Bridgeport, Conn. facility, which never began full commercial operation, suffered from this problem. Exposures during actual construction are comparable with other municipal utility projects like incinerators. Other risks are prevalent, however, during compliance testing and start-up. Potential problem areas are evaluated together with a risk assessment which should be detailed by the consulting engineer.

Protections specified in the construction contract and insurance policies are weighed. S&P considers what factors may contribute to cost overruns, project delays, or failure to meet performance standards and what participant has the legal responsibility to correct these problems. The ability of the corporate sponsor to stand behind completion guarantees is especially important. Comprehensive insurance to cover extra expense items not included in the construction contract is analyzed for possible exclusions. Properly identified project risks and adequate coverage are key elements in this evaluation.

Operating risk

High maintenance costs, inability to process waste up to design capacity, and failure to meet environmental emissions standards are some of the operational hazards that resource recovery facilities have faced. Specific problems over the years have included:

- —Corrosion of boiler tubes, grates, super heaters, etc.
- —Heavy residue deposits within the boilers
- —Crane breakdowns
- —Explosions
- —Grate failure
- —Inability to process waste up to design capacity

—Failure to produce a satisfactory product
—Failure to meet environmental emissions standards

S&P considers the level of redundancy built into the facility which may forestall the impact of some of these operating problems. Alternative boilers or processing equipment may provide flexibility during short-term outages.

In appraising the potential effects of both short and long-term outages, the disposal and energy contracts are reviewed in conjunction with the corporate operating agreement when present. Generally, communities retaining facility ownership assume a greater share of operating risk even with service contracts. Under either ownership option, however, additional costs incurred typically revert to the muncipal partner or may be covered by insurance.

In cases where the corporate operating guarantee is limited, S&P assesses the business interruption and other liability insurances in place to cover revenue shortfalls. Due to the potential for delay in recovering insurance payments, frequently large deductibles, and cancellation provisions, insurance may have marginal implications for credit quality.

A proven recovery process, economic and management experience all have a bearing on technological risk. If a private operator contracts to run a facility, S&P focuses on incentives to induce operating efficiencies. This may include sharing of energy revenues and profits between both public and private participants. If a plant is to be operated by a municipal entity, its experience in other similar municipal utilities such as electric or incinerator plants is reviewed. Overall, the public entity's philosophy in dealing with its enterprises is important. A fee schedule which provides strong economic incentives for all participants is viewed as a positive rating factor.

Environmental risk, due to new or changed air pollution control standards, is both a construction and operating hazard for resource recovery plants. The Resource Conservation and Recovery Act, presently under congressional review, may also be amended to include municipal solid waste residue as a hazardous material. This would require specialized disposal. A facility which is designed to accommodate possible equipment additions may serve to mitigate this exposure.

Risk assessment: facility operations

	Risk level		
	Low	Intermediate	High
Management	Full service Contractor Owner/operator	Experienced Municipal operator	Inexperienced/ Municipal/private operator
System	Operational history	Turnkey facility Operational history	No operational history
Rate structure (tip fee)	Adjustable tip fee (profit sharing)	Adjustable tip fee	Fixed tip fee (no profit sharing)

Construction and financing cost comparison

Location	Year finances	Year operational	Technology	Months to complete	Tons capacity	Unit cost ($/ton)[1] 1984 adjusted $ Construction	Total cost[2]	Financing sources Debt	(%)	Equity and other	(%)
Tulsa, Oklahoma	1984	1987	Mass-burn	30	750	$ 68	$105	$ 62,855	80	$15,900	20
North Andover, Mass.	1983	1986	Mass-burn	36	1500	$ 91	$145	$167,670	83	$35,260	17
North East, Maryland	1983	1985	Mass-burn	29	2010	$101	$140	$190,765	75	$63,738	25
Tampa, Florida	1983	1986	Mass-burn	36.5	1000	$ 91	$115	$115,600	100	$22,800 Equity	35
Lawrence/Haverill, Mass.	1982	1985	RDF	36	1300	$ 57	$ 84	$ 58,200	65	$8,000 Grant (UDAG)	

Source: Construction estimates based on Consulting Engineer's Reports.
[1]Unit cost adjusted to 1984 value—reflects 11% inflation for construction.
[2]Includes financing and costs of issuance.

Waste to energy technologies

Of the many ways to convert refuse into energy, mass burn and refuse derived fuel (RDF) are the most common and successful to date.

Refuse is delivered by truck directly to the facility's tipping floor in a mass-burn system. Large items, such as refrigerators and engine blocks are either diverted or sorted out. No other processing is applied. The refuse is deposited into a hopper leading to a series of traveling grates and finally to a boiler. The thermal energy output is primarily steam, used directly for heating or converted to electricity. There are more than 350 facilities of varying sizes in several countries utilizing this basic technology.

In an RDF system, the material is processed before being placed in the boiler. It is normally shredded, sorted to remove unprocessable materials, then shredded again to produce a fuel of a specified size. The reduction of non-burnable refuse increases the heat value per ton. RDF has been burned successfully in utility boilers in Madison, Wis. and in dedicated boilers in other locations. Where processing entailed more than shredding and sorting the waste, technological failures occurred. Bridgeport, Conn. and Akron, Ohio are examples.

Some of the key advantages and disadvantages of each technology are summarized here.

Comparison of RDF and mass-burn technologies*

Advantages

RDF	Mass-burn
Higher heat value of fuel entering the boiler (4600–7700 BTU/lb)	Simpler operating system
Greater energy revenues	Lower capital investment for processing equipment
Lower level of excess residue for landfill disposal	Process demonstrated in European plants
Potential for lower life cycle cost	

Disadvantages

RDF	Mass-burn
Processing equipment may require additional maintenance	Higher level of boiler corrosion
Materials handling problems	Lower heat value of fuel (3000–6000 BTU/lb.)
	Problems with particulate emissions

*International Conference on Resource Recovery Technology prepared for U.S. Department of Energy 1981.

Documentation requirements for rating

A. Legal documents
   Enabling legislation
   Waste control ordinances
   Bond ordinance
   Lease, mortgage, assignments
   Waste disposal contracts
   Energy contracts

B. Financial information
   Audited historical information for three years for both public and private participants
C. Feasibility study including sensitivity analysis
D. Enforceability opinions for disposal contracts, control ordinances

Waste stream and demand for service

Changes in waste composition and control laws or in the economic base of the service territory could have significant implications for credit quality. Each factor contributes to a facility's competitive standing in relation to disposal alternatives and influences economic feasibility. Assessing community support and commitment to the project in relation to these factors is key in the analysis.

Waste stream composition

The ultimate success or failure of a resource recovery facility depends on the quantity and quality of the waste it receives. The availability is dependent on the existing population and the rate of growth. Waste composition, and its energy value, differs depending on recycling laws and commercial and industrial activities. Heating value rises with increases in paper and plastics, and on average a 4500 BTU energy equivalent is expected per pound of municipal solid waste. S&P appraises the ability of the disposal fee structure to adjust to any loss of BTU value or decrease in the amount of waste disposed.

Economic base

A sound economy is more likely to maintain a steady flow of waste, increase demand for project output, and enable users to pay the proposed fee structure without stress. The evaluation of the service territory's economic base is similar to other debt analyses. Assessments of the economic viability of a community take the long-term perspective, focusing on income levels, economic growth, and employment trends.

The results of four sampling studies are summarized below. Energy value and output is seasonal as moisture content increases or declines. Because of these variations, sampling is essential in assessing the potential for energy value.

Municipal refuse composition (%)

Constituent	Weston	MRI	Envirogenics	G.E.
Paper	35.8	33.0	31.0	37.0
Glass	8.4	8.0	12.0	9.0
Ferrous metals	7.6	7.6	10.0	7.6
Nonferrous metals	0.6	7.6	10.0	0.8
Plastics, leather, rubber, textiles and wood	6.9	6.4	4.0	7.4
Garbage and yard waste	10.3	12.6	9.6	10.0
Moisture	28.8	23.0	20.3	25.0
Miscellaneous	1.6	1.8	3.1	3.2
Total	100.0	100.0	100.0	100.0

Waste stream control

The need for restrictive flow control legislation is important, but is no substitute for an economically viable project. Emphasis is placed on what flow control ordinances are in place, and the validity and enforceability of the law. Control legislation mandated by a state authority should be less susceptible to the antitrust charges which have risen when a public entity monopolizes an activity in the realm of the private sector. Public control of waste disposal, seen as an additional legal protection, centers on the need to contractually guarantee an adequate supply of waste over the life of the facility.

(continued on next page)

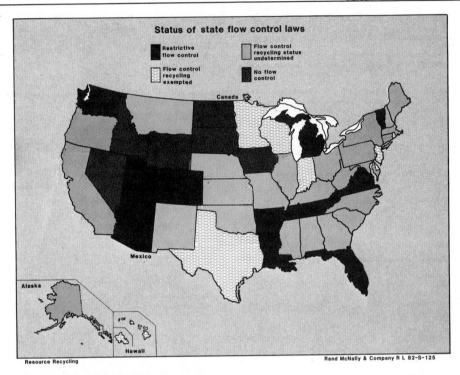

Status of state flow control laws

- Restrictive flow control
- Flow control recycling exempted
- Flow control recycling status undetermined
- No flow control

Canada

Mexico

Alaska

Hawaii

Resource Recycling

Rand McNally & Company R L 82–S–125

Economic and financial feasibility

Key elements reviewed in determining the economic feasibility of the resource recovery facility include the following:
—Construction and financing costs are reasonable with estimates derived from sound engineering practices.
—Solid waste required to obtain full operating capacity is reasonably available, given population growth and historical waste generation trends.
—Alternative disposal facilities will not pose any competitive threat to waste flow.
—Demand for energy, given certain assumptions about the availability and price of alternative sources, is consistent throughout the project life cycle.
—Facility and landfill options have a useful life corresponding to the life of the bonds.
—Management is capable of construction, start-up, and plant operation.
S&P partially relies on an independent consulting engineer's report to examine projected financial and operating results.
The income stream is evaluated for stability and balance between the revenues generated from tip fees and energy sales.

A wider margin of comfort to the bondholder is provided if annual revenues, excluding contributions from equity participants, or other subsidies, cover fixed and variable costs for the entire forecast period. Where projects have an operating history, historical coverage of current and future debt obligations in excess of sufficient levels is also viewed as a credit strength.
The review of financial projections involves analysis of underlying assumptions and the sensitivity to adverse change in energy prices or in the waste flow. At a minimum, the following sensitivity runs should be provided:
1) Base case, assuming best estimates for all key operating, waste, and energy assumptions.
2) A worst case operating scenario, assuming waste flows and plant operations at worst case levels
3) A worst case operating and energy sales scenario assuming case 2 operations and worst case energy prices and demand for service.
Additional scenarios may be required depending on individual project concerns.

Economic profile

Projected results for selected facilities based on 1987 'base case' projections are summarized in the chart below. Unit itemization includes revenue from all sources, expense projections including debt service, and an indicator of feasibility—net income. Five of the seven projects are expected to generate coverage of debt service in excess of sufficient levels. Both North Andover and Tulsa report 'break even' results because payment to the private operator is treated as a full expense item.

Accurate forecasts covering five to 10 years ahead are difficult due to uncontrollable economic circumstances and market risks such as fluctuations in energy prices. Economic and operating conditions vary for each project. In each case, however, the two revenue sources—tip fees and energy sales—may be subject to considerable exposure.
Tip fee formulas with payment based on fixed debt service requirements and adjusted to distribute profits between public

and private participants may be less susceptible to adverse economic conditions. Cash infusions of equity often serve to stabilize rate increases. For example, the North Andover structure applied this formula with additional cost reductions provided by the owner/operator in the form of revenues equating to tax benefits. Here tip fees are projected to be level, helping to maintain a competitive posture with alternative disposal options.

Revenues from energy sales are also vulnerable, especially in cases where electric sales contribute to a large portion of the income stream. Where electricity is sold to utilities with excess capacity or access to low priced hydro power, for example, pricing at the 'avoided cost' may not always be sufficient to sustain a viable solid waste facility. Pricing formulas structured with a fixed capacity charge in addition to a variable fee based on the utility's fuel cost may offset potential risks. In addition, if possible, stabilized electric price increases could serve to reduce tip fees in the critical early years of project operation. This may be beneficial both from an operating and credit perspective.

Issuer profile: economic summary
Selected credits—1987 base case projections

| | Unit itemization $/ton | | | | Forecast 1987–1992 % change | | | | | |
| | | | | | $/Ton | | $/Ton | | ¢ | |
Facility	Total revenue	Total expense	Income after debt service	Debt coverage (x)	Revs	Δ%	Tip fee	Δ%	Rev/kwh	Δ%
Pinellas, Florida	52.50	49.25	3.25	1.1	52.50	+12	35.32	−7	3.3	+26
Lawrence, Haverhill, Ma.[1]	56.30	51.15	5.15	1.27	56	+45	18.40	+40	2.3	+47
North Andover, Ma.[2]	58.83	58.83	—0—	1.0	52	+48	26.00	+3	8.5	+35
Saugus, Ma.[3]	48.00	34.60	13.40	1.89	48	+26	20.50	+25	6.1	+33
North East, Maryland	66.70	57.75	8.95	1.24	67	+54	35.00	+65	5.3	+47
Tampa, Florida[4]	136.00	130.70	5.30	1.1	136	+13	13.00	+10	6.4	+26
Tulsa, Oklahoma[5]	58.00	58.00	—0—	1.0	58	+27	26.25	+31	—	—

Source: Official Statements
[1] Average tip fee, includes spot non-contract sales.
[2] Projections include a $3.4 million annual equity contribution for owner operator
[3] Reflects average price for residential and commercial tonnage.
[4] Tampa Base Case $/ton are estimates of charges to the ultimate household user. This is not comparable to tip fee at the facility.
[5] Expense includes service fee and energy credit payment made to facility operator

Role of key players

Unlike most basic municipal bond financings, where the key partners in the transaction are limited to the municipal issuer and the trustee acting on behalf of the bondholder, waste to energy financings can have as many as five major categories of participants. Generally these include:

Project owner   The owner may be either a public entity (*i.e.* a municipality) or a corporation. While the public entity might enhance the creditworthiness of the project through a full or partial pledge of taxes, the corporate participant, as owner, may also provide guarantees for debt repayment. In addition, the corporation often constructs and operates the facility.

Equity partner   Investors from the private sector or the corporate owner may make a 20%–40% equity investment in the project in exchange for tax benefits. Arrangements may also call for corporate cash infusions once the project is operational. Equity contributions are typically made at the time of financing.

Municipal participants   Public entities commit to insure revenue for the project and support debt repayment by agreeing to bring waste for disposal and pay a fee ('tipping' or tip fee). These agreements in many cases hold even if the project is unable to produce energy.

Energy customer   Electric utilities or industrial customers commit to purchase electricity or steam generated from the project once operational. This provides another revenue source to meet debt service payment.

Trustee   Project debt is typically financed through traditional revenue bonds or industrial development bonds. All revenues related to the project and often the title to the facility are assigned to the trustee on behalf of the bondholders.

Conclusion

After a 10 year history in the United States marked by financial and technological failures, the future for resource recovery may improve as projects gain more operating experience. Uncertainties do exist however. Congressional legislation may limit the use of industrial development bonds which could increase financing costs. Changes in energy markets may make the cost of project output less competitive. Further, the operating experience of existing facilities may challenge assumptions about the reliability and feasibility of resource recovery technology. Irrespective of these potential developments, S&P's focus will remain on the prospect for long-term project feasibility as the greatest source of credit strength.

Diane R. Maurice, with the assistance of William Chew,
Maria Markham, Barry Genesen

Resource recovery in review

Each of the analyses that follows will examine a unique aspect of the growing resource recovery industry. Factors that can distinguish one credit from another are type of ownership, technology used, security facilities, and the size of waste disposal systems and its service territory.

The Pittsfield, Massachusetts solid waste facility, the first credit reviewed, is unique in many respects. The facility, which has the capacity to process 240 tons of waste per day, was one of the earliest mass burn units built and was designed to serve the needs of one community. Many other facilities are larger with capacity exceeding 2,000 tons per day and have a broader service territory.

Like many of the larger units since 1979, private equity supported initial projects costs for the Pittsfield project. Equity con-

tributions were made by a partnership in exchange for tax benefits and ownership rights. However, since the initial financing, the general partner has experienced financial difficulties. Should the situation worsen, the facility would have to retain a new operator. This exposure is mitigated, however, by the unconditional obligation of the city to make debt payments even if the partnership is forced into bankruptcy. In S&P's assessment, the economic feasibility of the project provides adequate incentive for the city to honor its commitment.

As shown in the Pittsfield analysis, S&P's analyses on the industry as a whole will focus on the prospect for long-term feasibility as the greatest source of credit strength.

CREDIT ANALYSES

Pittsfield, Massachusetts
Vicon Recovery Associates Resource Recovery Project

Rationale: The 'BBB+' rating on Pittsfield's solid waste revenue bonds is affirmed. The rating reflects the city's ability to meet its obligation to repay debt issued to finance the resource recovery project. This commitment continues even in the event the operating and facility lease agreement with the Vicon Recovery Associates partnership is terminated. The Vicon parent has experienced financial difficulties which could potentially jeopardize the lease. The economic viability of the project is a further consideration in the rating. Competitive disposal fees and steam rates are expected to continue throughout the life of the facility. These revenues are projected to continue to support project operating costs and debt obligations.

Legal protections: In 1979, the city of Pittsfield, through its Industrial Development Authority, sold $6.2 million in tax-exempt debt to finance a 240-ton per day mass-burn solid waste facility. A $2.7 million equity contribution from Vicon Construction Co., included in the original financing, was later supplemented in 1981 by $3.4 million in debt privately placed by Vicon's limited partners.

Bonds are secured under a lease agreement in effect for the life of the issue. The partnership leases the facility from the authority which acts as lessor on behalf of the city. Lease payments, which must always be sufficient to cover debt service, are derived from waste disposal fees from the city and steam sales revenues from Crane Paper Co. Both agreements are in effect for the life of the bonds and all revenue flows directly to the trustee. The indenture stipulates that lease payments are first allocated to replenish the debt service reserve (presently at maximum debt service), then for debt service payment, and finally for operating expenses which include payments to Vicon for facility management.

Should there be a default under the lease, the city is obligated under the indenture to make debt payments in accordance with an adapted tip. This is not a direct general fund committment however, although it is stronger than a non-adavalorem pledge. Because of this provision, bond counsel has opined that bankruptcy of the partnership should not affect the city's obligation to service the debt.

Under provisions of the disposal agreement, the city is obligated to pay a fixed fee of $11.59 for each ton of waste delivered. Even if the facility ceases to operate, the disposal fee must include sums sufficient to pay debt service. Taxes, operating expenses, and a management fee to Vicon adjusted annually by the consumer price index are also included as part of

this fee. During normal periods of operation, reduction of the city's payment occurs by adjusting for energy sales and revenue sharing provisions. This annual redistribution of excess revenues serves to keep disposal fees at competitive levels.

Steam sales to Crane Paper are conditioned upon facility performance. In the event of a six-month outage, the obligation under the steam contract ceases. Under normal operating conditions, Crane is required to purchase steam from the facility for 240 work days annually (five days a week).

Technological exposure: Vicon Recovery Associates, under the terms of a construction agreement, was unconditionally obligated to complete the 240-ton per day mass-burn facility by September 1980. Modifications to the facility design increased construction cost to $10.85 million from $4.46 million and extended the completion date to June 1981, 10 months after the original scheduled operation date. The limited partners assumed the additional cost of improvements which included the addition of another incinerator and emissions control equipment.

To date, the facility has demonstrated a 95%-98% availability with no unscheduled outages. To insure continued operating flexibility in the event of facility failure or interruptions in waste flow, equipment redundancy has been built into the facility. This includes three 120-ton per day furnaces and an auxiliary generator. There is also the capability to burn No. 6 fuel oil, if necessary, to insure continued steam generation should solid waste delivery temporarily cease.

Although this facility developed from a scaled-up model one-fifth its present size, many of the resulting initial operating problems have been resolved. Two problems are apparent, however. The first, related to potential extraordinary wear of the superheater, is expected to be minimized by keeping temperatures below 750 degrees Farenheit. In addition, underutilization has posed some difficulty in producing steam to desired levels. Related problems initially encountered with transmitting steam have been alleviated by Crane making the necessary modifications to its steam lines. As larger quantities of waste are delivered, the facility should be able to operate at levels greater than 53% of capacity demonstrated to date. This would reduce inefficiencies and increase steam output. Management is experienced and has been involved in constructing and operating the facility from its inception.

Operating data	(10 months)		
($000)	1981	1982	1983
Tons solid waste delivered:			
Contract-Pittsfield	36,973	49,722	48,674
Non-contract	4,741	12,468	19,673
Total	41,714	49,722	68,347
Disposal cost ($/ton)	13.13	13.07	11.95
Steam sales (¢/pound)	5.8	5.7	5.5
Income per ton ($)	1.56	3.64	4.26
Operating capacity (%)	32	38	53

Waste steam and demand for service: In April 1981 area landfills closed, eliminating most of the competition for available waste. At present, the nearest landfill which could vie for business is located outside a 30-mile radius of the facility. The city controls waste flow within its boundaries and has been able to deliver the 44,000 tons of waste annually to the facility as required under the disposal contract. Exposure exists, however, as 30% of the incoming waste is not under contract and could be diverted elsewhere should alternative disposal options develop. (With only contracted waste delivered, 'breakeven' operating results would barely be achieved.) However, average tip fees charged to haulers from outside the city are low at $12 per ton, and are expected to remain so.

Steam sales to Crane Paper are tied to the price of No. 6 fuel oil and discounted in determining the final fee for steam. Although steam production in 1983 at 233 million pounds was under the 300 million projected because of transmission problems and slight production declines at the company, rates at 5.5¢ per pound were low to moderate. Increases in tip fees to offset the decline in revenue which resulted were not substantial. In the future, demand for steam is expected to remain at stable levels. Retrofitting of Crane's equipment will facilitate acceptance of the project's output. Crane manufactures paper for the U.S. mint and other foreign currencies.

The economic base of the service territory is sound and will assist in continued demand for service. Employment has been maintained at fairly constant levels at Crane, despite job losses in manufacturing as a whole within Berkshire County. Much of the area's employment is tied to General Electric, and unemployment rates are generally above state averages. Pittsfield, the county seat, covers a 42 square mile area. It also serves as an economic center for western Massachusetts. The 1980 census reported a population of 52,000 representing a 9% decrease from 1970. However, economic indicators are strong. Per capita retail sales generally exceed state levels.

Pitsfield general fund operations*	—Fiscal—		
($000)	1981	1982	1983
Revenues	69,692	65,402	74,193
Expenses	70,005	65,014	72,676
Operating surplus (deficit)	(313)	388	1,517
Cash and investments	4,715	5,139	6,928
*Cash basis			

Economic and financial feasibility: As ultimate guarantor of the facility's debt service, the city would be able to assume this obligation, if necessary. Since 1979, the city's own bonded debt obligations have been reduced significantly. General fund operations offset by large and growing cash balances. Tax collections and growth in assessed valuation have been excellent and any problems associated with Proposition 2½ appear to have been mitigated. No future debt issuance is planned.

Financial performance of the solid waste facility has shown improvement in 1983, the second full year of project operations. Net income per ton of waste processed increased significantly, with disposal fees and steam prices held at competitive levels. Overall financial results have been below projections, however. Increases in disposal fees reflecting non-contract waste delivery were not substantial enough to offset revenue loss from unreal-

Financial Information	(10 months)		
($000)	1981	1982	1983*
Revenues			
Tip fees	431	811	816
Steam	662	1,192	1,317
Interest	202	128	87
Total	1,408	2,131	2,220
Expenses			
Operations & maintenance	926	1,103	1,154
Vicon fee	0	134	142
Total exp	926	1,237	1,296
Net revenues	482	894	924
Debt service	477	663	663
Coverage (x)	1.01	1.37	1.76
Debt service + note payment	699	820	873
Coverage (x)	0.69	1.09	1.06
*Unaudited results			

ized steam sales. Despite this, coverage of debt obligations has exceeded sufficient levels with excess revenues available to make subordinated payments due on the partnership's equity notes.

The system has maintained a sound balance sheet although cash availability may be constrained because of a slight lag in receivables collections. In 1982, the depreciation schedules were adjusted to a straight-line rather than accelerated method reflecting the extended useful facility life.

In addition to depreciation and tax benefits, Vicon receives repayment on its notes and has the option to purchase the facility for $1 at the end of the lease term. Incentives for the city to continue its participation in the project include the use of a low-cost method of waste disposal located 4.5 miles from the downtown area. Bankruptcy of Vicon is not seen as a deterrent to project operations as the city could find a replacement operator if necessary. The project, from the perspective of both public and private participants, is expected to continue to demonstrate economic and financial feasibility.

Diane R. Maurice, Patrick McCorry

Ohio Water Development Authority
Akron Recycle Energy System

Rationale: The rating on the $46 million Ohio Water Development Authority/Akron Recycle Energy System solid waste revenue bonds is affirmed at 'CC'. The rating is based on the plant's inability to meet operating expenses and debt service from operating revenue, even after turnkey plant modifications. An Environmental Protection Agency (EPA) grant received in October 1982 was used to remove the plant's troublesome source separation system and make other modifications to improve

performance. In the fall of 1983 a series of minor explosions in the boiler disrupted plant operations. The operator believes this problem was caused by powdered polymers in the industrial waste stream. The plant is currently operating on residential waste, pending a solution to this problem. A cooperative agreement between the authority and the city of Akron designates sufficient funds from the EPA grant to make the December 1984 coupon payment. In the absence of substantial rate increases

(continued on next page)

or subsidies, insufficient funds will be available for the June 1985 debt service payment.

Security: The bonds are secured by the net revenues of the Recycle Energy System consisting primarily of steam sales contracts with 33 steam users in the Akron central city area. Steam contracts provide for rates to be set at a level sufficient to cover operation and maintenance and to provide 1.5 times (x) coverage of annual debt service. Additional revenue is provided by tipping fees. Further security for the bonds was provided by a debt service reserve which was funded to maximum annual debt service from bond proceeds.

Operations: The Akron Recycle Energy System was designed to separate recyclable material from residential and industrial waste and to burn the remaining waste in semi-suspension boilers rated at 1,000 tons per day. Construction began in 1977 with completion projected for 1979. Due to major design and mechanical problems, the plant was not completed on time. The major problem was that the pneumatically operated density separation system, which was designed to recover recyclable material and increase the heating value of the combustible material, failed to operate effectively. The original bond issue was insufficient to cover required modifications. On June 1, 1982, the trustee assumed management of the system and later that month, the city, Ohio Water Development Authority, and the trustee entered a modification and operation agreement with Tricil Resources Inc. Contributions from the city and the authority plus a $19.7 million EPA grant received on Oct. 7, 1982 were directed to plant modifications ($11.4 million), an operating fee ($3.6 million), and debt service payments through Dec. 31, 1982 ($8.5 million).

Tricil replaced the pneumatic density separation system with a simple conveyer belt and performed other modifications to improve performance. After successfully operating at a 900 tons per day rate; the plant was certified by the city on Jan. 4, 1983 as ready for operation. For the nine months ended Sept. 30, 1983, the facility had a net cash shortfall of $4.7 million on revenue of $4.2 million. Debt service payments during this period were made from EPA grant funds. In the 1983 fourth quarter, a series of boiler explosions disrupted plant operations. Tricil believes that these explosions were caused by powdered polymers in the industrial waste stream. The plant burned natural gas to supply its customers in the last quarter of 1983. Currently the plant is operating on residential waste pending a solution to this problem. The total operating deficit in 1983 was $7.1 million. In addition to operating problems, only 60% of the steam which is generated can be sold due to high waste deliveries in the summer when customers do not need the steam for heating.

Rates: The system continues to face difficulties because of competitive limits which restrict increases in tipping fees and steam rates. Low tipping fees ($7.50 per ton) are limited because of available landfills. These landfills may be closed in the next several years, providing flexibility to increase these fees. Steam rates are currently competitive with oil and natural gas. The trustee estimates that a 100% increase in steam rates would be necessary to cover operation expense and debt service. Steam customers have stated that they will break their contracts and use alternative fuel if increases of this magnitude are enacted. As of Feb. 1, 1984, the trustee raised rates 18% as a short-term compromise, a level sufficient to cover operations and maintenance but not debt service. The city of Akron appointed a blue ribbon commission to recommend a solution to the system's problems. A preliminary report released May 1, 1984 indicated that the city considered the system a valuable community resource and the only feasible long-term waste disposal system. However, no specific plan for meeting debt service payments was recommended. Release of the final report, originally scheduled for June 1984, has been delayed due to continuing discussion among interested parties seeking a solution to the facility's short-term problems. These discussions are expected to continue, with no resolution likely until the fourth quarter of 1984.

Finances: Operating revenues, assuming no further unplanned shutdowns, are currently sufficient to cover operations but make no contribution to debt service. The project contingency fund has a zero balance. The debt service reserve fund, which was covenanted to be maintained at one-year maximum debt service of $4.320 million, has a balance of approximately $150,000. The modification fund which consisted of $24.4 million in City, Authority and EPA grant funds, has been exhausted; $1.68 million in Environmental Protection Agency grant funds held by the authority were used to make the June 1, 1984 debt service payment. Sufficient monies from this fund are being allocated by the authority and the city for the December 31 debt service payment. However, no funds will remain to make any further debt service payments.

Michael Maguire

Pinellas County, Florida

Rationale: S&P affirms the 'A' rating on the $160 million outstanding 1980 revenue bonds and the 'AAA' on the 1983 insured issue. The rating reflects the strong legal provisions, which include a rate covenant and fully funded reserves, in addition to the continued feasibility of the project. Increases in disposal fees anticipated in the next fiscal year are moderate and will enable the facility to retain its competitive position in relation to other disposal options. Future increases necessitated as a result of the system's 1000 TPD expansion will be held at moderate levels but place a slightly larger burden on system users. Minor plant modifications required to improve electric generating capacity will be complete within the next six months and bring operations to 85% of nameplate capacity.

Legal provisions: Series 1980 and 1983 bonds are secured by revenues of the system derived from operations of the solid waste facility. These include tip fees charged to haulers for waste disposal and revenues derived from the sale of the project's electric output. Further security for the earlier series is provided by a county *pledge of non-advalorem* tax revenues. County funds are available to replenish deficiencies in the 1980 debt service reserve. Other legal provisions include a county covenant to raise sufficient revenue from rates and other user fees derived from system operations which, when combined with renewal and replacement fund deposits, will cover annual debt 1.5 times. All reserves are fully funded to date, providing sufficient funds to cover maximum debt service 1.4x in any year. A privately placed loan to cover the taxable portions of the facility is repaid on a subordinated basis.

Florida Power Corp. under a take and pay purchase agreement, will acquire excess electric energy produced from the project. Payment is based on Florida Power's cost of replacement power. This agreement is in effect through the life of the bonds.

Construction schedule

Commence engineering	May 1984
Purchase boiler	August 1984
First refuse burned	August 1986
Turbine operating	October 1986
Acceptance test	January 1987
Commercial operation	February 1987

Technological exposure: Construction of the first phase of the mass burn facility was completed in May 1983, 100 days ahead of schedule. Although operations at this 2,000-ton-per-day plant were accelerated, excessive corrosion of the boiler tubes, and unanticipated moisture content in the solid waste has resulted in inefficient operating performance.

Electric output per ton of waste has not reached operating projections. Estimates of 500kwh/ton in 1980 were revised downward to 450 kwh with actual performance to date slightly below this level. Boiler modifications in November are expected to bring operations to 85% of capacity from 74%, and improve generation. Costs associated with these modifications are the responsibility of the construction contractor and are not expected to result in any increase in user fees. Minimal expense was incurred earlier this year with an unscheduled week-long outage and turbine failure.

Future operations will benefit from completion of the 1000tpd expansion. This will include the additon of a 25mw turbine bringing electric generation to 75.9mw. Additional operating flexibility will be provided by including other auxillary equipment.

Because this expansion is part of an operating facility, there is limited incentive on part of the contractor or the county to have the 33-month construction schedule accelerated. Operating revenues received upon completion of the expansion will accrue to the county rather than the operator. All milestones are expected to be met on schedule with little variance.

Solid waste delivery

				Population		
Projections—tons (000)			Actual	Projections—(000)		
Year	1980	1983		Year	1980	1983
1984	628	—	530	1980	848	753
1985	637	839		1985	872	826
1986	645	853		1990	926	899
1987	650	868				

Energy produced

				Energy revenues—unit cost			
						Mills	
Projections—(kwh 000)			Actual	Projections—(price/kwh)			
Year	1980	1983		Year	1980	1983	Actual
1983/							
1984	130		198	1980	36.9	35.8	36–37
1984	317	278		1985	41.9	35.7	
1985	326	278		1987	43.1	45.7	
1986	328	260					
1987	328	352					

Waste stream and demand for service: State legislation passed in 1975 subsequently modified, and adopted by the county in 1980, provided the mandate for county waste flow control. Although the county can legally compel all persons to utilize the solid waste system, because of limited options for disposal, they have not found it necessary to do so. The system expansion was designed to accommodate additional waste generation unanticipated at the time of initial construction. The consulting engineer's initial and revised projections of solid waste delivery support the assumption of increased waste generation. However, population growth projections and actual waste delivery are slightly below estimates.

Total production of electricity and related revenues are expected to be reduced based on the new estimates. Although FPC's demand for project output is still expected at the anticipated unit cost, because generation has been reduced there will be greater reliance on tip fee to produce sufficient project revenues.

Revenue sources

Projections		
(% of total revenues)	1980	1983
Tip fees	46	68
Electric sales	37.9	23

Economic and financial feasibility: Based on a year of actual operations, financial results are close to revised 1983 projections. Revenues derived from both tip fees and electric sales are meeting budgeted levels. Further, first-half results indicate the tip fee per ton is at a level competitive with the other disposal options in the area as well as on line with what was projected initially. Audited results for the prior two fiscal years are strong, reflecting a three-fold increase in operating income corresponding to plant start-up.

The first full year of financial performance will benefit from a $4.2 million revenue carryover from prior year operations. Without this cash infusion, debt service coverage would be well below the required levels. Tip fee increases of 36% are expected in 1985 bringing the actual cost to $26 per ton. It is not anticipated the use of prior year revenue to offset tip fee increases will occur in the future. Moderate pressure will be placed on system users in 1987 when the expansion is complete. Metal sales, while contributing minor contributions to the revenue stream, continue below projections.

Based on the sound economic base of the service territory, the limited options for waste disposal, and the consistent demand for electric output, the continued feasibility of this project is anticipated.

Finance status summary	1983–84	6 months	Budget
(unaudited)	projected	actual	(%)
($000)			
Income Item			
Electric revenues	9,476	5,633	60
Tipping fees	14,673	6,447	44
Metal sales	1,128	155	14
Fiscal 1983 revenue carryover	4,200	2,100	50
	29,477	14,335	48
Expenditure item			
O&M	11,132	4,144	37
Landfill	1,397	715	51
Debt service	16,948	8,474	50
Total	29,477	14,335	48.6
Refuse tonnage	839,400	395,200	52%
Average tip fee	17.50	16.30	—
Source: Pinellas County			

Diane R. Maurice, Maria C. Markham

Westchester County

Construction is complete at this facility with commercial operations expected approximately two months ahead of schedule. Although the design required certain plant modifications, the construction budget of $179 million is expected to be met. Acceptance testing in July resulted in consistent performance with both combustion and energy exceeding projected levels. Debt service has been capitalized through the anticipated final acceptance date.

Based on final acceptance tests, SO_2 and NOX emmissions were slightly higher than anticipated. The required revised, EPA permit is expected by mid-October. The State Division of Air expects no delay as the changes are minor and no intervening groups have been active since the project's inception.

Northeast Maryland

This facility was upgraded to three (750) ton per day units from a slightly smaller size in an effort to replicate the Westchester design. The resulting savings in engineering costs offset needed increases in construction capital and saved three months on critical construction path. Waste will be accepted for initial testing by September 15 with operations to begin during the first quarter of 1985.

Construction schedules

Facility highlights

	Westchester	North East Maryland	North Andover
Process lines (tons/day)	3 (750)	3 (750)	2 (750)
Daily capacity	2250	2250	1500

Major milestones:	Orig. Estimate	Months ahead	Orig. Estimate	Months ahead	Orig. Estimate	Months ahead
Commence Engineering	11/81	—	12/82	—	6/83	—
First refuse burn	7/84	+5	6/85	+9	1/86	+8
Begin acceptance tests	11/84	+4	10/85	+8	5/86	+9
Commercial operations	1/85	+2	12/85	+10	6/86	+7

Source: Signal Resco.

North Andover, Massachusetts

Construction at this facility is presently seven months ahead of schedule. Accelerated delivery schedules on boilers and the turbine have accelerated the overall project schedule. The design employed is similar to the Pinellas County facility. With 75% of the construction contracts committed, final completion is expected by the fourth quarter of 1985 with little variance in the fixed fee construction contract amount expected.

Diane R. Maurice, Maria C. Markham

CHAPTER 7
LEGAL ISSUES

BARRY S. SHANOFF
JANE C. SOUZON

Law Offices of Barry S. Shanoff
Silver Spring, Maryland

7.1 WASTE FLOW CONTROL

The viability and success of a solid waste disposal facility, whether it be a landfill or an incinerator, depends on a predictable stream of refuse. For landfills, waste flow used to be a simple matter of location and economics; it was cheaper to use the closest dump. As disposal facilities become more regulated and sophisticated, however, the owners and operators can no longer assume that their investment will pay off.

This section will discuss several different mechanisms used to ensure a steady volume of waste during the life of a waste facility. These mechanisms include market forces in the form of competitive tipping fees, private agreements and contracts, and public laws.

7.1.1 Competitive Tipping Fees

When every community had its town dump, charges, if any, for dumping refuse were nominal compared with the costs of hauling the trash to another dump. As the number of approved landfills dwindle, however, waste haulers may be faced with choosing among several facilities, all of which may be located outside of their service areas. Predictably, the hauler will choose the waste site that, considering distance, travel, time and tipping fees, is least costly.

Many factors go into the tipping fee charged by a disposal facility, including day-to-day operation and maintenance, insurance, site and design engineering, and permit costs. Such cost variables can result in site A charging half as much as site B.

High volume is not important in landfill operation if there is no hurry to bring the site to capacity. However, except in the case where a public agency operates a landfill without regard to profit, if the tipping fees are too high (driving business away) or too low (making the site a veritable "give-away"), the landfill will not generate sufficient revenues to pay for itself.

The pricing decision is more consequential for a waste-to-energy facility, which must have a relatively steady supply of waste to use as fuel. Sufficient volume may be obtained by dropping the tipping fees, but such a facility must set fees at a level that will create adequate revenues to operate and maintain the plant and to service a debt obligation.

7.1.2 Private Agreements and Contracts

One way to lock in a steady supply of waste is to contract with private entities such as waste haulers or generators, whereby the latter are obliged to provide a certain volume of waste to the facility within a given period. Such an arrangement presumably benefits both: the disposal facility (by giving it a measure of predictability and continuity) and the customer (by providing an assured avenue for disposal at a known price).

Most generators of solid waste do not create sufficient volumes of refuse to justify anything more elaborate than a service contract to have the waste collected and hauled away. The more attractive customer in the eyes of the disposal facility is the entity that collects and hauls waste for a large number of generators. This entity may be a private company looking to fix the cost of disposal for a period into the future, or it may be a unit of local government that operates or oversees the collection of refuse for its residents.

Where a municipality collects the waste with its own employees and equipment, it simply directs its drivers to the contracted disposal facility. The commitment may cover all the waste from a community or it may specify a minimum annual tonnage that the municipality is obligated to deliver. If the collection is performed by a private hauler under contract to the municipality, the contract must specify the disposal site to be used.

Municipal contracts have been effective in guaranteeing the supply of waste to a resource recovery facility. An example of this form of waste flow control is contained in the appendix to this chapter.

7.1.3 Legislative Controls

An increasing number of jurisdictions are relying on legislative moves to control the waste stream: state laws and local ordinances authorizing restrictions on disposal sites and collection practices. The approach is designed to eliminate competition in solid waste collection and disposal arrangements for the sake of the "public interest." To implement the program, the governmental unit customarily issues revenue bonds, signs contracts, and sets fees. Because of its obvious and intended anticompetitive effects, legislated waste flow control has led to court challenges on antitrust grounds wherever it is introduced.

In *Parker* v. *Brown*, 317 U.S. 341 (1943), the Supreme Court first addressed the question of whether the federal antitrust laws prohibit a state from exercising its sovereign powers to impose certain anticompetitive restraints. It held that a marketing program enacted by the California legislature to create price supports for raisins was exempt from challenge under the Sherman Act because the program "derived its authority . . . from the legislative command of the state." The Court stated:

> We find nothing in the language of the Sherman Act or in its history which suggests that its purpose was to restrain a state or its officers or agents from activities directed by its legislature. In a dual system of government in which, under the Constitution, the states are sovereign, save only as Congress may constitutionally subtract from their authority, an unexpressed purpose to nullify a state's control over its officers and agents is not lightly to be attributed to Congress.

In *City of Lafayette* v. *Louisiana Power & Light Co.*, 435 U.S. 389 (1978), the Supreme Court faced the question of whether the state action doctrine protected a municipality from federal antitrust liability. A plurality of four justices rejected the claim that the state action doctrine extended to "all governmental entities, whether state agencies or subdivisions of a State . . . simply by reason of their status as such." The justices nonetheless recognized that municipalities are instrumentalities of the state, and their actions may reflect state policy. The plurality thus held that "the Parker doctrine exempts only anticompetitive conduct engaged in . . . by [municipalities], pursuant to state policy to displace competition with regulation or monopoly public service."

The plurality observed that a state policy to displace competition could not be found "in the absence of evidence that the state authorized or directed a given municipality to act as it did." It concluded, however, that "an adequate state mandate for anticompetitive activities of cities and other subordinate governmental units exists when it is 'found from the authority given a governmental entity to operate in a particular area, that the legislature *contemplated* the kind of action complained of.' "

In *Community Communications Co. v. City of Boulder*, 455 U.S. 40 (1982), the Supreme Court ruled that broad home rule powers do not shield a municipality from antitrust liability for specific anticompetitive measures. The court rejected the city's state action defense, noting that the city failed to prove that its challenged restraint constituted either "the action of the State of Colorado itself in its sovereign capacity, . . . [or] municipal action in furtherance or implementation of clearly articulated and affirmatively expressed state policy."

How clearly must a state express its policy to displace competition? To what extent, if at all, must a state supervise a challenged activity within an area of traditional local government function? In its 1984–85 Term, the Supreme Court may answer these questions when it considers *Town of Hallie v. City of Eau Claire*. However, until the high court says differently, lower courts will continue determining whether the restraint in question is a reasonable and foreseeable exercise of delegated powers within the scope of the local government's authority. If the challenged restraints are reasonably related to a local government's express powers and reasonably designed to promote the state's aims within a designated field of regulation, the restraints will be linked to a "clearly articulated and affirmatively expressed state polity" to displace competition. [*See, e.g., Central Iowa Refuse Systems, Inc. v. Des Moines Metropolitan Area Solid Waste Agency*, 715 F.2d 419 (8th Cir. 1983) (ordinance requiring all solid waste generated in participating jurisdictions to be sent to defendant agency's landfill upheld as a "necessary or reasonable consequence" of engaging in the authorized activity of constructing a waste disposal facility with funds raised by revenue bonds); *accord., Hybud Equipment Corp. v. City of Akron*, 742 F.2d 949 (6th Cir. 1984).]

Another significant instrument in waste flow control is the creation of franchises where allowed by law. An example of a solid waste franchise administration mechanism is contained in the Appendix to this chapter.

Municipalities that engage in noncompetitive solid waste collection and disposal practices no longer face budget-breaking liability for money damages in federal antitrust suits. The Local Government Antitrust Act of 1984 (P.L. 98-544) limits successful plaintiffs to injunctive relief, that is, an order forbidding the city, town, sanitary district, or other unit of local government from engaging in the illegal actions. The law also extends the same protection to haulers and other private persons and firms who benefit from an exclusive municipal franchise. Roughly 300 antitrust suits were pending against municipalities when the law took effect. In these cases a plaintiff may still seek money damages unless the local government defendant shows that it would be unfair (considering, among other things, how far the lawsuit has progressed) to allow the plaintiff more than mere injunctive relief.

7.2 INTERSTATE COMMERCE

The Commerce Clause of the United States Constitution provides that, "The Congress shall have Power . . . To regulate Commerce . . . among the several States . . ." (U.S. Const. Art. I, Section 8, cl. 3). Implicit in this power is the concept that certain limitations exist on the power of the states to interfere with or burden interstate commerce. "Even in the absence of congressional action, the courts may decide whether state regulations challenged under the Commerce Clause impermissibly burden interstate commerce" [*Western and Southern Life Ins. Co. v. Board of Equilization of California*, 451 U.S. 648, 652 (1981)]. For the purposes of the Commerce Clause, local ordinances also are required to withstand at least the same degree of constitutional scrutiny as state laws. [See, *e.g., Huron Portland Cement Co. v. Detroit*, 362 U.S. 440 (1960).]

The precise limitations and parameters of the Commerce Clause do not appear in the Constitution, but have gradually emerged in the decisions of the Supreme Court. The broad purpose of the clause was expressed by Mr. Justice Jackson in *Hood & Sons, Inc. v. Du Mond*, 336 U.S. 525 (1949):

> *This principle that our economic unit is the Nation, which alone has the gamut of powers necessary to control of the economy, including the vital power of erecting customs barriers against foreign competition, has as its corollary that the states are not separable economic units. As the Court said in* Baldwin v. Seelig, *[citation omitted], "what is ultimate is the principle that one state in its dealings with another may not place itself in a position of economic isolation."*

This alertness to the evils of economic isolationism and protectionism was dispositive in a key high court decision: *City of Philadelphia v. New Jersey*, 437 U.S. 617 (1978). In *City of Philadelphia*, the Court concluded that movement of "valueless" wastes was "commerce" within the meaning of the Commerce Clause. "Just as Congress has power to regulate the interstate movement of these wastes, States are not free from constitutional scrutiny when they restrict that movement." (*Id.* at 622–623).

The case presented the Supreme Court with a New Jersey law that said:

> *no person shall bring into this State any solid or liquid waste which originated or was collected outside the territorial limits of the State, except garbage to be fed to swine in the State of New Jersey, until the commissioner (of the State Department of Environmental Protection) shall determine that such action can be permitted without endangering the public health, safety and welfare and has promulgated regulations permitting and regulating the treatment and disposal of such waste in this State.*

In summarizing the previous decisions of the Court, Justice Stewart (speaking for the 7–2 majority) concluded that a "virtually *per se* rule of invalidity has been erected" where simple economic protectionism is effected by state legislation. Although the New Jersey statute purported to protect the safety, health, and welfare of the citizens of New Jersey, the Court held:

> *The harms caused by waste are said to arise after its disposal in landfill sites, and at that point, New Jersey concedes, there is no basis to distinguish out-of-state waste from domestic waste. If one is inherently harmful, so is the other. Yet New Jersey has banned the former while leaving its landfill sites open to the latter. The New Jersey law blocks the importation of waste in obvious effort to saddle those outside the State with the entire burden of slowing*

the flow of refuse into New Jersey's remaining landfill sites. That legislative effort is clearly impermissible under the Commerce Clause of the Constitution.

Following the Supreme Court's decision in *City of Philadelphia,* state and federal courts have consistently invalidated town, county, and state enactments which, like the New Jersey law, attempted to discriminate against wastes solely on the basis of origin. [See *Browning-Ferris, Inc.* v. *Anne Arundel County,* 292 Md. 136, 438 A.2d 269 (1981) (county ordinance barring transportation or storage of hazardous, toxic, and special waste from areas outside county); *Washington State Building & Construction Trades Council AFL-CIO* v. *Spellman,* 518 F. Supp. 928 (E.D. Wash. 1981) (state statute adopted by voters initiative seeking effectively to bar storage and transportation of all nonmedical radioactive waste generated outside the state); *Dutchess Sanitation Service, Inc.* v. *Town of Plattekill,* 51 N.Y. 2d 670, 417 N.E.2d 74 (1980) (town ordinance broadly forbidding anyone other than those residing or conducting an established business in the town from dumping "garbage, rubbish, or other articles originating elsewhere than in the Town."); *Hardage* v. *Atkins,* 619 F. 2d 871 (10th Cir. 1980) (Oklahoma statute enacting comprehensive control over out-of-state industrial waste).] In *Shayne Bros. Inc., et al.* v. *Prince George's County, Maryland,* 556 F. Supp. 182 (D. Md. 1983), a county ordinance provided as follows:

> *No trucks or other vehicles shall transport from without the state trash or refuse to a dump or landfill within this County, unless authority has been granted by the Council and County Executive.*

Relying on *City of Philadelphia,* the district court struck down the ordinance, thereby rejecting the County's argument that the measure merely sought to reserve the limited landfill space within the county for the sole use of county residents and that such a purpose was a valid public policy. Instead, the Court concluded that "[s]imilar arguments were advanced and rejected by the Supreme Court in the *City of Philadelphia* case." The court concluded its analysis of the invalidity of the ordinance with the following comment:

> *In the concluding passage of his opinion in* City of Philadelphia, *Justice Stewart said the following [citation omitted]:*
>
> *'Today, cities in Pennsylvania and New York find it expedient or necessary to send their waste into New Jersey for disposal, and New Jersey claims the right to close its borders to such traffic. Tomorrow, cities in New Jersey may find it expedient or necessary to send their waste into Pennsylvania or New York for disposal and those States might then claim the right to close their borders. The Commerce Clause will protect New Jersey in the future just as it protects her neighbors now from efforts by one State to isolate itself in the stream of interstate commerce from a problem shared by all.'*
>
> *These observations apply with equal force in this case. If Prince George's County may today close its borders to trash and refuse from Virginia and the District of Columbia then the latter jurisdictions can likewise close their borders to out-of-state waste from Prince George's County or any other County in Maryland. The Commerce Clause protects all of these jurisdictions from efforts by adjacent states or counties to isolate themselves in the stream of interstate commerce or from a problem shared by all."*

In the face of these precedents, what arguments can local governmental officials make in justifying a ban on the disposal of trash originating in another jurisdiction? Perhaps things would be different if the state or locality were to own the landfill site(s); at least other legal issues would arise. These issues appear in two different lines of Supreme Court cases: (1) three recent cases that completely exempt from Commerce Clause review a state or local measure which is characterized as "market participation" instead of "market regulation," and (2) cases dealing with state measures seeking to preserve the state's natural resources for the use of its own citizens.

   1.   *Exemption as Market "Participant", Not Market "Regulator."*   What the Commerce Clause prevents are moves by individual states that interfere with the freedom of the marketplace, the freedom of private parties to buy from and sell to whomever they wish. If the state's action can be characterized, not as interfering with third parties' relationships in the marketplace, but as merely the state itself participating in the market, then the Commerce Clause may not apply.

   Hughes v. *Alexandria Scrap,* 426 U.S. 794 (1976) dealt with a situation where Maryland offered bounties to scrap metal processors for wrecked cars titled in Maryland, as a means of getting abandoned cars off the streets and vacant lots of the state. The bounties were available to both local and out-of-state metal processors, but out-of-state processors were saddled with providing considerably more documentation on the hulks. And the persons who supplied wrecks to the out-of-state processors, in turn, were obliged to provide more documentation, which meant they were more likely to supply the in-state processors instead. The Supreme Court said that a bounty system favoring in-state processors did not burden interstate commerce because Maryland, in effect, was choosing to "buy" from its own citizens.

In *Reeves, Inc. v. Stake,* 447 U.S. 429 (1980) the state of South Dakota operated a cement plant with a policy of selling cement only to state residents. The Court viewed this arrangement as market participation, not regulation, and thus exempt from review under the Commerce Clause.

In *White v. Massachusetts Council of Construction Employees,* 460 U.S. 204, 103 S. Ct. 1042 (1983), the Court ruled that a mayor's executive order was exempt from Commerce Clause scrutiny even though it arguably interfered more with private party contracting than the circumstances in the *Alexandria Scrap* and *Hughes* cases. The mayor of Boston had ordered that all construction projects financed completely or partly with city funds be performed by a work force at least half of which are city residents. In other words, the order dictated who the city's contractors could hire to do the labor, rather than merely determining to whom the city would directly give funds. Nevertheless, the court ruled that the city's program amounted to direct local participation in the market. [*See also County Commissioners of Charles County v. Stevens,* 299 Md. 203, 473 A.2d 12 (1984).]

These cases suggest that a state or local measure banning out-of-state waste might withstand a Commerce Clause challenge if the state or municipality operated its own landfill, on the theory that the governmental operator had merely chosen to "sell space" to only its own residents. In the *Reeves* case, however, the court hinted that the exemption that applied to selling cement and buying wrecked cars would not necessarily apply if the state or local measure dealt instead with natural resources—especially if the restriction could be characterized as "hoarding" those resources. Whereas the court in *Reeves* did not consider cement to be a natural resource, it did imply (by reference to *City of Philadelphia*) that landfill sites would be considered natural resources (447 U.S. at 443). Thus, to use the market participation exemption to justify a state or local measure banning importation of solid waste to a publicly owned landfill, one would have to make arguments like these:

1. The measure does not equal "natural resources." It does not regulate the selling of "space" (land). Instead, it controls the selling of "landfill services" to local residents only. These "services" are akin to cement, in that they are the end product of a natural resource (land), machinery, and human effort. Why can't a public agency limit the sale of this service to the people who pay for it with their taxes, just as a local jurisdiction limits fire protection service and public education to local residents? *See County Commissioners of Charles County v. Stevens, supra.,* and *Shayne Bros., Inc. v. District of Columbia,* 592 F.Supp. 1128 (D.D.C. 1984). In the *Charles County* case, the Maryland court pointed out that restrictions that apply only to publicly owned facilities do not "close the state's borders" to interstate commerce—haulers can still bring waste into the jurisdiction to deposit at privately owned facilities.

2. Even if by restricting access to landfill space one is choosing whom to sell "land," such property is not the natural resource the Supreme Court had in mind in the cases where it struck down limitations on the sale of natural resources to nonlocal residents. For example, in *Hughes v. Oklahoma,* 441 U.S. 322, (1978), the court said it would be a fiction to say that a state "owns" minnows. A better characterization, according to the justices, would be to say that the minnows have not yet been reduced to private owernship. On the other hand, land can truly be owned by a state or locality in a traditional legal sense. Moreover, land is different from the resources referred to in *Reeves,* such as minerals and natural gas, and this difference underscores a basic concern of the Commerce Clause expressed in *Reeves:* the fear that "Pennsylvania or Wyoming might keep its coal, the northwest its timber, and the mining states their minerals" [*Reeves Inc. v. Stake,* 447 U.S. 429, 443 (1979)]. These resources are unique to some states, so that if the "owner" states did not allow access to them by other states, the owner states would enjoy an abundance while other states, and the nation as a whole, would be deprived. By contrast, all states have some land, probably enough to meet their waste disposal needs if they but choose to use the land in this way. Thus, each state should use its own land for this purpose, especially where the state trying to ban out-of-state waste is a small one, or the state seeking to export waste has land in abundance.

3. The Court in *Hughes* appeared to reach its decision on the basis of a balancing test, rather than an automatic (*per se*) rule of invalidity. Further, the Court suggested that a balancing test, rather than a *per se* rule, had been the basis for its decision in the *Philadelphia* case. Therefore, the proponent of a state or local measure restricting out-of-state waste should at least be allowed to argue the necessity of the proposed restrictions.

7.3 FINISHED LANDFILL SITE CONTINUING LIABILITIES

Although this part of the handbook is devoted to nonhazardous wastes, the reader should consider the possibility that the "innocuous" materials disposed of today in a landfill may be declared "hazardous" tomorrow. The recent federal Superfund legislation is a case in point: Even though the placing of waste materials may have been "legal" at an earlier time, the owner of the dump site can be held responsible for cleaning up the wastes if they are later determined to be hazardous. No one can assure that Congress will not take similar actions in the future. Meantime, other sources of owner/operator liability may be based upon traditional common-law doctrines, such as nuisance or strict liability.

Both governmental and private owner-operators may seek protection against potential liability through insurance, surety bonds, and trust funds. Ultimately, either the owner/operator bears the costs as part of doing business or the facility users (through higher fees or surcharges) pay for the financial security.

7.3.1 Insurance

Not long ago it was virtually impossible to obtain insurance covering the types of long-range liabilities associated with solid waste disposal facilities. Too many unknowns existed: the term of coverage, which activities and occurrences were covered, and especially, the scope of persons and property affected. In the wake of RCRA (Resource Conservation and Recovery Act) and the superfund law, however, the insurance industry is rapidly moving to provide appropriate coverage. As more customers purchase insurance, carriers are becoming better able to purchase reinsurance and spread the risk of loss among a larger pool. Experience also helps the raters determine fair premiums.

Coverage for environmental restoration is becoming a necessity, albeit an expensive one. The alternative is to take the chance of losing the entire investment should something go wrong in the future. An added benefit from insurance is that the carrier will defend any lawsuit involving a covered claim, no matter how frivolous it may be.

7.3.2 Surety Bonds

Although surety bonds are generally more expensive than insurance, the coverage tends to be more certain. The concept of a surety bond is an agreement to answer for the payment of another person's debt, or for the performance of his obligation. The surety (usually an insurance company) is bound with the principal (the landfill owner/operator) on the obligation or debt. A surety contract in the form of a bond is a so-called bond on condition. The elements of such a surety contract include (1) the obligation to pay a stated sum of money, (2) a recital of the relationship of the parties and subject matter of the bond, and (3) the condition, or statement of circumstances and contingencies on which the bond becomes void (*e.g.,* payment by the principal of all damages and costs adjudged by reason of legal liability for landfill operation). The liability under the bond becomes fixed only upon nonperformance of the condition. Surety bonds have been used extensively in the past for all types of risks. As a result, the cost of such protection may be relatively easy to identify.

A sample surety (indemnity) bond is contained in the Appendix to this chapter.

7.3.3 Trust Funds

Trust funds are a self-insurance mechanism. Funding of the trust is usually accomplished by means of a surcharge on every ton of waste that a facility handles, earmarking these revenues for a special account that is tapped only for specified purposes. Accumulation of moneys in the fund amortize the anticipated costs over the life of the facility, thus creating a reserve for possible liability.

Sample trust fund documents are contained in the Appendix to this chapter.

APPENDIX 7.1   SAMPLE FRANCHISE ADMINISTRATION AND RATE AVERAGING PROCEDURE

SECTION I—SOLID WASTE FRANCHISE

Introduction

Under provisions of Pub. L. 1975 Chapter 326, the counties and the Hackensack Meadowlands Development Commission were mandated to prepare and implement 10-year solid waste management plans. An integral part of the implementation of a solid waste management plan is the establishment of techniques or procedures to direct the flow of the solid waste stream.

This exhibit is being prepared in response to Middlesex County's desire to direct the flow of solid waste to specific waste disposal facilities within the county. It is the intention of the county to

---

Appendix 7.1 appears courtesy of James C. Anderson Associates, Hainesport, New Jersey.

gain control of the flow of solid waste only, not to own or operate a disposal facility. The ownership and operation of disposal facilities is to remain within the private sector with the franchise controlling the type and quantity of waste disposed at each facility.

This exhibit is organized into four sections. In this section, Section I, various background and introductory aspects relating to the Solid Waste Disposal Franchise will be discussed. In Section II, the procedure for computing a uniform rate will be described. Section III of this exhibit describes the franchise administration procedure, and Section IV summarizes the proposed franchise procedure.

Background

In 1975, the State Legislature passed Chapter 326, formally known as the Solid Waste Management Act of 1975. This Act specifically mandated that each County Board of Chosen Freeholders was responsible for developing and implementing a Solid Waste Management Plan.

The Plan, among other items, inventoried sources of solid waste, inventoried existing solid waste facilities within the county, projected quantities of solid waste, and the estimated life expectancy of disposal facilities over the next 10 years. Also included in the Plan was a discussion on procedures for coordinating all activities related to the collection and disposal of solid waste.

Another, and most important activity considered, was the ability of the county to direct the flow of the solid waste to specific, designated solid waste facilities. Any solid waste management plan that proposes to use solid waste facilities, such as sanitary landfills, in an effective and efficient manner, and stipulates resource recovery facilities in the future, must provide provisions to control the solid waste stream.

Without the ability to direct the flow of solid waste, the annual loading rate of solid waste into a sanitary landfill would be variable making it nearly impossible or, at best, difficult to project the facility life expectancy. If sanitary landfills were to continue to be utilized with varying and uncontrolled waste deposition rates, it would be equally difficult to plan responsibly for future facilities.

Although important in planning for effective use of sanitary landfills, the ability to direct the flow of solid waste will create a positive atmosphere for resource recovery to occur, and is crucial in planning for the utilization of a resource recovery facility. The need for a guarantee of a specified amount of solid waste to a resource recovery facility cannot be overemphasized. Without a guaranteed supply of waste material, the resource recovery facility might not meet its financial obligations and, indeed, would be hard pressed even to obtain initial funding. The facility must produce projected quantities of energy or fuel to generate income and thereby meet its financial liabilities. Without guaranteed waste material, the necessary quantities of energy or fuel products cannot be assured.

Franchise Petition

Having recognized the vital importance of directing the flow of solid waste, the Solid Waste Management Plan recommends that a franchise be sought from the N.J. Board of Public Utilities.

The N.J. Board of Public Utilities is a commission charged with the regulation of certain businesses that are considered to be "affected with a public interest," and customarily are classified as public utilities. In the State of New Jersey, the collection and disposal of solid waste is considered to be a public utility. Pub. L. 1975, Chapter 326 states:

> Any solid waste facility constructed, acquired or operated pursuant to the provisions of this amendatory and supplementory act shall be deemed a public utility and shall be subject to such rules and regulations as may be adopted by the Board of Public Utility Commissioners in accordance with the provisions of the Solid Waste Utility Control Act of 1970.

This method of control was recommended over such other methods as long-term contracts with individual municipalities or a user charge system. The Plan concludes that the franchise method of gaining control of the waste stream is the best in terms of level of control and overall management of the solid waste stream.

The Plan recommends that the County Freeholders should petition the Board of Public Utilities requesting that they be granted a franchise to direct the disposal of solid waste within the District and cites the Board of Public Utilities' authority to grant a franchise through Section III of the Solid Waste Management Act:

> The Board of Public Utility Commissioners shall, after hearing, by order in writing, when it finds that the public interest requires, designate any municipality as a franchise area to be served by one or more persons engaged in solid waste collection and any solid waste management district as a franchise area to be served by one or more persons engaged in

solid waste disposal at rates and charges published in tariffs or contracts accepted for filing by the board; provided, however, that the proposed franchise area for solid waste collection or for solid waste disposal conforms to the solid waste management plan of the solid waste management district in which such franchise area is to be located, as such plan shall have been approved by the Department of Environmental Protection . . . .

Further support for the directing of the waste stream is found in the opinion of the New Jersey Attorney General furnished to Ms. Jerry English, Commissioner of the N.J. Department of Environmental Protection (DEP). The opinion was made following the request of the Solid Waste Administration ". . . to determine whether solid waste management districts, acting pursuant to solid waste management planning, have the authority to require that solid waste generated within the district be directed to specific disposal facilities." The reply, in part, by the Attorney General to this question was that, ". . . The Board of Public Utilities Commissioners may designate a solid waste management district as a franchise area . . ." and that ". . . in this manner the B.P.U. may exercise control over the destination of the waste stream."

The opinion concludes that, "It is therefore our opinion that the Solid Waste Management Act and the Solid Waste Utility Control Act establish the authority of the solid waste districts through their comprehensive planning to direct the flow of wastes to selected destinations. The exercise of administrative authority by the DEP can effectuate compliance with the district plans, and the B.P.U. can either directly franchise an area, or otherwise influence the marketplace through rate-setting in such a manner as to affect the flow of waste materials throughout the districts. [N.J.S.A. 13:1E-1 *et seq.*, N.J.S.A. 48:13A-1 *et seq.*, N.J.S.A. 48.2-1 *et seq.*] Therefore, through the combined abilities of the districts, the DEP and the B.P.U., solid waste generated within a district may be directed to specified waste disposal facilities."

Rate-Setting

The Board of Public Utilities has the requisite powers to control the destination of the solid waste stream and to set rates for solid waste disposal at the various facilities.

When a franchise is implemented, the rate-setting procedure will need to be carried one step further. It is anticipated that a uniform rate will be set and charged to the collector/hauler using disposal facilities in the county.

A uniform rate is considered necessary for several reasons:

1. To allow facility owners to upgrade their landfills to meet more stringent federal and state regulations.
2. To allow new sanitary landfills to meet RCRA Standards.
3. To guarantee a sufficient amount of waste to any proposed resource recovery facility.

If a uniform rate is not set, the collector/haulers that are directed to higher-priced disposal sites would be unfairly penalized. It would not be acceptable to direct certain collector/haulers to higher-priced facilities because the areas in which they collect are in close proximity to the more expensive disposal facility. Also, there would be no incentive for disposal facility operators to upgrade their facilities. Facility upgrading often requires a large amount of capital. This cost will then be passed on to the collector/haulers that use the facility. If a uniform rate is set, the cost of upgrading facilities, to meet more stringent environmental standards, will be evenly distributed among the users of all the facilities in the county.

Having firmly established the need for and the procedures available for obtaining a solid waste franchise, the method of administering and operating the franchise will be discussed.

SECTION II—RATE COMPUTATION PROCEDURE

Need for a Uniform Rate

A uniform rate for the disposal of solid waste at all facilities in Middlesex County is considered to be desirable for several reasons.

The primary reason for the implementation of a franchise is to control effectively the disposal of solid wastes in Middlesex County. To accomplish this goal, collector/haulers will be directed to certain disposal facilities. If there was no uniform rate, collector/haulers that are directed to higher-priced disposal facilities would be unfairly penalized. With a uniform rate, however, all collector/haulers will pay the same rate regardless of where they dispose of their wastes.

A uniform rate will also insure the efficient use of all disposal facilities. A uniform rate will limit overuse of sanitary landfills while guaranteeing a sufficient amount of waste to any proposed resource recovery facility. This insures that a high-technology resource recovery facility will not be bypassed for a low-technology landfill because of a higher tariff.

Finally, uniform rate will create an improved atmosphere for upgrading landfills to meet stringent environmental regulations. Landfill operators could, at present, be hesitant to upgrade their landfills due to the cost. The operators feel that this cost will have to be passed on to their customers and will, in turn, drive some customers away. With a uniform rate, however, the cost of environmental improvements and subsequent rate increase, if approved by the Board of Public Utilities would be recalculated into the uniform rate formula.

The Formula

A formula to establish a uniform disposal rate for solid waste must account for several varying factors. The intended uniform rate charged for disposal of solid waste at Middlesex County franchise disposal facilities will be based on a formula comprised of the following: the various Board of Public Utilities-approved tariffs at each disposal facility, daily quantity of each type of waste disposal at each facility, and administrative costs.

The mathematical formula for this rate is as follows:

$$(T_1^1 \times Q_1^1) + \frac{(T_1^2 \times Q_1^2) + \ldots (T_n^a \times Q_n^a)}{Q_1^1 \times Q_1^2 + \ldots Q_n^a} + \frac{C}{Q_t} = \text{Uniform rate}$$

where

T_n^a	= Current tariff for waste type a at facility n.
Q_n^a	= Semiannual quantity of waste type a disposed of at facility n.
Q_t	= Total quantity disposed of at all facilities.
C	= Administrative costs.
Uniform rate	= Expressed in dollars per unit of quantity.

Table 7.1 is a sample calculation based on current tariffs and estimated disposal volumes.

At many facilities different tariffs are established for different types or quantities of solid waste. For this reason, the formula must provide for the inclusion of several fees per facility. It is also necessary then to include in the formula the quantity of waste associated with each specific tariff. The above formula takes both of these items into account.

The costs associated with implementing and operating the franchise will also be included in the uniform rate. This cost will be termed the administrative cost and will include such items as: salaries for necessary personnel, printing fees, and the cost to the party administering the franchise. Another important factor to be considered is the accumulation of a positive or negative balance of cash flow. It is expected that a uniform rate will create cash excesses or deficits, as it is likely that quantities of waste will deviate from period to period. These inbalances will be negated with the reevaluation of the uniform rate by including it in the administrative cost component.

The last variable that requires consideration is how to express the quantity of waste. The typical units used throughout the solid waste industry are tons and cubic yards. An attempt, however, to incorporate both of these units into the formula is not feasible. Therefore, a decision must be made to use only one unit of quantity. A solution to this is to use volume (cubic yards) until one of the facilities installs scales. At this time, all the facilities should be required to install scales. The average rate would then be determined on a weight basis.

Reevaluation of the Uniform Rate

The uniform disposal rate will be reevaluated semiannually. Variables such as revisions to tariffs at any facility, administrative costs, and changes in the quantity of waste will be considered.

Consideration was given to both increasing and decreasing the frequency for which the uniform rate is recalculated. If the frequency of recalculation is increased, the data period over which actual quantities of disposed waste are recorded decreases. That is, if the uniform rate was recalculated monthly, the quantity of disposed waste would also be based on 1 month's records. Because of likely variations in any given month, a longer data period was determined to be more appropriate. For this reason, a semiannual reelevation of the uniform rate is more acceptable and less affected by a monthly variation.

There are, however, instances that would necessitate reevaluating the rate at intervals between the usual semiannual period. Occurances such as Board of Public Utility-approved tariff increases and the opening or closing of disposal facilities would necessitate reevaluation of the rate.

The uniform rate will be based on quantity data accumulated during the corresponding six (6) months of the previous year and the tariffs in effect on the first day of the current rate period. That is, the quantities used in the first 6 month's calculation will be based on the actual quantities recorded during the previous year's first 6 months. The tariffs used in the calculation will be those approved by the Board of Public Utilities and in effect on the first day of the rate period.

With any system, deficits or surpluses can be anticipated. These deficits or surpluses will be included as a component of the administrative cost factor contained within the formula. A uniform rate will evenly apportion the cost of solid waste disposal.

SECTION III—FRANCHISE ADMINISTRATION PROCEDURE

Introduction

After determining the formula to be used for the uniform disposal rate, a system must be designed to administer the franchise. The procedures that will have to be instituted include the following:

Administering body.
Collector/hauler registration.
Issuance of tickets to collector/haulers.
Disposal site transaction.
Purchase and return procedure for collector/haulers.
Redemption procedures for facility owners.

Administering Body

Consideration must be given to the type of organization to administer and oversee the franchise procedures. There are several possible alternatives: a department within the county government, an autonomous authority, an uninvolved third party (a fiduciary trust), or a combination of the three.

The administering body can expect to handle large amounts of information, as well as large amounts of cash. The selected administrator will be required to conduct the registration, direct waste to the proper disposal site, dispense tickets, arbitrate disputes, collect tickets, and reimburse disposal site operators.

The best method of administering the franchise will be determined before implementing the franchise.

Collector/Hauler Registration

Each collector/hauler utilizing disposal sites in Middlesex County will be required to register with the solid waste franchise before being admitted to any disposal site in the county. The information that will be required is:

Corporate name of the collector/hauler.
Number of trucks operated by the collector/hauler.
Capacity of each truck operated.
The municipality where each truck generally collects.
The daily quantity generally collected in each municipality.
The type of waste collected.
A means of identifying each truck (i.e., license number, truck number, etc.).

This information will be used for several purposes. First, it will be used to direct the waste to the proper disposal facility. The assigned disposal site will be based on where the waste is collected, and on the type of waste (i.e., residential, commercial, industrial, etc.). Second, the information will be utilized to identify trends in the solid waste stream in Middlesex County. Finally, the data will be used in planning for the future disposal of solid waste in Middlesex County.

Any changes in the collector/hauler's operation must be reported to the franchise. This is necessary to assure that waste loadings at disposal facilities remain equitable. Also, to insure that collector/haulers will not be denied admittance to a disposal facility if their operations change.

Issuance of Tickets to Collector/Haulers

Tickets will be used in lieu of cash to provide a means of record keeping and control over the waste stream. Collector/haulers will be able to obtain tickets for a particular month. The initial quantity of tickets printed will be determined from the information that the collector/haulers provide. Possession of these tickets will allow a collector/hauler to utilize a facility within the county. The tickets will contain the following information:

Color code for each disposal facility.

Collector/hauler name and/or numerical code.

Capacity of truck.

Type of waste.

Expiration date of ticket.

Serial code for tickets.

The tickets will be color-coded for each disposal facility and will be accepted only at that disposal facility. Disposal facility owners will not be able to redeem any tickets other than the tickets with their proper color codes.

The quantity of waste will be indicated on the tickets. The graduations will be determined by the information supplied by the collector/haulers to the franchise, typically they will be issued according to standard collection and transfer vehicle sizes (i.e., 20, 25, 31, 65 yd^3, etc.). The franchise will also issue tickets for odd-size vehicles and occasional users, such as loads from small construction projects.

The tickets will be issued based on calendar months. They will be accepted only during the month indicated. This system of monthly expiration, as opposed to longer durations, will provide the franchise a means of denying the use of disposal facilities to any collector/hauler that does not meet his financial obligation.

Tickets may be purchased by mail order or in person at the offices designated by the franchise. The purchase of tickets by mail will be satisfactory if the collector/haulers register and maintain a satisfactory financial balance. The purchase of tickets in person would necessitate an identification system to prevent their fraudulent purchase. This could be accomplished by issuing each collector/hauler an identification card to be used when tickets are purchased in person. This identification card would be presented each time the collector/hauler purchases tickets. It then becomes the responsibility of the collector/hauler to notify the franchise if the identification card is lost or stolen.

A numerical code will be used to prevent counterfeiting and to provide a means of control over the amount of tickets issued.

Disposal Site Transaction

One of the characteristics of the franchise system is to eliminate the transfer of cash at the disposal site. With this system, there will not be a need for any financial transactions at the disposal site.

Tickets corresponding to the truck's ownership and capacity will be presented upon entering the facility. The gate attendant will receive the ticket or tickets and validate the following information:

Correct color code.

Proper collector/hauler.

Verify the capacity.

Type of waste.

After the gate attendant is satisfied that the information on the ticket is correct, he will accept the ticket and cancel and retain the ticket.

Purchase and Return Procedure for Collector/Haulers

To initiate the program, the collector/haulers will be issued tickets for the first and second months' operation. However, before being issued tickets for the third month of operation, the collector/hauler must settle his account for the first month's billing. Accounts are settled by paying for used tickets or by returning unused ones.

Tickets may be purchased by the collector/hauler the month before, or during the month the tickets will be used. This assumes the collector/hauler has settled his account and returned previously unused tickets. Tickets purchased during the month they are good for will not have the expiration date extended. For example, tickets for March may be purchased during February or March, providing outstanding bills for January are paid. However, tickets for March purchased during March will not have their expiration date extended into April. This procedure then would continue throughout the life of the program. That is, to obtain tickets for any month, open accounts for the period 2 months prior must be settled.

The procedure for distributing the tickets will be developed during the final stage of the franchise implementation. Distribution, as discussed before, will take place at offices designated by the franchise, by mail or both.

Redemption Procedures for Facility Owners

Disposal facility owners will collect tickets from the collector/haulers that are directed to their facility. These accumulated tickets can be redeemed through the offices of the franchise for payment. The procedure for redemption will be as follows.

The facility owners will collect the tickets from the collector/haulers as they enter the facilities. At regular intervals the owners will submit an invoice to the franchise. This invoice may be submitted in person or by mail. Included with the invoice will be all of the tickets collected during the month that is being invoiced. After the invoice is received, the franchise will pay the owners within a specific time and in accordance with the approved rate as set by the N.J. Board of Public Utilities.

The facilities will be reimbursed only for tickets that are returned and tickets that are color-coded for their facility.

SECTION IV—SUMMARY

The granting of a franchise is seen to be important for these reasons:

1. To direct the flow of solid wastes generated within the county and imported from other countries and states to specified disposal facilities. This is necessary to make the most efficient use of present and future disposal facilities.
2. Assure that any proposed resource recovery facility will receive sufficient quantities of solid waste for the economical operation of the facility.
3. Encourage landfill operators to make environmental improvements in their landfills so as to provide environmentally sound solid waste disposal.
4. Encourage resource recovery through generation of energy or recycling recovered materials.

The Solid Waste Management Act establishes the procedures for meeting the above goals.

The Act states that solid waste disposal shall be considered a public utility, and gives the N.J. Board of Public Utilities the ability to regulate this utility. The Board of Public Utilities is also able to award franchises for the disposal of solid waste. The franchise method is the vehicle that will be used by Middlesex County to accomplish the previously mentioned goals.

Another aid in accomplishing the goals will be the establishment of a uniform rate. The formula for this uniform rate has been determined and a sample calculation is presented in Table 7.1. This uniform rate takes into account each type, quantity of each type, and tariff for each type of waste disposed of at each disposal facility.

The franchise administration procedure has also been described. Procedures that will need to be instituted are:

Administering body.
Collector/hauler registration.
Issuance of tickets to collector/haulers.
Disposal site transactions.
Purchase and return procedures for collector/haulers.
Redemption procedures for facility owners.

The system instinctively provides many checks and balances to prevent abuses of the system. Figure 7.1 presents a general flow diagram of the administrative procedure. Tickets will be issued based on the needs of the collector/haulers. The franchise, however, will know exactly how many tickets have been issued and, therefore, how many can be redeemed.

The development of this franchise system has taken many factors into account. As with a mechanical system, a start-up period will be necessary. During this period, fine tuning will take place and efficient operation and administration will result.

APPENDIX 7.2 SAMPLE INTERMUNICIPAL AGREEMENT RE: SOLID WASTE

This AGREEMENT entered into this ____ day of _____, 19__, by and between:

THE COUNTY OF _____, a municipal corporation of the State of _____, having an office and place of business in the City of _____, _____;

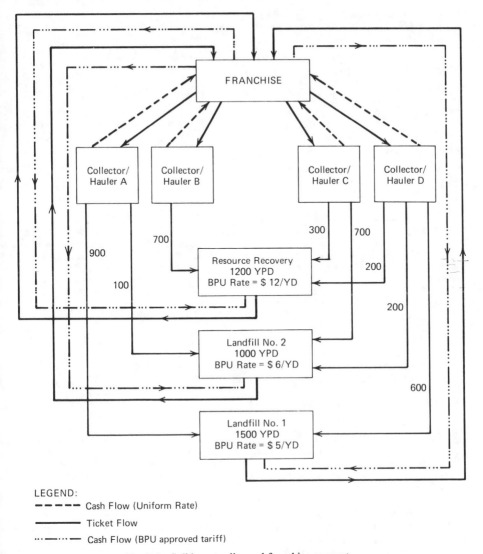

Fig. 7.1 Solid waste disposal franchise concept.

hereinafter "County", and

THE _____,
a municipal Corporation of the State of _____, having an office and place
of business _____

hereinafter designated "Participant":

WITNESSETH:

WHEREAS, the County, in 19__; pursuant to Resolution No. _____, undertook to investigate the problem of Solid Waste disposal in the County and to formulate environmentally sound, and economically viable solutions; and

WHEREAS, in 19__, pursuant to Resolution No. _____, and as a result of the aforesaid investigation the County Board of _____ (hereinafter "County Board") adopted a Plan for Solid Waste Disposal in the County pursuant to which the County undertook to assist municipalities with the disposal of Municipally Collected Solid Waste, and placed an emphasis on resource recovery; and

Table 7.1   Sample Rate Calculation

Type	Quantity $(yd^3)^a$	Tariff	$(T_n{}^a \times Q_n{}^a)$
Landfill No. 1			
Municipal	543,000	.75/yd^3	$ 407,250
Commercial/Industrial	345,000	1.00/yd^3	$ 345,000
Landfill No. 2			
Municipal	811,000	1.40/yd^3	$1,135,400
Commercial/Industrial	660,000	1.30/yd^3	$ 858,000
Landfill No. 3			
Municipal	709,000	.75/yd^3	$ 531,750
Commercial/Industrial	412,000	.75/yd^3	$ 309,000
Q_t =	3,480,000	Total =	$3,586,400

This sample rate calculation assumes that three landfills are accepting wastes. Table 7.1 identifies the landfills, type of waste, amount of water, and the product of the tariff and quantity. Another component of the uniform rate is the cost of administering the franchise. The estimated cost of administration is $75,000 per 6 months. This cost when divided by the total 6-month quantity of waste can now be included in the average rate. The final calculation is:

$$\left(\frac{\$3,586,400}{3,480,000 \text{ yd}^3} \right) + \left(\frac{\$75,000}{3,480,000 \text{ yd}^3} \right) =$$

$$1.03 \quad + \quad 0.022 \quad = 1.05/\text{yd}^3 \text{ uniform rate}$$

This is the uniform rate that will be charged, per cubic yard, regardless of waste type or quantity, to dispose of solid wastes at franchise disposal area.

^{a}The quantity reported is from data accumulated during the corresponding 6 months of the previous year.

WHEREAS, in 19__, pursuant to Resolution No. _____, the County Board designated two (2) Resource Recovery sites and authorized up to seven (7) County-owned and -operated transfer stations, at sites to be designated by the County Board, and reaffirmed the designation of eight (8) wastesheds approved by its Resolution No. _____; and

WHEREAS, in 19__, pursuant to Resolution No. _____, the County Board designated certain property on _____ in the City of _____ as a resource recovery site in lieu of the previous designation of a site in the Town of _____; and

WHEREAS, pursuant to the aforesaid Plan for Solid Waste Disposal, the County would arrange for the construction and operation of Transfer and Resource Recovery facilities and would assume responsibility for the disposal of Municipally Collected Solid Waste from the point at which it would be delivered into the County System; and

WHEREAS, the County has entered into an agreement with the City of _____ dated _____, 19__ relating to the construction and operation of a Resource Recovery Facility in _____; and

WHEREAS, the County is preparing to lease the aforesaid _____ site and is preparing to arrange for the financing, design, construction and operation of the aforesaid facility; and

WHEREAS, it is mutually understood that for the benefit of the Parties hereto and all municipalities in the County, in order to acquire the necessary land, to finance the construction and purchase of the necessary facilities and equipment and to eliminate wasteful duplication and uneconomical operations it is necessary that municipalities in the County enter into legally enforceable commitments to deliver substantially all Municipally Collected Solid Waste presently collected in the County into the County System and, for the availability of the County System (regardless of delivery), to pay specific minimum annual amounts to the County; and

WHEREAS, Participant seeks assurance that the County will, during the term of this Agreement, accept all of Participant's Municipally Collected Solid Waste, up to a specified maximum annual tonnage, and at a guaranteed maximum price per ton; and

WHEREAS, the parties have agreed to act in good faith and to take all necessary and appropriate actions, in cooperation with one another, to effect the purposes of this Agreement and enter into this Agreement pursuant to their respective lawful authorities.

NOW, THEREFORE, in consideration of the premises and of the mutual covenants and agreements herein set forth, and of the undertakings of each party to the other, the parties do hereby promise and agree as follows:

ARTICLE I
DEFINITIONS

101. "Aggregate Minimum Commitment" shall mean the sum of the then current Minimum Tonnage Commitments of All Participants for the System as a whole.

102. "Aggregate Maximum Allowance" shall mean the sum of the then current Maximum Tonnage Allowances of All Participants for the System as a whole.

103. "County Solid Waste Plan" shall mean the County-wide solid waste management and disposal program adopted by the County Board in resolutions _____, _____ and _____, and as such may hereafter be amended or modified in compliance with law.

104. "Fiscal Year" shall mean the fiscal year of the County.

105. "Minimum Commitment" shall mean the tonnage of Municipally Collected Solid Waste required to be delivered by Participant, pursuant to Section 205 below for the period during which such tonnage shall apply.

106. "Original Participants" shall mean each city, town or village within the County, duly executing an agreement in the form hereof prior to _____, 19__ and listed in Schedule "A" annexed hereto. In the case of each town, other than a town coterminous with a village, Original Participant will refer to the unincorporated portions of such town and shall not include any village lying wholly or partially within such town, unless the contrary is expressly stated in such agreement and in Schedule "A" hereof.

107. "All Participants" shall mean all municipalities in the County, whether or not an Original Participant, which enter into binding, long term, agreements for the delivery of Municipally Collected Solid Waste to the System.

108. "Person" shall mean any individual, firm, corporation, partnership, trust, governmental agency, or any other entity, or any group of such persons.

109. "Solid Waste" shall mean residential, governmental, industrial, and/or commercial refuse but shall not include human wastes; rendering wastes; demolition wastes; residue from incinerators or other destructive systems for processing waste, other than now existing individual building incinerators whose residue is presently collected as part of normal municipal collections; junked automobiles or pathological, toxic, explosive, radioactive material, or other hazardous wastes, which according to existing or future Federal, State or local laws, rules or regulations, require special handling in their collection and/or disposal.

110. "System" shall mean the County's overall, solid waste management and disposal system and every aspect thereof including, but not limited to, equipment, Transfer and Resource Recovery Facilities, and residue disposal sites acquired, constructed, or operated or to be acquired, constructed, or operated by the County or any agent, designee, or contractor in connection with the County Solid Waste Plan.

111. "Transfer Facility" or "Transfer Facilities" shall mean those Transfer Station(s) and/or Transfer System(s) that shall be established as determined to be necessary by the County, and in accordance with the provisions of Article V hereof, for the economical assembly and consolidation of Solid Waste for delivery to a Resource Recovery Facility or other disposal site.

112. "Full System Operation" or "Fully Operational" shall mean that date upon which the consulting engineer designated by the County certifies that the initial Resource Recovery Facility has undergone a shakedown consisting of a trial run and/or a test run, as those terms may be defined in, and as may be required by, the County's contract with its Resource Recovery Facility operator, and that such consulting engineer determines that the Resource Recovery facility is capable of processing not less than ____ tons of Solid Waste per seven (7)-day week, and that such engineer further determines that all Transfer Facilities deemed by the County to be necessary for the operation of such Resource Recovery Facility are likewise functional and capable of processing the appropriate tonnages.

113. "Municipally Collected Solid Waste" shall mean that Solid Waste, originating within the boundaries of the Participant, and collected or caused to be collected by Participant's employees and/or by Persons under contract to Participant for the express purpose of collecting, storing, transporting, and disposing of such Solid Waste.

114. "Point of Entry Into the System" shall mean that delivery site designated by the County for delivery of Solid Waste by Participant.

115. "Resource Recovery Facility" or "Recovery Project" shall mean any plant or facility capable of recovering resources, including without limitation, energy, from the disposal of Solid Waste. Resource Recovery Facilities shall not be deemed to include Transfer Facilities or any Source Separation facilities not located on a Resource Recovery Facility site and not operated as an integral part of a Resource Recovery Facility. The initial Resource Recovery Facility, the _____ facility, shall be of a type capable of producing steam and/or electricity as its end product, whether by mass burning or by other economical and demonstrably reliable means.

116. "Source Separation" shall mean the segregation and collection, prior to the Point of

Entry into the System, for the sole purpose of recycling, of individual components of Solid Waste, such as bottles, cans, newspapers, and corrugated containers. Source Separation shall not include the shipment of heterogeneous loads of Solid Waste out of the municipal boundaries of Participant or the shipment of any materials to any Resource Recovery facility not a part of the County System.

117. "Recycling" shall mean the use of a waste product as a raw material in the manufacture of the same or a similar product.

118. "Fees for Participants" shall have the meaning set forth in Sections 301, 302, 303, 304, and 305 below.

ARTICLE II
COUNTY UNDERTAKINGS AND DELIVERIES TO THE SYSTEM

201. *Construction of Facilities.* Subject to Section 802 below County will promptly arrange for the construction and operation of the _____ Resource Recovery Facility and for the establishment of suitable Transfer Facilities and shall use its best efforts to have the System Fully Operational by _____, 19__ and to adhere to the Schedule contained in Exhibit "B" annexed hereto.

202. *Commitment to Accept Solid Waste and Limitation on Fees.* Commencing on the date the system becomes Fully Operational and for the term of this Agreement, the County agrees to accept, at the Point of Entry into the System, and to dispose of the Municipally Collected Solid Waste of Participant, up to the maximum annual tonnage set forth in Section 205 below, for the fees set forth in Sections 301 *et seq.* below and further agrees that it will, during the term of this Agreement, never charge Participant more than the Maximum Fee set forth in Section 303 below.

203. *Notice of Start-Up.* The County will give the Participant sixty (60) days written notice prior to the date that it requires partial deliveries for testing and trial run purposes of any Resource Recovery or Transfer Facility. Participant will, upon such notice, deliver such quantities of Solid Waste as may then be required for such purposes. The County will also give Participant sixty (60) Days written notice prior to the date that the System becomes Fully Operational.

204. *Commitment to Deliver.* Participant agrees that, commencing on the date that the System is Fully Operational, and for the term of this Agreement, it will deliver, or cause to be delivered, all of its Municipally Collected Solid Waste into the System, up to the Maximum Allowance set forth in Section 205 below, but, in any event it will, for the term of this Agreement, deliver not less than its Minimum Commitment, set forth in Section 205 below.

205. *Maximum Allowance and Minimum Commitment.* For the period commencing on the date the System is Fully Operational and up to December 31, of the first complete calendar year of full operations, Municipal Participant may deliver, or cause to be delivered, into the System annually, not more than one hundred ten (110%) percent of the initial average tonnage for Participant set forth in Schedule "A" (Maximum Allowance) but shall, in any event, deliver, or cause to be delivered, into the System annually, not less than ninety (90%) percent of the initial average tonnage for Participant set forth in Schedule "A" (Minimum Commitment). Such initial Maximum Allowance and Minimum Commitment shall, unless thereafter adjusted in accordance with the provisions set forth below, constitute annual limits and requirements for the term of this Agreement. Subject to the provisions of Section 207 below, the Maximum allowance and the Minimum Commitment may, following the first complete year of full operations, be adjusted annually as follows:

(a) In the event that there is then sufficient uncommitted capacity in the System for such additional Municipally Collected Solid Waste, the Maximum Allowance may be adjusted by the County to an amount equal to 110% of the prior year's actual deliveries. In the event that in any year or years there is not sufficient uncommitted capacity in the System to permit the foregoing adjustment to be fully made with respect to All Participants then such adjustment of the Maximum Allowance shall be reduced by an amount proportionate to the ratio Participant's increase in actual tonnage in the prior year bears to the collective increase in the tonnage of All Participants, to the extent necessary to limit such adjustments to the available capacity.

(b) The Minimum Commitment may be adjusted by the County to an amount equal to 90% of the actual deliveries in the previous year provided, however, that the net effect of all such adjustments of the Minimum Commitments of All Participants shall not reduce the Aggregate Minimum Commitment below ____ tons per year. In the event that in any year or years the foregoing limitation precludes the full application of the adjustment then Participant's adjustment will be reduced by an amount proportionate to the ratio such Participant's decrease in deliveries in the prior year bears to the collective decreases of All Participants, to the extent necessary to achieve an Aggregate Minimum Commitment of ____ tons.

206. *Source Separation.* Participant shall have the unfettered right, prior to the Point of Entry into the System, to engage in Source Separation for its own account.

207. *Capacity and Additional Capacity.* The County will reserve ____ tons per annum of capacity in the initial Resource Recovery plant for municipalities since it is estimated that such reservation will be sufficient to accommodate all Municipally Collected Solid Waste in the County,

as such wastes are presently collected, for the term of this Agreement. The difference between the reserved ____ tons per annum and the then current Aggregate Maximum Allowances of All Participants may be committed in other short term agreements by the County or its agent. In addition, the County, or its agent, may, without restriction, commit all capacity available in such facility over ____ tons pursuant to other agreements.

When and if, in the judgment of the County, the aforesaid reserved capacity of the initial Resource Recovery Facility will be insufficient to meet anticipated requirements of All Participants, the County and All Participants shall, in good faith, work cooperatively with respect to such alterations of then existing facilities or construction of new facilities as may then be required and as may be determined to be economically viable and the County and All Participants shall, in good faith, enter into such additional agreements and undertakings as may be necessary and appropriate in relation to the handling of such additional Municipally Collected Solid Waste.

208. *Plan of Operations.* The County will, at least nine (9) months prior to the date the System becomes Fully Operational and after consultation with All Participants, promulgate and disseminate a Plan of Operations for the System. Such Plan of Operations shall, *inter alia,* deal with such matters relating to the operation, management and administration of the System as hours of operation, which hours of operation shall account for the regular collection schedule of All Participants, schedule and routing of deliveries, provisions for the deliveries of nonheterogeneous waste, issuance of authorizations to and regulation of delivery vehicles, measurements for quantity, quality and other characteristics, billing, rules, and regulations relating to the use of the System and such other items as may be appropriate.

ARTICLE III
CHARGES AND PAYMENTS

301. *Payment of Fees by Participants.* Commencing on the date the System is Fully Operational and for the term of this Agreement, Participant shall pay, on a monthly basis, the then current per ton Fee for Participants for all Solid Waste delivered or caused to be delivered into the System and shall, in any event, pay its minimum fee in accordance with Section 305 below. All charges and fees owed by the Participant pursuant to this Agreement shall proportionately constitute assessments against the real property within the boundaries of Participant, which may be extinguished monthly by their payment by Participant. The Fee for Participants shall be determined annually by the County in accordance with the criteria set forth in Sections 302, 303, and 304 below and in accordance with the provisions of Schedule "C" annexed hereto.

302. *Basis of Fees for Participants.* Subject to the provisions of Section 303 below, and subject to any other obligation to pay any interim fee for the usage of a portion of the System prior to the date the System is Fully Operational, the Fees for Participants shall be determined by the County, operating the System as an enterprise fund, such that the revenues to the County from the operation of the System, including without limitation, revenue from the sale of products and Fees from Participants shall be sufficient to reimburse the County for all of its expenses, costs, and obligations relating to the System including but not limited to design, construction, operation, maintenance, repair, replacement, alterations, enlargement, modifications, additions, capital, interest, maintenance of reasonable reserves, land and land use acquisition, insurance, administration, overhead, and alternate and/or emergency disposal and also including the costs of closing down, final grading, and capping the County's present landfill at _____. The Fees for Participants, from the date the System is Fully Operational and through December 31, 19__, shall not be greater than be $__.00 per ton. Thereafter the Fees for Participants shall be determined annually by the County on the basis of the criteria set forth above, subject to Section 303 below, and pursuant to the provisions of Schedule "C" annexed hereto.

303. *Maximum Fees.* From the date the System is Fully Operational Participant shall, during the term of this Agreement, never be required to pay more per ton than the annual Maximum Fee which shall be determined as follows:

(a) Through December 31, 19__ the Maximum Fee shall be $__.00 per ton.

(b) Thereafter the Maximum Fee for each succeeding year shall be calculated by adding to or subtracting from the then current year's Maximum Fee the amount obtained by multiplying such Maximum Fee by the average change, expressed as a percentage, of the following two indices as determined in the following manner:

   (1) From the Producer [Wholesale] Price Index for Durable Goods for the region including _____County, as determined and recorded by the U.S. Dept. of Labor, Bureau of Labor Statistics, as available on September 30, of the year in which the calculation is being made for the next fiscal year, subtract the amount of such Index as at the September 30, next preceding the date when such adjustment is being determined, the result being the change of such index expressed as a percentage;

   (2) The change, expressed as a percentage, for said period in the Consumer Price Index for the region including _____ County, as determined and recorded by said Bureau of Labor Statistics, shall be similarly calculated.

The change of each of the foregoing indices so determined shall be added together and then the sum of those numbers shall be divided by two to determine the average change, expressed as a percentage, of the two indices.

304. *District Charge.* In the event that in any year during the term of this Agreement a District Charge is required, in accordance with Section 302 and schedule "C," because of the limitations of Section 303 the County will levy upon a county-wide Solid Waste Disposal District a District Charge in an amount sufficient to make up the amount of such calculated limitation.

305. *Minimum Annual Fee.* Notwithstanding anything herein to the contrary and regardless of whether Participant delivers its required Minimum Commitment, commencing on the date the System becomes Fully Operational and for each year during the term of this Agreement, Participant shall pay the County at least The Annual Minimum Fee, payable in twelve equal monthly installments, during each year, which Annual Minimum Fee shall be determined by multiplying the then current Fee for Participants by Participant's then current Minimum Commitment, as determined pursuant to Section 205 above.

306. *Distribution of Surplus, If Any.* In the event that in any year during the term of this Agreement the revenues to the County on account of the System, not including revenues derived from District Charges, exceed all of the County's costs related to the System (which costs are generally described, without limitation, in Section 302 above), such difference shall be deemed a surplus in the year in which it was created. Following the determination by the County of the existence of a surplus such amount shall be distributed by the County to All Participants on a pro-rata basis determined by their actual tonnage deliveries in the year in which the surplus was created. Should such a surplus exist in the last year of this Agreement, it shall be first applied to closing down or disposing of the System, facilities, equipment, land and the like and the remainder, if any, shall then be distributed to Participants as set forth above.

307. *Uniformity.* The Fees for Participants, determined pursuant to Sections 301 and 302, shall be the same for All Participants provided, however, the County or its agent may from time to time and for limited periods accept into the System, Solid Waste from other sources so as to maximize the efficiency of the System and the revenues to be derived therefrom and may charge such rates therefore as the County determines will best achieve such objective.

ARTICLE IV
ACCOUNTING, WEIGH SCALES, RECORDS & LOCAL OPERATIONS

401. *Accounting and Audit.* The County will account for its administration of the System separate and apart from its general operations as an enterprise fund. The County will keep proper books of record and account, in accordance with generally accepted accounting principles, consistently applied. Such books and records, together with all documents and materials relating to the System (other than such as may be subject to legal privilege) shall, at all reasonable times, be subject to inspection by Participant. The County shall cause such enterprise fund to be audited annually by independent, certified public accountant(s) and shall provide a copy of such audit to Participant promptly upon receipt thereof.

402. *Weigh Scales and Records.* The County will arrange for and/or provide, weigh scales and/or other devices appropriate for determining the quantity, quality and other characteristics of all Solid Waste. The County will make and keep appropriate records of such measurements, which records shall be available to Participant at all reasonable times.

403. *Local Operations.* Upon the signing of this Agreement Participant shall not construct, enlarge, operate, or contract for any facility for the treatment and/or disposal of Solid Waste except as the County may expressly agree to in writing as being required on a temporary basis prior to the date that the System is Fully Operational. In addition, Participant shall take all such action as may be necessary to insure that all of its Municipally Collected Solid Waste will be delivered to the System once it is Fully Operational.

404. *Responsibility for Collection Costs.* Participant shall be solely responsible for and shall bear the total cost, expense and other obligation connected with the collection and transportation of Solid Waste prior to its Point of Entry Into the System.

ARTICLE V
TRANSFER FACILITIES

501. *Transfer Facilities.* The County and Participant shall cooperate in good faith with respect to the establishment of Transfer Facilities. The County agrees to establish, or arrange for the establishment of, Transfer Facilities such that the centroid of Participant's Municipal Solid Waste collection shall be within approximately _____ (__) minutes driving time from a Transfer Facility.

ARTICLE VI
FURTHER ASSURANCES

601. *Additional Actions.* The Parties shall, in good faith, during the term of this Agreement take all such actions as may be necessary or appropriate to carry out the purposes of this Agreement including, without limitation, the enactment of legislation, resolutions, and the like. In addition to the foregoing and without limitation thereof to the extent that any Fees to be paid by Participant shall or may be pledged in connection with the financing of any portion of the System,

the County and Participant shall each use its best, good faith efforts to defend, preserve and protect such pledge of such Fees.

602. *Assistance with Permits and Approvals.* The County and Participant shall use their mutual best, good faith efforts to obtain such agreements, approvals, licenses, permits, legislation, authorizations, and the like as may be necessary or appropriate in connection with the design, financing, construction, and operation of the System or as may be necessary or appropriate to carry out the purposes of this Agreement.

603. *Commitment to Pay Unconditional.* Participant agrees that so long as the County shall accept or shall be willing to accept Participant's Municipally Collected Solid Waste, up to the Maximum Allowance set forth in Section 205 above, and whether or not the System is then functioning as per design or functioning normally, Participant's obligation to make all aforesaid payments in the amounts required and at the time specified whether to the County or the County's designee, agent or contractor, shall be absolute and unconditional, and shall not be subject to any set-off, counterclaim, recoupment, defense (other than payment itself), or other right which Participant may have against the County, its designees, agents, or contractors or any other person for any reason whatsoever.

ARTICLE VII
MISCELLANEOUS

701. *Effect of Breach.* Each party specifically recognizes that the other is entitled to bring suit for injunctive relief, mandamus, or specific performance or to exercise other legal or equitable remedies to enforce the obligations and covenants of each Party hereto. In addition, it being recognized that the successful operation of the System—and therefore the ability of municipalities within the County safely, lawfully, and economically to dispose of their Municipally Collected Solid Waste—is dependent on All Participants fully living up to the terms and conditions of their agreements, it is agreed that All Participants shall be deemed third-party beneficiaries of all other Agreements entered into by All Participants.

702. *Assignability.* The County may assign or pledge this Agreement in relation to the financing of the System but no other assignment of this Agreement shall be authorized or permitted by either Party.

703. *Waiver Not to be Construed.* No waiver by the County or Participant of any term or condition of this agreement shall be deemed or construed as a waiver of any other term or condition, nor shall a waiver of any pledge be deemed to constitute a waiver of any subsequent pledge, whether of the same or of a different section, subsection, paragraph, clause, phrase, word, or other provision of this Agreement required of it under this Agreement or by law. The failure of either party to insist in any one or more instances, upon strict performance of any of the terms, covenants, agreements, or conditions in this Agreement shall not be considered to be a waiver or relinquishment of such term, covenant, agreement, or condition, but the same shall continue in full force and effect.

704. *Amendments.* This Agreement being for the benefit of All Participants, it may not be substantially amended without the concurrence of all other Participants and any such amendment shall be only by written agreement, duly authorized and executed. This writing represents the entire agreement between the parties and any modification or amendment shall be in writing and duly executed by the parties.

705. *Severability.* If any provision, paragraph, sentence, clause, or word of this Agreement shall, for any reason, be held to be invalid or unenforceable, the invalidity or unenforceability of such shall not affect the remainder of this Agreement and this Agreement shall be construed and enforced, consistent with its expressed purposes, as if such invalid and unenforceable provision, paragraph, sentence, clause, or word had not been contained herein.

706. *Duplicate Originals.* This Agreement may be executed in two or more counterparts, any of which shall be regarded for all purposes as duplicate originals.

707. *Insurance.* Schedule "D" annexed hereto and made a part hereof, contains the insurance coverage which each of the parties hereto have agreed to maintain with respect to matters relating to the performance of this Agreement.

708. *Notices.* All notices required hereunder to either party shall be in writing and sent by Registered Mail Return Receipt Requested to:

the County:

County of _____

County Office Building

[City] _____, _____ [State]

ATT: Commissioner of Public Works

with a copy to the County Attorney;

the Participant:

_____

_____

_____

ATT: _____

with a copy to _____.

ARTICLE VIII
TERM OF AGREEMENT

801. *Term.* It is the intention of the parties that this Agreement shall remain in effect and be binding until such time as the debt which shall have been created, by the issuance of bonds, in connection with the initial financing for the design and construction of the System shall have been discharged. It is presently estimated that such bonds will be issued on or before _____, 19__. It is also presently anticipated that such bonds will be issued for approximately twenty (20) year term. The parties are, however, cognizant that [reference to applicable state law] may presently preclude them from setting forth a fixed term greater than twenty (20) years. Accordingly, this agreement shall be binding as of the date of its due execution by the last of the Original Participants set forth in Schedule "A" and shall remain in effect for a period of twenty (20) years from such date. It is further agreed that should the provisions of [applicable state law] or any other provision of law be amended or enacted to authorize municipalities to enter into an agreement of the type herein for a period longer than twenty (20) years then, at such time, this Agreement shall be deemed amended such that the term of this Agreement shall be such longer period, but in no event longer than twenty-five (25) years from the effective date hereof. Notwithstanding the foregoing, and because the System to be financed on the basis of this Agreement is vital to the economic anxd environmental interest of All Participants, in the event there is no such amendment or enactment and in the event that the aforesaid financing cannot be amortized within the Term first stated above then on January 1, 19__ this Agreement shall be deemed renewed effective such date sufficient to amortize the aforesaid bonds and discharge the aforesaid debt.

802. *Termination.* Except as is expressly provided below, neither the County nor Participant shall have the right to terminate this Agreement for any reason whatsoever including breach or default in the obligations of either to the other and this agreement shall, for its term, remain in full force effect and may at all times be enforced by each upon the other at law or in equity.

(a) Notwithstanding the foregoing, and in order to provide further assurances to Participant, the Participant may, on _____ (__) days written notice to the County, mailed within _____ (__) days of the event(s) below listed, terminate this Agreement without fault or obligation if, but only if, and for no other reason whatsoever:

    (i)   The County has not, by _____, 19__, entered into Agreements in the form hereof providing for an Aggregate Minimum Commitment of _____ tons of Municipally Collected Solid Waste to be delivered to the System annually; or

    (ii)   By _____, 19__ the County has not obtained a commitment from the State of _____ entitling the County to utilize $_____.00 of State Environmental Quality Bond Act funds in connection with the financing of the System; or

    (iii) By _____, 19__ the County has failed to award a contract to a responsible bidder for construction of the _____ Resource Recovery Facility.

(b) Notwithstanding the provisions of Section 802 the County may, on _____ (__) days written notice, mailed within _____ (__) days of the event(s) below listed, terminate this Agreement without fault or obligation if, but only if, and for no other reason whatsoever:

    (i)   By _____, 19__ sufficient municipalities have not signed Agreements in the form hereof so as to provide for an Aggregate Minimum Commitment of _____ tons of Municipally Collected Solid Waste to the System annually; or

    (ii)   If any Original Participant listed in Schedule "A" hereof terminates its Agreement with the County pursuant to Section 802(a) above, and the County should then determine that such termination or terminations render it uneconomical to construct or operate the facilities and manage the System.

IN WITNESS WHEREOF, the Parties hereto have duly executed this Agreement the day and year first above mentioned.

THE COUNTY OF _____

By: _____

_____

By: _____

Authorized by the [recommending authority] of the County of _____ at a meeting duly held on the ____ day of _____, 19__.

APPROVED AS TO FORM AND
MANNER OF EXECUTION:

_____
Deputy County Attorney
The County of _____

Approved by the Board of _____ of the County of _____ at a meeting duly held on the   day of   , 19__.

APPENDIX 7.3   SAMPLE INDEMNITY BOND

SAFECO INSURANCE COMPANIES
SAFECO INSURANCE COMPANY OF AMERICA   Bond No. _____
GENERAL INSURANCE COMPANY OF AMERICA
FIRST NATIONAL INSURANCE COMPANY OF AMERICA
HOME OFFICE   SAFECO PLAZA, SEATTLE, WASHINGTON 98185

INDEMNITY BOND

KNOW ALL MEN BY THESE PRESENTS, That we, _____[Name of Landfill Operator]_____
_____, as Principal and SAFECO INSURANCE COMPANY OF AMERICA, a corporation organized under the laws of the State of Washington, and authorized to transact the business of surety in the State of _____, as Surety, are held and firmly bound unto _____[Project Sponsor; Landowner]_____ in the just and full sum of FOUR HUNDRED THOUSAND AND NO/100 DOLLARS ($400,000.00), for which sum, well and truly to be paid, we bind ourselves, our heirs, executors, administrators, successors, and assigns, jointly and severally, firmly by these presents.

WHEREAS, the said Principal and Obligee have entered into an Agreement dated _____, 19__ for Construction and Maintenance of   [Name of Landfill Project]
RECREATIONAL LAND FILL, _____City, State_____.

NOW, THEREFORE, THE CONDITION OF THIS OBLIGATION IS SUCH, That, if the said Principal shall indemnify the _____[Project Sponsor; Landowner]_____
_____ against any and all loss, cost, and expense, including labor and material which may be incurred in connection with the removal of buildings, structures, equipment and debris, the placing of the prescribed final cover, grading, seeding and landscaping necessary to render the landfill site suitable for forest preserve purposes, then this obligation to be void; otherwise, to remain in full force and effect.

PROVIDED, HOWEVER, This bond shall become effective on January 1st, 19__ and continue in force for a period of _____ (__) Years, or until the date of expiration of any Continuation Certificate executed by the Surety.

PROVIDED FURTHER, HOWEVER, That the Surety's liability under this bond shall in no event exceed an aggregate sum equal to the above written penal amount.

Sealed with our seals, and dated, this ____ day of December, 19__.

[Name of Landfill Operator]

By: _____
                [Title]

_____

COUNTERSIGNED   SAFECO INSURANCE COMPANY OF AMERICA

By: _____   By: _____
     , Resident Agent   Attorney-in-Fact

APPENDIX 7.4   SAMPLE PROVISIONS—LANDFILL ENVIRONMENTAL TRUST FUND

1. *Creation.*   There is hereby created by [Municipality] and ordered established with the Trustee a trust fund to be designated "Environmental Trust Fund–Municipal Landfill Project" (herein referred to as the "Trust Fund"). Moneys on deposit in the Trust Fund shall be used for those purposes specified in Sections 5-9.

2. *Disposal Surcharges.*   [Operator] shall collect from each user of the landfill a surcharge on each ton (or part thereof) of waste delivered to [Operator] for disposal. Such surcharges shall be the property of [Municipality], but shall be assessed and collected by [Operator]. Amounts collected by the Operator shall be paid, on behalf of Municipality, to Trustee for deposit in the Trust Fund.

   a.   The initial surcharge is established at _____ per centum (__%) of the user fee established from time to time under Section _____ of [local ordinance or operating agreement, etc.] The amount of such surcharge may be amended from time to time by the agreement of [Operator] and [Municipality], as they may find necessary to generate moneys to accomplish the purposes of this Agreement. Any such amendment shall be promptly communicated to the Trustee by [Municipality].

   b.   [Operator] shall deliver to the Trustee, for deposit in the Trust Fund, all surcharges collected from the period ending with the close of business on _____, 19__, and for each six (6)-month period thereafter. Delivery shall be made within two (2) months after the initial date and each subsequent six (6)-month period.

   c.   The _____ per centum collected hereunder shall be considered a part of the compensation rights of the [Municipality] per Section _____ of [local ordinance or operating agreement, etc.].

3. *Disbursement from the Trust Fund.*   The Trustee is authorized and directed to withdraw moneys from the Trust Fund sufficient to reimburse [Municipality] or any other person for any expenses directed, approved, or for which approval has been waived, by the Trust Administration Committee, pursuant to this Agreement. The Trustee is authorized and directed to issue its checks for each disbursement required.

4. *Trust Administration Committee.*   There is hereby created a Trust Administration Committee, comprised of three persons. One person shall be appointed by [Municipality]; one person by [Operator]; and one person by [other interested party]. The initial members of the Trust Administration Committee shall be _____ (Municipality); _____ (Operator); and, _____ (interested party). An appointing body may change its representative on the Trust Administration Committee, upon written notice to the Trustee and the other appointing bodies.

   The Trust Administration Committee shall meet within 30 days and at least once per year thereafter, at the offices of [Municipality]. Additional meetings shall be held at the written call of [Municipality] or any Committee member, at the time, date and place specified in the call, which meeting shall not be earlier than four days after the call.

   In the event of an emergency, [Municipality] or a Committee member may call a meeting of the Trust Administration Committee upon telephone notice to each member, which meeting shall be held not earlier than twenty-four (24) hours after such notice.

   Trust Administration Committee members shall serve without compensation.

   To be effective, any action or decision of the Trust Administration Committee shall be supported by a majority of all duly appointed members, unless a unanimous vote is otherwise provided.

5. *Providing Alternative Water Supplies.*   Moneys in the Trust Fund shall be used to pay the reasonable and necessary expenses of designing, developing, constructing, installing, and obtaining required permits or licenses for a system for delivery of an alternative water supply to the owner of an existing water supply which has been determined to be contaminated by leachate from the landfill. The Trust Fund shall be drawn upon to pay any costs or charges for use, operation or maintenance of such alternative systems.

   a.   For purposes of this paragraph, the following definitions shall apply:

   (1)   "Alternative water supply" means a supply of water of sufficient quantity and suitable quality to meet those uses to which the existing water supply had been put prior to its contamination.

   (2)   "Contaminated" means the presence, in water drawn from an existing water supply, of biological, chemical, physical, or radiological substances in such quantities and concentrations that use of such water for its previously used purposes would adversely affect human health or the environment.

   (3)   "Determined" means the entry of a final and unappealable order, after notice and the opportunity to be heard, by an administrative or judicial body, which finds that contamination has occurred and that such contamination has been caused, in whole or in part, by leachate from the landfill.

(4) "Existing water supply" means one of those wells or springs, shown on the map and two-page document marked Exhibit __ and attached hereto.

b. Trust Fund moneys shall be used for the above purposes at any time after the commencement of landfill operations, and shall be so available for a period of __ years following closure of the landfill by [Operator].

6. *Prevention of Groundwater Pollution.* Moneys in Trust Fund shall be used to pay the reasonable and necessary expenses of [Municipality] in preventing the escape of contaminants from the completed landfill area which may adversely affect groundwaters of the State of
_____.

a. The purposes of this Section shall be accomplished by:

(1) Monitoring leachate head levels at those places and at those frequencies specified in the "Groundwater Protection Contingency Plan," marked Exhibit __ and attached hereto.

(2) Obtaining samples of, and analyzing, leachate quality at those places and frequencies, and for those parameters, specified in Exhibit __.

(3) Establishing, operating, and maintaining a leachate recirculation system, as specified in Exhibit __.

(4) Applying for and obtaining such permits or licenses as are required for the discharge of attenuated leachate into subsoils or groundwaters of the State of _____, as specified in Exhibit __.

(5) Removal of leachate from the landfill in such volumes and at such frequencies as specified in Exhibit __.

(6) Disposal of leachate, and obtaining necessary permits therefor, according to Exhibit __.

b. Expenses for activities which will advance the general purposes of this section, but which are not included in subparagraph a, may be paid by moneys in the Trust Fund, but only upon the unanimous direction of the Trust Administration Committee to the trustee.

c. Trust Fund moneys shall be used for the above purposes commencing with the closure of the landfill by [Operator] and continuing for a period of __ years thereafter.

7. *Prevention of Surface Water Pollution.* Moneys in the Trust Fund shall be used to pay the reasonable and necessary expenses of [Municipality] in preventing the escape of contaminants from the completed landfill area which may adversely affect surface waters of the State of
_____.

a. The purposes of this Section shall be accomplished by:

(1) Maintaining a grass cover in, and on the slopes of, ditches on the landfill property which are used to drain precipitation run-off from the completed landfill area.

(2) Repair of erosion channels in, and on the slopes of, such ditches, through backfilling with low permeability soils and/or topsoils and recontouring the repaired area.

(3) Maintaining a grass cover on the exposed portions of the dam containing the sedimentation pond, and repair of erosion channels on the dam in the manner provided above.

b. The above purposes do not include activities or expenses designed to improve the appearance or aesthetics of drainageways, the sedimentation pond or the dam nor activities designed to alter the location, size, or configuration of such facilities for recreational needs.

c. Expenses for activities which will advance the general purposes of this Section, but which are neither included in subparagraph a, may be paid by moneys in the Trust Fund, but only upon the unanimous direction of the Trust Administration Committee to the Trustee.

d. Trust Fund moneys shall be used for the above purposes commencing with the closure of the landfill by [Operator] and continuing for a period of __ years thereafter.

8. *Groundwater and Surface Water Monitoring.* Moneys in the Trust Fund shall be used to pay the reasonable and necessary expenses of [Municipality] in evaluating the quality of groundwater or surface water which may be adversely affected in the event that contaminants should escape from the completed landfill area.

a. The purposes of this Section shall be accomplished by:

(1) Obtaining water samples from those wells and sampling stations, and at such frequency, as are specified in "Water Quality Monitoring Program," marked Exhibit __, and attached hereto.

(2) Analyzing the samples for those contaminant parameters, and at such frequency, as are specified in Exhibit __.

(3) Preparing reports on the results of the sampling and analysis.

b. Expenses for monitoring activities which will advance the general purposes of this Section, but which are not included in subparagraph a, may be paid by moneys in the Trust Fund, but only upon the unanimous direction of the Trust Administration Committee to the Trustee.

c. Trust Fund moneys shall be used for the above purposes commencing with the closure of the landfill by [Operator] and continuing for a period of __ years thereafter.

d. Trust Fund moneys shall be used for the above purposes commencing with closure of the landfill by [Operator] and continuing for a period of __ years thereafter.

9. *Stabilization and Maintenance of Earthen Cover.* Moneys in the Trust Fund shall be used to

pay the reasonable and necessary expenditures of [Municipality] in minimizing the infiltration of precipitation into, in promoting the run-off of precipitation from, and in minimizing the erosion of soils from, final earthen cover placed on complete landfill.

    a. The above purpose shall be accomplished by:

      (1) Maintaining a grass cover on completed landfill areas, through site preparation, re-seeding, fertilization, and top-soiling of barren or sparsely vegetated areas.

      (2) Repair of erosion channels, through backfilling with low permeability soils and/or top-soils and recontouring of repaired areas.

      (3) Repair of settlement depression areas, as necessary to promote surface run-off and prevent ponding of water, through backfilling with low permeability soils and/or topsoils and recontouring of repaired areas.

    b. The above purposes do not include activities designed to improve the appearance or aesthetics of completed areas nor to meet any planned recreational uses of the areas, such as establishment or trimming of trees, bushes or ornamental plants, the cutting of grass, or surface contouring for recreational needs.

    c. Expenditures for activities which will advance the general purposes of this Section, but which are neither included in subparagraph a nor excluded in subparagraph b, may be paid by moneys in the Trust Fund, but only upon the unanimous direction of the Trust Administration Committee to the Trustee.

    d. Trust Fund moneys shall be used for the above purposes commencing with closure of the landfill by [Municipality] and continuing for a period of __ years thereafter.

10. *Submission of Expenses by [Municipality]*. [Municipality] shall obtain the review and approval of the Trust Administration Committee before submitting any expenses to the Trustee for payment from the Trust Fund. The Committee may, by unanimous vote of all members, waive its review and approval of any particular class or category or activities or expenses. Any waiver shall be in writing, may specify a termination date, and shall be directed to the Trustee and [Municipality].

11. *Remainder to [Municipality]*. Any amounts remaining in the Trust Fund after payment in full of the fees, charges and expenses of the Trustee, and all other amounts required to be paid hereunder, shall be paid to [Municipality] upon the expiration or sooner termination of this Agreement as provided therein.

    IN WITNESS WHEREOF, the parties hereto have caused this Agreement to be signed by their duly authorized agents or representatives this __ day of _____, 19__.

                     [Landfill Operator]

                     By: _____
                           [Title]

Acknowledged:
[Trustee]   [Municipality]

By: _____   By: _____
    Officer   Officer

PART 2
IMPLEMENTATION ISSUES: SYSTEMS, HARDWARE, OPERATIONS

CHAPTER 8

COLLECTION OF RESIDENTIAL SOLID WASTE

H. LANIER HICKMAN, JR., Executive Director

Government Refuse Collection and Disposal Association, Inc. Washington, D. C.

8.1 INTRODUCTION AND POLICY OVERVIEW

The collection of residential solid wastes is provided to citizens to assure the protection of public health and environmental quality. From the time city officials of old first realized that they had to do something about the garbage piling up in the streets and gutters of their bustling burgs, it has been the public sector that has been responsible for "public cleansing."

Government-owned and -operated solid waste management systems are not the only entities that have handled municipal waste through the centuries, however. Early on, "trash men"—the first recyclers—made their way through a city's refuse to collect those articles that still had some economic value. In doing so, they accomplished two things: they made a profit, if they were successful, and they helped to keep their cities clean.

As the character and amount of solid wastes changed in our urban centers, the methods and techniques for collection also changed. Residential solid waste became bulky and more difficult to collect.

Commercial and industrial enterprises introduced wastes that were totally different from those of the homeowner. These changes brought about the evolution of compaction equipment and containerization, and the need for greater speed and capacity to collect and move waste through an urban center.

But perhaps the most important change was not in technology, but in institutional arrangements. The result is the partnership that exists today between government and industry that provides solid waste management services to citizens and businesses. Public and private sector solid wastes employees work side by side to collect the vast amounts of solid waste generated by homes, commercial enterprises, and industry. Generally speaking, the private sector has accepted the lion's share of commercial and industrial solid waste collection, whereas the public sector retains major responsibility for collecting residential solid wastes. It makes little difference which forces collect solid wastes, as long as the public interests are protected. Local government must assure the satisfactory delivery of collection services.

The commitment to assure that the public is served by an effective, efficient, and well-managed collection program has to be supported by a determination of what level and type of service is best for the community and its citizens. This can only be done with the involvement of the public. No matter how difficult or unpleasant it may be for the public official, the public must be involved in determining the level and character of the service to be provided. This requires that the responsible public officials, both elected and career, prepare the public for such involvement. The officials must do their homework so that they can provide the public with the information necessary for public input into this policy-making exercise.

We frequently hear the statement that "my collection service doesn't cost me anything!" This is, of course, absurd. It takes money to pick up solid wastes, and any community that has let its citizens think that the service is free has done those citizens a great disservice. The public has to know exactly what it costs to provide different levels and quality of services and what can and cannot be done with today's collection technology. The citizen has to understand that the more he wants the more it is going to cost. Citizens must understand that they are part of the solid waste management problem and therefore must be part of the solution. This is not an easy lesson to teach, and certainly not an easy lesson to learn, and certainly not one that most public officials—elected, appointed, or merit—really want to deal with. Nevertheless, the public must know; it must be educated; and it must be involved in the decision-making process.

The selection or determination of the level and type of service guides the delivery of services. In engineering terms, it is the technical specifications that describe the service to be provided regardless of whether the service is provided by government or private forces; that service should be provided based upon a clear description of level and type of service intended. The technical specifications depend heavily on the professional in the community responsible for solid waste management. With an adequate public policy on what the community wants, the job of the local solid waste manager becomes far easier.

Local government is responsible to see that their community is kept clean and the public health is protected. If state legislation is unclear in assigning this responsibility, then local government should work toward a clarification of that responsibility. Recent court decisions related to antitrust and local government make it essential that state legislation clearly assign the responsibility for the management of solid waste generated within the jurisdiction of a local government entity in the hands of local government. In turn, with that responsibility, local government can then assure adequate collection services and how those services will be delivered.[1]

8.2 MANAGING CHANGE IN A SOLID WASTE COLLECTION SYSTEM[2]

8.2.1 Introduction

Change is a constant factor in solid waste collection. Change is brought on by a number of pressures:

Political and public policy shifts.
Economic stresses.
The evolution of equipment technology

Regardless of what pressures bring about the demand for change, responsibility will fall to the manager of the solid waste management system to manage that change.

Naturally, the preferred reason for change is because the solid waste manager perceives the need to change technologies or change types of services to respond to changing times and need. Frequently, however, the pressures for change come from outside forces. Whatever the reason may be, a change will affect the citizens, the employees, the policy makers, and even the managers. Consequently, to effectively manage the changes in a manner to minimize disruption and negative response (change always brings resistance), the manager of change must prepare all involved parties for the changes.

8.2.2 Plan for Change

It should not be surprising that a change in service will be a politically explosive issue. Many solid waste managers are reluctant to suggest change because of the resistance by citizens, labor forces, and politicians. From the manager's perspective, these are rational reasons for resisting change. Such a change may necessitate renegotiation of a labor contract which frequently entails prolonged talks and disagreements. In addition, why go from a comfortable "known" to an uncertain "unknown", whatever the potential benefits? Finally, many managers are sensitive to public resistance to change and may be unsure how to develop an effective public relations program to sell the change.

However, it is possible to make a successful transition if a well-thought-out plan, which identifies major issues and an appropriate course of action, is developed and implemented. The plan needs to include:

A public information program to gain citizen participation and support.
Special provisions for the elderly and handicapped.
Meetings with the solid waste collectors and labor union to explain fully the proposed plan and its effect on collector employment.
Development of a plan that assures that no permanent employees will lose their jobs.
Thorough and timely briefings to elected officials to ensure that they have accurate information concerning the proposed change.

Finally, each of the components of change can only be successful if the solid waste manager prepares the necessary documentation and logistics to assure effective response to the normal resistance to change. Each of the components are complex in themselves and each must be thoroughly planned to assure success.

8.3 COST ACCOUNTING PROCEDURES FOR SOLID WASTE COLLECTION SYSTEMS[3]

8.3.1 Introduction

Management control of a solid waste system centers around its budgeting-accounting-reporting system. Generally accepted accounting principles require that, where a governmental activity derives a substantial portion of its revenues from user charges, the activity be accounted for on the same basis as the similar enterprise in private industry. In governmental accounting, the vehicle for accomplishing this objective is usually referred to as an enterprise fund. A clear and concise discussion of the enterprise accounting can be found in the publication, *Governmental Accounting, Auditing and Financial Reporting,* prepared by the National Committee on Governmental Accounting.

Even if a solid waste system is funded through general tax revenues, accounting concepts and principles should be utilized by management to the maximum extent possible.

8.3.2 Enterprise Fund Accounting

The principles that distinguish enterprise fund accounting from general fund accounting are:

Accrual basis of accounting.

Use of a self-balancing chart of accounts (i.e., assets, liabilities, equity, income, and expenses).

Classification of income and expenses as operating versus nonoperating, and classification of revenues as income versus contributions of fund capital.

Recording of fixed assets and recognition of depreciation as an operating expense.

Use of interfund billings to account for services rendered for and provided by other agencies of the same governmental unit.

For cost accounting principles to be truly a management tool, the solid waste manager must make every effort to assign all costs to various cost centers. This allows for a more realistic analysis of costs to be made. Further, by assigning costs and budgets to various line managers, the solid waste manager is able to assign responsibilities for budget control and fiscal management to those line managers. Through such responsibility, a cost accounting system can be used as a "management tool."

8.3.3 System Deficiencies

"True" or total costs of solid waste collection systems often are not properly accounted for. Simply stated, the reason for this dilemma is that the principles of enterprise fund accounting are not being adhered to in all instances. Examples of costs that should be charged to a solid waste collection operation, but often are charged to other departments, are as follows:

Capital costs (i.e., depreciation or principal portion of debt in lieu thereof) associated with the acquisition and/or construction of facilities and equipment (e.g., collection vehicles, garages, administrative services).

Interest cost of debt incurred to acquire and/or construct facilities and equipment.

Costs (i.e., labor, parts, oil, tires) of repairing and maintaining facilities and equipment.

Employee benefits, including pension contributions, for solid waste personnel.

Cost of temporary employees borrowed from other departments to fill a short-term need.

Overhead costs associated with the city or county executive and supporting staff agencies.

Costs associated with budgeting, accounting, and report activities.

Costs associated with billing and collecting user charges.

Liability and damage claims paid by a city or county which is self-insured.

Insurance premiums related to solid waste personnel, facilities, and equipment (e.g., personnel liability, fire, accident).

Obviously, the exclusion of such items from the costs of a solid waste collection department could understate significantly the result of operations. In turn, any analysis of operating costs or any comparison of such costs with other governmental entities, private enterprise, and/or alternative collection methods could lead to erroneous conclusions and improper decisions.

8.3.4 Summary

Financial management of a solid waste system involves a number of complex issues and principles. Depending upon the organization of a particular solid waste system, and the economic and political environment in which it functions, some of the principles may not be necessary or practical. However, all should be seriously considered in formulating alternative approaches to solid waste management. It should be obvious that financial management and operational management are inextricably linked. A decision in one area necessarily influences the available alternatives of the other. The solid waste manager must become familiar with and understand these financial concepts.

8.4. UNIONS AND SOLID WASTE COLLECTION[4]

8.4.1 Introduction

The growth of unionism and collective bargaining in state and local governments is a fairly recent phenomenon. Only 10 or 15 years ago, there existed hardly any state laws regulating public sector labor relations and consequently, from an historical perspective, the parties to bargaining have had only limited experience. By contrast, there has been comprehensive federal legislation protecting the collective rights of employees in the private sector since 1935 when Congress passed the Wagner Act. That Act guaranteed employees the right to organize and bargain collectively through representatives of their own choosing. The Act also defined unfair labor practices with respect to employees. However, the legislation specifically exempted from its coverage employees of the federal and state governments and political subdivisions of the states.

Although there have been several attempts to pass federal legislation governing bargaining rights of state and local employees, neither the Senate or the House of Representatives have ever reported such legislation out of committee.

One of the major obstacles is an overriding concern regarding the constitutionality of federal legislation designed to regulate labor relations of state and local governments. Unless this constitutional issue can be overcome, it is questionable whether federal legislation in this area can become a reality.

There are many reasons why collective bargaining in the public sector has lagged behind the private sector. Prior to 1945, the public employees were better off than their counterparts in private industry. Their job security was greater because, for the most part, government employees were not subject to periodic layoffs, which were a common practice in many private sector industries. In addition, widespread adoption of a civil service system based upon a merit system and reduced reliance upon a spoils system provided employees with job security. Furthermore, public employee wages were comparable with their counterparts in private industry whereas their fringe benefits were superior.

All of this began to change after World War II, when the unions in the private sector succeeded in winning round after round of wage increases and fringe benefits surpassing those received by public employees. In an effort to restore the balance, public employees, particularly at the state and municipal levels, turned to union leadership which was frequently militant.

What followed in the 1960s was a period that has been referred to as "the decade of the public employees." During this period, the unionization of employees in the private sector was stagnating, whereas the number of public employees who joined unions more than doubled.

8.4.2 Collective Bargaining in Residential Solid Waste Collection

Strikes

What are some of the basic differences between private and public sector collective bargaining? Probably the most significant distinction is the almost universal prohibition against public employee strikes. Only five states have given public employees a limited right to strike. Labor leaders argue that the absence of the strike weapon results in collective begging, rather than collective bargaining. However, that position is arguable. A bargaining atmosphere is often created in spite of the strike ban because of an implied or expressed threat of the union to strike, regardless of the law.

Supervisory Personnel

Another key difference concerns the status of supervisory personnel. In the private sector, supervisors are excluded from coverage of the Taft-Hartley Act. Consequently, employers need not recognize their organizations or bargain with them, although they may do so voluntarily. Consequently, few labor agreements are negotiated between supervisors and private employees.

In contrast, supervisory personnel in the public sector frequently bargain on a formal basis with the governments that employ them. This is partly due to the fact that the demarcation between management and nonmanagement employees in the public sector is much more obscure than in the private sector. In other words, public sector supervisors, even more than their private sector counterparts, are the proverbial "men in the middle." Often they perceive themselves as supervisors in name only and, consequently, seek to define their position more explicitly through unionism and collective bargaining.

Labor Relations Staff and Strategy

If government expects to meet the challenge of unionism, it must establish a labor relations staff function as a permanent part of the organization and develop the competence of the individuals who staff this function. All too often, the responsibility for collective bargaining is not fixed with any degree of certainty. If labor relations are not recognized as a distinct function, then the individuals who must assume this responsibility are generally expected to perform their normal duties as well; a situation that is less than satisfactory. These individuals may not be prepared by either training or background to deal with a skilled negotiator representing the employees.

The private employer generally orients a significant portion of management toward personnel and collective bargaining matters and maintains a well-ordered strategy concerning the labor relations function. Operation divisions are probably assisted by an industrial relations or personnel director who negotiates with one or more unions and provides advice on day-to-day contract matters. His future with the company depends on how effectively he manages his area of responsibility.

Establishing a separate labor relations function and holding its staff to high standards of performance is not enough. In the public sector, the political element must also be considered. Because the union may be a potent force, elected officials may permit the union to circumvent the collective bargaining process and make a direct appeal to the legislative body. This process is called an "end run" and, if successful, seriously undermines the effectiveness of the negotiating team. If collective bargaining is going to work in the public sector, the "end run" must be eliminated.

Motivation

There is also a difference in motivation between public sector management and managers in the private sector. The most important of these differences are:

1. The private manager is profit-motivated. He is well aware of the importance of retaining the right to manage from the point of view of the financial success of the business, and more importantly, in his personal interest in that success. This profit motivation is missing in the public sector.

2. Many companies have a continuous program of training supervisors to think as management and to direct the work force thereby emphasizing the importance of management to the success of the business. Such programs are frequently missing or sporadic in the public sector.

3. As noted earlier, there is a much sharper line of demarcation between management and nonmanagement employees in the private sector. Because of these differences in motivation, public managers sometimes fail to realize that it is their duty to represent management's viewpoint vigorously, not only during contract negotiations, but also in day-to-day dealings with the union. Frequently, supervisors do not want to be identified as management. Perhaps this has something to do with the civil service system and the practice of promoting from within.

It may be that employees who have spent a good share of their working career in nonmanagement roles have a difficult time when they move up the ladder and assume supervisory responsibilities. Consequently, they may be inclined to be a "good guy" with the people who work for them rather than try to understand and support management's side of an issue.

The question then is, what can be done to properly motivate public managers? The following suggestions make a great deal of sense:

1. Every effort should be made to impress upon public managers that it is their duty to represent and protect the interests of the governmental agency employing them, just as it is the duty of unions to represent public employees.

2. Negotiators in the public sector must recognize the importance of retaining the right to manage. This is critical if we are to retain the right to operate efficiently, to utilize technological change to reduce labor costs, and to avoid restrictive work rules.

3. All persons who hold supervisory positions should be considered part of the management team.

4. An attempt should be made to provide public managers with some form of financial reward for outstanding performance.

Hopefully at this point there will be a better appreciation for the need to gain labor support for any changes in the solid waste collection operation along with an attempt to highlight public sector labor relations and how it differs from the private sector. We look now at some of the steps that can be taken to more effectively manage change.

8.4.3 Managing Change

There are many factors that contribute to resistance to change. First, we should recognize that change cannot take place unless it occurs through people, and people have a natural tendency to resist change. That resistance can be caused by work-related factors, individual factors, or social factors.

But regardless of the underlying cause for the resistance, we must actively seek to reduce resistance to change by intelligently analyzing the forces for and against that change. The critical factor in the success of organizational change then, is how we go about managing change. To manage change, it is necessary to plan ahead to anticipate and resolve potential problems and constructively eliminate or mitigate those elements that would otherwise tend to reduce the effectiveness of the planned change.

What are some of the steps that can be taken to more effectively manage change? Some steps may seem obvious and rather basic, but they are nevertheless important:

1. *Is the Change Necessary?* To answer this question, the benefits as well as the costs involved in making the change. Change in areas where the returns are minimal should not be made.

2. *Is the Proposed Change the Correct One?* In almost all situations, there are alternative courses of action. The manager, in making a decision, should be sure that the intended change will correct the underlying problem or will return the greatest benefit at the least risk.

3. *Evaluate the Impact of the Change.* Prior to making a change, a manager should evaluate the impact of the change—on the problem it is intended to correct, on the people affected, and on the organization involved. This step is necessary to forecast the forces that may resist the change thereby enabling you to reduce or eliminate that force.

4. *What Are the Critical Factors in Gaining Acceptance of the Change?* As might be expected, those most directly affected by the change hold the key to its success. Unless their anxieties and concerns are addressed, and steps taken to resolve them where possible, even a sound change has little chance for a full degree of effectiveness.

For instance, when planning a change in a collection system, will that change necessitate special provisions for the elderly and handicapped and has the need been provided?

Will the change result in reduced staffing requirements and how will that be accomplished without laying off permanent employees? This should be of equal concern to the manager of a public or private collection operation.

If the organizational change will affect the customer, an effective information program that will advise the public of the need for, and benefit of the change will need to be prepared.

Are special interest groups such as environmentalists affected by the change and has a strategy to deal with their concerns been prepared?

The point here is that unless thorough consideration of a proposed change has occurred, including anticipated impacts, it is not possible to move ahead.

8.4.4 Summary

It is the role of managers to initiate and manage change. It must also be recognized that organizational and human forces act to resist change. But we need not be hesitant to initiate needed change, if we have properly prepared and are confident that the desired results can be attained.

It is equally clear that impending change will be difficult unless there is a strong commitment to the concepts from all levels of management. A concerted effort should be made to achieve greater participation and personal involvement on the part of first-time supervisors. This is most important as the conviction of their part of the inherent worth of a new operating procedure can greatly strengthen management's hand in gaining acceptance of the change.

By taking the time to meet personally with key personnel, both formally and informally, the manager can and should establish an atmosphere in which employees better understand the reason for and impact of the proposed change. The manager must make a special effort to determine as objectively as possible, what employee reaction will be to a proposed change. This requires that management actively listen to employee concerns and be willing to accept their constructive criticisms. If an accurate gauge of employee attitude can be attained, groundless fears can be dispelled and the proposed change explained in such a way that worker resistance can be minimized. This may require the manager to face unpleasant problems or listen to emotional outbursts from employees. However, it is best to handle these matters in a direct, open, and honest manner. The manager must recognize that it is a simple fact of life that individuals personally affected by a change will tend to look upon that change from the viewpoint of their own self-interest.

Initiating change can result in a frustrating, tension-filled, and painful experience. But it can also be challenging, worthwhile, and very satisfying. If governmental solid waste managers are not prepared to deal with the challenge of change, then our organizations will become less effective, or will fail to exist.

8.5 CONTRACTING FOR RESIDENTIAL SOLID WASTE COLLECTION[5]

8.5.1 Introduction

In many communities, the provision of collection services is done by contract forces. The assignment of this responsibility to the private sector does not negate the responsibility of local government to protect the public interest and see that those services are provided. Furthermore, as in the award of any contract, local government will still need to maintain oversight of the contract through a conscientious inspection program to assure satisfactory performance of the contract. Many communities also retain a portion of the community to be served with city forces as a means to assure comparisons of performance and effectiveness and to have a means for a continuation of services should a contractor default or experience labor disruptions.

8.5.2 Determining Type and Level of Service

Because of local government's historic responsibility for solid waste collection, it cannot wash its hands of the issue simply because it chooses to service the public by using private contractor services. The use of private contractor services merely puts the operational aspects of collection into the hands of another part of the team. Local government still has to determine the level of service to be provided.

The key to a successful contractor collection service is how local government defines the needs of the community and then manages those needs for the public good. Communities that decide to contract out residential collection services must take every step possible to define how that service is to be provided. Poorly conceived and managed contract services do a disservice to all parties involved. A poorly described scope of work makes it difficult for a contractor to compute a bid adequately. It makes it difficult to select a winner. And once the contract is awarded, it makes it very difficult to manage a contractor to assure delivery of the services bought.

The major elements of success in contracting out solid waste services are how the community goes about defining its collection objectives and how it states those objectives through a well thought-out set of bid documents and technical specifications.

The first step in this process is the commitment of local government to carrying out its responsibility to see that the community is clean and that the public health is protected. The acceptance of this responsibility means that the community will have to define how such services will be provided and upon what terms. It further means that community leaders will have to be assured that their objectives are met through a thoughtful and professional approach to the management of their solid waste program.

8.5.3 Technical Specifications

Once, for whatever the reasons may be, that local government decides to use contract collection services, they must assure that those services are provided consistent with what the public wants. Those wants must be translated into a set of technical specifications that clearly state the level and type of services to be provided.

It is essential that the specifications be developed so that they fully describe:

1. The community to be serviced, in sufficient detail that potential bidders fully understand what and whom they must service. A carefully prepared and thorough description of the area is essential to a complete scope of work. A discussion of past history and practices in solid waste management is helpful to the contractors bidding on the services. Maps, population counts and densities, and the location of facilities are all important to the potential contractor. A breakdown of normal and rush hour traffic patterns will help him understand how his vehicles will have to move through the city.

2. Possible routing, to guide the bidders further regarding how the community anticipates that

the services will be provided. The community should determine to some extent the configuration of the service it wants and do some homework on what makes sense as far as routing goes. Certainly there may be a need to indicate which portions of the city should be served on which days, if that makes a difference. If the community is going from a community-operated system to a contractor system, past routes and other operating information should be provided to assist the bidders.

3. The type and level of service that is to be provided, including any alternatives that would be acceptable. A community must clearly determine what it wants and needs in a collection service. Some communities that change to contract services fail to recognize and consider the variety of services that their community forces have provided. When they start to develop the scope of work, they should carefully review past services. As an example, a community should determine who is to provide:

a. Dead animal pickup.

b. Snow removal. City systems equip their collection trucks with the ability to plow snow. Will the contractor do the same?

c. Special clean-up programs. Many community-owned systems provide this service; if the contractor is to continue to do so, it must be included in the scope of work.

d. Emergency (e.g., storm, hurricane, etc.) service. What is the role of the contractor in these instances.

e. Pick-up at schools and public buildings, street sweeping, and so on. Many of these special services often are provided by city forces as part of the business of sanitation. Will the contractor be expected to do the same?

It is obvious that when a community decides it is going to contract out for "public cleansing," it had better do its homework on exactly what is to be cleaned up by the contractor.

4. How proposals will be evaluated and selections made. Anyone who has ever made a proposal for any type of work knows how important it is to have an understanding of how the proposals are to be judged. A community should, in its scope of work, identify the factors to be considered in judging proposals. Is the low bid the only factor considered in the evaluation of bids? Will past experience and past demonstrated capabilities be considered, and if so, what will be the relative weight of these factors? Will contractors be given an opportunity to best and final offers? The evaluation procedure should be numerical in its character, and it should be known to all who will be bidding.

5. How the local government solid waste manager will evaluate performance under the terms of the contract; what procedures will be followed when service is not consistent with the specifications; and how the manager will oversee and supervise the contractor in the performance of the contract. Any community that makes an award for contract services and then neglects to oversee that service is sticking its head in the sand. No intelligent institution that contracts out the construction of a building, highway, or bridge is going to neglect to oversee the construction of that facility. Why, therefore, would it not follow the same practice in contracting out solid waste management service? In overseeing that service, however, the community should clearly identify who will be overseeing; what will be the nature of the force that will be supervising; and what factors will be used to measure satisfactory completion of the service (missed stops, the number of complaints, responsiveness to unique occurrences, etc.). The determination and measurement of acceptable service must be defined in the scope of work.

6. Payment procedures, what financial requirements are expected of a contractor, and other financial matters that will affect the success of the contract. Government is notorious for slow payment for services and products received. This really is unfair to a vendor or contractor. Everyone has the problem of cash flow. A community should be sensitive to the fact that if it is slow in paying, a contractor might have to get commercial financing to continue to provide the service that the community is supposed to be paying for. Prompt payment is essential. The scope of work should commit the community to a payment schedule that is fair, and if it fails to meet that schedule, the community should pay an additional premium for its lackadaisical or cavalier approach to payment.

7. The rights of the community in regard to the continuation of services when a contractor defaults or for some reason is not able to perform according to the terms of the contract. Circumstances such as strikes may occur that prevent the contractor from meeting the terms of the contract. The document itself must deal with how the community will assure the protection of the public good in these instances. If the failure is one of a temporary nature (say several weeks) the community will have to find alternative collection methods on an emergency basis, or perhaps take over the operation of the system. How this is to occur and who is to pay for it must be described.

In the case of default, the contract document has to describe fully what will be the penalties placed on the defaulted contractor, what will be the disposition of the contract, and what is the role of the contractor's equipment and facilities in this endeavor. This is a very complex set of issues, but one that must be dealt with in the documents.

8.5.4   Summary

The above-listed items are major policy areas that must be addressed by a community and included in the specifications to assure fair and equitable treatment of a contractor and, at the same time, protect the public interest. As a community develops its contract documents, it can seek professional advice to assure that the bid documents are adequate to achieve the objectives of the community.

A bid document and the supporting specifications need not be overwhelming in either language or length. Certainly, the specifications related to the level of service, quality of service, and so on, can be written in terms that are simple and understandable. The long established guidance of KISS (Keep It Simple, Stupid) should be followed in writing bid documents and the technical specifications. Although it is desirable to have counsel review and assist in the development in the offering, the professional responsible for the documents and the project should not let counsel dictate the nature and content. The reader is reminded that an attorney is called a counsel, and in any dictionary the term "counsel" is defined as advice, not dogma.

It also seems advisable to avoid the common practice of cut and paste. Do not take an old bid document and attempt to adapt it. Although this may make the job easier, it does not assure that the resultant product will be on target for the particular project to be initiated. Previously used bid documents are certainly helpful as guidance in the development of the new bid document and specifications. However, the work should be original, with the specific objectives of the job in mind, and be prepared in the tone and language of the person(s) responsible for the project.

8.6   COLLECTION EQUIPMENT MAINTENANCE PROGRAMS[6]

8.6.1   Introduction

Any maintenance program must be built upon the bedrock of planned or preventive maintenance and must avoid at all costs a system based on demand maintenance. A responsible manager cannot depend on a system that permits a truck or piece of equipment to break down before repairs are made. Preventive maintenance can normally prevent most major breakdowns such as engine failures because systematic inspections will permit a manager to detect problems before they occur. In addition, preventive maintenance programs can extend the useful life of vehicles and equipment by the establishment of a planned maintenance program.

8.6.2   Planned Maintenance

Planned maintenance is accomplished by inspecting a vehicle systematically at regular intervals and by replacing, readjusting, tightening, repairing, or adding to it any part or system that shows a need for repair or adjustment during the inspection. Replacement is done also when the maintenance records on a vehicle indicate that an assembly or a component is nearing the end of its useful life.

Some advantages of planned maintenance are:

Fewer part failures, fewer emergencies, and fewer road calls.

Development of the records necessary to assist in preparing annual maintenance budgets.

Development of information to help determine the best specifications for new vehicles and equipment.

8.6.3   Components in a Maintenance Program

The following records should, at least, be part of a maintenance program.

A standard repair order.

A standard vehicle history jacket.

Standard codes for use with the repair order and the vehicle history jacket.

A liquid usage report.

Repair Order

The repair order, when properly filled out, will provide the information needed to complete maintenance records. This form should be used for all work performed so that the shop manager can allocate labor costs and parts costs to each vehicle and so that it is possible to determine what each vehicle costs to operate. All repair orders should be filed permanently in the individual vehicle jackets (on the computer bank if the operation has access to computer capability).

A standard repair order should provide the following advantages:

1. It will provide written orders to the shop personnel and will eliminate misunderstandings about the work to be accomplished.
2. It will be the document that authorizes the parts department to supply a required part to a mechanic.
3. It will provide an accurate record of the labor expended and the parts used for each repair.
4. It will provide a complete record of the work done at a given mileage (preferably given as hours of use) and on a specific date.
5. It will pinpoint the responsibility for the quality of the repairs done by listing the names of the mechanics who performed the work.
6. It will provide a means to check the productivity of the mechanics because it is possible to check work times against standard time allowances.
7. It will provide the date and the times for scheduling PM (preventive maintenance) intervals and inspections.
8. It will permit a check through all details of unscheduled shop visits and determine the reasons for them.
9. It will provide a running history of all mechanical work done on each vehicle along with the vehicle history jacket.

Vehicle History Jacket

The repair order probably will be the most important and useful document used in the shop by the mechanic and maintenance manager. The most important management document, however, will be the vehicle history jacket. This jacket should be designed to mirror the repair order so that the most important information placed on a repair order can be transferred quickly to the jacket. In that way, the jacket will highlight the critical maintenance actions on each truck, and will give a manager a quick and accurate picture of the truck's problems. The maintenance manager should study a truck's history jacket before he ever issues a repair order. The maintenance manager should determine if a complaint is a current one, if repairs have been excessive, or if maintenance has been performed at proper intervals. Based upon this information, the manager can determine what kinds of repairs should be made, if replacement is called for, and if planned repairs and/or replacements are cost effective. Ultimately, the jacket can be used to determine whether a vehicle should be replaced.

The vehicle history jacket should be kept for the life of a vehicle. More than any other record, the jacket is a management tool.

Codes for Maintenance

To simplify entries on the repair order, the manager should consider using a standard brevity code. The American Trucking Association has developed standard codes.

Liquid Use Record

The last record that should be mandatory in any simplified system should deal with liquids. It is essential to keep a daily record of fuel consumption, oil usage, and the usage of special liquids such as hydraulic fluid and antifreeze.

8.6.4   Other Factors to Consider

The four records mentioned can provide the essentials for a good maintenance system. There are other matters that should be considered in a maintenance program. For example, more complete records, such as: Purchase orders, purchase order logs, weekly preventive maintenance forms, quarterly or 250-hr preventive maintenance forms, annual or 1000-hr preventive maintenance forms, work to be done sheets, workmen's time tickets, daily truck cost reports, mechanical downtime reports, fuel pump readings reports, tire inventory reports, and similar items might be included. Equipment dealers can be of great assistance in planning preventive and planned maintenance.

8.6.5 Summary

Maintenance programs must be professionally managed. A simple system built upon planned maintenance should be the basis for any maintenance program. Keys to any planned maintenance program are:

Managerial interest and support.

Competent mechanics.

A simple system of procedures and records.

Standardization of components and fleets, if possible.

Attention to big dollar expenses.

8.7 OPTIMIZING THE PERFORMANCE OF COLLECTION SERVICES

8.7.1 Introduction

The provision of residential collection services are predicated upon the need to protect public health. The provision of public health protection, however, need not and should not be used as a justification for inefficient, unproductive, and ineffective service. These factors can result only in one outcome—an expensive system that fails to utilize the latest in work force and equipment utilization.

The keys to assuring the optimal collection system is a combination of equipment selection, maximum productivity, and effective routing. The selection of equipment is dependent upon the determination of type and level of service to be provided and which type of equipment can be the most productive within the configuration of the community to be served. In recent years little work has been done to define the variables and many characteristics that go together to determine a residential solid waste collection system.

Work in the 1970s by the federal solid waste management program did result in a number of findings that are still applicable today. This work however, was curtailed just at the time of the advent of more mechanized residential solid waste collection technologies. Consequently, there has been very little done in an analytical way to determine the efficiency and productivity of a fully automated one-man system. It seems reasonable, however, to expect that the analytical techniques developed to study the one- and two-man systems can be utilized to compare one-man automated, one-man semiautomated and two-man crews. Since the issue here is one of increasing productivity and optimal delivery of services, evaluations of three-man crews will not be considered.

8.7.2 Factors Affecting Productivity and Costs[7]

Work supported by the federal solid waste management program and conducted by ACT Systems identified a number of factors that when applied to a residential solid waste management system can result in optimizing the system. A summary of their findings follows:

Conclusions Regarding Crew Size

The productivity per crewman in terms of homes served and tons collected per collection hour is greatest with a one-man crew. On the average, the productivity of one two-man crew is less than the productivity of two one-man crews. Likewise, the productivity of one three-man crew is less than the productivity of three one-man crews.

The percentage of on-route productive collection time for one-man crews is significantly greater than the percentage of productive time for two- and three-man crews. For one-man crews, the on-route productive time is about 97%. For the two- and three-man crews, the on-route productive time is approximately 70%. There is no significant difference in the percentage of productive time between the two- and three-man crews.

In going to the route and in transporting the collected waste, only the driver is productive. All other crewmen, whether they ride with the driver or not, are nonproductive in these operational phases. With these phases consuming approximately 30% of the work day, then one-half and two-thirds of the man-hours of this effort are wasted for two- and three-man crews, respectively.

Conclusions Regarding Frequency of Collection

Increasing the frequency of collection from once a week to twice a week requires approximately 50% more crews and equipment than the once-a-week systems. The average number of homes served per week for a twice-a-week collection system is approximately two-thirds the number for a once-a-week collection system. Conversely, to decrease the frequency of collection from twice-a-week to once-a-week, requires approximately 33% fewer crews and equipment than the twice-a-week systems.

In terms of productivity factors, the twice-a-week collection systems served approximately 50% more homes per collection hour than the once-a-week collection systems. The weight collected per collection hour, however, was only 80% of the weight collected per collection hour by the once-a-week collection system.

Conclusions Regarding Storage Point Locations

The productivity of a backyard system in terms of homes served per collection hour and tons collected per collection hour, is approximately one-half the productivity of a corresponding curb or alley system.

Conclusions Regarding Incentive Systems

Collection systems operating under the task incentive system tend to work a smaller percentage of the normal work week than the standard day systems. The work effort of standard day collection systems has a tendency to expand into overtime operations. The collection production and productivity of the task incentive systems tend to be greater than the collection production and productivity of standard day systems.

Conclusions Regarding Storage Containers

The percentage of one-way items (bags and miscellaneous items) does have a significant effect on the system productivity. An increase in the percentage of one-way items reduces the time required to service a home, and conversely, increases the number of homes served per collection hour.

The weight per home per collection also affects the system productivity, and this effect is greater and opposite in direction to the effect of one-way items. An increase in weight per home increases the time required to service a home and decreases the number of homes served per collection hour.

Conclusions Regarding Productivity and Efficiency

Curbside is more productive and cost efficient than backyard service.
For the curb and alley systems:

Systems that have a collection frequency of twice a week tend to serve more homes per collection hour, but collect fewer tons per collection hour, than their once-a-week counterparts.

The larger crew sizes have a tendency to collect more tons per collection hour.

When productivity and cost efficiency are considered on a per crewman basis, there is a strong tendency for the smaller crew sizes to have the greatest productivity and best cost efficiency.

For backyard systems:

The system that uses the task incentive system has a greater productivity than the system that uses the standard day system.

There is no clear pattern between backyard systems regarding collection cost efficiency.

Conclusions Regarding System Costs

Regardless of the kind of equipment used, the initial cost of the equipment, or the number of days per week the equipment is being used, the daily equipment costs are of the same general magnitude for all systems. (*Note:* This concluson may or may not apply to a fully automated one-man system, given the costs of containers, etc. Readers should consider additional cost analysis to make a true comparison.)

The daily personnel costs were related directly to the crew size. For every system studied, using the study standardized cost data, the daily personnel costs were significantly more than the daily equipment costs. The manpower of equipment ratios averaged 1.4 for one-man crews, 3.0 for two-man crews, and 4.5 for three-man crews. The incremental effect of an increase in equipment costs of $1000 was small in comparison with an effective increase in labor costs per crewman of $0.50/hr.

Since daily personnel costs were significantly more than the daily equipment costs, cost reduction programs should look first in the area of personnel costs. Personnel costs can be lowered by

Table 8.1 Rank Order of Factors Affecting Residential Solid Waste Collection, Productivity, and Costs

Factor	Order for Productivity	Relative Magnitude of Effect	Order for Cost Efficiency	Relative Magnitude of Effect
Point of collection	1	58	1	52
Crew size (per crewman)	2	38	3	9
Frequency of collection	3	36	2	28
Incentive system	4	26	4	1
Percent one-way items (per percent)[a]	5	1	4	1

improving personnel productivity, by reducing the numbers of personnel or both. There was a strong tendency for personnel productivity to increase as crew size decreases.

Since incremental cost effects of an increase in equipment cost of $1000 were small in comparison with an increase in the effective labor rate of $0.50/hr; compromising equipment performance for the sale of a lower equipment cost appears to be counterproductive.

The ACT work also rank-ordered those factors that are most important to productivity and cost efficiency, and, although they may not be totally applicable to a specific system, the rankings clearly do demonstrate the conclusions that they established from their study (Table 8.1).

For the ranking, * to obtain the effect of a decrease of two crewmen, multiply the listed effect by 2. Only one- to three-man crew sizes can be used since these were the only ones studied.

For the ranking, ** to obtain the effect of more than 1% change, multiply the listed effect by the percent change. Due to the limited sample and nonlinearity of this function, a maximum of ±20% should be used.

For each of these factors, the direction to improve productivity and costs is, from less to better: point of collection (backyard to curbside), crew size (larger to smaller, but depends on the point of collection, amount of waste and distance between stops), frequency of collection (twice to once-a week), incentive systems (standard 8-hr day to task system), and percent one-way items (less to more, the impact is significantly greater with curbside collection than backyard).

8.7.3 Measuring Productivity in Residential Solid Waste Collection Systems[8]

It is not the speed at which collection crews work that affects productivity. Rather, as reviewed earlier, there are a number of nonhuman factors and subfactors related to the type and level of service provided and the equipment used to provide that type and level of service that affects productivity and costs. The most costly portion of collection, labor, and its productivity is greatly affected by that portion of the total work day actually devoted to productive collection activities. Consequently, increasing the productivity of work crews is not necessarily predicated upon making the crews work harder and faster, but the application of equipment and methodologies that will allow the worker to devote more time to actually collecting solid wastes.

Stearns, in a series of studies, developed several models that allow solid waste managers to measure the productivity of systems. The determination of productivity will necessitate that solid waste management managers conduct field studies and analyses of their current system to establish the basic data necessary to define the various characteristics that affect productivity.

Stearns determined that route-related subfactors are those that most directly affect productivity. These subfactors must be measured and compared when a solid waste manager is attempting to measure and improve productivity. Table 8.2 summarizes those subfactors.

A review of this table illustrates the compability of the Stearns findings with those of ACT Systems. These factors were utilized by Stearns to develop methodologies to measure and compare productivity.

Application of the methodology identifies the reasonable level of productivity which should be expected for the existing mix of men and equipment, and the collection situation, that is, a "fair day's work." Maximum utilization of the methodology is achieved by using the models to estimate expected productivity levels if alternative collection equipment and crew sizes are used in the same collection situation. Following categorization of the collection situation and identification of the optimal refuse collection vehicle for that situation, and subsequent application of local cost factors,

Table 8.2 Subfactor Characteristics Affecting Solid Waste Collection Productivity

Service Level Related	Route Related	Collection Methodology Related/ Climate Related
Collection Point	Containers	Crew size
alley curb	number	Collection procedures
curb	type	compaction policy (if any)
on-property	size	1 side of street/alley w/1 pass
	weight limitations	2 sides of street/alley w/1 pass
	Distance between collection stops	driver assistance to loader
Collection frequency	Quality of refuse per collection stop	Wind—ambient air temperature
once/week		Rain—snow and/or ice
twice/week	Haul distance to disposal site	
other		
Waste material collection	Maneuverability constraints	
type	vertical	
maximum dimensions	horizontal	
seasonal variation		
	Topography	
	collection route	
	haul route	
	Delays	
	traffic	
	parked cars and other obstacles	
	congestion at disposal site	
	container accessability to crew	
	compaction cycle	
	Road conditions	
	type	
	speed limits	
	load limits	

these accurate estimates of productive time requirements allow development of total service cost comparisons between existing and alternative collection crew size/equipment combinations. Thus, the most cost-effective residential solid waste collection systems for a specific collection situation and desired service level can thereby be readily identified and documented.

The model developed by Stearns to measure the productivity of two-man crews (rear loaders) is:

$$P = 0.001D + 0.16N + 0.09T + 0.03S + 0.02 \tag{8.1}$$

The model developed by Stearns to measure the productivity of one-man crews is:

$$P = 0.005D + 0.15N + 0.08T + 0.08 \tag{8.2}$$

where

P = Productive collection time required per stop, in minutes, including the driving time from the previous stop.
D = Distance between stops in feet.
N = Number of refuse containers at a service stop.
T = Total number of throwaway items (including paper or plastic garbage bags) serviced at a stop.
S = Number of services collected at each stop.

8.7.4 The Five-Stage Process to Improve Residential Solid Waste Collection Systems[9]

Shuster, of the federal solid waste management program, identified five steps which when combined with the measurement of productivity and a commitment by managers and policy officials can result in improved cost-efficiency and optimal delivery of collection services.

The five stages of the process are: (1) review existing policies and methodologies and the alternative of these, including institutional structure and objectives of the delivery system, (2) macrorouting, that is, determine the optimum assignment of the daily collection routes to existing or proposed processing and disposal facilities, (3) perform route balancing and districting to determine a fair day's work, to evaluate crew performances and costs for different policies and methods, and to divide the collection areas into equal workloads for each crew, (4) microrouting, that is, determine the path or route the collection vehicle is to follow as it collects waste from each service in a specified area, (5) implement changes.

The stages are generally performed in the order listed, with the exception that some of the policies and methodologies of stage 1 should be determined or revised after the route evaluations of stage 3 have been performed. Implementation, of course, must be considered in each stage.

Stage 1 is greatly dependent upon a political and policy commitment to take steps that might lead to change and perhaps community disruption during the period of change. A frank and analytical judgement of the current practices must be done so that a community will move forward and implement needed change. Stage 2 is very much an effort to determine current productivity and to determine and define the current practices followed in the delivery of collection services.

Route balancing (stage 3) is the process of determining the optimum number of services that constitutes a fair day's work and dividing the collection task among the crews so that they have equal workloads. Route balancing can be used to (1) estimate the number of trucks and men required to collect waste in a new or revised solid waste system, (2) develop or evaluate a bid price for a collection contract, (3) evaluate crew performances, as a whole or individually, (4) determine a fair day's work or a work standard necessary for task and wage incentive (bonus) systems, (5) balance or equalize the workloads among collection crews, or (6) determine the optimum size for new trucks or optimize the use of existing trucks.

Thus, route balancing is necessary if a new collection system (e.g., backyard to curbside collection) is going to take place, if the system is to be evaluated, or if a collection contract is up for bid.

Route balancing is accomplished by analyzing each component of time in the collection day, or how each spends its time. Adding these component times results in an equation (3) for the total time in the workday (Y):

$$Y = a + b + n(c_1 + c_2 + d) - c_2 + e + f + g \tag{8.3}$$

where

a = Time from garage to route.
b = Total collection time on route.
n = Number of loads.
c_1 = Time from route to disposal site.
c_2 = Time from disposal site to route.
d = Time at disposal site.
e = Time from disposal site to garage.
f = Time for official breaks.
g = Slack time: lost time due to breakdowns and other delays, incentive time, lunch time.

This equation is the basis for determining a fair day's work and for route balancing.

The data required for this analysis are (1) time and distance data related to the components of the collection day, (2) the number and type of services and where they are located, (3) the average amount of waste generated per service, including seasonal variations, and (4) basic equipment and labor cost data.

In designing a new collection system, it is necessary to apply Equation (8.4) to determine the appropriate number of services per crew per day that tells how many trucks and men are required. Reasonable values for variables a, c_1, c_2, d, e, and f are readily obtainable. For example, variable a, the time from the garage to route, is easily derived by considering the distance and route covered and the reasonable driving time to traverse it. Likewise, the number of formal breaks to be taken is a policy decision and is typically two 15-min breaks (in a task incentive system, crews frequently skip the breaks).

Variable n (number of loads per day) is based on the Equation (8.4) of the number of services per load (N):

$$N = \frac{x_x x_2}{x_3} \tag{8.4}$$

where

x_1 = Vehicle capacity (yd^3/load).
x_2 = Vehicle density capability (lb/yd^3).
x_3 = lb/service/collection.
N = Number of services per load.

Variable n is then determined by dividing the number of services that can be collected in the workday (calculated by the procedure which follows) by N and rounding up to the next whole number.

Variable b (total collection time on route) is a function of the number of services that can be collected per hour, or the on-route minutes per service. These values may be obtained in four ways: (1) conducting time-and-motion studies on the existing system or a similar one; (2) using regression equations developed from data on similar systems; (3) implementing experimental routes and trying different crews; or (4) utilizing the Stearns work with some modifications. Obviously, the values for variables n and b, and the time per service, vary seasonally as the amount of waste per service varies.

It makes a great deal of sense, therefore, for the values for the variables n, b, and the time per service be calculated for the peak, normal, and low waste generation periods.

The steps necessary to determine the number of services per crew per day (a fair day's work), the number of men and trucks required, and the system cost are described in the following procedure. These steps are based on Equations (8.1) and (8.2), and determine values for variable b (collection time), variable n (number of loads), and the number of services per crew per day. The time required for on-route collection and transport for each load is compared with the time that is left in the workday until the total time in the collection day is accounted for.

Step 1. Select the level of service, truck type and size, and crew size.

Step 2. Determine N (number of services per load) from Equation (2) using normal waste generation rate.

Step 3. Starting with Y, the total hours in the workday (e.g., 8 hr) from Equation (1), subtract variables a, e, and f, and add c_2. Then subtract the round-trip haul time per load ($c_1 + c_2 + d$) and the collection time per load [services per load (N) times the minutes per service]. Continue to subtract transport and collection times, load by load, until all the time in the workday is used up.

Step 4. Multiply the resultant number of loads (including partial loads if any) by the services per load (N) to get the total number of services per day for each crew.

Step 5. The number of trucks required is determined as follows (rounding the result to the next highest whole number).

$$\text{Trucks required} = \frac{(\text{total number services})}{(\text{services per truck per day})} \frac{(\text{collection frequency per week})}{(\text{number workdays per week})} \quad (8.5)$$

Step 6. Calculate the annual cost of a crew and truck by adding vehicle costs and labor costs:

Vehicle Cost = depreciation + maintenance + consumables + overhead + license fees and insurance

Labor Cost = salary of driver + salary of collector(s) + fringe benefits + indirect labor + supplies (e.g., gloves) + administrative overhead

Step 7. Evaluate the effects of peak and low generation periods on overtime and incentive time respectively by repeating steps 2 through 5 using peak and low generation rates.

Step 8. Multiply the cost per crew (from step 6) by the number of crews needed (from step 5), and add overhead expenses including overtime cost (from step 7) to obtain the total system cost. Divide the total cost by the total number of services to obtain the annual cost per customer.

Step 9. Repeat steps 1 through 8 for any other level of service, equipment, crew size, or other system alternatives being considered. Comparison of crew productivities, system slack, and total cost (and cost per customer) for each system alternative helps give a clear picture of which alternative is most acceptable.

The slack time, variable g, is built into this procedure by rounding to whole numbers and by using conservative estimates. For example, if the average number of loads is 2.3 (step 3), the number of trips is 3, giving a slack capacity of 70% for the last load for all trucks. Slack also results from rounding the number of trucks required up to the next whole number. For example, if the number of trucks required is 7.2 (step 5), then the actual number of trucks required is 8. If the number of services per truck per day is computed to be 650, based on 7.2 trucks (step 4), then the actual number of services per day for each of the 8 trucks is 585. This also means more slack in the number of loads. In this case, the actual length of the workday should be recalculated using 8 trucks and 585 services per truck.

Once the equitable number of services per crew (step 4) has been determined for each area, districting and microrouting can be performed to develop the individual routes. Districting is the process of dividing the collection area into equal workload sections according to the day of the week, and then dividing each daily section into specified routes for each truck, based on the equitable number of services per crew determined by the route-balancing procedure. Total collection and haul time should be reasonably constant for each route. Developing the daily routes may be done in conjunction with microrouting or before. When they are done together, a starting point is selected and a path or route is developed (continuous and concentrated in an area) until enough services to make a route are reached. This process is continued until the whole collection area is routed.

In districting and microrouting, natural boundaries should be used where possible for route boundaries. These include rivers, lakes, streams, mountains, valleys, railroads, highways, major roads, parks, cemeteries, hospitals, and other areas without services.

Routing can be done in a number of ways using complex computer approaches to more common sense techniques. The heuristic routing approach developed by the federal solid waste management program has broad application to many different-sized systems.

The heuristic approach to routing[10] is a relatively simple and expedient method for obtaining an efficient route layout that minimizes dead distances and delay times. The heuristic approach could also be called a pattern method of routing since it relies heavily on the application of specific routing patterns to certain block configurations. Admittedly, efficient routing requires both skill and aptitude. But guided by certain heuristic rules and patterns, and through experience, a router can readily develop the ability to scan a map rapidly and systematically and plot timesaving routes.

The heuristic rules for microrouting are:

Rule 1. Routes should not be fragmented or overlapping. Each route should be compact, consisting of street segments clustered in the same geographical area.

Rule 2. Total collection plus haul times should be reasonably constant for each route in the community (equalized workloads).

Rule 3. The collection route should be started as close to the garage or motor pool as possible, taking into account heavily traveled and one-way streets (see rules 4 and 5).

Rule 4. Heavily traveled streets should not be collected during rush hours.

Rule 5. In the case of one-way streets, it is best to start the route near the upper end of the street, working down it through the looping process.

Rule 6. Services on dead end streets can be considered as services on the street segment that they intersect, since they can only be collected by passing down that street segment. To keep left turns at a minimum, collect the dead end streets when they are to the right of the truck. They must be collected by walking down, backing down or making a U-turn.

Rule 7. When practical, steep hills should be collected on both sides of the street while vehicle is moving downhill for safety, ease, speed of collection and wear on vehicle and to conserve gas and oil.

Rule 8. Higher elevations should be at the start of the route.

Rule 9. For collection from one side of the street at a time, it is generally best to route with many clockwise turns around blocks.

Heuristic rules 8 and 9 emphasize the development of a series of clockwise loops in order to minimize left turns, which generally are more difficult and time-consuming than right turns; right turns are safer, especially for right-hand drive vehicles.

Rule 10. For collection from both sides of the street at the same time, it is generally best to route with long, straight paths across the grid before looping clockwise.

Rule 11. For certain block configurations within the route, specific routing patterns should be applied.

Implementation (stage 5) is accomplished only by a carefully planned effort.

The section on managing change points out a number of steps that must be utilized to ease the changing of a collection system. Those steps, when combined with the analytical efforts which are the basis for determining productivity and optimal use of human resources and equipment will result in a residential solid waste collection system designed for the community and its citizens that is efficient, effective, and the least cost for the level and type of service provided.

REFERENCES

1. D. Helsel, Local Government Policies for Solid Waste Collection (Proceedings, GRCDA 17th International Seminar and Equipment Show, 1979), in *Residential Solid Waste Collection*, GRCDA, 1982, pp. 1-1/1-6.

2. D. Kerton, Making the Change from Backyard to Curbside Collection (Proceedings, GRCDA 17th International Seminar and Equipment Show, 1979), in *Residential Solid Waste Collection*, GRCDA, 1982, pp. 5-1/5-9.

3. H. G. Larson, Cost Accounting Techniques for Solid Waste Collection Systems (Proceedings, GRCDA 18th International Seminar and Equipment Show, 1980), in *Residential Solid Waste Collection*, GRCDA, 1982, pp. 2-1/2-14.

4. D. Kerton, Labor Relations in Implementing Changes in Collection Services (Proceedings, GRCDA 18th International Seminar and Equipment Show, 1980), in *Residential Solid Waste Collection*, GRCDA, 1982, pp. 6-1/6-13.

5. H. L. Hickman, Jr., Precise Specs are Key to Successful Government/Private Refuse Contract, *Solid Wastes Management*, February 1982, pp. 54–58.

6. L. Gonyou, Refuse Equipment Maintenance (Proceedings, GRCDA 18th International Seminar and Equipment Show, 1980), in *Residential Solid Waste Collection*, GRCDA, 1982, pp. 9-1/9-9.

7. ACT Systems, Inc., *Residential Collection Systems, Vol. 1, Report Summary*, USEPA/OSW (SW-97c.1), 1974, 105 pp.

8. R. A. Stearns, Measuring Productivity in Residential Solid Waste Collection Systems (Proceedings, GRCDA International Seminar and Equipment Show, 1980), in *Residential Solid Waste Collection*, GRCDA, 1982, pp. 3-1/3-19.

9. K. A. Shuster, *A Five Stage Improvement Process for Solid Waste Collection Systems*, USEPA/OSW (SW-131), 1974, 38 pp.

10. D. A. Schur and K. A. Shuster, *Heuristic Routing for Solid Waste Collection Vehicles*, USEPA/OSW (SW-113), 1974, 45 pp.

CHAPTER 9
TRANSFER OF MUNICIPAL SOLID WASTE

LAURENCE T. SCHAPER

Vice President, Black & Veatch, Engineers–Architects

9.1 THE TRANSFER STATION

Transfer stations have been used for several decades, and the number of these stations has increased greatly in recent years because sanitary landfill sites are located greater distances from the collection areas. Transfer stations are most common in larger metropolitan areas. There is great variance in types, size, and degree of sophistication of transfer stations. In past years, especially in warmer climates, a large number of open-air stations were constructed. Many of the newer stations have buildings that enclose the transfer operation.

9.1.1 Potential Advantages

Transfer stations have gained widespread acceptance as a method of reducing transport costs, in addition to providing other advantages. Other potential advantages include the following:

Better haul roads for collection vehicles.
Greater traffic control.
Fewer trucks on the sanitary landfill haul route.
Improved landfill operating efficiency.

The transfer station can be an effective means of lowering overall haul cost when sanitary landfill locations are remote. The cost benefit results from the reduction in number of drivers required to transport the waste to the disposal site. When the entire collection crew also rides to the disposal site, the probability of savings from transfer haul increases. A careful analysis is needed to provide a true comparative evaluation of transfer haul versus direct haul in collection vehicles.

Transfer station roads are almost always paved and therefore provide a desirable driving surface. Sanitary landfill roads are difficult to maintain in good condition because they extend across fill areas to the working face. Poor road condition can result in damage to collection vehicles and delay to collection crews. Transfer station roads are easier to maintain in good condition and therefore reduce crew delay and collection vehicle damage.

People living along the haul route to a sanitary landfill often object to the large volume of truck traffic, as well as to the litter that sometimes results. Other concerns include congestion during rush hour traffic and the safety hazard to children going to school. A transfer system offers a potential solution to these concerns because storage at the station can provide the opportunity to keep transfer vehicles off the road during rush hour periods and at the time of heavy school traffic.

Transfer haul greatly reduces the number of trucks using the haul route to the sanitary landfill. Depending on the types of vehicles involved, the reduction in traffic can be in the range of 3:1 to 5:1. Thus, for every five trucks using a landfill road before transfer haul, the number can be reduced to as low as one vehicle with transfer haul. This can be a significant factor in reducing congestion on haul roads to a sanitary landfill.

L. W. Bremser, C. L. Hutchison, and F. E. Kirkpatrick assisted in preparation of this section.

195

Sanitary landfills located in congested areas sometimes have restrictions on which haul roads can be used by collection vehicles. These restrictions are difficult to enforce, especially when many haulers use the landfill. Road restrictions are easier to enforce because the limited number of transfer truck drivers can be controlled.

Improvements in sanitary landfill operation may occur when most of the waste is received in transfer trucks. The reduction in number of trucks that must be handled can result in a smaller working face and fewer traffic control problems. The smaller working face makes litter control easier, probably results in better compaction, and reduces the quantity of cover material used. Storage of solid waste either in the transfer station or in transfer trailers may make it feasible to reduce the operating hours at the sanitary landfill. A reduction in operating hours would result in lower disposal costs.

9.1.2   Types of Users

Transfer stations are constructed to serve varying types of users. Categories include the following:

Rural citizens
Urban public (automobiles and pickups).
Municipal and private collectors.

Stations to serve rural areas are frequently drop-box-type containers served by front-end loaders. The concept has been a successful approach to providing solid waste service to rural areas where collection at each house is not economically practical. The container size can be matched with the area served and the population density.

Urban citizens in many locations prefer to haul solid waste to a transfer station rather than pay for home collection service. Transfer stations to serve citizens must be planned to handle large volumes of waste delivered in automobiles and pickups. Peak usage frequently occurs during spring weekends when weather is favorable.

Many transfer stations are designed to handle municipal solid waste hauled by private or municipal collectors. The variety of collection vehicles using the facility may include open top trucks, vans, compactor-type collection vehicles, drop boxes, and so on. The types of vehicles using the facility will have significant impact on its capacity to handle solid waste in a given time period. For example, trucks that must be unloaded by hand occupy transfer station space for a long time. Many of the station design features must be tailored to accommodate the types of vehicles expected.

9.2   LOCATION

Transfer stations are sometimes perceived by the public as contributing noise, odors, dust, increased traffic, rats, flies, and litter. The alleged result of these environmental factors is lower property values. Most of the concerns can be alleviated through prudent site selection, good design and operation, and public education.

Extreme care should be used in selecting the transfer station location. Public acceptance will be greater if the selection process is carefully planned. Criteria for transfer station location include the following:

Near the collection area served.
Accessible to major haul routes.
Adequate land area to provide isolation.
Suitable zoning.
Served by utilities.

The economic benefits of transfer haul are maximized by locating the facility near the collection area served. The collection crew time for haul to the transfer station is nonproductive and should be minimized. Ideal locations are seldom available. Usually a compromise must be made between this and other site selection criteria.

Locating sites in proximity to main haul routes is critical to both public acceptance and economics. Transfer truck traffic is undesirable on residential streets or heavily used streets serving commercial areas. The interests of the general public are served best when major haul roads are utilized. The major roads reduce travel time that results in lower transfer haul costs.

The transfer station may change traffic patterns at the site entrance and on nearby streets. Traffic engineers should evaluate the traffic routes available to the collection and transfer haul vehicles. Where appropriate, traffic controls should be provided to assure safe and efficient traffic flow.

The amount of land required for a transfer station will depend on the volume of traffic to be handled. Sites that handle waste delivered in many small vehicles will need more land area to accommodate them. Storage of transfer trailers on site will increase the land area required. Additional land is needed if separate buildings are provided for maintenance and scale facilities. A significant land area will be needed to provide buffer areas around the site perimeter to reduce the off-site impact of noise and vehicle traffic. A general guide for the amount of land required is shown on Figure 9.1. Site-specific layouts for the individual site are necessary to determine exact requirements.

The transfer station location must conform to zoning and or special use permit requirements. Because of the traffic associated with a transfer station, zoning is either commercial or industrial. The suitability of a specific site is related to the extent of isolation and screening needed.

Utilities needed to serve the transfer station include water, sanitary sewers, storm drainage, electricity, and fuel for heating. Availability of these utilities should be considered in site selection. The cost necessary to obtain the necessary utilities is an important part of overall site development costs.

9.3 DESIGN CHOICES

Design choices include the type of station to be constructed, the extent of site development, and the need for other facilities such as scales, office space, and vehicle storage. The size of station needed may influence the design concept selected.

9.3.1 Station Concepts

There are three major station concepts and variations of each. They include:

Pit.
Direct dump.
Compaction.

The three concepts are illustrated on Figure 9.2.

Pit stations have gained popularity in recent years. Collection vehicles unload wastes into a large pit. The wastes are then pushed from the pit into an open-top transfer trailer by a tractor with a dozer or landfill-type blade. The pit provides storage of waste during peak periods. Some compaction of the waste, especially bulky items, is achieved by the tractor in the pit.

At direct dump stations, collection vehicles dump directly into open-top transfer trailers. Large hoppers direct the waste into the transfer trailers. Stationary or mobile clamshell equipment can be used to distribute the waste in the transfer trailer and can also accomplish some compaction of the waste. Very large transfer trailers are used since there is only minimal compaction of the waste in the trailers. The direct dump stations are inherently efficient because there is no intermediate handling required to transfer the waste from the collection vehicle to the transfer trailers.

Two types of compaction stations are commonly constructed. In smaller stations, the collection vehicles dump into a hopper from which the waste drops by gravity into a compactor which then packs the waste into the trailers. Other stations use a push pit. After waste from the collection vehicle is dumped into the push pit, a large hydraulically operated blade moves the waste to the stationary packer. The stationary compactor then packs the waste into the transfer trailer. Equipment for a push pit-type station is shown in Figure 9.3.

Each of the station concepts has advantages and disadvantages. A listing of these is provided in Table 9.1.

9.3.2 Sizing Transfer Facilities

Procedures for sizing transfer stations are not well established. The lack of widely accepted sizing procedures complicates comparison of station capacity and associated costs. The following formulas assume wastes are received in compactor-type collection vehicles. Modifications are necessary for stations handling smaller loads received in noncompaction trucks, pickups, and automobiles.

Station capacity can be controlled by either of two factors: (1) the rate at which wastes can be unloaded from collection vehicles or (2) the rate at which transfer trailers can be loaded. Therefore, two calculations must be made for most types of stations. The formulas are slightly different for different types of stations and are not applicable to stations where the tipping floor is used for

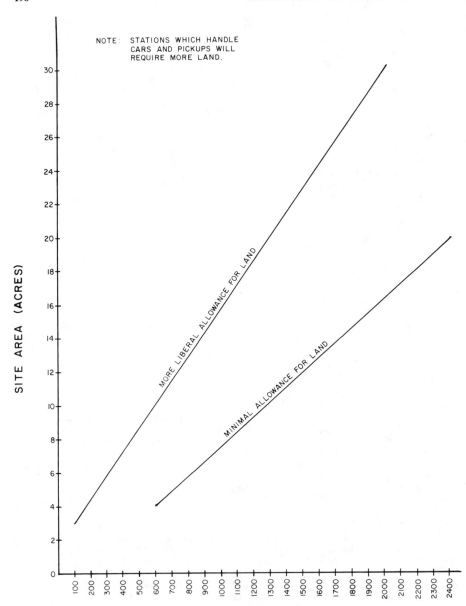

Fig. 9.1 Site area for transfer stations.

storage. Without the peaking factor, F, the equations provide a peak capacity. With the peaking factor, the equations provide station capacity with minimal queuing of collection vehicles.

For pit stations, the capacity based on the rate at which wastes can be unloaded from collection vehicles into the pit is calculated by Equation (9.1).

$$C = Pc \frac{L}{W} \frac{60\,Hw}{Tc} F \qquad\qquad (9.1)$$

TRANSFER STATION CONCEPTS

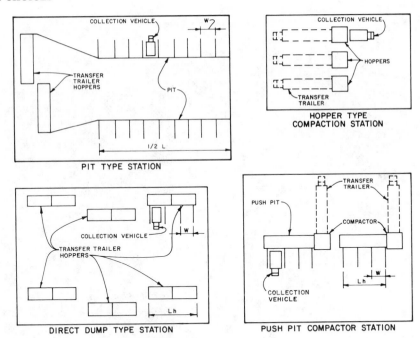

Fig. 9.2. Transfer station types.

where

C = Station capacity in tons per day.
Pc = Collection vehicle payload in tons.
L = Total length of dumping space in feet.
W = Width of each dumping space in feet.
Hw = Hours per day waste is delivered.
Tc = Time in minutes to unload each collection vehicle.
F = Peaking factor equal to the ratio of number of collection vehicles received during an average 30-min period to the number received during a peak 30-min period.

Capacity for the pit type station where the rate of loading into transfer trailers controls is calculated by Equation (9.2).

$$C = \frac{Pt\ N\ 60Ht}{Tt + B} \tag{9.2}$$

Where

C = Station capacity in tons per day.
Pt = Transfer trailer payload in tons.
N = Number of transfer trailers loading simultaneously.
Ht = Hours per day used to load trailers (empty trailers must be available).
B = Time in minutes used to remove and replace each loaded trailer.
Tt = Time in minutes to load each transfer trailer.

For direct dump stations, capacity can be calculated using Equation (9.3).

$$C = \frac{NnPt\ F\ 60Hw}{(Pt/Pc)(W/Ln)\ Tc + B} \tag{9.3}$$

Fig. 9.3 Equipment for a push pit-type station.

Table 9.1 Advantages and Disadvantages of Transfer Station Concepts

Advantages	Disadvantages
Direct Dump	
Lack of hydraulic equipment minimizes probability of a shutdown	If large amounts of uncompacted waste are received, maximum payloads are difficult to obtain
Extra handling of the waste is not required to load transfer trailers	Top-loading trailers are more difficult and time consuming to unload than enclosed transfer trailers with push-out blades
If the incoming waste is precompacted in collection trucks, maximum payloads in the transfer trailers may be achieved with a minimum of processing	Bulky items are not handled as easily as in a compactor trailer system or pit concept where compacting force is available
Drive-through arrangement for transfer vehicles can be easily provided	Peak deliveries must be handled by tipping floor storage or by additional direct dump areas and trailers

Table 9.1 (cont'd)

Advantages	Disadvantages

Pit

Advantages	Disadvantages
Convenient and efficient storage area for waste	Construction of receiving pit and purchase of the bulldozer require considerable capital investment
With uncompacted material crushed in the pit by the bulldozer, maximum loads are attainable without further processing	Top loading trailers are more difficult and time consuming to unload than enclosed-compactor trailers
Top trailers are less costly than enclosed-compactor trailers	
Peak loads may be handled easily with many incoming vehicles capable of being unloaded at the same time	
Drive-through arrangements for transfer vehicles can be easily provided	
Simplicity of equipment and operation minimizes possibility of complete station shutdown	

Compaction—Hopper

Advantages	Disadvantages
Maximum payloads can be obtained from compacted or uncompacted waste	No alternative way of loading trailers exists if compactor fails
Nearly all bulky material that can be placed in the hopper can be handled by the stationary compactor because of the large hydraulic force available	If most of the waste received is precompacted in collection trucks, the heavier enclosed trailer offers little advantage since maximum payloads can be achieved in lighter top loading trailers with top tamping.
The compaction equipment can handle light, fluffy types of waste	Limited hopper storage space causes queuing problems during peaks; therefore, more suitable for low-volume transfer stations
Trailers can be unloaded quickly and efficiently	Extra dead weight of the hydraulic ram ejection system and the required trailer reinforcement reduces legal payloads
	Procurement costs are more than for top loading trailers

Compaction—Push Pit

Advantages	Disadvantages
Pit provides some storage for peak waste loadings	Construction of push pit and purchase of hydraulic ram results in considerable capital expense
Maximum payloads can be obtained from compacted waste	No unloading into push pit possible when charging the stationary compactor
Nearly all bulky material that can be placed in the hopper can be handled by the stationary compactor because of the large hydraulic force available	No alternate way of loading trailer if compactor fails
Incoming waste usually receives minimum exposure because it is rapidly pushed into the enclosed trailers	Extra dead weight of the hydraulic ram ejection system and the required trailer reinforcement reduces legal payloads
Canvas or metal tops do not have to be dealt with when loading and unloading the transfer trailer because it is enclosed	Procurement costs are more than for top-loading trailers
Trailers can be unloaded quickly and efficiently	If most of the waste received is precompacted in collection trucks the heavier enclosed trailer offers little advantage since maximum payloads can be achieved in lighter top loading trailers with top tamping.

where C, Pt, F, Hw, Pc, W, Tc, and B are as previously defined, and Nn = number of hoppers, Ln = length of each hopper in feet.

Capacity for compaction stations (hopper type), based on the rate at which wastes are unloaded from collection vehicles, can be determined using Equation (9.4).

$$C = \frac{Nn\ Pt\ F\ 60Hw}{\dfrac{Pt\ Tc\ +\ B}{PC}} \qquad\qquad (9.4)$$

where symbols are as previously defined.

The rate at which transfer trailers can be loaded at a hopper-type compaction station is calculated using Equation (9.2).

Push pit-type compaction stations have capacity as determined by Equation (9.5). The equation is based on the assumption the hopper holds a quantity of waste equal to one transfer trailer load.

$$C = \frac{Np\ Pt\ F\ 60Hw}{\left(\dfrac{Pt\ W\ Tc}{Pc\ Lp}\right) + Bc + B} \qquad\qquad (9.5)$$

where C, Pt, F, Hw, Pc, W, Tc, and B are as previously defined, and

Lp = Length of push pit in feet.
Np = Number of push pits.
Bc = Total cycle time for clearing each push pit and compacting
        waste into the trailer.

9.3.3   Site Development and Ancillary Facilities

Site development can include on-site roads, parking, drainage, fencing, landscaping, fuel storage, and utilities. Ancillary facilities can include scales, office, and employee facilities. These complement the transfer station and should be carefully planned.

On-site roads must accommodate the expected traffic as safely and efficiently as possible. Where public dumping of waste is allowed, consideration should be given to providing separate access for cars and pickups and for collection trucks and transfer vehicles.

Site roads for each type of vehicle should have turning radii consistent with anticipated vehicle speed and turning characteristics. Ramp slopes should be less than 10%. Provisions for de-icing steep ramps should be considered in cold weather climates. Road surface design should be suitable for heavy vehicles.

Site drainage structures should be sized to handle peak storm flow to avoid disruption of station operation. An evaluation should be made of the potential for flooding, and, where appropriate, levees or other flood protection measures should be considered.

A transfer station site is usually fenced. The fence serves several purposes, including keeping the public away from the heavy traffic areas, providing security for the facilities, and assisting in control of blowing litter. Chain link-type fence is effective for these purposes.

Landscaping and screening can be especially important when stations are constructed in esthetically sensitive areas. Berms can be used to screen the vehicles using the station from view. Landscaping enhances the facility and makes the station appearance more pleasing. Plantings should be selected to require minimal maintenance. Evergreens provide screening throughout the year.

Fuel supply storage and dispensing facilities are frequently installed at transfer station sites to serve the transfer haul vehicles. The fuel can be stored in large underground tanks. Both gas and diesel fuel storage may be needed. Dispensing facilities require a significant area to accommodate the large transfer vehicles.

Utilities required for a transfer station commonly include telephone, water supply, electricity, natural gas, and sanitary sewers.

Water supply needs depend on the location and size of station. Possible uses include fire protection, dust control, potable water, water for sanitary facilities, and irrigation water for landscaping. Usually fire protection will set the maximum quantity needed. If necessary, on-site storage and pumps can be utilized to boost pressure and increase the flow available.

Needs for electricity include the maintenance shop, yard and building lighting, scales, equipment, fuel dispensing equipment, compactors, and other pumps or processing equipment. In some stations, the stationary clamshell used to distribute the load in transfer trailers is electrically operated.

Natural gas is used primarily for building heat. Sanitary sewers are required for the sanitary facilities and the washdown water. It is usually desirable to provide a sump or trap to separate large solids from washdown water prior to discharge to the sanitary sewer.

Requirements for office space at transfer stations vary greatly. The office space should be adequate for both administrative and clerical personnel. Additional space will be necessary if billing for collection and transfer services is handled at the transfer station. Space should be provided for files, employee records, and operation and maintenance manuals.

Employee needs for lunchroom, locker space, and showers should be considered. Where appropriate, these facilities should handle the transfer vehicle drivers as well as the transfer station personnel.

Scales are provided at many transfer stations and can service the collection trucks, transfer vehicles, or both. In some stations, scales are provided at the transfer station loading area to allow monitoring the weight of the transfer trailers as they are loaded. The number of scales needed to weigh incoming vehicles is a function of the traffic volume. The length and capacity of the scales should accommodate the largest and heaviest vehicles using the station. Where more than one scale is provided, one can be large enough to handle transfer vehicles and the others can be smaller to handle collection vehicles. Scale facilities are useful in monitoring the productivity of collection crews as well as providing the basis for fees charged.

There is a trend toward greater sophistication in scale controls. Frequently plastic cards are provided to regular station users. These cards can contain hauler identification, vehicle tare weight, and other pertinent data. Upon entering the scale area, the card is inserted into a reader and the information is printed onto a weight ticket. The gross weight and resulting net weight can be automatically calculated and printed on the ticket. Information regarding the type of waste, such as industrial, commercial, or residential, can be recorded.

Sophisticated scale systems frequently experience downtime due to equipment difficulties. Dust and temperature variations in the scale house may be hard on the equipment. Manual standby equipment can be provided for periods when the system is out of operation. Generally, the scale itself has a high degree of reliability.

The volume of traffic handled by a single scale is dependent on a number of factors such as whether the transaction involves fee collection, the type of equipment installed, and efficiency of operating personnel. A rough guide is that with most of the waste hauled in collection vehicles, a single scale can handle about 500 tons/day.

9.4 PROCESS OPTIONS

The transfer station can include processing equipment as part of the facility. Processing options include:

Baling.

Shredding.

The purpose of processing can be to prepare waste for transfer haul and subsequent disposal or it can be an initial step in resource recovery. The following discussion of baling and shredding pertains primarily to transfer stations. More detailed information on the processes is contained in other chapters.

9.4.1 Baling

Baling municipal solid waste has been practiced in the United States since the early 1960s. Potential advantages include reduced haul cost, a more controlled operation at the landfill, and lower cover material requirements (see Chapter 17).

The density of the bales is usually in the range of 1500 to 2000 lb/yd^3. Depending on the type of equipment utilized, bales may have wire ties or may not be tied.

Most stations using baling equipment provide for floor storage of the waste. A highloader is used to load waste onto conveyors that feed the baler. Large balers require little or no separation of wastes prior to baling. Bales are loaded onto trucks, barges, or rail cars for transport to the disposal site. Flat-bed semitrailer vehicles are commonly used for road haul. Flat-bed trailers are significantly lighter than other types of transfer trailers and so allow a greater legal payload for baled waste. The density and size of bales usually results in the need to haul only one layer of bales to achieve the legal payload. Because the bales have some loose material, it is necessary to cover the bales to prevent littering of the haul route.

Baling is adaptable to transport by barge or rail. The greater weight capacity of barges and rail cars makes them well suited for transporting high-density baled waste. Loading and unloading of baled waste can be done with a fork lift or a crane.

9.4.2 Shredding

Shredders installed at transfer stations require special design features (see Chapter 17). Floor or pit storage is required to receive and store the waste. Conveyors or cranes are utilized to move the waste to the shredder. The operator should have the opportunity to segregate items the shredder is not capable of handling.

Large shredder equipment requires adequate foundation design. Piling will be required in certain soil conditions. Design of shredder installations should consider safety since explosions can occur. Explosion prevention and control measures are discussed in Section 12.8.

Shredding equipment can be installed for several reasons. A shredder may be provided as a means of handling bulky wastes such as tree limbs, furniture, etc. In other cases, all of the waste is processed through the shredder. The heterogenous nature of municipal solid waste requires very large equipment. Conveyors serving shredders should be of heavy-duty design and capable of withstanding the severe service associated with handling municipal solid waste. Conveyor and shredder equipment requires routine and periodic major maintenance. Stations should include equipment and space necessary for such maintenance.

Transporting shredded waste commonly occurs in enclosed transfer trailers having a push-out blade for unloading. Shredded waste has good compaction characteristics, and a relatively small trailer is needed to obtain a legal payload. Trailers of 65 yd^3 capacity are common and a density of 615 lb/yd^3 (365 kg/m^3) will provide a 20-ton payload.

9.5 TRANSFER VEHICLES

Two kinds of transfer trailers are used depending on the type of transfer station.

Station Type	Trailer
Pit or direct dump	Top-loading trailers with walking bottom, moving chain, or push-out blade for unloading
Compactor	Rear loading enclosed trailers with push-out blade for unloading

Trailers are sized to obtain a legal payload. Variables include the type of material transported and the degree of compaction. For nonprocessed municipal solid waste, volumes typically vary from 65 yd^3 (50m^3) for compaction trailers to 125 yd^3 (96 m^3) for noncompaction trailers.

9.5.1 Compaction Trailers

Compaction trailers usually have a volume in the 65 to 75 yd^3 range (50–96 m^3). The push-out unloading blade can be powered by the tractor hydraulic system or by a separate engine provided for this purpose. A photograph of a typical compaction type trailer is shown on Figure 9.4.

Density in compaction trailers handling municipal solid waste is normally in the range of 400 to 600 lb/yd^3 (237 × 356 kg/m^3).

Compactor trailers are enclosed and reinforced to withstand stresses imposed by the compaction process. Because of the reinforced construction, compaction trailers are usually heavier than noncompaction trailers even though the volume is substantially smaller.

The hydraulically operated push-out blade used to unload the trailer is relatively simple and trouble-free. This unloading feature is one of the major advantages of compaction trailers.

9.5.2 Noncompaction Trailers

Noncompaction trailers are larger, but lighter, since they are not designed to withstand the stresses due to compaction. Both steel and aluminum are used as materials for construction.

Several systems have been used to unload the trailer. An early approach was to provide a net inside the front of the trailer with cables extending to the back. At the landfill, a tractor would hook onto the cables and pull the load from the trailer. A disadvantage of this concept was the time and expense associated with the tractor and operator required to assist in the unloading. More recent systems for unloading use moving chains, walking floor bottoms, or a push-out blade. Another unloading system is a hydraulic lift that tips the trailer to an angle adequate to discharge the waste. This type of equipment is shown on Figure 9.5.

Densities achieved in noncompaction trailers are in the general range of 275 to 400 lb/yd^3 (163–237 kg/m^3) for municipal solid waste. The lighter weight of the noncompaction trailers results in legal payloads that are usually 1 to 3 tons per load greater than with compaction trailers.

Fig. 9.4 Compaction-type transfer trailer.

Noncompaction-type trailers must have the entire top of the trailer open for loading. After loading, top doors close to prevent wastes from littering the roadway on the trip to the landfill.

9.5.3 Number of Vehicles Required

The number of transfer tractors and trailers required depends on the type of station, length of haul to the disposal site, hours of haul per day, quantity of waste, and number of spare vehicles provided. The station should have sufficient equipment to handle peak day requirements. It may be practical to maximize use of equipment and labor by working overtime on peak days.

Fig. 9.5 Unloading equipment for noncompaction trailers.

Table 9.2   One-Way Travel Time for Transfer Vehicles (Minutes)

Nontravel Activity	Trip Number	Trips Per Day			
		2	3	4	5
Warmup vehicle and connect trailer	1	15	15	15	15
Unload at landfill site	1	15	15	15	15
Break time	1	20	20	20	20
Change trailer	2	10	10	10	10
Unload at landfill site	2	15	15	15	15
Break time	2	20	—	—	—
Change trailer	3	—	10	10	10
Unload at landfill site	3	—	15	15	15
Break time	3	—	20	20	—
Change trailer	4			10	10
Unload at landfill site	4			15	15
Break time	4			—	20
Change trailer	5				10
Unload at landfill site	5				15
Shutdown and refuel		20	20	20	20
Total		115	140	165	190
Working day		480	480	480	480
Travel time (Working day— total nonproduction time)		365	340	315	290
Maximum one-way travel time per trip		91	57	39	20

The number of trips that can be made to the landfill each day is a function of the travel time to the site. A procedure to determine the number of trips per day is shown in Table 9.2. Travel time from the transfer station to the disposal site of 50 min would allow three round trips per day for each transfer vehicle.

Some stations, especially the direct dump type, may load trailers at a faster rate than tractors and drivers are available to take them to the disposal site. Excess trailers are needed to provide temporary storage of waste. The filled trailers are stored at the transfer station site and taken to the landfill as tractors and drivers become available.

9.6   MATERIALS-HANDLING EQUIPMENT

Types of materials-handling equipment vary with the station design. Examples are as follows:

Station Type	Materials-Handling Equipment
Pit	Track-type factor with landfill blade
	Stationary or mobile clamshell for load distribution in trailers
Direct dump	Stationary or mobile clamshell for load distribution in trailers
Compaction	Compactor and push pit equipment
Direct dump or compaction using floor storage	Rubber-tired front-end loader
Shredding or baling	The shredder or baler plus conveyors or traveling cranes

Depending on the size of station, a mechanical sweeper may be needed for station clean-up. Items of equipment for station maintenance are discussed in the maintenance section.

The function of the mobile or stationary clamshell is to distribute the load within the trailer. It can also provide compaction of the waste in the trailer. Good operator visibility is an important feature. When selecting the equipment, consideration should be given to both the horizontal and vertical operating range of the clamshell unit.

The track-type tractor used in a pit-type station should be large enough to break up and compact the waste and also to move it efficiently. A landfill-type blade is desirable for pushing large amounts of waste the length of the pit. Firefighting equipment on the tractor is needed due to fire hazards inherent in operating in the waste.

Conveyors handling municipal solid waste are subjected to severe operating conditions. Where conveyors are provided, they are usually essential to station operation, and failure renders the station inoperable. The conveyor width and general construction should be adequate to handle the variety of wastes received. The maximum conveyor incline depends on the type of equipment, but usually is approximately 30° from the horizontal.

Floor storage of waste generally requires rubber-tired front-end loaders. Considerations in equipment selection include the maneuverability required and the quantity of waste to be handled.

9.7 MAINTENANCE

Maintenance facilities should be tailored to the size and needs of each transfer station. Large stations may require extensive maintenance capability both for station equipment and transfer vehicles.

The transfer tractor fleet requires routine daily attention as well as major repairs and overhauls. On-site facilities should be provided for routine maintenance. More extensive maintenance may need to be contracted to repair shops staffed and equipped for major overhaul and transmission work.

Maintenance facilities can either be provided within the transfer station building or in a separate structure. The following checklist can be helpful in determining maintenance and service building needs:

Maintenance work area.

Spare parts and tool storage.

Maintenance equipment.

Underground storage and fuel supply.

Office for supervisory personnel and storage of operation and maintenance records.

Shower, restrooms, and locker facilities for maintenance crew.

Mechanical room for air compressor, HVAC (heating, ventilating, air conditioning) equipment, and water heater.

Communications system.

Drive-through bays are desirable for tractor maintenance. Each bay should be approximately 20 ft (6m) wide. Accessories for each bay can include electricity, grease, water, and air connections. A lift can be provided for each drive-through bay. The lifts are used primarily for preventive maintenance, which includes oil change, lubrication, and inspection.

Some bays can be dedicated to trailer maintenance. The dimensions of the bays should be 60 ft (18 m) long × 20 ft (6 m) wide. Each can be equipped with electricity, grease, water, and air connections.

A drive-through wash-down bay, 18 ft (5.4 m) wide and 70 ft (21 m) long, can be furnished for washing vehicles. The wash facility should be supplied with high-pressure hot water. Several types of washing systems are available.

A bay, 20 ft (6 m) wide × 60 ft (18 m) long, can be provided to change tires on tractors or trailers. The area reserved for tires would be equipped with storage racks, appropriate tools, and tire storage.

A separate partitioned welding bay can be provided. Much of the welding repair work will be on the trailers.

Storage space will be needed to house parts, special tools, and the parts and tool sign-out and control center. Consideration should be given to security of parts storage areas.

The maintenance equipment considered should include the following:

Forklift.

Portable arc welder.

Oxyacetylene welder and cutting torch.

Compressed air system and accessories.

Creepers.

Vacuum cleaner.

Battery charger.

Voltmeter.

Ammeter.

Vises.

Work benches.

Tire mounting equipment.

Electric drills.

Electric sander.

Soldering iron.

Propane torches.

Rivet gun.

Other hand tools.

When a wrecker converter is attached to the fifth wheel of a tractor, it becomes a wrecker that can be used to haul inoperable vehicles back to the transfer station.

A forklift can be used to remove and install heavy truck components and carry them to repair and cleanup areas. It also has the advantage over a monorail crane hoist in that it can move to all areas of the maintenance building.

A portable welder has a similar advantage in that it can be moved to the equipment requiring repair. The welder can be a heliarc type with a current selection of AC, DC+, and DC−. This provides the capability to weld most metals including aluminum and steel.

The air compressor system can include two compressors each capable of meeting the building air requirements. This provides a backup if one becomes inoperable.

Underground storage can be provided for engine oil, waste oil, and anti-freeze as well as for diesel fuel. This saves space in the building and provides better control of supplies.

An office area may be needed. Windows should look over the minor maintenance and preventive maintenance areas.

The employee facilities for the maintenance crew can include shower, restrooms, and locker facilities. The multipurpose room can be used as a lunchroom and for meetings with employees. Employee facilities can be estimated on the basis of one mechanic for six pieces of transfer haul equipment.

The mechanical room houses the compressed air system, water heater, and heating, ventilating and air conditioning equipment. Heating and air conditioning can be supplied to the office and employee facilities. The maintenance areas frequently are only heated.

An exhaust system should be provided to remove fumes. The welding bay will also require special ventilation.

The communication system can include a two-way intercom system. Telephones should be placed in the office and parts control areas.

9.8 COST ANALYSIS AND CASE STUDIES

9.8.1 Cost Analysis

A major incentive to construct a transfer facility is to lower total transportation costs. Cost savings associated with transfer haul are possible due to:

Reduction in mileage traveled by collection vehicles reduces operating cost.

Reduction in nonproductive crew time during the haul from the collection route to the disposal site may lower personnel costs.

Since less collection crew and vehicle time is used for travel to the disposal site, it should be possible to reduce the number of trucks and crews needed.

Transfer system cost estimates must be site specific and reflect cost levels appropriate for the individual situation. Procedures explained in the following text provide general guidance in calculating transfer cost.

Transfer costs can be categorized as follows:

Transfer station investment.

Transfer station operation and maintenance.

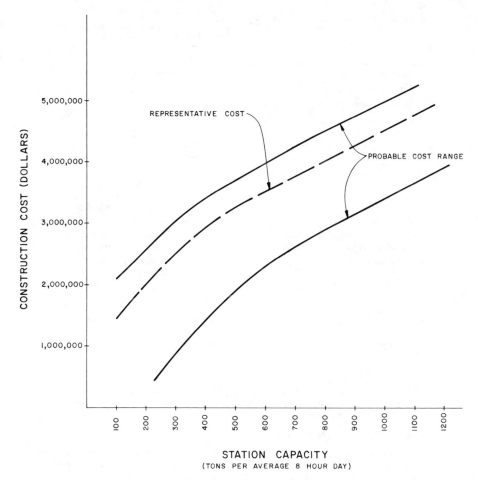

Fig. 9.6 Transfer station construction cost.

Transfer vehicle investment.

Transfer vehicle operation and maintenance.

The total transfer cost is the sum of the four categories.

The transfer station construction cost is an expenditure required prior to startup of the facility. It includes land, buildings, utilities, and site development. The land cost is proportional to the amount of land required which is related to station capacity. Figure 9.1 provides an indication of the acreage required. Since land can be sold at the end of the useful life of the transfer station, the annual cost can be considered equal to the annual interest on the land value.

Building and site development costs depend on variables such as:

Station capacity (facilities required for peak loading).

Foundation conditions (piling).

Architectural treatment (metal building—other).

Code requirements (fire control).

Dust and odor control.

Site work (availability of utilities).

Support facilities (scales, maintenance area, etc.).

Station costs are unique for each situation and must be refined after the design concepts are developed. Figure 9.6 was developed using cost data from many existing stations. Costs reflect an Engineering News Record Building Cost Index of 2220.

Table 9.3   Representative Labor Requirements

Position[a]	Station Capacity (Tons/8-hr day and 6-day week)			
	100	300	500	1000
Foreman		1	1	1
Equipment operator(s)	1	2	2	3
Scale operator(s)	1	1	2	3
Laborer(s)	1	2	3	4
Total	3	6	8	11

Annual costs for a transfer station should include the following:

Description	Annual Cost
1. Land	Annual interest on the land value
2. Structures and site development	Original cost ammortized over facility life
3. Equipment	Original cost ammortized over equipment life
4. Labor	Salaries plus fringe benefits
5. Utilities	
6. Building and site maintenance	
7. Equipment maintenance	
8. Insurance, taxes, etc.	
Total annual station cost[b]	————

[a]The Table does not include on-site tractor operators, mechanics, or janitors.
[b]The total annual station cost can be converted to cost per ton by dividing by the tonnage handled each year.

In addition to land, structure and site development, station cost must include the materials handling equipment discussed in Section 9.6. The purchase cost for this equipment will depend on the amount and types of equipment required.

Annual station operating and maintenance costs include labor, utilities, maintenance, insurance, and taxes. Representative labor requirements for several station capacities are shown in Table 9.3. When calculating labor costs, allowance should be included for fringe benefits which frequently are 25 to 35% of salary costs.

Maintenance costs will vary with the type of structure. An allowance can be made by using a percentage of the original construction cost.

Transfer vehicle cost is a function of the number of tractors and trailers required and miles traveled. The number of transfer vehicles required was discussed in Section 9.5.3.

Transfer haul includes the annual fixed cost of ownership of the tractors and trailers, licenses, taxes, insurance, and trailer repairs. In addition to fixed costs, certain costs vary with mileage traveled. Variable costs include fuel, oil, tires, and tractor maintenance and repairs. Both variable and fixed costs in terms of dollars per ton are shown in Table 9.4. The transfer haul cost curve shown in Figure 9.7 was developed from data on Table 9.4.

The total transfer cost is calculated by summing the initial transfer station and site development cost; the transfer station operation costs; the transfer haul fixed costs; and the transfer haul variable cost. All costs can be expressed in dollars per ton. The total transfer cost can then be compared with the cost of direct haul in collection vehicles. Collection vehicle haul cost is developed using a procedure generally similar to the transfer haul cost calculation shown on Table 9.4. The number of trips per day shown on the table would not be a consideration when calculating direct haul costs.

Favorable conditions for transfer include:

Long travel time between collection route and the disposal site.

Small collection vehicles.

Large collection crews traveling to the disposal site.

Table 9.4   Transfer Haul Cost

Equipment	Capital Investment Per Unit ($)	Estimated Service Life Years		Salvage Value ($)
Tractor	60,000	5		22,000
Trailer	43,000	10		5,500
Average Trips Per Day	2	3	4	5
Annual Fixed Costs	$	$	$	$
Capital cost per tractor	13,200	13,200	13,200	13,200
Licenses, personal property tax, and insurance	7,000	7,000	7,000	7,000
Allowance for spare tractors	3,300	4,000	4,600	5,300
Capital cost for trailer (one)	7,300	7,300	7,300	7,300
Allowance for spare trailers	5,800	8,800	11,700	14,600
Labor per unit	36,000	36,000	36,000	36,000
Trailer repairs	2,400	3,600	4,800	6,000
Overheads at 10% total annual fixed cost	7,500	8,000	8,500	9,000
	82,500	87,900	93,100	98,400
Cost per day (312-day year)	264	282	298	315
Cost per ton (avg. payload, 17 ton)	7.76	5.53	4.38	3.70
Representative one-way travel time (min)	77	45	31	23

Variable Cost Per Mile	
Fuel: 4 mpg	0.28
Oil, lub., service, wash, etc.,	0.01
Tires: 20,000 mi/set	0.30
Tractor repairs	0.15
Subtotal	0.74
Overheads at 10%	0.07
Total variable cost per mile	0.81

The decision to provide a transfer system depends on:

The availability of a suitable station site.
Funds for initial financing.
Need for the potential advantages associated with a transfer system.
Transfer system costs that are reasonable when compared with the cost of direct haul.

Generally, the decision to utilize a transfer system is tied to an anticipated reduction in costs. Lower cost is only one of several potential advantages. In some cases, the flexibility provided by a transfer station will justify higher costs than for direct haul in collection vehicles.

9.8.2   Case Studies

Determine the size and cost per ton for a pit-type transfer station capable of handling 600 tons/day. Assume review of operating system data reveals the following conditions. The average collection

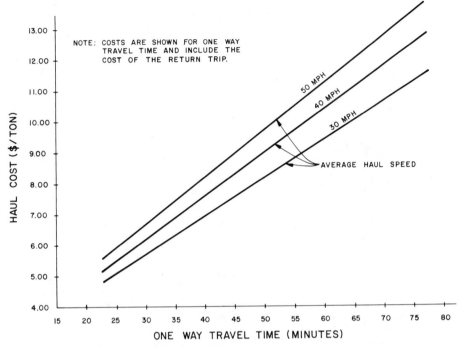

Fig. 9.7   Transfer haul cost.

vehicle load is 4 tons. Most collection vehicles arrive at the transfer station during a 2.5-hr period in the morning and a similar time period in the afternoon. Approximately 150 loads are delivered to the station. The peak arrival rate is 45 loads per 30-min period. The average vehicle unloading time is 10 min. The width of each unloading stall is 13 ft (3.9 m).

Solving Equation (9.6) for L provides the length of dumping space needed.

$$L = \frac{C\ WTc}{Pc\ 60HwF}$$

$$= \frac{(600)\ (13)\ (10)}{(4)\ (60)\ (5)\ (.33)} = 197\ \text{ft}$$

(9.6)

Since the vehicles unload on two sides of the pit, use 100 ft (30 m) for the length of the pit.

Check whether loading two transfer trailers at a time will result in the ability to handle 600 tons/day. Use Equation (9.7). Base the calculation on an average transfer trailer payload of 18 tons, loading time of 7 hr/day, 5 min to remove a full trailer and replace it with an empty one, and average trailer loading time of 20 min.

$$C = P_t\ \frac{N\ 60\ H_t}{T_t + B} = \frac{(18)\ (2)\ (60)\ (7)}{20 + 5} = 605\ \text{tons/day}$$

(9.7)

Conclude this meets the design requirement of 600 tons/day.

As a second case study, determine the total initial cost and the cost per ton for a direct dump type station having a capacity of 850 tons/day. Base the analysis on hauling the waste to a disposal site 25 miles (40 km) from the transfer station. Assume a one-way travel time of 38 min.

Using Figure 9.6, the initial cost can be estimated to be $4,000,000. The land required for the station is estimated from Figure 9.1 to be 9 acres (3.6 ha). At $60,000/acre, the site value would be $540,000. Using procedures shown in Table 9.2, it is determined that each vehicle will make four trips per day to the disposal site. The average transfer trailer load is expected to be 17.5 tons. Therefore, based on four trips per day, 13 over-the-road tractors will be needed. If two yard tractors are provided plus 10% spares, a total of 17 tractors will be required. A total of 32 trailers are needed. Of the 32 trailers, 13 will be on the road to the disposal site, 15 are needed for peak storage at the transfer station, and four are spares. As discussed in Section 9.6, two mobile clamshells will distribute the waste in the transfer trailers. The required initial investment can be summarized as follows:

Building and site development	4,000,000
Land	540,000
Transfer station mobile equipment	200,000
Transfer tractors (17 @ $65,000 ea.)	1,105,000
Transfer trailers (32 @ $47,000 ea.)	1,504,000
Total Initial Investment	$7,349,000

Next, determine the total transfer cost per ton.

Interest on land:

$$(540,000) (0.12) = 64,800$$

Amortization of the building and site development using 12% interest rate and a 20-year life:

$$\overset{0.13388}{(4,000,000) (crf\text{-}12\%\text{-}20 \text{ yr})} = \$535,520$$

Amortization of station mobile equipment and two yard tractors over 5 years with a 10% salvage value:

$$\overset{0.27741}{(330,000 - 33,000) (crf\text{-}12\%\text{-}5 \text{ yr})} + 33,000 (0.12) = \$86,350$$

Station annual costs for labor:
(See Table 9.3)

Position	Number	Annual Salary (each)	Salary ($)
Foreman	1	27,300	27,300
Equipment operator	3	23,600	70,800
Scale operator	3	16,700	50,100
Laborers	3	16,700	50,100
Subtotal	10		198,300
Fringe benefits allowance (30%)			59,490
Total labor cost			$257,790
Annual utilities allowance			$34,000
Building and site maintenance Allowance: (Use 2% of building and site development cost) (0.02) (4,000,000)			$80,000
Transfer station equipment maintenance (Use 3% of initial cost) (0.03) (330,000)			$9,900
Insurance allowance			$2,500
Summary of transfer station annual cost:			
Land			64,800
Structures and site development			535,520
Station equipment			86,350
Station labor			257,790
Station utilities			34,000
Building and site maintenance			80,000
Station equipment maintenance			9,900
Insurance allowance			2,500
Subtotal			$1,070,860

Express the cost in dollars per ton:

$$\frac{(850 \text{ tons})}{\text{day}} \frac{(6 \text{ days})}{\text{week}} \frac{52}{\frac{\text{weeks}}{\text{year}}} = 265,200 \text{ tons}$$

$$\frac{\$1,070,860}{265,200 \text{ tons}} = \$4.04 \text{ per ton}$$

The transfer haul cost can be calculated from Figure 9.7. Based on 38 min travel time and 40 miles (67 km)/hr average travel speed, the haul cost is $8.00/ton.

Therefore, the total transfer cost is $4.04/ton plus $8.00/ton or $12.04/ton. Expressed in terms of cents per one-way ton mile, the 1983 cost would be:

$$\text{ton} \frac{\$12.04}{25 \text{ mile (42 km)}} = 48 \text{ cents per one-way ton-mile (ton-km)}.$$

CHAPTER 10
SOURCE SEPARATION AND CITIZEN RECYCLING

ROBERT COWLES LETCHER

U. S. Department of Energy
Morgantown, West Virginia

MARY T. SHEIL, Administrator

Office of Recycling
N. J. Departments of Energy and Environmental Protection
Newark, New Jersey

10.1 PERCEPTIONS, ANALYSIS, AND STATUS
Robert Cowles Letcher

Background

With the exception of this chapter, the major focus of this handbook is waste: from how to handle it to how to dispose of it—even how to recover resources from it. This chapter is different: it addresses resources and it discusses source separation as an approach to managing them. As such, it is not an alternative among waste management programs; rather, it is an alternative to waste management itself.

Yet, despite this significant distinction, the combined problems of dwindling landfill capacity and increasing transportation costs are leading a small but growing number of waste managers to reconsider their longstanding antipathy toward source separation programs. The mere presence of this chapter in a handbook of this type testifies to the growing acceptance of source separation among waste managers. However, this transition has only just begun, with the preponderance of waste managers continuing to prefer "cheaper alternatives."

This situation places an additional burden on the presentation of this chapter. In addition to discussing programatic matters (the primary topic of other chapters of this work) the situation described here requires that the discussion begin with a justification of source separation itself. This additional discussion is required not only to provide a context in which the programatic information can be appreciated, but also to provide "ammunition" for arguments in support of source separation programs. And, at this stage of the diffusion of these programs, arguing in support of source separation must be considered to be part of the "job" by those who wish to move into source separation. The latter portion of this presentation addresses the more operational aspects of source separation. The reader who still considers such information to be "blue sky" is reminded that, "Anything that exists is possible."

Author's Note: The precepts and opinions in this presentation represent an independent analysis from the author's background in the subject and should not be construed as representing any official position of the U.S. Department of Energy.

Setting the Context

For decades, society's discarded materials have been disposed of as waste. However, since the early 1970s, a growing number of people have come to question the practice. To them, such materials are resources, not waste; and programs should be devised for their utilization, not their burial, and not even energy recovery. However, even though this concept is simply stated, it has proven to be difficult to implement.

10.1.1 Recycling Defined

Many waste managers associate the word "recycling" with resource recovery, but many other people ascribe other definitions to the term. To some, it is the stacking of newspapers and separating of cans and bottles from the trash. To others, it is the program that collects those materials. And, to still others, it is just another waste disposal program. These various interpretations indicate the relatively recent origin of the word. "Recycle" was not even included in the 1970 edition of *Webster's Collegiate Dictionary*. The lack of unanimity also indicates the difficulty of identifying just what recycling should be compared to, a difficulty complicated by the fact that discussions of recycling are generally rendered in irrelevant terms.

For the purpose of this discussion, "recycling" refers to a closed-loop system, the purpose of which is to optimize utilization of resources to the overall benefit of humankind while minimizing production of the waste. With this definition, recycling is cast as a systemic alternative to the "Produce-Consume-Dispose" (P-C-D) syndrome that governs the flow of materials in today's society and for which there exists no simple reference. Furthermore, while it is tempting to perceive recycling in terms of a P-C-D system, to do so would be inconsistent with the closed-loop nature of recycling. This inconsistency arises from the fact that each activity in a recycling system *both* consumes (the products of upstream activities) *and* produces (for consumption by downstream activities). As a result, *each* activity could properly be referred to as a "Consume-Produce" activity, and, as such, viewed as a "recycling" activity and a part of a recycling system.

From this perspective, the separate stacking of newspapers and the separating of cans and bottles, as well as the program that collects them, are *recycling* in the sense identified here. On the other hand, the "just another waste disposal program" definition is clearly incompatible with this definition. Subsequent sections will discuss the manner in which this particular incompatibility obstructs development of recycling.

10.1.2 Source Separation Programs Defined

Source separation programs can be considered an alternative to disposal and centralized recovery systems. Instead of mixing materials prior to discarding them, source separation programs are based upon the separation of materials at the place where they are discarded, according to categories established to expedite their re-use.

Source separation is the activity that marks divergence from the P-C-D material flow. As will be discussed in Section 10.1.10, it is a necessary yet insufficient condition for the existence of an efficient recycling system.

10.1.3 Recycling and the Waste Disposal Industry

The waste disposal industry appears not to think of recycling in the terms proposed here. As suggested above, that industry generally thinks of recycling (if it thinks at all of recycling) in terms of centralized recovery systems, or, when thinking of source separation programs, as a waste disposal option. Furthermore, waste disposal managers tend to be unfamiliar with the *crucial* marketing aspects of source separation programs, unfamiliar also with the secondary materials industry without which source separation programs cannot be successful. And it is useful to note that, with a few notable exceptions, source separation programs have tended to operate outside of the waste disposal/centralized recovery industry.

10.1.4 Implications for Both Concepts

The preceding sections describe a situation that is unique among chapters of this handbook. On the one hand, those involved in citizen-based recycling activities—groups such as Recycle Ann Arbor (Ann Arbor, Mich.), Garbage Reincarnation (Sonoma County, Calif.), and the Institute of Local Self-Reliance (Washington, D. C.), are not likely to consult a handbook on waste management. On the other hand, those who do consult this work are likely to be waste managers searching for better ways to dispose of their waste and unlikely to appreciate source separation as a recycling activity. The approach adopted here is directed toward that latter group, as well as toward those who regulate them and are familiar with regulating "waste disposal programs."

For this audience, an appreciation of recycling itself must be developed prior to discussing operational aspects of source separation programs. This appreciation requires that waste disposal programs be "put in perspective," so that source separation programs can be evaluated on their own merits instead of on their effectiveness as disposal programs. The following section creates that perspective.

10.1.5 Waste: Perceptions and Perspectives

The day-to-day activities of waste collection and disposal programs are limited almost exclusively to operations matters, i.e., "getting the job done." Even related planning efforts rarely address matters other than "How much, how to, and when?" Solutions to operational problems are usually sought through increased operational efficiency; little attention is given to the possibility that waste disposal might be part of the problem. Questions such as "Is there a better way?" are seldom asked; and, even if they were asked, those who ask them seldom possess the authority to act upon their own answers. In this section, the origins of this unquestioning attitude are discussed.

Historical Origins of Disposal Programs

Prior to the latter part of the 19th century, the "Waste not, want not" ethic motivated a conserving approach to material flows. Informal arrangements handled discarded materials; for example, farmers traveled to cities each spring to collect manure.[1] In fact, 70% of cities in the United States operated primary separation programs prior to 1920.[2] However, rapid industrialization during the late 19th and early 20th centuries outpaced the capacity of those informal arrangements to adapt. In the absence of an adequate processing system, discarded materials piled up in the streets and other informal "dumps." Eventually, those piles came to be perceived as public health hazards. Government officials responded by initiating programs intended to remove those offensive materials to locations safely distant from human activity.[3] Those public health programs gave rise to today's waste disposal programs.

10.1.6 The Institutionalization of Waste Disposal

Contemporary disposal programs continue to function in much the same manner as those original programs. However, they no longer do so for the same reason: neither the public health emergency nor the rapid industrialization that created it exist as such today. So, the question arises: Why are there still disposal programs? Two answers frequently put forth by the waste disposal industry are discussed below, along with a third answer that could be considered to be the "recycling position." The third answer has the added benefit of shedding light on the question of why other types of programs do not exist—it simply does more explaining.

Answer #1: Waste Disposal Programs Protect the Public Health by Removing Potential Threats.   The plausibility of this explanation derives from the fact that some of the same materials that created those public health hazards continue to be collected and removed before any hazard develops. This explanation benefits the waste disposal industry because it retains the aura of a public policy basis for such programs, an aura that tends to insulate disposal programs from strictly "bottom line" criticism. It has other flaws as well, the most serious being the absence of *a priori* evidence that would demonstrate that disposal is the best approach to *avoiding* the public health emergency. Furthermore, the narrow focus and action-orientation of emergency decision-making are not compatible with the more circumspect approach of true public policy-based decision making. The existence of recycling programs demonstrates that a more circumspect approach could generate programs other than disposal.

Answer #2: Disposal Is the Most Efficient Approach to Waste.   The plausibility of this explanation derives from the apparent sensibility of disposing of materials that are accorded no value. Its chief benefit to the disposal industry is that it creates a sense that disposal programs exist in response to a "given"—the existence of waste—much as the original programs existed in response to "given" public health hazards. This explanation is also flawed, primarily as a result of historical inaccuracy. The "no value" sense of waste is not a given: materials considered to be waste today were considered to be items of value prior to their becoming public health hazards. In fact, the existence of waste can be seen as a response to the existence of disposal programs, instead of the other way around. The circumstances which gave rise to disposal programs afforded little opportunity to consider the material value of offending materials—only time to react to them as public health hazards. Today's "no value" definition of waste can be seen as a natural adaption to the near-total reliance on programs which assign no value to the materials they process, a reliance that results from the paucity of alternative systems capable of reclaiming no-longer useful materials.

Answer #3: Waste Disposal Has Become Institutionalized. This explanation, which makes sense only *outside* the waste disposal industry, suggests that waste disposal has become a self-validating structure supported by an internally consistent and highly interconnected set of concepts, systems, definitions, and programs acting to thwart development of alternatives. This institutionalization can be seen at three levels which themselves are interconnected. In increasing order of generality, these levels are: individual waste generators, disposal programs, and the P-C-D system. Several factors that both effect and manifest this institutionalization are discussed below.

Institutionalization at the Level of the Individual Waste Generator. At this level, institutionalizing factors are most apparent. They include the absence of both a value assigned to materials at the point where they are discarded and information concerning the costs of this no-value assignment. Also important is the lingering sense of "uncleanliness" still attached to these materials, a sense that is grounded in the public health problems that started their institutionalization process in the first place. The reinforcement given to disposal by the P-C-D system also plays a role; an illustrative example being the increased use of throwaway packaging. To overcome these institutional factors, individual waste generators must be convinced of the value of their "wastes" as resources, and shown that such materials are safe to handle, and worth the inconvenience.

Institutionalization at the Level of the Disposal Program. At this level, institutionalizing factors are more complex, involving both people and institutional arrangements between them. These factors include the following:

1. *Financial Investments.* Large investments in single-compartment collection vehicles, landfill processing equipment, and even landfills themselves, create an economic "inertia" that impedes acquisition of equipment appropriate for the collection and processing of separated materials.

2. *Bureaucratization of Program Management.* The organization that has evolved to administer and operate disposal programs has created slots for people who possess skills appropriate to tasks associated with specifically those slots. Those people are usually professional managers and engineers, trained to identify and solve administrative and engineering problems and accustomed to the personal satisfaction of doing so. However, source separation programs involve people and institutional problems. Those involved in disposal programs are unlikely to conceive of problems to which source separation would be an appropriate solution. Furthermore, disposal programs are familiar to those involved, and they are likely to have a vested interest in continuing those programs. Similarly, a new program, calling for new skills, is likely to be perceived as a threat to the status of the people and entities involved in disposal programs—and resisted as such.

3. *Labor Factors.* Workers who actually perform the disposal operation are selected, trained—and treated—in accordance with the "no value" definition of the materials they process. In addition, wage rates and job assignments, themselves frequently formalized through contractual arrangements, are based upon the processing of worthless materials. The net effect is to create a situation in which workers have no interest in the materials they process, and even less interest in the possibility that source separation activities might introduce additional work for them.

4. *Performance Criteria.* Disposal programs have no *absolute* criteria for determining whether they are successes or failures. Only the relative performance criterion of *efficiency* exists. Thus, while *more* efficient disposal programs can be distinguished from *less* efficient disposal programs, there is no basis on which to determine whether the most efficient disposal program is efficient *enough*. This allows disposal programs to treat rising costs as "givens" without worrying about the impact of "bottom line" matters on program continuity. In this context, source separation programs are placed at a structural disadvantage: They are not efficient disposal programs. The situation is made even less favorable by the fact that source separation programs are not permitted to treat market variations as "givens" in the same sense.

5. *Political Considerations.* Because of the high visibility of waste disposal programs in the public eye, a tendency exists for risk-averse elected officials to support disposal programs strictly on the basis of those programs being "sure things"—immune to failure as long as increased costs are considered as "givens." Maintaining the *status quo* also avoids upsetting the political constituency that supports waste disposal programs. This constituency includes landfill owners who are usually prominent local businesspeople who operate on a "first name" basis with waste disposal program managers. This places the still-emerging source separation constituency at a disadvantage in its interactions with elected officials.

Institutionalization at the Level of the P-C-D System. At this level, institutionalizing factors are even more obscure, transcending the immediate context of waste disposal and involving the even larger and more pervasive institution that establishes the flow of materials within the system. That institution emphasizes one-way material flow and maximum throughput, continuous style changes, and other forms of planned obsolescence. The large industries that comprise the potential market for secondary (i.e., recycled) materials, the supply of secondary materials is neither sufficiently predictable nor adequately stable. As a result, those industries purchase their materials from other sources. The result of this business decision is a drastic reduction in the demand for recycled materials and the further entrenchment of the P-C-D system. This situation has itself become institutionalized by the following factors.

1. *Vertical Integration.* In response to their need to provide a predictable and stable supply of materials, large material-intensive industries have acquired ownership of properties and industries that generate the materials utilized in their particular processes. This arrangement, which internalizes the producer-consumer transaction at this stage of the material flow, creates the additional benefit for these businesses of a *controllable* supply of materials. Secondary materials suppliers face a difficult time competing with this arrangement since they must rely on the coordinated behavior of large numbers of people. The result for much of the secondary materials industry is erratic marketing opportunities—and low prices for *their* suppliers.

2. *Government Subsidies.* Firms involved in the extraction of virgin materials frequently receive tax incentives and other subsidies from various levels of government, with depletion allowances being one example. Such government involvement is widely acknowledged to introduce market distortions. In this case, those distortions manifest themselves by providing a financial incentive, in addition to those already attributed to vertical integration, for material-consuming industries to purchase virgin materials. The presence of a significant, industry-supported aluminum recycling system serves to demonstrate the significance of recycled material purchase support by other materials-consuming industries.

Implications for Recycling

The discussion in Section 10.1.5 raises questions about the influence of the waste disposal industry in evaluating the merits of source separation. The negative influence of structural factors in waste disposal has little to do with the merits of source separation. Furthermore, the discussion of performance criteria raises the question of whether the disposal industry is capable of evaluating its own programs. Ultimately, the discussion makes a case for "evaluating the evaluators"—for "removing the fox from the chicken coop." Such an evaluation is critical not only for the advancement of recycling, and source separation in particular, but also to deal with the growing problems faced by the waste disposal community itself. The absence of this evaluation has made it possible for the waste disposal community to create the problems they now face.

10.1.7 Benefits of Source Separation

Whereas Section 10.1.5 addressed limitations associated with evaluating source separation and recycling from a waste disposal perspective, this section addresses the benefits of source separation/recycling within their own, broader context. These benefits can be grouped under two headings: those attributable directly to source separation programs, and those attributable to the recycling system made possible by source separation programs. In each area, benefits can be associated with the substitution of source separation for other programs and to factors that arise from source separation itself. These areas are addressed separately below. Also, because of the ambiguous nature of centralized processing systems, and in recognition of their growing significance, the relationship between source separation programs and centralized systems is considered as a special case.

Benefits of Source Separation Programs

To the extent that individual members of the general public must choose whether to process their discards through either a source separation program or a waste disposal program, benefits of source separation can be considered in terms of the advantages generated by choosing the former instead of the latter. With consumers projected to spend $6 to $10 billion for collection and disposal of their discards by 1985,[4,5] the potential benefits associated with the choice are large. The immediate benefits of the source separation choice include the following:

1. *Extended Landfill Life.* Materials processed by source separation programs conserve landfill capacity. Such programs create an opportunity for program managers to manage landfill capacity as a resource itself. The recycling public appreciates this benefit: Survey results indicate that concern over dwindling landfill capacity is an important factor behind participation in source separation programs.[6] Extended landfill life also means that the costs and political headaches associated with new landfills can be postponed and potentially reduced. This benefit will take on additional significance as over one-half of the cities in the United States will run out of landfill capacity during the 1980s.[7]

2. *Reduced Waste Disposal Costs.* Materials processed by source separation programs

avoid or reduce several costs associated with current and future waste disposal programs. Source separation can postpone the added costs of transporting materials to more remote landfill sites. Construction of transfer stations can be deferred, and their capacity reduced in accordance with the amount of materials diverted through source separation programs. And, of course, no tipping fee is required for materials that do not arrive at the landfill.

3. *Direct Benefits.* In contrast to paying a "tipping fee" to dispose of waste, source separation programs are *paid* for by the materials they process. (Of course, for reasons discussed in Section 10.1.10, such payments are frequently low; and it is also true that markets for these materials are not guaranteed merely by the existence of available materials, again for reasons discussed in that section. However, such dislocations must be expected in a system undergoing development.) These payments offset programs costs. And, although program costs may not be entirely offset by those payments, in comparison to disposal programs that have been subsidized historically, source separation is the approach to "waste disposal" that comes closest to supporting itself.[8] Source separation, with its increased emphasis on material flows at the level of the individual, creates an opportunity for an unexpected benefit, but one which *is* closely connected with the origins of disposal programs. It is but a simple extension of such separation practices to include separation of materials for control of pests, especially in urban areas. Proper containment and storage of household materials can remove the food and harborage that facilitate the "lifestyle" of these pests. For example, organics require sealed containment whereas stacks of newspaper need only be stored on platforms elevated above the ground.[9] This practice could lead to reduced costs in pest control programs.

10.1.8 Benefits of the Recycling System

The benefits discussed in this section include those with which recycling programs are most frequently associated, as well as "spin-off" benefits that are not always recognized as being attributable to the recycling system. The former group includes conservation, pollution reduction, and related matters—factors that provide the primary motivation for participation in source separation programs.[10] Table 10.1 characterizes these benefits for the manufacturing of four commonly recycled materials from secondary materials instead of virgin materials. Other benefits of this type are illustrated by the following examples:

1. One Sunday edition of The New York Times consumes 62,000 trees.[11] Recycling 1 ton of newspaper saves trees.[12] However, only about 20% of all paper used in the United States is ever recycled.[13]

2. The United States imports 91% of its aluminum and throws away 1 million tons of it annually, worth over $400 million.[14]

3. There are no sources of tin within the United States, but 6 lb of ultrapure tin can be reclaimed from each ton of metal cans.[15]

4. Cullet (i.e., recycled and crushed glass) reduces the energy required to manufacture new glass by 0.5% for each 1% increase in its proportion in the cullet/raw material mixture.[16] Cullet lowers the temperature requirements of the process, reducing emission levels; several glass smelters have taken advantage of this fact to avoid the purchase of costly emission control equipment.[17]

5. Over 200 million tires are discarded annually in the United States.[18] If all tires were retreaded once before being disposed of, the demand for synthetic rubber would drop by one-third, and the tire disposal problem would be cut in half.[19]

6. Over 300 million barrels of oil were saved by recycling paper, metals, and rubber during 1979.[20]

Table 10.1 Benefits Derived from Substituting Secondary Materials for Virgin Resources (Percent Reduction)

	Paper	Glass	Steel	Aluminum
Enery	23–74	4–32	47–74	90–97
Air pollution	74	20	85	95
Water pollution	35	—	76	97
Mining wastes	—	80	97	
Water usage	58	50	40	

Sources: Refs. 49, 50, 51, 52, 53.

In addition to the conservation/environmental protection aspects, recycling is also a significant economic activity. In a recycling economy, source separation programs play a role restricted to extraction industries in the P-C-D system. As such, they generate functionally similar employment opportunities, ranging from extraction to manufacturing. One advantage that can be realized with source separation-based processes is that the associated economic benefits are available wherever valuable materials are discarded, and not restricted to locations possessed of raw material deposits. Some examples demonstrate these economic benefits (note that the final example describes an actual program):

1. For every 10,000 tons of waste materials recycled, 32.7 jobs are supported, compared with only 6.5 jobs if those same materials are used for landfill.[21]
2. For every job created by harvesting paper from trees, five jobs are created if the same amount of paper is recycled.[22]
3. Whereas recycled aluminum is priced at around 20¢/lb (450 g), solely on the basis of its value as a secondary material, it is worth up to 40¢/lb (450 g) if it receives some preliminary processing such as shredding and/or baling; up to 60¢/lb (450 g) if it is smelted into secondary ingots and sold to industry; and, it is worth over $1.00/lb (450 g) if it is fabricated into storm windows.[23] The added value at each step represents employment opportunities available locally.
4. A publicly created, for-profit corporation, Western Community Industries (WCI) of Fresno, California, manufactures over 800 tons of cellulose insulation each month from recycled newspaper. WCI employed 22 people in 1979, thereby contributing $300,000 in wages to the local economy. Sales for FY1980 were about $1.8 million. The whole project was established with a grant of $350,000.[24] This is much less than the capital cost associated with additional capacity required to burn this amount of paper in an energy recovery system. (See next section.)

In addition to describing the wide range of benefits generated by source separation programs and recycling, this discussion has illustrated some of the constraints placed on such programs by the waste disposal community. For example, a person in charge of a program designed to take advantage of all the potential benefits described above would be better entitled "Resource Manager" instead of "Waste Manager," and a wider range of skills would be required of the person who filled the position. Unfortunately, this is the sort of change that an institution is so capable of arresting. Nevertheless, as the example of WCI illustrates, these benefits have begun to be realized.

10.1.9 Source Separation Versus Centralized Resource-Recovery Process Systems

Centralized processing systems, often referred to by the waste disposal industry as "resource recovery" or even "recycling" systems, are being proposed by that industry as *the* solution to the interconnected problems of dwindling landfill capacity and increasing operational costs encountered by disposal programs. Yet these systems manifest the same characteristics as the problem-stricken programs they are intended to replace. For example, these systems continue to treat waste as a "given." Also, centralized processing systems continue reliance on engineering solutions and, thus, continue to insulate those involved in managing the systems from pressures beyond their area of professional expertise. Finally, no new performance criteria have been developed to evaluate centralized systems in comparison to other alternatives, such as source separation. In the absence of such criteria, it is difficult to interpret claims that centralized systems are "solutions" other than in terms of the disposal-efficiency criterion applied to other programs promoted by the waste disposal industry.

At this stage, however, one critical difference separates centralized processing systems from other waste disposal programs: centralized processing systems have not yet come into widespread use. With this in mind, the list of benefits attributable to source separation programs can be extended to include avoided disbenefits attributable to these so-called "resource recovery" facilities. Some of these avoided disbenefits are discussed in the following paragraphs.

Centralized Processing Systems Are Expensive. Current estimates fall in the $50 to $100,000/ton of daily capacity range.[25-27] Financing costs alone for a 2000 tons/day plant can approach $4.9 million.[28] Feasibility and prebonding studies can cost upwards of $150,000.[29] Taxpayers bear the risk, reducing their "borrowing capacity" available for other worthwhile projects for which there may be no alternatives. Source separation is an alternative to centralized processing systems. It requires relatively little investment; for example, a comprehensive source separation program for a city the size of Washington, D. C. could be established for approximately $3 million, including the cost of planning and public awareness,[30] whereas community recycling groups have demonstrated that source separation programs can be initiated on shoestring budgets and small grants. Furthermore, since materials collected through source separation programs do not require capacity at a centralized recovery facility—should such a facility prove to be a necessity—capital costs for the facility can be reduced significantly for every ton per day recycled.[31] Reductions of 20% are achievable even in today's disposal-oriented setting,[32] whereas even greater savings could be realized should recycling become more accepted and the systems that facilitate it better established.

Centralized Processing Systems Are Inflexible.   These facilities require a guaranteed supply of materials for the duration of their 20-year life-cycle and is usually a condition of the bonding arrangements that finance such plants. Unfortunately, such guarantees lock into place the waste generation patterns for which the facility was sized, and contributes to the controversy over "flow control." The creation of source separation programs involves shorter lead times—usually 2 years instead of the 7 years required for centralized facilities—and they can be modified to adapt to changing material flow patterns and markets.

Centralized Processing Systems Create Few Employment Opportunities.   Centralized systems are highly capital intensive, a serious shortcoming in the urban areas where most of these systems are under consideration. One proposed system of several such facilities would require an investment of $1.8 million for each job created, whereas another facility that has ceased operation generated only six jobs for its $3.5 million investment. In contrast, source separation collection and processing systems can generate one job for each $10 to $15,000 investment, with more sophisticated systems requiring somewhat higher investments in the $30 to $50,000 per job range. A 1000 tons/day recycling source separation program would employ several hundred people, most of them young and underskilled, another important consideration in urban areas where centralized recovery systems are targeted.[33]

Centralized Processing Systems Create Pollution Control Problems.   These include both air and water pollution control and the processing of hazardous wastes. Pollution control equipment, which represents a large portion of total plant costs, must control a wide range of substances, including the following: toxic gases from burning plastics and pesticides; heavy metals volatilized from noncombustibles; smoke, soot, and odors; organics in waste waters; and hazardous wastes in ash residues. Despite the expense of this equipment, it is not clear that adequate levels of control have yet been achieved, a decision that is made more difficult by the absence of emission standards for some of these pollutants (dioxins, etc.). Plant emissions are connected with employment as well. For example, a chemical plant that could be expected to generate the same level of pollution as a large centralized waste recovery plant, and face the wrath of the general public for doing so, would employ approximately 1000 people and generate 2000 to 4000 additional jobs in related low-polluting industries; the garbage plant would create only about 49 jobs.[34] *But*, source separation programs generate much less pollution than either the chemical plant or the garbage plant.

Centralized Processing Systems Recycle Inefficiently.   Just as source separation is an inefficient approach to disposal of wastes, centralized "recovery" systems, which are designed primarily as "trash compactors" are inefficient at recycling resources. Recycling is concerned with both energy and material flows. In the energy dimension, for example, if only one-half of the energy content of trees left standing as a result of paper recycling is credited toward recycling, a net energy savings is realized through recycling as compared with burning the same paper in an energy recovery system. This result applies to newsprint, printing and writing paper, tissue and sanitary paper, and corrugated cardboard; and it remains valid for distance of up to 1000 miles between the point of discard and the recycling mill.[35] As a further illustration, each representative ton of "garbage" that is recycled conserves 21.4 million BTU (5.4×10^6 kcal), whereas burning the same garbage releases only about 10 million BTU (2.52×10^6 kcal).[36] And recycled paper can itself be recycled, saving even more energy. Centralized recovery systems do no better in the materials recycling dimension. The high degree of purity required by markets that purchase secondary materials is extremely difficult to achieve once those materials have been mixed with "the trash." As a result, markets for mechanically separated materials are limited, and the prices offered are generally lower than those available to suppliers of higher-quality source-separated materials. And, like other aspects of centralized recovery systems, mechanical separation systems are expensive; for example, one glass recovery system currently in use cost $1 million after an initial estimate of $600,000.[37] In contrast, source separation programs require no complicated equipment, and what equipment is used is generally inexpensive in comparison.

The point that this discussion has attempted to illustrate is that source separation programs offer significant benefits relative to centralized recovery systems. These benefits suggest that source separation programs should be considered *before* a firm commitment is made to centralized systems. Finally, in response to the one significant argument put forth by the waste disposal community against source separation programs—that such programs are impractical because of the public's supposed unwillingness to participate in source separation—it is worth wondering what the effect on that "unwilling" public would be of a public relations program funded by the $65 to $85,000 saved by reducing the capacity of such a plant by just 1 ton/day, or whether even better results could be achieved by the straightforward payment for one year of the associated 9 to 12¢/lb (450 g) of delivered recyclables. The plausibility of these irreverant suggestions is just one further indication of the importance of evaluating all options before committing to a centralized "recovery" system.

10.1.10 Summary of Source Separation Program Incentives and Benefits

Section 10.1.7 has described many of the benefits of source separation programs. This topic constitutes an important aspect of any discussion of recycling activities because an important part of such activities lies simply in getting the program started. This task requires that a large number of widely different people be convinced that they should support recycling. These people include small numbers of top officials, large numbers of ordinary people, and several types of intermediaries. It is the benefits described in this section, along with the sense of real possibility that accompanies the operational information discussed in the remainder of this chapter, that will attract these people to the support of source separation programs.

Materials and Markets

The question of what to collect is one that rarely arises in waste disposal programs. It is a question that was answered during the genesis of such programs: any discards are included. Of course, limitations are sometimes applied, but they generally reflect operational considerations, not different answers to this question. In contrast, the question of what to collect is central to the operation of source separation programs. Its answer illustrates the purpose, as well as the interdependencies, of the recycling system of which source separation programs are a part. This section addresses the nature of the decision "what to collect," provides answers that have been adopted by operating programs, and addresses a number of related matters.

The Role of Markets

Source separation programs exist to provide for efficient utilization of material resources. Efficient collection and handling are simply means to that end. However, unlike disposal programs that control the "utilization" of what they collect (by paying for it), utilization in the source separation sense lies outside of the purview of source separation itself. Source separation programs must rely on external *markets* which also recognize value in what is collected and are, themselves, willing to pay for that value. In the absence of this type of demand, source separation programs would become little more than orderly disposal operations.

The important conclusion to be drawn from this is that source separation programs are demand-constrained. And, while they are supply-constrained as well—particularly in the current setting— the institutional factors discussed in Section 10.1.6 create a "tighter" constraint on the demand side. In practical terms, this means that source separation programs can collect only what markets will accept.

What these programs actually collect is determined by the strength of market demand (as measured by price and stability), the financial constraints under which the program operates, and the ingenuity of program leaders.

Specific Materials

Because of the dependence on markets, general statements regarding what source separation programs collect are difficult to make. The most commonly collected materials, however, are *newsprint, container glass, metal cans,* and *aluminum.* According to one survey, 76% of source separation programs collect newspaper, 16% collect container glass, and 14% collect some metal.[38] Specific information about these materials is provided in Table 10.2. Other materials (even items) can be collected as well. Some of these that have been successfully marketed are described in Table 10.3. Clearly, the question of what to collect must be made on a case-by-case basis in response to the specific realities of local markets.

Marketing Arrangements. In keeping with the flexibility of source separation programs, there are several approaches available for marketing what is collected. Of course, marketing arrangements must be compatible with other aspects of the program; in particular, how the materials are processed prior to marketing.

Table 10.2 Commonly Collected Materials

	Newspapers	Glass Containers	Metal Cans	Aluminum
Types of buyers	Paper mills, cellulose insulation manufacturers, dealers acting as intermediaries between programs, and end-use buyers	Glass container manufacturers	Scrap dealers and detinning plants	Aluminum industry-sponsored recycling centers, aluminum mills, scrap dealers
Prices	Prices tend to fluctuate in response to demand for products containing waste paper; tend to be higher during warmer months	Prices tend to be fairly stable, with a gradual tendency to rise; vary with color, mixing and contamination; higher prices often available for larger quantities	Prices tend to be stable due to limited market; lower prices for bimetal cans; steel cans priced at a rate which varies with #1 bundled scrap	Prices generally stable due to support of aluminum industry for recycling
Contamination	Must be free of nonpaper items and not mixed with other papers, such as magazines; wet paper is difficult to process	No ceramics or stones permitted; metal caps and rings generally allowed only if buyer has clean-up process at plant; garbage may cause load to be rejected	Excessive amounts of food residue may detract from buyer's willingness to purchase materials	Only pure aluminum accepted; fasteners must be removed from structural items
Material preparation	Paper mills require baling, while other markets may not accept bales without assurance of quality of contents	Crushing of clean glass saves money in transportation; however, crushing of dirty glass makes cleaning difficult; bulk loads preferrable	Some compaction, especially flattening, saves money in transportation; baling is not required; material should flow freely when agitated	Similar to metal cans

Table 10.3 Miscellaneous Collected Materials and Items

Material/Item	Description
High-quality office paper, computer paper, and cards	These classes of paper receive a high price in the market, and can provide a significant source of program income. They require tight quality control and must be contaminant-free, but high prices make the additional effort worthwhile. Computer paper must be free of "groundwood," which can be tested for by a simple test usually available through brokers.
Corrugated and kraft papers	This group includes cardboard boxes and brown paper shopping bags. They must be free of other papers, including especially waxed liners which are frequently encountered on cardboard. Prices for these types of paper are generally lower than for newspaper, but their value can be increased through bundling.
Mixed papers	This category of paper is not highly marketable or profitable, and brokers occasionally refuse even to accept such papers as donations during market downturns. However, this category of paper requires less effort on the part of the source separator.
Plastics	This category, which is commonly used in the waste management setting, is too broad for source separation, where classification must be based on reuse application. Some programs collect only plastic milk containers. Interest in PET is increasing, whereas the increasing price of petrochemical feedstocks will tend to make recycling of other classes of plastic more attractive.
Structural lumber	This material, cleaned of nails and with broken sections removed, can be marketed for "less-demanding" construction construction applications, such as toolsheds. Such materials can be sold to the general public at the landfill itself.
Tree trimmings	These materials can be cut up and set aside at the point of collection for sale to local citizens as firewood.
Organics	Food wastes, grass clippings, leaves, and other organics can be composted and sold to local plant shops, farmers, and gardeners.
Wine bottles	These items can be recycled directly as items if suitable arrangements can be made with nearby wineries. This "low-tech" opportunity for reuse can be taken advantage of by creation of new industries to perform the necessary cleaning.
Automobiles	These products can be utilized as a source of still useful parts that can be removed and inventoried in a manner similar to more traditional automobile parts outlets. Automobile tires can be mixed with asphalt to make a paving material.
Motor oil	Used motor oil can be reprocessed for several applications, including coolants and cutting oils. They can also be re-refined for use as lubricant. Automobile stations are often willing to participate as drop-off centers.

The most common marketing arrangement follows the accumulation of materials at a centralized location. When sufficient materials have been collected, or at intervals dictated by other factors such as availability of delivery trucks, materials are delivered to their market. This arrangement may qualify a program for higher unit prices, due to the larger volumes involved. In addition, some markets will deal only with large suppliers.

The advantage of size can be realized on an even larger scale through regional "marketing pools." The higher volume that such pools can generate can permit the members of the pool to achieve access to the most lucrative markets which generally require a high "entry level" supply. Furthermore, because a larger quantity of materials is being processed, the pool can more easily justify investments in equipment which upgrades the value of materials before marketing. Paper balers and glass crushers are examples of such equipment.

Other marketing approaches require less processing and storage of materials and, as a result, are more easily adapted to common mixed-waste disposal programs. One such approach is to unload recyclable materials directly from the collection vehicle to the market. However, this approach introduces additional transportation costs. Another approach that involves a minimum of change from in-place collection systems involves creation of a marketing entity. The purpose of this marketing entity is to act as an intermediary between the collection programs and outside markets. This "brokering" function allows collection programs to insulate themselves from the problem of marketing—a potentially significant problem in the case of municipal collection programs. In effect, the collection program continues to collect materials, and the marketing entity performs the marketing function.

Information about markets with which to make these arrangements can be obtained in several ways; for example, under "Scrap Dealers" and even "Recycling Centers" in the Yellow Pages of local telephone directories. In addition, numerous recycling-related organizations, including those listed in Appendix 10.2 can be contacted.

Contracts.   Because of the importance of the proceeds from sale of collected materials, source separation programs must make every effort to create the best possible financial arrangements with their markets. In this regard, three factors should be considered by the program: price, security, and flexibility. Different arrangements emphasize different mixes of these factors. A *"no contract"* arrangement allows a program to take advantage of seasonal and other fluctuations in demand by selling to the highest bidder. It suffers from the obvious drawback that marketing opportunities are unpredictable and insecure. A *"fluctuating rate"* approach ties the price of materials to the price of a reference commodity on a reference market—in the case of newspaper, for example, to the price of #1 News at a major paper exchange. Usually, this tie is expressed in terms of a fixed percentage of the reference price. This approach offers the advantage of a guaranteed market and occasional peak prices, but it suffers from the drawback of "bottomed out" markets. A *"floor price with escalator"* arrangement creates a guaranteed market, with a guaranteed minimum unit price. When prices climb above the "floor," the unit price is pegged to the reference price, as in the previous case but at a lower percentage. Thus, the protection against "down markets" comes at the expense of reduced prices during "up markets." Arrangements of this type generally require designation of the contracted market as the sole market for materials collected within the jurisdiction of the contracting program. These arrangements also tend to require larger quantities of materials, and are thus more accessible to larger programs and those that operate under some form of cooperative marketing agreement. In addition, the credibility that comes with being involved with a public agency is sometimes useful in convincing prospective markets to enter into such an arrangement.

Because of the high cost of storing large quantities of materials, most programs cannot afford to operate without a guaranteed market for the materials they collect.

Any formal marketing arrangement should address at least the following topics: price structure, duration, quality requirements, acceptable preparation, and terms of delivery. The duration of contracts that specify these matters is generally 1 year in length, as the variability in prices precludes longer-term arrangements. Quality requirements must be clearly specified, as the importance of meeting them cannot be overemphasized. Further, higher quality levels can sometimes justify higher prices. The preparation factor includes the nature of allowable processing within the acceptable quality specifications; for example, some markets will accept baled paper only if they are provided additional assurance that quality requirements on the paper itself are being met. Finally, the terms-of-delivery factor includes such matters as who pays for the delivery of materials to market, how often delivery is made and at whose initiation, and other similar matters. A sample contract is provided in Appendix 10.1.

Stimulating Markets.   Although the primary emphasis of source separation programs during their early stages must be directed toward existing markets, as these programs mature they can undertake the task of stimulating creation of new markets for products made with recycled materials. Such efforts not only further the purpose of source separation programs—utilization of materials that still possess resource value—but also tend to increase the price paid for the materials by their markets. This type of effort ranges from the creation of new industries that use recycled materials, such as Western Community Industries described in a previous section, to modification of existing systems to use products made from recycled materials. An important example of the latter is the growing number of government purchasing programs that specify a minimum "recycled fiber" content in office paper acquired for government use. The experience of the State of Maryland, for example, indicates that the use of recycled office paper can actually *save* money. Since the beginning of the Maryland program in 1977, over 270,000 reams of recycled bond have been purchased, along with substantial quantities of recycled corrugated boxes and paper towels. The price for bond paper has ranged from 20¢ *less* than the price of virgin paper to about 5¢ above that price. Overall, the State has saved $17,000 through these purchases.[39]

The establishment of programs such as this one involves several steps. One approach involves the following actions:

1. Provide for legal support of the purchase program, through either legislative or executive action. Recycling must be embraced as a matter of policy.

2. Develop purchase specifications that designate recycled fiber content and functional requirements. Since the purpose of such programs is to stimulate source separation programs, the recycled fiber content should be specified in terms of percent postconsumer waste; for example, Maryland's program specifies that recycled paper must include at least 80% postconsumer waste. References to "virgin only" should be deleted from existing specifications, unless function requires such specification. Also, guidelines should be established for purchasing such paper; for example, a "cost premium" may be established during the "phasing-in" period.

3. Identify current usage of recycled products. It is easier to begin the program by expanding on existing purchases than it is to create new ones.

4. Check on availability of specified products. The program should not be implemented until after adequate supplies of acceptably priced products have been identified.

5. Require certification of the products. The burden of proof that the products meet required "recycled content" specifications should be placed on the vendor.[40]

In closing this section, it is informative to note that purchasing programs such as these establish a direct relationship between the interests of purchasing agents and those of waste managers. The more successful the activities of the former, the greater the demand for the source separation services of the latter, which means reduced pressures on waste collection programs, reduced pressure on landfills, reduced capital requirements for centralized waste processing systems, and so on. The implication is that managers of waste collection and disposal programs, especially municipally operated programs, should, as a part of their normal activity, become actively involved in efforts to implement "recycled content" purchasing programs.

10.1.11 Summary of Materials, Markets, and Programs

Source separation programs can be designed to handle a wide range of useful materials. However, it is the marketability of these materials, not their collectibility alone, that determines what materials can actually be collected by such programs. This marketing aspect creates new responsibilities, as well as opportunities, for managers of waste-oriented programs. Such changes typify the movement toward the recycling perspective which gives purpose to source separation programs.

Program Models. The same value that motivates the existence of source separation programs also creates a wide range of operating models for such programs. The relative cleanliness of source separated materials contributes to creating these options. And, because of the inherent flexibility of the several options, programs can easily be switched from one model to another to adapt to changing program needs. In this section, several program models are discussed. The discussion draws heavily from the U.S. Environmental Protection Agency publication "Operating a Recycling Program: A Citizen's Guide."[41]

Drop-Off Centers. The simplest type of program is built around drop-off centers—facilities to which individual recyclers deliver their source separated materials. These centers can be located at shopping centers, grocery stores, churches, and other places convenient to recyclers; the location is relatively unlimited due to the generally clean character of many of the materials collected. Centers can also be located at landfills, creating alternative "disposal" options. These centers can be operated in two modes, attended or unattended, which are discussed separately below.

Attended Drop-Off Centers offer the advantage of having personnel available, on either a paid or volunteer basis, to assist in unloading and handling materials, as well as to insure that necessary quality control is maintained. In addition, attendants have the opportunity to educate people who visit the center and to "compliment" them for their participation. Programs of this type also create an opportunity for useful community-service work for youthful law offenders, senior citizens, and sheltered workshop members.

Unattended Drop-Off Centers offer the advantage of minimal overhead and coordination expenses. They can also operate on a 24-hr basis. However, they provide little opportunity for quality assurance activities and there is no assistance available for recyclers. What this approach does offer is an inexpensive way to introduce source separation practices into new communities, since they require little equipment other than drop boxes and they can be expanded upon as participation increases. Since participation is less convenient in this approach, however, low participation can be anticipated.

Buy-Back Programs. The drop-off center model can be expanded to include the concept of paying recyclers for the materials they deliver to the center. These so-called "buy-back" programs usually offer payments only for higher-value materials, such as newsprint and aluminum, while accepting other recyclable materials as donations. Unit prices for materials are generally tied to prevailing market prices.

Programs of this type illustrate, in explicit economic terms, the *value* of materials that would otherwise be discarded as waste, an important educational benefit. However, operation of such programs requires more work than simple drop-off programs (weighing, accounting, and associated paperwork), as well as additional equipment (scales, calculators, safes for money storage). Some of this additional paperwork can be avoided by limiting payments to community groups selected by the source separator, so that the number of payments is substantially reduced. One program that operates according to this model has returned over $100,000 annually to community groups in its area.[42] Of course, money paid to source separators is unavailable for balancing program costs, so this approach may not be feasible for all programs.

It is worth noting that programs of this sort have been operated by the aluminum industry for several years. By one count, 1300 industry-related recycling centers were paying source separators for aluminum in 1977.[43] One of the companies involved recycled over 2.9 billion aluminum containers during 1977, and paid out over $20 million for them.[44]

Collection Programs. Although the preceding models can lead to viable source separation programs, the model that is most likely to lead to high levels of participation involves some form of collection service. This service can take several forms, depending upon the number of passes involved in the overall collection of discarded materials, the number of materials collected, and the arrangements between source separators and the collection service. Although these three dimensions are interconnected insofar as decisions about programs specifics are concerned, it is useful to discuss the options on the basis of the number of passes.

Single-Pass programs generally operate using either racks mounted on normal collection vehicles or by towing a trailer. The "rack option" is useful for collection of a single material, usually newspaper. Racks are mounted low along the side of the collection vehicle, or high at the rear, with side-mounts being more common due to ease of fabrication, installation and, especially, loading/unloading. Capacity of these racks ranges from 1 to 1.5 yd³ (0.76–1.14 m³) or 500 to 900 lb (225–405 kg) of newspaper.[45] Provision must be made for handling materials in excess of that capacity. Such provisions can include intermediate transfer receptacles, "roving" transfer vehicles, and, as a last resort, disposal of excess separated materials along with the mixed waste.

The trailer option provides greater capacity than the rack option, and offers greater flexibility in selecting materials to be collected. However, trailers can prove to be difficult to maneuver, rendering this approach infeasible in many programs, and a third option is currently emerging to handle the single-pass, multimaterial programs that trailers are sometimes incapable of handling. This option involves the use of compartmentalized collection vehicles which are capable of storing both mixed waste and several types of source separated materials.[46] However, as mentioned in Section 10.1.6, the sizable investment in single-compartment collection vehicles creates a significant obstacle to the widespread usage of such vehicles.

Whichever collection mode is selected, the program must also provide for the unloading of separated materials. This can be accomplished most efficiently by designing an unloading and storage facility at the landfill (or transfer station) where the mixed waste is unloaded, thereby avoiding additional transportation costs. If racks are to be used, it is important to verify that they do not create ground clearance problems at the landfill.

Single-pass programs offer the advantage of avoiding the expense of additional collection routes, as well as the blanket coverage associated with mixed waste disposal programs. However, they fail to clearly distinguish between recyclable materials and "waste." Nevertheless, this type of program affords an opportunity for relatively easy transition to recycling-oriented programs.

Multipass programs collect separated materials apart from whatever mixed-waste collections are in operation. As a result, this approach offers greater flexibility in responding to the recycling demands of an area. Special purpose vehicles and trailers can be utilized, and modified or replaced as program needs evolve, so that multimaterial operation is greatly enhanced. Furthermore, since this collection is "separated" from the waste collection, the vehicles used in the collection can be incorporated into the process of promoting participation in the recycling program; for example, vehicles can be painted thematically and with the program logo.

Multipass programs create several options for arrangements between the program and its "suppliers." Subscription or call-in service creates an opportunity to begin a program without incurring the high cost of broad—and, initially, unnecessary—coverage. Collection services can then be performed with or without a fee, depending on the intent of the program and cost considerations. Another arrangement involves call-in service to neighborhoods that can demonstrate in advance a prescribed minimum level of participation. As the level of participation increases in such neighborhoods, service can be expanded from call-in to general service of the area, an expansion that can lead to participation of "uncommitted" recyclers.

Other Program Models. To this point, the discussion has focused on substitution of source separation activities for waste collection programs. However, the resource management perspective that underlies source separation leads to areas of program activity that have no counterpart within the waste management perspective. For example, the management of landfill capacity as itself a resource becomes a possibility within the source separation perspective. As a result, increased incentives are created to expand source separation activities beyond traditional residential waste collection boundaries. Three approaches for new program areas are discussed below.

Office Paper Recycling programs have been initiated by over 450 private and governmental office facilities.[47] Paper collected by these programs is of higher quality and value than most other papers collected through source separation. These papers include letterheads, dry copying papers, computer tab cards, and printouts. Office paper programs are generally designed to include some type of desk-top separation, with two important variants being separation according to particular classes of recyclable paper and separation of "mixed" recyclable papers. Collection can be effected along two lines. In one model, the office itself provides for collection and storage of separated papers at a centralized pick-up point. With this arrangement, the recycling program usually purchases the paper from the office at a negotiated price. In the other model, the recycling program provides for collection of papers within the office. In return for this service, the program usually accepts the paper as a donation. Participation in these programs is made more attractive to office management if the program provides paper shredding services.

Commercial Programs can also reduce pressures on landfills. Establishments such as restaurants, furniture stores, and automobile service stations usually pay to have their recyclable glass, paperboard, and oil taken away by waste haulers who deliver to landfills. These businesses can be attracted to source separation programs if they can be shown that switching will save them money. Their switching also conserves landfill resources.

Landfill Management practices can also be modified to conserve landfill capacity. The simple practice of creating a recycling station at the landfill creates one last opportunity for people to avoid "throwing away" resources. This approach can be facilitated by offering "dumping credits" in return for delivery of specified amounts of recyclable materials prior to visiting the landfill. Another approach is to provide a "high grading" service at the landfill face. "High grading" refers to the practice of hand sorting materials after they have arrived at the landfill. It involves four steps: (1) separation of useful materials and items, (2) sorting them according to their reuse potential, (3) upgrading them, and (4) selling them for reuse. During 1 year, one such program which emphasized metals, construction and demolition wastes, and compostables, sold more than $150,000 worth of materials removed from a 400 tons/day landfill.[48]

Program Model Summary. This discussion illustrates the range of program models upon which source separation programs can be based. These models include adaptations of mixed-waste collection programs, but also go beyond the "waste collection" perspective in an effort to implement the "resource management" perspective that underlies source separation activities. However, although these models are useful, the important criterion for such programs is how well they redirect material resources toward reuse. Success in this dimension depends upon how well a program responds to local conditions and how creatively it takes advantage of them—not upon how well it conforms to the program model on which it is based. Thus, unlike the institutionalized waste collection model, the models for source separation programs discussed here are important only as means to achieve that success.

10.1.12 Case Studies

The introduction to this chapter closed with the observation that, "Anything that exists is possible." Subsequent discussion has addressed source separation in primarily abstract terms. That discussion has, in effect, addressed issues of possibility: benefits that could be realized, institutional constraints that impede what could be, markets that would be involved, and models of how it would be done. But, that discussion was spared the task of *proving* that such programs are indeed possible by the actual existence of programs that generate the benefits described, deal with the constraints, find markets, and use models. This section describes five real programs to emphasize that existence; coincidentally, it also provides the kind of operational/programmatic information generally found in a handbook of this type.

Table 10.4   E.C. Ology Program Highlights

Multifaceted recycling program
   Curbside recycling services provided to three communities (El Centro, Albany, Kensington)
   Buyback services
   Drop-off
   Commercial collection
   Satellite drop-off programs
   Wine bottles (ENCORE) recovery programs
Weekly curbside collection provided on nonrefuse collection day
Program was designed to leverage the more cost-effective buyback service against the more costly and service-intense curbside program
Curbside program tonnages increased a reported 56% from 1978 to 1981 without a significant increase in collection costs
Extensive use of labor in overall program
Unique use of labor sources including, Comprehensive Employment Training Act (CETA), Work Incentive Program (WIN), Department of Rehabilitation, probation department referrals, volunteer workers, and contract labor
Use of program as a job skills training site
Unique computer system used for financial accounting of individual programs

Source: California Solid Waste Management Board, "Case Studies of California Curbside Recycling Programs," Vol. 1.

Description of Cases.   Among the most thorough evaluations of source separation programs is the set of case studies conducted by the California Solid Waste Management Board.[53] That set included five programs selected to represent the wide range of options available for source separation programs. The study itself focused on curbside collection programs. However, three of the programs selected on that basis offered "full service" recycling operations as well. The five cases were:

   E.C. Ology, City of El Cerrito, California.
   Palo Alto Recycling Program, City of Palo Alto.
   Downey at Home Recycling Program (DART), City of Downey.
   Fresno/Clovis Metropolitan Area Recycling Service (MARS), Cities of Fresno and Clovis.
   Recycle 3, Redwood Empire Disposal, located in city of Santa Rosa.

Tables 10.4 through 10.8 highlight these five programs. Table 10.9 identifies the assignment of key areas of responsibility. These exhibits illustrate the variety represented by the five cases; for

Table 10.5   Palo Alto Recycling Program Highlights

Combination of recycling services provided to the community
   Curbside recycling program
   Drop-off
   Composting project
   Wood recovery system
Activities other than curbside collection account for over 50% of revenues received from the sale of materials
The integrated recycling program diverts over 19% of the community's municipal solid wastestream
Equal sharing of gross revenues from the sale of collected newspaper with a community nonprofit organization
Storage receptacles (burlap bags) are provided to the participants
Participants can receive free dump passes as an incentive to recycle
Extensive in-kind services (labor, management, and public awareness) provided by the city
Employs one of the highest paid recycling staff in the state
Use of labor incentives for collection staff (profit-sharing and time incentives)

Source: California Solid Waste Management Board, "Case Studies of California Curbside Recycling Programs," Vol. 1.

Table 10.6   Downey at Home Recycling Team (DART) Program Highlights

Single-faceted recycling program
  Curbside recycling services
Collection of mixed recyclables (all materials placed in one container)
Minimal source separation required of participants
Bimonthly (every other week) collection
Recycling program operates through an agreement among the city, franchise hauler/collector, and processor/market
No processing of collected material by recycling operation—recovered materials sold to processor/market in "mixed form"
Use of processing line and hand-picking operation by processor/market to separate materials
25% of collected materials are not recyclable
Use of large, brightly colored storage receptacles provided by the operator
Minimal labor need (one collector)

Source: California Solid Waste Management Board, "Case Studies of California Curbside Recycling Program," Vol. 1.

example, curbside-only (DART) to "full service" (E.C. Ology), high level of city involvement (Palo Alto) to totally private operation (Recycle 3), direct "buy back" (E.C. Ology) to city billing and credits (MARS). These exhibits also illustrate that government tends to play an important role in successful programs.

Information concerning the areas serviced by each of these programs, as well as their collection rates, is presented in Table 10.10. Collection data are presented graphically in Figures 10.1 and 10.2. The service area data demonstrate that large population alone does not assure high participation, nor does high population density. The collection data show that newspaper constitutes the largest single class of material collected, with glass second. It is noteworthy that E.C. Ology collects both glass and wine bottles separately, demonstrating the ability of source separation programs to recover both "material value" and "manufactured value."

Public Awareness. As was suggested in Section 10.11, the success of a source separation program depends on the success of the effort to overcome the institutionalization of "waste," including resistance at the level of the individual. The case study programs addressed that resistance through "public awareness" activities. Such activities involve much more than publication of "date/time" information generated by waste disposal programs. The scope and variety of these activities are illustrated by the list presented in Table 10.11. Each of the programs undertook a

Table 10.7   Recycle 3 Program Highlights

Single-faceted recycling program
  Curbside recycling services
Development and usage of prototype collection vehicle
Use of specially-designed, standardized storage containers (three separate, interlocking, color-coded containers)
Good visibility of program vehicle and receptacles
No processing of collected material by recycling operation
All collected material sold by recycling operator FOB at its own yard, thus eliminating transportation and processing costs
Only program to be completely operated by a private operator—all administration, collection, marketing, processing, and publicity are handled either by the private refuse hauler or other private entities
Extensive media usage for public awareness
Minimal labor needs
  Incorporation of recycling labor needs into existing solid waste management operation
  Division of responsibility in major labor-intensive elements of recycling (collection, marketing, public awareness)
Use of labor incentives for collection staff (time incentives)

Source: California Solid Waste Management Board, "Case Studies of California Curbside Recycling Programs," Vol. 1.

Table 10.8   Fresno/Clovis Metropolitan Area Recycling Service (MARS) Program Highlights

Combination of recycling services provided to the community

    Curbside recycling service provided to: residential, single-family dwellings, multifamily dwellings (apartments) (provided on a limited basis)

    White office paper program

    Commercial, corrugated collection

Activities other than curbside collection materials account for over 41% of direct revenues received from the sale of materials

Serves the largest population (313,000) of any other curbside program in the state

One of the lowest metropolitan population densities (1600 persons/mi^2) served by a curbside program

Monthly curbside collection provided on nonrefuse collection day

Charge for recycling program included in refuse collection bill ($0.06/month per account)

Recycling incentive credit for regular participants in the curbside program ($0.40 per month credited to refuse collection bill)

Extensive use of labor (collection and processing)

Extensive economic commitment (subsidy) by the city to finance the program

Source: California Solid Waste Management Board, "Case Studies of California Curbside Recycling Programs," Vol. 1.

majority of the activities listed, with two employing nearly all of them. This list illustrates that public education and public involvement, not merely one-way information dissemination, play a critical role in successful programs. It is also noteworthy that the list of activities requires skills not generally required for the successful operation of disposal programs: community relations, communications, art, etc. This observation supports the observations made in Section 10.1.5 concerning "bureaucratization of program management" as an institutional obstacle to source separation programs.

    Collection, Processing, and Marketing.   The two critical operational aspects of source separation programs are collection and processing of materials. Table 10.12 provides information concerning the manner in which the five programs handled these two practical matters. This table further illustrates the variety of available options. Collection frequency ranges from weekly to monthly, representing different trade-offs between the reduced household storage requirements of the former and the reduced collection costs of the latter. Collection may be scheduled on the same day as refuse collection to reduce memory requirements placed upon participants in either collec-

Table 10.9   Program Responsibilities

Program/Location	Coord/Admin	Collection	Public Aware	Processing
E.C. Ology/El Cerrito	City	City	City	City
Palo Alto Recycling Program/Palo Alto	City	Private (fran hauler)	City	Private (fran hauler)
Downey at Home Recycling Team/Downey	Private (fran hauler)	Private (fran hauler)	Private and city	None[b]
Recycle 3/Santa Rosa	Private (fran hauler)	Private (fran hauler)	Private[a]	None[c]
Fresno-Clovis Metro Area Recycling Service (MARS)/Fresno-Clovis	City	City	City	City

[a] Recycle three contracts with a public relations firm for the public awareness activities.

[b] Mixed recyclables are collected by the program hauler; upon obtaining a full truckload of materials, they are taken immediately to market where they are sold and then separated, processed, and resold.

[c] After the recyclables are collected, they are immediately sold to one market that is located at the same place as the recycling program. Any processing required is performed by the market entity.

Source: Ref. 53.

Table 10.10 Service and Materials Information Fiscal Year 1980–1981

Prog/Loc	Total Pop.	Pop. Density[a]	One-family residence	Multifamily residence	Material	Curbside TPM	Other
E.C. Ology/ El Cerrito	43,000	na[b]	13,499	na	News	97.9	Buyback
					Ferrous	9.2	Drop-off
					Nonferrous	1.3	Satellite
					Glass	50.4	
					Wine bottles	1.5	Commercial
					Total	160.3[c]	
Palo Alto Recycling/ Palo Alto	57,000	na	15,500	8,500[d]	News	122.0	Drop-off
					Ferrous	11.5	Compost
					Nonferrous	3.0	Wood
					Glass	80.0	
					Cardboard	11.5	
					Total	228.0	
Downey at Home Recycling Team/Downey	82,662	6,503	19,812	13,876	News	35.7	None
					Ferrous	9.0	
					Nonferrous	0.8	
					Glass	27.2	
					Total	72.7	
Recycle 3/ Santa Rosa	83,205	3,000	23,000	na	News	147.00	None
					Ferrous	.25	
					Nonferrous	1.40	
					Glass	65.00	
					Total	213.65	
MARS/Fresno	313,200	2,240	84,000	34,500	News	90.3	Drop-off
					Ferrous	12.2	
					Nonferrous	.7	Wh paper
					Glass	61.8	Commercial
					Total	165.0	

[a]Population per square mile.
[b]Not available.
[c]Estimated from population and single family statistics provided in report.
[d]Figure obtained from "July 1981 Curbside Recap" published by E.C. Ology.
Source: Ref. 53.

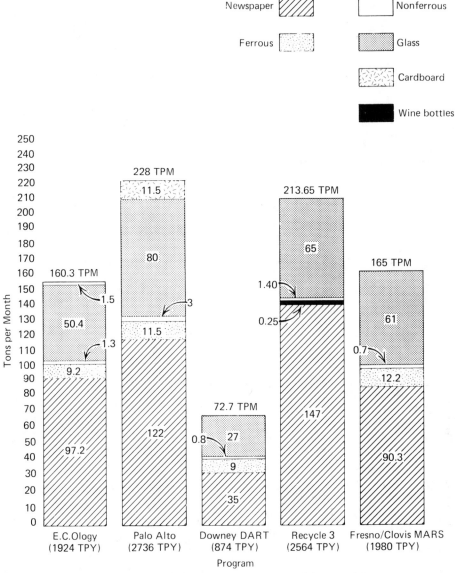

Fig. 10.1 Materials recovery, tons per month by material.

tion or to provide an opportunity for integrated recycling/disposal collection; alternatively, the source separation collection may be scheduled on a different day to avoid "mix-ups" with refuse collection or to establish a separate identity for resource-oriented programs. All the programs shared an effort to minimize collection costs by minimizing the number of people in a collection crew, but they differed in the equipment used to perform that collection. The variety in collection equipment illustrates the flexibility of programs to take advantage of available resources while still providing the necessary level of service. The same diversity is manifested in the processing equipment selected by the programs involved in that activity. As was pointed out in Section 10.1.10, this processing increases the market value of collected materials. However, two of the programs have chosen not to process their materials prior to marketing, thereby simplifying their operations. Of special note in this regard is the processing arrangement adopted by Recycle 3: materials collected by that program become the property of an independent processing/marketing firm, effective upon the return of the collected materials to Recycle 3's yard. Such an arrangement may be useful to other in-place "refuse" collection programs considering expansion into recycling services.

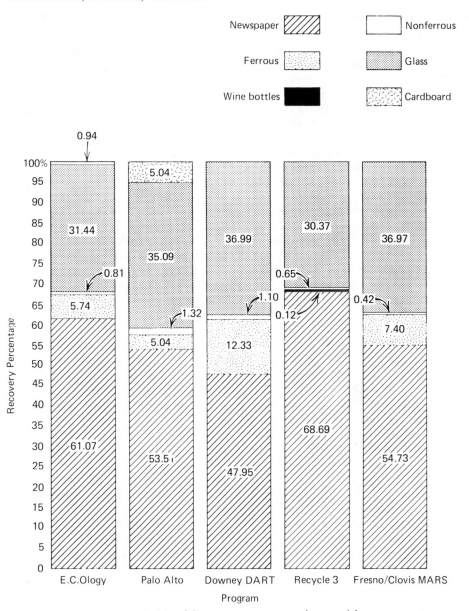

Fig. 10.2 Materials recovery percentages by material.

Program Economics. The final question addressed by the California study was the first question asked by the waste disposal community regarding source separation programs: How much does it cost? Constrained to address this question with an accounting system derived from principles similar to those that led to the institutionalization of waste in the first place, the study found that, under the least favorable set of accounting assumptions, program costs exceeded revenues for the four programs for which necessary data were available. In particular, this finding was based on a comparison of full operating costs, including in-kind services, with revenues received directly from the sale of collected materials. Table 10.13 provides the actual dollar figures upon which this finding was based, while Table 10.14 presents the same information on a unit-cost basis. Importantly, the latter presentation shows that the net unit cost of Recycle 3's curbside collection is actually *less than* the unit cost for collection, hauling, and disposal of refuse in the county where that program operates. Similar calculations were also performed on a "full-service" basis for the

Table 10.11 Public Awareness Activities

General approach
  Strategy and planning
  Log/slogan design
Generating active support
  City endorsement and community letter
  Community advisory committee
Media, print, and electronic
  Press releases
  Feature stories and editorials
  Public service announcements (PSAs)
  Newsletters and programs
Printed materials
  Brochures and fact sheets
  Posters
  Vinyl decals and stickers
  Displays
  Slide shows
  Reminder calendars
Community outreach
  Public surveys
  Door-to-door canvassing
  Billing inserts
  School, church, civic groups
  Workshops/seminars
  Program newsletter
  Mobile exhibit
Paid advertising
  TV and radio
  Yellow pages
  Newspaper and billboards
Other activities
  Fundraising for community groups
  Special events (e.g., parades)

Source: Ref. 53.

two programs that operated on that basis. Tables 10.15 and 10.16 present those calculations. Comparisons of corresponding entries for "full-service" and "curbside only" programs demonstrates a favorable impact on costs associated with expanded program services. Table 10.16 shows that, on this expanded basis, the Palo Alto program joins Recycle 3 in providing services at a cost advantage relative to local disposal programs.

Program economics were also evaluated for a less unfavorable set of accounting assumptions. For this second evaluation, full operating costs, excluding in-kind services, were compared with direct revenues (from sale of collected materials), indirect revenues (e.g., collection and disposal cost avoidance), grants, and other financial assistance. The results of this second evaluation are presented in Tables 10.17 through 10.20. As would be expected, this second approach "improves" the cost performance of these programs, with both Recycle 3 and Palo Alto showing absolute profits. Moreover, the unit cost of Palo Alto's curbside program drops below the local disposal cost, whereas the unit costs of both DART's curbside collection program and MARS' overall program drop to within about 30% of local disposal costs.

It is important to note that neither set of assumptions includes a mechanism for crediting benefits associated with recycling materials (as discussed in Section 10.1.7) to the program that collects them. This omission, which leads to underestimation of actual program benefits, is an accounting artifact, which suggests that the accounting results presented here are themselves "artifacts." These artifacts are joined by another: the omission of a debit charged to disposal programs for the opportunity cost associated with the resource value of disposed materials. And,

Table 10.12 Collection and Processing

Prog/Loc	Coll Freq	Same-Day Refuse Col	Collection Crew Size	Collection Equip	Processing Materials	Processing Equip	Processing Crew Size
E.C. Ology	Weekly	No	1/Vehicle	(3) 2-Ton collection trucks each modified to carry 6 bins	Yes	Forklift Baler Computer sys Alum shredder Magnetic separator Storage bins (1½–40 cy)	8 Ftime 9 Ptime[d]
Palo Alto	Weekly	Yes	1/Vehicle	(2) Mod van cabs and chassis (2) Midway trler and bin systems	Yes	Forklift (5)30 cy bins (5)40 cy bins (2)15 cy covered bins (1) Can crusher/separator/conveyor (1) Baler	2 Fulltime, addtl part time as needed
Downey at Home Recycling Team	Every other week	Yes	1/Vehicle	(1) Coll truck cab and chassis with 37 cy body and front loading bucket	No[b]	n/a	n/a
Recycle 3/Santa Rosa	Weekly	Yes	1/Vehicle with alternate[a]	(2) Evo-Eco Lodal coll trucks (1) Bin truck, used as needed	No[c]	n/a	n/a
MARS/Fresno	Monthly	No	1–2/vehicle (avg/.25)[a]	(3) ¾ Ton coll trucks (3) Midway trler and bin systems Used for other aspects: (3) ¾ t pickup (1) 1 t pickup (1) ¾ t flatbed (2) ½ t pickup	Yes	(1) Forklift (1) Baler (1) Mag can sorter/flattener (100) 3 cy bins (1) 50 cy dropbox	3 Fulltime

[a]The Downey program does not perform any processing of materials. However, they do have a processing/market contract; the collected mixed recyclables are taken directly to this market.
[b]The Recycle 3 program does not perform any processing of materials. The market/processing are located at the same location. When the collection trucks return to site, the materials become the property of the processor/market.
[c]On heavier routes, more than one collector is used per route. This need is on an infrequent basis.
[d]The entire E.C. Ology recycling program employs these numbers. Numbers include rehabilitation clients. Does not include another ¾ man year provided in-kind by city.
Source: Ref. 53.

237

Table 10.13 Program Cost Analysis—Comparison of Full Operations Costs[a] with Direct Program Revenues Curbside Programs Only[b] Fiscal Year 1980–1981

	Palo Alto	Downey	Recycle 3	MARS
Operating costs				
Labor and fixed overhead	158,400	44,445[g]	93,936	
Variable expenses				389,000[h]
(operating and maintenance)	40,700	18,362	33,143	
Public awareness	10,000	-0-	15,199	
In-kind services, ongoing				
(incl recycling incent)	111,000	-0-	-0-	38,400
Revenue sharing[c]	34,000	-0-	-0-	-0-
Amortization[d]	26,021	26,300	19,382	19,162
Subtotal	380,121	89,107	161,660	446,462
Direct Revenues				
Secondary materials sales	127,840	16,460	56,464	76,357
Recycling charge[e]	-0-	-0-	-0-	34,600
Subtotal	127,840	16,460	56,464	110,957
Annual net profit<loss>	<252,281>	<72,647>	<105,196>	<335,605>
Annual net profit<loss> per household[f]	< $16.28>	< $3.67>	< $4.57>	< $3.99>
Monthly net profit<loss> per household[f]	< $1.36>	< $0.31>	< $0.38>	< $0.33>

[a] Includes ongoing in-kind services.
[b] Palo Alto and MARS programs also provide other recycling services (e.g., drop-off, white paper, commercial collection, etc). Program cost analysis for their entire operations are presented in Table 10.14.
[c] Palo Alto pays one-half of gross revenues from sale of collected newspaper to California Association for the Retarded.
[d] Amounts determined by amortizing *state* purchased equipment and/or site improvements over a 5-year period.
[e] Fresno/Clovis Refuse Collection bill includes a $0.06 recycling charge.
[f] Based on service population, single family residences: Palo Alto—15,500; Downey—19,800; Recycle 3—23,000 and MARS—84,000.
[g] The figure was reduced from that quoted in the individual Downey report. Amortization figures were originally included in those operating cost figures. These costs are now included in another line item, only.
[h] Exact figures are not known; this figure is the most current one submitted by the City of Fresno. It is not possible to "breakdown" those costs attributed to the curbside only portion of the program. In all likelihood, the costs would be lower to operate only the curbside aspect.
Source: Ref. 53.

beyond these accounting artifacts, it is necessary to emphasize that the accounting itself was performed for programs operating in the adverse institutional context discussed in Section 10.1.5. Thus, the cost study discussed here raises the question of whether source separation programs are burdened by adverse costs, or merely by adverse cost accounting principles. This is a question from which early disposal programs were spared: The public health emergency that led to their creation superseded accounting. Supporters of source separation programs, and the recycling system as well, will be similarly spared from artificially adverse accounting principles—before a materials-related emergency requires it.

10.2 RECYCLING: A STATEWIDE PROGRAM FOR NEW JERSEY
Mary T. Sheil

Introduction

The State of New Jersey has set a goal of recycling, through material recovery programs, 1.3 million tons of materials annually from the municipal solid waste stream by 1986. To achieve this

Table 10.14 Unit Costs and Revenues—Curbside Programs Only[a] Fiscal Year 1980–1981

Programs	Recovered Materials	Coll Costs[b]	Coll Cost/Ton	Revenues[c]	Revenues/Ton[d]	Net Profit <Cost>/Ton[e]	County Disposal Costs[f]
Palo Alto Recycling Program	2741 TPY	$380,121	$138.68	$127,840	$46.64	<$ 92.04>	$47.00
Downey at Home Recycling Team	876 TPY	$ 89,107	$101.72	$ 16,460	$18.79	<$ 82.93>	$34.00
Recycle 3	2570 TPY	$161,660	$ 62.90	$ 56,464	$21.97	<$ 40.93>	$44.00
MARS	1971 TPY	$446,562	$226.57	$115,957	$58.83	<$167.74>	$50.00

[a] Palo Alto and MARS programs also provide other recycling services (e.g., drop-off, white paper, commercial collection, etc.). Unit cost analysis for their entire operations are provided in Table 10.18.
[b] Full operating costs (e.g., collection administration, processing, etc.), including in-kind services.
[c] Direct revenues, including recycling charge, if any.
[d] Revenues per material will greatly fluctuate; these figures reflect an average revenue per ton of material.
[e] Collection cost/ton minus revenues/ton.
[f] Refuse collection, hauling, and disposal costs per ton for the county in which each program is located. Estimates obtained from preliminary findings for the "1982 Garbage Crisis Cost Survey," SSWMB.
Source: Ref. 53.

Table 10.15 Program Cost Analysis—Comparison of Full Operations Costs[a] with Direct Program Revenues Palo Alto and MARS Programs Only[b] Fiscal Year 1980–1981

		Palo Alto	MARS
Operating costs			
Labor and fixed overhead		$203,690	
			$389,000[g]
Variable expenses (operating and maintenance)		40,700	
Public awareness		10,000	
In-kind services, ongoing (including recycling incentives)		111,000	38,400
Revenue sharing[c]		64,080	-0-
Amortization[d]		26,021	19,162
	Subtotal	$455,491	$446,562
Direct revenues			
Secondary materials sales		$258,600	$129,950
Recycling charge[e]		-0-	39,600
	Subtotal	$258,600	$169,550
Annual net profit<loss>		<196,891>	<277,012>
Annual net profit<loss> per household[f]		< $8.20>	< $2.34>
Monthly net profit<loss>per household		< $0.68>	< $0.20>

[a] Includes ongoing in-kind services.
[b] These two programs provide a wide variety of services in addition to curbside services. Materials are collected from these other program aspects; costs and revenues relative to the total programs are included.
[c] Palo Alto pays one-half of gross revenues from sale of collected material to California Association for the Retarded.
[d] Amounts determined by amortizing *state* purchased equipment and/or site improvements over a 5-year period.
[e] Fresno/Clovis Refuse collection bill includes a $0.06 recycling charge.
[f] Entire residential area included as curbside, and drop-off programs theoretically serve the entire community. Total estimated single family and multifamily residences: Palo Alto—24,000 households; MARS—118,500 households.
[g] Exact figure unknown at this time.
Source: Ref. 53.

goal, state legislation (N.J.S.A. 13:1E-92 *et seq.*) was enacted to provide a $4 to $5,000,000 annual recycling fund for a 5-year period. The fund is allocated as follows:

1. Not less than 45% of the fund must be set aside for grants to municipalities based on the tonnage of materials recycled annually.
2. Not less than 15% for education programs.
3. Not less than 20% for low-interest loans and loan guarantees to recycling businesses and industries.
4. Not more than 20% to municipalities, counties, and the state for program planning, implementation, and administration.

The effective date of the legislation was January 1, 1982, when a $0.12/yd^3 (766 m^3) recycling surcharge was placed on all waste landfilled in New Jersey. Annual revenues from the surcharge are placed in a state recycling trust fund (see Appendix 10.5 for typical programs). *Editor's Note:* The following syllabus is a precursor of the continuing expansion of the initial program and its database.

10.2.1 Background[56]

In 1975, the New Jersey Legislature amended the State Solid Waste Management Act in recognition of the need to develop a comprehensive state solid waste management program.

The amended act required the 21 counties, the Hackensack Meadowlands Development District, and the state to develop a 10-year solid waste management program. A major goal set forth by the legislature was the development of resource recovery programs for material and energy recovery.

Table 10.16 Unit Costs and Revenues—Palo Alto and MARS Programs Only[a] Fiscal Year 1980–1981

Programs	Recovered Materials	Coll Costs[b]	Coll Cost/Ton	Revenues[c]	Revenues/Ton[d]	Net Profit <Cost>/Ton[e]	County Disposal Costs[f]
Palo Alto Recycling Program	5,340 TPY	$455,491	$ 85.30	$258,600	$48.43	<$ 36.87<	$47.00
MARS	2,617 TPY	$446,562	$170.64	$169,550	$64.79	<$105.85>	$50.00

[a] These two programs provide a wide variety of services in addition to curbside services. Materials are collected from these other program aspects; costs and revenues relative to the total program are included.
[b] Full operating costs (e.g., collection, administration, processing, etc.), including in-kind services.
[c] Direct revenues, including recycling charge, if any.
[d] Revenues per material will greatly fluctuate; these figures reflect an average revenue per ton of material.
[e] Collection cost/ton minus revenues/ton.
[f] Refuse collection, hauling, and disposal costs per ton for the county in which each program is located. Estimates obtained from preliminary findings for the "1982 Garbage Crisis Cost Survey," SSWMB.
Source: Ref. 53.

Table 10.17 Program Cost Analysis—Comparison of Full Operations Costs[a] With Direct and Indirect Programs Revenues[b] Curbside Programs Only Fiscal Year 1980–1981

	Palo Alto	Downey	Recycle 3	MARS
Operating costs				
Labor and fixed overhead	$158,400	$44,445[j]	$ 93,936	
Variable expenses				$389,000[l]
(operating and maintenance)	40,700	18,362	33,143	
Public awareness	10,000	-0-	15,199	
Revenue sharing[d]	34,000	-0-	-0-	-0-
Amortization[e]	26,021	26,300	19,382	19,162
Subtotal	$269,121	$89,107	$161,660	$408,162
Direct revenues				
Secondary materials sales	$127,840	$16,460	$ 56,464	$ 76,357
Recycling charge[f]	-0-	-0-	-0-	39,600
Subtotal	$127,840	$16,460	$ 56,464	$115,957
Indirect revenues				
Landfill fee savings[g]	$ 8,890	$ 6,504	$ 15,402	$11,826
Reduced collection costs	37,500[i]	533	93,732[k]	12,367
Grant monies	30,400	26,840	19,728	22,700
Subtotal	$ 76,790	$33,877	$128,862	$ 46,893
Total, All revenues	$204,630	$50,377	$185,326	$162,850
Annual net profit<loss>	< 64,491>	< 38,730>	23,666	<245,312>
Annual net profit<loss> per household[h]	< $4.16>	< $1.96>	$1.03	< $2.92>
Monthly net profit<loss> per household	< $0.35>	< $0.16>	$0.09	< $0.24>

[a] Excludes in-kind services.
[b] Including one-fifth SSWMB grant award, landfill fee savings, estimated reduced collection costs.
[c] Palo Alto and MARS programs also provide other recycling services (e.g., drop-off, white paper, commercial collection, etc.). Program cost analysis for their entire operations are presented in Table 10.16.
[d] Palo Alto pays one-half of gross revenues from sale of collected newspaper to California Association for the Retarded.
[e] Amounts determined by amortizing *state* purchased equipment and/or site improvements over a 5-year period.
[f] Fresno/Clovis Refuse collection bill includes a $0.06 recycling charge.
[g] Figures will be highly variable due to differences in landfill fees throughout California.
[h] Based on service population, single-family residences: Palo Alto—15,500; Downey—19,800; Recycle 3—23,000; and MARS—84,000.
[i] Based on savings incurred from reduction of a one man refuse collection crew.
[j] Figure reduced from that quoted in the individual Downey report. Amortization figures were originally included in those operating cost figures. These costs are now included in another line item.
[k] Based on savings incurred from reduction of a two-man refuse collection crew.
[l] Exact figures unknown.
Source: Ref. 53.

By the end of 1980, most of the districts had completed their 10-year management plan, and the State Departments of Energy and Environmental Protection had consolidated their recycling programs in one office—the New Jersey Office of Recycling. The Office of Recycling with an advisory committee of environmentalists, government, and business and industry representatives drafted *Recycling in the 1980s*,[54] the state materials recovery plan.

The materials recovery program is an effort to develop a balanced solid waste management program in New Jersey. It is also a recognition by the State that no one waste management system will solve the problem. A statewide solid waste program must include source separation, energy and materials recovery systems, and landfills. Outlined in *Recycling in the 1980's* is a 5-year statewide program to recycle annually 25% of the municipal solid waste stream by 1986. (Estimates

Table 10.18 Unit Costs and Revenues—Curbside Programs Only[a] Fiscal Year 1980–1981

Programs	Recovered Materials	Coll Costs[b]	Coll Cost/Ton	Revenues[c]	Revenues/Ton[e]	Net Profit/ <Cost>/Ton[e]	County Disposal Costs[f]
Palo Alto Recycling Program	2741 TPY	$269,121	$ 98.18	$204,630	$74.66	<$ 23.52>	$47.00
Downey at Home Recycling Team	876 TPY	$ 89,107	$101.72	$ 50,337	$57.46	<$ 44.26>	$34.00
Recycle 3	2570 TPY	$161,660	$ 62.90	$185,326	$72.11	$ 9.21	$44.00
MARS	1971 TPY	$408,162	$207.08	$162,850	$82.62	<$124.46<	$50.00

[a] Palo Alto and MARS programs provide other recycling services (e.g., drop-off, white paper, commercial collection, etc). Unit cost analysis for these entire operations are provided in Table 10.20.

[b] Full operating costs (e.g., collection, administration, processing, etc.), excluding in-kind services.

[c] Direct and indirect revenues, including one-fifth SSWMB grant award, landfill fee savings, estimated reduction in collection costs (formulas provided in text).

[d] Revenues per material will greatly fluctuate; these figures reflect an average revenue per ton of material.

[e] Collection cost/ton minus revenues/ton.

[f] Refuse collection, hauling, and disposal costs per ton for the county in which each program is located. Estimates obtained from preliminary findings for the "1982 Garbage Crisis Cost Survey," SSWMB.

Source: Ref. 53.

Table 10.19   Program Cost Analysis—Comparison of Full Operations Costs[a] with Direct and Indirect Program Revenues[b] Palo Alto and MARS Programs Only[c] Fiscal Year 1980–1981

		Palo Alto	MARS
Operating expenses			
Labor and fixed overhead		$203,690	
Variable expense (operating and			$389,000[g]
maintenance)		40,700	
Public awareness		10,000	
Revenue sharing[d]		64,080	-0-
Amortization[e]		26,021	19,162
	Subtotal	$344,491	$408,162
Direct revenues			
Secondary materials sales		$258,600	$129,950
Recycling charge[f]		-0-	39,600
	Subtotal	$258,600	$169,550
Indirect revenues			
Landfill fee savings		$ 17,350	$ 15,840
Reduced collection costs		74,450	26,840
Grant monies		30,400	22,700
	Subtotal	$122,200	$ 65,380
	TOTAL, All Revenues	$380,800	$234,930
Annual net profit<loss>		$ 36,309	<173,232>
Annual net profit<loss> per household[g]		$1.51	< $1.46>
Monthly net profit<loss> per household		$0.13	< $0.12>

[a] Excludes in-kind services.
[b] Includes one-fifth SSWMB grant award landfill fee savings, estimated reduction collection costs.
[c] These two programs provide a wide variety of services in addition to curbside services. Materials are collected from these other program aspects; costs and revenues relative to the total programs are included.
[d] Palo Alto pays one-half of gross revenues from sale of collected material to California Association for the Retarded.
[e] Amounts determined by amortizing *state* purchased equipment and/or site improvements over a 5-year period.
[f] Fresno/Clovis refuse bill includes $0.06 recycling charge.
  Entire residential population included. Total estimated residences: Palo Alto—24,000; MARS—118,500.
[g] Exact figure unknown.
Source: Ref. 53.

indicate about 11.5 million tons of waste per year are landfilled in New Jersey, of this amount about 5.3 million tons is municipal waste.) Targeted recyclable materials include paper, metals, glass, plastics, oil, tires, food, and yard wastes.

The purposes of setting such a goal for 1986 was to:

1. Decrease materials flow to overburden landfills.

2. Develop organized source separation programs that offer the potential to provide a stable source of raw materials supply to secondary material industries.

3. Reduce the design capacity of energy recovery systems by 3000 to 4000 tons/day, for an estimated capital cost savings of $260,000,000.

4. Reduce solid waste management collection and disposal costs which are rapidly increasing because of the closing of landfills near population centers and requirements for stringent environmental landfill controls.

5. Conserve energy in the manufacturing process by at least 2 million barrels of oil equivalent per year.

Table 10.20 Unit Costs and Revenues—Palo Alto and MARS Programs Only[a] Fiscal Year 1980–1981

Program	Recovered Materials	Coll Costs[b]	Coll Cost/Ton	Revenues[c]	Revenues/Ton[d]	Net Profit <Cost>/Ton[e]	County Disposal Costs[f]
Palo Alto Recycling Program	5340 TPY	$344,491	$ 64.51	$389,800	$71.31	$ 6.80	$47.00
MARS	2617 TPY	$408,162	$155.97	$234,930	$89.77	<$66.20>	$50.00

[a]These two programs provide a wide variety of services in addition to curbside services. Materials are collected from these other program aspects; costs and revenues relative to the total program are included.
[b]Full operating costs (e.g, collection, administration, processing, etc.), excluding in-kind services.
[c]Direct and indirect revenues, including one-fifth SSWMB grant award, landfill fee savings, estimated reduction in collection costs (formulas provided in text).
[d]Revenues per material will greatly fluctuate; these figures reflect an average revenue per ton of material.
[e]Collection cost/ton minus revenues/ton.
[f]Refuse collection, hauling, and disposal costs per ton for the county in which each program is located. Estimates obtained from preliminary findings for the "1982 Garbage Crisis Cost Survey," SSWMB.
Source: Ref. 53.

The recycling goal of 1.3 million tons a year is limited by present market capacity and the available materials supply. The achievement of this recycling rate by 1986 requires improved collection systems, an increase in public participation, and an expansion of the secondary material markets.

Various methods to achieve the recycling goal were researched and analyzed by the advisory committee, a process that proved to be quite dynamic. The philosophies and viewpoints of the diverse constituencies that were represented often created much controversy before a final decision on an issue was reached. However, the democratic process and compromise strategies prevailed and the committee remained intact to complete a final report.

The following programs were recommendations that the committee voted to adopt:

1. Legislation that would impose a landfill disposal surcharge of $0.12/yd^3 (766DM^3) for a period of 4 years annually. The surcharge would decline to $0.06/yd^3 (766DM^3) in the fifth year (1986) and terminate at the end of that year. The purpose of the surcharge was to establish a recycling trust fund.

2. Procurement of recycled products by government and state funded agencies. Specifically a 50% recycled products procurement rate for all state purchases of paper and oil was recommended.

3. Restrictions on waste flow orders to prevent diverting recyclable and marketable materials to disposal facilities.

4. Implementation of tax incentives for the state's recycling industries.

5. Issuance of an Executive Order to require state agencies to recycle office paper.

6. Development of procurement practices by recycling businesses and industries that lend stability to recycling collection programs.

7. Leadership role by the container and beverage industries in the development and marketing of New Jersey's recyclable glass, plastic, and metal containers.

8. A leadership role by business and industry for a public relations and advertising program to promote recycling and purchase of recycled products.

9. An education and training program that would include technical assistance, development of school curriculums, advertising campaigns, and other promotional and education techniques to reach a wide audience.

10. The placement of recycling coordinators in the state's 21 counties to assist in the development of municipal and regional recycling programs.

10.2.2 Implementing the Recycling Plan

In September of 1980, the State Departments of Energy and Environmental Protection adopted *Recycling in the 1980's* as the official State recycling plan and incorporated it into the total solid waste management program for the State.

By December of 1980, the legislation recommended by the advisory committee was drafted and introduced in the State Assembly as Assembly Bill 2283. This bill established the Committee's proposed $0.12/yd^3 (766DM^3) recycling surcharge at all landfills in New Jersey. The bill passed unanimously in the legislature and was signed into law by the Governor in September 1981 (N.J.S.A. 13:1E-92 *et seq.*).[55] The surcharges are placed in a recycling trust fund for distribution as previously outlined in the Introduction.

10.2.3 Meeting the Challenge

Recycling in New Jersey holds great promise. The adoption of a statewide plan and the establishment of the State Office of Recycling gave many avid recyclers the incentive to expand and develop more stabilized local programs. It also brought about a recognition that recycling made "cents" in New Jersey as a way to address some of the rising costs of solid waste management.

As of 1980, New Jersey had listed about 250 recycling programs. Most of these programs received little municipal support and were generally volunteer citizen organized activities rising out of the environmental movement of the early 1970s. The percentage of material recovered compared with the population served in most towns was minimal, often reflecting a very low participation rate.

Local programs that are now evolving appear to have needed municipal support. Municipalities are enacting ordinances mandating recycling, developing curbside collection programs to increase public participation, and economically integrating recycling with collection of solid waste. Since January 1, 1982, the effective date of the Recycling Act, the number of recycling programs has increased from 250 to more than 400 operating in over 300 of the state's 567 municipalities. There are now 165 municipalities providing curbside recycling collection service. Municipalities are actively competing for their share of the New Jersey recycling market.

Two of the major weak links, however, in developing statewide recycling systems are the lack of private collectors offering curbside collection service of recyclables and the need to expand the consuming markets for certain types of materials.

10.2.4 Collection of Recyclables

The solid waste industry is beginning to recognize the value of recycling in keeping the cap on solid waste management cost. One company in northern New Jersey has put out a bid to contract with communities for recycling as well as solid waste collection services. A central New Jersey firm has surveyed its customers to determine their opinion on setting rates on a per can basis as well as customer interest in receiving recycling collection service. In addition, the New Jersey Waste Management Association, an association of haulers, is gearing up for a campaign to educate its members on the need to change waste management practices in the State.

The development of privately run collection routes requires new equipment purchases by industry as well as a recognition by municipalities that collection of recyclables will require contracts that not only benefit the community but also provide a reasonable profit for the collection industry. Municipalities must also accept recycling as a solid waste management system that will vary in costs and benefits depending on local market conditions, availability of materials, volume of materials collected relative to road miles of collection routes, and trash collection and disposal costs. Municipalities must view recycling not as a moneymaker but as a method to contain solid waste costs and to reduce waste to landfills. Recycling in New Jersey must be institutionalized as part of the solid waste management system (see Appendix 10.5).

10.2.5 Recycling—A Cost Avoidance Mechanism

New Jersey municipalities annually spend more than $150 million to collect and dispose of solid waste, a price tag that is rapidly rising because of landfill closures and environmental standards imposed on the operation of landfills. Two landfill operators that received environmental engineering design approval from the State Department of Environmental Protection in 1981 raised their rates an average of 220% to meet the increase in cost of proper disposal. [This was from an average of $0.95/yd^3 (766DM^3) to $2.10/yd^3 (766DM^3).] One town, the Town of Woodbury in Gloucester County, affected by this increase in landfill fees implemented a mandatory multimaterial recycling program and reduced by 50% its trips to the landfill. The 50% volume reduction, however, could not be attributed to the recycling rate which is about 35% of the municipal waste stream. Woodbury found that the type of materials collected for landfilling had a greater compaction ratio with cans, bottles, and dry paper goods removed. Thus, recycling netted for Woodbury a 50% reduction in disposal costs. In terms of dollars, this meant a savings of more than $1,000 per month in tipping fees for the town.

The savings achieved by each municipality as well as the investment required to begin a recycling program will vary depending on existing solid waste management systems in the community. The 567 towns in New Jersey also vary in population density. A small densely populated suburban town with its own equipment that collects trash at least 2 days a week can anticipate less expenditures for recycling than an expansive community with 1 day a week trash collection. There is no formula that can be set forth for every town to follow with the expectation that all towns will receive equal benefits and incur equal costs.

10.2.6 Market Expansion and Development

As previously mentioned, stronger markets in some areas are necessary to achieve the goals of the state plan. The market for recyclable aluminum is fairly stable in New Jersey. Glass, a previously stable commodity, is losing market demand to plastics. Paper, ferrous metals, plastics, and rubber require expansion or development to achieve the state recycling goal. Existing paper markets are substantial, but a greater commitment by industry to the utilization of New Jersey generated material is required. It is expected that the evolution of organized mandated municipal programs will help stabilize supply of materials and thereby encourage consuming markets to contract for domestic materials and expand processing capacity.

In an effort to stimulate the demand for recycled products, Assembly Bill 3342 was introduced in the legislature in June, 1983. This bill, if passed, will require the State Division of Purchase and Property to work with the Office of Recycling to review procurement specifications and to increase substantially the State's purchase of products made from recycled materials. In addition, the State Office of Recycling maintains and publishes a directory of all grades of paper, glass, and metals as well as other materials such as plastics, textiles, food waste, and asphalt. It is the most comprehensive listing of its type in the Northeast if not the country. The directory allows municipalities or private organizations to locate rapidly potential markets for the different materials they intend to recover.

The state also publishes a current listing of all municipal recycling programs in New Jersey. One primary value of this publication is that it enables recycling businesses that are interested in expanding into municipal recycling to identify potential customers. The Office of Recycling also assists recycling businesses in working with municipalities and has cooperated with industry in developing and designing new equipment to meet recycling needs.

10.2.7 Education—The Key to Success

The successful development of recycling in New Jersey requires a continuing, massive education program. The recycling program is planned essentially to affect almost every citizen in the State by gaining their cooperation to participate and to change "throw-away" habits.

Education is critical to achieving the goals of the program. The New Jersey Recycling Act provides about $650,000 a year for education programs [N.J.S.A. 13:1E-96(b)(5)]. The following programs have been implemented:

A noncredit seven-session evening course on recycling is held twice a year at colleges in various locations throughout the state. These courses are well attended and have provided training to more than 200 individuals representing 200 municipalities and businesses in the state.

Kindergarten through 12th grade recycling and litter abatement curriculum and training workshops for teachers.

Workshops and conferences for industry, local governmental officials, and the citizens of the state.

The expansion and development of brochures, how-to guides, and other education materials for public distribution.

A statewide recycling promotional and advertising campaign.

Grants to 75 municipalities and counties to provide information on recycling to local businesses and residents.

10.2.8 A Total Effort

New Jersey has the largest secondary materials industry in the nation, and much of the resources required to feed this industry must be imported from sources outside the state. There are large amounts of recoverable materials used for landfill every day in the state. An effective statewide materials recovery program requires the close cooperation of the general public, industry, and government. The public's role is to make a social commitment to participate in source separation programs, whereas private industry must not only participate but also expand markets for recycled materials. Finally, the government's responsibility is to create a favorable environment for recycling by removing the institutional and regulatory barriers that inhibit progress in the recycling. (*Recycling in the 1980s,* New Jersey Advisory Committee on Recycling, September 1980, pp. 8, 9.)

The state recycling plan is an ambitious undertaking that, if successful, will affect almost every citizen in the State. It is a program that encourages municipalities to restructure their solid waste management programs from collection and dumping, to collection, recycling, and only residual waste dumping. The quality and quantity of recyclable materials collected will depend on active citizen participation and cooperation through the source separation of materials.

The most important ingredient to success is the development and implementation of an expansive and continuous education program. Recycling must become part of the solid waste management system as well as an opportunity for all citizens individually and collectively to affect change in the State of New Jersey.

REFERENCES

1. J. Huls and N. Seldman, Developments in Recycling: Historic and Institutional Issues, presented to the 4th World Recycling Congress, April 1982; taken from *National Recycling News,* p. 7, (Fall 1982).

2. M. Melosi, *Garbage in the Cities 1880–1980,* Texas A&M Press; cited by Neil Seldman and Jon Huls in "100 Years of Garbage in the Cities," *BioCycle.*

3. Huls and Seldman, *op. cit.*

4. A. H. Purcell, The World's Trashiest People, *The Futurist,* pp. 51–59 (Feb. 1981).

5. N. N. Seldman, Institute for Local Self Reliance, "Economic Feasibility of Recycling," June 30, 1978; report to the Office of Technical Assistance, U.S. Dept. of Commerce.

6. A Survey of Mandatory Recycling Ordinances, *Resource Recycling,* Vol. I, No. 2, May/June 1982; adapted from a study performed by Robert M. Cooper.

7. N. Seldman, *op. cit.*

8. N. Seldman, Institute for Local Self Reliance. Testimony presented to the U.S. Environmental Protection Agency.

9. K. L. Cowles, John Muir Institute, personal communication.

10. A Survey of Mandatory Recycling Ordinances, *Resource Recycling,* Vol. 1, No. 2, May/June 1982; adapted from a study performed by Robert Cooper.

11. P. Grogan, The Future of Community Recycling, *National Recycling News,* p. 4, (Fall 1982).

12. J. Roumpf, "Sorting It Out," prepared for the California Office of Appropriate Technology.

13. M. W. Anderson, Garbage Reincarnation, Inc., "Philosophy and Goals of a Community Recycler," presented to the Western Division Conference of the National Association of Recycling Industries, August 1981.

14. A. H. Purcell, *op. cit.*

15. Maryland Recycling Directory, Sept. 1982.

16. *Resource Recovery,* Vol. I, No. 3 (July/Aug. 1982).

17. G. Easterbrook, Towns Do Make Passes at Guys Who Buy Glasses, *Waste Age* (June 1978).

18. Tires: The Basis for Rubber Recycling, excerpted from *Recycled Rubber Products* by the Rubber Recycling Division of the National Association of Recycling Industries in *Environmental Education Report,* p. 11 (Oct./Nov. 1982).

19. Recycling, Reuse, Repairs, *Science,* **202,** 34–35 (1978).

20. "Recycling: an Urban Frontier," U.S. Conference of Mayors, Nov. 1980.

21. Resource Integration Systems, Ltd., "Evaluation of Handling Stations in Waste Reclamation Systems," prepared for the Waste Management Advisory Board, Ontario, Canada, 1979, pp. 76–77.

22. R. Anthony, Fresno, Calif., Metropolitan Area Recycling Service, personal communication.

23. N. Seldman and J. Huls, Beyond the Throwaway Ethic, *Environment,* **23,** 29 (1981).

24. "Fresno: A Case Study of Economic Development and Resource Recovery," U.S. Conference of Mayors, Oct. 1982.

25. N. Seldman, Institute for Local Self Reliance, letter to Tina Hobson, U.S. Dept. of Energy, May 8, 1980.

26. "Recycling: an Urban Frontier," U.S. Conference of Mayors, Dec. 1980.

27. R. Anderson, Newsprint Collection Program Saves $40,000 During First 30 Months of Operation, *Solid Waste Systems,* p. 9 (Jan./Feb. 1977).

28. N. Seldman, Institute for Local Self Reliance, Citizen and Institutional Participation in Resource Conservation and Recovery, presented to the American Society of Civil Engineering Conference, Oct. 1979.

29. Sonoma County Community Recycling Center, "Garbage to Energy: The False Panacea," 2nd Ed., Sept. 1979, p. 38.

30. N. Seldman, Institute for Local Self Reliance, "Citizen and Institutional Participation in Resource Conservation and Recovery," *op. cit.*

31. D. B. DePasse, Waste Flow Control, *Resource Recovery,* Vol. I, No. 3 (July/Aug. 1982).

32. "Fresno: A Case Study of Economic Development and Resource Recovery," U.S. Conference of Mayors, Oct. 1980.

33. Sonoma County Community Recycling Center, *op. cit.,* p. 45.

34. Sonoma County Community Recycling Center, *op. cit.,* pp. 23–29.

35. P. Love, Middletown Associates, "Net Energy Savings from Solid Waste Management Options Summary," prepared for the Solid Waste Management Branch, Environment Canada, EPS3-EC-76-16, Sept. 20, 1976.

36. N. Seldman, Institute for Local Self Reliance, "Citizen and Institutional Participation in Resource Conservation and Recovery," *op. cit.*

37. Peter Love, *op. cit.,* pp. 19–20.

38. D. M. Cohen, U.S. Environmental Protection Agency, A National Survey of Separate Collection Programs, SW-778, 1979.

39. R. Keller, "Maryland's Program for Buying Recycled Paper," in Innovations, The Council of State Governments, 1980.

40. *Ibid.*

41. K. Mulligan and J. Powell, Portland Recycling Team, "Operating a Recycling Program: A Citizen's Guide," U.S. Environmental Protection Agency, SW-770, 1979.

42. R. Anthony and N. Seldman, Fresno: Building A Closed Loop System, *Compost Science/ Land Utilization*, p. 39 (Sept./Oct. 1980).

43. Office of Solid Waste, U.S. Environmental Protection Agency, "Fourth Report to Congress: Resource Recovery and Waste Reduction," SW-600, Aug. 1977.

44. T. Y. Canby, Aluminum, the Magic Metal, *National Geographic*, p. 204. (Aug. 1978).

45. U.S. Environmental Protection Agency, "Source Separation Collection and Processing Equipment: A User's Guide," SW-842, 1980.

46. R. Tichenor, The Combination Collection Experiment, *Waste Age*, p. 34, (Dec. 1980).

47. Office of Solid Waste, U.S. Environmental Protection Agency, Fourth Report to Congress: Resource Recovery and Waste Reduction, SW-600, Aug. 1977.

48. D. Knapp, *Resource Recovery: What Recycling Can Do,* Materials World Publishing, 1982.

49. R. Keller, "Maryland's Program for Buying Recycled Paper," in Innovations, The Council of State Governments, 1980.

50. J. Roumpf, "Sorting It Out," prepared for the California Office of Appropriate Technology.

51. M. Anderson, Sonoma County Community Recycling Center, private communication.

52. M. L. Renard, National Center for Resource Recovery, A Review of Comparative Energy Use in Materials Potentially Recoverable from Municipal Solid Waste, prepared for the Office of Renewable Energy, U.S. Dept. of Energy, DOE/CS/2016/12, March 1982.

53. California Solid Waste Management Board, Case Studies of California Curbside Recycling Programs, Vol. 1.

54. New Jersey Advisory Committee on Recycling, *Recycling in the 1980's,* September 1981.

55. New Jersey Statutes Annotated, 13:1E:92.

56. M. Sitgil, "Recycling, A Statewide Program for New Jersey," Proceedings, ASME Tenth National Solidwaste Conference, New York, N.Y., May 1982, pp. 421–428.

APPENDIX 10.1 SAMPLE CONTRACT TO SELL USED PAPERS*

This Agreement entered into this ———————— day of ———————— 1974, by and between WASTE PAPERS, INCORPORATED, a ———————— Corporation, with business offices at —— , party of the first part, hereinafter referred to as "Contractor," and THE COUNTY BOARD OF ARLINGTON COUNTY, VIRGINIA, a body corporate, party of the second part, hereinafter referred to as "County."

WITNESSETH:

That for and in consideration of the mutual promises and covenants herein contained, and Ten Dollars ($10.00) in hand paid, each to the other, receipt of which is hereby acknowledged, the parties hereto agree as follows:

1. Contractor agrees to purchase on a daily basis all waste newspapers collected on behalf of Arlington County, Virginia, either by County or its collection agents, and delivered daily to a mutually agreed upon receiving site located in Arlington County, Virginia; Alexandria, Virginia; or Washington, D.C.

2. Contractor agrees to accept delivery of all waste newspapers at its receiving site daily (Monday through Saturday, including all holidays except Christmas and New Year's Day) between the hours of ——— A.M. through ——— P.M.

3. It is mutually understood and agreed that County is not restricted as to either the minimum or maximum quantity of waste newspapers to be delivered to Contractor and that no adjustment shall be made on account of the moisture content of the waste newspapers due to inclement weather conditions.

4. It is mutually understood and agreed that waste newspapers shall be delivered to the receiving site in an "as picked up" condition and that no processing, bundling or baling of newspapers will be done by County or its collection agents; but that all processing, bundling, baling, transportation or service charges incurred after delivery of the waste newspapers to the receiving site shall be at the expense of Contractor.

5. Contractor shall deliver a certified weighing slip to the County or its agent at the time of delivery at the receiving site and such weighing slip may be verified by the County. The County reserves the right to challenge the weight as determined at the receiving site and to verify same at weighing scales located at the County Transfer Station. In case of discrepancy between weights determined at the receiving site and County Transfer Station, the weight determined at the County Transfer Station shall be used to determine the price for said waste newspapers.

6. Contractor shall pay to County on bi-weekly basis the amount due for waste newspapers

delivered to its receiving site. Payment shall be due and payable within ten (10) calendar days from the date of receipt of the last waste newspapers delivered during a biweekly period. Such payment shall be accompanied by an itemized list of daily receipts.

7. It is mutually understood and agreed that the price per ton (2,000 pounds) to be paid by Contractor to County shall be computed on the following basis:

The price per ton (2,000 pounds) of waste newspapers delivered during a calendar week (Sunday through Saturday) shall be determined by reference to *Official Board Markets* ("The Yellow Sheet") published by Magazines for Industry, Inc. and shall be the highest market value quoted in "Paper Stock Prices Per Ton" for "No. 1 News" in the market area of "Philadelphia" less a fixed charge of ten dollars and fifty cents ($10.50). The issue of *Official Board Markets* to be used in determining the price per ton to be paid by Contractor shall be the first publication date within each week. However, if no issue of *Official Board Markets* is published during the week, the last issue thereof published the preceding calendar week shall be used in determining the market value for the said week.

Notwithstanding anything to the contrary heretofore set forth, Contractor guarantees to purchase all accumulated and delivered waste newspapers at a minimum price of $10.00 per ton.

8. The initial term of this Contract shall be for a one-year period beginning _____ and County shall have the option of renewing the Contract for one (1) additional year under the same terms and conditions by giving a 30-day notice prior to the expiration date hereof.

9. Contractor shall not assign this Contract or any interest therein without the prior written consent of the County thereto.

IN WITNESS WHEREOF, _____

_____, has executed this Agreement on behalf of the County Board of Arlington County, Virginia, a body corporate, pursuant to a resolution of said Board, duly adopted on _____; and Waste Papers, Incorporated has caused this Agreement to be executed in its corporate name by its _____ and its corporate seal to be hereunto affixed, duly attested by its _____; said officers being thereunto duly authorized all as of the day, month and year first hereinabove written.

ATTEST: WASTE PAPERS, INCORPORATED

_____   By _____

ATTEST: THE COUNTY BOARD OF ARLINGTON
                                             COUNTY, VIRGINIA

_____   By _____

Source: Arlington County, Virginia, taken from a report written by P. Hansen, *Residential Paper Recovery: A Municipal Implementation Guide,* U.S. Environmental Protection Agency, SW-155, 1975.

APPENDIX 10.2 STATE RECYCLING ASSOCIATIONS

ALASKA

Alaska Recyclers Association
5650 Camelot Drive
Anchorage, Alaska 99508

CALIFORNIA

California Resource Recovery Assoc.
125 W. Swift Avenue
Clovis, California 93612
(209) 292-3247

COLORADO

Colorado Recycling Cooperative Assoc.
P.O. Box 4193
Boulder, Colorado 80306
(303) 444-6634

ILLINOIS

Illinois Assoc. of Recycling Centers
P.O. Box 48761
Chicago, Illinois 60648
(312) 470-0242 or (312) 769-6677

MICHIGAN

Michigan Recycling Coalition
P.O. Box 48107
Ann Arbor, Michigan 48107
(313) 665-6398

MINNESOTA

Recycling Assoc. of Minnesota
P.O. Box 30632
St. Paul, Minnesota 55175
(612) 646-2591

NEBRASKA

Nebraska State Recycling Assoc.
Route 2, Box 423
Kearny, Nebraska 68847
(308) 237-7339

NEW HAMPSHIRE

New Hampshire Resource Recovery Assoc.
P.O. Box 472
Concord, New Hampshire 03301

NEW JERSEY

New Jersey Recycling Forum
Park 80 Plaza West 1
Saddlebrook, New Jersey 07662
(201) 843-1450

OHIO

Federation of Ohio Recyclers
2801 Far Hills Avenue
Dayton, Ohio 45419
(513) 294-8080

OREGON

Association of Oregon Recyclers
P.O. Box 10051
Portland, Oregon 97210
(503) 228-5375 or (503) 227-1326

PENNSYLVANIA

Group for Recycling in Pennsylvania
P.O. Box 7391
Pittsburgh, Pennsylvania 15213
(412) 661-4447

VERMONT

Association of Vermont Recyclers
P.O. Box 965
Rutland, Vermont 05701
(802) 775-6482

WASHINGTON

Washington Citizens for Recycling
Curtis Chapman Express
P.O. Box 2449
Seattle, Washington 98111
(206) 621-8218

Washington State Recycling Assoc.
P.O. Box 569
Seattle, Washington 98111
(206) 363-8433

WISCONSIN

Wisconsin Coalition for Recycling
111 King Street, Room 31-32
Madison, Wisconsin 53703
(608) 256-0565

Source: National Recycling Coalition, New York, N.Y.

APPENDIX 10.3   TRADE ASSOCIATIONS OF INDUSTRIES WHICH PROCESS OR USE RECYCLED MATERIALS

Aluminum Association
818 Connecticut Ave., N.W.
Washington, DC 20006
(202) 862-5100

Aluminum Recycling Association
900 17th St., N.W.
Washington, DC 20006
(202) 785-0550

American Iron & Steel Institute
1000 16th St., N.W.
Washington, DC 20036
(202) 452-7100

American Paper Institute
260 Madison Ave.
New York, NY 10016

American Retreaders Association
P.O. Box 17203
Louisville, KY 40217

Assoc. of Petroleum Re-refiners
Suite 1111
2024 Pennsylvania Ave.
Washington, DC 20006
(202) 833-2694

Automotive Dismantlers &
Recyclers of America
1000 Vermont Ave. N.W.
Washington, DC 20005
(202) 628-4634

Can Manufacturers Institute, Inc.
821 15th St., N.W.
Washington, DC 20005
(202) 232-4677

Copper & Brass Fabricators Council
Suite 440
1050 17th St., N.W.
Washington, DC 20036

Fibre Box Association
5725 East River Road
Chicago, IL 60631

Glass Packaging Institute
1800 K. St., N.W.
Washington, DC 20006
(202) 872-1280

Institute of Scrap Iron & Steel
1627 K. St., N.W.
Washington, DC 20006
(202) 466-4050

National Assoc. of Recycling Industries
330 Madison Ave.
New York, NY 10017
(212) 867-7330

National Assoc. of Solvent Recyclers
1406 Third National Bldg.
Dayton, OH 45402
(513) 223-0419

National Assoc. of Wiping Cloth
Manufacturing
189 W. Madison Street
Chicago, IL 60602
(312) 726-0050

National Textile Processors Guild, Inc.
51 Chambers Street
New York, NY 10007
(212) 962-1183

National Tire Dealers &
Retreaders Assoc.
1343 L. St., N.W.
Washington, DC 20005
(202) 638-6650

Paperboard Packaging Council
1800 K. St., N.W.
Suite 600
Washington, DC 20006
(202) 872-0180

Rubber Manufacturers Assoc.
1901 Pennsylvania Ave., N.W.
Washington, DC 20006

Society of the Plastics Industry, Inc.
355 Lexington Avenue
New York, NY 10017

Solid Waste Council of the Paper Industry
1619 Massachusetts Ave., N.W.
Washington, DC 20036
(202) 797-5786

Technical Assoc. of Pulp &
Paper Industries
One Dunwoody Park
Chamblee, GA 30341
(404) 394-6130

Textile Fibers & By-Products Assoc.
144 Brevard Court
Charlotte, NC 28201

Tire Retread Information Bureau
P.O. Box 811
Pebble Beach, CA 93953

Source: National Recycling Coalition, New York, N.Y.

APPENDIX 10.4   STATE RESOURCE RECOVERY AGENCIES*

ALABAMA

Planning & Resource Recovery Section
Div. of Solid & Hazardous Waste
Department of Public Health
State Office Building
Montgomery, Alabama 36130
(205) 832-6728

ALASKA

Air & Solid Waste Management
Dept. of Environmental Conservation
Pouch O
Juneau, Alaska 99811
(907) 465-2635

ARIZONA

Bureau of Waste Control
Dept. of Health Services
1740 West Adams Street
Phoenix, Arizona 85007
(602) 255-1164

ARKANSAS

Dept. of Pollution Control & Ecology
P. O. Box 9583
8001 National Drive
Little Rock, Arkansas 72219
(501) 371-2130

CALIFORNIA

Resource Conservation & Development Div.
Solid Waste Management Board
1020 Ninth Street, Suite 300
Sacramento, California 95814
(916) 322-2649

COLORADO

Department of Health
4210 East Eleventh Avenue
Denver, Colorado 80220
(303) 320-8333

CONNECTICUT

Solid Waste Management Unit
Dept. of Environmental Protection
State Office Building
165 Capitol Avenue
Hartford, Connecticut 06115
(203) 566-5847

DELAWARE

Solid Waste Management
Dept. of Natural Resources &
Environmental Control
Edward Tatnall Building
P. O. Box 1401
Dover, Delaware 19901
(302) 736-4764

DISTRICT OF COLUMBIA

Dept. of Environmental Services
5000 Overlook Avenue, S.W.
Washington, D.C. 20032
(202) 727-5701

FLORIDA

Resource Recovery Subsection
Solid Waste Management Section
Dept. of Environmental Regulation
2600 Blair Stone Road
Tallahassee, Florida 32301
(904) 488-0300

GEORGIA

Planning & Resource Conservation
Unit, Environmental Protection Div.
Dept. of Natural Resources
3420 Norman Berry Drive
7th Floor, Scott Hudgens Bldg.
Hapeville, Georgia 30354
(404) 656-7404

HAWAII

Pollution Technical Review Branch
Department of Health
P. O. Box 3378
Honolulu, Hawaii 96801
(808) 548-6410

IDAHO

Solid/Hazardous Materials Section
Dept. of Health & Welfare
State House
Boise, Idaho 83720
(208) 334-4108

ILLINOIS

Div. of Land & Noise Pollution Control
Environmental Protection Agency
2200 Churchill Rd., Room A104
Springfield, Illinois 62706
(217) 782-9800

INDIANA

Solid Waste Management Section
State Board of Health
1330 West Michigan Street
Indianapolis, Indiana 46206
(317) 633-0176

IOWA

Air & Land Quality Division
Dept. of Environmental Quality
Henry A. Wallace Bldg.
900 East Grand Street
Des Moines, Iowa 50319
(515) 281-8927

KANSAS

Solid Waste Management Section
Dept. of Health & Environment
Topeka, Kansas 66620
(913) 862-9360

KENTUCKY

Division of Waste Management
Dept. of Natural Resources
& Environmental Protection
Pine Hill Plaza
1121 Louisville Road
Frankfort, Kentucky 40601
(502) 564-6716

LOUISIANA

Dept. of Natural Resources
P. O. Box 44396
Baton Rouge, Louisiana 70804
(504) 342-4506

MAINE

Bureau of Land Quality Control
Dept. of Environmental Protection
State House—Station 17
Augusta, Maine 04333
(207) 289-2111

MARYLAND

Maryland Environmental Service
Dept. of Natural Resources
60 West Street
Annapolis, Maryland 21401
(301) 269-3355

MASSACHUSETTS

Bureau of Solid Waste Disposal
Dept. of Environmental Mgmt.
1905 Leverett Saltonstall Bldg.
100 Cambridge Street
Boston, Massachusetts 02202
(617) 727-4293

MICHIGAN

Resource Recovery Division
Dept. of Natural Resources
Westland Plaza
Lansing, Michigan 48909
(517) 322-1315

MINNESOTA

Division of Solid Waste Management
Pollution Control Agency
1935 West County Road B-2
Roseville, Minnesota 55106
(612) 297-2734

MISSISSIPPI

Div/Solid Waste Mgmt. & Vector Control
State Board of Health
P. O. Box 1770
Jackson, Mississippi 39205
(601) 982-6317

MISSOURI

Solid Waste Management Program
Dept. of Natural Resources
P. O. Box 1368
1915 Southride Drive
Jefferson City, Missouri 65102
(314) 751-3241

MONTANA

Solid Waste Management Bureau
Dept. of Health & Environmental Sciences
Room A201, Cogswell Building
Helena, Montana 59620
(406) 449-2821

NEBRASKA

Litter Reduction & Recycling Program
Box 94877, Statehouse Station
Lincoln, Nebraska 68509
(402) 471-2186

NEVADA

Waste Management Programs Division
Dept. of Conservation & Natural Resources
Capitol Complex
Carson City, Nevada 89701
(702) 885-4670

NEW HAMPSHIRE

Governors Council of Energy
2½ Beacon Street
Concord, New Hampshire 03301
(603) 271-2711

NEW JERSEY

Office of Recycling
Dept. of Energy and Environmental
Protection
101 Commerce Street
Newark, New Jersey 07102
(201) 648-6295

NEW YORK

Division of Solid Waste
Dept. of Environmental Conservation
50 Wolf Road
Albany, New York 12233
(518) 457-6603

NORTH CAROLINA

Solid/Hazardous Waste Mgmt. Branch
Dept. of Human Resources
P. O. Box 2091
Raleigh, North Carolina 27602
(919) 733-2178

NORTH DAKOTA

Resource and Recovery Branch
Div/Environmental Waste Mgmt.
& Research, Dept. of Health
1200 Missouri Ave., 3rd Floor
Bismarck, North Dakota 58505
(701) 224-2366

OHIO

Office of Litter Control
Fountain Square
Columbus, Ohio 43224
(614) 268-6333

OKLAHOMA

Industrial & Solid Waste Service
Dept. of Health
P. O. Box 53551
1000 N.E. 10th Street
Oklahoma City, Oklahoma 73152
(405) 271-5338

OREGON

Recycling
Dept. of Environmental Quality
P. O. Box 17860
522 S.W. 5th Avenue
Portland, Oregon 97204
(503) 229-5913

PENNSYLVANIA

Resource Recovery Section
Bureau of Solid Waste Mgmt.
Dept. of Environmental Resources
Fulton Building
P. O. Box 2063
Harrisburg, Pennsylvania 17120
(717) 787-7382

RHODE ISLAND

Solid Waste Mgmt. Program
Dept. of Environmental Management
204 Cannon Bldg.
75 Davis Street
Providence, Rhode Island 02908
(401) 277-2808

SOUTH CAROLINA

Solid Waste Mgmt. Division
Dept. of Health & Environmental Control
J. Marion Sims Building
2600 Bull Street
Columbia, South Carolina 29201
(803) 758-5681

SOUTH DAKOTA

Office of Energy Policy
Capitol Lake Plaza
Pierre, South Dakota 57501
(605) 773-3603

TENNESSEE

Div. of Solid Waste Mgmt.
Dept. of Public Health
150 Ninth Avenue North
Nashville, Tennessee 37203
(615) 741-3424

TEXAS

Solid Waste Section
Dept. of Water Resources
1700 N. Congress Avenue
P. O. Box 13087 Capitol Station
Austin, Texas 78711
(512) 475-3187

UTAH

Bureau of Solid Waste Mgmt.
Div. of Environmental Health
Dept. of Health
P. O. Box 2500
150 West North Temple
Salt Lake City, Utah 84110
(801) 533-4145

VERMONT

Air & Solid Waste Programs
Agency of Environmental
Conservation
State Office Building
Montpelier, Vermont 05602
(802) 828-3395

VIRGINIA

Division Solid and Hazardous
Waste Management
Department of Health
109 Governor Street
Richmond, Virginia 23219
(804) 786-5271

WASHINGTON

Solid Waste Management Division
Department of Ecology
Olympia, Washington 98504

WEST VIRGINIA

Solid Waste Division
Dept. of Health

1800 Washington Street, East
Charleston, West Virginia 25311
(304) 348-2981

WISCONSIN

Solid Waste Recycling Authority
3321 West Beltline Highway
Madison, Wisconsin 53713
(608) 266-2686

WYOMING

Solid Waste Management Program
Dept. of Environmental Quality
Equality State Bank Building
401 West 19th Street, Room 3011
Cheyenne, Wyoming 82002
(307) 777-7752

---

*Copied from the *National Recycling Directory,* prepared for the Office of Recycled Materials, National Bureau of Standards by State of Florida, Department of Environmental Regulation, Resource Recovery Section.

APPENDIX 10.5   NEW JERSEY PROGRAMS: I. *PROGRAM A.*— MUNICIPAL CURBSIDE COLLECTION WITH A DROP-OFF CENTER*

Municipal profile:

Area	= 6 sq mi
Population	= 40,000
Municipal waste generated	= 25,000 tons (23,000 t) per year

Program—Recycling program is mandatory

A. COST ESTIMATES

    1. Start-up costs

        a. Equipment

Construction of a concrete and masonary storage area with 4 bins:	$18,000.00
Front-end loader:	10,000.00
Used 25 yd³ packer truck to store and transport newspaper to market:	5,000.00
(2) collection vans:	20,000.00
(2) glass collection trailers:	7,000.00
(2) 20 yd³ roll-off containers:	5,000.00
TOTAL EQUIPMENT COSTS:	**$65,000.00**

        b. Publicity

• postage, printing, brochures, calendars, etc.:	$5,550.00

    2. Annual Recurring Costs

        a. Labor

• (1) supervisor:	$16,000.00
• (4) driver/collectors:	50,000.00
• (2) laborers:	20,000.00
• Fringe Benefits at 25%	21,500.00
TOTAL LABOR COSTS:	**$107,500.00**

        b. Overhead (Fuel, maintenance, insurance, tires, spare parts, leasing charges): $10,500.00

TOTAL PROGRAM COSTS (1st Year):	$188,550.00

B. REVENUES

1. Newspapers = 1500 T (1364 t) per year [Municipality has contract with mill at an average price of $40/T]: $60,000.00
2. Glass = 580 T (520 t) per year (sold at an average price of $30/T): $17,400.00
3. Aluminum = 5.8 T (5.4 t) per year (sold at an average price of $500.00 T): $2,700.00
4. Magazines, office paper and other scrap = 260 T (236 t) (average price = $30/T): $7,800.00

TOTAL REVENUES: $87,900.00

C. COLLECTION & DISPOSAL SAVINGS PER YEAR

Landfill costs: 2,346 tons (2,128 t) × $5.45/ton: $12,800.00
Diverted mixed waste collection costs: 26,400.00†

TOTAL AVOIDED COSTS: $39,200.00

D. TOTAL REVENUES AND SAVINGS $127,100.00

---

*Source: Case studies of selected New Jersey Recycling Programs, 1984, N. J. Department of Energy, Office of Recycling.
†This represents 10 to 12 percent of the waste stream not collected for disposal.

II. *PROGRAM B*—MUNICIPAL CURBSIDE COLLECTION

Municipal profile:

Area	= 2.25 sq mi
Population	= 10,300
Municipal waste generated	= 5,000 tons (4,500 t) per year

A. ANNUAL COSTS OF TRASH COLLECTION AND DISPOSAL

1. Labor: $55,282,00
2. Fuel & maintenance: 11,587.00
3. Disposal fees [3,744 T (3,400 t) per year × $11.40/T]: 42,670.00

TOTAL COSTS: $109,539.00

B. ANNUAL COSTS OF COLLECTION OF RECYCLABLES

1. Labor: $39,944.00
2. Fuel & maintenance: 7,810.00

$47,754.00

3. Publicity: 2,000.00

TOTAL RECYCLING COSTS: $49,754.00

C. REVENUES FROM RECYCLABLES

1. Mixed Paper—750 tons (680 metric tons) per year (average price = $10/ton): $7,480.00
2. Glass—385 tons (349 t) per year (average price = $30/ton): $11,517.00
3. Aluminum—2.7 tons (2.5 t) per year (average price = $400/ton): $1,100.00

4. Ferrous metals—115 tons (104 t) per year (average price
   = $25/ton): $2,860.00

TOTAL REVENUES: $22,957.00

D. SAVINGS ON LANDFILL FEES
   1,228 tons (1,114 t) × $11.40/ton: $13,990.00
E. TOTAL ANNUAL REVENUES AND SAVINGS ON LANDFILL FEES $36,947.00
F. TOTAL ANNUAL COST OF RECYCLING PROGRAM
   (Cost-revenues and landfill savings) $12,827.00

G. TOTAL ANNUAL COST OF TRASH COLLECTION PROGRAM $109,539.00

Program Description:

Citizens are mandated to source separate all clean paper products, glass, aluminum, and other metals. Municipal trash collection twice a week was changed to the collection of recyclables once a week and trash once a week. No new investment was made for the recycling program and existing municipal facilities are used for storage of recyclables (when necessary) prior to marketing materials.

This program was primarily developed to decrease solid waste collection and disposal costs.

III. *PROGRAM C—DROP-OFF CENTERS IN URBANIZED/SUBURBAN REGIONAL AREA PROGRAM*

Voluntary program serving 75,000 population with 30 mi^2 area. Citizens must deliver materials to center. Investment is minimal and some volunteer labor is used.

A. COSTS OF RECYCLING DROP-OFF CENTER
   1. Equipment
      Center includes: Wooden newspaper shed, 10–15 drums (55 gal each)
      for glass and metal storage, fencing around center, dedicated
      municipal space.
      15 centers ($800.00 per center): $12,000.00
      1 used collection vehicle: 2,000.00

      TOTAL: $14,000.00

   2. Operation expenses
      Labor (volunteer labor also
      provided assistance): $12,000.00
      Fuel & maintenance: 4,800.00

                                                                        $16,800.00
      TOTAL COSTS: $30,800.00

B. REVENUES FROM SALE OF RECYCLABLES
   1. Newspaper—444 tons (403 t) per year (@ $25.00/ton): $11,100.00
   2. Glass—392 tons (356 t) per year (@ $30.00/ton):   11,750.00
   3. Aluminum—1.6 tons (1.5 t) per year (@ $400/ton):   640.00
   4. Steel & bimetal cans—48 tons (44 t) per year (@ $25.00/ton):   1,200.00

                                                                        $24,690.00

C. SAVINGS ON LANDFILL FEES
   886 tons (804 t) per year (recycled) × $5.60
   per ton tipping fee: $4,962.00

D. TOTAL LANDFILL FEE SAVINGS & REVENUES
   FROM RECYCLABLES $29,652.00

CHAPTER 11
LAND DISPOSAL

PHILIP R. O'LEARY

**University of Wisconsin
Madison, Wisconsin**

LARRY CANTER

**University of Oklahoma
Norman, Oklahoma**

WILLIAM D. ROBINSON

**Consulting Engineer
Trumbull, Connecticut**

11.1 LANDFILL DISPOSAL: THEORY AND PRACTICE
Philip R. O'Leary and Larry Canter

11.1.1 Definition and Background

Land disposal of solid wastes has been practiced for many centuries. In fact, over 2000 years ago the Greeks buried their solid waste without compaction. In the 1930s, New York City and Fresno, California, started compacting refuse with heavy equipment and covering it up; thus the term "sanitary landfill" was coined. This chapter addresses the practice of landfill disposal of solid wastes. Although not as publicly appealing as energy or materials recovery from solid wastes, nonetheless, landfill disposal typically remains the most economically viable disposal option in the majority of the United States. In fact, alternative energy and materials recovery programs require land disposal of significant quantities of a residue. This chapter contains information on the principles of landfill disposal with examples of regulatory guidelines, landfill development, including site selection, leachate control, methane control, landfill design and construction, site operations, on-site processing, site closure, and long-term care. A typical landfill is shown in Figure 11.1.

A sanitary landfill is traditionally defined as an engineered method of disposing of solid wastes on land in a manner that protects the environment, by spreading the waste in thin layers, compacting it to the smallest practical volume, and covering it with soil by the end of each working day.[1] This definition, modified from a definition first developed by the American Society of Civil Engineers, is the one usually quoted when sanitary landfills are discussed. It is instructive to examine the direct and indirect implications of the definition.

First, the definition calls sanitary landfills "an engineering method." This means a well-thought-out method based on scientific principles. Instead of simply finding a vacant piece of property that has a gully or hole big enough to accept loads of waste for several years, it means using data on waste generation quantities, locations, road capabilities, available cover soil, ground and surface water, land use, and so forth, to analyze sites for a landfill. It also means designing and constructing facilities that take advantage of natural conditions to protect the environment, and the use of methods that overcome natural deficiencies to insure environmental protection. The definition goes

Fig. 11.1   A typical landfill.

on to describe three components necessary to satisfy the requirements of a sanitary landfill: (1) spreading the wastes in thin layers, (2) compaction to the smallest practical volume, and (3) daily cover. These three steps are a necessary part of sanitary landfilling to reduce problems of settling, fires, fly and rat breeding, and the like, and to conserve landfill space by compacting to the smallest practical volume.

Indirect implications of the definition are related to some significant potential problems, including gas generation and migration and leachate generation and pollution, as well as considerations of aesthetics such as litter control and landscaping. Also implied is concern for designing the site to be compatible with other land use, such as building the landfill to be a park, or returning it to agricultural land.[2]

A properly sited, designed, and operated sanitary landfill offers several advantages when compared with other disposal alternatives. Examples of these advantages include provision of ultimate disposal (a necessity for any solid waste system), protection of public health and the environment, recovery of degraded land and/or construction of a ski hill, golf course, or park, lower capital and operating costs when compared with the cost of remedial clean-up actions at uncontrolled dump sites, and esthetic acceptability. The development of a sanitary landfill will have a high initial capital cost, may be subject to intense public opposition, and in order to locate on land that is environmentally suitable, long haul distances may be necessary.

11.1.2   Principles of Operation

As noted in the definition of a sanitary landfill, operations at a landfill include spreading the wastes in thin layers, compacting it into the smallest practical volume, and covering it on a daily basis. Landfill operations can be considered in terms of the trench and area landfilling methods. The method selected for a given site depends on the physical conditions involved and the amount and types of solid waste to be handled. In general, the trench method is used when the groundwater is low and the soil is more than 6 ft (1.8 m) deep. It is best employed on flat or gently rolling land. The area method can be followed on most topographies and is often used if large quantities of solid waste must be disposed of. At many sites, a combination of the two methods is used.[3]

The building block common to both methods is the cell. All the solid waste received is spread and compacted in layers within a confined area. At the end of each working day, or more frequently, it is covered completely with a thin, continuous layer of soil, which is then also compacted. The compacted waste and soil cover constitute a cell. A series of adjoining cells, all of the same height, makes up a lift as shown in Figure 11.2.[3] The completed fill consists of one or more lifts.

The dimensions of the cell are determined by the volume of the compacted waste, and this, in turn, depends on the density of the in-place solid waste. The field density of most compacted solid waste within the cell should be at least 1000 lb/yd^3 (595 kg/m^3). It should be considerably higher if large amounts of demolition rubble, glass, and well-compacted inorganic materials are present. The 1000-lb (595 kg) figure may be difficult to achieve if brush from bushes and trees, plastic turnings, synthetic fibers, or rubber powder and trimmings predominate. Because these materials normally tend to rebound when the compacting load is released, they should be spread in layers up to 2 ft (0.6 m) thick, then covered with 6 in. (15.2 cm) of soil. Over this, mixed solid waste should be spread and compacted. The overlying weight keeps the fluffy or elastic materials reasonably compressed.

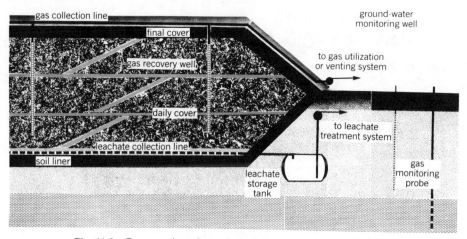

Fig. 11.2 Cross section of a typical sanitary landfill. (From Ref. 3.)

An orderly operation should be achieved by maintaining a narrow working face (that portion of the uncompleted cell on which additional waste is spread and compacted). It should be wide enough to prevent a backlog of trucks waiting to dump, but not be so wide that it becomes impractical to manage properly—never over 150 ft (45 m).

No hard-and-fast rule can be laid down regarding the proper height of a cell. Some designers think it should be 8 ft (2.4 m) or less, presumably because this height will not cause severe settlement problems. On the other hand, if a multiple lift operation is involved and all the cells are built to the same height, whether 8 or 16 ft (2.4 or 4.8 m), total settlement should not differ significantly. If land and cover material are readily available, an 8-ft (2.4 m) height restriction might be appropriate, but heights up to 30 ft (9 m) are common in large operations (and some are higher, for example, New Jersey Meadowlands; Fountain Avenue, New York City). Rather than deciding on an arbitrary figure, the designer should attempt to keep cover material volume at a minimum while adequately disposing of as much waste as possible.

Cover material volume requirements are dependent on the surface area of waste to be covered and the thickness of soil needed to perform particular functions. As might be expected, cell configuration can greatly affect the volume of cover material needed. The surface area to be covered should, therefore, be kept minimal.

In general, the cell should be about square, and its sides should be sloped as steeply as practical operation will permit. Side slopes of 20 to 30° will not only keep the surface area, and hence the cover material volume, at a minimum but will also aid in obtaining good compaction of solid waste, particularly if it is spread in layers not greater than 2 ft (0.6 m) thick and worked from the bottom of the slope to the top.[3]

Trench Method

In the trench method solid waste is spread and compacted in an excavated trench.[3] As shown in Figure 11.3, cover material, which is taken from the spoil of the excavation, is spread and compacted over the waste to form the basic cell structure. In this method, cover material is readily available as a result of the excavation. Spoil material not needed for daily cover may be stockpiled and later used as a cover for an area fill operation on top of the completed trench fill.

Cohesive soils, such as glacia till or clayey silt, are desirable for use in a trench operation because the walls between the trenches can be thin and nearly vertical. The trenches can, therefore, be spaced very closely. Weather and the length of time the trench is to remain open also affect soil stability and must be considered when the slope of the trench walls is being designed. If the trenches are aligned perpendicular to the prevailing wind, this can greatly reduce the amount of blowing litter. The bottom of the trench should be slightly sloped for drainage, and provision should be made for surface water to run off at the low end of the trench. Excavated soil can be used to form a temporary berm on the sides of the trench to divert surface water.[3]

The trench can be as deep as soil and groundwater conditions safely allow, and it should be at

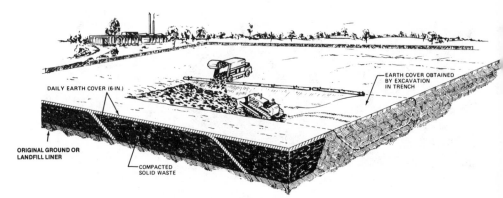

Fig. 11.3 The trench method of sanitary landfilling. (From Ref. 3.)

least twice as wide as any compacting equipment that will work in it. A typical depth is in the range of 15 ft (4.5 m). The equipment at the site may excavate the trench continuously at a rate geared to landfilling requirements. At small sites, excavation may be done on a contract basis.

Area Method

In the area method solid waste is spread and compacted on the natural surface of the ground, and cover material is spread over it and compacted. The operations during the area method are shown in Figure 11.4.[3] The area method is used on flat or gently sloping land and also in quarries, strip mines, ravines, valleys, or other land depressions.

Combination Methods

As noted earlier, the trench and area methods have relative advantages depending on the location and amount and type of solid wastes to be handled. For example, the trench method permits utilization of on-site material for cover and can be adapted for a wide variation in size of operation; adapted to a large variation in terrain condition; operated with minimum size of working face exposed; operated effectively during wet weather; designed for optimum drainage during filling operations; and efficient in land site use since borrow material is readily available. The area method can accommodate very large volume operations (large working face), and can be used where no excavation below original grade is feasible, that is, in constructing lifts on top of previous lifts.

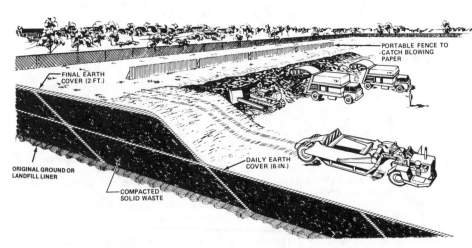

Fig. 11.4 The area method of sanitary landfilling. (From Ref. 3.)

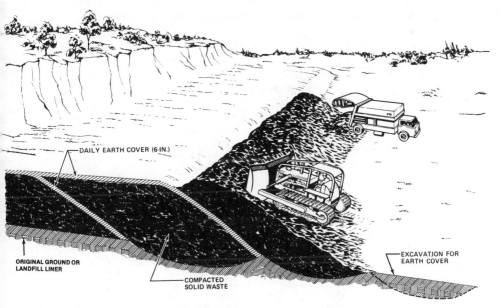

Fig. 11.5 The progressive slope method of sanitary landfilling. (From Ref. 3.)

Combinations of the two methods are possible, and flexibility is therefore one of sanitary landfilling's greatest assets. The methods used can be varied according to the constraints of a particular site.[3]

One common variation is the progressive slope or ramp method, in which the solid waste is spread and compacted on a slope. The operations in this method are shown in Figure 11.5.[3] Cover material is obtained directly in front of the working face and compacted on the waste. In this way, a small excavation is made for a portion of the next day's waste. This technique allows for more efficient use of the disposal site when a single lift is constructed than the area method does, because cover does not have to be imported, and a portion of the waste is deposited below the original surface.

Both trench and area methods might have to be used at the same site if an extremely large amount of solid waste must be disposed of. For example, at a site with a thick soil zone over much of it but with only a shallow soil over the remainder, the designer could use the trench method in the thick soil zone and use the extra spoil material obtained to carry out the area method over the rest of the site. When a site has been developed by either method, additional lifts can be constructed using the area method by having cover material hauled in.

Costs of Operations

Proper sanitary landfill operation is important for reasons of economy, efficiency, esthetics, maintenance, and environmental protection. From an economic standpoint, the operational cost is much higher than the design or site purchase costs. For example, a study of 17 landfills by the U.S. Environmental Protection Agency (EPA) determined that average capital and operating costs are proportioned as shown in Table 11.1.[2] These data indicate that landfill operating costs represent more than three-fourths of the total landfill disposal costs.

11.1.3 Biological and Chemical Processes

A sanitary landfill can be described as engineered burial of solid wastes that are subsequently degraded by soil microorganisms. Following burial, the microorganisms slowly degrade the organic portion of solid wastes to stable compounds.[4] As shown in Figure 11.6, it has been found that there are three distinct decomposition phases within landfills.[2] In the first stage (aerobic phase), the degradable solid wastes react quickly with the oxygen trapped during the landfilling process to form carbon dioxide and water. This is accompanied by heat and growth of the degrading organisms. The temperature will rise about 30°F higher than outside temperatures, and part of the carbon dioxide

Table 11.1 Average Landfill Disposal Costs Expressed as a Percentage of Costs Per Ton

Category	Item	Percentage
Initial capital cost	Land	4.3
	Development	4.3
	Stationary equipment	0.3
	Vehicles	12.8
	Subtotal	21.7
Operating cost	Labor	39.3
	Stationary equipment	1.3
	Vehicle operation and maintenance	27.5
	Administration and other	10.2
	Subtotal	78.3
	Total	100.0

Source: Ref. 2.

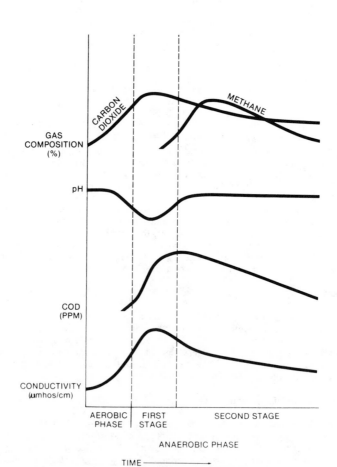

Fig. 11.6 Phases of solid waste decomposition. (From Ref. 3a.)

can dissolve within the water to form a weak acid that can dissolve other minerals. A simplified equation showing this aerobic decomposition is as follows:

$$6(CH_2O)_x + 5O_2 \rightarrow (CH_2O)_x + 5CO_2 + 5H_2O + \text{Energy} \qquad (11.1)$$
(organics in   (bacterial cells)
solid wastes)

After the oxygen is fully consumed, the second stage of decomposition (anaerobic phase with two stages as per Figure 11.6) begins. The organisms that grew off of the wastes when there was oxygen available can no longer live, whereas those organisms that can live without oxygen have not yet matured. The large organic molecules present in food, paper, and similar wastes are broken down into simpler molecules, including hydrogen, ammonia, carbon dioxide, and inorganic acids, and carbon dioxide levels reach a maximum of 50 to 90% of the gas generated. A simplified equation showing this organic acid production is as follows:

$$5(CH_2O)_x \xrightarrow[O_2]{no} (CH_2O)_x + 2CH_3COOH + \text{Energy} \qquad (11.2)$$
(organics in   (bacterial   (organic
solid wastes)   cells)   acid)

Complex organic solid wastes are degraded primarily by aerobic or facultative bacteria and fungi.[5] Facultative organisms are of importance in a landfill ecosystem since they can survive in both oxygen or nonoxygen environments. This is significant, since air does not penetrate a well-compacted landfill to any extent, and oxygen inside the fill is utilized rapidly by aerobic microorganisms as they decompose organic solid wastes. When this oxygen supply is depleted, decomposition by anaerobic facultative microorganisms begins and accounts for the degradation of most of the organic solid wastes in the landfill. Water-soluble organic acids are produced, through the hydrolysis of complex organic solid wastes, by these anaerobic facultative bacteria. These organic acids enter the water media and are then able to diffuse through the landfill soils where fungi and other bacteria aerobically metabolize this organic matter to carbon dioxide and water. Carbon dioxide produced inside a landfill can dissolve in groundwater, making the water weakly acidic.

In the third decomposition stage (second stage of anaerobic phase), methane-forming microorganisms utilize the carbon dioxide, hydrogen, and organic acids to form methane gas and other products. This is the final stage of the decomposition of the landfill, and the microorganisms work slowly but efficiently to decompose all available material. The organisms that produce methane require certain growth conditions such as water (the more the better), warm temperatures above 50 to 60°F (10 to 16°C), absence of oxygen or toxic materials, proper carbon-to-nitrogen ratios, and a near neutral pH. During the third stage, about 50% of the gas produced will be carbon dioxide; the other 50% will be methane. Total methane production depends on waste composition, but theoretically, about 6.5 ft[3] of methane gas can be produced from each pound of waste (0.5 m[3]/kg). A simplified equation showing methane production is as follows:

$$2\tfrac{1}{2}CH_3COOH \xrightarrow[O_2]{no} (CH_2O)_x + 2CH_4 + 2CO_2 + \text{Energy} \qquad (11.3)$$
(organic acid)   (bacterial acid)

Anaerobic methane bacteria occasionally can accumulate in large quantities in landfill systems, thereby discharging sizable amounts of methane gas through the soil. Aerobic bacteria can utilize a portion of this methane as it diffuses through the landfill; however, most of the methane is lost to the atmosphere.[4]

In addition to the biological decomposition processes in the landfill, chemical processes also occur. For example, ferrous and other metals may be oxidized.[3] Liquid waste products of microbial degradation, such as organic acids, increase chemical activity within a landfill. The changes in liquid phase pH can cause changes in solubilization, adsorption, precipitation, and ion-exchange patterns for many waste chemicals in sanitary landfills. Organic materials in sanitary landfills tend to undergo bacterial decomposition at variable rates. The decomposition products may become dissolved in the water phase or be released to the atmosphere as a gas, primarily either as carbon dioxide or methane. Inorganic materials tend to change under chemical reactions and may become associated with the water phase.

Food wastes degrade quite readily, while other materials, such as plastics, rubber, glass, and some demolition wastes, are highly resistant to decomposition.[3] Some factors that affect degradation are the heterogeneous character of the wastes, their physical, chemical, and biological properties, the availability of oxygen and moisture within the fill, temperature, microbial populations, and type of synthesis. Since the solid wastes usually form a very heterogeneous mass of nonuniform size and variable composition and other factors are complex, variable, and difficult to control, it is difficult to predict accurately contaminant quantities and production rates.

11.1.4 Environmental Protection Considerations

Properly sited, designed, and operated sanitary landfills represent a viable alternative to open dumps in the protection of public health and preservation of environmental quality. Landfills have advantages over open dumps in the following areas: (1) esthetics, especially litter; (2) flies and mosquitoes; (3) rats; (4) birds; (5) fires; (6) odor; (7) injury from scavenging; (8) decomposition gases; and (9) leachate.[2] Landfill esthetics include the screening of the daily operation from roads and nearby residents by berms, planting, or other landscaping. They include an attractive entrance with both good roads and easy-to-read signs. On the site itself, esthetic control means litter control, principally by the use of a fence to stop blowing paper and plastic, along with manual or mechanical pickup of the litter. The advantages related to flies and mosquitoes, rats, birds, fires, odor, and scavenging are focused on public health concerns. For example, harborage for flies and mosquitoes, rats, and birds is minimized in a well-operated landfill with daily cover.

Of major environmental concern relative to sanitary landfills are decomposition gases, mainly methane, and leachates. Gas can cause limitations in the usage of completed landfill sites; leachates have the potential for creating surface or groundwater pollution. Additional impacts from the development and operation of a sanitary landfill are highlighted in Table 11.2.[5] The table also includes actions that can be taken to mitigate the adverse impacts.

Table 11.2   Summary of Adverse Impacts and Counteractive Measures for Sanitary Landfills

Anticipated Adverse Impact	Actions Planned to Mitigate Adverse Impact
Public Health Esthetics	
Litter	Provide proper fencing Control working face area
Dust	Periodic watering
Odors	Assure prompt and consistent coverage of exposed wastes
Leachates	Diversion of runoff and drainage of precipitation incipient on the surface; if necessary, install underdrains and a collection/treatment system
Air quality impairment	Control dusts
Heavy equipment and collection vehicle movement	Provide proper traffic directors and spotters; assure adequate access roads
Methane gas generation	Install appropriate gas control venting system; minimize water infiltration to waste by drainage control
Local and Regional Biota	
Vegetation	Remove only the vegetation necessary for operations; install gas vents to preclude root-zone damage to adjacent vegetation
Animal life	Landscape finished landfill to reattract displaced native species; control leachates from entering water courses
Land and Land Use	
Visual unattractiveness	Plan cut and fill areas to minimize "desecration" appearances
Restricted land use	Plan for parks, golf courses, and open space
Social and Economic Environments	
Public opposition	Develop a comprehensive public relations/education program to promote and explain need for sanitary landfill and its operation; arrange for public meetings to air grievances—dispel aura of public powerlessness and promote participation in planning process
Cost increase	Incorporate discussions for landfill economics into public relations program

Source: Ref. 5.

Groundwater pollution from landfill leachates, particularly from the organics contained therein, is an important and long-term environmental issue. Robertson, Toussaint, and Jorque[6] conducted a study of organic compounds leached into groundwater from a landfill containing refuse deposited below or near the water table. Groundwater from wells within or near the landfill and a control well was sampled by modified low-flow carbon adsorption procedures incorporating all glass-Teflon systems to preclude introduction of extraneous organics. Column chromatography, solubility separation, and gas chromatography-mass spectrometry were employed for separation, identification, and quantitation of individual compounds in organic extracts. The groundwater was shown to contain low levels of many undesirable organic chemicals leached from the landfill. Of those compounds identified (over 40), most were chemicals commonly employed in industry for manufacturing many domestic and commercial products. The source of these compounds was apparently manufactured products discarded in the landfill, since it had not received appreciable wastes from industrial operations. Because the age of the refuse in the area studied was at least 3 years, the compounds identified were believed to be substances leached very slowly from the refuse and/or transported away from the landfill very slowly because of adsorption on aquifer solids. Potential long-term pollution of groundwater by industrial organic chemicals from landfills may be indicated by this study. Section 11.1.3 discusses leachate control in detail.

11.1.5 Guidelines: Federal and State

To encourage appropriate siting, design, and operation of sanitary landfills, the U.S. Environmental Protection Agency (EPA) has developed criteria for classification of solid waste disposal facilities and practices.[7] These criteria are for use under the Federal Resource Conservation and Recovery Act (RCRA) of 1976, as amended, in determining which solid waste disposal facilities and practices pose a reasonable probability of adverse effects on health or the environment. "Facility" means any land and appurtenances thereto used for the disposal of solid wastes. "Practice" means the act of disposal of solid waste. Facilities and practices that fail to satisfy the criteria will be considered open dumps. Key definitions are as follows[7]:

> *"Disposal" means the discharge, deposit, injection, dumping, spilling, leaking, or placing of any solid waste or hazardous waste into or on any land or water so that such solid waste or hazardous waste or any constituent thereof may enter the environment or be emitted into the air or discharged into any waters, including groundwaters.*

> *"Leachate" means liquid that has passed through or emerged from solid waste and contains soluble, suspended or miscible materials removed from such wastes.*

> *"Sludge" means any solid, semisolid, or liquid waste generated from a municipal, commercial, or industrial wastewater treatment plant, water supply treatment plant, or air pollution control facility or any other such waste having similar characteristics and effect.*

> *"Solid waste" means any garbage, refuse, sludge from a waste treatment plant, water supply treatment plant, or air pollution control facility and other discarded material, including solid, liquid, semisolid, or contained gaseous material resulting from industrial, commercial, mining, and agricultural operations, and from community activities, but does not include solid or dissolved materials in irrigation return flows or industrial discharges which are point sources subject to permits under the Federal Water Pollution Control Act, or source, special nuclear, or byproduct material as defined by the Atomic Energy Act.*

The key elements of the criteria as related to landfill disposal of municipal solid wastes are in Appendix 11.1.[7] Siting and operational criteria are enumerated for floodplains, preservation of endangered species, prevention of surface water and groundwater pollution, minimization of disease vectors and air pollution, and promotion of safety. Appendix 11.2 delineates maximum contaminant levels for determining whether solid waste disposal activities comply with groundwater protection criteria.[7]

In addition to the EPA criteria, each state has adopted sanitary landfill design and operational guidelines. The guidelines from the state of Oklahoma will be highlighted as an example. Twenty-seven evaluation factors are used for review of proposed and operating landfill sites. Inspections of operating sites involve the completion of an inspection report, a portion of which is shown in Appendix 11.3.[9] The 27 factors are described in Appendix 11.4 along with their reason for inclusion and criterion to be used to determine compliance.[8]

11.1.6 Landfill Development

Landfill development must be based on quantification of the land requirements for meeting present and future needs, selection of an appropriate site following the systematic comparison of candidate sites, and consideration and selection of design strategies to meet the solid waste disposal needs at the chosen site.

The development of a new landfill will generally include the following steps:

1. Service area delineation.
2. Estimate of refuse quantity.
3. Preparation of the preliminary design basis.
4. Site selection.
5. Preparation of a feasibility report (engineering report).
6. Application for regulatory agency permits (depending upon the state and/or local requirements this application may need to be filed earlier or possibly later in the process).
7. Filing of an environmental impact statement if required.
8. Preparation of detailed engineering plans.
9. Landfill site construction and initiation of operation.

Carefully and accurately completing each step is crucial to the overall success of the landfill project. Each step will be described in detail in the following sections.

11.1.7 Service Area, Waste Quantities, and Land Requirements

The area that the landfill serves will depend upon the nature of the entity developing the facility. A municipal government will be developing the landfill to serve all or part of its geographic jurisdiction. In contrast a private company will be developing the landfill to serve a market area that has been identified. The public entity may decide to offer service to entities outside its jurisdiction, in which case it must consider some of the same marketing factors that a private developer must evaluate. Both the public and private entity must decide which types of waste, residential, commercial, industrial, publicly or privately collected, and other classifications of waste, will be accepted at the landfill.

The next step is to estimate the quantity of waste that the landfill may receive. To make this estimate, population projection may be needed. These projections are usually available from local or state planning commissions and provide projections for at least 20 years. Similar projections may also be needed for commercial and industrial development. Using per capita waste generation rates or similar factors, future waste generation can be estimated. Past trends in waste generation may also be employed to predict future conditions assuming that the previously collected data are accurate and representative of a continuing trend. The complexity of this analysis will depend upon the nature and size of the service area and the need for accuracy. Developers of private landfills must also consider the nature of competitors, their prices, service market penetration, and possibly other factors.

The estimate of future solid waste quantities must also take into account actual or likely development of resource recovery facilities, and the impact on waste generation that recycling may have on future refuse quantities. An argument often put forth by opponents of landfills is that new landfills are not really needed if recycling is practiced. In some states, for example, Wisconsin, the entity developing the landfill must show that a need for the landfill will truly exist over at least a 10-year period. One element in this "needs determination" is the future potential impact of resource recovery activities on waste generation rates.

Preparation of the preliminary design basis requires several policy and technical decisions. A principal consideration is the desired landfill site operating life. A planned site life allows a better explanation of the facility to potential neighbors and is needed to make sound financial decisions. Ultimately the site life may depend on site restrictions that are not foreseen at this stage of landfill development but at a goal that should be established. Regulatory constraints may also specify the planned operating life of the landfill. A related consideration is the final use the site will be put to after the landfill has ceased operation. By planning for end use during site preparation, it is possible to better anticipate problems in developing the end use and also to reduce potential opposition by providing an attractive end use for the site.

The preliminary design basis should also identify any special factors or requirements that may be associated with the development of the landfill. This would include the density of uncompacted waste upon arrival at the landfill, expected volume reduction from compaction at the landfill, volume of cover in relation to the volume of solid waste, the expected design depth of the landfill (solid waste plus earth cover), and any waste characteristics that will result in the need for special handling.

The following equation can be used to calculate the landfill volume requirements:

$$V = \left[\frac{R}{D} \left(1 - \frac{P}{100} \right) \right] + CV \qquad (11.4)$$

where:

V = unit volume of compacted solid waste plus cover soil in the sanitary landfill (yd³/person/yr).
R = unit quantity of solid waste generated on an annual basis (lb/person/yr).
D = uncompacted density of solid waste as it arrives at the landfill (lb/yd³).
P = volume reduction achieved from solid waste compaction at the landfill; the reduction is a function of solid waste type, compaction prior to arrival at the landfill, type of compaction equipment, and number of passes of equipment over the solid waste during the compaction process (50 to 75% is a typical range).
CV = unit volume of soil cover; in a landfill with 8 ft of compacted solid waste and 2 ft of final cover there is one part of cover to four parts of solid waste; therefore, the equation can be rewritten to subsume CV as follows:

$$V = 1.25 \left[\frac{R}{D} \left(1 - \frac{P}{100} \right) \right] \tag{11.5}$$

To serve as an illustration of the use of this equation, determine the unit volume requirements for a city of 10,000 persons based on the following average factors:

R = 1825 lb/person/yr (5 lb/person/day times 365 days/yr).
D = 250 lb/yd³ (following transport to the landfill site in a packer truck).
P = 75% (heavy compaction equipment to be used with three to five passes over the solid waste).

$$V = 1.25 \left[\frac{1825 \text{ lb/person/yr}}{250 \text{ lb/yd}^3} (1 - 0.75) \right] = 2.3 \text{ yd}^3/\text{person/yr} \tag{11.6}$$

The area requirement can be calculated based on the following equation:

$$A = \frac{27\,(V)(N)}{(d)43{,}560} \tag{11.7}$$

where:

A = landfill area needed (acres/yr).
N = population.
d = landfill lift (compacted solid waste plus cover) (ft).
27 = ft³/yd³.
43,560 = ft²/acre.

To complete the above example for the city of 10,000 persons, and using a lift of 10 ft:

$$A = \frac{27 \text{ ft}^3/\text{yd}^3 \,(2.3 \text{ yd}^3/\text{person/yr}) \,(10{,}000 \text{ persons})}{(10 \text{ ft})\, 43{,}560 \text{ ft}^2/\text{acre}} = 1.4 \text{ acres/yr} \tag{11.8}$$

It should be noted that the above two equations do not include a "waste land factor." This factor, which accounts for the fact that a site is not 100% usable due to the need for access roads and other working areas, ranges from about 1.25 for the area method to 2.0 for the trench method. Therefore, for the above example, the land required for the area method is 1.8 acre/yr, and for the trench method 2.8 acre/yr.

Several "rules of thumb" have been developed for estimating the volume or area requirements for a sanitary landfill. Examples include: (1) 2 yd³/person/yr; (2) 2 acres/10,000 persons/yr (8 ft lift); and (3) 1.25 acre-ft/1000 persons/yr. Graphs have also been developed to allow quick estimates of the volume requirements for sanitary landfilling. Figure 11.7 displays volume requirements as a function of the solid waste production rate and density after compaction.[9]

11.1.8 Siting Procedures

The selection of a site is the first major step in the development of a new sanitary landfill. The importance of thoroughness in a carefully planned selection process cannot be overstated. Recognition not only of technical considerations, but also of environmental, economic, social, and political concerns is vital. A well-planned selection process will address all of these concerns.[2] The objectives of the site selection study is to find a site where solid waste disposal can be accomplished economically with minimal disruption of the environment. It must be acceptable to the public and satisfy the requirements of the applicable local, state, and federal agencies. Certain tasks must be completed before beginning the search for a new landfill site. The tasks described in the previous

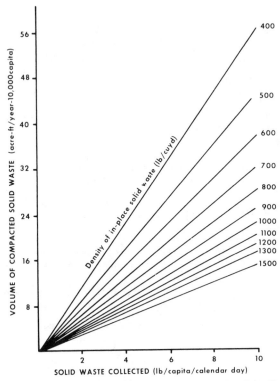

Fig. 11.7   Volume of solid waste after placement and compaction. (From Ref. 3.)

section, that is, service area delineation, estimate of solid waste quantity, and preparation of the preliminary design basis, must be well documented before initiating landfill siting.

A complete site selection process is a complex system integrating public opinion and involvement with existing policy, while evaluating environmental, safety, economic and engineering feasibility.[10] The relative importance of each of these factors is dependent upon the basic selection of objectives, namely, services to be provided by the facility and pertinent local, state and federal regulations and policies.[11] Typically, the overall site selection process should contain the following 10 steps:

1.   Developing site selection criteria.
2.   Identifying candidate sites best meeting these criteria.
3.   Initial review and evaluation of candidate sites.
4.   Selection of sites for final evaluation.
5.   Evaluation of regional awareness.
6.   Final technical evaluation and ranking of sites.
7.   Public involvement.
8.   Site selection.
9.   Public hearing.
10.   Review.

Site selection criteria are necessary to provide a base from which any analysis can be completed. This allows a site or site area to be evaluated with respect to specific environmental concerns as well as many of the technological aspects of the proposed sanitary landfill. Social, economic, and political constraints can also be added to the process by using the site selection criteria. Frequently the siting criteria is applied in two stages; a first stage where broad criteria are applied to a large geographic area, and a second stage where detailed evaluation of specific sites is undertaken.

A large quantity of information and data must be collected in order to thoroughly evaluate the potential landfill sites. Sources of information may include the following:

1. State environmental protection agency and EPA guidelines.
2. Local highway, land-use, and ownership maps.
3. Regional planning agency reports and maps.
4. U.S. Department of Agriculture Soil Conservation Service soils maps.
5. U.S. Geological Survey water supply papers and topographic, surface water resource, water table, and wetlands maps.
6. Well logs for existing public and private wells.
7. Existing records of waste generation and disposal rates.
8. Soil borings logs and monitoring well data for selected sites.
9. Detailed on-site surveys of sites identified as having a high potential for landfill development.

The nature and amount of data collected will depend somewhat upon the size of the planned landfill and the degree of potential environmental hazard and controversy that may be anticipated. Large controversial projects may require exhaustive study before a site is selected, whereas a small site at an isolated location requires less investigation. However, even a small site requires sufficient investigation to insure that potential operational and environmental requirements are anticipated.

A graphical approach for selecting potential landfill sites within a large geographic area is to prepare a series of overlay maps. As shown in Figure 11.8, a separate map is prepared for each major evaluation factor. The maps are prepared on transparent sheets with those areas having limitations for landfilling shaded the darkest. When the maps are placed one on top of the other the clear areas that show through have the fewest limitations for landfill construction. This approach also allows the easy preparation of visual aids that can be helpful during public meetings and other

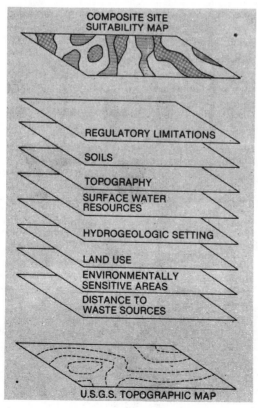

Fig. 11.8 Overlay of site criteria maps. (Haji-Djafari, et al., 1980.) (From Ref. 10a.)

forums that will be conducted during the siting process. The principal disadvantage is that each of the factor maps are given equal weighting in the decision process. This approach does allow the rapid screening of a large area for potential suitable landfill sites once the maps are prepared.

A numerical site-selection method that allows the weighting of the various evaluation criteria can also be employed. A grid system is established for the land area being evaluated. The land inside each element of the grid is rated for each individual evaluation factor. The individual scores are summed using the weighting formula shown below.

$$C = (W1 \times R) + (W2 \times S) + (W3 \times T) + (W4 \times W) + (W5 \times H) + (W6 \times L)$$
$$+ (W7 \times E) + (W8 \times D) \tag{11.9}$$

where:

W = Composite Suitability Score (0–5).
R = Regulatory Suitability Score (0–5).
S = Soils Suitability Score (0–5).
T = Topography Suitability Score (0–5).
W = Surface Water Resources Suitability Score (0–5).
H = Hydrogeologic Setting Suitability Score (0–5).
L = Land Use Suitability Score (0–5).
E = Environmentally Sensitive Areas Suitability Score (0–5).
D = Distance to Waste Sources Suitability Score (0–5), and where $W1$ through $W8$ are weighting factors with values of 0 to 1.0 for the respective suitability scores.

The rating for each parcel is displayed on a map. The level of detail will depend upon the size of the grid spacing, with smaller being more reliable. However, a smaller grid spacing also increases the amount of data that must be collected before the analysis is undertaken.

Each of these methods, graphical and numerical, offer an opportunity to assess a large area relatively quickly for potential landfill sites. After a list of potential sites is selected, the next step is to use more extensive criteria to evaluate in detail the relative suitability of particular candidate sites.

Examples of evaluation criteria are shown in Table 11.3 for rating soil suitability for trench-type landfills, Appendix 11.5 for rating various factors for suitability of chemical waste disposal, and Appendix 11.6, which lists site criteria required for the engineering report for a sanitary landfill by the Oklahoma State Department of Health.

11.1.9 Techniques for Comparing Candidate Sites by Specific Issues

Some systematic methodologies have been developed for comparing candidate sites in terms of specific issues. For example, the potential for groundwater pollution is a major concern of sanitary landfills, and should be a key issue in the site selection process. One example is a method called landfill site rating.[12] The landfill site rating method is based upon the experience gained by many individuals to establish objectively the more-favorable and less-favorable conditions for prevention of adverse groundwater impacts. Four key hydrogeological factors or variables are used. The following four factors are considered to represent the simplest and most easily determined and effective factors for a wide variety of applications: (1) distance from a contamination source to the nearest well or point of water use; (2) depth to the water table; (3) gradient to the water table; and (4) permeability and attenuation capacity of the subsurface materials through which the contaminant is likely to pass. The LeGrand-Brown methodology is illustrated in Appendix 11.7.[12] Some of the advantages and disadvantages of this methodology are as follows:[13]

Advantages:
Quantitative data can be obtained and compiled.
Site suitability can be quantified.
The level of site security or suitability can be related to a numeric rating.
It may be applied to define many categories of sites according to numeric rating.
A weighting factor can be assigned to a particular waste criterion.
Disadvantages:
Weighting factors are assigned primarily through the judgment of the user (subjective).
The ranking system must be verified and calibrated through case history testing.
This approach has not been extensively used by regulatory agencies.

Table 11.3 U.S. Department of Agriculture Soil Conservation Service Soil Limitations for Trench-Type Sanitary Landfills

Property	Slight	Moderate	Severe	Restrictive Feature
Flooding	None, protected	Rare	Common	Floods
Depth to bedrock (inches)	—	—	Less than 72	Depth of rock
Depth to cemented pan (inches)				
-thick	—	—	Less than 72	Cemented pan
-thin	—	Less than 72	—	Cemented pan
Permeability (inches/hours) (bottom layer)	—	—	Greater than 2.0	Seepage
Depth to high water table (ft)	—	—	+	Ponding
-apparent	—	—	0–6	Wetness
-perched	Greater than 4	2–4	0–2	Wetness
Slope (%)	0–8	8–15	Greater than 15	Slope
USDA Texture[a,b]	—	Clay loam, sandy clay, silty clay loam	Silty clay, clay	Too clayey
USDA Texture[b]	—	Loamy coarse sand, loamy sand, loamy fine sand, loamy very fine sand	Coarse sand, sand, fine sand, very fine sand, sand and/or gravel	Too sandy
Unified Classification[b]	—	—	OL, OH, PT	Excess humus
Fraction greater than[c] 3-in. (weight %)	Less than 20	20–35	Greater than 35	Large stones
Sodium absorption ratio (great group)	—	—	Greater than 12 matrix halite	Excess sodium
Soil reactions—pH	—	—	Less than 3.6	Too acidic

[a] If in kaolinitic family rate, one class better if experience confirms
[b] Thickest layer between 10 and 60 in.
[c] Weighted average to 60 in.
Adapted from *Nation Soils Handbook*, USDA Soil Conservation Service, 3–78.
Source: Ref. 2.

It may involve a high cost for site investigations; the owner of the disposal site generally bears this expense.

It requires skilled judgment and experience to use this system (for example, assigning weighting factors).

Final site selection should involve the use of a systematic methodology for comparison of the candidate sites (alternatives) relative to site selection criteria. This selection requires the use of trade-off analysis. Table 11.4 displays a conceptual trade-off matrix for systematically comparing specific alternatives relative to a series of decision factors. The following approaches can be used to complete the trade-off matrix[10]:

1. Qualitative approach in which descriptive information on each alternative relative to each decision factor is presented in the matrix.
2. Quantitative approach in which quantitative information on each alternative relative to each decision factor is displayed in the matrix.

Table 11.4 Trade-off Matrix

Decision Factors	Alternatives				
	1	2	3	4	5
Health risk					
Economic efficiency					
Social concerns (public preference)					
Environmental impacts Biophysical Cultural Socioeconomic					

3. Ranking, rating, or scaling approach in which the qualitative or quantitative information on each alternative is summarized via the assignment of a rank, or rating, or scale value relative to each decision factor (the rank, or rating, or scale value is presented in the matrix).

4. Weighting approach in which the importance weight of each decision factor relative to each other decision factor is considered, with the resultant discussion of the information on each alternative (qualitative; or quantitative; or ranking, rating, or scaling) being presented in view of the relative importance of the decision factors; and

5. Weighting-ranking/rating/scaling approach in which the importance weight for each decision factor is multiplied by the ranking/rating/scale of each alternative, then the resulting products for each alternative are summed to develop an overall composite index or score for each alternative.

Decision-making that involves the comparison of a set of alternatives relative to a series of decision factors is not unique in terms of its application to sanitary landfill site selection. This approach represents a classic decision-making problem that is often referred to as multiattribute or multicriteria decision-making. The conceptual framework for this type of decision-making appeared in the early 1960s in the U.S. Department of Defense. Since that time it has been applied to numerous situations requiring decisions, including those that involve consideration of environmental factors and impacts.

If the qualitative or quantitative approach is used for completion of the matrix as shown in Table 11.4, information for this approach relative to the decision factors (site selection criteria) can be found elsewhere in this chapter. It should be noted that this information would also be needed for the approaches involving importance weighting and/or ranking/rating/scaling. If the importance weighting approach is used, the critical issue is the assignment of importance weights to the individual decision factors, or at least the arrangement of them in a rank ordering of importance. One approach for importance weighting involves the use of an unranked paired comparison technique.[14]

Morrison[15] applied a comparison technique to the selection of a sanitary landfill site from three alternatives. The list of decision factors used are contained in Appendix 11.8.[15]

11.1.10 Public Involvement

For many who have located landfill sites, the subject of public attitude brings only negative reaction. The public, it seems, has a "NIMBY" (not-in-my-backyard) attitude. Public attitudes are sometimes ignored when locating a landfill with the rationale that there will be complaints no matter where the site is located. A well-planned siting program will include opportunities for public participation at appropriate times.[2] In addition to landfill site selection, public involvement can also be a component in the evaluation of alternative design strategies. The firm or agency designing the landfill should solicit input to the decision-making process from the individuals and groups who will be directly or indirectly affected by the location. Individuals and groups having an interest in the decision will include the site operators, the committee or board overseeing solid waste disposal, public and commercial refuse collectors, landowners and residents near the landfill site, local environmental groups, regulators at both the local and state level, elected public officials, and interested citizens in the region.

The basic purpose of public participation is to promote productive use of inputs and perceptions from private citizens and public interest groups in order to improve the quality of environmental decision-making.[17] Relevant citizen-oriented activities are variously referred to as citizen participation, public participation, public involvement, and citizen involvement. Public participation can be defined as a continuous, two-way communication process, which involves promoting full public understanding of the processes and mechanisms through which environmental problems and needs are investigated and solved by the responsible agency; keeping the public fully informed of the status and progress of studies and findings and implications of plan formulation and evaluation activities; and actively soliciting from all concerned citizens their opinions and perceptions of objectives and needs and their preferences regarding resource use and alternative development or management strategies and any other information and assistance relative to plan formulation and evaluation.[16]

If public participation is to be effective in the various stages of sanitary landfill site selection and design, the public participation program must be carefully planned. A good public participation program does not occur by accident.[16] Planning for public participation should address the following elements:

1. Delineation of objectives of public participation.
2. Identification of publics.
3. Selection of public participation techniques that are most appropriate for meeting the objectives and communicating with the publics. It may be necessary to explain techniques for conflict management and resolution.
4. Development of a practical plan for implementing the public participation program.

The delineation of objectives of public participation is an important element in developing a public participation plan. The objectives could be general or specific. Hanchey[17] suggested three general objectives that should be considered in the composition of a public participation program for specific planning or decision-making. These are referred to as: (1) the public relations objective; (2) the information objective; and (3) the conflict resolution objective. These public participation objectives are shown in Figure 11.9.

The identification of publics that would potentially be involved in the various stages of the decision-making process is another basic element in the development of a public participation program. Also of importance is a description of public participation techniques that are most effective in communicating with various publics. The general public cannot be considered as one body. The public is diffuse, but at the same time highly segmented into interest groups, geographic communities, and individuals. There are sets or groups of "publics" that have common goals, ideals, and values. Any one person may belong to several different sets of these publics since they may be professionally, socially, or politically oriented. The Venn diagram shown in Figure 11.10 illustrates the overlapping of some of these groups, and the fact that an individual may identify with one, a combination of two, or all three of the groups.[18] Two significant points may be drawn from Figure 11.10 in terms of communication:

1. Individuals are likely associated with various social, economic, and cultural orientations from which they draw their information and structure their values.
2. Multiple association thus allows the opportunity for multiple access to individuals as participants, clients, or critics in a planning process.

There are several ways of categorizing various publics that might be involved in a public participation program for sanitary landfill site selection and design. In targeting publics an attempt should be made to identify those persons who believe themselves to be affected by the decision outcome. The difficulty is that the degree to which people feel affected by a decision is a result of their subjective perception, that is, people the agency feels are most directly impacted may not be as concerned as someone the agency perceives as only peripherally involved. However, the starting point is always an effort to analyze objectively the likelihood that someone will feel affected by the study. Some of the bases on which people are most likely to feel affected are:[19]

1. *Proximity.* People who live in the immediate area of a landfill and who are likely to be affected by noise, odors, dust, or possibly even threat of dislocation, are the most obvious publics to be included.
2. *Economic.* Groups that have jobs to gain or competitive advantages to win, for example, landowners near the site versus the site landowner(s), are again an obvious starting point in any analysis of possible publics.
3. *Use.* Those people whose use of the area is likely to be affected in any way by the outcome of the study are also likely to be interested in participating. These include recreationists, hikers, fishermen, hunters, and so forth. In some cases these users are among the most vocal participants in a study.

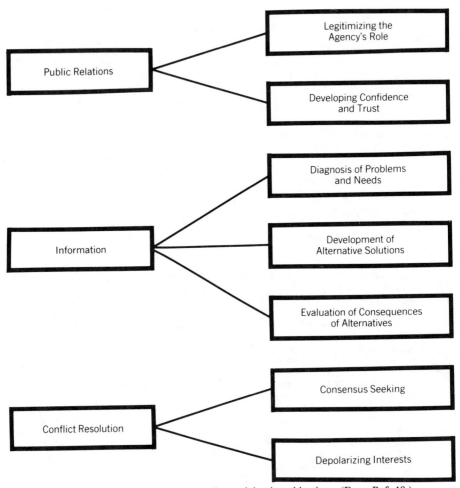

Fig. 11.9 Interrelationship of public participation objectives. (From Ref. 18.)

4. *Social.* Increasingly, people who see projects as a threat to the tradition and culture of the local community are likely to be interested in projects. They may perceive that an influx of construction workers into an area may produce either a positive or negative effect on the community. Or they may perceive that the project will allow for a substantial population growth in the area, which they may again view either positively or negatively.

5. *Values.* Some groups may be only peripherally affected by the first four criteria but find that some of the issues raised in the study directly affect their values, their "sense of the way things ought to be." Any time a study touches on such issues as free enterprise versus government control, or jobs versus environmental enhancement, there may be a number of individuals who participate primarily because of the values issues involved.

A critical element in planning a public participation program is associated with the selection of public participation techniques to meet identified objectives and publics. It should be noted that there are numerous techniques, and a well-planned public participation program will probably involve the use of multiple techniques over the lifetime of the decision-making process. The most traditional public participation technique is the public hearing, which is a formal meeting for which written statements are received and a transcript is kept. The public hearing is generally not an appropriate forum for public participation in the environmental impact decision-making process.[17] Classification schemes have been developed for public participation techniques in accordance with their function[20] and communication characteristics and potential for meeting stated objectives.[21]

Bishop[21] developed a structured classification scheme for 24 public participation techniques.

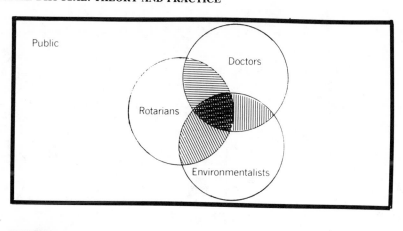

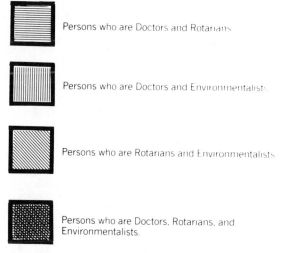

Persons who are Doctors and Rotarians

Persons who are Doctors and Environmentalists

Persons who are Rotarians and Environmentalists

Persons who are Doctors, Rotarians, and Environmentalists.

Fig. 11.10 Example of multiple public associations. (From Ref. 19.)

Table 11.5 displays the techniques in three groups: the first six listed techniques represent public forums, the following eleven listed techniques represent community contacts, and the final seven listed techniques represent interactive group methods. The communication characteristics of the 24 techniques are displayed in Table 11.5 in terms of the level of public contact achieved, ability to handle specific interests, and degree of two-way communication. A relative scale of low (L), medium (M), and high (H) effectiveness is used to delineate the communication characteristics of the 24 techniques. Table 11.5 also has information on the potential usefulness of each technique relative to meeting one or more of six objectives for a public participation program.

The following list of items represents some very practical ideas and suggestions that can be useful in organizing a public participation program:[17]

1. Coordinate the various federal, state, and local agencies that have interests and responsibilities in the same geographic or technical areas of the study. Develop formal agreements or informal relationships.

2. Develop lists of groups and citizens in the geographic area who have previously expressed interests or potential interests in the study.

3. Assemble a newspaper clipping file on project needs and previous history of the project or study.

4. Try to convey the attitude, "What can we do in this study to assist you in maintaining local environmental amenities and values."

Table 11.5 Capabilities of Public Participation Techniques

Communication Characteristics[a]				Impact Assessment Objectives					
Level of Public Contact Achieved	Ability to Handle Specific Interest	Degree of Two-Way Communication	Public Participation Techniques	Inform/ Educate	Identify Problems/ Values	Get Ideas/ Solve Problems	Feedback	Evaluate	Resolve Conflict/ Consensus
M	L*	L	Public hearings	X	X		X		
M	L	M	Public meetings	X	X		X		
L	M	H	Informal small group meetings	X	X	X	X	X	X
M	L	M	General public information meetings	X					
L	M	M	Presentations to community organization	X	X		X		
L	H	H	Information coordination seminars	X			X		
L	M	L	Operating field offices		X	X	X	X	
L	H	H	Local planning visits		X		X	X	
L	H	L	Planning brochures and workbooks	X		X	X	X	

Technique							
Information brochures and pamphlets	M	M	X				
Field trips and site visits	L	H	X				X
Public displays	H	L	X		X		X
Model demonstration projects	M	L	X		X	X	X
Material for mass media	H	L	X				
Response to public inquiries	L	H	X				X
Press releases inviting comments	H	L	X		X		
Letter requests for comments	L	H		X	X		X
Workshops	L	H		X	X	X	X
Charettes	L	H		X	X		X
Advisory committees	L	H		X	X	X	X
Task forces	L	H		X	X	X	
Employment of community residents	L	H		X	X		X
Community interest advocates	L	H		X	X	X	X
Ombudsman or representative	L	H		X	X	X	X

[a] L, low; M, medium; H, high.
Source: Ref. 2.

5. Disseminate study information through the news media (newspaper, radio, and television) and through regular publication of a newsletter. The mailing list should encompass all state and federal interests as well as local groups and individuals who have participated in previous meetings or shown interest in the study or decision.

6. Every third or fourth issue of the newsletter should contain a mail-in coupon for persons wanting to continue to receive the newsletter. Each issue should contain a coupon for suggestions or other persons or groups that should receive the newsletter.

Some practical suggestions for actually conducting a public meeting are as follows:[17]

1. Keep data presentations simple. The purpose of data presentations is to inform and not to confuse or disillusion.

2. An outline of any project or alternative should include a discussion of location, features, benefits and costs, and beneficial and detrimental environmental impacts.

3. Use simple visual aids. Slides of the actual area prove very beneficial.

4. Discuss project timing previous to the meeting and the anticipated timing of future required steps.

5. Discuss general concept, features, and requirements of benefit-cost ratio, or other relevant economic analysis.

6. Only those persons who can speak on general matters, commingled with engineering expertise, should be considered for meeting with the public. It is well known that not everybody has the ability to speak, answer questions, and perhaps debate while holding a specific image in front of the public. The ability to speak well is not the only trait on which the selection should be based. A person who can answer questions quickly and confidently from an audience in which the sentiment is mixed or opposed can establish a profound positive relation with the audience.

7. Avoid use of technical jargon or words that may be hard to understand, especially with local groups unaccustomed to engineering and ecological terminology.

8. Be familiar with the area.

9. Be earnest, sincere, and willing to work on problems with individual groups.

The difficulty in siting new landfills due to public opposition has given rise to the use of arbitration procedures in an effort to reach agreements between landfill developers and neighbors around the site. In the absence of formal negotiation procedures, and even with them, the siting of new landfills has been subject to extensive litigation. The arbitration procedures are patterned after labor arbitration procedures and involve the establishment of a formal procedure for the developer and the affected parties to articulate their viewpoints.

A formal or informal negotiation process proceeds until an agreement or impasse between the parties is reached. If an agreement is not reached, the views of each side are forwarded to an arbitrator. The arbitrator makes a decision regarding the conditions under which the development of the landfill can proceed. The specific procedures followed will depend upon the state laws and rights of appeal available.

11.1.11 Plan Preparation and Regulatory Approval

The successful development of a new landfill will depend heavily on the preparation of a technically sound plan for the construction and operation of the facility. This plan will usually be in the form of a detailed feasibility or engineering report, drawings showing the planned construction, a plan of operation, and possibly a closure plan. The specific documents and their detail will depend upon the regulations that apply to its development. Usually the regulatory review and approval process will involve the application of federal, state, and possibly local standards. Federal standards are summarized in Appendices 11.1 and 11.2. There is a wide variation in the content of state standards, but a typical standard for the items that must be addressed when developing a landfill is summarized in Appendices 11.3 and 11.4.

The design process is broken into five steps. They are:

1. Identify and establish goals.
2. Identify the design basis.
3. Prepare alternative designs.
4. Evaluate alternatives and select the best design.
5. Develop detailed design for the selected alternative.

Typical design goals should consider the following:

1. Meet all applicable regulations.
2. Protect the physical environment (groundwater, surface water, air quality).
3. Minimize operational nuisances (litter, dust, noise, fires).
4. Minimize dumping time for site users.
5. Minimize costs (initial, operation, total).
6. Protect worker and user safety.
7. Provide for maximum use of land when site is completed.
8. Maintain esthetic site.

The second step is to specify the landfill design basis. This will be a compilation of information derived from earlier studies and will include the following:

1. Environmental regulations.
2. Waste characteristics.
3. Site physical characteristics.
4. Geotechnical data.
5. Hydrologic and climatic data.
6. Transportation systems.
7. Site operational requirements.
8. Planned final use.

A portion of the information described above will have been tabulated before actual design commences and more information will need to be gathered during design. It is important that sufficient time and resources be available to gather the necessary information. For example, if an insufficient number of soil borings were conducted during the geotechnical evaluation, the danger exists that assumptions will be made during the design phase that cannot be justified.

The third step in developing plans is to prepare drawings of alternative layouts and operating procedures that can be employed on the site. The preparation of alternative plans promotes a more thorough evaluation of the most economical combination of features within the landfill. The final product of this step is a written description of the alternatives and basic drawings showing the major features of each alternative. This allows the landfill operators, the site owners, and possibly regulatory agency personnel to provide input and judgments with regard to the landfill development before the expense and time necessary to draw detailed plans are expended.

During step four, the alternatives are evaluated with regard to the goals and objectives that have been identified and trade-offs will be necessary. One alternative that may be best with regard to leachate management may have disadvantages for gas control or final end use of the site. If a large number of alternatives are being considered, it may be helpful to develop a numerical scoring system to rate each alternative plan with regard to attainment of the goals and objectives. This evaluation method allows for inclusion of many criteria in the evaluation processes. There may still be many concerns that are not quantified, and a numerical rating system should only be used to narrow the number of alternatives before selecting the one for which detailed plans will be drawn.

It is critical that the public participation activities effectively communicate the proposed plans to the public and, in turn, the process obtains feedback from the public for incorporation into the landfill design. If there is no input at this stage, the danger exists that after detailed plans are prepared, which is an expensive and time-consuming process, the individuals and groups having an interest in the landfill development may object. The site developer will find it much easier to modify his or her plans during the alternative evaluation process than after the detailed plans have been prepared. Also, there is a tendency to resist change in the plans once they have been drawn, and therefore modifications at the alternative evaluation stage are much easier to accomplish.

The final output from step 4 is a report describing the design basis, a description of the alternatives evaluated, and a detailed description of the design selected for implementation. Regulatory agency procedures generally specify that this landfill feasibility report be submitted for review and approval before detailed plans are prepared. Even if this approval process is not required, it may be a good strategy to inform the regulatory agencies about the proposed plan before detailed plan preparation is commenced. This approach allows the reconciliation of review agency questions before the detailed plans are prepared.

The fifth step in landfill design is the preparation of construction plans and operating specifications. The final product is a complete set of instructions for constructing, maintaining, operating, and closing of the landfill. This extends to the end of the long-term period, which may be

20 or 30 years after the landfill closes. Since the operating period for the landfill may be 10 or 20 years, the total plan may have a life of 30 to 50 years. The detailed design begins with the selection of landfill base elevations, top elevations, and slopes. Next, the volume of cover soils needed and quantity available are computed. Cut and fill balances are computed and a fill capacity is projected. Using hydrologic and climatic data, the water flow into and out of the landfill is calculated. This is accomplished with the water balance methods or computer models, which will be described in the next section. If the landfill base is below the water table, the quantity of groundwater which may enter the fill must also be estimated. Depending upon the quantity of percolate projected, engineering means for routing the leachate through the landfill may be necessary, and regulatory or environmental conditions may dictate that a leachate collection system be incorporated into the design. In a similar manner plans are developed to control methane gas migration, noise problems, litter, and other environmental and operational concerns.

Good design will ensure operational effectiveness of the site. Procedures must be incorporated into the design that ensure compatible procedures between different pieces of equipment and between landfill operating equipment and delivery of refuse to the site. Operational procedures must be incorporated into the design and should only be developed after consultation with the intended operator of the site. The operator must be provided with the necessary equipment to be effective in monitoring and managing the landfill. This includes: (1) scales for weighing incoming loads; (2) methods for determining the volume of the site that has been filled; (3) machinery records that identify use, maintenance, and fuel consumption; (4) an accounting system for costs and revenues at the landfill; and (5) data on waste quantities, soil, cover location, and safety records.

In addition to the interior design of the landfill, it will be necessary to develop plans for the land immediately adjacent to the landfill. This will include buffer zones, access roads, plants, berms to reduce noise, litter, and dust, and fences to prevent unauthorized access. The landfill entrance should be designed to provide an attractive and safe access to the site, including turnoff lanes from busy roads and signs to direct the traffic. A visual zone around the landfill should be established where the sometimes unsightly activities within the landfill are not visible. This will reduce the potential number of complaints regarding the operation of the landfill and enhance the landfill's acceptance by neighboring residents.

The landfill is usually depicted on a series of engineering drawings and accompanied by a narrative description and a technical report. A typical set of engineering plans will contain drawings showing the location, existing site conditions, base grade, engineering modifications, final site topography, phasing diagrams, geologic cross sections, site monitoring plan, site long-term care plan, and detailed drawings of gas vents, leachate collection systems, and other facilities. A design element checklist is shown in Table 11.6.

Generally, the landfill is constructed in such a fashion that modular development is accomplished. This phasing of the landfill construction allows each segment of the landfill to be operated independently. Building a landfill in this fashion allows the completed sections to be used immediately for other purposes and minimizes the amount of landfill surface that is left open and exposed to the elements. This will reduce runoff problems, erosion problems, and the possibility for litter and windblown paper problems. In a well-planned phase development, it is possible that the end use of the landfill, such as for hiking or cross-country skiing or use as a nature area, can begin while other areas within the landfill are still being utilized for disposal purposes.

Table 11.6 Landfill Design Element Checklist: Possible Engineering Site Modifications

Leachate Collection Systems

Leachate quantity	Leachate storage reservoirs
Collector pipe bedding to prevent clogging	Pumping facilities
Pipe materials to withstand heat and pressure	Continuous monitoring
Means to inspect and clean out	Modular integrity
Cross connections to provide alternative flow paths	Leachate recirculation during peak flow or for treatment
Positive drainage	

Leachate Containment

Material selection	Leak detection
Specifications	Subgrade preparation
Protection from puncture	Landfill capping
Construction techniques	Controlled release

Table 11.6   (Continued)

Methane Gas Control

Concentration of gases
Withdrawal system
Odor control

Recovery as resource
Protection of on-site vegetation
Monitoring

Perimeter Zone

Access control—fencing
Access control—signs
Screening berms
Screening vegetation
Screening fences

Windblown paper controls
Esthetics
Safety
Cover or topsoil stockpiling
Handle off-site drainage

Entrance Zone

Reasonable grades (7–10%)
Eliminate interference with access route
Need for merging lanes
Adequate site distances
Adequate turning radii

Adequate width
Paved with a permanent surface
Devices to keep mud from access road
Esthetics
Prevent straight-line view into site

Visual Zone

Designed to provide good image
Vehicle pull-off for viewing site

Traffic controls
Consider concurrent use

Interior Zone

Provide building for administration, weighing, equipment, maintenance, storage, and parking
Provide employee facilities
Examine layout of facilities for efficiency
Use pull-off areas to avoid congestion on access roads
Provide bulk containers
Fire controls
Adequate separation to property boundaries
Provide minimal gas venting
Set up monitoring program—gas and leachate
Moveable litter fences

Primary windblown paper controls
Cell construction in relation to wind direction
On-site surface water runoff
Erosion controls
Specification for access roads
Handling of cover materials
Equipment selection
Initial site improvement
Topsoil and cover stockpiles
Hours of operation
Utilities: water, electric, and telephone

Plan of Operation

Goals to be met
Handling of special wastes—dead animals, industrial, liquids, hazardous
Placement of cover
Maintenance of facilities, equipment, roads
Operation during variable weather conditions: winter, wet, warm weather, brisk winds
Fires
Salvage

Record keeping
Operating hours
Traffic routing
Staffing needs
Vector control
Dust control
Daily cleanup
Ban on salvaging

During the preparation of plans, the designer must remember that the landfill operator will ultimately determine the success or failure of the landfill's design. The plans should be prepared in a form that is clear and concise and easily read by persons who do not necessarily have a working engineering knowledge. Plans that are difficult to read will only be misinterpreted and not followed. In addition to the typical engineering plans, it is worthwhile to prepare isometric drawings of the landfill that show the various phases of construction and operation. A typical drawing, both plan view and isometric, is shown in Figure 11.11.

The timing for the application of state approvals will depend upon the applicable state regulations. The state's process may provide for obtaining a single approval for landfill construction and operation. Some states employ a multistaged approval process such as the following:

1. Initiation of state landfill siting arbitration procedure.
2. Submission of a feasibility (engineering) report to the state for approval.
3. Submission of detailed engineering plans to the state.
4. Final application for state landfill operating permits.

The usual procedure during this process is for the landfill developer to submit applications and plans to the agency who then reviews the merits of the proposal. If the agency finds the material acceptable, the process proceeds. More frequently, the agency will have additional questions which they will ask the developer to answer before proceeding. This process will take a number of months.

The need for additional permits such as for disturbance of environmentally sensitive resources, roads, utilities, and discharge permits will place a greater burden upon the landfill developer. Permits may also be needed from local agencies, other state agencies than the one dealing specifically with landfills, and federal agencies such as Corps of Engineers or Fish and Wildlife Service. Many of these approval and permitting procedures will require that the applicant present justification for the proposed plan at public meetings and hearings.

The National Environmental Policy Act and similar legislation enacted by many states requires that federal and state agencies prepare an environmental impact statement (EIS) before giving approval to projects that may have a significant impact on the environment. The purpose of an environmental impact statement is to disclose the nature of a proposed project, assess current and possible future environmental conditions, and to describe alternative courses of action. A typical environmental impact statement for a landfill would include the following:

1. The proposed action.
   a. Purpose and justification.
   b. Project description.
   c. Primary impact area.
   d. Relationship with laws, policies, and plans.
2. Existing conditions.
   a. Natural environment.
   b. Man-made environment.
3. Environmental impact.
   a. Topographic, geologic, and soils.
   b. Water quality and drainage.
   c. Ecological community.
   d. Land use, zoning, and socioeconomic functions.
   e. Esthetic.
   f. Health.
   g. Air quality.
   h. Noise.
4. Unavoidable adverse impacts.
   a. Disruption of agricultural lands.
   b. Increased traffic.
   c. Modifications of surface water drainage patterns.
5. Alternatives to the proposed action.
   a. No action.
   b. Alternatives to the project.
   c. Alternatives within the project.
6. Relationship between local short-term environmental uses and the maintenance and enhancement of long-term productivity.
7. Irreversible and irretrievable resource commitments with the proposed action.

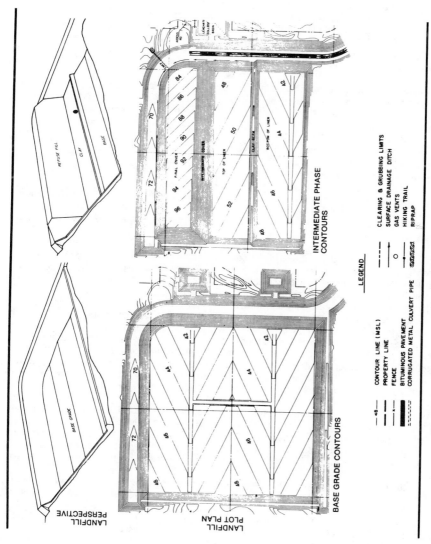

Fig. 11.11 Sanitary landfill construction plans. (From Ref. 17a.)

LANDFILL
PERSPECTIVE

LANDFILL
PLOT PLAN

BASE GRADE CONTOURS

BASE GRADE

INTERMEDIATE PHASE
CONTOURS

REFUSE FILL

CLAY

BASE

LEGEND

———48———	CONTOUR LINE (MSL)	
———▮———	PROPERTY LINE	
———x———	FENCE	
▬▬▬▬	BITUMINOUS PAVEMENT	
═══════	CORRUGATED METAL CULVERT PIPE	

———48———	CLEARING & GRUBBING LIMITS	
————————	SURFACE DRAINAGE DITCH	
o	GAS VENTS	
•——•——•	HIKING TRAIL	
▨▨▨▨	RIPRAP	

Obviously, large quantities of data must be tabulated when preparing an EIS. Frequently, the agency responsible for preparing the EIS will require the entity developing the facility, in this case the landfill owner, to prepare a preliminary document called an environmental impact report to provide a starting point from which the agency writing the EIS can proceed. After the agency prepares the EIS document, the document is the subject of one or more public hearings. Preparation of an EIS and the conduct of public hearings will usually take at least 1 year. Frequently, litigation concerning landfill approval by an agency will center around the need for or adequacy of an EIS.

The preparation of the required engineering reports and plans must proceed in harmony with the site identification process and the public participation program described in the two previous sections. All of this activity must be carefully coordinated. The resulting plan for landfill development must be technically sound, acceptable to the public, and capable of gaining regulatory agency approval. The time needed to guide a landfill plan from inception to the final receipt of regulatory approval may take several years. Long negotiations with neighbors during site selection, the preparation of an environmental impact statement, or litigation can slow the landfill's development further.

A key to successfully developing a landfill is to provide careful consideration of implementation, operational, and environmental concerns that may affect the long-term performance of the landfill. The following sections will describe the major technical considerations when developing and operating a landfill.

11.1.12 Leachate Formation and Control

Background and Definition

A principal environmental concern when constructing and operating a landfill is the formation of highly contaminated water that may emanate from the base of the landfill. This leachate is formed as water passes through the landfill. The escape of the leachate into either surface or groundwater systems can significantly degrade their quality.

The sources of water for leachate formation are primarily precipitation onto the operating landfill, infiltration through the cover of the completed landfill, and groundwater which may flow laterally from the geologic formation surrounding the landfill. To a lesser extent, water contained within the solid waste deposited in the landfill and surface runoff into the landfill from exterior areas will contribute to leachate generation.

The primary point for preventing leachate formation is at the landfill's surface. Water falling on the surface will eventually end up in one of three places: (1) running off the surface into the drainage systems that surround the fill, (2) infiltrating into the cover material and being evapotranspired back into the atmosphere, and (3) infiltrating into the landfill cover and seeping as deep percolation down into the solid waste below. Although a landfill has some ability to absorb water, it is frequently assumed during design that the quantity of water moving into the landfill as deep percolation will eventually result in an equal quantity of leachate being generated at the base. Thus the principal actions to limit leachate formation will be controls to prevent water from entering the landfill.

As the percolating water is moving through the landfill, it will react chemically and biologically with the solid waste. Biological reactions occur continuously in the landfill and, depending upon the stage of decomposition and the availability of oxygen, either aerobic or anaerobic decomposition will proceed. Concurrently with biological decomposition will be chemical decomposition, where leaching and other processes will result in the addition of contaminants to the downward-moving water.

Once the leachate reaches the bottom of the landfill or an impermeable layer within the landfill, it either will travel laterally to a point where it discharges to the ground's surface as a seep, or it will move through the base of the landfill and into the subsurface formations. The route that is followed will depend upon the particular hydraulic conductivities of the materials encountered. If the landfill is lined and a leachate collection system is placed at its base, the leachate collection system can recover the contaminated, downward-moving water. If no collection system is provided or the leachate escapes recovery by the collection system, the leachate will move below the fill and possibly enter the groundwater flow system. Depending upon the formations below the fill, this leachate can become a source of contamination for the aquifers underlying the landfill.

Leachate Composition

The composition of leachate will be highly dependent upon the stage of decomposition and the materials that are contained within the fill. Table 11.7 shows the range of compositions for a variety of leachates, and it can be seen that compositions vary widely. It is difficult to generalize as to the particular contaminant concentrations that a leachate will contain. Research on tests to simulate

Table 11.7 Leachate Chemical Characteristics

Parameters	Range Reported in the Literature (mg/l)[a]	Wisconsin MSW Leachates	
		No. of Values	Overall Range (mg/l)
T. Alkalinity (as $CaCO_3$)	0–20,850	50	4–10,630
Aluminum	0.5–41.8	7	ND–85.0
Antimony	NR	23	ND–2.0
Arsenic	ND–40	34	ND–70.2
Barium	ND–9.0	9	ND–2.0
Beryllium	ND	23	ND–0.08
BOD_5	9–54,610	876	67–64,500
Boron	0.42–70	2	4.6–5.1
Cadmium	ND–1.16	53	ND–0.40
Calcium	507,200	7	200–2,100
Chloride	5–4,350	98	2–5,590
T. Chromium	ND–22.5	42	ND–5.60
Hex Chromium	ND–0.06	3	ND
COD	0–89,520	108	62–97,900
Conductivity[b]	2,810–16,800	352	480–24,000
Copper	ND–9.9	41	ND–3.56
Cyanide	ND–0.08	27	ND–0.40
Fluoride	0.1–1.3	1	0.74
Hardness (as $CaCO_3$)	0.22-800	92	206–225,000
Iron	0.2–42,000	88	0.06–1,500
Lead	ND–6.6	46	ND–1.2
Magnesium	12–15,600	7	120–780
Manganese	0.06–678	19	ND–20.5
Mercury	ND–0.16	24	ND–0.01
Ammonia–N	0–1,250	28	ND–359
TKN		32	2–1,850
NO_2 and NO_3–N	0–10.29	36	ND–250
Nickel	ND–1.7	40	ND–3.3
Phenol	0.17–6.6	20	0.48–112
T. Phosphorus	0–130	92	0.16–53
pH[c]	1.5–9.5	432	5.7–7.66
Potassium	2–3,770	7	31–560
Selenium	ND–0.45	33	ND–0.038
Silver	ND–0.24	17	ND–0.196
Sodium	0–8,000	20	33–1,240
Thallium	NR	24	ND–0.32
Tin	NR	3	0.08–0.16
TSS	6–3,670	812	5–18,000
Sulfate	0–84,000	66	ND–1,800
Zinc	0–1,000	38	ND–162

[a] Refs. 24 to 28.
[b] μmho/cm.
[c] Standard Units

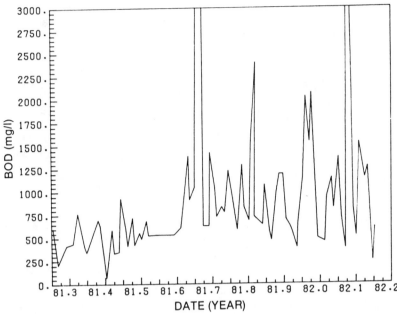

Fig. 11.12 Variations in leachate biochemical oxygen demand (BOD) over a 1-year period at a Wisconsin sanitary landfill. (From Ref. 24.)

potential leachate concentrations has been carried out by the EPA.[22] They concluded that test accuracy will be enhanced by mixing the solid waste with soil that is representative of the soil that will be used in the construction of the landfill. Additional research is currently under way by EPA in an effort to develop a standard procedure.

A study by Kmet and McGinley[23] of 16 Wisconsin operating landfills (see Table 11.7) provides additional evidence that leachate compositions can vary significantly between landfills and also at individual landfills over time. They summarized the results of reports submitted to the Wisconsin Department of Natural Resources by landfill operators who monitored leachate collection systems or leachate monitoring wells constructed inside the landfills. Nine of the 16 landfills reported having accepted industrial waste, and three had or were continuing to accept hazardous waste. The latter three serve a large metropolitan area and principally receive municipal solid waste.

Kmet and McGinley[23] also found significant variations in concentrations that occur over time. Figure 11.12 shows the marked variations in bio-oxygen demand (BOD) concentrations observed at one landfill. Similar results for pH and conductivity were observed. Chemical oxygen demand (COD) for 12 landfills is summarized in Figure 11.13.

The method of leachate sample collection is a possible source of variation in evaluating leachate concentrations. In those instances where the leachate is pumped directly from a collection system or pipe extended into the landfill, higher concentrations are expected than when the leachate is recovered from a groundwater monitoring well that is on the parameter of the landfill where dilution by groundwater is possible.

Certain factors may result in even higher than normally expected contaminant concentrations. Certain industrial wastes if introduced into the landfill may result in high leachate concentrations. Compositions will also change through the year as weather changes solid waste characteristics.

When assessing the potential characteristics of any leachate, the designer or landfill operator must be aware that the addition of hazardous materials to the landfill could have drastic effects upon the characteristics and potential ability to dispose ultimately of the leachate. The introduction of hazardous materials, either unknowingly or illegally, into a sanitary landfill may result in the generation of a leachate that may be classified as a hazardous waste and require special handling. Consequently, it is important that the composition of the waste entering the landfill be carefully monitored.

Recent attention has been directed towards the possible organic chemical contamination of leachate. The EPA has identified 114 priority pollutants for which tests were conducted at five landfills.[23] The results of this analysis are shown in Table 11.8. Ten of the priority pollutants were detected in over half of the samples analyzed. It should be noted that these landfills were selected

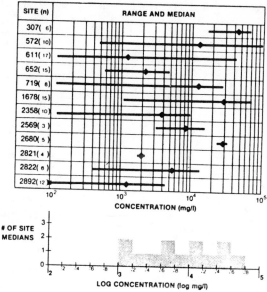

Fig. 11.13 Variations in leachate chemical oxygen demand (COD) at 12 Wisconsin sanitary landfills. (From Ref. 24.)

for evaluation due to their potential for having elevated concentrations of the priority pollutants and were not a random sample.

Physical modification of the solid waste may also impact leachate characteristics. Shredded refuse was shown[29] to generate higher initial contaminant concentrations than unprocessed waste. After a period of time, the concentration of contaminants declined sharply. This is in contrast to unprocessed solid waste where it was generally found that leachate concentrations declined slowly. Figure 11.14 shows an example of this phenomenon.

Operational procedures will also influence the leachate concentrations. Ham[29] placed solid waste in 4-ft (1.2 m)-deep test cells. Several of the cells were covered with earth, as is usually done at a landfill, while others were covered with shredded solid waste or left uncovered. Test cells that were not covered had high initial concentrations of chemical oxygen demand, but these concentrations rapidly declined when compared with those test cells where earth-covered material was placed over the cell.

Predicting Leachate Quantity

When designing a landfill and planning for its future operation, it is important to predict accurately the quantity of leachate that will be generated. The leachate quantity will be a principal consideration when predicting whether or not the groundwater flow system can safely attenuate the contaminants that may emanate from the base of the fill. If leachate collection is provided, the leachate quantity will dictate the design requirements for the leachate treatment or disposal system.

The amount of leachate that will be generated within the landfill will be a function of the amount of water that becomes available. Sources of water include:

1. Water falling directly on the solid waste during operations.
2. Water draining into the landfill from outside areas.
3. Water contained within the solid waste.
4. Percolation into the soil cover over the landfill, if one exists.
5. Water that may enter through the side walls of the fill from surrounding formations.
6. Water that flows up through the base of the fill such as would occur in a hydrogeologic discharge area.

Each of these potential water sources will affect the quantity of leachate generated and, whether controlled or not, must be accounted for in the projection of leachate quantities.

Table 11.8 Priority Pollutant Organics Detected in Municipal Solid Waste Leachate

Parameter[a]	No. of Samples Above D.L.	Analyzed	For Sites Where Detected Range PPB	Median PPB
Acid Organics (11)				
Phenol	3[b]	5	221–5,790	293
4-Nitrophenol	1	5	17	
Pentachlorophenol	1	6	3	
Volatile Organics (32)				
Methylene chloride	6	6	106–20,000	2,650
Toluene	5	5	280–1,600	420
1,1-Dichloroethane	3	5	510–6,300	570
trans-1,2-Dichloroethene	3	5	96–2,200	1,300
Ethyl benzene	3	5	100–250	150
Chloroform	3[b]	6	14.8–1,300	71
1,2-Dichloroethane	2[b]	5	13–11,000	
Trichloroethane	2	5	160–600	
Tetrachloroethane	2[b]	5	26–60	
Chloromethane	1	5	170	
Bromomethane	1	5	170	
Vinyl chloride	1[b]	5	61	
Chloroethane	1	5	170	
Trichlorofluoromethane	1[b]	5	15	
1,1,1-Trichloroethane	1	5	2,400	
1,2-Dichloropropane	1[b]	5	54	
1,1,2-Trichloroethane	1	5	500	
cis-1,3-Dichloropropane	1[b]	5	18	
Benzene	1[b]	5	19	
1,1,2,2-Tretrachloroethane	1	5	210	
Acrolein	1	5	270	
Dichlorodifluoromethane	1	5	180	
Bis(chloromethyl) ether	1	5	250	
Base-Neutral Organics (46)				
Bis(2-ethyl hexyl)phthalate	5[b]	5	34–150	110
Diethylphthalate	4[b]	5	43–300	175
Dibutyl phthalate	3[b]	5	12–150	100
Nitrobenzene	2[b]	5	40–120	
Isophorone	2	5	4,000–16,000	
Dimethyl phthalate	2[b]	5	30–55	
Butyl benzyl phthalate	2	5	125–150	
Naphthalene	1[b]	5	19	
Chlorinated Pesticides (19)				
Delta-BHC	1	5	4.6	
PCBs (7)				
PCB–1016	1	5	2.8	

[a] No. in parentheses represents total number of compounds analyzed in category.
[b] Includes suspect value near detection limit.

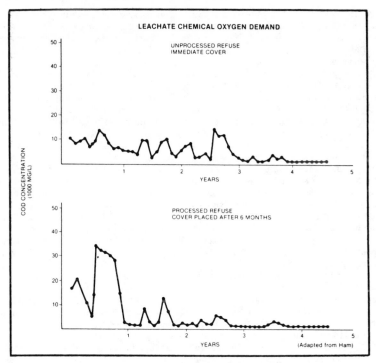

Fig. 11.14 Impact of waste processing and placement on leachate chemical oxygen demand (COD). (From Ref. 30.)

Most attention is given to the cover of the landfill when projecting the quantity of leachate that will be produced. The water balance equation in Figure 11.15 will predict the approximate quantity of water that will percolate through the cover of the landfill and into the underlying solid waste.[30] This method assumes that the landfill does not have a water-holding capacity and that all water passing through the landfill cover will eventually be leachate.

Examination of the individual elements of the water balance equation is instructive. Infiltration (I) into the landfill cover is the difference between precipitation (P) and runoff (R/O). Runoff (R/O) is estimated by applying runoff coefficients from the rational runoff calculation method.[30] When

$$PERC = P - R/O - AET - \Delta ST$$

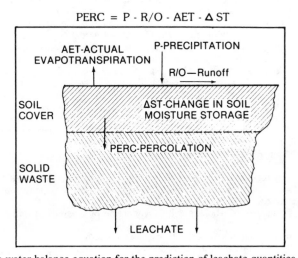

Fig. 11.15 The water balance equation for the prediction of leachate quantities. (From Ref. 31.)

estimating runoff, specific site conditions such as slope, soil, texture, and cover must be considered. Water infiltrating (I) into the cover eventually will be lost as actual evapotranspiration (AET) to the atmosphere or percolation (PERC) through the cover into the landfill. The change in soil moisture content over the period of analysis is ΔST. This method assumes that no lateral water movement occurs.

Potential evapotranspiration will limit the maximum quantity of water that actually will be evapotranspired (AET). Potential evapotranspiration can be estimated with the Thornthwaite and Mather method.[31] Actual evapotranspiration will be less than potential since insufficient water will be available for transpiration during dry periods of the year. During other periods of the year when precipitation exceeds potential evapotranspiration, percolation into the fill will occur if the water storage capacity of the landfill cover is exceeded. The results of an analysis of a hypothetical landfill for Columbus, Ohio, are shown in Table 11.9. As shown for this example, the water balance method, when computed manually, is usually done on a monthly basis.

A computer model for making similar predictions was developed by Perrier and Gibson.[32] This interactive computer program, designated HSSWDS, allows the input of daily precipitation and temperature records and handles the detailed calculations of evapotranspiration and runoff. HSSWDS has the capability to simulate the movement of water into and possibly through landfill covers that are constructed of different layers of soil material. The designer can attempt to reduce the amount of percolation that may pass through the landfill cover by simulating various combinations of layers and selecting the combination that allows the least percolation. This model has the added advantage of being able to easily incorporate sensitivity analysis to test the potential effect of variations in input parameters.

A comparison of the HSSWDS model results and field observations was conducted by Gibson and Malone.[33] This study compared actual leachate quantities collected from an experimental landfill test cell in Boone County, Kentucky, over a 3-year period with the HSSWDS model prediction. The HSSWDS model underestimated the leachate generation by 10%. Given the short duration and other limitations of the study in interpreting the results, the authors indicated that they concluded the HSSWDS model was providing satisfactory results for this site.

The HSSWDS model has subsequently been upgraded and renamed HELP.[34] This model has the additional features of being able to model lateral water movement through the landfill cover and also to predict the movement of leachate through the entire thickness of the fill.

The preceding models for predicting leachate generation imply that all water moving through the cover as percolation will eventually emanate from the base of the landfill as leachate. An alternative approach is to assume that no leachate will actually flow out of the landfill until the compacted solid waste reaches its field capacity for retaining water. The report by Fenn, Hanley, and DeGeare[30] indicated that the field capacity of solid waste to retain water is 20 to possibly as high as 35% by volume. After this volume is filled, it is assumed that all water entering the landfill as percolation through the cover will result in an equal quantity of water leaving as leachate. However, field experience has shown that the solid waste often does not reach field capacity before leachate generation begins. Channels of saturated flow are quickly formed in many landfills long before the entire fill is at field capacity. In these channels, water moves quickly to the bottom of the landfill and emanates as leachate.

Depending upon climate and depth of the landfill, the estimate of water-holding capacity may or may not have a significant effect on the calculation of leachate quantity. In a shallow landfill where precipitation is relatively heavy, the water-holding capacity of the landfill will have much less impact than in an arid region, especially if the landfill is deep. In the former case, the quantity of leachate generated will eventually approach the quantity of percolation through the cover; in the latter case, little or no leachate may be generated even over a long period of time.

Sources of water other than that which percolates through the cover can significantly add to the volume of leachate. Precipitation falling onto the active surface of the landfill will become contaminated and require special disposal practice. Limiting the active area can limit the quantity of leachate generated in this manner. The potential quantity of leachate will be approximately equal to the amount of precipitation falling onto the landfill working face. This assumes that good management practices prevent drainage from other areas of the landfill into the active area.

Water contained within the solid waste brought to the landfill may also be a significant factor. High-moisture sludges or liquids will add to the quantity of liquid retained by the landfill and may result in higher volumes of leachate. Predicting the quantity of water that will drain from sludges is difficult.

Site conditions may require that the base of the landfill is lower than the groundwater table. This can result in the controlled generation of additional quantities of leachate. More information regarding this manner of construction is described in Section 11.1.12.

In summary, the quantity of leachate will be a function of several factors as shown in Table 11.10.

The water balance equations described above can alternately be derived using different methods for predicting runoff, infiltration, and evapotranspiration. McBean[35] suggested several

Table 11.9 Water Balance for Hypothetical Landfill at Cincinnati, Ohio

Parameter[a]	J	F	M	A	M	J	J	A	S	O	N	D	Annual
PET	0	2	17	50	102	134	155	138	97	51	17	3	766
P	80	76	89	92	100	106	97	90	73	65	83	84	1,025
$C_{R/O}$	0.17	0.17	0.17	0.17	0.17	0.13	0.13	0.13	0.13	0.13	0.13	0.17	
R/O	14	13	15	14	17	14	13	12	9	8	11	14	154
I	66	63	75	58	83	92	84	78	64	57	72	50	872
I-PET	+66	+61	+58	+18	-19	-42	-71	-60	-33	+6	+55	+67	+106
NEG (I-PET)				(0)	-19	-61	-132	-192	-225				
ST	150	150	150	150	131	99	61	41	33	39	94	150	
ΔST	0	0	0	0	-19	-32	-38	-20	-8	+6	+55	+56	
AET	0	2	17	50	102	124	122	98	72	51	17	3	658
PERC	+66	+61	+57	+18	0	0	0	0	0	0	0	+11	213

[a]The parameters are as follows: PET, potential evapotranspiration; P, precipitation; $C_{R/O}$ surface runoff coefficient; R/O, surface runoff; I, infiltration; ST, soil moisture storage; ΔST, change in storage; AET, actual evapotranspiration; PERC, percolation. All values are in millimeters (1 in. = 25.4 mm).
Source: Ref. 30.

Table 11.10   Factors Affecting Leachate Quantity Generated

Factor	Effect on Quantity of Leachate
Precipitation onto landfill cover	Increase
Runoff over landfill cover	Decrease
Actual evapotranspiration	Decrease
Water-holding capacity of the waste	Decrease at least temporarily
Precipitation onto working face	Increase
Surface runoff into active area	Increase
Moisture in the waste	May increase
Groundwater intrusion into waste	Increase

methods, including infiltration models developed by Holtan[36] and Huggins and Monk.[37] McBean further noted that the use of water balance methods used to predict leachate should also include as part of the analysis the hydrologic characteristics of the runoff and that appropriate storm water detention facilities be incorporated into the landfill design. He further suggested that, in addition to predicting the quantity of runoff, some type of storm water runoff analysis should be conducted.

Techniques for controlling the quantity of leachate and disposal procedures will be described in later sections.

Surface Techniques to Limit Leachate Generation

The best approach to managing the potentially harmful effects of leachate is to prevent its formation to the greatest extent possible. Landfills that continuously generate leachate generally have covers that allow the entrance of precipitation. Other sources of water, such as the waste itself, precipitation onto the waste during operation, and surface drainage into the active area during operation, are sources of water that are temporary or can be controlled by good management practices. The entrance of groundwater into the site will be discussed later. The control of water movement through the landfill cover and into the waste is the principal point of intervention in the limiting of leachate generation.

Construction of a sanitary landfill cover out of materials that are totally impermeable to water is usually not practiced. Generally, the large land area will dictate that some type of soil cover be employed. Membrane covers and other nonsoil materials have been employed at some sites, but these landfills generally were located where appropriate soil material was unavailable or environmental concerns prompted extraordinary measures.

Inspection of the elements of the water balance equation provides insight into the opportunities for limiting percolation through the landfill's cover. Promoting runoff and actual evapotranspiration will reduce the quantity of leachate. Runoff can be increased by using soil materials that have a low permeability and infiltration rate. Increasing the slope also encourages the precipitation to run off, rather than infiltrate. A nonvegetated soil will also have higher runoff rate, but erosion may result. Typical values for runoff coefficients are shown in Table 11.11.[30]

The United States Department of Agriculture (USDA) Soil Conservation Service curve number method can be applied to assess various combinations of soil materials, slope, and vegetative cover to identify the optimum combination to promote runoff and reduce infiltration.

Promoting evapotranspiration requires identifying a relatively active growing plant that is com-

Table 11.11   Factors for Runoff

Surface	Slope	Runoff Coefficient
Grass—Sandy soil	flat–2%	0.05–0.10
	2–7%	0.10–0.10
	over 7%	0.15–0.20
Grass—Heavy soil	flat–2%	0.13–0.17
	2–7%	0.18–0.22
	over 7%	0.25–0.35

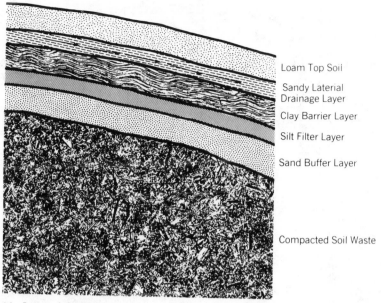

Loam Top Soil

Sandy Lateral
Drainage Layer

Clay Barrier Layer

Silt Filter Layer

Sand Buffer Layer

Compacted Soil Waste

Fig. 11.16 Layered landfilling cover system designed to promote lateral drainage of water infiltrating into cover. (From Ref. 34.)

patible with local climate and landfill development. The amount of actual evapotranspiration that occurs will be a function of the plant's ability to transpire water, weather conditions, and the water-holding capacity of the landfill cover soil. A thin soil cover will have a low water-holding capacity and therefore be less able to make water available to the plant roots during dry periods. Consequently, less water is actually evapotranspired. A low soil water-holding capacity also reduces the ability of the soil cover to retain water during precipitation events. When the soil pores become saturated, downward percolation speeds and water begins to enter the underlying solid waste. Consequently, a thicker cover soil with good water-holding capacity will promote evapotranspiration and reduce percolation.

Constructing landfill covers from several layers of soil that have marked differences in permeability and pore structure has been investigated. Certain combinations such as a fine textured soil overlying a coarse textured soil will retard the downward movement of soil water. Percolation is reduced by providing additional soil moisture retention for evapotranspiration or by routing the unsaturated water flow underground to an area where no solid waste is deposited. This approach is shown in Figure 11.16. This design feature can be analyzed with HELP.[34]

Occasionally there is a need to restrict percolation but no suitable soil material is available for the cover. One option is to place a membrane over the landfill. Soil would be placed on top of the deposited solid waste as a bedding material, the membrane is placed, and then a soil cover is placed on top. Additional information regarding membrane use is presented in Section 11.1.12 (Liner Systems: Soil and Membrane).

Design Strategies for Leachate Management

The development of a leachate management plan for a landfill presents a unique challenge to the designer. Many factors must be taken into account, including the surrounding land use, the hydrogeologic setting, the method of operation being proposed, and the options for the ultimate disposal of the leachate.

A very early consideration in the siting and design of a landfill is the hydrogeologic setting of the area. Depending upon the hydrogeologic conditions, one of the following approaches may be employed:

1. Natural attenuation landfill.
2. Lined landfill and leachate collection.
3. Zone-of-saturation landfill.

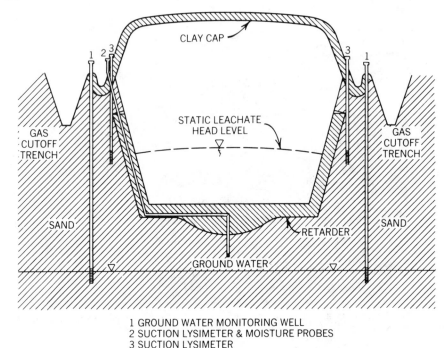

1 GROUND WATER MONITORING WELL
2 SUCTION LYSIMETER & MOISTURE PROBES
3 SUCTION LYSIMETER

Fig. 11.17 Zone-of-aeration leachate management system. (From Ref. 39a.)

A natural attenuation landfill is constructed without a liner. Leachate that is generated is allowed to seep below the base of the landfill into the underlying geologic formations. In choosing a natural attenuation approach, a judgment must be made as to whether harm will be caused to any nearby water resources, especially the groundwater, by the addition of a quantity of potentially contaminated leachate. A judgment also must be made as to the degree of waste strength reduction that may occur within the formation and thus provide natural attenuation of any percolating waters.

At a site where the formation is relatively permeable and the danger of seriously contaminating groundwater is high, it may become necessary to install a soil or membrane liner and possibly a leachate collection system. In the humid region, the installation of a liner at the base of the site implies that eventually leachate will accumulate and may need to be removed from the site through a collection system. This leachate will very likely require some type of treatment before ultimate discharge to a surface water course. Liners, leachate collection systems, and leachate treatment are described in greater detail in a later section.

The opportunities for natural attenuation are limited where the groundwater table is relatively close to the base of the landfill. An approach described by Glebs[38] combines the natural attenuation and linear retention techniques for leachate control (Figure 11.17). A liner is placed on the base and partway up the side walls of the landfill. Leachate is allowed to accumulate inside the landfill and rise to the point where it flows laterally over the top of the liner. Once outside the landfill, the leachate moves downward and is subject to natural attenuation mechanisms. To allow for the possibility of removing leachate, a collection system is installed on top of the liner. One mode of operation is to remove leachate until its contaminant concentrations decline, at which time the leachate is allowed to overflow and undergo natural attenuation.

In hydrogeologic settings where the groundwater table is quite near to the ground surface and the formation is moderately impermeable, designers have chosen to construct "zone-of-saturation landfills." At these facilities, leachate is contained within the landfill site by taking advantage of the particular hydrogeologic setting.

The landfill is constructed by excavating and placing the base below the elevation of the naturally occurring groundwater table. Water seeping into the open excavation through the sidewalls and base is collected by a system of pipes. Alternately, wells are installed to dewater the area immediately adjacent to the landfill. In either case, this results in a depression of the groundwater table and creates a hydraulic gradient toward the landfill from the area immediately surrounding the site. Any leachate escaping from the sides or base of the landfill after refuse is placed is carried by this hydraulic gradient toward the leachate collection pipes or wells.

The principal advantage of a zone-of-saturation approach is that landfilling can be accomplished in high-water-table soils without installing a liner. Installing a liner below the water table may require dewatering the entire region around the landfill. A disadvantage is that the quantity of contaminated water recovered is somewhat greater than would be recovered at a soil- or membrane-lined site since uncontaminated groundwater is also being drawn into the leachate collection system.

Choosing the best approach will depend upon site characteristics, the probable quantity of leachate, and the anticipated degree of natural contaminant attenuation. Additional factors when comparing the suitability of several sites are the cost of installing a liner system, the cost of disposing of larger quantities of leachate from a zone-of-saturation facility, and the potential risk of groundwater pollution if a zone-of-saturation facility does not perform as intended. The regulatory agency reviewing the plans for the landfill will very likely have specific standards that must be adhered to and may, in fact, not allow certain designs.

Natural Attenuation Mechanisms

Leachate reaching the groundwater will enter the flow system and be subject to the physical, biological, and chemical processes. The rate of movement will be predicted by Darcy's Law,

$$\bar{V} = \frac{I \times k}{n} \tag{11.10}$$

where

$\bar{V}$ = average linear velocity of the contaminant front (ft/day).
I = hydraulic gradient (change in head/travel distance, ft/ft [m/m]).
k = hydraulic conductivity, ft/day (m/day).
n = soil porosity, in.^{3}/in.^{3} (1/1).

The equation shown above indicates that the average linear velocity of contaminant transport will be directly proportional to the hydraulic gradient and the hydraulic conductivity. A steep gradient coupled with a high hydraulic conductivity can result in rapid transport of the contaminants away from the base of the landfill. Typical hydraulic conductivities for various formations are shown in Table 11.12.

Horizontal hydraulic conductivities may be five to ten or more times higher than the vertical conductivity. Therefore, it is necessary to access both the horizontal and vertical components of movement in order to describe accurately the direction of contaminant flow away from the landfill's base.

Table 11.12   Range of Values of Hydraulic Conductivity for Various Formations

Formation	k(cm/s)
Unconsolidated Deposits	
Gravel	10^{-1}–10^{-2}
Clean sand	10^{-3}–5×10^{-4}
Silty sand	10^{-5}–10^{-1}
Silt, loess	10^{-12}–10^{-3}
Glacial till	10^{-10}–10^{-4}
Unweathered marine clay	10^{-10}–10^{-7}
Shale	10^{-11}–10^{-7}
Rocks	
Fractured igneous and metamorphic rocks	10^{-6}–5×10^{-2}
Limestone and dolomite	10^{-7}–5×10^{-4}
Sandstone	10^{-8}–5×10^{-4}

Source: Ref. 39.

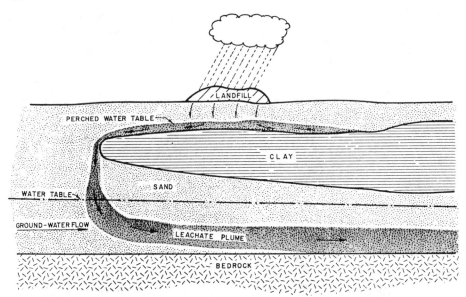

Fig. 11.18 Leachate plume movement away from a leaking landfill. (From Ref. 40a.)

Leachate entering the groundwater will generally follow flow lines. Irregularities in the formations may result in unpredictable routes for the movement of leachate away from the landfill. Several examples of leachate plume movement are shown in Figure 11.18.

Density effects may result in leachate moving in a direction different from that predicted by the groundwater flow lines below the landfill. A leachate which has a density that is greater than water will have the tendency to sink below the groundwater flow lines, whereas a leachate mixture containing certain lighter-than-water organic chemicals may float on the surface of the water table.

Some contaminants carried with the percolating leachate will react chemically with the soil in the formation. Depending upon solubility, the constituent may move very rapidly through the profile at approximately the same speed as the percolating water, or the contaminant may be greatly slowed almost to the point of no movement. Highly soluble inorganic chemicals such as the chloride ion will be highly mobile and not subject to chemical attenuation. A soluble compound such as nitrate may be subject to chemical and biological attenuation by being converted to nitrogen gas if the proper conditions exist. Other elements, such as the heavy metals, may be initially soluble in the percolating leachate but be subject to chemical reactions within the profile, which result in their being retained on the soil particles either through absorption or precipitation reactions. These reactions will reduce the concentration of the contaminant as the leachate moves through the formation (Figure 11.19).

Dispersion will result in a spreading out of the leachate plume as it moves away from the landfill. Figure 11.20 shows the effect dispersion has upon the leachate plume. Dispersion can be modeled using analytical, numerical, or stochastic models. The extent of dispersion will be principally a function of the formation, but the amount of dispersion is somewhat difficult to predict with a high degree of accuracy. The random walk model by Prickett, Naymik, and Longquist[40] is one method for predicting the amount of contaminant movement and dispersion that will take place down gradient from a landfill. Figure 11.21 shows the contaminant concentrations predicted by the model for a leachate plume.

Hydrogeologic Investigations

After preliminary data about the site have been collected and surface features catalogued, an extensive subsurface hydrogeologic investigation is usually conducted. Hydrogeologic studies are expensive, and therefore only the best candidate sites are studied.

A number of techniques are available for the subsurface evaluation. The most frequent approach is to do soil borings at a number of locations across the site. Various techniques for extending the borings into the ground include solid-stem augering, hollow-stem augering, and various types of rotary boring equipment. Soil samples are collected in various depths and analyzed for hydraulic conductivity, particle size distribution, and other characteristics. A number of the

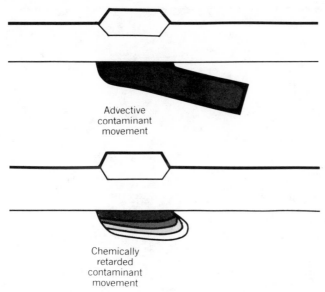

Fig. 11.19 Idealized leachate attenuation by chemical reaction with the underlying formation. (From Ref. 40.)

bore holes are often converted to monitoring wells and piezometers which can be used to identify the groundwater flow characteristics at the site.

Geophysical techniques may also be employed when evaluating new or existing landfill sites:[41]

1. General evaluation of geology and hydrogeology.
2. Evaluation of the extent and depth of existing refuse filling.
3. Locating plumes of contaminated groundwater.

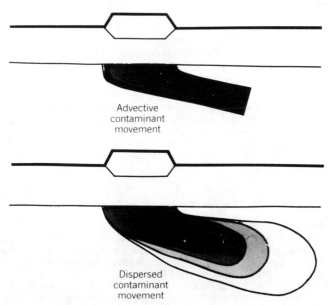

Fig. 11.20 Idealized leachate attenuation by dispersion within the underlying formation. (From Ref. 40.)

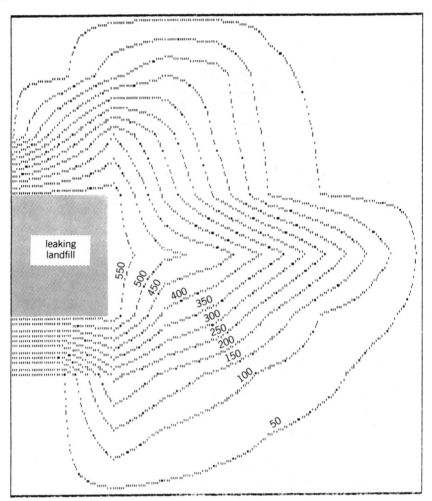

Fig. 11.21 Predicted maximum concentration (mg/liter) of chloride in the groundwater near a leaking landfill. (From Ref. 41.)

Borehole methods that are used in landfill investigations are single-point resistivity, self-potential, natural gamma, thermal neutron, and caliper logs. Data are collected from a probe that is lowered down boreholes. Geophysical logs are most often employed to assess lithology variations in the space between individual boreholes. In addition to borehole methods, four aerial methods are quite useful: ground-penetrating radar, electromagnetic, resistivity, and seismic refraction. Geophysical methods, when used in conjunction with conventional borehole techniques, provide a more detailed description of conditions than is possible from boring log interpretations alone.

The goal of the subsurface investigation is to ensure that the site is suitable to retain the solid waste. If the evaluation shows natural attenuation cannot be employed, then a liner or leachate recovery system may be necessary. As part of the investigation, soils which may be suitable for use as a liner will be sought. Also, potential cover soils are evaluated. A complex geologic formation will need a more extensive evaluation than one that is uniform.

Figure 11.22 is a soil-boring log for a 95-ft (28.5 m)-deep hole. Inspection of the log reveals that the top 2 ft of soil, using the Unified Soil Classification system, consisted of silts, very fine sands, silty or clayey fine sands, or micaceous silts (ML). Silty sands and silty gravelly sands (SM) were found between elevations 1362.9 and 1294.9 ft (409 and 389 m), with some changes in characteristics observed at elevation 1354.9 ft (407 m). The boring extended through three more soil layers before encountering bedrock at elevation 1274.4 ft (383 m). The detailed soil profile description can be interpreted with the key shown.

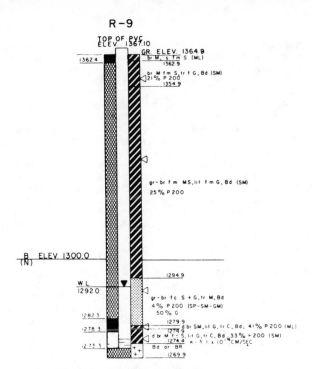

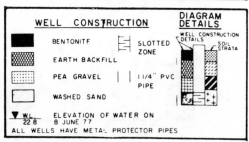

Fig. 11.22 Log for landfill site soil boring and observation well. (From Ref. 17a.)

Additional information shown includes percent (P200) of soil samples passing through a number 200 sieve and the hydraulic conductivity (K) of a sample collected at elevation 1274.4 (383 m). When boring was completed, the hole was converted to an observation well by the installation of PVC pipe which has a slotted lower end. The water elevation measured in the well was 1292 (408 m).

Utilizing the data collected from logs, several useful diagrams can be prepared. The first diagram is a map showing the water table elevations relative to the ground surface elevations. A typical map is shown in Figure 11.23. For this site, the water table is 60 ft (18 m) or more below the ground surface. The direction of groundwater flow is perpendicular to the groundwater elevation contour lines.

A second type of diagram is a cross section. This diagram is a side view of the soil and geologic formations under and near the landfill site. Figure 11.24 presents the results when boring R-9 and several other borings at the site are plotted to show a cross section. A dashed line shows the water table, a dotted line, the bedrock surface. Utilizing this diagram, a designer can easily visualize the setting in which the proposed landfill may be constructed. Soils with desirable or undesirable characteristics can be identified, and water table and bedrock conditions observed.

Liner Systems: Soil and Membrane

The need to protect the environment and regulatory agency requirements has resulted in the installation of liners at the base of many landfills. The liner's purpose is to limit the movement of leachate through the base of the landfill and into the underlying formations. Many materials and techniques have been tried in an effort to prevent leakage at a reasonable cost.

The liner must endure chemical and physical attack mechanisms. Many chemicals found in leachate have the potential to damage liner materials. Also, the liner must not fail structurally during installation or from the strain of the solid waste.

Liner materials include soils and, in particular, clay soil, admixed liners, flexible polymeric membranes, sprayed-on linings, soil sealants, and chemical absorptive liners.[42] The purpose of the liner is to prevent the movement of water and its associated contaminants through the base of the landfill and into the underlying formations. The liner may accomplish this in one of two ways. It may physically prevent the movement of the water and therefore the contaminants, or it may absorb any chemicals carried in the water that does move through the liner. Liners may be constructed on-site, such as when soil materials are placed and compacted, or they may be manufactured flexible membranes. Various types of liners are classified in Table 11.13.

Many landfills have had clay liners placed at their base. Clay minerals may be kaolinite, illite, or montmorillonite. A typical clay will contain one or more of these clay minerals and possibly will be mixed with other fine-grain soil materials such as silt. Clay minerals have a low hydraulic conductivity and therefore will significantly retard the movement of any leachate through them. Permeabilities for most soils which contain greater than 25% clay are in the range of 10–8 cm/sec to 10–5 cm/sec.

Clay liner thicknesses of 5 ft (1.5 m) or greater have been required at some sites. The success of the clay liner will not only depend upon its original characteristics, but also upon the method of

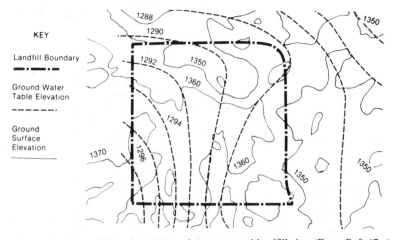

Fig. 11.23 Water table contour map for a proposed landfill site. (From Ref. 17a.)

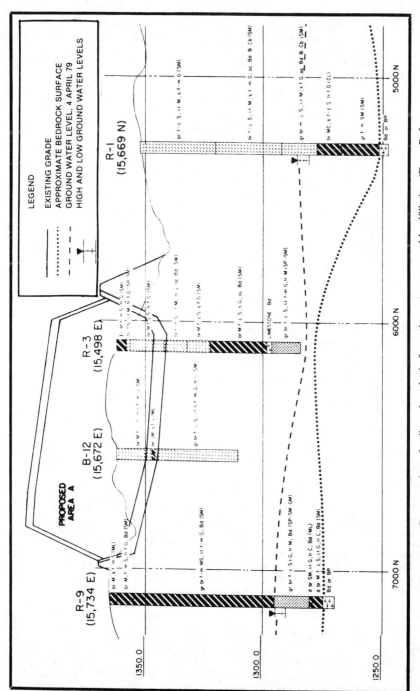

Fig. 11.24 Cross section of soils and geologic formations at a proposed landfill site. (From Ref. 17a.)

Table 11.13   Classification of Liners for Waste Disposal Facilities

A. By Construction:

On-site construction:
   Raw materials brought to site and liner constructed on site
   Compacted soil
   Mixed on site or brought to site mixed
   Sprayed-on liner
Prefabricated:
   Drop-in polymeric membrane liner
Partially prefabricated
Panels brought to site and assembled on prepared site

B. By Structure:

Rigid (some with structural strength)
   Soil
   Soil cement
Semirigid
   Asphalt concrete
Flexible (no structural strength)
   Polymeric membranes
   Sprayed-on membranes

C. By Materials and Method of Application:

Compacted soils and clays
Admixes, for example, asphalt concrete, soil cement
Polymeric membranes, for example, rubber and plastic sheetings
Sprayed-on linings
Soil sealants
Chemisorptive liners

Source: Ref. 42.

liner installation. Best results are achieved by placing several individual layers or lifts. Each lift is compacted before the next layer is placed. The degree of compaction achieved will be a function of the compacting equipment, the thickness of the lift, and the moisture content of the soil. Rollers and specialized equipment designed specifically for compaction will achieve better results than tracked equipment. Table 11.14 identifies various types of compaction equipment and their expected performance. Thinner lifts will be easier to compact than thicker layers. Specific recommendations with regard to the thickness of the layer will depend upon the particular soil material, but layers greater than 2 ft (0.6 m) will usually be difficult to compact. The soil moisture content is a significant factor. Soil that is either too dry or wet will be less than optimal for compaction.

Determining the suitability of a particular clay soil for use as a liner and the best means for constructing the liner must be based upon on an engineering evaluation and test program. The goal of this program is to obtain a liner with the lowest practical hydraulic conductivity. When the liner is under construction, careful inspection must be provided to ensure that the engineering standards are being achieved.

Admixed liners are formed-in-place liners. These include asphalt concrete, soil cement, soil asphalt, and bentonite clay liners. These are formed by mixing the lining material with the natural soil at the base of the landfill. The resultant mixture hardens or modifies the characteristics of the soil material to provide a low-permeability barrier. Each of these approaches has been employed successfully in the lining of impoundments.

Flexible polymeric membranes are manufactured materials that are 0.020 to 0.120 in. (0.51 to 3.0 mm) thick. The liner material is manufactured in rolls that are 48 to 96 in. (1.2 to 2.4 m) wide and hundreds of feet long. A uniform bedding material, such as sand, is placed at the base of the landfill prior to the installation of the liner. The particle size of the subgrade material usually should be less than 3/4 in. (19 mm). The base grade on the liner should be a minimum of 2% if gas release from materials beneath the liner is anticipated. Gas vents may also be necessary in order to adequately allow release. The liner is installed by unrolling the sheets of plastic and then using specialized equipment to form bonded seams between the individual sheets. The liner is usually extended up the side of the landfill to the ground surface where it is anchored.

Soil sealant and chemically absorptive liners are two other approaches that have been used on a limited basis for retarding the movement of the materials through the base of impoundments. These approaches are also being tested for use in landfills.

Moore and Roulier[43] have developed a series of formulas to predict the liner's efficiency in retaining leachate. Since a soil liner is not totally impermeable, a small quantity of leachate can be assumed to pass through it. This prediction allows assessment of the best combination of design features in order to minimize any leakage through the soil liner. No similar formation is available for membrane liners, but it can be assumed that the greater the head of leachate that is allowed to build upon the liner, the more likely that unacceptable levels of leakage will occur.

There has been extensive speculation and research into mechanisms by which landfill liners may fail. Concerns include weathering of the liner material, physical suitability, biological attack, and waste/liner incompatibility. In addition, there are concerns regarding the ability to install the liner properly and maintain it during the operation of the landfill.

Weathering of polymeric membranes can be a result of ultraviolet, thermal, and ozone attack.[42] Most membrane fabricators recommend the use of a soil cover over the membrane to protect it from temperature extremes and ultraviolet and ozone attack. Erosion and excessive drying followed by cracking are weathering processes that may degrade the eventual performance of a clay liner.

Forms of physical attack that liners must endure include high stresses from the weight of waste, membrane punctures and tears during installation, piping of clay liners, differential settling, hydrostatic pressure, abrasion, creep, and freeze-thaw cracking.

Polymeric liners may also be subject to biologic attack. This may include attack by microorganisms, plant roots, rodents, and other animals that may make their way onto the landfill site. Most polymeric resins are resistant to microbial attack, but additives to the membrane material may be attacked by fungi or bacteria. Plant roots, which may grow down through the membrane, or grasses which grow up through the membrane after it has been installed can cause punctures. Care must be exercised to ensure that this type of attack is prevented by limiting the location of plantings with roots which may extend to the liner and by removing all vegetation under the liner when it is placed. Limiting other forms of biological attack requires protecting the liner by barriers or some other means.

A concern that has received extensive research is the impact that the chemical constituents in leachate will have on the liner's performance. Laboratory and field studies have been conducted which show that particular liner materials, when exposed to certain chemicals in leachate, may reduce the performance of the liner.[44–47] The degree to which the liner's performance is reduced depends upon the nature of the chemicals and the specific properties of the liner. Several test methods have been developed that test the integrity of the liner material when exposed to chemical attack. These include pouch tests and membrane tests where municipal solid waste leachate and other types of liquid wastes are exposed to the material.[42] When designing a liner system for a landfill, it is generally recommended that in the absence of affirmative performance data, a test be conducted for liner compatibility with a liquid waste that it is being designed to retain.

Specifications for the construction of a soil liner should address the following:

1. Require low permeability to water and waste fluids.
2. Allow little or no interaction with the waste, which might increase permeability.
3. Provide absorptive capacity for pollutant species.
4. Provide strength initially and after contact with waste fluids to maintain slope stability.

When evaluating a particular soil for possible use as a landfill liner, several laboratory tests are usually performed. These include the Atterberg limits, permeability to water, permeability to leachate, and identification of soil strength characteristics. Also determined are soil moisture/density relationships which allow specification of the moisture content at which the greatest degree of compaction can be achieved. This detailed analysis is necessary in order to identify potential problems that may arise with a particular soil and to project reasonably the performance of a given soil liner. On the basis of this analysis, a recommendation for the liner thickness and compaction procedures can be specified.

The installation of a soil liner must be monitored carefully. Quality assurance procedures should include laboratory and field tests for density, hydraulic conductivity, and site-specific parameters. Checks should be conducted to insure that the required compaction effort is being provided. Lift thickness should be monitored.

The installation of membrane liners also requires a great deal of engineering study and analysis. The liner material must be suitable to withstand the chemical, physical, and other forms of attack

Table 11.14 Compaction Equipment and Methods

Equipment Type	Applicability	Compacted Lift Thickness, in. (cm)	Passes or Coverages	Dimensions and Weight of Equipment		
				Soil Type	Foot Contact Area, in.² (cm²)	Foot Contact Pressures, psi (MPa)
Sheepsfoot rollers	For fine-grained soils or dirty coarse-grained soils with more than 20% passing No. 200 mesh; not suitable for clean coarse-grained soils; particularly appropriate for compaction of impervious zone for earth dam or linings where bonding of lifts is important	6 (15)	Four to six passes for fine-grained soil; six to eight passes for coarse-grained soil	Fine-grained soil PI 30	5–12 (32–77)	250–500 (1.7–3.4)
				Fine-grained soil PI 30	7–14 (45–90)	200–400 (1.4–2.8)
				Coarse-grained soil	10–14 (64–90)	150–250 (1.0–1.7)
				Efficient compaction of wet soils requires less contact pressures than the same soils at lower moisture contents		
Rubber tire rollers	For clean, coarse-grained soils with 4–8% passing No. 200 mesh	10 (25)	3–5	Tire inflation pressures of 60 to 80 psi (0.41 to 0.55 MPa) for clean granular material or base course and subgrade compaction; wheel load 18,000 to 25,000 lb (80 to 111 kN); tire inflation pressure in excess of 65 psi (0.45 MPa) for fine-grained soils of high plasticity; for uniform clean sands or silty fine sands, use large-sized tires with pressure of 40 to 50 psi (0.28 to 0.34 MPa)		
	For fine-grained soils or well-graded, dirty coarse-grained soils with more than 8% passsing No. 200 mesh	6–8 (15–20)	4–6			

Equipment	Requirements	Compacted lift thickness, in. (cm)	Passes or coverages	Possible variations in equipment
Smooth wheel rollers	Appropriate for subgrade or base course compaction of well-graded sand-gravel mixtures	8–12 (20–30)	4	Tandem-type rollers for base course or subgrade compaction, 10 to 15 ton weight (89 to 133 kN), 300 to 500 lb per lineal in. (3.4 to 5.6 kN per lineal cm) of real roller
	May be used for fine-grained soils other than in earth dams; not suitable for clean well-graded sands or silty uniform sands	6–8 (15–20)	6	Three-wheel roller for compaction of fine-grained soil; weights from 5 to 6 tons (40 to 53 kN) for materials of low plasticity to 10 tons (69 kN) for materials of high plasticity
Vibrating baseplate compactors	For coarse-grained soils with less than about 12% passing No. 200 mesh; best suited for materials with 4 to 8% pass No. 200 mesh, placed thoroughly wet	8–10 (20–25)	3	Single pads or plates should weigh no less than 200 lb (0.89 kN); may be used in tandem where working space is available; for clean coarse-grained soil, vibration frequency should be no less than 1600 cycles/min
Crawler tractor	Best suited for coarse-grained soils with less than 4 to 8% passing No. 200 mesh, placed thoroughly wet	10–12 (25–30)	3–4	No smaller than D8 tractor with blade, 34,500 lb (153 kN) weight, for high compaction
Power tamper or rammer	For difficult access, trench backfill; suitable for all inorganic soils.	4 to 6 in. (10 to 15 cm) for silt or clay; 6 in. (15 cm) for coarse-grained soils		30 lb (0.13 kN) minimum weight; considerable range is tolerable, depending on materials and conditions

Requirements for compaction of 95 to 100% Standard Proctor, maximum density.
Source: Ref. 43.

described previously. In addition, the liner must be placed in an environment that is compatible with retention of long-term protective characteristics of the liner membrane. The subgrade, which serves to support the membrane, must have structural characteristics which are suitable for this purpose. In particular, the subgrade performance will depend upon

1. The loading imposed by the weight of the waste.
2. The subgrade characteristics and subsequent groundwater changes.
3. Slope instability.
4. Liner malfunctions.
5. Seismic activities.

Field and laboratory tests are available to evaluate each of these potential difficulties.[42]

Performance of the liner can be assessed with groundwater monitoring wells or lysimeters placed beneath the liner. Membrane liner coupons can also be employed to assess chemical and physical changes.

Leachate Collection System

Leachate will accumulate on top of the landfill liner. Usually a system of collection lines or sumps are placed at low points on top of the liner. The removal of leachate through the collection system will result in hydraulic gradients towards the collection points. If the leachate is not removed through the collection system, the pressure head will continue to build over the liner and possibly result in discharge through the landfill's sidewalls onto the ground surface, or it may increase to such magnitude that unacceptable quantities of leachate are forced under pressure through the liner. The success with which the collection system recovers the leachate will be a function of the hydraulic conductivity of the refuse, the hydraulic conductivity of any permeable materials placed over the liner and the configuration of the collection system.

The design of the hydraulic and structural components of the leachate collection system requires careful analysis. The system must be sized to carry the flow adequately and must also be installed in such a fashion that the collection lines can be periodically cleaned if they should become clogged. Generally, the collection pipe is placed in a trench that interfaces with the liner or has been constructed within the permeable blanket over the liner. In the latter case, the collection line would be set immediately on top of the liner. Generally, 6- or 8-in. (15.24 or 20.3 cm) polyvinyl chloride or other suitable material is employed for the pipe. The pipe must have structural integrity that is commensurate with the depth of the refuse that will be placed over it. Dietzler[48] describes techniques for specification of the structural design features.

The required hydraulic capacity of the leachate collection system will depend upon the peak quantity of leachate anticipated. The leachate collection system should extend completely around the perimeter of the landfill in order to limit any lateral flow away from the site. An interior grid is usually necessary to prevent unacceptable heads of leachate accumulating near the center of the landfill. The appropriate spacing of the leachate collection lines can be predicted using flow net calculations. A series of charts have been prepared by Matrecon[42] to aid in the hydraulic design of leachate collection systems.

Recent experience indicates that the landfill designer and operator must consider the pathway leachate will take as it flows through the landfill. The use of heavier textured soils as daily cover may interfere with the downward movement of leachate and has resulted in the appearance of leachate seeps on the side of the landfill. Operating procedures to overcome this problem have included importing sand onto the site for use as daily cover and removing the daily cover before beginning each day's operation. Auxiliary collection lines and recovery wells have been installed at landfills where daily cover or impermeable waste has prevented downward movement and caused leachate seeps.

Leachate Treatment and Disposal

Collected leachate must be handled and disposed of in a manner that will not cause pollution of nearby water courses or groundwater. Options for treatment and disposal include the following:

1. On-site biological or physical chemical treatment and surface water discharge.
2. On-site treatment and land application disposal.
3. Recycling of the leachate back into the landfill.
4. Discharge of the leachate to a municipal wastewater treatment system. The leachate may receive pretreatment prior to discharge.

The most frequently employed option is item 4. This choice relieves the landfill operator of the responsibility for a wastewater treatment and disposal system. If the landfill is located at a site that is some distance from a community, trucking of the leachate is necessary. If the haul distance is uneconomic or no wastewater treatment authority is willing to accept the leachate nor any facility operator is willing to accept the leachate, then another alternative must be found for leachate treatment and disposal.

Inspection of Table 11.7 indicates that leachate is a very high-strength waste. Average municipal wastewaters contain 150 to 200 mg/liter of BOD. Leachate contains far greater concentrations and, in addition, has many other constituents that are not common in municipal wastewaters. Leachate also has a pollutant strength that is much greater than many industrial wastes. Frequently, industrial wastes have 1500 to 2000 mg/liter BOD. Consequently, the discharge of leachate to municipal sewage treatment plants requires careful consideration. Loadings that exceed 5% of the hydraulic capacity of a conventional sewage treatment plant have been found to disrupt the performance of the plant.[49]

In addition to the high contaminant concentrations, seasonal flow and strength variations will complicate the design and operation of a municipal treatment facility receiving leachate or a plant constructed to handle only leachate. A report by Robinson and Maris[50] concluded the following:

1. The composition of leachate changes over the life of the site and therefore treatment methods that are appropriate initially may not be satisfactory later. In general, leachate from newly deposited waste is more amenable to biological treatment than are leachates from more established fill areas.

2. Generally, the leachates are deficient in phosphorus and possibly nitrogen; and therefore, without nutrient addition, biological treatment may be inhibited.

3. Anaerobic biological treatment can significantly reduce concentrations of contaminants, but the performance of the systems severely declines when temperatures fall below 20°C (68°F).

4. Physical chemical processes have not been shown to remove soluble organic matter efficiently from the leachates. Physical chemical techniques may, however, be necessary to remove toxic concentrations of specific chemicals.

5. Pilot-scale experiments have indicated that recirculation of leachate through the landfill has advantages in terms of both leachate control and accelerated stabilization of solid waste.

A report by Stegmann and Ehrig[51] reported effluent concentrations of less than 25 mg/liter BOD were achieved when organic loadings to aerated lagoons were less than 5100 lb BOD/million gallons. They noted that this was achieved only when sufficient phosphorus and nitrogen were available in the leachate to promote biological decomposition and the BOD-to-COD ratio was greater than 0.4.

Farquhar[52] recommended the following steps when designing a leachate treatment system:

1. *Requirements Analysis.*   A comparison between leachate properties and effluent standards determines the type and extent of contaminant removal required.

2. *Identify Constraints.*   The time-variant properties of the leachate must be assessed and accounted for in the design. This includes both seasonal fluctuations and long-term decaying trends as the site ages. Biological treatment processes are limited to what they can remove and the efficiency with which they can accomplish it.

3. *Collect Background Information.*   A review of previous work on related leachate treatment problems will provide guidance on effective treatment types and areas of potential difficulty.

4. *Conduct Treatability Studies.*   The potential contaminant removal level can be estimated. Nutrient supplement and other pretreatment activities such as pH neutralization and aeration can be evaluated at the time.

5. *Process Selection.*   If on-site treatment is necessary, processes can be reviewed beginning with the simplest and working into more complex systems as required. Off-site treatment should require consideration first.

6. *Identify Effluent Disposal Method.*   The ultimate means for discharge of all effluent streams including treated leachate, sludges, rejection liquids, and so forth, must be designed. These may be as difficult to handle as the raw leachate itself.

7. *Operation and Management.*   A program for the operation of the treatment system must be designed and implemented. Staff requirements, an analytical program, laboratory facilities, and contingencies in the event of changes in design conditions should be included.

The available treatment and disposal options must be carefully evaluated when developing a new landfill. The trend toward installation of lined landfill sites results in secondary leachate emission from the landfill, which, if not properly managed, may cause very serious environmental problems. The long-term cost of leachate treatment must also be accounted for during budget planning. Leachate may require treatment long after the landfill has ceased operation.

Groundwater Monitoring

The purpose of groundwater monitoring is to assess the success or failure of the landfill in retaining wastes. Leachate that is allowed to escape from the base of the landfill, either through a leaking liner or under natural attenuation conditions, will migrate downward to the water table and enter the groundwater flow system. The installation of wells or other monitoring devices provides an opportunity to detect contaminants escaping the landfill, assess the performance of the landfill, and give an early warning before contaminants are allowed to enter a public water supply or other valuable water resources.

Groundwater monitoring can range from very simple installations of two or three wells to elaborate networks or geophysical surveys. A typical groundwater monitoring well is shown in Figure 11.25. It consists of a 2-in. (5 cm)-diameter pipe, usually plastic pipe, to which a screen has been attached at the base. The pipe is installed in a 4-in (10 cm) or larger diameter hole with granular material placed around the pipe inside the hole to allow easy entrance of the water. The well is extended below the top of the groundwater table to a depth at which contaminants may reasonably be detected. The length of the screen will roughly define the aquifer thickness monitored. Determining the location, depth, and length of screen for monitoring wells being installed at a particular facility requires a thorough understanding of the subsurface hydrogeologic conditions. This work is usually the hydrogeologic investigation.

The complexity of the monitoring system will depend upon the nature of the geologic formation and, if the groundwater is contaminated, the magnitude of contaminant migration. At sites where uniform materials are found over the depth of the formation of interest, rather simple systems for monitoring groundwater can be employed. However, if the subsurface features are highly complex with varying formations and irregularities, a much more elaborate monitoring system will be necessary. More wells will be needed and probably will terminate at various depths.

Correctly identifying the monitoring points requires a thorough understanding of the groundwater flow system. The flow system will not only have lateral components, but also vertical components. Two approaches for monitoring groundwater quality at various depths within the formation are shown in Figure 11.26. The simplest approach is to install several adjacent wells that terminate at various depths. These are constructed in separate bore holes and have screens that are open to different elevations within the formation. An adaptation of this approach is to construct one large bore hole and to install two, three, or more wells within the hole that terminate at different depths. This reduces the cost of the bore hole construction, but increases the complexity of the well installation. Several devices have been developed that allow the collection of samples at various levels within a single long-screened monitoring well. This sampling device is lowered into the well to the depth desired and then a sample withdrawn at that depth. A refinement of the multiple well in the one-hole approach is to install a long pipe into which a number of small-screened collection ports have been installed. Attached to each port is a thin plastic hose from which a sample can be pumped at the ground surface.

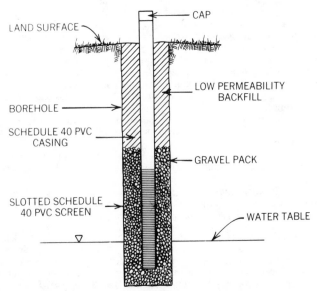

Fig. 11.25 Typical monitoring well screened over a large vertical interval. (From Ref. 40a.)

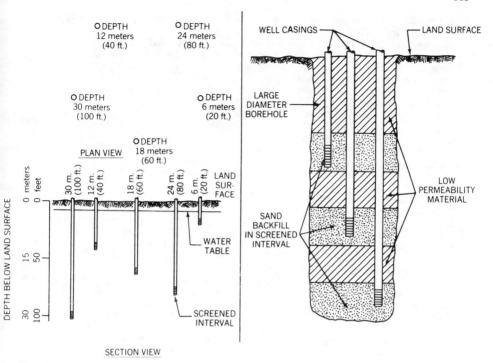

Fig. 11.26 Typical well cluster configurations for monitoring at multiple levels. (From Ref. 40a.)

Before removing water from the well, the depth at which the water is standing within the well is measured. This information allows the preparation of a topographic map of the groundwater table. This information can be used to chart the direction in which the groundwater flow is moving and allows better interpretation of the sampling results.

The methods employed for collecting samples will depend upon well type and depth from which the water is being drawn. Where the depth is greater than approximately 25 ft (7.5 m), a pressure pump or bailing system is needed to lift the water out of the well. At depths at less than 25 ft, these units or a suction pump can be employed.

It is generally recommended that a volume of water equal to two or three times the capacity of the well to hold water be removed before collecting a sample. This can be time-consuming if the well is of great depth or large diameter. A second problem which arises is that in many landfill situations, the wells are constructed in formations that yield only a small quantity of water and many hours are needed for the well to refill. To insure comparable results, the same type of equipment should be used for sample collection. Using a variety of types of equipment and procedures will almost certainly increase the variability in the results of the chemical analysis.

Suction lysimeters have been installed at sites where deep water tables or bedrock formations make the installation of monitoring wells difficult. The lysimeter consists of a porous ceramic collection cup into which soil water is drawn with a vacuum. This monitoring technique has two advantages over a monitoring well. Percolating water can be collected before it has had an opportunity to be diluted by groundwater, and a suction lysimeter allows early detection of leakage before the groundwater is contaminated. Installation of suction lysimeters requires special construction techniques and planning. Gravity lysimeters must be constructed under the fill at the time of the fill's construction. Suction lysimeters are installed before the landfill is constructed and possibly after construction if a hole can be drilled on an angle under the landfill.

Samples collected must be handled in accordance with standard practices for preservation and analysis. This includes keeping the samples cold after collection and, depending upon the type of analysis being conducted, field filtering of the samples before transport to the laboratory. Field filtering reduces the opportunity for chemical reactions occurring within the sample bottles between collection and analysis. Procedures for sample analysis are described in *Standard Methods for the Examination of Water and Wastewater*[53] and the *Annual Book of ASTM Standards: Water and Environmental Technology*.[54] Quality assurance procedures to ensure that samples are not being contaminated in the field and for assessing the accuracy and precision of the laboratory must be established and documented.

Geophysical techniques can be employed to identify the extent of buried refuse and to locate contaminant plumes. Table 11.15 lists the applicability of several aerial methods for identifying problems at disposal sites. These methods allow an investigator to survey a large area relatively quickly. The observations taken during the geophysical surveys are calibrated with monitoring well data.[41] Geophysical data can be used to guide the installation of additional monitoring wells in areas that are shown by remote means to have high concentrations of contaminated groundwater.

A groundwater monitoring program at a landfill results in the generation of a large quantity of data. Graphical representation of the data allows easy assessment of the groundwater quality changes from one sampling period to the next. An example is shown in Figure 11.27. Statistical tests should be employed to assess the significant changes observed within the data. Seasonal variations and other factors may result in fluctuations in groundwater quality that are naturally occurring. A number of statistical tests have been described by McBean.[56]

Groundwater Pollution Remedial Actions

Actions to abate groundwater pollution may become necessary if the groundwater monitoring system indicates unacceptable levels of contamination. Remedial actions are of basically three types with many variations and combinations possible within these types.

The first type of action is to cut off the source of groundwater contamination by limiting infiltration through the cover of the landfill and therefore reducing the amount of leachate from the base of the landfill. Techniques employed to abate pollution would be generally much more extensive and therefore more expensive than normally associated with the covering of a landfill. Activities may include the placement of a clay cover that is several feet thick or the installation of a synthetic membrane. These activities are undertaken to limit to the greatest degree possible infiltration into the landfill with the hope that percolation from the base of the landfill will be greatly reduced. If the landfill contains a large quantity of liquids, this type of activity will only become effective after a period of time has elapsed during which accumulated liquids are drained from the base of the landfill.

The second approach is to isolate the landfill by installing a clay cutoff wall or other impermeable material around the exterior of the landfill. This may also include attempting to limit vertical movement below the base of the landfill by injecting materials under the landfill that will retard downward percolation. A slurry wall is one example of this approach. A trench is dug to a depth that is below the zone of contaminated groundwater or to the top of an impermeable layer. This trench, which will be 2 to 3 ft (0.6 to 0.9 m) wide, is continuously filled with a slurry material that when dewatered becomes fairly impermeable. The goal is to construct a vertical wall that is relatively impermeable to leachate. Contaminated water is collected by wells or collection sumps on the interior of the slurry wall and given the appropriate waste treatment and disposal. Construction of the slurry wall requires careful techniques to ensure that caving of the trench does not result

Table 11.15 Applications of Geophysical Methods to Disposal Sites

Application	Radar	Electro-magnetics	Resistivity	Seismic	Metal Detector	Magne-tometer
Mapping of geohydrologic features	1[a]	1	1	1	—	—
Mapping of conductive leachates and contaminant plumes (e.g., landfills, acids, bases)	2[a]	1	1	—	—	—
Locations and boundary of buried trenches with metal	1	1	1	1	1	1
Location and boundary definition of buried trenches without metal	1	1	2	2	—	—
Location and definition of buried metallic objects (e.g., drums, ordnance)	2	2	—	—	1	1

[a]1. Primary method, indicates the most effective method; 2. secondary method, indicates an alternative approach.
Source: Ref. 55.

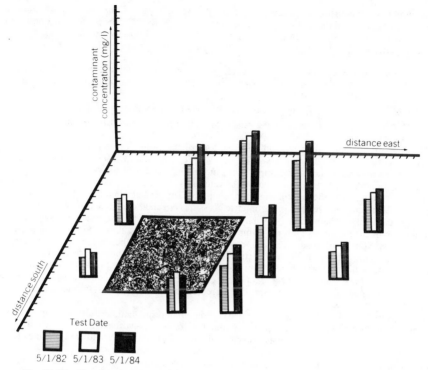

Fig. 11.27 Graphical presentation of groundwater monitoring data. (From Ref. 40a.)

in holes forming within the impermeable vertical wall. Slurry wall techniques are currently becoming much more refined as cleanup at Superfund sites gets under way.

A third approach is hydraulically to isolate and recover contaminated groundwater. This is accomplished by installing recovery wells. At some sites injection wells have also been installed to force the movement of the contaminated water toward recovery points. The recovery wells are drilled at strategic points that have been identified as having the optimum location for recovering the contaminants. In some instances, these locations have been determined through computer modeling. The principal objective is to recover as much of the contaminated water while limiting the incidental pumping of uncontaminated water.

For some sites, it has been determined that the optimum solution is to construct only injection wells whereby a pressure ridge is formed around the site and the contaminant plume is contained within that pressure ridge with no attempt made to actually recover the contaminants. Regardless of the particular technique employed, the principal goal is to stem the spread of the contaminants.

Recovered contaminated water may require some type of treatment before disposal, and therefore it is imperative that the quantities recovered are limited if at all possible. Possible alternatives include discharge to a municipal treatment system and installation of a physical/chemical treatment facility. An alternative approach is in situ treatment where chemical and biological treatment of contaminated water is provided while the water is still in the ground.

Remedial actions are frequently quite expensive, and in a number of cases, success has been difficult to achieve. In a few cases the groundwater pollution problems have been so serious and solutions so limited that the entire landfill was excavated and removed. This is obviously a very expensive remedial action.

11.1.13 Methane Gas Formation and Control

Background and Definition

Decomposition of solid waste will begin immediately upon placement in a landfill. Initially, aerobic decomposition will take place where oxygen is consumed to produce carbon dioxide gas and other by-products. Anaerobic decomposition begins when the supply of oxygen has been exhausted. During the first phase of anaerobic decomposition, carbon dioxide is the principal gas generated.

As anaerobic decomposition proceeds towards the second phase, the quantity of methane generated increases until the methane concentration reaches 50 to 60%. The landfill will continue to generate methane at these concentrations for 10 or 20 years, and possibly longer.

The generation of methane gas from a landfill is of concern for the following reasons:

1. Methane gas in confined spaces is explosive when concentration is between 5 and 15%. When the concentration is over 15%, insufficient oxygen is available for an explosion, but an extreme danger exists.
2. Methane gas can cause asphyxiation of personnel in a confined space.
3. Methane gas will kill vegetation by asphyxiation of the roots as it passes through the soil.

Methane gas is lighter than air and will rise toward the ceiling of an enclosed space. If the methane is in the 5 to 15% concentration range, a source of ignition will set off an explosion. Gas accumulations may be found on the upper story of buildings and near the ceilings of lower floors within these buildings. An additional reason for caution is that the methane gas concentration that occurs underground near the landfill can change rapidly. These rapid changes in concentration are generally associated with changes in barometric pressure. No clear guidance can be provided as to the maximum distance the methane gas may move underground. The distance the gas travels will be a function of the pressures built up within the landfill and of the soil formation surrounding the landfill. A relatively porous soil extending from the ground surface to below the base of the landfill will allow gas to move horizontally and vertically to the atmosphere. However, frozen ground, wet soil conditions, or impermeable layers may restrict the upper movement of the gas and cause it to flow horizontally for greater distances. McOmber et al.[57] have prepared a computer model that can account for these conditions and has shown to predict accurately steady-state methane distributions.

The principal components of landfill gas are carbon dioxide and methane. Other compounds are present and are listed in Table 11.16. One compound of concern is hydrogen sulfide, since it has an obnoxious odor and also adds to the asphyxiation danger of landfill gas. Volatilized organic chemicals are also of concern. Organic acids have been identified as the cause of extensive corrosion problems within gas collection systems and gas recovery systems. Volatile organic chemicals are

Table 11.16 Typical Landfill Gas Composition and Characteristics

Component	Component Percent (Dry Volume Basis)
Methane	47.5
Carbon dioxide	47.0
Nitrogen	3.7
Oxygen	0.8
Paraffin hydrocarbons	0.1
Aromatic and cyclic hydrocarbons	0.2
Hydrogen	0.1
Hydrogen sulfide	0.01
Carbon monoxide	0.1
Trace compounds[a]	0.5
Characteristic	Value
Temperature (at source)	41°C
High heating value	17,727 kJ/std cu m
Specific gravity	1.04
Moisture content	Saturated (trace compounds in moisture)[b]

[a]Trace compounds include sulfur dioxide, benzene, toluene, methylene chloride, perchlorethylene, and carbonyl sulfide in concentrations up to 50 ppm.
[b]Trace compounds include organic acids (7.06 mg/m³) and ammonia (0.71 mg/m³).
Source: Ref. 58.

also of potential concern from an air pollution standpoint and are just now beginning to be studied in detail.

In addition to the methane gas concentration in the landfill, the rate at which gas is being produced will be of interest when controlling gas or attempting to recover it for beneficial purposes. Several studies, which are summarized in Table 11.17, have predicted the rate and quantity of gas that ultimately will be produced within the landfill. Most of these estimates have been based upon the quantity of decomposable material contained within the refuse. Whether the estimated quantity of gas will ultimately be produced will depend upon other conditions such as moisture content and leakage of oxygen into the landfill.

Movement through Solid Waste and Soil

Gas released into the landfill atmosphere will be moved by two driving forces, diffusion and pressure gradient. Diffusion is the physical phenomenon that causes a gas to seek a uniform concentration throughout the volume within the landfill. Since there is resistance to the movement of the gas and the rate of generation varies throughout the landfill, this process will not result in a uniform concentration, but will cause the gas to move away from areas of higher concentration toward areas of lower methane concentration. The second driving force is gas pressure gradients that build up within the landfill. The gas will move from zones of higher pressure toward zones of lower pressure. Superimposed over these two physical phenomena will be changes in barometric pressure as weather conditions vary. An additional factor may be the introduction of large volumes of water into the landfill during precipitation events and freezing soil conditions. The net result is that the gas will move along routes that will allow it to escape from the landfill environment.

The possible escape routes include venting through the cover or movement through the side of the landfill and out into the surrounding soil formations. The rate of movement will depend upon the degree of concentration differences, the variations in pressure throughout the landfill atmosphere and surrounding formation, and the permeability of the solid waste and soil formation. The rate at which gas moves will be a function of the pressure gradient and the permeability of the materials through which it is moving.

The concentration of methane within the landfill gas will be a function of the stage of anaerobic decomposition. When rapid decomposition is taking place, concentrations in excess of 60%

Table 11.17 Summary of Reported Landfill Gas (CO_2 and CH_4) Production Rates from Municipal Refuse

Conditions	Basis	Gas Production Rate, l/kg (ft³/lb) Refuse as Received per Year
After 5 years in landfill	Theoretical	16 (0.25)
Literature and various data sources	Estimated	3.7 to 14 (0.06 to 0.23)
Lysimeter, calculated from 300 days CH_4 production	Measurement	4.9 (0.079)
Ave. range during active CH_4 production for solid waste only lysimeter	Measurement	4.9 to 6.1 (0.078 to 0.0098)
Lysimeter gas produced over 300 days	Measurement	0.21 (0.0034)
Lysimeter, maximum production rate observed	Measurement	8.1 (0.13)
Maximum production in lysimeter at "optimal" temperature and % H_2O	Measurement	32 (0.51)
Lysimeter with unusually high food waste content	Measurement	400 (6.4)
Pilot-scale landfill, low H_2O	Measurement	32 to 62 (0.51 to 1.0)
Test landfill, maximum and minimum observed production on per year basis over 3-years' monitoring	Measurement	16 to 41 (0.25 to 0.65)
Estimated during landfill pump tests, varies seasonally	Measurement	3.9 to 39 (0.063 to 0.63)
Estimated during landfill pump tests, both values given	Measurement	21 to 29 (0.33 to 0.47)
Theoretical extrapolation of short-term landfill pumping data	Measurement/ Theoretical	11 (0.17)
Testing of landfill, recalculated for 53% CH_4	Measurement	2.2 (0.035)

methane have been observed. As the gas containing high concentrations of methane begins to move through the landfill and out into the surrounding formations, its concentration will begin to diminish as it mixes with the surrounding soil atmosphere. In formations with low permeability, this mixing may be quite limited since the movement of gas is much slower than in higher permeability formations.

Changes in barometric pressure have been observed to have a significant influence on the rate of gas movement. It appears that during periods of falling barometric pressure, the gas that is trapped under the cover of the landfill flows more rapidly through the landfill sidewalls and out into surrounding formations. This results from a lowering of the atmospheric pressure within the surrounding formations as the barometer falls. Gas movement can be rapid under these conditions, with much higher concentrations of gas being observed within a matter of hours as barometric pressure falls. Wet or frozen soil may also restrict the release of the gas from the soil formation into the atmosphere. Changing barometric pressure, rainfall, and frozen ground may also cause the escaping gas to move in unpredictable or not previously observed directions.

Gas Venting and Collection

Two types of systems have been implemented for controlling and possibly recovering methane gas from landfills. Passive systems are structures that rely upon natural processes to vent the gas to the atmosphere or prevent its movement into undesirable areas. Active systems use blowers to withdraw the gas from the landfill or the surrounding formations. An adaptation of the active system is to pump air into the ground to block the further movement of methane gas.

Passive systems include the following:

1. Impermeable soil or clay cutoff walls.
2. Venting trenches.
3. Gas vents.

Each approach is shown in Figure 11.28.

The purpose of the cutoff wall is to retard the movement of the gas horizontally away from the landfill site. The cutoff wall may be used in conjunction with some type of venting system to release the gas to the atmosphere. One approach is to dig a trench along the outside edge of the landfill and to fill that trench with granular material into which the gas moves and then escapes vertically to the atmosphere. An alternative approach is to put in a passive gas collection system under the cover of the landfill. This system directs methane and other gases to a riser which vents into the atmosphere. In some cases, these risers are equipped with a flare that is ignited to burn off the methane gas. Passive venting approaches are suitable where the escape of gas in the surrounding formations does not present a high degree of hazard. They rely upon changes in barometric pressure and the physical processes of diffusion and pressure gradients to remove the gas from the landfill. Should the vent become blocked by moisture or frost conditions, the gas will seek other escape routes including movement into surrounding formations.

Active gas collection systems incorporate a mechanical means of withdrawing the methane gas from the landfill or surrounding soil formations (see Figure 11.29). These systems are usually installed where a higher degree of reliability is needed than can be accomplished with a passive collection system.

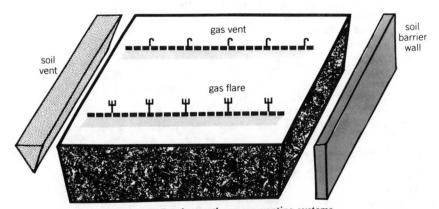

Fig. 11.28 Passive methane gas venting systems.

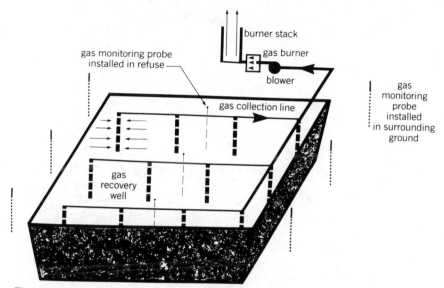

Fig. 11.29 Active (mechanical) method of gas withdrawal from surrounding formations.

A series of wells are placed into the solid waste or the adjacent formation. The gas collection wells are constructed from perforated pipe material. A typical gas collection well is shown in Figure 11.30. A 36-in. (1 m) diameter hole was drilled into the soil formation immediately adjacent to the landfill. A 6-in. (15 cm) slotted polyvinyl chloride pipe is installed in the hole and the hole is backfilled with a coarse aggregate. The collection wells usually extend to a depth that is below the base of the landfill or to a restricting layer that would prevent any movement of gas deeper into the ground than the base of the recovery well.

The individual wells are connected to a piping system and large vacuum pump or blower. Gas collected by this system is either vented to the atmosphere directly or sent through a burner facility for combustion prior to release to the atmosphere. The piping network is equipped with various valves and monitoring points so that the individual performance of each well can be observed and valves can be operated to adjust the amount of gas being recovered in each specific area.

The gas collection wells can be installed either in the solid waste or in the formation around the outside of the landfill. Where the protection of surrounding buildings is the principal purpose for installing the wells, it appears that installation in the soil formation exterior to the waste offers a slightly higher degree of protection than the installation of the wells in the waste. Wells installed in the waste may have a higher probability of plugging due to materials infiltrating into them from the refuse. However, the degree of protection provided to surround buildings will be strictly a function of the particular design developed. The spacing of the wells, the permeability of the solid waste, and the space available for installing the wells will all be factors that must be considered in the development of the best protection plan. Active gas collection systems that have been installed to protect surrounding buildings should also have appropriate instrumentation to monitor their performance.

Gas recovery wells that are installed solely for the purpose of recovering gas will possibly have a different design than those that are used strictly for venting gas from the landfill as a protective device. The recovery wells must be designed to limit the movement of air and therefore oxygen into the landfill. The drawing of air into the landfill will not only reduce the quality of the gas being recovered, but may also interfere with the anaerobic biological decomposition process and therefore reduce the quantity of methane gas being generated. The introduction of excessive amounts of oxygen into a landfill may also cause a fire to start from spontaneous combustion.

A means must be provided to monitor the performance of the passive or active gas venting or collection system. To monitor venting system performance, probes can be placed in borings around the landfill site. A typical probe is shown in Figure 11.31. Test equipment is employed to measure soil atmospheric pressure and the concentration of methane. Pressures in the soil atmosphere that exceed atmospheric pressure may be indicative of outward movement of methane gas. A network of gas monitoring probes from which pressure readings and gas concentrations are collected can provide the information necessary to map the migration of gas away from the landfill. Due to the possible rapid changes in the pattern of gas movement, frequent measurements may be needed to obtain an accurate assessment of gas movement.

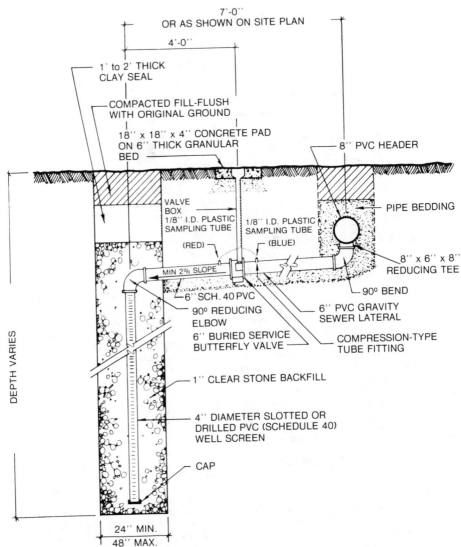

Fig. 11.30 Methane gas collection well. *Note:* Venting well design as shown has been developed for use outside of landfill boundaries. Modifications to this design for use within landfill boundaries would include: slip couplings on the well screen; flexible couplings on 6-in. (15.2 cm)-diameter lateral pipe; and flexible couplings spaced along the 8-in. (20.3 cm)-diameter header pipe. (From Ref. 59a.)

A second means of protection is to install methane gas detectors in enclosed structures that may be subject to infiltration of methane gas from the surrounding soil formations. These detectors will sound an alarm before the explosive limit of 5% methane gas is exceeded. These detectors require periodic calibration and testing to assure continued satisfactory performance.

Methane Gas Safety Guidelines

The following guidelines should be followed when working at a landfill in the presence of potentially dangerous gases:[59]

1. No person should enter a vault or a trench on a landfill without first checking for the presence of methane gas. The person should also wear a safety harness with a second person standing by to pull him or her to safety.

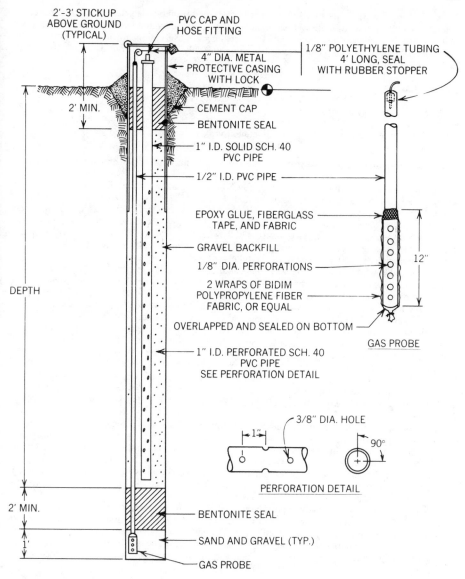

2'-3' STICKUP ABOVE GROUND (TYPICAL)

PVC CAP AND HOSE FITTING

4" DIA. METAL PROTECTIVE CASING WITH LOCK

1/8" POLYETHYLENE TUBING 4' LONG, SEAL WITH RUBBER STOPPER

2' MIN.

CEMENT CAP

BENTONITE SEAL

1" I.D. SOLID SCH. 40 PVC PIPE

1/2" I.D. PVC PIPE

EPOXY GLUE, FIBERGLASS TAPE, AND FABRIC

GRAVEL BACKFILL

1/8" DIA. PERFORATIONS

2 WRAPS OF BIDIM POLYPROPYLENE FIBER FABRIC, OR EQUAL

OVERLAPPED AND SEALED ON BOTTOM

12"

GAS PROBE

DEPTH

1" I.D. PERFORATED SCH. 40 PVC PIPE SEE PERFORATION DETAIL

3/8" DIA. HOLE

1"

90°

PERFORATION DETAIL

2' MIN.

BENTONITE SEAL

1'

SAND AND GRAVEL (TYP.)

GAS PROBE

TYPICAL GAS PROBE MONITORING DETAIL

NOT TO SCALE

Fig. 11.31 Typical gas probe monitoring detail.

2. Anyone installing wells in a landfill should wear a safety rope to prevent falling in the bore-hole. Open holes should be covered when they are left unattended.

3. Smoking should be prohibited on the landfill where drilling, excavating, or installation of equipment is taking place or where gas is venting from the landfill.

4. Collected gas from a mechanically evacuated system should always be cleared to minimize air pollution and any potential explosion or fire hazard.

5. Methane gas in a concentration of 5 to 15% is an explosive mixture. Gas accumulations should be monitored in enclosed structures to insure that explosive conditions are avoided and, if detected, appropriate response is taken to avoid a source of ignition and to vent the structure.

All personnel working on the landfill must be provided training regarding the danger posed by landfill gases. Personnel operating safety equipment around the landfill must be thoroughly trained in its use and have a clear understanding of the meaning of observations made with the monitoring equipment. Monitoring equipment must also be periodically calibrated to ensure continued accuracy in the results.

Energy Recovery

Three approaches have been adopted for recovering energy from landfill gas. They are:

1. Upgrading the gas quality to pipeline quality for delivery to utility distribution systems.
2. Use of landfill gas directly as a boiler fuel.
3. Generation of electricity by the operation of an internal combustion engine with landfill gas.

Typical landfill gas contains approximately 500 BTUs per standard cubic foot (4450 Kcal/m^3) of energy whereas pipeline-quality gas contains 1000 BTU/Scf (8900 Kcal/m^3). The energy content of landfill gas will vary widely in quality depending upon the performance of the gas collection system and the stage of decomposition within the landfill. Generally, the collection of gas for energy recovery purposes has been limited to large landfills with over 1 million tons of solid waste in place. Recent experience has shown that gas may possibly be economically recoverable from smaller landfills, especially where energy prices are relatively high.

A principal consideration in selecting the best approach for energy recovery at a landfill is the available energy markets and the price that can be received for the energy when compared with the capital investment needed to recover and possibly process the gas. An early step in developing an energy recovery system is to do a market analysis to determine available users of the recovered energy. Potential customers include pipeline and electric utilities and commercial, industrial, or institutional facilities that need a boiler fuel. Utilities are potential markets for a recovery facility which produces pipeline-quality gas or a facility which generates electricity. Electric utility distribution lines are usually accessible, but natural gas lines may not be installed in the vicinity of the landfill. The use of methane gas as a boiler fuel will require that the boiler either be at the landfill with steam piped to the user or the gas is piped directly to the end user, who will mix landfill gas with other fuel sources to operate a boiler. The challenge to the developer of the energy recovery system is to identify the combination of energy recovery facilities that satisfies the needs of the energy user.

Two important concerns to the energy customer, in addition to price, will be the timing of delivery of the energy and system reliability. Electricity generated from the landfill gas during on-peak demand periods will have a higher value to the electrical utility than electrical energy generated during off-peak periods. However, the generation of power only during peak periods will require the installation of a larger generating facility and therefore may not be economically practical over the life of the gas-recovery project. Seasonal variations in generation rates may not coincide with the season demands of the customer.

System reliability is extremely important when developing this type of energy recovery system. Much of the methane that is generated during periods of system failure may be lost to the atmosphere and therefore result in an economic loss. Depending upon the end use, contractual agreements may specify penalties for periods of system failure, or, if low reliability is anticipated by the energy customer, they may insist upon contractual terms that are not favorable for the recovery of the energy.

Upgrading the gas to pipeline qualilty requires removal of the carbon dioxide and other trace contaminants contained within the gas. A number of processes are available to do this and are listed in Table 11.18.

During initial development of several processes, extreme corrosion problems were encountered. Recently, techniques have been devised for removing these contaminants and also for reducing the corrosion damage.

Gas recovered for direct firing of a boiler requires less treatment before use. The removal of moisture is the primary concern. Equipment for detection of oxygen entry into the gas collection system is usually provided to minimize the danger of an explosion. Depending upon the needs of the energy user, the user may mix the landfill gas with other fuels to increase the output of the boiler.

The use of internal combustion engines to convert energy in landfill gas to electricity has a number of advantages including the following:

1. Electricity is a premium form of energy which is readily marketable.
2. There is a continuous demand for electricity.
3. Electrical distribution lines are usually located near landfills.

Table 11.18 Selected Operating Landfill Gas Processes and Uses

Process	Product (BTU/Scf)	Use
Pretreatment to remove water	500	Boiler fuel
Intake scrubber to remove free water	500	Boiler fuel
Triethylene glycol process to dehydrate	500	Boiler fuel
Chiller to remove water	530	Pipline to a refinery
Condenser to remove water	700	Natural gas pipeline
Cylindrical adsorber containing alumina gel, activated carbon, and molecular sieves to remove part of the carbon dioxide and other impurities	1000	Natural gas pipeline
Pretreatment to remove moisture, heavy hydrocarbons, and trace contaminants Selexol process to remove CO_2	1000	Natural gas pipeline
Pretreatment to remove moisture, hydrogen sulfide, and trace contaminants Molecular sieve to remove CO_2 and remaining contaminants	1000	Natural gas pipeline

Source: Ref. 60.

4. The use of methane and internal combustion engines has been successfully achieved at many sewage treatment plants, thereby providing a useful variety of information.

Landfill gas, when consumed in an internal combustion engine, can result in significantly more wear than would be experienced with conventional fuels. A water-wash flume scrubber or some type of pretreatment may be provided in order to reduce the maintenance needs for the engine. The installation of extensive pretreatment facilities may not be economical. Instead, a planned program of preventive maintenance on the internal combustion engine may be more feasible. A typical electrical generation facility is shown in Figure 11.32.

The future need for alternative sources of energy and the development of new approaches for the recovery and utilization of methane gas will encourage more landfill owners to investigate and possibly implement these systems.

11.1.14 Landfill Operations

All of the work that has gone into planning, designing, and developing a landfill will be jeopardized if proper site operations are not carried out. Operation is important for reasons of economy, efficiency, esthetics, maintenance, and environmental protection. An efficient operation minimizes

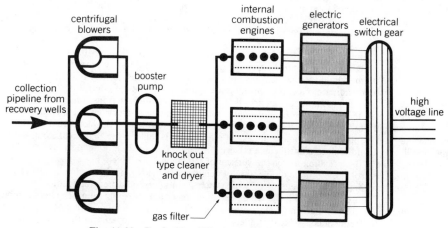

Fig. 11.32 Typical landfill gas electric generation facility.

environmental problems and reduces delays in delivery of waste to the site. A well-run site can be an asset to the community and can be operated in a manner that is compatible with other land uses. Good operation may also forestall some public opposition when attempting to develop future landfill sites.

The following are 12 rules developed by Fowler[61] specifically to guide landfill operation:

1. Start landfilling on high ground and work toward low ground.
2. First fill on the windward side of the site.
3. Spread each load of waste and thoroughly compact it using a heavy, wheel-type landfill compactor.
4. Never deposit wastes against the advancing wall of the excavation.
5. Once a load of soil has been picked up by an earth-moving machine, do not put it down until it is in its final resting place in the landfill.
6. Keep the active area as small as possible and fill upward to final grade as directly as possible.
7. Use the proper equipment and use it well within its capabilities.
8. Build interior haul roads high and drain them well.
9. Put interior haul roads on top of completed areas.
10. Keep intermediate waste slopes three horizontal to one vertical.
11. Keep surface and groundwater away from waste.
12. Keep trucks and equipment off all inactive areas.

The site operational plan is prepared as part of the plans and specifications for landfill construction. The operational plan provides the following:

1. A detailed list of operating procedures that are consistent with the design.
2. The operator's manual for building and running the site. It is important that the plan be in language that can be readily understood by the operating personnel, including the foremen, equipment operators, and other employees.
3. The technical basis for the landfill design and the procedures that will be employed during the life of the site.

When opening a landfill, the site must be prepared and several construction steps completed. Table 11.19 lists the typical steps that are necessary to place a landfill into operation.

Not all of the steps listed in Table 11.19 will be completed in order. Since many landfills are developed in phases, those facilities that will serve the landfill over its entire life are constructed initially. Excavation of fill areas, placement of liners and leachate collections systems, and the installation of gas management systems will probably be done in phases. Each phase will have the necessary construction completed before it is placed in service. During this site development stage, it is important that careful documentation be prepared. This includes the preparation of as-built plans and should include a pictorial record.

The operators of the landfill must be selected only after a careful hiring process and training program. The selection of good operators will help ensure that the plans for the landfill are properly carried out. Health records for the operators should also be maintained and record of safety training established. It is imperative that an organized safety program be carried out on a long-term basis.

The failure to control dust, fire, and litter at a landfill can foster public opposition to the further operation of the landfill. Dust control practices on site roads and possibly at certain key points within the landfill must be undertaken. These practices include paving surface roads that receive frequent use and spreading of calcium chloride on earth surface roads to control dust.

Procedures and equipment to control fires that may develop within the refuse must be available. These will include the availability of a water truck and operating instructions for the equipment operators should a fire break out. Also, arrangements should be made with the local fire department so that they can respond appropriately in case of a fire at the landfill. A fire that breaks out within the landfill may be difficult to control since covering it with refuse in attempting to smother the fire may only result in its spreading under the surface of the landfill.

Poor litter control is the principal objection of neighbors to many landfills. Unloading trucks at locations that minimize the movement of windblown materials should be practiced. Some site operators simply close the landfill if the winds get above a certain threshold point. Barriers can be installed that catch the windblown paper, but these barriers can become a maintenance headache due to the need for frequent cleaning. Portable screens can also be employed near the working face.

Erosion can be an unsightly and environmentally damaging problem at a landfill. It must be

Table 11.19 Landfill Site Preparation and Construction Steps

1. Clear site
2. Remove and stockpile topsoil
3. Construct berms
4. Install drainage improvements
5. Excavate fill areas
6. Stockpile daily cover materials
7. Install environmental protection facilities (as needed)
   a. Landfill liner with leachate collection system
   b. Groundwater monitoring system
   c. Gas control equipment
   d. Gas monitoring equipment
8. Prepare access roads
9. Construction support facilities
   a. Service building
   b. Employee facilities
   c. Weigh scale
   d. Fueling facilities
10. Install utilities
    a. Electricity
    b. Water
    c. Sewage
    d. Telephone
11. Construct fencing
    a. Perimeter
    b. Entrance
    c. Gate and entrance sign
    d. Litter control

controlled to prevent site damage and water pollution problems. Landfill areas that have reached final grade should be seeded as soon as possible, and areas at intermediate grades that will not be worked for a long time should be seeded or mulched. Providing and maintaining proper drainage, storm water diversion, and storm water detention structures will limit erosion problems.

The operations plan for the landfill should specify which soils are to be used as cover material, where they are to be obtained and how they are to be placed over the compacted waste. Cover materials used may be classified as daily, intermediate, and final; the classification depends on the thickness of soil used. Daily cover is usually 6 in. (15 cm), intermediate, 1 ft (30 cm), and final cover, 2 ft (60 cm).

At the end of each working day cover is placed over the compacted waste. This forms the top or end of a cell within the landfill. Subsequent cells are built up until the height of the landfill is achieved. Recent problems with leachate seeping from the side of landfills has prompted some operators to strip off a portion of the daily cover each morning before depositing more waste to allow the percolating leachate a downward escape route. Other operators have employed a sand blanket to accomplish the same purpose.

Final cover is placed and compacted after an area of the landfill is completed. In order to reduce permeability to greatest extent possible this cover should be compacted to the maximum extent possible. Figure 11.33 shows the relationship between compaction effort and the density of soil cover achieved. Increasing density reduces permeability. Usually a 6-in. (10 cm) layer of stockpiled top soil is placed over the cover material. This soil should not be extensively compacted so that vegetative growth can be promoted. Some landfill plans provide for a drainage or sand blanket to be placed between the cover material and the top soil.

11.1.15 Landfill Equipment Selection and Utilization

Various types of equipment will be needed to operate the landfill. Functions of this equipment are:

1. Waste movement and compaction.
2. Earth transport and compaction.
3. Support functions.

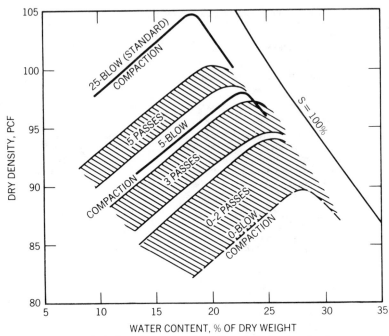

Fig. 11.33 Relationship, campaction effort, and density of soil cover.

The type and size of equipment selected will depend upon the size of the site, the nature of the waste, and the operator's preferences.

Waste handling equipment is principally machinery for spreading and compacting the waste material as it arrives. At high-volume landfills, dozers support the compactors by spreading the waste quickly. The compactors smooth the surface, break up the waste, and compact it.

When establishing the working face within the landfill, it is important that procedures are followed that result in little action movement of the waste. Trucks bringing solid waste to the landfill should deposit it at the toe of the working face in such a fashion that the operating personnel can simply spread the waste and then compact it without the need for hauling the waste a significant distance. The following procedures are frequently employed:

1. Wastes are deposited at the bottom of the working face.

2. The wastes are spread into a thin layer up the working face at a 20 to 30° slope.

3. The wastes are compacted by three to five passes of the compaction equipment or until the operator detects minimal settlement under the equipment.

4. On a daily or less frequent periodic basis, the compacted refuse is covered with earth. At some facilities where windblown paper and litter are less of a problem, wastes are deposited at the top of the working face and then spread downward.

The degree of compaction achieved within the landfill will directly affect the long-term capacity of the landfill and the future occurrence of settling problems within the fill. Achieving a high degree of compaction will result in a larger quantity of waste being disposed of within the planned volume of the landfill. Also, a good compaction assures that later problems with the landfill cover settling and disruption of drainage characteristics will at least be minimized. Even with good compaction, it can be expected that some surface settlement will occur.

The best compaction of the waste is achieved when it is spread in thin layers on as low a slope as practical. Figure 11.34 shows the effect layer or lift thickness has on the compacted density of the refuse. Generally, greater lift thicknesses are discouraged. Specially designed compaction equipment is frequently employed for this purpose. This equipment has steel-cleated wheels, and the units frequently weigh in excess of 50,000 lb. This equipment makes several passes over the wastes to achieve effective compaction. Figure 11.35 shows the relationship between the number of passes and compacted density. Actual results at a specific landfill will depend upon the waste and the particular pieces of equipment being employed. However, the relationship shown will generally be maintained.

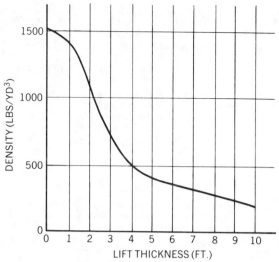

Fig. 11.34   Effect of layer (lift) thickness on compacted density.

When attempting to achieve the compaction, special care must be provided for the handling of bulky wastes, such as tires, tree stumps, brush, appliances, furniture, demolition wastes, and some commercial and industrial wastes. These materials are frequently difficult to compact and, if not properly handled, will cause later settlement problems. In addition to bulky waste, the operators must be on the lookout for wire and other materials which can become entangled in the equipment (see Chapter 17).

Inadequate compaction results in more settlement after closure; therefore, postclosure maintenance costs are increased. Some landfills have settled as much as 30 ft (9 m) due to inadequate compaction.

In addition to compaction and earth-moving equipment, each landfill needs service equipment to maintain roads, mow grass, sprinkle water for dust control, pump water, build drainage ways, install leachate pipes, and other tasks.

The road grader is used to maintain haul roads and side ditches and to plow snow if the landfill is within the snow belt. A sweeper is used for picking up mud and debris from paved areas and for cleaning the shop floor. Depending upon the amount of surface to be cleaned, the sweeper size can vary from a small three-wheeled unit weighting less than 200 lb (91 kg) up to a full-sized commercial unit.

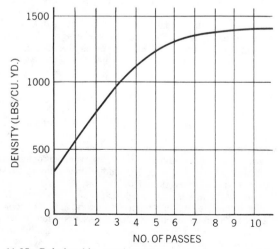

Fig. 11.35   Relationship, number of passes, and compacted density.

Water wagons also vary in size according to the facility. Some landfill operators are satisfied with a 200-gallon (5.6 m³) tank mounted on a pickup truck. Larger landfills use water wagons designed for use in constructing highway embankments. These units are mounted on off-road vehicle frames, have built-in pumps, and can haul from 5000 to 7000 gallons (140 m³ to 196 m³) of water.

Many landfills use additional utility machinery such as small rubber tire loaders, small track loaders, and backhoes, depending upon the special needs of the site.

Equipment accessories are particularly important. When specifying the equipment, the landfill manager must account for the weather, special operating problems that may occur at the landfill, and the diverse nature of the solid waste that the equipment must handle.

Special attention must be given to the control of fires that may occur on equipment. Automatic fire protection systems are available for many pieces of equipment and have been found to be a good investment.

11.1.16   On-Site Processing

On-site processing includes various types of resource recovery processing techniques such as recycling, shredding, and baling. The resource recovery techniques such as preparation of refuse-derived fuel (RDF) or incineration are described in other chapters of this book. Recycling may be planned with highly organized programs for the recovery of materials at the landfill or before the waste reaches the landfill. Shredding of the waste may precede placement of it into the landfill. Shredding may be the only operation performed on the waste or additional resource recovery processing steps can be employed. Baling of the solid waste may also precede its placement in the landfill, with the subsequent operations at the landfill termed balefilling. This section will address these topics.

Material Recovery/On-Site Recycling Operations

Salvaging usable materials from solid waste is laudable in concept, but it should be allowed only if the sanitary landfill has been designed to permit this method of recycling operation and appropriate processing and storage facilities have been provided.[62] All recycling proposals should be thoroughly evaluated to determine their economic and practical feasibility. Approaches to consider include:

1.   Operation of a manual separation area in conjunction with a baling facility. Materials recovered may include cardboard, newspaper, plastics, and ferrous metals.

2.   Imposing a mandatory source separation program whereby the waste generators are required to place in separate containers certain recyclable materials. These containers are hauled to the landfill in special compartments of the collection trucks or separate collection trips are made.

3.   Providing a voluntary drop-off area where waste generators may deposit materials that are recyclable.

The capital and operational costs of recycling operations at the landfill must be carefully evaluated. An important consideration is the availability and long-term viability of markets for the recovered materials.

Scavenging, sorting through waste that has been placed in the landfill, should not be allowed. Scavengers are too intent on searching to notice the approach of spreading and compacting equipment, and they risk being injured. Moreover, some of the materials collected, such as food, may be contaminated. Vehicles left unattended by scavengers interfere with operations at the working face.

Shredding

The concept of shredding (milling) solid waste for landfill disposal without strict requirements for daily cover is referred to as "millfilling." This process was initiated in France in the early 1950s, and in 1973 there were over 100 milling plants in the United Kingdom and on the European continent. A comprehensive study of the advantages and disadvantages of millfilling was conducted from 1966 to 1972.[63] It is estimated that over 100 landfill operations in the United States are currently receiving wastes that have been shredded.

Handling shredded solid waste at a landfill site is simple and convenient. It is merely dumped and graded as desired. Since shredded waste is relatively uniform in size and composition, compaction and settlement are even. The material provides excellent support for rubber-tired vehicles, and since glass objects are reduced to tiny pieces, damage to tires of landfill machinery is greatly reduced. The nuisance caused by blowing paper and dust produced by packing machinery is greatly reduced. The danger of fire in the landfill is greatly lessened.

Many vectors, especially rats and flies, are reduced in number or eliminated from the landfill.

Shredded solid waste is esthetically more pleasing than unprocessed waste; therefore, it is more readily accepted by the public.

The key advantages of shredding prior to compaction at a sanitary landfill are as follows:

1. The shredded waste can be compacted to a greater density than unshredded waste; thus, the life (or capacity) of a landfill site can be extended. The actual density of shredded waste on a wet-weight basis was found to be approximately 27% greater than the actual density of unprocessed solid waste given equal compaction.[63]

2. A given tonnage of shredded solid waste can be compacted more quickly than unshredded waste. For example, 100 tons of shredded solid waste could be compacted in about 70% of the time required for compaction of 100 tons of unshredded waste.[63] This would increase the cost-effectiveness of the sanitary landfill operations.

3. The requirement for daily cover is lessened. Cover should still be used, but some of the public health concerns prompting the need for daily cover are minimized.

4. There is a reduced fire hazard from shredded solid waste compared with unshredded solid waste.[63]

5. Covered unshredded solid waste cells never produced organics at as high a rate as did the covered shredded waste cells during initial stages of decomposition; however, the unshredded cells continued to produce organics at a fairly consistent rate throughout the 6-year study at Madison, Wisconsin. Thus, the shredded waste cells could be characterized as producing more leachate contaminants during initial stages of decomposition but less during later stages of decomposition than the unshredded waste cells.[63]

The primary disadvantage of shredding is the cost for shredder purchase and operation. However, if shredding can extend the useful life of a landfill site, the added costs may be justified.

Baling

Baling of solid waste can occur prior to its final compaction and coverage in a sanitary landfill. The baling operation can occur at transfer stations or at the landfill site. Baling involves compressing solid waste into a high-density block; then loading these blocks onto a truck, trailer, or railroad flat car; hauling them to a disposal site; and with a forklift, stacking them neatly in a trench and covering them with cover material.[64] The advantages of balefilling are as follows:[64]

1. Extends landfill useful life by increasing waste density by approximately 60%.

2. Improves cost-effectiveness of total waste collection and disposal by reducing haul time to distant disposal sites and/or providing transfer station to help keep packer trucks on their routes.

3. Reduces negative impact of disposal site by reducing litter, dust, odor, vermin, vectors, fires, traffic, noise, leachate, safety hazards, earth moving, and land settling.

4. Reduces cost and improves operating efficiency since less heavy equipment, personnel, and cover material are needed and operating standards can be upgraded.

5. Increases potential future use of disposal site by improving foundation-bearing factors; reduces waiting period for land stabilization.

6. Increases resource recovery opportunities by providing central processing facility for simplifying start-up of material separation and baling for storage or sale.

High-density baling and balefills offer unique environmental benefits. Settlement, noise, traffic, noxious leachate production, birds, vector attraction, combustible gas generation, fire, litter, dust, noise, vehicular air pollution, and odors are environmental problems common to most sanitary landfills. Solutions exist for reducing these problems, but they are expensive and local conditions such as lack of cover soil, high groundwater, and proximity to developed areas may make the solutions unacceptable. High-density baling can produce a fill with maximum density and minimum observed settling, low rates of combustible gas generation, dilute leachate, minimum landfill traffic (reduced noise, dust, and the chance of accidents), reduced labor and equipment requirements and costs, more efficient transport, and minimal littering problems.[64]

A stable, final landfill area can be provided almost immediately after the fill is completed. Vector (birds, rats, flies, and so forth) attracting hazards are significantly reduced at balefills because of the smaller working face and ease of providing complete daily soil cover. Results indicate that bale landfills are more esthetic and environmentally superior to conventional sanitary landfills.[64]

11.1.17 Operator Safety

Safety refers to freedom from the occurrence or risk of injury or loss. Many operations at a landfill disposal site could lead to injuries or losses if proper planning and precautionary measures are not invoked. Safety concerns at a landfill operation are related to both site employees and users. Employees are involved with equipment operations and materials handling. They may also be subject to fires and explosions. Users may be subject to unsafe conditions depending on the extent of allowed public access and scavenging.

Accidents are expensive, with the hidden costs often several times the readily apparent costs. Solid waste personnel work in all types of weather situations, with many different types of heavy equipment, with a variety of materials presenting diverse hazards, and in many different types of settings. The types of accidents possible at landfills include direct injury from explosion or fire, inhalation of contaminants and dust, asphyxiation due to workers entering a poorly vented leachate collection system manhole or tank, falls from vehicles, accidents associated with the operation of heavy earth-moving equipment, accidents from attempting to repair equipment while the engine is operating, exposure to extreme cold and heat, and traffic accidents at or near the site.[2]

Proper landfill operation depends on good employees. The number of employees needed depends upon the size and method of site operation. After computing the number of pieces of equipment, the number of equipment operators can be determined. If a scale is used, a scale operator is needed. Maintenance personnel will also be required, along with a laborer to assist when needed for directing traffic, picking up litter, assisting in machine repair, and other tasks. Large landfills will also need a supervisor.

In order to maintain an efficient landfill operation, employees must be carefully selected, trained, and supervised. The operation of a landfill is a complex task involving very detailed procedures and utilizing expensive and complicated equipment. Hiring procedures must insure that the employees selected will have the ability to operate the landfill properly under potentially adverse conditions. To insure efficient operation, employees must be properly trained in efficient, safe equipment use. Competent supervision will also insure that problems are corrected before they become major.

Responsibility for safety begins with dedication by top management. Some responsibilities include: providing employee training, providing safe equipment, enforcing all safety rules, supplying safety equipment, maintaining good housekeeping, promptly investigating accidents, insisting on participation in the safety program by employees, and observing landfill operation to detect potential safety problems. To serve as an example, Table 11.20 contains a list of safety pointers for landfill equipment operators.[65] In addition, protective equipment and clothing can be worn

Table 11.20 Safety Suggestions for Sanitary Landfill Equipment Operators

1. Check equipment before starting
2. Use steps and hand holds
3. Keep steps clean
4. Inspect area before moving
5. Operate from driver's seat
6. Wear seat belts
7. Never mount moving equipment
8. Authorized passengers only
9. Keep bucket or blade low
10. Check blind areas
11. Keep enough clearance
12. Avoid sidehill travel
13. Avoid excessive speed
14. Do not crush sealed containers
15. Go carefully over bulky items
16. Check work area
17. Park on level ground
18. Lower attachments to ground when parked
19. Never jump from equipment
20. Avoid leaving equipment unattended
21. Always have adequate lighting
22. Clean equipment before repairing
23. Remain in seat during equipment adjustments

Source: Ref. 65.

by employees, with examples including ear plugs, hard hats, safety gloves, safety glasses, and respirators.

Educational films and written material on safety at the landfills are available from the federal government as well as equipment manufacturers. Assistance in setting up a safety program is also available from insurance companies with worker's compensation programs, the National Safety Council, safety consultants, and federal and state safety programs.[66]

No burning of wastes should be permitted at a sanitary landfill, but fires occur occasionally. The use of daily cover should keep fire in a cell that is under construction from spreading laterally to other cells. All equipment operators should keep a fire extinguisher on their machines at all times, since it may be able to put out a small fire. If the fire is too large, waste in the burning area must be spread out so that water can be applied. This is an extremely hazardous chore, and water should be sprayed on those parts of the machine that come in contact with the hot wastes. The operations plan should spell out fire-fighting procedures and sources of water. All landfill personnel should be thoroughly familiar with these procedures. A collection truck occasionally arrives carrying burning waste. It should not be allowed near the working face of the fill but be routed as quickly as possible to a safe area, away from buildings, where its load can be dumped and the fire extinguished.[62]

11.1.18   Site Closure and Long-Term Care

When selecting, designing, and operating a landfill, it is important to have an end use in mind. This will foster more efficient site operation and may prove useful in gaining local support when describing the project to residents and officials during the site selection process.[2] It should be realized that long-term environmental planning and management is necessary for appropriate site closure following cessation of solid waste disposal at the site. This section addresses general closure considerations, long-term maintenance and environmental monitoring, landfill site and uses, and financial liability and planning for closure.

Closure Considerations

Often in the past, site closure and potential site closure costs have been ignored in the planning of landfills. Unfortunately, communities have later discovered to their chagrin that site closure can be very expensive and difficult to accomplish. Two basic goals need to be achieved. First, closure should minimize the need for further maintenance at the landfill site. Second, closure should place the landfill in a condition that will have the least possible detrimental environmental impacts in the future.

Table 11.21 identifies tasks that must be accomplished during site closure. Some regulatory agencies are requiring that the developer prepare a closure plan as part of the initial plans for the landfill. If a closure plan has not been developed, the tasks identified in the table under "Preplanning" must be completed. This includes specifying the final topographical contours for the landfill and establishing procedures for storm water removal.

A factor that is sometimes overlooked is the need for cover material to cap the landfill. A source of cover should be identified when the fill is designed. If additional cover material will be needed, it should be brought to the site while the landfill is operating. This will insure that cover is available when the landfill is closed, and the cost can be recovered from current landfill users. Another preplanning element is preparing a landscaping and vegetative cover plan for implementation upon closure. This is in addition to planning other features of the landfill such as gas vents, leachate collection facilities, or groundwater monitoring systems.

Long-Term Maintenance/Environmental Monitoring

The long-term maintenance of a closed landfill site will be a function of the ultimate site use. Many current landfills have gas and leachate collection systems that will require continuous attention after closure. Groundwater monitoring devices may also be incorporated into the design to check the performance of the leachate control system. Other site features that will require a degree of attention on a continuing basis are drainage control structures and erosion control features.

Methane gas control systems may be either active or passive. Passive systems allow the gas to escape to the atmosphere by natural means. The blower and pumps in the active system require periodic maintenance. In addition, the withdrawal pipes and collection lines may need condensate removed and repairs of damage caused by differential settling.

Leachate collection systems installed at landfills will require continuous attention, once the landfill is closed. The leachate collection system must be maintained to insure effective operation. This work may include annual leachate collection pipe cleaning, collection tank cleaning, and inspection and pump cleaning and repairs. Collected leachate needs to be disposed of in the appropriate manner. Failure to withdraw leachate could allow it to seep out of the side of the landfill and possibly contaminate groundwater. Records should be maintained that show the quan-

Table 11.21   Site Closure Checklist

Preplanning

Identify final site topographic plan
Prepare site drainage plan
Specify source of cover material
Prepare vegetative cover and landscaping plan
Identify closing sequence for phased operations
Specify engineering procedures for the development of on-site structures

Three Months before Closure

Review closure plan for completeness
Scheduled closing date
Prepare final timetable for closure procedures
Notify appropriate regulatory agency
Notify site users by letter if they are municipalities or contract haulers, and by published announcement if private dumping is allowed

At Closure

Erect fences or appropriate structures to limit access
Post signs indicating site closure and alternative disposal sites
Collect any litter or debris and place in final cell for covering
Place cover over any exposed refuse

Three Months after Closure

Complete needed drainage control features or structures
Complete, as required, gas collection or venting system, leachate containment facilities, and gas or groundwater monitoring devices
Install settlement plates or other devices for detecting subsidence
Place required thickness of earth cover over landfill
Establish vegetative cover

Source: Ref. 2.

tity of leachate removed. The leachate quantity will vary with the season of the year and should be carefully monitored, possibly with automated signaling devices to insure that it is being properly removed. The duration over which this must be practiced is somewhat uncertain.

The success of the leachate collection system or the effectiveness of natural attenuation of the soils in managing leachate can be assessed with a groundwater monitoring system. Regulations require that many landfills have a groundwater monitoring system incorporated into their design. The purpose of these wells is to evaluate the performance and design of the facilities provided for leachate control.

Monitoring of landfill gas and leachate control systems will provide valuable information about the landfill. Of prime importance is the detection of any problems as soon as possible and immediately implementing corrective actions. In this way the damage to the environment can be minimized and the cost incurred held to a minimum. Monitoring data can also be used to improve future landfill design. Also it may be found that certain expensive features were not effectively accomplishing the goals or could be accomplished at a much lower cost.

Often drainage control problems can result in accelerated erosion of a particular area within the landfill. Differential settling of drainage control structures can limit their usefulness and may result in failure to direct storm water properly off the site. In instances where erosion problems are noted or drainage control structures need to be repaired, proper maintenance procedures should be immediately implemented to prevent further damage. Failure to maintain the physical integrity of the landfill cover will promote additional infiltration into the landfill and eventually cause generation of larger leachate quantities. This will only antagonize the problems associated with leachate collection and disposal.

Table 11.22. Summary of Procedures for Planting Vegetation on Completed Sanitary Landfills

Vegetating Landfills with Limited Funds

1. Selecting an end use: Selection of the end use will depend on the amount of funds available. An open space or park will require less funds than an intensive use such as a golf course.
2. Determining depth of cover: Soil cover must be at least 2 ft (60 cm) for grasses and 3 ft (90 cm) for trees. On sites of less than 50 acres at least two test holes or pits per acre should be dug; on larger sites there should be at least one test per acre.
3. Establishing an erosion control program: Stabilize the soil as soon as possible. Mulching is recommended for areas with erosion potential. Terracing may be appropriate on steeper slopes. The progress of the cover growth should be monitored by a qualified expert.
4. Determining the soil nutrient status: Test for pH, nitrogen, potassium, phosphorus, conductivity, bulk density, and organic matter.
5. Determining soil bulk density: Test for bulk density. Compacted soil should be scarified and organic matter added to enhance the physical properties.
6. Amending soil cover: Incorporate fertilizer and organic matter into the top 6 in. (15 cm) of the soil.
7. Selecting landfill tolerant species: Make selection based on test plots or published results.
8. Planting grass and ground covers: The planting technique will be site dependent. Seed incorporation is generally recommended.
9. Developing trees and shrub growth: It is generally recommended that trees and shrubs be planted 1 to 2 years after the grass cover. Trees and shrubs will be at least as gas sensitive as the grass cover. If economically feasible a membrane barrier should be placed below each tree or shrub.

Vegetating Landfills Where Adequate Funds are Available

1. Constructing the landfill: Construct the landfill in a fashion that will encourage good drainage, limit erosion, minimize gas movement through the soil cover, prevent settlement, and provide a good seedbed.
2. Extracting gas: The most successful vegetative covers are those where a gas extraction system is installed.
3. Selecting gas barriers: Prevent gas from entering the root zone by covering the landfill with an impervious soil layer or synthetic membrane.
4. Selecting cover soil: Obtain good-quality soil. A loam soil should be provided where trees and shrubs will be planted. The soil should be tested for nutrients before use.
5. Spreading cover soil: The clay cover soil should be carefully placed over the compacted refuse. The top soil should be placed using methods and equipment that avoid compaction.
6. Providing proper soil depth: Trees and shrubs need 3 ft (90 cm) of soil, of which the top 8 in. (20 cm) is good-quality topsoil. Grasses need 2 ft (60 cm) of soil depth.
7. Locating areas unsuited for tree and shrub growth: Designate areas that are not suitable for trees and shrubs by physically inspecting the site for dead grass, anaerobic soil, high soil temperatures, and thin soil cover.
8. Selecting tree and shrub material: Choose species considering site end use, geographic location, type of waste in landfill, and the adaptability of the species to site conditions.
9. Planting and maintaining vegetation: Employ planting procedures that are specifically suited for the species. Watering will be important.

Source: Ref. 68.

Table 11.23   Costs for Landfill Development, Operation, and Long-Term Care

Category	Percent of Total
Predevelopment and site preparation	31
Site operation and maintenance cost (10 years)	35
Site closure cost	5
Long-term maintenance costs (20 years)	29
Total cost	100

Source: Ref. 69.

Table 11.24 Long-Term Care Fund Deposit and Payout Schedule for Landfills

		Fund Deposits			
Date	Site Life Year	Annual Payment to Fund on January 1	Fund Balance on January 1	Annual Interest Earned (13%)	Fund Balance on December 31
1/1/81	1	71,526[a]	71,562	9,303	80,865
1/1/82	2	78,718	159,583	20,746	180,329
1/1/83	3	86,590	266,919	34,699	301,618
1/1/84	4	95,249	396,867	51,593	448,460
1/1/85	5	104,774	553,234	71,920	625,154
1/1/86	6	115,251	740,405	96,253	836,658
1/1/87	7	126,776	963,434	125,246	1,088,680
1/1/88	8	139,454	1,228,134	159,657	1,387,791
1/1/89	9	153,399	1,541,190	200,355	1,741,545
1/1/90 through 3/31/90 closure period		—	1,741,545	56,600	1,798,145

[a] Fund deposits increased annually to reflect inflation 10% per year.

			Fund Payouts		
			4/1/90 Closure Payout $471,590		
Date	LTC Year	LTC Cost Estimate (current price)	Annual LTC Payout on April 1	Fund Balance on April 1	Annual Interest (Earned 13%)	Fund Balance on March 31
4/1/90	—	—		1,326,555	172,452	1,499,007
4/1/91	1	100,000	265,629[a]	1,233,378	160,339	1,393,717
4/1/92	2	90,000	262,972	1,130,745	146,997	1,277,742
4/1/93	3	80,000	257,129	1,020,613	132,680	1,153,293
4/1/94	4	70,000	224,988	928,305	120,680	1,048,985

4/1/95	5	60,000	192,846	856,139	111,298	967,437
4/1/96	6	50,000	160,705	806,732	104,875	911,607
4/1/97	7	40,000	141,421	770,186	100,124	870,310
4/1/98	8	37,000	143,896	726,414	94,434	820,848
4/1/99	9	34,000	145,451	675,397	87,802	763,199
4/1/00	10	31,000	145,879	617,320	80,252	697,572
4/1/01	11	28,000	144,938	552,634	71,842	624,476
4/1/02	12	25,000	142,350	482,126	62,676	544,802
4/1/03	13	22,000	137,794	407,008	52,911	459,919
4/1/04	14	19,000	130,905	329,014	42,772	371,786
4/1/05	15	16,000	121,259	250,527	32,569	283,096
4/1/06	16	13,000	108,375	174,721	22,714	197,435
4/1/07	17	10,000	91,702	105,733	13,745	119,478
4/1/08	18	7,000	70,611	48,867	6,353	55,220
4/1/09	19	4,000	44,384	10,836	1,409	12,245
4/1/10	20	1,000	12,245	0		0

[a] Annual long-term care (LTC) payout is equal to LTC cost estimate adjusted for 10% per year inflation from date of cost estimate (1/1/81) to beginning of individual year when LTC begins (4/1/90, 4/1/91, etc.)

Note: The above example assumes that LTC cost estimates (at current prices) decline from $100,000 for the first year to $1,000 for the last year. An alternate assumption could be a uniform expense estimate of $40,000 (at current prices) per year from 1996 to 2011. Use of this assumption would result in the following:

First annual payment to fund—$84,425
Fund balance on January 1, 1989—$2,054,608
LTC payout in 2011—$488,223

Quite obviously provisions beyond the 20-year LTC period will be necessary.
Source: Ref. 2.

It is also important that the site owner maintain the site in an esthetically acceptable manner. This may include periodically mowing the site and controlling the growth of weeds or other undesirable forms of vegetation. Depending upon the long-term use identified for the site, the amount of maintenance in this regard may be minimal.

Landfill Site End Uses

Completed sanitary landfill sites have been successfully used for parks and recreation, botanical gardens, residential and industrial development, parking areas, airport runways, and goods transfer yards.[67] The use of a completed sanitary landfill as a green area is very common.[62] No expensive structures are built, and a grassed area is established for the pleasure of the community. Some maintenance work is, however, required to keep the fill surface from being eroded by wind and water. The cover material should be graded to prevent water from ponding and infiltrating the fill.

The most commonly used vegetation is grass. Shrubs and small trees may be added where funds are available and will enhance the end use. Recommendations by Gilman, Flower, and Leone[68] for establishing vegetative cover on sanitary landfills are summarized in Table 11.22. Separate recommendations are provided for landfills with limited funds and those with adequate funding.

Completed landfills are often used as ski slopes, toboggan runs, coasting hills, ball fields, golf courses, amphitheaters, playgrounds, and parks.[62] Small, light buildings, such as concession stands, sanitary facilities, and equipment storage sheds, are usually required at recreational areas. These should also be constructed to keep settlement and gas problems at a minimum. Other problems encountered are ponding, cracking, and erosion of cover material. Periodic maintenance includes regrading, reseeding, and replenishing the cover material.

Light, one-story buildings are sometimes constructed on the landfill surface.[62] The bearing capacity of the landfill should be determined by field investigations in order to design continuous foundations. Foundations should be reinforced to bridge any gaps that may occur because of differential settling in the fill. Continuous floor slabs reinforced as mats can be used, and the structure should be designed to accommodate settlement. Doors, windows, and partitions should be able to adapt to slight differential movement between them and the structural framing. Roads, parking lots, sidewalks, and other paved areas should be constructed of a flexible and easily repairable material, such as gravel or asphaltic concrete.

A foundation engineering expert should be consulted if plans call for structures to be built on or near a completed sanitary landfill.[63] This is necessary because of the many unique factors involved—gas movement, corrosion, bearing capacity, and settlement. The cost of designing, constructing, and maintaining buildings is considerably higher than it is for those erected on a well-compacted earth fill or on undisturbed soil. Piles can also be used to support buildings when the piles are driven completely through the solid waste to firm soil or rock.[62] Some of the piles should be battered (angled) to resist lateral movement that may occur in the fill. Another factor to consider is the load imposed on the piles by solid wastes settling around them. Several peculiar problems arise when piles are used to support a structure over a landfill. The decomposing waste is very corrosive, so the piles must be protected with corrosion-resistant coatings. The fill underlying a pile-supported structure may settle and voids or air spaces may develop between the landfill surface and the bottom of the structure. Landfill gases could accumulate in these voids and create an explosion hazard.

Financial Liability

Closure and long-term care of a landfill are expensive. While some income may be expected through the sale or lease of the site and possibly through the sale of methane gas, this is not usually enough to pay all costs of closure and long-term care. The expenses incurred at closure and during a 20-year long-term care period, for example, can be a significant portion of the overall cost of operating a site. To serve as an example, the data presented in Table 11.23 show the relative expenditures over the life and long-term care period of a landfill being planned in Wisconsin.[2] Approximately 34% of the cost has been identified for site closure and long-term maintenance. The cost is equivalent to the site operation and maintenance cost for the planned life of the facility.

In Wisconsin, for example, a closure and long-term care financial responsibility program has been established. This is to insure that adequate funds will be available to close a landfill and provide for long-term maintenance of the facility's gas, leachate, and groundwater monitoring system.[2]

If a municipal government owns a landfill that is being phased out and has not set aside funds, then the current users of other solid waste facilities operated by the municipality or the general fund will probably have to supply the necessary money. When the landfill is operated by a private organization, the operators will have to obtain the necessary funds from other sources of revenue within their company. If the company cannot meet the closure and long-term care expenses, then real financial problems can arise.[69]

One approach which can be used to meet the financial liability for site closure involves the establishment of a long-term care fund. Table 11.24 summarizes the payments that a landfill authority could make into an account over the life of a site to finance closure and long-term care.[2] When the site has been completed, funds would be withdrawn to pay for closure, and then a sum would be withdrawn annually to pay for long-term care. This approach essentially is a savings account for the landfill authority. Several features should be noted. First, the payments into the account have been computed in such a way that the amount of money deposited is adjusted for inflation each year. The first year, $71,562 are deposited, and then, after assuming a 10% inflation rate, $78,718 are deposited the second year. These payments continue through the ninth year when they total $153,399. Each year interest is paid into the fund at a rate of 13%.

Payments from the fund commence when the landfill is closed and long-term care gets under way. This fund has been established in such a way that payments out of the fund occur only after the particular task has been completed. When the landfill was being planned it was estimated that the closure costs would be $200,000. However, since closure did not occur until 9 years after the cost estimate was made, the estimate was adjusted to reflect an annual rate of inflation of 10%. Therefore, when closure is completed, the landfill authority withdraws $471,590, which is equivalent to $200,000 after 9 years of 10% inflation.

Long-term care costs are estimated in a similar fashion. When the site was developed it was estimated that long-term care costs at current prices would be $100,000 for the first year and decline by $10,000 per year for the first 7 years after closure. The rate of decline then is reduced to $3,000 per year. These cost estimates are at 1981 prices and are adjusted for inflation as shown in column four of Table 11.24. At the end of each year of long-term care, the amount of money set aside is paid to the authority with the long-term care fund being exhausted at the end of the 20-year period.

This example illustrates one possible way a landfill authority can accumulate the necessary funds during the active life of the fill to pay for closure and long-term care. Many other alternatives have also been suggested such as net worth statements, trust funds, escrow accounts, or bonds.[1] Choosing a funding method requires careful study. Factors to consider include political sentiments within the community, the tax consequences to operators of private landfill facilities, and the desirability of tying up large sums of money where the operating authority appears to have sufficient financial resources already.

REFERENCES

1. Reindl, J., and O'Leary, P., Solid Waste Landfills Independent Study Course, 1981, University of Wisconsin—Extension, Madison, Wisconsin.

2. Reindl, J., Landfill Site Selection, Solid Waste Landfills Independent Study Course, 1981, University of Wisconsin—Extension, Madison, Wisconsin.

3. Brunner, D. R., and Keller, D. J., Sanitary Landfill Design and Operation, EPA Report No. SW-65ts, 1972, U.S. Environmental Protection Agency, Washington, D.C.

4. Hagerty, D. J., Pavoni, J. L. and Heer, Jr., J. E., Solid Waste Management, 1973, Van Nostrand Reinhold, New York, pp. 111–113.

5. Stearns, R. P., and Ross, D. E., Environmental Impact Statements for Sanitary Landfills, Public Works, Nov. 1973, pp. 63–66.

6. Robertson, J. M., Toussaint, C. R., and Jorque, M. A., Organic Compounds Entering Ground Water from a Landfill, EPA-660/2-74-077, Sept. 1974, U.S. Environmental Protection Agency, Washington, D.C.

7. U.S. Environmental Protection Agency, Criteria for Classification of Solid Waste Disposal Facilities and Practices, Federal Register, Vol. 44, No. 179, Sept. 13, 1979, pp. 53438–53468.

8. Oklahoma State Department of Health, Sanitary Landfill Standards—Compliance and Reason, Guidelines No. 2, May 1974, Oklahoma City, Oklahoma.

9. Anonymous, Empire State Study Defines Ideal Disposal Operation, Solid Waste Management/Resource Recovery Journal, Apr. 1972, p. 39 ff.

10. Canter, L. W., Site Selection for Hazardous Waste Disposal Facilities, Presented at the Conference on Ground Water Management: Quantity and Quality, Iowa State University, Ames, Iowa, Apr. 20–22, 1982.

11. Kiang, Y. H., and Metry, A. A., Hazardous Waste Processing Technology, 1982, Ann Arbor Science Publishers, Ann Arbor, Michigan, pp. 371–390.

12. LeGrand, H. E., and Brown, H. S., Evaluation of Ground Water Contamination Potential from Waste Disposal Sources, 1977, Office of Water and Hazardous Materials, U.S. Environmental Protection Agency, Washington, D.C.

13. Corbin, M., General Considerations for Hazardous Waste Management Facilities, Ch. 6 in The Handbook of Hazardous Waste Management, A. A. Metry, Ed., 1980, Technomic Publishing Company, Westport, Connecticut, pp. 158–192.

14. Dean, B. V., and Nishry, M. J., Scoring and Profitability Models for Evaluating and Selecting Engineering Products, *Journal Operations Research Society of America*, Vol. 13, No. 4, July–Aug. 1965, pp. 550–569.

15. Morrison, T. H., Sanitary Landfill Site Selection by the Weighted Rankings Method, Master's Thesis, University of Oklahoma, Norman, Oklahoma, 1974.

16. Canter, L. W., *Environmental Impact Assessment*, 1977, McGraw-Hill Book Company, New York, pp. 220–232.

17. Hanchey, J. R., Objectives of Public Participation, in *Public Involvement Techniques: A Reader of Ten Years Experience at the Institute of Water Resources*, Creighton, J. L., and Delli Priscoli, J., Eds., IWR Staff Report 81-1, 1981, U.S. Army Engineer Institute for Water Resources, Fort Belvoir, Virginia.

18. Bishop, A. B., Communication in the Planning Process, in *Public Involvement Techniques: A Reader of Ten Years Experience at the Institute of Water Resources*, Creighton, J. L., and Delli Priscoli, J., Eds., IWR Staff Report 81-1, 1981, U.S. Army Engineer Institute for Water Resources, Fort Belvoir, Virginia.

19. Creighton, J. L., Identifying Publics/Staff Identification Techniques, in *Public Involvement Techniques: A Reader of Ten Years Experience at the Institute of Water Resources*, Creighton, J. L., and Delli Priscoli, J., Eds., IWR Staff Report 81-1, 1981, U.S. Army Engineer Institute for Water Resources, Fort Belvoir, Virginia.

20. Schwertz, Jr., E. L., The Local Growth Management Guidebook, 1979, Center for Local Government Technology, Oklahoma State University, Stillwater, Oklahoma.

21. Bishop, A. B., Structuring Communications Programs for Public Participation in Water Resources Planning, IWR Contract Report 75-2, May 1975, Institute for Water Resources, Fort Belvoir, Virginia.

22. Garrett, B. C., Jackson, D. R., Schwartz, W. E., and Warner, J. S., Solid Waste Leachate Procedure Manual (SW-924) (Draft Technical Resource Document for Public Comment), March 1984.

23. Kmet, P., and McGinley, P. M., Chemical Characteristics of Leachate from Municipal Solid Waste Landfills in Wisconsin, presented at the Fifth Annual Madison Conference of Applied Research & Practice on Municipal & Industrial Waste, Department of Engineering and Applied Science, University of Wisconsin—Extension, Madison, Wisconsin, Sept. 22–24, 1982.

24. Clark, T. P., and Piskin, R., Chemical Quality of and Indicator Parameters for Monitoring Landfill Leachate in Illinois, Illinois Environmental Protection Agency, Joliet, Illinois, May 1976.

25. Chian, E. S., and DeWalle, F., Evaluation of Leachate Treatment, EPA-600/2-77-186a, USEPA, 1977.

26. Uloth, R., and Mavinic, J. S., EPA Procedures Manual for Groundwater Monitoring at Solid Waste Disposal Facilities, EPA/530/SW-611, Aug. 1977.

27. Myers, T. E., Chemically Stabilized Industrial Wastes in a Sanitary Landfill Environment, in *Disposal of Hazardous Waste, Proceedings of the Sixth Annual Research Symposium*, EPA-600/9-80-010, Mar. 1980, pp. 223–241.

28. James, S. C., Metals in Municipal Landfill Leachate and Their Health Effects, *American Journal of Public Health*, 67, 1977, pp. 429–432.

29. Ham, R. K., Decomposition of Residential and Light Commercial Solid Waste in Test Lysimeters (SW-190c), 1980, U.S. Environmental Protection Agency, Aug. 1978, Washington, D.C.

30. Fenn, D. G., Hanley, K. J., and Degeare, T. V., Use of the Water Balance Method for Predicting Leachate Generation from Solid Waste Disposal Sites, written for the Office of Solid Waste Management Programs, U.S. Environmental Protection Agency, 1975.

31. Thornthwaite, C. W., and Mather, J. R., Instructions and tables for computing potential evapotranspiration and the water balance, Centerton, New Jersey, 1957, pp. 185–311. Drexel Institute of Technology, Laboratory of Technology, Publications in Climatology, Vol. 10, No. 3.

32. Perrier, E. R., and Gibson, A. C., Hydrologic Simulation on Solid Waste Disposal Sites (HSSWDS), U.S. Army Engineer Waterways Experiment Station, Vicksburg, Mississippi (SW896), Sept. 1980.

33. Gibson, A. C., and Malone, P. G., Verification of the U.S. EPA HSSWDS Hydrologic Simulation Model, USAE Waterways Experiment Station, presented at the Proceedings of the Eighth Annual Research Symposium at Ft. Mitchell, Kentucky, Mar. 8–10, 1982.

34. Stanforth, R., Development of Synthetic Municipal Landfill Leachate, *Journal of Water Pollution Control Federation*, July 1979.

35. McBean, E. A., Surface Techniques to Control Leachate Generation, presented at University of Wisconsin—Extension Seminar, Sanitary Landfill, Gas, and Leachate Management, Dec. 1984.

36. Holtan, H. N., A Concept for Infiltration Estimates in Watershed Engineering, U.S. Department of Agriculture, Agricultural Research Service, 1961.

37. Huggins, L. F., and Monk, E. J., The Mathematical Simulation of the Hydrology of Small Watersheds, Technical Rep. 1, Purdue University Water Resources Center, LaFayette, Indiana, Aug. 1966.

38. Glebs, R., Alternative Design Strategies, presented at University of Wisconsin—Extension, Sanitary Landfill Planning Design, Feb. 1982.

39. Freeze, R. A., and Cherry, J. A., *Groundwater,* Prentice-Hall, Englewood Cliffs, N.J., 1979.

40. Prickett, T. A., Naymik, T. G., and Eonnquist, C. G., A "Random Walk" Solute Transport Model for Selected Groundwater Quality Evaluations, Illinois State Water Survey, Champaign, Illinois, 1981.

41. Montgomery, R., personal communications, 1984.

42. Lining of Waste Impoundment and Disposal Facilities, Matrecon, Oakland, California, U.S. EPA (SW-870), Sept. 1980.

43. Moore, C. A., and Roulier, M., Evaluating Landfill Containment Capability, EPA Eighth Annual Research Symposium, Ft. Mitchell, Kentucky, Mar. 8–10, 1982.

44. Haxo, H. E., Jr., Permeability Characteristics of Flexible Membrane Liners Measured in Pouch Tests, EPA Tenth Annual Research Symposium, Apr. 5, 1984.

45. Brown, K. W., Permeability of Compacted Clay Soils to Solvent Mixtures and Petroleum Products, EPA Tenth Annual Research Symposium, Apr. 4, 1984.

46. Daniel, D. E., Effects of Hydraulic Gradient and Method of Testing on the Hydraulic Conductivity of Compacted Clay to Water, Methanol, and Heptane, EPA Tenth Annual Research Symposium, Apr. 4, 1984.

47. Acar, Y. B., Organic Fluid Effects on the Structural Stability of Compacted Kaolinite and Montmorillonite, EPA Tenth Annual Research Symposium, Apr. 4, 1984.

48. Dietzler, D., Structural Design of Landfill Leachate Collection Systems, presented at the Sixth Annual Madison Conference of Applied Research and Practice on Municipal and Industrial Waste, sponsored by the University of Wisconsin—Extension, Sept. 14–15, 1983.

49. Boyle, J. S., and Ham R. K., Landfill Leachate Management Design Procedures, Residuals Management Technology, Madison, Wisconsin, 1984.

50. Robinson, H. D., and Maris, P. J., Leachate from Domestic Waste: Generation, Composition, and Treatment: A Review, Technical Report TR 108. Water Research Centre, Stevenage, Herts., Stevenage Laboratory, Mar. 1979.

51. Stegmann, R., and Ehrig, J. J., Operation and Design of Biological Leachate Treatment Plants, *Progressive Water Technology*, Vol. 12, Toronto, pp. 919–947. IAWPR/Pergamon Press, England, 1980.

52. Farquhar, G., Treatment and Disposal Alternatives for Leachate, presented at University of Wisconsin—Extension Seminar, Sanitary Landfill Gas and Leachate Management, Dec. 1984.

53. Standard Methods for the Examination of Water and Wastewater, 15th ed., APHA, Washington, D.C., 1980.

54. *1983 Annual Book of ASTM Standards, Section 11, Water and Environmental Technology,* Volume 11.01 Water (1), Philadelphia, ASTM, pp. 7–34.

55. Geophysical Techniques for Sensing Buried Wastes and Waste Migration, U.S. EPA, 1983.

56. McBean, E. A., Surface Techniques to Control Leachate Generation, Seminar, University of Wisconsin—Extension, Madison, Wisconsin, 1984.

57. McOmber, R. M., Verification of Gas Migration at Lees Lane Landfill, in Disposal of Hazardous Waste, Proceedings of the Eighth Annual Research Symposium, EPA-600/9-82-002, United States Environmental Protection Agency, Municipal Environmental Research Laboratory, Cincinnati, Ohio, Mar. 1982, pp. 150–160.

58. Ham, R. K., Gas Quantities and Chemical Characteristics, presented at University of Wisconsin—Extension Seminar, Sanitary Landfill Gas and Leachate Management, Dec. 1984.

59. Lutton, R. J., Regan, G. L., and Jones L. W., Design and Construction of Covers for Solid Waste Landfills, EPA-600/2-79-165, Aug. 1979.

60. Lofy, R., Gas Venting/Control Techniques and Facility Design, presented at University of Wisconsin—Extension Seminar, Sanitary Landfill Gas and Leachate Management, Dec. 1984.

61. Fowler, B., Landfill Operating Procedures, presented at University of Wisconsin—Extension Seminar, Sanitary Landfill Operation and Management, October 1984.

62. Brunner, D. R., and Keller, D. J., Sanitary Landfill Design and Operation, EPA Report No. JW-65TS, 1972, Washington, D.C.

63. Reinhardt, J. J., and Ham, R. K., Final Report on a Milling Project at Madison, Wisconsin, between 1966 and 1972, Vol. I, Mar. 1973, U.S. EPA, Washington, D.C.

64. Stone, R., Balefill Costs More Initially, But Can Pay Off in Public Acceptance, Lower Operating Dollars, *Solid Waste Systems*, Mar./Apr. 1977, pp. 16–18.

65. U.S. E.P.A., Training Sanitary Landfill Employees, Instructors' Manual, Course SW-43, c. 1, 1973, Washington, D.C.

66. Reindl, J., Sanitary Landfill Operation, Solid Waste Landfills Independent Study Course, 1981, University of Wisconsin—Extension, Madison, Wisconsin.

67. Aplet, J. H., and Conn, W. D., The Uses of Completed Landfills, *Conservation and Recycling*, Vol. 1, No. 3–4, 1977, pp. 237–246.

68. Gilman, E. F., Flower, F. B., and Leone, I. A., Standardized Procedures for Planting Vegetation on Completed Sanitary Landfills, U.S. EPA (EPA-600/S2-83-0055), Aug. 1983.

69. Reindl, J., Landfill Closure and Long-term Care, Solid Waste Landfills Independent Study Course, 1981, University of Wisconsin—Extension, Madison, Wisconsin.

11.2 LANDFILL WITH BALES[1]

William D. Robinson

11.2.1 Background

Landfill for solid waste disposal is rapidly becoming more costly and difficult. Escalating land prices, increased transportation distances, rising fuel costs, and increasing environmental restrictions are requiring local governments to employ more efficient methods for disposal of residential, commercial and industrial wastes.

New and improved technologies under development may offer a variety of solutions, but many communities must find acceptable solutions for the near term without commitments to huge bond issues at high interest rates. For such communities, solid waste baling can be a safe, effective, and financially affordable interim measure.

Baling works within the approved context of sanitary landfilling, minimizing or eliminating many objections to land disposal. It can reclaim otherwise useless land efficiently, economically and with esthetic enhancement (see Appendices 11.8 and 11.9).

Investment and operating costs are lower than for most other integrated disposal and resource recovery options and it is entirely compatible with source separation/recycling programs. System versatility can produce a variety of saleable baled metallic scrap and fibrous waste (paper, corrugated, etc.).

11.2.2 The Baling Process

There are two types of balers and they work on the same principle. Both types compact refuse under hydraulic pressures approaching 3000 lb/in² (20,700 KPa) or nearly 3.5 million pounds of force (1.6 × 10⁶ kg) on each load in the baling chamber. In comparing baling for landfill with traditional techniques, the first characteristics usually considered are in-place densities, bale weight, size, and density along with cover requirements. Although significant, some of these indices, such as bale weight and density, do not appear to be critical despite variations up to plus or minus 25% (Figure 11.36) because the same variable raw material compressibility applies in a comparison. In practice, landfill site life is prolonged on an average of about 60% with high-density baling and with cover material savings up to about 80%[2] (Figure 11.37) (see Appendix 11.8).

11.2.3 High-Density Balers

Three-stroke (@90° angles), high-density balers (Figures 11.38 and 11.39) are designed to handle large amounts of mixed (residential, commercial, industrial) or segregated refuse and can process over 800 tons per day (TPD) in two 8-hr shifts. The finished bales are compacted in three directions and of such density that wire tie or banding constraints are not required (Figure 11.40). This includes bulky items such as appliances and automobiles as well as secondary fibers.

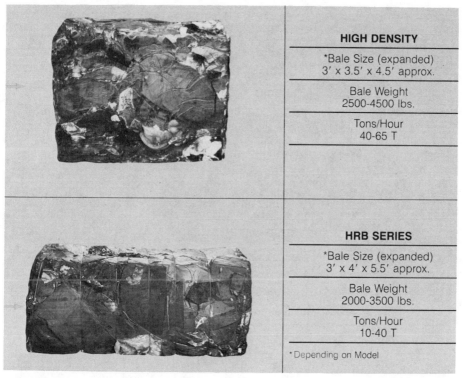

Fig. 11.36 Comparison of high-density (HD) and medium-density horizontal ram baler (HRB) bale characteristics.

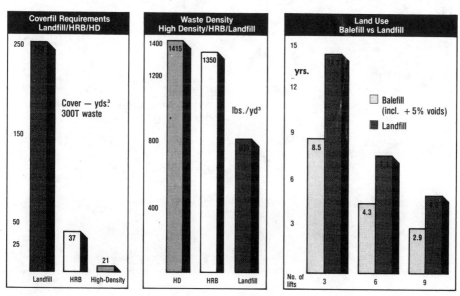

Fig. 11.37 Comparison of requirements for cover material, in-place waste density, and land use (site-life), balefill versus landfill. (Courtesy of American Hoist & Derrick Co.)

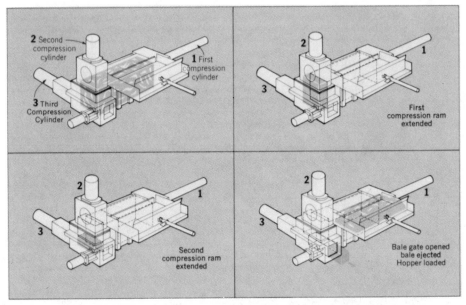

Fig. 11.38 High-density baler. (Courtesy of American Hoist & Derrick Co.)

Fig. 11.39 High-density baling system.

Fig. 11.40 Typical high-density bale.

11.2.4 Medium-Density Balers

These balers can process 10 to 40 tons of refuse per hour and are economically suitable for smaller communities and for producing mill and export size bales of aluminum, paper, corrugated, and the like.

They are single-stroke, usually horizontal ram balers (HRB) (Figure 11.41) producing bales that are wire-tied as they are ejected from the machine (Figure 11.42). Raw material is usually fed as received but with sorting out of any rogue material such as white goods, extraordinary oversize items, demolition debris, and the like. Preshredding (Figure 11.43) can be included for such material. Also, if a moderately higher density is desired and/or if an integrated materials recovery option justifies (present or future) shredding before baling, a system as shown by Figure 11.44 can be designed. This is the installation at the city of Omaha, Nebraska, that operated from 1977 until 1985 (see Chapter 17, Case History 17.5.5, and Ref. 3).

Fig. 11.41 Medium-density, single-stroke horizontal ram (HRB) baler. (Courtesy of American Hoist & Derrick Co.)

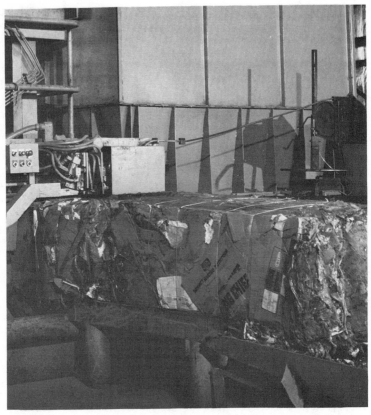

Fig. 11.42 Medium-density bales wire-tied as they are ejected from the press. (Courtesy of American Hoist & Derrick Co.)

Fig. 11.43 Horizontal tie baler/shredded material, with shredder in background (City of Omaha). (Courtesy of American Hoist & Derrick Co.)

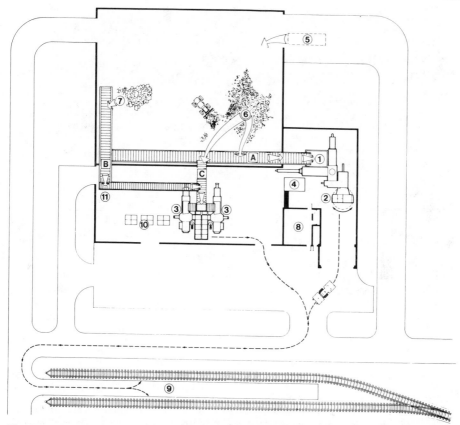

Fig. 11.44 Diagram, baling plant with high-density baling of raw refuse and medium-density baling of shredded material (City of Omaha). A, This is the main solid waste conveyor system that carries wastes from the second-floor dumping area to the high-density baler. B, Recyclables are carried on this system where a shredding system classifies materials. Once separated, they are baled in one of two HRB wire-tie balers, then sold to nearby consumers. C, Recyclables separated manually from the main dumping area travel directly to the HRB balers on this conveyor system. 1, American Solid Waste Systems High-Density Baler Model SWC-2528. 2, Revolving bale table for easy forklift access to pallets. 3, Two HRB automatic wire-tie balers. 4, Elevated control tower. 5, Incoming weigh station. 6, Main dumping floor and storage area. 7, Recyclable dumping floor. 8, Offices. 9, Rail spurs for loading solid waste bales onto rail cars. 10, Baled recyclables storage area. 11, Shredder and magnetic separator. (Courtesy of American Hoist & Derrick Co.)

In the HRB baler, refuse is fed directly into the charging box located on top of the baler. When it is full, a charging ram travels the full length of the charging chamber to densify the waste with a predetermined adjustable ram force. The finished bale is automatically tied with wire at predetermined intervals on its way out of the machine. It may then be stacked or loaded onto a waiting vehicle for transfer to the disposal site.

Figure 11.45 depicts a large, high-density baler plant and a smaller, medium-density facility is shown by Figure 11.46.

11.2.5 Transportation of Bales

Because of growing distances between urban areas and refuse disposal sites, transfer stations are becoming an increasingly necessary step in the transportation process. Solid waste is hauled by local collection trucks to a centrally located building where it is loaded onto transfer trailers that take it to the fill site. A transfer trailer with a single driver can hold the contents of several fully crewed collection trucks, and because of a higher payload-to-weight ratio, there are significant savings in labor, fuel, and collection vehicle maintenance.

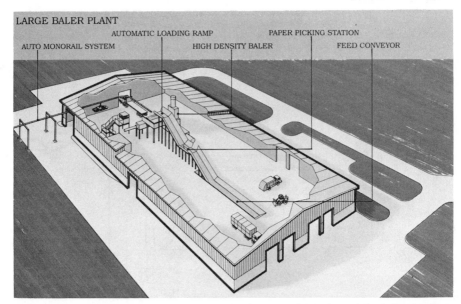

Fig. 11.45 Large plant for high-density bales. (Courtesy of American Hoist & Derrick Co.)

A "satellite" baling facility is an efficient transfer station in attaining maximum transportable loads due to orderly and extremely dense bales which can be handled automatically or by forklift at the baler. Transport to the balefill can be by standard flatbed trailer, railcar, or barge. Up to 16 high-density bales can be placed on a standard 40-ft (12 m) truck trailer, which is the equivalent of four or five average collection trucks, and any roadworthy flatbed truck combination will usually suffice. Expensive specialized hauling equipment is not required nor is it necessary to expose collection vehicles to the rigors of off-road driving at the fill site. Fewer drivers and vehicles are required with significant reduction in fuel consumption (Figure 11.47) and in fleet maintenance. Equipment does not have to operate in or on refuse, riding instead on top of dense bales.

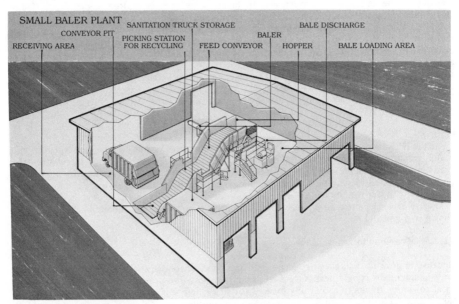

Fig. 11.46 Small baling plant for medium-density bales. (Courtesy of American Hoist & Derrick Co.)

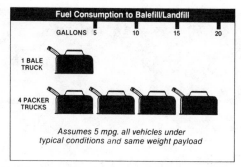

Fig. 11.47 Fuel consumption comparison, transportation to landfill, bales versus raw refuse. (Courtesy of American Hoist & Derrick Co.)

11.2.6 The Balefill

Bales are unloaded by a forklift and stacked in tiers usually three tiers high (Figure 11.48), although this number can be varied according to local preference and the type of forklift available. A standard bulldozer or loader covers the stacked bales in the usual manner with additional compaction unnecessary (Figure 11.49) with reduced labor and equipment requirements. It is also reported that because the bales are dense and stacked tightly together, they expand slightly into a solid mass after placement, reducing cover requirements by up to 50% with few if any problems with vermin, fire, or blowing debris.[4]

11.2.7 Approximate Capital and Operating Costs

There are approximately 50 balefill installations operating throughout the world, with about 40 of them in North America. Precise capital and operating cost data are not readily available for each of them, so typical costs have been assembled (1984).

Since solid waste baling installations usually do not operate on a 24-hr, round-the-clock basis (which is the usual rating index for incineration), the frame of reference for capital and operating cost must be qualified by the number of operating shifts and hourly or daily production.

A high-density baling plant without extraordinary site and land cost conditions designed for two 8-hr operating shifts to produce 800 tons would have a total installed capital cost up to about

Fig. 11.48 Forklift stacking bales at balefill.

Fig. 11.49   Balefill.

$4,000,000 ($5000.00/ton-day). This is the capital cost to be compared with the per ton-day capital cost for any alternative disposal method. These costs do not include operation of the balefill.

Typical owning and operating costs can range from $12.00/ton to $20.00/ton.

11.2.8   Summaries of Balefill Results and Testing of Bales as Foundation Material

An Evaluation of Solid Waste Baling and Landfilling

Appendix 11.8 presents the results of two evaluation projects funded by the EPA and performed by Ralph Stone and Co., Inc., Los Angeles, California. These programs evaluated the performance of a high-density baling system and a medium-density shred/bale system (the latter study was cofunded by the city of San Diego, California). In addition, the study evaluated both types of bales for the following performance criteria in landfilling:

Environmental factors of settlement, gas leachate and temperature.

Landfill characteristics of voids and expansion, densities, percolation, BOD5, NO_3, and sulfides.

Gas generation of methane, carbon dioxide, oxygen, hydrogen sulfide, and chimney effects.

Vector factors of rodents, flies, and birds.

Litter, dust and noise.

Benefits of solid waste baling.

An Engineering Study of Baled Solid Waste as Foundation Material

Appendix 11.9 presents the results of a study commissioned by the American Hoist & Derrick Co. and performed by Dr. Roger G. Slutter at the Fritz Engineering Laboratory in the Department of Civil Engineering at Lehigh University, Bethlehem, Pennsylvania. The objective of the report "was to determine some of the basic engineering properties of the material that could be used in evaluating foundation design for structures to be built on a landfill. In development of methods for obtaining bearing capacity, shear strength, permeability, density and leachate generation, the scope of our approach was extended to include a complete approach to the development of engineering properties and settlement analysis. The amount of data that we are able to provide in the report will show that a simple testing procedure can be used to determine the required information for foundation analysis.''

The report concludes with the following: ''Our experience has convinced us that the experience gained from bale sites will be far better than experience with conventional land sites.''

REFERENCES

1. Automated Baling From American Solid Waste Systems, Bulletin 5JR823, American Hoist & Derrick Co., St. Paul, Minnesota, 1982.
2. Decision Makers Guide SW500, USEPA (United States Environmental Protection Agency).
3. Omaha—The Nation's First Solid Waste Rail Haul Operation, Performance Report No. 1000-HD-PB-2, American Hoist & Derrick Co., St. Paul, Minnesota, 1977.
4. Stone, R., and Kahle, R., Evaluation of solid waste baling and landfilling, *The APWA Reporter* (American Public Works Association), **43**, No. 10 (1976).
5. Slutter, R. G., An Engineering Study of Baled Solid Waste, Lehigh University Department of Civil Engineering, Fritz Engineering Laboratory Report No. 200.74.562.1, Bethlehem, Pennsylvania, 1974.

APPENDIX 11.1 KEY ELEMENTS OF THE CRITERIA FOR CLASSIFICATION OF SOLID WASTE DISPOSAL FACILITIES AND PRACTICES[7]

Solid waste disposal facilities or practices that violate any of the following criteria pose a reasonable probability of adverse effects on health or the environment:

1. Floodplains.
   a. Facilities or practices in floodplains shall not restrict the flow of the base flood, reduce the temporary water storage capacity of the floodplain, or result in washout of solid waste, so as to pose a hazard to human life, wildlife, or land or water resources.
   b. As used in this section:
      (1) ''Base flood'' means a flood that has a 1% or greater chance of recurring in any year or a flood of a magnitude equalled or exceeded once in 100 years on the average over a significantly long period.
      (2) ''Floodplain'' means the lowland and relatively flat areas adjoining inland and coastal waters, including floor-prone areas of offshore islands, which are inundated by the base flood.
      (3) ''Washout'' means the carrying away of solid waste by waters of the base flood.
2. Endangered Species.
   a. Facilities or practices shall not cause or contribute to the taking of any endangered or threatened species of plants, fish, or wildlife.
   b. The facility or practice shall not result in the destruction or adverse modification of the critical habitat of endangered or threatened species.
3. Surface Water.
   a. A facility or practice shall not cause a discharge of pollutants into waters of the United States that is in violation of the requirements of the National Pollutant Discharge Elimination System (NPDES) under the Clean Water Act.
   b. A facility or practice shall not cause a discharge of dredged material or fill material to waters of the United States that is in violation of the Clean Water Act.
   c. A facility or practice shall not cause nonpoint source pollution of waters of the United States that violates applicable legal requirements implementing an areawide or statewide water quality management plan approved under the Clean Water Act.
4. Groundwater.
   a. A facility or practice shall not contaminate an underground drinking water source beyond the solid waste boundary or beyond an alternative boundary specified in accordance with paragraph b of this section.
   b. Only a state with a solid waste management plan approved pursuant to RCRA may establish an alternative boundary to be used in lieu of the solid waste boundary. A state may specify such a boundary only if it finds that such a change would not result in

contamination of groundwater which may be needed or used for human consumption. This finding shall be based on analysis and consideration of all of the following factors:

(1) The hydrogeological characteristics of the facility and surrounding land;
(2) The volume and physical and chemical characteristics of the leachate;
(3) The quantity, quality, and directions of flow of groundwater;
(4) The proximity and withdrawal rates of groundwater users;
(5) The availability of alternative drinking water supplies;
(6) The existing quality of the groundwater including other sources of contamination and their cumulative impacts on the groundwater; and
(7) Public health, safety, and welfare effects.

c. As used in this section:
(1) "Aquifer" means a geologic formation, group of formations, or portion of a formation capable of yielding usable quantities of groundwater to wells or springs.
(2) "Contaminate" means to introduce a substance that would cause:
  (i) The concentration of that substance in the groundwater to exceed the maximum contaminant level specified in Appendix 11.2, or
  (ii) An increase in the concentration of that substance in the groundwater where the existing concentration of that substance exceeds the maximum contaminant level specified in Appendix 11.2.
(3) "Groundwater" means water below the land surface in the zone of saturation.
(4) "Underground drinking water source" means:
  (i) An aquifer supplying drinking water for human consumption, or
  (ii) An aquifer in which the groundwater contains less than 10,000 mg/liter total dissolved solids.
(5) "Solid waste boundary" means the outermost perimeter of the solid waste (projected in the horizontal plane) as it would exist at completion of the disposal activity.

5. Disease.
a. Disease Vectors. The facility or practice shall not exist or occur unless the on-site population of disease vectors is minimized through the periodic application of cover material or other techniques as appropriate so as to protect public health.
b. As used in this section:
(1) "Disease vector" means rodents, flies, and mosquitoes capable of transmitting diseases to humans.
(2) "Periodic application of cover material" means the application and compaction of soil or other suitable material over disposed solid waste at the end of each operating day or at such frequencies and in such a manner as to reduce the risk of fire and to impede vectors' access to the waste.

6. Air.
a. The facility or practice shall not engage in open burning of residential, commercial, institutional, or industrial solid waste.
b. The facility or practice shall not violate applicable requirements developed under a state implementation plan approved or promulgated by the Clean Air Act.
c. As used in this section "open burning" means the combustion of solid waste without (1) control of combustion air to maintain adequate temperature for efficient combustion, (2) containment of the combustion reaction in an enclosed device to provide sufficient residence time and mixing for complete combustion, and (3) control of the emission of the combustion products.

7. Safety.
a. Explosive gases. The concentration of explosive gases generated by the facility or practice shall not exceed:
(1) Twenty-five percent (25%) of the lower explosive limit for the gases in facility structures (excluding gas control or recovery system components); and
(2) The lower explosive limit for the gases at the property boundary.
b. Fires. A facility or practice shall not pose a hazard to the safety of persons or property from fires. This may be accomplished through compliance with (6) and through the periodic application of cover material or other techniques as appropriate.
c. Bird hazards to aircraft. A facility or practice disposing of putrescible wastes that may attract birds and which occurs within 10,000 ft (3,048 m) of any airport runway used by turbojet aircraft or within 5000 ft (1524 m) of any airport runway used by only piston-type aircraft shall not pose a bird hazard to aircraft.
d. Access. A facility or practice shall not allow uncontrolled public access so as to expose the public to potential health and safety hazards at the disposal site.
e. As used in this section:

(1) "Airport" means public-use airport open to the public without prior permission and without restrictions within the physical capacities of available facilities.
(2) "Bird hazard" means an increase in the likelihood of bird/aircraft collisions that may cause damage to the aircraft or injury to its occupants.
(3) "Explosive gas" means methane (CH_4).
(4) "Facility structures" means any buildings and sheds or utility or drainage lines on the facility.
(5) "Lower explosive limit" means the lowest percent by volume of a mixture of explosive gases which will propagate a flame in air at 77°F (25°C) and atmospheric pressure.
(6) "Periodic application of cover material" means the application and compaction of soil or other suitable material over disposed solid waste at the end of each operating day or at such frequencies and in such a manner as to reduce the risk of fire and to impede disease vectors' access to the waste.
(7) "Putrescible wastes" means solid waste which contains organic matter capable of being decomposed by microorganisms and of such a character and proportion as to be capable of attracting or providing food for birds.

APPENDIX 11.2 MAXIMUM CONTAMINANT LEVELS FOR DETERMINING WHETHER SOLID WASTE DISPOSAL ACTIVITIES COMPLY WITH GROUNDWATER PROTECTION CRITERIA[7]

The maximum contaminant levels promulgated herein are for use in determining whether solid waste disposal activities comply with the groundwater criteria in item 4 in Appendix 11.1.

1. Maximum contaminant levels for inorganic chemicals. The following are the maximum levels of inorganic chemicals other than fluoride:

Contaminant	Level (mg/liter)
Arsenic	0.005
Barium	1
Cadmium	0.010
Chromium	0.05
Lead	0.05
Mercury	0.002
Nitrate (as N)	10
Selenium	0.01
Silver	0.05

The maximum contaminant levels for fluoride are:

Temperature[a]		Level
°F	°C	(mg/liter)
53.7 and below	12 and below	2.4
53.8 to 58.3	12.1 to 14.6	2.2
58.4 to 63.8	14.7 to 17.6	2.0
63.9 to 70.6	17.7 to 21.4	1.8
70.7 to 79.2	21.5 to 26.2	1.6
79.3 to 90.5	26.3 to 32.5	1.4

[a] Annual average of the maximum daily air temperature

2. Maximum contaminant levels for organic chemicals. The following are the maximum contaminant levels for organic chemicals:

	Level (mg/liter)
(a) Chlorinated hydrocarbons:	
Endrin (1,2,3,4,10,10-Hexachloro-6.7-epoxy-1,4,4a,5,6,7,8,8a-octahydro-1,4-endo, endo-5,8-dimethano naphthalene	0.002
Lindane (1,2,3,4,5,6-Hexachlorocyclohexane, gamma isomer	0.004
Methoxychlor (1,1,1-Trichloro-2,2-bis (p-methoxyphenyl) ethane)	0.1
Toxaphene ($C_{10}H_{10}Cl_8$-Technical chlorinated camphene, 67 to 69% chlorine)	0.005
(b) Chlorophenoxys:	
2,4-D (2,4-Dichlorophenoxy-acetic acid)	0.1
2,4,5-TP Silvex (2,4,5-Trichlorophenoxypropionic acid)	0.01

3. Maximum microbiological contaminant levels. The maximum contaminant level for coliform bacteria from any one well is as follows:
  a. Using the membrane filter technique:
      (1) Four coliform bacteria per 100 ml if one sample is taken, or
      (2) Four coliform bacteria per 100 ml in more than one sample of all the samples analyzed in one month.
  b. Using the five-tube-most-probable-number procedure (the fermentation tube method) in accordance with the analytical recommendations set forth in "Standard Methods for Examination of Water and Waste Water," American Public Health Association (13th Ed., pp. 662–688), and using a standard sample, each portion being one-fifth of the sample:
      (1) If the standard portion is 10 ml, coliform in any five consecutive samples from a well shall not be present in three or more of the 25 portions, or
      (2) If the standard portion is 100 ml, coliform in any five consecutive samples from a well shall not be present in five portions in any of five samples or in more than fifteen of the 25 portions.

4. Maximum contaminant levels for Radium-226, Radium-228, and gross α-particle radioactivity. The following are the maximum contaminant levels for Radium-226, Radium-228, and gross α-particle radioactivity:
  a. Combined Radium-226 and Radium-228—5 pCi/liter;
  b. Gross α-particle activity (including Radium-226 but excluding radon and uranium)—15 pCi/liter.

5. Maximum contaminant levels for other than health effects. The following are the maximum levels for odor, taste and miscellaneous contaminants:

Contaminant	Level
Chloride	250 mg/liter
Color	15 Color units
Copper	1 mg/liter
Foaming agents	0.5 mg/liter
Iron	0.3 mg/liter
Manganese	0.05 mg/liter
Odor	3 Threshold odor No.
pH	6.5 to 8.5
Sulfate	250 mg/liter
TDS	500 mg/liter
Zinc	5 mg/liter

APPENDIX 11.3 SANITARY LANDFILL INSPECTION REPORT*

NOTE

6 POINT ITEM—Corrected 48 hr or less.   4 POINT ITEM—Corrected 10 days or less.   2 POINT ITEM—Corrected 30 days or less.

* Applies only to Construction/Demolition sites.

		DEMERIT SCORE	PLEASE NOTE ITEMS CHECKED			REMARKS
			ITEM NUMBER 00	SL^a	C/D^b	
4.1.1 6.3.1	ACCESS ROAD All weather, heavy duty OR alternate site.	4	01			
4.1.2	EMPLOYEE FACILITY Shelter adequate. Sanitary facilities, etc., adequate.	2	02			
4.1.3	MEASURING PROCEDURE Adequate.	2	03			
4.1.4 6.3.3	FIRE PROTECTION Fires controlled. Adequate fire fighting equipment available.	4	04			
4.1.5 6.3.1	CONTROLLED ACCESS Fenced; with a gate. Operator on duty; depositing of wastes controlled.	6	05			
4.1.6 6.3.1	UNLOADING Traffic not congested. Unloading controlled.	2	06			
4.1.7 6.3.4	SIZE OF WORKING FACE Wastes spread, compacted and covered adequately.	2	07			
4.1.8 6.3.4	BLOWING LITTER Barrier confines blowing litter. Site policed adequately daily. *wastes spread regularly	6	08			
4.1.9 6.3.4	SPREADING AND COMPACTING REFUSE Spread and compacted in layers 2 ft or less in depth.	4	09			
4.1.10	DEPTH OF CELLS 8 ft or less in depth.	2	10			

APPENDIX 11.3 (Continued)

	DEMERIT SCORE	00	SL[a]	C/D[b]	REMARKS
					PLEASE NOTE ITEMS CHECKED / ITEM NUMBER
4.1.11 6.3.4	DAILY COVER *Weekly cover. 6 in. or more—compacted daily. No visible signs of refuse left on surface.	6	11		
4.1.12	INTERMEDIATE COVER 1 ft or more between lifts—compacted. No visible signs of refuse left on surface.	2	12		
4.1.13	FINAL COVER Cracked, eroded, uneven areas repaired promptly. 2 ft or more—compacted within 1 week. No visible signs of refuse left on surface.	6	13		
4.1.14	EQUIPMENT MAINTENANCE Repaired or replaced within 24 hr.	4	14		
4.1.15	SEWAGE SOLIDS OR LIQUIDS, HAZARDOUS WASTES Highly putrescible or readily decomposable excluded. Disposal procedure approved and followed OR excluded and alternate disposal method provided.	4	15		
4.1.16	DEAD ANIMALS, ETC. Adequate method of disposal at site.	2	16		
4.1.17	LARGE OR BULKY ITEMS Incorporated in site.	2	17		
4.1.18 6.3.3	BURNING No wastes burned.	6	18		
4.1.19	SALVAGE No scavenging. Salvage operation good—stored in separate area—clean.	6	19		
4.1.20 6.3.5	VECTOR CONTROL Control adequate.	4	20		
4.1.21 6.3.5	DUST CONTROL Control adequate.	4	21		
4.1.22 2.2 6.2.2	PLACEMENT IN GROUND AND SURFACE WATER At least 2 ft above high water table OR more when specified. Water pollution per se prevented.	6	22		

4.1.23 6.3.6	**DRAINAGE OF SURFACE** Run-off water diverted from fill. Graded and maintained to prevent ponding on fill.	4	23
4.1.24 6.3.6	**FINAL GRADING** *Vegetation not more than 12 in. in height. Graded to drain well—blends with surrounding area. Seeded adequately.	2	24
4.1.25	**ANIMAL FEEDING** Domestic animals excluded.	4	25
4.1.26	**ACCIDENT PREVENTION AND SAFETY** First aid supplies adequate. Active personnel trained in first aid and safety.	2	26
4.1.27	**OPERATIONAL RECORDS AND PLAN** Daily log and record kept—available.	2	27

*Oklahoma State Department of Health (March, 1979).

[a] SL, Sanitary landfill.

[b] C/D, Construction/demolition.

APPENDIX 11.4 SANITARY LANDFILL DESIGN AND OPERATIONAL GUIDELINES*

1. *Access Road.* Roads that provide access between public roads or highways and disposal sites shall be maintained so as to be passable in ordinary inclement weather. It is necessary that patrons of the sanitary landfill shall be able to enter the site and dispose of refuse in all sorts of weather. If the site is not available to them due to impassable access roads, it would be necessary to store the garbage and refuse at some point where it may create a public health problem. It is also known that when the access road is impassable, people will dump their loads of refuse at the entrance to the site or some other convenient location such as the roadside. This, of course, not only creates a public health menace and nuisance, but it also creates a traffic hazard as well as being unsightly. This standard shall be deemed to have been satisfied when:

     a.   The access road to the face of the unloading site shall have a base surface that will stand up under all traffic usage in ordinary inclement weather. It shall be so drained that water will readily run away from the road. It shall be wide enough for safe travel by large vehicles; or

     b.   An auxiliary unloading or disposal site for temporary use during inclement weather is available, and said site is accessible at all times during inclement weather.

2. *Employee Facility.* Suitable shelter and sanitary facilities shall be provided for personnel employed at the disposal site and those unloading vehicles at the site. Shelter is a desirable protection for the landfill employees as well as collection personnel unloading vehicles during inclement weather. Toilet and hand-washing facilities are desirable for good personal hygiene for landfill employees and collection personnel who are unloading vehicles at the site. It is also desirable to provide safe drinking water for these people. This standard shall be deemed to have been satisfied when:

     a.   A permanent or temporary shelter of adequate size has been constructed.

     b.   Safe drinking water, sanitary hand-washing facilities, and toilet facilities meeting the State Health Department requirements are provided.

     c.   Suitable lighting, heating facilities, and proper screening are provided.

3. *Measuring Procedure.* Provisions shall be made for measuring all refuse delivered to and disposed of in the sanitary landfill. Volume measurements (yd^3, m^3) are preferable; weights are acceptable. Measurement of all refuse received at a disposal site provides recorded data for planning and forecasting land requirements for future landfills, life expectancy of the present site, everyday operations, and earth space requirements. Measurement of refuse received provides a basis for establishing charges for the use of the landfill disposal site should charges be made. This standard shall be deemed to have been satisfied when:

     a.   The volume of each vehicle is duly estimated or actually measured and recorded in a record book. Quick reference tables for estimating volumes are desirable; or

     b.   Fixed or portable scales are available at the landfill site and are used continuously to determine the weights of refuse delivered by each vehicle and the weights are recorded in a record book; or

     c.   The landfill is cross-sectioned regularly to determine the volume in place in the filled portion of the sanitary landfill. This shall be accomplished a minimum of once each 3 months.

4. *Fire Protection.* Suitable measures shall be taken to prevent and control fires. Fires endanger life and property. Smoke and odors create nuisances to surrounding property owners, cause air pollution, endanger disposal personnel, and interfere with landfill operations. Fires on sanitary landfill sites cause them to revert to a status equivalent to open dumps. This standard shall be deemed to have been satisfied when:

     a.   An adequate supply of water under pressure is available at the site; or

     b.   A stockpile of earth is maintained close to the working face of the fill; or

     c.   A nearby organized fire department will provide immediate service whenever called.

     d.   Suitable fire extinguishers, maintained in working order, are kept on the equipment and in all buildings.

5. *Controlled Access.* Access to a sanitary landfill shall be controlled as to time of use and as to those authorized to use the site for disposal of refuse. An attendant should be on duty to control access. In most cases if public use of a sanitary landfill is allowed when no attendant is on duty, scavenging, burning, and indiscriminate dumping commonly occur. It then becomes necessary to divert men and equipment to police the area to restore sanitary conditions. When only authorized persons are permitted access to the site during operating hours, traffic and other accident hazards are minimized. This standard shall be deemed to have been satisfied when:

     a.   The landfill site is fenced; a well-defined entrance to the site and a gate that can be locked are provided.

     b.   Signs are placed at the entrace of the site. They shall clearly state the name of the site, the

---

*Oklahoma State Department of Health (March, 1979).

hours of operation, the kinds of solid waste excluded, other special instructions and the prices charged.

c.   An attendant is on duty at the site during the hours when the site is open; or there are signs within the site area that will direct vehicles to the proper unloading locations; or an attendant is on duty at the site during heavy use hours and adequate signs are posted;

d.   The waste deposited at the site is deposited in the right location in the desired manner; no wastes deposited in an indiscriminate manner.

6.   *Unloading.*   Unloading of refuse shall be restricted and controlled by the operator of the disposal site so as to minimize traffic congestion and to facilitate the handling of wastes. The operator is understood to be a ''person'' as defined herein. An orderly control of the unloading and systematic placement of refuse at the site, restricted to a small unloading area and coordinated with spreading and compacting operations, reduces work, minimizes scattering of refuse, and expedites unloading of collection vehicles and movement of traffic without undue congestion. This standard shall be deemed to have been satisfied when:

a.   Unloading of refuse is confined to as small an area as possible, takes place in an orderly and regular sequence, and does not unduly delay the spreading and compaction of the waste in the fill.

b.   Traffic congestion is at a minimum.

7.   *Size of Working Face.*   The working face of a sanitary landfill shall be confined enough to be easily maintained with available equipment. For efficient and satisfactory operation of a sanitary landfill the refuse must be confined to the smallest area possible so that the bulldozer or compactor will have the least amount of work to do, and the least amount of cover material will be required. To spread and compact refuse over a wide area may create more work than the machinery can keep up with and may consequently cause failure of the sanitary landfill operations. A recommended size varies from one and one-half to three times the width of a compactor vehicle. This standard shall be deemed to have been satisfied when:

a.   The working face is not much more than three times the width of the spreading-compacting vehicle.

b.   The spreading-compacting vehicle adequately spreads, compacts, and covers the wastes that are unloaded and the collection vehicles are able to unload promptly.

8.   *Blowing Litter.*   Blowing litter shall be controlled by providing fencing near the working area or by use of earth banks or natural barriers. The entire landfill site shall be policed regularly and unloading shall be performed so as to minimize scattering of refuse. The purpose of the sanitary landfill is to dispose of the refuse in a sanitary nuisance-free manner. Papers and other lightweight material blow away from the unloading area regardless of precautions. These materials create unsightly conditions, nuisances, and fire hazards if left to accumulate. Barriers and/or movable fencing help to prevent and limit this material from blowing into roads, ditches, and onto other property and to keep the appearance of the site area looking good. Fencing does not refer to perimeter fencing for the site but to some type of portable fencing in lengths that are easy to handle. This standard shall be deemed to have been satisfied when:

a.   Portable fencing and/or artificial or natural barriers placed near the unloading and spreading area of the fill to catch wind-blown paper and other light material does control this blowing litter.

b.   The portable fencing, barriers, and surrounding areas are policed daily and all scattered material is collected and placed in the fill.

9.   *Spreading and Compacting of Refuse.*   Refuse shall be spread and compacted in shallow layers. Any one layer shall not exceed a depth or thickness of 2 ft (0.6 m) of material after compaction is completed. The refuse should be compacted on a slope of 20 to 30° and worked from the bottom of the slope to the top. Successful operation of a sanitary landfill depends upon adequate compaction of the refuse. Solid wastes are bulky and displace large volumes. They need to be spread out and compacted so as to displace a smaller volume. Well-compacted solid waste allows the most judicious and economical use of the landfill site. Settlement will be excessive and uneven when the refuse is not well compacted. Such settlement permits the ingress and egress of insects and rodents and severely limits the usefulness of the completed area. Compaction is best initiated by spreading the refuse evenly in shallow layers and compacting each layer separately rather than placing the material in a single deep layer and attempting to compact it. Further compaction is provided by the repeated travel of landfill equipment, special compacting equipment, and collection vehicles over the covered portion of the fill. These procedures result in the greatest compaction and the least ultimate settlement, thus providing the most useful finished fill and best utilization of the capacity of the site. A 20° slope approximates a ratio of one vertical unit of measurement to three horizontal; a 30° slope, a ratio of one vertical to two horizontal. This standard shall be deemed to have been satisfied when:

a.   Additions of refuse are spread evenly by repeated passages of landfill equipment.

b.   Each layer is compacted thoroughly to a depth not greater than approximately 2 ft (0.6 m). Several layers may be placed in each cell.

10. *Depth of Cells in Fill.* The depth of the cells in a sanitary landfill, measured perpendicularly to the working surface of the fill area, shall be no greater than 8 ft (2.4 m) in thickness. The feasible vertical lift or height of a cell in the fill should be 8 to 16 feet (2.4 to 4.8 m) on the average; in larger operations higher lifts may be feasible. The total depth of a landfill is governed by the characteristics of the site, the desired elevation of the completed fill, and good engineering practice. Construction of a landfill in well-compacted cells of not more than 8 ft (2.4 m) each in depth minimizes settlement, surface cracking, release of odors, and provides internal fire protection when each cell is covered properly each day. Fills using cell lifts of less than 8 ft (2.4 m) do not generally make maximum use of available land, but provide for earlier reuse of the site. This standard shall be deemed to have been satisfied when:

    a.   The depth of all cells is 8 ft (2.4 m) or less as measured perpendicularly to the working face of the fill.

    b.   No health hazard, nuisance, or pollution results from the placement of one or more cells above other cells.

11. *Daily Cover.* A uniform compacted layer of at least 6 in. (15.24 cm) of suitable cover material shall be placed on all exposed refuse by the end of each working day. Daily covering of the refuse is necessary to prevent scattering of refuse, attracting flies and rodents, blowing litter, production of odors, fire hazards, and an unsightly appearance. Fly emergence generally is prevented by 6 in. (15.24 cm) of compacted soil. Daily covering divides the fill into "cells" that limit the spread of fires within the fill. This standard shall be deemed to have been satisfied when:

    a.   At least 6 in. of well-compacted cover material (measured perpendicularly to the surface of the compacted refuse) is placed daily to cover completely all refuse deposited that day in the fill.

    b.   After each day's operation there are no visible signs of refuse or materials associated with solid waste on the surface of the fill and the working face.

12. *Intermediate Cover.* On all but the final lift of a landfill, a layer of suitable cover material, compacted to a minimum uniform depth of 1 ft, shall be placed daily on all surfaces of the fill except those where operation will continue. Two feet (61 cm) or more of compacted soil cover might be wasteful of cover material in a landfill in which there is a clear intention to provide at least one additional lift. Under such circumstances a 1-ft (30 cm) layer of properly compacted and maintained cover will prevent health hazards, nuisances, and fire hazards and will hold out surface runoff. If more than 90 days are expected to pass before another lift is added, the area should be seeded to help prevent erosion. This standard shall be deemed to have been satisfied when:

    a.   At least 1 ft of well-compacted cover material (measured vertically to the surface of the landfill) is placed daily on all surfaces of each lift on which another lift will be constructed, except where 6 in. (15.24 cm) of daily cover are provided;

    b.   After each day's operation there are no visible signs of refuse or materials associated with solid waste on the surface of the fill.

    c.   The entire surface of the intermediate cover is inspected weekly, and all cracked, eroded, and uneven areas are repaired.

13. *Final Cover.* A uniform layer of suitable cover material compacted to a minimum depth of 2 ft (61 cm) shall be placed over the entire surface of each portion of the final lift not later than 1 week following the placement of refuse in that portion. A minimum final cover of 2 ft of compacted earth will prevent emergence of insects from the compacted refuse, minimize escape of odors and gases, and prevent rodent burrowing. This cover also provides an adequate bearing surface for vehicles, and sufficient thickness for the cover to remain complete and in a more or less unchanged condition in the event of settling or erosion. This cover should contain at least 6 in. (15.24 cm) of topsoil. This standard shall be deemed to have been satisfied when:

    a.   At least 2 ft of well-compacted cover material (measured vertically to the surface of the landfill) is placed within a week of the completion of each portion of the final lift.

    b.   Upon completion of the placement of the final cover there are no visible signs of refuse or materials associated with solid waste on the surface of the fill.

    c.   Until completion of the entire landfill site, the entire surface of the final cover is inspected monthly and all cracked, eroded and uneven areas are repaired.

14. *Equipment Maintenance Facilities.* Provision shall be made for the routine operational maintenance of equipment at the landfill site or elsewhere and for the prompt repair or replacement of landfill equipment. When the operational authority has adequate portable facilities for this purpose, this shall be deemed to be sufficient. Equipment breakdowns of a day or more result in the accumulation of refuse as in an open dump with all the attendant health hazards or nuisances. Systematic routine maintenance of equipment reduces repair costs, increases life expectancy, and helps to prevent breakdowns that interrupt landfill operations. In event of breakdown, prompt repair of equipment or immediate procurement of available standby equipment insures continuity of operations. Special advance arrangements for making major repairs and for providing standby equipment will materially reduce downtime. This standard shall be deemed to have been satisfied when:

a.   Adequate routine maintenance of sanitary landfill equipment is carried out. A record of maintenance will be available for inspection when desired.

b.   Inoperative equipment is repaired or replaced within 24 hr.

15.   *Sewage Solids or Liquids, and Other Hazardous Wastes.*   Sewage solids or liquids and other hazardous wastes shall not be disposed of in the sanitary landfill except when special permission by the Commissioner of Health has been given. Permission will be based upon the geology of the area as well as the kind of operation so that the ground or surface water will not be polluted. Sewage solids or liquids are infectious and create health hazards if not properly handled. Other materials, including oil sludge, waste chemicals, magnesium shavings, empty pesticide containers, and explosives, may also present special hazards. Unless properly handled, these wastes can be dangerous to landfill employees. When the design of the sanitary landfill includes special provisions for disposal of hazardous materials, they can be disposed of safely and need not be excluded. This standard shall be deemed to have been satisfied when:

a.   Suitable procedures for the disposal of sewage solids, liquids, or other hazardous wastes have been established and approved by the Commissioner of Health.

b.   Suitable procedures in a are followed for the disposal of these wastes; or sewage solids, liquids, or other hazardous wastes are excluded from the immediate sanitary landfill site, and an alternate method and/or place of disposal is provided.

c.   Pesticide containers may be received at a sanitary landfill for disposal provided the following procedures are observed:

(1)   Each container shall be rinsed at least three times with a volume of diluent equal to at least one-fifth (20%) of the capacity of the container. Each rinse should be added to the spray mixture or disposed as a hazardous waste material but not in a sanitary landfill unless the site is approved for such disposal.

(2)   Each container shall be punctured at least three times at each end and at least four times on the side surfaces. The punctures shall be equally distributed on each end and around the side surfaces. Each puncture shall create a hole of at least 0.5 in. (1.25 cm) as measured across the narrowest point of the puncture.

16.   *Large or Bulky Items* (also see Chapter 17).   Special provisions shall be made for the disposal of large, heavy, or bulky items not suitable for sanitary landfill operations. Some special method may be necessary for the disposal of such large items as car bodies, refrigerators, stoves, water heaters, large tires, some demolition wastes such as lumber, plaster, lathes, bricks, and concrete, also large tree stumps, trunks, and branches. At landfills with heavy equipment, such items generally can be handled routinely with other refuse; however, special provisions are necessary to incorporate large or bulky items into the fill at some small landfills. In other situations it may be necessary to provide separate sites for handling large or bulky items. This standard shall be deemed to have been satisfied when:

a.   All large or bulky items, both combustible and noncombustible, are incorporated into the fill at the sanitary landfill; set aside in a special area for resource recovery (recycling) processing; or buried in an area of the sanitary landfill site specifically designated and operated for such items; or

b.   Under adverse situations, such as storms and icy conditions, when excess woody material is produced in a locality for a short time, special permission may be granted by the State Health Department for burning such in a location remote from the sanitary landfill site. Each situation will be considered separately. Normal tree pruning and the like is not considered an adverse situation. Air pollution, water pollution, solid waste, and health nuisance laws and regulations will be observed; or

c.   A special site is planned and developed, separate from sanitary landfill sites, for such disposal (see Chapter 17).

17.   *Burning.*   No garbage or refuse shall be burned at the disposal site except in an approved incinerator. Burning of any material creates combustion by-products that can cause air pollution. Garbage cannot be burned without nuisance except in high-temperature incinerators. Any other method of combustion creates odors, air pollution, and fire and safety hazards. Such burning adversely affects public acceptance of the operation and proper location of future sanitary landfill sites. This standard shall be deemed to have been satisfied when:

a.   No solid waste materials are burned at the sanitary landfill site.

18.   *Salvage.*   No scavenging shall be permitted at the disposal site. Salvage and recycling operations, if carried on at the disposal site, shall be in a separate area and a planned operation. Any operation at the sanitary landfill site that interferes with the prompt sanitary disposal of refuse cannot be tolerated. Controlled salvage (recycling, resource recovery) operations can be permitted when conducted in view of the above fact. Improperly conducted salvage operations delay landfilling and create unsanitary conditions. The accumulation of salvaged materials at the disposal site often results in vector problems and unsightliness, which are detrimental to public acceptance of the operation. Scavenging (the culling over of unwanted material for something that may be

useful) is an unhealthy, esthetically objectionable practice that interferes with the orderly and efficient operation of a landfill. This standard shall be deemed to have been satisfied when:

a. No scavenging is permitted at the sanitary landfill.

b. No salvaging (recycling, resource recovery) operations are allowed at the sanitary landfill site; or any salvage operations are conducted according to a definite plan and in a sanitary, orderly, and dependable manner with minimum interference to landfill operations.

c. All salvaged materials are removed from the sanitary landfill site to a separate area or incorporated into the fill and covered by the end of each working day.

d. The salvaged materials are stored in a separate area in an orderly and sanitary manner and are removed to the salvage (recycling, resource recovery) market often enough to prevent overstocked and messy conditions at the site or at least every 2 months.

19. *Vector Control.* Vector control (rats, flies, insects, birds) shall be instituted whenever necessary in the judgment of the Commissioner of Health to minimize the transmission of disease through this route. Although operation of a sanitary landfill according to these standards will reduce insect, arthropod, rodent, bird, and other vector problems to a minimum, any lapse in proper operative procedures may result in attraction and rapid production of these possible carriers of disease and filth. Supplemental vector control measures may occasionally be necessary to prevent health hazards or nuisances. This standard shall be deemed to have been satisfied when:

a. There is proper maintenance of daily, intermediate and final cover; adequate spreading and compaction of refuse; adequate drainage; and compliance with other sanitary landfill standards so as to preclude the necessity for vector control measures.

b. Supplemental vector control measures are performed within 24 hr when they become necessary.

20. *Dust Control.* Suitable control measures shall be taken whenever dust is a problem. Excessive dust slows operations, creates accident hazards and esthetic problems, and may cause eye irritation or other injury and health problems to landfill personnel. Should there be nearby developments, dust could become an air pollutant and a health, public, and/or private nuisance. This standard shall be deemed to have been satisfied when:

a. Dust control measures are not necessary; or

b. Suitable measures are taken to control dust on the sanitary landfill site or on the access road whenever it becomes necessary.

21. *Placement in Groundwater.* The depositing of refuse in groundwater or within a minimum of 2 ft (0.6 m) of the highest groundwater table at the site is prohibited. Where necessary, the Commissioner of Health may require more than 2 ft (0.6 m) above the highest groundwater table at the site. In general, landfills should not be located at a site where groundwater will intercept the deposited material in the fill. To classify as a sanitary landfill, there should be a margin of safety by having a vertical distance of at least 2 ft (0.6 m) between the highest water table and the deposited refuse. Refuse often contains infectious material or other harmful substances that can cause serious health hazards or nuisances if permitted to enter groundwaters. The decomposition products provide soluble materials that also could cause problems if absorbed by groundwater. Streams could be polluted, and potable water could become contaminated. This standard shall be deemed to have been satisfied when:

a. Refuse is placed a vertical distance of 2 ft (0.6 m) or more above the highest groundwater table.

b. If so specified by the Commissioner of Health, refuse shall be placed at a distance greater than 2 ft (0.6 m) above the highest groundwater table.

22. *Drainage of Surface Water.* The entire site, including the fill surface, shall be graded and/or provided with drainage facilities to minimize runoff into and onto the fill, prevent erosion or washing of the fill, drain off rainwater falling on the fill, and prevent the collection of standing water. The final surface of the fill shall be graded to a slope of at least 1%, but no surface slope shall be so steep as to cause erosion of the cover. The natural surface drainage of an area should not be materially altered. Runoff from above the landfill and rain falling on the landfill may, unless diverted, leach into the fill and pollute surfacewater or groundwater with the leachate. The cover may be removed by erosion of the fill. Standing water may permit mosquito breeding and may interfere with the operation of the landfill. This standard shall be deemed to have been satisfied when:

a. All the runoff water from the area above the landfill is diverted from the fill.

b. The surface of the landfill is smooth and graded to a minimum slope of one foot in 100 ft (30.5 m) (1%).

c. The maximum slope of the sides on top of the completed fill is not greater than 1 ft (0.3 m) vertically in 2 ft (0.6 m) horizontally, the slope is adequately protected against erosion, and the bottom of the slope is protected against raveling (crumbling, breaking up, or caving off) and is so constructed as to provide either surface or subsurface drainage to prevent ponding.

d. Inspections are made weekly or more often for standing water on the site and on the access road, and all accumulations are eliminated promptly.

23. *Final Grading.* The completed fill shall be graded to serve the purpose for which the fill is ultimately planned. The surface drainage shall be consistent with the surrounding area. The finished construction shall not in any way cause interference with proper drainage of adjacent land nor shall the finished fill concentrate runoff waters into adjacent areas. Seeding of finished portions of the fill with appropriate vegetation to promote stabilization of the cover shall be performed. To promote the sanitary landfill as an acceptable refuse disposal practice, and to facilitate obtaining appropriate future sites, it is important that the fill not only be operated in an acceptable manner, but also that the completed landfill blend with its surroundings and, if possible, be utilized for some useful purpose. The final grade must not cause interference with the natural flow of groundwater on terrain so as to be a public nuisance or injure the surrounding properties. This standard shall be deemed to have been satisfied when:

a. The completed landfill site has been graded and constructed so as to blend with the surrounding land areas.

b. The completed landfill site is graded so as not to interfere with the proper drainage of adjacent land nor to concentrate runoff waters onto adjacent areas.

c. The completed landfill site has been seeded with appropriate vegetation so as to promote stabilization of the cover of the fill.

24. *Animal Feeding.* All animals shall be excluded from the site and garbage feeding of animals on the site is prohibited. Consumption of raw garbage by hogs is an important factor in the transmission of trichinosis in man, as well as trichinosis, hog cholera, and vesicular exanthema in hogs; therefore, hogs should be excluded from sanitary landfills. Domestic or wild animals will interfere with the landfill operations. Appropriate fencing will exclude animals, and prompt covering of refuse will make the site less attractive for gulls and other birds. This standard shall be deemed to have been satisfied when:

a. All domestic animals are excluded from the sanitary landfill site.

25. *Accident Prevention and Safety.* Employees shall be instructed in the principles of first aid and safety and in the specific operational procedures necessary to prevent accidents, including limitation of access. Accident precautionary measures shall be employed at the site. An adequate stock of first aid supplies shall be maintained at the site. The use of heavy earth-moving equipment, the maneuvering of collection trucks and other vehicles, and the explosive or flammable items that may be in the refuse create accident-prevention problems at landfills. The remote location of some landfills makes it particularly important that personnel be oriented to accident hazards, trained in first aid, and provided with first aid supplies. For reasons of safety, access should be limited to those authorized to use the site for the disposal of refuse. This standard shall be deemed to have been satisfied when:

a. At least one person trained in first aid is on duty during operating hours.

b. An educational program is maintained on safety and first aid.

c. Adequate first aid supplies are maintained at the site at all times.

26. *Operational Records and Plan.* A daily log shall be maintained by the sanitary landfill supervisor to record operational information, including the quantity of refuse received. The total solid waste program must be carried out in a systematic businesslike manner. Only with the proper daily records for background information can this be accomplished. With these records and a definite plan future solid waste disposal and land reclamation can be carried on. Daily operational information on the landfill site, quantity of refuse received, portion or portions of the landfill site used, any deviations from the plans and specifications, and other daily activities of interest, such as weather, equipment breakdowns, employee problems, fires, and so forth, should be included in the daily records. Copies of these records and/or logs should be filed with the regulatory agency. This standard shall be deemed to have been satisfied when:

a. Complete daily records and/or logs are kept; one copy to be filed with the local responsible agency or local health department and one copy to be kept and be available at the site; one monthly summary copy shall be forwarded to the Solid Waste Management Division, Oklahoma State Department of Health, at least each 3 months.

b. One copy of the plans and specifications is filed with the State Health Department; one copy filed with the local responsible agency and/or local health department.

APPENDIX 11.5 SAMPLE OF TECHNICAL SITE CRITERIA FOR CHEMICAL WASTE DISPOSAL[11]

Category	Site Characteristic	Tolerance/Suitability for Chemical Waste Disposal		Considerations
		Favorable Conditions	Limited or Unfavorable Conditions	
		Land		
Soils and topography	Topographic relief	Gently rolling terrain	Hilly, or near-steep slopes	Limited conditions will likely add to facility development costs
	Soils: composition, engineering and site development	Suitable soils for dike construction, building construction, and liner development	Poor foundation soils, unsuitable dike material; liner soils must be imported	
	Soils: slope, erodibility	Slopes (3 to 10%) to limit erosion potential	Slopes >10% resulting in a high erosion potential	The exact slope limit needs to be defined on a site-specific basis
	Soils: texture	Clay to silt or loam (very fine to medium grain sizes)	Fine sands to gravels (coarse grain sizes)	
	Soils: agriculture uses	Soils with lesser agricultural value	Prime agricultural land	
	Subsoils: composition	Suitable soils for dikes, buildings, and liner development	Poor foundation conditions, unsuitable for dike materials; liner soils must be imported	Cost is an important factor in this consideration
	Subsoils: permeability	Silt soils with high clay content and with low permeabilities ($\leq 10\text{-}7$ cm/sec)	Clean sands and gravels, with permeabilities $>10\text{-}5$ cm/sec	Here it is assumed that natural protection of low-permeability deposits are more favorable than higher-permeability deposits
	Subsoils: thickness	Thick deposits of low-permeable materials Few or no sand and gravel lens Uncompacted thickness no less than 4 ft (1.2 m)	Thin deposits of low-permeable materials underlain by large thickness of sand and gravel	Ideally, a site should be underlain by a good thickness of impermeable material; underlying sands and gravels are less favorable
Geology	Bedrock: depth	Bedrock covered by thick deposits of unconsolidated material	Bedrock at or near surface	

Category	Factor			
Water	Bedrock: subcropping formations	Shale or undisturbed very fine-grained sedimentary formation	Highly fractured limestone or dolomites; coarse-grained, permeable sandstone	The limitations introduced by this factor are dependent on the composition and thickness of the overlying unconsolidated material
	Bedrock: structural conditions	No major structural variations within an area	Areas of faulting, extreme fracturing, or severe folding	The limitations introduced by these factors are dependent on the composition and thickness of the unconsolidated material and are site specific
	Groundwater: unconsolidated formations	No connection with surficial or buried drift aquifers; low-permeable materials to bedrock	Underlain by surficial and buried drift aquifers of local and/or regional significance	Potential for polluting a usable aquifer is the primary concern here
	Groundwater: bedrock formations	Away from any recharge areas to major bedrock aquifers: no direct connection with a usable bedrock aquifer	On a major bedrock aquifer recharge area; direct connection between a drift and usable bedrock aquifer	
	Groundwater: flow direction	Local flow pattern	Regional flow pattern	
Man-oriented	Land use: forested	Areas where existing forest may serve as a buffer	Areas where significant amounts of existing forests may be removed are not as attractive	Significant removal of forests is an additional cost factor
	Land use: cultivated land	Minor removal of land from current cultivation	Areas where significant removal of prime agricultural land from cultivation is required	Significant removal of prime agricultural land from cultivation can be a local socioeconomic cost factor
	Land use: urban residential	Areas with little urban development	Areas with high urban development	Proximity to residences is considered less favorable
	Land use: extractive	Areas of no or low on-going activity	Areas currently being mined or actively used	Use of abandoned extractive areas is questionable and would require site-specific investigation
	Land use: pasture	Areas that are currently prime pasture lands	Significant pasturing activities	Extent of pasturing determined by site-specific investigation
	Land use: urban and nonresidential or mixed residential	Site specific	Areas with minimal commercial, industrial, or institutional development	These factors can be considered exclusionary (schools, hospitals, airports)

APPENDIX 11.5 (continued)

Category	Site Characteristic	Tolerance/Suitability for Chemical Waste Disposal		Considerations
		Favorable Conditions	Limited or Unfavorable Conditions	
	Land use: parks, wildlife preserves, recreation areas	Very limited	Site location in any of these land types	All federal, state, regional, county and local parks, preserves, historical areas are considered here; this factor is considered very limited area for a chemical waste disposal facility
	Land use: transportation	Good conditioned roads (≥9 ton) in area, lower traffic volume; near railroad	Roads in poor condition, high traffic volume, high-hazard roads	The absence of 9-ton roads is not exclusionary; however, upgrading of lesser roads may be a costly alternative
	Land use: historical, archeological	Dependent on site-specific details	Areas with confirmed historical or archeological significance	This area will require some interpretation since areas of possible archeological significance have been designated; services of a professional archeologist may be required
	Socioeconomic land availability	Land available for purchase; minimum amount of land owners involved	Land unavailable or must be acquired through legal means, numerous land owners involved	
	Location	Near the majority of the waste generators	Away from waste generators	Based on the waste generator/waste disposer relationship; this aspect can become matter of transportation economics
Natural conditions	Environmental: unique areas	Area of typical regional ecosystems	Areas of unique ecological sensitivity, for example, habitats of unique and/or endangered or threatened species	This aspect is extremely site specific
	Environmental: public health	Area where construction and operation will not adversely affect public health	Areas where dust, noise, fire, explosion may create a public health and/or safety hazard	Protection of the public is the primary consideration
Nondevelopment	Engineering suitability: electric	Adequate electric power is relatively available in site area		The economics of electrical transmission are a consideration

Engineering suitability: sewer	Site near interceptor sewer or wastewater treatment plant	Not required . . . but could be used to dispose of clarified effluent; this item becomes a matter of economics
Water		
Surface water: water bodies and water courses	Limited	Placement of facility on or near
Surface Water: floodplains, floodways	Limited	Placement of facility on or near
Surface water: wetlands	Limited	Placement of facility on or near wetlands
Drainage: natural	Areas where surface drainage exists and can be controlled	Areas of poor drainage or where ponding occurs; drainage areas requiring excessive engineered controls
Drainage: local watershed	Site location near a drainage divide where upstream surface area is small	Site location where upstream surface area is great and engineering precautions to handle runoff become costly
Air		
Ambient air quality	Good dispersive characteristics are important if the facility generates a discharge to the atmosphere	Dispersion is not expected to be an important consideration for land disposal facilities
Climatology		
Odor		These site characteristics are facility specific
Dust		Site and facility specific

APPENDIX 11.6 ITEMS TO BE INCLUDED IN THE ENGINEERING REPORT FOR A SANITARY LANDFILL

I. NARRATIVE REPORT
  A. General Information
    1. Name of site
    2. Person or agency owning and operating the disposal site
    3. Name of individual(s) who will be directing or supervising the disposal site
    4. Name of person(s) who will be responsible for the actual operation of the site
    5. Population and area proposed to be served:
      a. Communities
      b. Major commercial and industrial establishments
      c. Institutions
    6. Anticipated types and quantities of solid waste that will be received
    7. Estimate of the life of proposed site, number of years
  B. Specific Information about the Site
    1. Determine relationship to floodplains
    2. General geology of the area
    3. Soil profile to a depth 5 ft (1.5 m) below the lowest part of the disposal site
    4. Hydrology:
      a. Depth to water table and direction of flow
      b. Describe measurements to be taken to prevent pollution of groundwater
      c. Streams, lakes or ponds receiving surface drainage from the disposal site
      d. Describe measures taken to divert surface water drainage from disposal area and describe those measures taken to insure that the disposal site itself is well drained
    5. Cover material:
      a. Soil type(s)
      b. Quantities of cover material and topsoil needed for life of site
      c. Quantities of cover material and topsoil available at site
      d. How will deficiencies or excesses of cover material on soil be handled
    6. Equipment to be maintained at site—type, sizes, number
    7. Land use planning, current and master plan:
      a. Present zoning classification, if any
      b. Intended usage for completed site
      c. Land use of adjacent property
        1. current
        2. future
    8. Access roads to site:
      a. Type
      b. Who is to maintain them
    9. Location of airports within 2 miles (3.2 km)
    10. Operating procedures—describe those measures that will be taken to insure that the disposal site will operate in compliance with the standards for sanitary landfills as found in the regulations to the Oklahoma Solid Waste Management Act
    11. Monitoring—describe any methods to be used to monitor for any water pollution or gas production
II. MAPS AND/OR BLUEPRINTS AS FOLLOWS
  A. Map to Show Location to Communities, Highways, County Roads
  B. Vicinity Map with Legal Description of Site to Show the Following:
    1. Homes, buildings, irrigation used in the area, water wells, water treatment facilities, wastewater treatment facilities, and roads within 0.25 mi (0.4 km) of proposed site
    2. Rivers, creeks, dry or intermittent streams, canyons, ravines, lakes, ponds, marshes, rock outcropping, recharge zones, floodplains, or any other items of interest within 0.25 mi (0.4 km) of the proposed site
  C. Plot Map or Drawing (Scale no Greater than 1 in. = 200 ft.) of the Area Wanted Under Permit—Show the Following:
    1. Dimensions
    2. Locations of test wells, core holes, monitoring wells, and monitoring sites
    3. Original grades (contours—5 ft (1.5 m) intervals). Also show elevations of the area immediately adjacent to the site. (Landfills serving a population of less than 6000 can substitute contours of 10 ft (3 m) intervals.)
    4. Final grading (contours—5 ft (1.5 m) intervals)
    5. Surface drainage—diversion ditches, dikes, dams, etc.
    6. Fencing and gate(s)
    7. Access roads
    8. Proposed trenches and fill face areas
    9. Cover material borrow areas
    10. Employee and equipment shelters
  D. Cross Section of Proposed Methods of Operation and Proposed Cell Development

---

*Oklahoma State Department of Health (April, 1979).

APPENDIX 11.7 LANDFILL SITE RATING METHOD[12]

Key Hydrologic Factors

	Point Value	0	1	2*	3	4	5	6	7	8	9
Step 1 Determine distance on ground between contamination source and water supply. Record point value	Distance in feet	30	50	75	100	150	200	300	500	1000	2500 or more

*Where water table lies in permeable consolidated rocks (II in Step 4), no more than 2 (followed by ·) points should be allotted on distance scale.

	Point Value	0	1	2*	3	4	5	6	7	8	9
Step 2 Estimate the depth to water table. Record point value	Depth in feet of water table below base of contamination source more than 5% of the year	0	2	4	7	15	25	50	75	100	200 or more

*Where water table lies in permeable or moderately permeable consolidated rocks (II in Step 4), no more than 2 (followed by ·) points should be allotted, regardless of greater depth to water table.

	Point Value	0	1	2	3	4	5
Step 3 Estimate water table gradient from contamination site	Water table gradient and flow direction (related, in part, to land slope)	gradient greater than 2% toward water supply and is the direction of flow	gradient greater than 2% toward water supply but not the direction of flow	gradient less than 2% toward water supply and is the anticipated flow	gradient less than 2% toward water supply but not the anticipated flow	gradient almost flat	gradient away from all water supplies that are closer than 2500 ft (762 m)

365

Record point value

Step 4

Estimate permeability-sorption for the site of the contamination source

(1)

Thickness in feet of unconsolidated material over bedrock	Clean Coarse Gravel		Clean Coarse Sand		Clean Fine Sand		Sand with a Little Clay		Thin Layers of Sand and Clay		Clayey Sand		Even Mixture of Sand and Clay		Sandy Clay		Clay	
	I	II	I	II	I	II	I	II	I	II	I	II	I	II	I	II	I	II
100+	0A	0A	0A	0A	2A	2A	4A	4A	5A	5A	6A	6A	7A	7A	8A	8A	9A	9A
100	0B	0J	0B	0J	2B	2D	4B	3D	5B	4H	5D	4K	7B	5K	8B	6K	9B	6M
90	0B	0J	0B	0J	2B	1E	4B	3D	5B	4H	5D	4K	6B	5K	7C	5L	8C	6M
80	0C	0K	0C	0K	2B	1F	4B	3D	5B	4H	5D	4K	6B	4M	7C	5L	8C	5M
70	0C	0L	0C	0L	2B	1F	4C	3E	5C	4J	5E	4L	6C	4M	7D	4P	8D	5M
60	0D	0L	0D	0L	2C	1G	4C	2E	5C	3G	5E	3J	6C	4N	7D	4P	8D	5N
50	0D	0M	0D	0M	2C	1G	4C	2E	4D	3G	5F	3J	6D	4N	7E	4Q	8E	4R
40	0E	0M	0E	0M	1B	0S	4D	2F	4E	3H	5F	3K	6E	3J	6C	4Q	7F	4S
30	0E	0M	0E	0N	1C	0T	3B	2F	4F	3H	5G	2G	6F	2J	6G	2J	7G	3M
20	0G	0P	0G	0P	1D	0U	3C	1H	4G	2G	5H	2H	5H	2K	6H	2L	7H	3N
10	0H	0Q	0H	0Q	0Q	0V	2D	1J	3F	1J	4G	1J	5J	1K	6J	1L	6L	2M

(2)

| 0 | 5Z | 0Z | 5Z | 0Z | 5Z | 0Z | 5Z | 0Z | 5Z | 0Z | 5Z | 0Z | 5Z | 0Z | 5Z | 0Z | 5Z | 0Z |

I—Over shale or other poorly permeable, consolidated rock.

II—Over permeable or moderately permeable, consolidated rocks (some basalts, highly fractured igneous and metamorphic rocks, and cavernous carbonate rocks—also fault zones).

(1)—Suffix A means because of depth, bedrock is not to be considered, for example, a coastal plain situation.

(2)—Suffix Z means bedrock is at surface, that is, there is no soil.

Record point value

Point Value is determined from Matrix

For single type of unconsolidated material over bedrock, point value is determined by its thickness alone; for combination of unconsolidated materials, point value must be interpolated

Step 5

Add all point values determined in Steps 1 through 4 above

Record total point value

Total Point Value	0–5	6–7	8–13	14–20	21–25	26–32
Description of site in *Relative Hydrogeologic Terms* only (without regard to type of contaminant)	*VERY POOR to POOR* because one or more key factors must have values of less than 2		*FAIR* if no separate value is less than 2	*GOOD* if all separate values are 3 or greater	to *VERY GOOD* if all separate values are 3 or greater	*EXCELLENT* if all separate values are 3 or greater

	Special Site Identifier Suffixes		
	A	B	C
Step 6 Sensitivity of aquifer (choose appropriate category)	A permeable, extensive aquifer capable of contamination	Aquifer of moderate permeability not likely to be contaminated over a large area from a single contamination source	Limited aquifer of low permeability, or slight contamination potential from a source
	A	B	C
Step 7 Degree of confidence in accuracy of rating values (choose appropriate category)	Confidence in estimates of ratings for the parameters is high, and estimated ratings are considered to be fairly accurate	Confidence in estimates of ratings for the parameters is fair	Confidence in estimates of ratings for the parameters is low, and estimated ratings are not considered to be accurate
Step 8 Miscellaneous identifiers (add if appropriate)	A. *Alluvial valley*, a common hydrogeologic setting, especially important because of the general high permeability and prevalence of down-gradient water supplies. B. Designates property *boundary* when ground distance from a contamination site is to boundary rather than to a water supply. C. Special conditions require that a *comment* or explanation be added to the evaluation. D. Cone of pumping *depression* near a contamination source, which may cause contaminated groundwater to be diverted toward the pumped well. E. Distance recorded is that from a water supply to the estimated closest *edge* of an existing plume rather than to the original source of contamination. F. Indicates the contamination source is located on a groundwater discharge area, such as a *floodplain*, and would likely cause minimal groundwater contamination. M. *Mounding* of the water table beneath a contamination site—common beneath waste sites where there is liquid input or reduced infiltration capacity. P. *Percolation* may not be adequate—the permeability-sorption digit suggests the degree to which percolation may be a problem, a digit of 7 or more being a special warning of poor percolation. Q. Designates a "*recharge or transmission*" part of an extensive aquifer that is sensitive to contamination—may be suggested by a low rating on the permeability-sorption scale and A or B rating for Step 6. S. Indicates that the most likely water supply to be contaminated is a *surface stream*, rather than a well or spring.		
	COMPLETION OF NUMERICAL RATING		
Step 9 Completion of site numerical rating	The total point value determined in step 3 is recorded and then followed in sequence by the individual point values for the four key hydrogeologic factors: distance, depth to water table, water table gradient, and permeability-sorption; this is followed, in turn, by the special site identifier suffixes: aquifer sensitivity, degree of confidence, and miscellaneous identifiers; an example of a site rating with brief explanations and interpretations is shown below		

APPENDIX 11.7 (continued)

Step 9
Completion of site numerical rating

COMPLETION OF NUMERICAL RATING

The total point value determined in step 3 is recorded and then followed in sequence by the individual point values for the four key hydrogeologic factors: distance, depth to water table, water table gradient, and permeability-sorption; this is followed, in turn, by the special site identifier suffixes: aquifer sensitivity, degree of confidence, and miscellaneous identifiers; an example of a site rating with brief explanations and interpretations is shown below

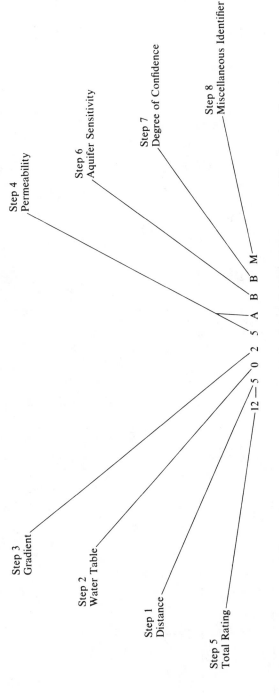

Explanation of sequence of digits and letters

12—Total point value as shown in Step 5
5—The first digit is rating for ground distance, Step 1
0—The second digit is rating for depth to water table, Step 2
2—The third digit is rating for water table gradient, Step 3
5—The fourth digit is rating for permeability-sorption, Step 4
A—Represents a closely defined position (5A) in permeability-sorption scale, Step 4
B—Represents sensitivity of an aquifer to be contaminated, Step 6
B—Represents degree of confidence or reliability of overall rating, Step 7
M—Indicates special conditions (mounding of water table in this case), Step 8

368

APPENDIX 11.8 DECISION FACTORS IN SANITARY LANDFILL SITE SELECTION[15]

A. Land Requirements
1. In-place refuse densities. One of the main goals in a landfill is to obtain maximum compaction. If the soil is spongy in nature or the cover material of poor compacting nature, there is a loss of compaction densities.
2. Cover material requirements. The site with available cover material would be best. The more cover material and the longer the distance it has to be hauled, the higher the operating cost of the site.
3. Per capita refuse production. The site or combination of sites should be picked so as to have sufficient capacity for the refuse generated.

B. Relative Location to Generating Areas
1. Time spent in hauling refuse (more important than distance). The truck and crew cost are usually expressed on an hourly basis. With today's labor cost, the time spent riding by the pickup crew could well mean the difference between profit and deficit on a collection system.
2. Highway systems available with ready access to and from the site. Time and money could be saved by the use of available highway systems.
3. The capacities of vehicles operating in the system. Where there is a long haul distance, a large-capacity truck is needed.
4. Utilization of transfer operations. A transfer system could be of great advantage if small-capacity trucks are already in use.

C. Relationship to Community Growth
1. Direction and magnitude of projected growth. The site should be located in the direction of the community growth.
2. Long-range area development. If there are long-range development plans for an area, it would be wise to place a fill in the area before zoning and prices go up.
3. Commercial and industrial development. If large commercial and industrial areas are planned, a site in the area could be used to decrease haul distance.

D. Utilities
1. Electrical power for lights and equipment.
2. Water supply for sanitary purposes, equipment washing, and fire protection.
3. Sewer service for sanitary wastes.
4. Telephone, radio communications. Items 1 through 4 are needed at any landfill, cost could be saved by locating a fill where there are already utilities in service.

E. Nuisances That Can Affect the Site
1. Traffic to and from site. At the entrance of a site, large trucks and autos pulling trailers will be entering and leaving. There should be easy access with well-marked instructions.
2. Noise of mechanical equipment. Because of the noise of the heavy equipment needed to operate a fill, the site should not be around quiet zones.
3. Dust is inevitable under certain weather conditions. Sites should be laid out so the dust stays inside the landfill boundaries.

F. Soil Conditions
1. Less-suitable soils can sometimes be improved. Poor soil areas or low areas could be built up and improved if good-quality topsoil is used.
2. Cover material may have to be brought to site. If the cover material is of poor compacting quality or of insufficient amount, cover may have to be brought in.

G. Groundwater
1. Location of groundwater table and proximity to surface. Put as much distance between the bottom of the fill and the water table as possible. Groundwater tables are seasonable. Leachate from the fill can contaminate water sources.
2. Leachate from fill. Have a thorough drainage system to handle and treat leachate and run-off to keep it from contaminating water sources.

H. Access to Site
1. Preferably over high-speed, unrestricted routes with easy on–off access in both directions.
2. Laid out to eliminate crossing of traffic and consequent tie-ups. At the entrance of a site, large trucks and autos pulling small trailers will be entering and leaving. There should be easy access with well-marked instructions. Waiting space on site for scales and parking space for employees' automobiles and stand-by equipment parking and waiting spaces should be close to but away from the site entrance. This would prevent traffic tie-ups, accidents, plus thefts from employees' cars and equipment.
3. Traffic controlled by signs and, if necessary, traffic control lights. The public is always

confused by the layout of landfills. Signs should be used to control and direct their every movement.

I. Legal Aspects
1. Jurisdiction, or lack of same, in any area for solid waste disposal. A fill should only be located inside of the operating agency's jurisdiction as to be able to maintain complete control.
2. State, county, and local laws. The fill should comply with all state, county, and local laws.
J. Public Opinion. Public opinion toward sanitary landfilling is generally negative and the term "sanitary landfill" is synonymous with open dump. A publicity campaign is money well spent.
K. Political Considerations
1. Political considerations may range from lack of political support to lack of authority. The lack of either one can mean the difference between the support a landfill gets.
L. Climatic Conditions
1. Wind
2. Rain or snow
3. Temperature. The number of days of wind, freezing temperature, rain, or snow should be known. The way a fill is planned, the kinds of roads needed, amount and placement of cover, and the types of shelters constructed depends on this type of data.
M. Ultimate Land Use. The final land use dictates the type and depth of final cover needed, and the required amount of compaction of the solid waste.
1. Parks and playgrounds
2. Industrial sites
3. Agriculture
N. Site Cost
1. Cost per acre. The true cost of a site should be calculated on the cost per usable acre of landfill site.

APPENDIX 11.9 EVALUATION OF SOLID WASTE BALING AND LANDFILLING*

Ralph Stone and Richard Kahle
Ralph Stone and Company, Inc.
Los Angeles, Calif.

Baling of solid waste has been proposed as a method of reducing transportation costs, improving landfill operations, extending landfill life, minimizing environmental impacts by reducing collection vehicle travel to distant landfills, facilitating waste transfer and resource recovery operations, reducing landfill volume requirements, reducing long-haul transportation and landfill costs, and providing structurally stable filled land. Baling consists of compressing solid waste in a mechanical baler to reduce the volume and obtain a dense bale suitable for transportation and landfilling. Baling as a technology has a long history of application in the scrap and salvage industries.

Two recent projects have evaluated baling. Funded by USEPA, System I was on a full-scale, high-pressure baling plant and landfill in St. Paul, Minnesota. System II, to investigate low-pressure baling combined with shredding, was cosponsored by USEPA and the City of San Diego, California. Wire ties were required for the latter.

11.9.1 Performance Evaluation, System I

The average number of high-pressure bales produced per hour in St. Paul was initially 18.1. After modifications to the baling chamber, the production rate increased to 22. Corresponding average daily bale production was 256 and 341, respectively, for two 8-hr shifts per day, five days per week.

The average weight per bale was 2820 lb (1282 kg), with a standard deviation of 110 lb (50 kg). Average daily quantities of solid waste processed at the two hourly rates were 25.5 tons (23,200 kg) and 31 tons (28,200 kg) per hour, respectively. The corresponding daily totals were 360 tons (328,200 kg) and 480 tons (436,400 kg). Less than 1% of the bales produced had to be rebaled due to imperfections.

Volume expansions at 1 hr, 1 day, and 1 week after production were 7.4%, 28.4%, and 24.6% respectively. This indicates that many void spaces in a bale landfill will be eliminated by subsequent expansion. Bales require solid wastes with a minimum moisture content of about 30% to remain stable.

*From *The APWA Reporter*, Vol. 43, No. 10 (October 1976). Based on a paper presented at the 1975 International Public Works Congress and Equipment Show, New Orleans, Louisiana.

The percentage of time that major equipment was idle was as follows: labor—30, conveyors—49, scale—38, baler—20, and bale transport trucks—29. Interference between the conveyor and baler caused the baler to wait for the conveyor 13% of the time. Since the baler is the key production unit, the conveyor interfered with production and accounted for a majority of the baler idle time.

The percent time that workers were idle was: gateman—55, loader operator—25, sorters—84, and truck drivers—31. Thus, improved labor task assignments could reduce labor idle time.

Reliability of the baling plant was 82%; the plant operated an average of 6.5 hr per 8-hr shift. Downtime includes one-half hour per day to clean waste from the conveyor and baler. Major causes of maintenance downtime include cleaning the conveyor and the scale platen, and malfunction of the platen.

Average time to load a transport truck was 41 min, and the time to unload at the landfill, 12 min. An additional 10 to 15 min were required to clean loose waste off the trailer and secure trailer curtains and tailgate.

Due to irregular dimensions and different sizes, bales placed in the landfill left an average void space of 6%. These void spaces were filled with loose solid wastes from the few broken bales or eventually with cover soil. About 2.2% of the bales broke during handling.

11.9.2 Performance Evaluation, System II

Average hourly bale production was 13, but the average per day was 44 due to intermittent pilot plant operation. At an average weight per bale of 1.8 tons (1636 kg), the plant actually processed 85 tons per day (TPD) (77,300 kg). These values are low because the plant was operated intermittently as a pilot test facility.

Average volume of the bales was 1.2 yd^3 (0.92 m^3). Since these bales were tied, no springback was measured.

No quantitative data were available. A review of available production data indicates that the plant produced at about 75% of its effective production rate.

Major causes of downtime include maintenance of the shredder, vertical bucket elevator, and the baler, and, for the shredder, replacement of hammer surfaces. The goal for satisfactory operating efficiency was shredding 1200 tons (1.1 million kg) between hammer changes. Labor productivity was about 1.5 man-hours per ton of solid waste baled.

11.9.3 Environmental Evaluation

A test cell was constructed containing 1 week's production of System I bales totaling 2179 tons (1.98 million kg). The test cell was 90 × 110 ft (30 × 33 m) at the base and 16 ft (5 m) deep. The test cell was monitored for settlement, gas, leachate, and temperature. The landfill operation was monitored for bale placement and landfill density, cover soil, litter, odor, and vectors.

11.9.4 Landfill Characteristics

Approximately 6% of the volume occupied by a three-bale depth of solid waste bales was joint void space. The joint void space varied from negligible in the bottom layer of bales (because of loading compression) to 12% in the top layer.

The range of densities per bale and baled solid waste for System II is much less than for System I, indicating the effect of negligible expansion due to the restraint of the bale tie wires. Baling achieves a minimum of 60% (System II) increase in effective landfill density over normal unprocessed solid waste when cover soil quantities are included.

Measurements of settlement on the test cell bales from System I showed minor expansion during the 1 year of field monitoring. Baling initially achieves the compaction density that could be expected in a sanitary landfill with conventional unprocessed waste after many years of settlement.

Test cell leachate was significantly lower in BOD$_5$, TDS, NO$_3$, and sulfides than at normal landfills. By comparison, leachate from a shredded waste fill (the Madison, Wisconsin, fill) was probably higher in total dissolved solids (TDS), but complete analyses were not found for this location. The percolation rate in a high-density balefill is slower than for traditional waste fills. Also, the percentage of rainfall percolating through is smaller than that of traditional fills.

11.9.5 Gas Generation

Gas samples were taken from the test cell at several points and levels. Carbon dioxide and methane levels increased with time, and oxygen levels decreased at all sampling points and depths. This trend was more pronounced at lower depths. Significantly, the greatest hydrogen-sulfide gas concentration measurement was less than that encountered in normal landfills. No firm conclusion can be drawn concerning gas generation rates, however, since no special tests were made. It is possible that the greater density in the balefill inhibited decomposition and gas generation, thus explaining the lower concentrations. But a "chimney effect" might also have occurred in the balefill, allowing gas to escape faster.

11.9.6 Vectors

Except for flies, birds were the only vectors observed. This contrasts with the normally greater vector problems of traditional fills. Rodents were never seen at the balefill. Significant numbers of flies were observed emerging from special fly traps on the surface. Fly larvae survive baler pressures. There is no reason to assume fly emergence per unit of area is greater at balefills; for a given volume of waste the surface area is less, and it seems reasonable to conclude that fly emergence for equal weights of raw solid waste is a less significant problem at balefills.

11.9.7 Litter, Dust, and Noise

Most of the litter at the balefill originated from: (1) bales which were broken as they were lifted off the transport truck and positioned, and (2) residual waste which was swept off empty transport truck beds. In general, baled litter remained near the working face; litter fences were not needed. There was far less litter at the balefill than at a conventional landfill.

The greater density of bales reduced the number of trucks required to haul the wastes. Hence, far less dust, noise, vehicular air pollution, and landfill traffic and safety hazards result at the balefill. Odors were not as great as at conventional landfills.

11.9.8 Costs

The total cost for bale landfill operation averaged $0.97/ton over a 21-month period—less than half the cost of a comparable, normal landfill.

Investment and operating costs are greater for the baler system, but the potential environmental benefits are also large. The System I plant in 1974 would cost about $1.3 million. Appurtenant transport and balefill facilities would be additional.

11.9.8 Benefits of Solid Waste Baling

Using a baler as part of a solid waste management system offers the following advantages:

Extends the landfill's useful life by increasing the in-place density by about 60%.

Improves the cost-effectiveness of local solid waste collection and provides a transfer station for better long-distance hauling and disposal.

Increases resource recovery opportunities by providing a central transfer facility that can incorporate materials separation and reclaimed product baling.

Reduces negative environmental impacts at the landfill including negligible settlement, and reduced litter, dust, odor, vectors, fires, traffic, earth addition, noise, pollution, and safety hazards.

Reduces the cost and improves the operating efficiency of the landfill by requiring less work equipment, personnel, and cover material while improving operating standards.

Increases the potential usability of the finished site by improving the foundation bearing values and reduces the landfill stabilization waiting time needed.

The first two potential advantages (longer landfill life and more efficient long-distance hauling) result from the great volume reduction achieved with high-density baling. The denser the solid wastes, the more that will fit into a transport vehicle and into the landfill. Effective landfill density for a traditional fill is 236 to 438 lb/yd^3 (140 to 260 kg/m^3) and for a balefill 1103 to 1416 lb/yd^3 (655 to 840 kg/m^3). Land costs are a relatively small part of total system costs even at $16,190/acre ($40,000/ha). The cost per unit of solid waste is relatively low—9 to 14 cents/lb (20 to 30 cents/kg) and are therefore not a major determinant in the decision to adopt baling. Transport cost is a major system cost, 7 to 9 cents/lb mi (15 to 20 cents/kg km) for significant hauling distances. Distances greater than 18 miles (30 km) can make a transfer and processing operation such as baling economical. In other words, the major saving from increased landfill life results from avoiding use of more distant sites. Additional savings of reduced long-distance transport costs via rail, barging or trucking also accrue.

Shredding, combined shredding, low-pressure baling, and incineration are other state-of-the-art volume-reduction techniques. Incineration is the most expensive alternative and can present air-pollution problems, but it does provide greater volume reduction and the potential for energy recovery. A comparison of shredding and shredding/baling with high-pressure systems shows a slight cost advantage for high-pressure baling. The final cost decision will depend on local solid waste management needs.

For equal weights, high-pressure baled solid waste requires only about 60% of the landfill space required by unbaled waste; the percent reduction in soil cover is even greater since baled refuse requires only 40 to 50% of the earth required for unprocessed waste for the same depth of soil cover.

Increased salvage opportunities are associated with all solid waste processing and transfer operations, baler-based or not. Corrugated paper and metals were hand-sorted from the waste stream at the St. Paul operation. The separated material was processed separately at the baler, and these reclaimed material bales were sold and reclaimed. After deductions for the sorters' labor, baling, and transport costs, this resource reclamation operation returned a profit.

High-density baling and balefills offer unique environmental benefits. Settlement, noise, traffic, noxious leachate production, birds, vector attraction, combustible gas generation, fire, litter, dust, noise, vehicular air pollution, and odors are environmental problems common to traditional solid waste sanitary landfills. Costly solutions exist for reducing these problems, but local conditions such as lack of cover soil, high groundwater, and proximity to developed areas may make the cure prohibitively unacceptable. In such cases, however, baling can upgrade an otherwise problem site. High-density baling can produce a fill with maximum density and minimum observed settling, low rates of combustible gas generation, dilute leachate, minimum landfill traffic (reduced noise, dust, and the chance of accidents), reduced labor and equipment requirements and costs, more efficient transport, and minimal littering problems. A stable, final landfill area can be provided almost immediately after the fill is completed. Vector (birds, rats, flies, etc.) attraction hazards are significantly reduced at operating balefills because of the smaller working face and the ease of providing a complete daily soil cover. The results indicate that bale landfills are more esthetic and environmentally superior to traditional sanitary landfills.

Baling is obviously feasible in large communities since economies of scale exist. The System I baling plant processed 360 TPD (328,000 to 436,000 kg). The System I plant served a city of 310,000 in competition with conventional landfills. Balers smaller than the System I baler are also available. One example of solid waste baling in a small community is Chadron, Neb., a community of 6000 residents. The Chadron baler processes 350 tons (320,000 kg) per month even though the baler capacity is 34 tons (31,000 kg)/hr. The baler has enabled Chadron to conserve a close-in landfill that results in a reported annual savings of $7000 due to reduced hauling time and cost over the previous 5 mi (8 km) distant landfill operation. Chadron also recovers metals (aluminum, copper, and ferrous), newsprint, and corrugated cardboard. Thus, the economic feasibility of solid waste baling may also apply to smaller communities. The feasibility of using baling should be evaluated for each community or region regardless of population size in order to minimize overall solid waste management system costs and related environmental impacts.

APPENDIX 11.10 ENGINEERING STUDY OF BALED SOLID WASTE AS FOUNDATION MATERIAL*

Roger G. Slutter
Lehigh University
Bethlehem, Pennsylvania

11.10.1 Introduction

Tests were conducted at Fritz Engineering Laboratory on fives bales of solid waste obtained from American Hoist Company. These bales were tested in compression by methods that would produce data sufficient to establish the properties of this material as a foundation material. The determination of this type of information for solid waste material has not been attempted prior to this by any investigator.

The engineering properties of landfill material has not been attempted because the material that has been placed in landfills by conventional landfill methods cannot be studied except in place after the landfill has been completed. Studies of foundations on completed landfill sites has been an expensive and time-consuming problem because virtually nothing is known about the material in the fill. The development of information has been the result of experience over a long period of time. Landfill material has been found to be a very nonuniform material when placed by conventional methods. The reasons for this are obvious when one considers the great effect that weather and climate have on the placement and compaction operations.

Source: Fritz Engineering Laboratory Report No. 200.74.562.1 Lehigh University, Bethlehem, Pennsylvania.

The important step that has been developed in the baling of solid waste is to produce a relatively uniform material that has properties that can be determined and held relatively constant throughout the year. The density of the material is uniform from bale to bale and can be kept constant throughout the year. Placement of these bales tightly in the landfill site so that expansion cannot take place will result in a uniform foundation material that can be evaluated by the determination of properties from laboratory tests. These properties can then be used in designing foundations for structures to be placed on the site.

Much of the experience with poor foundation conditions that has been encountered with landfill sites has been the result of using poor soil that does not compact as a cover material. This results in a situation in which the landfill material can never be uniform. The stacking of bales of fairly uniform size and shape tightly together will produce a more uniform foundation material than that obtained by the older methods. It has been found that the bales will expand if they are not confined. This property is important in two ways. First, it is important to move the bales into place in the landfill site soon after being produced to obtain the maximum density in place. Second, it is beneficial that they expand to fill the space between bales so that it is unnecessary to attempt to fill with soil between bales. The expansion will result in some reduction in density, but this will not change the properties of the material significantly if bales are stacked with only a small space between bales.

The tests conducted at Fritz Engineering Laboratory were conducted on bales that had already expanded during shipment from St. Paul, Minnesota, to Bethlehem, Pennsylvania. Therefore this testing program was completed on material that represents the expanded material in the landfill site. In fact the density achieved in a good operation should result in higher densities than those found in the bales after shipment to our laboratory. For this reason the results presented in this report are conservative.

11.10.2 Testing Program

The objective of this program was to determine some of the basic engineering properties of the material that could be used in evaluating foundation design for structures to be built on a landfill. In development of methods for obtaining bearing capacity, shear strength, permeability, density, and leachate generation, the scope of our approach was extended to include a complete approach to the development of engineering properties and settlement analysis. The amount of data that we are able to provide in this report will show that a simple testing procedure can be used to determine the required information for foundation analysis.

The material that was shipped to us expanded during shipment, so that the density of the material as we tested it was significantly less than the density as baled. In the following table the density that we found and the probable density immediately after baling are given:

Bale No.	Dimensions at Fritz Laboratory (in.)	Density as Found in lb/ft^3	Probable Density after Baling in lb/ft^3
1	41 × 40 × 43	41.6	60.8
2	38 × 38 × 38	50.4	66.7
3	40.5 × 42 × 49	45.5	63.5
4	47 × 39 × 59	40.0	57.1
5	48 × 38 × 54	40.4	63.9

The actual density when baled is not known for the first three but the density of the last two checks with data given by the American Hoist Company. A density between the two values given in the above table would be achieved in an actual landfill operation. A density near the higher value should be achieved in a good operation.

Bales 1 and 4 were tested in compression. In this test the unconfined compressive strength was determined following ASTM D2166 procedure. The bales were compressed in our largest testing machine to determine the load that would produce failure or 15% deformation. In testing Bale 1 we were unable to follow the ASTM D2166 procedure exactly because we had no idea what to expect. Therefore we are using the results from testing Bale 4 in this report. The results from Bale 1 were somewhat higher. The unconfined compressive strength was 4330 lb/ft^2 from Bale 4.

Bale 2 was tested by supporting the bale at the ends and loading in the center as a beam using 6 × 6 in. (15.2 × 15.2 cm) timbers for supports and loading point. This bale failed at a load of 15,000 lb (6802.8 kg). From this I estimated that the tensile strength of the material was approximately 500 lb/ft^2 (24 KPa). From this test we realized that the material had a dependable tensile strength but we were not satisfied with this method of measuring it.

We used Bale 5 to determine the tensile strength by the double-punch method. In this test we supported the bale on a steel punch 12 in. (30.5 cm) in diameter and pushed a punch of similar diameter into the top. The test was performed following the ASTM D2166 procedure and the tensile strength was computed as 520 lb/ft^2(25 KPa). This test has been successfully used for soils, concrete, rock, and various types of stabilized materials in our laboratory. The test on the solid waste worked out very well and we are confident of the results. The actual load that produced 15% deformation was 7860 lb (3565 kg) on the 12-in. (30.5 cm)-diameter punch.

All of the above tests reached the failure criterion due to deformation and did not actually fail when the deformation was carried to a 20% deformation. From these results we determined basic engineering properties as follows: Angle of internal friction = 19° and cohesion strength = 1310 lb/ft^2 (63 KPa). From these values and the unconfined compressive strength and tensile strength given above we were able to construct a Mohr Circle failure analysis. This analysis results in the following formulas for shearing strength and bearing capacity:

Shearing strength = $c + p \tan \phi$ where c is the cohesion strength given above, p is the overburden pressure at the point being considered in the fill, and ϕ is the angle of internal friction given above.

Bearing capacity = $1.35\gamma B + 8.8\ c + 4\ q'$ using the Terzaghi-Meyerhoff formula in which γ is the density of the material, B is the width of a footing, c is the cohesion strength, and q' is surcharge pressure.

It may be possible to develop a better formula for bearing capacity. However, with the limited data available at this time, the above formula is the best available.

Although the permeability test has not been completed, this will be done on Bale 3. We expect that our permeability test will verify the following data obtained from a report by American Public Works Association Research Foundation:

Density after Compaction lb/ft^3	Coefficient of Permeability ft/day
35.8	42.6
49.0	13.6
52.2	10.0
71.0	2.0

In attempting to conduct the permeability test on Bale 3 we found that a complete bale has channels of flow within it that were produced in baling. These channels opened due to expansion and allowed water to flow through the bale. The permeability of the bales of density equal to 49.0 lb/ft^3 would be classified as low permeability by comparison with natural soils. Because of the channels that apparently exist in a bale the material may actually behave more like medium-permeability natural soils. The flow channels would undoubtedly be closed by the tendency of the material to expand in an actual site because the bales are confined. For this reason the flow channels would not be detrimental in a dike or dam. Bales should be placed so that the joints do not line up when constructing dikes. At least three rows and preferably four rows should be used in this type of construction to provide sealing of all possible flow channels.

We also used Bale 3 to determine a time versus rate of settlement curve at a bearing stress of 1000 lb/ft^2 (48 KPa). These data can be used in the development of settlement characteristics of the material in conjunction with the density and permeability data.

Data on the effect of the variation of density on the bearing capacity, total settlement and the time versus rate of settlement for structures is not available at this time. The testing procedures outlined in this report will serve as a guide in developing more data for use in foundation design.

11.10.3 Analysis of Results

The results obtained in the tests of the bales of solid waste enable us to reach certain conclusions relative to their behavior compared with natural soils. The material is better than many of the poor soils found in nature. The value of the angle of internal friction is about the middle of the range for natural soils. The tensile strength is about the lower third point of the range for soils. The compressive strength of the material is in the range of medium to soft clays. The bearing capacity formula developed will give results comparable to the lower bearing capacity values used with medium clays.

The engineering properties of baled waste are better than one would consider for landfills placed by conventional methods. Often the cover material for landfills that has been used in the past is not as good as the bales of waste and is not easily compacted because of the rebound of the waste material. The greater density achieved by baling and the uniformity of the bales both contribute toward a better foundation material than previously achieved in landfills.

Obviously the material of the bales is similar to that placed in the older type of landfills and it has some of the same characteristics. These must be considered in the development of the land site for future use. Certain of the precautions used in regular landfills such as the use of a cover material must be followed. The minimum cover should be 2 ft (0.6 m) of suitable compactible material.

The use of a site for buildings will require the careful consideration of the settlement and the time rate of settlement. The settlement of a landfill with bales should be less than the best experience with other landfills having the same depth of fill. This means that the expected settlement is probably less than 10% of the depth within a period of 6 months to 2 years. The differential settlement should also be less than for conventional landfills. If the fill is wet, the amount of settlement may be slightly greater but the time of settlement may be less. Since each site is essentially different with regard to depth of fill, groundwater conditions, rate of placement of fill, and operating conditions, observations at the site are necessary to determine the actual rate and magnitude of settlement.

Immediately after the completion of filling operations the land can be used for recreation, agriculture, parking lots, forests and small airports. Some future fill will be necessary to maintain drainage and eliminate low spots. Flexible pavements should be used in the area until after the settlement has taken place. Flexible pavements for parking lots, roads and airports are probably the most economical and most easily repaired for long-term development as well.

The construction of buildings can proceed immediately if pile foundations are used. The driving of piling through a landfill is not a simple operation. However, with a shallow depth of fill it can be done. The best approach to this problem is probably to use the heavier "H" pile section with driving tips. With the benefit of the additional weight and driving tips, it should be possible to drive these relatively straight. Small buildings can also be constructed immediately if special foundation design is employed. Another method available for immediate construction is the use of compacted earth fill under the building.

Light commercial and industrial buildings can be built on landfills made up of bales after 2 years without piling or special foundation. Although the bearing capacity is rather low, it is of sufficient magnitude for economical construction. Large, spread footings are satisfactory and slab foundations are better suited if the type of building lends itself to this type of foundation. Long, narrow footings should be avoided because of the danger of differential settlement and tilting. Simple steel-frame buildings can successfully endure a differential settlement and tilting. Simple steel-frame buildings can successfully endure a differential settlement of up to $0.005L$ where L is the distance between two adjacent columns under consideration. These buildings are the most suitable for early construction on landfills. Other types of buildings require that the differential settlement be limited to $0.002L$ to $0.004L$. These buildings may be constructed also, but only after settlement experience for the site is available.

The entire problem of designing foundations for buildings on a landfill site requires a complete analysis of conditions and types of construction required at the time that the site is available. Structures that result in heavy column loads, loads that produce vibration in the columns, or columns with large moments will require pile foundations. Industrial buildings with cranes up to 5 tons and all commercial buildings of not more than three or four stories can be built on shallow footings. The time schedule for building is affected by groundwater conditions, depth of fill, and settlement experience on the site. The use of an overburden fill to speed up settlement can also be considered.

Our experience has convinced us that the experience gained from bale sites will be far better than experience with conventional landfill sites. The properties of bales are such that they can be used in building dikes, small dams, fills for caissons, and foundation mats for swamps. Low permeability and dependable engineering properties are the important factors that make their use possible. As developments in the use of the baling process occur and experience with the utilization of completed sites becomes available the use of baling will undoubtedly expand.

BIBLIOGRAPHY

1. H. Y. Fang and Hirst, T. J., A Method for Determining the Strength Parameters of Soils, 52nd Annual Meeting, Highway Research Board, Washington, D. C., January, 1973.

2. H. L. Hickman and Song, T. J., Sanitary Landfill Facts, United States Department of Health, Education and Welfare, Public Health Service Publication No. 1792, Washington, D. C., 1968.

3. T. W. Lambe and Whitman, R. V., *Soil Mechanics,* Wiley, New York, 1969.

4. G. A. Leonards, *Foundation Engineering,* McGraw-Hill, New York, 1962.

5. R. C. Merz and Stone, R., Landfill Settlement Rates, Public Works, Vol. 93, No. 9, September, 1962.

6. C. A. Rogus, Use of Completed sanitary landfill sites, *Public Works* **91,** No. 1, January, 1960.

7. G. B. Sowers and Sowers, G. F., *Introductory Soil Mechanics and Foundations,* 3rd ed., Macmillan, New York, 1970.

8. T. E. Winkler, Compaction, Settlement of sanitary landfills, *Refuse Removal Journal,* December, 1958.

CHAPTER 12

RESOURCE RECOVERY: PREPARED FUELS ENERGY AND MATERIALS

DAVID J. SCHLOTTHAUER

Henningson, Durham & Richardson, Inc.
Omaha, Nebraska

GEORGE E. BOYHAN

Waste Technologies International
Palm Beach, Florida

WILLIAM D. ROBINSON

Consulting Engineer
Trumbull, Connecticut

KENNETH L. WOODRUFF

Resource Recovery Consultant
Morrisville, Pennsylvania

JAY A. CAMPBELL

Henningson, Durham & Richardson
Alexandria, Virginia

GORDON L. SUTIN

Gordon Sutin Consultants Ltd.
Dundas, Ontario, Canada

DAVID G. ROBINSON

Waste Survey Consultant
Trumbull, Connecticut

E. JOSEPH DUCKETT

Schneider Consulting Engineers
Bridgeville, Pennsylvania

ANTHONY R. NOLLET

ROBERT H. GREELEY

Aenco, Inc.
Albany, New York

12.1 ENERGY RECOVERY OVERVIEW, PROCESSED FUELS
David J. Schlotthauer

The interest in energy and materials recovery from prepared solid waste fuels began in the late 1960s and early 1970s. This interest resulted as part of an effort to: (1) develop alternate methods to landfilling and reduce the quantity of waste being landfilled; (2) increase the recoverability and quality of recyclable materials (glass, aluminum, ferrous metals, etc.) from the solid waste; (3) develop a more energy-efficient method than mass burn incineration; and (4) develop a homogeneous fuel product derived from solid waste that can be burned in existing boilers thus reducing the cost of solid waste facilities.

The interest in prepared fuel systems was also helped by the enactment of the Resource Recovery Act of 1970. The Act provided funds for the demonstration, construction, and application of solid waste management and resource recovery systems. The energy crisis in 1973 played a major role in developing the interest in energy recovery from solid waste. With the increasing costs of fossil fuels, alternate energy sources were sought, of which solid waste was one.

Because of this increasing interest in energy and materials recovery from solid waste in the United States and Canada, research and pilot programs were undertaken to develop technology, systems, and plants to produce a homogeneous fuel from solid waste and at the same time recover recyclable materials. These programs led to a wide variety of systems and plants for this purpose; some of these facilities were successful whereas others were failures.

The initial stages of the development of the technology for the processing of and recovery of energy from solid waste used equipment from other industries such as the sugar cane, pulp and paper, lumber, and automobile shredding industries. Because these industries have been successfully processing wastes—bagasse from sugar cane stalks, bark and wood wastes from the lumber and paper industry, and scrap metals from automobile shredding—that had similar characteristics to municipal solid waste, the application of these technologies seemed to be a logical step. However, because of the heterogenous nature of municipal solid waste, it was soon recognized that some of this technology would have to be modified or abandoned and an alternate technology developed.

The concept of preparation and burning of prepared waste cellulose fuels was first used by the pulp and paper industry and sugar cane industry in the late 1930s for preparation of tree bark, woodchips, and bagasse for firing in spreader stoker-fired boilers. These materials were shredded with hammermill-type shredders (hogs). The shredded material was transferred to a storage facility and then fed to boilers for the production of steam and electrical energy.

The first facility that produced a prepared fuel from municipal solid waste and fired the fuel to recover energy was the St. Louis-Union Electric demonstration project.[1,2] This project was a joint venture between the City of St. Louis and the Union Electric Company to determine the feasibility of preparing a suitable fuel product from raw refuse which could be cofired in an existing utility boiler. The project, partially funded by the Environmental Protection Agency, was initially designed with a simple processing facility containing a shredder followed by a magnet for ferrous metal removal. The prepared waste (refuse-derived fuel [RDF]) was loaded into trucks and transported to a receiving and unloading facility at the power plant. The fuel was unloaded from the trucks and pneumatically conveyed to a storage silo. From the storage silo the RDF was fed to the boilers by another pneumatic conveying system. After approximately 1 year of operation, an air classification system was added to the process facility to improve the fuel product. A schematic of the St. Louis facility is shown in Figure 12.1.

The facility operated successfully until it was shut down. The operation showed that the processing of municipal solid waste to produce a fuel and the recovery of energy from this fuel was indeed viable.

Subsequent to the St. Louis project, the solid waste industry has taken numerous paths in developing technology and facilities for the production of a RDF and the recovery of energy and materials from municipal solid waste. During the 1970s, the processing plants became increasingly complicated and complex in an attempt to produce a higher quality RDF product and at the same time recover more and more materials such as glass, aluminum, and nonferrous metals from the waste stream.

Prepared fuel-processing systems were conceived, designed, built, and tested to produce RDF ranging in composition and particle size from 4.5 inches (114mm) down to a powdered product having a size less than 200 mesh. Other facilities were built to produce oil, gas, and char by the pyrolysis of the solid waste. Still other facilities were constructed to pelletize the RDF. All of these processes were efforts to produce a homogeneous fuel product that could be efficiently burned in conventional boilers, either alone or cofired with other conventional fossil fuels.

Processing plants added multiple stages of shredding, various types of screening, and air classification. As the plants became more complex, they also grew larger to handle larger quantities of solid waste. It was theorized that the larger the facility became, the more economical, on a dollars per ton basis, it would be to provide resource recovery. This all led to materials handling difficulties, higher operations, and maintenance costs and facilities that did not perform up to expectations. Some facilities were failures, other facilities required extensive modifications, whereas others are operating successfully.

Despite the fuel preparation systems, RDF sometimes had a lower quality than expected. The heating value was lower and the ash content higher. This was because the fuel still contained glass, sand, metals, and a high moisture content.

Problems were encountered in the feeding and firing of RDF in boilers. In existing suspension-fired units dump grates were added. The high ash content overloaded ash-handling systems. Furthermore, if not prepared properly, RDF was difficult to store, reclaim, and feed, with bridging in silos and excessive wear on feed mechanisms. Pneumatic transport systems, which would appear to be a good method for transporting the waste, had problems due to wear caused by the abrasive nature of unclean RDF. Boiler tube corrosion problems similar to those experienced in mass burn units occurred in units fired with RDF as a result of the high chlorine content of refuse fuel, likewise mass burn units.

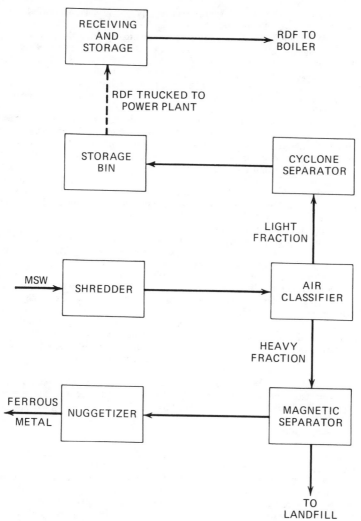

Fig. 12.1 Flow line diagram of the St. Louis demonstration project.

As citizens became more ecologically aware and interested in conserving natural resources, separation systems were devised and added to processing plants in an attempt to recover valuable material from the solid waste stream. Eddy current separators ("aluminum magnet") were developed to recover aluminum. Extensive systems, such as froth floating and optical sorters, were used to recover glass and separate it by color. Still other systems were developed to recover valuable nonferrous metals such as lead, copper, zinc, and brass. All of these processes added to the cost of the facilities and to the operational problems already encountered. The materials recovered were of low quality and often were not salable. This added to the economic problems by reducing the anticipated revenue from resource recovery facilities and adding to the capital cost.[3]

An additional problem, which was not originally anticipated, was encountered during the operation of the first resource recovery facilities. This was the problem of shredder explosions that were caused by gasoline cans, propane tanks, and other items in the waste stream. Safety precautions were added to the processing line to prevent explosions and reduce their severity if one did occur. Blowout vents, suppression systems, and explosion mats were added to shredders. Hand picking stations were added to the lines in an attempt to spot and remove potentially hazardous items. These precautions have added to the capital and operations costs of facilities.

Because of these operational problems, the resource recovery industry received a bad reputation, with people becoming skeptical about the viability of these types of facilities.[26] Some cities and municipalities were left with facilities that did not operate and some private companies abandoned their ventures in the solid waste business.

In fact, the economic viability of resource recovery projects, including mass burn, has been over estimated in many cases. Plant capital and O & M costs have increased above the anticipated costs made in early feasibility studies. The revenues anticipated from the recovery and sale of recovered material and from the energy conversion of RDF were overstated. Also, many of the facilities have not been able to operate at design capacity.* These factors have resulted in higher tipping fees or in tax increases to support the operation of the facilities.

Even with these major problems and the negative publicity (which parallels the historic struggle of mass burn here and abroad),[26] facilities designed and built to produce a RDF remain as a viable and economic alternative in the resource recovery industry. Successful plants have renewed the interest in prepared fuel facilities that are currently in the design and construction stages.

Facilities are using picking stations to remove hazardous materials before they enter the shredders. Also, trommels and disc screens are being added ahead of shredders to remove glass, grit, metals, and other abrasive materials. This reduces the wear on equipment and improves the quality of the fuel. If shredders are installed as the first stage, they can be coarse-type shredders (flail mills) for breaking open bags.*‡ Magnetic separators are used to remove ferrous metals prior to shredding and from the final RDF product.[3]

New facilities have been installing dedicated boilers specifically designed to burn RDF. Most of these units have been spreader stoker-fired boilers in which part of the fuel is burned in suspension and the heavier particles burned on a traveling grate.

The modification of existing boilers for cofiring still remains as an economic alternative. The use of existing boilers can greatly reduce the capital costs of energy recovery from prepared fuels. A good example of this is the General Motors Corporation Truck/Coach Division installation in Pontiac, Michigan as described in Chapter 16, Case History 1.

In evaluating the feasibility of RDF systems, the owner and engineer must closely compare boiler efficiencies, boiler island costs and emissions with the alternatives. Along with furnace fireside corrosion and deposits problems common to all, he/she must compare the operating problems and maintenance costs of the accessible front end RDF process system with those of say the mass-burn grates and bridge cranes. The owner and engineer should consider the installation of redundant processing lines[27] to improve the availability and reliability of any thermal processing facility. Also, the costs must be closely evaluated so that capital costs are not underestimated and revenues overestimated.

The following sections provide a general overview of prepared fuels resource recovery facilities. The operating history of existing plants is discussed to provide a general idea of what technology has and has not worked. Also, various types of combustion and process systems are discussed.

12.1.1 Dedicated Units

In the context of this discussion, these are facilities designed, built, and operated for a specific purpose, in this case, facilities dedicated solely to the processing of solid waste into a fuel product and the recovery of energy by the combustion of this fuel product. Dedicated prepared fuel-processing facilities and energy recovery facilities are generally designed and constructed at the same site. However, because of specific requirements and restrictions, such as having an energy customer in the center of a city but with limited space available, the processing facility may be located separately from the energy recovery facility. Dedicated units can be of all sizes and configurations but they will have one common purpose, the processing of and recovery of energy from solid waste.

An advantage of dedicated units is that the entire system and components have been designed and are operated solely for specific conditions and needs. A disadvantage of a dedicated unit is not realizing the cost savings resulting from the utilization of an existing utility or industrial combustion system modified for supplemental firing.

Dedicated prepared fuel energy recovery plants presently operating are shown in Table 12.1.

12.1.2 Modification of Existing Units

Existing industrial and utility combustion systems can be modified for the recovery of energy from prepared solid waste fuels. This is the most economical method for the recovery of energy from prepared fuels. The use and modification of existing equipment and facilities will generally reduce the capital costs of a prepared fuels (RDF) resource recovery facility.

The existing industrial and utility facility boilers that are best suited for modification to prepared

---

*Editor's note: Resources Recovery (Dade County) Inc. is a notable exception (See Sections 12.4.2 and 12.4.3).

*‡Editor's note: Although these precepts are becoming popular after Madison, Banyan-Dade, etc., there are skeptics for applications elsewhere with wider ranges of raw material.

Table 12.1 Dedicated Prepared Fuel Energy Recovery Facilities

Plant	Design Capacity (TPD)	Products
City of Akron, Ohio[a]	1000	Steam for process, heating and cooling, ferrous metal
City of Albany, New York	750 ton/shift	Steam for heating and cooling, ferrous metals, and ash
City of Hamilton, Ontario	500	Electricity and steam
Dade County, Miami, Florida	3000	Electricity, glass, ferrous metals, aluminum, nonferrous metals, and ash
Occidental Chemical Company, Niagara Falls, New York	2000	Electricity, steam, and ferrous metals
Refuse Fuels, Inc. Haverhill, Massachusetts	1300	Steam and electricity for a Lawrence, Massachusetts, industrial park
Eastman Kodak Co. Rochester, New York	120 RDF 114 sludge	Process steam

[a] Not operating since December, 1984, pending decision to rehabilitate following a major explosion and fire.

solid waste fuel firing are pulverized coal boilers and spreader stoker boilers designed for coal, bark, woodwaste, or other cellulose type fuels. These types of units generally have the required ash-handling system, air pollution control equipment, and soot blowers installed to facilitate the firing of RDF. Gas- and oil-fired boilers can also be modified for RDF firing, but they will require more extensive modifications and addition of new equipment.

During the evaluation and design of the conversion of an existing boiler or facility, the engineer and designer must consider the following basic requirements:[5]

1. The design of the furnace must be evaluated to determine if the furnace bottom is designed for ash removal and will have sufficient strength to withstand falling slag which may occur during the firing of RDF. Also, the furnace bottom on a suspension fired boiler will have to be modified for a dump grate. The dump grate is required to complete the combustion of the heavier organics. The furnace design should also be evaluated to determine that the heat release rates and gas velocities are acceptable.

2. The method of firing of the prepared fuel must be thoroughly evaluated. For existing suspension-fired or cyclone-fired boilers, the RDF will have to be pneumatically fed into the furnace for combustion. The location of the refuse entry ports must be reviewed with the boiler manufacturer. For stoker-fired boilers, the RDF will usually be fed into the furnace area above the existing fuel feeders by air-swept spouts. This will also require a design review by the boiler manufacturer. On stoker-fired units, the grate speed and design must be reviewed to insure that adequate burnout of the RDF will occur.

3. Tube spacing and gas velocities in the superheater and convections sections must be reviewed thoroughly. The tube spacing in the superheater and convection sections should be sufficient to minimize ash fouling and plugging and to permit effective sootblowing. The gas velocities through the superheater and economizer must not be excessively high because erosion of the tube surfaces may occur.

4. Sootblower location and spacing must be evaluated to permit effective cleaning of the tube surfaces and prevent bridging.

5. The air pollution control system must be adequate to handle the increased ash loading in the flue gas and still meet emission requirements. If the units do not have existing air pollution systems, new equipment for emissions control will be required.

6. An adequately sized ash handling system must be provided. On units with an existing system, the ash handling must be reviewed to determine if it has adequate capacity. If not, extensive modifications may be required to provide for the continuous removal of the ash. On gas- or oil-fired facilities major modifications will be required to add an ash handling system.

7. The plant and site arrangement must be reviewed thoroughly to determine if adequate space is available. The plant arrangement review should take into consideration the space required for an RDF feed system, which may include surge bins or silos, conveyor systems, pneumatic system, and dust-collection system; for a new ash-handling system or modifications to the existing system; for an air pollution control system; for control

modifications; and for additional auxiliary equipment and systems. The site review should take into consideration space for an RDF storage system (usually 2–3 days firing capacity), RDF processing system, RDF receiving and unloading facility if the process plant is off-site, and ash storage. The review of the site must also consider the additional truck traffic for delivery of either municipal solid waste or RDF and the impact this additional traffic will have on the surrounding area.

8. Steam pressure and temperature and flue gas temperatures of the existing boilers must be evaluated in terms of what impact the firing of RDF will have on boiler tube corrosion. Units with high steam conditions, such as 2400 psig, 1000°F (16,545 KPa, 538°C), generally are not suited for conversion to RDF firing because of the high flue gas temperatures required to produce this steam. Flue gas temperatures at the furnace exit which exceed 1600°F (871°C) and the resulting high tube metal temperatures have been attributed to the corrosion problems experienced on superheater tubes. Boilers with lower steam conditions, such as 600 psig, 750°F (4136 KPa, 398°C), are better suited for modification to RDF firing. The flue gas and tube metal temperatures are generally below the acceptable limits of 1600°F and 850°F (871°C and 454°C) respectively.

9. The boilers evaluation must include a study of whether the units will be cofired with RDF or will be modified for 100% RDF firing. Suspension-fired units being evaluated and designed for cofiring of RDF with coal generally limit the heat input of the RDF to 15 to 20%.

When evaluating the feasibility of modifying an existing boiler or facility, the manufacturers of the major pieces of equipment, such as boiler, ash handling, and air pollution control, must be included as a part of the overall effort. They can provide the expertise in evaluating the effects that the burning of RDF will have on their particular piece of equipment and in recommending what modifications, if any, will be required.

A summary of presently operating facilities that have been modified for recovering energy from the combustion of a prepared solid waste fuel is listed in Table 12.2.

12.1.3 Energy Recovery Methods and Products

The recovery of energy from prepared solid waste fuels can be accomplished by direct combustion in boilers. Energy from combustion is recovered in the form of steam or hot water. In the case where steam is generated, the steam can be used for district heating and cooling, for process steam by an industrial user, or can be expanded through a turbine generator for the production of electricity. Steam can also be expanded through a turbine generator with both electricity being produced and low-pressure steam being extracted for industrial or district heating use (cogeneration). Where hot water is produced during the combustion of RDF fuels, it is normally used for either district heating or industrial process.

Other methods for the recovery of energy consist of the generation of methane gas, oils, and chars (charcoal) from the pyrolysis or anaerobic digestion of the prepared solid waste. These products then can be burned in existing facilities to generate steam, hot water, or electricity or can be further refined to produce pipeline-quality natural gas or other petroleum-based products.

The pressure, temperature, and load or demand requirements of an existing district heating system or industrial customer will generally govern the design requirements of a resource recovery facility. This usually requires that the steam be generated at low-pressure saturated conditions or reduced in pressure and desuperheated if high-pressure, superheated steam is generated. The flow demands of the district heating system or industrial user often restrict or control the size of the resource recovery facility and are not usually compatible with the seasonal variations in the solid waste supply. Also, the location of the resource recovery facility is restricted because it has to be near the steam or hot water customer. Facilities designed for the production and sale of steam or hot water are usually in the smaller size range, 300 TPD or less.

Table 12.2 Modified Existing Facilities

Plant	Capacity (TPD)	Products
City of Ames, Iowa	200	Electrical generation and ferrous metals recovery
City of Madison, Wisconsin	400	Electrical generation and ferrous metals recovery
Baltimore Gas & Electric Co., Crane Station (cyclone unit)	400	Electricity generation from RDF prepared by Baltimore County, Maryland

Resource recovery facilities for electrical generation have more flexibility in design in that (1) the location of the plant is not restricted by the need to be near the energy users; (2) the size of the facility is governed by the solid waste stream and not the customer of the energy product; and (3) the load variations of the facility can more readily match the seasonal variations in the solid waste supply. Depending on the size of the facility and technology used the steam produced for electrical generation is in the 600 psig, 600 to 750°F (4136 KPa, 315 to 399°C), range. Facilities for the generation of electricity can be larger in size, up to 2000–3000 TPD.

Cogeneration facilities have the same advantages of electrical generation facilities with the exception of the flexibility in location. Because of the need to be near the steam customer, the location of a cogeneration facility is restricted. When designing a cogeneration facility, the engineer and designer have various types of systems from which to choose. For example, the engineer must decide on whether to use a backpressure turbine or a full condensing turbine. If a full condensing turbine is selected the engineer must then decide on whether to use an automatic (controlled) extraction or a floating pressure (uncontrolled) extraction turbine. The selection on the type of turbine generator to be used will be governed by the steam requirements of the customer. Also, in the selection of the turbine generator the size of the unit is usually selected to provide for a full generation capability in case the industrial steam customer shuts down. Figures 12.2, 12.3, and 12.4 show various cogeneration cycles.

The recovery of energy using pyrolysis or anaerobic digestion (bioconversion) of prepared solid waste is still in the experimental and developmental stages. The products from these processes—methane gas, oils, and chars—are of low quality. The methane gas has a low heating value, 300–500 Btu/ft^3 (11,025 to 18,375 kJ/m^3). The gas can be refined to pipeline quality, but by expensive processes. The oil products from pyrolysis are low grade having a heating value of 10,000 to 15,000 Btu/lb (23,000 to 34,800 kJ/kg). The chars produced will usually have a high ash content as a result of the glass, sand, and metals in the processed waste.

12.1.4 Cofiring

Cofiring is defined as the simultaneous firing (combustion) of two different fuels in the same furnace or combustion chamber. In this case, cofiring refers to the simultaneous combustion of a prepared solid waste fuel with either oil, coal, natural gas, or sludge. RDF has been used, tested, and cofired in utility boilers, industrial boilers, and cement kilns since the early 1970s. The first facility to cofire RDF was the St. Louis–Union Electric RDF demonstration project. During the testing at the St. Louis project, RDF was fired as a supplemental fuel with pulverized coal in a suspension-fired boiler. Since that time RDF has been, or is currently being fired in over 15 different facilities. Table 12.3 lists some of these facilities.

The Ames plant currently cofires all RDF in a suspension-fired boiler. During the initial operations of the plant, RDF was also cofired in two of the utility plant's stoker-fired boilers. Because the

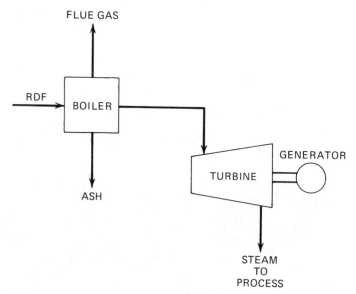

Fig. 12.2 Cogeneration topping cycle with backpressure turbine.

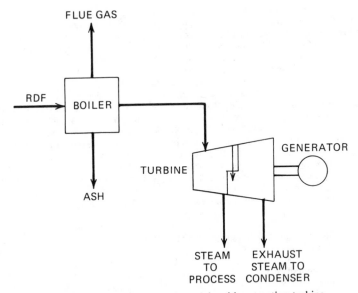

Fig. 12.3 Cogeneration topping cycle with extraction turbine.

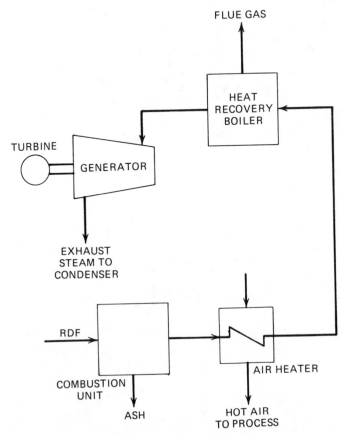

Fig. 12.4 Cogeneration bottoming cycle with drying process and electric generation.

Table 12.3 Facilities Designed for Cofiring Refuse-Derived Fuel

Plant	Firing Method	Type of RDF	Fuel with RDF
Ames, Iowa	Suspension and spreader stoker	Shredded	Coal
Baltimore County, Maryland[c]	Suspension and cement kilns	Shredded	Coal
Bridgeport, Connecticut[a]	Cyclone furnace	Powdered	Oil
Chicago, Illinois (Southwest Plant)[a]	Suspension	Shredded	Coal
Lakeland, Florida	Suspension	Shredded	Coal
Madison, Wisconsin	Suspension	Shredded	Coal
Milwaukee, Wisconsin[a]	Suspension	Shredded	Coal
Rochester, New York[a] (Monroe County)	Suspension	Shredded	Coal
St. Louis, Missouri[a]	Suspension	Shredded	Coal
Wilmington, Delaware	Suspension	Shredded	Coal
Hagerstown, Maryland[c]	Spreader stoker	dRDF	Coal
Erie, Pennsylvania[b]	Spreader stoker	dRDF	Coal
Westbury, England (Blue Circle Group)	Cement kiln	Shredded	Coal
Columbus, Ohio	Spreader stoker	Shredded	Coal
Houston, Texas[c] (Gulf Coast Portland Cement Company)	Cement kiln	Shredded	Gas
Woodstock, Ontario[c] (Cement Lafarge Ltd.)	Cement kiln	Shredded	Coal/gas
General Motors Corp. Truck Coach Division Pontiac, Michigan	Spreader stoker	Shredded	Coal

[a]Shutdown.
[b]Proposed or in start-up.
[c]Test facility.

suspension-fired boiler is more efficient than the stoker-fired boilers and because it can burn all of the RDF produced, the stoker-fired boilers are no longer used.

Besides Ames, four other municipal plants cofire municipal wastes. They are: Baltimore County, Maryland (Crane Station); Columbus, Ohio; Madison, Wisconsin; and Westbury, England. Baltimore, Bridgeport, Chicago, and Milwaukee are currently shutdown for financial and technical reasons.

The primary methods of cofiring RDF with other fuels are in full suspension or spreader stoker fired boilers. In a suspension-fired boiler, the RDF is pneumatically introduced into the furnace by locating the refuse ports (burners) above or among the existing coal burners. The primary combustion occurs when the RDF is in suspension. The heavier particles fall to the bottom of the furnace. To facilitate complete burnout of the RDF, dump grates are required at the bottom of the furnace. Cofiring RDF in a suspension-fired boiler generally requires the RDF to be processed to less than 1.5 in. in size.

For spreader stoker-fired boilers, shredded RDF is burned in semisuspension where the prepared fuel is fed through several fuel distributors. The lighter particles burn in suspension while combustion of the heavier particles is completed on the traveling grate. The prepared fuel can be fed to the boiler by either mechanical feeders and air swept spouts or by a pneumatic conveying system. RDF up to 4 to 4.5 in. (100 to 114 mm) in size is acceptable for cofiring RDF in stoker fired units. Densified RDF (dRDF)* may also be cofired in spreader stoker-fired boilers by blending the dRDF with lump coal and feeding the mixture onto the grate or by separately feeding the dRDF.

Another method of cofiring-prepared solid waste fuels is in cement kilns. The supplementing of gas or coal in kilns has been successfully tested in at least four plants.[7] One of these plants in England commercially uses RDF to supplement coal. In the facilities tested to date the shredder RDF, usually less than 1.5 in. (38 mm) in size, is pneumatically injected into the kiln at a point above the existing burners. Flue gas velocities are maintained at a high flow rate to keep the prepared fuel in suspension (see Chapter 18, Sections 1, 2, and 3).

The cofiring of prepared fuels with coal, oil, or gas requires that the ratio of RDF to the other fuel be limited. By limiting the ratio of refuse to the other fuel, the steam production fluctuations will be minimized, the steam temperature can be better controlled, the particulate collection efficiency can be maintained by controlling normal excess air requirements, and potential for corrosion can be reduced.[5] Experience at existing facilities has shown that the ratio of prepared refuse firing should be limited to 15 to 20% of the total heat input on suspension fired units. For spreader stoker-fired units the ratio should be limited but the limit is not well established. Existing spreader stoker units have cofired RDF at a ratio of up to 60%. Based on testing with cement kilns, ratios of up to 30% of the energy input to the kiln have shown satisfactory results. Above this point, problems have occurred with temperature control and lower-quality cement being produced.

Boiler performance is slightly reduced when supplementing a fossil fuel-fired boiler with pre-pared refuse. Experience at existing facilities has shown that the boiler efficiency is reduced from 0.5 to 3% depending on the ratio of the RDF to primary fuel and boiler load. As a result of higher ash loadings, the collection efficiency of precipitators may also be reduced. The cofiring of RDF with coal has shown to reduce SO_2 emissions because of the lower sulfur content in the fuel and reduced NO_x emissions because of the higher moisture content which reduces flame temperature.[5]

Some problems have been encountered when cofiring RDF with other fuels. The major problem is with potential tube metal corrosion resulting from the formation of acid gases. This is due to the higher chlorine content of RDF. Additional problems reported at existing facilities include in-creased ash slagging and fouling due to higher ash content and lower ash fusion temperature; increased ash-handling problems due to higher ash loading, unburned combustibles, and metals and glass in the fuel; and wear and pluggage problems on pneumatic fuel transport systems.

Although there have been problems with units cofiring RDF, facilities have successfully demon-strated that refuse can be fired in industrial and utility boilers and cement kilns as a supplement to conserve other fuels. Supplementing fossil fuels with prepared waste fuels has four potential advantages. These advantages are:

1. Reduced solid waste disposal problems.[8]
2. Reduced total plant SO_2 emissions.[8]
3. Lower total fuel costs.[8]
4. Existing boilers and cement kilns can be utilized thus reducing the capital cost of prepared fuel facilities.

In considering the feasibility of converting an existing boiler or kiln for supplemental fuel firing, the engineer must conduct a detailed study. The study must include the boiler manufacturer for evaluation of the effects on boiler performance and the air pollution control equipment manufac-turer for evaluation of precipitator performance. The engineer must also consider the effects on ash handling systems, siting requirements, type of processing systems required, and plant emissions. The most suitable boilers for conversion are those that already have ash handling and precipitators installed.

A study is currently being developed to provide guidelines for RDF cofiring in utility boilers.[6] This study, when completed, should provide valuable information to assist in the making of deci-sions when considering and designing projects for RDF firing.

12.1.5 Codisposal

Codisposal refers to the combined disposal of one type of waste with another type of waste. In this case the disposal of wastewater treatment plant (sewage) sludge and prepared municipal solid waste will be discussed.

Sewage sludge generally contains 95 to 98% moisture before concentration. The moisture content is dependent upon the wastewater treatment process. Various sludge concentration pro-cesses can reduce the moisture content to 70 to 80%. At this point, the sludge is considered to be

---

Editor's note: dRDF (densified Refuse Derived Fuel) also has the ASTM designation RDF-5.

autogenous, that is, the heating value of the sludge is equivalent to the heat required to vaporize the moisture in the sludge.

Drying systems are required to dewater the sludge to the point where it has some value as a fuel. The usable heat from the dry sludge has to offset the cost and/or energy to prepare it, or it has a negative energy or cost impact on an energy recovery system. Because of the possibility of the negative costs and the high cost of drying systems, research and testing have been done in the codisposal of sewage sludge with prepared solid waste by either burning the sludge directly with the solid waste or by using waste heat from solid waste facilities to dry the sludge and then burn it with the solid waste. Since these types of codisposal systems do not reduce the cost of solid waste disposal, the real benefit of codisposal is the offsetting of costs incurred for the disposal of sludge and the positive impact it may have on encouraging resource recovery.

Codisposal has been tried numerous times in the past 30 to 40 years through incineration. From the 1940s through the 1960s, various methods of codisposing sludge with solid waste were tested. This included the predrying of sludge and burning it in combination with refuse, spraying the sludge on the refuse before incineration, and spraying sludge directly into the incinerator. These practices have been discontinued since the operation of incinerators in this country has generally been abandoned due to the high operating costs and air pollution regulations. More recent developments in the codisposal of sludge with solid waste involves the combustion in various types of furnaces and pyrolysis systems. The following discussions review the codisposal of sludge with prepared solid waste fuels. The codisposal of sludge in mass burn units will be discussed in Sections 13.11, 13.12, and 13.13 of Chapter 13.

There are numerous types of systems for the codisposal of sewage sludge and prepared municipal solid waste. However, each system will consist of preparation and storage systems for the sludge and refuse, feed systems, furnaces or combustion units, heat recovery units, air pollution control equipment, and ash removal and handling systems. Some codisposal facilities may also require a supplemental fuel supply.

The selection of the preparation, handling, and storage systems for sludge and RDF will depend on the type of furnace and combustion units selected. Some combustion units may require that the sludge be dried prior to being fed to the unit and the RDF may have to be finely shredded with the removal of metals, glass, grit, and other nonorganics. The finely shredded RDF and dried sludge may be mixed together prior to feeding into the combustion units or may be fed separately into the unit. This would be typical of a suspension-fired boiler. Other systems, such as spreader stoker-fired boilers, may only require a coarse RDF with the sludge in its concentrated form (70 to 80% moisture) being mixed with the RDF just prior to being fed to the combustion unit. Other types of combustion units such as pyrolysis, fluidized bed, and multiple hearth units may require yet another type of preparation. The engineer and designer must closely coordinate the design of the preparation, handling, and storage systems with the design and operation of the combustion systems.

The storage and handling systems for a concentrated sludge may consist of truck receiving and unloading station, storage and holding facility, and sludge pumping or conveying system. If the facility is near the waste treatment plant, the sludge could be pumped or conveyed to the on-site storage and holding facility. For a dried sludge, the storage and handling systems may consist of truck unloading facility, pneumatic or mechanical conveying system, storage silos, and a mechanical or pneumatic dry sludge feeding system. The storage, handling, and feed systems for RDF are discussed in detail in Section 12.9 of this chapter.

The combustion of prepared solid waste fuel and the combustion of sewage sludge are both well developed as separate technologies. The development of the technologies for the combustion of these two wastes together (codisposal) is relatively recent. Currently, there is one successful, commercially operating combustion system for the codisposal of prepared solid waste and sludge. (Eastman Kodak, Kodak Park, Rochester N.Y., 100 TPD RDF @ 1½ in. (38 mm) and plant sludge burned in suspension.) However, several facilities have been designed and constructed as pilot plants and commercially operating facilities.

One commercial-size facility in Duluth, Minnesota, the Western Lake Superior Sanitary District, has installed two fluidized bed reactors with a waste heat recovery boiler to codispose of sludge with shredded refuse. Each reactor has a 20-ft inside diameter refractory lined combustion zone and is designed for balanced draft operation. The reactors are provided with both a fluidizing air blower and an induced draft fan to maintain a slight negative pressure on the freeboard zone of the reactor. Each reactor is designed to burn approximately 136,000 lb/day (61,608 kg/day) of sludge solids together with either 375 gallons/hr (1.40 m³/day) of No. 2 fuel oil or 160 tons per day of prepared solid waste. The sludge consists of 20% dry solids and 65% volatiles and has a 9835 BTU/lb (22,817 kJ/kg) heating value on a dry basis. The solid waste is shredded to a nominal 1.5 in. (38 mm) material with ferrous metal removal and air classification. The prepared fuel is pneumatically injected into the reactor and has a 30% moisture content, 65% combustible content and fired heating value of 5500 BTU/lb (12,760 kJ/kg). Waste heat produced by each reactor is passed through a heat recovery boiler generating 45,000 lb/hr (20,385 kg/hr) of 280 psig (1931 KPa) saturated steam. The process system is illustrated in Figure 12.5.

The RDF processing facility, fluidized bed reactors, and flue gas treatment system have experienced operating difficulties which have necessitated design modifications and repairs to the equipment. After design modifications were made, the plant operated in the co-combustion mode for 3 months (May–July, 1982) until a shredder explosion closed the RDF processing line. (The RDF processing facility was abandoned in 1983.) The plant was still closed down as of July, 1983, because of slagging problems in the precipitators.

Another type of combustion system that has undergone pilot testing is the use of a multihearth furnace. This combustion system consists of hearths stacked one above the other. Prepared solid waste is fed at the third level while prepared solid waste mixed with sludge is fed in at the first level. A rotating arm progressively rakes each level working the combustible material and ash down to the lowest hearth for removal of ash. As the sludge is passed down through each level, it is dried until it ignites. This system has been extensively tested in Contra Costa County, California. A diagram of the test facility is shown in Figure 12.6. The test facility did not have any heat recovery boiler, however, one could be installed for the production of steam.

During the testing of the plant, it was determined that the unit was most efficient when operated as a starved air incinerator. The unit then required a special afterburner to burn the combustible off-gases at a temperature of 1400°F (760°C). On the basis of this testing, it was determined that large commercial-size units could feasibly be constructed for the codisposal of RDF and sludge. To date no commercial facilities using this technology are in operation.

Other types of combustion systems, such as stoker-fired and suspension-fired boilers and pyrolysis systems, may be adapted for the cocombustion of prepared solid waste and sewage sludge. To date no known testing or pilot operation of these systems has been attempted. Problems that may be encountered with these systems are (1) fuel-feeding difficulties resulting from mixing wet sludge with prepared refuse, (2) poor combustion of the fuel in the furnace due to wide fluctuations in moisture content, (3) unburned sludge coming out with the ash, (4) odor problems from insufficient combustion temperatures to destroy the odors [1400 to 1600°F (760 to 871°C) are required to control odors from sludge], and (5) corrosion problems in the combustion units and downstream equipment.

The codisposal of prepared solid waste and sewage sludge and the recovery of energy from this codisposal can be accomplished by a method called bioconversion. Bioconversion systems recover methane gas from the anaerobic digestion of the prepared solid waste and sludge and burn the methane gas to generate process steam or electrical energy. The anaerobic digestion process has been used for many years at wastewater treatment plants to digest sludge. However, anaerobic digestion of prepared solid waste and sludge is currently in the developmental stage.

An experimental 50 to 100 TPD "proof of concept" solid waste facility is located in Pompano Beach, Florida. Solid waste is shredded, screened, and air-classified with the resulting light fraction from the air classifier being blended with sewage sludge, water, and nutrients. The mixture is anaerobically digested in mechanically agitated tanks to produce methane gas. For more information on bioconversion, see Section 12.3.7 of this chapter and Chapter 18.

12.1.6 Economics and Case Histories

Over 30 facilities have been designed, constructed, tested, and operated since the early 1970s for the recovery of energy and material from prepared solid waste. Some of these facilities were designed and operated as test or demonstration projects, such as the St. Louis–Union Electric plant. Other facilities from 200 to 3000 TPD, such as the Ames, Iowa, plant, and the Dade County, Florida, resource recovery plant (world's largest) are in successful commercial operation, whereas others have been temporarily or permanently shutdown because of technical and economic difficulties.

Table 12.4 is a list of 21 selected facilities. The table provides a brief description of the process, design capacity, cost, and status of each facility.

To provide general information on the development and progress of energy recovery from prepared solid waste fuels, case histories of five of the facilities listed in Table 12.4 are described below. The information provided is not intended to give every detail of the facilities, but to provide a general overview of the operating success and problems and the designs that have been incorporated into various facilities.

CASE I. Bridgeport, Connecticut

The most refined prepared fuel product is the powdered RDF produced at the Resource Recovery System facility at Bridgeport, Connecticut. The facility was designed to operate at a nominal capacity of 561,000 tons/year based on a waste input of 1800 TPD for 312 operating days/year.[11]

The RDF processing plant consisted of two process lines designed to process at a peak of 75

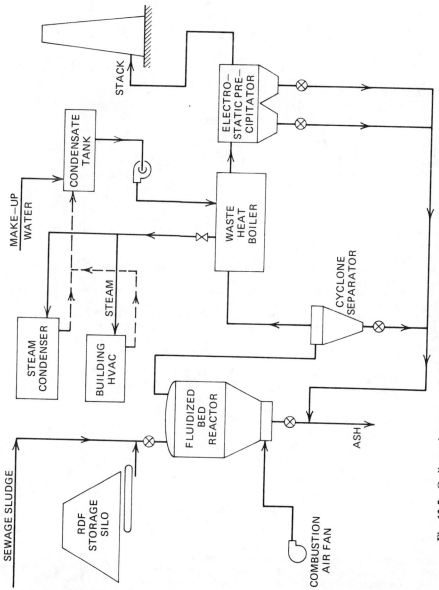

Fig. 12.5 Codisposal system at Duluth, Minnesota, utilizing a fluidized bed reactor.

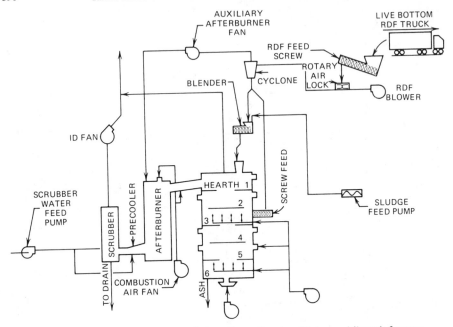

Fig. 12.6.   Codisposal system at Contra Costa County utilizing multihearth furnace.

tons/hr. Each line had primary and secondary trommels. The main line had a shredder (spare line a flail),* air classifier, magnetic separator, ball mill, and product screen to produce the powdered RDF and recover ferrous metals, glass, and aluminum. The RDF was pneumatically conveyed to storage silos and then loaded into trucks for transporting to a nearby power plant. The design of the Bridgeport facility was based on the development of the powdered RDF process at a pilot plant in East Bridgewater, Massachusetts. A schematic flow diagram of the process is shown in Figure 12.7.

The powdered RDF was reported to have the following characteristics:[11]

Higher heating value	7500–8000 BTU/lb (17,400–18,560 kJ/kg)
Particle size	Less than 50% minus 200 mesh
Moisture content	Less than 5%
Ash content	Less than 15%
Sulfur content	Less than 1%
Density	25–50 lb/ft³ (396–792 kg/m³)
Percent RDF recovered per ton of MSW	40–50%

At the power plant, the RDF was cofired with a low-sulfur residual oil in a cyclone-fired furnace. Since the boiler was originally designed for coal firing, the required electrostatic precipitators were already installed. The only major modifications required to the power plant and boiler consisted of installing an unloading and storage facility, a pneumatic injection system to the cyclone burners, and a wet bottom ash system. It was anticipated that the boilers could cofire the RDF at rates up to 50% of the required thermal input when operating at or above 80% load.

Shakedown and startup of the processing plant commenced in May, 1979, and continued intermittently to process solid waste.[11] RDF firing in the boilers did not start until November, 1979.

Due to the proprietary nature of the process, information has not been published or made available on the operations of the processing facility. Because of environmental constraints on sulfur emissions the firing rates of the RDF were limited to 2 to 10% of the boilers' heat input. Other than this firing limitation the testing of cofiring the RDF is reported to have no adverse effects on

---

*The main process line had a large 2500 HP (1865 kW) shredder. The second line had a flail but the second line was never operated.

Table 12.4 Prepared Fuels Energy Recovery Facilities[a]

Facility	Process Description	Design Capacity (TPD)	Capital Cost ($ Millions)	Status as of October, 1985
Bridgeport, Connecticut	Shredded, air-classified, ball-milled prepared fuel, powdered fuel fired in utility boiler; ferrous metals recovered	1800	53	Shutdown
Wilmington, Delaware	Shredded, air-classified prepared fuel; RDF fired in utility boiler; ferrous, nonferrous, and glass recovered	1000	72.3	Operational
Dade County, Florida	Basic shredding and peak load pulping, air-classified prepared fuel; fuel fired in plant boilers; electrical generation; ferrous and nonferrous metals; aluminum and glass recovered	3000	165	Operational
Lakeland, Florida	Shredded prepared fuel; RDF cofired with coal in utility boiler; ferrous metals recovered	300	5	Intermittent operation
Chicago, Illinois (Southwest)	Shredded, air-classified prepared fuel; RDF fired in utility boiler; ferrous metals recovered	1000	19	Shutdown
Ames, Iowa	Shredded, screened, air-classified prepared fuel RDF cofired in utility boilers; ferrous metals recovered	200	6.3	Operational
Baltimore County, Maryland[b]	Shredded, prepared fuel; RDF used in utility cyclone boiler, and testing in cement kilns; ferrous metals and glass recovered.	1200	11	Operational
Duluth, Minnesota	Shredded and air-classified prepared fuel; RDF and sludge fired in fluidized bed boiler; steam generated for heating and process; ferrous metals recovered	400	19	Shutdown
St. Louis, Missouri	Shredded and air-classified prepared fuel; RDF cofired in utility boiler (demonstration facility)	300	N/A	Shutdown
Albany, New York	Shredded prepared fuel; RDF fired in nearby dedicated boiler; steam generated for heating and cooling; ferrous metals recovered	750 (per shift)	28.2	

Table 12.4 (continued)

Facility	Process Description	Design Capacity (TPD)	Capital Cost ($ Millions)	Status as of October, 1984
Hempstead, New York	Wet pulped prepared fuel; RDF fired in spreader stoker boilers; electrical energy generated; glass, aluminum, and ferrous metals recovered	2000	130	Shutdown
Monroe County, New York	Shredded and air-classified prepared fuel; RDF cofired in utility boiler; glass, nonferrous, and ferrous metals recovered	2000	62.2	Shutdown pending decision for modification or dedicated energy recovery addition
Occidental Chemical Corp., Niagara Falls, New York	Shredded and air-classified prepared fuel; RDF fired in dedicated boilers; steam used in industrial plant; electrical energy also generated. Ferrous metals recovered.	2000	100 +	Operational
Akron, Ohio	Shredded and air-classified prepared fuel; RDF fired in semisuspension boilers; steam generated for heating and cooling; ferrous metals recovered	1000	80	Shutdown pending decision to resume regular operations following pit explosion.
Columbus, Ohio	Shredded prepared fuel; RDF cofired in stoker-fired boilers; electrical energy generated; ferrous metals recovered	2000	175	Operational
Richmond, Virginia	Screened and shredded prepared fuel; RDF cofired in an industrial boiler; glass, aluminum, and ferrous metals recovered	400	1.85	Operational
Madison, Wisconsin	Shredded, screened and air-classified prepared fuel; RDF cofired in utility boiler; ferrous metals recovered	400	2.5	Intermittent operation
Milwaukee, Wisconsin	Shredded and air-classified prepared fuel; RDF cofired in utility boiler; ferrous metals recovered	1000	30	Shutdown
Hamilton, Ontario	Shredded prepared fuel; RDF fired in spreader stoker boiler; electrical energy generated; ferrous metals recovered	500	9	Operational
Westbury, England	Shredded prepared fuel; RDF cofired in cement kiln	N/A	N/A	Operational

Table 12.4 (continued)

Facility	Process Description	Design Capacity (TPD)	Capital Cost ($ Millions)	Status as of October, 1984
Eastman Kodak Co., Kodak Park, Rochester, New York	Plant waste single-stage shred to nom. 1½ in. (38 mm) mill air sweep classified; full suspension cofiring with plant sludge	100 TPD RDF Sludge	—	Operational
Haverhill, Massachusetts	Shredded and screened RDF, ferrous recovery	1300 TPD	99.5	Operational
General Motors Corporation, Truck/Coach Division	Shredded and air classified RDF cofired with coal	RDF: 225 TPD	N/A	Operational

[a] Resource Recovery Activities, *City Currents*, October, 1984, U. S. Conference of Mayors, Washington, D. C. and author's update.
[b] A process plant only and shreds for landfill. It has process systems to prepare various RDF types for test and began continuous production of RDF for sale/use in cyclone boilers at C. P. Crane Station of Baltimore Gas & Electric Co. in early 1984.

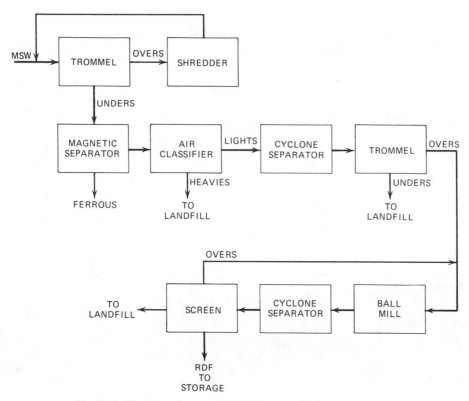

Fig. 12.7 Flow line diagram of the Bridgeport, Connecticut, facility.

the boiler operation. The project operations were suspended in 1980 due to operational and financial difficulties.

CASE II. *Dade County, Florida*

The world's largest and probably the most sophisticated facility for the production of a RDF and recovery of energy and materials is the Dade County facility in Miami, Florida. The facility has a design capacity of 3000 tons/day and includes an optional process that adds water to a portion of the solid waste to produce a fuel product.

The RDF processing plant is divided into two functions. One is for dry processing waste that is classified as trash, essentially commercial and industrial types of waste, along with household garbage. The process consists of shredders, screens, air classification, and magnetic separation. The shredded light fraction from the air classifier goes directly to fuel storage while the heavy fraction goes to the wet systems for further processing.

The wet area of the processing facility handles the remaining waste received. The wet processing system consists of a trommel, four hydrapulpers, and a classification and dewatering system. The RDF product is combined with the light shredded fraction from the dry process area. The reject material from the hydrapulpers is conveyed to a material recovery area where ferrous and nonferrous metals, aluminum, and glass are recovered.[25] A simplified schematic of the processing facility is shown on Figure 12.8.

The fuel product (RDF) is fed either directly from processing or from storage to four dedicated spreader stoker boilers. Each boiler has a rating of 185,000 lb/hr (83,805 kg/hr) at 625 psig (4309 KPa) and 750°F (407°C). The steam generated in the boilers drives two 37.5 mw turbine generators. The electrical energy is sold to a local utility.

The Dade County facility began operations in January, 1981. Since that time, the plant has processed waste at rates in excess of 3000 TPD. No major operating problems have been reported with the processing facilities. During shakedown, there were problems with the ash-handling system and boiler fuel feed system along with superheater corrosion that required modifications to the boiler superheater tubes.

CASE III. *Ames, Iowa*

The first commercial RDF facility to be constructed and operated in the United States is the Ames Resource Recovery Facility in Ames, Iowa. The plant also has the distinction of being the oldest operating resource recovery facility in the United States.

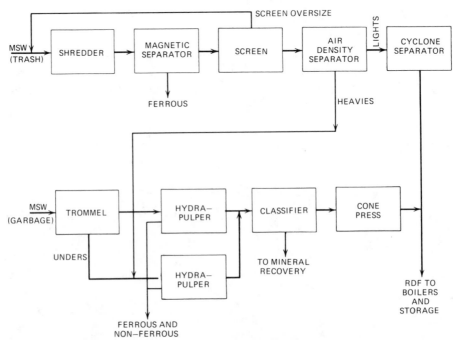

Fig. 12.8 Flow line diagram of the Dade County, Florida, facility.

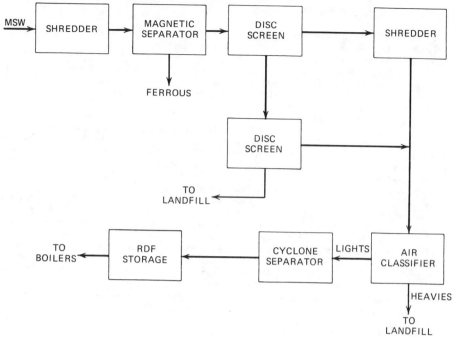

Fig. 12.9 Flow line diagram of the Ames, Iowa, facility.

The Ames facility has a design capacity of 200 TPD. The processing facility, which was patterned after the St. Louis demonstration plant, consists of one process line with two stages of shredding, two disc screens, air classifier, magnetic separation, and an aluminum, glass, and nonferrous metal separation system.[24] The light fraction from the air classifier is pneumatically transported to a storage silo located at a nearby power plant. The heavy fraction was to be further processed for recovery of aluminum, glass, and nonferrous metals. Figure 12.9 is a schematic of the Ames process line.

RDF is mechanically reclaimed from the storage silo and is pneumatically transported, fed to, and cofired by either of three boilers in the power plant. Two of the boilers are spreader stoker units and have been routinely cofired with up to 50% RDF on a heat-input basis. The third boiler is a suspension-fired unit with RDF being cofired with coal at a rate of up to 20% on a heat input basis.

The RDF produced at Ames has the following characteristics:

Higher heating value	6000 BTU/lb (13,920 kJ/kg)
Product size	Less than 1.5 in. (38 mm)
Ash content	10%
Moisture content	23%
Percent RDF produced per ton of MSW	60–70%

Since the start-up of the plant in September, 1975, the facility has undergone modifications to improve operations. Two disc screens were added to the process line to improve the fuel product and to minimize wear on the pneumatic transport systems by removing glass, sand, and grit. This modification has greatly improved the RDF quality and reduced wear on the pneumatic system. A dust-collection system was also added to the processing facility. In the spring of 1978, the suspension-fired boiler was modified to add a dump grate to improve the combustion of the heavy organics. The materials recovery system is no longer operated because of operational problems. The only materials recovered and sold are ferrous metals. The air classifier heavies and disc screen rejects are being landfilled.

No operating problems with the suspension-fired boiler has been reported since the dump grate was added. The two stoker-fired boilers are no longer operated because the suspension-fired boiler is more efficient and because it can burn all RDF produced.

CASE IV. Niagara Falls, New York

The Niagara Falls facility, known as the Occidental Chemical Corp. Energy from Waste facility, has a design capacity of 2200 TPD.[13] The plant is a cogeneration facility in which both electrical energy and steam are sold from the combustion of RDF.

The RDF processing system consists of a 1600-ton receiving and storage pit that utilizes a multiram movable floor to feed a surge hopper feeding three process lines. Each process line consists of a primary shredder and air classifier. The light fraction from the air classifier is pneumatically conveyed to a 5000-ton capacity storage bin. The heavy fraction is conveyed to a two-stage magnet for ferrous metal recovery.

The RDF storage bin is divided down the middle and has a single, variable position auger on each side to reclaim RDF. This is similar to storage bins used in the wood industry for reclaiming wood chips and hogged fuel. From the storage bin, the fuel is conveyed by either of two conveyors which feed two live bottom surge metering bins, one in front of each of the plant's boilers. The RDF is fed through eight air-swept distributors on each boiler.

Each of the two spreader stoker-fired boilers is designed to produce 300,000 lb/hr (135,900 kg/hr) of 1250 psig/750°F (8618 KPa/399°C) steam when burning RDF. The boilers are also designed to burn coal or oil as a back-up fuel to assure the availability of steam to the plant's customers. The boilers also supply two 25-MW automatic extraction turbine generators. The low-pressure extraction steam is sold to two nearby steam customers and the electrical energy sold to the local utility.

The facility began operation in August, 1980, and problems were reported with the fuel storage reclaim augers and surge-metering bins. The reclaim augers in the storage bins did not meet the discharge requirements. The single screw auger has been replaced with a double screw to increase capacity. The surge-metering hoppers have undergone adjustment to promote material movement to the outside of the bin and with a limited height to which the bin can be filled (about half capacity) to prevent bridging. The front-end processing system has undergone modification to improve the fuel quality.

During initial operation there was significant slagging on the refractory walls above the grates. By removing the refractory, adjusting the underfire air, and reducing feed from the two outside distributors, the problem has been solved. Also rectified was a high ash content in the fuel causing excessive wear on the grates and overloading the ash-handling system. Superheater tube erosion reportedly caused by a combination of sootblower operation and a large quantity of highly erosive ash has been corrected. Also, corrosion of the side and rear-wall water tubes has been remedied, and the boiler performance goals have been met or exceeded.

A schematic of the RDF processing system is illustrated by Figure 12.10. The following are the characteristics of the RDF fuel:

Higher heating value	5400 BTU/lb (12,528 kJ/kg)
Moisture content	23.2%
Ash content	17.8%
Particle size	90% less than 4 in. (101 mm)
Percent RDF produced per ton of MSW	80%

CASE V.   Madison, Wisconsin

The Madison facility is a third-generation system designed, constructed, and operated by the City of Madison. The processing plant has a design capacity of 250 TPD with a single 50-ton/hr processing line.

The processing line consists of a flail mill for bag breaking and coarse size reduction, a magnet for ferrous metal removal, a trommel screen for removal of glass, nonferrous metals, and other noncombustibles, and a secondary shredder for final RDF production. The secondary shredder has an air-swept, pneumatic takeaway system for conveying the RDF to stationary packers where the RDF is loaded onto trucks for shipment to a nearby power plant. A schematic of the process system is shown on Figure 12.11.

The RDF is transported to the nearby power plant for cofiring in either of two suspension-fired pulverized coal boilers. The boilers are each rated at 425,000 lb/hr (192,525 kg/hr) steam flow at 1250 psig/950°F (8619 KPa/510°C).[14] RDF is unloaded into either of two 18-ton surge storage silos in the RDF receiving building. Each bin serves one boiler. The RDF is fed from the silos by the two twin augers onto individual conveyors which discharge the material into lump breakers and air-lock feeder. A pneumatic transport system feeds the RDF directly to the furnace for suspension firing. The boilers were modified by the installation of dump grates, two new fans to supply underfire and overfire air to the grates, and air-cooled fuel nozzles.

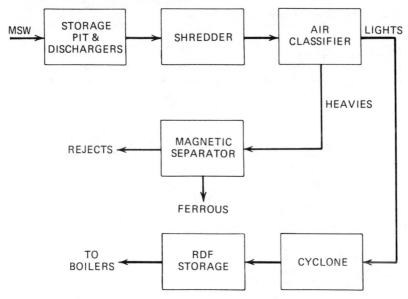

Fig. 12.10 Flow line diagram of the Niagara Falls, New York, facility.

The processing plant began processing solid waste in January, 1979. The facility began firing RDF in June, 1979. The processing plant has been operating smoothly since the start-up. Throughput capacities as high as 400 TPD during an 8-hr shift have been achieved. The cofiring of the RDF has had no adverse effect on the boilers. Some clinker formation has occurred above the grate but this is easily removed by lowering the grate and rodding the clinker. The efficiency of the boilers has been reduced by about 1.5%. A dust collection system has been added to the RDF receiving and storage system.

The characteristics of the RDF are as follows:[14]

Higher heating value	6100 BTU/lb (14,152 kJ/kg)
Ash content	10–12%
Moisture content	18–22%
Particle size	91% less than ¾ in. (1.9 mm)
Percent RDF produced per ton of MSW	54%

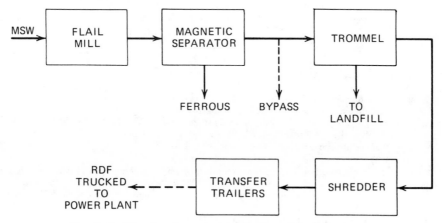

Fig. 12.11 Flow line diagram of the Madison, Wisconsin, facility.

Table 12.5   ASTM Refuse-Derived Fuel Classifications

ASTM Classification	Description
RDF-1	Municipal solid wastes used as fuel in as-discarded form
RDF-2	Municipal solid wastes processed to a coarse particle size with or without ferrous metal separation
RDF-3	Municipal solid wastes processed to a particle size such that 95% by weight passes through a 2-in. (50 mm) square mesh screen and from which most metals, glass, and other organics have been removed
RDF-4	Municipal solid waste processed into a powdered form, 95% by weight passes through a 10-mesh screen and from which most metals, glass, and other organics have been removed
RDF-5	Municipal solid waste that has been processed and densified (compressed) into the form of pellets, slugs, cubettes, or bricquettes
RDF-6	Municipal solid waste that has been processed into a liquid fuel
RDF-7	Municipal solid waste that has been processed into a gaseous fuel

As can be seen from the list of facilities in Table 12.4, some of them have had operational or financial problems. This has caused many municipalities and cities to prematurely consider alternate means of resource recovery or abandon current plans.*

12.2   PROCESSED REFUSE FUEL TYPES

David J. Schlotthauer

Engineers and designers often specify variations in the physical sizes and characteristics of RDF to meet their particular requirements and objectives. To provide an industry-wide standard for the classification of various types of RDF, the American Society for Testing and Materials (ASTM) has developed seven categories into which the types of RDF can be classified. The ASTM classification numbers and descriptions are listed in Table 12.5.

Municipal solid waste used as a fuel in its as-discarded form (RDF-1) would be burned in mass-burn type units without any processing† other than the removal of nonprocessible and oversize items such as white goods, large quantities of wire and cable, etc. Larger facilities which burn RDF-1 include the 1600 TPD Chicago Northwest Waste-to-Energy Incinerator, the 1500 TPD Saugus Facility, the 2000 TPD Pinellas County Facility, the 720 TPD Harrisburg plant, the Westchester County N.Y. and the city of Baltimore, Maryland plants @ 2250 TPD. Refer to Chapter 13 for a more detailed discussion of mass-burn energy recovery.

RDF classified as RDF-2 would be produced in facilities that have processing equipment for the screening, shredding, and removal of ferrous metals from raw municipal solid waste. The fuel produced from this type of facility may have the following specifications:

Product size	95% less than 4–4.5 in. (100 mm)
Moisture content	15–25%
Ash content	12–20%
Higher heating value	4500–5500 BTU/lb (10,440–12,760 kJ/kg)
Density	3–7 lb/ft^3 (48 kg/m^3–111 kg/m^3)
RDF yield (RDF/Raw MSW)	70–80%

A processing line for the production of RDF-2 may have the configuration shown in Figure 12.12. Fuel classified as RDF-2 would be burned in spreader stoker-fired combustion units. Existing facilities designed to produce and burn RDF-2 are the 2000 TPD Niagara Falls Facility, the 1000 TPD Akron Facility, the 750 ton/shift at Albany, the 225 TPD at General Motors Truck Coach, the 1300 TPD at Haverhill, the 2000 TPD at Columbus, and so on.

Editor's note: A realistic bottom line type appraisal of waste-to-energy projects is presented in Standard & Poors *Credit Week*, issue of October 15, 1984 in ''A Resource Recovery Review,'' pp. 50–54 and as presented at their symposium on Municipal Bond Financing, New York, NY, October 15, 1984. Refer to Chapter 6, Appendix 6.2.

†*Editor's note:* A recent exception is the Gallatin, Tenn., 200 TPD rotary combustor steam plant which employs a modest front-end process system for fuel beneficiation and material recovery.

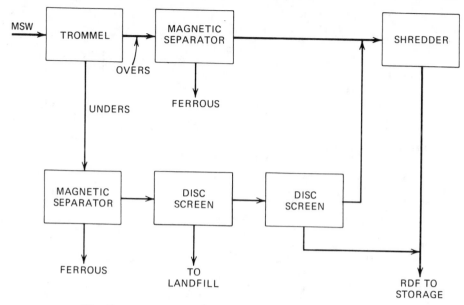

Fig. 12.12 Typical processing line for the production of RDF-2.

Solid waste processed to produce RDF-3 requires additional shredding and screening to reduce the fuel size and remove glass, grit, and other inorganics. The processing facility for this may have two stages of shredding and air classification. Alternate technologies, such as hydropulpers, may also be used to produce this type of fuel. A typical process system for the production of RDF-3 is shown in Figure 12.13. Existing facilities that produce and burn RDF-3 are the 3000 TPD Dade County Facility, the 200 TPD Ames Facility, the 400 TPD at Madison, the 400 TPD at Baltimore Gas & Electric, and the 120 TPD at Eastman Kodak. This classification of fuel may be burned either in spreader stoker-fired units or in suspension-type units equipped with dump grates. A typical fuel specification for RDF-3 may be as follows:

Product size	95% less than 1.5 in. (38 mm)
Moisture content	12–20%
Ash content	15–20%
Higher heating value	5000–6000 BTU/lb (11,600–13,920 kJ/kg)
Density	2–6 lb/ft³ (32–95/m³)
RDF yield (RDF/Raw MSW)	55–65%

Processed RDF classified as RDF-4 is produced by using systems similar to those used to produce RDF-2 and RDF-3 and then by drying and grinding the refuse to a uniform powdered consistency. This type of RDF would be in the form of a free-flowing powder with a high-bulk density and having a uniformly higher heating value. This provides a fuel that allows for complete and rapid combustion and allows for a higher degree of control over the combustion process. RDF-4 could be burned in suspension units using pulverized coal-type burners or in cyclone-fired furnaces. The only process developed to date to produce RDF-4 is the ECO-Fuel II process developed by Combustion Equipment Associates, Inc. A typical specification for RDF-4 is as follows:

Product size	95% less than 10 mesh
Moisture content	1–5%
Ash content	10–20%
Higher heating value	6500–7500 BTU/lb (15,080–17,400 kJ/kg)
Density	25–35 lb/ft³ (396–555 kg/m³)

RDF-5, which is a densified refuse derived fuel (d-RDF), is produced by first processing the raw municipal solid waste with systems similar to those used to produce RDF-2, 3, and 4. The product

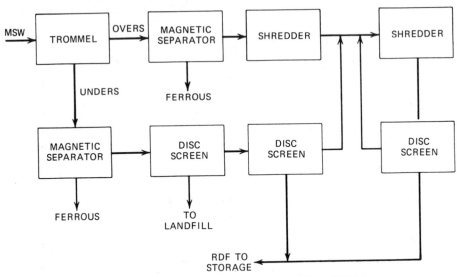

Fig. 12.13 Typical processing line for production of RDF-3.

is then further processed into pellets, cubes, or bricquettes. This provides a fuel that has a high density and is easily stored and handled. RDF-5 may be burned in spreader stoker boilers. This process is experimental with no known plant operating on a continual, commercial basis to produce or burn RDF-5 (see Section 12.8).

RDF-6 and RDF-7, being classified as a RDF produced from solid waste processed into a liquid or gas, is produced by the processing of waste by the method called pyrolysis. Pyrolysis is defined as the thermal decomposition of matter in the absence of oxygen. Depending on the system design, waste fed into the pyrolysis system can range from raw, unshredded refuse (RDF-1) to a d-RDF (RDF-5). The output from pyrolysis can be liquid (oil), gaseous (methane), a powdered char, or a combination of these products. The gases produced from pyrolysis are of the low to medium BTU class with a heating value of 200 to 500 BTU/scf (1763 to 4400 kcal/m³). The oils produced are a low-grade product having a heating value of 10,000 to 15,000 BTU/lb (23,200 to 34,800 kJ/kg). Numerous pilot and demonstration plants using pyrolysis have been built and tested to date. The only commercially operating pyrolysis facility using solid waste is in Japan.

Process systems for the production of a fuel from municipal solid waste are discussed in detail in Sections 12.4–12.9.

12.3 METHODS OF COMBUSTION OR ENERGY RECOVERY OF PROCESSED FUELS

David J. Schlotthauer

Currently there are seven methods for the combustion or recovery of energy from RDF. These are:

1. Combustion in a spreader stoker-fired boiler
2. Combustion in a suspension-fired boiler
3. Combustion in a fluidized bed
4. Combustion in a cyclone furnace
5. Combustion in a cement kiln
6. Pyrolysis of the RDF with the combustion of the resulting gases, oils, and/or chars
7. Bioconversion of the RDF with the combustion of the resulting methane gas

The methods of energy recovery from prepared solid waste that are considered to be proven are combustion in spreader stoker- and suspension-fired boilers (see Chapter 6, Appendix 6.2). The other methods are presently in the pilot or demonstration stages and are considered as experimental. With the exception of combustion in cement kilns, all of the above methods can be used to generate low-pressure/saturated steam for district heating and process use or to generate higher-pressure/superheated steam for use in a turbine generator to produce electrical energy. The high-pressure/superheated steam can also be used in an extraction turbine to generate electrical energy and supply low pressure steam for industrial use (cogeneration).

Many factors must be considered when evaluating or designing a facility to burn or recover energy from prepared fuels. The following sections will discuss some of these considerations as well as the above combustions and energy recovery methods.

12.3.1 Spreader Stoker Firing

The principle of spreader stoker firing was developed in the early 1930s to improve the combustion of coal and to permit the effective burning of low-ranking coals. In the latter part of the 1930s, the principle of spreader stoker firing was employed on cellulosic fuels such as bagasse and wood fibers. It was not until 1972 that spreader stoker firing was used for the combustion of a RDF. The first significant application of the spreader stoker firing of RDF was at the Hamilton, Ontario, Solid Waste Reduction Unit.

Spreader stokers consist of multiple feeder-distributor units (spreaders) and a grate system (stoker). Spreader stokers use a combination of suspension burning, in which the lighter portion of the fuel burns above the grate, and a thin-burning fuel bed, where the heavy portion burns on the grate. Because of this method of combustion, spreader stoker units are often referred to as a semisuspension system. The feeder distributors consist of a fuel hopper, a feeder mechanism, and a distributor. These units are generally mounted on the front wall of the boiler. The hopper feeder and distributors can be integrated into a single unit or mounted separately with the feeder mechanism at a raised elevation and the distributor mounted on the boiler wall just above the grate. The integrated units are common on coal-fired boilers, while separating the feeder from the distributor is common practice on units firing cellulose-type fuels, such as RDF.

Feeders for stoker-fired boilers may be of several different types. They all regulate the flow of fuel in proportion to the boiler load. Reciprocating plate or ram, variable-speed chain or belt conveyor and vibrating types of feeders are typically used for coal firing. A rotating drum-type feeder is used for bagasse firing. All four types of feeders can and have been used for RDF firing. Also, RDF boilers sometimes use a screw auger-type feeder.

The distributors uniformly distribute the fuel into the furnace and on the grate. There are two basic types of distributors—mechanical and pneumatic. The mechanical distributor is essentially a rotating drum with two to four blades of the overthrow or underthrow type. The pneumatic distributor, or an air-swept spout, uses air to sweep the floor of the spout and convey the fuel onto the grate. Motorized rotary dampers control the air in each distributor and high-pressure air jets fan the flow of air and fuel from the spout. The mechanical distributors are generally used for coal firing whereas both types of distributors can be used for cellulose fuels. The air-swept spout is the primary choice for RDF fuels.

The grate for spreader stoker boilers provides a supporting surface for the combustion of heavy fuel particles and a high-resistance method for metering underfire combustion air. The grates may be of three different types: continuous-cleaning, stationary, and intermittent-dumping. The continuous-cleaning type grates can be further classified as either traveling, reciprocating, or vibrating.[17] The majority of the spreader stoker fired boilers in operation today use continuous-cleaning, traveling grates. For this type of grate, the direction of travel is toward the front of the unit. The traveling grates are variable speed with an operating range from 2 to 20 ft/hr (0.6 to 6.0 m/hr).[18]

The ash from the grate discharges into an ash hopper located at the front of the boiler. A hopper is also provided under or at the rear of the grate for the collection of ash which sifts through the grate. The ash from the grate discharge (bottom ash) and ash from the siftings hopper can be handled dry or wet. With dry ash handling, the ash can be conveyed by either a pneumatic system, a vacuum system, or a mechanical conveyor to a storage silo. At the silo the ash is conditioned with water when loading the ash into a truck for disposal. The pneumatic and vacuum systems are not considered practical for RDF units because pieces of metal or other large unburned items may plug these systems.

With a wet ash-handling system, the bottom ash and siftings discharge into a water-filled hopper beneath the boiler. The ash can then be removed by a hydraulic sluice system or a drag chain conveyor. The hydraulic sluice system uses an ash pump or venturi (jet pump) to convey the slurry to a silo where it is dewatered and hauled to a landfill or to convey the slurry to a settling pond. The drag chain conveyor system utilizes an inclined section of the conveyor to dewater the ash before it discharges onto another conveyor or directly into a truck for disposal. Because of the possibility of wire, metals, and pieces of unburned material in the bottom ash the wet system using a drag chain conveyor should be considered for RDF firing units. If a wet-type ash handling is used, the ash should be removed continuously and not stored for any great length of time because of the cementitious properties of the ash. Likewise, if a dry system is considered the ash should be removed on a continuous basis because of the potential for fires and explosions due to hot embers and unburned material remaining in the ash.

The combustion air for stoker-fired boilers is provided by a forced draft system. The air, generally preheated, is divided into underfire air and overfire air. The underfire air is supplied to the combustion zone through a plenum under the grate. The plenum is sectionalized and contains dampers to control the air supply. The high-pressure drop across the grate ensures even distribution of the air supply over the entire grate area. The overfire air is supplied to the combustion zone through multiple overfire air nozzles above the grate. The nozzles can be located on the side walls or front and rear walls of the furnace. The overfire air provides the required turbulence and mixing to ensure burnout of suspended particles and volatile matter. The amount of overfire air will depend on the quality of the fuel and the amount of excess air required for combustion. RDF stoker-fired units should be capable of providing from 15 to 25%[15] of the total combustion air required as overfire air. The quantity of excess air for RDF firing is typically between 40 to 50%.

Spreader stoker-fired units are provided with boilers for the recovery of the thermal energy released during combustion. The boilers generally consist of a furnace section, superheater section, boiler bank, and economizer section.

The construction and arrangement of the furnace section may vary from manufacturer to manufacturer. They generally utilize an all-welded membrane water-wall type construction. A portion of the heat released during combustion is absorbed by water circulating through the tubes. Some furnace designs have refractory installed on the tubes in the area above the grate to protect the tubes from erosion and corrosion. Most boilers, however, have bare metal tubes throughout the furnace section. The upper, rear wall of the furnace generally contains an arch to help distribute the combustion gases through the convection section of the boiler and to protect a pendant superheater from radiant heat. Furnace sections must be liberally sized to provide adequate residence time to complete combustion of the fuel and to reduce the gas temperatures before entering the convection section of the boiler. Also, the gas velocity through the furnace must be low to minimize the entrainment and carryover of the fuel. For RDF-fired spreader stoker boilers, the furnace section is generally designed for a gas velocity of 15 to 17 ft/sec (4.5 to 5.1 m/sec), a maximum furnace exit gas temperature of 1600°F (871°C) with 1400°F (760°C) being preferred, and a volumetric heat release rate of 12,000 to 15,000 BTU/hr·ft^3 (4.4 to 5.5 × 10^5 kJ/m^3·hr).

The superheater sections, when used, provide the final heat to the boiler outlet steam and are used to superheat the saturated steam generated in the steam drum. Depending on the final steam temperature desired, either two-stage or single-stage superheaters are used. The single-stage superheaters are provided on low-temperature boilers, generally below 600°F (315°C) on RDF-fired units, and use a terminal-type spray attemperator to control final steam temperature. Two-stage superheaters are provided on high-temperature applications, generally above 600°F (315°C) for RDF-fired units, and utilize a spray attemperator between the two stages to control the final steam temperature. For RDF spreader stoker-fired boilers other considerations in the design of a superheater are tube spacing, gas velocity, and tube metal temperature. To avoid fouling, plugging, and erosion problems in RDF-fired boilers, the tube spacing and gas velocity in the superheater should be conservative, approximately 6 in. (152 mm) side spacing center-to-center, and a maximum velocity of 30 ft/sec (9 m/sec). To minimize corrosion on the outside surface of the superheater tubes, the steam temperature limitation is required. Current practices limit the steam temperature to approximately 750°F (399°C) and tube metal temperature to 850°F (454°C).

The boiler bank is the section of the boiler where saturated steam is generated. This section generally consists of two drums connected with a rear and forward bank of tubes. The upper drum (steam drum) contains steam separators and secondary scrubbers for controlling the solids and moisture content in the steam. The lower drum (mud drum) is where solids that remain from the boiling of the water are concentrated. As with the superheater section, the tube spacing and gas velocities in the boiler bank are critical. For fuels, such as RDF, which tend to have a fouling and abrasive-type ash, the tube spacing should be wide to minimize plugging and to maintain low velocities. A tube side spacing of 5 in. (120 mm) center-to-center and a maximum velocity of 30 ft/sec (9 m/sec) are generally specified for RDF-fired units.

The economizer section is provided on boilers to add heat to the incoming feedwater and reduce the temperature of the exiting flue gas. The tube spacing and gas velocities through economizer sections on RDF-fired boilers should be similar to gas velocities and spacing in the boiler bank.

To provide effective cleaning of tube surfaces in spreader stoker-fired boilers, sootblowers are generally installed. The sootblowers can use either high-pressure steam or air as the blowing medium. Recent boiler designs have incorporated rappers and sonic systems to try to clean units that fire fuels with slagging and fouling characteristics. Because of the abrasive nature of RDF ash, care must be taken when selecting and locating sootblowers.

Regenerative or tubular air heaters are generally provided on spreader stoker-fired units to preheat combustion air. The air heaters are located in the gas stream after the economizer. Spreader stoker boilers burning RDF generally utilized tubular-type air heaters to preheat the combustion air to approximately 350°F. The flue gas temperature exiting the air heaters must be controlled and maintained above the acid dewpoint of the gas. If the flue gas temperatures are allowed to cool too low, acids may condense in downstream equipment resulting in corrosion. For RDF-fired units, the flue gas temperature is usually maintained between 400 to 450°F (204 to 232°C).

Air pollution control equipment is required on spreader stoker-fired units to meet local, state, and federal emission requirements. Depending on the type of fuel, plant location, and size, the air pollution control equipment required will consist of either precipitators, baghouses, or scrubbers or a combination of these units. For RDF-fired units, precipitators have been accepted in most areas. Some states, such as California, Connecticut, and New Jersey are requiring the installation of either wet or dry scrubbers.

The RDF for spreader stoker-fired units require only minimal processing. The processing system can consist of a single shredder, magnetic separation, and screening to produce a RDF of a nominal size of 4 in. (100 mm). The fuel should be 90 to 95% less than 4 in. (100 mm) in size with minimal streamers and a maximum size of 6 in. (150 mm). This type of fuel is classified as RDF-2 per ASTM standards.

The RDF is fed at a controlled rate to the boilers from live bottom metering bins which generally have a storage capacity of from 15 to 20 min. The metering bins are located at an elevated position in front of the boilers. The metered RDF is fed through downspouts to the air-swept spout distributors. A typical spreader stoker, RDF-fired boiler is shown on Figure 12.14. A spreader stoker boiler for cofiring is shown on Figure 12.15.

Spreader stoker units can be designed for 100% RDF firing or for cofiring with coal, oil, or gas. On units that would cofire RDF with either oil or gas, the oil or gas burners are generally located above the RDF distributor spouts to prevent overheating of the grates. New units can be designed for cofiring and most existing stoker-fired units can generally be modified for cofiring. The modification of existing units will generally result in a substantial capital cost savings to projects.

The largest RDF spreader stoker-fired units in operation to date are the 300,000 lb/hr (135,900 kg/hr) units at the Occidental Chemical Corp. facility. The limiting factor in unit size is generally the design of the grate. The largest single unit available today is approximately in the 400,000 lb/hr (181,200 kg/hr) size range.

The advantages to RDF-fired spreader stoker units include:[18]

1. Faster boiler response than for mass burn.
2. Units can be designed for cofiring.
3. Higher thermal boiler efficiency because of lower excess air requirements.
4. Boiler island and overall plant costs are generally lower than for mass burn units.
5. Potential for sewage sludge codisposal.

These types of units have the following disadvantages:

1. The need to build, own, and operate a prepared fuel system is required.
2. Because of the required processing facility the overall facility horsepower requirements may be higher than for mass burn units.
3. Depending on the process, some combustible material may be lost in processing.

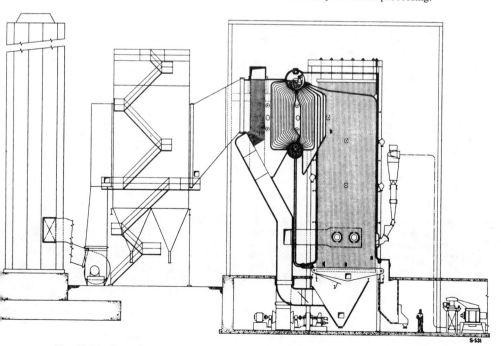

Fig. 12.14 Spreader stoker RDF-fired boiler. (Courtesy of Babcock and Wilcox.)

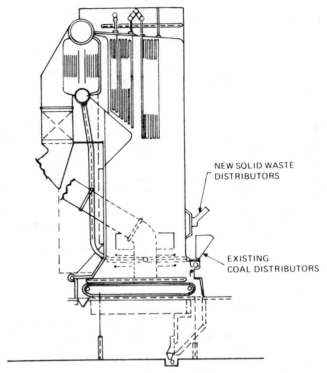

NEW SOLID WASTE
DISTRIBUTORS

EXISTING
COAL DISTRIBUTORS

Fig. 12.15   Spreader stoker boiler cofiring RDF and coal. (Courtesy of Combustion Engineering.)

The above discussions provide general guidelines and "rules-of-thumb" that the engineer and designer can use when considering prepared fuel-fired spreader stoker units.

12.3.2 Suspension-Fired Units

Suspension firing, the complete, rapid burning of a solid fuel suspended in the combustion zone, was first tried in utility boilers during the early 1920s with the development of pulverized coal equipment and firing systems. The application of suspension firing of a prepared solid waste fuel was first employed at the St. Louis demonstration facility in 1972. The facility cofired RDF with coal in a utility boiler.

The boilers on suspension-fired units are very similar to the boilers on spreader stoker-fired units. They both utilize a waterwall-cooled furnace section. Both types of units can contain a superheater section, a boiler section, and economizer section. Modern suspension-fired units (pulverized coal) generally utilize a single drum design, with primary and secondary superheater section and a reheater section and are generally designed for steam conditions up to 2400 psig (16,548 KPa) and 1000°F (538°C). Suspension-fired boilers also use combustion air heaters and precipitators, baghouses, scrubbers, or a combination of precipitators or baghouses and scrubbers to meet emission requirements.

The major differences between spreader stoker-firing and suspension-firing are the requirements for fuel preparation and method of firing the fuel. Because the average retention time for suspension firing is generally less than 2 sec,[8] the fuel requires a high degree of processing to control the particle size and to reduce the moisture content. On pulverized coal-fired units, coal is reduced in size, generally 85% minus 200 mesh, and dried in several types of pulverizers or mills. For the suspension firing of RDF the fuel should have a size of 90 to 95% less than 1.5 in. (38 mm). This will generally require two stages of shredding. Also, to reduce the moisture content and minimize erosion in the RDF fuel-handling system, municipal wastes are screened, magnetically separated, and/or air-classified to remove the majority of the metals, sand, glass, and grit in the waste.

Suspension-fired boilers employ two methods of firing—horizontal firing and tangential firing. On horizontally fired units, the burners are located in rows on the front and/or rear walls of the furnace. To provide the turbulence required for complete combustion, the burners impart a rotational motion to the flame by introducing the fuel and primary air tangentially to the burners and by using adjustable vanes in the secondary air portion of the burner.

Tangentially fired boilers use burners located in the corners of the furnace. The burners direct the fuel and combustion air along a line tangent to a small circle laying in a horizontal plane within the furnace.[18] This section imparts a cyclonic motion to the flame. The fuel nozzles can be tilted to raise or lower the flame within the furnace.

In suspension units firing coal, the coal is pulverized and is then pneumatically conveyed directly to the burners using heated primary air. For RDF-fired units, the prepared fuel is either mechanically or pneumatically conveyed from the processing system to a fuel storage facility. From the fuel storage facility, the RDF is metered and fed pneumatically to the individual burners on the furnace. Either heated combustion air or ambient air can be used for conveying the RDF.

Within the furnace, the fuel must be entrained in the vertically rising gas stream or it will fall to the ash pit. Suspension-fired units are designed for a furnace gas velocity of approximately 20 ft/sec (6 m/sec)[8] to maintain the fuel in suspension. Since RDF has a larger size distribution and higher moisture content than coal, some of the fuel will not remain in suspension. Dump grates are added to the bottom of the furnace as a means to provide additional residence time to complete combustion. Combustion air nozzles are generally provided above and below the grate to ensure complete combustion of the heavier particles.

Because of the high gas velocities through the furnace, suspension-fired units produce a higher percentage of fly ash than spreader stoker units. This requires that the downstream particulate control and ash-handling equipment be sized to handle the higher loading. For pulverized coal-fired units, approximately 70 to 80% of the coal ash is carried over as fly ash. When RDF is fired alone or cofired with coal, the ash carry over will be slightly lower than this depending on size distribution and moisture content of the RDF.

The design criteria for the furnace and convective sections for RDF suspension-fired units are similar to those on spreader stoker-fired units. The furnace should be conservatively sized to provide a maximum furnace outlet temperature of 1600°F (871°C) with 1400°F (760°C) preferred. The maximum steam temperature is generally set at 750°F (399°C) to minimize tube metal corrosion. The convective sections should have wide tube spacing and have a maximum gas velocity of 30 ft/sec (9 m/sec) to minimize turbulence, tube erosion, and ash fouling. The tubes should be arranged in-line to permit effective cleaning with retractable sootblowers. Precautions should be taken to minimize or eliminate corrosion in downstream pollution control equipment by maintaining the flue gas temperature well above the acid dew point. Current practice is to design the units for a flue gas exit temperature of 400 to 450°F (204 to 232°C). Boilers that cofire RDF with coal may be designed for higher steam pressures and temperatures because corrosive effects from the combustion of RDF may be reduced by the coal ash. Suspension units designed for cofiring of RDF with coal are normally limited to firing 15 to 20% RDF on a heat input basis. The design parameters for RDF firing should be closely reviewed with the boiler manufacturer. Figure 12.16 illustrates a typical RDF suspension-fired boiler.

The ash-handling systems for suspension-fired units are similar to those of spreader stoker-fired units. The bottom ash is either handled by a hydraulic sluice system or a mechanical dewatering-type drag chain conveyor. The fly ash can be handled with pneumatic-type systems or mechanical systems such as screw or drag chain-type conveyors. As with spreader stoker-fired units, the design of the ash-handling system must receive careful attention and evaluation.

Although it is feasible to fire 100% RDF in suspension units there are no facilities currently utilizing this method. All suspension-fired units currently burning RDF are doing so by cofiring the prepared fuel (baseload) with coal (or oil) in existing utility or industrial boilers. The use of existing facilities is more economical thereby enhancing the feasibility and viability of a project.

Suspension firing of prepared fuels has numerous advantages over other types of combustion systems. These include faster boiler response rate, smaller grate, higher efficiencies, capability of modifying existing units for cofiring thus reducing costs, potential for lower boiler tube metal corrosion because of the homogenous nature of the fuel, and capability for codisposal of sewage sludge.[16] However, suspension firing may have higher fuel preparation equipment costs and higher auxiliary power requirements because of the additional processing required.

12.3.3 Fluidized Bed Units

Since the initial development in Germany in the 1920s, fluidized bed technology has been commercially applied in the oil refinery and steel industries.[19] The technology is used in the oil refineries to extract more gasoline from crude oil and in the steel industry for ore roasting. More recently, there has been an increasing interest in the application of fluidized bed boilers for the generation of high-pressure and -temperature steam in utility and industrial power plants. This recent interest is primarily due to the ability of the fluidized bed boilers to burn low grades and wide varieties of fuels and to control NO_x and SO_2 emissions.

The application and interest in the fluidized bed technology is world wide and is considered to

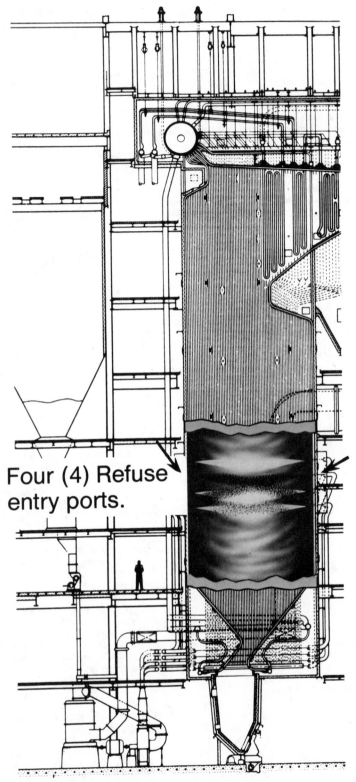

Four (4) Refuse entry ports.

Fig. 12.16 Suspension-fired boiler for cofiring RDF and coal. (Courtesy of Combustion Engineering.)

have passed the proof-of-concept stage. The technology is still undergoing refinement. However, units are commercially available in the 10,000 to 600,000 lb/hr (4530 to 271,800 kg/hr) steam capacity range using technology that has undergone extensive testing and development.

In fluidized bed combustion, the fuel is burned in a turbulent bed of inert material, such as sand, limestone, or ash. The bed is kept suspended and turbulent in the lower section of the combustion chamber by air that enters through a distribution system below the bed. The fluidizing action created in the bed provides the turbulent action required for good combustion and make the bed of materials act like a liquid.

Fluidized bed combustion units have many configurations. The units can be of the fire tube or water tube design with the combustion chambers integral or separate from the heat-recovery unit. The combustion chambers can have single-stage or two-stage beds. The beds can also be dilute-phase beds or dense-phase beds or a combination of both. Figure 12.17 shows a schematic of a fluidized bed boiler. Even with the wide variation in units available and being developed, they all contain the following basic elements.

Fuel and bed material feed system
Air distribution system
Fluidized bed section (combustion chamber)
Heat-recovery system
Ash removal and bed material recirculation system
Emissions control system

The fuel feed systems are either an underbed or overbed design. In the underbed design, the fuel is generally fed pneumatically to the combustion chamber. Multiple feed points are required to

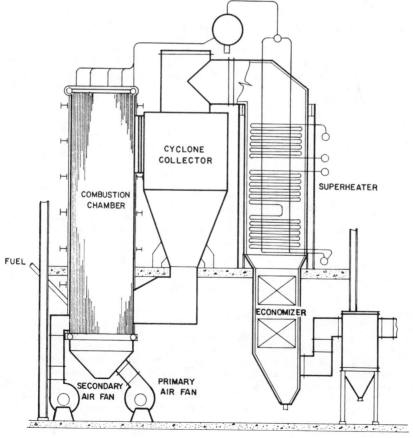

Fig. 12.17 Schematic of a fluidized bed boiler.

provide proper distribution of the fuel. As combustion chambers become larger, the problems of providing even fuel distribution increase. Current underfed units provide points for every 9 to 20 ft^2 (0.84 to 1.86 m^2) of the bed area. The overfed units can utilize conventional feeder units found on spreader stoker-fired boilers or can use pneumatic- or gravity-type systems. Multiple feed points are also required, however, the distribution of the fuel is easier.

The bed material feed and makeup systems are required to supply new material and also to vary the height of the beds to control the load. The feed and makeup can be incorporated into the fuel feed system or the bed material recirculation system. Also, separate pneumatic or gravity feed systems can be employed.

The design of the air distribution system is critical to the operation of a fluidized bed unit since it provides the required fluidizing action and combustion air. The design of the distribution systems can range from perforated plates to multiple layers of sparging nozzles in the bottom of the combustor. Some units contain nozzles that supply air tangentially to the bed to provide in-bed circulation. A zoned air plenum beneath the bed supplies and controls the air distribution to the various zones of the chamber.

The fluidized bed section (combustion chamber) contains the distribution plates, bed of fuel, and inert material, and is where the combustion of the fuel occurs. As previously noted, the combustion chambers can have a two-stage bed system or dilute phase bed section with dense bed section. In the two-stage bed design, the lower bed is the combustion or burning bed and the upper bed is used for desulfurization. A hybrid concept utilizes a dense bed where the combustion reaction occurs and a dilute phase bed where the fine particles are recirculated. The combustion chambers are either water cooled or refractory lined. On most units, each combustion chamber is sectionalized to provide flexibility in controlling the units.

Heat-recovery systems are used on fluidized bed units to control temperature in the combustion zone as well as recovery of thermal energy. A heat sink, in the form of in-bed tube surface or waterwall tubes, must be installed in the combustion zone to control temperatures. The combustion temperatures are generally maintained in the 1500 to 1600°F (816 to 871°C) range. The heat-recovery sections can be integral to the combustion section or can be completely separate. Separate heat-recovery sections and combustion sections are generally provided on facilities that utilize an existing boiler or on units that have multiple combustion chambers. The heat-recovery units can be designed to generate hot water for district heating, steam for utility turbine generators or industrial uses, or to heat air to drive a gas turbine. Also, energy recovery from fluidized bed units can be in the form of cogeneration where hot flue gases from the combustion are passed through a gas turbine and the waste heat from the gas turbine is passed through a heat-recovery boiler.

Ash removal and bed material recirculation systems are required on fluidized bed units to ensure good combustion characteristics and to improve SO_2 absorption and limestone utilization. The bottom ash and spent or reacted bed material must be removed to maintain the height and quality of the bed material. This is accomplished by the ash removal system. Ash and bed material are extremely hot, therefore, a cooling system, generally using feedwater, is required. The bottom ash removal systems can be either mechanical or pneumatic and are incorporated into the fly ash-handling system. The recirculation systems reinject a portion of the ash and bed material that escapes with the flue gas. The material is collected in a cyclone and reinjected into the combustion chamber.

The control of particulates from fluidized bed units is accomplished by the use of cyclone and either baghouses or electrostatic precipitators. The ash may contain larger-than-normal quantities of carbon and calcium compounds and lower sulfur-bearing compounds that can cause operational problems in precipitators and baghouses. The high carbon content can cause fires in a baghouse, whereas the calcium compounds and low sulfur content can cause resistivity problems in precipitators.

In conjunction with the use of a bed material such as limestone, the SO_2 emissions from fluidized bed units can be reduced greatly. In most cases, depending on the fuel, the use of downstream SO_2 scrubbers is not required. As a result of the low combustion zone temperatures, the formation of NO_x is greatly reduced.

Fluidized bed units offer the advantages, over conventional forms of combustion, of (1) less volatilization of alkali compounds, (2) relative insensitivity to the quantity and nature of ash in the fuel, (3) smaller furnace volumes, (4) ability to fire a wide variety of fuels, (5) the ability to remove SO_2 in the combustion chamber with the use of limestone for the bed material, and (6) low NO_x emissions. The disadvantages of fluidized bed combustion are that the technology is relatively new and is still evolving, and that because of the use of an abrasive bed material many units have experienced erosion on surfaces within the bed.

The utilization of fluidized bed units for the combustion of prepared solid waste fuels is not readily established. There have been only two known units that have used RDF in fluidized bed units. These are the Western Lake Superior Sanitary District Co-disposal Facility in Duluth, Minnesota, and a district heating plant in the City of Eksjö, Sweden. The Duluth facility is not currently burning RDF because of operational problems. The Eksjö facility, which started up in

1979, has been operating satisfactorily.[19] With careful design and future testing, the use of fluidized bed combustion can and probably will become a viable alternative for combustion of prepared solid waste fuels.

12.3.4 Cyclone Furnace Firing

The cyclone furnace was developed primarily to improve on and to overcome some of the problems pulverized coal-fired boilers experienced with low-grade, high-ash coals. The first commercial cyclone furnace boiler was installed at a power plant in 1944.[17]

The cyclone furnace is essentially a refractory-lined, water-cooled, horizontal chamber for the firing and combustion of the fuel, with the cyclone furnace separate from the boiler furnace section. The unique feature of the cyclone furnace is that during the combustion process high heat-release rates are developed along with a high gas temperature, often exceeding 3000°F (1650°C).[17] In addition to coal, the cyclone furnace is capable of firing many other types of fuels including oils, gases and by-products of coal, petroleums, wood, and prepared solid wastes. A cyclone furnace is illustrated in Figure 12.18.

The fuel, along with approximately 20% of the combustion air (primary air), is introduced tangentially into the burner end of the cyclone furnace. High-velocity secondary air, at approximately 300 fps (90 m/sec) is supplied tangentially to the main section of the furnace. The action of the fuel, primary and secondary air, imparts a cyclonic motion to the combustion process causing the fuel and ash particles to be thrown to the walls of the furnace by centrifugal force. This action causes the high turbulence required for rapid combustion. Also, the fuel particles are held in a thin layer of slag on the furnace wall. The incoming secondary air scrubs the fuel held by the slag,

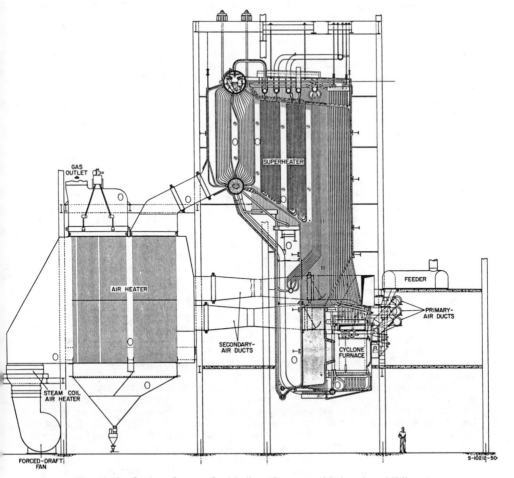

Fig. 12.18 Cyclone furnace-fired boiler. (Courtesy of Babcock and Wilcox.)

supplies the required oxygen to burn the fuel held in the slag, and removes the products of combustion.

The relatively small size of the cyclone furnace, 6–10 ft in diameter, results in low heat absorption rates within the furnace itself. This, in combination with the high heat-release rates, results in the very high temperatures developed. These high temperatures ensure complete combustion and also melt the ash into a slag within the furnace chamber.

The hot combustion gases are discharged from the cyclone furnace into a boiler furnace. The boiler furnace and convective sections are very similar in design to the spreader stoker and suspension-fired boilers. The boilers consist of a waterwall-cooled furnace section followed by a convective section consisting of superheaters, reheaters, boiler bank, and economizers depending on the particular application. Combustion air is preheated by either regenerative or tubular air heaters.

The cyclone furnaces are arranged on either a single wall of the boiler furnace or on opposite walls of the boiler furnace. The number and location of the cyclones depends on the size of the unit.

The cyclonic section of the combustion process results in most of the ash being formed into a slag within the furnace. The molten slag drains from the cyclone furnace through a slag tap and discharges to a slag tank at the bottom of the boiler furnace. The slag is quenched in the tank and is either periodically or continuously removed via a hydraulic sluicing system or drag chain conveyor. The fly ash, approximately only 20 to 30% of the total ash, is collected in downstream ash hoppers and air pollution control equipment such as precipitators.

The advantages of a cyclone furnace-fired boiler over a pulverized coal suspension-fired boiler are the cyclone unit does not require the finely pulverized coal thus reducing power requirements, a major percentage of the ash is removed in the furnace section, the low quantity of fly ash minimizes tube erosion, the cyclone furnaces operate with only 10–15% excess air, and the units are well suited for multiple fuel firing. The cyclone furnace has one potential problem when burning fossil fuels. As a result of the high combustion temperatures, the formation of NO_x is greatly increased.

There are two facilities that have utilized cyclone furnaces for the recovery of energy from the combustion of a prepared solid waste. One was at a utility boiler in Bridgeport, Connecticut. A powdered RDF (RDF-4) produced at a nearby facility, was cofired at rates of up to 10% on a heat input basis in the boiler with oil. This facility is no longer in operation with RDF-4 because of problems with a sophisticated processing system. The cyclone firing, however, was considered successful.

C.P. Crane Station of Baltimore Gas & Electric Co. has burned 400 TPD of RDF-3 in a cyclone furnace since early 1984 with ¾ in. × 0 (20 mm × 0) coal. The RDF is furnished by the Baltimore County shredding plant operated by National Ecology, Inc., at Cockeysville, Md. sixty days of successful test firing led to the contract.

12.3.5 Pyrolysis

Pyrolysis is a process whereby material, such as coal, wood, heavy oils, solid waste, is thermally decomposed in an oxygen-deficient atmosphere. Pyrolysis requires the addition of heat or is an endothermic reaction, whereas incineration releases heat or is an exothermic reaction. Depending on the type of process used, pyrolysis produces a gaseous or liquid product and a carbonaceous char.

Commercially, pyrolysis has been used for years for the production of charcoal, methanol, and turpentine from wood, and coke from coal or heavy petroleum oils. More recently pyrolysis processes have been developed for the production of liquid and gaseous fuels, oils, and methane, from coal and other materials such as solid waste. In the mid-1970s, several pyrolysis projects were implemented to use prepared solid wastes as a feedstock.

The composition and yield of the pyrolysis process products can be varied by controlling operating parameters such as pressure, temperature, time, feedstock particle size, and the use of catalysts and auxiliary fuels. High-temperature processes, greater than 1400°F (760°C), are used to produce primarily gaseous products such as hydrogen, methane, carbon monoxide, and carbon dioxide. Low-temperature processes, from 850 to 1300°F (454 to 740°C), produce primarily liquid products such as oils, acetic acid, acetone, and methanol. In all pyrolysis processes, a carbonaceous residue (char) is produced.

The types and quality of products may be controlled by utilizing oxidizing agents, such as air, oxygen, and water, or reducing agents, such as hydrogen or carbon monoxide. The addition of hydrogen, hydrogeneration, or hydrogasification produces higher-quality oils or methane gas having a higher heating value approaching that of natural gas. The addition of air for partial oxidation produces a fuel gas that is diluted by nitrogen and has a low heating value. The addition of water may produce carbon monoxide and hydrogen that can be converted to natural gas substitutes.

The quality of the char residue is greatly dependent on the quality of the feedstock. In the pyrolysis process, chemicals and residues such as sulfur, metals, glass, and ash remain in the char. If the feedstock contains significantly high quantities of these materials, the char may become contaminated and unusable.

The types of equipment and systems developed for pyrolysis have been numerous. Systems have been developed as continuous or batch feed-type processes. The types of equipment utilized can range from refractory-lined rotary kilns, vertical shaft-type reactors, regenerative towers, gasifiers, to feedstock processing systems. Some systems use a portion of the gases or oils produced to sustain the pyrolytic reaction while others use auxiliary fuels such as natural gas or coal. Still other systems blend the char with the incoming feedstock to enhance the reaction. In all cases, the feedstock, whether coal, wood, or solid waste, is processed to produce a uniform material, is heated to drive off water, is fed to a pyrolytic converter or reactor, and is heated further to drive the reaction. The pyrolytic gases or liquids and chars that evolve are collected and either used directly or processed to produce higher-quality fuel products.

The use of a prepared solid waste as a feedstock for pyrolysis has met with limited success. There are currently two facilities, one in Japan and one in France, that are operational. To date, the application of pyrolysis in the United States has been unsuccessful both technically and economically. Two, 200 TPD demonstration facilities, one in South Charleston, West Virginia, and one in San Diego, California, have been shutdown. The pilot plant in South Charleston was designed to yield a medium BTU gas. The San Diego plant was designed to produce a high BTU liquid fuel. Two commercial facilities, one in Baltimore, Maryland (Landgard), and one in Florida (Disney World) have also been shutdown. Problems encountered at all facilities have mostly been due to insufficient design controllability of the quality of the fuel product and other technical problems such as corrosion. Even with these failures, companies are still investigating and testing the feasibility of utilizing pyrolysis for the recovery of energy from prepared solid waste.

For pyrolysis systems using prepared solid waste as a feedstock, the waste (RDF) must generally be finely shredded and screened to produce a high-quality material that has the majority of the glass, sand, grit, and metals removed. A typical feedstock may be RDF-3 or RDF-4.

The primary advantage of a pyrolysis system is that refuse can be transformed into gaseous or liquid fuel products that can be utilized by a wide variety of customers and in conventional boilers. However, the major limitation is that the technical and economic feasibility of a large-scale facility has yet to be proven.

12.3.6   Cement Kilns

The manufacturing of portland cement began commercially in the United States in the 1870s. The process used in cement manufacturing is very energy intensive. As a result of this, the process has great potential for energy utilization from the combustion of prepared solid waste fuels. To date the use of RDF as a fuel or fuel supplement has been very limited. In the United States, the combustion of RDF in the cement manufacturing process has been limited to test firing. In England, one facility occasionally utilizes RDF as a fuel supplement.

To understand better how RDF may be used in the cement manufacturing process, it is important to understand the process itself. The cement manufacturing process consists of basically five steps: (1) grinding and mixing of raw materials, (2) drying, (3) calcining, (4) clinkering, and (5) pulverizing. The thermal portion of the manufacturing process—steps 2, 3, and 4—requires large quantities of energy. In the drying stage, the raw materials are dried at temperatures of approximately 200 to 250°F (93 to 121°C). The calcining stage utilizes temperatures in the range of 1000 to 1500°F (538 to 815°C) to dehydrate the raw material and to burn off the carbon dioxide from the limestone. Clinkering, the next step, is the chemical reaction in which the calcined raw materials are hardened into granules of calcium silicates, calcium aluminates, and calcium ferrites. Clinkering, sometimes called burning, occurs at a very high temperature, in the range of 1800 to 2700°F (982 to 1482°C).

All three stages of the thermal process can occur in the same piece of equipment, a rotary kiln. Approximately 80% of the cement produced in the United States is produced using this process. The remaining portion is manufactured at new facilities which separate the drying process from the clinkering process. In the facilities incorporating all three stages into one kiln, the wet or dry mixture of raw material is placed in the back end of the kiln and travels countercurrent to the flow of hot gases through the three stages. The clinkered material exits the front, or combustion end, of the kiln. The composition of the raw material and temperatures requires careful control to account for any ash that may be in the fuel.

The heat required for the process is generally supplied by suspension firing of the fuels through burners at the front end of the kilns. Over 60% of the cement in the United States is produced in units firing pulverized coal. Other facilities use gas, oil, petroleum coke, or electrical power. The current trend is to convert existing cement plants to coal as the primary fuel. Figure 12.19 illustrates a typical rotary kiln.

In kilns that burn fuels with an ash content, a majority of the ash falls to the bed of the kiln and is combined with the cement clinker. The ash that remains in suspension (fly ash) and fine cement powders that are entrained in the gas stream exit the back end of the kiln and are collected in cyclone separators and/or precipitators or baghouses. The fine ash collected is sometimes mixed with the raw material and refed through the kiln.

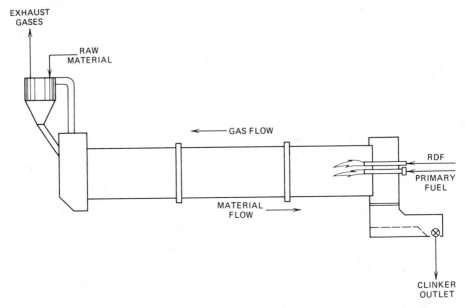

Fig. 12.19   Rotary cement kiln.

Since the raw material contains a high percentage of calcium oxide, the kiln acts as a scrubber for the removal of sulfur oxides and chlorine compounds. This reduces some of the pollutants that could potentially exit the kiln and permits the use of higher-sulfur fuels. In limited quantities, sulfur and chlorine compounds are acceptable in the cement product.

In addition to the energy recovery potential of firing RDF in cement kilns, the features of the ash being entrained in the final product and the scrubbing effect offer the advantages of reducing residue disposal requirements and minimize the need for expensive flue gas scrubbers, respectively. Even with these advantages there has been very little progress in using RDF as a fuel supplement. This is in part due to the wariness of the cement industry, the lack of long-term operating data, and the availability of RDF near cement plants.

There have been over eight portland cement companies that have test-fired RDF as a supplementary fuel in kilns. Four facilities, one in England, one in Canada, and two in the United States have conducted extensive testing with RDF.[21] The tests have shown that RDF, up to 30% on a heat input basis, can be fired in kilns with no major problems. The RDF is pneumatically fed to a burner located above the existing burner. However, the tests indicated that a highly refined and finely shredded fuel is required to minimize contamination of the cement with glass, metals, and unburned carbon. Since the kilns are refractory lined and because of the scrubbing action within the kiln, there should be no corrosion problems as has been experienced in boilers firing RDF.

With the increasing interest in developing resource recovery facilities, the firing of RDF in cement kilns can offer another viable alternative for the recovery of energy from prepared fuels. Several major points must be considered, however, when evaluating the use of RDF in kilns. These include:

1. Fuel-handling properties
2. Temperature control
3. Effect of combustion on kiln coatings, clinker chemistry, and air emissions
4. Changes in physical properties of cement
5. Economics
6. Location of cement plants with respect to sources of RDF

Chapter 18 discusses refuse fuels in the portland cement industry.

12.3.7   Bioconversion

Bioconversion is the use of an anaerobic digestion process to produce a fuel, methane gas, from prepared solid waste. In the anaerobic digestion process, bacteria and enzymes attack the organic fraction of waste converting the carbon, hydrogen, and oxygen to methane and carbon dioxide.

The technology for anaerobic digestion of wastes has been used for a long time. The use is employed mainly in wastewater treatment plants to digest sludge and produce energy for plant operations. It was not until the late 1970s that this process was used on prepared solid waste.

A typical bioconversion process may consist of four main systems. In the first system, the front-end system, raw municipal solid waste is processed through various stages of screening, magnetic separation, and shredding to produce a prepared solid waste product that is compatible with the digestion process. The processing system would be very similar to those utilized for the production of a RDF-3 fuel. The process could be either a dry or wet system.

In the second system, the feed and digestion system, the prepared solid waste is blended with nutrients and recycled liquids and then fed into anaerobic digesters. Sewage sludge can also be mixed with the prepared waste to help codispose of both types of wastes. The digesters are essentially large tanks equipped with agitators, air lock feed systems, air seals, effluent removal system, and gas collection and metering system. In the digesters, anaerobic bacteria and faculative bacteria breakdown and decompose the organic matter, producing methane and carbon dioxide gases. The process requires elevated temperatures, in the 130 to 140°F (54 to 60°C) range and that the material be held in the digester for 5 to 7 days. Steam is generally used to control the temperature.

The third system consists of the effluent treatment system. For bioconversion systems to be effective, utilization of the effluent from the digesters is essential. The digester residue is dewatered using either cone presses, vacuum filters, or centrifuges. The liquid is recycled for use as a make up to the feed slurry. The resultant sludge, occupying 25 to 30% of the original feedstock volume, is relatively stable and can be burned to provide energy to operate the system, landfilled, or possibly used as a soil conditioner.

The gas recovery system, the fourth system, consists of gas collection and metering equipment. The gas produced during the anaerobic digestion process contains approximately 55 to 60% methane and 40 to 45% carbon dioxide. The gas may be upgraded to produce pipeline-quality gas or may be burned in existing equipment, such as boilers or gas engine generator sets to produce steam or electrical energy.

There are no commercially sized facilities using bioconversion of prepared solid waste in operation today. A 50 to 100 TPD "proof-of-concept" facility in Pompano Beach, Florida, has been in operation since November, 1978. The facility shreds, screens, and air classifies the municipal solid waste. The light fraction is blended with sewage sludge, water, and nutrients and is then anaerobically digested in tanks for 5 days to produce a medium BTU gas. A schematic of the system is shown on Figure 12.20. Chapter 19 discusses biological processing in detail.

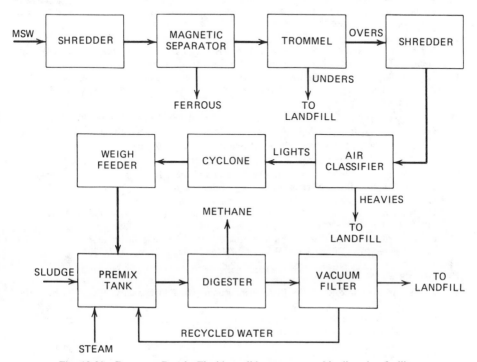

Fig. 12.20 Pompano Beach, Florida, solid waste anaerobic digestion facility.

REFERENCES

1. G. E. Dreifke, D. L. Klumb, and D. J. Smith, "Solid Waste as a Utility Fuel," Union Electric Company. Presented at American Power Conference 35th Annual Meeting, May 10, 1973, 22 pp.

2. E. K. Dille and D. L. Klumb, "Recycling Solid Waste for Utility Fuel and Recovery of Other Resources," Union Electric Company. Presented at 1973 Frontiers of Power Technology, 17 pp.

3. J. Reason, "RDF: Now you see it, now you don't," *Power*, December 1982, pp. 82–83.

4. H. Alter and J. J. Dunn, *Solid Waste Conversion to Energy*, Marcel Dekker, Inc., New York, 1980.

5. J. H. Fernandes and G. J. Prohazka, "Solid Waste and Coal Firing in Industrial Boilers," Combustion Engineering Inc. Presented at 5th Energy Technology Conference, Feb. 26–March 1, 1978, 8 pp.

6. D. E. Fiscus, et al, "RDF Cofiring in the Electric Utility Industry." Presented at ASME Solid Waste Processing Conference, June 10, 1983, 28 pp.

7. E. J. Duckett, "Technical and Marketing Issues Affecting the Use of RDF as Kiln Fuel," Schwartz & Connolly, Inc. Proceedings of the International Conference on Prepared Fuels and Resource Recovery Technology, New York, NY, Feb. 10–13, 1981, pp. 413–427.

8. W. H. Pollock, "Supplementing Coal with Solid Waste Fuels," Combustion Engineering, Inc. Presented at American Power Conference, Apr. 21–23, 1980, 12 pp.

9. Resource Recovery Activities, *City Currents*, October, 1984, U.S. Conference of Mayors, Washington, D. C.

10. R. E. Schwegler and H. L. Hickman, "Waste to Energy Projects in North America-A Progress Report," GRCDA Reports, Feb. 1981.

11. F. Hasselriis, "The Greater Bridgeport, Connecticut Waste-to-Power System," Combustion Equipment Associates, Inc. Proceedings of 1980 National Waste Processing Conference, pp. 435–445.

12. N. J. Weinstein and R. F. Toro, *Thermal Processing of Municipal Solid Waste for Resource and Energy Recovery*, Ann Arbor Science, Ann Arbor, Michigan, 1976.

13. B. Schwieger, Refuse-Derived Fuel Wins Out Over Oil and Coal at New Industrial Power Plant, *Power*, May 1978, pp. 33–40.

14. R. J. Vetter, Madison's Energy Recovery Program the Second Generation. Department of Public Works, City of Madison, Wisconsin. *Journal of Resource Management and Technology*, January 1983, pp. 11–16.

15. R. S. Rochford and S. J. Witkowski, "Considerations in the Design of a Shredded Municipal Refuse Burning and Heat Recovery System," Babcock and Wilcox. Presented to ASME 8th National Waste Processing Conference, May 7–10, 1978, 15 pp.

16. W. H. Pollock, "Generating Steam from Refuse in Industrial Boilers," Combustion Engineering. Presented at International District Heat Association 70th Annual Conference, June 18–20, 1979, 8 pp.

17. *Steam—Its Generation and Use*, The Babcock and Wilcox Co., 37th Edition, 1963.

18. *Combustion-Fossil Power Systems*, Combustion Engineering, Inc., Rand McNally, 1981.

19. J. Makansi and B. Schwieger, Fluidized Bed Boilers, *Power*, August 1982, pp. S-1–S-16.

20. L. M. Pruce, Waste Pyrolysis: Alternative Fuel Source, *Power*, June 1978, pp. 101–103.

21. C. R. Willey, "Co-firing Fluff RDF and Coal in a Cement Kiln," Maryland Environmental Service. Prepared for Conference of Waste-to-Energy Update 1980, April 15–16, 1980, 22 pp.

22. D. K. Walter, "Methane Production from Municipal Wastes in a 100 Ton-per-Day Plant." U.S. Department of Energy. Symposium Papers—Energy from Biomass and Wastes IV, Jan. 21–25, 1980, pp. 497–508.

23. C. H. Perron, "Refuse Conversion to Methane," Waste Management, Inc. Symposium Papers—Energy from Biomass and Wastes IV, Jan. 21–25, 1980, pp. 509–526.

24. J. C. Even, Ames, Iowa: A Successful Solid Waste Management Story, *NCRR Bulletin*, December 1979, pp. 87–92.

25. L. Nelson, Resource Recovery-Dade Plant Running at Full Capacity, *World Wastes*, May 1983, pp. 10–12.

26. L. R. Galese, Wheelabrator Finds Turning Garbage into Energy One Heap of a Problem, *Wall Street Journal*, Thursday, Nov. 16, 1978, p. 14.

27. H. G. Rigo, "Simulation as a Resource Recovery Plant Tool," Proceedings, ASME 10th National Waste Processing Conference, New York, 1982.

12.4 FUEL PROCESS SYSTEMS

George E. Boyhan

Background

The design of fuel process systems includes the preparation of raw solid waste as received in order to recover materials and prepare a refuse-derived fuel (RDF) suitable for burning with the option of generating steam and/or electricity. The first significant use of prepared solid waste began in the 1970s, and since then a total of 24 plants have been built, of which 14 are still in operation in the USA. The processes vary considerably and the sizes range from 100 (91 t) up to 3000 TPD (2722 t) of solid waste.

Most employ the so-called "dry process," which is a mechanical preparation of the incoming raw solid waste to recover saleable materials and to use the remainder as RDF for production of energy either at the plant site or at a remote utility installation. A typical installation includes mechanical size reduction by primary shredding followed by magnetic separation of ferrous materials and screening. A second-stage shredder or air classification can follow. The purpose is to introduce a fuel relatively uniform in particle size with a moisture content of approximately 30% and an ash content of less than 15%.

Most of the plants in operation were completed in the 1970s and are presently operating at 60–80% capacity. The largest plant operating at rated capacity has been Dade County, Florida, using a combination wet/dry process that started full operations in January, 1982. This is the largest solid waste plant in the world, with a rated capacity of 18,000 tons (16334 t)/week or 3000 TPD (2722 t). Fourteen plants presently in operation in the United States are shown in Table 12.6. Three of them produce RDF for sale to utilities with coal-fired boilers, with the RDF normally amounting to 10–20% of the total heat input. At 20% of total heat input of coal, the RDF requirement is about 40% by weight and 115% by volume. Coal and RDF are not compatible in conveyor handling and feeding systems, thus a system designed exclusively for coal is not suitable for RDF. Eleven of the plants produce RDF for combustion in dedicated boilers fired with prepared refuse fuel. A dedicated boiler is one designed for exclusive firing of RDF, usually a two-drum waterwall type with suspension and grate burning (spreader stoker).

There are ten plants that have been closed for a variety of reasons and they are shown in Table 12.7. The reason for failure in most cases was either technology or economic deficiency. Most of these plants chose to produce RDF for sale to reduce capital costs by eliminating the steam and

Table 12.6 Refuse-Derived Fuel Preparation/Utilization Plants, U.S.A.

Year Started	Plant	Rated TPD	Process	ASTM RDF Type
1970	Eastman Kodak Rochester, New York	120	D	3
1975	City of Ames, Iowa	200	D	3
1976	Baltimore County, Maryland	1200	D	2,3
1984	(Baltimore Gas & Electric Co.)	400	D	3
1979	City of Madison, Wisconsin	400	D	2
1979	City of Akron, Ohio[a]	1000	D	2
1976	General Motors Corporation Truck Coach Division Pontiac, Michigan	225	D	2
1980	City of Albany, New York	750	D	2
1980	Occidental Chemical Co., Niagara Falls, New York	2000	D	2
1981	Dade County, Florida	3000	D/W	3
1982	Wilmington, Delaware	1000	D	3
1982	City of Lakeland, Florida	300	D	2
1982	City of Henrico City, Virginia	400	D	2
1984	Refuse Fuels, Inc. Haverhill, Mass.	1300	D	2

D = Dry process.
D/w = Dry/wet process.
[a]Not operating pending decision to rehabilitate following a major explosion and fire, December, 1984.

Table 12.7 Prepared Fuel Plants Closed, USA.

Year Started	Plant	Rated TDP	Process	RDF Use
1974	Baltimore City, Maryland ('Landgard' Pyrolysis)	2010	D	3
1977	Chicago SW, Illinois	1000	D	3
1977	Milwaukee, Wisconsin (Americology)	1600	D	1
1978	Hempstead, New York	2000	W	3
1979	Bridgeport, Connecticut	1800	D	1
1979	E. Bridgewater, Massachusetts	300	D	1
1979	Duluth, Minnesota	400	D	3
1979	Lane County, Oregon	500	D	2
1981	Disney World, Florida	100	D	3
1984	Monroe County, New York	2000	D	1

Dry = dry process; W = wet process.
1 = Sale to utility; 2 = burned off site; 3 = burned on site.

electric power generation plant. Arrangements were made with local utilities to buy the fuel and contracts were signed. The contracts were generally onerous in the extreme with total responsibility and liability on the RDF producer. They found both real and imagined reasons for not using it. It was argued, for example, that RDF was not homogeneous, had too much dirt and metals, caused erosion and corrosion of the boiler tubes and excessive ash with lower efficiency, and so on. As a result, plants with very large investments, such as Chicago S.W. Supplemental Fuel and Americology Milwaukee, were suddenly without markets for RDF and were forced to close.

In the case of Baltimore (Southwest Facility), the pryolysis technology simply did not work and the original supplier lost millions of dollars. In the case of Disney World, the pyrolysis plant was too small and proved uneconomical.

The case of Hempstead, New York (a wet system), was a combination of several issues. There is no question that the system performed and the plant was able to operate at rated capacity. The plant was entangled in a political fight for the office of U.S. Senator from New York and a questionable issue on Dioxin finally forced the plant to close by mutual agreement of the town and the builder.

12.4.1 Dry Process

Solid waste as received is usually fed to hammermill-type shredders with either horizontal or vertical shafts. The purpose is to reduce the incoming material usually to about 4 in. (102 mm) for further processing, such as magnetic separation and screening. Occasionally, secondary shredding and air cleaning is employed for sizing, cleaning, and removal of nonferrous materials. Scalpers and/or trommels as well as conveyors are used extensively. Conveyors can be belts, aprons, drags, vibrating pans, screws, or a combination of these. The goal is to produce a homogeneous RDF with particle size normally under 3 in. (76 mm) and free as possible from all metals as well as glass, sand, and grit.

There are presently 14 plants in operation in size from 100 to 2200 TPD (900 to 1800 t/day) (rated capacity) employing various techniques. A few are selling their RDF and several have gone through the expected start-up difficulties encountered in any new plant. The economics vary from plant to plant. For example, in New York State, Monroe County chose to sell RDF to the local utility whereas the City of Albany sends RDF to two dedicated boilers owned and operated by the State of New York in the Center-city. Niagara Falls (Occidental Chemical Company) burns RDF (plant and local municipal solid waste) in dedicated boilers for steam and power generation on the premises.

Explosions have occurred in several installations with costly damage, and four fatalities (three in one episode). Inspection and separation are now common at most installations to screen incoming material. The actual physical layout and installation of the shredders has become extremely important to reduce or eliminate the danger of explosion damage to buildings, equipment, or injury to personnel.*

Editor's note: Experienced, knowledgeable system designers now employ effective preventive and control measures that virtually relegate shredder explosion episodes to routine operating experience with little or no damage, downtime, or injury. See Sections 12.13 and 12.14.

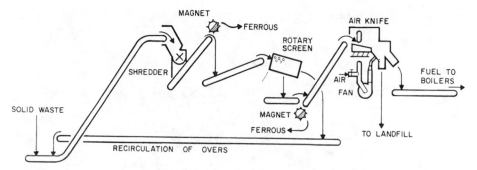

Fig. 12.21 "Dry" system: Front-end separation.

Ferrous metals separated directly after the shredders can be sold in the local market. However, the price obtained is relatively low since the shredding operation frequently imbeds extraneous materials into the metal. If further processing is undertaken, the value of the ferrous increases appreciably.

Figure 12.21 shows a typical diagram for the dry process with a trommel (rotary) screen after the shredder to separate oversize material that is recirculated back to the shredder. Also, an air knife is incorporated for extracting the nonferrous to produce an RDF free of metals. Figure 12.22 shows an outdoor dry system installation including bag houses for dust control. A combination trommel screen with a ⅜ in. (9.5 mm) hole front section for removal of sand, grit, and glass followed by a 3 in. (76 mm) hole section for size control, with overs recycled as above, can also be employed. The sand, grit, and glass can be further washed to produce a merchantable product (also see Sections 12.7, 12.10, and 12.11).

12.4.2 Wet Process

In this case, the incoming solid waste goes through a first-stage trommel screen with holes in the order of 2.5 in. (63 mm) creating two streams—"overs" and "unders." The "overs" include most of the combustibles while the "unders" include virtually all the sand, grit, and glass (99% +). Each stream of solid waste is sent to separate hydrapulper systems. These are very large rotor-hammer-type agitators @ 1200 hp (895 kW) in which water is added for macerating the solid waste into a slurry that passes through perforated bottom extraction plates and is pumped to cyclones via a dump tank. By centrifugal action, the cyclones arranged in a primary-secondary configuration remove and separate the organic and nonorganic fraction. The nonorganic or heavy fraction consisting primarily of small metal pieces, ceramic, stone, bone, sand, grit, and glass is conveyed to the minerals plant for further processing. The separated organic fraction goes to an agitated chest prior to dewatering. The hydrapulpers have an opening about 1 ft² (914 cm²) for removal of large pieces, particularly metals that are conveyed to a permanent rotating magnet for separation of ferrous and nonferrous. The nonferrous is subjected to a hand sort for recovery of flake and cast aluminum, copper, brass, zinc, and bronze. The current quantities and prices for extracted by-products are shown by Table 12.8. Figure 12.23 shows the hydrapulper system.

The slurry from the agitated tile storage chest is pumped to a two-stage dewatering system to produce an RDF with a moisture of approximately 50%. The only installation with a wet system only is the 2000 TPD plant in Hempstead, New York, which operated at rated capacity for over a year before closing.

Figure 12.24 shows a typical flow diagram for a wet system with the separate "overs" and "unders" hydrapulpers.

12.4.3 Combined Dry/Wet System

Both the dry system and the wet system have advantages and disadvantages. The dry system produces a lower moisture fuel with higher BTU value but the fuel is not as clean or homogeneous as the wet system. The wet system produces a very homogeneous fuel but has a higher moisture content and a lower BTU value. Combining the two has advantages. The Dade County installation in Florida uses both wet and dry systems. Each has separate infeed lines.

Figure 12.25 is a simple combined system with the "overs" from the rotary trommel screen and the "heavies" from the air knife as well as all metals sent to the hydrapulpers.

At Dade, domestic garbage, constituting about half the input, is sent to both systems whereas trash, commercial and industrial waste is sent to the dry system. It should be noted that there is a fair amount of comingling. The resulting RDF from each, when combined, produces a homoge-

Fig. 12.22 "Dry" subsystem equipment: Two-line drum magnets, conveyers, and shredder dust collection dry cyclones and baghouses (outdoor installation).

neous fuel with a moisture of approximately 40% and a BTU value as fired of approximately 4500 BTU/lb (2493 kcal/kg). This fuel has a particle size less than 2.5 in. (63 mm) and is relatively free of metals, glass, sand, or grit. It burns extremely well in standard boilers with traveling grate spreader stokers, and can be stored indefinitely without any apparent degradation or smell. The plant demonstrates a combined dry/wet system, front-end separation and burning RDF in standard boilers.

Selection of a system depends on the solid waste profile, as well as the quantity of mixed solid waste (MSW), whether trash and industrial-commercial waste are to be handled, as well as the quality and quantity of by-products and the market for steam or power. Environmental considerations and other factors such as location and transportation also play a part.

Table 12.8 Quantities and Prices, Recovered Materials

| By-Product | Price (July 1983) | | Quantity Sold per Week |
	Delivered	Net to Plant	
Ferrous	$60/G ton	$31/N ton	800/1000 tons
Aluminum			
MRP		$340/ton	30 tons
Cast		$600/ton	10 tons
Can		$940/ton	5 tons
Brass, bronze, zinc		$700/ton	3 tons
Copper		$1000/ton	1.25 tons
White goods (unshredded)	$45/G ton	$15/N ton	200 tons
Heavy cup (small size, nonferrous)		$340/ton	1 ton
Coins		$1000/week	$1000/week
Ash		$2.50/3.50 per ton	1900 tons
Glass and grit		$2.50/3.00 per ton	1200 tons

12.4.4 Energy Output Comparison

The energy output can be measured by the amount of steam generated from a fixed amount of solid waste. It clearly depends on the percentage and nature of combustibles and moisture in the incoming waste.

Table 12.9 indicates the comparison. In all cases, 100 tons/hr of solid waste are assumed coming in. In the case of the wet system, water is added at the time of separation. The amount of recovered materials is the same in all three cases; however, the total amount going to landfill is lower when the wet system is used.

12.4.5 Characteristics of Dry/Wet Systems

Both the dry and wet systems as well as the combined systems are based on the principle of removing as much of the noncombustible material as possible from the solid waste before it is burned in the boilers as well as achieving reasonably uniform sizing.

1. *Fuel Characteristics.* The RDF produced by any of the three systems compares very favorably with other fuels used for steam generation, such as bark, bagasse, and other agricultural and industrial wastes. Table 12.10 summarizes these characteristics. If solid waste is not subject to presort or preparation, it may be difficult to burn efficiently because of the highly variable nature of the fuel.

2. *Feeding Methods.* Fuel handling is quite similar whether it is dry, wet or combined. In all three cases, the fuel is homogeneous and easy to handle. It can be controlled and metered quite accurately and does not require turning and mixing on a stoker.

RDF is autogeneous and requires no supplementary fossil fuel. It can be fired in standard utility coal boilers in combination with coal, bark, or other fuels. The percentage to date with coal has usually been less than 20% of the total fuel calorific value, primarily due to utility demands. In any case, automatic control is easily accomplished. Table 12.11 summarizes fuel feeding alternatives.

3. *Boiler and Grate Design.* Boilers for RDF are standard units used for coal, bark, agricultural fibers, and industrial wastes. The grates are simple, standard continuous types, with easy-replacement parts and standard metallurgy. All boiler components, including furnace walls, generating sections, economizers, and air heaters are conventional.

The ash removal system is also standard and can be supplied by a number of vendors. Spare parts and repairs are neither costly nor time consuming. There are also many experienced operators available for these standard boilers and their associated systems and components.

12.4.6 Market for RDF Fuel

A large number of RDF plants built in the 1970s opted to save money on capital cost by producing an RDF to be used by others, especially utilities. At the time, it appeared to be an excellent idea

Fig. 12.23 "Wet" system hydrapulpers.

that would permit relatively low-cost disposal/process plants and an assured commitment for the sale of the RDF. Furthermore, the reasoning followed that burning RDF would assist a utility in reducing sulfur emissions.

The experience with this approach has usually been less than satisfactory. Large and expensive plants, i.e., Baltimore "Landgard," Chicago Southwest Supplemental Fuel, Americology Milwaukee, Monroe County, NY, and Bridgeport are closed due to technology deficiencies or lack of RDF market. A notable and successful exception is RDF from Baltimore County being burned in a cyclone boiler at Baltimore Gas & Electric Co. RDF as a market fuel continues to face great difficulties due to a number of factors:

1. Low value set by utilities
2. Total liability on RDF producer
3. Utility intransigence

RDF is simply not attractive to a public utility, and following are some of the reasons:

1. The utility normally burns one kind of fossil fuel in a dedicated boiler. It has all necessary equipment to do so and with trained personnel. To introduce a new fuel in the system, even in the very small percentage of only 10 to 20%, introduces unknown elements in the operation of the dedicated boiler.
2. There is no economic advantage to the utility in the use of RDF. Actually, the utility has very high pressure-temperature boilers operating at a high efficiency. The RDF can actually decrease efficiency and perhaps more importantly increase maintenance.
3. Utilities in general have excellent housekeeping. Introducing RDF aggravates housekeeping and ash handling can increase costs significantly.
4. If the RDF has not been properly prepared, it could contain metals, sand, and grit and other

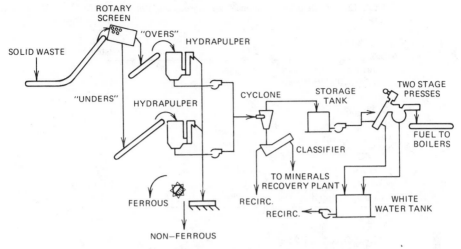

Fig. 12.24 "Wet" system front-end separation.

contraries that can seriously damage boiler tubes, stoker grates, ash handling, etc. Maintenance may increase appreciably when using solid waste fuel.

5. Transportation of RDF is expensive due to its low density and the need for compaction equipment as well as a truck fleet and both load and unload facilities.

Why should the utility burn RDF? Possibly for public relations or as an accommodation. Actually, burning RDF should be considered an added utility service. Clearly, the most important reasons are to decrease imported oil (national security) and to reduce the balance of payments.

The dedicated solid waste plant generating electricity can consume a portion of the energy generated for in-plant requirements and export the excess to the utility. By law (Public Utility Regulatory Policies Act PURPA), the utility must accept the excess.

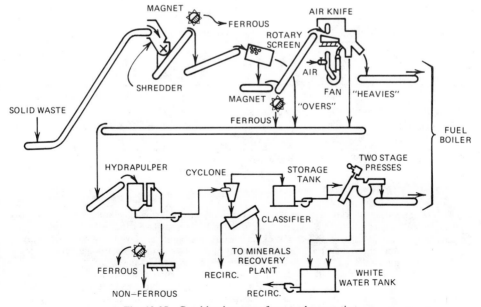

Fig. 12.25 Combined system front-end separation.

Table 12.9   System Performance Comparisons Per 100 Tons of Raw Waste

	Dry System[a]	Wet System[a]	Combined System[a]
Solid waste input (55% Combustibles)	100	100	100
Water added	—	25	12.5
Total input	100	125	112.5
RDF produced	83	108	95.5
Recovered material	15	15	15
Residue to landfill[b]	2	2	2
Ash from boilers[c]	15	15	15
Total to landfill	10–20	10–20	10-20
Steam generated	288	274	281

[a] Tons
[b] Non processibles.
[c] Most of the ash is sold.

Table 12.10   Comparison of Fuel Characteristic/Firing Methods

	Lb/ft^3 Density	BTU/lb Oven Dry	BTU/lb As Fired	% Moisture	% Ash[a]
Unprepared solid waste	5–18	6600	4620	25–50	25–50
Dry RDF	5–10	7600	5260	30–32	15[b]
Wet RDF	24	7600	3710	50–52	15[b]
Combined RDF	20	7600	4,000–5,000	37–43	15[b]
Sugar cane bagasse	27–34	8,000–10,000	4000	45–55	2
Bark (Ohio)	24–26	8,500–9,500	3600	45–60	3
Coal (Kentucky)	75–95	11,000–14,000	10,000–13,000	3–20	5–12
No. 6 fuel oil	55–65	17,500–19,000	17,500–19,000	1	0.5

[a] Ash includes non-combustibles
[b] Ash referred to incoming solid waste, % by weight.

Table 12.11   Fuel Feed and Combustion Control Comparisons

	% of Total Fuel	Particle Size	Grate Turning	Feeding Method	Combustion Air Excess
Dry RDF	100	2–4 inches	no	G&P	50–60
Wet RDF	100	under 2″	no	G&P	50–60
Combined RDF	100	under 2″	no	G&P	50–60
Fired with Coal	10–20	under 2″	no	P	—

G = Gravity or hydraulic ram; P = pneumatic.

12.4.7 RDF Storage

Since the generation of RDF will not always coincide with the demand, storage provisions are necessary and Figure 12.26 shows simple and effective RDF storage. Furthermore, solid waste is normally delivered over 5 or 5.5 days a week during daytime hours only, while a power plant must operate on a 24-hr basis 7 days a week. It is therefore necessary to provide storage between the RDF-producing facilities and the power plant, even if both are at the same site.

A variety of live storage bin designs have been tried at high cost immediately ahead of the boilers with mixed results. A dependable RDF supply to the boilers is essential. The best storage is a simple floor/bunker with adequate crane and loader handling with remote control, variable speed reclaiming conveyors. (See Section 12.9.5.)

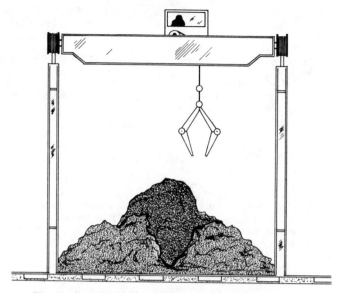

Fig. 12.26 RDF Concete bunker storage with bridge crane.

12.4.8 By-Product Recovery

A resources recovery facility, whether dry, wet, or a combined system can maximize by-product recovery. Markets permitting, most metals, including ferrous and nonferrous, and various types of aluminum as well as glass, sand, and grit can be recovered, cleaned, and successfully merchandised. Bottom ash and fly ash from the boilers has been sold to cement companies and over $50,000/year in coins has been recovered.

Table 12.12 compares the recovery capacities of typical 2000 TPD (1814 t) solid waste plants. Depending on the solid waste profile and the market for recovered materials, the income from this source may vary upward presently (early 1984) from a minimum of $2/ton of solid waste received up to $4/ton. These by-products will increase significantly in value in the future.

Table 12.12 Potentially Saleable Recovered Material, 2000 TPD Plant

Recovered Material	Resource Recovery RDF Plant (TPD)
Ferrous metal	100
Nonferrous metal	4
Glass	120
Sand and grit	40
Ash for sale	240
Ash to landfill	Minimal (120–200)

12.5 PROCESS AND MATERIALS-HANDLING SYSTEMS AND EQUIPMENT: SHREDDING AND RECEIVING SYSTEMS

William D. Robinson

12.5.1 Background

"I wish the guy that designed this thing had to make it work!" How many times has this lament been heard from plant operators who have had to make unexpected, sometimes prolonged and costly retrofits to suit their particular conditions in order to function with any degree of success.

This is usually because many designers of incinerator systems and consultants who evaluate them have not had operating experience or the perspicacity to recognize what works and what does not, as evidenced by the less than expected (or failed) performance of many incinerator plants, *RDF or mass burn.* The tendency to repeat mistakes continues but with less frequency, although there is a continuation of misplaced priorities in RDF systems such as shredder hammer wear and horsepower along with questionable system concepts and component selection. These mistakes include the flail mill and rotary shear in primary size reduction service with a determination to minimize power requirements for a critical function. This approach jeopardizes entire downstream efficiencies and continuity of operation unless thoroughly effective preprocess front-end separation is provided by air drum, screening, tipping floor scrutiny, or highly disciplined source separation and collection.

12.5.2 Typical RDF Dry Process Components and Systems

Table 12.13 and Figs. 12.27 through 12.34 show process components and systems, RDF characteristics, and their applications. Figure 12.27 illustrates a simple low horsepower system with a high RDF/raw material yield ratio using a rotary shear for primary reduction of a fairly narrow range of raw material to a coarse product size with limited hourly output. A magnet removes ferrous metals, but all other extraneous material remains—glass/grit, nonferrous metals, etc. Figure 12.28 depicts another simple system for high RDF/raw material yield ratios, 85% and above, but with higher horsepower for primary shredding of a wide range of raw material to a nominal 4 to 6 in. (10 to 15 cm) topsize with high average throughput and followed by magnetic ferrous removal only.

The penalties for the simplicity of each of these systems, however, can be significant if downstream handling and furnace designs are not conservative and have trouble with:

RDF containing glass and grit with the potential for materials handling abrasive wear

Increased ash

Potentially troublesome furnace fireside conditions for tube erosion and deposits

Excessive topsizes that foul RDF storage, distribution, and burner control

Insufficient removal of heavy and sometimes large (long) ferrous material along with firm plastics (jugs, etc.) which extrude through grate openings, fouling any part of the system, including grates, hopper bottoms, and ash handling

Figure 12.29 shows a system concept using a high-horsepower, heavy-duty shredder with installed power of 20 to 30 hp (15 to 22 kW) per ton-hr. and capable of accepting a wider range of raw material. Primary shredding reduces about 90% of it below a desired topsize followed by ferrous magnetics and a screen with two hole sizes for entry section removal of glass and grit, followed by drop out of RDF sizes, say 4 in. (10 cm), with discharge overs to discard. This arrangement has an RDF/raw material yield ratio of 50 to 60% but may not require such scrupulous presorting (except for flammable liquids/explosives), depending on the shredder design (vertical shaft mills usually require more pre-sorting). If the trommel overs are returned to the primary shredder in closed circuit for reshred, RDF yield ratios upwards of 80% are attainable.

Figure 12.30 shows a raw refuse front end trommel feeding a shredder. It drops out most of the sand, grit, bottles (glass), and cans, with the overs going to the primary shredder. Options for front-end materials recovery are most effective and the RDF quality is improved with regard to ash content, but the yield and BTU/lb (MJ/kg) will decrease if a significant quantity of the -4 in. (-10 cm) combustible in the undersize is not returned to the RDF fraction. In any case, downstream abrasive wear is reduced (especially the shredder) and the aforementioned benefits to furnace operation are realized.

A word of caution, however, regarding a common misconception, that is, that the reduction in shredder throughput (sizing) will vary directly with the ratio of the overs-to-unders split by weight. Instead, any such effect will be closer to the ratio of overs-to-unders bulk densities. For a 50/50 split by weight and with bulk densities of 7 lb/ft^3 (114 kg/m^3) and 21 lb/ft^3 (342 kg/m^3), the reduction in shredder load will not be 50%.

Front end trommel:

	Split by Weight (%)	Bulk Density
Overs	50	7 lb/ft^3 (114 kg/m^3)
Unders	50	21 lb/ft^3 (342 kg/m^3)

Approximate reduction in shredder load: $\dfrac{7 \text{ lb/ft}^3}{21 \text{ lb/ft}^3} = 33\%$.

Table 12.13 Typical Dry Process Components and Systems with RDF Characteristics and Applications

Figure Number	Process	Product Size Range	Yield, RDF/Raw Material	Firing Method	Other Comments
12.27	Rotary shear + magnet	+ 8–12 in. strips (20–30 cm) 4–8 in. (10–20 cm) topsize	90–95%	Mass burn, fluid bed, pyrolysis, if no further processing	Frequent and erratic stop/start/reverse interruptions; maintenance and capacity uncertainty. In shakedown stages
12.28	Shred + magnet	90% 6 in. (15 cm)	+ 90%	Spreader stoker	RDF retains glass/grit and topsizes
12.29	Shred + magnet + screen	90–95% 4 in. (10 cm) 60–70% 3 in. (7.5 cm)	50–60% if screen overs discarded	Spreader stoker	50–60% glass/grit and most onerous topsize removed if screen overs discarded
12.30	Trommel + shred + magnet	90% 6 in. (15 cm)	50–60% if trommel combustible unders discarded	Spreader stoker	90% materials recovery, cans and glass + grit removal; reduced downstream wear.
12.31	Trommel + shred + magnet + screen	90–95% 4 in. (10 cm) 60–70% 3 in. (7.5 cm)	50–60% if raw trommel combustible unders discarded	Full suspension, spreader stoker	
12.32	Flail + magnet + screen + shred + air knife	90% 2 in. (5 cm) 60–70% 1 in. (2.5 cm)	50–60%	Full suspension, spreader stoker	50–60% Glass/grit removal
12.33	A. Shred, magnets, screens, air knife B. + Screen overs to reshred	A. 90% 4 in. (10 cm) 75% 2 in. (5 cm) B. 90% 2 in. (5 cm) 75% 3/4 in. (1.9 cm)	85–90%	Spreader stoker or full suspension	80–85% material recovery, primary shred option to 2-in. (51 mm)
12.34	dRDF preparation	(Typical pellet cubette sizes)	40–50%	Spreader; mass burn; coal-mixed then pulverized	90% material recovery; clean, dry and fine pre-pelletizing preparation required

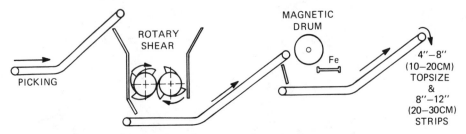

Fig. 12.27 Rotary shear and Fe magnet.

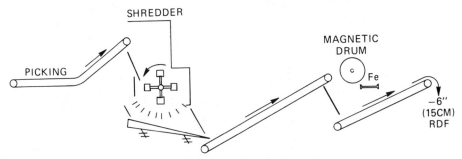

Fig. 12.28 Shredder and magnet.

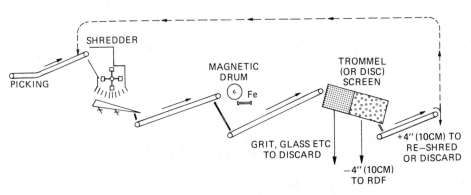

Fig. 12.29 Shredder, magnet, and screen.

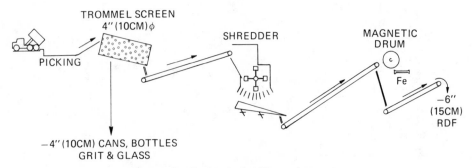

Fig. 12.30 Trommel, shredder, and magnet.

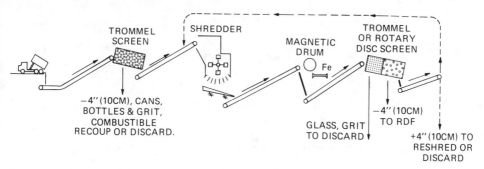

Fig. 12.31 Trommel, shredder, magnet, and screen.

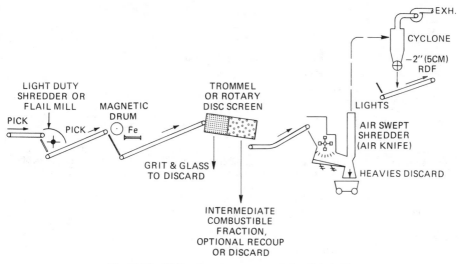

Fig. 12.32 Flail, magnet, screen, and shred/air knife.

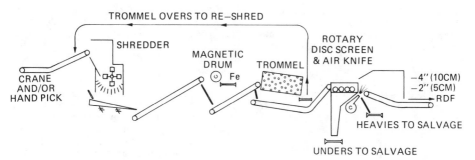

Fig. 12.33 Shred, magnets, screens, air knife.

Figure 12.31 shows a system that performs much the same as that described by Figure 12.30 and Table 12.13 except that the RDF product size will be more uniform with a smaller nominal topsize, higher heat value, lower ash, and less abrasive potential but with about the same RDF yield ratio.

The system in Figure 12.32 reflects the fixation for minimal horsepower in primary size reduction (not much more than bag breaking) which must rely upon RDF sizing mainly by screening (trommel or rotary disc) and secondary shredding rather than primary shredding, with the glass/grit undersizes and the intermediates to discard, and the screened overs going to an air-swept second-

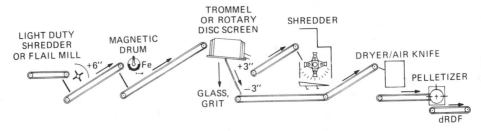

Fig. 12.34 dRDF Preparation: Flail, magnet, screen, dryer/air knife, and pelletizer.

ary shredder (air knife) with lights (RDF)/heavies separation. This technique has RDF yield ratios of 50 to 60% and requires careful preprocess sorting.

Figure 12.33 shows a highly versatile RDF dry process system with regard to:

1. RDF product size versatility and low ash
2. Near maximum:
   a. RDF yield and heat value
   b. Front-end materials recovery and cleanliness
3. Wide range raw material
4. On-line availability and maintenance

Figure 12.34 shows preparation of dRDF which requires selected raw material for careful processing of the several pellet/cubette sizes regardless of the firing method. This is required not only to maintain adequate integrity of the pellets with regard to cohesion for acceptable transport, storage, and retrieval, but also to assure availability and continuity of pellet/cubette production with acceptable extruder maintenance. The process system must assure:

Uniform feedstock product size

Minimal extraneous material (bits of wire, glass, stone, etc.)

Acceptable feedstock moisture levels

Section 12.8 discusses dRDF production and utilization in detail.

12.5.3 Shredding and the Air-Classifier Anomalies

In the evolution of MSW process systems, air classifiers were employed in early efforts, but they have not proved capable of providing the desired range of fractions without additional separation of product sizes. Classification by air is now relegated mostly to an "air knife" (Figure 12.33) or a fluidized air bed screen cleaning concept in an effort to recover the "small lights" combustible fraction to increase RDF yields, with a collateral benefit of dropping out more of the remaining undesirable glass and grit. Likewise, air sweeping of a secondary (or primary) shredder essentially can do the same (Figure 12.32).

Largely unrecognized has been the contribution of primary shredders to the downfall of the original air-classifier concept. Excessive troublesome topsizes (textiles, plastics, metal banding, etc.) would fly along with the desired lights, so attention was focused mainly on screening out not only the combustible oversizes but combustible smaller sizes as well in the effort to eliminate grit and glass (also see Section 12.11).

The penalty is a reduction in RDF yield with significant increase in requirements for raw material delivery, storage and handling, processing, and residues disposal (Figure 12.29 without reshred option).

The aversion to higher shredding horsepower has delayed any serious effort to remedy excessive topsize production, that is, innovative redesign of mill internals with more shear interface, smaller grate openings (or anvil clearances), and larger mills if necessary. [Recent use of shear-channel grates and four-corner hammer design appears to be a successful step in this direction. (See Section 12.5.8.)] Of course, the trade-offs are higher shredding power consumption and increased wear. But it will be worth it if increased yields become necessary in order to avoid:

Limitations imposed by raw material availability/flow control.

Doubling of ancillary services such as raw material collection and delivery, in-plant handling, and process residue handling and disposal.

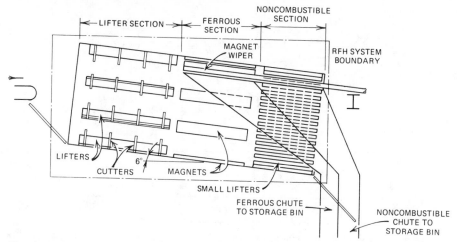

Fig. 12.35 Rotary drum, mass-burn raw material preparation, and material recovery. Figure courtesy of National Recovery technologies, Inc., Gallatin, Tennessee.

12.5.4 Size Reduction: Key Factors

Size reduction is required to prepare RDF for full or partial suspension burning, and certain pyrolysis concepts. Also, there is nascent interest in moderate pretreatment (without size reduction) of raw feed for mass burning* units (Figure 12.35 and 12.36) to provide:

A more homogenous fuel bed for improved control of firing rate/heat release and combustion efficiency.[1]

Removal and/or dispersal of concentrations of low-melting-point materials troublesome to grates and which agglomerate ash and siftings into "clinkers."

Removal of cans and larger ferrous material to minimize ash system jamming and to recover ferrous and aluminum.[2]

This discussion, however, will focus on size reduction for full and partial suspension during applications.

Background

Until recently, size reduction, particularly shredding, has had less technical evaluation than any other of the solid waste-processing technologies. There has been no established body of knowledge other than euphoric and questionable proprietary information from manufacturers and limited accumulations of scattered and sometimes defensive operating plant data (mass-burn installations are also guilty of the latter).

However, by means of federally funded study grants in the 1970s, research institutions, universities, and individuals began evaluating shredder performance, and the literature has become replete with publications formally reporting the analysis and testing of operational and laboratory-size shredders. These reports are characterized by ponderous utilization of materials process theory, mathematical modeling, and statistical analysis in proposing guidelines for comparing performance and shredder selection, including EPA[3] and NTIS[4] published manuals.

As scholarly as these exercises appear, they present an analytical overkill distracting to the system designer, in the opinion of this author. There are also a few misplaced priorities and questionable criteria, for example,

Shredder motors tend to be excessively undersized. *Example:* 665 hp (496 kW) is recommended for a throughput of 66 tons (60 Mg)/hr with a nominal topsize, X_{90}, of 4 in. (10cm). Most manufacturers would not guarantee less than 1000–1250 hp (746–942 KW).

---

*Raw feed size reduction is not an objective of this technique. It is rather a materials removal/recovery process to enhance combustion efficiency with a collateral recovery of saleable materials.

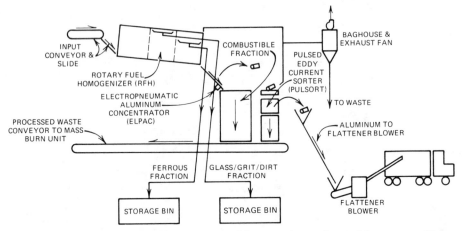

Fig. 12.36 Flow diagram, mass-burn raw material preparation, and materials recovery. Figure courtesy of National Recovery Technologies, Inc., Gallatin, Tennessee.

A preoccupation with characteristic particle size X_0 (63.2% passing)

Absence of thorough raw material identification[9] and related comparisons in test programs and theoretical analysis

Omission of worrisome adverse effects of removing hammers

Inconsistent and indecisive comparisons of single- and two-stage shredding, power requirements, process system complications, etc.

System designers may now call upon manufacturers and operators of resource recovery plants for the most current and relevant operating experience despite the fact that each installation has had to adapt to its own set of conditions which inevitably require retrofit with customized operating and maintenance procedures peculiar to their needs. These sources will likely continue to be the most reliable and useful reference base.

There are three basic options for size reduction:

1. *Shredders.*   Swing hammer-type hammermills, horizontal and vertical shaft
2. *Flail Mill.*   Lightweight swing hammermill for very coarse reduction (a bag breaker)
3. *Rotary Shear.*   Slow turning (<100 rpm) with circular cutter discs

12.5.5 Shredders

Advantages:

Better product size control, from coarse to fine
Wide range material input with less presorting
Heavier construction less vulnerable to shock/explosion damage

Disadvantages:

Higher power requirements
Rotor windage requires dust control along with effective explosion containment and venting
Higher capital cost

Tables 12.14, 12.15, and 12.16 are application/selection charts for horizontal shaft mills. (Although not shown, similar data for vertical shaft mills are available from the manufacturers.)

There are three basic types of horizontal-shaft, solid waste, swing hammer-type shredders and two basic vertical shaft designs.

The horizontal shaft types are:

Topfeed, single-direction rotor rotation
Topfeed, reversible rotor rotation
Controlled feed, single-direction rotor rotation

Table 12.14 Selection Chart: Shredder and Motor Sizes with Estimated Operating Costs[a]

Shredder Size (Hammer Swing Diameter × Feed Width)	Inches (cm)	42 × 60 (106 × 150)	60 × 60 (150 × 150)	60 × 80 (150 × 200)	72 × 100 (180 × 252)	96 × 100 (216 × 252)
Types of Material		Average Continuous Throughput, Tons (2000 lb)/hr				
Packer truck/residential		25	40	60	100	150
Bulky: residential/commercial		—	20	25	45	60
Motorsize for nom 4 in. (10 cm) nominal topsize	hp	600	800	1250	2000	3000
(About 20 hp or 15 kW per ton/hr)	kW	450	600	950	1500	2250

Estimated 1984 shredder operating costs, not including labor
(For conveyers and other process equipment, add 25 to 50% of shred cost)

		Power	
	Wear Parts	Electric	Diesel
Packer truck/residential	$0.75/ton	8 kWh/ton	1 gal/ton
Bulky: residential/commercial	$1.00/ton	12 kWh/ton	1.5 gal/ton

[a]Figure shredder operating and maintenance labor at about $2.00/ton-hr. design capacity, including fringes, as a fixed cost.

Table 12.15   Selection Chart Horizontal Shaft Shredders for Various Types of Refuse[a]

Types of Refuse	Recommended Shredder
1. *Residential.* These are materials collected by typical city packer-type trucks and generally include garbage can material, loose and in plastic bags; cardboard and paper cartons; cans; bottles; clothing; small or table top appliances; toys; auto tires; smaller crating; chairs and furniture to kitchen chair size; miscellaneous electrical and plumbing hardware; tree and lawn trimmings; and similar size and weight materials.	Standard type shredder top feed
2. *Commercial and Bulky Trash.* This is the larger, heavier appliances and furniture; stoves, refrigerators, washers, dryers; doors; beds; springs; mattresses; truck tires; larger tree trimmings; dunnage and similar size and weight materials. Maximum gauge of metal objects should not be heavier than that in the usual household appliances.	Heavy-duty top feed or controlled feed shredder
3. *Heavy Trash.* This would be demolition wastes; masonry; lumber; metal siding; roofing; auto parts; logs; railroad ties; and heavy industrial wastes up to the weight and gauge of automobile parts.	Controlled feed shredder

[a]The classifications are typical and serve to help select a suitable shredder and power unit to do an approximate volume of various waste materials.

Table 12.16   Shredder Types, Horizontal Shaft

Types of Shredders

1. *Standard Shredders.* Horizontal-shaft hammermills with top- or side-feed chutes, standard disc and hammer rotor assembly, and grate circle cage for product size control. Approximate product nominal topsize -6 in. (-2 in. minimum) (-15 cm, minimum -5 cm) with characteristic particle size, X_0 (63.2% passing) 2–3 in. (5–8 cm).

2. *Controlled Feed-Type Shredder.* Same as standard shredder above except with compression feed device for positive feed and more uniform power load. Long and short hammers optional, with a cutter bar standard on these shredders. Product size nominal -6 in. (-2 in. minimum) (-15 cm, minimum -5 cm) topsizes.

3. *Reversible Centerfeed Shredders.* Same as standard shredder above except with limited top-feed opening for smaller feed and secondary shredding only, product size nominal -2 in. (or smaller to 3/4 in.) (-5 cm or smaller to 20 mm). Not recommended for primary shredding. Rotors with replaceable tip wear caps and shear channel grates are available for most shredders of all styles.

4. *Flail Mill.* For flail mill service, a standard top-feed shredder is recommended, but with internals (rotor, hammers, discharge, etc.) designed for coarse bag breaking reduction. This affords easy optional field modifications if a wider range of output characteristics is eventually required, i.e., if raw infeed material, system requirements, etc., change beyond the limited capability of the conventional, marginally powered lightweight flail mill or modified fixed hammer paper/corrugated shredder that are not recommended for primary service.

The vertical shaft types are:

Topfeed, swing hammer, reversible rotor rotation
Topfeed, ring grinder, single-direction rotation

Each shredder has operating characteristics peculiar to its style.

12.5.6   Shredder Operating Characteristics

Horizontal Topfeed Single Direction Rotation (Figure 12.37)

These mills have the in-feed entering mostly by free fall over a breaker plate/chute inclined between 60° and vertical. Shortcomings are:

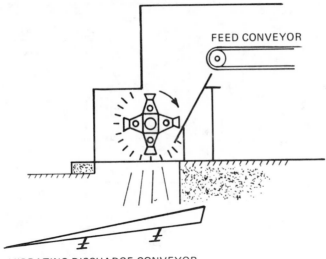

Fig. 12.37 Standard top-feed shredder, single-direction rotation.

Frequent difficulty penetrating the hammer circle with a resulting rejection tendency that contributes to increased retention time, orbital inventory, hammer and rotor disc wear

"Shrapnel" throwback requiring effective containment hoods and curtains, with "off limits" restricted movement of personnel

High-shock loads on mill and motor rotors and bearings, couplings, and alignments

Uneven discharge flow rates, that is, slugs or surges that can reduce processing efficiencies down stream

Adverse rotor windage conditions that can aggravate housekeeping and significantly contribute to explosion severity[10] and distribution (see Section 12.14).

Horizontal Topfeed Reversible Rotor Rotation (Figure 12.38)

Mills of this type have the infeed entering by free fall as close to the rotor center line as possible with the option of reversing the direction of rotor rotation. Its cross-section profile is essentially symmetrical. The principal advantage claimed is reversible rotor rotation, usually on a day-to-day basis, with a more even division of hammer wear and less frequent manual turning of hammers.

The rejection tendency and adverse rotor windage factors described previously also apply to this style shredder but with greater severity because: (1) more material tends to fall in the upward hammer path and (2) the entry throat is longer.

Furthermore, when material already bouncing off the hammers is suddenly forced into the hammer paths along with incoming feed, shock loads are increased notably. The theory that rejected material impinging on incoming material provides an initial reduction by attrition in midair is not confirmed in practice. It is not very effective, even with friable material.

Hammer wear advantages claimed for reverse rotor rotation and adjustable breaker plates are not valid because:

Downtime and labor for manually turning hammers in a shredder is not significant (some operators change out an entire set in less than 3 hr)

Hammer life is equivalent

Self-sharpening hammers are a myth; there is no way to compensate for rounding-off of hammer and grate edges

Controlled-Feed, Single-Direction Rotation (Figure 12.39)

The controlled or force feed-type shredder has the infeed entering by gravity slide down a feed chute at a minimum angle of 45° assisted as required by a crawler apron feeder directly above the feed chute (or a drum roller) at the mill throat. For residential material, it can be positioned above and clear of the raw feed as it slides into the throat. For difficult and bulky material, it can be

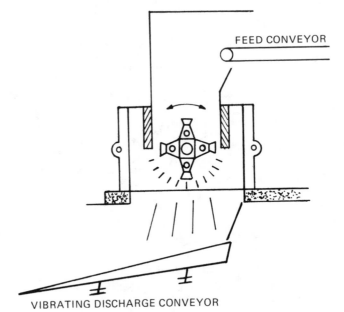

Fig. 12.38 Reverse-rotor rotation top-feed shredder.

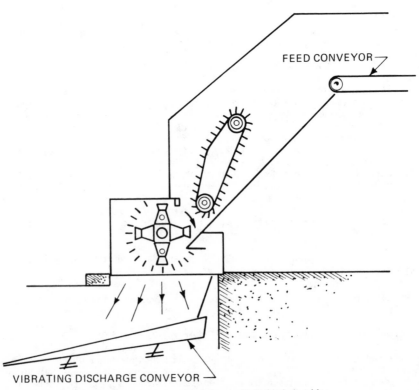

Fig. 12.39 Compression (controlled) feed shredder.

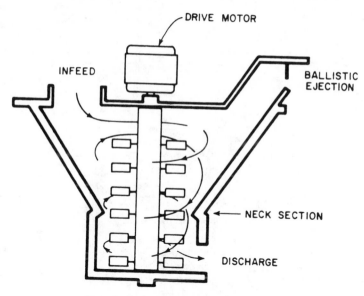

Fig. 12.40 Vertical hammermill/shredder.

positioned to grip and force the material into the downward hammer path tangential to an anvil plate. Advantages are:

Minimal rejection, retention time, and shrapnel back-fling
Lower shock and power consumption peaks
Minimal adverse rotor windage, tending to direct windage down and out instead of back and out the feed opening

Vertical-Shaft, Swing-Hammer, Reversible-Rotor Rotation (Figure 12.40)

Infeed material must penetrate a horizontal plane of rotating hammers, turn 90°, and take a helical path downward, exiting via a somewhat tangential discharge vector that can load up one side of the discharge conveyer. Reduction is by attrition between hammer tips and anvils spaced around the inner shell. Unlike most horizontal shredders, rotor windage is claimed to be mostly downward and out the discharge opening, and is easier to control. The larger mills will have a top-mounted direct drive motor that would be vulnerable to serious damage from a severe detonation-type explosion (see Sections 12.13 and 12.14). Deflagration-type explosions, however, would be less worrisome.

Vertical Shaft Ring Grinder (Figure 12.41)

This style vertical shaft shredder turns more slowly (about 300 rpm) than the vertical shaft hammermill (600 rpm) but is similar with regard to the material flow pattern, attrition reduction, rotor windage, and product size control. The drive motor is vertically mounted and drives the rotor at the bottom through a bull gear and pinion. Major maintenance of the rotor shaft bottom thrust bearing has not been uncommon (see Chapter 16, case history II).

Ejection/Kick-Out Traps

Usually ineffective are rejection devices (for difficult or nonreducible material) such as kick-out panels, skip ejection, or ballistic trajectory escapes. They are: (1) overly selective, or (2) reject too much material, or (3) do not function at all.

An exception is the remote-control pneumatic escape trap (Figure 12.42) successful in horizontal scrap shredders and actuated by the operator when knocking/vibration is felt and/or heard from the protected control cubicle beside and above the shredder, a heresy in certain misguided refuse shredding system design practice (with regard to explosion protection).

An explosion vent opening was completely blocked by rejected material and contributed to a damaging explosion in a large landfill shredder in 1982.

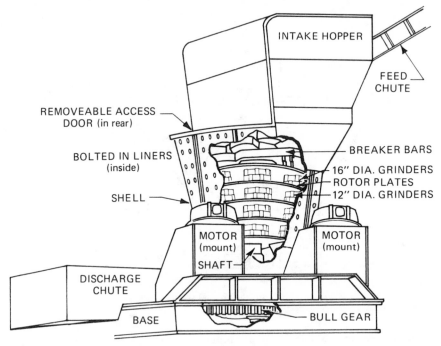

Fig. 12.41   Vertical-shaft ring grinder.

12.5.7 Design/Operating Factors Common to Topfeed Shredders

The adjustable breaker plate feature that provides clearance adjustment between hammers and breaker plate, purportedly compensating for hammer wear, is not effective in the solid waste RDF application. In any topfeed free-fall entry-style mill, hammer sharpness is much more critical regarding product size, fragmentizing ability, and grabbing hold of certain smooth, rounded surface items that otherwise might bounce and roll above the rotor.

In fact, sharp new edges can overgrab (until a small radius is worn on) pulling in more material than the mill can ingest and thus stall it. For this reason, some operators order their replacement new hammers with a small radius. In this connection, many system designers, operators, and research teams do not realize that experienced operators seldom buy hammers, grates, and other wear parts from the original equipment (shredder) manufacturer. There are many established professional steel foundries that can furnish almost any wear part, including a complete rotor assembly as a replacement or as original equipment in the shredder manufacturers mill if the buyer believes it to be a superior design (see 12.5.8).

As emphasized throughout this chapter, the vagaries of nominal topsizes continue to plague primary size reduction in the production of RDF, leading to a rationale of discarding the topsizes (screen overs) with recovery of acceptable RDF sizes. This trend now elevates characteristic particle size, X_0 (63.2% passing), to a more significant index in estimating regressive RDF yield potentials.

Shredding with Fewer Hammers: Benefits and Penalties

The principal benefits are higher throughput and lower power consumption but with an increase in topsize and percentage along with an increase in characteristic and bottom particle size. A disadvantage is the reduction of rotor inertia (flywheel effect), with more frequent stalling at choke-feed thresholds. The rotor inertia dissipates faster than the motor torque-speed response can pull through, and an extremely onerous and a time-consuming stall delay may follow, for example:

Manually clearing a stuffed, tightly packed mill

Waiting for the motor to cool down

A potential revenue loss

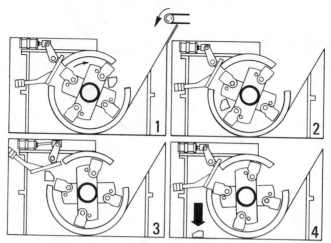

Fig. 12.42 Remote-controlled shredder ejection trap for large uncrushables.

Likewise, removing hammers can impose an expensive and aggravating operating maintenance burden. Even with a full complement of hammers, rotor arms (or discs) are exposed to abrasive wear that must be resurfaced frequently if not daily. *Surfaces exposed by hammer removal will wear significantly faster,* and unlike hammers, this work must be done inside the mill. The following section describes some remedies.

12.5.8   Recent Improvements in Shredder Design

Rotor Wear

Replaceable wear caps for rotor arms (discs) (Figure 12.43) is standard practice in metal scrap shredding for these reasons, and it will likely become standard procedure for solid waste shredders. The advantage of no-weld discard-type hard hammers (550 brinell) are also gaining favor, and, along with replaceable cap rotors, wear surface welding in or out of the mill will be an absolute minimum.

Rotor Windage

Rotor windage has finally been recognized as a major concern and points to an effective remedy of designing "scroll-like" mill cavities with limited throat openings and a containment baffle (upper breaker bar) to guide windage mostly down and out the bottom for primary shredding.

Test sampling by the author[5] and others[6] indicates rotor windage air patterns as shown by Figure 12.44. Test data were obtained by traversing above and below the shredder rotor with pitot tube and anemometer equipment. Although cursory, the results strongly suggested changes in shredder internal design that ultimately led to more recent and significant investigation by Galgana[7] via smoke model testing. The resulting test data were incorporated in the shredder design specification for the Albany, New York, RDF Process Plant with operating retrofit modifications by Nollet[8] to correct deviations from design by the low-bid manufacturer.

This effort has been a significant contribution to state-of-the-art improvement in shredder design and application regarding rotor windage vis-à-vis explosions management, dust collection, and housekeeping. Figure 12.45 illustrates some of the recent shredder design improvements.

Hammers and Grates

Depending upon the operator's preferences and the foundry casting them, hammer design features have shown marked improvement in the last several years, much of it to meet the requirements of the scrap processors.

The most notable improvement is the no-weld discard concept suitable for most shapes and with Brinnel hardness over 500.[9] The fear of such hard hammers breaking or cracking has proved unfounded and hammer life of 12,000 tons (coarse shred ~6 in.) with no welding is being reported, with a welcome improvement in maintenance procedures and economics.

The following is a description of recent design developments that have significantly improved

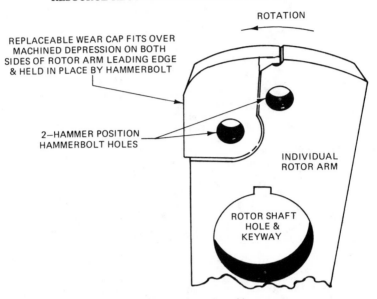

Fig. 12.43 Shredder rotor replaceable wear caps.

the performance of shredders as presented by Mr. Terry Francis, the inventor and reported in *Recycling Today* magazine, December 1983:

> *Adaptation of the swing hammer type hammermill rock crusher to the task of breaking up ductile material (e.g., refuse and scrap metal) has been successfully accomplished. There are between 300 and 500 of these machines in existence throughout the world and the various parts are of a design best suited to the breaking and sizing of brittle materials. For solid waste service, however, no account was taken until recently of the fact that most solid waste materials are ductile and require a positive cutting action. This oversight is evident even though the machines have been re-named shredders.*

> *Blunt faced hammers, capable only of impacting on brittle material were used and mated with random opening grates with no consideration of ripping, shredding, or cutting the material being processed. When the blunt hammers round off, efficiency drops. This combined with grate wear causes enlargement of shredded particle size. In breaking brittle material, wear is slower and not proportional to the tonnage processed, as the hammer is only an impacting device. Nor is grate wear proportional because the grate is only a sizing screen. In processing ductile material, co-action between a moving device (hammer) and a stationary device (grates, liners, or anvils) is necessary and wear is proportional to reduction of the material (less sand, grit, and glass abrasion factors).*

> *An appropriate title for this discussion could be 'co-acting hammer and grate design for shredding ductile materials.' Of prime importance for this application is a grate and ham-*

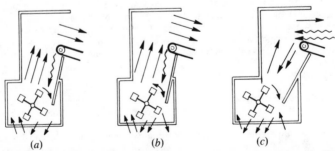

(a)   (b)   (c)

Fig. 12.44 Typical rotor windage vectors, velocities to .25 ft/sec (8 m/sec). (*a*), Single direction rotation, wide throat; (*b*), reversible rotation, wide open throat; (*c*), single direction rotation, medium throat and windage directional baffle.

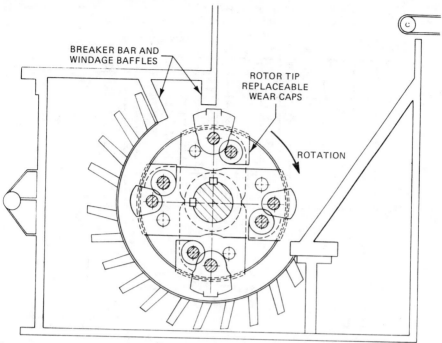

Fig. 12.45 Rotor windage direction baffles and rotor tip replacement wear caps. Courtesy of Hammermills Inc.

mer configuration aligned so that the hammers pass directly over the openings in the grates. When the hammers pass only over the grate openings, projections can be raised on the grates between the openings to act as an interface of cutting edges with materials which the hammers are carrying on their faces, and thus present a definite cutting action. If the shredder design is such that it requires short hammers in the rows where passage is over the projections, short hammers can be made by trimming down worn out long hammers for this purpose. However, a rotor designed with hammers only over the openings is more efficient and uses fewer hammers. Grates with forward slant openings can be used to gain production though particle size will increase and the cleanliness of shredded metals will decrease a little. Figure 12.46 illustrates these features.

This grate design will greatly enhance shredder performance, i.e., shear action, topsize control, throughput, etc., but with reduced hammer life (and higher power consumption). Elements of hammer design for ductile material vary from those for brittle material. They are:

> *The angle of the cutting face*
> *The elongated hammer pin pivot holes*
> *Working end notches*
> *Hammer thickness*
> *Four cutting edges*
> *Hollow center*

The typical shredder hammer is shown in Figure 12.47, and the advanced design features described above are shown in Figure 12.48, illustrating the differences.

The attack angle of the face of a typical hammer for brittle material can vary from 0 to 25°, most commonly in the 15° range. For shredding of soft, ductile material, the cutting face has a 45° angle to give the point greater penetrating ability and allowing the ripping of ductile material. The sharp point design also relieves unnecessary weight from the hammer as the tip does not extend as far up into the body of the hammer. This fact enhances its manufacturability.

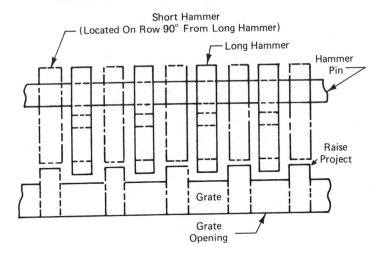

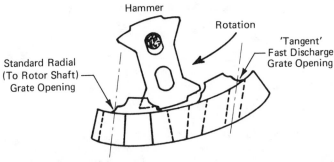

Fig. 12.46 Four-cutting-edge hammer and shear-channel grates. Courtesty of Adirondack Steel Specialties, Inc. and *Recycling Today* magazine.

The elongated hammer pivot hole is important in the design of hammers for soft/ductile material to protect the shredder itself. When a hammer is loaded normally on its working face, (Figure 12.49, sketch 1) it swings back or possibly attempts to rotate 360°. In either instance, a hammer can reach a point where its center line is forward of a line through the centerline of the rotor shaft (Figure 12.49, sketch 2). At this time, a load can be wedged between the bottom of the hammer and the grate (Figure 12.49, sketch 3). With brittle material, the load would be crushed with no problem. However, with ductile material, wedging can occur between hammer bottom and the grate, causing excessive stress with possible cracking of grates, bending the hammer pin, or bearing damage (Figure 12.49, sketch 4). By elongating the pin hole, the hammer is free to give way, precluding such

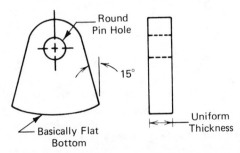

Fig. 12.47 Typical shredder hammer. Courtesy of Adirondack Steel Specialties, Inc. and *Recycling Today* magazine.

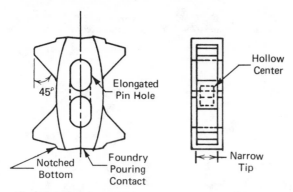

Fig. 12.48 Four-cutting edge, double-ended hammers. Courtesy of Adirondack Steel Specialties, Inc. and *Recycling Today* magazine.

possible damage. Since the work of the hammer is at its face and not the opposite end, efficiency is not reduced.

Elongating the pin hole also allows weight reduction of the hammer and improves its manufacturability. The notched hammer bottom permits increasing the quantity of wear metal available on the hammer. The swing radius of the hammer is the distance from the furthest point of the opposite pinhole to either cutting tip on the hammer. However, if this radius is in a smooth unbroken line, the sharp angle of the bottom will cause the shredder to reject material. This is especially true of machines which feed material tangentially to the periphery. Hammers for brittle materials have flat or very shallow radius bottoms to prevent this action. By notching the full radius of a hammer for soft/ductile material (municipal solid waste, etc.), the problem is avoided and maximum wear alloy can be added to the bottom of the hammer, providing a primary cutting face which increases wearability and hammer life. The more steeply angled bottom also increases the manufacturability.

Decreasing the thickness of the angle of the cutting tip of a refuse shredder hammer relates directly to increased angle of the cutting tip and enhances the sharpness of the point for more efficient shredding. This design would decrease the efficiency of a hammer for friable material because the impacting face would be narrower and present less impacting area. Narrowing the tip enhances the durability of the refuse shredder hammer because the thinner sections can be more efficiently heat treated to maximum hardness for long life.

The combination of a shorter, narrower tip and elongated pinholes allows the shredder hammer to be manufactured in a double ended design for smaller diameter shredders (under say 72"), providing four cutting edges instead of two in conventional designs. The weight of the double ended refuse shredder hammer is not much more than the common single ended design. It offers significant saving in purchase cost, hammer turning/replacement procedures, and the per-edge wear life.

The hollow center design concentrates the weight of the hammer at the two working ends, increasing striking force where it is needed for greater penetration. Without the hollow center, the hammer would spin more easily and less effective.

12.5.9 Flail Mills (Figure 12.50)

Advantages are:

Low horsepower
Lower rotor windage
High throughput
Low cost and space requirements

Disadvantages are:

Careful, extensive presorting required
Coarse product with limited size control requiring extra processing for product quality control

Note: For clarity, rotor/loads shown only in a 6–o'clock position.
Actually, hammer/workload positions occur anywhere in grate arc.

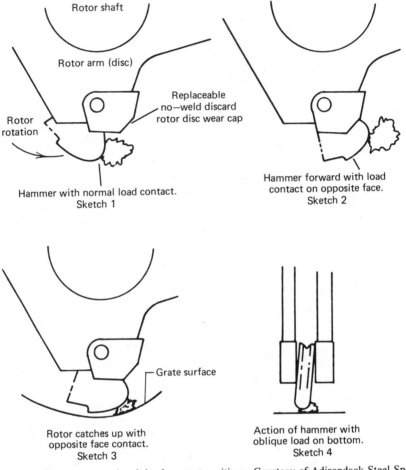

Fig. 12.49 Various hammer/work load contact positions. Courtesy of Adirondack Steel Specialties, Inc. and *Recycling Today* magazine.

Lightweight construction/resistance to pounding, explosion, etc.

Questionable ability to operate without excessive interruption at continuous average rates above 30 tons/hr without careful and continuous presorting and source separation.

12.5.10 Rotary Shear* (Figure 12.51)

Advantages are:

Low horsepower < 100 hp (75 kW)
Negligible rotor windage
Small space requirement

---

*A more complete discussion of the rotary shear is presented in Section 12.6.

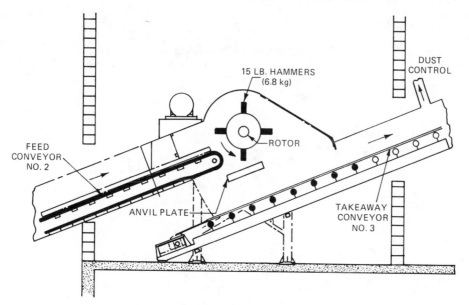

Fig. 12.50 Flail mill.

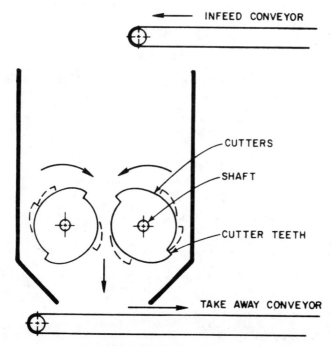

Fig. 12.51 Rotary shear.

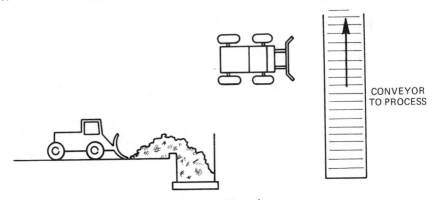

Fig. 12.52   Floor tip.

Disadvantages are:

Coarse, stringy product with limited size control
Erratic stop/go and reverse-to-clear feed patterns, with continuous production uncertainties

The importance of product size control, mainly topsize, in size reduction cannot be overem-phasized if vexing tradeoffs are to be avoided:

Downstream process inefficiency (magnets, screens, etc.)
Excessive screening
Lower RDF yields (without topsize reshred)

12.5.11   Front-End Raw Material Receiving Systems

Early front-end receiving systems for transfer station, on-site shred for landfill, or RDF/material recovery plants were floor tip and dozer/loader push (or lift) onto a feed conveyer, Figure 12.52. This simple cost-effective design practice has continued along with variations of it in the search for arrangements suitable for the delivery logistics and specific process technologies.

In any event, a by-pass to landfill is usually required for the contingencies of inadequate storage resulting from:

Process system downtime
Extraordinary delivery surges due to poststrike or paralyzing weather followed by accelerated collection.

Some of the adjuncts to the basic floor tip are shown in the following illustrations:

Floor Tip with Push Pit (Figure 12.53).   This arrangement offers a relatively small additional (live) storage capacity of say 250 yd^3 with metered feed to the process by rams. A hydraulic knuckle-boom loader remotely controlled by the operator in an adjacent control room is available to remove undesirable items. The bucket can be laid against the face of the material as it drops off onto the feed conveyor to meter the feed and prevent avalanching or cantilevered slug drop-off. Plants at Odessa, Texas, and Cockeysville, Maryland, employ this system or a variation of it.

Floor Tip with Reciprocating Ram Live Bottom Pit (Figure 12.54).   The objective of this ar-rangement is to supplement floor tip with a live bottom pit and conveyer arrangement, providing increased receiving and storage capacity along with controllable metering to the shredder feed conveyer. Pit bridging and slug feed from ram-faces onto feed conveyers requires an hydraulic picking/levelling crane and bucket remotely controlled by the operator. In at least one installation, incoming raw material is deposited only on the forward push ram area at the feed conveyer loading end. This system has been in use at plants in Akron, Ohio, and Niagara Falls, New York.

Deep Pit with Apron Conveyer Live Bottom (Figure 12.55).   Additional pit storage was an objective of this design with truck tip into the pit, but bridging/tunnelling led to erratic conveyer loading ranging from trickle to slug feed. Also, with a depth of burden upwards of several feet

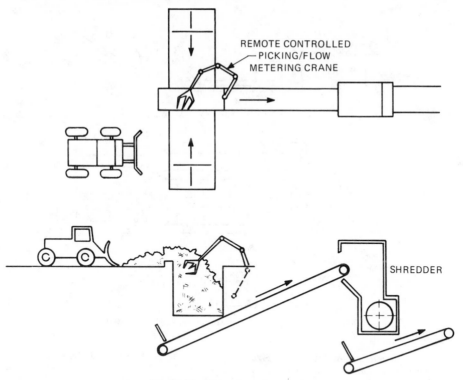

Fig. 12.53 Floor tip and push pit.

covering the conveyers completely, the conveyer drives can stall. A remedy has been to reduce the depth of burden and load only the first 25–30% of the horizontal bottom conveyers at their discharge end. This design is employed at Hamilton, Ontario.

Truck Dump, Ram Push, and Storage Pit with Bridge Crane (Figure 12.56). Increased (pit) storage is an objective of this arrangement which includes a large vibratory conveyer loaded by the crane bucket, the purpose of which is to shake down slugs and evenly distribute the stream of raw material for feeding to a shredder from a steel belt apron conveyer. An arrangement of this type has operated successfully at a shred and barge load transfer station in London, England, beginning in the early 1970s.

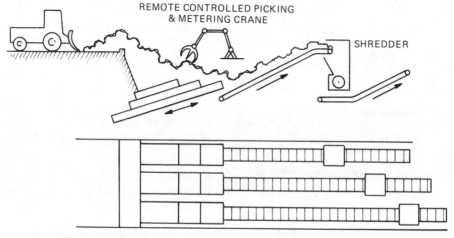

Fig. 12.54 Live-bottom pit: Reciprocating rams and apron conveyers.

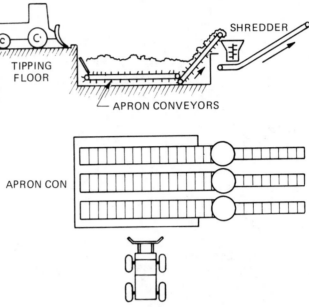

Fig. 12.55   Deep pit with apron conveyer live bottom.

Truck Dump to Shallow Pit Storage (Figure 12.57). This arrangement avoids more costly deep pit/bridge crane storage if delivery/storage logistics permit. It is in common use, with a variety of raw feed conveyer arrangements possible.

All of these arrangements require and/or employ sorting or picking at collection, tipping, or conveyer picking stations. Likewise, mass-burn plants require the same degree of surveillance/removal of undesirable material before or after passage through the furnace and onerous removal from the ash system.

12.5.12   Front-End Receiving Conveyers and Burden Depth Control

Background

Raw material feed to the system is critical because process equipment requires steady, uniform feed rates to perform efficiently, that is, uninterrupted throughput of most material delivered to the plant without excessive hand picking or segregation. Even with the presorting required by most incinerator plants, a difficult burden is placed upon the raw materials receiving and handling system.

The vagaries in material delivery to the receiving conveyer include slug dumping from trucks, payloader, or grapple, followed by avalanche or slug discharge from the receiving conveyer often causing excessive interruption of feed to a shredder while it digests its erratic loading. The result is average hourly production less than expected and poor downstream process efficiency.

Burden Depth Control

The experience to date with methods to optimize input flow characteristics has been mixed. This includes levelling devices (Figure 12.58, sketches A, B, and C). Sketch A shows trapeze-type levelers (fair to good), sketch B illustrates rigid (or hinged) strike-off baffles (poor), and sketch C shows a steep slope conveyer roll-back (a failure).

Receiving Conveyer Arrangements

A popular feed conveyer arrangement consists of multiple belts (Figure 12.59, sketch A). However, the transfer points require constant vigilance despite differential (metering) conveyer speeds which have not been especially effective.

When a vibratory conveyer discharges directly into the shredder (Figure 12.59, sketch B), a uniform feed to the mill can be obtained. However, a large vibrating mass at that elevation requires

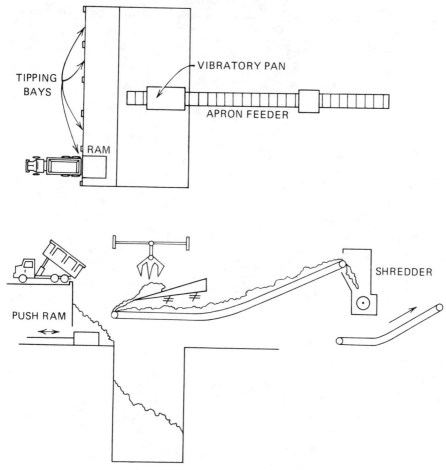

Fig. 12.56 Trunk dump, ram push, and storage pit with bridge crane.

careful structural design and it is vulnerable to dislocation and damage if adequate explosion protection (venting, suppression, etc.) is absent. It is also vulnerable to damage from shredder shock forces and "missile" pounding and access for maintenance is more difficult.

Feed conveyers can be steel or synthetic belts, or vibratory pans. If belt conveyers are inclined too steeply (over 25°), a dangerous rollback or slideback can occur with certain items such as tires, pallets, etc. In fact, pallets can tumble back, turning on the corners in an upright position, endangering any picking station alongside or at the bottom.

Vibratory conveyers must be designed for combinations of raw material mass, moisture, and irregular shapes that may not move, emphasizing the importance of the most thorough raw material evaluation possible in the early system design stage (see Section 12.12). Most conveyer manufacturers experienced in the solid waste application can design properly with regard to amplitude, frequency, stroke, and angle of decline if they have sufficient material identification. Also, the possibility of establishing resonance with the natural frequency of substructures, adjacent or nearby masses must be considered. If this occurs, it is usually necessary to change the vibratory mode without rendering the conveyer ineffective, otherwise it must be replaced by a belt.

Another common feed conveyer arrangement (Figure 12.59, sketch C) has the inclined section discharging material directly into the mill after having to turn an obtuse angle. It has several disadvantages:

Longer items must extend further out over the conveyer head pulley (discharge) before cresting over into the feed chute and bridging in the mill throat above the rotor.

Items with uneven weight distribution (refrigerator, etc.) can crest over with a tendency to whip, tumble, and bridge. Center of gravity/force vectors explain this.

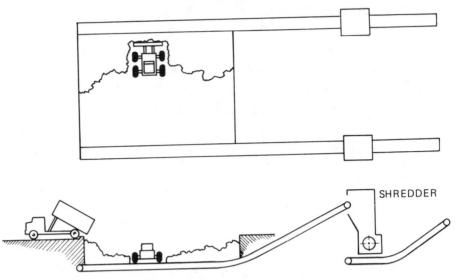

Fig. 12.57 Truck dump, shallow pit storage.

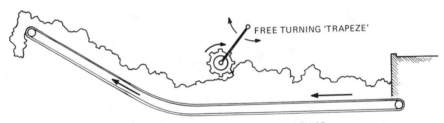

SKETCH 'A' TRAPEZE TYPE LEVELLING DEVICE

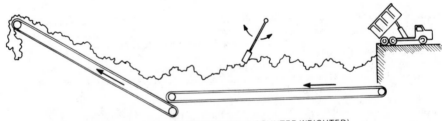

SKETCH 'B' FIXED (OR PIVOTED & COUNTER WEIGHTED)
STRIKE—OFF BAFFLE.

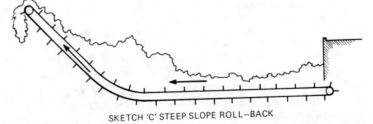

SKETCH 'C' STEEP SLOPE ROLL—BACK

Fig. 12.58 Raw material feed conveyor levelling devices.

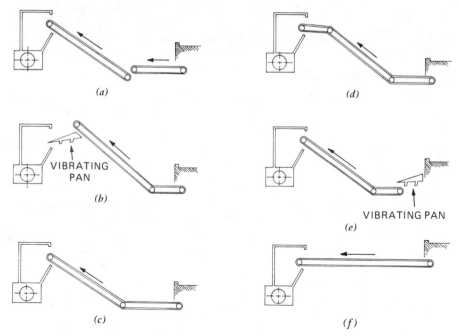

Fig. 12.59   Receiving conveyer arrangements.

Opening the mill throat in the flow direction to minimize bridging can be a poor trade-off. It increases rejection tendency, "shrapnel" kick-out, and adverse rotor windage.

A single continuous steel belt of the "bend-back" type (Figure 12.59, sketch D), with vertical feed hopper sides and vertical skirt boards (sides), has performed very well if the tipping floor loader avoids pile or slug loading and maintains a continuous, uniform, and acceptable depth of material.

Figure 12.59, sketch E, depicts a vibrator loading section discharging to the inclined feed conveyer. This arrangement can help provide a uniform continuous feed stream at an acceptable height if the vibrating feeder is constructed to withstand the shock loads with minimal downtime and again, if the tip loader is careful.

Figure 12.59, sketch F, is the simplest and least costly arrangement with minimum potential trouble points, slopes and turning angles, transfer points, etc., but must also rely on the integrity of the loader.

12.5.13   Shredder Discharge Conveyers

These can be steel or flexible belts, or steel vibratory pans. Selection for reliability of flow depends upon the types and combinations of shredded material with regard to impact, abrasion, as well as mass, moisture, and shapes.

Also to be considered are downstream separation systems in which the vibratory pan tends to even out the flow, an advantage to any process equipment. On the other hand, it adds a transfer point whereas a single flat belt can run directly to the next operation if the stream is already smooth. Flat belts, however, require more attention (edge condition, tracking, etc.) and a disputed trade-off is a combination of troughed belt followed by a vibratory pan.

In general, materials handling/process systems for solid waste perform best with straight line flow and few transfer points, and again, the importance of uniform continuous material flow cannot be overemphasized.

12.5.14   Summary

Design and Operating Improvement Needs

Background.   Preparation and suspension burning of prepared waste cellulose fuel is not a recent development and municipal solid waste fits into the cellulosic category but with an added burden of wide-ranging and unpredictable extraneous materials.

The concept was first used in the late 1930s by the pulp and paper industry and the sugar cane growers for firing tree bark, bagasse, and furfural in spreader stoker fired boilers. The "hogged fuels" systems usually consist of single-stage shredding in a swing hammer mill (commonly called a "hog" in those industries) with removal of tramp iron by a trap in the mill and by magnet.

RDF Processing: Quality and Handling

The adaptation to municipal solid waste (MSW) seemed to be logical but with additional preparation requirements not entirely recognized at first which presented operating problems, but which have been overcome in some cases, for example:

Shredders that can produce an acceptable topsize fraction (minimal long stringy, etc., material)

Rotor windage adversities

Screening devices for removing undesirable top and bottom sizes (but with marginally acceptable RDF/raw material yield ratios sometimes)

Abandonment of air classification unable to provide wider and controllable separating ranges in favor of a simpler lights/heavies air knife "cleaning" kind of separation used by scrap metals processing, etc.

Recognition of the shredder explosion potential but with a few useless control nostrums still practiced (overdependence on ineffective vent designs and placement despite adequate theoretical vent ratios, indicating a lingering lack of full comprehension of the phenomenon)[10]

A trend away from magnetic belts toward electromagnetic drums

Improved flat belt conveyers that maintain a uniform full width distribution of processible material for optimum downstream magnet and air separation and fuel distribution functions.

Control of RDF distribution at the boilers has finally been recognized for its vital importance, especially with mechanical systems. After unsatisfactory experience at the beginning, second-generation designs show marked improvement, but further improvement in RDF processing for better material-handling characteristics will be a continuing goal requiring improved process equipment design/performance such as:

Shredder product size control for wider input ranges

Screen efficiencies and wider material range tolerance

Mainstream magnet systems with a trend away from magnetic belts toward drums

Avoiding troughed conveyer design disadvantages of concentrated, uneven streams by trending toward improved flat belts or troughed belt/vibrating pan combinations.

Raw Material Handling and Downstream Process/Utilization Factors

Management of raw material receiving and delivery to the mainstream process lines can be improved with more effective logistics for surveillance and sorting raw material for hazardous and unprocessible items, including increased use of hydraulic picking/loading cranes.

Areas with well-disciplined citizenry and collection crews or quasi-captive collection and process combinations, especially in light or nonindustrial areas, usually assure an easily digested ("baby food") raw material delivery with minimal process difficulties.* But for "major league garbage" from undisciplined urban and industrial areas, equipment and system design performance and versatility is fully tested.† This applies to mass-burn techniques as well, very few of which can handle and process wide-range material, requiring costly transport, handling, and landfill.

Improvement in performance of raw material receiving and feeding systems in regulating raw material feed stream profiles (depth, width, and speed) for increased picking station and equipment efficiencies is needed, that is, better control of:

Differential conveyer speeds between loading and feed conveyers at transfer points

Metering at transfer points between receiving/loading conveyer and the mainstream feed conveyer

Stream height levelling devices

Presently, most of these combinations are not very effective except at low rates, usually below production targets.

---

*Ames, Madison, Pompano, Recovery 1, Pinellas, Pittsfield, and so on.
†Bridgeport, Resources Recovery (Dade County), Albany, and so on.

Mass-burn plants (especially small scale) usually require presorting by well-disciplined residents, collection crews, scale house surveillance, or the most difficult and time-consuming pit/bucket picking, removal, and disposal. Otherwise, constant vigil and disruptive handling and removal from the furnace and ash discharge/disposal system is continuous and an impediment to production.

RDF quality and handling betterment factors apply also to pneumatic conveying and delivery to burners of finer RDF, say ≲ 1 1/4 in. (32 mm) for full suspension burning. Stray topsizes can foul any part of the system—elbows, rotary air lock feeders, burner nozzles, and so on. Likewise, erosion from glass and grit plagues storage and air transport systems as well as boiler fireside surfaces (erosion and deposits).

It is astonishing how undesirable extraneous material can escape removal during all of the shredding, screening, magnetics, and blowing to appear in the finished product.

Avoiding such contamination is possible and has been overcome in some cases, but trade-off penalties must be justifiable such as higher process horsepower, lower RDF/raw material yields, increased process monitoring, and maintenance.

In addition to the effect of RDF physical quality on fuel transport, delivery, and combustion systems, it is likewise an important factor in RDF storage and retrieval systems, and it appears that this is an area needing more attention and improvement—perhaps less passion for automation and a turning back to a simple bulk piling with bucket (Figure 12.26) or loader retrieval* if the "bin system" problems persist.

A somewhat similar evolution ensued years ago (1920s) in the electric utility industry with the advent of pulverized coal firing that led the way to steadily increasing combustion efficiencies, steam pressures, temperatures, and burgeoning unit generating capacities with steadily declining costs not available with traditional mass-burning techniques.

It started with the "bin system" where incoming raw coal was stored in bunkers, fed to pulverizers, and then stored in bins and reclaimed for pneumatic delivery to the burners. House-keeping was an enormous chore considering the pervasive pulverized coal dust during storage and retrieval, and severe coal dust explosions were not uncommon.

A better way was urgently needed that led to a simpler "unit system" where raw coal goes directly from storage to a pulverizer and directly to an individual burner or burner bank in a totally enclosed and sealed delivery train. A similar approach in RDF preparation and firing might be well worth pursuing to eliminate problems of RDF storage and retrieval with improved processing techniques, especially shredding, which must produce RDF quality that permits simpler and more reliable distribution and firing techniques.

For larger plants upwards of say 1000 TPD, an idea of merit is the "preprep" satellite transfer station where raw material sorting and the potentially more difficult initial processing is performed followed by a more efficient haul to the boiler site. A finishing stage there would lend itself to a unit system concept, thus avoiding the enormity of most collection truck delivery logistics and super-plant size with equipment too closely coupled and stuffed into a single "all the eggs in one basket" area totally vulnerable to any operating (or other) mishap.

Several somewhat similar smaller versions of this remote process satellite concept are presently operating successfully but without the multiple preprep transfer station logistics. Albany, Madison, and to some extent, Ames, are in this category.

REFERENCES

1. E. J. Sommer, Jr., "Effects of Material Recovery on Waste-to-Energy Conversion at the Gallatin, TN, Mass Fired Facility," Proceedings, ASME 11th National Waste Processing Conference, Orlando, Florida, June, 1984.

2. G. R. Kenny, "A Simplified Process for Metals and Non-Combustibles Separation from MSW Prior to Waste-to-Energy Conversion," Proceedings, ASME 11th National Waste Processing Conference, Orlando, Florida, June, 1984.

3. E.P.A. "Engineering Design Manual for Solid Waste Size Reduction Equipment," 600/S8-82-028, Municipal Environmental Research Laboratory, January, 1983.

4. N.T.I.S. "Significance of Size Reduction in Solid Waste Management," No. PB83154344, U.S. Dept. of Commerce, Springfield, Virginia, January, 1983.

5. W. D. Robinson, Hammermills, Inc., unpublished data and correspondence, re: prototype trials, Cleveland, Ohio, 1966.

6. B. Crawford, National Center for Resource Recovery Prototype Test Facility, Washington, D.C., circa 1975.

---

*Resources Recovery (Dade County) Inc., successfully employs this simpler system (see Section 12.5.4).

7. Unreported test/development work by R. R. Galgana, Chief Mechanical Engineer, Smith and Mahoney, Engineers, Albany, New York.

8. A. R. Nollet, President, Aenco, Inc., Contract Operators, City of Albany, "Answers" RDF Process Plant.

9. W. D. Robinson, "Shredding Systems for Mixed Municipal Solid Waste," Proceedings, ASME Seventh National Solid Waste Processing Conference, Boston, Massachusetts, 1976.

10. W. D. Robinson, "Solid Waste Shredder Explosions—A New Priority," Shredder Explosions Seminar, Aenco, Inc., New Castle, Delaware, 1979.

12.6. PROCESS AND MATERIALS HANDLING EQUIPMENT: ROTARY SHEAR SHREDDERS, DESIGN AND OPERATION

Kenneth L. Woodruff

12.6.1 Background and Description

Rotary shear-type shredders have been used for several years in the scrap-processing industry and more recently for municipal solid waste. The unit is characterized by two parallel counterrotating shafts (turning toward each other), with a series of cutter discs keyed to them. The shafts turn at two different speeds, usually at a ratio of approximately 2:1. Figure 12.60 is a sketch of a shear shredder showing the cutter discs mounted on the shafts. The number of teeth per disc varies from one to six depending on the application. Two or three teeth per disc are common.

Shear shredders are available with drives rated from 5 hp (3.8 kW) to 1000 hp (750 kW) and with capabilities ranging from 0.5 to 60 tons (500 kg to 55 Mg) per hour. Usually up to 100 hp (75 kW), the units are electrically driven through a gear reducer. Above 100 hp (75 kW) a hydraulic drive system is usually employed. In the hydraulic drive approach, each shaft is powered by a radial piston motor through a gear reducer to provide the speed differential. A typical unit will have one

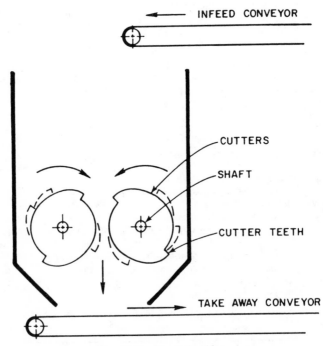

Fig. 12.60 Shear shredder.

Fig. 12.61   Typical shear shredder installation.

shaft turning at 20 rpm and the other at 40 rpm. Each hydraulic motor is powered by an electrically operated high-pressure pump. These pumps operate at constant speed and direction with all shaft speed and directional variations accomplished by changing the flow rate and direction of the hydraulic fluid. Normal operating pressure is about 1000 psi (6900 kPa), however pressure relief valve settings are usually in the 2500–3000 psi (17,240–20,700 kPa) range to allow for shredding of difficult materials. When material fed to the unit is not shredded at this high pressure, the unit will go into self-cleaning mode through one of several sensing methods.

Feed material to the unit drops into the top and is directed toward the center of the counter rotating shafts. The material is reduced in size through a shearing or tearing action of the cutter teeth turning at different speeds. Shredded material is pulled through the unit and drops from the bottom onto a conveyor (Figure 12.61). In the event a nonshreddable item enters the unit or jamming occurs, the hydraulic control system is designed to reverse the cutters to release the item and then resume forward operation. If jamming continues, the machine must be shut down for removal of the object. The system of self-reversing operates on the basis of a pressure-sensing device, zero-speed switch, or proximity switch, depending on the particular manufacturer.

12.6.2   Operating Experience

Operating experience with the slow-speed, high-torque shear shredder on municipal waste, though limited, appears to be satisfactory. Although output particle size is not positively controlled as in a horizontal shaft hammermill shredder, lower power consumption and improved safety of operation are benefits of this type of unit. Power consumption for a shear shredder operating on municipal waste is generally 10–20% less than for a hammermill shredder designed to operate at the same feed rate and output particle size. Actual installed horsepower is generally 20% less. Although a shear shredder utilizes 6 to 8 kWh/ton, it is not subject to a high-demand charge due to its lower installed horsepower and the fact that the smaller motors do not start up under any significant load as is the case with a hammermill.

Due to the slow speed of the cutters and the shredding action employed (shearing rather than impact) the units are less prone to explosions. Volatile liquids in containers fed to the shear shredder present a fire hazard rather than an explosion hazard. The container is merely cut or torn, allowing the contents to drain rather than be volatilized by the airflow as in a hammermill. Sparks

Fig. 12.62   Sample cutter disc arrangement.

are less likely to be generated by the shearing action. Explosion danger does exist, however, for ordnance such as mortar shells or other munitions. Due to the lack of air flow through the shear shredder, less dust is generated by this type of unit. Also the shredder itself is a low noise operation. However, the hydraulic power pack is high noise and requires indoor installation or isolation. Noise data indicates levels in excess of 100 dBA adjacent to the power pack of a 400 hp (300 kW) unit.

12.6.3   Operating and Maintenance Costs

The primary maintenance cost is associated with cutter disc replacement (Figure 12.62). Cutters are usually fabricated from 4140 or 8620 steel plate. They may be heat-treated to provide improved wear characteristics, but this limits their weldability. Experience indicates that maintenance costs of shear-type shredders operating on municipal waste are $1.00 to $2.00/ton, depending on output particle size and the degree of flexibility allowed in output size fluctuations. Several operators producing a 4 to 6 in. (100 to 150 mm) product report maintenance costs of about $1.50/ton. This cost is comparable to that experienced by hammermill operators. The difference is that hammermill maintenance costs contain a greater proportion of labor than for shear shredders. The cutter discs are the primary cost. Hammermill expenditures are spread more uniformly throughout the year, whereas shear shredder costs are large periodic expenditures. Although annual maintenance costs are similar, the lack of daily maintenance allows it to be operated more scheduled hours per day.

12.6.4   Applications

Several hundred shear shredders are operating on various scrap materials, including pallets, drums, batteries, tires, railroad ties, paper, plastics, white goods (major appliances), wire, furniture, demolition waste, and so on; only a few installations have operated on municipal waste. These include Banyan-Dade Resource Recovery in Dade County, Florida; Charleston, South Carolina; Chemung County, New York; Oklahoma City, Oklahoma; and Deadhorse, Arkansas. As more experience is gained in the operation of the units and in the use of various alloys and configuration of cutter discs, this type of shredder may gain wider acceptance. The results of the New York State ERDA sponsored program at Chemung County, New York, will be of great benefit to the industry.

12.6.5 Shear Shredder Manufacturers

Iowa Manufacturing Company
Cedar Rapids, Iowa 52402
319-399-4800

MAC Corporation
Saturn Shredders
Grand Prairie, Texas 75050
214-790-7800

Mitts & Merrill
Saginaw, Michigan 48601
517-752-6191

Morgardshammer AB
Centro-Morgardshammer (Canada) Inc.
Rexdale, Ontario M9W 5Y4
416-675-2662

Officine Meccaniche
Pierangelo Colombo

24045 Fara Gera D'Adda (Bergamo), Italy
Telex 340148 OMPC-1

Shred Pax Corporation
Bensenville, Illinois 60106
312-595-8780

Shredding Systems, Inc.
Wilsonville, Oregon 97070
503-682-3633

Triple/s Dynamics
Dallas, Texas 75223
214-821-9143

Williams Patent Crusher & Pulverizer Co.
St. Louis, Missouri 63102
314-621-3348

BIBLIOGRAPHY

G. R. Darnell and W. C. Aldrich, *Low-Speed Shredder and Waste Shreddability Tests*, EG&G Idaho, Inc., Idaho Falls, Idaho, April, 1983.

12.7. PROCESS AND MATERIALS HANDLING EQUIPMENT: SCREENS FOR SOLID WASTE PROCESSING

Kenneth L. Woodruff

12.7.1 Background

The development of the use of screens as classification devices for the processing of mineral ores has been reported as early as 1556 by Georgius Argicola. Rotary screens, or trommels, have been more widely used than any other type of movable screens. However, since the early 1900s rotary screens have been largely replaced in the mineral-processing industry by vibrating screens. In the waste-processing industry, three types of screens are employed—vibrating, rotary, and disc. However, rotary and disc screens are the most widely used due to the nature of the material screened.

12.7.2 Vibrating Screens

Vibrating screens are equipped with flat decks that are sometimes inclined to facilitate movement of the material along the screening surface. The deck of wire cloth or punched plate is powered by an electric motor and drive mechanism designed to transmit a side-to-side, lengthwise or vertical vibrating motion. A screen with a lengthwise or reciprocating motion tends to move the material upward and in the direction of material flow (Figure 12.63). A side-to-side or gyratory motion tends to move the material on a spiral down the screen while the material never leaves the screen surface (Figure 12.64). A vertical motion throws the material upwards, which allows it to contact the

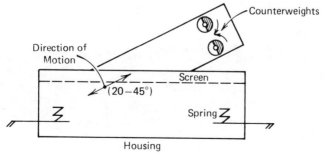

Fig. 12.63 Reciprocating vibrating screen.

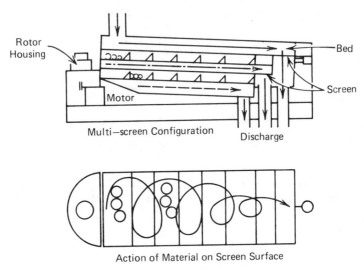

Multi—screen Configuration

Action of Material on Screen Surface

Fig. 12.64   Gyrating vibrating screen.

inclined screen deck at a different location each time (Figure 12.65). The inclination of the screen determines the residence time of the material.

Flat deck vibrating screens of all types have had poor experience and hence little use in solid waste processing. Blinding has been a major problem contributing to the poor separation experience. Several facilities including East Bridgewater, Massachusetts, El Cajon, California, and Monroe County, New York, have had poor experience. In addition, the National Center for Resource Recovery test program in Wilmington, Delaware, and Washington, D. C. experienced difficulty. The only successful application of vibrating screens in solid waste processing is the screening of relatively fine fractions of material that tend to be concentrated streams, such as glass-rich and aluminum fractions for fines removal. In addition, this type material tends to be either wet screened (spray bars are mounted over the screen decks) or dried prior to screening to reduce blinding.

12.7.3 Trommel Screens

The use of trommel screens (rotary screens) in solid waste processing was initially proposed and tested in England. In 1972 the use of a trommel as a refuse preprocessing device was proposed in the United States. As a result, a pilot test program was conducted to develop supporting data for the design and specification of a trommel for use as a preprocessing device in a prototype materials recovery facility. That facility, Recovery I, is owned and operated by Waste Management, Inc. in New Orleans, Louisiana.

Since the mid-1970s, additional experimental work has been conducted in both England and the United States. Several more recent resource recovery facilities presently in operation utilize trommel screens (Figure 12.66). These include Resources Recovery (Dade County), Florida, Madison, Wisconsin, and Haverhill, Massachusetts.

The Recovery I trommel has been in operation since 1976 and much operating data has been

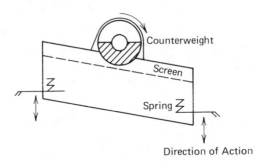

Fig. 12.65   Vertical motion vibrating screen.

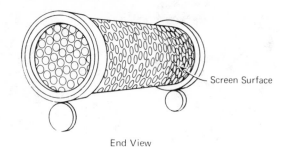

End View

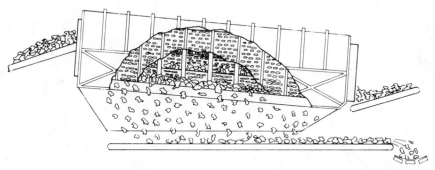

Fig. 12.66 Trommel (rotary) screen.

published. On the basis of experience at that facility, operating at the rate of 60 tons/hr, several resource recovery system developers currently include trommels in their processing schemes (Figure 12.67).

Trommels may be used prior to shredding as at New Orleans to remove material already smaller than the shredder output, thus saving energy, reducing wear and maintenance on the shredder. In addition, removing glass prior to shredding eliminates pulverization of glass and subsequent contamination of RDF by glass fines.

Trommels used after shredding aid in glass and grit removal to produce a higher-quality fuel containing a lower ash content than is possible without screening. However, screening before shredding results in a higher efficiency of glass removal then after shredding, since glass fines tend to become imbedded in and stuck to paper.

In addition to more efficient glass removal prior to shredding, aluminum cans may be concentrated in whole form by screening. Reynolds Metals Company has received U.S. Patent 4387019 for an Aluminum Can Reclamation Method using two stages of trommeling. This system has been demonstrated since 1980 at Houston, Texas, and more recently at Salem, Virginia. Another U.S. Patent No. 4095956, that has been issued to Holmes Bros., Inc., Syracuse, N.Y. utilizes two stages of trommeling to produce a high-quality RDF. In addition to Recovery I and the Houston, Texas, facilities utilizing trommels, Resources Recovery (Dade County) currently uses a trommel as a preprocessing device. Hempstead, New York, and Bridgeport, Connecticut, also included trommels as preprocessing devices, although those facilities presently do not operate. Haverhill, Massachusetts, utilizes trommels effectively after shredding for RDF production as does Madison, Wisconsin.

Typically, trommels used for waste processing are 9 to 12 ft (2.7 to 3.7 m) in diameter. Length varies depending upon capacity and retention time desired. Both the New Orleans and Resources Recovery (Dade County) units are 45 ft (13.7 m) long with capacities in excess of 60 tons/hr. Drive mechanisms include trunnion-tire and sprocket-chain arrangements. Maintenance requirements of trommels are minimal, with screen plate requiring replacement after 5 or more years. Routine service includes lubrication and periodic removal of rags and other items that tend to blind openings, and this may be required on a daily basis. Generally, trommels require less cleaning than disc screens.

Trommels are efficient separatory devices with a long operating history. They are reliable and have wide application, including raw waste, processed fractions, and incinerator residue. As low-

Fig. 12.67 Typical trommel installation.

maintenance devices, they have wide application, although in some systems they have been replaced by disc screens.

12.7.4 Disc Screens

Disc screens are employed in similar applications as trommels. The principal reason for their acceptance is the fact that they occupy less space and require less capital cost than a comparable capacity trommel. This type of screen was originally used in the forest products industry to separate chunks of frozen and oversize pieces from woodchips. Since 1978, their use in solid waste processing has been expanding.

Disc screens consist of a series of horizontal shafts perpendicular to the direction of material flow. The shafts have staggered disc arrangements resulting in openings through which undersize pieces may pass (Figure 12.68). Discs are typically lobed or star-shaped. As the discs turn, the lobes carry material across the surface. They are powered hydraulically, which allows for variable speed control. In the event of blinding or blockage, shaft stoppage monitored by a shaft motion indicator signals the programmable controller resulting in reversal of the shafts to effect a self-cleaning action.

Facilities utilizing disc screens include: Ames, Iowa, Baltimore County, Maryland, Oklahoma City, Oklahoma, Dade County, Florida, and Lakeland, Florida. Uses include preprocessing prior to shredding as well as shredded refuse screening and RDF fines removal.

Problems experienced by disc screens include wrapping of long, stringy material around the shafts, especially in coarse screening applications. This tends to blind the screen and ultimately create a blockage that is not removed by the reversing, self-cleaning action. Shutdown and cleaning is required at least once per shift. Another problem is the wear of the discs that are usually fabricated from mild steel plate. Depending on the tonnage processed, lobed or star-wheeled discs may need replacement annually due to wear. This might be avoided by fabrication of the discs from abrasion-resistant steel plate at somewhat higher initial cost. Due to blinding problems and the necessity for frequent cleaning, one system vendor plans to use trommels in subsequent systems rather than disc screens despite initial capital cost savings.

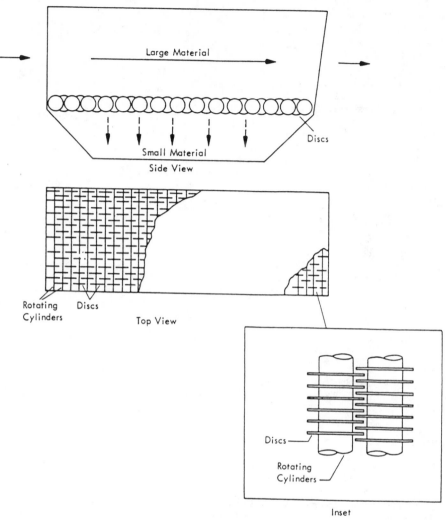

Fig. 12.68 Disc screen.

12.7.5 Summary

Trommels and disc screens are the most widely used screens in waste processing. Disc screens appear to be at least as efficient as trommels, while being less costly initially. However, disc screens have experienced more frequent downtime and operational problems due to blinding than have trommels. Both are likely to become more widely utilized in the industry as the desire to recover more materials of higher quality increases.

12.7.6 Representative Installations

Ames, Iowa	Disc screens
Lakeland, Florida	Disc screens
Haverhill, Massachusetts	Trommels
New Orleans, Louisiana	Trommel
Houston, Texas	Trommels
Salem, Virginia	Trommel
Resources Recovery (Dade County), Florida	Trommel and disc screen
Banyan-Dade, Miami, Florida	Disc screens
Baltimore County, Maryland	Disc screen

Oklahoma City, Oklahoma	Disc screens
Madison, Wisconsin	Trommel
Ontario, Canada	Trommel
Eastbourne, U.K.	Trommel
Doncaster, U.K.	Trommel
Steuenage, U.K.	Trommel
Byker, U.K.	Trommel

12.7.7 Solid Waste Processing Screen Manufacturers

Gruendler Crusher and Pulverizer Co. St. Louis, Missouri 63106 314-423-0600	Trommels
The Heil Company Milwaukee, Wisconsin 414-647-3439	Trommels
Koppers Company, Inc. Sprout-Waldron Division Muncy, Pennsylvania 717-546-8211	Trommels
Rader Companies, Inc. Memphis, Tennessee 38117 901-365-8855	Disc screens
Rotex, Inc. Cincinnati, Ohio 45223 513-541-1236	Vibrating screens
Triple/s Dynamics Dallas, Texas 75223 214-821-9143	Trommels and vibrating screens
Williams Patent Crusher & Pulverizer Co. St. Louis, Missouri 63102 314-621-3348	Trommels

BIBLIOGRAPHY

H. Alter, J. Gavis, and M. Renard, Design Models of Trommels for Resource Recovery Processing, in *Proceedings of the 1980 National Waste Processing Conference*, American Society of Mechanical Engineers, New York, New York, pp. 361–371.

J. F. Bernheisel, P. M. Bagelman, and W. S. Parker, Trommel Processing of Municipal Solid Waste Prior to Shredding, in *Proceedings of the 6th Mineral Waste Utilization Symposium*, Chicago, Illinois, May 2–3, 1978, Illinois Institute of Technology.

Cal Recovery Systems, Inc., Trommel Screen Research and Development for Applications in Resource Recovery, DOE Contract No. DE-AC03-79CS20490, June 1981.

J. A. Campbell, et. al., Trommeling Proves Effective Step in Many Resource Recovery Systems, *Solid Waste Management*, January, 1981, pp. 34–35, 63.

R. L. Chrismon, J. F. Bernheisel, and P. M. Bagelman, *Trommel Initial Operating Report, Recovery I*, TR 78-3, National Center for Resource Recovery, Inc., October, 1978, 26 pp.

L. F. Diaz, G. M. Savage, and C. G. Golueke, Trommel Screening, in *Resource Recovery from Municipal Solid Wastes, Volume I, Primary Processing*, CRC Press, Inc., Boca Raton, Florida, 1982, pp. 109–126.

E. Douglas and P. R. Birch, Recovery of Potentially Reusable Materials from Domestic Refuse by Physical Sorting, in *The Technology of Reclamation*, University of Birmingham, Birmingham, England, April 7–11, 1975.

D. E. Fiscus, et. al., Evaluation of the Performance of the Disc Screens installed at the City of Ames, Iowa Resource Recovery Facility, in *Proceedings of the 1980 National Waste Processing Conference, Washington, D.C.*, May 11–14, 1980, ASME, pp. 485–495.

R. M. Hill, Rotary Screens for Solid Waste, *Waste Age*, **8**, 33–37 (1977).

Materials Recovery System, Engineering Feasibility Study, National Center for Resource Recovery, Inc., Washington, D. C., December, 1972.

New Orleans Resource Recovery Facility Implementation Study: Equipment, Economics and Environment, National Center for Resource Recovery, Inc., September, 1977, pp. 1–9.

W. S. Parker, Application of Trommeling to Prepared Fuels, in *Proceedings of the International Conference on Prepared Fuels and Resource Recovery Technology*, Argonne National Laboratory, Argonne, Illinois, April, 1981, pp. 101–116.

W. S. Parker, Trommel Technology, *Waste Age*, November, 1981, Vol 16, No. 7, pp. 43–45.

P. K. Patrick, "Waste Volume Reduction by Pulverization, Crushing and Shearing," The Institute of Public Cleansing, 69th Annual Conference, Blackpool, England, June 5–9, 1967, pp. 21–25.

Problems at Recovery I, *Waste Age*, **8**, 2 (1977).

G. Savage and G. J. Trezek, Screening Shredded Municipal Solid Waste, *Compost Science*, January/February, 1976, pp. 7–11.

A. F. Taggart, *Handbook of Mineral Dressing*, John Wiley & Sons, Inc., New York, 1927, pp. 7-24–7-34.

"Tests to Investigate the Use of Rotating Screens as a Means of Grading Crude Refuse for Pulverization and Compression Treatment," Greater London Council, Public Health Engineering Dept., London, 1966.

"There is Nothing New!" (Illustrations reproduced from the July 1924 edition of *My Magazine*). *Solid Wastes, Monthly Journal of the Institute of Solid Wastes Management*, **LXVI**, 536–537 (1976).

Trommel Performance at Nominal Design Conditions, prepared for USEPA Office of Research and Development, Cincinnati, Ohio, NCRR, 1978.

J. L. Warren, The Use of a Rotating Screen as a Means of Grading Crude Refuse for Pulverization and Compression, *Resource Recovery and Conservation*, **3**, 97–111 (1978).

E. Williams, *Review of the Literature of the Use of Trommels in Waste Processing and Resource Recovery*, National Center for Resource Recovery, Inc., Report No. DOE/CS/20167-6, July, 1981.

K. L. Woodruff, Preprocessing of Municipal Solid Waste for Resource Recovery with a Trommel, *Transactions of the Society of Mining Engineers*, **260**, 201–204 (1976).

K. L. Woodruff and E. P. Bales, Preprocessing of Municipal Solid Waste for Resource Recovery with a Trommel—Update 1977, in *Proceedings of the 1978 National Waste Processing Conference*, American Society of Mechanical Engineers, New York, pp. 249–257, May 1978.

K. L. Woodruff and E. P. Bales, Why Start with Shredding . . . , *The American City and County*, October, 1978.

12.8 DENSIFIED REFUSE-DERIVED FUEL (dRDF)*

Jay. A. Campbell

12.8.1 Background

The production of densified refuse-derived fuels (dRDF)† as a substitute for coal in industrial or institutional spreader-stoker boilers is a developing waste-to-energy alternative. The early production experience at several pilot plants and continuing operations at eight densification facilities in the United States and abroad will be reviewed. Equipment requirements, performance, and product properties are summarized from the production of over 1400 tons (1300 Mg) dRDF. Experiences in storage and handling are also discussed. Projection of capital and operating costs for a densification module are provided.

Over 25 combustion tests have been conducted utilizing various sizes and quantities of dRDF in controlled blends with coal in several types of stoker-fed boilers. Observations and results on the boiler capacity, efficiency, and emissions are reviewed briefly.

The emphasis of early waste-to-energy demonstration and commercial projects in the United States was on the preparation and firing of RDF in suspension-fired boilers burning pulverized coal. These types of boilers, which are typically large with steam-generating capacities generally over 400,000 lb/hr (180 Mg/hr) steam, offer a potential for combustion of large quantities of waste. However, the plants tend to be concentrated near larger cities and, due to economies of scale and the nature of utility rate structuring, their cost of primary fuel tends to be lower. The pressure to minimize fuel costs is also less than that for industrial boiler operators.

The industrial and institutional boiler population, including those with steam-generating

---

*This Report is reprinted from the Proceedings of the 1981 International Conference on Prepared Fuels sponsored by Argonne National Laboratories and with their permission. Mr. Campbell prepared this report for the National Center for Resource Recovery and for presentation at that conference.
†dRDF (densified refuse derived fuel) has an ASTM designation of RDF.5.

capacities of 66,000 to 240,000 lb/hr (30 to 110 Mg/hr) steam includes some 8500 units.[1] These boilers are widely dispersed in both large and small communities; and, because of their small size, they cannot benefit from discounts for quantity, purchases, and deliveries of the coal. Regulations on sulfur dioxide emissions are also forcing these operators to either buy expensive low-sulfur coal or invest in costly air pollution control equipment.

For these reasons of location, size, and fuel specifications and costs of this smaller power boiler market, there appears to be a good match to the availability, locations, and fuel characteristics of waste-derived fuel. To minimize the modifications to the existing feed and combustion systems, dRDF rather than fluff RDF burned in suspension-fired boilers would be attractive. The goal then is to produce a dRDF with physical and combustion properties comparable to the "lump" or stoker coal normally burned in these industrial-scale boilers.

Production and test work on dRDF accomplished by the National Center for Resource Recovery (NCRR) from 1976 to 1979 for the U.S. Environmental Protection Agency and Department of Energy has been reported both in final project reports [2,3,4] and on specific topics or in abbreviated versions in other papers.[5,6,7] This discussion will include results and observations from each of these NCRR projects, as well as summaries of other activities in the production, handling, and combustion of dRDF both in the United States and abroad.

12.8.2 Production Technology Status

dRDF is obtained by compacting the light organic fraction of solid waste, that has been processed to remove noncombustibles, into densified particles. The term dRDF as used here refers to mixed residential or commercial wastes densified in pelletizing, cubing, extrusion, or briquetting devices. Typical products from these various equipment types are shown in Figure 12.69. Densification of biomass feedstocks, although of growing interest, is not considered here.

Densification Equipment

Pelletizing of RDF is accomplished with a hard steel die perforated with a dense array of holes 0.025 to 1.25 in. (6 to 32 mm) in diameter. As the die and an inner set of pressure rollers rotate against each other, the feedstock is forced through holes, and the pellets broken off at random lengths up to 1.5 in. (38 mm). Cubetting is a modification of pelletizing in which the dRDF is produced in the form of cubes, generally 1.25 in. (32 mm) square.

Extrusion as a means of producing dRDF employs a screw or reciprocating plunger to force the feedstock under pressure through a die. Large-diameter cylinders 2 to 80 in. (50 to 200 mm) are produced, and binding agents such as pitch or paraffin are typically added to increase integrity. (Artificial fireplace logs, for example, are manufactured by an extrusion process.) The larger cylinders place the products outside the size range acceptable for stoker-fired boilers.

Briquetting involves the compaction of feedstock in cavities between two rollers. A material is produced similar in size and shape to charcoal briquettes. Although some experimental work has been conducted in the briquetting of refuse, problems with feeding and binding have been experienced. Briquetting of RDF is not practiced commercially at this time.

The mechanism that bonds compacted RDF is not fully understood. With wood and agricultural feedstocks, natural binders in the form of lignin are present. For other materials and in most briquetting operations, binders such as paraffin, lignins, or bettonite (clay) are added. For densification of residential waste feedstocks, there has been no reported need or use of binders. Apparently the presence of moisture and lignin, waxes, and greases in the waste combine to bind the particles.

A densification system is typically made up of several component parts: (1) a feeder, typically a screw-bottom bin, to meter the feed to the system; (2) a conditioner or mixer, usually a screw conveyor where binders or other additives may be mixed with the feedstock; (3) the densifier with associated dies, rollers, drive system, and base structure; and (4) a cooler, usually a conveyor carrying a thin bed of dRDF through which air is drawn to remove excess heat or moisture and thereby harden the product. Figures 12.70 and 12.71 are illustrations of densifying equipment.

Table 12.17 provides a partial listing of densification equipment manufacturers. A summary of some of the larger currently operating and notable older dRDF production facilities in the United States and United Kingdom are noted in Table 12.18.

12.8.3 Densification Equipment Performance and Problems

As in many instances where equipment from other industries has been applied to solid waste, the performance of the densification equipment has not equalled performance on traditional agricultural and forest product residue feedstocks. The following discussion of solid waste densification will be viewed from the perspective of the differences in feedstock and application. Lacking

Fig. 12.69 dRDF pellets, densified refuse-derived fuel. (Courtesy National Ecology, Inc.)

extensive operational experience, such a viewpoint provides a good base for understanding observed performance or identifying needs for improved feed preparation and delivery systems.

Note that whereas nearly all of the densification efforts to date have been with pilot and
experimental plants, most of the densifiers have been full size. Thus, the observations and data on
throughput, reliability, and feedstock/product relationships are valid for assessing commercial-
scale performances, capabilities, and limitations. Also note that these data and relationships should
apply to most types of densifiers, although experience has been exclusively with pelleting and
cubetting-type machines.

Throughput

Processing capacities of 8–12 tons/hr are common for pelleting traditional feedstocks with the
equipment size and configurations utilized at NCRR, and several other waste densification plants.
However, maximum capacities of only 5 tons/hr and sustained feedrates of only 2 to 4 tons/hr have
been achievable with RDF feedstocks without frequent blockages or deterioration of dRDF quality.[2] Rather than a single explanation for the shortfall, there seem to be several interrelated factors,
including equipment configuration and condition, feedrate stability, and feedstock properties such
as moisture, density, and particle size. The impact of these factors will be discussed below.

Equipment Configuration and Condition

Densifier capacity and product characteristics are affected by the die and roller configuration (die
hole diameter, taper and thickness, and roller diameter and surface covering). Lacking experience
with waste feedstock, for the first-generation plants these components have been selected based on
the manufacturer's intuition. Today there is still insufficient documented operating experience with
a variety of hardware configurations or feedstocks to improve much on such a selection process.

Prepared feed material in

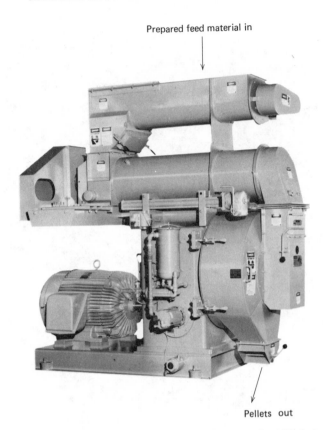

Pellets out

Fig. 12.70   dRDF pelletizer. (Courtesy California Pellet Mill Co.)

The impact of the condition of the dies and rolls on throughput was evident at NCRR in changes in throughput rate and the frequency of jams on the densifier with increasing wear on dies and rolls. When a new die with sharp corners at the hole inlet or new rolls with unworn corrugations were first installed, the feedrate could be increased by 40%; blockages and stalls of the machine were also reduced. After processing less than 110 tons (100 Mg), the abrasive feed blunted the die and roll surfaces; again, only lower throughputs could be maintained without jams.

Feedstock Size

In traditional densification applications, the feedstock is reduced to a top size, equal or smaller than the diameter of the die hole opening. In all waste-densification applications to date, the particle size has been larger than the die aperture. This would lead to greater power consumption and, thus, reduced capacity, as material must be sheared by the die and rolls as it is forced into the holes. In addition, materials particularly resistant to size reduction such as plastics and textiles have been observed to wind on augers and impellers and jam the rolls, stopping the machine.

In exploring sizing approaches at both NCRR and Warren Spring Laboratory (WSL), it was found that the products from hammermill and knife shredders varied both in the proportion of top-size materials as well as particle and bulk properties.[2] Conclusions from the investigations suggested that two or more stages of hammermilling were necessary to achieve the same size control as could be obtained with a single pass through a knife shredder. In addition, the knife mill product tended to be flatter, crumpled less, and have a higher bulk density than the hammermill product. Recognized limitations of a knife mill applied to solid wastes are susceptible to damage from tramp metals and wear from abrasive fines.

Moisture

The moisture content of most traditional pelletizer feedstocks is controlled by steam conditioning or drying to a level of 12 to 15%. NCRR experience suggests a comparable range of moisture of 12

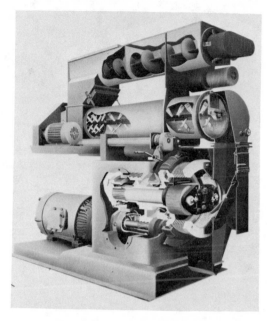

Fig. 12.71 Interior of pelletizer.

to 20%, that results in formation of pellets of good quality. Moisture content between 20 and 30% resulted in generally acceptable but variable pellet quality. With moisture content over 30%, pellets are difficult to form and have little sustained integrity. In all cases, it was noted the hardness and durability improved with cooling.

The first use of dryers to lower and control moisture content of the solid waste feedstock has been in commercial plants in Doncaster and Eastbourne, U.K., that began operating in 1980. Results have not been reported.

Bulk Density

Agricultural and forest product feedstocks tend to have fairly consistent bulk densities in the range of 8 to 12 lb/ft^3 (128 to 192 g/dm^3). This compares with densities ranging from 2 to 5 lb/ft^3 (32 to 80 g/dm^3) for waste fuels. Such lower and more variable densities lead to difficulty in introducing shredded waste into the working area of the densifier at a uniform rate. The feed augers tend to fluff or compress the feed or, at lower densities, the feed tends to bridge temporarily in the inlet section to the die. Both conditions lead to feed surging, a condition apparent in the fluctuating densifier power consumption noted in making power measurements.

Table 12.17 Selected Densifier Equipment Suppliers, 1981

Firm	Address
California Pellet Mill	1114 E. Wabash Ave. Crawfordsville, Indiana 47933
Agnew Environmental Products	P. O. Box 1168 Grants Pass, Oregon 97526
Papakube Corporation	1211 Rockhurst Dr. San Diego, California 92120
Ferro-Tech	467 Eureka Rd. Wyandotte, Michigan 48192
Sprout-Waldron	Koppers Co., Inc. Muncy, Pennsylvania 17756
Bühler-Miag	4515 Willard Ave. Chevy Chase, Maryland 20015
Trans Energy Systems	1605 116th Ave., N.E. Bellevue, Washington 98004

Table 12.18 Summary of Selected dRDF Production Facilities, 1981

Operator/Location	Operating Period	Plant Type/Capacity	Front-end Processing	Densified Product	Report Reference[a]
Vista Chemical Co., Los Gatos, California	1974–	Prototype, 6–8 ton/hr (5–7 Mg/hr)	Trommel, shred, air classify, trommel	Ring extrusion, ¼–1½ in. (6 mm–38 mm) dia.	—
Leigh Forming Co., Easton, Pennsylvania	1977–	Prototype, 10 ton/hr (9 Mg/hr)	Shred, air classify	Ring extrusion, ⅝ in. (16 mm dia.)	—
NCRR, Washington, D.C.[b]	1974–1979	Pilot plant, 10 ton/hr (9 Mg/hr)	Shred, screen, air classify, shred	Ring extrusion, ½–1.0 in. (13 mm and 25 mm dia.)	2, 3, 4
National Recycling, Ft. Wayne, Indiana[b]	1971–1973	Prototype, 6–14 ton/hr (5–13 Mg/hr)	Shred, air classify, screen	Cuber, 1¼ in. (32 mm dia.)	8
Maryland Environmental Service, Cockeysville, Maryland	1976–	Commercial, 10 ton/hr (9 Mg/hr)	Shred, air classify, trommel, shred	Ring extrusion (2) ½ in. (13 mm dia.)	9
U.S. Navy, Jacksonville, Florida	1980–	Prototype, 6 ton/hr (5 Mg/hr)	Flail, trommel, classify	Ring extrusion, ¾ in. (19 mm dia.)	10
Doncaster, U.K. (municipal)	1980–	Commercial, 11 ton/hr (10 Mg/hr)	Trommel, air classify, shred, dry	Ring extrusion, ⅝ in. (16 mm dia.)	—
Buhler Miag, Inc., Eastbourne, U.K.	1980–	Commercial, 10 ton/hr (9 Mg/hr)	Shred, trommel, air classify, shred, dry	Ring extrustion ¾ in. (20 mm dia.)	—
Byker, U.K. (municipal)	1980–	Commercial, 33 ton/hr (30 Mg/hr)	Shred, trommel, air classify, shred	Ring extrusion, ⅝ in. (16 mm. dia.)	—

[a] References are found after Section 12.8.
[b] Shutdown.

466

Power Consumption

The power consumption of the densifier is a function of the die and roller size and condition, feedstock properties (particle size, composition, and moisture), and throughput rates (average and surges). With a 2.5 ton/hr feedrate, nominal wear on a 1/2 in. die and for a typical MSW waste feedstock, the measured power requirement was 48 hp ton · hr (82 kW or 36 kW/Mg · hr).[2] For an office waste (high paper) feedstock of similar size and moisture, power consumption over the same ranges of throughput was 73 hp/ton · hr (55 kW/Mg · hr).[4] The reason for this difference is not known.

Both of these figures are higher by a factor of 2 to 3 than estimates of 22 hp/ton · hr (16.4 kW/ Mg · hr) made by the densification equipment supplier based on experiences with other feedstocks.[11]

Maintenance

Wear on the die and press roller is the most significant component of maintenance. Although the wear patterns are similar to those observed for traditional feedstocks, they occur at a much faster rate with RDF. This is due both to the abrasive characteristics of cellulosic materials and the presence of abrasive inorganic fines (glass and grit) in the feed. Projections based on experience by NCRR suggest that a die life of 3850 tons (3500 Mg) and roller life of 1870 tons (1700 Mg) could be expected for a low-ash content (10 to 12%) waste feedstock. Measurement of wear on a variety of densifiers and components, over a long duration and for a range of feedstock characteristics, is required to develop reliable wear rates and maintenance costs for waste fuels densification.

12.8.4   dRDF Properties and Characteristics

For utilization of dRDF as a supplemental fuel in stoker boilers, there are several physical characteristics that have a bearing on handling, transport, storage, feeding, and combustion. These include dRDF size, particle density, bulk density, fines content, and integrity. Typical values for these characteristics for dRDF sampled at several production and boiler facilities are summarized in Table 12.19 and are discussed below. Data on dRDF properties such as moisture content, ash content, chemical composition, and heat content are also shown. Care should be taken in comparing or interpreting these data since the sampling and analysis procedures may not have been consistent.

dRDF Size

The dRDF dimensions (cross section and length) have an impact on handling through relation to packing density and to burning rate in the boiler. Most past and current commercial and test production has been with particles from 1/2 to 3/4 in. (13 to 20 mm) diameter or 32 mm². Lengths are random, and range from one to two times the diameter. Difficulty with particles larger than these dimensions result from reduced bulk density and an increased tendency to bridge.

dRDF Particle Density

More dense particles indicate higher mechanical strength and resistance to abuse. With denser particles, flow is improved in and out of storage, and the ballistic behavior through the stoker system would more closely approximate particles of coal. Particle density of coal is on the order of 85 lb/ft³ (1.3 g/cm³). Table 12.19 indicates pellet densities 10 to 20% below that for coal.

dRDF Bulk Density

Due to the lower calorific value for dRDF (typically one-half to one-third that of coal), a larger mass and volume of dRDF must be fired to maintain the same boiler loading compared with coal alone. A higher bulk density is therefore desired to minimize the effects of increased fuel volume on the stoker feeder's capacity and combustion bed thickness. Higher bulk density will also alleviate possible problems in material flow during storage, retrieval, and firing. The bulk density ranges in Table 12.19 reflect variation in dRDF size and particle density as well as density variations related to the feed composition.

dRDF Fines Content

For coal, fines are specified as screened, minus 0.2 in. (6 mm) material. The definition of fines for dRDF has not been standardized, but in the cases where it is reported in Table 12.19, it has generally been defined as 1/8 in. (3 mm) less than the pellet diameter. Fines can accentuate

Table 12.19 Summary of dRDF Properties Results on Dry Weight Basis

dRDF Sample Source and Description	NCRR, As Produced from MSW (Ref. 2)[a]	NCRR, As Produced from MSW (Ref. 2)	NCRR, As Produced from Office Wastes (Ref. 3)	NCRR, Post Storage (Ref. 13)	Teledyne, Post Storage (Ref. 13)	PLM, As Produced (Ref. 14)
Number of samples	>25	>10	>10	4	13	Unknown
Diameter (in., mm)	½, 13	1.0, 25	½, 13	½, 13	½, 13	32 × 32 × 30
Length (in., mm)	⅝, 15.9	—	¾, 20	—	—	—
Pellet density (lb/ft³)	1.01	0.76	1.13	—	—	—
Bulk density (g/m³, kg/m³)	35.9, 575	27, 432	42.9, 687	33.6, 538	32.6, 522	28, 450
Fines content (%) −⅜ in., −9 mm	14.9	9.6	2.6	46.7 (−6 mm)	27.9 (−6 mm)	—
Moisture content (%)	22.9	22.8	21.5	29	31.4	15
Ash content (%)	23.4	22.8	9.6	30.7	13.8	10
Heating value (BTU/lb, MJ/g)	7485, 17.4	—	7744, 18.0	6754, 15.7	8131, 18.9	7012, 16.3
Carbon (%)	42.6	—	46.0	40.6	45.7	—
Nitrogen (%)	0.77	—	0.17	0.59	0.37	—
Sulfur (%)	0.48	—	0.13	0.43	0.23	0.29
Chlorine (%)	0.57	—	0.16	0.31	0.38	—

[a]References are found after Section 12.8.

problems of material flow or segregation and result in dusting in handling and feed systems. Although some fines may be desirable for obtaining a mix of suspension and grate burning (for coal a 30% fines level is often specified), higher levels of dRDF fines could increase slagging on the boiler tubes and could overload air pollution control equipment. The tolerable level of fines will vary with the boiler fuel, equipment, and operations at a specific site. It should be noted that fines content of the dRDF product can be controlled by screening after production.

Integrity

The integrity of the dRDF product is defined as its ability to sustain handling without breaking or losing mechanical strength. Fracture planes are created in production of the pellets by pieces of plastic or textile that bond poorly to adjacent materials. Observations during storage and handling by NCRR[2] and Systec[13] further suggest that the susceptibility to breakage is a function of moisture and temperature. Neither test methods nor units expressing relative integrity are as yet refined and defined for dRDF, although they have been and are being explored.[2,12] The dRDF particle density property is probably the best interim indicator of integrity.

Care in handling and storage after production is important to prevent delamination and disintegration leading to increases in fines content and reduced particle and bulk density. Such a loss in integrity can be seen in Table 12.19 in comparing the properties for dRDF as produced at NCRR and after handling and storage as fired in the test burn in Erie, Pennsylvania.

dRDF Ultimate Analysis

The moisture, ash, and chemical contents and heating values reported in Table 12.19 for the various dRDF products are a function of the composition and preparation of the waste feedstock and are not properties directly affected by the densification process (although the reverse is not true, as was seen in discussions on the densification process and equipment). From the standpoint of combustion, it is clearly desirable to produce a dRDF with the highest possible heating value by minimizing moisture and ash to levels consistent with that required by densification.

12.8.5 Storage and Handling

The experience accumulated thus far with densified fuels has shown, as expected, that in addition to being easily handled and transported, they alleviate the problems of material spillages, dusting, and bridging common with fluff forms of RDF.

However, NCRR experience with storage of more than 1000 tons of dRDF from mixed municipal waste indicated that problems with degradation, composting, and oxidation can occur over a longer term. After 10 months in a covered, 6 to 13-ft (2 to 4 m)-high pile, several seams of smoldering pellets surfaced at the edges of the pile and ignited plastic sheet covers. The pellets themselves did not ignite, although the seams of pyrolyzed, charred pellets ran deep into the pile. Excavations revealed both very wet (> 130% moisture) and dry areas (< 4% moisture), interspersed with material of average, as-produced (20%) moisture contents. Such conditions indicated that significant moisture transfers had also been occurring. Both the wet and oxidized materials exhibited significant pellet degradation (increased fines).

The reasons and mechanisms for formation of the oxidized seams are not clear. It is felt that inadequate cooling of the pellets after production, use of an unventilated cover, the long storage period, and interim movement of the stored material all contributed to the problems. Similar problems with oxidation of dRDF in storage have apparently not been observed elsewhere, and it is felt that with proper storage conditions and/or for reasonable periods, the problem of oxidation would be eliminated and degradation would be held to a lower, more tolerable level.

12.8.6 Densification Costs

As part of the investigations of densification technology supported by EPA, the economics of dRDF systems for small communities (100 to 200 TPD) were explored.[2] In the study, the capital and operating costs for a densification module were estimated and used to back-calculate what feed system preparation costs (plant and equipment) could be afforded, given a range of fuel values and landfill cost avoidance.

The reader is directed to the EPA report for detailed discussion of the approach and result of this analysis. For the purpose of the discussion here, only the costs developed for the densification module are cited.

Table 12.20 provides a list of equipment and costs for a 7.9 ton/hr (7.2 Mg/hr) capacity densification module: two each ring extrusion pelletizers with nominal 1-in. (25 mm) holes. Such a system operated two shifts per day (14 hr) and 250 days per year would produce 25,000 tons annually. Table 12.21 provides the operating and maintenance costs for a two-shift operation. Table

**Table 12.20   Densification Module Equipment Cost Detail (Ref. 2)
Two Densifiers—1979 Costs**

Item	Cost
Equipment	
Conveyers (4), feed and product	$ 32,700
Densifiers (2), w/surge bin, spare die, and rolls, motor	184,600
Pellet cooler	14,500
Pellet screener	6,200
Motor control center	5,000
Freight and taxes	17,000
Installation	65,000
Building allocation (807 ft^2, 75 m^2)	20,000
Engineering	27,600
Contingency	17,300
Total capital cost	$389,900

References are found after Section 12.8.

**Table 12.21   Densification Module Operating and Maintenance Cost Detail (Ref. 2) Two
Densifiers, 14 hr/day, 250 day/yr; 7.9 ton/hr (7.2 Mg/hr) Nominal Throughput, 27,720 ton/yr
(25,200 Mg/yr)—1980 Costs**

	Annual	Unit Throughput
Labor (annual basis)	$ 30,000	$1.19
Materials, supplies	12,000	0.48
Utilities	32,250	1.28
Maintenance		
Dies	50,400	2.00
Rollers	32,250	1.28
Miscellaneous (densifiers, conveyers, cooler, screen)	15,900	0.63
	$172,800	$6.86/Mg ($6.24/ton)

References are found after Section 12.8.

**Table 12.22   Densification Module Capital and Operating Cost
Summary (Ref. 2)—1980 Costs**

Item	Annual
Capital Costs	
Total cost	$389,900
Annual 8%, 10 years	58,485
Unit cost	2.34/Mg ($2.13/ton)
Operating Costs	
Labor	1.19
Materials, supplies	0.48
Utilities	1.28
Maintenance	3.91
Total Operating Costs	6.86/Mg ($6.24/ton)
Total Capital and Operating Costs	$9.20/Mg ($8.36/ton)

References are found after Section 12.8.

12.22 summarizes these costs yielding a total capital and operating cost of $8.36/ton ($9.20/Mg). Processing of the feedstock is presumed to include sizing to less than 1 in. (25 mm), and controlling ash content to less than 12% (reflecting removal of most inorganic fines). Drying of feedstock is not suggested unless the average moisture content of the feedstock is above 25%.

12.8.7   dRDF Combustion Experience

Table 12.23 provides information on six test firings of dRDF in the United States from the earliest reported tests at the Fort Wayne Municipal Power Co. in 1972.[8] Several are noteworthy for their duration or more detailed documentation of results.

More than 20 other tests (not shown) have been conducted (listed in Refs. 2 and 15). Most were of short duration (several hours) and generally yielded only observations on performance of the fuel-handling system, and indications of significant changes in boiler responses, burnout, and stack emissions.

In December, 1977, the EPA sponsored combustion trials using dRDF at the Maryland Correctional Institute in Hagerstown. During the tests, 285 tons (259 Mg) were burned over 230 hr at various blends of dRDF and coal. No particular problems in material handling were observed.[16]

Stack capacity, along with SO_2, was found to decrease while chloride emissions increased. The Hagerstown program served as a feasibility field test to the larger scale test burn sponsored by EPA in a 150,000 lb/hr (68,000 kg/hr) spreader stoker-fired boiler in Erie, Pennsylvania in early 1979.

Just over 1700 tons (1545 Mg) of dRDF from the National Center and Teledyne National production facilities were burned at Erie in volumetric blends of coal to dRDF of 1:1, 1:2, and 1:4. The tests spanned 400 hr of cofiring and 230 hr of coal-only firing. A full set of measurements and data was made by Systems Technology on material handling, boiler performance, and environmental control.[13]

The results of the Erie tests demonstrated that cofiring of coal and dRDF had minimum impact in the performance of the power plant test boiler. Weathering and subsequent deterioration of the dRDF from multiple handlings, transportation, and storage (for periods of 5 to 18 months) were significant. Nearly 500 tons (455 Mg) of the dRDF fired to the boiler contained 50% unpelletized material (fines) with the other 1200 tons (1090 Mg) of dRDF having an average of 30% fines. These material conditions led to some problems of channeling, erratic flow, and bridging of the coal/dRDF blends in the fuel bunkers, but had no noticeable influence in the operation of the stokers or fuel distribution in the furnace.

Boiler performance as measured by efficiency was reduced only 2 to 3% at the 1:2 coal to dRDF blend. Derating of the boiler was not experienced with substitution of the dRDF for coal (in part due to excess capacity in the design). While higher-ash pellets (20 to 30% ash) tended to form ash clinkers, lower-ash pellets (12 to 15%) did not exhibit this tendency. There was increased slag in the lower walls of the furnace but it generally fell away during operation or during scheduled maintenance. Fouling of the tubes was not observed. Testing on samples of slag and ash showed no corrosion problem.

From an environmental performance standpoint, the particulate emissions rate and precipitator performance was unchanged from firing coal only. Comparing gaseous emissions from combustion of coal alone with gaseous emissions from combustion of the blends, increases were noted in lead, cadmium, zinc, and chromium; hydrocarbons and carbon monoxide remained constant and sulfur dioxide decreased.

Several short-duration test burns were conducted in early 1979 at the Pentagon power plant near Washington, D.C. using pellets prepared from office wastes.[3] Firing trials were conducted with baseline coal, and blends of 20, 40, and 60% dRDF by volume. Unfortunately, the tests were run with a reduced (40 to 50% of rated capacity) and fluctuating boiler load. Although bridging and channeling were again evident in the fuel bunker, they were less than that observed at Erie; this was probably due to the condition of the coal rather than that of the pellets. Segregation of the dRDF and coal, that occurred during loading of the bunker, resulted in uncontrolled fluctuations in the ratio of dRDF and coal fed to the boiler. At the 60% volumetric blend, this segregation resulted in higher-volume feeds, that exceeded the stoker capacity and caused hot spots in the fuel bed in areas of high dRDF concentration. These problems were not evident with the 20 and 40% blends.

The amount and handling of bottom ash was no different than with coal. Particulate emissions actually decreased, while the gaseous emissions displayed the same trends evident in the Erie tests.

Short-term, low-tonnage firings have been tested to date in the United Kingdom with small industrial boilers using reciprocating stokers.[17] The conclusions were that with few modifications dRDF could be used in existing combustion systems and conform with environmental regulations. Significant data are expected over the next several years from the commercial dRDF firing programs in Doncaster, Byker, and Eastbourne in the United Kingdom.

In Sweden, the PLM Company has test fired up to several hundred tons in several 1 to 5-day tests.[14] The results of these tests are yet to be reported.

Table 12.23 Summary of Selected dRDF Combustion Tests

Location	Date	Boiler Type/Size	dRDF Description/Blend	Report Reference
Ft. Wayne, Indiana	1972	Underfeed stoker; 40,000 kW	40 ton, 1.5 in. × 1.5 in. cubes (36 Mg, 38 mm × 38 mm cubes); 3:1 coal:dRDF by volume	8
Chanute, A.F.B., Illinois	1975	Traveling chain grate, gravity overfeed; 35,000 lb/hr (16 Mg/hr) steam	148 ton, 1 in. pellets (135 Mg, 28 mm pellets); 1:1, 0:1 coal:dRDF by volume	—
Hagerstown, Maryland	1977	Traveling grate, spreader stokers (2); 27 Mg/hr, 60–75,000 lb/hr (27–33 Mg/hr) steam	277 ton, ½ in. pellets (252 Mg 13 mm pellets); 1:1, 1:2, 0:1 coal:dRDF by volume	16
Washington, D. C.	1979	Underfeed multiple retort; 70,000 lb/hr (32 Mg/hr) steam	124 ton, ½ in. pellets (113 Mg 13 mm pellets) (office waste); 4:1, 3:2, 2:3 coal:dRDF by volume	3
Erie, Pennsylvania	1979	Traveling grate spreader stoker; 150,000 lb/hr (68 Mg/hr) steam	1386 ton, ½ in. pellets (1260 Mg 13 mm pellets); 1:1, 1:2, 1:4 coal:dRDF by volume	13
Wright-Patterson A.F.B.	1980–1981	Traveling grate, spreader stoker; 80,000 lb/hr (36 Mg/hr) steam	Proposed 7920 ton/yr, ½ in. pellets (7200 Mg/y 13 mm); pellets 1:1 variable coal:dRDF by volume	—

References are found after Section 12.8.

REFERENCES

1. T. Divitt, P. Spaite, and L. Gibbs, "Population and Characteristics of Industrial/Commercial Boilers in the U.S.," a report by PEDCO Environmental under Contract 68-02-2603 to the EPA Office of Research and Development, Research Triangle Park, North Carolina, 1979.

2. J. A. Campbell and M. L. Renard, "Densification of Refuse Derived Fuels: Preparation, Properties and Systems for Small Communities," final report EPA Grant 804150, National Center for Resource Recovery, Inc., Washington, D. C., December, 1980.

3. J. A. Campbell, Final Test Report, "Demonstration dRDF Burn at the GSA Pentagon Power Plant," Contract DOE-ES-76-C-01-3851, Task 8, October, 1979.

4. Z. Khan and M. L. Renard, "The Use of Waste Oils to Improve Densified Refuse Derived Fuels," Contract DOE-ES-76-C-01, 3851, Task 5, Washington, D. C., October, 1979.

5. H. Alter and J. A. Campbell, The Preparation and Properties of Densified Refuse Derived Fuel, in *Thermal Conversion of Solid Waste and Biomass Symposium Proceedings*, American Chemical Society, Washington, D. C., September, 1979.

6. J. M. Arnold and D. C. Hendrix, "Initial Mass Balance for Production of Densified Refuse Derived Fuel," Technical Report RR77-3, National Center for Resource Recovery, Inc., Washington, D. C., December, 1977, 6 pp.

7. C. C. Wiles, "The Production and Use of Densified Refuse Derived Fuel," Proceedings, Fifth Annual Research Symposium, Land Disposal and Resource Recovery, Orlando, Florida, March, 1979.

8. H. I. Hollander and N. F. Cunningham, "Beneficiated Solid Waste Cubettes as Salvage Fuel for Steam Generation," Proceedings, 1972 National Incinerator Conference, American Society of Mechanical Engineers, New York, 1972, pp. 75–86.

9. K. R. Sheppard, "Fuels in a Waste-to-Energy System: The Teledyne Experience," presented at the International Conference on Prepared Fuels and Resource Recovery Technology, Nashville, Tennessee, February 12, 1981. (In press.)

10. F. C. Hildebrand, "Navy Experience in the Conversion of Solid Wastes into Energy," Naval Facilities Engineering Command, Alexandria, Virginia, June, 1979.

11. T. Reed and B. Bryant, "Densified Biomass: A New Form of Solid Fuel," SERI-35, Solar Energy Research Institute, Golden, Colorado, 1978, 30 pp.

12. P. A. Vesilind, Duke University, private communication regarding research project on dRDF properties and test procedures funded by EPA through ASTM Committee E-38 to begin at Duke in January 1981 (1980).

13. M. Kleinhenz, "Coal: dRDF Demonstration Test in an Industrial Spreader-Stoker Boiler," Final Report EPA 68-03-2426 by Systems Technology, Zenia, Ohio, July, 1980.

14. B. Hansen, Process for the Recovery from Household Waste of Solid Fuel (dRDF) Alternatively Secondary Paper and Their Reuse, in *Recycling Berlin '79*, K. J. Thome-Kozmiensky, Ed., Vol. 2, Springer-Verlag, Berlin, 1979, pp. 937–941.

15. Proposed Draft document for GSA Office Waste Removal and Procurement of dRDF for Use as Supplemental Fuel in GSA Operated Boilers, Final Report DOE-ES-76-C-01-3851, Task 8A, 1980.

16. H. G. Rigo, G. Degler, and B. T. Riley, "A field Test Using Coal: dRDF Blends in Spreader Stoker-Fired Boilers," Draft Interim Report EPA-68-03-2426 by Systems Technology Corporation, Xenia, Ohio (to be issued).

17. P. R. Birch, "Pilot-Scale Production and Firing of Densified RDF," Warren Spring Laboratory Report for The British Department of the Environment, 1979, 7 pp.

12.9 REFUSE-DERIVED FUEL STORAGE, RETRIEVAL, AND TRANSPORT

Gordon L. Sutin

12.9.1 RDF Storage, Retrieval, and Transport

Systems for storage, retrieval, and distribution of refuse-derived fuel (RDF) are critical to the success of solid waste prepared fuels technology, and it is in these areas where major challenges have been encountered in the evolution of system reliability. RDF, whether coarse or fine, is difficult material to handle. It is highly abrasive and may contain stringy materials, wires, metal objects, pieces of rope, etc., along with an unpredictability confounding to any materials-handling equipment. RDF prepared by several front-end processing stages can significantly reduce the percentage of problem components, but the quantity of combustible fuel is also reduced and process residue landfilling requirements are increased. For example, a midwestern plant reports an RDF yield less than 60% of the processed throughput in producing a refined nominal 1½ in. (38 mm) product. By contrast, an eastern facility with but single stages of shredding and magnetic separation produces a 4 in. (100 mm) product with a yield of 90% and over. Maximum RDF yields will be vital if solid waste flow control and availability of raw material become critical.

In any case, the prudent design professional will assume that even in highly processed RDF, there will remain sufficient undesirable materials such as plastic streamers, rags, metal, and wire to influence the choice of a storage and retrieval system. There are two basic concepts in waste-to-energy systems where storage of RDF may be required as described in the following Sections.

12.9.2   Remote Steam Plant and RDF Transport

When the front-end process is in a geographical location remote from the boiler facility, it is necessary to store RDF for transport to the boiler plant. This can be conveniently accomplished with transfer trailers commonly used for transfer haul of raw unprocessed MSW (Figure 12.72). If the front-end process is sized to process each day's raw waste in one or two shifts, and when boiler operation is continuous, loaded trailers may be stored at the processing site and delivered to the boiler facility over the 24-hr period. On arrival at the boiler plant, the trailers may be left to serve as storage until the RDF is needed. Alternatively, trailers may be unloaded immediately (Figure 12.73), in which case a bulk storage system is necessary at the steam plant. This alternative is in successful operation in at least one facility (Albany, New York) where RDF is produced during one shift, delivered to the steam plant, in packer trailers, and discharged immediately into a concrete bunker with a storage capacity of 2 days supply of RDF.

The most commonly available trailers for transfer of RDF are compacter trailers and live-bottom trailers. Compacter trailers are fully enclosed units into which RDF is loaded by stationary compacter into the rear of the trailer and unloaded from the rear by a ram, which is part of the trailer equipment. Live-bottom units, which are less costly, are loaded from the top, with no compaction, and are unloaded by means of push blocks on the floor of the trailer. Although capital investment is lower, load capacity is also lower. Also, maintenance costs of the unloading system will be higher. One must analyze and carefully compare capital costs, operating and maintenance costs, highway load limits, and trailer size. Whether or not trailers should provide storage at the steam plant depends on space available for parking trailers, an analysis of capital costs of extra trailers, and comparison with alternate approaches to bulk storage at the steam plant.

12.9.3   Processing Facility and Steam Plant, Same Site

Bulk storage is a necessity if the RDF processing system is capable of producing the daily requirement in one or two shifts per day. This is usually the case because shredder size is more often

Fig. 12.72 Typical RDF transport (and/or temporary storage) truck.

Fig. 12.73   RDF transport truck unloading at remote steam plant.

dictated by "worst case" rotor size/feed opening requirements than anything else and is usually much higher than maximum boiler firing rate. Front-end systems have two or more processing lines, each capable of producing 50 to 75 tons (45 to 68 t) of RDF per hour while overall plant consumption is generally in the range of 40 to 70 tons (36 to 64 t) per hour. Larger plants will usually have three processing lines.

If RDF bulk storage is not provided, front-end equipment must be run 24 hr/day with disproportionately high front-end operating costs. To allow for maintenance of process lines, each line must be arranged so that maintenance personnel are protected while other lines are operating which mandates RDF storage. Despite operational difficulties with storage and retrieval in early applications, retrofit and experienced-based design breakthroughs are eliminating this problem area as recent experience demonstrates.

During the early evolution period, several systems adapted from other industries have been tried with varying degrees of success.

12.9.4   Atlas Storage and Retrieval System

Atlas Bins have had a long history of use in the storage and retrieval of waste products. The equipment had its origins in the forest products industry, handling items such as wood shavings, bark, sawdust, hogged fuel, and wood chips. During the mid and late 1960s, the Atlas concept was adapted for use in the storage and retrieval of bagasse and copra—both very difficult materials to store and retrieve due to stringiness and abrasiveness.

The concept behind the Atlas Bin is comparatively simple. Fuel is fed into the bin through an opening in the top of the bin onto a rotating slide, which distributes the material around the bin. The retrieval system (Figure 12.74) is described as follows in Atlas System Corporation literature:

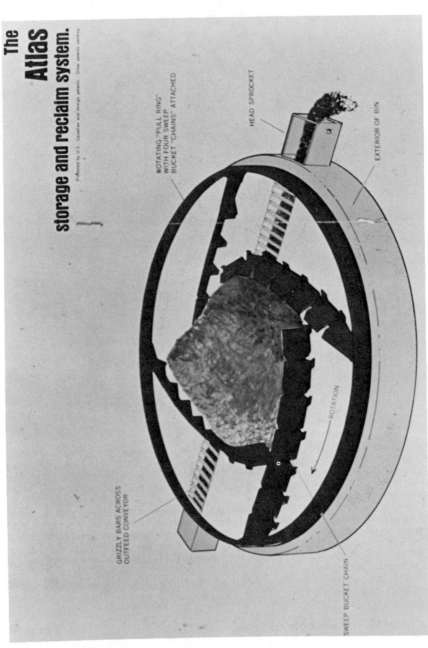

Fig. 12.74 RDF retrieval system used in the Atlas Storage Bin. (Drawing courtesy of Atlas Systems Corporation.)

Recovery of the bulk material from storage is accomplished by chains of sweep buckets. From three to six sweep chains are used, depending on bin diameter and volume of flow required. Each sweep chain is fixed at one end to a powered rotating "pull ring" encircling the storage area. The other end is free or trailing.

As the pull ring rotates around the periphery of the bin, the sweep chains automatically trail toward the center. The sweep buckets contact the stored material at the outside of the pile, and as the pull ring continues to rotate, the buckets fill and the material is swept through the grizzly openings onto an outfeed conveyer recessed in the floor. The conveyer then delivers the recovered material at a uniform and controlled rate to the next plant or mill operation.

The unique action of the trailing chains provides continuous and automated position adjustment so that the scrapers feed from the outside of the pile under all conditions. Bridging of compacted materials does not interfere with the reclaim process.

The first major installation of an Atlas Bin in the municipal solid waste industry was in 1971 when a 600-ton (544 t) capacity unit was installed to supply fuel to two 300 ton (272 t)/day spreader stoker-fired boilers in Canada.

Early Atlas Bin installations exhibited problems that the manufacturer has solved by improved design. The modified equipment now provides a steel cone-shaped center core to avoid the build-up of a hard core of RDF in the center of the bin. Atlas strongly recommends very hard floor finishes to avoid early damage to the floor due to the abrasive qualities of the sand and glass found in RDF. Bridging can still occur between the center core and the outside walls, but judicious design of bin wall slope minimizes this problem. And, of course, more refined RDF is less likely to cause any problems with entangling of rags or streamers in the buckets. Sophisticated control systems are now recommended to ensure that all drive motors equally share loads, and that the bin outfeed rate can be adjusted to meet steam demand.

The Atlas Bin has been shown to be capable of providing a reliable, uniform flow of fuel at variable rates. It should be noted, however, that there is a time delay between demand for a fuel rate change and actual delivery of the revised rate at the boiler face.

12.9.5 Miller Hofft Bin and Retrieval System

Systems supplied by Miller Hofft Inc. employ entirely different techniques. The Miller Hofft Bin (Figure 12.75) is rectangular in shape with sides that converge toward the top to avoid (or at least minimize) bridging problems. Loading of the bin is by means of a traversing conveyer high in the bin that distributes material in a triangular pile along the length of the concrete floor.

Immediately above the floor are twin counterrotating feed screws spanning the bottom width and traversing the entire length of the bin to discharge onto a horizontal conveyer parallel to the bin side. Screws have variable-speed capability to vary discharge rates. For extra large installations, screws can be applied from each side to discharge into a protected trough in the center of the bin floor.

Experience has shown that the abrasive and compacting characteristics of RDF can create high wear on the screw flights, and stringy material may tend to accumulate, requiring manual cleaning. These routines will be less frequent with more uniform RDF.

12.9.6 Concrete Bunker Bulk Storage

In at least one plant, the use of a pit and traveling crane has been adapted for storage and retrieval of RDF to assure continuity of RDF delivery (Figure 12.76). Two cranes (or one crane and a payloader) are required for back-up/repair time. Such an installation, although very positive, requires careful attention to pit and crane (or loader) styles to ensure long life and capacity. RDF in the bunker can have a density range of 3 to 15 lb/ft³ (48 to 190 kg/m³), and bucket size and lifting speed is important along with traversing speed to assure adequate feed to a return-type burner distribution system.

If not enclosed, the pit may require dust control, and there should be sufficient crane time for relocating RDF in the pit to allow for incoming RDF. The pit and crane system does not lend itself to small plants (< 500 TPD) or to direct feed of RDF to the boilers. It has been quite successful, however, with a return type fuel distribution system in a large plant.*

Having a small, live surge storage bin feeding the burner distributors and integral with a "return-type" burner feed system is recommended when pit and crane are used for storage and retrieval of RDF. Operation of pit and crane must be on a continuous 24-hr basis with proficient

---

*Resources Recovery (Dade County) Inc.

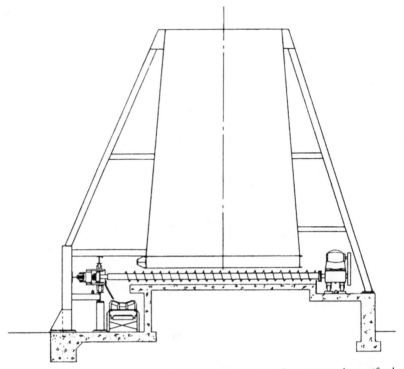

Fig. 12.75 Miller Hofft bulk storage and retrieval system showing transversing outfeed screw conveyers. (Drawing courtesy of Miller Hofft Inc.)

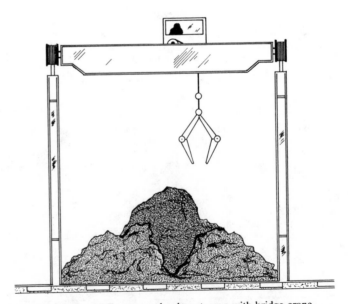

Fig. 12.76 RDF concrete bunker storage with bridge crane.

crane operation if optimal utility-type steam electric generation/revenue is to be achieved. Labor costs may be higher than for live bin storage and retrieval systems in plants with passive or marginally critical outputs (ample landfill by-pass, or standby alternatives).

12.9.7 Floor Bulk Storage

A system for storage and retrieval of RDF that has often been discussed but has not as yet been utilized, is the storage of RDF on the floor of a building fed by conveyers from the front-end process. (The City of Lakeland, Florida, utilizes a modification: excess RDF is returned to the raw material receiving floor.) RDF would be retrieved from the storage by means of rubber-tired tractors with large push blades or buckets. There is little doubt about positive retrieval of the RDF. Material feed, however, would be uneven and a surge storage system before the boilers would be necessary. As with pit and crane systems, operating labor is required on a 24-hr basis. Dust could also be a problem with, perhaps, a risk of fire. Special consideration must be given to feeding RDF into the building and reclaiming RDF with tractors without creating extreme working conditions for the tractor operators.

12.9.8 Surge Storage

Ideally, a surge storage bin to provide an immediate source of RDF close to the boilers should be included. This surge storage should respond quickly to boiler demand and must be able to feed fuel evenly over a wide range of firing rates from zero to full boiler capacity.

Surge bin capacity is usually quite small, in the range of 5 to 10 min fuel supply. The size depends on the time required to begin feed into the surge bin when the fuel level sensors demand.

12.9.9 Miller Hofft Surge Bins

Units provided by Miller Hofft for surge storage (Figure 12.77) are similar to the RDF bulk storage and retrieval systems, except for one fundamental difference. Rather than traversing screw conveyers moving along the length of the bin, the entire bin floor is covered with multiple parallel screw conveyers having multiple variable speed drives. Each motor drives three screws so that multiple discharge locations are available along the length of the bin. Thus, it is possible to feed directly into the individual boiler feed chutes and to more closely control the rate of RDF flow into each chute as required by the boiler. As expected, a problem with wear and jamming of the screw conveyers is possible but recent improvements to screw design and alignment have minimized such possibilities. This type unit has the advantage of providing several discharge points that can be directed into individual boiler feed chutes, with increased versatility in firing rate control.

12.9.10 Sprout Waldron Surge Bins

Designed to provide continuous loosening of the stored RDF and variable discharge from two discharge points, the Sprout Waldron Bin (Figure 12.78) uses screw conveyers for both functions. The bin is rectangular in cross section with a triangular hopper bottom. Several parallel vertical screws are designed to continuously move material from the bottom of the bin upwards and then down along the outside of the bin. This serves not only to keep the RDF loose but also to relieve vertical loads on horizontal discharge screw conveyers at the bottom. Again, improved screw design and alignment minimizes the wrapping of rags and plastic streamers on the vertical screws if such rogue material escapes size control. When large metal objects escape magnetic separation, jamming of the discharge conveyers can occur, sometimes damaging the screw conveyers and the bin discharge enclosures. In early installations, these systems were used only to the height of the bottom hopper, with the upper portion of the bin being left unused. Thus, where 1 to 2-hr storage had been anticipated in design, as low as several minutes storage was available. As with other equipment incorporating screw conveyers, the greatest likelihood of success is when there are no streamers or metal objects in the RDF.

12.9.11 Moving By-Pass Surge Storage Systems

An approach that has been taken by Rexnord in at least one plant is the transport of material past the boiler face in a trough with openings feeding the boiler feed chutes. The RDF is moved along the trough by means of a drag conveyer having replaceable wooden flights. The material drops through the openings in the trough into small rectangular bins with a variable-speed conveyer at the bottom to feed RDF into the boiler fuel chutes. Information about the success of this approach is not available at this time. It is likely that with stringy materials, one can expect some difficulties with build-up of material at the trough openings and subsequent jamming or breakage of the

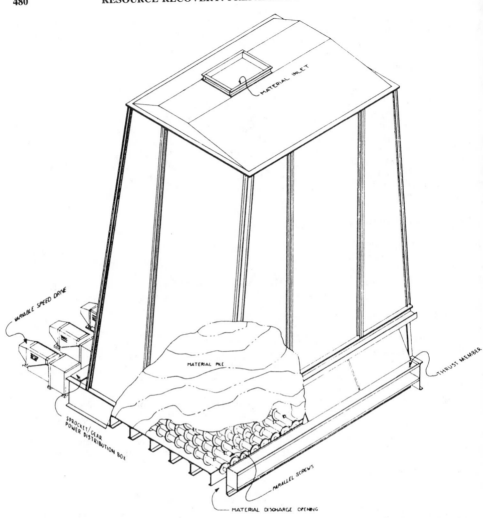

Fig. 12.77 Miller Hofft surge storage and retrieval system with multiple outfeed screw conveyers. (Drawing courtesy of Miller Hofft Inc.)

wooden conveyer flights. No doubt, if the RDF particles are consistently small and contain no streamers, rags, or metals, one can expect fewer problems.

12.9.12   Hooper Live-Bottom Bin

Originally used for storage and retrieval of bark and wood waste, two 100-ton (91 t)-capacity Hooper bins and discharges were installed in one North American experimental plant for handling RDF.

The discharger retrieval system (Figure 12.79) located at the bottom of the bin consists of a series of oscillating cross bars driven by hydraulic rams. The cross bars are shaped to push RDF on the forward stroke and to slide under the RDF on the return stroke. Little operating data is available but in discussion with plant operators, it was learned that the units were most effective when bins were between one-third and one-half full and that modifications were required. When used for bark and wood waste, the bins have been installed using multiple parallel dischargers. This allows individual feeds to separate boiler chutes from one bin. Is it not known, however, whether this can be successfully accomplished with RDF. It is likely that unless the RDF is very fine and uniform, jamming and plugging difficulties would be encountered. The designer considering this

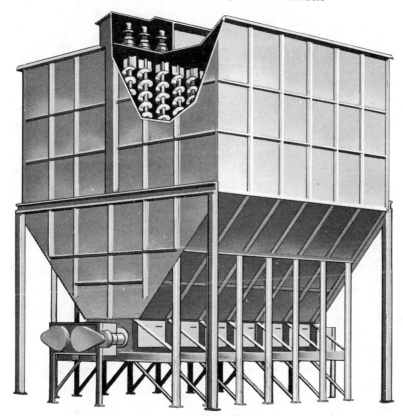

Fig. 12.78 Sprout Waldron surge storage and retrieval system with vertical RDF loosening screws and horizontal discharge screws. (Photograph courtesy of Sprout Waldron.)

equipment for RDF needs to be satisfied with its ability to perform with the anticipated RDF quality.

12.9.13   RDF Distribution and Feed

The previous section has pointed out the early material-handling difficulties that were encountered with RDF. The characteristics that caused these difficulties also caused problems in feeding the RDF into boilers.

After retrieval of RDF from storage, it must be fed into multiple chutes that carry the fuel to the burners. The number of chutes on each boiler can vary from two to eight, depending on boiler width. The boiler operator should have reliable control of the rate of feed in each chute and should be able to obtain an immediate response when a change of firing rate is needed.

The use of multiple discharge points from the surge bin as offered by the Hooper, Sprout Waldron, and Miller Hofft storage and retrieval systems presents one approach that is now successful. In one installation, eight individual outfeed points go directly to eight boiler feed chutes from a Miller Hofft surge bin. In another installation, two outfeeds from a Sprout Waldron surge bin feed directly into two boiler feed chutes with automated adjustment of feed rates. In both cases, it is apparent that the highly variable physical characteristics of the RDF required early modification, and prompt remedial action has been a major accomplishment.

A widely used technique for distribution of RDF into multiple feed chutes is the swinging spout (Figure 12.80). These units are available to provide divisions of the RDF stream into two, three, or four feed chutes. RDF is fed into the top of the spout. The bottom of the spout oscillates back and forth in a pendulum action over the top of each of the boiler feed chutes. Powered rollers at the connection between the chutes are designed to avoid hang-up of rags on the divider and to avoid bridging of material across any one feed chute. With careful design and correct timing, the feed rate of RDF can be reasonably well balanced between each chute. The system, however, does not

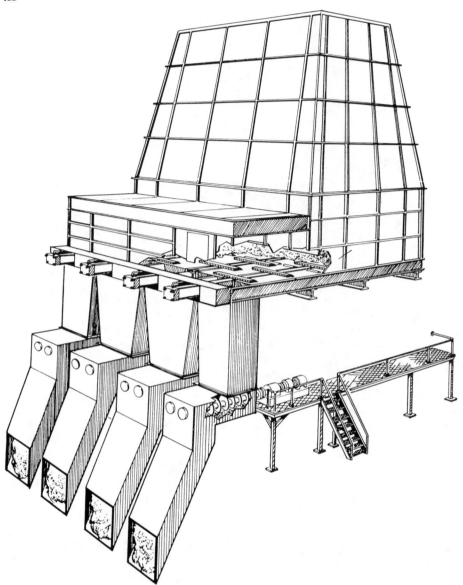

Fig. 12.79 Hooper Storage bin and SR discharger showing individual feeds to boiler feed chutes. (Drawing courtesy of S.W. Hooper and Co. Ltd.)

provide wide range control of feed rates in individual chutes. It is important that access be provided to the chutes for routine inspection and preventive maintenance. Swinging spouts, although often used, may not provide all the flexibility that would be desirable.

RDF processing quality control and materials handling systems will require continued review in the effort to assure optimum storage, retrieval, and boiler feed performance.

Summary

It can be stated that storage and retrieval systems have been a major problem area and that most have had difficulty in reliably and continuously handling this obstinate material. Fewer problems can be expected if the RDF has been processed to eliminate oversize, stringy, and abrasive materials.

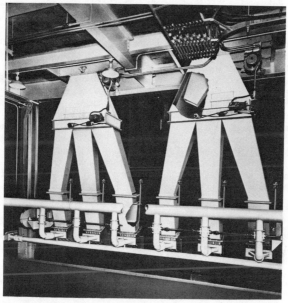

Fig. 12.80 Swinging spouts with cutaway view to show internal mechanism and arch breaker rollers. (Photograph courtesy of Detroit Stoker Company.)

In choosing equipment for storage and retrieval of RDF, the designer should make every effort to determine not only the exact characteristics of the RDF *but the raw feed as well*! Particular attention must be paid to rags, streamers, wires, hard plastics, and metals. It must also be recognized that RDF is rarely uniform, particle size will vary, moisture will vary, and unexpected materials will appear from time to time. Heaps of RDF can be fluffy or very dense and usually cannot squeeze through small spaces or be sliced. Storage and retrieval of RDF has been significantly improved to the point of acceptable reliability when the prudent designer anticipates potential trouble points and is ready with reasonable remedies.

12.10   RECOVERED MATERIALS; SPECIFICATIONS AND MARKETS
Kenneth L. Woodruff

12.10.1   Introduction

Resource-recovery facilities designed and operated to produce a prepared fuel (RDF) and recover saleable materials must have long-term markets available to accommodate the materials and assure a continuing revenue stream. Revenues produced by the sale of recovered materials at a resource recovery facility are generally 10 to 20% of the total. The balance of the revenue is derived from fuel/energy sales and tipping fees. To justify the capital investment and assure the viability of the materials recovery approach, knowledge of the markets and the marketability of products to be recovered is mandatory. Since waste composition varies and markets vary by location throughout the country, evaluating materials markets is a site-specific requirement. A general discussion of markets and market requirements will be presented for the following materials: ferrous, glass, aluminum, paper and corrugated, other miscellaneous materials (also see Chapter 15, Section 15.2).

12.10.2   Ferrous Metals

Ferrous metals contained in municipal solid waste may generally be divided into two categories, bulky ferrous and containers. Bulky ferrous includes white goods, appliances, and other miscellaneous materials, whereas containers consist of food, beverage, and other types of cans. Bulky ferrous is generally separated during processing and may be sold "as is" to a scrap processor or may be shredded and subsequently magnetically separated to produce a clean, ferrous fraction suitable for use by the steel industry. Further processing of bulky ferrous is generally only economically viable in the larger resource recovery facilities (1000 TPD and up). Smaller facilities generally can locate a local scrap processor (preferably one with a shredder) that can efficiently handle the bulky scrap material.

Container ferrous may be sold to detinners, the steel industry, foundries, or for copper recovery. All markets are site specific and must be examined. ASTM Specification E702-79, reproduced here in Figure 12.81, describes Municipal Ferrous Scrap. This has been widely accepted in the industry. It should be noted that the market price obtained is dependent upon the cleanliness of the product. Municipal ferrous scrap is usually sold in a loose form.

Due to relatively recent interest by the steel industry in the recovery of tin-free bimetal steel cans, it may now be possible to recover bimetal cans as a separate ferrous product. The detinners are not really interested in having tin free cans as part of the detinning feed since their processing yields no tin and the aluminum top consumes the detinning solution with no resulting benefit. Some facilities that utilize screening of raw waste prior to shredding and subsequently recover container ferrous before shredding, can readily employ a hand-sorting operation to separate bimetal cans from other steel containers. It has been reported that bimetal cans have a value of up to $100/ton, whereas other container ferrous generally has a value no higher than $65/ton. However, bimetal cans usually must be baled prior to shipment to the end user, thus increasing processing costs.

12.10.3 Glass

Glass may be recovered from mixed solid waste (MSW) in two forms, color-sorted and color-mixed. Color-mixed glass recovered for the manufacturing of new containers is generally prepared by froth flotation. A specification for this material, ASTM Specification E708-79, Standard

Standard Specification for
MUNICIPAL FERROUS SCRAP[1]

This standard is issued under the fixed designation E 702; the number immediately following the designation indicates the year of original adoption or, in the case of revision, the year of last revision. A number in parentheses indicates the year of last reapproval. A superscript epsilon (ϵ) indicates an editorial change since the last revision or reapproval.

1. Scope

1.1 This specification covers the chemical and physical requirements of municipal ferrous scrap that is intended for use by such industries listed as follows:

1.1.1 Copper industry (precipitation process),

1.1.2 Iron and steel foundries,

1.1.3 Iron and steel production,

1.1.4 Detinning industry, and

1.1.5 Ferroalloy industry.

1.2 Questions concerning material rejection, downgrading, and retesting based on failure to meet the requirements of this specification shall be dealt with through contractual arrangements between the purchaser and the supplier.

2. Applicable Document

2.1 *ASTM Standard:*

E 701 Methods of Testing Municipal Ferrous Scrap[2]

3. Definitions

3.1 *municipal ferrous scrap*—ferrous waste that is collected from industrial, commercial, or household sources and destined for disposal facilities. Typically, municipal ferrous scrap

consists of a metal or alloy fraction, a combustible fraction, and an inorganic noncombustible fraction that includes metal oxides.

3.2 *total combustibles*—materials that include paints, lacquers, coatings, plastics, etc., associated with the original ferrous product, as well as combustible materials (paper, plastic, textiles, etc.) which become associated with the ferrous product after it is manufactured.

3.3 *metallic yield*—the weight percent of the municipal ferrous scrap that is generally recoverable as metal or alloy.

4. Chemical Requirements

4.1 Municipal ferrous scrap shall conform to the requirements as to chemical composition for the respective end uses prescribed in Table 1.

4.2 The chemical requirements listed in Table 1 are based on melt analyses except where noted.

---

[1] This specification is under the jurisdiction of ASTM Committee E-38 on Resource Recovery and is the direct responsibility of Subcommittee E38.02 on Ferrous Metals.

Current edition approved Nov. 5, 1979. Published January 1980.

[2] *1983 Annual Book of ASTM Standards*, Vol 11.04.

Fig. 12.81 ASTM Specification E-702-79, Municipal Ferrous Scrap. (Reprinted with permission from the Annual Book of ASTM Standards, Copyright, ASTM, 1916 Race Street, Philadelphia, Pennsylvania 19103.

5. Physical Requirements

5.1 Municipal ferrous scrap shall conform to the physical properties for the respective end uses prescribed in Table 2.

6. Test Methods

6.1 Determine the physical and chemical requirements of municipal ferrous scrap in accordance with Methods E 701.

TABLE 1 Chemical Requirements

Element	Copper Industry (Precipitation Process)	Iron and Steel Foundries	Iron and Steel Production[A]	Detinning Industry[B]	Ferroalloy Production
	Composition, %				
Phosphorus, max	...	0.03	0.03	...	0.03
Sulfur, max	...	0.04	0.04	...	...
Nickel, max	...	0.12	0.08	...	...
Chromium, max	...	0.15	0.10	...	0.15
Molybdenum, max	...	0.04	0.025	...	...
Copper, max	...	0.20	0.10	...	0.20
Aluminum, max	...	0.50	0.50	4.00[E]	0.15
Tin	...	0.30 max[D]	0.30 max	0.15 min[F]	0.30
Lead, max	...	0.03	0.15	...	...
Zinc, max	...	0.06	0.06	...	...
Iron (metallic), min	96.0	...	...	...	...
Silicon, max	...	...	0.10	...	...
Manganese, max	...	...	...	...	0.35
Carbon, max	...	...	...	...	0.6
Titanium, max	...	...	...	...	0.025
Total combustibles, max	0.2[C]	4.0	4.0	...	0.5[G]
Metallic yield, min	...	90.0	90.0	...	90.0

[A] Experience has shown that material which has been incinerated probably will not meet these requirements.

[B] A minimum of 95 weight % of the material delivered shall be magnetic. Nonmagnetic material attached to the original magnetic article may be included in the minimum requirement.

[C] The scrap shall be appropriately processed (for example, by burning, chemical detinning, etc.) to be virtually free of combustibles.

[D] For steel castings, the requirement for tin content is 0.10 max %.

[E] Not based on melt analyses due to aluminum losses during melting; to be determined by a method mutually agreed upon between the purchaser and supplier.

[F] Refer to sections on magnetic fraction and chemical analysis of tin in Methods E 701. Normal separation of white goods and heavy iron yields tin contents equal to or greater than 0.15 weight %. Lesser tin contents would impact severely the value of the scrap to detinners.

[G] The scrap shall be appropriately processed (for example, by burning, chemical detinning, etc.) to be virtually free of combustibles.

Fig. 12.81 (Continued)

Specification for Waste Glass as a Raw Material for the Manufacture of Glass Containers, is indicated in Figure 12.82. Although glass meeting this specification has been produced and glass container manufacturers have used it, marketing of froth-floated glass is difficult since its color-mixed nature makes it only useful in certain types of glass manufacturing and in limited quantities.

The glass container industry really desires color-sorted glass and can utilize it in large quantities. Optical sorting equipment has been demonstrated and is installed in several facilities, however, it is expensive and inefficient. Each piece of material of a closely sized fraction must be examined and a binary separation made. Hence more than one stage of separation is required for color sorting and contaminant (refractories) removed. Optical sorting is applicable only to relatively large pieces, coarser than 0.24 in. (6 mm). Another method that may be utilized is hand sorting of whole or nearly whole bottles removed from the primary screening of MSW. Glass recovered in this manner requires further processing prior to its use as cullet to assure its cleanliness. This includes removal of caps, labels, and other contaminants and also crushing to produce a minus 0.75 in. (19 mm) product. Processing may be performed prior to shipment of the glass or by the purchaser of the glass. For example, Owen-Illinois has installed glass cullet-processing systems throughout the country that are capable of accepting color-sorted glass from resource recovery facilities as well as from source-separation programs. Other uses for glass from MSW include: brick manufacturing, glass wool production, manufacturing of masonry block, use as an aggregate substitute, and structural fill. Several resource recovery facilities produce mixed glass fractions that are sold for use in one or more of these applications. However, it must be noted that the market value of waste glass for these uses may be as little as $1.50/ton, whereas cullet for container glass manufacturing may have a value of $50 to 55/ton. Specifications for waste glass for use in other than structural fill applications have to be developed. Requirements must be determined on a case-by-

Standard Specification for

WASTE GLASS AS A RAW MATERIAL FOR THE MANUFACTURE OF GLASS CONTAINERS[1]

This standard is issued under the fixed designation E 708; the number immediately following the designation indicates the year of original adoption or, in the case of revision, the year of last revision. A number in parentheses indicates the year of last reapproval. A superscript epsilon (ε) indicates an editorial change since the last revision or reapproval.

1. Scope

1.1 This specification covers particulate glass (cullet material, recovered from waste destined for disposal, smaller than 6 mm intended for reuse as a raw material in the manufacture of glass containers.

2. Applicable Documents

2.1 *ASTM Standards:*
C 162 Definitions of Terms Relating to Glass and Glass Products[2]
C 169 Methods for Chemical Analysis of Soda-Lime and Borosilicate Glass[2]
C 429 Method for Sieve Analysis of Raw Materials for Glass Manufacture[2]
E 688 Methods of Testing Waste Glass as a Raw Material for Manufacture of Glass Containers[3]

3. Definitions

3.1 *flint glass cullet*—a particulate glass material that contains no more than 0.1 weight % Fe_2O_3, or 0.0015 weight % Cr_2O_3, as determined by chemical analysis.

3.2 For definitions of other terms used in this specification, refer to Definitions C 162.

4. Representative Sample

4.1 The following requirements qualify the glass lot to be used for direct use in soda-lime glass container manufacturing. Sample should be prepared and examined in accordance with Methods E 688.

NOTE 1—A preponderant proportion of glass cullet will be soda-lime bottle glass, the glass cullet having a composition as follows, as determined by Method C 169.

Oxide	Composition, Weight %
SiO_2	66 to 75
Al_2O_3	1 to 7
CaO + MgO	9 to 13
Na_2O	12 to 16

NOTE 2—All percents referred to in this specification are weight percents.

5. General Requirements

5.1 The sample shall show no drainage of liquid and be noncaking and free flowing. A moisture content of less than 0.5 weight % is required to meet the free-flowing characteristics of a cullet that is predominantly of smaller particle size, 1.18-mm (No 16) sieve or smaller.

5.2 *Screen Size*—No material shall be retained on a 6-mm (¼-in.) screen. Material not exceeding 15 weight % shall pass through a 106-μm (No. 140) screen.

5.3 *Organic Materials*—The total content of organic materials, as measured in accordance with Section 6 shall not exceed 0.2 weight % of dry sample, except for color-mixed glass where the content of organic material may exceed 0.2 weight %. However, a content of organic material greater than 0.2 weight % must be held within a tolerance of ±0.05 weight %, with a maximum organic limit of 0.4 weight %.

5.4 *Magnetic Materials*—The total magnetic materials shall not exceed 0.05 weight % of dry sample weight for flint glass and 0.14 weight

---

[1] This specification is under the jurisdiction of ASTM Committee E-38 on Resource Recovery and is the direct responsibility of Subcommittee E38.05 on Glass.
Current edition approved Nov. 30, 1979. Published January 1980.
[2] *1983 Annual Book of ASTM Standards,* Vol 15.02.
[3] *1983 Annual Book of ASTM Standards,* Vol 11.04.

Fig. 12.82   ASTM Specification E-708-79 Standard Specification for Waste Glass as a Raw Material for the Manufacture of Glass Containers. (Reprinted with permission from the Annual Book of ASTM Standards, Copyright, ASTM, 1916 Race Street, Philadelphia, Pennsylvania 19103.)

---

case basis between buyer and seller. ASTM Specification E850-82, Standard Practice for Use of Process Waste in Structural Fill, is included as Figure 12.83.

12.10.4   Aluminum

Aluminum from MSW is the most marketable and highest-value component of solid waste. However, it is found in two different forms that affect its marketability. First, food container aluminum consisting of beverage cans, food cans, and foil are the types most desired by the primary aluminum manufacturers. This is the type of aluminum generally purchased by recycling centers and can

% for colored glass of dry sample weight in accordance with Section 6.

5.5 *Permissible Color Mix for Color Sorted Glass Cullet by Weight*:

5.5.1 *Amber Glass Cullet*:

> 90 to 100 % amber
> 0 to 10 % flint
> 0 to 10 % green
> 0 to 5 % other colors

5.5.2 *Green Glass Cullet*:

> 50 to 100 % green
> 0 to 35 % amber
> 0 to 15 % flint
> 0 to 4 % other colors

5.5.3 *Flint Glass Cullet*:

> 95 to 100 % flint
> 0 to 5 % amber
> 0 to 1 % green
> 0 to 0.5% other colors

5.5.3.1 Percents above 0.1 weight % of Fe_2O_3 or 0.0015 weight % of Cr_2O_3, or both, as determined by chemical analysis shall be considered mixed color glass. These limits are consistent with industry experience on raw material.

5.5.3.2 Flint glass cullet may contain up to 1 weight % emerald green or 10 weight % Georgia green, or a combination within the limits: 1 % Georgia green = 0.1 % emerald green.

5.6 *Other Inorganic Material* (such as non-magnetic metals or refractories)—As measured, material larger than 850-μm (No. 20) screen size shall not exceed 0.1 % of the dry sample weight. Material smaller than 850-μm screen size shall not exceed 0.5 % of the dry sample weight.

5.6.1 *Refractories*—Based upon U.S. series screen size and sample weight, the following refractory particle limits shall apply for each screen fraction as stated below.

+20 mesh	1 particle per 18-kg (40-lb) sample
−20, +40 mesh	2 particles per 450-g (1-lb) sample
−40, +60 mesh	20 refractory particles per 450-g (1-lb) sample

5.6.2 *Nonmagnetic Metals*:

+20 mesh	1 particle per 18-kg (40-lb) sample

Upon failure to meet the previously stated specification limits, retesting is permissible.

6. Sampling and Testing

6.1 Sampling and testing shall be in accordance with Methods E 688.

Fig. 12.82 (*Continued*)

be readily recycled back into beverage containers. All other forms of aluminum—pots and pans, lawn furniture, appliances, etc. (miscellaneous extrusions and castings)—are alloys that are unsuitable for mixing with container aluminum scrap if the highest value is to be obtained.

ASTM Specification E753-80, Municipal Aluminum Scrap (Figure 12.84), provides for six grades of aluminum recovered from MSW. Grade 1 is essentially aluminum cans with each subsequent grade containing more contaminants and alloy constituents as found in other types of aluminum present in MSW.

All the primary aluminum manufacturers have buy-back programs for aluminum cans recovered by a resource recovery facility FOB the facility. However, most want only cans whereas others will accept other food container aluminum with the cans (foil and food cans). Primary aluminum manufacturers do not usually wish to purchase the other aluminum forms, nor do they usually purchase a mixture of the container and other aluminum; such outputs generally must be sold to the secondary aluminum industry.

During the past several years, market prices for container aluminum have ranged from $0.29 to 0.48/lb, while aluminum mixtures or streams of other aluminum have ranged from $0.10 to 0.45/lb. Sale of aluminum mixtures and lower grades are generally on lesser terms than is the container (higher grade) aluminum.

12.10.5 Paper and Corrugated

Both newspaper and corrugated paper (cardboard boxes) may be recovered from mixed municipal waste, however only in limited amounts once they have been comingled with household waste. Of the two, corrugated is the more readily recovered since it is generally found in large concentrations in commercial waste. As a result, it is generally much cleaner and drier than corrugated in typical household waste. Resource recovery facilities can direct loads of commercial waste that may have

Standard Practice for
USE OF PROCESS WASTE IN STRUCTURAL FILL[1]

This standard is issued under the fixed designation E 850; the number immediately following the designation indicates the year of original adoption or, in the case of revision, the year of last revision. A number in parentheses indicates the year of last reapproval. A superscript epsilon (ϵ) indicates an editorial change since the last revision or reapproval.

1. Scope

1.1 This practice specifies procedures for evaluating certain process wastes and similar construction materials to be used in structural fill and it specifies precautions necessary where a potential for pollution of the ground or surface water, or both, exists. This practice lists some of the test methods useful for predicting and evaluating physical in-place characteristics of the wastes and those of their resultant engineering and environmental properties specifically related to the material's structural integrity and to the protection of ground and surface water.

2. Applicable Documents

2.1 *ASTM Standards:*

C 593 Specification for Fly Ash and Other Pozzolans for Use with Lime[2]

D 698 Test Methods for Moisture-Density Relations of Soils and Soil-Aggregate Mixtures, Using 5.5-lb (2.49-kg) Rammer and 12-in. (304.8-mm) Drop[3]

D 1633 Test Method for Compressive Strength of Molded Soil-Cement Cylinders[3]

D 2049 Test Method for Relative Density of Cohesionless Soils[3]

D 2166 Test Methods for Unconfined Compressive Strength of Cohesive Soil[3]

D 2434 Test Method for Permeability of Granular Soils (Constant Head)[3]

D 2664 Test Method for Triaxial Compressive Strength of Undrained Rock Core Specimens Without Pore Pressure Measurements[3]

D 2850 Test Method for Unconsolidated, Undrained Compressive Strength of Cohesive Soils in Triaxial Compression[3]

D 3080 Method for Direct Shear Test of Soils Under Consolidated Drained Conditions[3]

D 3987 Test Method for Shake Extraction of Solid Waste with Water[4]

2.2 *Other Documents:*

Department of Interior, Earth Manual (Second Edition, 1974.[5]

Corps of Engineers', Soils Testing Manual 1110-2-1906.[6]

Bureau of Public Roads/Highway Research Board, Classification of Subgrade Materials.[7]

Resource Conservation and Recovery Act, Federal Register Sept. 13, 79, and May 19, 80.[8]

3. Significance and Use

3.1 This practice is intended for inorganic process waste that can be used in structural fill and similar applications.

3.2 This practice is not intended to limit the flexibility of design in the use of these waste materials. All demonstrated sound engineering procedures that result in appropriate in-place physical and environmental properties are acceptable.

NOTE—Sound engineering practice should be especially considered for materials whose solubility loss can impair stability.

3.3 This practice includes a number of test

---

[1] This practice is under the jurisdiction of ASTM committee E-38 on Resource Recovery and is the direct responsibility of Subcommittee E38.06 on Construction Materials from other Recovered Materials.
Current edition approved June 25, 1982. Published August 1982.
[2] *1983 Annual Book of ASTM Standards*, Vol 04.01.
[3] *1983 Annual Book of ASTM Standards*, Vol 04.08.
[4] *1983 Annual Book of ASTM Standards*, Vol 11.04.
[5] Available from Government Printing Office Book Stores. 219 S. Dearborn St. Room 1463, Chicago, IL 60604.
[6] Available from U.S. Corps of Engineers, Waterways Experiment Station, P.O. Box 631, Vicksburg, MS 39180.
[7] Available from The Transportation Research Board, The Joseph Henry Building, 2100 Pennsylvania Ave., Wshington, D.C. 20037.
[8] Available from U.S. Government Printing Office, Washington, D.C. 20402.

Fig. 12.83 ASTM Specification E-850-82 Standard Practice for Use of Process Waste in Structural Fill. (Reprinted with permission from the Annual Book of ASTM Standards, Copyright, ASTM, 1916 Race Street, Philadelphia, Pennsylvania 19103.)

---

a high corrugated content to be dumped to one side of the tipping area so that the material may be more readily sorted. This is a typical approach used at many transfer stations that recover corrugated.

Newspapers are difficult to recover from MSW in a clean or dry enough form to have value in recycling, unless they are bundled or bagged prior to disposal by the homeowner. As a result, unless there is a requirement to bundle or bag waste newspapers prior to disposal, they are usually better left in the mixed waste to become part of the fuel/energy output.

Recovered newspaper and corrugated must be clean and free of tramp materials. It may be shipped in baled or loose form, although the loose form is usually only applicable if shipped short methods that are particularly applicable to the types of materials contemplated for use. It shall be understood that the selection from among these and other tests and specifications for the properties required for any construction use are the responsibility of the design engineer.

4. Definitions

4.1 *cemented materials*—materials consisting of one or more substances that develop hardness by chemical reaction after placement of the materials within a structure.

4.2 *cohesive materials*—materials in which attractive forces exist between the particles that confer upon the materials sufficient shear resistance to provide properties as defined in 4.9.

4.3 *effective permeability*—the permeability of a total structure results from the materials of construction or suitable construction techniques such as capping, impermeable layers, drainage, etc.

4.4 *granular materials*—materials consisting of individual particles of the same or of various sizes and shapes, but usually less than 10 mm and more than 0.075 mm in their principal linear dimension.

4.5 *leachate*—liquid that has percolated through, or passed over, a solid waste or other medium and ¹ as extracted, dissolved, or suspended materials from it.

4.6 *permeate*—a leachate produced by causing a liquid to flow through a porous solid.

4.7 *process waste*—inorganic by-product materials such as mine tailings, culm piles, coal-processing conversion, and combustion wastes, cement and limekiln dust, phosphogypsum, and chemically treated compositions made from these wastes or waste mixtures. In certain cases, process wastes may produce concentrations of constituents in the leachate that exceed environmental standards. Special provisions are included to accommodate this class of material.

NOTE—Some wastes, for example, phosphogypsum and uranium tailings, contain low concentrations of radioactive minerals. Use of such wastes will be in accordance with regulatory requirements.

4.8 *runoff*—a leachate produced by causing a liquid to flow essentially over the surface of a structural fill.

4.9 *structural fill*—a fill consisting of any material or composition that, by reason of its physical and chemical properties, can be used

Fig. 12.83 (*Continued*)

to fill a void or occupy a volume in a fill and also contribute to the mechanical integrity of the fill. (Examples of such uses include, but are not limited to certain landfills, landfill liners, road bases, embankments, and earthen dams.)

5. Material Properties

5.1 The properties of the process waste defined under 4.7 cover both cohesive and cemented materials and granular material. Cohesive and cemented materials have been divided into two classes. Class 1 relates to process waste that, either as produced or after conditioning prior to placement, has properties conforming to those normally required for virgin in-place borrow materials such as those conventionally used for construction of compacted embankments. Class 2 material relates to process waste that requires curing or aging after placement in order to develop physical properties in situ that meet the structural requirement of the intended use.

6. Methods of Test

6.1 Table 1 lists representative test methods that are recommended to evaluate the structural and environmental properties of the process waste or waste composition. Applicable limits for the numerical value of these properties will vary depending on design requirements and are determined by accepted engineering practices and regulatory requirements.

6.2 Testing of Class 2 materials should be performed after curing and aging. Tests of Class 2 waste should be performed on specimens that have been compacted, cured, and aged to duplicate in-place properties as accurately as possible. Cured undisturbed specimens removed from the field may also be used.

7. Construction Practice

7.1 Construction of a project using the process waste should conform to standard practices as used with conventional structural materials. Methods such as described in the *U.S. Department of Interior Earth Manual, Corps of Engineers' Soil Testing Manual*, and Department of Transportation publications are suitable for this purpose. During construction, visual inspection should be made to ensure that the in-place material is uniform in consistency, homogeneous, and free from large voids.

7.2 Materials having a predicted in-place

permeability $\leq 1 \times 10^{-7}$ cm/s and properties meeting the requirements of Section 3004, Subsection 250.45-2, of the proposed regulations under the Resource Conservation and Recovery Act are suitable for landfill liners when placed a minimum of 5 ft (1.52 m) above the historical high-water table.

8. Special Provisions

8.1 When process wastes that are used as constuction materials produce concentrations of constituents in the leachate, as determined by Method D 3987 or other required methods, that exceed one hundred times the EPA National Interim Primary Drinking Water Standards (100 $\times$ DWS), the following provisions should apply:

8.1.1 Structures that are made of materials having a predicted effective permeability of $>1 \times 10^{-5}$ cm/s shall have an environmentally acceptable underdrain and permeate collection and disposal system.

8.1.2 Structures that are made of materials having a predicted effective permeability of $\leq 1 \times 10^{-5}$ cm/s do not require leachate collection and treatment systems provided that surfaces exposed to the environment are sloped so that 90 % of the incident precipitation runs off these surfaces, and provided the material is placed a minimum of 5 ft (1.52 m) above the historically high water table where the subsoil has a permeability $>1 \times 10^{-5}$ cm/s. The test method shall be determined by the engineer.

8.1.3 Where concentrations of constituents in permeate or surface runoff exceed environmentally acceptable limits, such leachates shall be collected and treated (within the available technology) prior to discharge.

TABLE 1   Representative Methods Recommended for Evaluation of Performance Characteristics of Process Wastes Suitable for Construction Use

Characteristics	Test Methods
Cohesive and cemented materals:	
Moisture-density (laboratory)	D 698
Unconfined compressive strength:	
Class 1	D 2166
Class 2	C 593, D 1633, D 2166
Triaxial shear strength:	
Class 1	D 2850
Class 2	D 2664
Permeability (falling head)	Corps of Engineers Earth Manual-1110-2-1906
Granular materials:	
Moisture-density (laboratory)	D 698, D 2049
Direct shear strength	D 3080
Triaxial shear strength	D 2850
Permeability (falling head) (constant head)	Corps of Engineers Earth Manual-1110-2-1906, D 2434
Environmental properties[4]:	
Leachate extraction	Method D 3987
Leachate testing procedures	Annual Book of ASTM Standards, Part 31

[4] Methods under jurisdiction of ASTM Committee D-19.

The American Society for Testing and Materials takes no position respecting the validity of any patent rights asserted in connection with any item mentioned in this standard. Users of this standard are expressly advised that determination of the validity of any such patent rights, and the risk of infringement of such rights, are entirely their own responsibility.

This standard is subject to revision at any time by the responsible technical committee and must be reviewed every five years and if not revised, either reapproved or withdrawn. Your comments are invited either for revision of this standard or for additional standards and should be addressed to ASTM Headquarters. Your comments will receive careful consideration at a meeting of the responsible technical committee, which you may attend. If you feel that your comments have not received a fair hearing you should make your views known to the ASTM Committee on Standards, 1916 Race St., Philadelphia, Pa. 19103.

Fig. 12.83   (Continued)

distances to a local waste paper processor or dealer. Recovered paper or corrugated shipped to an end user must be baled.

In summary, only limited amounts of paper and corrugated may be readily recovered from MSW by hand-sorting in a clean enough form to be saleable for recycling. Experience indicates that corrugated paper may be recovered largely from loads of commercial waste at both transfer stations and resource-recovery facilities. Sale of recovered paper and corrugated to a local waste paper processor usually requires no further processing; the material may be shipped loose. However, long distances or end user markets require baling. Corrugated may require shredding (hogging) prior to baling. If the market value of recovered paper and corrugated is less than $30–35/ton, leaving it in the waste to be processed for its energy content will generally be its best use.

Mechanical systems for recovering a paper-rich fraction suitable for recycling in the form of

Standard Specification for
MUNICIPAL ALUMINUM SCRAP (MAS)[1]

This standard is issued under the fixed designation E 753; the number immediately following the designation indicates the year of original adoption or, in the case of revision, the year of last revision. A number in parentheses indicates the year of last reapproval. A superscript epsilon (ε) indicates an editorial change since the last revision or reapproval.

1. Scope

1.1 This specification covers municipal-refuse originated aluminum alloy scrap (MAS), not source-separated, that is recovered from industrial, commercial, or household wastes destined for disposal facilities.

1.2 Municipal aluminum scrap (MAS) covered by this specification is suitable for use by the following industries:

1.2.1 Secondary aluminum smelters,

1.2.2 Primary aluminum producers,

1.2.3 Aluminum scrap dealers,

1.2.4 Iron and steel industry,

1.2.5 Foundries,

1.2.6 Nonintegrated aluminum producers, and

1.2.7 Independent aluminum fabricators.

2. Applicable Documents

2.1 *ASTM Standards:*

D2013 Method of Preparing Coal Samples for Analysis[2]

E 11 Specification for Wire-Cloth Sieves for Testing Purposes[3]

E 34 Methods for Chemical Analysis of Aluminum and Aluminum Alloys[4]

E 101 Method for Spectrographic Analysis of Aluminum and Aluminum Alloys by the Point-to-Plane Technique[5]

E 122 Recommended Practice for Choice of Sample Size to Estimate the Average Quality of a Lot or Process[3]

E 227 Method of Optical Emission Spectrometric Analysis of Aluminum and Aluminum Alloys by the Point-to-Plane Technique[5]

E 276 Test Method for Particle Size or Screen Analysis at No. 4 (4.75-mm) Sieve and Finer

for Metal Bearing Ores and Related Materials[4]

3. Classification

3.1 This specification covers two classes based on fines content (7.2) and six grades based on chemical composition of MAS material, as listed in Table 1 (see also 7.2).

4. Description of Terms

4.1 *combustible material* (organic)—material that is measured by weight loss of a dried sample input after heating to red heat in an open crucible in a vented furnace. Combustibles include both loose organics and organic coatings.

4.2 *loose combustible material* (organic)— loose combustible organics (LCO) that consist of, but are not limited to, nonmetallic materials such as paper, rags, plastic, rubber, wood, food wastes, and yard or lawn wastes, etc., which are not permanently attached to noncombustible objects. The LCOs are defined as material larger than 12 mesh (U.S. Standard Sieve). A determination of LCOs is best done by sampling the material and handpicking, handcleaning, and visually identifying the materials described previously.

4.3 *moisture percent*—liquid content, as determined by weight loss when sample material

---

[1] This specification is under the jurisdiction of ASTM Committee E-38 on Resource Recovery and is the direct responsibility of Subcommittee E 38.03 on Nonferrous Metals.
Current edition approved Sept. 2, 1980. Published November 1980.
[2] *1983 Annual Book of ASTM Standards,* Vol 05.05.
[3] *1983 Annual Book of ASTM Standards,* Vol 14.02.
[4] *1983 Annual Book of ASTM Standards,* Vol 03.05.
[5] *1983 Annual Book of ASTM Standards,* Vol 03.06.

Fig. 12.84 ASTM Specification E-753-80 Municipal Aluminum Scrap. (Reprinted with permission from the Annual Book of ASTM Standards, Copyright, ASTM, 1916 Race Street, Philadelphia, Pennsylvania 19103.)

---

pulp have been found to produce a very low-quality material that is a small fraction of the paper contained in the raw waste. Due to the cost of processing and recovery and the typical low value of mixed paper fiber, it is generally more economical to utilize paper for its energy content. Recovery by hand-sorting when market prices are high is usually the best approach.

12.10.6 Other Miscellaneous Materials

A number of other materials may be recovered by resource-recovery facilities. Experience is varied and depends somewhat on the processing method employed as to whether a saleable product can be recovered.

is dried to a constant weight at $110° \pm 5°C$.

4.4 *recovery*—the percent material recovered after an assay using the procedures prescribed in this specification.

5. Ordering Information

5.1 It is recognized that variations in the MAS may occur due to the heterogeneous nature of the solid waste stream. The grades indicated are intended as a means for the purchaser and the seller to establish the value and quality of the MAS.

5.2 MAS shall be considered to be of a particular grade if the value for each element specified, as obtained by the test method agreed upon between the purchaser and seller, does not exceed any of the limits for that grade.

6. Chemical Requirements

6.1 The MAS shall conform to the requirements as to chemical composition prescribed in Table 1.

7. Physical Requirements

7.1 *Density*—The density for MAS is not specified and shall be agreed upon between the purchaser and the seller.

7.2 *Fineness*—MAS shall contain not more than the amount of minus 12 mesh (U.S. Standard Sieve) material, described in 7.2.1 and 7.2.2.

7.2.1 *Class A* material shall contain not more than 1 weight % fines.

7.2.2 *Class B* material shall contain not more than 3 weight % fines.

7.3 *Loose Combustibles*—MAS shall contain not more than 2.0 weight % of loose combustible material.

7.4 *Moisture*—MAS shall contain not more than 0.5 weight % of moisture.

7.5 *Metal Recovery*—A minimum metal recovery of 85 % is required.

Fig. 12.84 (*Continued*)

7.6 *Magnetics*—The presence of free magnetic material is not specified and shall be as agreed upon between the purchaser and seller as part of the purchase contract.

8. Sampling

8.1 Sampling shall be in accordance with the procedures described in Annexes A1 or A2. Either procedure may be used, as determined by agreement between the purchaser and the seller.

8.1.1 Annex A1 covers sampling at the point of origin.

8.1.2 Annex A2 covers sampling at the point of receipt.

9. Test Methods

9.1 Determine the properties enumerated in this specification in accordance with the following:

9.1.1 *Fineness*—Annex A3.

9.1.2 *Moisture*—Annex A3.

9.1.3 *Metal Recovery*—Annex A3.

10. Rejection and Rehearing

10.1 Material that fails to conform to the requirements of this specification may be rejected. Rejection should be reported to the producer or supplier promptly and in writing. In case of dissatisfaction with the results of the test, the producer or supplier may make claim for a rehearing.

11. Shipping

11.1 MAS shall be shipped in rail cars, trailers, or other containers as agreed upon between the purchaser and the seller. The shipping equipment shall be sufficiently watertight to prevent the MAS from becoming wet during shipment.

Mixed other nonferrous metals (copper, brass, zinc, lead, etc.) as well as coins may be recovered from MSW. Several facilities employ wet processes consisting of jigging and heavy media separation to produce a mixed nonferrous fraction which may be sold to a scrap processor or secondary metals smelter. In addition, coins may be sorted from the nonferrous mixture which may be sold to the U.S. Mint. Facilities have reported coin recovery in the range of $0.05 to 0.40/ton of waste processed.

Other materials that may be recovered include various types of plastics. For example, clean polyethylene terephthalate (PET) soft drink bottles and high-density polyethylene (HDPE) milk bottles may be recycled. Various plastics must be sorted by type and cannot be mixed if they are to be recycled. Hence, plastic recovery requires hand-sorting of the various plastic containers. Once separated, they must be processed further to be sold as recyclable material. The final form required for sale to an end user is a clean granulated form. Depending on facility location, it may be possible to sell the various plastic containers to an intermediate processor who prepares them for sale to an end user. Usually the plastic must be baled for shipment to an intermediate processor. Recently quoted prices for baled material are $0.05 to 0.08/lb, while final products may have a value of $0.20 to 0.22/lb.

Some facilities have proposed the recovery of textiles for sale as rags. Although clean rags are in demand as wiping cloths for various industries, these generally must be sorted by type of fabric,

TABLE 1 Chemical Requirements

Element[A]	Composition, Maximum % Allowable					
	Grade 1	Grade 2	Grade 3	Grade 4	Grade 5	Grade 6
Silicon	0.30	0.30	0.50	1.00	9.00	9.00
Iron	0.60	0.70	1.00	1.00	0.80	1.00
Copper	0.25	0.40	1.00	2.00	3.00	4.00
Manganese	1.25	1.50	1.50	1.50	0.60	0.80
Magnesium	2.00	2.00	2.00	2.00	2.00	2.00
Chromium	0.05	0.10	0.30	0.30	0.30	0.30
Nickel	0.04	0.04	0.30	0.30	0.30	0.30
Zinc	0.25	0.25	1.00	2.00	1.00	3.00
Lead	0.02	0.04	0.30	0.50	0.10	0.25
Tin	0.02	0.04	0.30	0.30	0.10	0.25
Bismuth	0.02	0.04	0.30	0.30	0.10	0.25
Titanium	0.05	0.05	0.05	0.05	0.10	0.25
Others (each)	0.04	0.05	0.05	0.08	0.10	0.10
Others (total)	0.12	0.15	0.15	0.20	0.30	0.30
Aluminum	balance	balance	balance	balance	balance	balance

[A] By agreement between the purchaser and the seller, analysis may be required, and limits established for elements or compounds not specified in this table.

ANNEXES

A1. METHOD FOR COLLECTION OF A SAMPLE OF ALUMINUM SCRAP, RECOVERED FROM MUNICIPAL SOLID WASTE, AND ITS PREPARATION FOR ANALYSIS

A1.1 Scope

A1.1.1 This method describes procedures for collection of a sample of shredded aluminum metal scrap recovered from municipal refuse, and the shredding, mixing, and secondary sampling of the metal for analysis.

A1.2 Summary of Method

A1.2.1 A selected size, gross sample of shredded, nonferrous metal scrap is taken from the metal recovery system conveyor belt in increments. Sample increments are taken at timed intervals from a full cross section of the conveyor while it is stopped or by briefly taking the total flow at the discharge of the conveyor while it is moving.

A1.2.2 The quantity of gross sample may be further reduced by mixing, cone-and-quarter sampling, and riffling.

A1.3 General Precautions

A1.3.1 In solids sampling, each step must be designed to eliminate accidental classification by size or gravity. Different sizes usually have different analyses.

A1.3.2 The increments obtained during the sampling period shall be protected from changes in composition due to exposure to the weather.

A1.3.3 Plan the sampling arrangement to avoid contamination of the increments with foreign material.

A1.3.4 A satisfactory sampling arrangement is one that takes an unbiased sample at the desired degree of precision of the constituent for which the sample is to be analyzed. The weight or volume of the

collected sample is compared with that of the total lot to assure a constant sampling ratio.

A1.3.5 It is preferable that the nonferrous metal scrap be weighed and sampled at about the same time. If there is a long lapse in time between these two events, both the purchaser and seller should give consideration to changes in moisture during this interval and the consequent shift in relationship of moisture to the true content at the instant when ownership of the nonferrous metal scrap transfers from one to the other.

A1.3.6 Samples and subsamples shall be collected in such a manner that there is no unmeasured loss of moisture of significant amount. The samples shall be weighed before and after drying or other operations to measure all significant weight loss. The material balances shall be adjusted accordingly.

A1.4 Selection of Gross Sample Size

A1.4.1 Choose the gross sample size by methods given in Recommended Practice E 122, whenever practicable. The chief difficulty for implementing this practice can be that insufficient information concerning possible variation is available. This information should be gathered with practice. Due to the heterogeneity in size and type of material comprising municipal solid waste, the choice of a large sample is desirable.

A1.5 Taking a Gross Sample

A1.5.1 In order to obtain complete representation of materials in a gross sample, it is desirable that the sample increments be withdrawn from the full cross section of the stream. The best possible increment is

Fig. 12.84 (*Continued*)

either a full cross section removed from a stopped conveyor belt or the total flow at the discharge of the moving conveyor taken during a suitable interval of time.

A1.5.2 The choice of sample size can be estimated using Recommended Practice E 122. It is imperative for a given degree of precision that not less than the minimum size and number of sample increments be collected from a lot (see Table A1.1).

A1.5.3 *Number of Gross Samples*—For quantities up to approximately 20 tons, it is recommended that one gross sample represent the lot. Take this sample in accordance with the requirements prescribed in Table A1.1.

A1.5.4 *Distribution of Increments*—It is essential that the increments be distributed throughout the lot to be sampled. The taking of increments shall be at regularly spaced intervals.

A1.6 Sample Preparation

A1.6.1 Cone and quarter the sample until approximately 2 ft^3 (0.06 m^3) remains. Pile the material to be sampled into a conical heap and then spread out into a circular cake. Divide the cake into quarters, take two of the diagonally opposite quarters as the

sample, and reject the two remaining quarters. Collect the two quarters taken as the sample and repeat the procedure of coning and quartering until the desired size is obtained.

A1.6.2 Divide the sample into approximately equal parts. Take one half (1 ft^3) (0.03 m^3) for use in the melt test (see Annex A3). Divide the sample by riffling until the analytical sample is obtained. (Typical riffles can be found in the apparatus section of Method D 2013.)

A1.6.3 Store the prepared analytical sample in a covered, labeled, corrosion-resistant metal can or plastic container until needed for chemical analysis.

TABLE A1.1 Number and Weight of Increments for Sampling

	Top Size, in. (mm)		
	⅝ (15)	2 (50)	6 (150)
Minimum number of increments	15	15	15
Minimum weight of increments, lb (kg)	2 (1)	6 (3)	18 (9)

A2. SAMPLING AT POINT OF RECEIPT

A2.1 Sampling During Unloading

A2.1.1 *Sample Size*—Take a representative quantity of approximately 1 yd^3 (0.76 m^3) from each car or truck of shredded aluminum scrap received.

A2.1.1.1 *Car Sample*—Take two shovels (No. 2 size) of aluminum scrap from the top, middle, and bottom of opened doorway area of car in two locations of doorway area as shown in Fig. A2.1 to obtain six samples. Take an additional 18 samples as shown in Fig. A2.1. Place all material sampled into a suitable receptacle for the total sample from the car. Adequately identify the sample container. Weigh and record the sample from the car.

A2.1.1.2 *Truck Sample*—Take two shovels (No. 2 size) of aluminum scrap from upper, middle, and lower areas of shredded material starting at the rear of the truck. Starting at the truck rear, take samples every 8 ft (2.4 m) at locations one third the distance from the side of the truck. Place all material sampled into a suitable receptacle for the total sample of the truck. Adequately identify the sample container. Weigh and record the sample from the truck.

A2.1.2 *Reduction of Sample Size*—Using riffle sampling equipment[6] or coning and quartering (or equivalent method), reduce the size of sample to approximately 1 ft^3 (0.03 m^3) by putting through the riffle sampler five times. Retain one half of the split sample on each pass through the riffle sampler until a fine sample of approximately 1 ft^3 is attained. Carefully bag and identify the sample for the assay and retain a duplicate sample. Retain the duplicate sample until assay is completed and accepted. If assay is not ac-

cepted, then the duplicate sample may be used for settling the claim.

A2.1.3 *Identity Ticket*—It is recommended that the following information be included on the ticket:
A2.1.3.1 Supplier,
A2.1.3.2 Car number or truck identification,
A2.1.3.3 Net weight of car or truck,
A2.1.3.4 Date unloaded and sampled, and
A2.1.3.5 Initial sample weight.

A2.2 Off-Specification Shipments

A2.2.1 *Sampling*—If shipments are received as off-specification material or are suspected of being off-specification, sample the shipment in the doorway of the car or truck using core sampling equipment,[6] or other acceptable procedure, from at least ten different locations prior to car unloading. Take at least 2 ft^3 (0.06 m^3) of representative sample. Weigh and record the sample weight using a scale accurate to within ±0.1 lb (0.05 kg) or ±0.5 %, whichever is the more precise.

A2.2.2 *Reduction of Sample Size*—Using riffle sampling equipment or coning and quartering (or equivalent procedure), split the sample into equal parts 1 ft^3 (0.03 m^3) remains in each of the last split fractions. Weigh each split fraction, identify, and retain duplicate sample until acceptance or rejection of shipment.

A2.2.3 *Identity Ticket*—It is recommended that

---

[6] Core sampling equipment, available from W. S. Tyler Co., or equivalent, has been found suitable for this purpose.

Fig. 12.84 *(Continued)*

the following information be included on the ticket:

A2.2.3.1 Supplier,

A2.2.3.2 Car number or truck identification,

A2.2.3.3 Net weight of car or truck,

A2.2.3.4 Date unloaded and sampled, and

A2.2.3.5 Initial sample weight.

A2.3 Recommended Testing Frequency

A2.3.1 *New Sources*—Test all new sources of supply on an "every shipment" basis until a total of 1 000 000 lb (450 000 kg) have been received on-specification, at which time, the supplier is considered to be an established source.

A2.3.2 *Established Sources:*

A2.3.2.1 Randomly sample shipments from established source and test a minimum of 20 % of all shipments.

A2.3.2.2 A single shipment from any established or new source of supply that fails to meet the agreed-upon limits within 20 % on any individual factor may require that the source be treated as a new source (see A2.3.1).

A2.3.2.3 Any established or new source failing to meet the agreed-upon limits within 10 % on any two or more factors may require treatment as a new source (see A2.3.1).

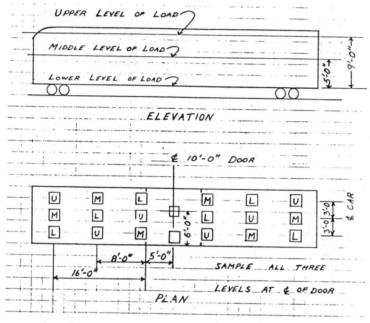

NOTE—All samples consist of two No. 2 shovels from each location sampled.

FIG. A2.1 **Sample Location Chart for 50-ft Railway Car**

A3. ASSAY AND ANALYSIS PROCEDURE

A3.1 Significance

A3.1.1 This is a laboratory procedure used to produce an analytical specimen, determine the percent of metal recovery by remelting, and determine the amount of contaminants present.

A3.2 Determination of Physical Requirements

A3.2.1 Take approximately one third of the sample, weigh, and record the weight.

Fig. 12.84 (*Continued*)

A3.2.2 Dry this one-third sample for 2 h at 110 ± 5°C and record the weight of the dried material.

A3.2.3 Remove the magnetic materials and separately record the weight of the magnetic materials removed and the remaining nonmagnetic materials.

A3.2.4 Screen the sample over a 12-mesh (U.S. Standard) sieve using the Dry Screening Procedure of Method E 276. Record the weight of the sieved material and the material remaining on the screen.

A3.3 Determination by Assay and Chemical Analysis

A3.3.1 Take the remaining two thirds of the original sample weight and place in one or more crucibles using a minimum sample size of about 225 g.

A3.3.2 Prepare a salt flux comprised of 47.5 % NaCl, 47.5 % KCl, and 5 % cryolite. (If cryolite is not available, sodium fluoride may be substituted.)

A3.3.3 Mix a minimum of 2 parts by weight of salt flux to 1 part metal (a salt-to-metal ratio of 3:1 is preferred).

A3.3.4 Place the crucible containing the sample in a gas-fired furnace to obtain a sample temperature of 815 ± 55°C (1500 ± 100°F). Stir occasionally with a graphite rod until all metal is melted. Hold for 15 min.

A3.3.5 *Alternative Procedure A for Pouring Sample:*

A3.3.5.1 Decant salt and sludge and save for removal of metal fines.

A3.3.5.2 When entire sample is melted in one crucible, pour remaining sample into a mold.

(a) *Samples over 450 g*—If sample weighs more than 450 g, pour analytical sample into appropriate mold.

(b) *Samples Under 450 g*—If sample weighs less than 450 g take the entire sample of metal recovered and remelt in a suitable laboratory furnace.[7] Melt the sample at 675 to 735°C (1250 to 1350°F) and pour into an analytical sample mold.

A3.3.6 *Alternative Procedure B for Pouring Sample:*

A3.3.6.1 Allow metal and salt to cool to room temperature while still in the crucible.

A3.3.6.2 Invert the crucible to remove metal; break crucible, if necessary, to remove all metallics.

A3.3.6.3 Mechanically separate salt from metal and pick out any metallics that may be entrapped in the flux.

A3.3.7 Weigh all metal. Wash with water and then dilute nitric acid (0.1 *N*), rinse, dry, and weigh.

A3.4 Calculations

A3.4.1 Calculate the percent recovery on the basis of the as-received sample weight, before drying, screening, and magnetic separation as follows:

$$\% \text{ recovery} = \frac{R}{S} \times 100$$

where:
R = weight of metal recovered in A3.3.7, g, and
S = weight of sample in A3.3.1, g.

A3.5 Spectrochemical Analysis

A3.5.1 Determine the composition of the sample using an appropriate spectrochemical technique.

NOTE—It should be recognized that if the metal recovered in the assay is analyzed, the results for magnesium will be lower than the actual magnesium content of the MAS. This is due to a reaction of the assay flux with magnesium. If there is reason to suspect that the magnesium content of the MAS is higher than the maximum specified, analysis should be made on a duplicate sample of the MAS, which is remelted either under a salt flux mixture or in a small induction furnace.

---

[7] Laboratory furnaces, available from Jelrus and Lindberg, or equivalent, have been found suitable for this purpose.

The American Society for Testing and Materials takes no position respecting the validity of any patent rights asserted in connection with any item mentioned in this standard. Users of this standard are expressly advised that determination of the validity of any such patent rights, and the risk of infringement of such rights, are entirely their own responsibility.

This standard is subject to revision at any time by the responsible technical committee and must be reviewed every five years and if not revised, either reapproved or withdrawn. Your comments are invited either for revision of this standard or for additional standards and should be addressed to ASTM Headquarters. Your comments will receive careful consideration at a meeting of the responsible technical committee, which you may attend. If you feel that your comments have not received a fair hearing you should make your views known to the ASTM Committee on Standards, 1916 Race St., Philadelphia, Pa. 19103.

Fig. 12.84   *(Continued)*

otherwise they have a very low value. No facilities are known to recover and sell rags at the present time, but the availability of local markets may warrant consideration of their recovery. Prices may range from $0.08 to 0.20/lb depending on fabric type and condition.

Flyash from the combustion of RDF may be sold to cement producers as an additive for cement manufacturing. It may be added up to several percent of the feed. Several cement plants currently purchase this material for several dollars a ton, which greatly reduces the disposal needs of the resource-recovery facility.

12.10.7   Conclusion

In conclusion, there are numerous materials that may be recovered from municipal solid waste by resource-recovery facilities. However, most may only be effectively recovered by hand sorting rather than by fully mechanized system. Markets must be determined on a site-specific basis by contact with nationwide firms that purchase various materials, local scrap processors, local end users, and local recycling centers. In many cases, recovery by hand sorting is a cost-effective and efficient method of recovery. Recovery of recyclable materials in those cases serves two purposes: (1) it removes valuable materials that otherwise would be incinerated or landfilled with little or no resulting benefit while generating revenue for the facility and (2) it creates new jobs, especially for those in the lower echelons of the economic ladder, those who are most impacted by the uncertainties of the times. As reliable, cost-effective mechanized systems for materials recovery are developed and demonstrated, they may be installed in future facilities and existing facilities may be retrofitted.

REFERENCES

H. Alter and W. R. Reeves, *Specifications for Materials Recovered from Municipal Refuse*, EPA—670/2-75-034, U.S. EPA, Cincinnati, Ohio, May, 1975, 120 pp.

American Society for Testing and Materials, *Resource Recovery and Utilization*, Technical Publication 592, ASTM, Philadelphia, Pennsylvania, 1975, 200 pp.

Annual Book of ASTM Standards, Volume 11.04, Pesticides; Resource Recovery; Hazardous Substances and Oil Spill Response; Waste Disposal; Biological Effects, American Society for Testing and Materials, Philadelphia, Pennsylvania, 1983.

Y. M. Garbe and S. J. Levy, *Resource Recovery Plant Implementation: Guides for Municipal Officials, Markets*, SW-157.3, U.S. EPA, 1976, 47 pp.

Market Locations for Recovered Materials, Report SW-518, U.S. Environmental Protection Agency, 1976, 81 pp.

Plastic Beverage Container Division, the Society of the Plastics Industry, New York, New York. 10017.

Resource Recycling, Journal of Recycling, Reuse and Waste Reduction, published bimonthly by Resource Conservation Consultants, Portland, Oregon 97210.

K. L. Woodruff, H. Alter, A. Fookson, and B. Rogers, Analysis of Newsprint Recovered from Mixed Municipal Waste, in *Resource Recovery and Conservation*, Elsevier Scientific Publishing Co., Amsterdam, Vol. 21976, pp. 79–84.

12.11   RECOVERED MATERIALS-EQUIPMENT AND SYSTEMS
Kenneth L. Woodruff

12.11.1   Introduction

As discussed in Section 12.10, Recovered Material Specifications and Markets, various markets are available for materials recovered from municipal solid waste. However, extracting them in a saleable form, meeting user quality requirements is necessary to assure a continuing market as well as revenue to the facility. This section discusses the equipment and systems that have been demonstrated and are available.

12.11.2   Air Classifiers

Typically solid waste-processing systems begin with some type of size reduction and/or sizing system, that is, shredders, trommels, or disc screens. Size reduction and screening devices have been discussed previously. To facilitate both materials recovery and refused-derived fuel (RDF) production following size reduction, air classification may be employed.

During the early 1970s when a number of waste-processing systems were being developed in the United States, it was believed that air classification was the second processing step after shredding. Shredded refuse was air-classified to produce a light, organic-rich fraction (RDF) and a heavy noncombustible-rich fraction to be processed further for materials recovery. Air classification technology was demonstrated on a commercial scale at various locations by Combustion Power Co., the National Center for Resource Recovery (NCRR), the EPA-sponsored City of St. Louis Project, and the U.S. Bureau of Mines, among others.

Through actual operating experience, it was found that shredded refuse could not be efficiently separated into two fractions by air classification. The major problem was the carryover of glass fines into the light fraction (RDF). Furthermore, a significant portion of the combustibles contained in the waste reported with the heavy fraction when the units were operated to minimize glass and aluminum carryover to the light fraction. The most well-documented experience with the air classification of shredded waste was the St. Louis Facility operated by the city to produce RDF for sale to Union Electric Co. Initially, shredded, magnetically separated refuse was shipped to Union Electric for use as supplemental fuel. Extensive wear and abrasion problems, as well as boiler slagging necessitated the examination of further refuse processing. It was believed that the implementation of air classification would greatly reduce the glass and ash content of the refuse fuel. An air classifier was installed, however the RDF glass content was not significantly reduced and abrasion problems in RDF handling, as well as boiler slagging problems continued. Through analysis of the shredded refuse and RDF, it was found that the bulk of the glass, because it had been pulverized by shredding, reported to the air classifier light fraction. In fact, ash content of the refuse fuel did not significantly decrease. Originally ash ranged from 22 to 24%; following air classification, ash content was reduced only to about 18%. Of this amount, approximately half was glass.

Some processing system developers recognized the deficiencies in the St. Louis operation and proceeded to either develop more efficient air-classification processes or to modify the size reduction process by including prescreening to minimize glass pulverization and hence RDF contamination (also see Section 12.5.4).

Unfortunately, there were those developers that believed that rather than modify the size reduction step and/or improve the air-classification system, further downstream processing of the RDF was the answer. Facilities using this approach were Ames, Iowa, Milwaukee, Wisconsin, and Chicago Southwest. These plants all found that although screening of the light fraction (RDF) removed a portion of the glass fines, significant amounts of glass were still present, and significant additional amounts of combustible materials were lost to the fines fraction. Boiler slagging and abrasion problems continued. Of these facilities, it should be noted that Ames is the only one still operating.

Facilities utilizing the modified size reduction approach to reduce the amount of glass pulverization include Madison, Wisconsin; Miami, Florida; New Orleans, Louisiana; and Baltimore County, Maryland. All these facilities continue to operate. The Monroe County, New York, and Tacoma, Washington, facilities use improved air-classification systems. Several other RDF facilities including Akron, Ohio, and the Hooker Chemical Co., Niagara Falls, New York, facility originally included air classification, but have essentially abandoned it. Since these facilities utilize semisuspension fired dedicated boilers, the operators have learned to live with the high ash-content fuel. Along these lines, the Albany, New York, facility uses no air classification at all, simply burning shredded, magnetically separated refuse with a spreader stoker. However, early tests of the Aenco air-swept rotary drum on raw refuse or after shredding (97% lights/3% heavies split) shows promise.

On the basis of experience with municipal solid waste (MSW) air-classification systems over the past 15 years, the use of an air classifier is dependent upon the end use of the RDF. An air classifier may not produce a low ash-content RDF from shredded refuse. Preprocessing of the refuse prior to shredding to remove glass and grit is the preferred approach to produce a high-quality RDF with air classification. An alternative, screening of RDF after air classification, is not as effective in glass and ash reduction.

If RDF is to be fired on-site in a dedicated boiler and minimal materials recovery is desired, air classification may not even be of benefit to the system; however, if RDF is to be densified or must be of very high quality for specialized combustion equipment or certain bioconversion processes, air classification of prescreened shredded material should be employed. (*Editor's note:* Section 12.5.2 adds to this discussion.)

The following list indicates vendors of air-classification systems for solid waste. Figures 12.85–12.90 depict the general configuration of air-classification systems demonstrated and available for MSW processing.

SOLID WASTE AIR CLASSIFICATION SYSTEM
VENDORS

AENCO, INC. A Cargill Company Albany, New York 12205 518-869-3651	Rotary drum type
Iowa Manufacturing Company Cedar Rapids, Iowa 52402 319-399-4800	Rotary drum type
MAC Equipment, Inc. Sabetha, Kansas 66534 913-284-2191	Vertical type
Rader Companies, Inc. Memphis, Tennessee 38117 901-365-8855	Vertical and horizontal type
Triple/s Dynamics Dallas, Texas 75223 214-821-9143	Vibrating type

12.11.3 Ferrous Metal Recovery

Ferrous metals, primarily food and beverage containers, may be readily recovered from shredded or screened MSW by magnetic separation. However, to produce a clean, high-quality product that may be sold, an effective cleaning-type ferrous recovery system is required.

In general, magnetic separation requires a size-reduced or sized fraction so that recovery may be optimized. This means that magnetic separation of prescreened undersize fractions, shredded

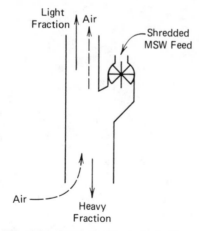

Fig. 12.85 Vertical column air clasifier.

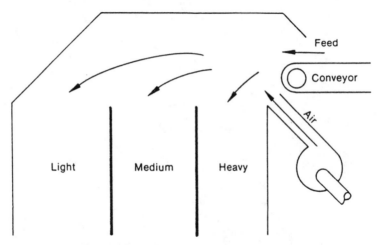

Fig. 12.86 Horizontal air classifier (air knife).

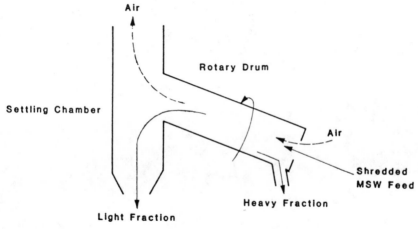

Fig. 12.87 Rotary drum air classifier, shredded refuse.

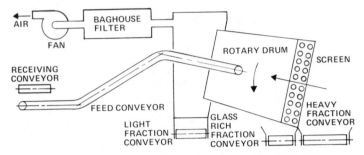

Fig. 12.88 Rotary drum air classifier, raw or shredded refuse.

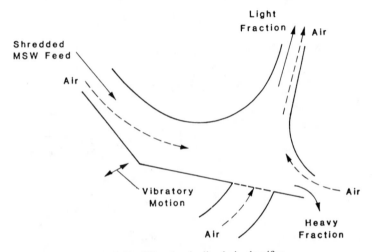

Fig. 12.89 Vibrating inclined air classifier.

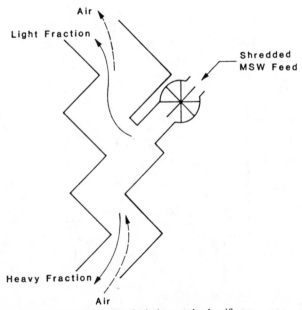

Fig. 12.90 Vertical zig-zag air classifier.

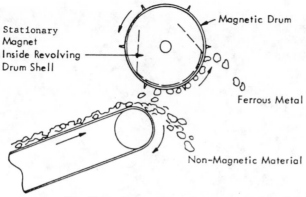

Fig. 12.91 Drum magnet separator.

MSW, or air classifier heavy fractions will result in high levels of ferrous metal recovery. Typically, manufacturers of magnetic separators are willing to guarantee at least 95% recovery. Although this appears attractive, the problem is that a standard belt- or drum-type magnetic separator in achieving a high recovery tends to carry over tramp material, entrained material, and other contaminants which render the ferrous product unsaleable. The ferrous output from standard separators may be only 80 to 90% ferrous metal. Hence, although these magnetic separators provide excellent recovery efficiencies they provide poor separation efficiencies. Figures 12.91–12.93 depict typical drum- and belt-type magnetic separators. It should be noted that both are available in permanent and electromagnetic types and although permanent magnets generally offer a stronger, more uniform magnetic field, electromagnets provide the advantage of being able to be turned off, allowing operational flexibility.

Through much experience in solid waste processing, it has been found that at least two stages of magnetic separation are required to produce a saleable ferrous product from municipal waste. This is a minimum. In addition, it may be desirable to include an intermediate size reduction or shredding step followed by air classification to produce a really "clean," 98 or 99% ferrous metal product. Several magnetic separator vendors have developed specialized magnetic separation systems for solid waste that feature multistage magnetic separation. The idea is to maximize recovery with the first stage and maximize separation efficiency in the second. Figures 12.94–12.96 depict these systems, one consisting of multiple drums and the other a belt-type unit with multiple magnets. Both methods allow for the material to be picked up and dropped and then picked up again. During the second pickup, contaminants are allowed to drop free of the metal. The belt type has been widely employed in the waste industry.

Facilities wishing to produce an even cleaner metal product utilize air separation and/or shredding prior to secondary magnetic separation. Design of a ferrous metal recovery system for a resource recovery facility cannot merely include a simple single-stage magnetic separator. A good knowledge of the market(s) to which material is to be sold is necessary with the ferrous metal recovery system designed accordingly. A number of resource-recovery facilities have ceased ferrous metal recovery due to a claim of lack of markets. Low-quality material will always be the first to be rejected in a market downturn, whereas high-quality product will always have a market even in bad times, although the price may be depressed.

The following list indicates vendors of solid waste magnetic separation equipment.

SOLID WASTE MAGNETIC SEPARATION
EQUIPMENT MANUFACTURERS

Dings Magnetic Separator Co.
Milwaukee, Wisconsin 53246
414-672-7830

Eriez Manufacturing Company
Erie, Pennsylvania 16514
814-833-9881

Stearns Magnetics, Inc.
Cudahy, Wisconsin 53110
414-769-8000

Fig. 12.92 Drum magnetic separator in operation.

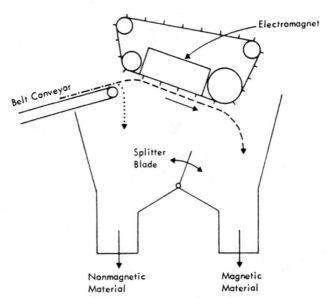

Fig. 12.93 Belt magnetic separator.

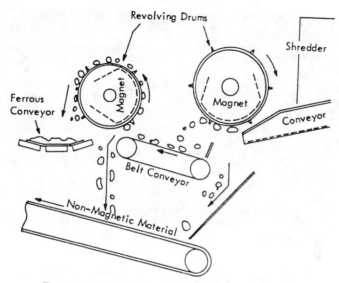

Fig. 12.94 Double-drum magnetic separation system.

12.11.4 Nonferrous Metals Recovery

The principal nonferrous metal component of MSW that is desirable for recovery is aluminum. Other nonferrous metals are generally recovered in a mixed form and sold to scrap processors, as is. In some instances, coins may be recovered from the mixed nonferrous fraction for sale to the U.S. Mint. Coin recovery rates of $0.05 to 0.40/ton have been reported.

Aluminum in the form of beverage cans is the most highly valued aluminum component of solid waste. Other aluminum, miscellaneous extrusions and castings, such as lawn furniture, pots and pans, and so on, also have a market, however *the highest value may be obtained for municipal aluminum scrap when beverage containers are kept separate from other forms of aluminum.*

Although mechanical aluminum recovery systems have been demonstrated, including heavy media separation and various types of eddy current separation devices, hand-sorting is still the most efficient means of aluminum recovery.

Several facilities, including Ames, Iowa, New Orleans, Louisiana, and Monroe County, New York, have utilized eddy current devices for aluminum recovery; all have been discontinued as a result of inefficiency. The only eddy current system operating successfully is the NRT Preburn™ system in Gallatin, Tennessee. It processes 200 ton/day of raw waste prior to incineration. Reynolds Metals has developed and patented (U.S. Patent 4,387,019) a system for the recovery of aluminum cans from mixed MSW based on differential screening by trommels, density separation,

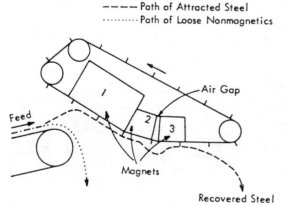

Fig. 12.95 Three-stage belt magnetic separation system.

Fig. 12.96 Three-stage belt magnetic separation system installation.

and hand-sorting of aluminum cans. The system has been demonstrated to be cost effective and has operating experience at Houston, Texas, and Salem, Virginia. Several other facilities also recover aluminum by screening to enhance hand-sorting; all operate successfully due to the high value of the recovered product. It has been reported that aluminum recovered by this method is nearly the same as source-separated aluminum cans and it commands a very high price.

Other aluminum constituents of MSW may also be recovered by hand-sorting. The most effective recovery occurs prior to size reduction since fewer, smaller pieces must be sorted. Following shredding, pieces of aluminum larger than 2 in. (51 mm) may be hand-sorted provided the stream has been concentrated by screening. Otherwise the aluminum reports with the mixed nonferrous fraction which must be sold to a scrap processor at a reduced market price.

Nonferrous metals in total, rarely exceed 1% by weight of the waste stream. However, aluminum generally comprises about 70 to 80% of the nonferrous metals stream. Because of its extremely high value in comparison with other materials that may be recovered from waste ($200 to $1000/ton of metal), nonferrous metals recovery, specifically aluminum, is the largest revenue source of a resource recovery facility, exceeded only by energy sales and tipping fees. (*Editor's note:* Successful pulsed eddy current aluminum recovery, mostly cans, is employed by the NRT, Inc., mass-burn preprocess system at Gallatin, Tennessee. Figure 12.36 in Section 12.5 illustrates this system.)

12.11.5 Paper Recovery

The most successfully demonstrated method of newspaper and corrugated paper recovery from MSW involves the segregation of these materials from the mixed waste, either on the tipping floor or on the primary feed conveyer to the processing system. This type of recovery is routinely practiced at solid waste transfer stations. Newspaper and/or corrugated recovered in this manner is subsequently baled for shipment to a waste paper processor, or in some instances, may be shipped directly to an end user.

Several system developers and vendors have demonstrated mechanical systems designed to recover paper fiber from mixed MSW. *However, it has been found that the fiber that is recoverable is of low quality with limited uses. Considering the cyclical waste paper market, as well as generally strong energy markets which develop as part of most resource recovery projects, it would appear to be more beneficial to utilize the bulk of the waste paper contained in MSW for its energy value with only minimal hand separation of paper from the heavily concentrated commercial waste loads delivered to the facility.*

Mechanical systems generally consist of multistage screening, shredding, and air separation to concentrate the paper. Removal of film plastics may be accomplished by flash-drying the paper, thereby melting and densifying the plastic so that it can be air-separated and/or screened from the paper. Another alternative is to pulp the paper fiber which may then be screened from the nonpulpable plastic. Such mechanical systems have been developed and demonstrated by AENCO, Flakt, Forest Products Laboratory, NCRR, ORFA Corporation, and University of California.

Perhaps the fiber recovery system most well known is the Hydrasposal System developed by the Black-Clawson Co. As demonstrated at Franklin, Ohio, for a number of years, the system successfully recovered a low grade of pulp from MSW using a wet pulping process. The fiber was used in the manufacturing of roofing shingles, as it did not meet higher-grade pulp specifications.

Once recovered in dry form, paper or corrugated may be readily baled for shipment to market using one of several types of balers. These are readily available and are widely used in the scrap-processing industry. They are available in fully automatic, semiautomatic, or manual tie models. The type used depends on the volume to be baled and end-user size and density requirements. It should be noted that hogging (shredding) of the paper or corrugated may be required prior to baling if shipped directly to an end user, rather than to an intermediate processor.

12.11.6 Glass Recovery

Since glass may comprise up to 10% by weight of the MSW stream delivered to a processing facility, recovery is desirable. To recover glass for reuse in the manufacturing of containers, two methods have been demonstrated; however both have had limited success.

The processes follow metal and organic material removal. The first method involves optical sorting of coarse glass pieces larger than 3/16 in. (4.75 mm). Closely sized glass fractions are visually inspected piece by piece. A binary separation is made on the basis of the amount of light transmitted to a photocell. Pieces to be rejected trigger a pneumatic device which blows it from the stream. This method of glass separation suffers from very high cost and poor levels of refractory material removal. Facilities at Franklin, Ohio, Hempstead, New York, and Doncaster, England, have all employed optical sorting.

The second method, froth flotation, produces a color-mixed fraction that has more limited marketability. However, operationally it is more reliable than color sorting and it is essentially applicable to the entire glass content of MSW since it requires fine particles (less than 10-mesh). Froth flotation affects a separation on the basis of surface chemistry of the glass particles. The technique is routine mineral-processing technology. The glass-rich fraction is ground and sized and slurried with water. Chemicals are added to the slurry that selectively make the glass particles hydrophobic and allow air bubbles to become attached when the mixture is aerated. The air bubbles float the glass particles to the surface where they are scraped off, washed, and dried. This method is effective and produces a saleable product. However the operation is expensive and markets are limited due to the color-mixed output. Demonstration of the technique has occurred in New Orleans, San Diego, and Monroe County, New York. The only facility presently in operation is at the New Castle County Delaware Reclamation facility.

The only other method that has been demonstrated for the recovery of glass suitable for reuse in the manufacturing of containers is hand sorting. Banyan-Dade Resource Recovery in Dade County, Florida, hand-sorted glass containers from the raw refuse primary screen -5 in. $(-12.7$ cm) undersize fraction. It was found that 25 to 30% of the incoming glass could be recovered in this manner. The glass was sold as recovered (color-sorted) for further processing (removal of caps, labels, and crushing) prior to its use as cullet. However, the economics of this recovery method dictate that the operation must be located nearby a cullet user to minimize transportation costs.

Other uses of glass include brick, concrete aggregate, foamed glass, ceramic tile, terrazo tile, building panels, glass wool, slurry seal, and glasphalt. All are reportedly able to consume waste glass, but in varying quantities. Most of these uses require a lower-quality product than does the glass container industry, however substantially lower prices must be paid. Glass recovery systems that have been used to produce glass for this type of market include air tables, stoners, wet gravity tables, and heavy media systems. At least two stages of processing are required to produce a 93 to 95% glass fraction. Although the market value of the product is higher, the cost of recovery is such that widespread recovery of glass for the container industry is unlikely for some time.

12.11.7 Plastics Recovery

Recovery of plastics from MSW has been extremely limited. Due to their high heating value, they are generally left in the fuel fraction for energy recovery purposes. However, if recovered in saleable form, which requires separation by type as well as contaminant removal and granulation, they have a high market value (in excess of $100/ton depending on type and degree of processing). Plastics that may be most readily identified and recovered from the waste include high-density polyethylene (HDPE) (milk bottles) and polyethylene terephthalate (PET) (soft drink bottles).

The only method presently available to recover these materials is hand-sorting following one or more stages of screening of raw MSW. Since these materials are easily recognizable, they are easily recovered. However, clean-up of the recovered material once it has been mixed with other wastes is necessary to produce a saleable product. Further processing may include washing to remove labels and contaminants followed by granulation. Owens-Illinois received U.S. Patent 4,379,525 for a Process for Recycling Plastic Container Scrap, which describes such a system.

Several European firms have demonstrated systems that recover film plastics, primarily poly-ethylene bags. Both Flakt and Sorain-Cecchini recover film plastics for recycling. The material is recycled back into plastic bags.

Due to the growing use of plastic containers, replacing both glass and metal, interest in the recovery of plastics will continue to grow. Although all plastics have high energy value as fuel, they have even a higher value if recovered and sold for recycling purposes.

Editor's note: As of this writing (early 1985), ASME, ASTM, the U.S. EPA, Argonne National Laboratories, Battelle, private industry, and certain states (California, New York, New Jersey, etc.) are working cooperatively in a concerted effort in evaluating the impacts of trace amounts of dioxins which are appearing in most recent incinerator stack tests. An objective is to establish operating and test protocols leading to establishment of realistic guidelines for the industry. In general, plastic content, moisture, combustion efficiencies, furnace temperatures, etc., are thought to be principal precursors in the formation of dioxin emissions, and there are proponents of plastics removal from raw municipal, industrial, and commercial wastes prior to incineration for this reason alone. As part of the 1984 amendments to the Resource Conservation and Recovery Act (RCRA), enacted on November 8, 1984, Congress has mandated the U.S. E.P.A. to report to Congress on the risk of dioxin emissions and operating practices appropriate to controlling the emissions.

12.11.8   Ash Processing for Metals and Aggregate Recovery

Processing of municipal waste combustion residues was initially examined by the U.S. Bureau of Mines. Incinerator residues were processed to recover metals and glass. A pilot plant was operated in College Park, Maryland, for a number of years and numerous publications have resulted. Due to the relatively low recoveries and poor quality of materials when compared with raw refuse processing, this technology has not been widely implemented. Although several incinerators recover low-grade ferrous scrap, only three known facilities in the United States are processing combustion residues to recover other products.

The Albany, New York, facility (Answers Project) includes ferrous metals, mixed nonferrous, and aggregate recovery. The system employed includes magnetic separation, multistage screening, and grinding. The metal fractions are sold for further use, while the aggregate may be used in fill, stabilized roadbase, lightweight concrete, and asphalt aggregates.

The Pinellas County, Florida, mass burn facility includes an ash-processing system using mineral processing equipment as demonstrated by the Bureau of Mines. Operating results have not yet been made available.

Resources Recovery (Dade County), Florida, recovers metals and glass in a front-end processing system. As a result, its ash has been found to be suitable for use as an additive in the manufacture of portland cement. This further reduces the amount of residue to be landfilled.

Processing of ash for materials recovery has been limited due to poor recoveries and poor quality of recovered material. In certain areas due to specific market availability, it has application; otherwise, it is generally more beneficial both from a materials recovery, as well as an enhanced fuel quality standpoint to recover materials prior to combustion.

Advantages of growing importance in the processed fuel concept with front-end materials recovery include not only the reduction in overall ash quantities going to landfill (approximately 10% by weight of raw refuse versus 25 to 40% for unprocessed mass burn), but the significant reduction in heavy metals, concentrations of landfill leachate contamination of groundwater and trace organic emissions levels.

REFERENCES

American Society for Testing and Materials, *Resource Recovery and Utilization*, ASTM publication 592, ASTM, Philadelphia, Pennsylvania, 1975.

D. Bendersky, et al., *Resource Recovery Processing Equipment*, Noyes Data Corporation, Park Ridge, New Jersey, 1982.

L. F. Diaz, et al., *Resource Recovery from Municipal Solid Wastes, Volume I, Primary Processing*, CRC Press, Inc., Boca Raton, Florida, 1982.

E. Hainsworth, et al., *Energy from Municipal Solid Waste-Mechanical Equipment and Systems Status Report*, EG&G Idaho, Inc., Idaho Falls, Idaho, EGG-PSE-5974, March, 1983.

F. R. Jackson, *Recycling and Reclaiming of Municipal Solid Wastes*, Noyes Data Corporation, Park Ridge, New Jersey, 1975.

Materials and Energy from Municipal Waste, Office of Technology Assessment, Congress of the United States, Washington, D. C., July, 1979.

S. C. Schwarz and C. R. Brunner, *Energy and Resource Recovery from Waste*, Noyes Data Corporation, Park Ridge, New Jersey, 1983.

12.12   RAW MATERIAL QUANTITY AND COMPOSITION: A FINAL CHECK
David G. Robinson
William D. Robinson

12.12.1   Quantification Survey

Background

Too often, a solid waste management program is launched on the basis of old and often inaccurate information, be it for collection, land disposal, source separation/recycling, or centralized energy/materials resource recovery.

Michael Farber (a resident of Hamden, Connecticut) who advocates incentives for source separation/recycling, reviews critical planning inaccuracies via his "reader comment" reply in the *New Haven Journal Courier* of March 22, 1984, in response to reporter Tony Doris' series titled "The Garbage Pileup." The following is an excerpt from Farber's comments:

> *As the primary researcher for the Schenectady County Planning Department in upstate New York, I spent two years studying the problems of garbage management, and accumulating information based on the current writings and interviews with industrial, academic and government experts. I would like to make some additions and emphasize alternative solutions, particularly recycling, which will put the high-tech garbage-to-energy approach into a new light, revealing it to be fraught with uncertainties, dangers and exorbitantly large costs.*

> *When planning for a resource recovery plant, it is essential to know the approximate amount of garbage in a region which will need to be processed. Yet, there is a complete lack of tangible and reliable data available upon which to base the necessary planning. Typically, landfill operators are often in non-compliance with state regulations. They neglect to record and even falsify the amount and kind of refuse which they allow to be dumped on their property. It is the rare operator who even has scales at his landfills, much less scales that are accurate.*

> *In Doris' article we are told Connecticut generates garbage at a daily rate of 4 pounds (1.8 kg) per person. This statistic should not be stated with confidence. Estimates throughout the country range from 2.5 pounds (1.1 kg) per person to 9.0 pounds (4.1 kg).*

> *The President's 12th Council on Environmental Quality estimated a generation rate of 4.5 pounds (2.0 kg) per person per day in 1980 with an expected increase of 20 percent by 1984. In New York City, the almost 25,000 tons per day (tpd) averages out to a daily amount of 6–8 pounds (2.75–3.6 kg) per person. My own studies at Schenectady, a mixed urban and rural county, suggest a generation rate of approximately 8 pounds per person. As you can see, estimates do vary widely. How can a garbage-to-energy plant be planned without even this basic information? Much more effort and research needs to be conducted in this area.*

There are also examples of multiple feasibility studies performed for the same locality, county, or region by various consultants as politics and patronage changes along with whatever urgency is perceived at the time to "Do something about the solid waste disposal problem."

It is astonishing how these layers of studies can vary for essentially the same conditions, especially the quantity and quality evaluation segments. Therefore, it behooves the responsible entity to ascertain the realities *before* "lift-off" by cutting through any veil of cut-and-paste puffery and outright guessing in prior studies. The following will lead the way to the realities.

Overview of Current Practice

With nearly excessive concentration on other vital resource recovery/disposal issues, there finally appears to be recognition of a neglected yet critical facet of solid waste management: *The assessment of waste generation.* Historically, consultants and municipal officials too often rely upon demographic/socioeconomic analysis techniques to obtain per capita waste generation factors to approximate expected waste quantities. Such approximations have been the basis for facilities sizing, system design, projected energy/materials recovery revenues, financing, required landfill space, and the solid waste management plan itself.

A responsible study of solid waste management options must be based upon hard data! Too much off-the-shelf modelling, austerity-budgeted superficial studies, and approximations of per capita generation has led to misinformed decisions. The result is disposal/recovery facilities improperly sized and designed with costly consequences. *Example:* During the first year of operation (1983), a large new solid waste-to-energy steam/electric plant has found it necessary to add capacity because it just cannot handle the load without additional redundancy. Another had to perform major surgery in redesigning furnace and ash-handling systems. Still another is encountering significant stoker grate shut down during early operation, purportedly due to deleterious industrial wastes.

A first step in any solid waste management feasibility study is a field survey of available wastes, and it can be complicated by several factors:

1. The study area may be a fairly large region with many landfills.
2. The landfills may be privately operated and not amenable to allowing either placement of scales or survey personnel to record incoming data.

The importance of weight/volume/composition data cannot be overemphasized as part of a comprehensive field survey, and certain states—New Jersey is one—may require it before a permit can be issued for a disposal/recovery system.

Although prior feasibility studies include quantity and *maybe* composition factors for current and projected availability of solid wastes, such studies must be carefully reviewed as implementation time approaches. This review must not only ascertain the credibility of the original (or latest prior) study, but it must also adjust to changed (or changing) conditions just before "take-off." The following sections will present actual survey conditions, problems, and solutions to assist those responsible for field surveys of solid waste quantities and character prior to project implementation, or for updating waste management planning.

12.12.2 Presurvey Planning

1. *Preparation.* The program manager can easily be overwhelmed by what appears to be a confusing tangle of tasks and responsibilities, field coordination, and meetings. From the outset, the planner must be:

An organizer.
A diplomat.
A politician.
A cheerleader.
A spokesman.

He will find himself addressing such diverse groups as:

Engineers.
Mayors.
Waste haulers.
The public.

The political aspect includes promotional effort to encourage cooperation and coordination, and the manager may choose to publicize the program before it starts. This approach can avoid considerably tedious and often unsettling explaining to haulers at the scale locations who are often suspicious and ask "Who, What, and Why?"

Regardless of the scope—a town, city, county, or regional authority—the planner must undertake a number of tasks when the program begins. Most important is a program schedule, and it must include:

2. *Program Duration.* Days, weeks, months, etc.
3. *Survey Frequency.* How many times a year (or other time frame).
4. *Seasonality.* Effects on quantities and character, including weather (moisture), yard and rubbish clean-ups, holidays, and so on.
5. *Weather Patterns.* Inclemency deterrence, road conditions, and the like, indigenous to the area.

As trite and hackneyed as it sounds, but with the same inevitability as death and taxes, waste generation and character varies by the month, the week, day to day, hour by hour, and so on. A longer program can develop more accurate flow rates for a particular time of year, and an ideal duration would be 5 weeks and about four times a year. For a corroborative last look before starting final facilities design and construction, a minimum duration should be no less than 4 weeks.

Budgetary restraints are often the limiting factor in program frequency and duration. A week-long program will not provide reliable data and can be a misleading waste of time.

12.12.3 Survey Scope

Background

This step will define the extent of the survey and its costs for equipment and manpower. First, all landfill locations and types of waste received must be identified regardless of what is indicated by permits and/or personnel. Some sites may be permitted for oversize bulky wastes only, whereas others may be permitted for residential wastes only, and a few may be privately owned and operated by industry for their own particular discards. In any event, it is vital that all wastes be accounted for and sufficient time must be allotted to the following salient factors in the survey plan.

Location of All Disposal Sites

Disposal sites are often privately owned and operated, and the operators may be reluctant to divulge any information regarding their facilities, especially the quantity and types of materials collected and received. If such is the case, and survey personnel are denied access to the site, then off-site but near-by observation should be attempted.

Location of Truck Scales in the Area

This can also be time consuming, although local or state transportation departments usually maintain records of existing certified (or uncertified) scales. Another source may be the state Department of Weights and Measures. The necessity for such information is in the event that the disposal site does not have a platform scale and portable units cannot be placed there.

The number of disposal sites will determine the number of scales and personnel requirements. Each disposal site should be examined for multiple entrances and exits, and all points of access to the landfill must be monitored during the program. The hours of daily operation must be known so that all activities are included. Weighing/survey personnel should be present during all hours of operation each day to assure an accurate day-of-the-week load profile.

Direct contact with private haulers is a high priority; their cooperation is critical to the success of the program. They are private businesses and this realization must temper the relationships. Private haulers may covet their operating data to the point of not divulging anything deemed inimical to their competitive position or client/customer relationship. Once an understanding of mutual benefit is established, it is wise to not only solicit load data and patterns but any other relevant and useful advice they may have. They can help you avoid spinning your wheels, guessing, and lost time. Figure 12.97 illustrates a questionnaire useful for this purpose.

Orientation Meeting with All Program Participants

An effective beginning is an initial contact by a letter requesting their assistance and cooperation and outlining the program, its scope, purpose, and mutual benefits along with pertinent details. If possible, a meeting of all involved haulers is most effective when all questions and concerns can be examined and clarified. This may not be possible, in which case individual personal contact must be attempted. In any case, such an approach can be impressive to the extent of demonstrating the integrity and determination in the program and above all the necessity and, again, the mutual benefit. These familiarization procedures should be complete several weeks before scale and site locations, access points, and best ways to get there are needed.

Identify Haulers Serving Study Area: Residential, Commercial, and Industrial

A list of haulers can usually be provided by the municipalities within the study area. Residential hauling services are more easily profiled since they are usually provided via municipal contract.

Commercial waste collection, however, is often an unregulated laissez faire enterprise involving a number of competing haulers among which routing and collection schedules vary widely. This particular area of waste generation and collection is the most difficult to assess.

With the exception of the large agglomerate collection networks such as Browning Ferris Industries, Waste Management, Inc, and so on, the smaller independent commercial haulers' routing may not be as systematic, although computerization and optimization is infiltrating the latter group rapidly. For example, a hauler may service a number of commercial accounts in the study area, but may continue his collection activity outside the study area in a contiguous region in which final pick-ups and disposal take place. The opposite is equally onerous. This is a fairly

Haulers Survey

Your name: —————————————   Organization: —————————————

Landfill used: —————————————   Earliest load to landfill: ————————— A.M.

Number of trucks used for daily collection: ——————————————————————————

Collection schedule:

	Mon.	Tues.	Wed.	Thurs.	Fri.	Sat.
Number of loads to the landfill:						

Number of commercial accounts: ————————————

Number of industrial accounts: ————————————

Number of institutional accounts: ————————————

Number of residential accounts: ————————————

Truck Information

License Plate #	Truck #	Cubic Yards	Empty Weight	# Gallons in Gas Tank	Truck Type*

Can you make your routing available for each truck each day? ———————————

Comments:
*Truck type:

FL = Front loader
RL = Rear loader
T = Trailer
RO/C = Roll-off compacted
RO/O = Roll-off open top
SL = Sideloader
D = Dump

Fig. 12.97   Hauler's questionnaire. (Courtesy of Sanders & Thomas, Engr's.)

common occurrence and can create significant errors for the program planners, since commercial wastes often comprise up to 30% of the waste stream.

More often than not, the problem of quantifying commercial wastes cannot be totally resolved. Some relief, however, does occur albeit unintentionally, by the fact that most resource recovery concepts are not capable of handling the unpredictably difficult types of commercial/industrial wastes (textiles, bulkies, chemical solids, and the like) and do not wittingly accept it. *Nonetheless, accounted for or not, someone, somewhere, and at sometime must dispose of it. The special problems with commercial and industrial waste requires knowledgeable approaches that, with few exceptions, have been woefully lacking and ignored in overall planning, especially in design and implementation of resource recovery (see Chapter 17).*

Table 12.24 Truck Parameters for Portable Scale Selection

Type	Load Capacity	Size
Transfer trailer	50 ton	70 × 10 ft (21 × 3 m)
Roll-off open top	50 ton	70 × 10 ft
Roll-off (compacted)	25 ton	30 × 10 ft (9 × 3 m)
Rear or front loader	25 ton	30 × 10 ft

Source: Courtesy of Sanders & Thomas, Engr's.

Commercial haulers are often licensed by municipalities, and this can be an additional source in developing a profile of the total collection area. The same briefing meeting and contact procedures described previously are also recommended for this sector.

Implementation: Portable Scales, Type and Quantity

There are several types of scales available for the weighing program. Scales can be rented but are not always easy to locate. When selecting the type of scale, the largest truck to be weighed must be ascertained. It is often a transfer tractor-trailer rig that can range up to 70 ft (21 m) in length, and its largest permissible load must be assumed. Table 12.24 shows truck parameters useful when selecting a scale. Portable scales are available in several types and sizes and from several manufacturers although renting locally is the norm.

1. *Axle Scales.* Consist of two units designed to weigh one axle at a time at each end and simultaneously. They are the least expensive and easiest to erect and place. They are, however, the most time consuming to use (Figure 12.98).
2. *Tandem Set.* Portable scales for tractor-trailers consisting of two sections 25 to 30 ft (7.5 to 9 m) in length and also suitable for front- and rear-loading compactors and roll-offs (Figure 12.99).

Scales Set-Up Requirements

Portable scales do not require a pit, but ramps for access are required. Rental of one scale-set is usually sufficient for most surveys and ancillary equipment may include a read-out device and printer. The scales can be direct-reading manual balance beam or automatic electronic with digital readout. If heavy traffic at the scale is anticipated, automatic electronic is the choice and an electrical source is required. If a standard 115 V outlet is not available, then a small gasoline-driven generator or a 12-volt battery source with suitable conversion/regulation must be obtained. Figure 12.100 is a recommended portable scale construction and placement.

Field scale installation requires construction and possibly earth-moving equipment for placing the scales and ramping up to them. Pitless portable scales may stand from 1 to 2 ft (30 to 60 cm) high and usually require ramps of earth. Rental construction equipment can include one or two items such as a backhoe, payloader, bulldozer, crane—whichever is most suitable—and a few timbers.

Fig. 12.98 Portable scales: Axle type.

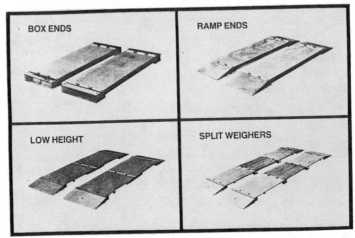

Fig. 12.99 Portable scales: Tandem set.

Estimate of Requirements for Personnel, Equipment, and Installation

Survey preplanning and scope determine the number and type of portable scales and weigh monitor personnel. If experienced judgment for this is not available from the survey team, be it a consultant or the municipality, reliable advice is available from the scale rental agency, which will require information on location, terrain, traffic, and truck types. This procedure can be an informal discussion in their yard with a likely possibility that they will provide scale set-up supervision and personnel. Also, they can assist with local authorization/permits if required. Temporary-hire personnel outside the survey organization—scale monitors, verification load checkers, and so on—usually is available through a local state employment agency and help-wanted ads.

Cost Estimates, Waste Quantification Survey

Equipment and personnel requirements are costed on the basis of survey duration—weeks, months, and so on, and include:

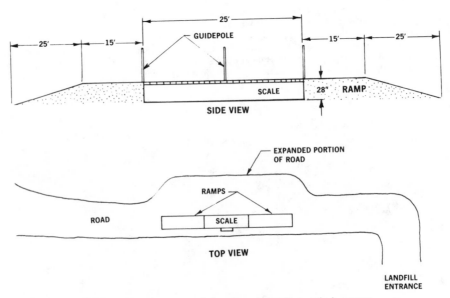

Fig. 12.100 Recommended scale construction and placement.

Equipment rental costs.

Wage rates for internal and temporary personnel.

Days/week.

Temporary office expense.

Travel and living expense of outside consultant.

Consultant's overhead expense.

For a municipality with a population of 25,000, typical per capita commercial activity and a little light industry, a waste quantification survey by consultants may range between $40,000 and $60,000 and include one or two consultant personnel and two or three outside local hires. If the municipality conducts its own survey, funding would be according to local cost allocation procedures.

12.12.4   Quantification Survey Work Tasks

Weighing Program

A.   With Permanent or Portable Scales on Site.   When preplanning is complete, the program scope may be more precisely defined. The required number of scales will have been determined along with a field personnel roster. If scales already exist at the desired locations, remaining field tasks would be confirmation of the survey schedule with the sites and commencing data collection activity, and to repeat, a strong reminder to field personnel that weigh sites must be monitored during *all* hours of operation.

All trucks tare weights must be logged (verified if at all possible) although they may have already been submitted with the hauler's questionnaire (Figure 12.97). Scale personnel should be thoroughly briefed on the questionnaire/log, and specific points of orientation should emphasize:

Load source verification.

Does all incoming material come from the survey area and if not, a best estimate of the percent.

Familiarity with the usual incoming collection vehicle types for log identification. Figure 12.101 illustrates the various types in common use.

Truck identification by fleet or registration number (or other symbol). Occasionally, numbers or other identifying symbols are obscured and it may be necessary to apply temporary, numbered adhesive stickers.

Thorough familiarity with scale operation and routing when temporary portable scales are required. This includes truck approach, weigh, and exit protocols as well as readout regimen, power source connections if available (a manual balance beam scale otherwise), and above all, safety precautions. Figure 12.102 is a typical scale data log.

B.   With Off-Site Scales or No Scales.   Quite often, alternative next best or even last resort tactics are required because ideal conditions do not exist, that is, no scales at a disposal site with constraints that preclude use of portable units (unavailable, permission, terrain, etc.). If an existing scale can be located reasonably near by and arrangements for its use are possible, including diverting the hauler truck traffic to it, more than one field observer may be required for that particular data source in an attempt to prevent by-passing the scale. If this is not possible, the log at the no-scale site (Figure 12.103), must be compared with the off-site scale log entrys (Figure 12.102) to determine the number of by-passed trucks.

C.   Truck By-Pass Contingency.   There are two important requirements for this procedure to be reasonably accurate:

1.   Average load bulk densities must be compiled from scale data for each truck type and applied to unweighed by-pass trucks. Table 12.25 shows average bulk density data for various truck types.

2.   The survey observer at the tip site must record estimates of load size (volume) during unloading (pushout blade stroke, truck size, etc.) (Figure 12.104).

Experienced observers are a distinct advantage but quite often, casual-type temporary hires are unavoidable and sufficient briefing, persistent surveillance of alertness, tardiness, truancy, and disciplining are major task responsibilities of the program manager at the beginning to eliminate chronic offenders.

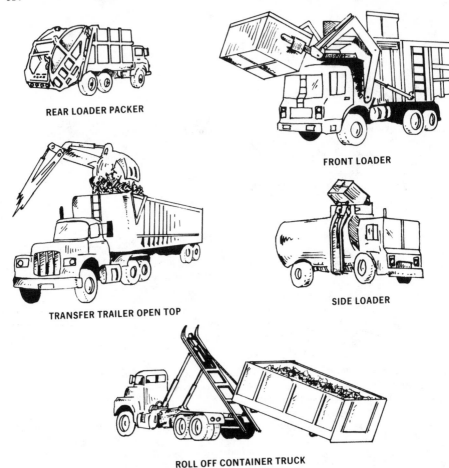

REAR LOADER PACKER

FRONT LOADER

TRANSFER TRAILER OPEN TOP

SIDE LOADER

ROLL OFF CONTAINER TRUCK

Fig. 12.101 Typical collection and transfer vehicles.

D.   Load Source Verification.   When a load of suspicious origin enters a disposal site or if the entire disposal site operation is suspect, a load verification procedure may be justified. "Suspicious origin" refers to unauthorized material "smuggled" in from outside the jurisdictional or study area and can consist of partial or entire loads.

Load source verification is a crude and somewhat indirect method for gathering evidence and is simply a matter of culling through a particular load(s) for any identifying material. This can be addresses on mail, packages, dunnage, or material specific to a certain source. Multiple collection stops are often involved and it is necessary to estimate the strategic places to look, that is, incremental collection points in a pile or in the laid-down ribbon. The procedure usually requires an extra observer or two, and it is especially helpful if the team has walkie talkies. Figure 12.105 shows suggested sample points. It is similar to waste characterization except for scope, purpose, and sampling.

12.12.5   Quantification Survey Summary Report

Waste Totals by Type and Season

All data and relevant operating procedure and conditions must be displayed for the following solid waste categories:

Residential.
Commercial.
Industrial.

For each category, the following information should be included:

Source, transport mode/truck type, and haul identification.

Truck traffic and loads via generation curves (or charts) for the period of the program, (Figure 12.106).

Anticipated seasonal fluctuation (Figure 12.107) due to weather, landscape/agriculture, holidays (festive, vacation, plant shutdowns, and so on), and trash clean-up periods.

Daily and weekly totals and averages by type and source are also required. Table 12.26 and 12.27 present typical examples of such data (a 4- to 6-week period is desirable).

At this time, cursory verification of salient factors is worthwhile if at all possible:

Who owns the delivery vehicle?

Affirmation of truck numbers (see Section 12.12.5).

Load verification.

Collection Vehicle Traffic Volume

A. Traffic Patterns and Condition of Roads. Data recorded during the survey is fundamental in constructing a profile of waste vehicle traffic, including public perceptions, that in turn is useful in reviewing adequacy of road structure and design, traffic systems, and so on, for new resource recovery facilities, transfer stations, or land disposal sites; that is, the adequacy of:

1. Existing road conditions including shoulders.
2. Traffic signals at nearby intersections.
3. Queuing or holding areas for waste vehicles during scale/tip rush hours.
4. Strike or weather interruption contingency planning.

Tables 12.26 and 12.27 show not only daily and weekly collection loads, but also indicate collection/delivery traffic patterns for the study period, and Figure 12.108 is a graphic representation of the vehicle traffic.

B. Effect on Recovery/Disposal Facilities Design and Operations. Collection vehicle traffic volume information is also necessary for:

1. Determination of need for transfer station(s).
2. Vehicle maintenance planning: collection/delivery and materials handling.
3. Planning resource recovery/transfer station maintenance and downtime.
4. Raw material storage and retrieval design and operations.
5. Process load and output scheduling.
6. Design of:
   a. Scale facilities.
   b. Waste delivery approach ramps and entries.
   c. Tipping patterns and protocol.
   d. Exits and by-pass of unacceptable loads (escape scale check), disabled vehicles, and so on.
   e. Residue disposal (25 to 40% by weight of rawfeed for mass burn up to 25% for prepared fuel, RDF, and so on).

12.12.6 Waste Composition Survey

Background and Purpose

Most raw waste surveys concentrate on quantities with insufficient attention to waste composition, especially physical size characteristics. For proposed land disposal, transfer station, or source separation/recycling projects, differences of opinion on this subject have hardly been noticeable, but for systems of energy/materials resource recovery, the debate continues.

The necessity for characterization has been debatable to the extent that many proponents of resource recovery appear to believe that as long as a basic core of residential material, that is, 30 gallon (113 dm^3) garbage cans, plastic bags, and so on amounting to at least 80% of the total, is reasonably certain, characterization is not necessary because sorting out undesirable, difficult

Date _____   Scale Record   Recorded by _____

_____ Owner

Truck Type & No.	Yd³	Gross Weight	Empty Weight	Net Weight	Area of Collection	% Study Area Waste	Net Weight	Type of Trash

RL   = Rear loader
FL   = Front loader
SL   = Side loader
D   = Dump
RO/OT = Roll-off open top
RO/C = Roll-off compacted

I   = Institutional
R   = Residential
B   = Brush
WG   = White goods
OBW = Oversize bulk waste
IND   = Industrial

Weather _____
Starting time _____

Fig. 12.102 Typical scale data log. (Courtesy of Sanders & Thomas, Engr's.)

_____ Date   Recorded by

Area of Collection	Owner	Truck No.	Yd³	Truck Type	Description of Trash

RL	= Rear loader	I	= Institutional
FL	= Front loader	R	= Residential
SL	= Side loader	B	= Brush
D	= Dump	WG	= White goods
RO/OT	= Roll-off open top	OBW	= Oversize bulk waste
RO/C	= Roll-off compacted	IND	= Industrial

Fig. 12.103 Landfill (without a scale) survey log. (Courtesy of Sanders & Thomas, Engr's.)

Table 12.25 Average Bulk Densities for Various Collection and Transfer Vehicles

Vehicle Type	Bulk Density Range
Compactor Collection Vehicle	400–600 lb/yd^3 (238–335 kg/m^3)
Open dump, pick up loose in truck	100–200 lb/yd^3 (60–115 kg/m^3)
Stationary compactor/trailer	400–600 lb/yd^3 (238–355 kg/m^3)
Stationary compactor/roll-off	400–600 lb/yd^3(238–355 kg/m^3)
Open trailer, tamped	300–350 lb/yd^3(177–207 kg/m^3)

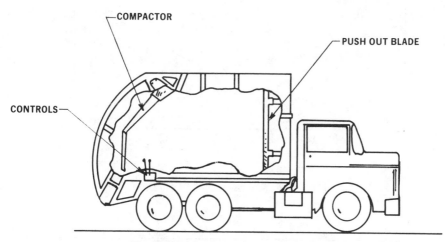

Fig. 12.104 Anatomy of a packer truck.

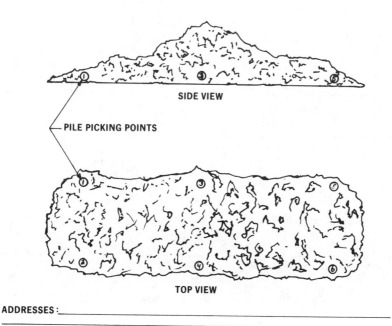

Fig. 12.105 Load source verification sample points.

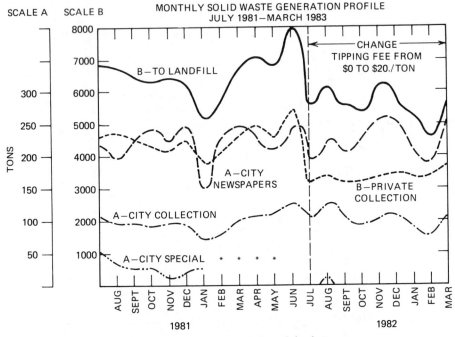

Fig. 12.106 Graphic total, 5-week loads.

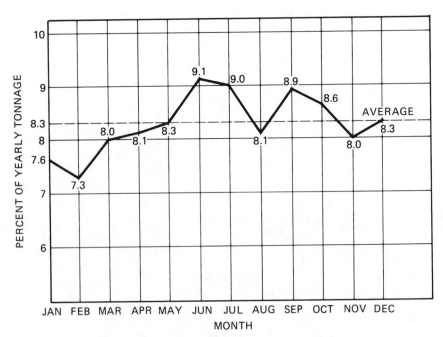

Fig. 12.107 Expected solid waste seasonal variations.

Table 12.26 Average Daily Waste Generation

	TPD$_6$[a]	TPD$_7$[b]
9/13–9/18	94.0	80.0
9/21–9/25	104.0	89.0
9/27–10/2	98.0	84.0
10/4–10/9	105.0	90.0
10/11–10/16	96.0	82.0
Average	99.0	85.0

[a] 6 = 6-Day/week.
[b] 7 = 7-Day/week.
Source: Courtesy of Sanders & Thomas, Engr's.

Table 12.27 Typical One-Week Waste Collection Totals

Week 1	Residential (lbs/tons)	Indust./Commercial (lbs/tons)	OBW (lbs/tons)	Brush (lbs/tons)	Total (lbs/tons)
9/13	107,785/54	75,206/38	34,230/17	12,070/6	229,291/115
9/14	87,184/43	89,812/45	54,502/27	11,660/6	243,158/121
9/15	79,020/40	66,811/33	9,780/5	9,390/5	165,001/92
9/16	80,240/40	69,462/35	7,460/4	13,900/7	171,062/83
9/17	92,225/46	62,580/31	3,130/1.5	7,080/3.5	165,015/82
9/18	120,208/60	31,880/16	690/.5	1,210/.5	153,988/77
Total	566,662/283 (50.2%)	395,751/198 (35.1%)	109,792/55 (9.7%)	53,310/28 (4.7%)	1,127,515/564

Source: Courtesy of Sanders & Thomas, Engr's.

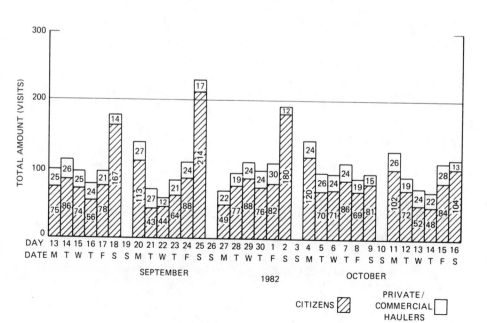

Fig. 12.108 Landfill vehicle traffic.

521

material can be improvised after plant start-up, that is, during collection, at a transfer station, the receiving scale, on the tipping floor, or by pit crane, and so on.

This is often the rationale of incineration advocates who usually believe they can accommodate rogue materials—that they can tolerate the aggravation to already irregular and unpredictable fuel beds, live with whatever fireside corrosion and deposits that may develop, and wrestle the unwieldy noncombustibles through ash-handling systems, and it appears that stack emissions is now in this picture.

These are penalties for ignoring the nature of the raw feed material in advance of system/plant design, planning for collection, and delivery logistics, and so on. Likewise, most prepared fuel advocates have ignored preproject evaluation of waste composition and find it necessary to reject and/or remove undesirable residential, commercial, and industrial materials to facilitate materials handling and processing (*Editors note:* An exception is Resources Recovery (Dade County) Inc., Miami, Florida. They are processing mixed residential, commercial, and industrial material as required by contract).

Proximate and ultimate chemical analyses of municipal solid wastes have been analyzed and chronicled very competently by well-known investigators.[1-3] There is also an ample backlog of generic type and quantification data (paper, glass, wood, and the like) as presented in proceedings of the ASME Solid Waste Processing Division Conferences. Too often, such sources are considered sufficient for feasibility studies due to aforementioned convenience and cost advantages in lieu of a waste composition study concurrent with a quantity survey (if there even is one) for a specific area.

In addition to the objectives described above, there are other singular advantages:

1. Obtaining specific composition data necessary for reasonable decisions in selecting a disposal concept, especially energy and materials recovery programs, their design and subsequent operation. Municipal solid waste (MSW) composition has a profound effect on refuse fuel inherent characteristics (heat values, moisture, ash), materials recovery, precombustion preparation (including mass burn tipping floor or pit mixing, ash-handling, RDF preparation), and provision for by-pass of unacceptable materials.

2. Anticipation of a possible mandate for presorting and removal of materials deemed contributory to toxic or deleterious airborne emissions such as Dioxin, lead, and acid gas.[4,5] Likewise, there is a growing concern for heavy metals leachate contamination of ground water (mainly lead and cadmium) in landfills for incinerator ash.[6,7]

3. An indication of the effect of source separation programs (bottle bills, citizen recycling, and so on) upon the profile of raw material to be received.[8]

Recognition of physical size characteristics as an important element in feasibility evaluation has finally surfaced in two excellent waste characterization reports. These are the Central Wayne County, Michigan,[9] and the Santa Clara County, California,[10] studies that are detailed in papers presented at recent ASME National Waste Processing Conferences.

The purpose of a waste composition survey is aptly summarized in these papers. Hollander, Stephenson, Eller, and Kieffer state in the Central Wayne County paper that:

> For any solid waste project, there must be a basis for economic analysis, design and subsequent operation to meet prescribed expectations and goals, be it a new program or a retrofit/optimization of an existing one. Any system designed to handle and process solid waste must have the flexibility to cope with the infinite variations in its component combinations.

12.12.7 The Sorting Program

Background

This section will present:

1. Details of a sorting method that produces the desired composition data for a specific study area.
2. Composition data from several areas of the United States.

Again, the necessity and validity of such data is seldom challenged when required by state or county regulation either in an environmental impact statement or project logistical planning. Field sampling is the most reliable method of obtaining a current profile of waste composition in an area, and many such exercises have been performed using a variety of procedures and laboratory analysis techniques. The procedures and their scope vary as much as the waste materials surveyed and data can be rendered inaccurate or irrelevant by faulty or incomplete methods by ignoring:

Yard wastes that can be high in moisture, causing inefficient incineration or ignition loss.

Slugs of low-melting-point metals that foul grates and windbox linkage.

The incidence of difficult items that "inadvertently" find their way into the collection, that is, white goods, bed springs, carpets, and so on.

Sorting/Sampling Procedure and Requirements

Ideally, this program should be conducted several (four or five) times during the year, but for the one-time "Final Look" the shorter time span over say 4 to 6 weeks will reveal significant information that, with proper response, will improve the probability of the project's success. It is recommended that an entire truckload should be sorted and sampled for each sampling run, with thorough documentation of its collection route.

Figure 12.109 is a suggested routing log and should include the following:

The truck identity: number, owner, driver.

The truck route and type of collection stop.

Time of pick-up and weather.

Truck weights: gross, tare, and net.

Type of waste.

The truck should be weighed before and after the collection run and each driver should be given a map of scale locations and the sorting site. Ideally, one truckload from each day of collection should be delivered to the sorting/sampling site during a typical week. Weigh tickets should be obtained by each driver for verification of the full load and tare weights, and in addition, routes from which sorting loads originate should be preselected and should also be:

Diverse by neighborhood character (socio-economic).

Diverse in typical waste types (residential, commercial, industrial).

Preferably municipal for maximum cooperation and control (city hall leverage, and so on).

Preferably once-a-week pick-up for better overall representative samples (twice weekly can skew the sample if say week-end yard/garage/trash clean-up is missed). If the schedule is twice-a-week, the second sample pick-up should be made.

Equipment, Manpower Requirements, and Sorting Facility Layout

The sorting facility should be a covered area, preferably enclosed, of at least 600 ft^2 (560 m^2) and the equipment should include:

1. *Equipment.*

A dial portable platform scale with a maximum capacity of 500 lb (227 kg) with ¼ lb (114 g) graduated increments.

Twenty heavy-duty, round 32-gallon (121 dm^3) plastic trash containers equipped with detachable platform casters for easy mobility (Rubbermaid Brute).

Long-sleeve coveralls for each sorting crew member.

Four to six snow shovels for use in handling the fines in the later sorting stages.

Two or three rakes and two to four pushbrooms.

Four to six small hand cultivators (or garden rakes).

Twelve to fifteen pairs of heavy-duty leather gloves.

One hundred 4-mil 32-gallon (121 dm^3) trash bag liners for constituent samples.

Fifty to sixty heavy corrugated cardboard boxes for sample storage.

Tape and identification tags for samples.

Several small hand-held magnets for sorting metals.

One 20 yd^3 (15.3 m^3) rental roll-off container for disposal after sampling.

2. *Manpower.* The crew size should be three men per ton of total material to be sampled. Each sorting crew member should wear heavy work boots.

3. *Sorting Facility Layout.* The portable scale should be placed as close as possible to the receptacle used for deposition of sorted material. Likewise, sample boxes should be placed adjacent to the scale for ease of sample recovery and packaging. Figure 12.110 shows a recommended facility layout with regard to equipment arrangement, load placement, and sampling area.

Time: _____

Collection company: _____   Truck no: _____

Truck gross weight: _____   Driver: _____

Truck net weight: _____   Full tank capacity: _____
Gallons level: ¼   ½   ¾   full

Collection in: _____ City township

_____ District

_____ Development/center

Residential area (indicate first
and last street collected) _____

From _____ to _____ Street/Avenue

_____   _____

_____   _____

_____   _____

_____   _____

_____   _____

_____   _____

_____   _____

_____   _____

_____   _____

<u>Comments</u>

Weather:

Bulky waste: (Anything unusual)

Collection frequency per week: one, two, three?
(circle one)

Fig. 12.109 Collection truck routing log. (Courtesy of Sanders & Thomas, Engr's.)

4. *Sorting Procedure.* After weighing, the collection truck should deposit its load in the center of the sorting area. The sorting crew members will then circle the load while pulling one or two of the caster-mounted plastic trash drums (Figure 12.111).

Initially, the sorters are assigned constituents thought to be in greatest abundance (a different one for each sorter). The assignments can be changed quickly if it becomes apparent that the predominant constituents are different from the original assignments. When a drum is filled, it is wheeled to a weigh-ready storage area adjacent to the scale, and the sorter returns to the sorting pile with an empty drum to repeat the cycle.

The crew chief and his helper weigh the filled drums and record the net weights on a weighed-sample data sheet (drum tare weights have been recorded previously) (Figure 12.112). The scale

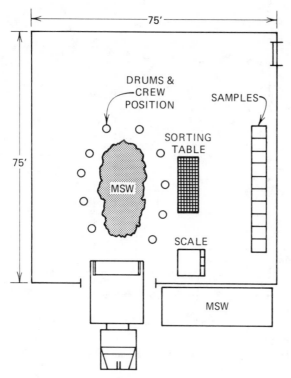

Fig. 12.110 Sorting facility layout. (Source, Reference 9.)

Fig. 12.111 Portable sampling drums. (Photo source, Reference 9.)

RECORDER _____

DATE _____ TRUCK IDENT. _____ WASTE SOURCE _____

TIME _____ TRUCK GROSS WT. _____ TRUCK NET WT. _____ PAGE ____ OF ____

CATEGORIES	PER-CENT 100	TOTAL WT	BARREL NO. (NET WEIGHT)												
1. NEWSPRINT															
2. OTHER PAPER															
3. CORRUGATED															
4. TEXTILES/GARMENTS															
5. PLASTIC (FILM)															
6. PLASTIC (RIGID)															
7. FOOD WASTE															
8. WOOD LUMBER															
9. YARD WASTE															
10. SWEEPINGS															
11. FERROUS															
13. NON FERROUS															
14. GLASS															
15. BRICK															
16. OBW															
REMARKS															

RAW REFUSE CHARACTERIZATION SAMPLING

Fig. 12.112 Raw refuse characterization sampling log. (Source, Reference 9.)

Fig. 12.113 Oversize bulky waste.

"chief" selects representative pieces from each drum for the laboratory constituent sample bag. After weighing and sampling, each drum is emptied and returned to active storage.

5. *Oversized Materials.* Oversized residential, commercial, or industrial items (Figure 12.113) are removed for separate classification, data logging, and photographing (mattresses, springs, rugs, furniture, bales of corrugated, paper and textiles, dunnage/pallets skids, tires).

6. *Metals.* Sorting of ferrous and nonferrous metals can be accomplished initially by random collection. During the sorting period, sorters can empty the drums of metallics and pass a hand-held magnet over the spread-out contents. Further identification of nonferrous metals if required can be accomplished by visual or more sophisticated minerals beneficiation and/or laboratory techniques.

7. *Miscellaneous.* All containers—jugs, bottles, cans, drums—that are unopened and appear suspicious are removed unopened as potential hazards, logged-in, and handled separately for proper disposal.

Assorted small particles, 1 to 4 in. (25 to 100 mm) "residues" that remain as the sorting pile diminishes can increasingly accumulate and slow the sorting process. To mitigate this phenomenon, snow shovels and a sorting table with 1-in. (25 mm) square holes in the surface are employed to screen out the fines. Figure 12.110 depicts this 4-ft (120 cm)-high table and its location.

8. *Sorted Constituents.* Although municipal solid wastes contain a multitude of materials, a typical and common constituent presort selection listing expedites the sort/logging procedure. A realistic generic listing follows:

Combustible	Noncombustible
Newsprint	Ferrous
Other paper	Nonferrous
Corrugated cardboard	Stone/masonry
Textiles	Glass
Film plastic	Grit and sand
Rigid plastic	
Food waste	
Wood waste	
Yard waste	
Sweepings	

The tabulation of probable items found in each category is listed in Figure 12.114.

12.12.8 Laboratory Analysis

Background and Methodology

During the sorting program, samples can be taken for analyses of the 10 combustible constituents, and the scale operator is responsible for picking small representative samples from the barrels of sorted wastes. During daily sorting activities, he scans the sorted waste for samples representative of a particular constituent. Until approximately 3 ft³ (85 dm³) are accumulated in two 4-mil plastic bags. Combustible samples should be obtained on each day of the sorting program and tagged with weight, type, source, and date data. Each double-bagged sample is then placed in its own 4 to 5 ft³ (113 to 142 dm³) heavy-duty corrugated box and sealed with plastic tape on all edges. The boxes are also numbered and the contents identified, ready for transport to a laboratory.

The combustible constituents are tested for the following:

1. Moisture.
2. Proximate and ultimate analysis for:
   a. Moisture and ash free.
   b. Dry basis.
   c. As received, or determined lab value.
   d. As received, adjusted for moisture.
3. Lead.
4. Chlorine and chlorides.
5. Heat value.

Oversize items are not included in such laboratory analyses for obvious reasons, that is, special preanalysis processing requirements in addition to the fact that most resource recovery concepts in

Analytical Basis

Analysis and evaluation can be on a composite basis.

1. *Newsprint.* Newspaper—black and white, colored ads, handouts, store handouts, comics, shredded newsprint.
2. *Other paper.* Magazine, filler, notebook covers (paper), wrappings, bags, computer cards.
3. *Corrugated cardboard.* All cardboard packaging materials.
4. *Textiles and garments.* Leather, belts, hats, shoes, stockings, clothing, underclothing, shirts, undershirts, shorts, panties, pants, bras, slips, foundation garments, gloves, jackets, sweaters, coats; long and short rubber shoes, rubber hoses, rubber sheets. Materials may be fur, leather, cloth, or plastic. (Do not include plastic sheet.)
5. *Plastic* (film). Plastic wrap, trash liners, kitchen savers, food savers, cigarette pack (covers), and plastic hoses.
6. *Plastic* (rigid). Molded toys, toothbrushes, kitchen helpers, knives, forks, spoons, utensils, scrubbers, containers (metal lids excluded). Cocktail stirrers, plastic pipe, imitation metal.
7. *Food wastes.* All commissary food waste, and so on.
8. *Wood items.* Processed, that is, lumberyard stock, furniture, toys, broomsticks, fencing boxes, wooden bowls, utensils.
9. *Yard waste.* Grass, bedding plants, tree and bush trimmings, bundled and tied brush.
10. *Sweepings* (misc.). Undefinables—leftovers after categories have been sorted out. Under a 2-in. top size.
11. *Ferrous metals.* To be identified by use of a hand magnet.
12. *Nonferrous metals.* (Excluding aluminum, brass, copper, lead, and the like.)
13. *Aluminum.* Pots, pans, utensils, tubing, cans (beer, soda), raw stock material.
14. *Glass.* Do not break; if unbroken: bottles, jars, dishes, window glass, panes, eye glasses, glass insulation, marbles. Weigh and photograph.
15. *Stone/Masonry.* Cement, plaster, rock wall, ceramic. Weigh and photograph and set aside for possible landfill material.
16. *OBW and oversize bulky waste.* Unique items, tar, asphalt, bicycles, lawnmowers, loose yard brush, and timber.

Fig. 12.114 Probable items in each constituent category. (*Source:* Ref. 9.)

need of raw material analysis make every effort to avoid such raw material. Chapter 17 discusses oversized and bulky wastes in more detail with the growing awareness that it can no longer be ignored and neglected.

Table 12.28 shows a basis upon which typical composite analyses are determined. Laboratory personnel create a composite sample for each constituent the method for which follows ASTM E38.01 Protocol (ASTM E-38 Committee on Resource Recovery, 01 Energy Subcommittee), and as presented by Tables 12.29 and 12.30. This data can be used for design of a waste-to-energy facility.

Comparison, Proximate and Ultimate Analysis Data

Table 12.31 compares data from a large eastern military base with equivalent data from the surrounding county. It reveals that moisture and ash content is higher for the municipal/county waste than for the military base that shows lower volatile content. This appears to corroborate the contention that commercial, industrial, or military wastes (a combination of both) are higher in heat value than municipal. A reason for this difference is higher percentages of plastics, corrugated, dunnage, and so on.

Values for Lead

Values for lead were obtained from constituents believed most likely to contain it as follows:

Corrugated.	Rigid plastic.
Newsprint.	Textiles.
Other paper.	Yard wastes.
Film plastic.	Sweepings.

Table 12.28   Typical Waste Analysis (Free of Oversize)

Constituent	Total Weights (with moisture added) (lb)	Percent of Total Weight (%)
Newsprint	1,961	6.3
Other paper	11,108	35.6
Corrugated cardboard	5,387	17.3
Textiles	1,243	4.0
Plastic film	1,277	4.1
Plastic rigid	1,004	3.2
Food waste	1,218	3.9
Wood waste	2,009	6.4
Yard waste	1,093	3.5
Sweepings	848	2.7
Ferrous	1,302	—
Aluminum	324	—
Nonferrous	493	13.0
Glass	1,885	—
Brick	52	—
Total	31,206	100

Composite Proximate Analysis		Composite Ultimate Analysis	
Moisture	24.6%	Moisture	24.6%
Volatile matter	52.1%	Carbon	29.9%
Fixed carbon	6.0%	Hydrogen	3.9%
Ash	17.3%	Nitrogen	0.34%
HHV Per Pound	5455 BTU	Oxygen	23.6%
		Sulfur	0.12%
		Chlorine	0.14%
		Ash	17.3%

Analysis showed that lead in the military base waste was about 55 ppm, which is lower than a figure of about 350 ppm in the sorted municipal waste.[11]

Editors'/Authors' acknowledgment: Much of the information and data in this chapter was made available by Sanders & Thomas, Engineers, Pottstown, Pennsylvania and Reference 9.

REFERENCES

1. E. R. Kaiser, Chemical analysis of refuse composition, Proceedings of the Winter Annual Meeting, ASME, 1965.

2. H. I. Hollander and J. K. Kieffer, Developing analytical procedures for reproducible determinations of thermo-chemical characteristics of RDF—an ASTM program, ASME 1978 Winter Annual Meeting.

3. W. R. Niessen and S. H. Chansky, The nature of refuse, Proceedings, ASME National Incinerator Conference, Cincinnati, Ohio, 1970.

4. California Air Resources Board, Air Pollution Control at Resource Recovery Facilities, 1983 report.

5. S. C. Remis, The environmental aspects of resource recovery facilities—a position of the Department of the Public Advocate," New Jersey Division of Public Interest Advocacy, May, 1984.

6. P. Giordano, A. Behel, E. Lawrence, J. Solleau, and B. Bradford, Mobility in soil and plant availability of metals derived from incinerated municipal refuse, TVA Division Agricultural Development. *Sci. Technol.* Vol 17, No. 4, pp 193–197 (1983).

Table 12.29 ASTM Method References

1.	Heating value	ASTM E711-81: Standard Test Method for Gross Calorific Value of Refuse-Derived Fuel by the Bomb Calorimeter
2.	Forms of chlorine	ASTM E 776-81: Standard Test Method for Forms of Chlorine in Refuse-Derived Fuel
3.	Sample preparation	ASTM D 829-81: Standard Method of Preparing RDF-3 Laboratory Samples for Analysis
4.	Ash	ASTM D 830-81: Standard Method of Test for Ash in the Analysis Sample of Refuse-Derived Fuel
5.	Total moisture	Proposed EDS-6: Standard Method of Determining Total Moisture (Single Stage) in RDF-3 Samples
6.	Preparation for lead	Proposed EDS-21: Standard Test Method for Preparing Refuse-Derived Fuel Samples for Analysis of Metals
7.	Volatile matter	Proposed EDS-24: Standard Test Method for Volatile Matter in the Analysis Sample of Refuse-Derived Fuel-3
8.	Analysis for lead	Proposed EDS-25: Standard Atomic Absorption Spectrophotometric Method for Metals
9.	Carbon, hydrogen, and nitrogen	Proposed EDS-17: Alternate Procedure for the Determination of Carbon, Hydrogen and Nitrogen in the Analysis Sample of Refuse-Derived Fuel (A Perkin Elmer 240 B Elemental Analyzer was used)
10.	Sulfur	LECO Sulfur Analyzer
11.	Fixed carbon	Calculation 100% (% moisture + % ash + % volatile matter)
12.	Oxygen	Calculation 100% (% Moisture + % ash + % H + % C + % N + % Cl)

7. C. L. Allison, Environmental quality considerations, residuals disposal, Resource Recovery Seminar, University of Wisconsin, Department of Engineering Extension, Madison, Wisconsin, February, 1984.

8. A. R. Nollet, ASME 11th National Solid Waste Processing Conference, Orlando, Florida, June 1984. Discussion of the paper, G. R. Kenny, and E. J. Sommer, Jr., A simplified process for metal and noncombustible separation from MSW prior to waste to energy conversion, in postconference written discussion volume.

9. H. I. Hollander, J. W. Stephenson, V. L. Eller, and J. K. Kieffer, A comprehensive municipal refuse characterization program, Proceedings, 9th ASME National Waste Processing Conference, Washington, D. C., 1980.

10. J. C. Glaub, G. M. Savage, and T. M. Henderson, Comprehensive waste characterization on a quarterly basis, 11th ASME National Waste Processing Conference, Orlando, Florida, June, 1984.

11. A. C. Salas, D. F. Lewis, and D. A. Oberacker, Waste-to-energy facilities, a source of lead contamination, in Proceedings of EPA Seventh Annual Research Symposium on Municipal Solid Waste: Resource Recovery, Philadelphia, Pennsylvania, March, 1981.

12.13 HEALTH AND SAFETY: HEALTH ASPECTS

E. Joseph Duckett

Background

As reflected in the preceding sections of this chapter, recovery of materials and fuels from solid wastes requires the use of several types of processing equipment. Like all industrial equipment, the processes employed in resource recovery each can pose health or safety hazards to workers unless the equipment is properly designed, installed, and operated. For a resource recovery facility, an assessment of these potential health or safety hazards is complicated by the wide range of substances than can be presented in municipal solid wastes and the similarly wide range of processing options available.

Many of the safety and health concerns in a resource recovery facility are no different than for

Table 12.30 Proximate and Ultimate Laboratory Data Log (Courtesy of Sanders & Thomas, Engr's)

As-Received Basis—by Weight (adjusted for moisture)
Proximate Analysis

Constituent	% Moisture	% Volatile Matter	% Fixed Carbon	% Ash	Higher Heating Value
Newsprint	23.0	68.5	6.93	1.54	6565
Corrugated cardboard	17.0	71.4	9.58	1.95	6594
Other paper	30.6	58.8	6.80	3.80	5430
Textiles	28.0	65.0	5.86	1.09	6948
Film plastic	26.2	66.9	2.55	4.32	12285
Rigid plastic	12.9	85.7	0.03	1.36	15744
Wood wastes	22.3	63.3	12.5	1.97	6699
Food wastes	72.4	22.4	3.33	1.81	2613
Yard wastes	47.3	26.6	2.77	23.3	3031
Sweepings	28.7	22.4	3.84	45.1	2046

Ultimate Analysis[a]

Constituent	% Moisture	% Carbon	% Hydrogen	% Oxygen	% Nitrogen	% Sulfur	% Chlorine	% Ash
Newsprint	23.0	37.0	3.53/6.11	34.8/55.2	<0.10	0.16	—	1.54
Corrugated cardboard	17.1	39.0	4.98/6.90	36.7/51.9	<0.10	0.25	—	1.95
Other paper	30.6	30.7	4.26/7.69	30.4/57.5	<0.10	0.12	0.12	3.80
Textiles	28.0	39.7	4.17/7.30	23.0/47.8	3.92	<0.10	0.09	1.09
Film plastic	26.2	54.2	8.86/11.8	5.42/28.7	0.30	0.04	0.68	4.32
Rigid plastic	12.9	71.2	11.6/13.0	1.11/12.6	0.10	<0.10	1.72	1.36
Wood wastes	22.3	39.2	4.73/7.23	31.1/50.9	0.67	<0.10	—	1.97
Food wastes	72.4	14.0	2.00/10.1	8.89/73.2	0.63	<0.10	0.26	1.81
Yard wastes	47.3	15.7	1.83/7.12	11.3/53.3	0.59	<0.10	—	23.3
Sweepings	28.7	13.9	1.69/4.90	9.27/34.8	0.88	0.13	0.38	45.1

[a]These tests will be performed for only the following constituents: Other paper, film plastic, rigid plastic, food wastes, textiles, sweepings.

Table 12.31 Comparison of Proximate and Ultimate Analysis Data (as-received basis) (Courtesy of Sanders & Thomas, Engr's)

Type of Analysis	Military Base	Surrounding County[a]
Proximate		
Moisture (%)	24.6	29.5
Volatile (%)	52.1	40.6
Fixed carbon (%)	6.0	7.9
Ash (%)	17.3	22.0
Higher heating value (HHV) (BTU/lb)	5455 BTU/lb	4650
Ultimate		
Moisture (%)	24.6	29.5
Carbon (%)	29.9	25.58
Hydrogen (%)	3.9	3.59
Oxygen (%)	23.6	18.7
Nitrogen (%)	.34	0.24
Sulfur (%)	.12	0.17
Chlorine (%)	.14	0.13
Ash (%)	17.3	22.0

[a]Data obtained from ASTM Standard E-38.01 Resource Recovery Committee, Energy Subcommittee.

many other industries. For example, the needs for—and methods of—protecting against falls, tripping, electrical shocks, and injuries from contact with the moving parts of machinery are common among most industries. There are, however, at least three safety and health concerns that are especially applicable to resource recovery facilities. These are concerns for explosion protection, dust control, and noise control. Over the past 10 years, much has been learned about the nature and control of the potential hazards posed by explosions, dust exposures, and noise at solid waste processing plants. The purpose of this section is to review what has been learned and to present some approaches that have been employed to protect the health and safety of resource recovery workers.

12.13.1 Explosion Protection

The presence of flammable and/or explosive materials in solid wastes raises concern for explosion protection, particularly during size reduction (that is, shredding) of the wastes. Several major resource recovery facilities have experienced explosions (not always in shredders), and there is understandable interest in explosion protection within the resource recovery industry. A 1983 survey of 123 resource recovery plants in the United States concluded that, among operating plants, almost 20% of the nonscheduled outages were due to explosions or fires.[1]

Examples of flammable and/or explosive materials that can be present in solid wastes include gasoline, solvents, propane, and even discarded military ordnance. One processing facility reported finding a World War II antitank land mine in the waste stream en route to a shredder; the mine was manually removed before shredding.[2] Other plants have reported finding automobile fuel tanks, propane bottles, and cans of industrial solvents.

Within a refuse processing plant, there are several areas where explosions have occurred. These include shredders, dust hoods, pneumatic ducts, cyclones, and storage bins. The common feature among these areas is that they are all places where flammable/explosive materials can accumulate and where there is an opportunity for ignition. It is important to note that fires and explosions have occurred in refuse-processing plants that were not engaged in resource recovery (for example, plants for shredding refuse prior to landfill disposal) as well as in plants that process conventional fuels (for example, coal dust explosions in ball mills) or conventional forms of scrap materials.[3-7]

Definition

An explosion can be defined as a chemical reaction or change of state occurring rapidly and resulting in the release of high temperature and a large quantity of gas. In simplest terms, an explosion is an extremely rapidly burning fire that occurs in an enclosed space and causes a rapid rise in pressure within that space. There are two specific types of explosions of interest to refuse processing: deflagrations and detonations. In a deflagration, the flame front propagates at a subsonic velocity and the pressure rise within the enclosure (for example, shredder) is relatively slow (taking on the order of 0.1 to 1.0 sec) compared with the virtually instantaneous pressure rise during a detonation. Supersonic propagation of the flame front is characteristic of detonations, and, consequently, these explosions are very difficult to control. Detonations in refuse shredders are probably due to discarded dynamite or military ordinance, and it is fortunate that relatively few shredder explosions have been of this type.[8]

History

Within the resource recovery industry, most of the interest in explosion protection has focused on shredders. This is because shredders are often the first processing step and because shredders provide the conditions (enclosed space and metal-to-metal sparks) that can lead to accumulation and ignition of explosible gases or vapors.

A 1976 study of shredder explosions in refuse-processing plants was conducted by the Factory Mutual Research Corporation (FMRC).[8] The FMRC study documented over 100 explosions that had taken place at refuse-processing facilities (primarily shred-landfill operations) over a 3-year period. The study excluded such incidents as "pops" due to aerosol cans and reported that three of the 100 explosions had resulted in injuries (nonfatal) and five had caused more than $25,000 in damage. The FMRC determined that most shredder explosions are due to flammable vapors and gases rather than dusts and that most of the exlosions were deflagrations rather than detonations.

Since 1976, there have been several highly publicized explosions, not all of them traceable to shredders. From the viewpoint of human loss, the worst occurred in December, 1984, in the raw refuse receiving area at the Akron, Ohio, waste-to-energy plant, fatally injuring three workers. Purportedly, the shredders were not involved. Another occurred in November, 1977, at the East Bridgewater, Massachusetts, resource recovery plant. This explosion killed one worker.[9] The exact causes of the explosion have never been documented, but it took place in a cyclone designed to de-entrain pulverized refuse-derived fuel (RDF) from a pneumatic conveyor. Oddly, the plant was not in operation at the time of the explosion.

A December, 1978, explosion at the Milwaukee resource recovery plant injured one worker seriously (he was hospitalized for 4 days and then released) and reportedly led to $1 million in repairs.[10] Like the East Bridgewater incident, this episode failed to fit the expected pattern of shredder explosions. Much of the accumulation of flammable vapors (gasoline, in this case) was attributed to the inlet hood and dust-control shroud and to the noise control enclosure surrounding the shredder, rather than to the shredder housing itself. Also, the source of ignition was external to the shredder; it was a radiant heater above the infeed conveyor.

In 1982, explosions occurred at the Albany, New York, and Rochester, New York, resource recovery plants. The most notable aspects of both of these 1982 explosions were that, in both cases, there were no serious injuries and damage to equipment was minimal.[11] Both plants had been equipped with explosion-venting systems and with other forms of explosion protection.

Other explosions that can clearly be traced to shredders have been reported in Brevard County, Florida; New Castle County, Delaware; Lane County, Oregon; Chemung County, New York; Duluth, Minnesota; Anchorage, Alaska, and Toronto, Ontario. The Toronto plant experienced three shredder explosions in its first year of operation.[12] It is noteworthy that the Toronto facility receives as much as 90% of its wastes from industrial sources. In the Toronto plant, the shredder is housed in a structure separated from the rest of the plant and there are no workers within the shredder room. There have been no injuries reported from the Toronto shredder explosions. The Toronto plant has undergone several modifications to prevent accumulations of flammable vapors in the shredder and to detect materials in the waste stream.

Prevention and Protection

To prevent and protect against shredder explosions, a variety of approaches have been suggested and implemented. In 1983, many of these approaches were published in "Guidelines for Shredder Explosion Protection" developed by ASTM Committee E-38 on Resource Recovery.[13] Other sets of guidelines are available from individual shredder manufacturers.

Among the approaches to explosion protection are:

Employ management controls to prohibit acceptance of high risk wastes (e.g., gasoline tanks, industrial solvents) in the processing plant.

Visually screen incoming wastes on the tipping floor or along a conveyor, and remove potentially explosive materials (gasoline cans, propane tanks, etc.) from the waste stream before processing.

Use automatic gas detection sensors to monitor concentrations of flammable vapors near the shredder and to signal when a dangerous level [e.g., the lower explosibility level (LEL) for the flammable gas] is being approached.

Fig. 12.115 Shredder vented through the roof. (Courtesy Aenco, Inc and 'Energy Answers,' Albany, N.Y.)

Locate shredders either outdoors, in a structure separated from the rest of the plant, or in a location suitable for venting to the outdoors.

Provide an explosion venting system using blow-off panels (with retention devices) or vent doors that would open in response to a pressure rise within a shredder (or other piece of equipment), thus directing the pressurized gases outside and relieving the pressure. Figure 12.115 shows a typical shredder explosion vent extending upward through the roof.

Use suppression agents such as halogenated hydrocarbons which, when discharged in response to a rise in pressure within a shredder (or other enclosed space), chemically extinguish the incipient explosion.

Use either a continuous or a gas detector-activated water mist to dissipate the heat generated by combustion of explosive vapors and to minimize flame propagation.

Provide radial plates at the end of the shredder rotor shaft to prevent accumulation of flammable gases near the ends of the shaft.[12]

Maintain an inert atmosphere within the shredder (or other enclosed space) to reduce the oxygen content to a level unable to support combustion.

Continuously purge explosible vapors by maintaining a large air flow through a shredder or other enclosure.

Mechanically separate the wastes (e.g., in a trommel or air classifier) as a primary step to facilitate identification and removal of potentially dangerous items.

Establish the area surrounding the shredder as "off limits" to personnel (both workers and visitors) during operation of the shredder.

For employees who might be exposed to shock waves or flying debris in the event of an explosion, provide protections such as blast mats, heavy-duty revetments, and high-strength windows.

Locate the plant control room away from areas where explosions might occur.

Install industrial fire protection equipment, including automatic sprinklers, to control fires generated following an explosion.

For a designer or operator of a solid waste processing facility, the choice among the listed approaches to explosion protection is complicated by issues of reliability, compatibility, cost, and practicality.

Protection Reliability

At present, there are no fail-safe approaches to shredder explosion protection. For example, neither waste management controls nor visual screening of the waste stream will be 100% effective in keeping explosible materials out of the plant. Similarly, suppression systems are not 100% effective; explosions have occurred in shredders equipped with suppression devices. Although automated systems for detecting flammable vapors are commercially available, their reliability and usefulness in the harsh environment of a shredding facility has not yet been documented.

One major reliability issue that continues to confound attempts to develop fail-safe explosion protection systems is the difficulty of designing explosion vents for shredders. Standard explosion venting guidelines such as those published by the National Fire Protection Association (NFPA) are usually based on tests with simple structures such as spherical or cylindrical pressure vessels and room-shaped explosion bunkers.[14] The NFPA standards are intended for use in a variety of applications and are therefore based on a general set of test conditions. Shredders represent a more complicated and severe explosion environment because of the effects of rotor windage, internal obstructions (shaft, hammers, breaker plates, trash throughput, etc.), and peripheral equipment such as inlet hoods and discharge gratings and conveyors. In combustion explosions, these effects can escalate the rate of pressure rise and reduce vented flow rates, thereby decreasing venting effectiveness.

Under the sponsorship of the U.S. Environmental Protection Agency (EPA), a series of shredder explosion tests were conducted in 1981 to determine whether the NFPA venting standards can be effectively applied to shredders and, if not, what new standards should apply.[15] The tests were conducted by Factory Mutual Research Corporation on a full-scale mock-up of a typical horizontal hammermill shredder. A near stoichiometric propane-air mixture was used to represent a worst-case condition for deflagration explosions involving flammable gases and vapors.

The results of the explosion tests suggested that:

1. The existing NFPA venting guidelines do not adequately address the venting requirements of a refuse shredder.
2. Explosive overpressures as high as 15 psi (105 kPa) may occur even when vent ratios (i.e., the ratio of vent cross-sectional area to shredder enclosed volume) are larger than those required by NFPA are used.

The results of the EPA-Factory Mutual Tests were discouraging because they suggested that venting alone could not provide full protection against a worst-case explosion (deflagration) in a shredder. The tests did, however, result in the publication of a set of venting guidelines that are more closely directed to the needs of shredder designers and operators than the existing NFPA standards.[15] Under these guidelines, the vent area, A_v, required to maintain explosive overpressures within a shredder damage threshold, P_m, is calculated by the following equation:

$$A_v = 0.13 V^{2/3} P_m^{-0.435} (5 + 0.034 v_H) \tag{1}$$

Where V is the internal shredder volume, v_H is the hammer velocity in feet per second, and P_m is in units of psig. The above equation has not yet been completely verified nor has it been incorporated into a nationally approved standard.

Compatibility of Protective Measures

The compatibility issue arises in at least two cases. First, in the case of visual screening, the effectiveness of the inspection may be improved by stationing workers along the shredder infeed conveyor, where there is usually a lower burden depth than on the tipping floor itself. This approach conflicts, however, with the idea of keeping workers as far away from the shredder as possible. The workers who were injured in the above referenced explosion at the Milwaukee plant had been stationed along the infeed conveyor immediately preceding the shredder. They were efficient (but not 100% effective) at spotting and removing hazardous or oversize items but they were also vulnerable to the blast from a shredder explosion.

One potential remedy for this dilemma is the use of remote TV cameras to view the wastes on the infeed conveyor. The conveyor could then be stopped when a dangerous item was observed. The effectiveness of this approach for identifying explosive materials has not been quantitatively evaluated but, for reasons of worker protection alone, the use of remote visual monitors has been adopted at several refuse-processing plants, often in combination with direct visual screening in the tipping area.

A second compatibility issue arises when explosion venting and suppression systems are to be installed together. Unless the venting system is designed to be activated at a higher pressure than the suppression system, it is possible for the vents to relieve overpressures just enough to prevent the triggering of the suppression system or to allow the suppression agent to be vented before the flame is extinguished.

Cost of Protective Measures

The provision of an explosion and fire protection system can be a major cost for a waste-processing plant. Both initial costs and operating costs are important. For venting and isolating the shredder, initial costs would be of most concern. For suppression systems, the costs of maintenance and recharging are most important.

Initial costs for the installation of vents for a 1000 hp shredder are in the range of $200,000 to $300,000; costs for installation of a suppression system are in the range of $40,000 to $60,000. The cost of a typical recharge for the suppression system would be approximately $6000.

In addition to the initial costs of explosion protection, it is important to recognize and provide for the costs of inspecting and maintaining the protective systems. Suppressive systems, gas-detection sensors, and even explosion vents can be rendered inoperative unless they are properly maintained. In at least two explosion incidents, it has been reported that explosion vent doors did not deploy, apparently because they had become obstructed.[12] Similar problems can result if pressure transducers become damaged or gas sensors become fouled. It is estimated that the annual costs of maintaining a venting/suppression system for a 1000-hp shredder are in the range of $1500 to $2500.

Protection Trade-offs

The practicality issue manifests itself in several examples. If, for some reason, a solid waste-processing plant must accept industrial wastes, the likelihood of keeping industrial solvents out of the plant diminishes. The use of a continuous water spray has been reported to be beneficial for auto shredder and shred-landfill operations, but this approach may be impractical where it is important to keep the wastes dry—as in plants producing RDF.*

The use of processes other than shredding as primary processing steps may be impractical for resource recovery facilities where the subsequent process flow, or the recovered product specifications, require size reduction of the wastes. Inerting may be impractical because of the difficulty of providing an airtight seal around the areas where explosions might occur. Finally, the use of an air-purging system may require air flows too large to be considered practical in a refuse-processing facility.

Conclusion

To conclude this section on explosion protection, it appears that there is no single approach that can be relied upon for full protection. Many refuse-shredding facilities now employ waste surveillance, explosion venting, suppression systems, and shredder isolation to reduce the risks of injuries and damage from explosions. Individually, none of these approaches provides absolute protection, but in combination, however, these and other approaches can provide acceptable explosion protection for solid waste-processing plants.

12.13.2   Dusts

Background

Dusts are generated during the processing of refuse and these dusts may contain chemical substances and microorganisms of health significance. The concentration, particle shape, size distribution, chemical composition, and microbiological characteristics of these dusts have been investigated at pilot plants and at full-scale resource recovery facilities.

Definitions and Exposure Standards

The U.S. Occupational Safety and Health Administration (OSHA) has established standards for occupational exposure to dusts. These standards limit the 8-hr time-weighted exposure to inert or nuisance dusts to 0.0066 g/ft^3 (15 mg/m^3) for airborne particles of all sizes and to 0.0022 g/ft^3 (5 mg/m^3) for dusts of "respirable" size (generally, dusts less than 5 μm in diameter).[16] Several years ago, total dust concentrations exceeding the OSHA 8-hr exposure standard had been reported for resource recovery pilot plants and for a small recovery plant not equipped with dust controls.[17-19] Concentrations of respirable dusts in these plants did not exceed OSHA standards except in areas where workers would not be exposed for a full 8-hr day (e.g., areas adjacent to a shredder). More

---

Editor's note: This has not proved to be of much concern vis-a-vis actual RDF moisture increase. Refer to Sections 12.4 and 12.14.

recent sampling programs conducted in full-scale plants equipped with dust controls have reported both total and respirable dust concentrations well within OSHA standards.[20,21]

Most of the dusts in resource recovery plants are fibrous organic materials (e.g., cellulosic fibers).[19] Chemical analyses of the dusts at resource recovery plants have revealed nothing especially alarming. Reported concentrations of trace metals (lead, cadmium, chromium, etc.) have been less than 1% of their respective threshold limit values.* Asbestos fibers have been reported present in the air of plants *not* equipped with dust controls but the reported concentrations have been below the threshold limit value. For plants equipped with dust controls, sampling has not yet confirmed the presence of asbestos fibers.[22]

12.13.3 Microbiological Aspects

Background Uncertainties

Interest in the microbiological aspects of dusts at resource recovery plants can be traced to the historical public health concern for sanitary disposal of municipal solid waste and to the potential presence of pathogens in the waste stream.[23-25] Despite such traditional interest, there is little reliable information on relationships between solid waste processing and transmission of infectious diseases.[26] Epidemiological studies have been unable to identify any significant differences between sanitation workers and control groups for rates of skin infections or gastrointestinal disorders.[27,28] Comparing respiratory disease rates among worker populations, one U.S. study reported no significant differences between sanitation workers and controls, whereas another has reported that the chronic bronchitis rate among Swiss refuse collectors is twice the national rate.[29] Neither of these respiratory disease investigations was designed to account for smoking habits. To date there have been no thorough epidemiological studies of the health experiences of persons working in, or living around, resource recovery plants. There have also been no reports of any outbreaks of disease, or any particular health problems, among recovery plant workers.

Current Studies

Despite the lack of evidence of specific health problems among resource-recovery workers, there are a sufficient number of unknowns about wastes to warrant concern and inquiry. Within the past several years, at least seven studies of dusts within resource recovery plants have been conducted. These studies have revealed that the dusts at solid waste-processing facilities do contain airborne bacteria and fungi, some of which are potential pathogens. To date, however, no viruses have been recovered from the air within a resource recovery plant, although at least three separate studies have attempted to do so.

The two most extensive studies of microbiological aerosols at resource recovery plants were carried out by the Ames Laboratory under the sponsorship of the U.S. EPA and U.S. Department of Energy and the Midwest Research Institute (MRI) under the sponsorship of the U.S. National Institute for Occupational Safety and Health.[22,30] The Ames Laboratory study covered several years of microbiological aerosol sampling at one resource recovery plant (the Ames, Iowa, plant). The MRI work was a survey of dust (and other) conditions within six full-scale operating resource recovery plants.

The results of several studies of microbiological aerosols are summarized in Table 12.32. The table lists not only the levels reported within resource recovery plants but also the results reported for other types of waste-processing facilities, references 31 through 39.

Comparison of bacterial concentrations recorded in and around resource recovery plants with concentrations recorded in other settings (including ambient air) is inhibited by the fact that the reported levels of bacteria have been measured by different sampling techniques under varied conditions. The fact that the reported levels of microbiological aerosols within some resource recovery pilot plants appear to be higher than levels recorded in ambient air or within other types of waste-processing facilities is not surprising, given the fact that recovery plant samples have generally been taken indoors and in plants not equipped with dust control systems. The results of the Ames and MRI studies provide data on the effect of a dust control system on the levels of airborne bacteria within a resource recovery plant. The controlled concentrations have been reported to be two to three orders of magnitude less than the uncontrolled levels, thus bringing the controlled concentrations within the range of reported levels for factories and other types of waste-treatment facilities.

---

*Threshold limit values (TLVs) are established by the American Conference of Governmental Industrial Hygienists. Each TLV is defined as the maximum airborne concentration of a substance to which a worker may be exposed for a working lifetime (8 hr/day, 40 hr/week) without adverse effect. For lead, cadmium, chromium, and asbestos, the OSHA standard for 8-hr exposure is higher than the corresponding TLV.

Table 12.32   Reported Concentrations of Airborne Microorganisms

| | Concentration (organisms per m^3) | | |
Location	Total Aerobic Organisms	Fecal Coliforms[a]	Reference
Urban air	500–6000	NR	31, 32
Schools/offices	3500	NR	32
Factories	4000	NR	32
Sewage treatment plants	45,000	NR	33–36
Spray irrigation	2000–30,000	NR	37, 38
Refuse trucks	500–90,000	70–120	18, 29
Incinerators	2000–30,000	NR	39
Resource recovery pilot plants	Up to 10,000,000	up to 18,000	17, 18, 19
Resource recovery full-scale plants	4000–80,000	100–400	22

[a]NR = Not Reported.

Interim Remedial Measures

The dust controls employed at the resource recovery plants studied by Ames Labs and MRI were reasonably conventional. For example, a typical system included maintaining the plant itself under slightly negative pressure, providing ventilation hoods at all major transfer points between conveyors (or from conveyors to processing equipment) and ducting of ventilation exhaust air to a bag house for dust removal. The use of covered conveyors is also employed at many plants as a means of reducing dust generation.

Missing: Reliable Sampling and Health Standards

Two major factors complicate the study of microbiological aerosols and make it much more difficult to draw firm conclusions about the results of microbiological sampling than it is to interpret the results of physical or chemical analyses of dusts. These two factors are the lack of standardized and reliable sampling techniques and the lack of health effects standards against which to judge the significance of airborne microbiological concentrations. The first of these factors, the lack of sampling techniques, has been partially remedied by the publication of an ASTM standard for sampling microbiological aerosols at waste-processing facilities.[40] The second factor, the lack of definitive health effects standards, appears to be far from resolution and will require studies to determine infective dosages for selected microorganisms via the inhalation route. At present and for the foreseeable future, there are no quantitative standards for occupational exposure to airborne microorganisms. It does appear, however, that the use of conventional ventilation and dust-cleaning equipment is an effective means of reducing microbiological aerosol concentrations to those commonly found in other types of waste treatment facilities.

12.13.4   Noise Control

The U.S. Occupational Health and Safety Administration (OSHA) establishes regulatory limits for the noise levels to which workers can be exposed. The limits apply to the frequency and intensity of the sound pressure level as well as to the duration of the noise. For an 8-hr exposure, OSHA has set a maximum time-weighted exposure of 90 decibels on the "A" scale of frequency distribution (90 dBA).[16]

The MRI survey of resource-recovery plants revealed several cases in which the noise levels at certain locations within a plant exceeded 90 dBA. In fact, individual noise levels well over 100 dBA (116 dBA near a ball mill) were recorded at several plants.[22] Table 12.33 presents examples of noise levels at selected locations in the plants surveyed by MRI, and Appendix 17A, Chapter 17, includes a complete shredding plant noise level survey for the city of Omaha, Nebraska.

Although such levels indicate that the equipment within a refuse processing facility can create relatively high noise levels, these levels themselves do not necessarily represent a serious occupational health hazard. This is because few, if any, resource recovery plant employees are actually

Table 12.33 Reported Noise Levels in Resource Recovery Plants[a]

Location	Noise Level (dBA)
Tipping floor	85–90
Shredder infeed	85–90
Primary shredder	96–98
Magnetic separator	90–96
Secondary shredder	91–95
Air classifier fan	95–120
Shop	78
Control room	70
Offices	67
Maintenance laborer	89
Shredder operator	83
OSHA 8-hr Std.	90
OSHA 4-hr Std.	95

[a]Source is Ref. 22.

exposed to excessively high noise levels for extended periods of time. When the MRI researchers monitored the actual personal exposures of workers, they found that the time-weighted 8-hr exposure of the workers seldom exceeded the 90 dBA OSHA limit.

The MRI researchers noted that many refuse-processing plants are inherently noisy, not only because of the use of heavy processing equipment but also because such equipment is often installed within bare concrete (not sound absorbing) walls and because the equipment is often arranged to provide multiple vertical working planes, thus creating an additive effective on noise levels.

Noise-control measures for resource-recovery plants are essentially the same as those employed in other types of manufacturing or processing industries. Dampening and isolation techniques such as the use of noise suppressors, noise enclosures, and acoustic insulation are available. Although it is often not cost effective to retrofit an existing plant for noise control, new plants should be able to be designed for noise control without adding excessively to design or construction costs. For existing plants, the key to controlling noise exposures is to assure that individual workers are not assigned to high-noise areas for extended periods of time. This may be accomplished either by rotating assignments throughout the workday or by providing periodic "quiet periods" in a sound-insulated room in between periods of exposure to high-noise areas.

12.13.5 Conclusion

To conclude this section on safety and health hazards, it is interesting to review what is known about the actual experiences of workers in resource-recovery plants. Public documentation of safety and health records for such workers has not been extensive but, from the evidence available, some preliminary conclusions can be made.

Occupational safety records reflect the frequency and severity of injuries from accidents such as falls, cuts, sprains, and electrical shocks. For many years, it has been reported that the frequency of injuries among refuse collection workers is among the highest for any occupational group.[27] Unlike refuse collection workers, most employees of resource-recovery plants do not come into direct physical contact with the wastes themselves. Consequently, resource-recovery workers are less likely than collection workers to incur the back sprains, cuts, and other injuries associated with lifting and handling municipal solid wastes.

The safety records of resource-recovery employees report the types of accidents more often associated with industrial work: falls, welding flashes, burns, bruises, injuries from moving machinery, and so on.[22] There are also sprains, pulls, and cuts reported but these are due to working with machinery (e.g., replacement of shredder hammers) rather than to handling the municipal wastes themselves. The MRI study of workers from six resource-recovery plants suggested that the most frequently occurring injuries were cuts, bruises, and eye injuries. Because of the relatively small number of resource-recovery workers, it has not been possible to develop reliable statistics for the rates of accidents and injuries among such workers but it does appear that the types of injuries are more similar to those found among industrial workers than to those reported among refuse collection workers.

Occupational health records report the types and frequencies of illnesses among workers. Unlike safety records, there is very little health data on refuse collection workers, much less resource-recovery workers. After reviewing studies of the overall relationship between solid waste and disease, one author concluded that none of the reported data permits a "quantitative estimate of any solid waste/disease relationship." This same reviewer also concluded, however, that the strongest circumstantial evidence suggesting a solid waste/disease relationship involves infectious diseases.

At least two resource recovery plants are known to have maintained detailed health records on employees and the results from one plant have been published.[41] Generally speaking, the conclusion to date seems to be that working in a resource recovery plant is not significantly different from working in other similar industrial or municipal jobs (e.g., street maintenance or boiler operation) in terms of types and number of illnesses reported, days off work, etc. The records do not, of course, provide conclusive epidemiological information, and it is impossible to say with certainty that there is *no* effect from working in a resource-recovery plant.

REFERENCES

1. Resource Recovery Overview, *Engineering News Record*, January 6, 1983, p. 18.

2. A. R. Nollet and E. T. Sherwin, "Causes and Alleviation of Explosions in Solid Waste Shredders," presented at the 1979 Convention of the GRCDA, San Diego, August, 1979.

3. W. H. Dedrick, "Explosions in Grain Dust Milling Operations," Pennsylvania State University Department of Mechanical Engineering, 1917, as referenced in W. D. Robinson, "Solid Waste Shredder Exlosions—A New Priority," presented at a seminar on Explosion Protection for Solid Waste Shredders, New Castle, Delaware, June 26–28, 1979, sponsored by Aenco, Inc.

4. *Proceedings, International Symposium on Grain Dust Explosions*, Kansas City, Missouri, October 4–6, 1977, Grain Elevator and Processing Society, Minneapolis, Minnesota, 1977.

5. Coal Pulverizer Blows Up, *Washington Post*, November 14, 1978, p. A-12.

6. W. Coco, Worker Seriously Hurt in Bristol-Myers Blast, *Newark Star-Ledger*, February 12, 1979.

7. W. D. Robinson, Solid Waste Shredder Explosions: What Do They Have in Common, *Solid Wastes Management Magazine*, **22**, 46–48 (1979).

8. R. G. Zalosh, S. A. Wiener, and J. L. Buckley, *Assessment of Explosion Hazards in Refuse Shredders*, Factory Mutual Research Corp., Norwood, Massachusetts, 1976 (prepared for U.S. Energy Research and Development Administration).

9. M. Kramer, Hydrogen May Have Been Factor in Blast, *Brockton (MA) Enterprise*, May 1, 1978, p. 1.

10. American Can Company, "December 28 Explosion at Americology," memorandum made available to public dated February 12, 1979.

11. A. R. Nollet, personal communication, September, 1982.

12. N. R. Ahlberg and B. Boyko, Explosions and Fires—Ontario Centre for Resource Recovery, in Proceedings, ASME 9th National Waste Processing Conference, Washington, D. C., May, 1980.

13. American Society for Testing and Materials (ASTM), Proposed Guide for Shredder Explosion Protection, ASTM, Philadelphia, Pennsylvania, 1983.

14. National Fire Protection Association (NFPA), *Explosion Venting*, Boston, Massachusetts, No. 68, 1954.

15. R. G. Zalosh and J. P. Coll, *Determination of Explosion Venting Requirements for Municipal Solid Waste Shredders*, Factory Mutual Research Corporation, Norwood, Massachusetts, 1982.

16. U.S. Occupational Safety and Health Administration, *OSHA* Safety and Health Standards: *General Industry*, Title 29, Chapter XVII, Section 1910.1000. U.S. Department of Labor, Washington, D. C., January, 1976.

17. L. F. Diaz, et al., Health Aspect Considerations Associated with Resource Recovery, *Compost Science*, Summer 1976, pp. 18–24.

18. D. E. Fiscus, et al., *St. Louis Demonstration Final Report: Refuse Process Plant—Assessment of Bacteria and Virus Emissions*, U.S. EPA, Cincinnati, Ohio, 1977.

19. E. J. Duckett, J. Wagner, R. Welker, B. Rogers, and V. Usdin, Physical/Chemical and Microbiological Analyses of Dusts at a Resource Recovery Plant, *American Industrial Hygiene Association Journal*, **40**, 908–914 (1980).

20. L. Lembke and R. Kniseley, Coliforms in Aerosols Generated in a Municipal Waste Processing System, *Applied and Environmental Microbiology*, **40**, 888–891 (1980).

21. S. Z. Mansdorf, M. Golembiewski, C. Reaux, and S. Berardinelli, Environmental Health and Occupational Safety Aspects of Resource Recovery, in *Proceedings of the Seventh Annual Mineral Waste Utilization Symposium*, Chicago, October 21, 1980, sponsored by IITRI and the U.S. Bureau of Mines, Illinois Institute of Technology Research Institute, Chicago, 1981.

22. S. Z. Mansdorf, M. A. Golembiewski, and M. W. Fletcher, *Industrial Hygiene Characterization and Aerobiology of Resource Recovery Systems*, Midwest Research Institute, Kansas City, Missouri, 1981, for National Institute of Occupational Safety and Health, Morgantown, West Virginia.

23. W. L. Gaby, *Evaluation of Health Hazards Associated with Solid Waste/Sewage Sludge Mixtures*, EPA Report No. 670/2/75-023, Cincinnati, Ohio, 1975.

24. R. W. Peery, "Public Health Implications of Staphylococcus and Streptococcus at a Landfill," Masters Thesis, West Virginia University, Morgantown, West Virginia, 1971.

25. M. Peterson, Soiled Disposable Diapers: A Potential Source of Viruses, *Am. J. Public Health* **64**, 912–914 (1974).

26. T. G. Hanks, *Solid Waste/Disease Relationships: A Literature Survey*, U.S. Department of Health, Education and Welfare, Cincinnati, Ohio, 1967.

27. J. A. Cimino, Health and Safety in the Solid Waste Industry, *Am. J. Public Health*, **65**, 38–46 (1975).

28. E. M. Sliepcevich, "Effect of Work Conditions Upon the Health of the Uniformed Sanitationmen of New York City," Doctoral Dissertation, New York University, New York, 1955.

29. G. Ducel, et al., The Importance of Bacterial Exposure in Sanitation Employees When Collecting Refuse, *Sozial Praventivmedizin*, 21, 136–138 (1976).

30. R. N. Kniseley, Microbiological Air Quality of the Ames Municipal Solid Waste Recovery System, *Quarterly Report of the Ames-DOE Laboratory for the Period January 1–March 31, 1979,* prepared for U.S. Department of Energy, Office of Health and Environmental Research, Ames Laboratory, Ames, Iowa, 1979, pp. 9–10.

31. J. F. Hers and K. C. Winkler, Eds., *Airborne Transmission and Airborne Infection*, John Wiley & Sons, New York, 1973.

32. E. A. Glysson, et al., The Microbiological Quality of Air in an Incinerator Environment, Proceedings of the 1974 ASME National Incinerator Conference, Miami, Florida, pp. 87–97.

33. A. P. Adams and J. C. Spendlove, Coliform Aerosols Emitted by Sewage Treatment Plants, *Science*, **169**, 1218–1220 (1970).

34. J. L. Hickey and P. C. Reist, Health Significance of Airborne Microorganisms from Wastewater Treatment Processes—Part Two: Health Significance and Alternatives for Action, *Journal WPCF*, **47**, 2758–2773 (1975).

35. J. O. Ledbetter, I. M. Houck, and R. Reynolds, Health Hazards from Wastewater Treatment Practices, *Environmental Letters*, **4**, 225–232 (1973).

36. M. R. Pereira and M. A. Benjaminson, Broadcast of Microbial Aerosols by Stacks of Sewage Treatment Plants and Effects of Ozonation on Bacteria in the Gaseous Effluent, *Public Health Reports*, **90**, 208–212 (1975).

37. H. T. Bausum, et al., *Bacterial Aerosols Resulting from Spray Irrigation with Wastewater*, U.S. Army Medical Bioengineering Research & Development Laboratory, Frederick, Maryland, 1976.

38. C. A. Sorber, et al., A Study of Bacterial Aerosols at a Wastewater Irrigation Site, *Journal WPCF*, **48**, 2367–2379 (1976).

39. M. L. Peterson and F. J. Stutzenberger, Microbiological Evaluations of Incinerator Operations, *Applied Microbiology*, **18**, 8–13 (1969).

40. American Society for Testing and Materials (ASTM), "Standard Practice for Sampling Airborne Microorganisms at Municipal Solid Waste Processing Facilities" (E 884–82), ASTM, Philadelphia, Pennsylvania, 1983.

41. T. V. Sprenkel, Health and Recycling, *Waste Age*, **10**, 74–81 (1979).

12.14 HEALTH AND SAFETY: IMPLEMENTATION

Anthony R. Nollet and Robert H. Greeley

12.14.1 Background and Scope

This chapter reviews the salient and frequent health and safety problems encountered in the solid waste industry, with emphasis on the safety of refuse-derived fuel (RDF) plants. It offers the best current (1984) opinions of the authors, but cannot be considered the definitive treatment of all hazards and safety problems to be encountered. Each author has 13 years of hands-on experience in operating two solid waste resource-recovery plants which processed in excess of 1.6 million tons of solid waste. These plants experienced 58 explosions, about 10 fires, and a number of accidents. *While we believe that our recommendations are sound, the reader is cautioned that this is no guarantee or warranty of the efficiency of our observations or recommendations. Our recommendations represent personal professional experience, and do not necessarily represent the opinions of AENCO, Inc. or its parent company, Cargill, Incorporated.*

We do not discuss the handling of toxic or hazardous substances except as such substances may incidentally be found in household, commercial, and industrial solid waste. Recommendations are limited to the handling and processing of nonhazardous solid waste.

Although there is reference to relevant hazards in collecting and landfilling solid wastes, most discussion relates to the hazards of operating resource-recovery plants that produce RDF. Recommendations are oriented more toward the operational aspects of health and safety, as opposed to the specific design of equipment, although we do suggest design criteria.

The Vagaries of Solid Waste

Almost everything that mankind produces will one day be discarded, and the material entering the waste stream represents standards of living, consumption, and life-styles. Some items are inherently or potentially hazardous: guns, knives, certain drugs, explosives, volatile flammable liquids, high-pressure containers, military ordnance, etc. Other items are relatively harmless, but in combination with other substances, can become hazardous. For example, used oil mixed with high-nitrogen fertilizer can be detonated.

Those who collect, landfill, or process solid waste must anticipate the behavior of those who discard wastes.

Actual examples illustrate this precept:

1. A retired Army Colonel, having worked on its development, kept a WW II antitank land mine in his home as a souvenir. When he passed away, the estate discarded it. An alert "picker" at a solid waste resource recovery plant removed the object from a processing line about 30 sec before it would have been shredded. Because the mine was equipped with a Munroe Charge, it could have destroyed the shredder, and perhaps the shredding plant.

2. In springtime, many suburban householders discard last season's stale gasoline (and save the can) for mowers or snowblowers by pouring it into a glass or plastic container and capping for discard. This creates potential danger because it may be broken during the compaction cycle of a collection vehicle, thus causing a fire, may be broken and ignited while being landfilled, thus causing a fire, may be broken and ignited while being landfilled thus causing a fire under a dozer, may be shredded, causing an explosion in the shredder system, or cause a fire or explosion in the pit, charging chute or furnace of an incinerator.

3. A trap shooter refills his own shells, with smokeless powder from 12-lb cans. As the new season approaches he may decide that his smokeless powder is damp so he discards it, thus creating a fire hazard in the collection vehicle and at the landfill, an almost certain deflagration if the container is shredded, or a deflagration if the container is fed to a mass-burn incinerator.

4. The high school and university terms end in June, and the chemistry laboratories are cleaned. Someone has not replaced a stopper on a large bottle of picric acid, and the solution may have dried out and crystallized, resulting in an impact-sensitive explosive. Or a student may have deliberately concocted explosives and left a sample in the laboratory. One of the most severe explosions we ever experienced was accompanied by identifying papers (before and after the shredder) that were from the chemistry laboratory of a noted university. But, as the dean of chemistry insisted, the explosion could not have been caused by their discarding a chemical. He informed us that the university had rules forbidding creation of explosives in the university laboratory, and discarding them.

Although this chapter deals with nonhazardous wastes, we must reiterate that hazardous material will occasionally be encountered in so-called nonhazardous wastes.

12.14.2 Safety Rules and Practices

RDF plants may incorporate shredders, flails, shears, trommels, disc screens, air classifiers, conveyors, mineral jigs, froth flotation tanks, magnets, bag house filters, explosion protection systems, mobile equipment such as front-end loaders and yard mules, compactors, and compactor trailers. Auxiliary equipment will probably include a manlift, a fork lift, and various lifting devices.

Personnel will eventually be hired and trained to operate and maintain this equipment. In most cases, system and equipment design changes will be required to operate and maintain the plant safely. Ideally, plant operators should work closely with the design engineers during design and construction of the RDF plant, but this does not usually occur.

Operators of newly built RDF plants are usually under time constraints to begin operations as soon as possible. They must insist on being given sufficient time to insure that it can be operated safely. Following is a short safety checklist for operators of RDF plants.

1. *Noise Survey.* With all mobile and stationary plant equipment in full operation, there should be a comprehensive noise survey by professionals. It may be that drastic plant modifications will be required to reduce noise levels in specific locations, or it may be that ear protection will always be required at some work stations. The solutions ultimately adapted must conform with OSHA requirements despite enforcement relaxation via deregulation, and the like.

2. *Work Stations.* There will be locations in any RDF plant in which people should not work during operations. For example, in plants that shred as the first processing step, it is inevitable that containers of gasoline or other flammable substances will be ruptured with delayed ignition. The liquid volatile may be conveyed with the shredded waste and/or volatilize with widespread distribution. Eventually a spark (probably in the shredder) will ignite the vapor and the resulting explosion and/or fire may spread rapidly through all plant equipment downstream from the shredders. It is obvious that personnel should not be allowed to work in or around equipment located downstream from the shredder. This means that periodic inspection of equipment in these areas should not occur while material is being fed to the shredders. Good remote-sensing instrumentation, such as television cameras, can monitor operations in these prohibited areas.

3. *Accessibility and Maintainability.* Each machine will eventually have to be replaced during the life of the plant. In some cases, replacement must be accomplished by rental cranes or other lifting equipment. But many heavy parts will require frequent replacement, for example, heavy shredder grate bars and liner plates and built-in lifting equipment should be provided for such items *before* plant operations begin.

4. *Jam Prevention.* One of the most hazardous and unpleasant jobs in an RDF plant is the clearing of jams in flow of material. Jams in the shredder are particularly difficult and dangerous to clear. Interlock circuitry that stops the infeed conveyors can prevent most shredder jams. If they are not already installed, the operator should consider retrofitting zero-speed sensors on all conveyors after the shredder. Sensors for high level above conveyor transfer points is also desirable.

5. *Space Heaters and Lights.* Since it is possible for explosive vapors to accumulate anyplace in an RDF plant, it follows that there should be no open flame, electric sparks, or hot wire space heaters.

6. *Receiving Conveyers.* Most RDF plants involve pushing raw material into recessed receiving conveyors. It is well to suspend ribbons from the ceiling at a position so that when the hanging ribbons contact the cab of the front-end loader, the operator knows that he must stop to avoid driving into the conveyer pit, Figure 12.116.

7. *Ladders and Manlifts.* Most RDF plants involve high ceilings, and often with equipment that requires maintenance located close to the ceiling. If built-in stairs and platforms are not provided, it may pay to consider purchase of a manlift. Also, high-quality OSHA-approved ladders in several lengths will be required.

8. *Fire Bottles.* Fire bottles should be installed at strategic locations. They should be inspected at least monthly to confirm their presence and condition. We recommend that special fire bottles be provided for all electrical control rooms. Such special fire bottles should be filled with a chemical that does not damage electrical equipment.

9. *Fire Hydrants and Sprinkler Systems.* We recommend that all RDF plants be equipped with overhead automatic sprinkler systems, and that fire hydrants with wall-mounted hoses and good quality nozzles be installed at strategic locations. There is a tendency to steal brass nozzles, and we therefore favor plastic nozzles. The hoses are not meant for professional fire fighting. They are meant to be used by plant personnel to extinguish small fires. Any serious fires should be handled by a trained fire department.

10. *Lifting Slings, Come-Alongs, and Hoists.* An adequate supply of these items must be procured before they are needed, and personnel must be trained in their proper operation, inspection, and maintenance.

A preliminary plant safety manual should be drafted before plant operations begin. At a minimum, the manual should include duties for each job station and should cover what each job station should do in the event of defined emergencies. It should also cover general safety items, such as instructions for lifting.

A plant safety director should be appointed before plant operations begin, and his duties should be well-defined. First aid equipment must be placed in strategic and accessible locations. Designated personnel must receive formal first aid training. A company doctor should be retained. He should have a strong background in industrial medicine.

Contact must be made with the nearest paramedical unit, and with the nearest hospital. Acci-

Fig. 12.116 Tipping floor warning ribbons.

dent and disaster planning must be complete and incorporated in the safety manual. Paramedical personnel and fire department officials should be invited to conduct a physical survey of the plant, and they should be provided with a plan sketch of the plant. Self-contained or water-supplied eye wash stations must be established at strategic locations.

12.14.3 Personnel Safety

The need for a strong continuing safety program cannot be overstated. Following is a partial checklist of items to be considered:

1. *Mandatory Protective Clothing.* All personnel working in an RDF plant should be required to wear OSHA-approved safety boots, hard hats, leather gloves, and safety glasses with side shields. We recommend that all employees be provided with uniforms that are not flammable. Welders must be equipped with leather coats and an approved welder's mask. Annual expenditures of about $500/employee will be incurred if company uniforms are issued. We recommend work uniforms to eliminate the possibility of workers wearing highly flammable or dirty clothing that might spread disease.

2. *Preemployment Physical Examinations.* All employees should be given a thorough preemployment physical examination to document any baseline conditions. The physical examination should include spinal and chest X-rays, a thorough hearing test, and blood analysis. The purposes of the physical examination are to be established that the prospective employee is capable of doing the work and also establish incontrovertible preexisting conditions in the event of future disability claims.

3. *Optional Protective Equipment.* Dust will exist in RDF plants, and parts of these plants may have objectional noise, which, while not exceeding OSHA standards for 8-hr exposure, may nevertheless prove annoying to employees. OSHA-approved dust masks and ear plugs should be made available to employees who request them. In particularly dusty locations, "Air-Hats" with self-contained battery-powered filters may be desirable. Additionally, when workers must clean a baghouse filter system, it may be necessary to equip them with Scott Air Packs or equivalent breathing apparatus. After the plant is in operation, we recommend the NIOSH-certified professional's test for total dust quantity at each station, and also for heavy metals. We do not know of any case in which dust generated in RDF production plants has been found to be hazardous, but an analysis of the dust quantity and toxicity is nevertheless recommended.

4. *Supervisory Briefings of Employees.* Any resource recovery plant is inherently dangerous, and it is essential that the plant operator thoroughly train all employees to avoid injury. The tipping floor will usually have a supervisor to direct incoming collection vehicles and one or more

front-end loader operators. The tipping floor is often small or the pit full and incoming waste is stacked high. The supervisor must be particularly alert to stay clear of the many vehicles operating on the tipping floor. He should be equipped with standard safety equipment, a dust mask, and a highly visible orange vest.

5. *Picking Station.* Most well-designed RDF plants are equipped with an on-line picking station located adjacent to the conveyer feeding the primary shredder, or the primary presort (air drum or trommel) equipment. The pickers are usually instructed to remove material that might cause an explosion (unidentified, unopened containers of possible flammable material), items that might damage the primary size reduction equipment (usually large pieces of metal), or items that might cause the primary reduction to jam (usually large quantities of stringy material that may wrap around the rotor). The pickers are usually equipped with potato rakes and knives (used to cut away stringy material). Some plants are equipped with hydraulically operated claws at picking stations, to remove undesirable material, but the authors believe that such devices are not often necessary* and may cause too much delay in removing unwanted materials. The pickers should be instructed to remove as much unwanted material as possible without stopping the infeed conveyer. Whenever they must walk onto the conveyer, they must definitely activate the emergency STOP switch that is required on all conveyers. Pickers should never puncture sealed containers, for they may contain hazardous chemicals under pressure. The possibility of being cut, bruised, or sprayed is always present at the picking station, and the pickers must be instructed to work safely—even at the expense of lost production. There is one special hazard at the picking station: If a long piece of cable or a long piece of stringy material such as a garden hose is strung along the infeed conveyer, then when one end of the cable contacts the shredder rotor, it may become entwined on a hammer. If this happens, then the entire cable is suddenly accelerated to the tip speed of the hammer—usually about 300 ft/sec (90 m/sec). If the other end of the cable has not yet passed the picking station, then the entire cable may whip violently and conceivably decapitate a picker. It is essential that the pickers cut cables, hoses or similar items into lengths that are shorter than is the distance from the shredder to the picking station.

6. *Control Room.* Although there are few hazards to the control room operator, he can inadvertently set up accidents for others. For example, if he sees an object that must be picked, he usually stops the conveyer and asks the pickers to remove it. The pickers should not walk on the conveyer until the pickers have activated their emergency conveyor STOP switch. They should not rely on the control room operator to keep the conveyer STOPPED. The control room operator should be equipped with television monitors, and he must be especially alert to avoid jams of material, or to detect such jams before they become difficult to clear. If material is flowing into a machine, then material should be flowing out of it. If not, the operator must immediately cut the flow of material into the machine, as a jam is probably beginning.

7. *Conveyers.* All conveyors must be equipped with an emergency STOP cord along the full length of the conveyer convenient to locations where anyone may be. Belt conveyers are especially hazardous. Typically they move at high speed, and if an employee's clothing is caught between a moving belt and a conveyer roller, the employee may be pulled between the belt and the roller, causing death or serious injury. Generally, employees should be forbidden to work adjacent to moving rubber belts. If it is essential that they work near a moving belt (to track the belt), the employee must be especially careful not to wear loose clothing. A second employee should stand by the emergency STOP cord or switch, using a "buddy" system to protect the other employee. Some plants are, unfortunately, equipped so that conveyers can be started remotely, often from the control room. We believe that belt conveyers should be wired so that they *cannot* be started remotely without the active cooperation of a worker near the conveyers whose job is to ensure that the conveyer is clear before it is started.

8. *Lockouts.* All employees working on machinery that can be started electrically should be equipped with an individual padlock, and they should not work on such equipment until its electrical switch has been locked out using their own padlock. During routine maintenance of the entire plant, management may elect to lock out blocks of designated equipment until the maintenance has been completed.

9. *Clearing Jams.* Plugged flow in RDF plants is inevitable. It is astonishing how mixed solid waste can compact, either raw or shredded, with shredder jams being extremely onerous. Although management wants uninterrupted production, safety precautions must be maintained, and proper equipment lockout steps followed. Injuries that occur as a result of clearing jams can result from restarting equipment in an effort to clear the jam without taking a head count to be certain that all persons who were working on it are in safe locations. Plant management must personally be

---

Editor's note: If residential oversize, industrial, and commercial trash must be processed, a knuckle boom crane is an advantage, i.e., Resource Recovery (Dade County). Likewise, push-pit receiving (Odessa, Baltimore County) uses them for restraint of avalanching onto the feed conveyor.

involved in all jam clearing to insure personnel safety. Jams that occur between the shredder discharge grates and the shredder discharge conveyer can be particularly difficult and dangerous. In some cases, the jammed material packs very densely. Those working on the bottom of the jam are seriously endangered if the mass of material above them should suddenly break loose and bury them. Generally, the clearing of jams is a long and slow process. Attempts to speed it up are usually counterproductive and often unsafe. The authors believe that *the clearing of jams in RDF plants is perhaps the most dangerous job in the plant*, and we consider that investment in equipment to detect and prevent jams is well worth it.

12.14.4   Raw Material Presort

Size reduction in RDF processing employs one of five types of mill: a horizontal hammermill, a vertical hammermill, a horizontal shear shredder, a flail, or a wet pulper. Horizontal hammermills force the material to pass through grate bars, whereas vertical mills crush the material between the hammers and the outside shell of the mill. A very lightly constructed horizontal hammermill that is not equipped with grate bars is sometimes referred to as a flail mill, which really describes bag breaking coarse reduction only. Generally, it is desirable to remove certain classes of items before they enter the above mills (flails and shears) requiring extensive presorting either on the tipping floor, source separation, collection constraints, etc., and preferably all three, that is:

> "Gross," dense, high-tensile metal items that may damage mill internals.
>
> Material that might cause fire or explosion, such as containers of flammable liquid, Figure 12.117.
>
> Large quantities of stringy material that might wrap around the mill rotor and jam it, Figure 12.118.
>
> Abrasive material such as glass that causes high wear of mill parts.
>
> Mattresses, springs, sofas, tires, etc., unsuitable for flails, Figure 12.119.

Mechanical sorting will not remove all such items, neither will hand-picking on the tipping floor and at the picking stations. Without well-disciplined source separation and collection, primary size reduction may be required to withstand a wide range of material input.

To reduce abrasion, it would be desirable to cast and heat-treat hammers for minimum abrasive wear. However, abrasion-resistant steel will have a very high hardness and is usually brittle. Excessively hard hammers or grate bars will certainly break if they encounter a dense metallic object such as a die block or a crankshaft. However, hammers in a typical mill can go up to 550 Brinell, with little danger of cracking, with optimal maintenance cost on a no-weld discard basis. To shred stringy material, it would be desirable if the mill were equipped with narrow, relatively sharp

Fig. 12.117   Containers of flammable liquid, ordnance, etc.

Fig. 12.118 Undesirable long stringy material.

Fig. 12.119 Typical residential oversize material.

knives. However, such knives would wear quickly and would certainly break if they encountered glass and dense metal. The hammers in primary solid waste mills are typically 5 to 6 in. (13 to 15 cm) in width and weigh some 200 lb (90 kg). Yet in some cases, the raw feed can be up to 50% paper which could be shredded using narrow, sharp, and very hard knives, if little or no grit, glass, or tougher material is present. Because of the unpredictability of solid waste raw material, primary size reduction must adapt to such vagaries unless the raw material undergoes maximum sorting and separating.

1. *Trommels and Screens.* To minimize certain downstream maintenance and to enhance materials recovery efficiencies, raw refuse screens (trommel or rotary disc) are employed as the initial process step before size reduction and only the "overs" are processed. Typically, preshred-

ded screens are equipped to drop minus 4.5 in. (11.4 cm) material and to pass plus 4.5 in. (11.4 cm) material to the primary mill. This procedure certainly removes most of the glass in the incoming material and also removes much of the small pieces of metal. It also removes small-size combustibles, which may be a prohibitive penalty with regard to RDF/raw material yield, MSW availability, etc. Removal of the glass certainly reduces wear of shredder hammers and grates and reduces RDF slagging characteristics. The trommel (or disc screen) overs will contain unsorted oversize items, including containers of flammable liquids, etc., and the use of trommels or screens will not reduce shredder explosion probability. Likewise, they will not permit major redesign of shredder internals or their hardness beyond 550 Brinell.

   2. *Air Classifiers.* Normally, air classification following size reduction (shredding, etc.) is not a factor in presort or evaluating the probabilities of explosions in RDF production, and air classifiers in their usual applications are discussed in Section 12.5. Air classification of incoming raw refuse as a first process step, however, suggests a new direction for innovation in process equipment, system design and plant safety.[3]

An air-swept rotating drum, Figure 12.120, for raw refuse which separates flyable lights from the rest of it presents several important innovations:

   Virtual elimination of the principal source of explosions in size reduction, i.e., containers of flammable liquids, detonating solids, and deflagrating chemicals.

   An essentially clean, low-ash combustible fraction via more efficient size reduction in a knife-type shredder with significantly harder hammers, grates, anvils, liners, and minimal maintenance.

   Separation of recoverables and undesirable material into smaller gross fractions for cleanup, recovery of smaller combustibles, and saleable secondary material.

Although there is no system of this type yet in operation, the success of the development prototype along with the apparent need for improvement of RDF process systems assures an initial installation.

12.14.5   Raw Material Surveillance

Most incinerator plants cannot process all incoming solid waste. Some loads must bypass the plant for landfill. Such loads include demolition debris, bundles of long stringy material such as leather

Fig. 12.120   Raw refuse rotary drum air classifier.

strips, metal and wire, drums of liquid, and material large enough to jam in a hopper, ash systems, and so forth. In practice, at least 20% of the general waste collection must bypass most RDF plants and be landfilled directly. Inevitably, however, unprocessable items will arrive at the plant and are removed at the scale, on the tipping floor, at on-line picking stations, or by remote control grapple controlled from the control room.

1. *Tipping Floor.* In most RDF plants, waste is dumped on a tipping floor and is pushed into receiving conveyers using front-end loaders (FELs). A few RDF plants were designed without tipping floors and the waste is deposited directly into receiving pits. Such plants are more difficult to operate if there is no opportunity for raw material surveillance. Well-managed RDF plants constantly remind the FEL operators to spot and remove objectionable material. Usually such plants are equipped with roll-off containers or dumpsters for such items. The FEL operator cannot be expected to see all objectionable items. In two plants, the FEL operator spreads the incoming material on the tipping floor and several inspectors walk through the spread-out load to manually remove objectionable items. This primitive system can slow production to an unacceptable level. The authors recommend that the FEL operators remove as much objectionable material as they can see without delving and do not recommend that inspectors walk through spread-out waste. In practice, good FEL operators will remove about 1.5% (by weight) of the waste that is dumped on the tipping floor.

2. *On-Line Picking Stations.* On-line picking stations can be quite effective in removing objectionable material from the waste stream before it is processed. The picking stations should be located at an horizontal section of a metal pan conveyer, and the floor of the picking station should be approximately level with the conveyer pans. A lip no more than 6 in. (15 cm) high should be provided at the conveyer side at the picking station, Figure 12.121. Burden depth should be as low as possible, consistent with a good production rate, and the conveyer should be run at the fastest practical speed, say 60 ft/min (18 m/min) for metal pan types. A 7-foot (2.1 m)-wide conveyer traveling at 60 ft/min (18 m/min) will have an average burden depth of about 6 in. (15 cm) when the feed rate is about 90 tons (82 t)/hr. It also helps if the pickers can observe the fall of waste from the receiving conveyer to the infeed conveyer (where pickers are usually stationed), enabling them to see more of the undesirables. However, the pickers will not see all objectionable material, and occasionally, very large pieces of metal will escape their notice as will containers of volatile fluid such as gasoline unless the feed rate is reduced to a very low level, say 25 tons (23 t)/hr. The picking station should be equipped with a switch to stop the conveyer if a difficult-to-remove item appears. Conversely, the pickers can remove fairly large material without stopping the belt by using potato rakes or similar rakes. There should be two pickers at each station, or the picking station should be manned by one picker who can be observed from a control booth. This is in the interest of safety to prevent a picker from being dragged to the processing equipment. If the first processing step is a

Fig. 12.121 Raw material on-line picking station.

shredder or flail mill, suitable baffles and shelter for the picking station and adjacent areas must be provided.

Shredder, flail, and infeed hoppers must be enclosed to contain the release of high-speed ballistic objects, and in addition, baffles must be provided to deflect explosion blast waves.

3. *Control Room.* A few RDF plants rely on control room operators to spot objectionable material and remove it, using remote-controlled hydraulic grapples. We do not recommend this procedure, because the control room is usually located too far away from the conveyer for the operator to spot objectionable material. Additionally, control room operators are usually too busy monitoring other instruments and equipment to pay close attention to the waste itself. The use of grapples is questionable. The authors believe that grapples are useful only for very large heavy items. Most objectionable items can easily and quickly be removed by hand, and even very heavy items can be removed manually in relatively short periods of time. We think that grapples merely slow down production.*

12.14.6 Explosion Protection

Commercial, industrial, and residential waste occasionally contains material which, if shredded, will almost certainly result in explosions. There are three general classes of such items: self-oxidants, vaporized flammable liquids, and dust. Self-oxidants include cans of smokeless powder, sticks of dynamite, military ordnance, and so on. Vaporized liquids include gasoline, propane, paint thinner, and aerosol cans pressurized with hydrocarbons. Explosive dusts include flour, coal, and similar carbon-bearing powders.

Explosions of self-oxidants cannot be prevented or suppressed. The others can be alleviated by heavily constructed shredders, conveyers, and hoppers and by providing adequate vents, properly placed and sufficiently strong to divert forces generated. Such vents must be designed to contain ballistic objects propelled by the hammers in the mill while allowing gases to escape.

Explosion of vapor or dust mixtures may be mitigated by water fog suspended in the shredder and other process enclosures. Many such explosions may be suppressed by pressurized blanketing chemical systems. In all cases, however, venting is required.

Mechanical sorting to remove explosives effectively prior to size reduction (shred, flail, etc.) would be the ideal safeguard. Unfortunately, no mechanical device to remove explosive material automatically prior to size reduction has ever been installed in a production plant, although a 150-ton (136 t)/hr prototype rotary air drum was tested some years ago with encouraging results. Records of early attempts to shred refuse are not clear although two shredder manufactures claim that early versions of their mills were used to process "garbage" near the turn of the century and Racine, Wisconsin, used a portable diesel-driven hammermill at an expiring landfill for over a year in 1954.

During the late 1950s and early 1960s, at least 22 composting plants were built in the United States alone. Nearly all of these plants employed some form of size reduction and were economic failures. If serious explosions occurred in these plants, the information did not make its way into the technical literature.

In late 1960, the Public Health Service sponsored a solid waste shredding plant in Madison, Wisconsin, for leachate, vectors, compaction, and no-cover landfill testing. Although this plant experienced several relatively minor explosions circa 1971, the facts concerning these explosions were not well-publicized.†

Since 1971, approximately 100 plants have been contracted for or built which shred the solid waste as the first processing step, and several have suffered explosion damage severe enough to discontinue operations. The Environmental Protection Agency (EPA) did not appear to include shredder explosions in their studies agenda until the 1970s, when such a program was vastly underfunded in our opinion.

The Energy Research and Development Agency (ERDA), now the Department of Energy (DOE), funded a study of explosion episodes by Zalosh[1] of Factory Mutual Research Corporation in 1975. This study surveyed all plants that shredded solid waste and was published in early 1976. It showed that at the end of 1975 approximately 8,295,000 tons of solid waste had been shredded. In all, 97 explosions were recorded, and 69 of these caused significant damage, but little if any serious injury.

Editor's note: An interesting variation to this precept is the 3000 TPD Resource Recovery Plant in Miami, Florida. Pedestal-mounted, manned hydraulic cranes not only very quickly remove undesirables but load the feed conveyers efficiently. It receives and processes a wide range (residential, industrial, and commercial) of waste.

†*Editors's note:* Madison, Wisconsin, has a very well-disciplined source separation and collection system which delivers a relatively innocuous supply of raw material to the present (and original) RDF production plant.

In 1980, EPA's Office of Research and Development, Municipal Environmental Research Laboratory (MERL), funded a modest experimental program with Factory Mutual. Dr. Robert Zalosh, who supervised the experiments, discovered that pressures in the order of 20 psig (138 KPa) could be experienced in top-feed shredder feed hoppers when propane was detonated in this region. This compared with a theoretical 6 psig (41 KPa). The unexpectedly high pressures were attributed to "worst case," highly efficient fuel/air mixing caused by rotor windage turbulence.

During the past two decades, at least 20 bulky trash shredders have been installed at incinerators and transfer stations. Not unexpectedly, all have had explosions, but no reported fatalities or serious injuries; some of them have been closed down for economic reasons until such time as the concept may have to be resumed under future logistical mandate of diminishing landfills. Chapter 15 discusses this concept in detail.

The authors were both involved in the design and operation of the New Castle County, Delaware, Shredding Plant, from 1972 through 1980. Initially, the plant was equipped with 1/4-in. (6 mm)-thick steel explosion vents extending through the roof above the shredder feed hopper. These vents were destroyed by an explosion in 1974 that did $250,000 worth of damage. Reinforced 1-in. (25 mm)-thick steel vents were retrofitted to withstand 20 psig (138 KPa). After these vents were installed and the shredder discharge chutes similarly strengthened, expensive explosions were greatly reduced. The plant experienced six explosions before the strong vents were installed, and total damage was about $325,000. Thereafter, 42 explosions occurred after the reinforced vents were installed and total damage was less than $50,000. A water-fog system was also installed and very likely reduced damage also.

A purpose of this section is to examine the causes of shredder explosions, to suggest means to alleviate (but not eliminate) the explosion problem, and to suggest broad guidelines for planning what to do when an explosion occurs. It should not be construed, however, as a preference for shredding of solid waste as the first step, and we suggest several prudent measures if shredding is required. It is ironic that several fatalities in resource recovery are not attributable to shredding and that more injuries are from other causes.* Property damage is another matter, however, and constant vigilance must be maintained.

Explosions in or near size reduction equipment occur for these reasons:

1. Sparks are produced constantly inside the shredders, flails, and shears (to a lesser extent) but most of them do not have enough heat energy to ignite a gas or dust deflagration.[4]

2. High-tensile scrap metal, however, may be heated to a glowing red if retained long in a mill and can be a source of ignition—likewise sudden rupture (instantly exceeding its elastic limit, exemplified by the hot ends of a coat hanger broken by rapid back and forth bending by hand).

3. Shredders, flail mills, and rotary shears will certainly open containers of volatile liquids, and very likely rupture acetylene bottles.

4. Burning/glowing material (heated by hot metal and fanned by rotor windage) is a source of ignition and/or fires.

5. Windage caused by rotor rotation creates turbulent conditions, effectively mixing vapors and dusts with air for maximum distribution and deflagration combustion efficiency.

6. The intimate mixing of incoming waste that occurs in a shredder may create an explosion from two or more innocuous materials.

Three general classes of material can cause shredder explosions:

1. *Self-Oxidants.* Explosions of dynamite and military ordnance will usually cause a *supersonic shock wave* in the vicinity of the explosion. Venting, microfog, or detection/suppression systems are all ineffective against this type of explosion. The only method to reduce damage is to provide extra-heavy equipment that may be exposed to such episodes. Compared with the most common vapor deflagrations, these more severe detonation explosions are somewhat rare.

Explosions caused by smokeless powder, on the other hand, are believed to propagate subsonically. Although they will not be attenuated by water microfog or by detection/suppression systems, venting definitely will reduce damage.

2. *Vapors and Gases.* The most common deflagrations in solid waste shredders are caused by vaporized volatile liquids such as gasoline, paint thinner, cases of alcohol-based perfumes, propane, and so on. All that is required is mixing of these vapors with air to reach the lower explosive limit (LEL). Fortunately, the flame front of the most commonly encountered vapor/air mixtures and explosive gas/air mixtures will propagate subsonically, and venting, microfog and

---

Editors note: In December, 1984, an explosion and fire in the raw refuse receiving area at the Akron, Ohio, waste-to-energy plant fatally injured three workers. Purportedly, solvent soaked sawdust was the cause and the shredders were not involved.

detection/suppression are all effective in reducing damage from such explosions. Flame fronts for a few mixtures propagate supersonically such as hydrogen/air and acetylene/air mixtures.

If the shock wave is supersonic, damage can be limited only by materials strong enough to withstand such pressures, and it may be impractical to design for the worst case. The exact relationship between the pressure behind a shock wave and the pressure ahead of the wave is given by Equation (1):

$$\frac{\text{Pressure behind wave}}{\text{Pressure ahead wave}} = \frac{2\,(\gamma)\,2}{(\gamma + 1)\,M} - \frac{(\gamma - 1)}{(\gamma + 1}\tag{1}$$

where:
γ = The ratio of the specific heat of a gas at constant pressure divided by the specific heat of the same gas at constant volume.
M = The Mach number of the shock wave.
In the case of air at standard sea level conditions, $\gamma = 1.4$, so that for most shredder installation, Equation (1) reduces to Equation (2):

$$\frac{\text{Pressure behind wave}}{\text{Pressure ahead wave}} = 1.17M^2 - 0.17\tag{2}$$

Thus, for a shock wave propagating at Mach 4 (about 4400 ft/sec) (1320 m/sec) the pressure behind a shock wave would be 18.55 times the pressure ahead of the wave. If the wave is propagating in air at sea level, the pressure behind the shock wave would be 18.55 × 14.7 > 229 psia (> 1580 KPa). Such a pressure would result in a force of about 1.1 million pounds (0.50 MKg) on a 6 × 6 ft (1.8 × 1.8 m) plate.

It is not likely that such pressure will occur in practice in solid waste shredders because there is a run-up distance required before a flame-front velocity becomes supersonic, and it is doubtful that there is sufficient unobstructed distance in most shredder installations for such events. However, covered conveyers leading into or out of shredders may be sufficiently long and sufficiently similar to a shock tube to permit supersonic flame-front propagation to develop with resulting high pressures behind the shock wave.

We believe that suspended water microfog particles will slow the velocity of flame-front propagation so that supersonic velocities may not be achieved. Microfog may even extinguish the flame front in some cases.

3. *Dusts.* Explosions caused by dusts have occurred since the early days of mechanical flour milling. During 1977–1978, a series of disastrous explosions in export grain elevators seriously crippled the ability of the United States to export grain. Dust particles smaller than 200 mesh (0.075 mm), if properly mixed with air, are especially explosive. Cornstarch is a commonly discarded substance that is almost certain to explode if properly mixed with air in the presence of an ignition source. One box of cornstarch is probably sufficient to create a damaging explosion in a typical solid waste-processing installation. In his comprehensive book *Dust Explosions and Fires*,[2] K. N. Palmer tabulates the explosive characteristics of some 300 dusts, many of which are quite common—such things as flake aluminum, cellulose, calcium silicide, coconut shell, coal, egg-white, iron pyrites, magnesium, malt, barley, paper, potassium sorbate, resin, rubber, soya protein, sugar, titanium (ignites in carbon dioxide), tobacco, zinc, yeast, and zirconium (which also ignites in carbon dioxide) are included in Palmer's table. Pressures in excess of 100 psi (690 KPa) and more have been reported by Palmer in connection with dust explosions in closed vessels. Flame propagation speeds of up to 4400 ft/sec (1350 m/sec) or Mach 4 are reported. Maximum rates of rise in pressure of from 5000 to 20,000 lb/in.[2] sec (345 to 1380 bar/sec) are not uncommon.

Explosions that propagate supersonically or explosions that result in very high rates of rise in pressure will almost certainly not be relieved much by vents or by detection/suppression systems. It is likely that the propagation velocity might be somewhat attenuated by suspended water microfog particles. Furthermore, it is probable that water microfog particles will tend to adhere to dust particles, somewhat reducing the likelihood of ignition to begin with.

A possible dust hazard can exist outside the shredders, that is, inside the building housing the shredder. Secondary dust explosions are not uncommon. Such secondary explosions occur as follows: A primary explosion dislodges dusts that may have accumulated in a building and suspends these dusts in areas within the building to form an explosive dust/air mixture. Then, either the flame from the primary explosion, or another source, ignites the secondary dust/air mixture, creating an explosion that is often more serious than the primary explosion. This indicates the need for dust collection and relatively good housekeeping in any solid waste facility.

12.14.7 Remedial Measures: Explosions in Resource-Recovery Plants

1. *Visual Inspection.* The importance of visual inspection of incoming wastes cannot be overemphasized. The wastes should be inspected when: (1) loaded into the truck, (2) discharged

from the truck, (3) pushed by a loader, (4) at conveyer transfer points with a minimum fall of 6 ft (1.8 m). An inspector stationed at the "garbage fall" has a reasonable chance of seeing explosive material in the falling stream. We estimate that vigilant visual inspection may remove one-third of the explosive items in the waste stream in a high-capacity plant ($\leqq$ 1000 TPD). A low-capacity plant (< 800 TPD) using a low-burden depth on conveyors may allow the detection and removal of two-thirds of the explosive items. Visual inspection will seldom detect gasoline in a plastic bleach bottle or military ordnance inside a plastic bag. But it certainly is a step in the right direction.

2. *Public Awareness.* The necessity for an advertising campaign to allay public fears of environmental/nuisance factors (truck traffic, odors, fire, emissions or hazards, etc.) for any resource recovery plant is debatable. Instead, public relations, such as educational and awareness programs to dispel unfounded concerns with industry, schools, civic groups and citizens, generally is wise.

3. *Explosion Vents.* THE FIRST AND BEST MECHANICAL METHOD TO ALLEVIATE EXPLOSIONS IN TOP-FEED, FREE-FALL SHREDDERS IS TO PROVIDE ADEQUATE VENTING, although such vents will be relatively ineffective against explosions that propagate supersonically. Uniform vent design criteria are not yet practiced, although ASTM has recently (1983) suggested a vent ratio minimum of 1 ft²/15 ft³ (0.09 m²/0.42 m³). Others recommend allowing 1 ft² of venting area for each 30 ft³ (0.9 m²/0.84 m³) of space to be protected. We provided about 1 ft² for each 20 ft³/(0.09 m²/0.56 m³) of protected space at the New Castle County plant. Particular care must be taken to provide vents that open easily with less than 1 psi (6.9 KPa). This has been considered difficult while still containing ballistic objects. A fairly simple and effective method, however, is shown by Figure 12.122.

The authors believe that explosion vents should be a vertical extension of the shredder feed hopper and should extend through the roof of the building in which they are installed (Figure 12.123). Vents should provide straight-through flow and not change direction, as the momentum of high-velocity products of combustion makes it difficult, if not impossible, to redirect the gases without high stress and possible damage. Moreover, the vents should expand in cross-sectional area, with the largest opening at the point of discharge.

4. *Water Microfog.* Factory Mutual Research Corporation in conjunction with the U.S. Bureau of Mines has developed a quenching distance theory for maximum distance between metal plates at which an explosive mixture injected between the plates cannot be ignited. For most solvent/air mixtures, this distance has been determined to be 0.08 in. (0.2 cm), or less. No criteria had been developed for aqueous microfog as an energy absorber, but such spacing of droplets requires a theoretical minimum of 3.6×10^6 tiny particles/ft³ (0.28 m³) of space protected plus that required by make-up air. New Castle found that minimum quantities of water to meet the above criteria were not adequate to absorb the heat generated by burning a stoichiometric gasoline/air

Fig. 12.122 Ballistic containment/vent discharge design.

Fig. 12.123   Straight-through explosion vent.

mixture, and water was increased to a total of 4 gpm (15 L/min) to protect 16,000 ft³ (448 m³) of space, Figure 12.124 space, or about 2 × 10¹² particles/min.[5]

It may be desirable to provide microfog protection in ducts, plenums, and so on, downstream from size reduction equipment to prevent distribution of combustible vapor or dust when followed by delayed ignition deflagration.

5. *Detection/Suppression Systems.*   These systems usually consist of four or more pressure sensors. If any two sensors simultaneously detect a rise in pressure, say, 1 psi (6.9 KPa), a number of high-pressure bottles containing a fire suppressant are discharged at high velocity into protected spaces. It is desirable to provide a slow-release (10-sec) bottle to discharge into any ductwork (such as dust collectors) located downstream from the mills. These systems effectively extinguish the flame front before it can travel far and are effective for gas/air or dust/air mixtures that are subsonic. Suppression systems are not effective for detonations and supersonic front velocities.

Suppression systems are effective for gasoline/air explosions and other commonly encountered explosive/air mixtures. The sensor pipes leading from the protected spaces to the pressure sensors tend to plug with waste, which renders them useless. Therefore, these pipes must be cleaned daily, and more often, if necessary, in specific installations.

6. *Explosive Vapor Detection Systems.*   The authors recommend installation of wide-band explosive vapor detectors downstream from any size reduction equipment and as close as possible to its discharge. When the detectors identify a lower explosive limit (LEL) in the range of 20%, the infeed conveyer should stop automatically, and the mill rotor deluged with water for about 10 sec to

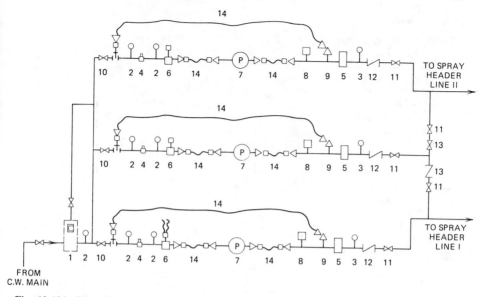

Fig. 12.124 Microfog spray system piping and layout. 1, Pressure reducing valve; 2, pressure gage; 3, pressure gage; 4, strainer; 5, strainer; 6, solenoid valve; 7, high-pressure pump and motor; 8, pulsation damper; 9, pressure relief valve; 10, valve—globe; 1, valve—needle barstock; 12, valve—horizontal check; 13, valve—vertical check; and 14, hose. *Note:* 1, pump capacity = 4 gpm @ 1000 psi; 2, operating pressure = 650 psi.

prevent friction sparks. There is little data as yet to validate this procedure, but the authors recommend it nevertheless.

7. *Personnel Protection.* Wherever possible, protective steel barricades, preferably multiple protection barricades, should be provided for employees who work in spaces that might be exposed to "shrapnel" ricochet in normal operation or to explosion flame/shock wave. Electrical control panels should be moved as far from process fire and explosion potentials as possible, consistent with operational requirements.

At New Castle, two shredder control panels were provided: one near the mills for start-up and for lock-out during maintenance, the other in the control room for remote control after mill start (control-transfer before feed start from the control room).

12.14.8 Postexplosion Procedures

1. *Generally allow discharge conveyers to keep running.* Postexplosion fires will often occur, and it is wise to move burning material out of the plant. Most fans should be turned off, especially fans associated with air classifiers, but this must be studied for each plant, as the best solution is clearly site-specific.

2. *Evacuate the Building.* This building should be evacuated immediately, and the *workers assembled and mustered in a safe predesignated place. Eager workers may wish to fight the fires; THEY MUST NOT DO SO EXCEPT UNDER CLOSE SUPERVISION.* Certain flames (alcohol, etc.) may not be visible. If inhaled, invisible flames can be fatal. Toxic gases may be generated. If personnel are missing at the muster, the manager or foreman should direct a *calm* search, using gas masks if available. Management should remain calm, and, if all workers are accounted for, the crew should generally do nothing for a short time pending calm instructions. Firefighting by the work force should be only under the very close supervision of management using a rehearsed plan whenever possible.

3. *Activated Sprinkler Systems.* Self-actuating sprinkler systems should be installed wherever possible. If manually activated systems are used, they should be activated as soon as possible, consistent with personnel safety.

4. *Call the Fire Department.* An automatic or semiautomatic alarm to the nearest fire department should be considered. Automatic-dialing telephones are relatively inexpensive and should also be considered.

REFERENCES

1. R. G. Zalosh, S. A. Wiener, and J. L. Buckley, Factory Mutual Research Corporation, "Assessment of Explosion Hazards in Refuse Shredders," prepared for the U.S. Energy Research and Development Administration under Contract No. E (49-1) − 3737, April, 1976.

2. K. N. Palmer, *Dust Explosions and Fires*, Chapman and Hall, London, England, 1973 (Halsted Press, Division of John Wiley & Sons, Inc., New York).

3. A. R. Nollet and E. T. Sherwin, "Air Classify First, Then Shred," Proceedings, ASME National Waste Processing Conference, Chicago, Illinois, May, 1978.

4. W. G. Courtney, "Frictional Ignition in Underground Coal Mining Operations," U.S. Bureau of Mines, Pittsburgh Research, P.R.C. Report No. 1762, June, 1980.

5. A. R. Nollet, E. T. Sherwin, and A. W. Madora, "An Approach to Energy Attenuation of Explosive Wastes in Processing Equipment," proceedings at the Sixth Mineral Waste Utilization Symposium, Chicago, Illinois, May, 1978.

CHAPTER 13

RESOURCE RECOVERY: MASS BURN ENERGY AND MATERIALS

MIRO DVIRKA

William F. Cosulich Assoc., Engineers
Woodbury, New York

13.1 MASS BURN ENERGY RECOVERY OVERVIEW

Energy recovery from mass burning incinerator systems has been practiced as far back as the 1930s and 1940s starting with hot water heating coils used in secondary chambers of refractory lined incinerators. Flue gas slipstream boilers were introduced in the early 1950s to generate low-pressure steam for heating and for partial electrical power generation.

Total heat recovery for low-pressure (250 psig) (1724 KPa) steam power generation was implemented in 1952 to 1954 at the Merrick (Town of Hempstead, N.Y.) incinerator plant using batch-fed refractory furnaces with tandem waste heat convection boilers.

The breakthrough in continuous gravity feed incinerator furnace design at the New York City Betts Avenue incinerator plant in the mid-1950s eventually led to the use of waterwall continuous feed furnaces developed primarily on the European continent in the 1960s while incinerator energy recovery in the United States was, for all practical purposes, abandoned. The disparity of fossil fuel, gas, and labor costs in Europe and in the United States made the use of maintenance-intensive waste heat boiler systems noncompetitive in the United States while the development of energy recovery systems in Europe continued at a rapid pace.

The implementation of the 1973 Clean Air Act eventually caused the almost total abandonment of incineration due to the high cost of required air pollution control retrofits in favor of cheaper waste disposal methods—landfills. Only after the mid-1970's worldwide oil crisis was the interest in incineration with energy recovery renewed, although it was hampered by new regulatory requirements and reluctance in public acceptance.

13.1.1 Dedicated Units: Boiler Types

The cost of incineration without revenue-producing energy recovery remains high and could be justified only in rare instances where development of new landfills is extremely costly or impossible. Because of new regulations (PURPA) requiring previously disinterested private and quasi-governmental utilities to purchase electrical energy from small producers (such as incinerator plants) at a price equal in most cases to avoided cost, the design of incinerator plants with or without partial energy recovery has changed to designs optimizing the energy recovery potential within certain constraints.

The optimization of energy recovery systems requires dedicated incinerator-boiler units where municipal solid waste (MSW) incineration is coupled directly with heat recovery. Two basic concepts used in the design of energy recovery incinerator systems are refractory furnaces followed by convection waste heat boilers and waterwall furnaces with convection boiler sections.

Each concept has its advantages and disadvantages and the selection of either system depends on specific conditions and overall economics.

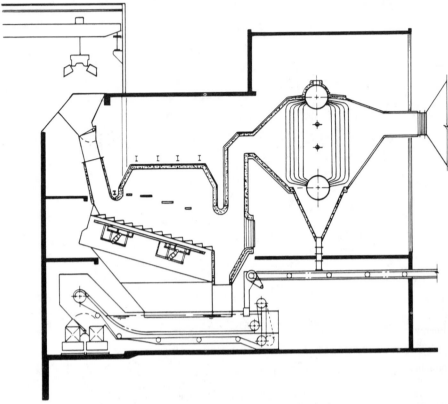

Fig. 13.1 Refractory furnace, convection boiler incinerator.

Refractory Furnace—Convection Boiler

This concept (Figure 13.1) is preferred for incineration of low calorific value, high-moisture-content wastes because of low heat flux through the furnace walls and the inherent capability of maintaining adequate combustion temperatures even with very low-quality fuel.

The "flywheel" effect of heat-storing refractory enclosures is quite helpful in maintaining steady combustion conditions, even with substantial short-term variations in fuel quality. Recommended for incineration of refuse with high heat value (HHV) of 3500 BTU/lb (8.14 MJ/kg) or less, the refractory furnace concept has been used successfully with higher HHV values refuse as well.

The low heat flux to the furnace enclosure requires higher excess air level, generally 130 to 170%, to keep the hot face of the refractory enclosure below 1800°F (982°C) slagging temperature of ash. Due to the excess air level and absence of radiant heat-absorbing, water-cooled walls, the heat recovery efficiency of the system is lower compared with waterwall furnaces.

Approximate excess air levels in refractory furnaces are given in Figure 13.2 as a function of furnace gas temperature and calorific value of refuse. The heat recovery efficiency is affected primarily by latent heat of water vapor in the flue gas, generally proportional to the HHV of the waste fuel as indicated by the family of curves in Figure 13.3.

Water-Cooled Wall Furnace Design

This concept (Figure 13.3) offers higher heat recovery rates than comparable-size refractory furnaces but somewhat lower reliability of operation. Early designs, using low-quality refuse burning in unlined waterwall chambers, had to be modified to mitigate corrosion of the waterwall tubes and to reduce the heat flux to the water-cooled walls that caused "cold furnace" operation resulting in high levels of unburned hydrocarbons in the flue gas.

Lining of the waterwall tubes with castable refractories greatly reduced the corrosion problem and, to some extent, the "cold furnace" syndrome; nevertheless, the sizing of the furnace envelope

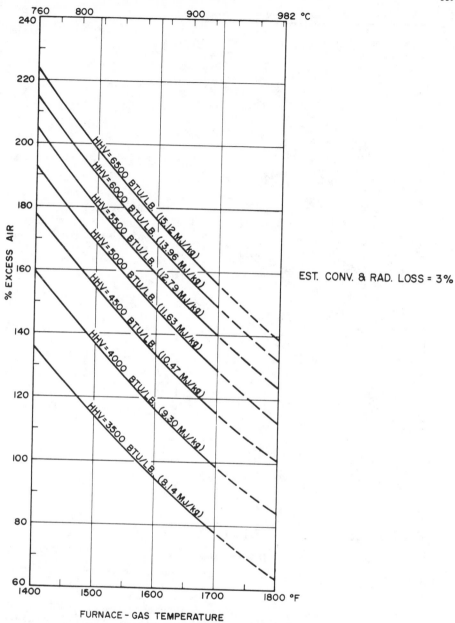

Fig. 13.2 Refractory furnaces: Excess air as a function of gas temperature and HHV.

and the control of the heat flux to the water-cooled walls remains a critical aspect of the waterwall furnace design. Figure 13.4 shows a typical waterwall furnace and convection boiler arrangement.

Although the excess air level in a refractory furnace is set to yield safe furnace wall temperatures and thus the optimum operating point, a waterwall furnace can be operated at varying levels of excess air, resulting in varying boiler efficiency.

Accurate evaluation of the combined waterwall-convection boilers' efficiencies is a tedious, largely empirical process. An estimated efficiency of a MSW fired waterwall boiler system is given in Figure 13.5 as a function of the excess air levels and the HHV of refuse. Although a range from 50 to 110% is shown, most systems operate in the 70 to 100% range.

The efficiency curves in Figure 13.5 are based on a flue gas temperature of 500°F (260°C). Because of the possibility of corrosion of downstream air pollution control systems due to the

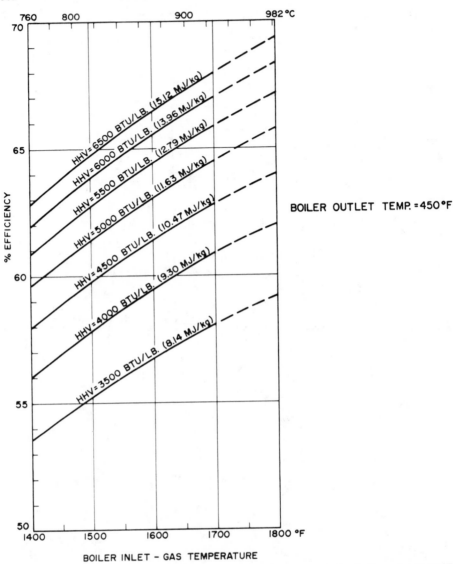

BOILER OUTLET TEMP. = 450°F

BOILER INLET – GAS TEMPERATURE

Fig. 13.3 Refractory furnaces: Approximate waste heat boiler efficiencies; boiler outlet @ 450°F.

presence of hydrogen chloride, few systems use flue gas temperatures below 450°F (232°C), while some may use temperatures as high as 550°F (288°C). Approximate efficiency correction for lower or higher flue gas temperature is shown in the same Figure 13.5.

13.2 EXISTING UNITS AND RETROFITS

The existence of refuse incinerator plants, most of them decommissioned because of failure to update obsolete air pollution control systems, can make a retrofit to energy recovery systems attractive and economically viable. Such retrofits must be done with caution and with extensive planning.

    Although it may be possible to save most of the building structure including the refuse storage pit, the process equipment, starting with refuse-handling cranes through the furnace feeders or hoppers, furnaces with stokers, residue-handling equipment, and so on, is seldom usable either due to its condition after an extended shutdown time or outdated technology. In rare instances, it may be possible to use existing refractory furnaces and update ancillary equipment including refuse-

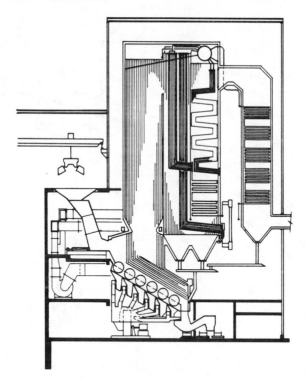

Fig. 13.4   Typical waterwall furnace convection boiler systems arrangement.

handling cranes, residue-handling conveyors, air pollution control systems, and adding convection boilers in lieu of fly ash settling chambers or gas spray cooling chambers.

As a rule, each retrofit project must be studied on its own merits. Because of a great variety of old incineration designs, there are no easy standard procedures that can be applied.

13.3   MASS BURN ENERGY PRODUCTS

Useful products of energy recovery systems are steam and electricity.

13.3.1   Constraints

Because of variations in the fuel (waste) quality, a precise control of the combustion process and furnace and gas temperatures is not possible, eliminating the application of direct high-pressure, high-temperature water-generating boilers. Where high-temperature water district heating systems or industrial processes are the best potential energy users, hot water is usually generated via heat exchangers with steam as the heating fluid.

Medium-temperature hot water and low-pressure steam applications are limited to the incineration of industrial-type wastes or biomass (wood chips, wood, bagasse) without chlorine compounds where there is no danger of low-temperature fireside chloride tube corrosion.

Similarly, the upper limit of the steam temperature must be carefully selected to avoid liquid-phase salt fireside tube corrosion and corrosion by other compounds present or formed from various compounds found in municipal solid waste stream. A rule-of-thumb steam temperature range is 300°F (149°C) to 750°F (399°C) with gas temperatures entering the superheater banks at 1500°F (816°C) or less.

Higher-temperature steam generators have been designed and built in all instances at the cost of frequent superheater tube replacement. Where higher steam temperatures are mandated by the energy user, the superheaters are generally located after several banks of saturated steam generating convection boiler tubes to reduce the gas temperatures to 1400°F (760°C) or lower, depending on the actual superheated steam temperature.

A generalized estimate of energy yield in the form of steam flow or power rate for a range of energy recovery systems is not possible because of a number of possible combinations of system-

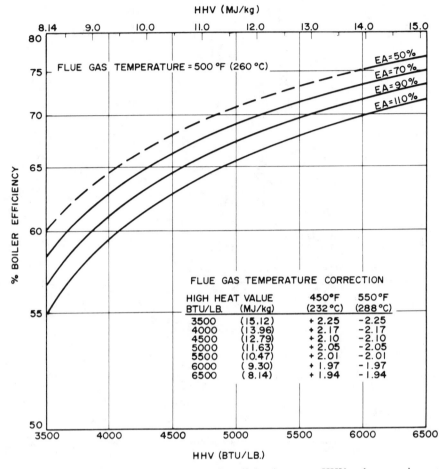

Fig. 13.5   Waterwall boiler systems: Boiler efficiencies versus HHV and excess air.

specific variables that must be evaluated for each application. Steam or power generation rates often given by overzealous salespersons "on the spot" must be viewed with utmost caution.

To facilitate the estimates of energy yields from a given system designed to service any energy user with specific energy-form requirements, three elementary energy cycles are given as a guide to a quick approximation of anticipated energy conversion potential.

13.3.2   Steam Generation

Interdependent Energy Balance Factors

The elementary steam generation cycle in Figure 13.6 shows the interdependence of various phases of the cycle, most of them project specific.

The energy balance at the boiler outlet yields the following equation:

$$W_\phi = \frac{EB(Q)/100}{h_1(1 - BD/100) - h_\phi + h_b(BD)/100} \tag{1}$$

where
W_ϕ = Feedwater mass flow LB(or Kg)/unit time.
EB = Boiler efficiency (%).
Q = Total heat input/unit time.
h_1 = Enthalpy of steam at boiler outlet.
h_ϕ = Enthalpy of boiler (economizer) feedwater.
h_b = Enthalpy of blowdown.
BD = Blowdown flow (% of feedwater flow).

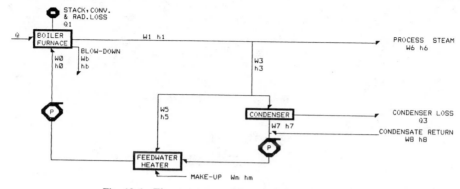

Fig. 13.6 Elementary steam generation cycle diagram.

The stack, convection and radiation losses are given by:

$$Q_1 = Q(1 - EB/100) \tag{2}$$

The net boiler outlet flange steam flow is

$$W_1 = W_\phi(1 - BD/100) \tag{3}$$

The energy balance at the feedwater heater gives an elementary relationship

$$W_\phi h_\phi - W_3 h_1 - W_8 h_8 - W_m h_m = \phi \tag{4}$$

$$\text{and } W_\phi - W_3 - W_8 - W_m = \phi \tag{4a}$$

where
W_3 = Steam for feedwater reheat (lb kg/unit time).
W_8 = Process condensate return (lb kg/unit time).
h_8 = Enthalpy of process condensate return.
W_m = Makeup water flow (lb kg/unit time).
h_m = Enthalpy makeup water.

Posing $W_m = W_6 - W_8 + W_\phi(BD/100)$, where W_6 is the process steam flow, and solving the elementary balance equations we can write that

$$W_3 = W_\phi(1 - BD/100) - W_6 \tag{4b}$$

Also, from Equation (4)

$$W_3 = \frac{W_\phi(h_\phi - BD/100 \times hm) - W_8(h_8 - hm) - W_6 hm}{h_1} \tag{5}$$

and posing Equation (4) = Equation (5)

$$W_6 = \frac{W_\phi[h_1 - h_\phi - BD/100(h_1 - hm)] + W_8(h_8 - hm)}{h_1 - hm} \tag{6}$$

Where the condensate return is given as percentage of the process steam flow, Equation (6) can be written as

$$W_6 = \frac{W_\phi[h_1 - h_\phi - BD/100(h_1 - h_m)]}{h_1 - h_m - R/100(h_8 - h_m)} \tag{6a}$$

Example 1: Given boiler heat input $Q = 75,000,000$ BTU/hr (equivalent to a 200 TPD incinerator processing 4500 BTU/lb waste); boiler efficiency $EB = 64\%$; blowdown $BD = 3\%$ of feedwater flow; feedwater enthalpy $h_\phi = 208$ BTU/lb (240°F); makeup water enthalpy $h_m = 28$ BTU/lb (60°F); process condensate return $R = 50\%$ of process steam flow; condensate enthalpy $h_8 = 148$ BTU/lb (180°F). Find actual process steam flow at 250 psig, saturated.

250 psig ≈ 264.7 psia, $h_1 = 1202$ BTU/lb, $h_b = 381.5$ BTU/lb

Feedwater flow (Equation 1)

$$W_\phi = \frac{0.64(75)10^6}{1202(0.97) - 208 + 0.03(381.5)} = 49,516 \text{ lb/hr}$$

Stack radiation and convection losses

$$Q_1 = 75(10)^6(0.36) = 27(10)^6 \text{ BTU/hr}$$

Process steam flow (Equation 6a)

$$W_6 = \frac{49,516[1202 - 208 - 0.03(1202 - 28)]}{1202 - 28 - 0.5(148 - 28)} = 42,617 \text{ lb/hr}$$

Feedwater reheat steam flow

$$W_3 = 49,516 (0.97) - 42,617 = 5414 \text{ lb/hr} \tag{6b}$$

The above equations are valid for a case where all generated steam can be sent to the user as a base load for a given process.

Where the user's steam demand fluctuates so that part of the steam flow must be wasted or diverted to a condenser, the boiler energy and flow balance will be computed using Equations (1), (2), and (3). With known process steam flow, the extraction steam flow is simply

$$W_3 = W_1 - W_6$$

and the direct steam flow to the feedwater heater is given by

$$W_5 = \frac{(W_1 - W_6 - W_\phi)h_\phi + W_8 h_8 + W_m h_m}{h_\phi - h_1} \tag{7}$$

The portion of the extraction steam diverted to the condenser is

$$W_7 = W_3 - W_5$$

Example 1a: In Example 1 set the process flow
$W_6 = 35,000 \text{ lb/hr}$.
The extraction flow will be
$W_3 = 49,516 (1 - 3/100) - 35,000 = 13,030 \text{ lb/hr}$.
$W_1 = 48,030 \text{ lb/hr}$.
The direct steam flow to the feedwater heater is

$$W_5 = \frac{(13,030 - 49,516)208 + 17,500(148) + [17,500 + 0.03(49,516)]28}{208 - 1202} = 4495 \text{ lb/hr}.$$

The condenser steam flow will then be
$W_7 = 13,030 - 4495 = 8535 \text{ lb/hr}$.

13.3.3 Power Generation

Interdependent Power Generation Factors

Figure 13.7 shows an elementary power generation cycle diagram. As in the case of the steam generation cycle, the feedwater flow and throttle steam flow is computed from Equations (1) and (3).

The properties of extraction steam from an ideal cycle (enthalpy and moisture content) at a given extraction pressure, used for feedwater reheat, are found in the Mollier chart following the constant entropy line starting with the throttle steam conditions (Figure 13.8).

Given the turbine efficiency, and assigning approximately 1% for mechanical losses of the turbine, the true extraction enthalpy is

$$h_3 = h_1 - (h_1 - h_3')(ET/100 - 0.01) \tag{8}$$

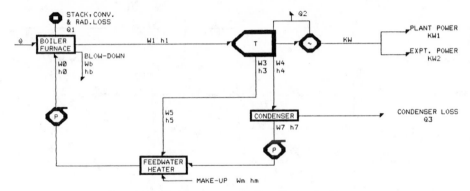

Fig. 13.7 Elementary power generation cycle diagram.

where h_3' is the ideal cycle enthalpy and ET is the turbine efficiency. The required extraction flow $W_3(= W_5)$ with enthalpy $h_3(= h_5)$ is given by

$$W_3 = \frac{W_\phi h_\phi - W_1 h_7 - W_8 h_8 - [W_\phi(BD)/100 - W_8]hm}{h_3 - h_7} \tag{9}$$

The turbine exhaust flow is given by

$$W_9 = W_1 - W_3$$

and the turbine exhaust enthalpy for a given exhaust pressure (vacuum) is computed in the same way as the enthalpy of the extraction flow.

The net turbine-generator set energy output is

$$KW = \frac{\{W_1 h_1 - W_3[h_3 + 0.01(h_1 - h_3')] - W_4[h_4 + 0.01(h_1 - h_4')]\}EG/100}{3413} \tag{10}$$

where EG is the efficiency of the generator set (generally about 97–98%).

The mechanical losses of the turbine-generator set can be approximated as

$$Q_2 = 0.01[(h_1 - h_3')W_3 + (h_1 - h_4')W_4] + 3413\,KW(1 - EG/100) \tag{11}$$

In the absence of turbine manufacturer's data, the turbine efficiency can be conservatively estimated from an empirical curve in Figure 13.9 based on the anticipated net power output.

13.3.4 Cogeneration

Application Criteria and Output Calculations

In many instances the energy market in the form of steam is more attractive in terms of overall economics than an electric power generation alternative but the steam demand is too small or fluctuating to the extent that the use of the steam output as a baseload for the user is not possible if the resource recovery plant must dispose of a given quantity of waste. In such a case, a cogeneration concept may offer a better solution than a steam generation system where part of the steam flow would have to be wasted or passed through a condenser to minimize the use of excessive makeup water flow.

Approximate energy output of a cogeneration system is given by a modified Equation (9), that is,

$$W_3 = \frac{W_\phi h_\phi - W_1 h_7 - W_8 h_8 - [W_\phi(BD)/100 - W_8 + W_6]h_m + W_6 h_6}{h_3 - h_7} \tag{12}$$

The values W_ϕ and W_1 are given by Equations (1) and (3), respectively, and electrical power output is given by Equation (10).

A schematic cogeneration diagram is shown in Figure 13.10.

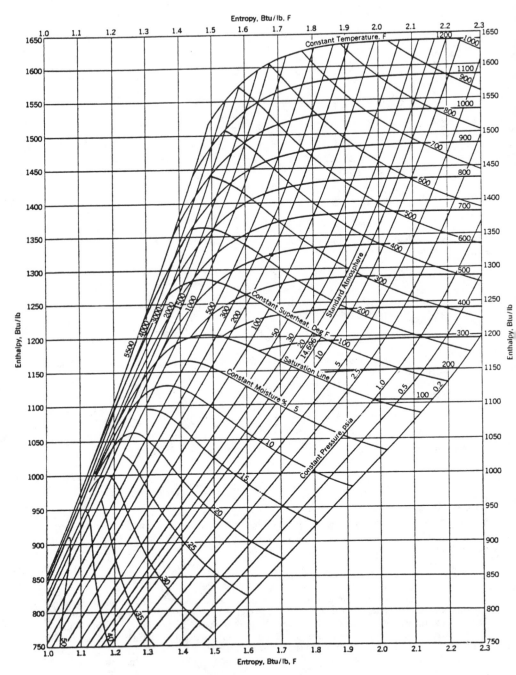

Fig. 13.8 Mollier chart for steam.

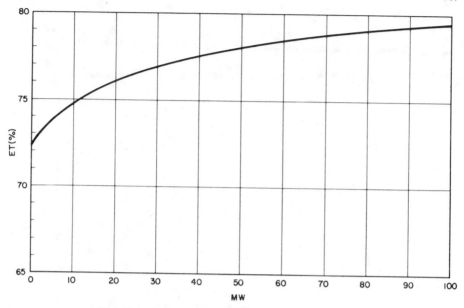

Fig. 13.9 Approximate turbine-generator set efficiency.

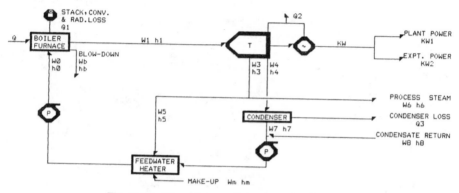

Fig. 13.10 Elementary cogeneration cycle diagram.

13.4 CODISPOSAL

Codisposal of sewage sludges and solid waste can be an economically viable alternative to land disposal due to increasingly stringent constraints imposed by environmental control agencies on landfills.

Codisposal systems can be segregated into two main concepts: (1) Coburning (in suspension) of dewatered and predried sludge above grate-fired refuse, (2) coburning of dewatered sludge in mass-fired refuse incineration furnaces layered above refuse in furnace feed.

13.4.1 Coburning (in suspension) of Predried Sludge above Grate-Fired Refuse

Coburning of partially dewatered sludge has been practiced using various degrees of sludge dewatering by mechanical and other means.

The system in Figure 13.11 consists of a separate waterwall combustion train designed for municipal refuse sludge with 26% solids content. The calorific value of municipal refuse could vary from about 3500 to 4500 BTU/lb. The sludge received from a sewage treatment plant at a consistency of about 5 to 10% solids is dewatered by centrifuges to approximately 26% solids content

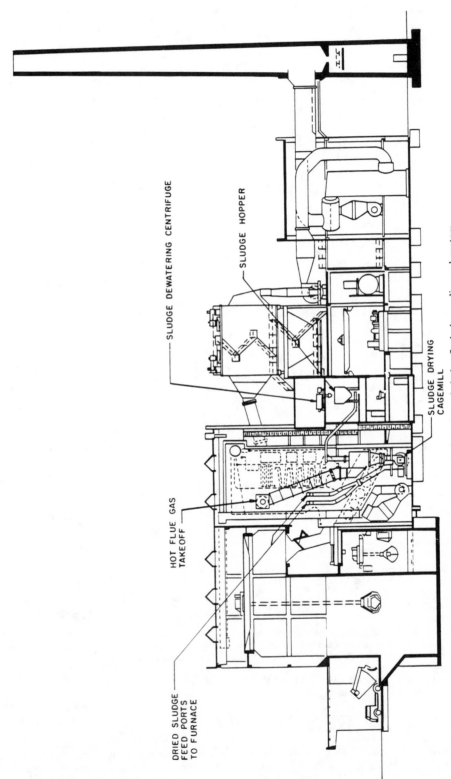

SLUDGE DEWATERING CENTRIFUGE

SLUDGE HOPPER

HOT FLUE GAS TAKEOFF

SLUDGE DRYING CAGEMILL

DRIED SLUDGE FEED PORTS TO FURNACE

Fig. 13.11 Mass burning furnace and sludge flash dry codisposal system.

before the flash-drying process. The drying of the sludge takes place in a combination mini-hammermill–material-handling fan developed and used originally for milling and drying of lignite and brown coal. The sludge is dried by a flow of hot flue gas entering the system on the suction side of the fan and is then discharged into the furnace together with the flue gas and evaporated moisture. To insure intimate contact between the sludge particles and the hot flue gas, the sludge is fed into the flash dryer via a steam atomizing nozzle using approximately 0.2 tons of steam per ton of centrifuged sludge, using approximately .18 to .22 tons of steam per ton of centrifugal sludge depending on the dewatering efficiency of the centrifuges and characteristics of the sludge.

The critical design considerations are the furnace and flue gas temperatures and the sludge-to-solid-waste ratio. A momentary drop in calorific value of refuse could result in substantial drop in furnace temperature, incomplete sludge drying, and deposits of partially burned sludge particles on boiler tubes. This potential problem can be alleviated by careful monitoring of the furnace temperatures and adjustment of sludge feed rate. A more positive remedy would be automation of the combustion process with a feedback to the sludge feed system.

Alternate methods of sludge conditioning include rotary kiln flue gas dryers, multiple hearth flue gas dryers, indirect steam dryers, and more sophisticated processes, such as the Carver-Greenfield "separator-dryer."

As most of the sludge is burned in suspension, the amount of airborne fly ash is substantially higher than with only municipal solid waste mass-burning and direct coburning systems having lower gas velocities in the boiler, larger tube spacing but somewhat larger overall boiler size.

13.4.2 Coburning Dewatered Sludge, Layered with Refuse in Furnace Feed

Figure 13.12 shows a direct coburning system where wet sludge is burned directly in a refractory furnace. A 20% ± solids consistency sludge is fed in a thin layer, resting on top of the bed of refuse. In the 30 to 45 min residence time in the furnace, the sludge first dries, then burns as it progresses through the furnaces. This concept offers a high degree of reliability when combined with fully automated combustion process control system.

Several systems to incinerate pretreated sludge together with unprepared municipal solid waste have been used since the mid-1950s with various degrees of success. Two basic problems have been encountered in the operation of the codisposal systems: the preconditioning of the sludge to proper consistency where it can be handled efficiently by a conventional incinerator furnace, and the combustion efficiency of older incinerator systems as such.

It was found that a sludge cake with 25 to 50% solids concentration would have to be reduced to a particle size of about ¾ in (19 mm) or smaller in order to be properly incinerated on an incinerator stoker. It was also found that sludge of this consistency is difficult to precondition and that reagglomeration of the cake particles is a major impediment to efficient combustion. Agglomerated cake is usually burned only on the surface, the inner part remaining unburned and insulated by the crust of the cake.

In later work it was found that sludge dewatered to 20% solids content or less will mix with unprepared refuse much more readily when charged into a refuse-feed hopper of a continuous-feed furnace, resulting in dispersion and better contact of the sludge and refuse particles and minimizing the potential of reagglomeration of the sludge cake. The approximate combustibility limits are indicated in Figure 13.13. The sludge cake consisting of 50%+ solids, however, can be readily incinerated in a conventional incinerator furnace, together with typical municipal refuse either as suspension-fired fuel fraction or stoker-fed fuel, depending on its consistency.

As to coburning of 20% solids sludge with unprepared refuse, the process must include a high-combustion-efficiency stoker such as a reverse-reciprocating, double-reciprocating (kascade) or drum roller type that can provide gentle agitation and thorough mixing of the burning fuel bed. Earlier experience shows that traveling, low-step, single-reciprocating or rocking-type stokers will not yield satisfactory results due to either the lack of fuel bed agitation or, as in the case of rocking grates, large open grate area resulting in the loss of unburned sludge particles to residue and siftings handling systems.

Where the free moisture fraction of the partially dewatered sludge is introduced into the refuse incinerator furnace in the form of vapor or together with the sludge, the sludge feed rate and the combustion process itself must be carefully monitored and controlled to prevent excessive suppression and fluctuation of furnace temperatures.

In addition to the high-combustion-efficiency stokers, the combustion volume of the furnaces must be increased to provide for adequate reaction time increased by the presence of water vapor in the flue gas. Where power generation is contemplated, automated combustion control systems are mandatory.

All systems handling mixed, partially dewatered or dried sludge and typical municipal refuse are subject to limitations given by the specific characteristics of the two dissimilar components of the final "fuel" mix. The combustion limits of municipal refuse have been determined by experience and are given by proportions of refuse components not exceeding 60% ash, 50% free moisture, and

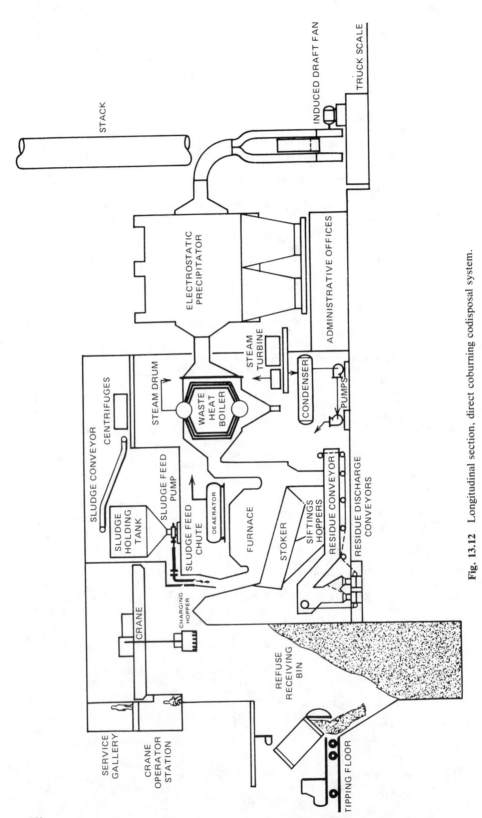

Fig. 13.12 Longitudinal section, direct coburning codisposal system.

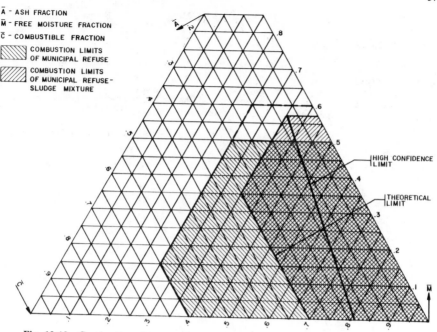

Fig. 13.13 Combustion characteristics of municipal refuse and refuse/sludge mix.

not less than 25% combustibles. Some sources claim a combustion limit up to 60% free moisture. However, in most instances, auxiliary fuel is used where the moisture content exceeds 50%.

13.5 FIELD-ERECTED UNITS: SYSTEMS AND SIZING

Dedicated energy recovery plants consist generally of two or more individual boiler units with single-unit capacities ranging from approximately 200 tons/day (TPD) to 1000 TPD. Although theoretically there is no upper limit, practical and economic considerations usually restrict the number of individual units to three or four; yet a five-unit plant is not necessarily to be ruled out.

The sizing of individual boiler units and the determination of their number is extremely site specific. It is dictated either by the potential energy market or waste disposal requirements or, in some instances, by both and must be based on mechanical availability of a given system.

Statistical data indicate a mechanical availability of proven systems of 85 to 90%. Adding approximately 5% for unscheduled downtime, 80 to 85% overall availability is achievable, yielding a scheduled downtime of 4 to 8 weeks per year for major maintenance and overhaul work and about 2 to 3 weeks for unforeseen downtime.

Where a constant energy yield or a constant daily plant capacity are mandated, the plant sizing would require redundancy equal to at least the capacity of one unit. For instance, in a two-unit facility, each unit would be sized for peak capacity: one running and one in a standby mode.

In a three-unit plant, each unit would have a rated capacity of 50% of the total demand; a four-unit plant would consist of units each rated at 33.3% of the total demand, and so on.

As a rule, it is desirable to plan a resource recovery facility to provide a baseload to an energy consumer with its own redundant capacity so that the plant can be sized to satisfy the peak waste disposal capacity given most often by seasonal variations in waste generation rates. Using the availability factor of 0.8, the unit and the plant peak capacity would be 125% of the average demand, an amount that should comfortably cover most typical seasonal variations in waste generation rates.

The established plant "peak" capacity dictates the sizing of all subsystems and components—receiving and storage pit, refuse charging crane duty cycle, residue-handling and storage systems, steam- and power-generating equipment, and the air pollution control systems. In some cases the subsystems and components must be sized to provide additional redundancy, especially where a particular system services more than one individual boiler unit.

13.6 RAW MATERIAL RECEIVING AND STORAGE

13.6.1 Pit Bunker Sizing

Comparing solid waste-fired energy recovery plants with fossil fuel, in particular coal-fired utility plants, it is apparent that the "fuel" storage and handling facilities are inordinately large in relation to the energy yield of the plants. If the calorific value of MSW is one-third to one-half of that of coal and its bulk density is about one-fourth of that of coal, the storage and handling facilities for a plant with a given energy yield firing typical municipal solid waste will be about 10 to 12 times as large as in a coal-fired plant.

Statistical data by Scott and Holmes[1] give the density of refuse stored in a bunker (storage pit) for approximately 2 days as a function of refuse depth.

An empirical equation representing the density of settled refuse at a depth D_i is

$$W_r = \frac{\ln D_i}{0.0117} + 320 \text{ lb/yd}^3 \tag{13}$$

$$\left(W_r = \frac{\ln D_i}{0.019721} + 250 \text{ kg/m}^3 \right)$$

with D_i in ft (m).

The average density of stored refuse from the top of the pile to a depth D_1 is, by integration,

$$W_r = \frac{\ln D_1}{0.0117} + 234.53 \text{ lb/yd}^3 \tag{13a}$$

$$\left(W_r = \frac{\ln D_1}{0.019721} + 139.14 \text{ kg/m}^3 \right)$$

for a level pit as shown in Figure 13.14, or

$$W_r = \frac{85.515(D_2 \ln D_2 - D_1 \ln D_1)}{D_2 - D_1} + 148.8 \text{ lb/yd}^3 \tag{13b}$$

$$\left(W_r = \frac{50.734[D_2(\ln D_2 + 1.1881) - D_1(\ln D_1 + 1.1881)]}{D_2 - D_1} \text{ kg/m}^3 \right)$$

for a pit with triangular surcharge. Figure 13.15 illustrates these typical refuse storage pit configurations.

The above equations are a good approximation for typical MSW with about 25% free moisture fraction $\bar{M}$. Since the MSW density varies primarily according to the moisture content, a density correction factor must be applied to the density computed in accordance with Equations (13), (13a), and (13b). Approximate values of the correction factor are given in Table 13.1.

The sizing of the storage pit is given by the refuse density using a value corresponding to the refuse with the lowest moisture content expected to be processed and by the plant capacity allowing for no-delivery days. For a 24 hr/day, 7 days/week operation, the pit capacity should be a minimum 2½ days and preferably 3 days of the nominal daily plant capacity.

The width/depth ratio of the pit should be, according to a good practice, about 0.6 or smaller if permitted by site and other considerations, although greater ratios are not uncommon in actual installations. The length of the pit is given generally by the number of tipping stations determined by the planned number and size of the MSW delivery vehicles. The reinforced concrete structures of the storage pits must be watertight to prevent infiltration of groundwater and degradation of refuse quality. The provisions for drainage of the pit, on the other hand, combined with elaborate sumps and sump pump arrangements, are ineffective. Any drainage system at the base of a pit usually becomes inoperable within a few days of the pit fill-up.

13.6.2 Oversized Material

The problem of oversized items, often found in MSW, is best handled with guillotine-type hydraulic shears rather than shredders if there is not very much of it. Since the size reduction to small particles is unnecessary and undesirable in a mass burning system, hydraulic shears will usually do the work at lower energy and maintenance cost than a hammermill for 25 TPD or less.

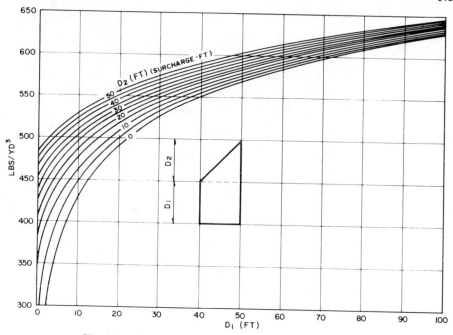

Fig. 13.14 Refuse storage pit design, average refuse density.

13.6.3 Fire and Ventilation

Ventilation of the storage pit area is best accomplished with air intakes using dust-laden and odorous air as combustion air for the combustion process. As a rule, the air intakes should be located as high above the pit as is practically possible to prevent clogging of inlet screens by airborne dust, fiber, and paper. The capture velocity of the air intake openings should be in the range of 500 to 1000 ft/min (2.5 to 5.0 m/sec), preferably closer to the lower value.

Fire protection of storage pits deserves special attention. Unlike other storage areas, the use of sprinkler systems is to be avoided. Since a pit fire generally starts at an isolated point, it is best observed and controlled from the crane operator's station using a high-capacity water gun. Sprinkler systems create a dense vapor which, combined with smoke from smoldering refuse in the pit, reduces the visibility to a point where effective fire-fighting efforts may be seriously hindered.

More extensive fires are best controlled from fire stations located strategically at the tipping floor area.

13.7 RETRIEVAL AND FURNACE FEED

13.7.1 Crane Design Criteria

Because of the heterogeneous nature of refuse and large-volume storage requirements, flat floor storage and refuse-handling bulldozer–conveyor-type retrieval systems are impractical in plants with a capacity of about 100 TPD and larger, although they do offer a better opportunity to segregate bulky objects unfit for the furnace feed systems. Crane and grapple arrangements are used exclusively due to their capability in handling heterogeneous materials. Smaller plants in a capacity range from 100 to 150 TPD in the past have used one-directional monorail or trolley-type cranes, since the storage pit can be narrow and elongated, thus allowing the crane grapple to cover the better part of the pit width.

Larger-capacity plants exclusively use bridge crane systems where the bridge travel covers the entire length of the storage pit and the trolley travel covers the width of the pit. The bridge-mounted crane operator's stations offer good visibility of the entire storage pit area and furnace charging hoppers. Nevertheless, from the point of view of safety and workplace environment, they are undesirable. The air conditioning and ventilation systems are generally ineffective in a dusty and odorous environment. More importantly, the access and egress from a bridge-mounted operator's station is limited and can result in injuries during storage pit fire episodes. Remote-control station-

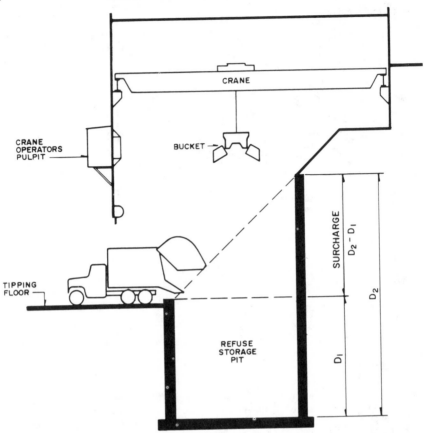

Fig. 13.15 Refuse storage pit configuration.

ary operator's stations with isolated access and egress are a standard design practice. Poorer field visibility, inherent to a stationary operator's station, can be compensated for by bridge positioning lights and trolley travel automation. Mirrors can enhance the view of the furnace charge hoppers. The remote control of the crane subsystems (bridge, grapple, trolley) has been greatly simplified in recent installations by the application of radio signal control systems, thus eliminating the need for complex remote control wiring.

13.7.2 Crane Feed Cycle Design Criteria

Fully automated computer-controlled crane systems, based on a "grid pick-up" concept, can be used successfully only in isolated instances where the delivered refuse is of a uniform consistency and density. In most cases, however, the wide variation in as-received refuse characteristics mandates visual judgment by the crane operator to provide as uniform as possible feed stock to the furnaces by stockpiling and mixing of waste streams of different consistencies.

The bridge crane design criteria must take into consideration the reliability factor comparable to cranes designed for steel mill duty (class F) which is the most stringent in the industry except for nuclear power plant requirements. Industrial, or modified industrial-duty cranes, are no longer acceptable in energy recovery plants.

Table 13.1 Pit Design: Moisture Correction Factor for MSW Density

$\bar{M}$	0	0.05	0.10	0.15	0.20	0.25	0.30	0.35	0.40	0.45	0.50
f_w	0.68	0.72	0.77	0.83	0.91	1.00	1.12	1.25	1.40	1.58	1.79

$\bar{M}$, Moisture, lb/lb of refuse.
f_w, Density correction factor.

The duty cycle of a refuse-handling crane should conform to minimum requirements established by experience yielding a charging time of 30 to 35%, a restocking time of 35 to 50%, and an idle time of 15 to 35%. All bridge and trolley motors should be sized for at least 115% of the design load and the hoisting and grapple motors should have a rating of at least 125% of the design load.

The grapples are generally of a clamshell-tine or orange-peel type, depending on the configuration of the furnace feed hoppers. Commonly used clamshell grapples with sheave and cable control mounted on the crane trolley require frequent closing-opening cable replacement. The tendency of the grapple to rise while in the closing mode yields a payload generally below that of the true rated capacity of the grapple, a factor that must be taken into consideration in the duty cycle design.

Electrohydraulic grapples, with an opening and closing mechanism mounted directly on the grapple, offer a greater actual payload but require a higher degree of maintenance. Electromechanical grapples, featuring a sheave and cable operating system mounted directly on the grapple assembly, offer a higher degree of reliability and an optimum payload characteristic combining the advantages of older mechanical and electrohydraulic grapple systems.

Since the entire operation of a resource (energy) recovery plant depends on the reliability of the charging cranes, no plant should be designed without adequate crane redundancy. Where one crane can satisfy the duty cycle requirements, a standby crane of equal capacity should be provided. Where two cranes are required by the duty cycle, each having at least 50% plant capacity rating, a standby crane of equal capacity is mandatory.

13.8 STOKER AND FURNACE DESIGN

The mass-burning combustion system design is based on the composition, quality, and characteristics of the fuel (solid waste), in this case of a largely cellulosic nature. Unlike some other solid fuels (coal), solid waste has high volatile and low fixed carbon content and relatively low calorific value. The behavior of such a fuel with frequently varying properties requires special attention in selecting proper design criteria for a site-specific facility.

13.8.1 Combustion Process Equations

Solid waste combustion characteristics are best evaluated by their proximate analysis, giving a combustible fraction $\bar{C}$, free moisture fraction $\bar{M}$, and ash and inert fraction $\bar{A}$, and the ultimate analysis of the ash and moisture free combustible fraction.

In the absence of actual calorimeter data, the HHV of the combustible fraction is closely approximated by the D.L. Wilson[2] equation.

$$HHV = 14,096(C_o) + 64,679(H - 0/8) + 3982(S) - 6382(C_i) + 2137(0) - 1040(N) \; BTU/lb$$

$$[HHV = 32.765(C_o) + 150.3425(H - 0/8) + 9.256(S) - 14.8346(C_i) + 4.967(0) - 2.4174(N) \; MJ/kg] \tag{14}$$

With different concentrations of industry, socioeconomic, and geographic conditions, the HHV of municipal-type solid waste can vary from about 3500 BTU/lb (8.1355 MJ/kg) to as much as 6500 BTU/lb (15.1 MJ/kg).

The elements given in Equation (14) are in lb/lb (kg/kg) of the fuel (waste). Because of usually low sulfur concentrations in solid waste, the sulfur term is often omitted in the basic combustion computations, as is the nitrogen term, which may be of importance only in high temperature applications. The carbon designation C_o indicates carbon in substances of organic origin whereas C_i (generally about 1.2% of total carbon) is the inorganic substance carbon.

Dry stoichiometric air required for theoretical complete combustion is

$$W_a' = 4.31[2.667C + 8(H) + S - 0]\bar{C} \; lb/lb \; (kg/kg) \tag{15}$$

of refuse.

The products of stoichiometric combustion are

$$W_g' = \underbrace{3.667C}_{CO_2} + \underbrace{9(H_2) + 0.0135W_a' + \bar{M}}_{H_2O} + \underbrace{0.768W_a' + N}_{N_2} + \underbrace{1.998S}_{SO_2} \tag{16}$$

The term $0.0135W_a'$ in the H_2O summation represents moisture content in combustion air at 60% relative humidity often assumed as a typical condition.

For the computations of the composite gas constant and enthalpy, the reader is referred to standard published data.[3,4] To facilitate chain computations desirable in computer-assisted design, a good approximation of the composite gas enthalpy at atmospheric pressure and at different temperatures is obtained from a series of specific heat equations in Table 13.2.

Table 13.2 Specific Heat Cp of Gases at Atmospheric Pressure

Gas	$Cp - BTU/lb/°R$ (MJ/kg/°K)
O_2	$0.2226 + 3.8842T^{-1/3} - 61.64665T^{-2/3} + 242.16T^{-1}$ $(0.0009314 + 0.01336T^{-1/3} - 0.174309T^{-2/3} + 0.5629T^{-1})$
CO_2	$0.21363 + 6.57977T^{-1/3} - 95,4797T^{-2/3} + 334.4432T^{-1}$ $(0.0008938 + 0.02263T^{-1/3} - 0.26997T^{-2/3} + 0.7774T^{-1})$
H_2O (vapor)	$2.044 - 34.11T^{-1/3} + 238.519T^{-2/3} - 543.7228T^{-1}$ $(0.008552 - 0.11733T^{-1/3} + 0.67442T^{-2/3} - 1.26385T^{-1})$
CO	$-0.52307 + 39.358T^{-1/3} - 652.3333T^{-2/3} + 4459.7565T^{-1} - 10918.74T^{-4/3}$ $(0.00218852 + 0.1353734T^{-1/3} - 1.8445T^{-2/3} + 10.3665T^{-1} - 20.8641T^{-4/3})$
N_2	$-0.636674 + 45.55635T^{-1/3}\ 773.796T^{-2/3} + 5456.452T^{-1} - 13828.82T^{-4/3}$ $(-0.002664 + 0.1567T^{-1/3} - 2.188T^{-2/3} + 12.68322T^{-1} - 26.42482T^{-4/3})$
Air (dry)	$-0.301323 + 29.6T^{-1/3} - 503.71536T^{-2/3} + 3485.9662T^{-1} - 8588.92T^{-4/3}$ $(-0.00126 + 0.1018155T^{-1/3} - 1.42428T^{-2/3} - 8.103T^{-1} - 16.412T^{-4/3})$

$T = °R(°K)$

The enthalpy of the gases at a temperature T_g above a given (ambient) baseline temperature T_a is

$$hg = \int_{Ta}^{Tg} Cp \, dT \tag{17}$$

For example, the enthalpy of O_2 above the baseline temperature of 530°R (294.44°K) is

$$ho = 0.2226Tg + 5.8263Tg^{2/3} - 184.94Tg^{1/3} + 242.16lnTg - 521.9346 \text{ BTU/lb} \tag{17a}$$

$$(ho = 0.0009314Tg + 0.02004Tg^{2/3} - 0.52293Tg^{1/3} + 0.56291lnTg - 0.8824 \text{ MJ/kg})$$

Excess air, required to maintain a desired combined combustion gas temperature is given by

$$Wea = (HHV - H'_g - dHHV)/ha \text{ lb/lb (kg/kg)} \tag{18}$$

of refuse,

where

    H'_g = Enthalpy of products of stoichiometric combustion in BTU/lb (MJ/kg) of refuse at the given temperature.

    $dHHV$ = The fraction of higher heat value allocated to losses (convection, radiation, and residue).

    h_a = The unit of enthalpy of air at the desired gas temperature T_g.

The excess air percentage is given as

$$EA = \frac{Wea}{W'_a} (100) \tag{18a}$$

The total products of combustion are

$$Wg = W'_g + Wea \text{ lb/lb (kg/kg)} \tag{18b}$$

of refuse and the conversion to volume is

$$Qg = Wg \, Rg \, Tg/p \tag{18c}$$

where

Qg = Gas volume per lb(kg) of refuse.

Rg = Gas constant.

p = Atmospheric pressure in lb/ft^2 (kg/m^2).

    Statistically typical combustible fraction $\bar{C}$ of municipal-type solid waste consists, with minor variations, of 49.7% carbon, 7.05% hydrogen, 3.94% nitrogen, and 39.31% oxygen. Approximate HHV of this combustible fraction is 9100 BTU/lb (21.152 MJ/kg). Assuming this composition, the set of combustion equations can be simplified to yield

Dry stoichiometric air $W'_a = 6.449\check{C}$ lb/lb (kg/kg) of refuse

60% relative humidity stoichiometric air $W_a = 6.537\check{C}$ lb/lb (kg/kg) of refuse

Products of stoichiometric combustion $W'_g = 7.537\check{C} + \check{M}$ lb/lb (kg/kg) of refuse

Total air $Wa = 6.537\check{C} + W_{ea}$ lb/lb (kg/kg) of refuse

Total products of combustion $Wg = 7.537\check{C} + \check{M} + Wea$ lb/lb (kg/kg) of refuse

The term $7.537\check{C}$ consists of $1.822\check{C}$ of CO_2, $0.723\check{C}$ of H_2O, and $4.992\check{C}$ of N_2, all in lb/lb (kg/kg) of refuse.

The enthalpy equation of the products of stoichiometric combustion of the typical composition refuse is reduced to

$$
\begin{aligned}
H'_g = &\, (-1.13112345\check{C} + 2.044\check{M})T \\
&+ (322.115\check{C} - 51.16734\check{M})T^{2/3} \\
&+ (-11{,}592.908\check{C} + 715.5567\check{M})T^{1/3} \\
&+ (27{,}454.852\check{C} - 543.7228\check{M})\ln T \\
&+ (207{,}100.43\check{C})T^{-1/3} - 123{,}635.204\check{C} + 938.82\check{M} \\
&\quad \text{BTU/lb of refuse}
\end{aligned}
$$

The condensed typical waste enthalpy equations can be used in a number of instances where the variations of the waste quality is related primarily to the moisture and inert (ash) fractions. Where the ultimate analysis of the combustible fraction $\check{C}$ is substantially different, a set of project-specific condensed equations can be easily derived in each case.

13.8.2 Stoker Design

The design of stokers in terms of surface burning rates, combustion efficiency, and related required effective areas is dependent on a number of variables, including fuel (waste) quality factor given by its HHV and fuel (waste) bed correction factor based on primary (underfire) air flow and its temperature, physical fuel (waste) characteristics, the stoker combustion efficiency, and the fuel bed depth itself.

The general "primary" burning rate F_A, dependent directly on the HHV of the refuse, is

$$
\begin{aligned}
F_A &= K_L\, Ca^{1/3}/3 \ \text{lb/ft}^2/\text{hr} \\
(F_A &= K_L \times 6.35449 \times Ca^{1/3}/3 \ \text{kg/m}^2/\text{hr})
\end{aligned}
\tag{19}
$$

where

K_L = Waste quality factor.

Ca = Stoker capacity in lb/hr (kg/hr).[5-10]

For waste with a HHV B below approximately 3000 BTU/lb (6.972 MJ/kg) the K_L factor is computed from

$$
\begin{aligned}
K_L &= 0.155973 \ \sqrt{B} \\
(K_L &= 3.235 \ \sqrt{B})
\end{aligned}
\tag{20}
$$

and for waste with a HHV B equal or higher than 3000 BTU/lb (6.972 MJ/kg) from

$$
\begin{aligned}
K_L &= 0.0010933B + 5.207 \\
(K_L &= 0.47035B + 5.207)
\end{aligned}
\tag{20a}
$$

The above empirical equations are represented graphically in Figure 13.16.

The actual burning rate is

$$
F'_A = F_A \frac{G_A}{W'_a}\left(\frac{\gamma}{a}\right) av. \ \text{lb/ft}^2/\text{hr} \ (\text{kg/m}^2/\text{hr})
\tag{21}
$$

The term G_A/W'_a is the ratio of actual primary air flow G_A and required stoichiometric air W'_a. Although the actual air flow G_A could theoretically be higher than W'_a, good practice indicates a maximum ratio of 1.0.

In many instances, especially in the case of large-capacity furnaces, the G_A/W'_a ratio is limited by manufacturer's criteria of heat release per unit area of the stoker surface, ranging generally from 300,000 BTU/ft²/hr (9235.5 MJ/m²/hr) to 350,000 BTU/ft²/hr (10,774.8 MJ/m²/hr).

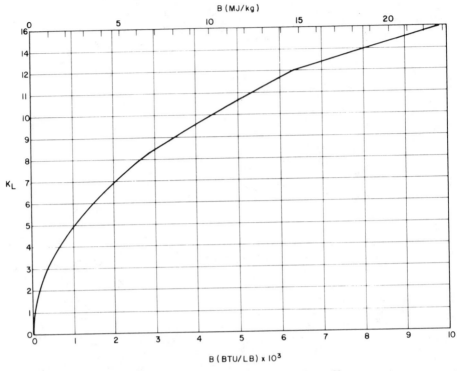

Fig. 13.16 Primary refuse burning rate factor K_L.

G_A/W_a' ratios higher than 1.0, where permitted by the maximum stoker area heat release, can result in excessive entrainment of ash into the gas stream, often causing serious slagging of the furnace walls at elevated gas temperatures, fouling of the superheater and boiler convection sections, and ultimately overtaxing of the air pollution control systems.

The term $(\gamma/a)av$ is the combined fuel (waste) bed depth correction factor, including correction for primary air temperature C'/C, stoker combustion efficiency a, fuel bed depth D, and fuel (waste) characteristics factor α'/d combining the influence of fuel "particle size" and other properties such as the ratio of volatile matter versus fixed carbon in the combustible fraction $\bar{C}$.

The primary air correction C'/C is 1.0 for ambient temperature and empirically approximately 1.29 for air preheated to about 320°F (160°C). A wide range of values for higher temperatures is not available and extrapolations should be based on actual test data specific to various stoker systems.

The fuel bed depth correction term is

$$\frac{\gamma}{a} = 6.045 \frac{C'}{C(a)} [1 - e^{-D\alpha'/d}] \tag{21a}$$

$$\left(\frac{\gamma}{a} = 6.045 \frac{C'}{C(a)} [1 - e^{-3.28084D\alpha'/d}] \right)$$

or, for ambient air

$$\frac{\gamma}{a} = \frac{6.045}{a} [1 - e^{-D\alpha'/d}] \tag{21b}$$

$$\left(\frac{\gamma}{a} = \frac{6.045}{a} [1 - e^{-3.28084D\alpha'/d}] \right)$$

The values of stoker combustion efficiencies are given in Table 13.3 and the waste characteristics factor is

$$\frac{\alpha'}{d} = \frac{\ln [a/(a - 0.1654)]}{1.75} \tag{21c}$$

Table 13.3   Stoker Combustion Efficiency Factors

a	Type of Stoker
$1 + 0.2D$ $(1 + 0.656D)$	Traveling (chain) grate
$1 + 0.1635D$ $(1 + 0.5364D)$	Reciprocating (low step)
$1 + 0.127D$ $(1 + 0.41667D)$	Reciprocating (high step), rocking
$1 + 0.08925D$ $(1 + 0.2929D)$	Kascade, reverse reciprocating, rotary drum

D = ft (m).

The different types of stokers are shown schematically in Figure 13.17. Equations (21a) and (21b) are valid for a fuel bed of a uniform thickness. For a fuel bed with the thickness varying from D_1 and D_2, the average bed thickness correction is

$$\left(\frac{\gamma}{a}\right)_{av}\Bigg]_{D_1}^{D_2} = \frac{1}{D_2 - D_1}\left[\sum_0^{D_2}(\gamma/a) - \sum_0^{D_1}(\gamma/a)\right] \tag{21d}$$

$$\left(\left(\frac{\gamma}{a}\right)_{av}\Bigg]_{D_1}^{D_2} = \frac{0.3048}{D_2 - D_1}\left[\sum_0^{D_2}(\gamma/a) - \sum_0^{D_1}(\gamma/a)\right]\right)$$

Curves giving the values of $\sum_0^{D_i}(\gamma/a)$ for most common types of stokers are shown in Figure 13.18.

The average fuel bed correction for a triangular-shape fuel bed is

$$\left(\frac{\gamma}{a}\right)_{av} = \frac{1}{D}\left[\sum_0^D(\gamma/a)\right] \tag{21e}$$

$$\left(\left(\frac{\gamma}{a}\right)_{av} = \frac{0.3048}{D}\left[\sum_0^D(\gamma/a)\right]\right)$$

and for a trapezoidal-shape fuel bed

$$\left(\frac{\gamma}{a}\right)_{av}^1 = \frac{L_1(\gamma/a)_D + L_2(\gamma/a)_{av}}{L} \tag{21f}$$

where
$(\gamma/a)_D$ = A uniform bed thickness factor.
$(\gamma/a)_{av}$ = As given by Equation (21e).

The fuel bed profile parameters are indicated in Figure 13.19.

Equation (21d) permits the computation of bed depth-correction factors for multiple primary air plenum sections and, with the use of Equation (21), determines the primary air flows through each plenum and stoker section.

The application of Equation (21f) best approaches the conditions found in practice and will yield somewhat higher overall burning rate for wider furnaces as compared with narrow and long tunnel-type furnaces.

The minimum stoker width can be approximated using Rank Index and fuel input rate per unit width of the stoker recommended by American Boiler Manufacturers Association (ABMA)[11] adjusted for typical waste characteristics.

The maximum capacity of a given stoker width $W < 20$ ft (6 m) is

$$Ca = \left[\frac{(20 - W)^{1.63093}}{-15.49497} + 10.908\right]\frac{W(10)^6}{B} \text{ lb/hr} \tag{22}$$

$$\left(Ca = \left[\frac{(6.096 - W)^{1.63093}}{-2.2318} + 10.908\right]\frac{.003459\,W(10)^6}{B} \text{ kg/hr}\right)$$

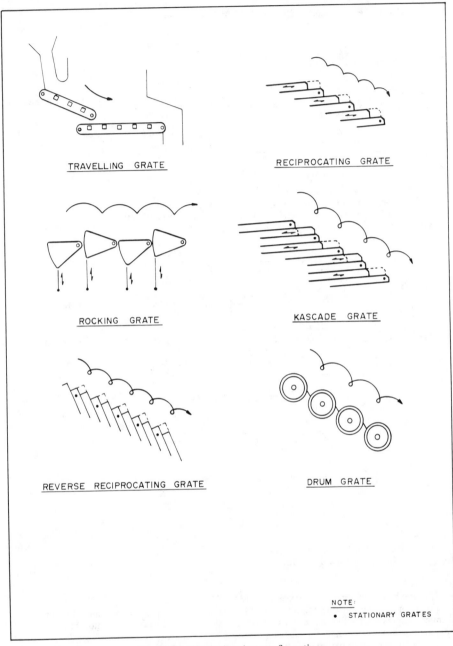

TRAVELLING GRATE

RECIPROCATING GRATE

ROCKING GRATE

KASCADE GRATE

REVERSE RECIPROCATING GRATE

DRUM GRATE

NOTE:
• STATIONARY GRATES

Fig. 13.17 Typical stoker configurations.

where
W = Actual width in ft (m).
B = HHV of the fuel in BTU/lb (MJ/kg).

For stoker widths $W > 20$ ft (6 m), the equation takes a form

$$Ca = \frac{10.908(10)^6 \, W}{B} \text{ lb/hr} \qquad (22a)$$

$$\left(Ca = \frac{37{,}730.77 \, W}{B} \text{ kg/hr}\right)$$

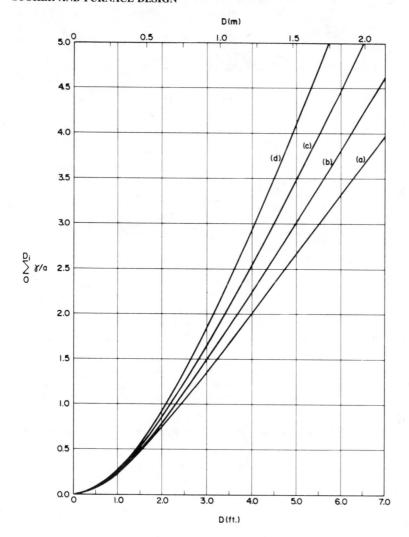

Fig. 13.18 Stoker burning rates and bed depth correction.

(a) - TRAVELLING GRATE   (c) - HIGH STEP RECIPROCATING, ROCKING GRATES
(b) - LOW STEP RECIPROCATING GRATE   (d) - KASCADE, REVERSE RECIPROCATING, DRUM GRATES

Equations (22) and (22a) are valid for higher capacity furnaces and higher calorific value fuel (waste). For lower furnace capacities and lower BTU values, the stoker width computed by Equations (22) or (22a) may yield a narrow and long furnace configuration.

To prevent such undesirable results, the stoker length/width ratio shall not exceed 2.

Figure 13.20 shows refractory furnace configurations, and Figures 13.21 through 13.24 give approximate stoker burning rates based on furnace capacities and HHV of the fuel with L/W ratios not exceeding 2 and maximum stoker surface heat release of 325,000 BTU/hr (10,774.8 MJ/m²/hr).

13.8.3 Furnace Design

Rigorous analysis of combustion chambers is one of the most difficult tasks in combustion systems design, even with constant quality fuels and ideal combustion conditions. With variable composition and quality fuels, such as MSW, an exact analysis of the combustion process and combustion chamber conditions becomes essentially impossible.

A number of more or less complex methods have been used with results suggesting that the accuracy of simplified methods, based on sound judgment and experience, can yield predicted combustion chamber conditions in close agreement with the conditions observed during actual

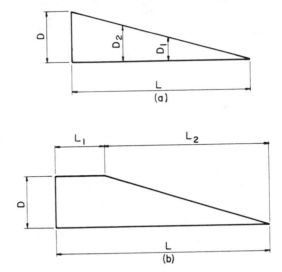

(a) TRIANGULAR BED
(b) TRAPEZOIDAL BED

Fig. 13.19 Stoker burning rates: Profile parameters for fuel bed thickness correction.

operation. The most important parameters taken into consideration in the evaluation of a combustion chamber performance are the combustion volume, emissivities of flame, combustion chamber enclosure and of the products of combustion, and the quantities and distribution of the combustion air.

Combustion volume must be sufficient to permit the completion of the combustion process within the primary and secondary combustion chambers for multiple chamber systems, or within the main combustion chamber of single chamber furnaces.

Combustion volume sizing is given by the combustion intensity of the fuel. For typical municipal refuse with HHV varying primarily with its free moisture content, the combustion intensity is expressed as

$$I = 0.005913B^{7/4} \text{ BTU/ft}^3/\text{hr} \tag{23}$$

$$(I = 6462.26B^{7/4} \text{ MJ/m}^3/\text{hr})$$

where
B = HHV of the "as received" fuel. The minimum required combustion volume is

$$V_C = \frac{Ca\,B}{I} \tag{23a}$$

with Ca as capacity in lb/hr (kg/hr).

In practice, the upper limit of combustion intensity I for solid fuels is 20,000 BTU/ft³/hr (743.84 MJ/m³/hr).

Elementary heat balance dictates the total enthalpy of the gas leaving the combustion chamber as

$$Hg = Q - \Delta Q \tag{24}$$

where
Q = Total fuel heat input per unit time.
ΔQ = Sum of heat losses consisting of convection and radiation losses from the furnace enclosure and grates, and heat lost in the inert and unburned residue or carbon.

The temperature of the gas T_3, corresponding to the total enthalpy Hg, is dependent on the heat exchange between the radiant flame, hot face of the furnace enclosure, and the gas itself. As stated

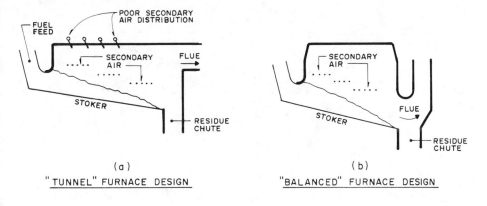

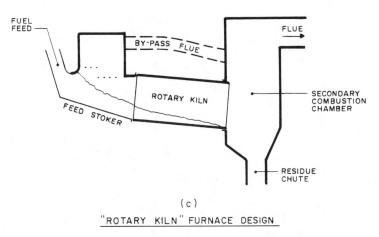

Fig. 13.20 Basic refractory furnace configurations.

above, a rigorous analysis with variable fuel composition is hardly possible; nevertheless, a simplification of the heat-exchange process concept will yield approximate results that are within a range satisfactory for the intended purpose.

The radiation from the fuel bed to the furnace enclosure in a vacuum is approximated by

$$Q_1 = Ag\sigma \, E_f[(T_1/100)^4 - (T_2/100)^4]\mathscr{F} \tag{25}$$

where

$$Ag = \text{Area of the fuel bed.}$$
$$E_f = \text{Flame emissivity.}$$
$$\sigma = \text{Stefan-Boltzman constant.}$$
$$\mathscr{F} = \text{View factor.}$$
$$T_1 \text{ and } T_2, \text{ respectively} = \text{Absolute flame and enclosure walls' temperatures.}$$

For a heat source surrounded completely by an absorbing and reradiating enclosure the view factor $\mathscr{F}$ is equal to 1. The emissivity of the flame varies with the carbon/hydrogen ratio and is approximated by an empirical polynomial expression.

$$E_f = 0.7058(10)^{-4}(C/H_2)^5 - 0.2607(10)^{-2}(C/H_2)^4 + 0.033375(C/H_2)^3 \tag{25a}$$
$$- .018(C/H_2)^2 + 0.4837(C/H_2) - 0.2098$$

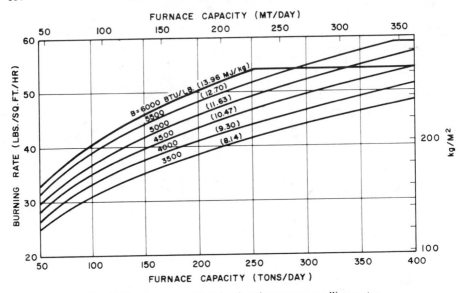

Fig. 13.21   Approximate stoker burning rates; travelling grates.

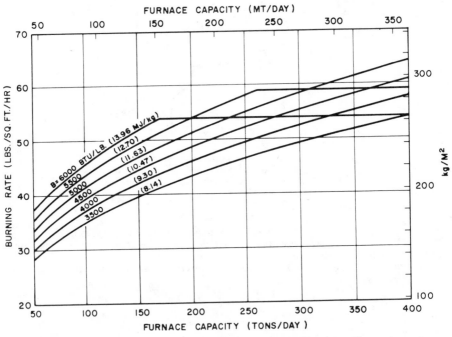

Fig. 13.22   Approximate stoker burning rates; low step reciprocating grates.

The convection heat exchange between the gas and the furnace enclosure is roughly

$$Q_2 = 2950 \, Ae \, \frac{(T_2 - T_3)}{T_3} \text{ BTU/hr}$$ (25b)

$$\left(Q_2 = 3.11 \, Ae \, \frac{(T_2 - T_3)}{T_3} \text{ MJ/hr} \right)$$

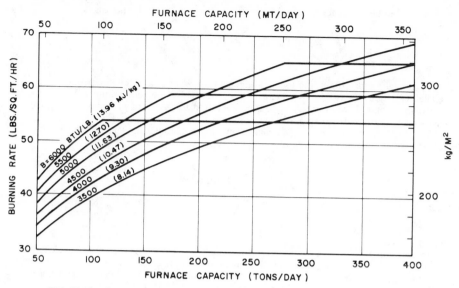

Fig. 13.23 Approximate stoker burning rates; high step reciprocating grates.

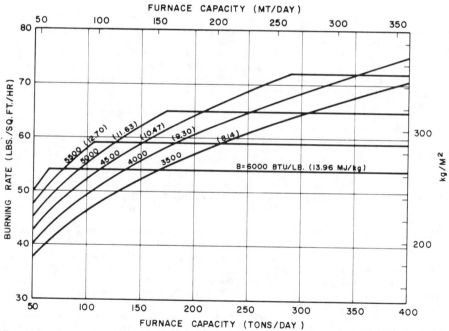

Fig. 13.24 Approximate stoker burning rates for Kascade reverse reciprocating and rotary drum grates.

where

Ae = The furnace enclosure area.

T_2 and T_3 = Absolute temperatures of the enclosure walls and gas, respectively.

The radiant heat transfer from the fuel bed (flame) to the gas is given by

$$Q_3 = A_g \, \sigma \, E_f \left[\alpha'_{G1} \left(\frac{T_1}{100} \right)^4 - E_G \left(\frac{T_3}{100} \right)^4 \right] \tag{25c}$$

and the radiant heat transfer from (to) the furnace enclosure to (from) the gas is

$$Q_4 = A_e \, \sigma \, E_e \left[\alpha''_{G1} \left(\frac{T_2}{100} \right)^4 - E_G \left(\frac{T_3}{100} \right)^4 \right] \qquad (25d)$$

where

E_e = Emissivity (absorbtivity) of the furnace enclosure walls at absolute temperature T_2.

The derivation of gas emissivity (absorbtivity) E_G and combined emittances α'_{G1} and α''_{G1} is beyond the scope of this work. The reader is referred to published literature dealing with the radiation heat transfer between gases and solids.[12,13]

The "gas beam dimension," L_G, which is one of the variables influencing the emissivities of CO_2 and H_2O, is approximated by

$$L_G = \frac{1.7 \, WLH}{WL + LH + WH} \qquad (26)$$

where

W = Furnace width.
L = Furnace length.
H = Furnace height.

Having simplified the problem based on the assumption that within the furnace enclosure there are three distinct mean temperature levels (flame, walls, and gas), and that the major component of ΔQ is the radiation and convection heat loss from the furnace enclosure, the net heat balance takes the form

$$Q_1 - Q_2 - Q_3 - Q_4 - \Delta Q = \phi \qquad (27)$$

If one of the temperatures is selected as a fixed value, then the other two temperatures can best be determined by successive approximation assigning a series of values to the second temperature and computing the third temperature for each case.

In the case of refractory furnaces, the critical temperature is the temperature of the hot face of the refractory walls T_2 which must be kept at or below 1800°F (982°C) to minimize the slagging of the walls. Computing the gas enthalpy by Equation (24) and assigning a series of values to the gas temperature T_3, the flame temperature T_1 is calculated so that the heat transfer rates satisfy the equilibrium Equation (27).

To satisfy Equation (27) applied to a refractory enclosure furnace, the highest system temperature is T_1 (flame) and the lowest temperature is T_3 (gas).

In a water-cooled wall furnace, however, the gas temperature T_3 is generally higher than the temperature T_2 of the hot face of the walls, whereas the flame temperature T_1 is, by necessity, the highest. This condition is due to a high heat absorption rate of the water-cooled walls, which is considered part of the "heat loss" Q for the purpose of the equilibrium Equation (27).

If the total area of the water-cooled walls is not carefully proportioned and the heat flux to the walls not adequately controlled, the gas temperature T_3 can easily fall below adequate combustion level, generally 1400°F (760°C), resulting in poor combustion conditions, partial extinction, and emissions of CO and unburned hydrocarbons.

The potential of a "cold furnace" syndrome is best demonstrated by comparing combustion volume requirements for coal- and solid waste-fired furnaces. Using 12,000 BTU/lb (27.9 MJ/kg) coal and 4500 BTU/lb (10.46 MJ/kg) refuse, the respective combustion intensities are 20,000 BTU/ft³/hr (743.84 MJ/m³/hr) and 14,620 BTU/ft³/hr (543.75 MJ/m³/hr). For the same total heat input per unit time, the refuse-burning furnace volume has to be 36.8% larger than the coal-fired furnace. Assuming a furnace with a square base and a height twice the size of the base, the wall area of the refuse fired furnace is 23.2% larger. By the same reasoning, a furnace firing refuse with a HHV of 3500 BTU/lb (8.14 MJ/kg) will have a wall area 65.2% larger than the wall area of a coal-fired furnace of the same heat input rate. Obviously, the increased rate of heat absorption by the larger water-cooled wall areas can result in excessive chilling of the products of combustion unless reduced by refractory lining to slow down the heat absorption rate.

The physical configuration of a solid waste fired furnace and the secondary air distribution are as important as the actual sizing. Narrow and long refractory furnaces with high flue takeoffs, popular in the late 1950s and mid-1960s, are no longer representative of a good combustion design practice. The uneven temperature distribution caused by the elongated furnace profile and unbalanced heat transfer result in overheating and slagging of portions of the side walls while other parts of the furnace are relatively cool. Slagging and differential expansion of the refractory enclosures cause their rapid deterioration and the need for frequent replacement.

Although the uneven temperature distribution can be mitigated to some extent by judicious distribution of secondary air, the unbalanced heat radiation between the fuel bed and the enclosure walls remains a problem.

Applying the rule of maximum length/width ratio equal to or smaller than 2, and improving the residence time and mixing of the gases within the furnace cavity by the use of bottom flue takeoff, results in a furnace configuration superior to most other combustion concepts, especially with low calorific value fuels as illustrated by sketch b in Figure 13.20 which compares the basic refractory furnace configurations.

A careful secondary air distribution is a prerequisite for a good combustion system performance with any furnace configuration concept. The introduction of the air through the roof arch of a furnace (sketch a in Figure 13.20) negates the efficient use of the furnace cavity as a true combustion volume. The air chills the gaseous products of partial combustion rich in unburned hydrocarbons and CO with internal energy too low to complete the combustion process. To achieve as complete combustion as possible, the secondary air must be introduced into the luminous portion of the flame above the fuel bed so that the remaining part of the combustion process can occur within the furnace cavity. To accomplish good mixing of the combustion air (oxygen) with the burning products of partial combustion rising from the fuel bed, the air must be supplied through a series of closely spaced small nozzles under a pressure that gives a jet penetration of 70 to 80% of the furnace width.

A special case of a refractory-type furnace is the rotary kiln system (sketch C in Figure 13.20). The feed stoker feeds fuel (waste), predried and ignited in the primary "drying" chamber, into a refractory lined rotary kiln where the fuel (waste) is tumbled and mixed while burning.

The system yields an excellent residue burnout efficiency; nevertheless, the combustion process itself is not necessarily the best.

Because it is impossible to supply the necessary combustion air in the kiln where it is most needed, the kiln usually operates at high temperatures with reducing or only slightly oxidizing atmosphere, resulting in cracking of hydrocarbon molecules and formation of soot in addition to high emissions of CO, both difficult to reignite in the secondary combustion chamber. A bypass flue from the primary to the secondary combustion chamber is sometimes provided to complete the combustion of gases emanating from the initial drying and ignition of the incoming waste.

The configuration of large water-cooled wall furnaces is quite different in concept. To take maximum advantage of radiant heat transfer, the furnaces are generally tall and narrow or short. Figure 13.25 shows commonly used types of waterwall furnace shapes.

The frequently used long rear arch-type furnace (sketch a in Figure 13.25) is a carryover from high fixed carbon fuel boiler design. The volatile fuel fraction is driven off shortly after ignition on the front end of the stoker. The fixed carbon is burned off at a much slower rate at the rear section of the stoker and the low rear arch takes maximum advantage of the radiation heat transfer from the incandescent fuel bed.

When applied to high volatile content fuel, such as solid waste (85% of combustible fraction), the long rear arch is of no value. It could be detrimental to the burnout efficiency of the system by excessively cooling a fuel bed with very low residual carbon fraction unless most or all of the fuel bed combustion is completed on the front end of the stoker. The unit area stoker heat-release limitation, in such cases, becomes meaningless. Due to the configuration of the lower portion of the furnace, the primary air is usually introduced through the front and rear walls rather than through the side walls.

The open-pass furnace design (sketch b in Figure 13.25) allows for a more uniform burning rate over the entire length of the stoker by proper primary air flow distribution and eliminates the possibility of localized high heat-release rates and related hot spots. The secondary air can be supplied through side wall nozzles as in the case of refractory furnaces.

Refractory lining of the waterwall tubes has a double purpose of protecting the tubes from corrosion in the luminous flame zone due to alternating reducing and oxidation atmosphere, and reducing the heat flux to the walls, thus mitigating the potential of the "cold furnace" syndrome.

The type of refractory lining and its thickness depends on its heat transfer coefficient and the heat balance computations.

13.9 WATER-COOLED ROTARY COMBUSTOR

Combining the rotary kiln and water-cooled wall concepts resulted in the development of the proprietary "O'Connor Combustor" consisting of a rotary kiln constructed of a membrane type water-cooled shell followed by a radiant waterwall secondary combustion chamber and a typical convection boiler section (Figure 13.26). The primary combustion air is admitted to the rotary combustor via perforated plates joining the water tubes of the "membrane" shell and the secondary air is introduced in the radiant waterwall secondary chamber.

Although the introduction of the primary air into the rotary combustor appears to overcome the disadvantages of a refractory lined rotary kiln, actual operation reveals that the operating temperatures within the rotary combustion itself are quite high for the type of fuel in question. Temperatures in the range from 2500°F (1371°C) to 2800°F (1538°C) are apparently not uncommon and are conducive to NO_x formation, which could cause a problem with respect to air pollution control

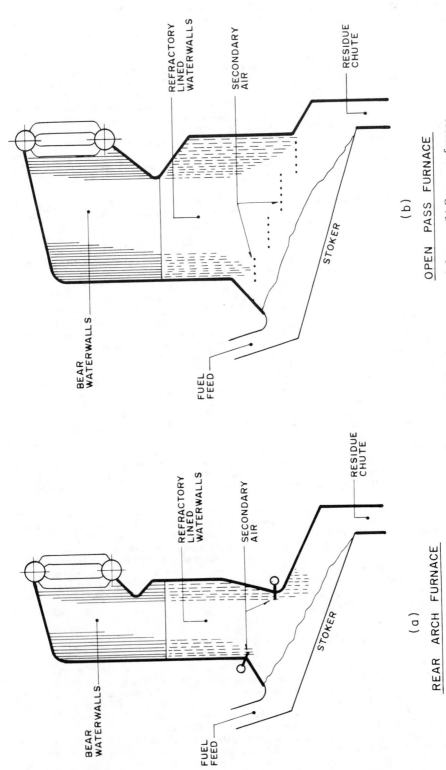

REAR ARCH FURNACE

(a)

OPEN PASS FURNACE

(b)

Fig. 13.25 Typical waterwall furnace configurations. (a) Rear arch furnace. (b) Open pass furnace.

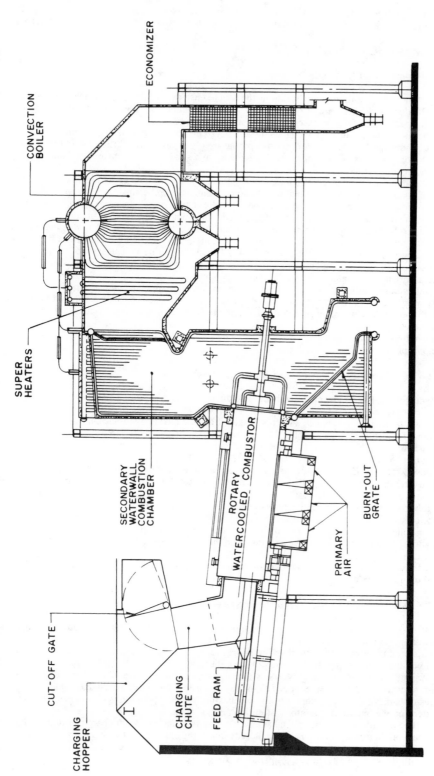

Fig. 13.26 Water-cooled rotary combustor.

589

regulations. The feed system used with the "O'Connor Combustor" inherently limits the sizing of the waste feed stock, requiring elimination of oversized objects that could jam the feed ram(s).

13.10 SMALL-SCALE "MODULAR" UNITS

13.10.1 Combustion Concepts

The term "modular," although applicable in the strict sense of the word to any field-erected or preassembled system consisting of two or more identical combustion trains (modules), has been adopted for partially or fully preassembled combustion systems.

Another distinction is often made labelling some systems as "starved air" or "controlled air" concepts, where the first definition describes partial pyrolysis process and the second definition means fully oxidizing incineration process equivalent, on a small scale, to custom-designed field-erected incinerator furnaces.

Both "starved air" and "controlled air" concepts use similar system components and have a number of common characteristics contrary to some manufacturers' claims.

Developed primarily as an allegedly inexpensive alternative to full-scale energy recovery incinerators, the modular systems are generally limited, with some exceptions, to an individual unit capacity up to 50 TPD, taking advantage of a loophole in the Clean Air Act regulations exempting incinerators under that capacity from the emission requirements imposed by regulations for new stationary sources. Some states have closed this legal gap by requiring all incinerators within their jurisdiction to comply with the emissions limitations set for larger systems, whereas other states allow substantially more lenient criteria for small units.

The ambiguous wording of the present regulations makes it possible, in many instances, to build a facility with three or four combustion trains of 50 TPD capacity, each of which is allowed to emit twice to four times the amount of pollutants than a plant of the same total capacity consisting of two parallel trains.

Minor differences between the "starved air" and "controlled air" system will be discussed where appropriate.

13.10.2 Raw Material Receiving and Storage, Modular Units

Consistent with the low capital cost trend of the modular systems, the storage of waste is mostly on a flat floor at ground level. Because of storage depth limitations, the floor storage area must be substantially larger than an area of a deep storage pit of an equivalent storage capacity.

Assume, for instance, a 100 TPD (90.7 t/day) facility requiring a 3 days' storage capacity. Using an average stockpile depth of 7 ft (2.13 m), the average waste density is approximately 400 lb/yd^3 (138.7 kg/m^3) and the floor loading is 933 lb/yd^2 (296 kg/m^2).

To store 300 tons (272 t) of waste, the required storage area is about 643 yd^2 or 5788 ft^2 (538 m^2).

On the other hand, using a 20-ft (6.1 m)-deep pit with a 20-ft (6.1 m) triangular surcharge (Figure 13.15), the average density of the stored waste is about 545 lb/yd^3 (189 kg/m^3). The average loading is 5450 lb/yd^2 (1729 kg/m^2) requiring an area of only 110 yd^2 or 990 ft^2 (92 m^2).

In the case of floor storage, additional free area must be provided for maneuvering of front-end loaders or bulldozers used for charging of the furnaces. The storage pit concept may or may not require additional covered tipping area depending on the overall on-site traffic considerations, dust and odor control, and aesthetics.

The comparative economics of the two alternates are generally very site specific and must be analyzed on a case-by-case basis.

13.10.3 Raw Material Retrieval and Feed Systems

Unlike large-scale, or even small custom-designed units, the modular systems usually use floor level feed hoppers and feed rams, resulting in a cyclic charging rather than a continuous charging characteristic of tall gravity feed chutes. The floor level hoppers are charged by front-end loaders or small bulldozers retrieving the fuel (waste) from the stockpile.

The frequency of the charging cycles is entirely dependent on the ability of the bulldozer operator to retrieve and charge the required quantity of waste, as well as on his judgment in mixing of wastes of variable quality. Inherently, the combustion system operation is also dependent to a great extent on the bulldozer operator's attention and judgment.

From the point of view of operational safety, floor storage and bulldozer charging requires the operator to work in a dusty, odorous, and generally unhealthy environment with possible exposure to accidental fires in the storage area, in contrast to the clean controlled environment that is possible with the storage pit and crane operation concept.

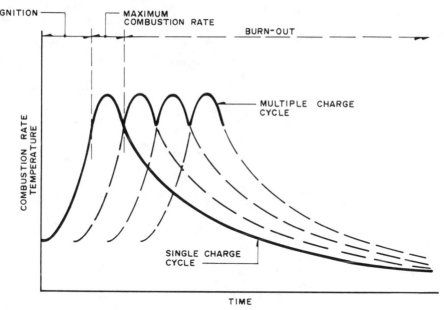

Fig. 13.27 Combustion rate cycle, modular units.

13.10.4 Combustion Systems

As mentioned earlier, the combustion design concept of the modular units is based on partial pyrolysis of the feed stock in a primary chamber or reactor and on the combustion of the products of partial pyrolysis in a secondary combustion chamber, with the assistance of auxiliary burners to assure reignition of the unburned gases and to maintain adequate combustion temperatures.

The combustion process cycle of the original small-capacity batch-feed systems, from which the semicontinuous feed systems were developed, were characterized by a single-charge combustion intensity curve, shown in Figure 13.27, consisting of an ignition period, maximum combustion or reaction rate period, and long burnout period. Superimposing multiple charge cycles within a given period of time results in a cyclic combustion rate and temperatures shown schematically in the same figure.

The "starved air" concept tends to reduce the peak combustion rates to some extent but, because of the cumulative effect of the burnout periods, results in incompletely burned residue inherent, in any event, to a pyrolytic process. Typically, the residue from a semi-continuous "starved air" system can have a carbon content as high, or in excess of, 30% by weight of the dry residue.

The "controlled air," or fully oxidizing, units tend to show more pronounced maximum combustion rate peaks but yield better combustion efficiency with respect to residue quality. Nevertheless, a combustion efficiency comparable to stoker-fired units cannot be achieved because of generally low combustion efficiency of interior rams and the vibrating type of hearth or similar devices which are intended as low-cost substitutes for an efficient stoker design.

An example of a typical modular system with internal rams is shown in Figure 13.28. The useful life of modular units is limited to about 3 to 5 years or slightly longer depending on the quality of construction of the furnace modules and ancillary systems, such as the ram feeders, the internal rams, residue conveyors, waste heat boilers, and so on.

An important factor contributing to a relatively rapid deterioration of the refractory lining of the furnace sections is the cyclic operation with wide temperature variations and the combustion atmosphere which often changes rapidly from reducing to oxidizing mode.

13.10.5 Emissions Control, Modular Units

When applied to incineration of low-ash-content wastes, such as institutional waste, wood, and cardboard in a batch-feed, single-charge mode, the modular units can meet fairly stringent emission limitations averaged over the entire single-cycle duration without elaborate air pollution control

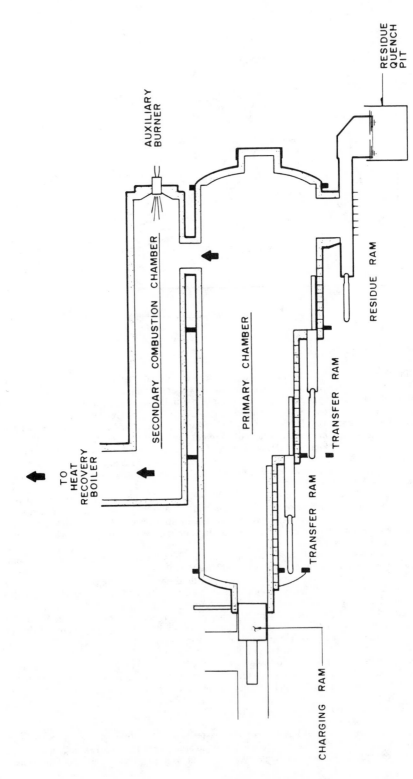

Fig. 13.28 Typical modular furnace configuration.

systems. When used, however, for a semicontinuous operation with high-ash-content wastes, such as mixed municipal solid wastes, the modular systems must be equipped with air pollution control devices (bag filters, electrostatic precipitators) to meet most state or federal regulations.

13.10.6 Application Constraints

The adaptation of modular construction to mass burning systems to replace custom-designed field-erected or partially preassembled units must be viewed with utmost caution. Developed originally as small units for industrial, institutional, and commercial applications, their operational character-istics do not generally satisfy the requirements for continuous mass burning installations with a rated capacity over 50 TPD.

Low combustion efficiencies that translate into reduced effective energy revenues of resource recovery systems, and substantially increased operating and maintenance costs due to limited useful life of the modular units (compared with the usual life of the bond issues used to finance such plants), often result in high life-cycle costs. It is, therefore, imperative to base a system selection on overall life-cycle costs rather than on often seemingly attractive initial capital costs.[13,14]

Editor's note: For additional capital and operating cost information covering small-scale and modular incinerator energy recovery data, illustration, and references, refer to Chapter 16, its Appendix 16A, and references.

REFERENCES

1. P. J. Scott and J. R. Holmes, The capacity and principal dimensions of refuse storage bunkers in modern incinerator plant, Proceedings, 1974 ASME National Incinerator Conference, Miami, Florida.

2. D. L. Wilson, Prediction of heat combustion of solid wastes from ultimate analysis, *Environ. Sci. Technol.,* **6,** p. 19, December (1972).

3. W. Keenan and R. Kaye, *Gas Tables,* Wiley, New York, 1966.

4. National Bureau of Standards, *Gas Tables,* Washington, DC, 1978.

5. R. H. Essenhigh, Burning rates in incinerators, Proceedings of 1968 ASME National In-cinerator Conference, New York, NY.

6. J. E. Williamson and H. M. Twiss, Multiple-Chamber Incinerator Design Standards for Los Angeles County, Technical Publication, Engineering Division, Air Pollution Control District, County of Los Angeles.

7. M. Kuwata, T. J. Kuo, and R. H. Essenhigh, Burning rates and operational limits in a solid fuel bed, Proceedings of 1970 ASME National Incinerator Conference, Cincinnati, Ohio.

8. Incinerator Institute of America, *IIA Incinerator Standards,* 1968.

9. M. Dvirka, Practical application of incinerator burning rate equations, Proceedings, 1976 ASME National Solid Waste Processing Conference, Boston, Mass.

10. American Boiler Manufacturers' Association, Recommended Design Criteria for Stoker Fir-ing of Bituminous Coals, October, 1962.

11. W. H. McAdams, *Heat Transmission,* McGraw-Hill Book Co., New York, 1968.

12. W. M. Rohsenow and J. P. Harnett, *Handbook of Heat Transfer,* McGraw-Hill Book Co., New York, 1964.

13. Argonne National Laboratory, *Thermal Systems for Conversion of Municipal Solid Waste, Small Scale Systems: A Technology Status Report,* D.O.E. No. ANL/CNSV-TM-120, Vol. 3.

14. U.S. EPA Office of Water and Waste Management, *Small Modular Incinerator Systems with Heat Recovery: A Technical, Environmental and Economic Evaluation-Executive Summary.*

CHAPTER 14

RESOURCE RECOVERY: AIR POLLUTANT EMISSION AND CONTROL

WALTER R. NIESSEN

Camp, Dresser & McKee, Inc.
Boston, Massachusetts

Introduction: Importance

The decision by an industry or a governmental jurisdiction to implement a combustion-based process as a key part of their solid waste management program is usually regarded as an environmentally beneficial act relative to ground and surface water contamination, land use, and traffic impacts (with reduced hauling). If energy recovery is planned, fossil fuel energy conservation credits are also perceived. However, these environmental benefits are often reduced or even totally overwhelmed in the public view by the impacts, real or imagined, of air pollutant emissions.

Looked at objectively, the air emissions of waste incineration systems are often unimportant when compared with power plant, industrial, automotive, fugitive, and "natural" (for example, volcanos, the resinous exhalation of conifers, etc.) emissions. Incinerator emissions do have the characteristic, however, of extending the sphere of influence of any proposed facility to relatively long distances and, as discussed below, include Dioxin pollutant species (such as the polychlorinated dibenzo-*p*-dioxins and dibenzo furans) which generate extraordinarily deep concern in the general public. Thus, whether important or not in some objective framework, those desiring to implement incineration systems must forthrightly address potential air emission problems from three standpoints, each to be discussed in this chapter:

1. The regulatory framework. What must I do to analyze an emission problem, deal with it, and secure a permit to build and operate?
2. The emission estimate. How much of each pollutant will be generated by waste combustion?
3. The control strategy. What air pollution control methods (a combination of installed hardware and operating technique) will adequately control emissions to acceptable levels?

14.1 REGULATORY CONTEXT—FEDERAL

Although all states and many local jurisdictions have statutes and/or regulations affecting emissions from incineration systems, most replicate the requirements promulgated by the federal government. Furthermore, the sampling methods developed and recommended by the federal government have been adopted by most states. Finally, although in many states it is the state agency that issues the construction and operating permits relative to air emissions, the states exercise those powers at the sufferance of the federal government regarding the emission standards, the test methods, the air quality impact that is acceptable, and, to a degree, the permitting process itself (especially the need for public involvement). Thus, any consideration of the regulatory context in which incineration systems must be evaluated necessarily begins at the federal level.

14.1.1 National Environmental Policy Act (1969)

The passage of the National Environment Policy Act (NEPA) in 1969 had a very major impact on waste incineration. It would be another year until the Clean Air Act of 1970 would be signed into law with its specific emission limitations and enforcement machinery. NEPA, however, requires an in-depth assessment of air impacts by permitting agencies, and it often mandates an opportunity for public review and comment that significantly publicizes and makes vulnerable the permitting process. NEPA-like legislation in many states makes it even more likely that any new incineration system will receive close scrutiny with particular focus on air pollution aspects.

14.1.2 Clean Air Act of 1970 and Amendments

The Clean Air Act of 1970 and its amendments affect the incinerator air permitting process in several ways:

1. *Emission Limitations.* These specify the maximum emission rate of the criteria pollutant particulate matter (the other criteria pollutants, sulfur and nitrogen oxides, carbon monoxide, and hydrocarbons, are not limited directly) emitted by incinerators (1) burning more than 50 tons/day (TPD) of municipal solid waste (MSW) and (2) constructed after August 17, 1971. The emission limit, known as the New Source Performance Standard (NSPS), is contained in Title 40, Part 60, Subpart E, of the Code of Federal Regulations and specifies a maximum particulate emission of 0.08 grains per dry standard cubic foot (gr/dscf) corrected to 12% CO_2. Note, however, that this standard does not apply to chemical waste or hazardous waste incinerators which are constrained by special regulations promulgated under the Resource Conservation and Recovery Act (RCRA).

   For industrial waste incinerators or incinerators burning MSW containing large fractions of industrial waste, the Clean Air Act may also influence design relative to pollutants whose emissions are limited in accord with the National Emission Standards for Hazardous Air Pollutants (NESHAPs). These NESHAP pollutants presently include benzene, vinyl chloride monomer, and lead and may be extended to include acrylonitrile, ethyl benzene, ethylene oxide, and formaldehyde. NESHAPs *emission* limitations are derived by back-calculation using computer diffusion modeling techniques such that the ambient air quality change that is judged to result from the facility operation is acceptable.

2. *Impact Limitations.* These act to limit emissions based on the projected impact that operation of the incinerator is forecast to have considering the existing air quality in the area surrounding the proposed incinerator site and the magnitude of the annual pollutant emission rate. Particularly if the surroundings are exceptionally clean (for example, a pristine forest) or exceptionally dirty [for example, a heavily industrialized area where violation of the National Ambient Air Quality Standards (NAAQS) may occur], the projected impact of the incinerator may be unacceptable without making extraordinary efforts to control emissions. Note that impact-related incinerator emission limitations may arise for pollutants other than particulate.

A major source proposing a modification that will cause a significant net increase in emissions and is located in an attainment or unclassifiable area must obtain a Prevention of Significant Deterioration (PSD) permit before construction can be initiated (40 CFR 52.21).

In general, a PSD permit application must contain the following basic components:

A Best Available Control Technology (BACT) determination and assessment.

An analysis of existing ambient air quality.

An impact assessment demonstrating that emissions from the new source in conjunction with other nearby sources will not cause a violation of ambient air quality standards or PSD increments.

An assessment of the source's impact on air quality related values including soils, vegetation, and visibility.

Air quality impacts associated with indirect growth created by the new source.

The control technology requirement required for new sources is the application and evaluation of BACT, defined as follows:

An emission limitation based on the maximum degree of reduction for any regulated contaminant emitted from or which results from any regulated facility which the Department on a case-by-case basis, taking into account energy, environmental, and economic impacts and other costs determines is achievable for such facility through application of control of production processes and available methods, systems, and techniques for each such contaminant.

The BACT requirements are intended to ensure that the control systems incorporated in the design of a proposed facility reflect the latest in control technology used in a particular industry in keeping with local air quality, energy, economic, and other environmental considerations. Compliance with NSPS or applicable state emission standards may indeed by deemed application of BACT. Nevertheless, an evaluation of the proposed air pollution control system is required, including an analysis of alternative control systems "capable of a higher degree of emission reduction."

Once compliance with PSD increments has been determined, compliance with NAAQS must be demonstrated. Total air quality levels are determined by combining the projected concentrations with existing background levels. Background levels can be estimated through modeling of existing sources, from existing ambient measurements, or by some combination of both.

The EPA has defined a set of impact levels used to determine whether a major new source or modification will "significantly" affect an area. In general, the reviewing agency does not analyze the impact of a major new source beyond the point where its contribution is seen to fall below those significant concentrations.

On January 16, 1979, the EPA promulgated as a final rule the Emission Offset Interpretative Ruling. This rule describes the requirements for preconstruction reviews that apply to major new or modified sources affecting areas where the NAAQS are not being maintained (nonattainment areas). Such sources would be subject to the following requirements: (1) meet an emission limitation which specifies the lowest achievable emission rate (LAER) for such source; (2) certify that all existing major sources owned or operated by the applicant in the state are in compliance with all applicable emission limitations and standards; (3) obtain emission offsets such that there will be reasonable progress towards attainment of the applicable NAAQS; and (4) demonstrate that the emission offsets provide a net air quality benefit in the affected area [not applicable for volatile organic compounds (VOC) or NO_x emissions]. LAER is defined as: "that rate of emissions which reflects the most stringent emission limitation in the State Implementation Plan (SIP) of any state, or control which has been achieved in practice by comparable sources."

3. *Operational Limitations.* These act to limit emissions by establishing operating guidelines with which the plant must comply. Such guidelines could include (minimum) scrubber pressure drop, afterburner (minimum) temperature, combustor (minimum) oxygen concentration, and/or (minimum) gas residence time.

14.2 REGULATORY CONTEXT—STATE AND LOCAL

Federal NSPS regulations for large incinerators are minimums and may be made more stringent by states or local jurisdictions. No common definition for an "incinerator" is used at the local level but all states regulate both opacity and particulate emissions.[1] These regulations are summarized in Tables 14.1 and 14.2. An approximate scale between differing emission units is presented in Table 14.3.

14.3 AIR POLLUTANT UNCONTROLLED EMISSIONS

Estimation of the basic emission rate of air pollutants (prior to reduction by the air pollution control equipment) is a key step both in the design process and in the evaluation of the potential environmental impact of incinerators. The uncontrolled emission rate is, in most cases, a complex function of waste composition, hardware design characteristics, and real-time operating conditions in the unit. Still, extensive analysis of emission data has resulted in the development of useful average values or functional relationships with which to estimate emission rates and/or to predict the impact of proposed system or waste changes on emissions. These "emission factors" are discussed below.

The air pollutants emitted from waste combustion processes include inorganic particulate matter; combustible solids, liquids, and gases [carbonaceous soot and char; carbon dioxide; "hydrocarbons"; and specialized classes of carbon-hydrogen-oxygen-nitrogen-halogen compounds such as benzene-soluble organics (BSO), polycyclic organic matter (POM), polyhalogenated hydrocarbons (PHH) including the polychlorinated dibenzo-*p*-dioxins, dibenzofurens, and biphenyls (PCB)—all of which are suspected to have a significant carcinogenic action]; specific chemical pollutants where emission levels are largely related to fuel chemistry (sulfur oxides, hydrogen chloride, trace elements, radioactive elements); and the nitrogen oxides where emission levels are related to a wide variety of combustor and fuel-related interactions.

14.3.1 Inorganic Particulate and Comparison of Firing Methods

Inorganic particulate is differentiated from total particulate as an acknowledgment of its refractory nature (that is, emission rates cannot be reduced by better combustion), and to establish clearly its primary source: the ash content of the fuel or waste that is burned. In principle, inorganic particu-

Table 14.1 Opacity Regulation for New and Existing Commercial and Industrial Incinerators

State	Value	Unit	Process Conditions	Validity
			Regulation	
Alabama	60	% Opacity	3 min discharge/60 min	
	20	% Opacity	All other times	
Alaska	40	% Opacity		Installed before 7/1/72
	20	% Opacity		Installed after 7/1/72
Arizona	Exempt		0.5 min discharge/60 min	
	20	% Opacity	All other times	
Arkansas	No. 3	Ringelmann	5 min discharge/60 min	
	No. 1	Ringelmann	All other times	Built after 7/30/73
California				
Colorado	20	% Opacity		
Connecticut	40	% Opacity	5 min discharge/60 min	
	20	% Opacity	All other times	
Delaware	20	% Opacity	3 min discharge/60 min	
D.C.	20	% Opacity	2 min discharge/60 min	Existing
	Prohibited		All other times	Existing
Florida	20	% Opacity	≤50 TPD, 3 min discharge/60 min	
	Prohibited		All other times	
Georgia	20	% Opacity	All other times	Installed after 1/1/72
	40	% Opacity	6 min discharge/60 min	Installed after 1/1/72
	40	% Opacity	All other times	Installed before 1/1/72
Hawaii	40	% Opacity		
Idaho	No. 2	Ringelmann	3 min discharge/60 min	Built before 4/1/72
	No. 1	Ringelmann	3 min discharge/60 min	Built after 4/1/72
Illinois	30	% Opacity	All other times	
	30–40	% Opacity	1 min discharge/60 min	
Indiana	40	% Opacity	15 min discharge/60 min	
Iowa	40	% Opacity		
	40	% Opacity	3 min discharge/60 min during breakdown, etc.	
Kansas	20	% Opacity		
Kentucky	20	% Opacity		
Louisiana	No. 1	Ringelmann	All other times	
	No. 1	Ringelmann	4 min discharge/60 min	
Maine	No. 1	Ringelmann		
Maryland				
Massachusetts	No. 1	Ringelmann		
Michigan	40	% Opacity	3 min discharge/60 min	
	20	% Opacity	All other times	
Minnesota	20	% Opacity		
Mississippi	40	% Opacity		
Missouri	No. 1	Ringelmann		Built after 2/10/72
	No. 2	Ringelmann		Built before 2/10/72
Montana	10	% Opacity		
Nebraska	20	% Opacity		
Nevada	20	% Opacity	1 min discharge/60 min	
New Hampshire	No. 1	Ringelmann	3 min discharge/60 min	
New Jersey	No. 2	Ringelmann	3 consecutive minutes	
	No. 1	Ringelmann	All other times	
New Mexico	No. 1	Ringelmann	2 min discharge/60 min	
New York				
(state)	40	% Opacity		Built before 1/26/67
	20	% Opacity		Built after 1/26/67
(city)	No. 1	Ringelmann	3 min discharge/60 min	

Table 14.1 Continued

State	Value	Unit	Process Conditions	Validity
			Regulation	
North Carolina				
North Dakota	No. 3	Ringelmann	4 min discharge/60 min	
	No. 1	Ringelmann	All other times	
Ohio	60	% Opacity	3 min discharge/60 min	
	20	% Opacity	All other times	
Oklahoma	No. 1	Ringelmann	All other times	
	No. 3	Ringelmann	5 min discharge/60 min	
Oregon	40	% Opacity	3 min discharge/60 min	Built before 4/1/70
	20	% Opacity	3 min discharge/60 min	Built after 6/1/70
Pennsylvania	20	% Opacity	3 min discharge/60 min	
Puerto Rico	20	% Opacity	All other times	
	60	% Opacity	1 min discharge/60 min	
Rhode Island	20	% Opacity	3 min discharge/60 min	
South Carolina	No. 1	Ringelmann	3 min discharge/60 min	
South Dakota	20	% Opacity	3 min discharge/60 min	
	60	% Opacity	3 min discharge/60 min	
Tennessee	20	% Opacity	5 min discharge/60 min	
Texas	30	% Opacity	5 min average	Built before 1/31/72
	20	% Opacity	5 min average	Built after 1/31/72
Utah	No. 1	Ringelmann		
Vermont	40	% Opacity	6 min discharge/60 min	Built before 4/30/70
	20	% Opacity	6 min discharge/60 min	Built after 4/30/70
Virginia	20	% Opacity		
Washington	20	% Opacity	3 min discharge/60 min	
	>20	% Opacity	15 min/8 hr	
West Virginia	No. 1	Ringelmann		
Wisconsin	20	% Opacity		Built after 4/1/72
Wyoming	20	% Opacity		

late can arise from mechanical degradation of refractory or oxidation and flaking of fireside metal surfaces, but the great majority of the emission results from the carryover of mineral matter introduced with the fuel or waste. In unusual conditions, where combustion air is drawn from an area with a high dust loading, a portion of the emission could be associated with the air supply.

A desk-top assessment of the fraction of the inorganic matter in a fuel that appears in the flue gas is often difficult because of the complexity of the processes involved. Thus, empirical estimation methods are necessary unless the system design is such that essentially all solids that are introduced must leave via the flue gas stream. This latter situation is often true for systems burning industrial gaseous or liquid wastes. For conventional gaseous and liquid fossil fuels, combustion systems seldom include air pollution control for particulate abatement. When burning liquid or gaseous waste streams, however, sampling or estimation of inorganic content is an appropriate design step to assure that no control system is needed. Then, consideration can be given to either precombustion removal of the solid matter or to the installation of a flue gas particulate air pollution control device.

Unlike liquid or gaseous fuels which may contain little inorganic matter, almost all solid wastes and fuels have a substantial ash content. Since these latter fuels are seldom susceptible to pretreatment to reduce emissions, the designer must give careful attention to design features and operating practices to minimize the emission rate. Even with such precautions, it is likely that emission control will be required.

For *suspension burning,* the maximum inorganic particulate emission rate corresponds to the situation where all noncombustible solids are swept from the combustion chamber. This is largely the case for burners fired with pulverized coal, shredded and air-separated municipal refuse, sawdust, rice hulls, and other suspension firing systems. In such systems, a portion of the fly ash settles in the bottom of the primary furnace or in following chambers or boiler sections. The degree to which settlement reduces emission rate depends on the flue gas properties, on the characteristic dimensions and weight of the fly ash particles (settling rate), and on the velocity of the gas stream relative to the chamber dimensions (transit time).

Table 14.2 Particulate Emission Limitations for New and Existing Incinerators—By States

State	Value	Units	Corrected To	Process Conditions	Validity
				Regulation	
Alabama	0.1	lb/100 lb charged	12% CO_2	>50 TDD	
	0.2	lb/100 lb charged		≤50 TPD	
Alaska	0.3	gr/scf	12% CO_2	≤200 lb/hr	
	0.2	gr/scf	12% CO_2	200–1000 lb/hr	
	0.1	gr/scf	12% CO_2	>1000 lb/hr	
Arizona	0.1	gr/scf	12% CO_2		
Arkansas	0.2	gr/scf	12% CO_2	≥200 lb/hr	
	0.3	gr/scf	12% CO_2	<200 lb/hr	
California	0.3	gr/scr	12% CO_2	Typical of the 43 APCDs	
Colorado	0.1	gr/scf	12% CO_2		Designated control areas
	0.15	gr/scf	12% CO_2		Other areas
Connecticut	0.08	gr/scf	12% CO_2		Built after 6/1/72
	0.4	lb/1000 lb	50% excess air		Built before 6/1/72
Delaware	0.2	lb/hr		100 lb/hr	
	1.0	lb/hr		500 lb/hr	
	2.0	lb/hr		1000 lb/hr	
	5.0	lb/hr		3000 lb/hr	
Florida	0.08	lb/hr	50% excess air	≥50 TPD	Built after 2/11/72
	0.1	gr/scf	50% excess air	≥50 TPD	Built before 2/11/72
Georgia	0.1	gr/scf	12% CO_2	≤50 TPD— type 0,1,2 waste	New (Built after 1/1/72)
	0.2	gr/scf	12% CO_2	≤50 TPD— type 3,4,5,6 waste	New (Built after 1/1/72)
	0.2	gr/scf	12% CO_2	Type 0,1,2 waste	Existing before 1/1/72
	0.3	gr/scf	12% CO_2	Type 3,4,5,6 waste	Existing before 1/1/72
	0.08	gr/scr	12% CO_2	≥50 TPD	New (Built after 1/1/72)
Hawaii	0.2	lb/100 lb charged			
Idaho	0.2	lb/100 lb charged			
Illinois	0.08	gr/scf	12% CO_2	2000–60,000 lb/hr	
		gr/scf	12% CO_2	≤2000 lb/hr	Built before 4/15/72
	0.1	gr/scf	12% CO_2	≤2000 lb/hr	Built after 4/15/72
Indiana	0.3	lb/1000 lb gas	50% excess air	≥200 lb/hr	
	0.5	lb/1000 lb gas	50% excess air	<200 lb/hr	
Iowa	0.2	gr/scf	12% CO_2	≥1000 lb/hr	
	0.35	gr/scf	12% CO_2	<1000 lb/hr	
Kansas	0.2	gr/scf	12% CO_2	<200 lb/hr	
	0.2	gr/scf	12% CO_2	200–20,000 lb/hr	
	0.1	gr/scf	12% CO_2	>20,000 lb/hr	
Kentucky	0.2	gr/scf	12% CO_2	≤50 TPD	
	0.08	gr/scf	12% CO_2	>50 TPD	
Louisiana	0.2	gr/scf	12% CO_2		
Maine	0.2	gr/scf	12% CO_2		
Maryland	0.1	gr/scf	12% CO_2	<2000 lb/hr	Built after 1/17/72
	0.03	gr/scf	12% CO_2	>2000 lb/hr	Built after 1/17/72
	0.3	gr/scf	12% CO_2	<200 lb/hr	Built before 1/17/72
	0.2	gr/scf	12% CO_2	>200 lb/hr	Built before 1/17/72
Massachusetts	0.1	gr/scf	12% CO_2		Existing
	0.05	gr/scf	12% CO_2		New
Michigan	0.65	lb/1000 lb gas	50% excess air	0–100 lb/hr	
	0.3	gr/scf	12% CO_2	>100 lb/hr	

Table 14.2 Continued

State	Value	Units	Corrected To	Process Conditions	Validity
				Regulation	
Minnesota	0.3	gr/scf	12% CO_2	<200 lb/hr	Existing before 8/17/71
	0.2	gr/scf	12% CO_2	200–2000 lb/hr	Existing before 8/17/71
	0.1	gr/scf	12% CO_2	>2000 lb/hr	Existing before 8/17/71
	0.2	gr/scf	12% CO_2	<200 lb/hr	New (Built after 8/17/71)
	0.15	gr/scf	12% CO_2	>2000 lb/hr	New (Built after 8/17/71)
Mississippi	0.2	gr/scf	12% CO_2	Design capacity	
	0.1	gr/scf	12% CO_2	New sources near residential areas	
Missouri	0.2	gr/scf	12% CO_2	≥200 lb/hr	
	0.3	gr/scf	12% CO_2	<200 lb/hr	Existing before 9/5/75
	0.1	gr/scf	12% CO_2	≤200 lb/hr	Existing before 9/5/75
Nevada	3.0	lb/ton charged		<2000 lb/hr	
	Variable	$E = 40.7 \times 10^{-5} C$	C,E = lb/hr	>2000 lb/hr	
New Hampshire	0.3	gr/scf	12% CO_2	≤200 lb/hr	
	0.2	gr/scf	12% CO_2	>200 lb/hr	
	0.08	gr/scf	12% CO_2	>50 TPD	Built after 4/20/74
New Jersey	0.2	gr/scf	12% CO_2	<2000 lb/hr	Type 0,1,2,3, waste only
	0.2	gr/scf	12% CO_2	All others	
New Mexico	Only opacity regulations			<50 TPD	
	0.08	gr/scf	12% CO_2	>50 TPD	New (Built after 8/17/71)
New York	0.5	lb/100 lb charged		>2000 lb/hr	Built between 4/1/62 and 1/1/70
	0.5	lb/100 lb		≤2000 lb/hr	Built between 4/1/62 and 1/1/68
	Variable (e.g., 0.3)	lb/hr		≤100 lb/hr	Built after 1/1/68
	Variable (e.g., 3.0)	lb/hr		@1000 lb/hr	Built after 1/1/68
	Variable (e.g., 7.5)	lb/hr		@3000 lb/hr	Built after 1/1/70
North Carolina	0.21	lb/hr		0–100 lb/hr	
	0.4	lb/hr		@200 lb/hr	
	1.0	lb/hr		@500 lb/hr	
	2.0	lb/hr		@1000 lb/hr	
	4.0	lb/hr		≥2000 lb/hr	
North Dakota	Variable	lb/hr		@100 lb/hr @1000 lb/hr @3000 lb/hr	
Ohio	0.1	lb/100 lb charged		≥100 lb/hr	
	0.2	lb/100 lb charged		<100 lb/hr	
Oklahoma	Variable	lb/hr		@100 lb/hr @1000 lb/hr @3000 lb/hr	
Oregon	0.3	gr/scf		≥100 lb/hr	
	0.2	gr/scf		>200 lb/hr	Built before 6/1/70
	0.1	gr/scf		>200 lb/hr	Built after 5/1/70
Pennsylvania	0.1	gr/scf			
Rhode Island	0.16	gr/scf	12% CO_2	<2000 lb/hr	
	0.08	gr/scf	12% CO_3	≥2000 lb/hr	
South Carolina	0.5	lb/10^5 Btu		B10 mm Btu/hr	
South Dakota	0.2	lb/100 lb charged			
Tennessee	0.2	% of charge		≤2000 lb/hr	
	0.1	% of charge		>2000 lb/hr	
Texas	Variable	lb/hr		@1000 lb/hr @3000 lb/hr	

Table 14.2   Continued

State	Value	Units	Corrected To	Process Conditions	Validity
				Regulation	
Utah	0.08	gr/scf	12% CO_2		
Vermont	0.1	lb/100 lb charged			
Virginia	0.14	gr/scf	12% CO_2		
Washington	0.1	gr/scf	7% O_2		
West Virginia	8.25	lb/ton		≤200 lb/hr	
	5.43	lb/ton		>200 lb/hr	
Wisconsin	0.2	lb/100 lb exhaust gas	12% CO_2	5–4000 lb/hr	Built after 4/1/72
	0.3	lb/1000 lb exhaust gas	12% CO_2	≤500 lb/hr	Built after 4/1/72
	0.5	lb/1000 lb exhaust gas	12% CO_2	>500 lb/hr	Built before 4/1/72
	0.6	lb/1000 lb exhaust gas	12% CO_2	≤500 lb/hr	Built after 4/1/72
Wyoming	0.2	lb/100 lb charged			

Table 14.3   **Approximate Conversion Factors between Particulate Regulation Bases**

	gr/scf @ 12% CO_2	gr/scf @ 50% EA	lb/1000 lb of flue gas @ 50% EA	lb/1000 lb of flue gas @ 12% CO_2	lb/100 lb of charge
gr/scf @ 12% CO_2	1	0.8795	1.563	1.789	0.833
gr/scf @ 50% EA	1.137	1	1.777	2.034	0.9471
lb/1000 lb of flue gas @ 50% EA	0.640	0.5629	1	1.145	0.533
lb/1000 lb of flue gas @ 12% CO_2	0.559	0.4916	0.873	1	0.465
lb/100 lb of charge	1.20	1.055	1.876	2.147	1

EA = Excess air.

An approach to estimating the potential for fallout makes use of Stokes' Law (and its various extensions into the "transition" regime for larger particles), although the result will tend to overestimate the fallout rate. The complications tending to lower the measured settling efficiency below the calculated value are:

Nonsphericity of particles.

Particle reentrainment due to turbulence. This can be minimized by use of baffles (creating a low turbulence zone or, if appropriate, a wet, water-sluiced chamber floor).

Nonuniformity of particle density: the coarser particles tend to be lower in density.

Particle-particle interactions ("hindered settling").

Fallout of the very coarse particles prior to the chamber under analysis, leaving a less easily settled dust in the gas stream.

In general, settling velocities for homogeneous spherical particles can be calculated from the relation

$$u_t = \sqrt{\frac{4d(\rho_s - \rho_a)g}{3\rho_a C_x}} \qquad (1)$$

where
u_t = Terminal settling velocity (m/sec).
d = Particle diameter (m).
ρ_s = Density of particle (kg/m³).
ρ_a = Density of gas (kg/m³).
g = Gravitational constant (m/sec²).
C_x = Drag coefficient (dimensionless).

In the streamline flow region (small particles), the drag coefficient is inversely proportional to the Reynolds number N_{Re}:

$$C_x = \frac{24}{N_{Re}} = \frac{24\mu}{du_t\rho_a} \tag{2}$$

where

μ = Viscosity of gas (kg/m · sec⁻¹).
N_{Re} = Reynolds number $(du_t\rho_a)/\mu$, which is dimensionless.

In the streamline flow region, combination of Equations (1) and (2) leads to Stokes' Law:

$$u_t = \frac{d(\rho_s - \rho_a)g}{18\ \mu} \tag{3}$$

For practical purposes, the streamline flow region stops at about N_{Re} = 0.5 to 1.0. Thus, for 200°C gases and particulate densities of 2.5 g/cm³ (2500 kg/m³), Equation (2), and thus Stokes' Law, is valid only up to diameters of about 90 μm. Above this region, we must use relationships other than Equation (2) to characterize the dependence of C_x on N_{Re}. For larger particles, the relation of Schiller and Nauman[1a] indicates

$$C_x = \frac{24}{N_{Re}} [1 + 0.15N_{Re}^{0.687}] \tag{4}$$

This relation is valid for $0.5 < N_{Re} < 800$.

Unlike the combination of Equations (1) and (2), the combination of Equations (4) and (1) no longer gives an explicit relation for u_t, but rather requires a trial and error solution. As an example, application of these relations to particles of density 2.0 and 3.0 g/cm³ in hot air at two temperatures leads to the settling velocities shown in Table 14.4. Note that for large particles, the effect of temperature is not very great.

In *mass-burning*, the fuel or waste is moved through the combustion chamber on a grate. The introduction of a portion of the combustion air through the grate provides a mechanism whereby a portion of the ash can be fluidized and carried off with the flue gases. This mechanism is favored by fuels or wastes with a high percentage of fine (that is, suspendable) ash, by high underfire air velocities, or by other factors that induce a high gas velocity through and over the bed. Secondary inducements to emission include the inadvertent deflection of overfire air jets onto the bed. Agitation of the bed and the volatilization of metallic salts. Grate systems designed with large air passages (such that a substantial fraction of the fine ash is dropped out) tend to reduce the emission rate.

Ash Content

Ash particles may be entrained when the velocity of the gases through the fuel bed exceeds the terminal velocity of the particles as calculated using Stokes' Law [Equation (3)]. Undergrate air velocities in municipal incinerators typically vary from a minimum of 0.05 sm³ · sec⁻¹ · m⁻² of grate area to 0.5 sm³ · sec⁻¹ · m⁻². On the basis of the terminal velocity of ash particles (Table 14.4), it is, therefore, expected that particles up to 70 μm will be entrained at the lowest velocities and up to 400 μm at the highest (at a mean temperature of 1100°C). The postulated mechanism of particulate entrainment is supported by the observed range of particle sizes of fly ash, the increases in particulate emission with increases in refuse ash content, the increases in particulate emission with undergrate air velocity, and the similarity in the chemical analyses of fly ash and those of the ash of the principal constituents of refuse.

Evaluation of the particle size distribution of fly ash[2] shows 70 wt % of the particles are smaller than 250 μm. This is consistent with the calculations of terminal velocities. The data available, however, are insufficient to test the expectation that the maximum particle size of the fly ash emitted increases with undergrate air velocity.

Table 14.4 Air Pollutant Generation—Calculated Settling Velocities in Hot Air (Spherical Particles)[a]

Diameter (μm)	Settling velocity (m/sec)			
	(Density = 2.0 g/cm³)		(Density = 3.0 g/cm³)	
	at 200°C	at 800°C	at 20°C	at 800°C
30	0.03	0.03	0.06	0.03
40	0.06	0.03	0.09	0.06
50	0.12	0.06	0.15	0.09
60	0.15	0.09	0.24	0.12
70	0.21	0.12	0.34	0.18
80	0.27	0.15	0.43*	0.24
90	0.37*	0.21	0.49	0.30
100	0.43	0.24*	0.55	0.37*
200	1.13	0.85	1.55	1.22
400	1.92	1.65	2.56	2.29
400	2.68	2.50	3.54	2.41
500	3.38	3.35	4.45	4.54
600	4.08	4.21	5.33	5.64
700	4.75	5.03	6.16	6.71
800	5.36	5.85	6.98	7.74
900	6.00	6.64	7.89	8.75
1000	6.58	7.41	8.50	9.72

[a]The asterisk denotes the largest particle for which Stokes' Law was used; Schuller-Nauman extension [Equation (1)] was used for all larger particles.

The dependence of the particulate emission factors on the ash content of refuse is most striking. The very high ash content of refuse in Germany, particularly during the winter in areas where a high fraction of the residential furnaces are coal-fired, accounts for the unusually high emission rates demonstrated by the data of Eberhardt and Mayer[3] and Nowak.[4] These findings show that the ash content of the refuse is a major factor in determining emission rates, and that the percentage of the ash carried over ranges mostly between 10 and 20% of the total. The actual percentage carried over for a particular incinerator will, of course, depend on other factors, such as the underfire air rate. Consideration of the particle terminal velocities indicates that the above conclusions apply only to fine particle ash (finer than 400 μm).

Underfire Air Rate

A systematic study of the effects of underfire air, secondary air, excess air, charging rate, stoking interval, and fuel moisture content on the emission rate from an experimental incinerator by the Public Health Service (PHS) led to the conclusion that the velocity of the underfire air was the variable that most strongly influenced particulate emission rate. The data on 25 and 50% moisture fuel were correlated[5] by

$$W = 4.35 V^{0.543} \tag{5}$$

where
W = The emission factor expressed in units of kilograms of particulate per ton of refuse burned.
V = Underfire air rate (in sm³ · sec^{-1} · m^{-2} of grate area).

The range of undergrate velocities studied ranged from 0.01 to 0.5 sm³ · sec^{-1} · m^{-2} of grate area. Subsequent field evaluation by the PHS[6] of emission rates indicate that the effect of underfire air rate was less pronounced than that predicted by the above equation and suggested that, for the two municipal incinerators tested, the effect was small for undergrate velocities below 0.18 to 0.2 sm³ · sec^{-1} · m^{-2} of grate area but significant above that rate.

Walker and Schmitz[7], however, correlated their test results on three municipal incinerators with the relationships developed by the PHS on the experimental incinerator. The results showed general agreement with the predicted slope but scattered ±20%. This scatter is not surprising in view of probable differences in refuse composition, furnace, size, and other variables.

Although reduction in the underfire air rate can reduce the emission from the furnace, there is usually[6,7] an attendant reduction in burning rate. An economic analysis of the trade-off between furnace capacity and air pollution control equipment costs would be needed to optimize the underfire air rate. There is, of course, a minimum underfire air requirement.

Incinerator Size

Larger incinerator units seem to have slightly higher emission rates, but the effect of size on emission factors has not been established quantitatively. Part of the increase is due to the higher burning rates and, hence, higher underfire air rates associated with large units. However, if emission on large and small units is compared at equal underfire air velocities, a residual effect of size is found. Possibly the higher emission rates for the larger size is a consequence of the higher natural convection currents encountered in large units.[2]

Burning Rate

For reasons similar to those presented in the preceding paragraph, it is expected that higher emission factors will be encountered at higher burning rates. Rehm[9] cites that reductions in the rate of burning to 75% of rated capacity have resulted in as much as a 30% reduction in the furnace emission from that at full capacity; however, insufficient data are available for a quantitative relationship to be established.

Grate Type

Stoking has been observed to increase particulate emission rates. This is particularly evident from the tests on Plant No. 76[2] and the PHS field tests[6] on the effect of underfire air rate, where the stoking required at the lowest air rates led to a significant increase in emission. Walker and Schmitz,[7] however, report results that suggest that the effect of grate design on emission is secondary to the effect of underfire air rate or ash content. The results of their tests on the emission from furnaces operated with a two-section traveling grate, a rocking grate, and a reciprocating grate led them to conclude that the differences in emission factors were primarily due to the differences in underfire air velocities used in the different units.

To assess the effect of grate type on emission factors, data on the different types of units were compiled.[2] The emission rates from reciprocating grate stokers were seen to be significantly higher than those from other grate types. It is probable that higher rates in the reciprocating grate units are due to a combination of greater stoking, higher underfire air rates, and larger furnace sizes than in the other units. Another factor is the difference in grate openings, which can result in large differences in the amount of fine ash that can sift through the grate; lower emission rates would be expected from grates (for example, rocking grates) that have large amounts of fine ash sifting through the grate.

Volatilization of Metallic Salts

Although a major fraction of fly ash from municipal incinerators seems to have been entrained from the fuel bed or formed by cracking of pyrolysis products (soot), trends in the emission rate from small incinerators cannot be explained solely by an entrainment mechanism.[8] For example, the sizes of the particulate in the stack discharge data reported by Rose and Crabaugh[8] are mostly in the 0 to 5 μm range. On the basis of the physical shape as determined from microscopic examination, and on chemical analyses of the noncombustible, they concluded that particles were formed by volatilization and recondensation of metallic salts. This finding, however, is inconsistent with the data on large units for which particles of 0 to 5 μm usually constitute a small fraction of the total emission. In municipal incinerators, the oxidation of the metals is known to be significant, but the volatilization of the salts formed is only of importance for trace constituents; the majority of the oxides are mechanically entrained from the bed.

14.3.2 Combustible Particulate

The incomplete combustion of fuels containing carbon can result in the formation of a wide spectrum of chemical species. The simplest, carbon itself, can contribute importantly to the opacity of the effluent, due both to the refractive index and color of the particles and to the typically small particle size (thus increasing the light-scattering power for a given mass loading).

In burning carbon-bearing wastes, conditions of high temperature and low oxygen concentration can lead to the formation of soot. The mechanisms responsible for soot formation include preferential oxidation of the hydrogen and thermally induced dehydrogenation. Whatever the cause, soot formation is a problem to the incineration system:

1.   Dark, high-optical-density flue gas emissions which, though less than the applicable mass emission limits, may violate opacity restrictions.
2.   Soot is a sign of poor combustion conditions and, most likely, is accompanied by high carbon monoxide emissions.
3.   If metal surfaces (for example, the fire side of boiler tubes) are exposed to an atmosphere that swings from oxidizing to reducing conditions (the regions where soot is formed are often reducing), rapid metal wastage occurs.
4.   If an electrostatic precipitator is used for air pollution control, the presence of carbon lowers the resistivity of the dust and may lower collection efficiency. In fabric filter control devices, capture of the slow-burning char may greatly decrease bag life.
5.   Unfortunately, once soot is formed, its slow burning rate makes subsequent control efforts difficult.

The control of soot burnout can best be understood by examining the kinetics of combustion of carbonaceous particles. For spherical particles, review of the considerable research on this topic[10] suggests the following:

$$q = \frac{p_{O_2}}{1/K_s + 1/K_d} \tag{6}$$

where
q = The rate of carbon consumption (g · cm^2 · sec^{-1}).
p_{O_2} = Partial pressure of oxygen (atm).
K_s = Kinetic rate constant for the consumption reaction.
K_d = Diffusional rate constant. Both kinetic and diffusional resistances to reaction are thus seen to influence burnout rate.

For (small) particles of diameter d (centimeters) typical of soot, the diffusional rate constant at temperature $T(K)$ is approximately given by:

$$K_d = \frac{4.35 \times 10^{-6} T^{0.75}}{d} \tag{7}$$

The kinetic rate constant is given by:

$$K_s = 0.13 \exp\left[\left(\frac{-35,700}{R}\right)\left(\frac{1}{T} - \frac{1}{1600}\right)\right] \tag{8}$$

where
R = The gas constant (1.986 cal/g mol K).

For a particle of initial diameter d_0 and an assumed specific gravity of 2, the time (t_b) in seconds to completely burn out the soot particle is given by:

$$t_b = \frac{1}{p_{O_2}}\left[\frac{d_0}{0.13 \exp\left[\left(\frac{-35,700}{R}\right)\left(\frac{1}{T} - \frac{1}{1600}\right)\right]} + \frac{d_0^2}{5.04 \times 10^{-6} T^{0.75}}\right] \tag{9}$$

14.3.3   Total Particulate

The total particulate emission factor is the sum of the inorganic particulate and the combustible (soot) particulate. As can be seen from the previous discussion, the uncontrolled emission rate is a complex function of waste composition and system design and operating conditions.

For mass-burning systems with generally vigorous stoking of the refuse bed (for example, rotary kilns and reciprocating grates) and somewhat typical refuse composition, the uncontrolled particulate rate is approximately 30 lb/ton of refuse fired (13 kg/t). Less vigorous stoking (as for a traveling grate) results in uncontrolled rates of approximately 15 lb/ton (7.5 kg/t). For the very low primary furnace gas flow rates typical of the starved-air modular combustion unit, uncontrolled emission rates of approximately 3 lb/ton (1.5 kg/t) are reported.

For full suspension burning in incinerators of a prepared refuse-derived fuel (RDF), the uncontrolled emission rate will approach the fine ash content although perhaps 15 to 30% of the fine ash and a much larger fraction of the coarser fragments of glass, metal, rock, and so on, will fall out in the primary furnace and/or settle in the boiler flues. Although a comprehensive set of data is lacking, an uncontrolled rate of about 120 lb/ton fired (60 kg/t) appears reasonable.

For semisuspension burning in incinerators of a prepared RDF, the uncontrolled emission rate will also depend importantly on both waste and system design-operating characteristics. Although a comprehensive set of data is lacking, an uncontrolled rate of about 60 lb/ton fires (30 kg/t) appears reasonable.

14.3.4 Carbon Monoxide (CO)

CO is produced by the incomplete combustion of the pyrolysis products of solid or liquid wastes, from the char in a refuse bed, or as an intermediate combustion product. The oxidation kinetics of CO to CO_2 have been studied by several investigators and, although there are differences in the rate constant reported, a reasonable estimate can be made of the times required to complete the combustion. The rate expression by Hottel et al.[12] can be used to calculate the kinetics of oxidation in a combustion chamber for the condition where the CO and oxygen (air) are intimately mixed. The kinetic expression for the rate of change of CO mole fraction(f_{CO}) with time is given by

$$-\frac{df_{CO}}{dt} = 12 \times 10^{10} \exp\left(\frac{-16,000}{RT}\right) f_{O_2}^{0.3} f_{CO} f_{H_2O}^{0.5} \left(\frac{P}{R'T}\right)^{1.8} \tag{10}$$

where

$f_{CO}, f_{O_2},$ and f_{H_2O} = The mole fractions of CO, O_2, and water vapor, respectively.

T = The absolute temperature (K).

P = The absolute pressure (atm).

t = Time in seconds.

R = The gas constant (1.986 kcal/kg $\cdot$ mol $\cdot$ K).

R' = Also the gas constant, but in alternate units (82.06 m $\cdot$ cm^3/g $\cdot$ mol $\cdot$ K).

In reviewing the reaction rate expression, it is instructive to note the dependence upon the mole fraction of water vapor. This results from the participation of hydrogen and hydroxyl (OH) free radicals in the complex sequence of reaction steps involved in CO oxidation.

To the extent that the combustion chamber may be considered as an isothermal reactor, the decline in CO concentration over an interval of t (sec) from an initial mole fraction $(f_{CO})_i$ to a final mole fraction $(f_{CO})_f$ is given by

$$\frac{(f_{CO})_f}{(f_{CO})_i} = \exp(-Kt) \tag{11}$$

where

$$K = 12 \times 10^{10} \exp\left(\frac{-16,000}{RT}\right) f_{O_2}^{0.3} f_{H_2O}^{0.5} \left(\frac{P}{R'T}\right)^{0.8}$$

Other workers have concluded that Equation (10) gives unrealistically high rates at low temperatures (for example, in afterburners operating below 1000°C). A more conservative result is given in Morgan[13]

$$-\frac{df_{CO}}{dt} = 1.8 \times 10^{13} f_{CO} f_{O_2}^{0.5} f_{H_2O}^{0.5} \left(\frac{P}{R'T}\right) \exp\left(\frac{-25,000}{RT}\right) \tag{12}$$

The emission factor of CO from an incinerator is, thus, strongly related to the temperature and oxygen concentration experienced by the various packets of combustion gases as they traverse the system. Of particular importance for mass-burning systems is the adequacy with which the CO-rich plume of gas arising from the gasification zone of the grate is mixed with combustion air by the overfire air system. For full suspension burning systems, the intense turbulence in the system minimizes CO formation and expedites its destruction. The semisuspension system falls in between but can be expected to be closer in emission characteristics to the suspension burning system.

In view of the complex mix of refuse and system characteristics that influence CO generation and, importantly, the sensitivity to operating practices, it should not be surprising that reported CO emissions vary widely, even for the same unit. Data[2] from pre-1970 refractory and waterwall mass-burning units, taken most usually with Orsat testing apparatus, showed an average emission factor of about 35 lb/ton (17.5 kg/t). Improvements in gas sampling and analysis techniques, in incinerator hardware, and in operator skill have considerably reduced average emissions but not necessarily variability.

Modular combustion units (MCU) are reported[14] to show a CO emission rate of about 0.5 lb/ton (0.25 kg/t). A rotary kiln, mass-burning unit, showed for 35 tests an average rate of 6.4 lb/ton (3.2 kg/t) with a standard deviation (SD) 135% of the mean. Three mass-burning waterwalls showed an average emission factor of 14.8 lb/ton (7.4 kg/t) with a SD of 102% of the mean. Yet, a recent compilation of data[11] suggests that 1.9 lb/ton (0.95 kg/t) is a "somewhat conservative estimate." This author tends to agree with the authors of Ref. 11 but only for modern (post-1975), well-operated waterwall plants. Data for dry RDF-fired systems are lacking but rates similar to or slightly lower than the rates for mass-burning units are expected. Wet-process RDF firing may give somewhat higher CO emission rates due to the relatively low flame temperature.

14.3.5 Nitrogen Oxides (NO$_x$)

Nitric oxide (NO) is produced from its elements at the high temperatures obtained in furnaces and incinerators. At lower temperatures, NO formation is limited by both equilibrium (which favors dissociation to the elements) and kinetics. Although only a small portion of the NO further oxidizes to nitrogen dioxide (NO$_2$) within the furnace, oxidation does take place slowly after leaving the stack at the temperatures and high oxygen concentrations of the ambient atmosphere. The air quality impact of the nitrogen oxides (referred to collectively as NO$_x$ and reported as NO$_2$) arises from their participation in atmospheric chemical reactions. These reactions, especially those stimulated by solar ultraviolet light (a photochemical reaction), produce a variety of oxygenated compounds that account for the visibility reduction and eye irritation associated with smog.

In combustion systems, nitrogen oxides arise through fixation of nitrogen from the combustion air with oxygen (thermal generation). Also, NO$_x$ is formed by oxidation of nitrogen entering the system bound in the fuel (fuel nitrogen generation). At very high temperatures, the dominant source of NO$_x$ is thermal generation, but at lower temperatures, fuel nitrogen mechanisms dominate. The keys to the distribution among these mechanisms are the equilibrium and kinetic relationships that control the process.

As for CO emissions, NO$_x$ emission data scatter widely. A recent data summary[11] suggests an emission rate of about 1.6 lb/ton (0.8 kg/t) for a modern waterwall, mass-burning incinerator although other data compilations[2] suggest about twice this value. Niessen[15] discusses the influence of fuel nitrogen and overall incinerator heat release rate effects on NO$_x$ emission rates. Data (nine tests) on a U.S. MCU (Modular Combination Unit) plant[14] showed an average NO$_x$ emission factor of 4.6 lb/ton (2.3 kg/t).

14.3.6 Sulfur Oxides

Many waste streams and fossil fuels contain sulfur. The sulfur can be present in any or all of its many oxidation states from S^{-2} to S^{+6}. Of particular interest relative to air emissions is the sulfur appearing as sulfides (organic and inorganic), free sulfur, or sulfur appearing in organic or inorganic acid forms. In each of these cases, the sulfur can be expected to appear in the flue gases as sulfur dioxide or trioxide. A small portion of the sulfur that exists as inorganic sulfates in the fuel or waste (e.g., gypsum-calcium sulfate) may be released by reduction reactions, especially in mass-burning situations.

Depending upon the chemical composition (alkalinity) of the mineral ash residues, a portion of the sulfur oxides may be lost from the flue gas by gas-solid reaction. Also, some sulfur may remain with the ash. Typically, however, these losses are relatively small, and in excess of 95% of the sulfur (other than that appearing as inorganic sulfates in the fuel or waste) will be found in the flue gases from suspension-fired combustors,[16] and up to about 70% for mass-burning systems,[17] based on analogy with coal-burning plants. Based on an average municipal refuse sulfur content of about 0.1 to 0.13% (as fired), an SO$_2$ emission factor of about 2.0 to 2.6 lb/ton (1.0 to 1.3 kg/t) would be expected for mass-burning systems with, perhaps, slightly greater emissions for MCU and slightly lesser emissions for RDF-fired units due to the difference in the exposure of the flue gases to alkaline fly ash.

The proportioning of sulfur between the dioxide or the trioxide forms depends on the chemistry of sulfur in the fuel, the time sequence of temperature and composition of the fuel gases, and on the presence or absence of catalytic ash material. Although cold-end chemical equilibrium considerations and excess oxygen concentrations favor oxidation to the trioxide, reaction rates are slow, and generally only 2 to 4% of the sulfur appears as the trioxide. Higher proportions of trioxide result from the burning of organic sulfonates, some heavy metal sulfates (which dissociate to SO$_3$ and an oxide), or to the "burning" of wastes such as discarded automobile batteries, which contain free sulfuric acid.

Sulfur oxides have importance as a pollutant due both to their health effects (especially in combination with respirable particulate matter) and to their corrosive effects on natural and man-made materials. Within the combustion system itself, sulfur trioxide will react with water vapor to form sulfuric acid, which has a dew point considerably above that for pure water. Indeed, to prevent serious corrosion (for example, in the stack) from sulfuric acid, combustion system cold-end temperatures should be limited to a value safely above the sulfuric acid dew point.[15]

14.3.7 Hydrochloric Acid

Chlorine appears in waste streams both in inorganic salts (for example, sodium chloride) and in organic compounds. In the combustion of many industrial wastes, and importantly, in municipal solid wastes, a substantial quantity of organic matter containing chlorine may be charged to the furnace. In the combustion environment (usually containing hydrogen in considerable excess relative to the chlorine), the organic chlorine is converted, almost quantitatively, to hydrogen chloride: hydrochloric acid.

Sources of organic chlorine include the following:

Compound	Chlorine (wt %)	Uses
Polyvinyl chloride	59.0[a]	Bottles, film, furniture
Polyvinylidene chloride	73.2[a]	Film
Methylene chloride	82.6	Solvent
Chloroform	88.2	Anesthetic
DDT	50.0	Insecticide
Chlordane	59.0	Insecticide

[a] Pure resin.

The importance of hydrogen chloride emissions from combustion sources depends on the quantity in the fuel, but is usually small. Of importance to system designers, however, is the high solubility of hydrogen chloride in scrubber water (or in condensed "dew" on cold-end surfaces) with well-demonstrated rapid acid attack and chloride corrosion of metal surfaces.

There are wide variations between measured and estimated HCl emission factors in the published literature. In concept, one may develop such factors from refuse composition (ultimate analysis) data or from direct-stack gas sampling, but problems are inherent in either procedure. Changing refuse composition from locale to locale and with time (both seasonally and as fundamental, long-term changes in waste character occur) leads to significant differences in both estimated and measured factors.

In stack measurements, it is difficult from a sampling and analysis standpoint to identify HCl *per se* in the stack gas. The presence of finely divided fumes of inorganic chlorides and acid gases (SO_2, SO_3, organic acids) confuses the data—especially when only limited mechanisms, for example, a glass wool plug, are used to remove particulates ahead of the impinger set.

Other analytical problems occur in the analysis of refuse. The total refuse chlorine content has been shown to be about one-half organic (yielding HCl on combustion) and one-half inorganic (not yielding HCl and, thus, not relevant to this discussion).[18]

Lastly, several researchers[19,20] have reported that 20 to 40% of HCl is absorbed by alkaline particulates within the incinerator flues. The majority of the data[20] (and earlier Swedish data) suggest 35 to 40% (approximately 37.5%) removal can be expected by such absorption.

A compilation of data on HCl emission from resource-recovery facilities from the literature is presented in Table 14.5.

An average U.S. emission concentration for HCl ranges between 100 and 150 ppm. This corresponds to a concentration of organic chlorine in refuse of 0.2 to 0.3% and approximately a two-thirds reduction prior to emission due to fly ash absorption.

14.3.8 Micropollutants

Modern sampling and, more importantly, modern analytical techniques have shown the existence of a wide variety of inorganic and organic compounds in incinerator effluents. In view of the diversity and largely uncontrolled nature of the waste itself and the complexity of the "chemical reactor" represented by the incinerator, this result is not unexpected. Yet, the verification that incinerator effluents contain substances with known or suspected carcinogenic toxic or other hostile properties (even if only in picogram quantities) gives pause to elected policymakers, to sanitation officials and to the general public who are considering the implementation of a waste combustion system. Also, such knowledge provides a powerful weapon to the "adversary" for any proposed solid waste combustion project. The public's fear of adverse health effects is a particu-

Table 14.5   Reported HCl Emission Factors

Reference Number	Comment	Average Emission Factor	
		lb HCl/ton refuse[a]	ppm[b]
21	Range 30.3–265 ppm	—	116
22	Range, U.C. 50–300 ppm	—	—
23		—	197
18	Refuse analysis for *organic* Cl	1.78	46
24	Refuse analysis	5.33	138
25	Range, 63–177 ppm	—	110
26	Range, 5–1260 ppm	—	—
20	Average of 11 tests	—	252
27	Range 10–60 ppm (after scrubber)		N/A
28	Range, 142–262 ppm	4.0	166
29	Refuse analysis for total Cl	8.03	2.8
29	Per Ref. 5[c]	4.02	104
30	Projected refuse organic Cl	4.02	104
2	Assuming 0.3% Cl in refuse	6.00	155
2	Per Ref. 5[c]	3.00	78

[a] Based on reported reuse chlorine content.
[b] Presumes 37.5% removal by fly ash absorption for estimates based on waste analysis. Concentration equivalents for mass emission factors based on typical refuse and represent concentrations at the point of discharge.
[c] Assumes organic chlorine is one-half of total refuse chlorine content.

larly effective rallying point in the 1980s, which this author would characterize as the decade when hazardous waste issues dominated the national environmental concern.

The challenge to the policy-maker and technologist related to micropollutant emissions as of this writing comes from many sides:

1. *What Are the Emission Rates?*   Sampling and analysis techniques are still evolving. Reported data often scatter over orders of magnitude and are often disassociated from the relevant waste, hardware system, and operational characterizations which might allow patterns in the data to emerge.

2. *What Is the Health Impact?*   The diversity of the chemicals involved and the high cost and long time required to carry out health effects research makes it hard to be sure what emission rate represents a significant health hazard and what rate is only an interesting benchmark of the advancement of man's analytical skill.

3. *How Much Potential Impact Constitutes an Acceptable Risk?*   Clearly, as analytical techniques have improved, our policy-makers are being forced to grapple in the public arena with the difficult matter of accepting a finite risk of adverse health impact . . . else be frozen into nonaction on almost all fronts by a "zero-risk" requirement. The lack of reliable epidemiological data at the subchronic effects level, of health data bases for the affected population, and of a well-accepted computational method for risk assessment compounds the policy problem.

Heavy Metals and Other Elemental Emissions

Several elements with significant health effects for several or all of their valence states (especially mercury, lead, arsenic, beryllium, and cadmium) are found in municipal waste and are emitted with the fly ash from a municipal incinerator. Clearly, the amounts emitted will be scaled by three factors: (1) the amounts in the waste, (2) the process variables that either boil/sublime or elutriate the compounds of the elements such that they leave the furnace suspended in the gas stream, and (3) the efficiency of the air pollution control system (especially regarding fine-particulate removal).

The amount of the heavy metals in the waste varies. Only lead, which appears in both the solder on "tin cans" and electronics, in discarded auto batteries, and in plumbing scrap, might be thought of as present in relatively large quantities and with some reliability. Still, average lead emission rates[11] are only about 0.012 lb/ton (0.006 kg/t).

Typical emission rates of other elements[11] are:

Element	lb/ton × 10^{5a}	kg/tonne × 10^5
Arsenic	4.4	2.2
Barium	18.0	9.0
Beryllium	<0.0056	<0.0028
Cadmium	38.0	19.0
Chromium	12.0	6.0
Copper	38.0	19.0
Mercury	640.0	320.0
Nickel	5.8	2.9
Zinc	1800.0	900.0

[a] Based on a stack particulate emission rate of 0.34 lb/ton.

It should be noted that the primary chamber of some of the smaller MCUs are operated (in the starved-air mode) at, say, 1300°F (700°C) where volatilization of all but mercury compounds is quite limited and, also, particulate carry-over is low. One would expect significantly lower elemental emissions from such plants.

Hydrocarbons, PNH, PCB

As for CO, hydrocarbons are found in incinerator effluents as a consequence of incomplete combustion. The preponderance of these materials are low-molecular-weight hydrocarbons and partially oxygenated species such as aldehydes and organic acids. A small but important fraction includes the polynuclear aromatic hydrocarbons (PNH) and PCBs, which may be significant because of their carcinogenic activity. Whereas the PNH materials are common products of incomplete combustion, the PCBs in the effluent are most likely the unburned residual PCBs initially present in the waste—a man-made pollutant. Present restrictions on the manufacture and distribution of PCBs strongly suggest that emissions in the future will be lower than those reported in the literature.

"Hydrocarbon" emissions in older incinerators, as for CO, are relatively high. Data[2] on older (pre-1970) refractory plants showed about 2.3 lb/ton (1.15 kg/t) and on waterwall units built in the late 1960s of about 1.6 lb/ton (0.8 kg/t). More recent data[11] suggest about 0.12 lb/ton (0.06 kg/t) for modern, waterwall, mass-burning systems; perhaps about twice as much for the modern MCU systems and slightly less for RDF systems might be expected.

Data on PNH are limited and scatter widely. An average value of about 1×10^{-5} lb/ton has been recommended.[10] Data for PCBs are even less common but a conservative value of 1.3×10^{-4} lb/ton (0.65×10^{-4} kg/t) is suggested based on tests at three U.S. waterwall units.[11]

Polychlorinated-p-Dioxin and Polychlorinated Dibenzofuran Emissions

In the late 1970s and early 1980s advances in analysis techniques revealed trace quantities of chlorinated dibenzo-p-dioxin and dibenzofuran compounds associated with the fly ash of municipal incinerators. Although the quantities found were expressed in the dimensions of nanograms per gram of particulates, this discovery generated great interest because of the known ability of these compounds at very low concentrations to affect liver enzyme activity and other health-related bodily functions.

Of particular concern to incinerator owner/operators around the globe is the question of the significance of polychlorinated dibenzo-p-dioxin (PCDD) and polychlorinated dibenzofuran (PCDF) emissions. At this writing, the data available on PCDDs and PCDFs is scarce, incomplete, and for the most part incohesive. Most of the PCDD/PCDF emissions data reported in the United States are from samples taken in the stack, and for the tetrachloro dibenzo-p-dioxin (TCDD) cogener and/or the 2,3,7,8-TCDD isomer. In Europe, the data are predominantly from samples of collected fly ash with the emission of key PCDD/PCDF cogeners reported separately. Most data do not include the information necessary to convert emission data to consistent units for direct comparison with other published information.

Understanding the fundamental reason(s) for the emission rate differences would assist in answering the following important questions:

Which emission rates best reflect the "true" case and should be used to evaluate the public health effects and risks of RR facilities?

What guidance as to: (1) acceptable wastes, (2) design features, (3) operating strategies should be given to existing and new incineration systems to limit/control PCDD/PCDF emissions?

What changes in sampling and/or analysis protocols should be considered in the future to assure that future data relative to PCDD/PCDF emissions best reflect the "true" case?

Although sampling and analysis methods could account for the differences between reported U.S. and European PCDD/PCDF emission data, it is also possible that the differences arise due to differences in generation rates, inherent (combustion environment) control, and/or removal by the air pollution control (APC) device.

During incineration, PCDD/PCDF compounds have been thought to be generated by one or more of the following:

Burning MSW components that contain trace levels of PCDD/PCDF.

Dimerization or condensation reactions of two or more molecules of man-made precursor chemicals (such as chlorinated phenols).

Partial oxidation of single-molecule precursor chemicals, such as PCBs.

Halogen attack of basic aromatic structures derived from lignins present in wood, woody vegetable residues, etc.

Within the incinerator environment, PCDD/PCDF could be generated in more than one region. These dioxins/furans brought in with the MSW will be released as vapors (volatized) in the burning bed, experiencing the full incineration system combustion history prior to emission.

Also likely to be generated on the burning bed are those PCDD/PCDFs formed by condensation reactions of chlorophenols or similar precursors present in the MSW. Such precursors (present in low concentrations) would be greatly diluted on entering the bulk flue gas flow, significantly reducing the reaction rate.

PCDD and PCDF formation by partial oxidation of single-molecule precursors (such as PCBs) can occur anywhere along the flue gas path. There are relatively small quantities of PCBs in MSW, and those present are likely to occur with some irregularity. This suggests that, in view of the widespread and relatively consistent PCDD/PCDF emissions rates reported for MSW incinerators, the PCB generation path may be relatively unimportant except on an episodal basis.

PCDD/PCDF formation by halogenation of basic aromatic structures could also occur in a distributed manner through the incineration system. One might speculate, for example, that some PCDD/PCDFs are formed by the attack of partially pyrolyzed or graphitized char particles by HCl or other chlorine sources present in the flue gas due to the incineration of chlorinated plastics and other halocarbon wastes or, perhaps, from reactions with inorganic chloride salts. The relative importance of this mode of PCDD/PCDF formation is uncertain. Investigators of such phenomena in the early 1960s detected neither PCDD nor PCDF in the off-gas from PVC combustion. Assuming that such reactions could occur, however, they would most likely take place in the refuse bed in zones where chlorocarbons are burning. There, the concentration of reactant species is highest and reducing conditions prevail (which limits competing oxidation reactions).

Lignin, the class of compounds found in woody vegetable matter, has been suggested[32] as the most probable starting material for PCDD/PCDF compounds. Lignins are complex polyphenolic, three-dimensionally branched network polymers. Experiments[32] wherein paper (largely delignified) and then wood were burned in a small fluidized-bed laboratory reactor fed an air-HCl mixture produced PCDD/PCDF compounds from the wood but essentially none from the paper.

The results of the fluid bed experiment, data from other investigators in both laboratory and field combustion situations where lignin-bearing materials were being burned, and the worldwide appearance of PCDD/PCDF in incinerator effluents suggests strongly the role of a common, ubiquitous precursor such as lignin (rather than unusual man-made chemical precursors such as PCBs and chlorophenols) as a principal source of PCDD/PCDF compounds. This is not to exclude other precursors or generation pathways (including de novo synthesis) but rather to suggest that the primary PCDD/PCDF source may be the reaction of lignin (and/or pyrolyzed fragments of lignin polymers) with appropriate chlorine donors.

There are many sources of lignin in municipal waste, with paper made from groundwood pulp (newsprint), wood and residential garden waste, and foodstuffs the most prominent. Lignin constitutes almost 30% of common coniferous and deciduous trees and comprises an even higher percentage of materials such as bamboo, rice hulls, and peanut shells. A large fraction of the lignin is removed in the process of converting wood to fine paper. Thus, a high paper content in refuse does not necessarily equate to a high lignin content. It is difficult to compare quantitatively lignin sources in European and U.S. refuse, although southern European waste appears to have a higher percentage of food and yard waste.

Sources of chlorine must also be identified to complete the formation scenario. Polyvinyl chloride (a common plastic) appears to be a possible donor-precursor but investigators believe that

Table 14.6 PCDD/PCDF Emissions from European/U.S. Incineration[a]

Compound (Mean)	Olie, Netherlands	Benfenati, Italy	Redford, U.S.	Gizzi, Italy	ADL, U.S.	Swiss Federal Office, Switzerland
Reference	36	Unpublished	37	38	39	40
Tri-CDD	NR[b]	—	65	NR	NR	NR
Tetra-CDD	610	—	32	250	175	26
Penta-CDD	2,419	—	—	467	451	72
Hexa-CDD	3,263	—	82	642	766	164
Hepta-CDD	3,367	—	38	581	766	158
Octa-CDD	1,900	—	13	705	175	323
PCDD	11,559	2,563[c]	230	2,645[c]	2,333	743
Tri-CDF	NR	—	1,529	NR	NR	NR
Tetra-CDF	1,562	—	447	700	260	147
Penta-CDF	2,704	—	—	743	541	179
Hexa-CDF	4,787	—	306	804	906	123
Hepta-CDF	2,614	—	38	880	641	82
Octa-CDF	581	—	3	590	80	54
PCDF	12,248	3,773[c]	2,323	3,717[c]	2,428	584

[a] μg/tonne, i.e., μg of PCDD or PCDF per 1 metric ton of waste burned.
[b] NR, Not reported.
[c] Eliminating samples 4–5 and 18.

direct formation from PVC is limited. The most obvious source of chlorine is HCl from oxidation of the PVC. Inorganic chlorides have been shown by some investigators[33] to act as a chlorine donor to lignins present in paper mill black liquor.

Studies[34] summarizing recent U.S. data indicate an average HCl concentration of 150 ppmv in the flue gases. HCl emissions in northern Europe are significantly higher (reflecting a reportedly higher PVC content in their refuse). Recent data[35] show 500 ppm for the Amager plant in Copenhagen, 620 ppm in Dusseldorf, and 975 ppm for the Issy plant in Paris—an average of 665 ppm. Based on the large dilution with water vapor and the relatively smaller plastics content of southern European (Italian) refuse, one might expect lower HCl concentrations in the flue gases from incinerators burning Italian refuse.

A recently published analysis of PCDD and PCDF data[31] suggests a correlation of generation rate with the concentration of chlorine in the flue gases as estimated from the chemical equilibrium of the reaction of HCl and oxygen to form chlorine and water vapor. Emission data for the various PCDD and PCDF cogeners are summarized in Table 14.6.

14.4 CONTROL TECHNOLOGY

The minimum control of air pollution from municipal incinerators with a furnace capacity greater than 50 TPD (45.36 t/day) targets the attainment of the federal NSPS for such units of 0.08 gr/dscf corrected to 12% CO_2. Equipping the units to do better than this for particulate and/or to control other pollutants (especially HCl, SO_2, NO_x, and CO) may reflect stringent state regulations or the unfortunate (but common) location of the unit in industrialized metropolitan areas which may be designated (by the federal EPA) as a nonattainment area for one or all of the listed pollutants. Thus, the incinerator owner is often forced to consider the Emission Offset Ruling and PSD regulations which trigger BACT review and, perhaps, installation of BACT systems.

In general, very low net particulate emissions can be achieved by application of conventional (if costly) flue gas cleaning equipment. Also, emissions of combustible pollutants (CO, hydrocarbons, PNH, POM, and so on) may be reduced significantly by a combination of good system design (especially the provision of a well-located overfire air supply) and thoughtful operation and maintenance. Acid gas control is feasible and has been demonstrated in Europe, Japan, and to a limited extent in the United States. Operating data on the acid gas control systems (wet and dry scrubbers) is inconsistent but suggests that this element of air pollution control adds significantly to capital and operating cost and reduces unit availability.

14.4.1 Particulate Matter

Several particulate control systems have been considered for review under BACT, including electrostatic precipitators (ESP) with various control efficiencies, fabric filters, a dry scrubber in line with an ESP, and a dry scrubber in line with a conditioned fabric filter.

The electrostatic precipitator is the most common flue gas cleaning device used on MSW incinerators and coal-fired utility boilers for particulate control. Fabric filters have been used on utility boilers but to a much lesser degree on MSW incinerators and can still be considered to be in the developmental stage from a long-term operations point of view. Use of wet and dry scrubbing systems on utility and industrial boilers is not uncommon; however, their application on MSW incinerators has been very limited. Operating experience with wet scrubbing systems on MSW plants has been largely unsuccessful due to the severe corrosion problems from the saturated flue gas and the costs associated with wastewater treatment and disposal. Dry scrubbing with a baghouse or ESP (as the particulate collector) is presently in operation on a few incinerators, and it appears to be a viable technology for combined particulate and acid gas control. Long-term reliability has yet to be demonstrated.

These devices have been historically applied to various combustion processes and all systems have either demonstrated or were designed to achieve control efficiencies in compliance with NSPS.

Alternative Control Systems

A. Electrostatic Precipitator.   Collection of particles is accomplished in an ESP by producing an electrical charge on the particle with a corona discharge; the particle is then attracted to collecting surfaces of opposite polarity. The collected dry particles are mechanically removed from the collecting surfaces and fall to a collecting hopper of the ESP. Some major MSW incineration plants equipped with ESPs are located in Braintree, Massachusetts; Harrisburg, Pennsylvania; Chicago NW, Illinois; Dade County, Florida; Albany, New York; Nashville, Tennessee; Norfolk, Virginia; Saugus, Massachusetts; Hooker Chemical, Niagra Falls, New York; and Montreal (Des Carriers), Quebec. Throughout the United States and the world MSW combustion facilities are successfully utilizing ESPs for the efficient control of particulate matter.

The use of ESPs to control particulate emissions from municipal incinerators has proven to be acceptable on the basis of many years of operating experience. Also, ESPs, when compared with other control technologies discussed hereafter, have, in general, proven to be most cost-effective, operationally reliable, and the most frequently applied technology available for controlling particulates from major combustion sources, especially municipal incinerators.

Several factors affect the collection efficiency of an ESP system. These include:

Collection plate area per unit gas flow rate (specific collection area).

Gas velocity through the collector.

Size distribution and electrical properties (resistivity) of the particles.

Number, width, and length of gas passages.

Electrical field strength and degree of high-tension sectionalization.

Particle in-field residence time.

Gas temperatures and humidity.

It is possible to attain the desired control efficiency by altering the design specification for each of the above parameters. Furthermore, several positive characteristics of an ESP are important when considering their use in controlling incinerator emissions. These include:

High overall collection efficiency and reliability.

Relatively low power requirements.

Ability to accommodate flue gas temperatures in the range of 250 to 600°F (121 to 316°C), with the low limit dictated by the gas dew point.

Minimal change in collection efficiency over a wide range of particle size.

Minimal fire hazard potential.

However, when incinerating municipal solid waste, the potential operational problems typically encountered with ESPs are:

Variations in flue gas temperature and fly ash resistivity.

Condensation of corrosive flue gas constituents.

Variation in incinerator processing (and emission) rate.

Deviations between design and actual gas flow rates.

Improper operation and resulting equipment deterioration.

The major adverse effect of these problems is to lower the collection efficiency. Corrosion problems can be alleviated through construction to avoid air leakage into the ESP and insulation and external heating of the unit to avoid cold spots where condensation can occur. Corrosion is normally not a problem as long as flue gas temperature remains above 250°F (121°C). Also, at high flue gas temperatures, fly ash resistivity is more predictable and particulate collection efficiency is enhanced.

B. High-Energy Wet Venturi Scrubber.   Collection of particles in a venturi is accomplished through inertial impaction of the particles on droplets of water. The collection efficiency of the particles is proportionately related to the water droplet size. Due to the abundant concentration of submicron particulate in the combustion gases of resource-recovery plants, a very high energy input is needed to form the small water droplets required for the high particulate removal efficiencies needed to meet NSPS regulations. Disadvantages of the venturi option are:

Materials of construction must be corrosion resistant, hence costly.

Flue gas will be saturated with resultant highly visible stack "steam plume."

Relatively poor plume rise due to low plume temperature (unless stack gas reheat is utilized, which consumes energy).

Additional wastewater treatment facilities required.

A wet sludge product must be disposed.

Excessive energy-power costs will be required for the venturi to meet NSPS. It is estimated the venturi will operate at approximately a pressure drop of 100 in. w.c.

Due to these disadvantages, the ventury alternative is not considered a viable means of particulate control for incinerators.

C. Fabric Filters.   The use of a fabric filtration system (baghouse) for controlling particulate emissions from combustion sources is an evolving technology and its application to municipal incinerators has been mostly limited to plants of 100 TPD or less. Operating experience in large mass-burning plants has been largely unsuccessful due to deterioration of the bags from flue gas temperature excursions, to blinding of the bags from variable moisture content in the flue gas, and to corrosion of the bags and metal components from acid gas condensation.

Collection of particles is accomplished through filtration by multiple tubular fabric filter media (bags). The bags are contained in multiple modular units which comprise the total baghouse system. Initial collection forms a thick porous cake of collected particulate on the bags. This cake then acts as the filter collection device, with the bag serving to support the particulate cake. As the cake builds, the pressure drop across the baghouse increases and the cake must be removed. Baghouses for incinerators are categorized according to the method used for cleaning the bags: reverse air and pulse jet. The two systems gently clean the bags by reversing the air flow through the bags. The reverse air baghouse utilizes an external centrifugal fan and the pulse jet uses high-pressure compressed air.

The major advantage of the fabric filtration system relative to the ESP is the improved control efficiency for particulate removal. A fabric filter is not as dependent on gas flow rate and composition as is the ESP, and it will accept surges in gas flow and particulates with no significant increase in particulate emissions. Resistivity of fly ash for low-sulfur fuels, which can be a problem with ESPs, is not a problem with fabric filter. Fabric filtration systems also have a greater control efficiency for smaller particle sizes (including the fumes bearing an enriched fraction of the heavy metals) and show some ability to control condensible organics and acid gases as compared with ESP systems. Problem areas include:

Susceptibility to smolder perforation or fires.

Blinding or clogging of the filter fabric if the dewpoint is reached.

Loss of structural integrity for high temperature surges above 550°F (288°C).

Cementation in a humid, low-temperature gas stream.

Short and uncertain bag life because of limited operating experience.

Individual bag failures can be limited when the fabric and cleaning methods are properly selected. Gas velocity, pressure drop, and the air-to-cloth ratio (volume flow rate of flue base per unit area of bag surface) are important operating parameters that must be considered in designing a system. However, predicting the build-up of filter cake and the resulting increase in pressure drop remains a difficult task that introduces uncertainties into baghouse design.

Fabric filters have not been thoroughly demonstrated on refuse incineration plants in comparison to with ESPs. To date, there have been only four full-scale mass-burning facilities utilizing fabric filters as the sole control device, two in Switzerland, one in East Bridgewater, Massachusetts, and one in Gallatin, Tennessee. The East Bridgewater plant showed severe corrosion of metal and fabric surfaces due to acid condensation at cold spots during system shutdowns. Operating experience at one of the Switzerland plants is reported to be satisfactory following considerable system modifications. Gallatin had severe problems with bag burn-out and is converting to an ESP. Recently, special-design baghouses have been utilized on several small municipal plants utilizing modular combustion units (MCUs). These plants include Windham, Connecticut, and Auburn, Maine.

At these facilities, 304 stainless steel mesh bags were used to eliminate the possibility of bag damage due to fires. However, blinding of the bags has been a recurring problem due to variation in the moisture content of the flue gas. At the Windham plant, blinding of the bags was so severe that the suction from the I.D. fan collapsed the bags.

Thus, the major uncertainty of fabric filtration systems, as compared with ESPs, is their reliability, which is dependent on bag life and other variables and can result in unacceptably high O & M costs.

D. Dry Scrubber-Electrostatic Precipitator System. An efficient particulate control system with the ability also to control acid gases can be effected when a dry scrubber and an ESP are placed in tandem. A dry scrubbing system manufactured by Teller Environmental Systems, Inc., includes a quench reactor tower and dry venturi agglomerator followed by an ESP or a baghouse. Systems offered by D.B. Gas Cleaning Corporation and Niro Atomizer Inc. are similar, except that they do not include the venturi agglomerator.

Dry scrubbing systems control acid gases through chemical neutralization. This process is accomplished by contacting the flue gases with a fine alkaline slurry spray. The spray can be generated with atomizing nozzles or rotating discs. A description of the process is as follows:

The fine droplets of either calcium- or sodium-based alkaline slurry dry quickly when contacted with the furnace gases in the quench reactor tower. The acid gases react while the H_2O is being evaporated. The evaporation process cools the flue gases and usually extinguishes glowing sparks.

Large-size particles drop to the bottom of the scrubber.

Smaller-size particles exit the dry scrubber and enter the particulate control device. If a baghouse is utilized as the particulate control device, the unreacted reagent will be retained in the thick dust cake on the bags. The unreacted reagent acts to neutralize further the acid gases (approximately 5%) as the flue gases pass through the cake with the reagent. An ESP will not further provide acid gas control.

The disadvantages of a dry scrubber are as follows:

Reagent cost (recycle of reagent is often practiced to maximize chemical utilization).

Minimal design and operation/maintenance experience.

Susceptibility of clogging spray nozzles and/or quench tower wall chemical accumulation.

Additional space requirements.

Additional solids disposal (containing soluble salts).

Decrease in plume dispersion due to gas cooling.

The main advantage of a dry scrubbing system (in comparison to conventional particulate control devices) is the ability to remove efficiently both particulate and acidic gaseous elements (SO_2, H_2SO_4, HF, HCl) from the gas stream. Other advantages include:

No visible steam plume (as with a wet scrubber) when ambient temperature is above approximately 40°F (4.4°C). This temperature will vary due to relative atmospheric humidity.

No corrosion problems as compared with conventional wet scrubber. Materials of construction are normally mild carbon steel.

Lower pressure drop as compared with wet scrubber.

Operating experience with dry scrubbing technology is limited. The only system in MSW service in the United States is the Teller Dry Scrubbing System in operation since 1979 at Framingham, Massachusetts. The system uses a baghouse for final particulate control, and the alkaline dust on the bags appears to have acted as a protective coating against bag smolder perforation.

There are a few European and Japanese incinerators that utilize dry scrubbing systems. A 1983 survey of dry scrubbing systems without the venturi agglomerator on MSW incinerators in Europe[41] concluded that the dry scrubbing process is still in the experimental stage based on visits to: Hamburg-Stellinger Moor, Oberhausen, and Dusseldorf. Two of the installations were pilot plants. The major problems observed at these installations were:

Fouling of the reactor tower with alkaline chemicals.

Clogging and frequent replacement of the atomizing slurry nozzles.

Frequent maintenance of chemical- and residue-handling systems.

Much higher than anticipated chemical usage.

Poor system reliability, that is, acid gas absorption efficiencies have not been achieved on a sustained basis.

In contrast to the reported European experience, Teller Environmental Systems claims to have 55 operational installations using their dry scrubbing technology. Of these, 28 are industrial applications and 26 are on municipal incinerators in Japan, and the one U.S. installation in Framingham, Massachusetts. Operating data on six of these installations shows control of particulate emissions to less than 0.01 gr/dscf and acid gas removal efficiencies of greater than 95% on HCl and greater than 85% on SO_2. The Framingham facility has had few unscheduled shutdowns and rebagging was required only once after 2.5 years of operation. High system availabilities have also been claimed at the most recent Japanese installations.

Chain deslaggers are now used by Teller on the quench reactor to minimize the growth of lime deposits on the walls. Spray nozzles use auxiliary air cooling to prevent baking of lime on the nozzle. Easy disengagement of the nozzles allows cleaning while in operation. Monthly cleaning of the nozzles and replacement of nozzles after 6 months is necessary due to abrasion. The upflow reactor lengthens the drying time in the quench reactor and helps prevent a wet bottom or sludge accumulation in the reactor.

In summary, although recent facilities appear to be operating successfully, dry scrubbing has been (and still is) an evolving technology that as yet has not demonstrated long-term reliability on refuse combustion facilities.

Performance Comparison of Alternative Systems

The control systems outlined above are all capable of meeting the particulate emission requirements established by state and federal NSPS guidelines. For the ESP, the particulate control efficiency can be upgraded by adding additional fields and by reducing the gas velocity (though a penalty of increased capital investment and higher electrical cost during operation is incurred).

The particulate loading is routinely brought to or below the 0.08 gr/dscf (corrected to 12% CO_2) level of NSPS using the ESP. Data from Baltimore, Maryland, and Nashville, Tennessee,[11] show that 0.018 to 0.025 gr/dscf is probably the best estimate of long-term emission levels. Although data are limited, these levels also appear to be within the range of the fabric filter.

14.4.2 Carbon Monoxide and Hydrocarbons

As noted in the discussion of pollutant generation, the emission rates of these (combustible) pollutants are very strongly related to the characteristics of the combustion environment that is maintained within the incinerator. The single most important process variable is the mean temperature; low values (less than 1200°F or 650°C) indicate that combustion will be quenched in many eddies of gas. Niessen[3] correlated CO emission (kg/t) with T_{AVE}, the average furnace temperature (K), for a refractory MSW incinerator in Italy as:

$$CO(kg/t) = 29.96 - 0.02333\ T_{AVE} \tag{13}$$

Furthermore, the control of combustible pollutants is influenced by the effectiveness with which the burning zone furnace gases are mixed with combustion air. In general, this is best accomplished with overfire air jets which both add air (oxygen) and also introduce a strong momentum flux to stimulate mixing.[2,15]

14.4.3 Oxides of Nitrogen (NO_x)

The installation of nitrogen dioxide (NO_x) emission control technologies on combustion units is in various stages of development. The U.S. EPA is sponsoring pilot programs in conjunction with major utilities for post-combustion NO_x control on fossil fuel fired boilers. NO_x emissions can also be controlled in the combustion process.

NO_x emissions from combustion processes result from an oxidation of nitrogen in the combustion air and in the materials incinerated. Formation of NO_x is highly dependent on temperature, pressure, and residence time in the combustion unit. A lower, uniform temperature and uniform mixing of air and burning refuse generally eliminates high oxygen concentration gradients and the sharp temperature gradients that are conducive to nitrogen oxide formation. The amount of NO_x released from a specific source is therefore a strong function of the design and operating characteristics of the particular combustion unit.

NO_x emissions from refuse incinerators have been controlled, to about 40% of their original values, by injection of ammonia (NH_3) into the primary furnace. Reaction of a portion of the ammonia with HCl led to the condensation of ammonium salts in the economizer of the boiler with consequent fouling problems.

14.4.4 Acid Gases

The application of a control device to remove any one of these pollutants (SO_2, HCl, H_2SO_4, and HF) would be effective in reducing a major percentage of all of these emissions. Federal authorities have not established sulfur dioxide or acid gas NSPS for MSW incinerators, and based on recent NSPS reevaluation, do not plan to impose any further emissions standards on MSW combustion facilities. Therefore a case-by-case review is required to determine an allowable emission rate for SO_2 and acid gases. The following analysis for SO_2 has been prepared to satisfy the impact assessment criteria within the scope of a typical BACT evaluation.

Alternative Control Systems

The control alternatives that would reduce acid gas emissions include wet scrubbers and dry scrubbers usually followed by an ESP or baghouse. Electrostatic precipitators and baghouses alone are not normally considered to be effective.

A. **Fabric Filters.** In general, fabric filters are a particulate control device and are not designed to control acidic emissions from combustion sources. Some manufacturers have claimed a slight (approximately 10%) removal efficiency with fabric filters due to the absorption of acids by the alkaline dust coating the filter bags.

B. **Dry Scrubber–ESP.** The use of a dry scrubber and electrostatic precipitator is a relatively new control technology for application in sulfur dioxide removal. The control mechanism considered in this application was described in the control of suspended particulates. The operative removal mechanism is the interaction that occurs between the caustic material and acid gases within the flue gas stream. The removal efficiency as quoted from performance data furnished by a system vendor would be 65% for sulfur dioxide and 80% for other acid gases (HF, HCl, and H_2SO_4).

C. **Wet Scrubber–ESP.** The use of wet scrubbers and electrostatic precipitators for controlling the emission of acid gas has been applied to utility power plants with increasing frequency over the last 5 to 10 years. The mechanism relies on the physical and chemical interaction of acid gases with the available absorbent materials. Based on typical performance specifications presented by a system vendor, the approximate removal efficiency for SO_2 is quoted at 75%. The removal efficiency for other acid gases is 90%.

D. **Dry Scrubber–Baghouse.** The final alternative control system proposed is a dry scrubber followed by a fabric filter (baghouse). The technology of dry scrubbing is relatively new and the control efficiencies as quoted by system suppliers have not been substantiated by long-term operating data. According to vendors, the estimated control efficiency for SO_2 would be 70% and 92% for other acid gases.

Performance Comparison of Alternative Systems

The data indicate that the highest degree of control for SO_2 and acid gases are achieved by the dry scrubber/fabric filter system and the combination of an ESP and a wet scrubber system.

14.4.5 Micropollutants

Combustible Pollutants

As noted for CO and "hydrocarbons," control of combustible pollutant emissions would be expected to improve as the general characteristics of the combustion environment improved: higher and more uniform temperatures and both air and momentum (mixing energy) addition to attain largely uniform gas compositions with a sufficiency of oxygen. Unfortunately, there is an incomplete data base to quantify these relationships relative to the significant combustible micropollutants. See, however, the several papers on the topic of dioxin emissions in the 1984 ASME National Waste Processing Conference held in Orlando, Florida.

Toxic Metals

Mercury emissions are minimally affected by particulate control devices and essentially 100% of the mercury compounds in the waste are volatilized and emitted. Limited control may occur in fabric filter installations due to the possible absorption of the mercury vapor onto the finely divided cake of captured dust.

The emissions of other volatile toxic metals (lead, cadmium, zinc) will arise due to their volatization and subsequent condensation as submicron fume particles which, owing to their small size, escape collection. The rate of release will be importantly related to the temperature of the burning refuse bed, a process variable that is not readily controlled. For these fume-type emissions, the fabric filter control technology may be more effective than the ESP but there is insufficient data available to support this hypothesis.

Other metal emissions (nickel, chromium, iron, beryllium) leave the furnace as relatively "macroparticulate" lifted from the burning bed by the underfire airflow. Since control of underfire air flow within certain limits is vital to realizing acceptable furnace burning rates, there are, clearly, limits to the reduction of these emissions by process optimization. Generally, the normal air pollution control system dust collection characteristics are sufficient to adequately control these emissions.

REFERENCES

1. L. Weitzman, Environmental regulations governing incinerator design and operations, *CPA*, March, 1984.

1a. L. Schiller and A. Naumann, *Z. Ver. Dent. Ing.*, **77**, 318 (1933).

2. W. R. Niessen, S. H. Chansky, E. L. Field, A. N. Dimitriou, C. R. La Mantia, R. E. Zinn, T. J. Lamb, and A. S. Sarofim, Systems Study of Air Pollution from Municipal Incineration, NAPCA, U.S. DHEW, Contract CPA-22-69-23, March, 1970.

3. H. Eberhardt and W. Mayer, Experiences with refuse incinerators in Europe, in Proceedings of 1968 National Incinerator Conference, ASME, New York, 1968, pp. 142–153.

4. F. Nowak, Erfahrungen an der Mullverbrennungs—anlage Stuttgart, *Brennst-Warme-Kraft*, **19**, 71–76 (1967).

5. R. L. Stenburg, R. R. Horsley, R. A. Herrick, and A. H. Rose, Jr., Effects of design and fuel moisture on incinerator effluents, *JAPCA*, **10**, 114–120 (1966).

6. R. L. Stenberg, R. P. Hangebrauck, D. J. Von Lehmden, and A. H. Rose, Jr., Field evaluation of combustion air effects on atmospheric emission from municipal incinerators, *JAPCA*, **12**, 83–89 (1962).

7. A. B. Walker and F. W. Schmitz, Characteristics of furnace emissions from large, mechanically stoked municipal incinerators, Proceedings of 1964 National Incinerator Conference, ASME, New York, 1964, pp. 64–73.

8. A. H. Rose, Jr. and J. R. Crabaugh, Research findings in standards of incinerator design, in *Air Pollution* (Mallette, Ed.), 1st ed., Reinhold, New York, 1955.

9. F. R. Rehm, Incinerator testing and test results, *JAPCA*, **6**, 199–204 (1957).

10. M. A. Field, D. W. Gill, B. B. Morgan, and P. E. W. Hawksley, *Combustion of Pulverized Coal*, British Coal Utilization Research Assn., Leatherhead, Surrey, England, 1967.

11. W. L. O'Connell, G. C. Stotler, and R. Clark, Emission and emission control in modern municipal incinerators, Proceedings, 1982 ASME National Waste Processing Conference, New York, 1982.

12. H. C. Hottel, G. C. Williams, N. M. Nerheim, and G. Schneider, Combustion of carbon monoxide and propane, in 10th Symposium (Int'l) on Combustion, Combustion Institute, Pittsburgh, 1965, pp. 111–121.

13. A. C. Morgan, Combustion of Methane in a Jet Mixed Reactor, D. Sci. thesis, MIT, Cambridge, Massachusetts, 1967.

14. H. G. Rigo, J. Raschke, and S. Worster, Consolidated data base for waste-to-energy plant emissions, 1982 ASME National Waste Conference, New York, 1982, p. 305.

15. W. R. Niessen, *Combustion and Incineration Processes, Applications in Environmental Engineering*, Marcel Dekker, New York, 1978.

16. E. S. Grohse and L. E. Saline, Atmospheric pollution: The role played by combustion processes, *JAPCA*, **8**, 255–267 (1958).

17. H. F. Johnstone, *Univ. Ill. Eng. Exp. Station Bull.*, **228**, 221 (1931).

18. H. I. Hollander, J. K. Kieffer, V. L. Eller, and J. W. Stephenson, The comprehensive municipal refuse characterization program, Proceedings, 1980 ASME National Waste Processing Conference, Washington, DC, 1981.

20. E. R. Kaiser and A. A. Carotti, Municipal incineration of refuse with 2 percent and 4 percent additions of four plastics: Polyethylene, polyurethane, polystyrene and polyvinyl chloride, Proceedings, 1972 ASME National Incinerator Conference, 1972.

21. J. W. Stephenson and V. L. Eller, The quest for incinerator air pollution control, Proceedings, 1980 ASME National Waste Processing Conference, Washington, D.C., 1980.

22. A. J. Teller, Dry system emission control for municipal incinerators, Proceedings, 1980 ASME National Waste Processing Conference, p. 581 et seq., Washington, D.C., 1980.

23. R. A. Olexsey, H. M. Freeman, and P. A. Brailey, Estimated ground level concentrations of pollutants from waste-to-energy facilities, Proceedings, 1980 ASME National Waste Processing Conference, p. 589 et seq., 1980, Washington, D.C.

24. EPA Resource Recovery Task Force and L. S. Wegman Co. Summary Report—Comprehensive Solid Waste Management Plan for Refuse Disposal of Material and Energy Resources, New York DEC Contract CSWP-30, June, 1977.

25. C. B. Bozeka, Nashville incinerator performance tests, Proceedings, 1976 ASME National Waste Processing Conference, Boston, Massachusetts, 1976.

26. D. A. Vaugn, P. D. Miller, and W. K. Boyd, Fireside corrosion in municipal incinerators versus PVC content of the refuse, Proceedings, 1974 ASME National Incinerator Conference, Miami, Florida, 1974.

27. J. A. Jahnke, J. L. Cheney, R. Rollins, and C. R. Fortune, A research study of gaseous emissions from a municipal incinerator, *JAPCA*, **27**, 747 (1977).

28. P. A. Brailey, Significance of Chlorine Emissions from the Thermal Processing of Municipal Solid Waste, EPA IERL Internal Report, March 25, 1979.

29. R. Rollins and J. B. Homalya, Measurement of gaseous hydrogen chloride emissions from municipal refuse energy recovery systems in the United States, *Environ. Sci. Technol.*, **13**, 1380 (1979).

30. Study of Municipal Solid Waste Quantity, Composition, and Fuel Characteristics—Summer 1980, Report by SCS Engineers for Port Authority of New York and New Jersey, February, 1981.

31. W. R. Niessen, Production of polychlorinated dihenzo-*p*-dioxins (PCDD) and polychlorinated dibenzofurans (PCDF) from resource recovery facilities, Proceedings, 1984 ASME National Waste Conference, Orlando, Florida, 1984.

32. Personal communication, Dr. K. Olie, University of Amsterdam (April 1983).

33. B. Ahling and A. Lindskog, Emission of chlorinated organic substances from combustion, in *Chlorinated Dioxins and Related Compounds: Impact on the Environment*, proceedings of workshop held at Instituto Superiore di Sanita, Rome, Italy, October 22–24, 1980 (O. Hutzinger, R. W. Frei, E. Merian, and F. Pocchiari, Eds.), Pergamon Press, New York.

34. Comments on the Need for the HCl Control Devices on New Resource Recovery Facilities in the State of New Jersey, prepared for the Port Authority of New York and New Jersey by Camp Dresser & McKee, July 8, 1981.

35. U. S. Environmental Protection Agency, Refuse-Fired Energy Systems in Europe: An Evaluation of Design Practices, Battelle Columbus Labs, Ohio, November, 1979.

36. K. Olie, J. Lustenhouwer, and O. Hutzinger, Polychlorinated dibenzo-*p*-dioxins and related compounds in incinerator effluents, in *Chlorinated Dioxins and Related Compounds: Impact on the Environment*, proceedings of a workshop held at the Instituto Superiore di Sanita, Rome, Italy, October 22–24, 1980. *Chemosphere* **9**, 501–522 (1980).

37. S. Redford, T. Haile, R. Lucas, Emissions of PCDDs and PCDRs from combustion sources, presented at The International Symposium on Chlorinated Dioxins and Related Compounds, Arlington, Virginia, October 25–29, 1981.

38. A. Gizzi, D. Reginato, P. Benfenati, J. Fanelli, Polychlorinated dibenzo-*p*-dioxins (PCDD) and polychlorinated dibenzofurans (PCDF) in emissions from an urban incinerator 1. Average and peak values. *Chemosphere*, **11**, 577–583 (1982).

39. Municipal Incinerator Emission Estimates, report to O'Brien & Gere Engineers, Inc. by Arthur D. Little, Inc., Cambridge, Massachusetts, March, 1981.

40. Swiss Federal Office for Environmental Protection, Environmental Pollution Caused by Dioxins and Furans from Commercial Refuse Incineration Plants, Bern, June, 1982.

41. M. Dvirka, Technology of Dry and Wet Scrubbing of Gases from MSW Incinerators, prepared for the Port Authority of New York and New Jersey, WF Cosulich Assoc., P.C.

CHAPTER 15

MARKETING RESOURCE RECOVERY PRODUCTS

RIGDON H. BOYKIN and
BERNAYS THOMAS BARCLAY

Chadbourne & Parke
New York, New York

CALVIN LIEBERMAN

Magnimet Corporation
Toledo, Ohio

15.1 ENERGY

Rigdon H. Boykin and Bernays Thomas Barclay

15.1.1 Energy Marketing Principles

The energy products of a resource recovery project must be marketed effectively to capture their full potential value and thereby to protect the project's long-term viability. Too often these projects have been sited, sized, and operated primarily to process the maximum amount of the waste resource available while avoiding environmental impacts as much as possible. The thinking here is: "We'll sell the steam and electricity for whatever we can get, and cover the rest of the costs of the plant with the tipping fee."

This may or may not be a legitimate set of objectives for a particular project. It relegates energy to a by-product status, however, where, as the project gains a momentum of its own, the potential value of the energy revenue stream may unintentionally be limited through the request-for-proposals (RFP) process. Restricting the project's marketing opportunities in this way can be hazardous to its health. Overdependence upon tipping fees can easily lead to increased competition for the waste resource, or to escalating (and unpopular) subsidies of the project, and early failure.

The return from a vigorous energy marketing effort, on the other hand, may contribute impressively toward forestalling these potential problems. Indeed, it may be an important factor in the owner's decision to go forward with the project in the first place. A difference of one-tenth of a cent in the price per kwh paid for the project's electrical output can mean a difference of $160,000 to $400,000 *per year* in electric revenues for large projects. Put another way, the difference of a tenth of a cent per kwh could reduce tipping fees by 30 to 50 cents per ton.

To the extent possible, therefore, the various steps of the decision to go forward with a project should, within applicable physical and environmental constraints, be marketing decisions. The developer's objective should be to maximize the energy product revenue streams and to minimize the tipping fee. This, then, is the first of the principles that should be applied to develop an effective energy marketing strategy: Plan, from the outset, to make money. Following this threshold decision, several additional principles follow in a commonsense order.

1. *Prepare Thoroughly.* There can be no substitute for vigorous, detailed, exhaustive preparation. Knowing the project's technical attributes, knowing the laws applicable to energy generation and sales, and knowing the potential purchasers will require a substantial investment of time and effort. The extent of the marketing team's preparation and knowledge will be the major determinant of long-term success.

2. *Explore the Market Creatively.* With superior preparation, the developer need not miss any opportunities to market the project's energy output. Innovative thinking will be required to unearth some of these opportunities, and "brainstorming" can pay off. The project's highest and best energy value will rarely be found in the "standard offers" made by an electric utility to its customer/generators. To realize that value, the developer may have to look for markets beyond the local utility, for two or more purchasers, for purchasers who are the utility's customers, or for customers of other utilities.

3. *Negotiate Aggressively.* Clear objectives and thorough preparation arm the developer for aggressive negotiation with the target purchaser. Although the developer must remain flexible, persistence, together with a willingness to use all of the resources available to obtain the developer's desired goals, will be necessary to translate the project's potential into a satisfactory contract.

The foregoing principles may be more effectively described in the framework of marketing tasks. The developer may find that it will in fact be useful to impose a task-type management structure upon the marketing effort. Since the problems involved are sufficiently complex, however, the effort should be a structured one regardless of the management technique employed. The sale of energy products will depend upon the following tasks:

1. Familiarization with the legal and political environment.
2. Identification and investigation of potential steam customers.
3. Identification and investigation of potential electricity customers.
4. Negotiation of energy sales contracts.

We will discuss the most important considerations that apply to these tasks later in this chapter. It is appropriate at this point, however, to develop an overview of their significance and scope in marketing the project's energy products.

Legal Familiarization

A thorough understanding of the seller's rights and potential liabilities under the law is essential for properly assessing market opportunities and effectively negotiating energy sales contracts. All too often resource recovery projects fail to fully realize their potential value because the developers attempt to arrange for and negotiate the sale of their energy products without this understanding. Such an oversight will be especially costly today in light of recent legal developments that have improved the bargaining position of most resource recovery projects vis-à-vis utilities and other potential customers and that make this an excellent time to negotiate for the sale of energy.

Familiarizing oneself with the applicable legal environment may also generate new marketing ideas. Moreover, a degree of expertise in this area can avoid the loss of precious time and opportunities that may occur if the developer must struggle to determine whether a particular marketing idea is legally viable.

An understanding of the legal rights of a developer of a resource recovery project must begin with Sections 201 and 210 of the Public Utility Regulatory Policies Act of 1978, commonly known as PURPA (16 U.S.C. §§ 796 (17-22) and 824a-3 (1978)). This law, among other things, requires electric utilities to purchase electricity from qualifying generation facilities at a reasonable rate, which for most purchases will be a rate equal to the utility's full "avoided costs." Avoided costs will be discussed in detail later in this chapter, but they are the incremental or marginal costs that the utility avoids by buying power from, for example, a resource recovery project, rather than generating it or purchasing it elsewhere.

The developer should also become familiar with the rules issued by the Federal Energy Regulatory Commission (FERC) which implement Sections 201 and 210 of PURPA (18 C.F.R. § 292.100 *et seq.*); the preamble to those rules (45 Fed. Register 12214, Feb. 25, 1980; 45 Fed. Register 24126, April 9, 1980); and the FERC's policy statement regarding enforcement of the rules (23 FERC ¶61,304, May 31, 1983). The Federal Power Act (16 U.S.C. § 791 *et seq.*), and the Public Utility Holding Company Act (15 U.S.C. §§ 79a-79z(6)), may also provide major pieces of the federal regulatory puzzle, especially for noncogenerating projects with a capacity in excess of 30 mw.

At the state level, there are a myriad of different laws and regulations that may apply to a resource recovery project. First, the developer must look at the particular state law establishing regulatory jurisdiction over utilities. This law will not only provide useful insight into the environment in which the utilities will be operating and/or negotiating, but it also may provide important information as to the extent to which a project can market its energy without subjecting itself to extensive state regulation as a "public utility." The developer must also determine whether there are any state energy laws (for example, a "mini-PURPA") that deal specifically with arrangements for small power production facilities. At this writing, mini-PURPA laws have been enacted in Alabama (Ala. Acts No. 83-574, July 27, 1983); California (Cal. Pub. Util. Code § 2801 *et seq.*, West

1982); Connecticut (Conn. Gen. Stat. § 16-243a, 1982); Indiana (Ind. Code Ann. §8-1-2.4, Burns 1982); Iowa (Iowa Code §§419.1(2)(a) and 476.34-.38); Kansas (Kan. Stat. Ann. §66-1, 184, 1980); Maine (Me. Rev. Stat. Ann. tit. 35, §2321 *et seq.*, 1982); Minnesota (Minn. Stat. Ann. § 216B.164, West 1983); Montana (Mont. Code Ann. § 69-3-601 *et seq.*, 1981); New Hampshire (N.H. Rev. Stat. Ann. § 362-A, 1983); New York (N.Y. Pub. Serv. L. § 66-c, McKinney's 1981); North Carolina (N.C. Gen. Stat. § 62-156, 1982); Oregon (Ore. Rev. Stat. § 758.500 *et seq.*, 1981); Texas (Tex. Rev. Civ. Stat. Ann. art. 1446c §§ 3(c) (1) and 16A, Vernon 1982); and Vermont (Vt. Stat. Ann. tit. 30 § 209(a) (8), 1982). Finally, whether or not a mini-PURPA law exists, the state utility regulatory authorities have been required by FERC to implement the federal rules under PURPA, and many have adopted regulations or issued orders in accordance with the FERC requirements. These will be of obvious and immediate interest to developer.

It is often important as well to understand the state laws, regulations, and procedures that govern a utility's ability to collect from its customers, through retail rates, the costs it incurs in generating and purchasing energy. This insight is essential to understanding the risks and potential benefits facing a utility in its negotiations for the purchase of energy from resource recovery projects. In most cases, the utility's primary concern is not how *much* it pays, but whether it will be able to *recover* those payments in its next retail rate case before the state regulatory agency. To the extent that the developer can design the sale of energy to avoid or reduce these perceived risks, a utility will look at the project in a more favorable light.

Some states also have laws regulating the manufacture and sale of steam. These laws are too particularized to permit meaningful generalization here; rather, the developer will have to evaluate them on a case-by-case basis.

The variety of applicable laws, as well as widespread uncertainty among practitioners regarding the interaction of federal and state regulatory jurisdiction in the area, may confuse and overwhelm all but the most experienced developer. Complicating matters even further is the fact that much of the applicable law has not been tested in the courts, either because the law is recent or because the participants in most of the energy transactions in the past—the utilities—rarely sue each other. The result is that different, and often contradictory, interpretations of these laws abound. The utilities' interpretation may seldom mesh with the seller's point of view.

These complications increase the potential for disagreement in the marketing and negotiating process, and require that the project developer be familiar with the substantive and procedural rules of the state or federal regulatory agencies administering the applicable laws. If the parties are unable, or unwilling, to agree upon an interpretation of their legal rights and obligations, it may become necessary to use available dispute resolution procedures before the appropriate regulatory agency. Although such litigation may be expensive and/or time-consuming, it is essential to establish a firm position and to be willing to take these disputes to a regulatory authority, especially in those instances when a utility is taking an unreasonable or abusive position.

Political Familiarization

The political environment of the state in which the project is located is also vital to the effective marketing of its energy products. Political factors and key relationships should be assessed in the earliest prefeasibility studies of the project. If the political environment appears to be unfavorable, the project developer should examine the opportunities for improving the situation. Resource recovery projects are never politically benign.

Political relationships are also a significant factor in any effort to obtain prompt and responsive action by potential purchasing utilities, or expedited remedial action by a regulatory agency. Litigation alone will rarely provide as satisfactory a solution. In addition, some marketing opportunities may only come to fruition if an applicable law is amended through the political process to provide all parties the comfort they require.

Identification and Investigation of Potential Steam Customers

The search for a good steam customer may be a difficult one and should start in the earliest stages of a project. The substantial potential benefits of acquiring a steam purchaser make a thorough search worthwhile. The fuel efficiency of a resource recovery project that cogenerates steam and electricity can be up to 82%, versus the 35% (or much less) experienced with straight generation of electric energy. With these efficiency gains—and associated revenue enhancement—in mind, the developer may find that sizing, siting, and operations planning for a project should await the outcome of steam market investigations.

Moreover, if the installed electrical generating capacity of the project will exceed 30 mw the resource recovery project will experience additional benefits by enlisting a steam customer. Under PURPA and the related FERC rules, small power production facilities that do not cogenerate steam and electricity are only exempt from FERC regulation under the Federal Power Act and the Public Utility Holding Company Act if their capacity does not exceed 30 mw. They do not qualify for

PURPA benefits at all if they exceed 80 mw. A bona fide cogeneration facility, however, qualifies under PURPA and is exempt regardless of its generating capacity.

Initial decisions regarding the project site are critical since steam quality cannot economically be maintained if it must be transported over long distances. One or two miles, depending on the intended steam use, is generally considered to be the limit. Moreover, the steam needs of many potential buyers fluctuate widely, depending upon the weather and the user's operational cycle, as well as the business cycle. Therefore, the developer may need to search for ways to supply the complementary steam needs of two or more customers, rather than to try and meet the needs of any one buyer.

Alternatively, the developer may be able to move the market to the project. Especially if the political groundwork has been attended to, a developer may be able to work with the local government to offer an attractive package of tax, labor and energy considerations to an otherwise remote industrial steam (and/or electric) user in need of an expansion or relocation opportunity.

As the developer locates potential steam purchasers, the development of the project's technical capabilities should begin. What are the viable siting options? Will the plant be sized and operated to run continuously? Will it operate on weekends? To what extent can supplementary firing systems or fuels be used economically to permit control over steam quantity and quality? The answers to these and other technical questions will have a significant impact upon the value of the project's steam product to potential purchasers.

The best source for these questions is likely to be the potential steam customer. The developer will only be able to develop this market on a customer-specific basis. Of course, the availability of the waste resource will permit a preliminary estimate of steam production capability sufficient to narrow the market survey to steam users of the appropriate size. After that, however, the developer must investigate a potential user's specific needs and requirements. This information may then be used to establish project specifications and requirements for the RFP process.

Identification and Investigation of Potential Electricity Customers

An effective marketing approach also requires thorough research of the potential market for the project's electricity product. Generally speaking, potential purchasers of electricity from resource recovery projects include investor-owned electric utilities, municipal utilities, electric cooperatives, and industrial, commercial, and institutional end users, including government facilities. The most readily apparent of these will be the local electric utility, whether investor-owned, municipal, or cooperative. Indeed, PURPA requires any electric utility* to offer to purchase the project's output, or any portion of it made available to them.

Market research should not be restricted to the local area, however. Nothing in PURPA requires that the project's electricity be sold to the local utility, or that all of it be sold to any single utility. If the state's laws are not unduly restrictive, the developer is free to shop anywhere on the grid, subject to the constraints imposed by (1) availability of transmission capacity, (2) willingness of the utility to wheel, (3) line losses, and (4) transmission charges.

If the project is located within the service area of a dominant utility, the price available for the project's electricity from any other potential local purchasers may be closely tied to that utility's rate structure. In this case, if the dominant utility's costs are too low to support an adequate price for the project's sales, the price available from other local purchasers may be no better, and the developer should examine opportunities for transmitting, or "wheeling," electricity to a more distant purchaser who can offer a more favorable price. Indeed as a practical matter, unless a special situation is presented, an initial survey should always be made of all the utilities, including municipals and cooperatives, which are connected to the project's local utility.

In some cases, the local utility's retail rate structure is such that its customers may be willing to pay a better price for the project's electrical output than would the utility. This can occur where the utility's retail rates (which, simply stated, recover the utility's average costs plus a rate of return on investment) exceed its avoided costs.

Sales to end users (that is, nonutilities), however, typically encounter two problems: First, since PURPA does not exempt such sales from retail rate regulation by the states, the seller may subject itself to regulation as a "public utility" under state law. Second, potential purchasers may be unwilling to jeopardize their existing relationship with the electric utility, especially if they will still have to rely on that utility for essential emergency or backup power.

The first problem has been resolved in a number of states by new legislation or rules expressly permitting unregulated sales to a limited number of end users, "affiliates" (partners, investors) of

---

*An electric utility for purposes of PURPA means any person (including corporations and other legal persons as well as natural persons) or state agency that sells electric energy; it includes the Tennessee Valley Authority but does not include any federal power marketing agency (16 U.S.C. § 796(22)).

the seller, or purchasers "at or near" the project site. (See, for example, Maine Stat. Ann. Ch. 36; New York Pub. Serv. L. §§ 2(2-d), (3), (13), (16) and (22), 1981.) In other states, existing laws may be interpreted to the effect that such limited sales do not bring the seller into the category of a "public" utility. In still others, clarifying actions by either the legislature or the utility regulatory authority may be necessary.

Where it exists, however, the fear of utility reprisal is not so easily overcome. The regulatory agency may provide some assistance. If persuasion and/or agency assistance does not resolve the issue, purchasers will be reluctant to experiment with the security of their electrical supply, and the developer may have to build backup reliability into the project to make the sale.

In the event that preliminary market analysis indicates that a sale to a utility is the best opportunity for the project, a great deal of research remains to be done. First, the specific target utility or utilities must be selected. Then the developer must vigorously explore ways to increase the target utility's purchase rate structure. Although the developer could accept the utility's published purchase rate offer at face value (if one exists), few projects can afford this luxury. This is true even if the utility's offer is purportedly based upon its "avoided costs." Calculation of avoided costs, as will be discussed later in this chapter, is frequently plagued by self-interest, lack of effort, and theoretical and arithmetical error. As noted earlier, if the elimination of any of these problems results in even a tiny increase in the purchase rate, the potential impact of that increase upon the project's long-term liability can be substantial.

Negotiation of Energy Sales Contracts

Since PURPA requires utilities to purchase electricity from qualifying resource recovery projects at the utility's "avoided costs," you may ask: What is there left to negotiate about? The short answer is, everything. First of all, many of the state jurisdictions have not implemented PURPA properly and have not established appropriate purchase rates for all of their utilities. Second, the FERC rules expressly state that qualifying facilities and utilities are not required to transact their business as provided for in the rules if they *agree* to other terms and conditions. These terms may, among other things, include a different price. The FERC rules, therefore, essentially provide an incentive for utilities to negotiate, and serve to backstop the parties if they fail to agree on different terms.

These negotiations frequently consist of a resource recovery project trading down from a full avoided cost rate in exchange for utility concessions to meet the project's other special needs for which the FERC or state rules do not provide. Where third-party or project financing is involved, for instance, it is often said that the developer may have to give in on price in order to obtain a long-term contract, since the utility's credit rating, secured through a long-term contract, may provide the lender's security in such an arrangement. Price concessions may also be necessary to obtain a special rate design suited to the project's financing requirements. Whether any particular price concessions are required in a particular case will depend upon the state rules for these transactions, and upon the preparation and skill of the negotiator.

As a practical matter, negotiations with potential purchasers of steam and electricity do not begin in earnest until after the RFP phase has been completed and a vendor selected for the project. At that time, in-depth research should be conducted with respect to those markets that were identified as most favorable in the preliminary surveys, and negotiations should begin as soon as possible. If two or more good marketing opportunities have been identified, it is probably unwise to pursue negotiations with only one at a time. Negotiations, especially with utilities, can require an inordinate amount of time, and may break down or stall for reasons completely beyond the developer's control. In such an event, the project can continue to move toward closing if other potential sales opportunities are well developed. Not infrequently, it becomes apparent to both parties during the course of negotiations that their mutual interest lies in having the utility assist the project in marketing its power to a third party. Only if negotiations with that other party are already being pursued can such a deal progress without undue delay.

Later in this chapter we will discuss negotiations in greater detail, including the essential elements of the power sales contract. Before addressing those issues, however, we will review in greater depth the applicable federal law, and several energy valuation techniques that may be available to the developer.

15.1.2 Federal Energy Law Affecting Marketing Considerations

The effects of Section 210 of PURPA (16 U.S.C. § 842a-3) are by far the most pervasive of any laws applicable to the marketing of energy for resource recovery projects. It should be recognized, however, that PURPA § 210 deals only with the sale of electricity and only with its sale to electric utilities for resale by them. It was expressly designed by Congress to encourage the development of cogeneration and small power production facilities in the face of the electric utilities' historic reluctance or refusal to deal with these facilities. The remedies provided by PURPA Section 210 (1) require utilities to buy power from and sell power to qualifying facilities, at rates and under terms and conditions that are just, reasonable, and nondiscriminatory, and (2) provide statutory exemptions that free qualifying facilities from the jeopardy of comprehensive financial and operational regulation by FERC and state authorities.

To enforce these and other requirements of the law, the FERC has prescribed regulations that apply to most aspects of transactions between qualifying facilities and utilities. The remainder of this part discusses the effect of these rules upon the sale of electricity by a resource recovery project.

Qualifying Status

To qualify for the benefits of Section 210 of PURPA, the resource recovery project must satisfy the rules' criteria for either a small power production facility (18 C.F.R. § 292.203(a)) or a cogeneration facility (18 C.F.R. § 292.203(b)).

To qualify as a small power production facility, (1) the project's primary (that is, 75%) energy source must be biomass, waste, renewable resources, geothermal resources, or any combination thereof,* (2) the project must not be more than 50% owned by any electric utility or combination of utilities or their subsidiaries, and (3) the nominal installed capacity of the project together with all other facilities that are owned by the same person, use the same energy resource, and are located at the same site must not exceed 80 mw. Although there are some additional regulatory concerns for small power production facilities that exceed 30 mw (for example, limited FERC regulation pursuant to the Federal Power Act and potential Securities Exchange Commission regulation under the Public Utility Holding Company Act), these are not likely to present insurmountable hurdles to development of such large facilities. See, for example, *Energy Conversions of America, Inc.*, 21 FERC ¶ 61, 329, Dec. 23, 1982.

To qualify as a cogeneration facility, the project first must produce electricity *and* "useful thermal energy" (such as heat or steam) through the "sequential use of energy," and the thermal energy must be "used for industrial, commercial, heating or cooling purposes" (18 C.F.R. § 292.202(c)). There are no size limitations on cogeneration facilities, but the projects must still meet certain operating, efficiency, and ownership standards. Most resource recovery projects that cogenerate, for example, will operate on a "topping cycle." In other words, their energy input is first used to produce electricity, and the reject heat from power production is then used to provide useful thermal energy. To qualify, this type of cogenerator must operate in such a manner that its useful annual thermal output is at least equal to 5% of its total energy output (18 C.F.R. § 292.205(a) (1)). In addition, if the cogenerator was built after March 13, 1980, and uses any natural gas or oil, FERC's efficiency standards require that the facility's electrical output plus one-half of its thermal output exceed approximately 45% of the total input of natural gas and oil, on an annual basis (18 C.F.R. § 292.205(a) (2)). Finally, the cogeneration facility, like the small power producer, must not be more than 50% owned by any electric utility or combination of utilities or their subsidiaries.

It is not necessary to obtain any certification of qualifying status from FERC; rather, the project owner or developer is merely required to furnish notice to FERC of the project's qualifying status, including with the notice certain information about the project (18 C.F.R. § 292.207(a)).

The rules do provide a method for obtaining certification, however, if the developer should so desire. The procedure requires the owner or developer to file an application for certification with the FERC that provides enough information about the project to permit FERC to make a determination of status (18 C.F.R. § 292.207(b)). The FERC must act upon such an application within 90 days or it is deemed to have been granted (*Id.*). The rules provide an opportunity for other interested parties to present their objections for 30 days after notice of the application is published. As a practical matter, if the FERC staff members have any questions regarding the application, they will usually contact the applicant by telephone for clarification. The staff is helpful and typically seeks nothing more than a complete and entirely accurate statement of the facts.

Although most purchasing utilities merely require the seller's representation that it is a qualifying facility, some insist on compliance with the letter of the law. The rules provide that a utility is not required to purchase energy from a qualifying facility with a design capacity in excess of 500 kW until 90 days after it receives notice that the facility is a qualifying facility, or 90 days after the facility files an application for certification with the FERC (18 C.F.R. § 292.207(c)). Consequently, the developer should request FERC certification no later than the early stages of energy sales negotiations, in order to eliminate the possibility that the potential purchaser may challenge the project's qualifying status and so cause a costly delay.

Finally, it is important to note that certification need not await actual construction and opera-

---

*"Biomass" is defined as "any organic material not derived from fossil fuels," and "waste" is defined as a "by-product material other than biomass" (18 C.F.R. § 292.202(a) and (b)).

tion of the project. It is often applied for, and granted, before construction of the project begins. Supplementary FERC approval should be requested before making any alterations or modifications to the actual project that differ from the original representations made in the notice of, or application for, certification (18 C.F.R. § 292.207(d) (2)).

Right to Contract

A qualifying facility owner or operator has an absolute right to agree to transact business with a utility at any price and on any terms satisfactory to the parties, and still be exempt from regulation as a utility (18 C.F.R. § 292.301(b)).

Utility Cost Data

The rules require regulated utilities to submit extensive cost data to their state regulatory authorities (18 C.F.R. § 292.302). Nonregulated utilities are required to keep this data available for public inspection. The state may require different or additional data if it determines that avoided costs can be calculated from such data (*Id.* at 292.302(d)).

Any data submitted by a utility pursuant to this section of the rules may be reviewed by the state regulatory authority, and if reviewed, must be justified by the utility (*Id.* at (e)).

Utility Obligation to Purchase

The FERC rules require that any electric utility must offer to purchase any electricity made available to it by a qualifying facility, whether through a direct interconnection or through a wheeling arrangement with another utility or utilities (18 C.F.R. § 292.303(a)). Thus, a resource recovery project is not necessarily tied to the purchase rate structure of its local utility. If the project can arrange for the transmission of its electricity to a distant utility, that utility is then obligated to offer to purchase it (*Id.* at (d)). The purchasing utility may adjust the price it pays in order to account for line losses, but not for transmission charges (*Id.*).

Electric utilities are also required to make such interconnections with the qualifying facility as may be necessary to permit the purchase of electricity from the facility (18 C.F.R. § 292.303(c)). An exception to this requirement is provided if such interconnection would subject an otherwise nonjurisdictional utility to FERC regulation under the Federal Power Act (*Id.* at (c) (2)). This exception is of extremely limited applicability. The qualifying facility is generally responsible for any expenses incurred by the utility in establishing this interconnection, at least to the extent that they exceed the corresponding costs the utility would have incurred if it had not purchased the facility's electricity, but had generated an equivalent amount itself or purchased it elsewhere (18 C.F.R. § 292.101(b) (7)).

Rates for Purchases

An electric utility must offer to purchase the qualifying facility's power at a rate equal to the utility's "avoided costs" (18 C.F.R. § 292.304). Avoided costs are the incremental costs the utility would have experienced if it had generated an equivalent amount of electric energy and capacity or purchased it elsewhere. In other words, the higher a utility's cost structure, the greater the cost the utility can avoid and the more it must offer to pay for electricity. Since the utility and the qualifying facility are free to agree to any other price, however, the "avoided cost" rule serves more to promote negotiation than it does to impose an inflexible standard.

Under the FERC rules, the qualifying facility may choose either of two basic transactions with the purchasing utility. First, it may provide the utility with electric energy as and when it wishes to do so ("as available" energy), in which case the utility's avoided costs will be determined as of the time of delivery . Under this arrangement, the qualifying facility has no legal obligation to provide the utility with any energy. Under the second option, the qualifying facility may provide the utility with electricity pursuant to a "legally enforceable obligation" for a "specified term," in which case the facility may choose at the outset of the obligation whether to have the purchase rate be based on the utility's avoided costs as determined at the time of delivery, or as estimated at the time the obligation is incurred (18 C.F.R. § 292.304 (d)). If the seller undertakes a legally enforceable obligation to deliver, and the parties agree to base the purchase rate on avoided costs as estimated at the beginning of the undertaking, that rate will satisfy FERC's rule regardless of whether the utility's actual avoided costs at the time of delivery of the electricity turn out to be above or below the rate agreed upon by the parties (18 C.F.R. § 292.304 (b) (5)).

Whether these rules require utilities to enter into long-term contracts with qualifying facilities is a subject of dispute, with utilities generally taking the position that they are not so obligated. State regulatory agencies have handled the issue in various ways but long-term contracts are usually available. In fact, long-term contracts are now more common than the short-term arrangements.

System Emergencies

If an emergency exists on the utility's system, the utility's acceptance of deliveries by the qualifying facility may be discontinued if they would contribute to the emergency. Alternatively, the facility may be required to provide power to the utility during emergencies if a contract so provides, or if ordered to do so by FERC under emergency provisions of the Federal Power Act (18 C.F.R. § 292.307).

Implementation

FERC requires that its rules be "implemented" by state agencies and nonregulated utilities (18 C.F.R. §292.401). The various state authorities have undertaken this task with varying degrees of enthusiasm and success. A few have done nothing of substance. The state regulatory authority inevitably will play an important role in the developer's marketing effort; however, an attempt to catalog the widely disparate treatment given qualifying facilities by those authorities is beyond the scope of this chapter. As a general rule, it may at least be said that the developer should endeavor to stay informed of any state proceeding in which the FERC rules are being implemented, and should participate in such proceedings when possible. Precedents that may be applicable to the project's marketing effort will probably be established in each one.

15.1.3 Energy Values

The price of energy from resource recovery projects will generally be based on its value to potential purchasers, and not the cost of producing it. The marketing effort, therefore, is only begun when potential customers are located: The difficult task of discovering the value of the energy products to those customers still remains. This uncertain and complex inquiry is made even more difficult by the fact that potential purchasers—whether industrials or utilities—consider information on the costs of their alternative energy sources to be sensitive. Moreover, there are no rules of thumb: The value of the energy product will be peculiar to the circumstances of the individual purchaser. This section is aimed, therefore, at setting out some practical methods of identifying and evaluating those circumstances in order to assist a project developer in arriving at a price for the project's energy output.

Energy and Capacity

The value of a resource recovery project's energy product to any particular energy user (whether steam or electricity) will either be as energy only, or as energy and capacity. The "energy" value reflects the purchaser's direct benefits or savings from receiving the energy product without any consideration of whether or not this energy will continue to be available in the future. Thus, this value may be measured by the variable costs of the alternative energy source that the purchaser is able to displace.

The "capacity" value, however, does involve considerations of continued availability over a period of time. If the purchaser is confident that an energy product of satisfactory quality will remain available from the project on a firm basis, it may be able to avoid or defer the costs of new investments in additional (steam or electric) generating capacity and/or the "demand" (or capacity-related) premium for a long-term firm purchase contract with another supplier. With respect to steam purchasers, the project's energy product will have "capacity" value to the extent that it enables the steam user to retire or mothball a boiler, to forestall investment in an additional or replacement boiler, or to take advantage of other efficiencies due to enhanced enthalpy (heat and pressure characteristics). Similarly, the project's electric output will be of value either as energy, (kWh) or as energy and capacity (kW), whether the purchaser is an electric utility or an end user.

Energy value may be determined on the basis of the purchaser's current variable costs of fuel, operations, and maintenance per unit of energy at the time of the purchase. Typically, however, the frequent revisions and updates necessitated by such a short-term approach will fail to provide the certainty and stability that a project needs. Therefore, the developer may find it preferable to determine the energy value on the basis of a reasonable term of years. These estimates may be based on projected changes in the components of and/or the composite value of energy costs, or may be tied to an index such as the fuel portion of the Regional Producers' Price Index. The forecasting and analytical problems involved will probably be familiar ones to most developers. Moreover, the nature of the analysis of variable costs lends itself to a convenient per-unit pricing mechanism.

Capacity value, if any, can properly be determined *only* with a long-term analysis of the purchaser's needs, plans, and options. Here it is necessary for the developer to exercise diligence and expertise in order to discover and assess the full value of the project's capacity. The task may involve the estimation of the likelihood, timing, and cost of discrete future construction events, and the impact upon those events of purchases from qualifying facilities. Even more often it involves the selection and analysis of an appropriate proxy to represent likely future scenarios. (The "proxy unit" approach to valuation is discussed in detail below.)

The determination of capacity value is where negotiations most frequently break down. Whether due to tactical considerations or to a failure to appreciate the possibilities, potential purchasers usually deny that the project's output can have any capacity value. Therefore, extensive research and expert presentation of financial and economic analyses will probably be necessary to establish capacity value and persuade the purchaser to pay for it.

Capacity value is often essential to the project's economics. Capacity credits, where established, in many cases equal or exceed energy prices. Unfortunately, some state regulatory agencies are reluctant to do the analysis necessary to question the utilities' representations that they need no additional capacity. This reluctance is usually justified by a short-term view, and in the short-term it will almost always be correct. Since utilities (and other purchasers as well) do not plan to be capacity deficient, few will find themselves in such a position at any given time. In the longer view, however, retirements, load growth, and other factors usually dictate that some new capacity will be required.

Once the purchaser and/or state regulatory agency is persuaded to take the appropriate long-term perspective, the developer still faces the problem of translating this long-term need for capacity into a current price for the project's power. In general, the requirement must be estimated as a dollar value to the purchaser and then discounted back to its present value. To take a simple example: If 25 mw of reliable power can be provided by a project to a utility on a long-term (for example, 15 years) basis, the availability of that power may permit the utility to meet its projected load growth in such a fashion that, instead of building a new 100 mw station for $100 million in 10 years, it can defer construction for an extra year. If financing that $100 million for that year would have cost $15 million, the present value of the project's capacity will be approximately $129/kW. ($15 million discounted at 15% for 11 years, divided by 25,000 kW.) This kind of payback potential obviously justifies the developer's close attention to the accurate determination of a project's capacity value.

We turn now to the practical considerations involved in establishing value for steam and electricity output in particular circumstances.

Steam

There is, unfortunately, little that can be said of a generic nature that will guide efforts to establish steam value. Steam sales will inevitably be made, if at all, to an end user for industrial or commercial processes or for heating or cooling functions. Given the competitive sensitivity of cost and operational information, as well as basic tactical common sense, a potential steam customer can be expected to be reluctant to divulge the true potential value of the energy product to his operation. The developer will have little ability to establish potential steam value without such disclosure. (PURPA has nothing to do with sales of steam. Most state regulatory agencies do not regulate steam sales, either. Where they do, care must be taken either to avoid such regulation or to abide by the state's rules.) His marketing efforts, therefore, will essentially take place as negotiations from the outset.

It is important to begin such negotiations as early in the project's development as possible—preferably in the conceptual stage. At that point, the steam product's quality, quantity, and delivery systems can most economically be tailored to meet a purchaser's specific requirements. By the same token, the purchaser can be induced at this point to think imaginatively about profitable uses of a new steam source. The more the purchaser requests in terms of product and delivery characteristics, the clearer will be the developer's idea of value to that consumer.

Determining energy value for the steam product is usually not a difficult problem from the developer's standpoint. If the purchaser is currently generating its own steam, its avoidable fuel costs will be the largest component of energy value for the project's steam product. If the purchaser is currently purchasing its steam from another source, its costs under those arrangements are indicative of energy value for the project's steam, taking into account the comparable quality, quantity, delivery, and reliability aspects of the two sources. Some uncertainty regarding the purchaser's present and future projected fuel mix may arise, but once the avoided fuel is known, only a straightforward calculation should be required to determine the energy value. This calculation would be based on the amount of fuel that the purchaser would have to consume to produce steam at the same specific pressure and temperature as the steam provided by the project, given the purchaser's feedwater quality and prevailing conditions of temperature and barometric pressure.

Determining steam capacity value is by far a more difficult task. It may be easier if the purchaser's contemplated use of the steam is a new one, however, so that the purchaser may be free to develop specifications and performance standards at the same time as the project is developing them. Moreover with respect to such new uses, the developer's expert estimates of the purchaser's capital costs of alternatives may be as valid as those of the purchaser itself.

In the absence of a new or expanded use, however, quantifying the savings that the purchaser can realize from the project as a reliable source of steam may involve estimates of: (1) the purchaser's need to switch fuels away from its existing sources (for example, from oil/gas to coal or wood); (2) the need that the purchaser would otherwise have faced to modify its existing steam generation facility (for example, to replace or improve environmental controls or other equipment); (3) the degree to which a supplemental steam load enables the purchaser to achieve new efficiency levels—either fuel purchasing economies or operating economies; (4) the ability of the purchaser to remove its existing facilities to utilize space more effectively and/or sell the salvageable equipment; and many other issues. One or more of these considerations could support a significant capacity value for the project's steam. Developing reasonable estimates of these benefits may be quite difficult, however, and often depends on the willingness and openness with which the purchaser will discuss these issues.

Electricity—To Electric Utilities

As noted in Section 15.1.2 above, the FERC rules under PURPA Section 210 require electric utilities to offer to buy power from a PURPA-qualifying facility constructed after November 9, 1978, at a rate equal to the utility's "avoided costs" (18 C.F.R. § 292.304(b) (2)). Avoided costs are not the utility's average costs, but are defined as "the *incremental* costs to an electric utility of electric energy or capacity or both which, but for the purchase from the qualifying facility or qualifying facilities [in the aggregate], such utility would generate itself or purchase from another source" (18 C.F.R. § 292.101(b) (6) (emphasis added)). Incremental operating costs for utilities are almost always higher than average costs. In other words, a utility turns on last those generating plants which are most costly to run.

The mechanism by which the definition of aided costs may be translated into a rate for utility purchases of power from the resource recovery project, however, is not provided by the FERC rules. The most the rules do in this regard is list a number of "factors" which "shall, to the extent practicable, be taken into account" (18 C.F.R. § 292.304(3)). These factors include:

1. Detailed cost information supplied by the utility to and reviewed by the state regulatory authority pursuant to the requirements of FERC's regulations at 18 C.F.R. § 292.302(b) (c) and (d).
2. Availability of energy from the qualifying facility during peak periods, including:
   a. Ability of the utility to dispatch the facility.
   b. Reliability of the facility.
   c. Terms of any contract including duration, ease of termination and sanctions for noncompliance.
   d. Extent to which scheduled outages of the facility can be usefully coordinated with scheduled outages of the utility's facilities.
   e. Ability of the facility to separate its load from its generation during system emergencies.
   f. Individual and aggregate value of qualifying facilities' energy and capacity on the utility's system.
   g. Smaller capacity increments and shorter lead times available through bringing new qualifying facilities on-line.
3. Savings in line losses and the firmness of the qualifying facility's capacity.

Avoided cost rates are thus left to be determined by state regulatory authorities, utilities, and qualifying facilities. The states have in most cases made some effort to establish a standard baseline rate, but these rates generally apply only to very small projects and even then reflect only short-term considerations. At this writing, only a few states have fixed rates that give additional credit to long-term sales by a qualifying facility. Most have opted instead to encourage producers to negotiate with utilities for this consideration and to bring their disputes to the regulatory agency for resolution.

The state proceedings in which avoided costs are determined are often special cases established solely for that purpose. Alternatively, or sometimes as a supplementary update to the special proceedings, avoided costs may be considered in the context of a utility's general request for retail rate increases.

The developer, however, cannot afford to depend on periodic ratemaking efforts by the regulatory agency to establish the value of the project's electrical output over time. First, there is no guarantee that state regulators will remain interested in updating electric values over the life of the project. More importantly, long-term project financing, which is necessary for many projects and highly desirable for others, will be virtually impossible to obtain when the project's major revenue stream is left subject to such regulatory risk.

Another reason that a series of short-term redeterminations of avoided costs by the agency should not be substituted for a true long-term rate is that such a scheme systematically undervalues a long-term firm power sale, even if some capacity credit is given. Theoretically, the first facility to provide power to a utility permits the utility to avoid its highest incremental costs. As more producers come on-line, the utility is able to avoid less and less expensive generation. Each time avoided costs are redetermined, those projects already selling to the utility will have lowered its incremental costs. If those projects are depending on the redetermination for *their own* rate, however, their production will have served to lower the rate they receive.

Therefore, it will probably be necessary to secure a rate with a long-term power sales contract. As a practical matter, the utility's avoided costs will be the ceiling for the price term of such a contract. Notable exceptions to this rule have occurred in Oregon and New York, where rates which are purportedly above avoided costs during some periods for some utilities have been mandated by the state legislatures.

Thus, until a long-term contract has been executed, it behooves the developer to participate in any state proceedings that have been established for the purpose of determining avoided costs for utilities that are potential purchasers. Even if the state agency is only considering short-term standard rates, the developer should realize that these rates, and the mechanism or formula by which they are calculated, will establish the point of departure for determining the value of the project's electric production in a long-term contract. To the extent that the developer can bring the appropriate factors to light in the agency's ratemaking process, therefore, the price for the project's power will ultimately be affected.

A. Avoided Cost Calculation Methods. Leaving aside minor differences, there are about four or five basic methods being employed to calculate or estimate avoided costs. The actual application of any of these approaches typically requires a substantial knowledge of utility operations, financial analysis, and statistics, as well as a current data base of utility cost information. Expert consultants are frequently employed, therefore, both in regulatory proceedings as well as in direct negotiations with a utility.

Nevertheless it is important for the developer to gain a basic understanding of the various avoided cost methodologies in order to permit more focused market research and, by the same token, to be alert to opportunities that might not otherwise be apparent. We turn, then, to some examples of these methodologies and a brief discussion of some of the advantages and potential problems of each.

a. Long-Run Approaches. Long-Run Differential Revenue Requirement. The long-run differential revenue requirement (LRDRR) approach to estimating avoided costs employs sophisticated computer modeling techniques that are designed to simulate utility operations in the face of various financial and economic conditions. First, the LRDRR planner runs this computer model on the basis of various projections, including the cost of purchases and the cost of new capacity, but *not* including purchases from new qualifying facilities, to produce a "base case" estimate of the total cost incurred by the utility in meeting its projected demand over the planning horizon. The planner then makes predictions or assumptions about the amount and kind of cogeneration and small power production that will be available from qualifying facilities over the same period,* and "plugs in" that capacity and energy as fully available, modeled at zero cost to the utility. Other inputs to the model are then adjusted to reflect changes that a prudent system planner would make to the utility's generation expansion plan in light of the availability of this new power to the utility. In fact, yet another computer model employing linear programming to optimize the generation expansion planning adjustments can be used for this purpose. Finally, the simulated operations and financial modeling program is run a second time incorporating these purchases from new qualifying facilities. By comparing the total cost under the base case with the total cost under the scenario incorporating these alternative power sources, the planner determines the cost the utility is expected to avoid as a result of the assumed levels of such new power.

The theoreticians and computer scientists are more enamored of the LRDRR approach than any other. Regulators also like it because it can produce a specific answer in terms of dollars without the "labels" of energy and capacity. Thus, the regulators can impose policy considerations in dividing the total into energy and capacity rates that will, they hope, induce qualifying facilities to behave optimally.

Despite the LRDRR's theoretical and policymaking elegance, however, it encounters serious problems in application. The first of these is that the method is enormously data-hungry. The LRDRR is based on an in-depth long-run revenue requirement analysis that, in turn, requires such detailed inputs as: load forecasts into the distant future; assumptions about fuel price escalation,

---

*These predictions or assumptions are typically sized in 10 mw or 50 mw blocks (or "decrements"), depending upon the size of the utility, to facilitate the ratemaking process.

future construction costs and technological advances in the industry; predictions of heat rates, available off-system sales, and purchases; assumptions regarding the cost of money, and, even more elaborate, the marginal cost of money; and many other inputs as well. Furthermore, once these assumptions and forecasts are developed and plugged into the program, they are locked in stone—the analysis has been made and a value derived. If the planner should decide that any of the inputs should be changed or updated, the entire program would have to be run again before a more accurate value could be derived.

In short, the calculation of an LRDRR is an expensive game that few can afford to play. Moreover, all of the existing computer models on the market today have operating deficiencies that usually render them inadequate to the task of accurately modeling a utility and producing valid avoided cost numbers. Despite its cost, the precision of the answer far outruns its accuracy.* For these reasons, the LRDRR is generally not a useful tool for negotiations between the developer and the utility, even in its less sophisticated variations. An exception to this rule might occur, however, if a recent base case run were already available and acceptable to the parties, and if the project were large enough to justify a readjustment of the utility's generation expansion plan—even if only by a matter of months on the timing of a new generation addition. In this event, it might be cost-effective for the parties to run the LRDRR to measure the precise capacity credit due to the project.

The Proxy Unit Approach. The starting point for this method is the lifetime capital and running costs of a generating unit that a utility actually plans to build. A hypothetical proxy unit—one that the utility might reasonably build in the future—may also be used. Another alternative is to refer to the utility's planned or reasonably available unit power purchases. (A unit power purchase is one in which the utility purchases the output of a particular generating station owned by another.) Whichever proxy is used, this information is then deflated (adjusted for inflation only, but not discounted) to comparable financial values in the ratemaking period, and certain operating factors (such as expected life) may be converted to compare costs on an equivalent per-unit basis with the anticipated operation of qualifying facilities. The total costs of the proxy unit are then used as the starting point for total avoided costs.

The validity of this approach is based upon the assumption that both the utility, deciding to plan the unit, and the state power plant siting board, deciding to certify the unit as consistent with the public convenience and necessity, performed an LRDRR analysis. In other words, the approach assumes that the utility has examined its requirements and selected the least-cost alternative to meet them. If the utility's generation expansion plan is prudent, the proxy values—indeed the highest cost unit the utility plans to build—should be considered as the *floor* on an estimate of total (energy and capacity) avoided costs.

The proxy unit approach is useful due to its simplicity and the fact that, in most applications, it is based upon an actual unit planned by the utility. It can reasonably be expected to produce at least rough justice over time, at a cost that makes it accessible to parties in negotiation, as well as in the hearing room.

One potential disadvantage of the proxy approach is that it is easily misapplied. Some regulators, for instance, assume that the approach is only useful if it can be shown that construction of the selected proxy unit will actually be deferred or avoided. It must be stressed that the validity of this approach does not depend upon the answer to such a question. Properly applied, the approach should produce the same result whether or not the proxy unit used will actually be deferred or avoided. If questions of the timing and manner in which a unit will be avoided are pursued in detail, the proxy unit approach will evolve into a type of LRDRR inquiry, and will tend to lose its most attractive characteristics of simplicity and economy.

b. Short-Run Approaches. Marginal Running Costs. Numerous states have adopted, and many utilities have proposed, avoided cost rates that reflect only the utility's running costs on the margin—the utility's variable costs of producing its marginal (last or next) kilowatt hour. This is generally due to a myopic view of the utility's capacity needs or a lack of enthusiasm for the development of qualifying facilities, or both. Here the burden will be heavy on the developer to discover and negotiate a basis for capacity credits. First, however, close attention should be given to the running-cost rate that has been offered, to insure that it captures the utility's true avoided energy costs. A number of questions may profitably be asked regarding a running-cost based rate:

1.   Does it reflect marginal costs, or is it really an average cost rate? Utilities too often report average costs to the state regulatory agency as their "avoided" costs. Unless these reports are seriously questioned, they may pass. Another variation of this mistake is to report average "fossil fuel" costs as avoided. Fossil fuels include coal, which lowers the average cost dramatically from

---

*The Maine Public Utilities Commission recently concluded a proceeding using LRDRR for a single utility. The case required nearly 3 years just to model the energy aspects. At the end of the hearings, the hearing examiner recommended abandonment of the LRDRR and the use of the proxy unit approach instead. The legal and consulting fees for the various parties to this proceeding exceeded $1,000,000.

the cost of other fossil fuels (gas and oil), which are actually used at the margin. A similar error is to average the incremental costs over too wide a band on "decrement" of the generation mix.* In other words, if the cost to the utility of serving its last 50 mw of load is $0.08/kWh, but the cost of serving its last 200 mw of load is $0.04, the utility might report an avoided running cost of $0.04 even though it has no qualifying facilities on line or on the planning horizon with more than 50 mw of capacity.

2. Do the rates take into account the possibility that the utility may, by virtue of the addition of qualifying facilities to its source mix, be able to sell additional power to other utilities at a profit?

3. Do the rates account for increased savings in running cost due to reduced purchases from other utilities? The addition of qualifying facility power may not only permit a reduced level of purchases from other utilities; it also may permit purchases at a lower cost. Utilities typically sell to each other at a rate that splits the difference between the incremental running cost of the purchasing utility and the running cost of the seller's unit providing the power for sale. The addition of new qualifying facility generation to the purchasing utility may reduce its incremental running cost (by allowing it to reduce its most expensive generation), which will lower the rate it pays for its purchases from others.

4. Are the rates generated by a computer forecasting model that has been validated against historical data? Several very popular models have not yet been validated and may have major technical flaws in some applications.

5. Do the marginal "running costs" include operations and maintenance costs? These are often overlooked. They can run between 2 and 4 mills (tenths of a cent)/kWh.

6. Are line losses considered? Line loss consideration at the subtransmission level can add 3–4% to the rate. At the primary distribution level it can add 7% or more.

Marginal Running Costs Plus the Capacity Cost of a Peaker Unit. This short-run approach seeks to recognize the capacity value of the qualifying facility generation by combining the marginal running costs of a utility with the annual carrying charges on a peaking unit (typically a gas turbine generator). The peaker methodology is often resorted to when a utility has no firm plans for constructing new generation capacity (so that no easily identifiable unit can be deferred or avoided), or in regulatory jurisdictions that have adopted marginal-cost-based retail ratemaking. This makeshift approach has some theoretical validity only if the carrying costs of a new peaker are used. Frequently, however, the market price of peaker capacity is used as a basis for the calculation. This is a misapplication of the peaker theory which will result in a lower capacity value whenever the utility has excess capacity. Since utilities almost always have excess capacity in the short run, the "market price" mistake will consistently undervalue long-term avoided capacity costs.

The peaker unit approach can be useful for the purposes of negotiations since it can provide the parties with readily available data and a simple method of calculating a price. Its ability to produce an overall (energy plus capacity) rate that approximates true avoided cost value, however, is suspect. At best, under current conditions, the use of new peaker costs as representative of capacity value may add 1.5¢ per peak period kilowatt hour to the rate. If the market value of a peaker is used, however, it may only contribute a third of that amount.

Pool or Market Prices. Where there is a tightly integrated pool of nonaffiliated utilities, or a holding company power pool, the sister utilities may agree that their avoided energy costs are equal to the marginal running costs of the pool itself. The same questions that apply to the marginal running cost rates discussed above may be asked here as well.

Designating the pool's capacity deficiency charge as the avoided capacity cost of the pool's utilities is supposedly consistent with this approach. The capacity deficiency charge is the per kilowatt charge a utility pays into the pool if its load and reserve responsibility exceeds its generation capability. Since the capacity deficiency charge rarely will have any relation to the generation expansion plans of pool members, however, the developer generally will not receive full value for the project's electrical output if this capacity credit is accepted.

B. Escalators. Short-run approaches are often important in the context of establishing a point of departure for the estimation of long-run avoided costs. In other words, they may serve as a floor for the value of the project's electrical output to a utility. In some circumstances, however, they

---

*A decremental rate structure seeks to recognize that the first facilities to supply the utility with power enable it to avoid more expensive generation than those facilities that come on-line at a later date. Calculating an avoided-costs-based rate for various decrements of power supplied to the utility is theoretically sound, and is not usually a difficult task. In at least two states, however, serious disputes have occurred in establishing which particular qualifying facilities are eligible for each decrement's rates. Unfortunately, no rules have yet been established to provide guidance as to how these disputes will eventually be resolved.

may be the best a developer can cost-effectively negotiate, given constraints on time and other resources.

In this event, to provide a long-run revenue stream that presents acceptable risks to the project owners, the financial backers, and the utility, it is worthwhile to avoid the regulatory risks of periodic agency redeterminations by contractually tying this year's short-run rates to an index, formula, or escalator that will automatically redetermine the rates. Where energy and capacity rates are involved, two indices may be used. In addition, the indices may be set to change (for example, from the price of oil to the price of gas) at some point in the contract term.

One goal in selecting an index or escalator is to find an appropriate estimator of the level and variability over time of those factors that determine a utility's marginal costs. Fuel costs and the utility's cost of capital are, therefore, likely candidates. Another goal is to select an index that will be understandable to financial backers (for example, the GNP deflator or the prime rate), so that they will be able to discount the financial risks appropriately and provide capital at an appropriate cost. The developer's imagination and judgment and the course of negotiations will undoubtedly produce any number of additional acceptable options.

C. Other Valuation Issues. In the course of determining the full value of the project's electricity, a number of possibilities may be presented that do not fit neatly under any of the topics discussed so far. They arise with sufficient frequency, however, to merit brief discussion.

a. Transmission Capacity Credit. It is possible that a project's output, together with that of other facilities, may permit the utility to avoid transmission capacity costs, whether or not it receives credit for avoided costs of generation capacity. At least one state (New York) has ruled generically that a utility's marginal transmission capacity costs are equal to its avoided transmission capacity costs. In other situations, an expert may be able to demonstrate that the project's output at its particular location will permit the utility to defer or avoid a specific transmission capacity investment, such as a substation or line reinforcement. In such an event, the project should receive the value it will provide the utility.

Any time a project will be producing power during peak periods, and an equivalent or greater amount of power will be consumed by an adjacent or nearby end user, the possibility of transmission capacity credit should be vigorously explored. Peak period loads drive utility investment in transmission (as well as generation) capacity. To the extent the project relieves the utility from the need to bring in peak power from elsewhere on the grid to serve an end user near the project, incremental transmission capacity investment may be avoided.

b. Hydroelectric Plant Relicensing. A utility with hydroelectric plants should be queried in some depth concerning the status of its FERC licenses for these plants. Since these licenses must be renewed on occasion, two events can take place that may affect the utility's capacity construction plans. First, the utility can lose the license (and the hydroelectric capacity) to another developer, in which case it may need new capacity earlier than it had otherwise planned. Second, FERC may require as a condition for retaining the license that the utility develop additional capacity at the site. This may provide an actual "proxy unit" for use in capacity credit negotiations, because the utility will have decided, if it chooses to retain the license, that its incremental costs are greater than those of the additional development (or at least that such development is its least-cost alternative).

c. Scheduled Maintenance Coordination and Dispatchability. If the project, and others like it, are able to schedule their maintenance outages in cooperation with the utility's needs, a capacity value may thereby be afforded the utility. The state of Maine, for example, deems that such coordination is worth 1% on top of the otherwise applicable avoided-cost rate.

Another value enhancement may be provided if the project is to some extent "dispatchable" by the utility (that is, the project will, within reasonable operating limits, obey utility requests to increase or decrease its output). In this event, the power from the project is more valuable to the utility as it enables the utility to schedule its sources as economically as possible. Once again, Maine deems this value to be equal to 3% of the otherwise applicable rate, and forbids the utility from dispatching a project in such a way as to reduce its total output. Certain additional metering and reporting facilities may be required to earn the credit, but they are cost-effective.

d. Wheeling. If the project's power is transmitted or wheeled to a remote utility at the project's request, the value of the output to the distant utility must be netted against wheeling charges, which are set by the FERC, and against transmission losses. Generalization about specific costs is difficult, but wheeling charges of 1 to 3 mills/kWh are not unusual, and losses may range between 2 and 10%.

Many utilities vigorously resist requests to wheel, even where sufficient transmission capacity exists. In order to get the value available from another utility, the developer may have to call upon the receiving utility for assistance. As a last resort, there are certain wheeling requirements in some nuclear generation operating licenses that may be insisted upon, and certain refusals to wheel may also be found to be violations of the antitrust laws. *See Otter Tail Power Co.* v. *United States*, 410 U.S. 366, 93 S. Ct. 1022, 1973.

Electricity—To End Users

In those instances where the state's law and regulatory scheme do not unreasonably interfere, direct sales of electricity by the project to one or more end users may provide a more attractive marketing opportunity than would sales to a utility. These direct sales offer several benefits, even where the price does not exceed the purchase rate a utility would pay for the project's output. First, sales to the customer at a percentage of the otherwise applicable retail rate may provide a satisfactory return to the project in their own right. In addition, a retail-rate-indexed price may be expected to reflect general changes in regional cost levels over time. A retail-rate-indexed price also may tend to behave less erratically than a utility's avoided-cost-based purchase rate as set by the regulatory agency. Furthermore, electricity sales may accompany steam sales to the same customer, in a load-following type of operation that maximizes the benefit of the transaction to both parties. And if the project is a municipal one, the purchase of power at a discount from retail can provide an anchor in the community for an industrial customer.

To understand the value of the project's electricity to end users generally, it will be necessary to analyze the retail rates otherwise available to such users from the serving electric utility. These rates are usually complex affairs, structured such that one rate or "tariff" is applicable to all customers in a particular service classification, yet the tariff applies to each customer in the class in such a fashion that no two may have the same average cost of electric service on a per-unit basis. The tariff may involve time-of-use (peak, off-peak and shoulder) pricing, minimum billing, stepped or "block" rates which decline (or increase) with variations in usage, seasonal rates, a fuel cost adjustment clause, and any number of other special provisions, adjustments, and conditions that affect the customer's cost of service.

In the final analysis, however, most rates for large electric consumers consist of energy-related charges, billed on the basis of kilowatt hours consumed, and separate capacity-related charges, billed on the basis of kilowatts of demand at the point of the customer's greatest demand during the billing cycle. For most large industrials, the energy-related charges comprise the largest part (two-thirds or more) of their total electricity costs.

To the extent that an electric customer can reduce its kilowatt-hour purchases from a utility by purchases from the project, the project will have an energy value. If the project is to have a capacity value to the end user, however, it will probably be necessary to relieve the customer's retail demand charges in proportion with its reduction in energy charges. The availability of this capacity value will usually be necessary if the project is to be paid a price for its electricity by the end user that exceeds the purchase rate available from a utility. The mere availability of this capacity value, of course, does not ensure that the user's rate will be higher than the utility's.

Regardless of the attractiveness of a direct-sale arrangement, there are important obstacles that must be overcome to permit each party to reap its respective benefits. In fact, whether such sales can be made at all will probably depend upon the cooperation—willing or otherwise—of the serving utility. First, the potential customer may not be willing to purchase electricity from the project unless backup or supplementary power is available from the utility at reasonable (that is, nonpunitive) rates. This reluctance may be due to various concerns. For example, a purchaser large enough to be a viable candidate for the electrical output of a resource recovery project is likely to have a fairly constant electrical demand, and therefore may be unwilling to risk a loss of power due to operational difficulties at the project. Another purchaser may simply need more power than the project can deliver on a regular basis.

For obvious reasons, most utilities do not offer special rates to accommodate the needs of customers who intend to purchase the bulk of their power requirements elsewhere. Such a special rate may be necessary, however, because the typical utility retail rate to large users contains a "ratchet" that requires the customer to pay a minimum demand charge based upon a percentage (for example, 80%) of its highest 15-min demand in the past year or longer. Thus, a customer normally drawing 13 mw from the project and a supplementary 2 mw from the utility might have to pay the utility a demand charge for 80% of 15 mw, or 12 mw for a year or more, if the project were to experience even a brief outage while the purchaser was operating. Thus, unless a special rate is available, these demand charges could easily destroy the economic value of purchases from the project.

Although the FERC rules under PURPA require utilities to offer supplementary, backup, maintenance, and interruptible power at reasonable rates to a qualifying facility "upon request" (18 C.F.R. § 292.305(b) (1)), it is not clear that federal law requires the utility to offer these rates to *customers* of qualifying facilities. In this case, assistance from the state regulatory authority pursuant to state law and ratemaking policy will be vital to persuade or require a utility to accommodate the situation. Even if the utility were to establish such a new rate voluntarily, the regulatory agency (assuming the utility is regulated) will have complete authority to approve or disapprove the rate. Even if the FERC rules are construed to require utilities to offer special rates to purchasers from qualifying facilities, the rules also permit state regulatory authorities to waive this requirement upon a finding that it would impair the utility's ability to serve other customers or would place an "undue burden" on the utility. For non-state-regulated utilities, FERC may waive the requirements [Id].

The next major difficulty facing direct sales agreements is the fact that most utilities embody their rate or tariff provisions for large customers in a contractual agreement executed by the customer, and the term of that contract may be for several years. If the project's output is to be of any value to an existing utility customer, therefore, the utility may have to be persuaded to release the customer from its contractual obligation. If this becomes an obstacle, the regulatory agency may not be willing to provide any effective relief.

Finally, when the technical and legal considerations are solved, the largest obstacle to marketing electric power to an existing utility customer may be the comfort factor. Despite the genuine dissatisfaction with utility rates expressed by many large customers, most come to appreciate and depend upon the reliability and responsiveness that are characteristic of good utility service. Overcoming the potential purchaser's fear of change, which may include a fear of utility reprisal, may, in fact, be the most difficult marketing task facing the developer with regard to such direct sales arrangements.

15.1.4 Negotiating a Power Sales Contract

Most of the developer's marketing efforts will be brought together in the negotiation of an energy sales contract. This section is intended to provide some guidelines for the negotiating process itself, including a discussion of key provisions of an electric-power sales contract.

Basic Rules

The keys to successful negotiation of a power sales contract, regardless of the purchaser, are the same as the basic elements of any effective negotiation strategy: Good books have been written on the subject of negotiation generally. As applied to the present subject matter, however, the important rules include:

1. *Preparation.* Know the law. Know what the project is technically capable of delivering. Know the target purchaser's needs and existing costs. Know how the local and state political and regulatory system works and know the players in those systems.

2. *Objectives.* Know what will be required in terms of price, timing, and other special conditions, to make the project work. Know not just what will permit it to be built and financed, but what is necessary to make it a *success* over its expected life.

3. *Flexibility.* With a view to rules 1 and 2, be willing and eager to innovate. Despite the fact that a regulated utility may be involved, there are no hard or fast rules that have been established for these transactions. Everything is negotiable. The more possible solutions that are placed on the table, the more likely it is that a solution will be found. There is always a deal there to be made.

4. *Carefulness.* Neither the utility nor the regulatory agency has any obligation to protect the interests of the project. Any public service obligation the utility may have runs only to its customers, not to its suppliers. Therefore, do not assume anything. Clear away ambiguities in communications immediately upon receiving them. Both sides expect that they are dealing with professionals, or will act as if they are. Meet deadlines, and meet them under the exact terms in which they have been given. Confirm oral communications in writing. Keep records and notes. Read and understand every word of a proposal and have several others do the same.

5. *Persistence.* Work hard to keep the purchaser's attention focused until the deal is done. The project will have competition for that attention.

Sources of Information About Utilities

The strength of the developer's negotiating position will be enhanced by the effective acquisition and use of information as to the utility's costs, operations, load forecasts, generation mix, future construction plans, power pool arrangements, regulatory climate, and related matters. This information can be difficult both to gather and to comprehend, as the available data will often be inconsistent, misleading, erroneous, obsolete, and/or cryptic. Even when it is accurate and complete, it may be difficult to evaluate without the assistance of an expert consultant.

With this in mind, let us review some of the most promising sources of information about utilities. Centralized sources of data concerning municipal utilities and cooperatives do not abound. One of the best is the *Electrical World Directory of Electric Utilities*, published annually by McGraw-Hill, Inc. This includes useful information respecting the costs, generation mix, sales of various types, interconnections, and key personnel of virtually every electric utility in the United States, its possessions and Canada, including, in the United States, 235 investor-owned companies, 1739 municipal systems, 939 rural electric cooperatives, and 70 public power districts, among others (1982–83, 91st ed.).

Moody's Public Utility Manual (Moody's Investors Service; R. Hanson, Ed.) also contains much detailed information concerning investor-owned utilities' sales, costs, generation mix, future plans, and problems. Utilities have been heard to complain that this information is obsolete by the time it is published, but it provides a substantially valid basis for comparison in an initial survey.

The FERC regulations require each regulated utility to provide extensive cost information to the state regulatory authority at least every 2 years, and to keep the information available for public inspection [18 C.F.R. §292.302]. Nonregulated utilities (that is, municipal utilities and electric cooperatives) are also required to maintain the information for public inspection. Unfortunately, enforcement of these regulations has been spotty at best. Even where information has been forthcoming, it has rarely been validated by any cross-checking mechanism.

It is usually informative with respect to these and other sources to obtain previously published reports as well as the most recent ones. A comparison will often highlight trends and/or gross inconsistencies in the data that would not be apparent from a single report.

Another useful source in the past has been the comprehensive report filed by most investor-owned utilities pursuant to Section 133 of PURPA (16 U.S.C. § 2643, 1978). These reports, which can be obtained from the FERC in Washington, D. C., were filed in 1980 and 1982 and provide detailed information regarding the utility's costs of service. FERC is phasing out these reports, and at this writing there is no assurance that they will continue to be required in the future.

The Section 133 report can be voluminous, but not all of it is necessary for marketing research purposes. Subpart C of the report, which contains marginal cost information and power pool data, should be sufficient for the developer's needs. Two of the most useful sections of Subpart C are the breakdown of operating costs by generating unit and the list of planned generation additions and retirements.

Each investor-owned utility also files with FERC a comprehensive annual report of operations, known as the FERC Form 1. This document is generally filed in April or May. It is available from the FERC and from the utility directly.

The utility and its power pool, if any, make numerous reports at the state level that can be illuminating. The first avenue to pursue is the utility's most recent rate increase filing. The testimony and exhibits justifying a rate request provide comprehensive cost and planning data for which the utility is immediately accountable. There may also be other kinds of regulatory proceedings in which the utility has made recent reports: Typical examples are investigations as to need for revised fuel purchase practices or greater conservation measures, and proposals for new transmission or generation plant. A brief chat with staff members at the agency will alert the developer to these proceedings, and copies of the utility's submissions are usually available.

Indeed, where time and resources permit, the developer should seriously consider intervention and participation in such proceedings. This is the arena where a utility can be required to respond to specific questions and be held accountable for the answers. Participation may also assist the developer to build a better working relationship with the regulatory agency and its staff.

Finally, check with all of the state's energy regulatory agencies to collect copies of their annual and special reports. All such agencies justify their budgets with these reports, which often contain valuable information concerning the state's energy situation and prospects. Since the electric utilities are usually a large component of the state's energy picture, they are typically the focus of many of these reports, as well as the source of the underlying data they contain.

Tactical Considerations in Dealing With Utilities

A. Utility Objectives. The utility's overriding concern is that the price it pays for power from the project, together with any other costs it incurs in the transaction, must be recoverable through retail rates charged to its customers. The state regulatory agency determines whether a utility's purchases have been prudent, and thus merit passthrough to retail rates. Thus, the utility may need some comfort from the agency before it agrees to an innovative pricing scheme.

The utility's second concern is to obtain the best possible deal for its ratepayers. Utilities do not earn a return for their shareholders on purchased power expenses. To the extent that the price is below calculated avoided costs, or when the utility realizes a benefit that is not necessarily included in the avoided-cost calculation (this may occur most frequently with respect to transmission capacity benefits), it will have an incentive to negotiate. Otherwise, utility negotiators are businessmen who have no incentive to accommodate the project's special needs or problems.

A third concern of the utility is reliability. This includes safety, preserving the operational ability of the utility's system to serve its customers, and insuring that the project meets its contractual obligations to deliver full value in terms of reliable capacity.

B. Dispute Resolution. Disputes during negotiation that cannot be resolved by the parties may be taken to the regulatory authority for interpretation, mediation, or a more formal action resulting in an order. Often an informal discussion of a problem with key staff at the regulatory agency can produce a potential solution that would not otherwise have occurred to the parties. Occasionally one party may need some regulatory assurance as to its position, in order to accept a proposal that might otherwise be seen as too risky. It may be useful in some cases for the utility and the producer to agree to disagree, in order to bring their dispute before the agency and, through the agency's order, to gain formal approval of a particular transaction. Any formal agency action will be governed by the law and regulations of the state.

If the matter is one of the project's rights under PURPA or FERC rules, and if the developer fails to get satisfaction from the state agency, it may petition the FERC to order implementation of the FERC's rules (16 U.S.C. § 824a-3(h) (2) (B)). If FERC fails to initiate an enforcement action within 60 days, or issues notice of its intention not to act, the question may be taken to a Federal District Court (*Id.*).

Another dispute-resolution methodology appropriate in some situations is to use the political process to obtain an amendment to the state law that will assure the parties of the security of a particular arrangement. One such amendment would be, for example, a guarantee of passthrough to retail rates of utility costs under purchase contracts based on estimates of avoided cost. Even where a statutory change is not possible, political pressure may be applied to the utility or the regulatory agency to help resolve particular problems. Municipal projects are often in a particularly good position to be able to call upon political resources in this manner.

The best dispute resolution method may be to have another marketing opportunity available. If the local utility refuses to negotiate in good faith, it may be far more cost-effective to sell power to a more remote municipality, for instance, than to litigate PURPA rights against a utility that is highly experienced in the field of regulatory law. Unlike the project developer, the utility is not greatly concerned about the cost of litigation or the time required to seek an administrative and judicial solution.

An alternative market possibility may be of little use to the project, however, if it is not developed until after the first alternative has eventually come to loggerheads. Therefore, it is a commendable practice to explore all of the best opportunities simultaneously to the extent resources permit, and to lose none until the deal is done.

Elements of the Long-Term Power Sales Contract

A. Price—Basic Mechanisms. The price will be the most important item on the agenda for negotiation. The (avoided cost) basis for calculating the project's value has been discussed earlier in this chapter. The parties must also agree, however, upon a mechanism by which value can be translated into a contract price. Although the variants may be legion, the essential mechanisms include:

1. A rate, or percentage of a rate, that is determined from time to time by the regulatory agency.
2. A flat rate, for example, $0.06/kWh for 15 years. This rate may or may not relate to avoided costs.
3. A specified rate for each year of the contract, for example $0.06/kWh for deliveries in 1984, $0.067 in 1985 to 1987, $0.079 in 1988.
4. A rate adjusted periodically, and automatically, according to changes in an escalator, index or formula.
5. A "levelized" rate.

These basic mechanisms and their variations may be combined in innumerable ways to suit the needs of the parties. With one exception, they are fairly self-explanatory, and in many cases the particular mechanism selected will closely reflect the method by which the parties have agreed (or failed to agree) on the value of the project's output.

One rate design method, however, deserves special mention at this point. The "levelized" rate has become the rate-of-choice among project developers wherever it has been made available. This rate is set above the utility's avoided costs in the initial years of a long-term contract, but at a level that is estimated to be sufficiently below the utility's expected avoided costs in the latter years of the contract, so that the utility will eventually be repaid the amounts advanced, with interest. In other words, if the parties agree, for instance, that the utility's avoided costs will rise from 4.6 cents/kWh in year 1 to 16.3 cents/kWh in year 15, the levelized rate for a 15-year contract might be set at 10.4 cents/kWh.

If the parties so desire, the level of this rate may be determined arithmetically, by discounting agreed-upon future avoided costs back to find their cumulative present value, and then finding the annuity value that (at an interest rate equal to the discount rate used) produces the same cumulative present value. In this case, assuming constant project production levels over the term of the contract, the utility will be made whole without any adjustment.

The purpose of a levelized rate, however, is to assist the developer in meeting the typically front-loaded costs of a new project, including the cost of debt. Depending upon project-specific cost considerations, therefore, the developer may actually need a rate that is even higher than the arithmetically-derived level in the first years of the contract. In return, the rate level in the latter years will be different from, and more deeply discounted than, the expected avoided costs.

Such a modified levelized rate may be a useful tool for a project-financed resource recovery project, to maintain stability in tipping fees. If a high rate is paid for electricity until the debt service on the project is retired, the electric revenue stream may be reduced thereafter without unduly affecting tipping fees.

It is important to remember that the levelized rate is at bottom only a device for borrowing money from the utility and later repaying it. The implicit interest rate for this service is usually the utility's cost of long-term capital, and the developer may find that money is available less expensively elsewhere. Moreover, the amount of money effectively "borrowed" depends, assuming constant production levels, upon the underlying avoided cost to which the payment is being compared. It is vital, therefore, that the avoided cost itself be negotiated, and not just the level of the special payment. If a level rate is agreed upon but the underlying avoided costs are left open, for example, for periodic determination by the regulatory agency, the developer cannot be sure how much will be borrowed, or if it will be repaid by production. Utilities invariably insist on full repayment well before the end of the contract term.

In application, any levelizing mechanism will probably require a bond or other assurance that the utility will be repaid, with appropriate interest, if the project fails before early overpayments are compensated for.

B. Price—Additional Considerations. A number of rate design features are available in addition to blends of the basic mechanisms. The most common of these is the time-of-day (TOD) rate, which differentiates between deliveries of power during the utility's peak demand periods (typically 8 AM to 8 PM weekdays other than holidays) and deliveries during off-peak periods. Where TOD rates are used, it will often be the case that per-kilowatt-hour capacity credits are paid only on peak period deliveries. This may be advantageous for projects where production can be more heavily loaded into peak hours or, conversely, where maintenance downtime can be loaded primarily into off-peak periods. Another twist on the TOD methodology is seasonally varied rates, which may differentiate, for instance, between the winter peak rate and the summer peak rate. This should be considered carefully by a resource recovery project where, for instance, the energy value or quantity of the resource may vary predictably by season.

Some utilities (and regulatory agencies) like to determine avoided costs, translate them into rates, and then establish an adjustment mechanism for the rate based upon the utility's "fuel cost adjustment clause." This adjustment is a standard feature of utility retail rates, whereby a large portion of the utility's costs of fuel are billed to the customer as the utility actually incurs them (usually with a 2-month lag), as part of the kilowatt-hour rate. It is said that its use in an avoided-cost rate (as an adder on non-fuel-based avoided costs) may protect the producer and the utility from unforeseeable swings in utility fuel costs, which are usually a major component of avoided costs.

If such an adjustment is considered at all, however, extreme caution must be exercised by the producer. The fuel adjustment clause varies with the utility's *average* cost of all fuel (and purchased power). Thus, it is far more likely to be affected by fluctuations in the price of baseload fuel, such as coal, than by the price of fuel at the margin (for example, oil or gas), which is the major determinant of the utility's avoided costs. For most utilities, a fuel cost adjustment index may be developed to track the cost of the utility's marginal fuel. If so, it would be preferable to the retail rate fuel clause.

Avoided capacity costs may sometimes translate to a per kilowatt, per annum rate paid monthly or quarterly. More often, however, they are converted to an energy-based, or per kilowatt-hour rate. This process can become the negotiator's equivalent of an art form. Assume, for instance, that avoided capacity costs are determined to be $100 per kilowatt-year. This figure may arguably be increased for avoided capacity line losses, depending upon the voltage level at which the project's deliveries are to be made. A 4% line loss adjustment would bring the capacity value to $104 per kilowatt-year. Then, if the basis for the determination of capacity value is, for instance, the cost of a new utility peaking unit, the $104 will usually be divided by the utility's peak period hours (say, 3570) to arrive at a capacity credit of $0.02913/kWh at peak.

This relatively straightforward calculation may not be fair to the project, however, since it would require production at full rated capacity during *all* of the utility's 3570 peak hours in order for the project to be compensated for its full capacity value of $104. An actual peaker on the utility's system would run only about 900 hr per year or less: Its value is as *capacity*, whether or not it is required actually to produce energy in any given year. Should the $104 then be divided by only 900 hr to develop a peak period rate? If the project then actually produces during four times the peak hours that a utility peaker would, should it be paid the peaker value four times over? The answers to these questions will probably be found by negotiation, but they are the kind of questions that intrigue the state's regulators as well.

Similar questions arise whenever the basis for capacity credits is the purchaser's avoidance of the costs of a particular generating plant (whether actual or hypothetical) or firm power purchase. The LRDRR approach to determining avoided costs, on the other hand, does not distinguish energy from capacity values, or peak from off-peak values: It therefore permits an even freer discussion of policy toward encouraging optimal behavior by producers.

Generally speaking, the project will stand to benefit anytime its projected capacity factor is greater than the capacity factor used to translate capacity value into energy-based rates.[*] Where this is the case, the project will have the ability and an incentive to produce at a higher level than that which is expected of it, in order to earn a premium on its capacity credit. By the same token, it may permit the utility to avoid additional costs not otherwise recognized by a flat per-kilowatt rate.

C. Term of the Agreement. The term of the agreement will be dictated by the expected life of the project, financing requirements, and the need to provide power for a period long enough to earn a capacity credit. To earn the greatest possible capacity credit, however, developers often contract to provide firm power for as long a term as the utility will accept. This is a practice that must be viewed with some concern. The contractual liability for default on an obligation to provide long-term capacity can be enormous. In Idaho, for example, the liability of a 25 mw facility that stops producing after 15 years of a 30-year contract can be about $40 million. Certain provisions in Idaho's rules, *if* incorporated into the contract, might limit liability to only $20 million; but even if insurance is available to cover such a liability, the premium necessary to insure such an exposure cold be a significant expense line item for the project. Moreover, at this writing, insurance policies covering risks of this nature are generally available for terms of no more than 5 years.

D. Delivery. Energy sales agreements are commercial contracts. The specification of a place where delivery will be made is a necessary and material term of the agreement. The point of delivery can be specified in words or by a line diagram appended to the agreement.

E. Metering and Payment. When they are inaccurate, electric meters always run slow: They never run fast. In other words, if the meter is wrong it will be to the project's detriment, not that of the utility. Inspection and adjustment of meters to a condition of zero error costs approximately $100 per meter. On a 10 mw project selling power at $0.06/kWh, however, an error of 2% can cost the project $80,000 to $100,000/year. It is thus cost-effective to insist upon regular metering inspections and upon billing true-ups based upon the results of the inspections.

In a true-up or billing adjustment provision, both the period of inaccuracy and the amount of the inaccuracy must be accounted for. In case of disputes, the parties may refer to back-up meters or the project's steam production records for the period in question. If the period of inaccuracy is unknown, it is common to use half the period since the last inspection and adjustment.

Payment is usually made monthly, but some utilities want to establish a complicated scheme of vouchers, statements, and notices that can result in eventual payment 2 or 3 months after the power is actually delivered. The utility should instead be encouraged to read the meters itself, calculate how much it owes, and submit payment along with a brief statement of the calculation. Meters should be read monthly and payment should follow within 30 days. Adjustments due to correction of meter inaccuracies should be paid immediately. Late payments should be made with interest at least equal to the utility's cost of capital.

F. Interconnection. Interconnection provisions—especially those drafted by the utility before the final project specifications are completed—can be hazardous to the project's financial health. The developer must take care that no control over the generating facility itself is given to the utility under the guise of interconnection requirements. Reasonably necessary interconnection

---

[*] Capacity factor, expressed as a percentage, is the ratio of the unit's average output in kilowatts over a period of time to its nominal rated capacity in kilowatts. Put another way, it is the ratio of its actual kilowatt-hour output during some time period (usually a year) to its potential kilowatt-hour output for that period. Resource recovery projects may have capacity factors in excess of 90%, whereas typical utility system-wide capacity factors are in the range of 65 to 70%. Since utility capacity is that which is avoided, utility capacity factors should arguably be used to translate capacity values to per kilowatt-hour rates.

equipment may be prescribed by the utility according to the dictates of prudent electrical practice, including considerations of economy and efficiency. As a practical matter, however, the project's engineers should agree with the utility's field engineers to an itemized table of equipment and specifications, all of which may be included in a schedule appended to the contract. The utility should be permitted reasonable inspections of, but no authority to change or disapprove, the generating facility itself.

Changes in the interconnection equipment should be requested by the utility only if necessary to its operations. The project should pay only for actual interconnection costs directly incurred (not allocated overhead), of interconnection equipment, installation, and maintenance. "Maintenance charges" and other "fees" should, therefore, be avoided. Any interconnection equipment purchased by the project should be *owned* by the project.

In addition, where the project is receiving electrical service from the utility, the project should only be responsible for costs in excess of those costs of interconnection that would have been incurred by the utility to provide such service. Service-related interconnection costs are recovered by the utility in its retail charges.

G. Utility's Obligation to Accept Power Delivered to It. Except for emergency or hazardous conditions, the project's Power Sales Agreement ought to be a take-if-tendered contract: PURPA requires a utility to purchase any electricity made available to it by a qualifying facility. Nevertheless, some agreements invite abuse on this point by permitting the utility to refuse to take power, for example, (1) whenever it inspects or maintains its equipment or system regardless of the reasonableness or necessity of such inspections, (2) whenever it can get cheaper power elsewhere, and (3) whenever it otherwise deems such disconnection or curtailment to be "necessary or convenient." This is usually without any liability to the utility and often without any provision for restoring acceptance.

These dangerous provisions may easily be avoided by the careful negotiator. The problem may not always be an obvious one, however. One industrial producer in a southern state signed an agreement that allows the utility to refuse to accept power if taking it would require the utility to reduce the output of any of its own "intermediate-level" (oil-fired) generating plants. This provision not only violates the intent and letter of PURPA, it is doubly abusive because the utility involved *only has* intermediate-level generating plants.

On the other hand, either party must be free to disconnect if it "sees sparks," or to prevent a reasonably foreseeable hazard. Best efforts should be used to remedy the situation and restore delivery and acceptance as soon as possible.

H. Land Rights. Each party may need a license to enter the other's property for reasonable access to its equipment for construction, operation, maintenance, replacement, and to perform its other obligations under the contract.

I. Assignment. A limitation on either party's ability to assign to another its rights and obligations under the agreement is common practice and, especially where long-term capacity is being sold, may be necessary to the utility. It is possible to agree in advance, however, that certain transactions in the nature of assignment will not require further permission of the other party. These may include, among others, transfer of limited partnership interests to a limited or general partner, or transfer of the contract from the developer or contract operator to the owner.

J. Indemnification. Indemnification occurs when one party reimburses the other for losses the other suffers as a result of being found to be legally liable for injury or damage to a third party. If injury or property damage occurs to third persons due to operations under the contract, the utility, having deep financial pockets, will probably be sued. In anticipation of this, the utility will want the project to agree to indemnify it in the event it is found liable. There is no reason, however, why the project should not receive the same consideration from the utility.

Indemnification provisions are often convoluted and confusing, so they should receive particularly close attention by the negotiator. A point that should not be overlooked is that neither party should agree to indemnify the other against the consequences of the other's own negligence. The courts may hold that such an agreement has taken place if the parties promise to indemnify each other against "all" claims, losses, etc.

Another indemnification problem that may arise is that a municipal owner may have no power under the applicable state law to provide a general contractual indemnification. In this event, the developer must also look to the state law for a solution.

K. Insurance. Various types of liability, workers' compensation, auto, and operational insurances may be insisted upon by a utility. Before agreeing to provide performance bond or other type of assurance to back up the sale of capacity, however, the developer should be certain such a surety can be obtained without crippling the economics of the project. Experience to date indicates that many sureties will talk but few will write such a policy unless the developer has substantial assets in addition to the project, with which assets the bond may be collateralized.

L.   *Force Majeure.* Parties should be excused from performance, but not from obligations to make payments, if they are prevented from performing by acts or events of significant magnitude and seriousness that are beyond their reasonable control. Notice, explanation, and best efforts to remedy the situation should be provided for. Issues may arise whether work stoppages, shortages of fuel, or economic hardship will constitute a *force majeure* event. Each should be analyzed in light of the particular position and outlook of the parties.

M.   Arbitration. Whether the parties wish to incorporate an arbitration provision in the contract may be a matter of policy for either party and/or a matter of legal requirement for a municipal project. The state regulatory agency may provide an ideal dispute resolution forum for this purpose, but the parties cannot by contract confer upon the agency any jurisdiction it does not otherwise have. The arbitration provision may thus require the parties to submit a dispute under or concerning the agreement to the agency for resolution, but it cannot require the agency to accept it.

N.   Notices and Records. The agreement must state precisely how and to whom notices of various types shall be given. Different persons may be designated for billing matters, operational and safety matters, and other matters. In addition, each party should be required to maintain accurate records of all transactions under the agreement, and to make them available to the other party upon request and upon reasonable notice.

O.   Events of Default. Many power sales contracts have been executed specifying no particular events that will constitute default or breach of the contract, but providing that if there is a default, the contract will simply be terminated. This may, in part, result from the utility's refusal to consider that it might ever default on a contract, in part from the utility's eagerness to be rid of the contract if the producer commits even the slightest sin, and in part from careless draftsmanship. The project developer should not permit it to result from his own oversight.

Events of default can be simply stated: They may include failure to perform any obligation, including obligations to inspect, maintain, pay, give notice, and so forth. They should not automatically result in termination: The party in default should be given notice and reasonable opportunity to cure the default. The time allowed to cure may vary with the nature of the problem. If cure is not effected within the allotted time, the nonbreaching party should have the *option*, and not the obligation, to terminate the contract.

A default in the agreement to provide capacity may occur, for instance, if the generating facility experiences a prolonged outage or is no longer capable of performing at its contract-rated capacity. Here, depending upon the situation, provision may be made for forfeiture of some or all capacity credits, including some already paid, by way of liquidated damages. Termination need not result, and sales of "energy only" may continue.

P.   Termination. Aside from termination at the end of the contract term or as a result of unremedied default, an abbreviation of the relationship may be provided for under certain other justifiable conditions:

1.   Some states require that newly executed power sales contracts be submitted to the state regulatory authority for approval, disapproval, or modification. In this situation, either party should have a window of opportunity (30 days) to terminate the contract without liability in the event that the agency attempts to impose any conditions that are unduly burdensome. For the utility, this may include a condition that less than all of the utility's costs under the contract may be passed through to its retail customers. For the project such a condition might be the requirement that all financial and accounting records of the owner be made available to the agency or any other person.

2.   If for any reason beyond its control, including a change in the applicable law, the project loses its status as a qualifying facility exempt from utility-type regulation by FERC or the state regulatory authority, the project should have an option to terminate without liability. The consequences of the project assuming utility status will clearly not have been contemplated by the parties to this kind of transaction: They may be so serious that the benefit of the bargain could easily be lost.

3.   Under some conditions, the project developer may want the ability to "buy out" the contract and terminate it. This is provided for in the applicable regulations of the state of New Hampshire. It would allow a producer to reimburse the utility with interest for any amounts

overpaid to date due to a levelized rate, and renegotiate a new contract, presumably at a higher rate.*

Occasionally a utility may attempt to insist upon a right to terminate the contract if the law changes in such a manner that utilities generally are no longer required to offer to purchase power from cogenerators and small power producers (that is, if PURPA and similar state laws are repealed). Alternatively, the utility may want to amend the contract if the law (or the applicable FERC rule) changes to permit utilities to offer to pay less than avoided costs. This kind of provision should be resisted vigorously, as it is likely to render the project unfinanceable.

The project developer should similarly resist any provision that purports to place the provisions of the contract or the operation of the project under the jurisdiction of any regulatory authority, or that requires amendment of the agreement to conform with any order of such an authority. It is possible that a regulatory body may, in the public interest, and under extraordinary circumstances, be able as a matter of law to require alteration of the private contract between the producer and the utility. This, however, would be a far cry from amending rules and regulations. It would involve questions of considerable constitutional gravity, and would require a final, nonappealable order, directed at specific parties, and requiring a particular amendment to the agreement. If such an order were issued, however, no contractual provision would be necessary for the parties to be subject to it.

15.2 RECOVERED MATERIALS: A VIEWPOINT OF THE PRIVATE SCRAP PROCESSOR

Calvin Lieberman

15.2.1 Choices in Strategic Planning

Most management guides for materials recovered from the MSW stream provide a list of possible extracted materials, indicate the type and the quantities that can be expected as a percentage of the waste feed, provide a general description of their condition or form, and outline the potential markets or possible purchasers of the product. Since we are concerned with "management," we should not ignore the need for managerial skills long before the question of marketing these materials arises. At this latter stage, the "manager" has but two options: Can he sell the product in the form, condition, and quantities he is committed to produce, and if so, where?

Following, are three verities to be considered at all times:

"Herein lies a hard-earned truth: Whatever the treasure in trash, it isn't the materials, or the energy one might get out of it—it's mainly what one can collect for just getting rid of the stuff." (The tipping fee.)[1]

"Perhaps the paramount message which should flow from this document is this: *Markets First: specifications determine technology.* Market availability and specifications determine not only the basic components of a recovery system, but also the specific manner in which those components are designed and operated."[2]

"Lack of markets to absorb secondary materials is a major deterrent to massive levels of recycling. Looking at the volume of secondary materials that are now recycled, one might assume that greater supplies would easily stimulate greater demand. Unfortunately, this is not the case. Inventory levels for many scrap wastes are already excessive. Substituting a ton of scrap iron (steel) from the municipal waste stream for a ton of recycled industrial scrap will solve no problems. We need a market that can take both."[3]

There are several approaches that must be considered. More often than not, the reason for removing anything from the process stream is a protective measure rather than for recovery in order to minimize downstream process problems with troublesome and/or extraneous material. Once this selection process is completed, whether for philosophical, political, public relations, or the bias of consultants, the operator may be locked into an inflexible system incapable of adapting to realistic market requirements and changes, or even corrective system technical change.

A community should choose to dispose of wastes by the least expensive means for volume reduction to conserve landfills and to satisfy environmental regulations, or it can modify this intention and extract energy and materials from the system.

---

*This may only be appropriate where the project continues to sell to the same utility or at least in the same state (so that the capacity that has been depended upon is not lost) *and* where the parties recognize, as in New Hampshire, that the rates upon which the contract was initially based are interim rates that are not fully representative of the utility's avoided costs.

The energy option can be steam, electricity, or both (cogeneration).

It may extract noncombustibles to enhance the combustion of either prepared RDF (refuse-derived fuel) or the mass-burn process and also minimize maintenance.

Or, it may extract the metallics from the burned waste stream if a market exists.

This series of choices involves substantial investment in equipment and operational expense. Management must then recognize that there are two important factors that will affect its choice: (1) The nature of the refuse stream in the particular location of the facility. (2) The end product it desires to produce, both the energy derivative as well as the materials to be recovered, if any.

Because of the many variables as well as the position of these variables in a flow model, the computer becomes a valuable tool. It is possible to develop mathematical models for real phenomena that predict "what if" results. Because the "what if" approach is so important in devising a multimillion dollar system, it should be applied. This is common methodology practiced by major firms to plan distribution of their products. For example: Should a fruit drink processor locate warehouses in Chicago, Denver, Los Angeles, and Dallas if the product is made in California? Or, would it be less expensive to also install a processing plant in Chicago; or, should distribution be made solely through wholesalers in various locations; or would it be less expensive to piggy-back trailers from one or two locations; or should the company utilize its own fleet of trucks and drop-ship loads to either wholesalers or to its own warehouses, or directly to retailers. In addition, what is the probable impact of freight rate changes on any system chosen to use its own equipment, common carriers, or combinations. But first and foremost, management will put great stress upon pin pointing its markets; what areas it serves and/or will serve; what demographic changes in the population it serves will affect the market; what the competition is doing, and so on. In short, it starts with the marketplace and then works backwards to determine its optimum future course.

15.2.2 Identifying and Evaluating Markets

The reader will recognize the necessity of exploring the markets for materials he is thinking of removing. The same holds true for planning the sale of energy, be it RDF, steam, and/or electricity. This discussion, however, is focused upon materials, recognizing, however, that the type of energy production may affect the choices for recovery of materials.

It is not sufficient to examine "markets" merely at the current level. The factors that will have an impact upon the marketplace itself must also be considered. Since ferrous metals are the largest metallic percentage in the waste stream, mainly tin cans (really steel cans), observe what is happening in the entire world, not just the United States. This exemplifies what is meant by an impact upon the marketplace. Reference here is to the demand for scrap iron/steel. A recent updated study by the Robert Nathan organization[4] indicates that the United States has approximately 700 million tons of potential steel scrap and that figure is growing yearly. As the steel industry curtails its operations and capacity, the demand for scrap has declined sharply. If national steel mills are successful in purchasing slabs from England, and/or any other country (and the trend may be irreversible), the demand for scrap will decline further. Since steel cans from MSW are at the lowest possible level of usable scrap regarding contamination and universal demand, it must compete with high-quality scrap metal in a declining market.

While the technology of recovering materials exists, and the methodology has improved, the driving force for these techniques might be misdirected; that is, applying the myth that the mere accumulation of materials will somehow insure that a market will develop. The market is the engine that should determine whether or not materials should be recovered, and if so, to what extent.

Although it might seem irrelevant to examine possible flow charts in the discussion of the impact of markets upon materials to be recovered, the interrelationship of markets and the choice of systems is ignored at great risk. In planning a solid waste disposal system, the least expensive but scientific approach is the application of a model or formula. If energy is to be produced by thermal reduction, the methodology of recovering materials may be limited in scope by the particular thermal technique. If noncombustibles must be removed prior to combustion, a certain marketplace must be explored. If however, metallics can only be recovered from the ash (mass burn), a different marketplace must be explored. Thus, the impact of markets and/or the projection of possible changes in current conditions will affect the choice of equipment, operations, and even the wisdom of removing the metallics for saleable recovery. Regarding recovered material specifications, the particular requirements of scrap purchasers and their markets are critical. Are you able and willing to produce such products?

1. *Steel Mill Remelting.* Unburned, shredded, and hydraulicly compressed bundles in size 24 × 24 × 60 in. maximum (61 × 61 × 152 cm) with density not less than 75 lb/ft^3 (1144 kg/m^3) containing less than 5% dirt, plastics, and loose organic matter, free from copper and other nonferrous metals except those used in making the cans. Aluminum tops shall not exceed 1% by weight.

2. *Ferro-Alloys.* Loose, free-flowing, shredded into size not over 6 in. (15 cm) or under ½ in.

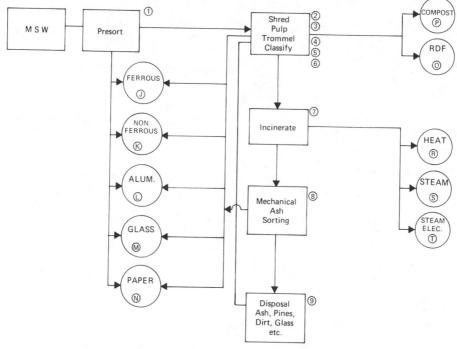

Fig. 15.1 Materials recovery flow chart.

(13 mm), minimum density 35 lb/ft³ (534 kg/m³), free of organics, nonferrous material, dirt; but may be burned or unburned.

3. *Detinning.* Free-flowing unburned cans; not baled or crimped to the extent chemical detinning is infeasible. Dirt, loose paper, plastics, rags, and other nonmetallics not to exceed 5% by weight; not over 5% to exceed 6 in. (15 cm) and not more than 3% to be smaller than ½ in. (13 mm), in large volume only.

4. *Electric Furnace: Rebar Only.* Loose, shredded, after incineration, under 2 in. (5 cm) in size but containing no fines, dirt, nonferrous material—limited volume.

In constructing a flow model, refer to Figure 15.1. Using ferrous materials as an example (it represents the largest percentage of metallics in MSW), we can begin the "what if" exercise. Select "B," removal after incineration, from Figure 15.2, FERROUS RECOVERY VS NO FERROUS RECOVERY. We can then utilize Figure 15.3, PROJECTION OF EXTRA COST VS REVENUE (ferrous flow line "B"), to determine the cost of equipment, operations, and related costs. There is now but one variable in our model. We can arrive at costs versus revenue based upon current markets, but the projection is invalid if it stops there and does not factor in possible market changes as well as the risks involved, and a model is a must. While amortization cost of the equipment will remain constant, the operational costs will escalate with inflation. As market prices can range from zero through the current price, if a contract exists, only the model can account for these possibilities when other materials are included in the formula. This is basic in planning a resource recovery system and should be considered when examining various materials to be recovered.

The planner needs insight into the marketplace. It is not enough to merely consider the supply and demand aspect, even though it is fundamental in regulating the market for materials. Secondary materials (and that is precisely what the system will produce) bear a special relationship to virgin materials unlike any other product. Because of environmental and political factors—that is, tax incentives and disincentives, demographic changes, legislation involving mandated source separation, or litter and deposit laws, inequitable freight rates, historical emphasis on virgin versus secondary, competition with the private tax paying sector, and so on—a "moving target" is the norm.

15.2.3 Evaluating Raw Material Supply and Recovery Technologies

The amount and quality of waste will also help determine which system you select—batch-type furnace, moving-grate waterwall unit (mass burning), production of RDF, wet pulping, and so on.

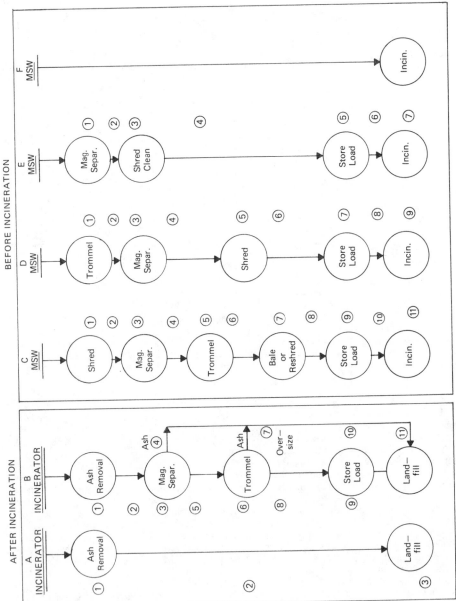

Fig. 15.2 Ferrous recovery (mostly cans) versus no ferrous recovery.

646

PROJECTION——EXTRA COST vs REVENUE——FERROUS——FLOW LINE "8"

(For illustration only "B" is shown here. Apply same procedure to C, D, or E, and/or for any other materials to be recovered.)

COST PER YEAR

ITEM	CAPITAL COST (Amortized 10–15 Years)	MAINTENANCE	POWER	DUST CONTROL EQUIPMENT	PERSONNEL	ADD'L BLDG. AREA	Sub. Totals
Conveyors:							
2							$
4							
5							
7							
8							
10							
Equipment:							
3							
6							
9							
Sub. Totals							

GRAND TOTAL _____ $

* Projected revenue from sale of ferrous: Year ____ $ ____

Total extra cost to recover the ferrous: $ ____

Projected gain or loss, per year. $ ____ + or –

Affect upon MSW processed per year: Gain or Loss _____ + or –
                                    Waste _____ ** TPY

RISK FACTORS TO BE CONSIDERED:

Volume of ferrous obtainable.
  Affect of deposit legislation—cans removed and not in the MSW stream.
  Affect of steel can replacement by aluminum, paper, or plastics.
Changes in market price.
Changes in demand for this product by consumers.
NOTE: Reduction of landfill volume by removal of cans is minimal.

* F O B Site.
** Tons Per Year

Fig. 15.3 Projection of extra cost versus revenue (ferrous flow line B).

647

You can choose between burning for reduction, for steam, for heat, for production of electricity, or any combination.

Assuming that incineration, whether for reduction only or energy recovery, is chosen, there are many factors affecting model formulation. Residential, commercial, and industrial waste varies widely throughout the United States. It is affected by the life-style of the population and the degree of urbanization with regard to the types of material generated, that is, cans, bottles, plastics, metallics, local flora, paper, and so on, raising the following questions:

Should you remove metallics and other materials to produce fuel and save wear and tear upon the equipment and/or to enhance processing? Should you burn the waste and extract the metallics from ash? Has the raw material survey provided meaningful data as to the capacity of the system for current and projected volume during the estimated lifetime? What are the capital costs of equipment that is used mainly to recover materials? Are the materials removed merely and hope-fully to obtain revenue from their sale?

The emphasis is upon the necessity for raw material analysis and availability and market research BEFORE you choose your equipment or design. These questions again underscore the importance of constructing a model to test various systems and costs prior to construction. Note that the impact of the market will help you choose. If there is a market for unburned cans in your area, properly prepared by removal of contaminants, and you install such processing equipment, will the additional tonnage lower the value of these cans in a limited marketplace? It is a kind of "regression analysis" or working backwards when materials recovery is considered. It is determin-ing if there is a market, and what specifications are to be met. Should you process the material to meet these specifications and can you do it? Should you explore having a processor of scrap do so? Will market changes in demand or price entail excessive risks? It is an investment decision that must be justified. The decision is accomplished with a model utilizing the variable to determine system capital investment along with owning and operating costs, via a series of "what if" analyses.

15.2.4　Evaluating Risks

To reduce risk, you can design bypass in your waste flow system. If sorting is feasible in a high market for materials, you sort. But if there is no market temporarily or permanently, you allow the materials to pass through for disposal. If the cost of conveyors and/or equipment allowing these options is expensive and marginally useful, avoid that installation, in which case the system design must permit future retrofit to provide the bypass option, and so on. The objective is not to commit the operation to markets that do not materialize or may be temporary, but to adapt to attractive opportunities when they develop.

15.2.5　Recovered Materials Quality/Saleability

Any discussion of markets for the sale of materials recovered from MSW must ask if the product quality is saleable, and will changes in demand render it marginally profitable or a loser?

Considering only recoverable materials—metallics, fibers, plastics, or glass—who will buy your product? Ultimately, it must be a secondary processor and/or consumer who will remelt it, or reprocess it, resulting in raw material for making new products. The consumer, whether it be a steel mill, foundry, paper processor, plastic manufacturer, must also produce saleable products, and is also vulnerable to the vagaries of the marketplace. The manufacturer of autos purchases steel for fenders and wants quality at low prices that meets his specifications. Fenders cannot be made from steel that is brittle or has poor surface qualities for accepting and retaining enamels. With the drive for zero defects, and penalties for selling unsafe products, quality control is vital. Rejects are expensive, forcing rigid control clear back to the processor of scrap iron (or iron ore), through the mill, and up to the final manufacturer.

In practice, aluminum can stock has not been made from 100% aluminum cans; likewise, 100% tin cans (really steel cans) cannot make new tin can steel stock. Recent developments in the USA and Japan, however, may lead to more direct conversion for aluminum cans.

Therefore, knowing that the materials in MSW are the most abundant discards, the most contaminated, and the least valuable, can these materials be made saleable? The answer is, some-times yes, sometimes no, but in any case, recovered materials must compete in the marketplace with virgin materials and also with cleaner scrap that is source-separated, graded, processed to specifications, and collected, avoiding the MSW stream.

15.2.6　Disincentives in Resource Recovery

A decision to produce electricity via RDF usually enhances the opportunity to recover materials in the RDF process stream, but despite PURPA, which mandates a market for electricity to utilities on a least-avoided cost basis, the revenue can be disturbingly uncertain due not only to generating irregularities but also to fickle fossil fuel prices and depressing effects of nuclear generation and hydroelectric power on avoided costs. Furthermore, bottle bills, litter laws, and source-separation programs are disincentives to materials recovery, especially if industry support of source-separation proliferates.

There can be little doubt that electric utilities will continue to seek relief from the federal mandate (PURPA) to purchase electricity despite a contention by resource recovery entrepreneurs that capital investment for increasing network capacity will be reduced. A comparison of capital, owning, and operating costs to produce steam (if there is a market) and electricity or burning for reduction and disposal only are becoming well known as compiled from the track record of successful plants on stream,[5] those in start-up, and of course, the failures (refer to Section 12.4 and Table 12.4).

15.2.7 Engineering with Unpredictable Raw Material

Engineers (likewise physicians, economists, lawyers, and certainly bankers) are most comfortable when dealing with behavioral and physical predictability. They rely upon known characteristics of matter to react as expected when the flow of material falls within the limits of their known characteristics. However, MSW contains infinite combinations of an infinite variety of materials including organics, metals, plastic, sludge, and so on. To illustrate, certain shredders are capable of reducing autos, rock, and the like, but must be modified to handle MSW. Feed material characteristics determine the design, and this applies as well to all types of incinerators, with or without energy recovery. The ASTM (American Society for Testing and Materials) through its E38 Committee on Resource Recovery tried for several years to establish standards for raw MSW. Although it has established standards and testing procedures for recovered materials, developing any more than the most cursory identification/standards for raw solid waste is elusive and sometimes illusory. The only certainty is that raw MSW will vary by the minute, hour, day, week, and by season, weight, makeup, moisture content, hazards, and value.

Unfortunately, we approach a "catch-22" situation. For a facility to perform as planned, it must have a minimum flow of raw material to justify the increased capital expenditures and operating costs required by the decision to recover resources. The result has been that other disposal alternatives to diminishing landfill availability have become secondary to resource recovery. If there are other alternatives to land disposal, is the cost of resource recovery materials and/or energy justifiable? Could disposal be accomplished on a smaller scale and at a lower net cost to the taxpayer?

15.2.8 Raw Material Flow Control: A Word of Caution

Recovering or recycling is the principal pretext in efforts to control MSW. This leads to the dilemma of deciding whether it is better to burn paper rather than recycle it. A monopoly created by flow control legislation mandates burning the paper for energy. However, if private haulers are allowed to source-separate materials before using the disposal facility, the waste materials markets are well served. Resources are recovered without severe constraints upon an energy producing system. Chapters 2 and 7 discuss these questions in more detail.

Most of this flow control legislation stems from the Akron, Ohio, ordinance required by the bond underwriters, to "insure a supply of raw materials" for producing steam. It has been alleged, however, that it really was the desire to assure more tipping fees, which are unlimited, and not merely to assure a supply of refuse. The plaintiffs in the case stressed the violation of antitrust laws, which the district court denied on the theory that a monopoly was justified under police power and was in the public's interest. The court did not address the major issue involving the economic justification for the entire system versus other forms of disposal.

Neither did the court consider the original cost of $56 million versus the final total of $80 million, required for retrofit. Nor was the court asked to consider the justification to sell steam at a cost above that which could be obtained by the steam customer from other sources. Since courts normally avoid second-guessing administrative projections or decisions involving economics, granting the monopoly without restraint is surely a license to proceed without restraint.

But note what happened since that decision was rendered. The Supreme Court of California (normally a pacesetter) accepted the argument of poor economics as a justifiable reason for a community to ban a nuclear plant. If this is an impending new theory, and the economic theory is applied, how many recovery systems could pass this test? (Chapter 7 discusses this in detail.)

When legislation is written that controls the flow of all waste to a facility, without allowing the removal of materials for recycling by the private nonsubsidized sector prior to disposal, it reinforces the argument that the facility needs more revenue because of its redundancies. The tipping fee continues to be the escape value for projects that make the wrong logistical, technical, and market decisions.

15.2.9 Markets for Recovered Materials: The Hard Facts

The scrap industry also has problems and promise in considering markets. Fortunately the problems are well known and traditional: the fact that scrap is purchased and not sold; the law of supply and demand (it operates very well for metals and paper); that specifications determine technology; that markets differ radically geographically; that each consumer has his own specifications; that recovered materials are considered secondary materials and must compete with virgin materials; that legislation affects markets; and that long-term contracts are the exception (weeks/months and not years). Unfortunately, planners of MSW facilities too often believe that somehow if materials are recovered in quantities, a market will develop. The fact is that if and when quantities do increase, the market sags and often disappears.

At the very outset of our analysis, we must ask: Are we designing a solid waste system to produce materials and/or energy or to dispose of refuse? The reflex answer, of course, is "both," which then raises other vital questions:

What are the specifications by consumers for materials we hope to recover?

What condition will they be in when extracted from the waste stream?

What processing will be necessary before a buyer will accept them, either for further processing or for direct consumption?

What quantities are being consumed in this area, where are the markets, what is the net FOB basis or price we can hope for?

What will it cost in capital and processing to make the product?

Do we have to remove materials to produce energy anyway, and if so, what is the least cost even if they must be discarded for landfill or removed by others for a fee?

What has been the total demand for the materials during the past 5 years and what effect will our production of more have on price and/or demand?

What factors may have an impact upon potential consumers in the way of competition from foreign sources, or substitute products, or substitute materials?

What possible effect upon our supply will legislation have, such as litter or deposit laws; or banning of certain products, or volunteer programs; or source-separation legislation?

Can a combination of energy recovery/disposal succeed without extracting materials?

What is the experience of other systems throughout the country and the world in recovering these materials (or not doing so)?

Have we contacted private industry to seek their interest and advice as to the acceptability of the materials we might produce?

What might affect the volume of these materials in the waste stream and how would that affect the capacity projection of the equipment we must install?

If metallics must be removed prior to incineration, there are several methods for ferrous recovery. If refuse derived fuel is to be produced, several options are:

1. Shred, separate magnetically, trommel, air-classify, then burn.
2. Trommel, separate magnetically, air-classify, shred, then burn.
3. Sort, shred, separate magnetically, reshred, then burn.

For mass-burn concepts after ash removal for processing, options are:

1. Separate magnetically, trommel.
2. Trommel, separate magnetically, shred, and so on.

Each operation is a cost center, and process selection must consider any impact on RDF quality in the RDF system. But for mass burn, the choice is based entirely on market requirements (for an inferior product).

A careful examination of the recoverable/saleable materials in raw MSW is critical. Of the metallics, the ferrous component is the largest, and there are four markets for it (mainly tin cans): melting for steel, melting for ferro-alloys, copper precipitation, and detinning with ultimate melting for steel.

1. *Melting for Steel.* The total quantity of cans, not detinned, and melted for steel is insignificant and usually represents public relations efforts and not a viable sustained market. At best, it is marginal and risky. But that status may change as minimills, making rebar, experiment with this material. Figure 15.4 is a typical basic steelmaking flow chart.

GENERAL FERROUS FLOW CYCLE

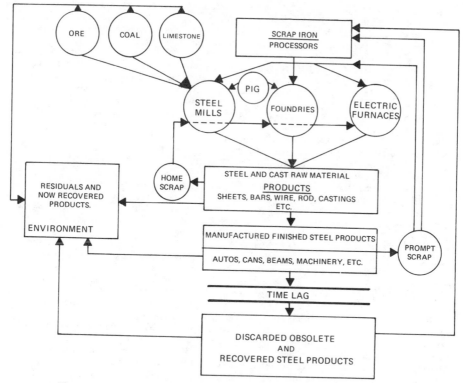

Fig. 15.4 Flow diagram, basic steelmaking, including ferrous scrap input.

2. *Melting for Ferro-Alloys.* If your facility is within a reasonable freight rate haul to this market, mainly around the St. Louis area, it is worth considering. You might obtain a reasonable long-term contract, up to 3 years, based upon a formula price related to market quotations for #2 Bundles in a trade publication. However, you must meet specifications for cans with organics removed, and in the size, shape, and density the consumer will accept.

3. *Copper Precipitation.* This is a dying market that exists mainly in the area west of the Mississippi River. As copper mining declines, and as pollution regulations become more stringent, this will be much less a factor (market) as small as it is now.

4. *Detinning.* Again, like ferro-alloy consumption, there is a limited market for cans prior to burning if properly prepared with limited organics, free from extraneous material, in proper size and density, in large volume, and within reasonable freight rate haul/cost to the purchasing process plant. However, it appears that only one MSW facility in the country at this date is selling cans under contract for this purpose.

5. *Nonferrous Metallics.* (Aluminum cans are discussed separately.) The usual run of other metals will consist of contaminated and mixed coppers, brasses, zinc, lead, stainless steel, and aluminum noncan items. This mixture will appear in small quantities in any system utilizing magnetic separation for the ferrous fractions or if it has a float separation system. If a trommel is used in the system to remove dirt, glass, and fines, you might interest a nearby scrap processing company to purchase the mixture at some price. Making this material saleable to a consumer requires expertise, equipment and trained personnel not justifiable for MSW systems. With perhaps one exception (Resources Recovery, Inc., Dade County, Miami, Florida), there is no plant successfully producing separated metals which can meet the specifications of the consuming industry.

6. *Aluminum Cans.* There is a relatively small fraction of aluminum cans in the typical raw MSW stream. It can be traced directly to the success of an incentive approach where consumers (can manufacturers, beverage industry, and so on) of aluminum have excelled in establishing a buy-back program. Because of a tremendous public relations campaign, and active participation by scrap processors and volunteer groups, relatively few cans are discarded in the waste stream.

Although this has successfully proved the desirability of source separation, it refutes the early projections for revenue from aluminum in MSW. The primary aluminum industry has also funded development of equipment (mostly eddy current technology) to extract aluminum from processed waste but with little success, with perhaps one exception [The NRT (National Recovery Technologies) Pre-Burn Process System, Gallatin, Tennessee].

7. *Glass.* To match the success of the aluminum campaign, glass producers and affiliated industries have also developed and encouraged collection by volunteers for sale at the local level. Using catchy circus-related equipment design, plus monetary inducements, designed to compete with aluminum cans, the tonnage of recycled glass has grown from 6,995 tons in 1970 for just one company, Owens-Illinois, to 232,858 tons in 1981. The effect of this campaign has implications also for the design of MSW facilities. Many systems have discovered that the installation of trommels is desirable to save downstream abrasive wear, boiler slagging, and volume reduction for landfilling, by removing glass fractions prior to processing waste. As this campaign grows, just as in aluminum, projected revenue from glass may encounter severe resistance. Although glass is useful as cullet to make new glass, the requirement for color-sorting makes it feasible only by hand-picking and in volunteer operations. No equipment to date has successfully been developed (despite much research and expense) that can economically color-sort glass with precision and dependability.

8. *Plastics.* Recovering plastics for revenue is unfeasible. Although plastic recycling continues, it is in source separation of postconsumer items, such as plastic bottles prior to mixing with the waste stream. Plastic products manufacturers recycle their own useable waste successfully. The technology to sort plastics property (either thermosetting or thermoplastic) from MSW other than hand-picking is in the distant future, unless intense and continuing investigation of the precursors of Dioxin formation and incinerators precludes burning it by law.

9. *Paper.* As previously stated, there is a dilemma: Shall we burn it for energy or shall we recycle its fiber to make more paper products? Since most systems being considered today include energy recovery, we should examine only what might be acceptable for recycling and removed from the tipping floor, and so on, before burning. This does not include the impact of source separation in communities that gather paper separately in their collection system. Paper from that source has market value depending upon the geographic location, proper sorting, economical gathering, and transportation costs. Some systems have experimented with retrieval of paper from a tipping floor or conveyor before moving the waste to the furnaces or pulper. It is not feasible for any large-volume system, considering the time, the labor, the equipment necessary and the danger of injuries to personnel. Once gathered and dumped, contamination has taken place. If a market does exist, it will act as the magnet and paper will not flow into the system. Schools, scouts, and other groups will do the job, mainly with volunteers.

Conclusion

For communities already operating a resource recovery facility and frustrated by the reality of marketing problems, they still have the option to stop. They can ignore the fact that the system may have been "sold" to the public by projected but unrealized revenue to be obtained from the sale of recovered materials. They can still scale down or remove sophisticated and expensive equipment and operations which process items that finally come to rest in a landfill. Or they may pay to have the recovered materials, unsaleable, removed by private industry for disposal in a landfill. The offsetting result is that they can reduce landfilling costs as well as operating costs, when materials removal was necessary in their energy-producing systems. Business makes these hard decisions regularly.

For those communities that have not made financial or contractual commitments for a system scaled up in capacity, size, and cost, on the premise of revenue from materials to be extracted, they are most fortunate. They have a wealth of data from multimillion-dollar projects that have succeeded and those that have failed.

Note: Section 15.2 expresses the personal views of the author and does not necessarily represent the position of the Institute of Scrap Iron and Steel.

REFERENCES

1. P. T. White, The fascinating world of trash, *National Geographic*, April, 1983, pp. 4–15.
2. Y. M. Garbe and S. J. Levy, Energy products and markets, in Resource Recovery Plans and Implementation Guides for Municipal Officials, U. S. EPA 1976 (SW 157.3).
3. R. L. Lesher, Bulletin, National Center For Resource Recovery, Inc., Washington, D.C., October, 1971.
4. R. Nathan, (Robert R. Nathan Associates, Inc.), Iron and Steel Scrap: Its Accumulation and Availability Updated and Revised to December 31, 1981, for Metal Scrap Research and Education Foundation.
5. L. Nelson, Resource recovery: Does it work? *World Wastes*, May 1983, p. 10.

CHAPTER 16
ENERGY FROM REFUSE IN INDUSTRIAL PLANTS

WILLIAM D. ROBINSON

Consulting Engineer
Trumbull, Connecticut

FRED ROHR

Consulting Engineer
Oak Brook, Illinois

16.1 BACKGROUND

Energy recovery from combustion of industrial wastes has been practiced since boilers and incinerators came into being. The first sawmills with steam engines used cut-off waste and sawdust for fuel. Since the Industrial Revolution, burning solid and liquid wastes for steam production has been standard procedure in the forest products industries (pulp and paper, furniture, and other wood products) and in others such as sugar extraction (bagasse) and food producers. Such by-products are essentially homogeneous waste fuels rather than random mixed refuse.

Industrial manufacturing industries usually have more random mixtures of discarded combustibles which makes handling and burning more difficult. As a result, development of systems for these random mixtures has lagged behind systems for the more homogeneous by-product wastes. The principal motivation for development of improved energy-producing refuse firing equipment has always been economics and emissions control.

Increasing concern for energy cost, conservation, and the advances in air pollution control techniques has renewed interest in the concept. This discussion is intended to review the classification of combustion and heat recovery equipment available today, the value of refuse energy, and its cost of recovery.

16.2 INDUSTRIAL WASTES AS BOILER FUEL

The prospects for using plant solid waste from most industrial production facilities for energy recovery is encouraging. The material is usually dry and contains little garbage or other wet material, as in municipal waste having lower heat values.

There are, however, a few large industrial facilities that have chemical process as well as "hard" manufacturing operations (see Case Histories) that produce waste mixtures resembling the somewhat presorted ("Baby Food") raw refuse at many municipal incinerator facilities with regard to varieties of materials (corrugated, paper, glass, textiles, plastics, and so forth). Moisture variation is the principal difference in intrinsic characteristics in these instances.

In a typical industrial plant, solid wastes are collected in containers located at convenient locations near the points of generation. In areas of metal working and fabrication, the wastes may be all chips or cutoffs that are frequently collected in containers designated specifically for the purpose and sold for scrap. Foundries also have a solid waste problem that is mainly spent sand and core butts. Manufacturing plants producing large quantities of noncombustible waste material such as metal also generate waste paper and wood. In a large facility, the quantity of combustible material may be large enough to justify an on-site heat recovery plant or to provide separate containers for collection and delivery to an energy recovery facility elsewhere.

Most combustible solid wastes are produced in areas where production materials (chemicals, parts, and so forth) are unloaded from pallets, sacks, or shipping containers, and such continuous flow usually provides continuous quantities of waste material. Another source of solid waste in manufacturing plants is the demolition material from modification and/or expansion programs. This material is usually sporadic, with random quantity and composition, but most often the material is broken concrete and old lumber.

Liquid wastes may be available in quantities that are worthwhile investigating for fuel value. The sources may be a variety of cutting oils, coolants, or hydraulic or engine oils, and only an analysis will provide the basis for including this material as a fuel for heat recovery. The oils can be used as auxiliary fuel for boilers or incinerators but they must have a minimum of corrosive producing contaminents. In any case, all liquid or fume wastes considered for combustion must be evaluated in terms of emissions control requirements.

A survey of all plant wastes is necessary to determine the actual types of materials available for energy recovery and the rates of generation. Since plant wastes are usually collected in containers and logged on a volumetric/frequency basis, it is necessary to convert all data to a weight basis. Guessing at the density of various materials and extrapolating the data can result in significant error. The most effective way to conduct a survey is to weigh and inspect the containers for a period of time to establish a level of confidence for composition and average weights. It may also be found that the wastes will have a cyclic pattern of generation, depending on rates of plant production, vacation shutdowns, and plant cleanup periods. The minimum as well as maximum rates of generation are important to the planning of an energy recovery system. Projections of future rates of generation due to increased plant production, expansion, or transfer of similar wastes from other plants are also necessary for planning the size of the recovery system and its hours of operation.

With essentially the same sampling and analysis difficulties as for mixed municipal waste, determination of the heat value of industrial solid waste is somewhat uncertain since the type of material and composition can vary from one load to the next.[1] The closest approximation is the average of cardboard, wood, plastics, rubber, and so on. When specific types of material, that is, wood, identical plastics, paper, and the like, appear consistently in dominant quantities, heat value reference tables[2] can establish or confirm the values for the particular project with reasonable certainty.

During the 1960s when the incinerator industry was developing rapidly, an industry manufacturers' association, The Incinerator Institute of America (IIA), was formed. They classified wastes into types 1 through 6, Figure 16.1. These designations may be considered a categorical reference only and should be used as fuel analysis criteria for furnace design with caution.

16.3 INDUSTRIAL INCINERATORS

16.3.1 Background

Incinerators for the destruction of refuse came into general use in the late 1800s. Smaller units were incorporated into buildings using what was termed "builders hardware," which the contractor could purchase in the form of doors, grates, a screen for the top of the chimney, and a set of installation drawings. The usual design consisted of a single chamber lined with firebrick which was incorporated into the building by the architect. Larger free-standing units were also developed at an early date with various single or multichamber arrangements for drying of garbage and burning of material to which coal was sometimes added to increase the heating value.

16.3.2 The Early Los Angeles County Excess Air Refractory Furnace

Designs gradually improved, with the most rapid development occurring after World War II through the 1950s and with capacities up to about 5000 lb/hr (2275 kg/hr). The early air pollution problems in Los Angeles led to more restrictive air pollution codes along with research on incinerator design and operation. The result of this research was *Design Standards for Multi-Chamber Incinerators*, published by the Los Angeles County Department of Air Pollution Control in 1960.[3] Although manufacturers were not universally in agreement on these design standards, new equipment could not be installed unless it conformed to the guidelines. Opinions differed greatly at that time as to designs necessary to minimize air pollution. New York City also developed guidelines that were different from those issued in California.[4]

The basic outlines of the California units are shown in Figure 16.2. Typically, the design of this incinerator is a refractory lined natural draft type with stationary grates, and either a preassembled or field erected unit in which no mechanical equipment is used for forced underfire or overfire air.

Design calculations for an excess air incinerator includes determination of the quantities of air flow necessary for combustion and cooling, usually about 150% excess air. For each incinerator design, capacity, and type of waste burned, cursory operating data such as furnace temperatures and excess air can be determined for waste fuels of various BTU values as shown in Figure 16.3.

INCINERATOR INSTITUTE OF AMERICA *Essentials for*

The I.I.A. Incinerator Standards have purposely dealt only with basic incinerator designs. Careful consideration must be given to the following:

(1) Collection and method of charging the refuse. (Can be charged at side, end, top, or at floor level.)

(2) Ample areas around the incinerator for charging, stoking, and ash handling, as well as general maintenance.

(3) Adequate air supply to the incinerator room at the stoking and charging levels.

(4) The effect which any air conditioning equipment, ventilating fans, etc., may have on the air supply or the draft available from the draft producing equipment.

classification of wastes to be incinerated

Type	Type of Waste Description	Principal Components	Approximate Composition % by Weight	Moisture Content %	Incombustible Solids %	B.T.U. Value/lb. of Refuse as Fired	B.T.U. of Aux. Fuel Per Lb. of Waste to be included in Combustion Calculations	Recommended Min. B.T.U./hr. Burner Input per lb. Waste
*1	Rubbish	Combustible waste, paper, cartons, rags, wood scraps, floor sweepings; domestic, commercial, industrial sources	Rubbish 100% (Garbage up to 20%)	25%	10%	6500	0	0
*2	Refuse	Rubbish and garbage; residential sources	Rubbish 50% Garbage 50%	50%	7%	4300	0	1500
*3	Garbage	Animal & vegetable wastes, restaurants, hotels, markets; institutional, commercial, and club sources	Garbage 100% (Rubbish up to 35%)	70%	5%	2500	1000	3000
4	Animal solids and organic wastes	Carcasses, organs, solid organic wastes; hospital, laboratory, abattoirs, animal pound, and similar sources	100% Animal and Human Tissue	85%	5%	1000	1800	8000 (5000 Primary) (3000 Secondary)
5	Gaseous, liquid or semi-liquid wastes	Industrial process wastes	Variable	Dependent on predominant components	Variable according to wastes survey	Variable according to wastes survey	Variable according to wastes survey	Variable according to wastes survey
6	Semi-solid and solid wastes	Combustibles requiring hearth, retort, or grate burning equipment	Variable	Dependent on predominant components	Variable according to wastes survey	Variable according to wastes survey	Variable according to wastes survey	Variable according to wastes survey

*The above figures on moisture content, ash, and B.T.U. as fired have been determined by analysis of many samples. They are recommended for use in computing heat release, burning rate, velocity, and other details of incinerator designs. Any design based on these calculations can accommodate minor variations.

Fig. 16.1   IIA (Incinerator Institute of America) waste classifications.

As air pollution codes become tighter after the Clean Air Act, design limitations of existing units began to limit their use. Many installations could no longer meet the newer codes and either shut down, added scrubbers, and/or converted to a "controlled air" design.

16.3.3   Controlled Air Designs*

Dual Chamber, Excess Air

Small-scale, modular incinerators, Figures 16.4 and 16.5, were developed in the 1950s and have been successful in solid and liquid waste applications. They are usually operated in an excess air mode in the lower chamber with flame combustion, followed by a secondary afterburner chamber

*Editor's note: This discussion interprets "controlled air" as meaning regulation of underfire and overfire air for either excess air or "starved air" operating modes. ASTM committee E38, many consultants, and much of the industry curiously restrict the definition to a pyrolytic "starved air" misnomer.

(5) Adequate draft (negative pressure) to assure safe operation and complete combustion at reasonable temperatures. Draft producing equipment should be adequate to handle all theoretical and excess air required at not less than .35″ of water column for Class III incinerators, .45″ for Class IV incinerators, and .25″ for Class VI incinerators.

(6) The location of the top of the chimney or stack to ventilation intakes, and penthouses or other obstructions.

(7) The immediate environments to determine the advisability of the use of auxiliary equipment such as fly ash collectors or washers, pyrometers, secondary burners, draft gauges, smoke density indicators, etc.

(8) Current local codes and ordinances.

incinerator selection chart

CLASSIFICATION	BUILDING TYPES	QUANTITIES OF WASTE PRODUCED
INDUSTRIAL BUILDINGS	Factories Warehouses	Survey must be made 2 lbs. per 100 sq. ft. per day
COMMERCIAL BUILDINGS	Office Buildings Department Stores Shopping Centers Supermarkets Restaurants Drug Stores Banks	1 lb. per 100 sq. ft. per day 4 lbs. per 100 sq. ft. per day Study of plans or survey required 9 lbs. per 100 sq. ft. per day 2 lbs. per meal per day 5 lbs. per 100 sq. ft. per day Study of plans or survey required
RESIDENTIAL	Private Homes Apartment Buildings	5 lbs. basic & 1 lb. per bedroom per day
SCHOOLS	Grade Schools High Schools Universities	10 lbs. per room & ¼ lb. per pupil per day 8 lbs. per room & ¼ lb. per pupil per day Survey required
INSTITUTIONS	Hospitals Nurses or Interns Homes Homes for Aged Rest Homes	8 lbs. per bed per day 3 lbs. per person per day 3 lbs. per person per day 3 lbs. per person per day
HOTELS, ETC.	Hotels—1st Class Hotels—Medium Class Motels Trailer Camps	3 lbs. per room & 2 lbs. per meal per day 1½ lbs. per room & 1 lb. per meal per day 2 lbs. per room per day 6 to 10 lbs. per trailer per day
MISCELLANEOUS	Veterinary Hospitals Industrial Plants Municipalities	Study of plans or survey required

NOTES

Do not estimate more than 7 hours per shift for Industrials.

Do not estimate more than 6 hours operation per day for Commercial Bldgs., Institutions, and Hotels.

Do not estimate more than 4 hours operation per day for schools.

Whenever possible an actual survey of the amount and nature of refuse to be burned should be carefully taken. The data herein is of value in estimating size and determining class of incinerator where no survey is possible and also to double-check against an actual survey.

classification of incinerators

CLASS I
Portable, packaged, completely assembled, direct fed incinerators, having not over 5 cu. ft. storage capacity, or 25 lbs. per hour burning rate, suitable for Type 1 or Type 2 Waste.

CLASS IA
Portable, packaged or job assembled, direct fed incinerators, 5 cu. ft. to 15 cu. ft. primary chamber volume, or 25 lbs. per hour up to but not including 100 lbs. per hour burning rate, suitable for Type 1 or Type 2 Waste.

CLASS II
Flue fed incinerators, with more than 2 sq. ft. burning area, suitable for Type 1 or Type 2 Waste. (Not recommended for industrial wastes).

CLASS III
Direct fed incinerators with a burning rate of 100 lbs. per hour and over, suitable for Type 1 or Type 2 Waste.

CLASS IV
Direct fed incinerators with a burning rate of 75 lbs. per hour or over, suitable for Type 3 Waste.

CLASS V
Municipal incinerators.

CLASS VI
Crematory and pathological incinerators, suitable for Type 4 Waste.

CLASS VII
Incinerators designed for specific by product wastes, Type 5 or Type 6.

Fig. 16.1 (*Continued*)

I.I.A. incinerator standards

MAXIMUM BURNING RATE LBS./SQ. FT./HR.
OF VARIOUS TYPE WASTES

CAPACITY Lbs./Hr.	LOGARITHM	#1 WASTE FACTOR 13	#2 WASTE FACTOR 10	#3 WASTE FACTOR 8
100	2.00	26	20	16
200	2.30	30	23	18
300	2.48	32	25	20
400	2.60	34	26	21
500	2.70	35	27	22
600	2.78	36	28	22
700	2.85	37	28	23
800	2.90	38	29	23
900	2.95	38	30	24
1000	3.00	39	30	24

Figures calculated as follows:

MAXIMUM BURNING RATE LBS. PER SQ. FT. PER HR. FOR TYPES #1, #2, & #3 WASTES USING FACTORS AS NOTED IN THE FORMULA.

B_R=FACTOR FOR TYPE WASTE × LOG OF CAPACITY/HR.
#1 WASTE FACTOR 13
#2 WASTE FACTOR 10
#3 WASTE FACTOR 8
B_R=MAX. BURNING RATE LBS./SQ. FT./HR.

I.E.—ASSUME INCINERATOR CAPACITY OF 100 LBS/HR, FOR TYPE #1 WASTE

B_R=13 (FACTOR FOR #1 WASTE) × LOG 100 (CAPACITY/HR.)
13 × 2 = 26 LBS./SQ. FT./HR.

weight and conversion tables

	Lbs. Per Cu. Ft.
Garbage (70% moisture)	45
Rubbish	10
Loose Paper	5 to 7
Scrap Wood and Saw Dust	12
Shavings (loose)	6 to 8

	Cu. Ft.
Garbage Can 18"x24"	3.6
Garbage Can 16"x22"	2.00
Bushel	1.25
Barrel (U.S. standard)	4.00
55 gal. Drum	7.00
1 Gallon	0.134

Fig. 16.1 (Continued)

with auxiliary fluid fuel firing as required to maintain 1600°F (871°C) to consume smoke and odors. Minimal particulates emission is effected by changes in direction and velocity of the flue gas stream, including orifice or weir-type velocity changes induced in the breeching between the lower and upper chambers, and so forth. A variety of ram feed/automatic grate configurations are in use along with continuous ash discharge where required.[5]

Rotary Kilns

Rotary kilns, Figure 16.6, normally operate with excess air for solid and liquid wastes and sludges and are most adaptable to applications with burning rates over 2 tons/hr with unprocessed as-received solid waste. Chapter 13 describes rotary kiln design factors in detail.

Starved Air

These designs, Figure 16.7, were also a development of the controlled air, small-scale, modular incinerator and employ a somewhat peripatetic form of pyrolysis. Commercially available in the 1960s, both excess air and starved air modular units were not universally accepted because of higher cost than the older brick-set, excess air types. As air pollution codes become tighter, however, the added cost of pollution control devices on the older designs made the controlled air units more economically attractive along with on-site incineration due to rapidly increasing landfill costs and constraints.

A starved air operation depends upon control of the operating conditions in two chambers. The lower chamber is intended to operate at low interior gas velocities to reduce turbulence and entrainment of particulate matter and to limit air for combustion which is also intended to control the operating temperature. Although manufacturers differ somewhat on the methods of control, the intent of the design is to avoid active flame combustion in the large (pyrolizing) lower chamber by restricting the air supply to less than stoichiometric, thus the term starved air. The effluent gas from the lower chamber is mixed with additional air in the small upper chamber (or stack) to complete the combustion reaction. In any case, staged thermal reduction takes place in controlled air units—both excess and starved air.

Whatever the furnace type, energy recovery is usually available via heat exchangers in the flue

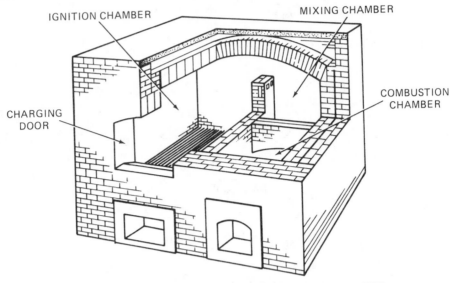

CUTAWAY OF A RETORT MULTIPLE-CHAMBER INCINERATOR

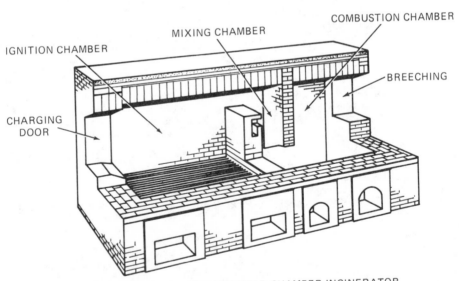

CUTAWAY OF AN IN-LINE MULTIPLE-CHAMBER INCINERATOR

Fig. 16.2 The early Los Angeles County incinerator designs.

gas discharge ducts (Figure 16.6) or a waste heat boiler, firetube or watertube, Figures 16.8 and 16.9.

An advantage claimed for controlled air packaged incinerators is that they can meet most pollution control particulate regulations without additional control devices. It should be remembered, however, that codes for emission levels are less restrictive under the clean air act for small units, 50 TPD and under, although some states have imposed more stringent requirements.

Controlled air incinerators are available in a variety of equipment options and sizes. Charging rams and stokers are standard, and the requirement for control of air flow into the lower chamber along with continuous ash discharge is also available. The upper limit capacities for this type equipment is unrestricted except for the shipping size of components, and stock models over 2000 lb/hr (910 kg/hr) are available from several manufacturers.

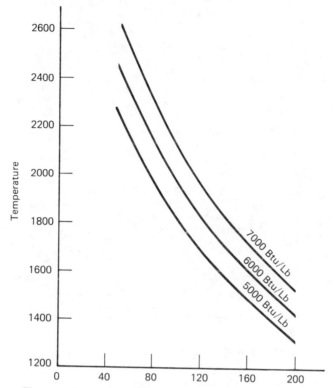

Fig. 16.3 Incinerator furnace temperature versus excess air.

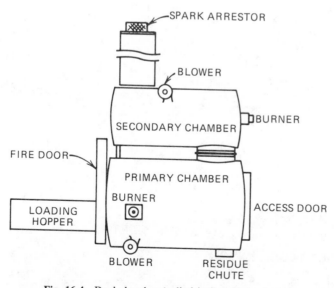

Fig. 16.4 Dual chamber (cylindrical) modular design.

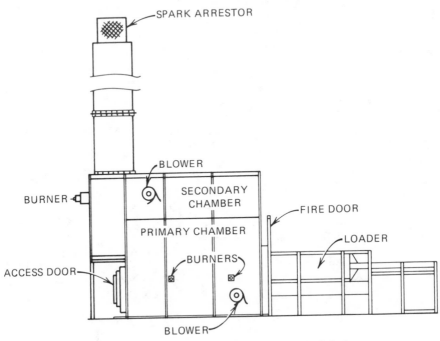

Fig. 16.5 Dual chamber (rectangular) controlled air design.

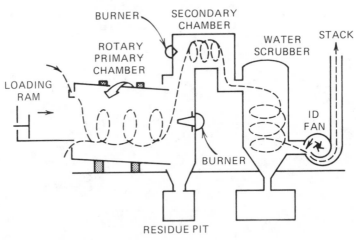

Fig. 16.6 Rotary kiln.

A critique of design, operating, and maintenance characteristics of small-scale incinerator units in any application are discussed in detail in Chapter 13, and in Reference 5.

16.4 ENERGY RECOVERY METHODS

16.4.1 Background

Although refuse-fired municipal district heating plants were in operation at the turn of the century, energy recovery from small excess air incinerators was not widely practiced due to unfavorable economics. When this type of incinerator reached its highest level of development before the enormous price increases of fluid fuels in the 1970s, the installed cost of a waste energy recovery

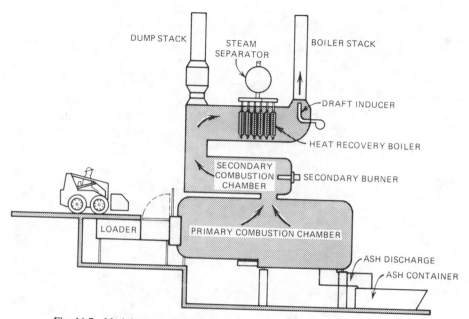

Fig. 16.7 Modular controlled air design with gas duct heat exchanger boiler.

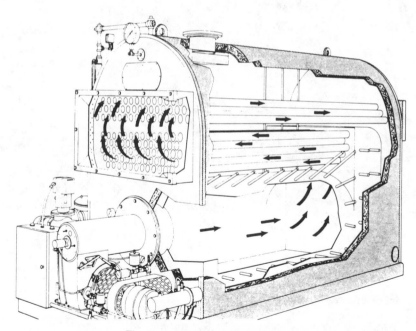

Fig. 16.8 Firetube waste heat boiler.

Fig. 16.9 Watertube waste heat boiler.

boiler was high compared with small packaged gas or oil-fired units. The owning and operating costs were low enough to preclude utilization of "waste heat" energy recovery in most cases.

Today, it is virtually an economic necessity to reclaim the energy in high-temperature waste gases with heat recovery boilers.

16.4.2 Utilization Choices: Steam, Hot Water, Hot Air, KW

Generation of steam is usually the most effective method of recovery of heat from incinerators. Less transfer medium is needed for steam compared with hot water because of the latent heat associated with phase change. Also, equipment, piping, and so on are readily available and compatible with most plant distribution systems. Unlike district heating or institutional uses (hospitals and so on), the potential for hot water systems is far less; likewise, requirements for hot air.

Generation of electricity is an attractive adjunct and viable in small- to medium-sized plants via cogeneration discussed in Chapters 12 and 13.

Steam can be generated in heat recovery boilers at nearly any pressure normally found in industrial plants. If possible, it is usually best to generate steam at the plant distribution pressure and feed the main piping since a constant heat sink is advantageous. If the main steam distribution system does not have sufficient summer requirements, or if the boiler supplies a branch low-pressure system for winter heating only or intermittent summer use, the boiler will be off the line part of the time and the economic advantages of the concept will be limited.

16.4.3 Boiler Types

Firetube Waste Heat Boilers

A compelling reason that firetube boilers (Figure 16.8) are frequently used with smaller and medium-sized industrial incinerators (2000 lb/hr, 910 kg/hr) is that they are less costly in smaller sizes than watertube boilers because of their simpler construction. From a design standpoint, the horizontal firetube boilers normally used with incinerators have several basic differences compared with watertube boilers.

Construction consists of a horizontal tube bundle supported at each end by a tube sheet. The bundle and tube sheets are enclosed in a cylinder or shell. Hot combustion gases flow through the inside of the tubes in a one- or two-pass configuration. Steam is generated outside of the tubes and rises to the upper portion of the horizontal shell. Due to the limited space within the shell, steam accumulation and purity can be less than in the watertube boilers, in which the drum or separator can be of larger size. Firetube boilers can be constructed, however, with a separate steam drum. The water circulation pattern in the firetube boiler is also not defined as well as a watertube boiler, in which the water is directed through the tubes. The lack of a circulation pattern can create conditions in which steam is not carried away rapidly enough in some locations within the shell causing blanketing of the outside of the tubes by the steam film. If the steam is not carried away rapidly enough during periods of high heat flux, or high rate of heat input to the tube surface, the tube can overheat and fail because the film prevents water from cooling it. For these reasons, firetube boilers are generally designed for lower rates of heat flux than watertube boilers. Properly installed and applied, however, firetube boilers are a satisfactory energy recovery method for smaller incinerators. The minimum controls should include means for monitoring and controlling maximum and minimum combustion gas temperatures entering the boiler, steam pressure demand, an induced draft fan damper, and controls for regulation of boiler draft with incinerator secondary chamber pressure, modulating feedwater level control, and water level alarms.

Watertube Waste Heat Boilers

Depending upon the economies of size and the type of installation, watertube heat recovery (Figure 16.9) may be required. Watertube heat recovery can be:

A heat exchanger in a hot flue gas duct (Figure 16.7).

A separate watertube boiler with auxiliary firing option.

A waterwall furnace.

The surface area of the tubes exposed to the hot gases is more effective than in the case of the firetube boiler and the tubes can be finned for greater efficiency. Soot blowers are usually included with watertube boilers for incinerators. Finned tubes reduce the surface area required and the overall size of heat exchangers, but are considerably more susceptible to fouling and are not easy to clean.

Waterwall Furnace Units

Waterwall furnace boiler units, Figure 16.10, have had a long and successful operating history with fossil fuels and they have been mass-fired with municipal refuse in European incinerators with mixed results for many years. Although smaller-sized industrial units have been available during the past 20 or so years, a major advantage of waterwalls in the primary chamber of an incinerator is that the heated surface absorbs more heat from radiation than from convection. Higher efficiencies are possible with a waterwall and convection combination, but combined maintenance may render it a marginal advantage. A disadvantage of the waterwall mass burn furnace is that the entire unit must come off the line if either the boiler or stoker is inoperative and requires greater plant redundancy.

Further and detailed comparison of waterwall furnace designs with the combination of refractory furnace and waterwall convection package is discussed in Chapter 13.

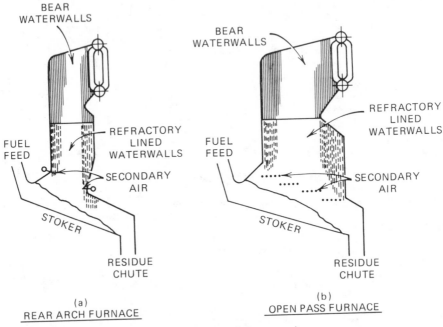

Fig. 16.10 Waterwall furnace units.

16.5 OPERATING AND MAINTENANCE FACTORS

16.5.1 Waterside Tube Failure

Waterside tube failures are not uncommon in watertube incinerators and they occur for a variety of reasons. One reason is improper feedwater treatment which allows scale to build up, thus insulating the inside of the tube from the water, and failure results from overheating. Other problems leading to tube failures relate to temperature excursions above the designed heat flux and boiler feedwater distribution. Boiler controls should include a high-temperature limit cutoff to help prevent over-heating and the means to take the boiler off-line if necessary.

16.5.2 Fireside Tube Wastage

Fireside tube wastage is likewise not uncommon in refuse-fired boilers. Fireside corrosion and deposits are probably the most vexing operating/maintenance problem in the entire concept of incineration of solid wastes along with lower furnace grate warping, cracking, plugging, wear, and replacement.

Industrial wastes have the advantage of a lower propensity for these continuing problems because the raw material contains fewer precursors of such problems. Chlorinated plastics, glass, sand/grit, and the like are examples. Tube wastage from acid gas, slagging from glass, and low-melting-point materials is usually less than with municipal refuse mixtures.

Medium temperature hot water and low pressure steam applications are usually more compat-ible with industrial wastes without chlorine compounds where there is little danger of low-temperature (dewpoint) fireside chloride corrosion of tubes, and life is easier. Similarly, the upper limit of steam temperature must avoid liquid-phase fireside tube corrosion, and corrosion by other compounds present or formed from various compounds if supplemental municipal refuse is involved.

16.5.3 Refractory Linings

Such linings tend to deteriorate somewhat frequently due not only to periodic abrasion, but mainly to erratic temperature control and excessive cycling between reducing/oxidizing atmospheres and occasional dew point acid corrosion.

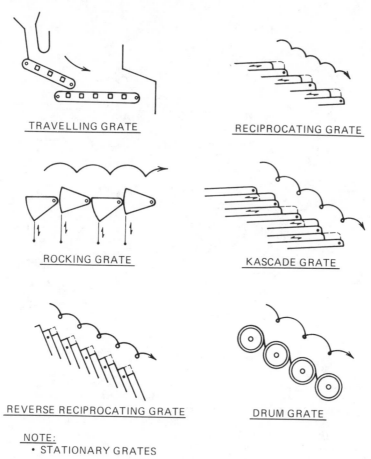

TRAVELLING GRATE

RECIPROCATING GRATE

ROCKING GRATE

KASCADE GRATE

REVERSE RECIPROCATING GRATE

DRUM GRATE

NOTE:
• STATIONARY GRATES

Fig. 16.11 Stoker types.

16.5.4 Stokers

Stokers are usually reciprocating or vibratory types, Figure 16.11. Insufficient agitation/mixing of fuel and air contributes to the reducing/oxidizing cycling pattern, with consistently effective control of air sometimes unattainable. Grate warping can be a result of these exigencies along with abrasive wear from glass/grit siftings and more difficult emissions control.

16.5.5 Ram Feed

Ram feed, Figure 16.12, can experience the same air control and mixing difficulty with similar reducing/oxidizing cycling, temperature excursions, and jamming. A tendency to compact the fuel bed contributes to this.

16.5.6 Ash Removal

Ash removal systems for continuous ash discharge have always required significant attention and maintenance due to jamming and stalling, with frequent replacement of damaged and worn parts from abrasion of competitively priced components often unsuitable for the task.

Regardless of the incinerator size (Chapters 12 and 13) or its application to either industrial or municipal waste, the following comments also apply.

Although dry ash systems are simpler and considerably less expensive with lower operating and maintenance costs than wet (quench) types, they usually create excessive and pervasive dust that is difficult to contain with sprays and conditioners. It aggravates materials handling and housekeeping and requires inordinate attention. However, dry and hauling costs are considerably lower.

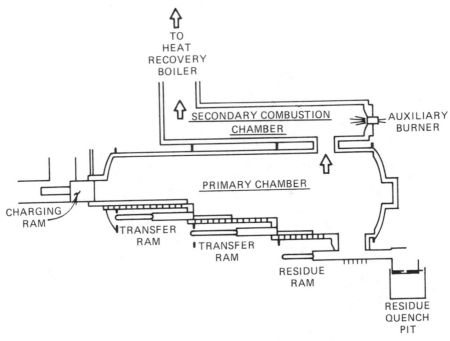

Fig. 16.12 Ram-type furnace feed.

Wet systems are significantly more expensive but are somewhat "cleaner" in operation with housekeeping usually confined to water and wet fines spills and leaks. Operating, maintenance and ash hauling costs are significantly higher however.

Common to any ash handling concept is the growing requirement for mixing the various size gradations of ash before disposal. These are the coarser bottom ash (a plurality), and the collected fly ash and grate siftings. This is because the fly ash and siftings usually have higher concentrations of heavy metals (lead, cadium, zinc). By mixing, there is a diluting reduction of the concentrations, and it appears that mixing may become a universal requirement in the continuous effort to minimize any deleterious effects in land disposal of incinerator ash.

16.5.7 Feedwater Treatment

All boilers need feedwater treatment. Tube failures have occurred at many incinerator installations because treatment was incorrect, allowing excessive scaling. Scaling causes overheating of boiler tubes and subsequent failure because the heating surface is insulated on the cold side and is insufficiently cooled by the boiler water.

The method of feedwater treatment should meet the criteria of the boiler manufacturer but should be determined by a water consultant on the basis of the analysis of the water supply. The water consultant should also recommend a frequency of blowdown.

16.6 INDUSTRIAL SOLID WASTE INCINERATION

16.6.1 Concept Choices

Mass-burn units are usually more suitable for smaller on-site applications ($\leq$100 TPD) when oversize items and dunnage are minimal and size reduction is not necessary or justifiable. Small mass-burn units are much more commonplace.

For larger and load-leading systems, fuel preparation has the advantages of accepting larger-sized combustible material with faster response to load changes via firing rate heat release control, and is useful as a base load fuel supplement to coal or oil.

Mass Burn

Mass burning is a term that describes the process of burning materials as-received on a hearth or grates by feeding the materials to the incinerator in unprocessed bulk form, with sorting out of any

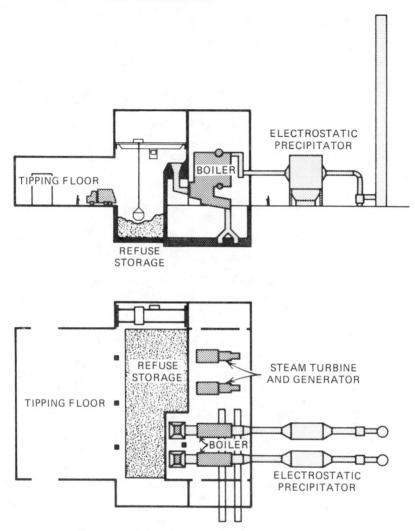

Fig. 16.13   Refuse-fired steam electric generating facility.

troublesome materials such as oversize or unburnable items but with no preparation before burning. However, a number of industrial installations have installed equipment for crushing oversize and cumbersome items to permit feeding and burning with the other material and to improve ash-handling operations. Historically and at present, mass burn in smaller size ranges continues to be most common for industrial on-site incineration.

Recent advances in the state-of-the-art for excess-air waterwall furnaces and in waste heat boiler technology are popularizing these designs for larger industrial and municipal refuse-fired installations. Individual furnace units have been designed for 1000 TPD and with continuous operation for generation of steam and electric power, as shown by Figure 16.13. A typical plant includes a tipping floor, storage pit and crane, continuous feed charging chutes, improved grate systems and hardware, residue conveyers, ash-processing systems, and air pollution controls. Chapter 13 discusses mass-burn technology design factors in detail with comparisons of the various options.

Prepared Fuel

Preparation of refuse as an industrial boiler fuel is an outgrowth of the utilization of by-product fuels in certain industries, Figure 16.14. Partial suspension firing (spreader stoker) of moderately

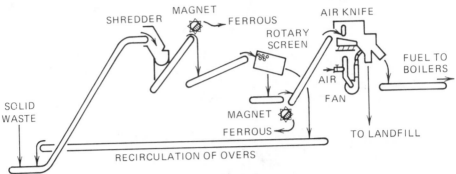

Fig. 16.14 A large prepared fuel energy recovery system.

prepared bagasse (sugarcane waste) and forest industry waste material from pulp (wood bark and the like) and paper mills has been very successful for years, and it was a logical progression to apply the same techniques for burning cellulosic refuse (paper, wood) in municipal refuse.

As air pollution regulations became progressively more restrictive, large industrial plants producing over 100 TPD of refuse found it necessary to modify incinerators with improved air pollution controls or switch to an alternate disposal method. As a result, many large and older units were abandoned. Eventually, burgeoning landfill costs and constraints, along with exhorbitant fuel costs, led to renewed interest in prepared waste fuels with energy and materials recovery.

The purpose of prepared waste fuels is to homogenize mixed raw material via established minerals processing techniques to obtain more uniform fuel, optimize combustion control and efficiency with superior burnout, and obtain reduced ash quantities.

Despite some early failures, several successful prepared fuel (refused-derived fuel, RDF) plants have been operating for the past several years, including the world's largest mixed refuse-fired steam-electric plant in Miami, Florida; the Eastman Kodak unit firing RDF and sludge in full suspension at Kodak Park in Rochester, New York; and the combination RDF and coal spreader stoker-fired unit at General Motors Truck/Coach Division in Pontiac, Michigan (see Case Histories).

Contrary to a not uncommon misconception, RDF process system maintenance must now be considered routine—likewise, the boiler/stoker combinations which are considerably less troublesome than the more complicated mass burn lower furnace/stoker/boiler combinations. As a result, overall maintenance and operating costs are no more than equivalent, with routine downtime and high capacities for RDF systems. RDF storage permits process equipment maintenance during full boiler capacity on-line operation without interruption.

The commonalities in small-scale (≦100 TPD) industrial incinerators and small-scale incinerator systems for municipal refuse (≦500 TPD) are significant, and two comprehensive studies including detailed case histories are in References 5 and 6.

16.7 INDUSTRY AS THE PURCHASER OF REFUSE ENERGY

In addition to on-site energy recovery by industry from its own waste, cooperative ventures whereby a municipality (or a consortium of towns) generates energy or RDF from waste for sale to a nearby industry have been an outgrowth of the energy-from-waste trend. Tables 16.1 and 16.2 list some of these for both "excess air" and "starved air" small-scale systems (Reference 5, Argonne National Laboratory).

Larger installations upwards of 200 TPD of similar cooperative arrangement are also operating, such as Refuse Fuels, Inc., Haverhill, Massachusetts; Pittsfield, Massachusetts; and Saugus, Massachusetts.

The reverse arrangement also exists at Niagara Falls, New York, where the Hooker Chemical waste-to-energy project receives municipal waste from surrounding areas for use (along with its own waste) in generating steam and electricity for plant use.

16.8 INDUSTRIAL COGENERATION

16.8.1 Background

At present, there is a renewed interest in the technology of combined generation of power and heat recovery called cogeneration. For many years, when numerous coal-fired boilers were operating in

Location	Number of Modules	Capacity (TPD), Each Module	Date of Start-up, Past or Projected	Capital Cost ($10^6)	System Vendor[a]	Energy Market
Norfolk, Virginia	2	180	5/67	2.2	A&E	Naval station
Braintree, Massachusetts	2	200	3/71	2.85	Riley	Weymouth Art & Leather Co.
Portsmouth, Virginia	2	80	1976	4.5	A&E	Naval shipyard
Waukesha, Wisconsin[b]	2	60	1971	1.7	A&E	Unknown
Hampton, Virginia[c]	2	100	1980	10.3	J.M. Kenith Co.	NASA-Langley
Pittsfield, Massachusetts[c]	2	120	7/81	1.0	Vicon/Enercon	Crane Paper
Gallatin, Tennessee[c]	2	100	3/82	9.8	O'Connor	R.R. Donnelly, TVA, others
Prudhoe Bay, Arkansas	1	120	1982	Unknown	Basic Env. Eng.	Unknown
Collegeville, Minnesota[c]	1	65	1982	2.4	Basic Env. Eng.	St. John's University
Harrisonburg, Virginia	2	50	1982	Unknown	A&E	
Glen Cove, New York (codisposal)	1	200 (MSW) 10 (sludge)	1982	22 + 12	A&E	City of Glen Cove
Bannock County, Idaho	2	100	1983–84	9	Clark-Kenith Co.	FMC, Idaho Power
Lassen County, California	1	96	1984	7.15	Bruun & Sorrenson	Lassen Community College
Fayettesville, Arkansas	1	150	1984–85	Unknown	Brunn & Sorrenson	University of Arkansas
Savannah, Georgia	2	220	1984–85	28	Katy-Seghers	American Cyanamid
Pasagoula, Mississippi	1	150	1984–85	5.9	Sigoure Freres	Thiokol
Islip, New York	2	255	1984–85	30–40	O'Connor-Penn. Eng.	Long Island Lighting Company
Davis County, Utah	2	250	1984–85	33	Katy-Seghers	Hill AFB
Rutland, Vermont	2	120	1984–85	11	Vicon	Central Vt. Public Service
Dutchess County, New York	2	200	1984–85	Unknown	O'Connor-Penn. Eng.	IBM, utility
Burlington, Vermont	2	60	1984–85	8	A&E	University of Vermont
Claremont, New Hampshire	1	200	1984–85	18	Clark-Kenith	Central Vt. Public Service
New Hanover County, North Carolina	2	100	1984–85	13.84	Clark-Kenith	W.R. Grace Plus Elec
Washington County, New York	2	200	1984–85	Unknown	Vicon	CIBA-GEIGY
Delaware County, Pennsylvania	1	50	1984–85	2	N/A	County Geriatric Center
Tri Cities, California	4	120	1984–85	25	Vicon	

[a] A&E indicates that the system was designed by an architectural and engineering firm and the equipment provided by various vendors.
[b] Retrofitted in 1979 (for $10^6 · 3.9).
[c] Case studies of these installations appear in Reference 5.

Table 16.2 Selected Data on Small-Scale U.S. Systems Using the Starved Air Design, as of July 1983

Location	Number of Modules	Capacity (TPD), Each Module	Date of Start-up, Past or Projected	Capital Cost (10^6)	System Vendor	Energy Market
Siloam Springs, Arkansas[a]	2	10.5	6/75	Unknown	Consumat	Allen Canning Co.
Blytheville, Arkansas[a]	4	12.5	8/75	Unknown	Consumat	Chrome Plating Co.
Groveton, New Hampshire	2	12	Unknown	Unknown	ECP	Groveton Paper Mill
North Little Rock, Arkansas[b]	4	25	8/77	1.45	Consumat	Koppers
Salem, Virginia	4	25	9/78	1.9	Consumat	Mohawk Rubber
Jacksonville, Florida[c]	1	48	1978	Unknown	SEE	Unknown
Osceola, Arkansas	2	25	1/80	1.1	Consumat	Crompton Millsa
Genesee, Michigan	2	50	2/80	2	Consumat	Unknown
Durham, New Hampshire	3	36	9/80	3.3	Consumat	University of New Hampshire
Auburn, Maine	4	50	4/81	3.97	Consumat	Pioneer Plastics
Dyersburg, Tennessee	2	50	8/81	2	Consumat	Colonial Rubber
Windham, Connecticut	4	25	8/81	4.125	Consumat	Kendall Co.
Crossville, Tennessee	2	30	12/81	1.11	Env. Services Corp.	Crossville Rubber
Cassia County, Idaho	2	25	1982	1.5	Consumat	J.R. Simplot
Batesville, Arkansas	1	50	1982	1.2	Consumat	General Tire & Rubber
Park County, Montana	2	36	1982	2.321	Consumat	Yellowstone Park
Waxahachie, Texas[d]	2	25	1982	2.1	Synergy/Clear Air[d]	International Aluminum Co.
Miami Airport, Florida	2	30	1982	Unknown	Synergy/Clear Air[d]	Miami Airport
Portsmouth, New Hampshire	4	50	1982	6.25	Consumat	Pease Air Force Base
Red Wing, Minnesota	2	36	1983	Unknown	Consumat	S.B. Foote Tanning
Cattaraugus County, New York[d]	3	37.5	1983	5.6	Synergy/Clear Air[d]	Cuba Cheese
Miami, Oklahoma	3	36	1983	3.14	Consumat	B.F. Goodrich
Oswego County, New York	4	50	1983	Unknown	Consumat	Armstrong Cork
Pasagoula, Mississippi	3	50	1983–84	6+	Consumat	Unknown
Oneida County, New York	4	50	1983–84	Unknown	Unknown	Griffis AFB
Tuscaloosa, Alabama	1	240	1984	13	Consumat	B.F. Goodrich
Fitchburg, Massachusetts	1	200	1984	Unknown	Consumat	Paper companies
Fergus Falls, Minnesota	3	29	1984	3	Control	Mid-America Dairymen, Inc.

[a] System now shut down and equipment removed.
[b] System now shut down and equipment removed, but Consumat is to supply new equipment.
[c] System now shut down.

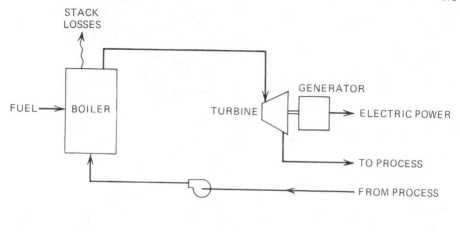

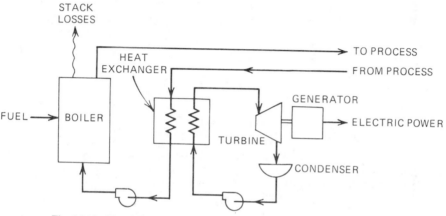

Fig. 16.15 Topping (back pressure) and condensing cogeneration cycles.

industrial plants, it was not uncommon for the powerhouse to generate superheated steam at 600 psi (41 bar) to drive turbines for power generation and to use steam at a lower pressure for plant process or heating. The economics of cogeneration became unfavorable at some plants when standby or "dump" gas fuel came into general use, but some energy-intensive industries such as pulp and paper continued the practice. The present renewed interest is the result of exhorbitant fossil fuel prices, incentives in the form of tax credits, accelerated depreciation for equipment cost and a guarantee (PURPA) of "fair" market price for electricity sold back to local utilities (see Chapter 2 and Chapter 15, Section 15.1).

16.8.2 Technology and Systems

The advantages of cogeneration, however, are not clear-cut and the identification of technical, economic, and legal problems has been the object of much study. From a technical standpoint, there are two basic methods of producing electric energy or shaft horsepower from a refuse-fired incinerator or boiler, Figure 16.15. Using a topping cycle as one example, steam would be produced by the boiler at a high pressure, generate power at a turbine, and exhaust to a lower pressure for process or plant heating. Pressure options are varied with a popular upper limit of about 600 psi (41 bar) and a temperature of 700°F (371°C), which requires a superheater. If the system backpressure is 150 psi (10 bar), reflecting the distribution pressure of many plants, a turbine would require about 55 lb of steam/kwh (25kg/kwh). Two incinerator boilers of 2000 lb/hr (910 kg/hr) of refuse would produce about 14,000 lb (6400 kg) of steam per hour resulting in generation of 250 kwh. For a straight condensing cycle, steam consumption could be as low as about 20 lb/kwh (9 kg/kwh), but capital and operating costs would be greater.

Small (modular) incinerators, however, are not known for consistent steam flow while turbines operate best at steady-state conditions. One possibility is to utilize the steam-generating capability of incinerator heat recovery with a steam engine. Steam engines are available today with direct connected generators and are designed for industrial use. The modern engine is low-speed, multi-cylinder, and does not require superheated steam or pressure as uniform as required by turbines.

Engines are somewhat more efficient at lower pressures than turbines and at an inlet pressure of 600 psi (41 bar) saturated and 150 psi (10 bar) out, the engine would require only about 40 lb (18 kg) of steam per kilowatt as opposed to 55 lb (25 kg) for a turbine. Inlet pressures as low as 150 psi and backpressures of 12 psi (.80 bar) to vacuum are feasible. The disadvantage of an engine is that it is more costly. At present pricing, however, an engine or turbine would probably have a more favorable return on investment than the incinerator plant without a generator.

The costs of turbines and engines involve comparisons of equipment pricing for various combinations of boiler and piping pressure ratings, exhaust steam pressure and availability, generators of synchronous or induction design, switchgear, and power connections. Chapter 13 also discusses cogeneration mainly for municipal refuse energy recovery.

16.8.3   Regulatory Factors

The decision with regard to cogeneration, nonetheless, is essentially one of economics rather than of technology, and the renewed interest in the economies of cogeneration is related to the national energy strategy and legislation.

As a result of the energy crisis in the 1970s, Congress enacted the National Energy Act of 1978. This Act has five constituent parts:

1. Power Plant and Industrial Fuel Use Act (Pub. L. 95-620).
2. Energy Tax Act (Pub. L. 95-618).
3. Public Utility Regulatory Policies Act (PURPA) (Pub. L. 95-617).
4. National Gas Policy Act (Pub. L. 95-621).
5. National Energy Conservation Policy Act (Pub. L. 95-619).

Cogeneration systems, including gas turbines and diesel engines intended to produce process heat and electric power, are governed by 1 through 5 above, the Crude Oil Windfall Profit Tax Act (COWPTA), and the depreciation and investment credits of the Economic Recovery Tax Act of 1981 (ERTA). The provisions of all of the above legislation were intended to reduce emphasis on scarce fuels (oil and gas) and stimulate investment and use of alternative sources of energy.

Incinerator and refuse-fired steam-producing boilers normally would not be governed by item 1, which covers individual units of 100 million BTU/hr (1.05 MJ/hr) and greater using oil or natural gas.

One of the most important sections of the National Energy Act is item 3, the Public Utility Regulatory Policies Act (PURPA). This legislation requires that electric utilities purchase electricity from qualified cogenerators at rates reflecting the utilities' avoided costs. Section 210 of PURPA requires the Federal Energy Regulating Commission (FERC) to establish policies to encourage use of cogeneration. FERC finalized its standards for cogenerators on March 20, 1980, after public hearings. These standards were then to be implemented by each state regulatory commission for regulated utilities and by each nonregulated utility. In the state of Illinois, the area served by Commonwealth Edison is governed by Rider 4 (Parallel Operation of Customer's Qualifying Generating Facilities effective September 7, 1981), which provides for cogeneration compensation and charges.

Firms seeking to employ the cogeneration option should consider not only the services of a consulting engineer/specialist, but also sufficient legal assistance to determine which regulations and tax credits apply. It is also necessary to keep current because an administration may change certain existing provisions or propose new legislation. Chapter 2 and Chapter 15 discuss regulatory factors in detail.

16.8.4   Economic Factors

The collection, processing, and disposal of liquid and solid wastes from manufacturing operations has generally been a costly process and most firms have devoted considerable time investigating means to reduce this cost.

The low cost of energy in the form of gas, oil, and coal in the past has discouraged the development of systems to recover energy from refuse in North America. Historically, plant engineers have always considered methods to recover lost heat only to be dissuaded by the higher capitalized cost of recovery versus low costs of fossil fuels.

Times have changed, however, and methods are available to recover energy from industrial waste with a return on investment attractive to most managers.

The capital cost can vary considerably depending on the installation requirements of the facility. In some cases, an incinerater with a boiler can be installed in an existing building with modifications, or a new building may be required. A common and important index for approximate cost comparisons is dollars per ton per 24-hr day as applied to municipal incineraters. In that case, the incinerater is given a 24-hr continuous rating whether or not the unit is actually in operation for that period of time. For example, a 1 ton/hr unit would have a rating of 24 TPD. On that basis, incineration equipment with heat recovery can cost from $45,000/ton-day to $80,000/ton-day if electric generation is included and if installed at a prepared site. A complete installation including a new building with the incinerater equipment and necessary auxiliaries can cost approximately $70,000 to $90,000 per ton/day based on the requirements for two 2000 lb/hr (910 kg/hr) units. These are plant and equipment costs only.

16.8.5 Operation and Maintenance Cost Factors

The two most important operating cost savings from incineration of plant waste, both solid and liquid, are: (1) lower hauling and disposal cost and (2) the value of the recovered steam, hot water, or warm air. The cost, installation, and maintenance of incineration and boiler equipment is of such magnitude today that incineration only without recovered energy does not show sufficient return on investment to be worthwhile where existing land disposal of raw solid wastes is not in crisis and such costs are not prohibitive.

Normally, the methods of in-house collection and handling of plant waste would not be greatly influenced by the installation of a new central system for energy conversion. The only new cost item that may be necessary is the use of a truck to transport refuse containers from various plant areas to the waste processing center if that service is now being performed by an outside contractor.

The calculation of the value of recovered steam is largely the determination of the cost of boiler fuel that is to be replaced by the solid and liquid wastes. This cost for fuel is only a portion of the total cost of steam production at an industrial plant. Other costs such as boiler plant operating and maintenance personnel and boiler repair would normally continue and not be influenced significantly if a waste/energy system is placed in operation.

Operating expenses that must be subtracted from the above cost savings will include labor, material handling equipment, auxiliary fuel, purchased power and demand charge, maintenance, and ash hauling, with total operating labor about two men per ton/hr burning rate.

Auxiliary fuel requirements would depend upon the availability of waste oils from manufacturing and the type of incineration/boiler equipment. Where an ample supply of waste oil is available, only a standby oil system would be required plus a gas supply for pilot ignition of the burners.

Maintenance cost can be estimated at 2 to 5% of total capital cost depending upon grate, refractory, boiler repairs, preprocessing, and so on. By volume, residual ash will be 5 to 10% of the raw waste infeed and can vary from 25 to 50% by weight (if quenched).

16.8.6 Operating Cost Summary

The following, including tables and curves, is from a U.S. EPA Compendium of Technical, Environmental and Economic Evaluations of Small Incinerator Systems with Heat Recovery.[6]

This evaluation of the net operating costs of operational industrial and municipal incinerators with heat recovery occurred when the units were not operating at optimum conditions and have been extrapolated to optimum conditions. To determine the net operating costs of the industrial facilities, the assumptions made are shown by Table 16.3.

The resulting computational data are summarized in Figures 16.16 and 16.17 where the curves A through F represent possible operational modes. In the development of these figures, it was assumed that the refuse would be generated only 5 days/week. The 7-day operational mode is burning a 5-day/week refuse generation over 7 days of burning.

At 100% of rated capacity, the net operating costs for 15 of 21 shifts per week in municipal systems are nearly the same. As seen in Figure 16.17, the net operating cost per unit of refuse feed rate is $9/ton ($10/Mg) or less for the capacity range between 50 and 100 tons/5 days (45 and 90 Mg/5 days).

With refuse feed rates in the range of 25 tons/5 days (27.5 Mg/5 days) and above, industrial facilities will yield a positive balance, or revenue, in the net operating cost computation when they operate at 100% of rated capacity but an actual cost in the net operating cost computation when they operate at 50% of rated capacity. This cost must be compared with the cost of alternative waste disposal methods and fuel sources to determine the economic feasibility of a proposed system.

Table 16.3   Projected Optimum Operating and Maintenance Costs, Industrial Plants in the 25 TPD (27.5 Mg/day) Range, 1978 Dollars

1. Average annual salary, including benefits	$20,800
2. Auxiliary fuel: Natural gas	$2.50/Mcf ($.088/m³)
3. Electric power	$.035/kwh
4. Water	$.91/1000 gal. ($.24/m³)
5. Volume ratio: Wet ash residue/raw refuse feed	0.10
6. Residue disposal cost	$4.40/ton ($4.00/Mg)
7. Interest rate	12%
8. Depreciation	7 yr
9. Heat value, raw refuse	7500 BTU/lb (7.91 mJ/kg)
10. Recovered energy value	$3.28/M BTU ($.0031/mJ)

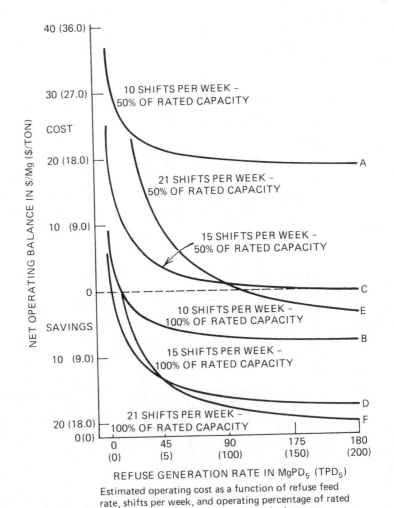

Estimated operating cost as a function of refuse feed rate, shifts per week, and operating percentage of rated capacity for industrial small modular incinerators.

Fig. 16.16   Estimated operating costs versus daily tons, small modular industrial energy recovery incinerators.

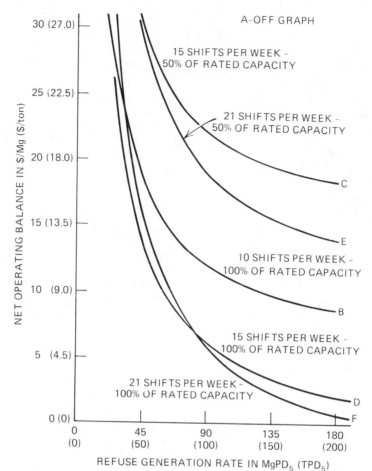

Estimated operating cost as a function of refuse feed rate,
shifts per week, and operating percentage of rated capacity
for municipal small modular incinerators.

Fig. 16.17 Estimated operating costs versus daily tons, small modular municipal energy recovery incinerators.

16.9 CONCLUSIONS

Many industrial plants generate sufficient quantities of solid and liquid wastes to recover energy in the form of steam, hot water, or air to show an adequate return on investment. An inventory of the various classifications of plant wastes and their rate of generation is the first step in the feasibility study of a heat recovery system.

Processing and heat recovery equipment is available in various types and can be obtained in packaged systems meeting a broad range in design and capacity.[5] The PURPA legislation has made industrial cogeneration a strong incentive for waste heat energy recovery.

A successful energy recovery installation requires careful equipment selection based upon performance and reliability "track records" and design of an adequate and sensible building or enclosure for raw material receiving and storage, furnace charging, ash handling, feedwater treatment, and utility services.

A list of manufacturers of packaged and modular incinerator/heat recovery systems is shown in Table 16.4.

Table 16.4   Representative Municipal Waste-To-Energy System Vendors

Mass-Burning Starved Air

Burn-Zol
P.O. Box 109
Dover, New Jersey 07081
(201) 361-5900

Clean Air, Inc.
226 W. 20th Street
Ogden, Utah 84844
(801) 399-3710

Comtro Division
Sunbeam Inc.
180 Mercer Street
Meadville, Pennsylvania 16335
(814) 724-1456

Consumat Systems, Inc.
P.O. Box 9574
Richmond, Virginia 23228
(804) 764-4120

Econo-Therm
1132 K-Tel Drive
Minnetonka, Minnesota 55343
(612) 938-3100

Environmental Control Products
P.O. Box 15753
Charlotte, North Carolina 28210
(704) 588-1620

Kelley Co., Inc.
6720 N. Tentonia Avenue
Milwaukee, Wisconsin 53209
(414) 352-1000

Lamb-Cargate
P.O. Box 440
1135 Queens Avenue
New Westminster, British Columbia, Canada
(604) 521-8821

Scientific Energy Engineering, Inc.
1103 Blackstone Building
Jacksonville, Florida 32202
(904) 632-2102 or 399-1111

Simonds Co.
P.O. Box 1443
Auburndale, Florida 33823
(813) 967-8566

Synergy Systems Corp.
1211 S.W. 10th Street
Boca Raton, Florida 33432
(305) 391-6850

U.S. Smelting Furnace Co.
P.O. Box 217
Belleville, Illinois 60222
(618) 233-0129

Washburn and Granger
85 5th Avenue
Paterson, New Jersey 07075
(201) 278-1965

Mass-Burning Excess Air

Babcock and Wilcox
P.O. Box 2423
N. Canton, Ohio 44726
(261) 494-7610

Basic Environmental Engineering, Inc.
21W161 Hill Street
Glen Ellyn, Illinois 60137
(312) 469-5340

Browning-Ferris Industries (VKW System)
Energy Systems Division
Fanin Bank Building
P.O. Box 3151
Houston, Texas 77001
(713) 870-8100

Combustion Engineering, Inc.
900 Long Ridge Road
Stamford, Connecticut 06902
(203) 329-8771 or 688-1911

Clark-Kenith, Inc.
2200 Century Parkway
Suite 300
Atlanta, Georgia 30345
(404) 329-1441

Detroit Stoker Co.
1510 East First Street
Monroe, Michigan 48161
(313) 241-9500

Katy Seghers Incino Systems
   (Seghers System)
Fulton Iron Works Co.
3844 Walsh Street
St. Louis, Missouri 63116
(314) 752-2400

International Incinerators, Inc.
P.O. Box 19
Columbus, Georgia 31902
(404) 327-5475

Morse Boulger
1431 Ford Rd.
Cornwells Heights, Pennsylvania 19020
(215) 638-2700

O'Connor Combustor Corp.
107 Music City Circle, Suite 203
Nashville, Tennessee 37372
(615) 883-0078

Ogden Martin Systems, Inc.
140 E. Ridgewood Ave.
Paramus, New Jersey 07652
(201) 599-2400

Signal RESCO
Liberty Lane
Hampton, New Hampshire 03842
(603) 926-5911

Vicon Recovery Systems, Inc.
10 Park Plaza, P.O. Box 100
Butler, New Jersey 07405
(201) 492-1776

Waste Management, Inc. (Volund System)
3003 Butterfield Road
Oak Brook, Illinois 60521
(312) 654-8800

APPENDIX 16.1 TWO 200 TPD COMPOSITE PLANT DESIGNS FOR A STARVED AIR SYSTEM AND FOR AN EXCESS AIR SYSTEM

Editor's Note: The material in this appendix is excerpted from R. Hopper, *Thermal Systems for Conversion of Municipal Solid Waste,* Volume 3, Argonne National Laboratories, July, 1983.

Factors in Estimating the Costs of a Proposed Facility

To assist in estimating the capital and operating costs of a proposed facility, two 200 TPD Composite Plant Designs for a Starved Air System and for an Excess Air System are presented as follows, with exceptions in the small-scale range, under say 200 TPD. It is interesting that the capital cost of excess air systems has generally remained higher than that of starved air systems. (Personal communications with plant personnel where waste-to-energy systems are installed as of July 1983.) To a great extent, this is because: (a) excess air systems tend to use a pit-and-crane approach to waste receiving and charging, as opposed to a less capital-intensive approach of a receiving floor and front-end loader; (2) starved air systems tend to use less expensive boilers; and (3) as previously mentioned, excess air systems may require larger and more expensive air pollution control equipment.

On the other hand, starved air systems have generally had higher operating costs. The starved air approach usually requires a receiving floor and a front-end loader, resulting in constant tire maintenance and upkeep. If fin-tube boilers are employed, the system must be shut down weekly for cleaning of the boiler tubes which in turn requires the weekly use of auxiliary fuel for shutdown and start-up. The throughput of starved air systems is typically 70 to 75% of design capacity, compared with 80 to 85% for most excess air systems.

Whether or not the typically lower O & M cost of a traditional excess air system makes up for its typically higher capital cost is something that can be determined only on a project basis. That assessment depends greatly on such factors as the reliability needed for a project, the quality of energy to be produced, and the type of financing.

Probably the single most significant factor in any economic comparison of systems and design approaches is the long-term cost of equipment repair and/or replacement. This cost category is significant because all the systems generally have similar O & M costs.

Not only should a project stay within its budget in procuring a resource-recovery system, it will also need to consider the system's long-term reliability and cost of operations. It must decide whether or not spending additional money for design and construction will lead to significant cost savings over the life of the facility, and if the need for system reliability will be adequately met.

There is very little operating experience to date among small-scale waste-to-energy systems to help determine such trade-offs. Also, most of the small-scale systems installed to date have had to compete with inexpensive landfilling and/or have been procured from the lowest bidder (that is, on the basis of capital cost only), and thus have not necessarily been designed to achieve the best long-term "life-cycle" costs. In this regard, most of the small-scale systems in the United States appear to have absorbed essentially equivalent costs for long-term equipment repair and replacement that is, there do not appear to be any distinct repair/replacement cost differences among European-derived or traditionally styled, excess air systems, the newer, "unconventional" excess air systems, and the starved air systems.

To assist in estimating the capital and operating costs of a proposed facility, Table 16.5 summarizes cost estimates for two hypothetical 200 TPD (design capacity) facilities: an excess air system and a similar starved air system. Construction is assumed to begin in January, 1983, and commercial operation to begin in January, 1984 (allowing for a 2-year construction period). Many of the cost categories are project specific and have considerable variations between projects.

CASE HISTORIES

Case History 1: General Motors Corporation, Truck and Bus Group, Pontiac Central Manufacturing and Assembly Plant

This case history is presented through the courtesy of the General Motors Corporation,[7] and The Riley Stoker Corporation.[8]

Raw Refuse Handling and Processing

Figure 16.18 is a diagram of the refuse processing, storage and retrieval system that prepares refuse-derived fuel (RDF) for partial suspension (spreader stoker) firing with coal in Riley Boiler No. 8. The design capacity is 200,000 lb/hr steam (90,720 kg/hr) @ 160 psi (1110 KPa) saturated.

Table 16.5 Capital and Operating Costs, Two Composite 200 TPD Energy Recovery Systems

Cost Item	Excess Air System[b]	Starved Air System[c]
Capital Costs ($)[a]		
Facility design, engineering, building, and site preparation	5,304,000	3,520,000
Electrical equipment, plumbing, etc.	900,000	700,000
Incineration and heat-recovery systems	5,724,000	3,000,000
Two ash-removal systems	864,000	(inc. above)
Electrostatic precipitators	961,000	700,000
Controls, two cranes, and scale	1,447,200	—
Controls, scale, four front-end loaders	—	280,000
Cost/design ton	76,000	41,000
Steam line (10^3 ft)	400,000	400,000
Total capital costs	15,600,000	8,600,000
Financing/Debt Service ($)[a]		
General-obligation bond financing		
Total bond issue[d]	15,912,000	8,772,000
Total annual debt service[e]	1,869,018	1,030,356
Debt service/ton	32.00	19.74
Annual Operating Costs ($)[a,h]		
Labor ($)[a]		
Plant manager (1)	35,000	35,000
Plant operators (4)	91,200	91,200
Helpers (4, excess air only)	91,200	—
Ash truck drivers (2)	31,920	31,920
Crane operators (4, excess air only)	91,200	—
Front-end loader operators (8, starved air only)	—	182,400
Maintenance mechanics (2)	45,600	45,600
Janitor/groundskeeper (1)	15,960	15,960
Secretary/scale operator (1)	15,960	15,960
Subtotal (19, each system)	418,040	418,040
Fringe benefits (30%)	125,412	125,412
Total labor	543,452	543,452
Supplies and chemicals	50,000	100,000 [i]
Maintenance and equipment replacement (2.5 × total capital cost for excess air, 3 × total capital cost for starved air)	390,000	258,000
Contract vehicles (dump trucks) for ash disposal	50,000	50,000
Water ($/yr, @ 20¢/$10^3$ gal)[j,k,l,m]	4,375	3,443
Fuel ($/yr)		
Auxiliary fuel for start-up	576[n]	78,300[o]
Front-end loaders	—	7,830[p]
Electricity	168,192[q]	150,336 [r]
Residue disposal transportation[s]	28,908	29,839
Other utilities ($/yr)	10,000	10,000
Insurance ($/yr)	50,000	25,000
Accounting ($/yr)	10,000	10,000
General and administrative ($/yr)	100,000	100,000
Operations and maintenance		
Total O & M ($/yr)	1,410,111	1,376,665
Per throughput ton ($)	24.15	26.37

Table 16.5 Continued

Cost Item	Excess Air System[b]	Starved Air System[c]
Landfill costs for unprocessable waste		
$/yr	0	37,200 [f]
$/throughput ton of waste	0	0.71
Steam revenues[u]		
$/yr	2,184,160	2,004,480
$/throughput ton of waste	37.40	38.40
Required net disposal fee ($/ton) if plant is financed with general-obligation bonds	18.75	8.42

[a] All dollar amounts for capital costs and financing/debt service expressed in January, 1983, $; annual operating costs expressed in January, 1985, $ (calculated by adding 10%/yr, for each of 2 years, to January, 1983, $).

[b] Based on capital cost estimates provided by the Clark-Kenith, Inc., for building a facility similar to the Hampton, Virginia, facility, escalated to account for inflation, with minor cost modifications made by Battelle.

[c] Based on capital costs for the Portsmouth, New Hampshire, facility provided by Consumat, escalated to account for inflation, with minor cost modifications made by Battelle.

[d] Assumes a 1.02 bond-build-up factor to account for bond-issuance costs.

[e] Assumes a bond interest rate of 10% and a 20-year amortization period.

[f] Assumes a 1.4 bond-build-up factor to account for bond-issuance costs, and a 40% equity contribution by the leverage lessor.

[g] Assumes a bond interest rate of 13% and a 20-year amortization period.

[h] Based on a throughput capacity of 200 TPD (365 day/year) design capacity for both systems. However, an 80% availability is assumed for the excess-air system—or an excess-air total of 58,400 ton/year; and a 71.6% availability is assumed for the starved-air system (assuming shutdown 2 day/week to clean fin-tube boiler tubes—this availability might be higher if another type of boiler were used, though boiler cost might also be higher)—or a starved-air total of about 52,200 ton/year.

[i] In particular, tires for front-end loaders.

[j] Boiler feedwater calculated by multiplying amount of steam per pound of waste (2.75 lb for excess-air system, 2.4 lb for starved-air system) by annual throughput of waste (see note h), then dividing by 8.34 lb/gal.

[k] Makeup water for blowdown calculated by multiplying boiler feedwater by 0.05.

[l] Water for personal consumption assumed to be 100 gal/day per person (19 each for the excess-air and starved-air systems) for 365 day/year.

[m] Ash dragout assumes no water consumption because quench water comes from boiler blowdown.

[n] Calculated as follows: 12 hr × 8 start-ups/year × 10^3 BTU/hr × $6/$10^3$ BTU.

[o] Calculated as follows: 250,000 BTU/ton of waste × 52,200 ton/year of waste × $6/$10^6$ BTU.

[p] Calculated as follows: 0.10 gal/ton of waste × 52,200 ton/year of waste × $1.50/gal—assuming one loader is running constantly. (Excess air systems do not use front-end loaders.)

[q] Based on 36 kwh/ton of waste × 58,400 ton/year—of waste × 8¢/kwh. Amount of electricity used by the excess air system assumes that 15% of the steam produced is used in-plant for driving pumps, etc.—if steam is not so used, more electricity would be required. Excess air systems typically require more electricity or in-plant steam, compared to starved air systems, because of their larger and more energy-intensive air-pollution control equipment and their use of a crane for waste-charging instead of a front-end loader.

[r] No existing starved air systems utilize any of their steam for in-plant uses.

[s] Calculated for excess air as: 1 gal/ton × 58,400 ton/year of throughput × 33% ash generation × $1.50/gal. Calculation for starved air is identical except that throughput is 52,200 ton/year.

[t] Calculated as follows; 58,400 ton/year throughput (excess air) less 52,200 ton/year (starved air) = 6,200 ton/year of waste × $6/ton assumed landfill cost.

[u] Calculated as follows: Annual throughput of waste times the pounds of steam per pound of waste (2.75 lb for excess air systems, 2.4 lb for starved air systems), times $8/$10^3$ lb of steam. For excess-air systems, in-plant steam use (0.4125 lb/ton of waste) is also subtracted. No existing starved air systems use steam for in-plant purposes.

The refuse fuel is industrial solid waste collected from various General Motors (GM) plants in the vicinity of Pontiac, Michigan. The characteristics of the prepared RDF and the coal are given in Table 16.6. (The refuse has remained relatively constant in its constituent makeup over the refuse firing years.) The processing plant has a capacity of 224 short tons (200 tons) per day and provides RDF to Riley Unit No. 8 as well as to another unit at the plant site. Approximately 112 short tons (100 tons) of coal and refuse in a single-shift operation. On a yearly basis, this amounts to approximately 22,400–28,000 short tons (20,000–25,000 tons) per year of coal.

Designed for combined coal and refuse firing, the system started up in 1973 and fired coal only until late 1975 when combined refuse/coal firing commenced.

Raw material is delivered by truck to the process/power complex (Figure 16.19) and deposited on a tipping floor (Figure 16.20). A payloader (Figure 16.21) loads the process line conveyer feeding a Williams shredder, a Rader Air Density Separator, and A Pneumatic Conveying System (Figure 16.22) that transports the RDF to an Atlas storage bin (Figure 16.23).

Fuel Burning Equipment Design

As shown in Figure 16.24, Riley Unit No. 8 was originally designed to handle either coal firing or combined refuse/coal firing. It includes a traveling grate spreader stoker with front ash discharge, with a design provision to burn a 70:30 refuse to coal ratio on a heat input basis.

Fuel Burning Operating Experience

Experience has shown that the optimum ratio of refuse to coal is 60:40 by weight (46:54 by heat input). This allows for a safe operating margin whereby the coal feed rate can be moderately increased without a drastic loss of steam output if refuse is unavailable. The 60:40 refuse-to-coal weight ratio is used during weekend operation and a 50:50 refuse-to-coal ratio is maintained during

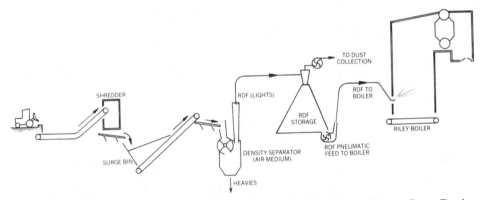

Fig. 16.18 Diagram, RDF processing, storage, and retrieval system, General Motors Corp., Truck and Bus Group, Pontiac Central Manufacturing and Assembly Plant.

Table 16.6. GM Truck and Bus Group Fuel Characteristics

General characteristics of raw refuse (design):	From various GM manufacturing plants: wood (42%), paper (33%), cardboard (23%), rubber and plastics (2%)
Analyses of prepared refuse (design):	C (41.5%), O (34.2%), S (0.5%), H (5.9%), ash (6.7%), water (11.2%), 7500 BTU/lb (17,445 kJ/kg) as fired
Analyses of coal (design):	C (71.44%), O (12.6%), S (0.98%), H (5.21%), N (1.69%), ash (8.08%), 12,250 BTU/lb as fired; ash fusion temp = 2700°F; 45 Hardgrove grindability
Prepared refuse (actual):	7000 BTU/lb (16,282 kJ/kg) as fired
Coal (actual):	0.8% sulfur (present allowable limit)

Fig. 16.19 Delivery of industrial solid waste to process building, General Motors Corp., Truck and Bus Group, Pontiac Central Manufacturing and Assembly Plant.

Fig. 16.20 Industrial solid waste on tipping floor, General Motors Corp., Truck and Bus Group, Pontiac Central Manufacturing Assembly Plant.

Fig. 16.21 Refuse payloader feeding process line conveyer, General Motors Corp., Truck and Bus Group, Pontiac Central Manufacturing Assembly Plant.

Fig. 16.22 Refuse shredder and air density separator, General Motors Corp. Truck and Bus Group, Pontiac Central Manufacturing and Assembly Plant.

Fig. 16.23 Atlas storage bin for RDF General Motors Corp. Truck and Bus Group, Pontiac Central Manufacturing and Assembly Plant.

weekday operation as an additional conservative measure when there is a higher average steam demand.

Steam Generating Operating Features and Experiences

The summary of design process conditions at maximum continuous rating (MCR) steam load are given in Table 16.7. A unit is operated continuously, 18–20 hr/day (average refuse burning hours availability), in generating steam. There has been a steady increase in operating time per year on refuse. Last year the goal of 65% of working days for refuse firing was met, at an average of 112 short tons (100 tons) per day per unit. On Riley Unit No. 8, there have not been any appreciable slagging problems (especially in the upper firing chamber) since refuse firing was initiated in 1976. Screen tube section cleaning by the water-blasting method has been required at an average rate of only once per year on this unit. Flue gas flow paths are appropriately sized for refuse firing and the furnace chamber is sized properly such that the products of combustion and any refuse carry-over do not have time to cool below their fusion temperatures and subsequently fuse before reaching the upper tubes.

Since refuse is fired, the unit is rated as an "incinerator," thereby requiring a mechanical collector plus a wet scrubber for emission control. In comparison to coal burning, flyash from

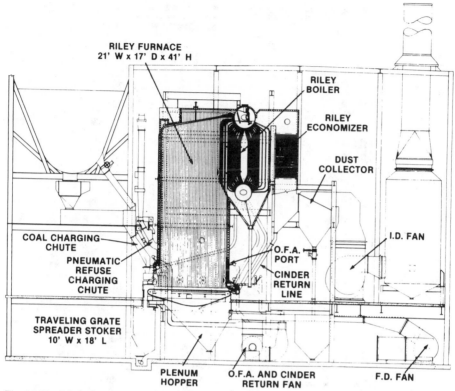

Fig. 16.24 Riley Boiler Unit #8, Plant 2 GMC Truck and Bus Group, Pontiac Central Manufacturing and Assembly Plant.

refuse/coal is lighter and finer (similar to talcum powder). Flyash escaping capture in the mechanical collector has previously resulted in I.D. (induced draft) fan wear, including erosion to the fan wheel and blades and abrasion to the I.D. fan inlet box.

The fan housing has been rebuilt three times, most recently with the addition of a ceramic-type tile welded to the housing, to deter abrasion. The latest retrofit of tile has been very successful in minimizing expenditures for metal replacement in the fan housing and no further material changes have been made to the fan housing in the past 3 years.

A stainless steel stack was retrofitted, following initiation of refuse burning. The stack remained intact for 5 years before needing replacement. GM personnel estimate an improvement to 8–20 years of life of the second stainless steel stack, since operating know-how has been gained.

The predicted collection efficiency of the mechanical collector at MCR steam load is 94% when firing coal and 92% when firing a 60:40 refuse-to-coal ratio by weight. Actual tests performed in recent years indicate an actual efficiency at 75% of MCR steam load, of 93% on coal firing, and 87% on 60:40 refuse/coal (by weight) firing. In summary, the mechanical collector provides nearly maximum efficiency in removing coal burning particulate and does quite well when firing a refuse/coal mix. However, as the refuse-to-coal ratio increases, collector efficiency may decrease.

In summary, the mechanical collector provides nearly maximum efficiency in removing coal burning particulate and does quite well when firing a refuse/coal mix. However, as the refuse-to-coal ratio increases, collector efficiency may decrease.

GM Truck and Bus Group Future Plans

GM plans to continue with this refuse burning practice. It should be emphasized that GM has evaluated the trade-offs of savings in fuel costs and landfill costs versus increased capital and operating costs (caustic chemicals for wet scrubber, and so forth) and has determined that a positive return exists with refuse firing based on existing fuel costs and landfill costs. This advantage could disappear should fuel and/or landfill costs decrease significantly, and GM could easily return to coal firing only on the existing equipment.

Table 16.7 Boiler Design Conditions at MCR Steam Load

Unit	Refuse Coal[a]	Coal
Metric Units		
Steam flow (kg/hr)	90,718.47	90,718.47
Sat. steam press. (KPa)	1110.09	1110.09
Outlet superheater pressure (KPa)		
Sat. steam temp. (°C)	188.33	188.33
Fuel flow	300 (R)	
(tons/day)	75 (C)	120
(tons/hour)	12.5 (R)	
	3.2 (C)	10
(kg/hr)	11,339.80 (R)	
	2,834.95 (C)	9,071.84
Air flow (kg/hr)	127,459.46	114,305.28
Excess air (%)	50	38
Heat input (MW)	55.95 (R)	
	22.45 (C)	71.80
Fuel heat content (kcal/kg) (as fired)	1,890 (R)	
	3,087 (C)	3,087
Furnace heat release (W/M³)	181,245.58	177,102.83
Furnace heat release (W/M²)	239,780.41	235,363.4
Grate heat release (W/M²)	2,303,153.9	2,192,728.7
Overall unit efficiency	75.64	80.88
English Units		
Steam flow (lb/hr)	200,000	200,000
Sat. steam press. (psi)	161	161
Outlet superheater pressure (psi)		
Sat. steam temp. (°F)	371	371
Fuel flow	300 (R)	
(tons/day)	75 (C)	120
(tons/hour)	12.5 (R)	
	3.2 (C)	10
(lb/hr)	25,000 (R)	
	6,250 (C)	20,000
Air flow (lb/hr)	281,000	252,000
Excess air (%)	50	38
Heat input (Btu/hr)	187.5×10^6 (R)	
	76.6×10^6 (C)	245×10^6
Fuel heat content (Btu/lb) (as fired)	7,500 (R)	
	12,250 (C)	12,250
Furnace heat release (Btu/ft³/hr)	17,500	17,100
Furnace heat release (Btu/ft²/hr)	76,000	74,600
Grate heat release (Btu/ft²/hr)	730,000	695,000
Overall unit efficiency	75.64	80.88

Source: Data from General Motors Corp. Truck & Bus Group.
[a]R, refuse; C, coal.

Summary

Operating performance, costs, and savings are summarized in Tables 16.8, 16.9, and 16.10.

Case History 2:   Eastman Kodak Company

Background[9]

This industrial waste-to-energy plant started up in 1970 and has operated successfully without major interruption ever since. The following is a description of the installation, operation, and subsequent modification of a system for preparation of plant refuse and sludge for full suspension firing in a combustion engineering VU40 boiler (Figure 16.25) for a maximum steaming rate of 150,000 lb/hr @ 400 psig, 550°F (2758 KPa, 288°C). The information presented is excerpted from technical papers in the proceedings of national conferences of the American Society of Mechanical Engineers (ASME), Waste Processing Division, as authored and presented by Kodak engineers.[9,10]

Kodak Park is an industrial complex in Rochester, New York, where the center of photographic technology is located. It is a 2,000 acre complex where 30,000 people are employed. This great concentration of people and facilities has all the problems of a city including the problem of solid waste disposal.

A study of waste disposal requirements was begun in 1964 and after consideration of various schemes, it was decided that a suspension fired system would best meet the needs for burning general plant wastes and industrial waste treatment plant sludge. Some chemical wastes are also disposed of in this facility. All waste fuels are prepared and then conveyed into the boiler furnace combustion chamber (Figure 16.26).

As an early effort in this landmark system concept, several materials handling problems occurred with the storage bins and conveyors, with space limitations also a factor. Boiler problems, however, have been of a minor and routine nature.

Table 16.8   Total Loads of Refuse Processed, 1976 through March, 1984

Total in-plant loads of refuse processed	30,193
Total outside loads of refuse processed	14,182
Total loads of refuse	45,005

Table 16.9   Overall Plant Performance, Totals for 1976 through March, 1984—General Motors Corp. Truck & Bus Group

Total tons of refuse processed	193,000
Cubic yards, landfill space conserved	1,690,000
Equivalent tons of coal conserved	111,000
Average value of coal	$5,288,000
Landfill savings, truck & bus division	$2,537,000

Table 16.10   Refuse Plant Statement 1982 and 1983 Budget Year Comparison—General Motors Corp. Truck & Bus Group

	1982	1983
Outside credits	$ 15,000	$ 19,000
Coal savings	847,000	972,000
Landfill savings	319,000	364,000
Total savings	$1,181,000	$1,355,000
Total operating costs	$ 653,000	$ 699,000
Profit	$ 528,000	$ 656,000

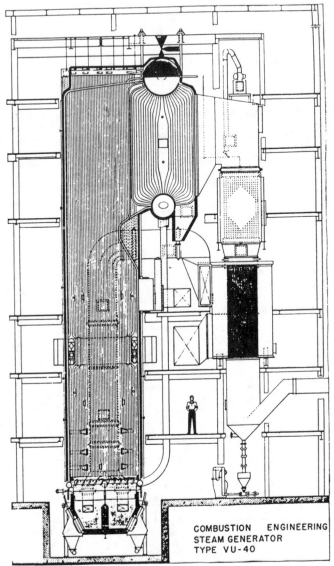

COMBUSTION ENGINEERING
STEAM GENERATOR
TYPE VU-40

Fig. 16.25   Combustion Engineering Co. Type VU40 steam generator, Eastman Kodak Co.

This facility (Figure 16.27) was a new system at a new location, and it was necessary to hire new personnel to develop training programs for all employees involved. The area was a desirable location because it had an existing incinerator, an ash silo, a steam distribution system, and a 360 ft. (110 m) chimney.

The decision was for a solid waste disposal system sized for 314 ton per day (285 MTD) of general plant refuse and industrial waste water treatment plant sludge. The general refuse portion of the system was sized for 180 ton per day (163 MTD) of paper, wood, plastics, rags, garbage, etc. with a heat value of approximately 7300 BTU/lb (4044 Kcal/Kg). The ultimate analysis is shown in Table 16.11. The sludge system was sized for 134 ton per day (122 MTD) of 30 per cent solids sludge or 114 ton per day (103 MTD) of 20 per cent solids sludge. Table 16.12 shows sludge characteristics.

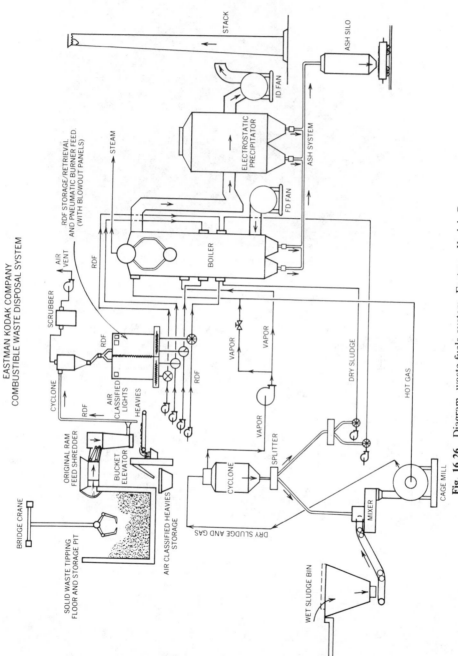

EASTMAN KODAK COMPANY
COMBUSTIBLE WASTE DISPOSAL SYSTEM

Fig. 16.26 Diagram, waste fuels system, Eastman Kodak Co.

Fig. 16.27   New waste-to-energy facility, Eastman Kodak Co.

Disposal System

The solid waste disposal system selected was a suspension-fired water wall boiler. This system was chosen because Kodak Park has its own steam distribution and power producing facilities and can justify heat recovery. We believed much better combustion would be attained with shredded refuse. Use of an electrostatic precipitator was required to meet air pollution regulations and to obtain a dry flyash. The system would be required to dispose of both sludge and general plant refuse, and the systems in common use were considered inferior.

Boiler Operation

The boiler is equipped with combustion control equipment normally associated with a power boiler in addition to boiler permissives required for firing waste fuels. Boiler start-up is accomplished with No. 2 oil ignitors and No. 6 main oil burners as prescribed by manufacturers' warm-up requirements. When the boiler load exceeds the oil fire ball capability, shredded refuse may be fired into the boiler. Normally, the refuse is started on one refuse burner at a time and the oil flow to the boiler reduced to maintain approximately the same load. Prior to burning sewage sludge, the boiler exit temperature must be above 900°F (482°C) in order to have sufficient hot flue gas to dry the sludge in the flash-drying system.

**Table 16.11   Ultimate Analysis,
Industrial Solid Waste Fuel—Eastman
Kodak Co.**

C	40
H	5.9
O	34.2
Cl	1.5
S	0.5
Ash	6.7
Moisture	11.2

Heat value: 7300 BTU/lb (4044 KCal/kg)

Table 16.12   Dry Industrial Waste Treatment Plant Sludge—Eastman Kodak Co.

Moisture	15%
Volatile matter	49.5%
Fixed carbon	3.88%
Ash	31.62%
Heat value: 4170 BTU/lb (2310 kcal/kg)	

The sludge burning requirement is such that the sludge-drying system must be operated on a continuous basis. When sufficient refuse is not available to maintain the boiler, main oil is fired in conjunction with sludge. Prior to refuse firing on a continuous basis, the boiler was operated on main oil and sludge in conjuction with the sludge flash-drying system. During approximately 9 months of operation in this manner, it was found the boiler could operate from 70,000 lb/hr (31,752 kg/hr) to 150,000 lb/hr (67,890 kg/hr) or steam load and still maintain the boiler exit tempeature near 1000°F (538°C) to operate the sludge-drying system. This is accomplished by selected use of four soot blowers in the boiler bank area for cleaning the boiler tubes. Operation in this manner allowed the boiler to be used to fire waste fuels in addition to using it as a power boiler to meet the plant steam load demanded.

Start-up of the boiler and operation in firing waste fuels pointed out several problems and operating criteria not originally anticipated. Coordination with the manufacturers resolved the main problem within a short period.

The vapor return plenum chamber refactory, located on the back of the boiler just above the burner level, was not installed with sufficient heat resistivity and was replaced with a castable insulating refractory.

The economizer and air heater is equipped with a shot cleaning system to clean the gas side of the tubular surfaces. Shot carry-over problems were experienced in the breaching leading to the precipitator. This problem was resolved by installing false tubes and an egg-crate-type baffle below the air heater to reduce the shot fall velocity, thus preventing carry-over.

Inspection of the boiler, air heater, economizer, and precipitator after operation on waste fuels indicates no signs of deterioration or corrosion due to waste fuels firing.

Ash Handling

The bottom ash is removed from the boiler by the use of air cylinder-operated dump grates to ash hoppers. Ash is dumped manually. There are provisions for future installation of clinker grinders and automated removal of the ash conveying line. The ash is pneumatically conveyed to the ash storage silo.

As stated earlier, sludge and oil firing resulted in clinker formation on the boiler walls. Large falling clinkers caused the grates to bend and proved difficult in passing clinkers through and into the hoppers. Much labor was required in reducing the clinkers down to a size for conveying. Since the bottom ash was very abrasive, it has been found necessary to replace wear backs in the ash conveying line elbows at 6-month intervals.

Some problems were also caused from the ash-conveying system pulling flue gas into the conveying line, resulting in vapor condensation and line plugging. Other flyash problems encountered in the precipitator hoppers during removal were burning and clinker-like ash formations. A new ash grate with tight shutoff was installed to prevent air infiltration. More frequent ash removal and improved combustion have minimized this.

Operator Training

Since the solid waste disposal system was a new concept with very little knowledge and experience available from our operational aspect, a complete training program was formalized. Each operator was given a 2-month on-the-job training period on boiler operation and feedwater testing in Kodak Park's power plant prior to start-up of the solid waste disposal boiler. Approximately 25 hr of classroom instruction was presented to each operator along with a training manual containing the operating principles of all systems and equipment, operator maintenance, operating procedure, and start-up and shut-down procedures. In addition, a great deal of training and supervision was required during the initial start-up and operation of each system. Operators are also required to attend classes and obtain a stationary engineer's license from the city of Rochester in order to operate a high-pressure boiler.

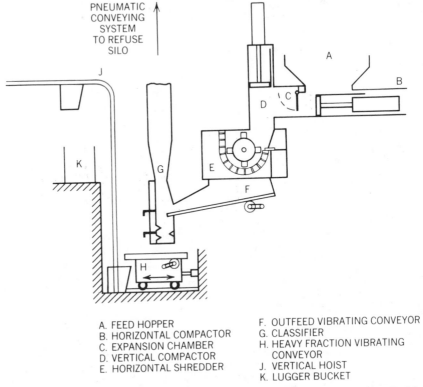

A. FEED HOPPER	F. OUTFEED VIBRATING CONVEYOR
B. HORIZONTAL COMPACTOR	G. CLASSIFIER
C. EXPANSION CHAMBER	H. HEAVY FRACTION VIBRATING
D. VERTICAL COMPACTOR	CONVEYOR
E. HORIZONTAL SHREDDER	J. VERTICAL HOIST
	K. LUGGER BUCKET

Fig. 16.28   Diagram modified RDF fuel preparation system installed in 1980, Eastman Kodak Co.

Experience to date indicates that refuse and sludge can be burned in suspension. Following modifications of materials handling equipment and location, the system is working extremely well.

Shredder Experience[10]

In the spring of 1980, a horizontal shredder, complete with an infeed and outfeed system as represented in Figure 16.28, was placed in service at the Kodak Park solid waste disposal facility. This shredder has become the primary shredder for preparing solid waste for incineration in the suspension-fired boiler. The vertical shredder, which had been in service since 1970, was then relegated to service as an operational backup.

This new shredding system represents Kodak Park's efforts over the past 10 years to improve its solid waste preparation system. Before discussing the details of this shredder installation, a brief review of the operating experience and modifications that have evolved will be presented.

The original shredder for fuel preparation at the Kodak Park Solid Waste Disposal Plant was selected in 1969. Specifications required that a single shredder have the ability to shred typical plant solid waste to a 2-in. (50 mm) nominal size. The horizontal shredder manufacturers contacted at that time felt that both a primary and a secondary shredder were required to obtain this shred quality. A manufacturer of vertical shredders felt that their shredder could meet the specifications. After shredding tests with this equipment produced the desired shred quality, a vertical shredder driven by two 400-hp motors was purchased. The vertical shredder produced an acceptable shred quality which could be handled by our boiler refuse feed system. However, after operating at full capacity for several months, we began to experience considerable downtime for shredder maintenance.

Refuse Infeed Experience

Refuse is delivered from a receiving pit by an overhead crane into a feed hopper. This solid waste consists of paper and plastic sheets, small rolls of paper and plastic, office waste, corrugated materials, fiber drums, and plastic pieces. Moisture content is low (5–10%) and some material is quite bulky. These characteristics caused the waste material to remain stationary on the original flat conveyor located below the feed hopper.

The infeed system was then modified to provide a positive feed of material into the vertical shredder. A horizontal ram was installed in the bottom of the feed hopper that pushed refuse into a pair of converging pinch belt conveyors. These conveyors discharged refuse into a downwardly curved duct connected to the shredder inlet. Two manually controlled vertical hydraulic rams were positioned in the curved duct so that the shredder operator can extend them to break up any refuse bridging that might occur in the shredder inlet.

Shredded Waste Conveying Experience

The original shredded waste conveyor was a series of seven enclosed belt conveyors that carried this material to a large elevated cylindrical live-bottom storage silo. This system was plagued with several problems: the storage discharge would jam, the shredded material would not convey, the transfer points would plug, refuse would foul the drives, and extremely dusty conditions were experienced. Fires in the conveyors were a common experience, sometimes occurring on all seven conveyors simultaneously.

After unsuccessful corrective measures, the belt conveyor system was removed and replaced with a classifier and pneumatic conveying system. The pneumatic conveying system and classifier operated successfully. Problems were experienced with the heavy fraction conveying system since the classifier discharge was cramped and lower than the shredder. A drag conveyor was installed beneath the shredder to convey the heavy fraction on a flat pan to the inlet hopper of a vertical bucket conveyor (Figure 16.28). The drag conveyor experiences problems with heavy materials jamming the sprockets and chains. Similar problems are encountered with the bucket conveyor, which carries this material vertically to a lugger bucket at grade level.

Horizontal Shredder Installation

A 60 × 60 in. (1.5 × 1.5 m) horizontal top-feed shredder with an 800 hp motor was installed on one end of our incoming refuse pit in 1969 to shred wood scrap, which was mostly truck and container loads of dunnage, predominantly pallets and crates. It was never possible to feed large slugs of wood scrap continuously this way due to excessive infeed hang-up on the infeed conveyor and in the mill throat opening, which is approximately 48 × 60 in. (1.2 × 1.5 m) at the rotor. Also, the 4 × 8 in. (10.2 × 20.4 cm) grate openings frequently produced a shredded product topsize fraction containing splinters which would plug up the pneumatic feed system. Closing the grate opening would increase retention time and the probability of fires and wood waste was no longer shredded.

In 1977, we began looking into the feasibility of using this horizontal shredder for fuel production. Our shredding experience now indicated that the 2 in. (50 mm) shred size was not necessary and material up to 4 × 4 in. (100 mm × 100 mm) topsize could be handled in the pneumatic conveying system without difficulty.

We tested the solid waste shred quality from the horizontal shredder with various shredder modifications such as reduced grate opening size, additional stationary grate bar shear anvils, and split (tandem) hammers. Reducing the grate opening was the only modification that significantly reduced the shred size to an acceptable level. To adapt the horizontal shredder to continuous refuse preparation with bucket loading from the pit, it was necessary to reposition the shredder by rotating it 180°. This change provided ready access to the shredder hammers and allowed the horizontal infeed ram to be installed at a lower elevation.

The 8 × 4 in. (100 × 100 mm) grate openings were reduced to 3 × 4 in. (76 × 100 mm) by welding 2-in. (50 mm)-thick steel bars into the grate openings. This change reduced shredder throughput and increased the horsepower requirement. We are able to shred 10–12 tons (9.1–10.9 t) of refuse per hour, but the crane operators must judiciously feed the shredder; otherwise the drive motor can be stalled.

We find that a finer shred occurs when the shredder is loaded at or near motor design current. When motor current is allowed to drop due to reduced infeed charging, the shred size increases and the incinerator conveying system occasionally plugs with this material. Shredder hammers made of manganese steel (300 to 350 Brinell hardness) have experienced 1200 to 1500 hr of operation before replacement was necessary. We have recently changed to harder TI (450 to 500 Brinell hardness) steel hammers; therefore we cannot comment on their operating life as yet.

The shredder became operational as our primary shredder in April, 1980, and has processed approximately 40,000 tons of material through November, 1981. The shred quality has been acceptable for conveying through 5-in. (12.5 cm)-diameter pneumatic lines into the incinerator. Only one type of material, small pint- and quart-size plastic bottles, does not shred to an acceptable size when hammer wear progresses beyond a certain stage.

Shredder Infeed System

After our successful experience with ram-feeding refuse into the vertical shredder, we sought to improve this technique further when designing the horizontal shredder infeed system. We designed a system consisting of a feed hopper mounted above a horizontal compactor. The horizontal compactor discharge is attached to an expansion chamber whose size was dictated by the restraints of the infeed hopper and shredding locations. This chamber is connected to the inlet of a vertically positioned compactor ram mounted on the horizontal hammer mill shredder inlet (Figure 16.28).

During normal operation, the ram of the vertical compactor is not required and remains stationary in the retracted position. The ram of the horizontal compactor pushes refuse in the expansion chamber. The incoming refuse charge then pushes the accumulated refuse in the chamber into the chute of the vertical compactor where this material falls by gravity into the hammer mill shredder.

If bridging occurs in the vertical chute, the operator activates the ram of the vertical compactor, pushing refuse into the shredder inlet. When material that is difficult to feed must be shredded, the horizontal and vertical compactor rams can be operated in sequence automatically so that the material is continually force-fed from the infeed hopper to the shredder inlet.

This infeed system has performed satisfactorily in feeding refuse to the shredder. However, two problems have arisen that were not encountered with the vertical shredder installation. We have experienced fine dust emissions and metal scrap projectiles flying out of the shredder through the horizontal compactor and infeed hopper with high velocity into the refuse pit area. This is typical of top-feed, horizontal-shaft, swing-hammer shredders and the projectiles were a safety hazard for the operators. As an immediate correction, we installed a hinged ½-in. (13 mm)-thick steel plate in the expansion chamber with small clearances on the vertical sides and 6 in. (153 mm) of clearance at the bottom. With the hinge located at the top of the plate, the refuse ahead of the horizontal compactor ram pushes the plate forward to enable refuse to enter the chamber and the vertical chute into the shredder. This hinged-plate scheme has not affected our refuse-feeding capability and has eliminated our projectile problem. Based on visual observation, we feel that this hinged plate has also reduced fine dust emissions from the shredder.

On one occasion, fine dust emissions from the shredder accumulated in the windings of the open construction shredder motor. This dust accumulation ignited and began to smolder. The motor was opened and the dust was removed. No damage occurred in the motor windings. To alleviate this problem, dust-stop filters were installed on the motor air intake openings. In addition, a dust-tight partition was installed between the feed hopper area and the shredder motor area.

Shredded Waste-Conveying System

The shredded waste-conveying system consists of a tightly enclosed inclined vibrating conveyor located below the air-swept shredder which discharges into an air-separation column in the basement. A horizontal vibrating conveyor resting on tracks to permit horizontal movement is positioned beneath the classifier to convey the heavy fraction. The discharge lip of the horizontal conveyor is positioned over the edge of the bucket of a vertical hoist (Figure 16.28). The vertical hoist lifts this bucket to its dumping elevation, where its contents are dropped into a lugger located at grade level.

In operation, the shredded material exiting the shredder falls onto the pan of the vibrating conveyor to be carried into the air separator. The tight enclosure around the conveyor allows the vacuum present in the classifier to pull air through the shredder, thereby providing shredder heat removal and an assist in moving the fluffy refuse. The classifier causes the light refuse fraction to be swept vertically upward off of the vibrating conveyor along with the conveying air for delivery to the refuse silo. The more dense heavy fraction falls onto the vibrating heavy fraction conveyor where this material is carried horizontally to the bucket of the vertical hoist.

A television camera aimed at this bucket monitors the heavy fraction level. When the shredder operator observes a high level on the television monitor, he starts the vertical hoist cycle. In this cycle, the horizontal conveyor vibrator is stopped. This conveyor is then retracted to clear the lip of the hoist bucket and the hoist bucket begins its vertical travel. At the top of its vertical travel, the bucket is tipped and the heavy-fraction material drops into a load lugger. After emptying, the hoist bucket is returned to its position beside the horizontal conveyor. This conveyor is then repositioned over the lip of the hoist bucket and the conveyor vibrator is started.

While the horizontal conveyor is retracted with its vibrator out of service, heavy-fraction material continues to fall into it. After completion of the dump cycle, the horizontal conveyor begins vibrating and this accumulated material is conveyed to the hoist bucket. A timer is actuated when the hoist cycle begins and an alarm serves to notify the operator when the dump cycle extends beyond a present time.

The tight enclosure around the inclined vibrator conveyor is critical to smooth conveying of the shredded waste. On one occasion, a conveyor side shield became damaged and the operator removed it. This produced a "leak" and a reduction in refuse flow to the classifier, causing a build-up on the conveyor pan and providing a spot for hot metal particles to ignite the refuse. We also feel that maximum air flow through the shredder is critical for reducing dust emissions and cooling the shredder hammers and grates.

Summary

This review of our experience and the approach taken in resolving problems should be of interest to those concerned with solid waste preparation. The technology which has evolved in developing our particular solid waste preparation system has been expensive and time-consuming. The commitment of the management of Eastman Kodak Company to provide a reliable solid waste disposal operation has been the key to our success.

Case History 3:   John Deere Dubuque Works

Background[11]

Start-up in 1977 with a total construction cost of approximately $1,400,000, this waste-to-energy installation serves the Dubuque, Iowa, manufacturing plant of John Deere, Inc., a world leader in the agricultural equipment industry, by providing alternatives to the uncertainties of escalating energy and waste disposal costs.

Desiged by Perkins & Will, consulting engineers, Chicago, Illinois, and the plant engineering staff of Mr. George Hellert, this energy-efficient facility (Figure 16.29) is designed to sort, shred, reclaim, and incinerate solid and liquid wastes from manufacturing operations and to generate steam to heat the equivalent of 200,000 ft^2 (18,600 m^2) of Deere's manufacturing areas.

Plant Operation[12]

Trucks enter the 14,000 ft^2 (1302 m^2) building and empty compactor boxes of 30 yd^3 (23 m^3) capacity onto the floor (Figure 16.30). The oversize refuse is sorted and fed into a slow-turning, coarse-shredding, preburn size-reduction unit, and the remainder is fed directly into the incinerator-charging hoppers (Figure 16.31) by John Deere front-end loaders (Figures 16.32 and 16.33). Three modular incinerators, each rated at 2000 lb/hr (910 kg/hr) solid waste, are equipped with hydraulic ram chargers, automatic ash discharge, and waste heat boilers. Controls automatically regulate supplies of combustion air, solid waste, and spent coolant oils. The system was originally designed to dispose of 35 tons of plant wastes and up to 500 gallons (1892 DM^3) of waste oils and coolants each production day. The system's steam generation capacity is a maximum of 20,000 lb/hr (9091 kg/hr) saturated @ 50 lb/in.^{2} (345 KPa).

REFERENCES

1. W. R. Niessen, A. F. Alsobrook, "Municipal and Industrial Refuse: Composition and Rates," Proceedings, ASME National Incinerator Conference, New York, New York, 1972.

2. *Marks Standard Handbook for Mechanical Engineer,* Eighth Edition, 1978 McGraw-Hill, New York, pp. 7–13.

3. Los Angeles County, *Air Pollution Engineering Manual,* LA-AP40.

Fig. 16.29   Waste-to-energy plant, John Deere Dubuque Works.

Fig. 16.30 Tipping floor, John Deere Dubuque Works.

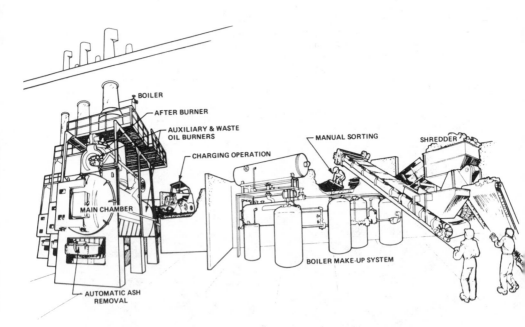

BOILER

AFTER BURNER

AUXILIARY & WASTE OIL BURNERS

CHARGING OPERATION

MANUAL SORTING

SHREDDER

MAIN CHAMBER

BOILER MAKE-UP SYSTEM

AUTOMATIC ASH REMOVAL

Fig. 16.31 Diagram, waste-to-energy system, John Deere Dubuque Works.

Fig. 16.32   Loader feeding preburn coarse shredder.

Fig. 16.33   Loader feeding incinerator.

4. City of New York Local Law 14.

5. Argonne National Laboratory, "Thermal Systems for Conversion of Municipal Solid Waste, Volume 3: Small-Scale Systems, A Technology Status Report." DOE No. ANL/CNSV-TM-120, Vol. 3.

6. U. S. EPA Office of Water and Waste Management, "Small Modular Incinerator Systems with Heat Recovery: A Technical, Environmental, and Economic Evaluation-Executive Summary." SW 797, November, 1979, Washington, D.C.

7. Data and graphics courtesy of Mr. L. Y. Lee, Plant Engineering Programs, General Motors Corporation Technical Center, Warren, Michigan.

8. C. Shafer, "Wood and Waste Burning in Industrial Plants," Proceedings, Purdue University Industrial Fuels Conference, Lafayette, Indiana, September, 1982.

9. R. L. Merle, M. C. Young, G. R. Love, "Design and Operation of a Suspension Fired Industrial Solid Waste Disposal System for Kodak Park," Proceedings, ASME 7th National Waste Processing Conference, Boston, May, 1976.

10. W. R. Matthews, "Preparation of Industrial Solid Waste into Boiler Fuel with a Single Horizontal Shredder," Proceedings, ASME 10th National Waste Processing Conference, New York, May, 1982.

11. Brochure, Perkins & Will, Consulting Engineers, Chicago, Illinois, 1977.

12. "Tracks," January/Feburary, 1978, A Publication for Employees of John Deere Dubuque Works (Iowa) & John Deere Davenport Works (Iowa).

CHAPTER 17

RESIDENTIAL, COMMERCIAL, AND INDUSTRIAL BULKY WASTES

WILLIAM D. ROBINSON

Consulting Engineer
Trumbull, CT 06611

17.1 INTRODUCTION

The predominant practices in bulky waste disposal have mostly remained crude and casual despite a number of technologically successful process plants in operation during the 1970s and at present. Those that closed had become victims of not only the international oil cartel's "world price" explosion, but also an accelerating and higher priority need for alternatives to rapidly diminishing residential garbage landfills and their burgeoning environmental constraints. A few of these bulky waste process plants never became fully operational due to poor planning and/or design.

Where shredders were employed, early explosion episodes created a lasting paranoia, obscuring subsequent successful preventive measures that have allayed such fears among knowledgeable system designers (see Chapter 12, Section 12.14, and References 1 and 2).

Although bulky waste disposal has always had a somewhat lower priority, more sophisticated process and disposal methods can no longer be ignored, hence the following case histories, including successful past and present efforts. The solid waste industry today is mostly unaware of this experience that can form the basis for improved and necessary programs.

17.2 NATURE OF THE WASTE

Bulky, oversize, trash, rubbish[3] are all descriptive and valid terms to describe the more difficult and larger types of waste. It is also referred to as OBW, an ambiguous reference which means "oversize burnable waste" to some and "oversize bulky waste" to others. When New York City operated 13 large incinerators, veteran city engineers coined the term to describe oversized burnable waste and the acronym spread throughout the industry but became misinterpreted when referred to as oversized bulky wastes. The aberration has survived, however, and OBW is now an occasional reference to bulky waste, burnable or not. Disposal methods are as follows:

1. Landfill as received.
2. Process for land disposal.
3. Process for resource recovery.

17.2.1 Residential Bulky Waste

Included in this category, Figure 17.1, are household furniture and "white goods" appliances such as stoves, washing machines and refrigerators, mattresses and springs, rugs, TV sets, water heaters, tires, lawn mowers, auto parts, tree & brush debris, and so forth.

Collection is usually at curbside by open dump truck during specified trash cleanup periods and comprises approximately 10 to 20% by weight of total urban and suburban residential solid waste on average.[4]

Fig. 17.1　Residential bulky waste.

17.2.2　Commercial Bulky Waste

Much of this, Figure 17.2, is packaging and containers in a wide range of sizes, including corrugated cardboard, and wood boxes, fiber, plastic and steel drums usually under 40 gallons (0.15 m^3), loose and bundled paper (office, printouts), bundles of textiles and plastics, bales of corrugated and paper, furniture and equipment, and flat and wire banding.

Collection is by rear loader, open dump truck, roll-off container, or containerized side or front loader.

17.2.3　Industrial Bulky Waste

This category, Figure 17.3, includes dunnage, including crates, cartons, pallets, skids; large and small steel, fiber, and plastic drums; bales and rolls of paper, plastics, and textiles; miscellaneous metal boxes, tubing, rod, punchings, and skeleton; wire, rope, and metal banding; and paper, textile, and plastic streamers.

Collection is usually by roll-off container or containerized side or front loader, and when combined with residential and commercial bulky waste can comprise from 25 to 40% of the total disposable wastes in industrialized municipal areas (see Chapter 12, Section 12.12).

17.3　PRESENT DISPOSAL STATUS

17.3.1　Background

There has never been any significant change in bulky waste disposal methods despite a number of promising and technologically successful processing efforts in the late 1960s and into the 1970s that fell victim to mainly nontechnical problems. Open burning has mostly been eliminated, but the typical historic treatment of bulky wastes at landfills has continued, Figure 17.4:

1. Landfilling as received.
2. Cursory volume reduction by crunching or mauling by bulldozers with (or without) track-mounted mauling shoes, Figure 17.5.
3. Work-over by ribbed or spiked wheel landfill machines, Figures 17.6 and 17.7.
4. Contract removal by scrap processors or by scavengers.

Collection, transport, and disposal of bulky wastes have always been a nuisance, and the foregoing disposal practices have been least expensive and complicated. This also applies to the less bothersome residential and commercial material during the past 20 years when well over 100

Fig. 17.2 Commercial bulky waste.

incinerators were closed[5] due to nuisance exploitation and inability to comply with progressively restrictive emissions codes. Also effective was a determined effort by the land disposal industry to convert most incinerators to transfer stations.

Ironically, landfill availability is diminishing rapidly not only with intensified restrictions on leachate monitoring and control, but a broader definition of hazardous solid waste. The result is an apparent permanent trend toward recycling and incineration with energy recovery. Mandates are also emerging for "BACT" (expensive "best available control technology") not only for ash residues to landfill but also for dioxin and acid gas emissions. It is noteworthy that most major national collection/landfill agglomerates are also waste/energy incineration vendors now.

Even without such restraints, the resource recovery alternatives (mass burn and most processed fuel incineration concepts) will reject most bulky waste anyway. Notable exceptions operating today are Resources Recovery (Dade County) Inc., Miami, Florida, and the City of Glen Cove, New York, and both are described in this chapter.

Exceptions notwithstanding, the improvement of handling and disposal techniques for bulky wastes has been relegated to a holding pattern in this era of permanently higher costs for solid and fluid fossil fuels, *and it is likely that this stagnation will continue until worsening landfill availability and crisis conditions soon force the issue.*[11,12]

Fig. 17.3   Industrial bulky waste.

17.4   BULKY WASTE PROCESS EXPERIENCE

17.4.1   Background

With perhaps the aforementioned exceptions, there are no high-volume (> 20 TPD) bulky waste-processing facilities presently operating daily and in continuous service in the United States, although upwards of 10 have performed successfully during the past 20 years (this does not include baling waste paper, corrugated, or metals, which are not in the disposable category and are sold).

Almost without exception, the reasons for discontinuance have been economic or logistical and not technical. In most cases, the process did an acceptable job of reducing volume and increasing product bulk density to enhance storage, transportation, landfilling, materials recovery, or combustion.

A fair evaluation of why these operations succumbed reveals that they were a little early, that is, it was still possible to dispose of bulky wastes without processing at less cost and much less bother. The inevitable crisis, however, is fast approaching in many areas via last-ditch landfill

Fig. 17.4 Typical treatment of bulky waste at a landfill. (Courtesy of Caron Compaction Co.)

Fig. 17.5 Bulldozer with track-mounted demolition pads. (Courtesy of Caron Compaction Co.)

Fig. 17.6 Spiked compaction wheel. (Courtesy of Caron Compaction Co.)

Fig. 17.7 Compactor wheel landfill machine. (Courtesy of Caron Compaction Co.)

finality and the inability of most current resource recovery practice to handle such material. The bulky waste processes included:

1. Shredding by heavy-swing hammermills (> 300 rpm, < 1200 rpm).
2. Guillotine shears.
3. Heavy-duty scissor-type shear.
4. Heavy-duty baling press.

This interlude did not include the fairly recent low-speed rotary shear because it is still in a period of evaluation in continuous daily service and was not involved in any of the installations, although the concept had existed previously in Europe where most processed oversize material is much less difficult than in North America.

A rather comprehensive "Consumer Reports"-type test program funded by the U.S. Department of Energy and performed by EG&G, Inc.[6] on reduction of assorted bulky waste compares four commercially available rotary shears. The report indicates apparent capabilities to reduce a wide range of materials. However, probabilities for uninterrupted operation in continuous daily service could only be extrapolated with the usual vagaries of scaling up, modelling, and so forth, to be expected.

17.5 BULKY WASTE PROCESSING CASE HISTORIES

17.5.1 City of Harrisburg, Pennsylvania

Background and Plant Description

The mass-fired waterwall incinerator was an A & E Project that started up in early 1972 and included a large bulky trash shredder, Figures 17.8 and 17.9. It was driven @ 900 rpm by a 2000 hp (1.5 × 10^5 kgf-M/sec) single-stage condensing steam turbine @ 3600 rpm with a spray pond for condenser cooling water, and with turbine throttle steam @ 260 psig, 530°F (18 bar, 276°C). A direct drive train consisted of a heavy-duty Lufkin Reduction Gear and Falk Steelflex Couplings, Figure 17.10. The typical torque-speed characteristic of a steam turbine (zero speed, maximum torque) are favorable in a heavy-duty shredder application, Figure 17.11. Surprisingly, the direct drive gear reduction train was essentially trouble free in such incessant high-shock service.

Two Martin Waterwall Furnace/Stoker units @ 375 TPD were installed with electrostatic precipitators, although there was no attainable steam market at the time (the emissions tail wagging the furnace dog) despite the fact that Pennsylvania Power and Light Co. operated a district steam heating system in the city several miles away. Steam not used by the shredder and turbine driven plant auxiliaries was condensed on the roof (most of it).

The shredder was a Hammermills, Inc., compression feeder type, 60 in. (152 cm) hammer swing diameter, 80 in. (203 cm) wide with 150 lb (68 kg) hammers @ 500 brinell and a capacity of 25 TPH. It was installed in a partially covered concrete vault adjacent to the receiving pit with a bulky waste receiving area on the tipping floor, Figure 17.12. Raw material, Figure 17.13, was accumulated and fed to the shredder by a shredder operator and the payloader operator, each having other duties as well.

Operating History

The shredder installation performed reliably on a daily basis from 1972 until 1978, when a connecting steam line to the existing district steam system was complete. The enormous increases in fossil fuel prices of the 1970s along with renewable energy credit payments was a windfall that justified the expensive pipeline and helped vindicate the earlier waterwall decision. Unfortunately, the high water rate (steam consumption) of the shredder turbine and its load swings would be difficult to control along with the newly imposed load-leading requirement of the district heating system, so the shredding operation was discontinued. Nonburnable, bulky waste that escaped detection and managed to get through the furnace/ash systems was piled behind the plant, Figure 17.14, until bulky waste bypass of the plant could be arranged. In addition, a higher-priority (than bulky waste shred) steam drying sludge facility was built adjacent to the incinerator but it has not yet become fully operational. Nonetheless, the shredder performed well for a period of 6 years.

Most recently (1984), the nearby Steelton Works of Bethlehem Steel Corp. has signed a long-

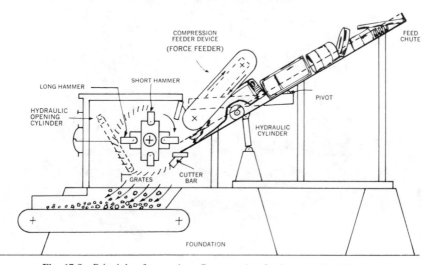

Fig. 17.8 Principle of operation: Compression feeder-type shredder.

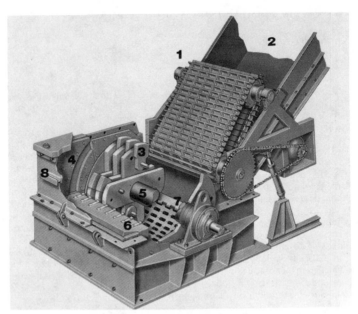

Fig. 17.9 Cutaway view, compression feeder-type shredder. 1. Compression feed device pivots up or down to grip or crush large items for controlled feed. Variable-speed feed track reversible. Reduces shock loads on mill, motor, etc. 2. Feed angle guides material into cutting wedge beneath hammers, minimizes hammer back-fling, stalling, power, etc. 3. Double-edge hammers, long and short, permit two life cycles. 4. Replaceable wear liners. 5. "Spider"-type rotor with staggered hammer rows provides shredding across full width of mill. 6. Deep section tangential grate openings for fast discharge minimum plugging. 7. Notched reversible cutter bar provide shear interface with hammers. 8. Hydraulic quick-opening rear quarter for fast access.

Fig. 17.10 2000 hp steam turbine reduction gear shredder drive train, Harrisburg, Pennsylvania.

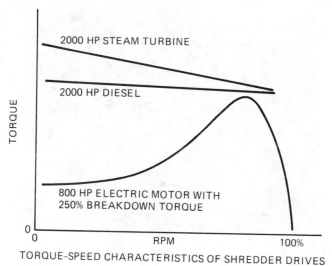

Fig. 17.11 Comparison of shredder drive torque-speed characteristics: Electric motor, diesel engine, steam turbine.

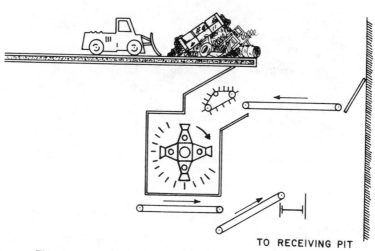

Fig. 17.12 Diagram of Harrisburg bulky waste shredder installation.

term contract with the Harrisburg incinerator for purchase of 100,000 lb/hr (45,500 kg/hr) of steam @ $10.00/1000 lb (455 kg). This timely development is permitting rehabilitation of an otherwise troubled[7,8] mass-burn incinerater but without a useful and reliable feature . . . the bulky waste shredding system.

Shredder explosions were not frequent and the shredder design with its concrete "cell" isolation contained explosions with no personal injury or significant downtime. Until it was replaced by a special rubber belt system, the piano hinge, outboard-roller, steel belt, shredder discharge conveyer was the most troublesome element in the installation and required nearly constant attention.

Ironically, the most damaging and expensive explosion and fire by far ($950,000) occurred in the furnace charging chute, causing a major pit fire with extensive structural steel and crane damage. This occurred shortly after the shredder was shut down when a drum of solvent[8] eluded observa-

Fig. 17.13   Bulky waste receiving floor, shredder feed area, Harrisburg, Pennsylvania.

Fig. 17.14   Unshredded nonburnable oversize pulled from furnace ash system, Harrisburg, Pennsylvania.

tion in the pit. (The shredder compression feeder would have passively ruptured the drum before it entered the shredder.)

During its operating tenure (1972 to 1978), the bulky waste shredder operating and maintenance costs averaged from $1.50/ton to $2.50/ton including labor, and its production average was 50 to 60 TPD of a very wide range of residential, commercial, and especially industrial bulky waste with a tipping fee from $8.00 to $12.00/ton.

17.5.2   City of Chicago, Illinois, Goose Island

Background and Plant Description

Built on the site of an old incinerator (1929 to 1934), this plant was a bulky waste shredding transfer station, Figure 17.15, that operated successfully from late 1970 until 1981. Goose Island is a small natural island in the Chicago River at about 1200 North, a few miles from the downtown loop area. Despite about 10 years of useful and successful operation, the service was eliminated for economic reasons, mainly the costs of collection due to skyrocketing truck operating costs, that is, exhorbitant fuel prices along with accelerating maintenance and labor costs in a highly inflationary period. Although shredding costs also increased, collection cost factors and logistics were the underlying reasons.

Fig. 17.15 City of Chicago "Goose Island" bulky waste shredding transfer station. (Courtesy Public Works Magazine.)

The shredder was a Hammermills, Inc. 60-in. (152 cm) diameter hammerswing, 80-in. (203 cm) wide topfeed design, Figures 17.16 and 17.17, 900 rpm with an 800-hp (600 kw) electric motor @ 250% breakdown torque and a direct connected "donut" coupling. With the assistance of city engineers and the consulting engineers, Perkins & Will, slowly transversing rabble arms were added to the upper rear wall of the shredder hood to minimize occasional bridging of long items.

Completed in late 1970, the facility capital cost, including site work, building, and equipment, was approximately $1,200,000 and financed by city garbage and refuse disposal bonds.[9]

Operating History

The city generated approximately 300 TPD of bulky waste[9] including demolition debris of which about 60 TPD of the most difficult material was delivered to the plant by truck (Figure 17.15). Material dumped on the tipping floor, Figure 17.18, was checked for hazardous and uncrushable material (I-beams, die blocks, and the like) by the payloader and roving/shredder operator, Figures 17.18 and 17.19. White goods and other salvage metals were removed when convenient and sold to a local scrap processor (@ $45.00/ton in 1975).

Shredded material was transported outside the building by belt conveyer, Figure 17.20 to a vibratory live-bottom circular bin for truck loading, Figures 17.21 and 17.22. This storage bin was not effective (plugging) so the bottom was removed and trucks loaded by payloader from piles on the ground and weighed again, Figure 17.23, then directed to one of the city's incinerators.

Space limitations prevented the inclusion of magnetic separation of shredded ferrous following the shredder, and this material could be recovered for sale only after magnetic separation from incinerator ash.

17.5.3 Resources Recovery (Dade County) Inc., Miami, Florida

Background and Plant Description

Designed, built, and operated for Dade County by Resources Recovery (Dade County) Inc., this is the largest refuse-fired steam electric plant in regular operation. Since early 1982, it has processed upwards of 3000 TPD, producing refuse-derived fuel (RDF) and recovering materials from a very wide range of residential, commercial, and industrial raw material, including bulky waste. Delivery is by city, county, and private haulers.

Designed as a dual fuel system, RDF preparation can be by an auxiliary wet-pulping process, Figure 17.24, and mainstream "dry"-shred RDF process systems, Figure 17.25, with from 6 to 10% by weight of total ash residue with minimal heavy metals (lead, cadmium, and so on) remaining for landfill.

Normally, most RDF production for steam/electric generation is by the dry shredders, Figure 17.26. Bulky trash (up to 50% of total by weight) is separated at delivery or by manned hydraulic

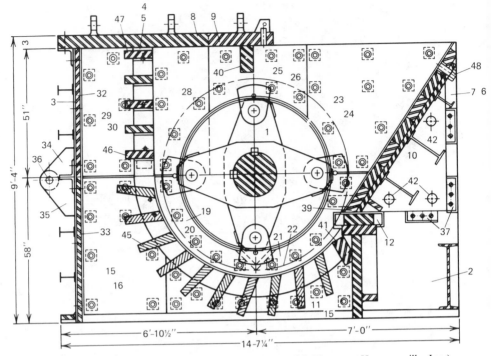

Fig. 17.16 Top feed-type shredder at "Goose Island." (Courtesy Hammermills, Inc.)

Fig. 17.17 "Goose Island" shredder feed conveyer and hood.

Fig. 17.18 Bulky trash on the tipping floor, "Goose Island."

Fig. 17.19 Checking raw feed for uncrushable and hazardous material.

Fig. 17.20 Shredded material conveyed to outdoor truck loading storage bin. (Courtesy Perkins & Will, Engineers, Chicago, Ill.)

Fig. 17.21 Shredded material outdoor storage truck loading bin. (Courtesy Perkins & Will, Engineers, Chicago, Ill.)

Fig. 17.22 Loading truck with shredded material from storage bin.

Fig. 17.23   Weighing outgoing shredded material. (Courtesy Public Works Magazine.)

Fig. 17.24   Dade County RDF auxiliary pulping equipment. (Courtesy Resources Recovery, Dade County, Inc.)

knuckle boom cranes (if mixed with garbage), Figure 17.27, for separately scheduled shredding. There are four 190,000 lb/hr (86,360 kg/hr) spreader stoker boilers @ 600 psig, 700°F (41 bar, 371°C) serving two 37.5 mw turbo-generators. The wet-process RDF system beneficiates recovered "recyclable" materials (ferrous, aluminum, and other nonferrous, glass, and most ash) for enhancement of saleability and market continuity, Table 17.1.

Two large compression feeder-type shredders (Figures 17.9 and 17.26), 66-in. (168 cm) diameter hammerswing × 104 in. (264 cm) wide are driven by 1250-hp (936 kw) electric motors @ 900 rpm and direct-connected flex couplings with 250% breakdown torque. The shredders are fed by heavy-duty "NICO" apron feeders with inboard tractor-type sealed bearings and chains, and the feeders are loaded from the tipping floor by operator-manned Barko pedestal-type hydraulic cranes (Figure 17.27), with optional payloader assist. The shredder operator is in a protected control room alongside and above the shredders for optimum sensory perception (see, hear, feel) supplementing instrumentation and controls. In addition, there is surveillance by the crane operator and a roving tipping floor director and/or a payloader operator.

Each shredder is equipped with a 20,000 CFM (cubic feet/minute) dust collection system with explosion relief panel-type dust pickup points at the shredder discharge, with relief panels throughout the downstream ducts, plenums, cyclone collectors, and baghouses, Figure 17.28. Each shred-

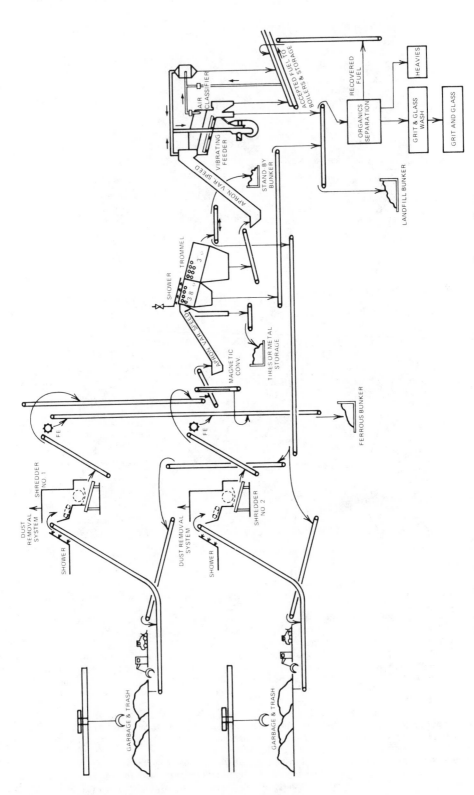

Fig. 17.25 Dade County dry shred RDF and bulky waste process system. (Courtesy Resources Recovery, Dade County, Inc.)

Fig. 17.26 Dry RDF production shredder for residential, commercial, industrial, and bulky wastes, Dade County, Florida (Courtesy Resources Recovery, Dade County, Inc.)

ding line includes two stages of magnetic ferrous separation followed by screening and air knife cleaning (Figure 17.25).

Operating History

The shredders and RDF preparation systems have also performed well when processing bulky wastes, with operating and maintenance cost well within acceptable limits of about 12 kwh/shredded ton, 10,000 ton/hammer, and with over 95% availability.

There has been no downtime, personal injury, or significant damage caused by shredder explosions nor has extremely difficult rogue material (wire or plastic bales, steel plate) caused extraordinary maintenance or downtime. Judicious system design (semioutdoor), equipment selection, and

Fig. 17.27 Barko "Knuckle-Boom" hydraulic loaders. (Courtesy Resources Recovery, Dade County, Inc.)

Table 17.1 Recovered Materials, Resources Recovery (Dade County) Inc.

By-Product	Price, 1984 ($)	Quantity Sold per week
Ferrous	65.00/Ton	800/1000 Tons
Aluminum, MRP	400.00/Ton	30 Tons
Aluminum, Cast	760.00/Ton	10 Tons
Aluminum, Can	1000/Ton	5 Tons
Brass, bronze, zinc	700.00/Ton	3 Tons
Copper	850.00/Ton	1.25 Tons
White goods	75.00/Ton	150 Tons
Heavy cup (small size nonferrous)	340.00/Ton	1 Ton
Coins	1000/Week	—
Ash and flyash	2.50 to 3.50/Ton	1900 Tons
Glass and grit	2.50 to 2.50/Ton	1200 Tons

disciplined operating management appear to account for this (see Chapter 12, Section 12.14 and References 1 and 2).

17.5.4 City of East Chicago, Indiana

Editor's note: The following is exerpted from a description by Mr. Fred Rohr, P.E., Vice President, Perkins and Will Engineers, Chicago, Illinois, that appeared in the March, 1978, issue of *Public Works Magazine*. Mr. Rohr presently heads his own consulting engineering firm in Oak Brook, Illinois.

Fig. 17.28   Shredder dust collection system.

Background and Plant Description

Every year, Americans discard millions of tons of tires, mattresses, couches, and kitchen appliances. Municipalities contribute to the bulky waste load with tree trunks, demolition rubble, and concrete from street and sidewalk improvements.

Most cities haul such refuse to landfill dumps and bury it, but diminishing landfill space now demands another solution. The option selected was a new bulky refuse processing and shredding facility which has been operating since early 1977. The refuse is then easily managed for incineration or landfill.

The municipality that buries its bulky refuse without shredding is discarding valuable material. Bulky waste contains from one-fourth to one-third ferrous metal recoverable by shredding. By comparison, household garbage only has about 5 to 8% salvageable ferrous, mostly "tin" cans.

Bulky waste is more difficult to shred than ordinary refuse. Plants built to handle household garbage normally cannot handle the heavy items in bulky material, yet only a few municipalities have developed facilities to handle bulky waste materials.

The Solid Waste Center includes a new vehicle maintenance garage with office space, an animal facility for receiving, holding, and disposing of dead animals, the shredding plant, and the incineration plant. The latter is a modern 450 TPD facility equipped with a scrubber for air pollution control.

The shredding system, Figure 17.29, reduces 25 tons of bulky refuse an hour to pieces no larger than 6 in. (15 cm). It was designed to handle three specific types of bulky waste materials. Combus-

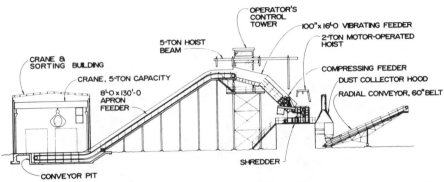

Fig. 17.29 Shredding system sketch, East Chicago, Illinois. (Courtesy Public Works Magazine.)

Fig. 17.30 East Chicago, Indiana, outdoor shredding system. (Courtesy Public Works Magazine.)

tible items such as mattresses, sofas, and wood materials are shredded and conveyed to the incineration receiving pit. Shredded metallics from kitchen appliances, washers, dryers, and other white goods are collected for salvage of the ferrous metal. Broken concrete from sidewalk and street renovation is crushed by landfill.

The shredder plant was approved by the Board of Sanitary Commissioners of the city and financed by a bond issue for the plant, small animal facility, and other needed projects. Design work began after selecting machinery and offering bid documents only to bidders manufacturing shredders equipped with compression feeders. Bids were opened in February, 1974, and Jeffrey Manufacturing Company, a division of Dresser-Industries, in Columbus, Ohio, was awarded the contract to furnish and install the "913 Scrap Hog" shredder, feeder, all conveyors, and the dust-handling system.

Using this equipment as a basis of design, plans and specifications were prepared for bidding the building, foundations, and utilities.

An objective was to reduce downtime and maintenance with a shredder large enough to minimize jamming and overloading. Equipment selection safety, and layout were important considerations, and to reduce cost, the equipment was located outdoors next to the 8000 ft^2 (743 m^2) sorting building, Figure 17.30. This greatly enhances shredding safety factors as well.

The sorting building is 34 ft (10 m) high inside so that large dump trucks can raise their beds to discharge their loads of broken furniture, wheels and tires, mattresses, or crumbling blocks of concrete. Electrically operated 16-ft-high (4.9 m) rolling aluminum doors located at both ends of the building allow delivery to each side of the "T" conveyer that runs the width of the building. A bridge crane for sorting the heaviest materials runs the length of the building, Figure 17.31.

Fig. 17.31 Bulky waste Receiving Floor, East Chicago, Indiana. (Courtesy Public Works Magazine.)

Trucks dump refuse on the floor, where the crane can select material for loading onto the 8-ft-wide (2.4 m) conveyer belt situated 2 ft (0.61 m) below the surface of the main floor.

Operating History

As the material accumulates, it is sorted and shredded by category. The conveyor belt moves the material out of the building upward to the vibrating feeder which is elevated 47 ft (14 m) above grade. The feeder controls the flow of material under the compression feeder where it is flattened and/or gripped for controlled feed into the shredder opening. The compression feeder keeps the shredder opening from jamming by regulating the flow of material.

The shredder, driven by a 2000 hp (1500 kw), 4160-volt electric motor, has 34 hammers, each weighing up to 150 lb (68 kg) mounted on a 6-ft-diameter (1.8 m) rotor in four rows—two rows of long hammers and two of short. These hammers are free to rotate a full 360° on hammerpins, which keeps wire, textiles, and the like from wrapping around them. The refuse is sprayed with water to minimize dust and the possibility of fire along process lines.

The shredder is isolated on a stabilizing concrete foundation 17 ft (5 m) above grade, to minimize vibration. The foundation module contains three equipment rooms. The first room contains electrical switch-gear and the motor control center. The second houses the hydraulic pump set that controls the force of the compression feeder, opens the top cover of the shredder, and operates the "pinpuller" for removing the hammers from the rotors during maintenance. This room also contains the lubrication system for the shredder bearings. In the third room, there is an air compressor for operating the impact wrenches used for maintenance.

Shredded material passes through grate bars onto a steel apron conveyor and combustible batches are dropped into the adjacent incinerator receiving pit by a radial stacker belt conveyor. By means of a traversing belt conveyor, ferrous scrap can be discharged outside by recycling, and crushed demolition rubble can be piled and hauled away to landfill, Figure 17.32.

The shredder operator directs operations from an enclosed control tower located above the equipment. From this point, he controls the material moving along the feed conveyor and the vibrating feeder, regulating speeds to maintain uniform flows. The raw feed conveyor also has automatic controls that slow its speed, or stop it if the shredder motor becomes overloaded.

The shredding operation is protected by an explosion suppression system that "senses" the start of an explosion when pressure inside the shredding module rises slightly. It releases suppressive agents that inhibit the rise in pressure and retard the explosion. A carbon dioxide fire extinguishing system safeguards the dust-collecting ducts and cyclone collectors, Figure 17.33.

Fig. 17.32 Ferrous and Noncombustibles for recycling, East Chicago, Indiana. (Courtesy Public Works Magazine.)

Fig. 17.33 Shredder dust collection system, East Chicago, Indiana. (Courtesy Public Works Magazine.)

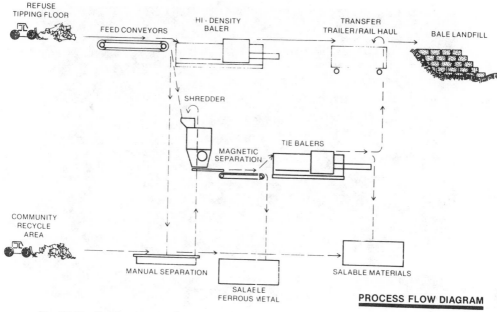

Fig. 17.34 Omaha shred/bale process flow diagram. (Courtesy Harris Press and Shear.)

Another alarm system monitors carbon monoxide emissions in the sorting building. Occasional explosions have not caused serious damage, downtime, or personal injury. As of early 1985, the plant operates five mornings per week.

17.5.5 City of Omaha, Nebraska, Solid Waste Recycling Center

Background and Plant Description

In operation since 1975, this combination shredding and baling system, Figure 17.34, was designed by Henningson, Durham & Richardson to process up to 600 TPD of mixed solid waste for anticipated future area growth, with a present raw material availability of about 300 TPD.

There are two 40 TPH process lines that produce bales for truck (or rail haul) to landfill. One line makes high-density bales from raw refuse as received, and the other shreds mixed raw material, Figure 17.35, including bulky waste, followed by two-stage magnetic drum ferrous recovery and a horizontal tie baler, Figure 17.36. The two lines are interconnected for optimum versatility (Figure 17.34), and the system design provides for future RDF production by the addition of process components to supplement the shredder and magnet modules.

The shredder is a Hammermills, Inc., compression feeder type (Figure 17.9), 60 in. (152 cm) diameter hammerswing, 80 in. (203 cm) wide with 150 lb (68 kg) hard hammers @ 500 Brinnell and a capacity of 50 TPH of mixed solid waste @ 900 rpm with a 1250 hp (932 kw) electric motor and 250% breakdown torque, Figure 17.37.

The baling presses are by Harris Press and Shear, Inc. (American Solid Waste Systems Division of Amhoist).

Operating History

Operating procedures have depended considerably upon seasonal variations in the characteristics of the raw material, mainly moisture levels and quantities of organic material, such as landscaping/yard wastes, land clearing, and the like.

During the fall and winter seasons when raw material is dryer, mainstream processing produces high-density (untied) bales with raw material as received. The shredding line handles excess mixed material and separated industrial and commercial wastes including scrap metals high in saleable recyclables. The integrity of the high-density bales (how well they hold together) is adequate and they are less troublesome.

Conversely, mainstream processing during spring and summer is by the shredder/tie baler line because of the higher integrity of the shredded and tied bales when moisture and organics are high.

Fig. 17.35 Mixed bulky and residential solid waste, Omaha, Nebraska.

Fig. 17.36 Horizontal auto-tie baler (foreground) and compression feeder shredder (background), Omaha, Nebraska.

The high-density untied bales are approximately 3 × 3.5 × 4.5 ft (0.9 × 1.0 × 1.4 m) weighing 2500 to 4500 lb (100 to 1800 kg). The tied bales of shredded material are approximately 3 × 4 × 5.5 ft (0.9 × 1.2 × 1.6 m) weighing 2500 to 4000 lb (1000 to 1600 kg).

Shredder power consumption averages vary from about 7 kwh/ton for residential/commercial material, and about 12 kwh/ton for bulky wastes, with hammerwear at 8000 to 10,000 tons of production per set. Although there have been no shredder explosions, there has been an occasional low-velocity deflagration—a passive sheet of flame mostly contained by the shredder module with easily and quickly quenchable small fires and with little or no operating interruption. Appendix 17.1

Fig. 17.37 1250 hp (932 kW) compression feeder-type shredder, Omaha, Nebraska.

includes Tables 17.2, 17.3, and Figure 17.55 that present a shredder product size analysis and a noise level curve.

The baling presses have experienced no extraordinary maintenance that has been essentially routine. It was found, in fact, that a reduction in hydraulic pressure did not significantly effect bale density/integrity in their particular operating range, with a collateral benefit of a significant reduction in wear plate maintenance.

The plant operates single shift 5 days/week with a full-time operating and supervising staff of 16, with six additional temporaries on a seasonal basis. Annual production rates are from 65,000 to 85,000 ton/year, depending upon material availability.

Total operating cost, including baling/shredding transport and landfill is about $16.50/ton, approximately half of which is the cost of operating the process station. Landfill is from 8 to 10 miles (13 to 17 km) away.

Ironically, the plant is scheduled to cease operations and close down completely in 1985 for economic reasons, with conventional landfill costing from $4.50 to $5.00/ton and presumably no immediate concern for land availability.

17.5.6 City of Glen Cove, New York, Codisposal/Energy Recovery Facility

Background and Plant Design

Designed by William Cosulich Assoc., Inc., Consulting Engineers, Woodbury, New York, under a conventional A/E Public Works Procurement Program with municipal ownership, this facility, Figure 17.38, is operated under private contract. Design commenced in 1974, and the plant started up in 1982 and includes a Mosely Bulky Trash Shear installation, Figure 17.39, for up to 30 TPD. Total plant capital cost was $24 million.

The 250 TPD Codisposal System, Figure 17.40, employs two 125 TPD mass-burn refractory furnaces with convection boilers, each for 112.5 TPD of municipal solid waste and 12.5 TPD waste-activated sludge (20% solids). Steam is generated at 600 psig SAT (41 bar SAT), 34,000 lb/hr (15,500 kg/hr) per boiler to produce 2500 kw, about half of which supplies plant requirements and the remainder available for sale to the Long Island Lighting Co. The solid waste tipping fee is presently $26.00/ton and electricity revenue is $.06/kwh (September 1984).

Operating History

Daily delivery of residential, commercial and industrial bulky waste varies from 20 to 35 TPD (10 to 15%) of total material received and is reduced to manageable size, Figure 17.41, by the hydraulic

Fig. 17.38 Glen Cove, New York, codisposal plant. (Courtesy Waste Age Magazine, Nov. 1983.)

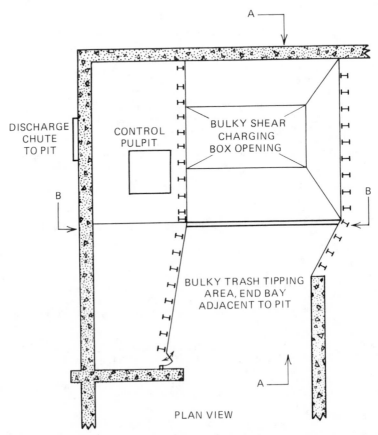

Fig. 17.39 Bulky trash guillotine shear, Glen Cove, New York. (Courtesy Wm. F. Cosulich Associates.)

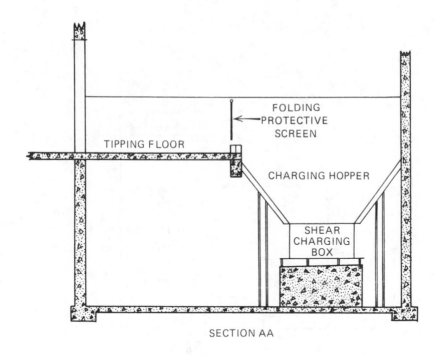

SECTION AA

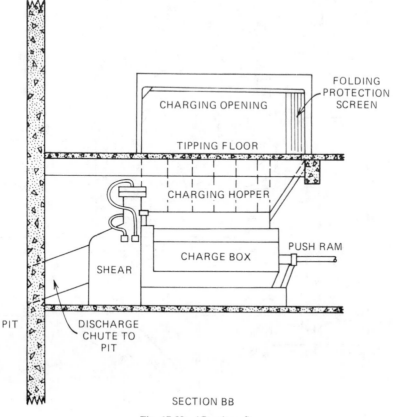

SECTION BB

Fig. 17.39 (*Continued*)

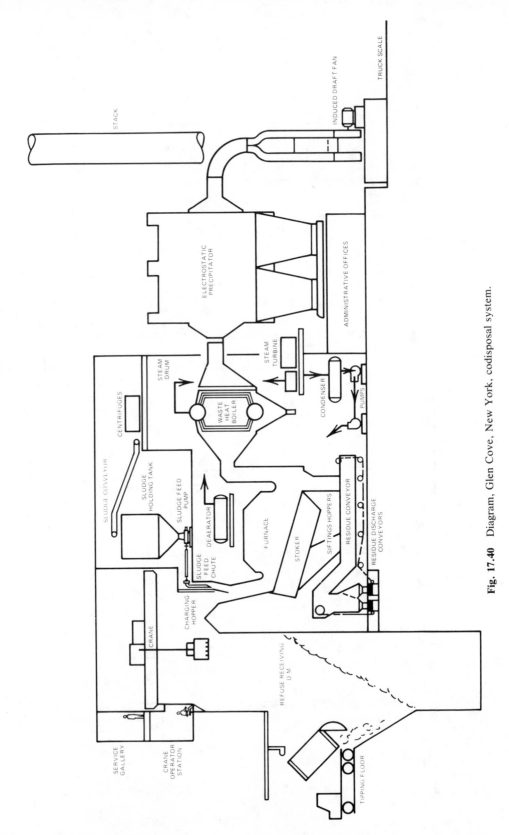

Fig. 17.40 Diagram, Glen Cove, New York, codisposal system.

Fig. 17.41   Sheared bulky waste, Glen Cove, New York.

guillotine shear. Bulky material is delivered to an end-receiving bay, dumped, and then pushed by payloader into the shear charging box, after which the charging box lid closes automatically to compact the admixture. A ram moves the material forward beneath the shear blade in incremental steps, and the shear blade descends to slice the load into pieces which are chuted down into the receiving pit. Operating cost for shearing the bulky waste has not yet been developed and/or released.

17.5.7   City of Montreal, Quebec, Canada

Plant Description and Background

The incinerator started up in 1971, designed for 1200 TPD, and consists of four 100,000 lb/hr (45,500 kg/hr) steam boilers @ 250 psig, 550°F (17 bar, 288°C) of the Von Roll Mass Burn Grate, Furnace and Boiler Design, fabricated and installed by Dominion Bridge Ltd. of Toronto. Included is a Von Roll Bulky Waste Shear, which can be described as a hydraulic multiblade, top-feed type with bottom pivot scissor action, Figure 17.42. It was installed at the furnace feed chute level, Figure 17.43, fed by the crane and with discharge to the receiving pit, similar to a demonstration unit, Figure 17.44.

Operating History

It is reported that the bulky waste shear operated on a regular basis until a reduction in the quantity of larger bulky materials delivered to the plant reduced the shear's operating hours. Burnable unsheared oversize material can sometimes be fed into the feed chute and on through the furnace.

17.5.8   City of Kyoto, Japan

Plant Description and Background

A bulky waste size reduction and materials recovery station, Figure 17.45, adjacent to mass-burn incineration facilities was designed and built by Mitsubishi Heavy Industries and started up in late 1972. Since then, approximately 21 other shredding stations have been built in Japan, about half of which recover materials, with the residues going to nearby incinerators. Often, the shredding facility is adjacent to the incinerator and is not an integral part of the incinerator building. The remaining shredding operations are for landfill.

Figure 17.46 is a flow diagram of the shredding and materials recovery system. The Kyoto installation is the largest of the bulky waste reduction stations in Japan. It includes a large Hammermills, Inc. 74-in. (188 cm) diameter × 104-in. (264 cm) wide compression feeder-type shredder (Figure 17.9), 2000 hp (1500 kw) @ 720 rpm, with 40 TPH bulky waste average capacity. The shredder and heavy-duty "NICO" inboard roller chain receiving/infeed conveyer, Figure 17.47, were manufactured under license by Mitsubishi in Japan, and the recovered materials storage bin system is shown by Figure 17.48.

Dimensions

Width	13 ft.
Length	26 ft.
Height	13 ft.
Weight	40 tons
Feed opening	11 x 12 ft.

Technical Data

Drive	electric-hydraulic
Power requirements	50 HP.

Capacity

Crusher volume	700 cu. ft.
Cutting width	1 ft.
Capacity	4250-7000 cu. ft./hr.

1   Inlet

2   Discharge

3   Hydraulic Drive

4   Cutting Edges

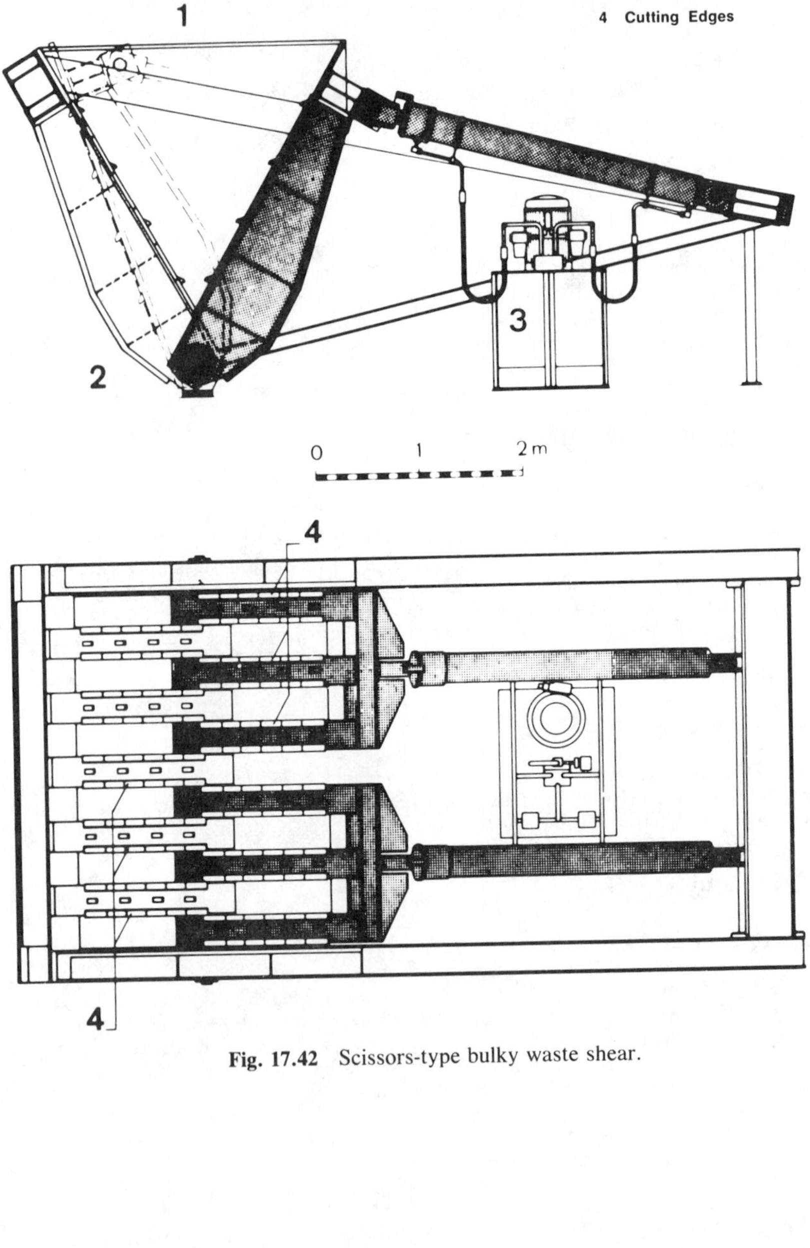

0   1   2 m

Fig. 17.42   Scissors-type bulky waste shear.

Fig. 17.43 Von Roll bulky waste shear feed opening at furnace charging level.

Fig. 17.44 Sheared bulky waste; Von Roll shear.

Fig. 17.45   Bulky waste shredding station adjacent to incinerator.

Operating History

There is little operating data from this bulky waste facility other than a reported 40% of the total feed to the incinerator has been processed through the bulky waste system at times, although much of it was not oversized. Purportedly, this was due to improved furnace performance with shredded residential "production" material.

It has also been reported that a preponderance of straw mats, often wet, frequently plasters or sticks to the shredder internals and requires cleaning at regular intervals.

17.5.9   City of Ansonia, Connecticut

Plant Description and Background

The bulky trash shredding plant, built with federal, state, and local funding, was started up in 1975 to reduce oversize material delivered to the Ansonia incinerator consisting of two 100 TPD refractory furnace incinerators. The incinerator was started up in 1968.

A steady increase in oversize residential, commercial, and industrial bulky trash led to the construction of the shredding facility which consisted of a separate building adjacent to the incinerator pit and included magnetic drum separation of ferrous metals.

Raw oversize material was dumped on a tipping floor and fed to a steel belt conveyer by a payloader, Figure 17.49, for delivery to the shredder, which was a Williams 600, 60-in. (152 cm) diameter hammerswing × 60-in. (152 cm) wide, 1200 hp (900 kw) @ 900 rpm. Integral with the shredder top hood was a ballistic ejection opening and chute for uncrushable material, Figure 17.50.

Operating History

An explosion in 1977 damaged part of the roof structure, and roll-up doors, with no damage to the incinerator. The shredder was essentially intact with the exception of contiguous hopper and conveyer parts. After repairs, operation was resumed and the incinerator plant continued to operate daily.

17.5.10   City of Tacoma, Washington

Plant Description and Background

Started up in 1972, this bulky trash shredder, Figure 17.51, was installed outdoors on a hillside at a city landfill under a federal EPA prototype installation study grant. The oversize material contained

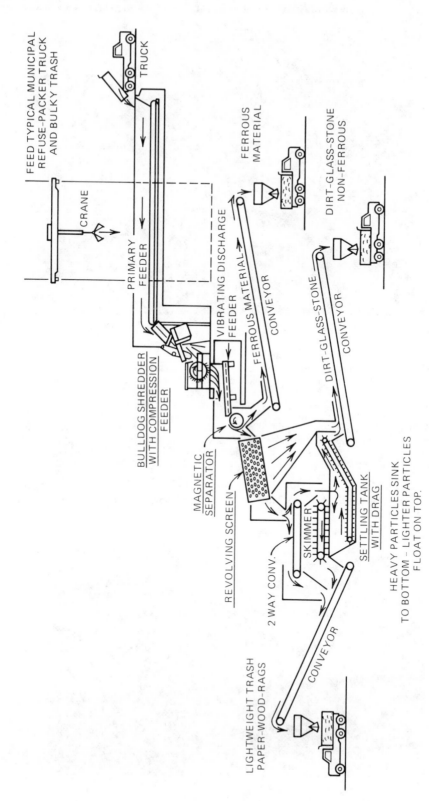

Fig. 17.46 Bulky waste shredding and materials recovery system, Kyoto, Japan.

FEED TYPICAL MUNICIPAL REFUSE-PACKER TRUCK AND BULKY TRASH

TRUCK

CRANE

FERROUS MATERIAL

DIRT-GLASS-STONE NON-FERROUS

PRIMARY FEEDER

VIBRATING DISCHARGE FEEDER

BULLDOG SHREDDER WITH COMPRESSION FEEDER

FERROUS MATERIAL CONVEYOR

MAGNETIC SEPARATOR

DIRT-GLASS-STONE CONVEYOR

REVOLVING SCREEN

2 WAY CONV.

SKIMMER

SETTLING TANK WITH DRAG

HEAVY PARTICLES SINK TO BOTTOM – LIGHTER PARTICLES FLOAT ON TOP.

LIGHTWEIGHT TRASH PAPER-WOOD-RAGS

CONVEYOR

Fig. 17.47 Raw bulky waste receiving conveyer and adjacent storage bin, Kyoto, Japan.

a preponderance of assorted wood wastes along with municipal items—mattresses, furniture, and so forth. Delivery trucks dumped onto a concrete apron at both sides and the end of the shredder infeed conveyer that was fed by a payloader. The shredder discharged directly into a dump truck underneath it, Figures 17.52 and 17.53, which shuttled between the shredder and the working landfill cell. There was no separation of shredded materials for magnetic ferrous recovery.

The shredder was a Williams 680 topfeed style, 60-in. (152 cm) hammerswing diameter × 80 in. (203 cm) wide, with 800 hp (597 kw) @ 900 rpm. Design capacity was reported to be 30 TPH with a 6 in. nominal topsize. Operation was 5 days/week, as required by the availability of material.

The installation operated as a bulky trash shredder by the city of Tacoma until 1978 when the Boeing Engineering and Construction Co. and the city started up a 400 TPD prototype RDF production plant on the site using the existing bulky trash shredder for primary reduction of mixed city solid waste, Figure 17.54. The facility was designed by Boeing[10] but has experienced limited use due to uncertain RDF markets.

Fig. 17.48 Recovered materials storage bin system, Kyoto, Japan.

Fig. 17.49 Bulky waste shredder receiving/feed system, Ansonia, Connecticut.

Fig. 17.50 Bulky waste shredder, Ansonia, Connecticut.

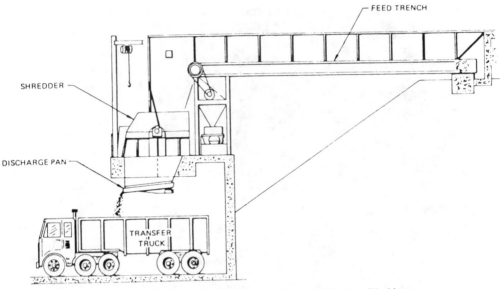

Fig. 17.51 Bulky waste shredder system diagram, Tacoma, Washington.

Fig. 17.52 Bulky waste shredder Hillside installation, Tacoma, Washington.

Fig. 17.53 Bulky waste shredder discharge to truck, Tacoma, Washington.

Fig. 17.54 Tacoma bulky waste shredder adapted to RDF production system.

As an outdoor installation with careful raw material surveillance, there were few explosions with insignificant damage, and operations were considered quite satisfactory, with no extraordinary difficulties.

17.6 ABORTED BULKY WASTE PROCESS PROJECTS

17.6.1 Background

Several large mass-burn incinerator plants were designed and built since the early 1970s with provision for size reduction of bulky wastes. This would permit accepting such material on a regular basis for minimal difficulty with blocked furnace charging chutes, stoker efficiency, and burnout, with no wrestling of nonburnable oversize through the ash removal system.

In each case, the size reduction effort was abandoned soon after plant start-up for a variety of reasons.

17.6.2 Summary of Aborted Projects

Case 1: A large, mass-fired incinerator plant in the mid-Atlantic states included a large swing hammer-type shredding system for municipal bulky wastes that was never placed in regular active service, purportedly for the following reasons:

1. Sufficient quantities of oversized material were not readily available to justify operating the large system, even on an intermittent or as available basis.
2. Excessive dust in the shredding area within the building.
3. Bothersome shock and vibration reverberations in the plant offices adjacent to the shredding area.

The incinerator continues to operate successfully but bulky trash is not delivered to the plant and is diverted to landfill.

Case 2: A mass-fired incinerator in the Northeast included a large bulky trash shredder installed at the furnace charging level and was fed by the pit/crane bucket, with shredded material chuted down into the pit. Although the region generates large quantities of bulky waste, the shredding system was never used regularly. It was removed and sold within a few years, reportedly for the following reasons:

1. Picking the bulky material from the garbage receiving pit was sometimes difficult. Also, feeding the shredder trash hopper with a large swinging cable-operated bucket resulted in time-consuming diversions from furnace charging and pit distribution/mixing routines.
2. The possibility of pit fires by shredding undetected containers of flammable fluids. Such an occurrence demonstrated the possibility.
3. Unshredded burnable oversize in the pit could be worked through the charging chutes and furnace without undue disturbance and nonburnables were not always excessively troublesome in the ash discharge system.

Case 3: A large, mass-fired incinerator in the Southeast has abandoned use of a large rotary shear-type shredder for bulky waste installed at the furnace charging chute level when the plant was built. It has been reported that feeding and uninterrupted operation were not possible without excessive labor, and continued operation could not be justified.

17.6.3 Analysis of Aborted Bulky Waste Process Projects

An important common factor in each case was the imprudent location of the shredder. Bulky waste processing as an adjunct to mass-burn incineration should be at or near grade level and remote from the main receiving/storage operations, office, and the like. Receiving materials handling and storage should be separate for minimal interference with production garbage operations.

Glen Cove, East Chicago, Ansonia, and Harrisburg can be considered correct in this respect, although Harrisburg was overly cramped with a bothersome pit discharge location for shredded material.

For processed fuel (RDF) installations, oversize material should be received, stored, and processed separately from residential garbage, and the Miami, Florida, installation of Resources Recovery (Dade County) Inc., is an example of good practice in this respect.

REFERENCES

1. W. D. Robinson, "Solid Waste Shredder Explosions—A New Priority," Shredder Explosions Seminar, AENCO, Inc., Div. Cargill Corp., Wilmington, Delaware, June, 1979.

2. W. D. Robinson, Proceedings, ASME Eighth National Waste Processing Conference, Discussions, Chicago, Illinois, 1978, p. 239.

3. ASTM, Thesaurus of Resource Recovery Terminology (H. I. Hollander, Ed.), by the ASTM Committee E-38 on Resource Recovery, ASTM Special Publication 832.

4. W. D. Robinson, Bulky Waste Processing Second Opinion Feasibility Reports for Government and Industry, by Hammermills, Inc., 1966 through 1982.

5. R. J. Alvarez, Status of incineration and generation of energy from thermal processing of MSW, Proceedings, ASME Ninth National Waste Processing Conference, Washington, D.C., May, 1980, and supplemental discussions volume.

6. G. R. Darnel and W. C. Aldrich, Low Speed Shredder and Waste Shreddability Tests, EG&G Idaho, Inc., for U.S. Department of Energy, Contract No. D.E.-Aco7-761Do1570.

7. P. B. Beers, "At Large" column, *The Harrisburg Sentinel*, November 4, 1983.

8. W. D. Turner, Thermal systems for conversion of municipal solid waste, in *Mass Burning of Solid Waste in Large Scale Combustors: A Technology Status Report*, Vol. 2, Argonne National Laboratories, NTIS DE84-002305.

9. M. Pikarsky, Commissioner of Public Works, City of Chicago, Chicago looks to refuse grinding, in *Public Works Magazine*, September, 1970, p. 82.

10. D. E. O'Neil and J. A. Bronow, Solid waste management development for the city of Tacoma, Washington, SAE-ASME-AIAA-ASMA-AIchE Intersociety Conference on Environmental Systems, San Diego, California, 1973, ASME-Publication 73-ENAS-34.

11. Fred Musante, "Tons of Trash Suffocating City," *Bridgeport Post*, May 15, 1985, p. 18.

12. "Spring Trash Cleanup Cancelled due to Garbage Crisis," public notice to Huntington (Long Island) N.Y. residents in the weekly trade paper *The Penny Saver* and as reprinted in *Waste Age Magazine*, May, 1985, p. 21.

APPENDIX 17.1 OMAHA SHREDDER PRODUCT SCREEN ANALYSIS AND NOISE LEVEL SURVEY

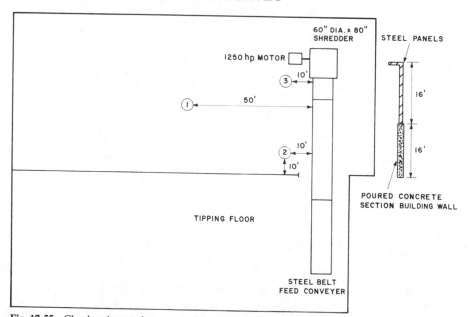

Fig. 17.55 Check points, noise survey, City of Omaha shredder system (couresy of Hammermills, Inc.).

Table 17.2   **Shredded Product Screen Analysis and Bulk Density Data, Hammermills, Inc. Model 6080 [60 in. (1.5 m) diameter × 80 in. (2 m) wide]**

"Bulldog" shredder 1250 @ 900 rpm; grate openings = 9 in. (22.8 cm) × 7 in. (17.8 cm)

Product screen analysis, average of three 28 lb (12.7 kg) samples:
- 98% are 10 in. (25.4 cm)
- 95% are 6 in. (15.2 cm)
- 90% are 4 in. (10.2 cm)
- 80% are 2 in. (5 cm)

Product bulk density at shredder discharge: Average of three samples @ 1 yd^3 approximately 95 lb/yd.^{3}

Shredder infeed: Approximately 80% household garbage can material + approximately 20% household and industrial oversize (mattresses, rugs, furniture, appliances, corrugated dunnage, and so forth).

Shredding rate during test: Approximately 40 TPH (80% of 50 TPH rated hourly average capacity).

Table 17.3   **Shredder Noise Survey, Bulldog Shredder Model 6080 @ 1250 hp**

Octave band	Locations[a] (Refer to Figure 17.55)					
	1	2	3	1A	2A	3A
16,000	50	50	50	50	54	52
8,000	50	50	50	50	62	61
4,000	63	63	59	64	66	67
2,000	66	67	63	66	70	73
1,000	69	72	63	69	74	78
500	71	73	70	72	83	81
250	72	72	71	75	88	91
125	80	78	75	81	88	90
63.0	81	82	82	84	94	96
31.5	90	93	94	92	97	98
All pass	93	93	94	93	100	103
A	73	77	76	75	83	84

[a] 1A through 3A: Shredding mixed municipal and industrial solid waste, approximately 80% packer, 20% bulky at approximately 40 TPH (approximately 80% capacity). 1 through 3: Running empty.

CHAPTER 18

REFUSE FUELS IN THE PORTLAND CEMENT INDUSTRY (INCLUDING TIRES AND AUTO SHREDDER RESIDUE)

DAVID WATSON

Blue Circle Industries
London, England

HEINRICH MATTHEE

Dyckerhoff Engineering
Wiesbaden, West Germany

WILLIAM D. ROBINSON

Consulting Engineer
Trumbull, Connecticut

18.1 EXPERIENCE IN ENGLAND*

David Watson

18.1.1 Refuse Versus Other Fuels—Technical Factors

Refuse is one of a number of combustible waste materials that can be used as fuels. There are various ways of classifying and comparing such unusual fuels. Calorific value alone is not an adequate parameter.

Physical characteristics are important. They determine the method and equipment for handling and preparation, and the capital and operating costs (Table 18.1). Refuse physical properties and composition vary according to country and districts within countries, and depend on the local arrangements for refuse collection. Also, there may be seasonal variations.

The main factors affecting kiln performance and the appropriate refuse usage rate are:

Percentage and size of incombustibles in the prepared refuse.

Content of K_2O, Na_2O, S, Cl, and so on.

Refuse moisture content and size.

Waste gas volume per net kcal combustion heat release.

Flame temperature.

---

*This section is a reprint of a paper presented by Dr. Watson at the 1982 International Cement Seminar, Chicago, Illinois, and titled "England's Blue Circle Burns Refuse; Practical Operating Experience with Waste Materials as Kiln Fuel—A Panel Discussion." By permission of *Rock Products Magazine*, sponsor of these annual seminars.

Table 18.1   Refuse Physical Characteristics and Kiln Application

Type	Examples		Possible Methods of Use
Brittle,	Charcoal fines, petroleum coke,	Finely ground	To burning zone
low ash	some lignites, rubber char	As small lumps or grit	To riser pipes of s.p. kilns, to precalciner vessels
Brittle,	Colliery minestone, coal washery tailings,	Finely ground	To burning zone, to riser pipe of s.p. kilns,
high ash	oil shales high carbon P.F.A., B.F. dust		to precalciner vessels, with raw meal feed, with slurry feed
Tough, nonbrittle, dense or large size	Domestic refuse, peat, woodchips, oil palm shell, rubber tires, acid battery cases	Pulverized or shredded	To burning zone, to riser pipe of s.p. kilns, to precalciner vessels
Tough, nonbrittle, but small size and/or low density.	Rice husks, chopped straw, sawdust	As received	To burning zone, to riser pipe of s.p. kilns, to precalciner vessels
Fluid	Waste oils, oil re-refining residue organic chemicals, acid tars	As received heated as necessary	To burning zone, to riser pipe of s.p. kilns, to precalciner vessels, to R.M. dryer furnace

Combustion gas volume is an important parameter because if it is too unfavorable versus the main, conventional kiln fuel, it can cause loss of kiln output. The clinker output of many kilns is limited by gas velocity/dust entrainment in the kiln, or pressure drop through the preheater, or ID fan capacity. Table 18.2 shows that refuse is toward the bottom-end of this league table, and is dependent on the as-fired moisture content.

The other main thermodynamic factor affecting kiln performance is flame temperature. If reduced too much by a waste fuel it can affect heat transfer rate in the kiln, kiln thermal efficiency, and output.

Note from Table 18.3 that refuse, if kept reasonably dry, is comparable with palm oil shells and lignite. Note also that rice husks, with a calorific value of only 3700 gross kcal/kg (6660 BTU/lb), have a higher flame temperature than many coals.

18.1.2   Development of Blue Circle's Interest

In about 1970, three main factors combined in the United Kingdom to stimulate Blue Circle Industries' (BCI) interest in refuse.

1. Some municipal authorities, particularly in the larger cities, were starting to run out of local landfill sites and were having to resort to the much more costly method of refuse disposal via purpose-built incinerators or haulage to distant tips.
2. The price of conventional cement kiln fuels was starting to escalate.
3. BCI had a number of redundant wet-process kiln plants.

The first studies were aimed toward using redundant kiln plants as incinerators. Schemes were investigated which used the rotary kiln as a dryer and the clinker cooler for incineration. Grate clinker coolers have similarities with grate incinerators and some coolers had often been used to destroy unwanted confidential papers. Our Westburn Works manager experimented with incineration of bagged domestic refuse in the grate cooler of his 1000 TPD kiln.

We carried out a short trial using the 150-ft (46 m) kiln at our Barnstone experimental plant as a straight incinerator, passing once-pulverized refuse through the high-speed peg mill, normally used for coal, and firing it as a P.F. (pulverized fuel) flame. This was followed by a similar, longer, and larger-scale trial on the redundant 230-ft (80 m) rotary kiln–Miag calcinator plant at our Rodmell Works.

Table 18.2 Comparing volume of products of combustion with heat value of conventional and refuse fuels. *Note:* As-fired moisture is included in combustion products volume and is used in estimate of net heat values. Five percent excess air is assumed in all cases.

	CHARCOAL FINES, LOW MOISTURE	WASHED COAL REJECTS	TYPICAL BITUMINOUS COAL	LIGNITE, LOW MOISTURE	RICE HULLS	LIGNITE, HIGH MOISTURE	PALM OIL SHELLS	RESIDENTIAL REFUSE LOW MOISTURE	RESIDENTIAL REFUSE HIGH MOISTURE	PREPARED PEAT 20% MOISTURE
FUEL MOISTURE, AS FIRED	10	8	14	10	8	30	10	10	30	20
% ASH AS RECEIVED	6.0	70.0	10.0	5.0	19.0	3.9	9.8	43.0	33.5	4.0
GROSS HEAT VALUE, KCAL/Kg AS RECEIVED (BTU/LB)	7000 (12600)	1930 (3474)	6100 (11000)	5400 (9720)	3700 (6660)	4200 (7560)	4660 (8400)	2600 (4600)	2000 (3600)	4080 (7344)

(CUFT/1000 BTU)
2.0 — 18.0
1.5 — 13.3
9.0
1.0

M³/1000 KCAL

739

Table 18.3 Comparison of adiabatic flame temperatures, conventional and refuse fuels. *Note:* Five percent excess air and a combustion air temperature of 328°C (622°F) are assumed in all cases.

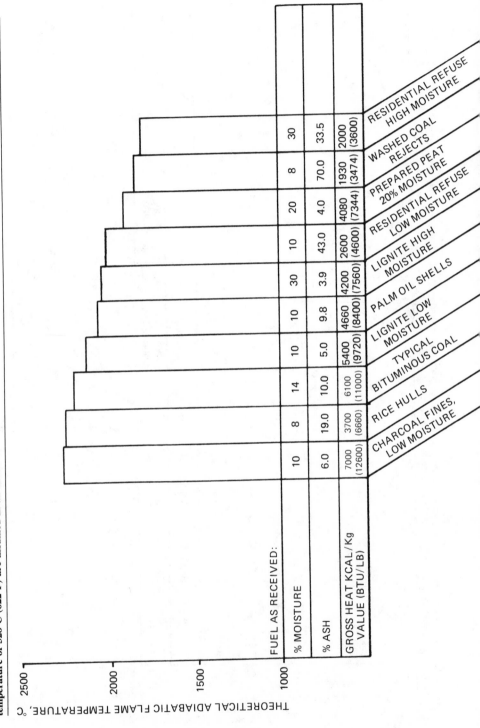

THEORETICAL ADIABATIC FLAME TEMPERATURE, °C

FUEL AS RECEIVED:	CHARCOAL FINES, LOW MOISTURE	RICE HULLS	TYPICAL BITUMINOUS COAL	LIGNITE LOW MOISTURE	PALM OIL SHELLS	LIGNITE HIGH MOISTURE	RESIDENTIAL REFUSE LOW MOISTURE	PREPARED PEAT 20% MOISTURE	WASHED COAL REJECTS	RESIDENTIAL REFUSE HIGH MOISTURE
% MOISTURE	10	8	14	10	10	30	10	20	8	30
% ASH	6.0	19.0	10.0	5.0	9.8	3.9	43.0	4.0	70.0	33.5
GROSS HEAT KCAL/Kg VALUE (BTU/LB)	7000 (12600)	3700 (6660)	6100 (11000)	5400 (9720)	4660 (8400)	4200 (7560)	2600 (4600)	4080 (7344)	1930 (3474)	2000 (3600)

740

The results of this work showed that it should be technically feasible to convert redundant cement kilns to refuse incinerators, but at that time such an operation would not have been profitable at the locations that we had available. However, by that time other experimental work and economic evaluations did indicate that it would be profitable at some plants to use refuse as a supplementary fuel for a producing cement kiln, and so that was the course that we pursued.

18.1.3 Resumé of Blue Circle's Experience

During 1970 to 1971, refuse that had been coarsely pulverized by a Municipal Council for landfill purposes was transported to the Barnstone experimental plant where it was further processed in a vertical shaft crusher, blended with the kiln coal fuel in various proportions as required, and fired into the 150-ft (46 m) cement kiln via the high-speed, peg-type coal mill. The experiment was continued over several days encompassing a range of operating conditions, during which samples and measurements were taken to determine relevant clinker chemistry and process engineering data. The experiment was successful and the data were evaluated to establish technical and economic scale-up criteria.

In 1974, domestic refuse was processed through a vertical-shaft primary crusher, followed by a horizontal shaft, gridded secondary, at the city of Bath. The product was taken to our Westbury Works where it was pneumatically conveyed and separately fired as supplementary fuel in the 500-ft (152 m), 1,000 TPD, coal-fired kiln. The techniques for pneumatic conveying and separate firing were developed, scale-up from the 1971 Barnstone operation was confirmed, and technical and cost data were obtained on the performance of the crushers.

The Shoreham Works trials, carried out over a period of several months, were undertaken primarily to determine the maximum refuse addition rate consistent with not prejudicing cement quality.

In this and some of the other trials, atmospheric emissions via the main stack were sampled and analyzed in detail by BCI and by an independent government laboratory to check that no unusual pollutants were being emitted and that acidic gases were being substantially retained. The Dunbar Works trial was carried out to evaluate the application of the process to a Lepol kiln plant. It showed that with a Lepol process the tolerable addition rate of refuse would be very limited if the nodule properties were marginal, unless the refuse was subjected to benefication and ash reduction.

A trial was undertaken at the Plymstock four-stage suspension preheater plant which showed that refuse could readily and successfully be used as a supplementary fuel within the normal constraints for that process.

Following the agreement of a long-term contract between the West Wiltshire County Council and Blue Circle, a commercial refuse plant was constructed and commissioned at Westburn Works in 1977. Domestic municipal refuse is received by road vehicle during 5 days a week. It is pulverized on a 5-day/week, two-shift basis and fed pneumatically to the kilns (2 × 1000/1200 TPD clinker, wet process, coal fired) continuously 7 days per week. A schematic flowsheet is shown in Figure 18.1.

The pulverizing plant initially installed the Mark I equipment; two lines each comprising primary pulverizer, rotary screen, and shredder were not adequate for the arduous duty required of it and never met required performance guarantees.

The remainder of the plant and process, however, performed well. BCI engineers, therefore, designed and specified a replacement pulverizing plant. This Mark II pulverizing equipment was installed in 1979, and the original equipment was removed. The Mark II pulverizing plant again comprises two separate lines. Each line has as the primary pulverizer a gridded, horizontal-shaft, impact crusher with ballistic rejection tower for "uncrushables." For secondary crushing one line has two gridded, swing-hammer crushers in parallel, whereas the other line has one larger, gridded, swing-hammer crusher of different manufacture. Both lines have fully met all the specified requirements of product quality, output, reliability, and operating cost.

The Westbury plant has processed nearly 200,000 tons (181,440 t) of refuse to date and is operating profitably.

18.1.4 Current Developments

An ongoing activity at the Westbury plant is a series of trials to determine the optimum materials for such regularly wearing parts as the crusher hammers and the rotary valves and bends on the pneumatic conveying pipelines. Also, to increase further the refuse capacity and profitability of the Westbury plant, screens are being installed to screen off some of the finer material for landfill. This will reduce the ash and moisture contents of the refuse used as fuel and thereby will enable more to be burned in the kilns.

A further activity is the recovery of methane from quarries already filled with refuse. Trial boreholes have been sunk in one works quarry to determine gas quality and quantity. The results

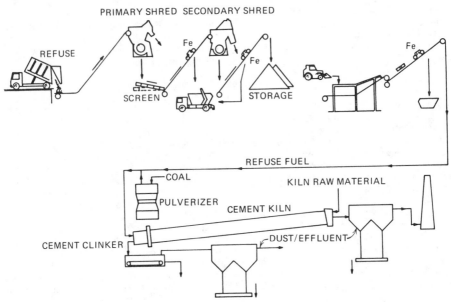

Fig. 18.1 Flow diagram: cement kiln firing a combination of prepared refuse and coal.

were satisfactory and permanent boreholes, pumps, pipeline, and so on, have been installed to enable the gas to be used as partial replacement for the normal kiln coal fuel. This plant will come on-line in December, 1982.

Following successful, manually operated trials, permanent handling equipment has been set up to add scrapped rubber tires into the kiln backend chute-riser pipe at one of the BCI's four-stage suspension preheater plants. This operation came on-line in the summer of 1982. Schemes are being drawn up for this fuel-saving technique to be applied at other plants.

Our research division has in hand a program to develop a fluidized-bed process for gasification of refuse to make a hot producer gas that could be used to fire cement kilns or other high-temperature processes. The advantage of this approach is that most of the ash should be retained by the bed. Also, we are to install a fluidized-bed combuster to produce hot air for raw materials drying at one of our plants. It is designed to be fired with coal initially but is readily adaptable to pelletized refuse at a future date.

18.1.5 Questions and Answers

Watson: Considering the question of why the U.S. companies haven't pursued waste fuels, the reason why Blue Circle and other European companies in 1970 were looking around for all kinds of unusual fuels, and the Americans weren't, is that we were paying relatively high prices for fuel, and your gas prices made us all very envious.

Q: Could the Westbury plant accept the whole tires?

Watson: I don't think so. It wasn't designed to shred tires, which is a difficult business anyway, and I think they'd probably stall the crushers. It wasn't designed to accept tires.

Q: With the insufflation pipe at the one o'clock position in the kiln hood, what was the direction of rotation of the kiln?

Watson: The one o'clock position was the position furthest from the clinker lying in the kiln, so it wasn't on the feed side, it was on the opposite side, whichever rotation that might be.

Q: What is the average BTU-per-pound content of solid waste?

Watson: Assuming you mean domestic municipal garbage, the content will vary according to the country in which you're operating. In the United Kingdom it is between 3000 and 4000 BTU/lb. If you don't let it hang around and get rained on, the average moisture content is about 30%. In some more dry countries it can be as low as 20%.

Q: What about brass, copper, aluminum, etc., in the waste? Does it affect product clinker?

Watson: There isn't much to start with, and we haven't detected any quality effects whatsoever. If a machine comes into the tipping area, and it looks like a washing machine, it never finds its way into the pulverizer, because the guy pulls (the metal) off and puts it in his pocket. That's fair enough . . . we also have fixed up a metal trapping device, and that, too, is operated with some enthusiasm by the operators.

Table 18.4 Analysis of Combustible Waste Materials

Analysis	Tires	Shredder Waste	Sludge Asphalt	Activated Bentonite
Moisture (%)	0–1	13–16	—	4.8
Volatiles (%)	83	53	95	48
Ash (%)	16	47	3–5	51
S (%)	1.3–1.7	0.7	16	1–3
Cl (%)	0.09	0.7	0.01	0.03
Fe (%)	12–15	10–12	0.6	n.d.
Zn ppm	10,000–26,000	5,500–26,500	3,300	n.d.
Pb ppm	100	1,750–13,000	1,700	n.d.
CV : kcal/kg	6,500–7,000	3,500	3,955	4,800
MJ/kg	27.2–39.3	14.6	16.5	20.1

18.2 EXPERIENCE IN WEST GERMANY*

Heinrich Matthee

18.2.1 Background: Tires

In Germany, the annual amount of used tires is about 350,000 tons. Of these, about 54% are dumped. In the United States, a figure of 200 million old tires has been mentioned, equal to some 12 million tons/year. The problem in dumping these tires is that they will not rot, they take up large amounts of space due to their shape, and they may give rise to hygienic problems.

Despite all these problematic features of old tires, you will look at them differently when considering that their calorific value is 11,600 to 12,900 BTU/lb (27 to 30 MJ/kg). Then the perspective really changes. During 1978, as part of our research program for various alternative combustible waste materials (Table 18.4), we embarked on an investigation of the fuel substitute potential of used tires, in order to replace part of the lignite dust, bituminous coal, and fuel oils used in most of our kilns at that time.

We thought about ways of introducing the tires as fuel into the process. From various considerations and burning duration studies, it was decided to use the tires as fuel for partial precalcination by feeding them into the kiln inlet chambers. Tests were made, reducing the tires in size by shredding or cutting. It was found that shredded rubber and other waste rubber chips could be burned without detrimental effects.

However, the steel inlays of the radial tires that predominate in the German tire market made this a rather costly process due to the high wear exerted on the cutting tools. Then it was thought about using the complete tires, and we started to throw smaller units into the inlet chamber, hoping for the best. As they did not show up in the clinker cooler, they must have burned out completely. The tires became successively bigger, and eventually we even introduced complete truck tires of up to 175 lb (80 kg) weight. The burning time of the tires is estimated at 0.5 to 1.5 min. So we decided to start a pilot test at our Amöneburg plant in Wiesbaden where we operate two preheater kilns of 2500 TPD each.

It is now more than 3½ years ago that we started using tires as a fuel, and we have not encountered any insurmountable operational problems since. The use of tires provides an average of around 16 to 18% of heat energy. The total use amounted to about 40,000 tons (36,287 t) of tires in 1980 and 45,000 tons (40,823 t) in 1981. You might ask now, why are we not converting more kilns to the use of tires? The answer is that the supply of used tires is not unlimited and that the cement plant under consideration has to be strategically placed in the market for used tires.

At our Amöneburg kilns, tires are taken from the storage by means of a forklift and placed on a storage roller conveyor (Figure 18.2), followed by a weighbelt feeder and a roller conveyor (Figure 18.3) before the tires reach the kiln inlet chamber and the kiln, passing a double-flap gate (Figure 18.4).

One of the areas where we expected problems to occur was the place where the tires fall onto the brick lining in the kiln inlet chamber. However, as the tires are elastic, the rebound absorbs

---

*This section is a reprint of an article written by Dr. Matthee that appeared in the April, 1983, issue of *Rock Products Magazine,* and titled "Discarded Rubber Tires Provide 16 to 18% of Heat Energy at Dyckerhoff Plant, but Raw Meal Mix Must Be Adjusted."

Fig. 18.2 Loading tires unto roller feed conveyor. (Reprinted, with permission, from *Rock Products Magazine,* October 1980, Fig. 82.)

most of the kinetic energy, and no changes through erosion have been noted so far. Although higher oxygen concentrations are measured due to false air entering, the kiln operation can still be controlled without difficulties. When feeding truck tires, it sometimes can occur that CO peaks develop due to spontaneous combustion, although so far, without negative influence on the operation.

There was an influence of the tires on the chemistry of the clinker; the Fe_2O_3 content rose by 0.4%, resulting in a reduction of the silica ratio by 0.1 and of the lime saturation factor by 0.5. The raw material mixture has to be adjusted accordingly to reach standard quality of the clinker and cement. There was also a slight decrease in the relative brightness of the cement. That means that for certain special cements with low iron content the use of tires is not feasible.

In one of our plants, there was noted a beneficial influence by the introduction of tires. The sulfur content of the tires facilitated the reduction of circulating dust loads in the kiln which ran with a raw mix that initially had a large oversupply of alkalis in relation to sulfur. The majority of trace metals is incorporated in the clinker, only minor amounts being emitted with the exhaust gas dust through the electrostatic precipitator. The benefits of this process are best illustrated by the cost of heat energy in the kiln. Disregarding any specific plant influences, the energy costs per heat unit when burning tires may be reduced by up to 50% compared with lignite.

18.2.2 Miscellaneous Shredder Wastes

In addition to tires, we also use waste rubber chips in our Beckum plant. Larger pieces of rubber, plastic, or wood are reduced to manageable size with the aid of a small shredding machine. Due to the flexibility of the conveying and feeding system, most of our tests using quite a variety of waste fuels are carried out here. The kiln in question is a 3,000 TPD FLS kiln, while the preheater is a

Fig. 18.3 Tire weigh belt feeder assures constant tire feed to kiln. Reprinted, with permission, from *Rock Products Magazine*, October 1980, Fig. 86.)

four-stage Dopol. Normally, the use of rubber chips was accomplished without any major difficulties. Only when chlorine-rich foam rubber was introduced were blockages in the preheater noted, due to the volatization of the chlorine components and their deposition as alkali-chlorides.

18.2.3 Auto Shredder Wastes

Our survey for combustible waste materials revealed that in Germany some 300,000 tons/year of waste are "produced" during the destruction of used cars by shredding and then by the ensuing extraction of metallic scrap. The material consists of textile fibers, paint remainders, timber, plastics, and rubber, and is partially soaked with engine oil. The dumping of such materials without endangering the groundwater is an extremely costly procedure, the average cost being 60.-DM/ton.

Fig. 18.4 Tires enter kiln via chute and air-lock feeder (average rate: one tire every 2 min). (Reprinted, with permission, from *Rock Products Magazine,* October 1980, Fig. 87.)

At a calorific value of about 6300 BTU/lb (14.6 MJ/kg), the potential of this material as a substitute fuel became attractive.

Due to the rather limited knowledge of the composition of auto shredder waste, its burning behavior, and its influences on the production process, tests have been carried out in our Beckum plant on the basis of a preliminary permit, commencing in the last quarter of 1981. During the test, the kiln plant used predominantly pulverized hard coal in addition to auto shredder residue. The shredder waste contributed about 9.5% of the heat supply.

Storage and handling problems were caused by the low bulk density (19 lb/ft^3, 0.3 kg/dm^3) and by the content of paper and foil particles that were blown from the belt conveyor by gusts of wind, leading to contamination of the plant area. This was prevented by the subsequent installation of side guards.

The use of auto shredder waste has not unexpectedly increased the concentrations of Al_2O_3 and Fe_2O_3 in the clinker by about 0.1%, and consequently reduced the silica and alumina modules. These changes are acceptable, although the increase in the Fe_2O_3 content would mean a darker color of the resulting clinker and cement.

18.2.4 Asphaltic Sludge

We also have experimented with other combustible waste materials. One of these is sludge asphalt, a waste product derived from the reprocessing of used oils.

The highly viscous sludge contains most of the contaminants of the oil, but also is mixed with 15 to 70% sulfuric acid, depending on the refining process. This admixture is making the handling of sludge asphalt a rather expensive process, especially when metallic pumps and pipes are used. However, at a calorific value of 7200 BTU/lb (16.5 MJ/kg), sludges with a low remaining content of sulfuric acid still are an interesting fuel substitute.

18.3 EXPERIENCE IN NORTH AMERICA

William D. Robinson

18.3.1 Background

Despite a notable resurgence in waste-to-energy solutions to the problems of declining landfill space and increasing generation of municipal, commercial, and industrial solid waste, there does not appear to be a compelling interest in the use of RDF (refuse-derived fuels) by the Portland cement producers in the United States and Canada as yet.

Although considerable prototype trials and testing began in the 1970s[1-4] and are being continued by several large producers in tracking relevant state-of-the-art and world energy factors, the prospect for continuous firing of supplemental RDF in the near future is doubtful.

18.3.2 Factors in a Discouraging Outlook

Unlike their counterparts in the higher-fuel-cost nations of the world including Europe and Japan, there is little enthusiasm for firing any supplemental refuse fuel, that is, municipal solid waste, tires, and auto shredder residue. Interestingly, this parallels the reluctance of the investor-owned electric utilities to embrace refuse fuels after considerable testing in each industry. With the latter, it has ironically been further exacerbated in the United States by the Public Utility Regulatory Policies Act (PURPA), which requires electric utilities to purchase electricity from qualified cogenerators and small power facilities (80 mw or less)* and might well eliminate direct use of RDF by utilities in their boilers with a few exceptions (cyclone firing, and so on). They will prefer buying kilowatts from outside dedicated generating facilities under PURPA unless the least-avoided-cost provision becomes unfavorable (operating cost anomalies, etc.).

In an industry as capital intensive and competitive as Portland cement production, the capital and operating costs to adapt to supplemental refuse fuels appear to override any lower energy cost advantage. Requirements for space, equipment, labor, product quality and emissions monitoring/control, and the not-in-my-backyard syndrome account for this.

However, a favorable factor in firing supplemental waste fuels in cement kilns is the typical temperature-time profile from 1000 to 2700°F (550 to 1500°C) and 1.5 to 3 sec which can mitigate the burgeoning concern for deleterious and toxic emissions, including dioxins, from any waste-burning stack. Such concern may well change stoker-fired mass-burn incinerator techniques in which the least controllable function in the entire operation is the combustion process itself, with unpredictable and highly variable temperature-time gradients and combustible-to-air ratios.

18.3.3 Scrapped Auto Shredding Residues

Background

The near-term outlook for disposal of residue from shredding scrapped automobiles by other than the usual method (dumping at a landfill) is not encouraging, and the outlook for land disposal is one of ever increasing costs for strict control of leachates from all but the most innocuous materials.

As of today, there appears to be no viable alternative operating anywhere in the United States. This includes

1. Supplemental fuel for cement kilns.
2. Incineration, disposal only.
3. Incineration with energy recovery.
4. Recycling as a saleable commercial fuel, construction filler material, agricultural mulch, and so.

Despite today's exorbitant domestic energy costs and diminishing landfill space with rising costs, none of these alternatives yet appear to be logistically and economically feasible.

Cursory test burning in commercially available modular incinerator equipment indicates the following:

Limited autogenous combustion requiring an auxiliary heat source.

50% weight reduction.

75% Volume reduction.

Acceptable emissions control will require filtration/scrubbing.

Incineration, Disposal Only

There are nearly 200 auto-shredding yards in North America, and, typically, each has one shredding line for one-shift production. When the market is good, each might produce 300 to 500 TPD of ferrous product (400 to 700 auto hulks) and generate approximately 60 to 100 tons (54 to 91 t) of

*See Section 2.5.

disposable residue during its daily one-shift operation, 5 days/week. On this basis, an on-site incinerator facility with adequate pollution controls for 24 hr/day, 5 day/week operation would require a capital investment of about $2.0 to 2.5 million.

For one-shift operation, such an incinerator would become even less feasible costwise ($3 to 5 million), sizewise, and operationally due to the deleterious thermal and chemical stresses of daily start-up and shut-down via temperature cycling from cold to hot, and the likelihood of auxiliary oil or gas burner fuel costs.

In addition to the capitalized fixed costs, there would be the as yet uncertain operating costs of labor, maintenance, utilities, and auxiliary fuel as well as the trucking and dump fees for ash residue.

With tipping fees approaching (or over) $30.00/ton at the larger municipal solid-waste-to-energy plants becoming commonplace, the daily disposal cost to an auto shredder operator would be prohibitive for raw residue.

Incineration with Energy Recovery

Although energy recovery options are available, that is, steam, hot water, and/or electricity, the cost/benefit ratio will be marginal at best, even if the yard operation itself can use such energy or if it is adjacent to an industrial user of steam which is the most profitable energy by-product. Although PURPA now can assure an electricity market for the excess, and electric generation for plant auxiliary power (violent shredder load swings would preclude on-site electric power generation for the shredder @ 2500 to 4000 HP [1875 to 3000 KW]) is technically feasible, the sophisticated complication of power generation on-site would not likely be compatible with a single-shredder operation. In any case, energy recovery of any kind usually requires utility levels of reliability, continuity of service, and labor skills higher than most operators might find justifiable at present.

Joint Venture

A more likely energy recovery approach might be a joint venture plant strategically located for residue delivery by the partners and with an industrial energy customer nearby or for kilowatt generation only under PURPA. The economies of scale and the necessity for specialized labor and management skills would lend themselves to a centralized operation. In addition, consideration could be given to thermal destruction of sludge and liquid wastes from area industries on a fee basis.

The need for such disposal facilities is urgent, especially for materials considered even slightly hazardous or toxic. However, such an operation would present siting and, permitting contingencies, which could be tedious, expensive, and insurmountable in some cases, or just plain not worth it. The approach, however, deserves a closer look.

18.3.4 Conclusion

It appears that the use of supplemental waste fuels in Portland cement production in North America is in approximately the same economic holding pattern as synfuels—it all depends upon the price and availability trends of preferred fossil fuels: coal, oil, and natural gas.

REFERENCES

1. E. J. Duckett and D. Weiss, RDF as a Kiln Fuel, Proceedings, ASME 9th National Waste Processing Conference, Washington, D. C., May, 1980.

2. C. S. Weinberger, Evaluation of Secondary Shredding to Enhance RDF Production as Fuel for Cement Kilns—A Research Test, Proceedings, ASME 9th National Waste Processing Conference, Washington, D. C., May, 1980.

3. J. D. Dorn, Uses of Waste and Recyclable Material in the Cement Industry, presented at IEEE Cement Industry Technical Conference, Tucson, Arizona, May, 1976.

4. R. Segal, Kiln Fuel: Rice Hulls, Wood Chips, Proceedings, 19th International Cement Seminar, Chicago, Illinois, December, 1983, Sponsored by *Rock Products Magazine*.

CHAPTER 19
BIOLOGICAL PROCESSES

DONALD K. WALTER

Office of Energy from Municipal Waste
U. S. Department of Energy
Washington, D. C.

JAMES L. EASTERLY

Meridian Corporation
Falls Church, Virginia

ELIZABETH C. SARIS

Science Applications, Inc.
McLean, Virginia

THEORY AND PRACTICE
Donald K. Walter
19.1 BACKGROUND

All of today's fossil fuels were produced over millions of years by natural processes. About two-thirds of the world's coal was formed from plants which grew during the Carboniferous Period from 280 to 350 million years ago. Heat and pressure drove off the volatile matter and left relatively pure carbon. Some of the plant material was converted to a methane/carbon dioxide mixture that is still found in coal seams today. Other organic material was converted to oil and natural gas by actions that are not fully understood. Natural gas probably was produced by the organisms which produced the methane/carbon dioxide gas in coal. Oil could have been produced by organisms that today produce a high body weight of oil, although heat and pressure may also have been the dominant system.

Biological systems range from commercial to highly experimental. For example, today the organics that have been buried in our landfills are being digested by organisms to produce methane/carbon dioxide (called digester gas for the remainder of this chapter) gas mixtures. This gas is being purified by removal of the carbon dioxide and being used as synthetic natural gas (SNG). We are researching to control and accelerate other biological processes and to add systems that convert the biological product to a useful fuel more rapidly.

The biological or biochemical systems generally are slow. They require reaction times that are measured in days. However, these systems also operate at mild conditions, that is, temperatures under 160°F (70°C) and at atmospheric pressures. In addition, the biochemical systems produce specific products such as methane, carbon dioxide, and ethanol. In contrast, the thermal systems require very short reaction times that are measured in seconds. However, these systems operate at very harsh conditions, that is, temperatures of 1000°F (540°C) or more and pressures from atmospheric up to 3000 psi (200 atm). In addition, the products are highly variable over short periods of time. The combustion option in thermochemical systems is described in Chapter 12. As the available oxygen is limited in a thermal system, the products switch to combustible gas, liquids, and char.

Biochemical systems require that the organic substrate, to be converted, be biodegradable, that is, capable of being readily decomposed by bacterial action. Many of the synthetic organics, such as plastics, textiles, and so on, are not biodegradable. In addition, the inorganic materials in municipal solid waste (MSW) and the dirt and rock in manures are not biodegradable. More seriously, the lignin in wood, agricultural waste, and other plant matter is quite resistant to biodegradation and shields the cellulose from attack by bacteria. Thus, plant matter is a poor substrate for biochemical processes unless some pretreatment system is used to separate or solubilize the lignin. For example, wood in a MSW digester is basically untouched while paper fiber is digested. In the remainder of this paper, organics will be used to indicate biodegradable and plant matter and plastic to indicate nonbiodegradable organics.

Organic compounds (carbohydrates) typically consist of a linked carbon atom and water molecule. For instance, xylose (a 5-carbon sugar) consists of 5 carbon and 5 water molecules, glucose (a 6-carbon sugar) 6 carbon and 6 water molecules, and cellulose is formed by glucose molecules linked by oxygen atoms into long chains. As the molecules become longer, they tend to stray from the precise carbohydrate formula and become more resistant to bacterial attack. Therefore, biological processes tend to prefer sugar and starch as feedstocks. These, unless they are a waste from another product or process, tend to be more valuable for other uses.

The usual method to measure the efficiency of biochemical processes is by volatile solids conversion. Volatile solids are measured by the percentage loss in solids when a sample is heated to 825°F (440°C).

There are many potential biochemical systems to convert organics to useful fuels and products. The three most developed (anaerobic digestion, fermentation, and composting) are discussed below.

19.2 ANAEROBIC DIGESTION

19.2.1 Introduction

Anaerobic digestion occurs in the absence of air. The controlled systems depend upon air-tight containers into which the organic material is introduced in a slurry form. To date the systems depend upon the slurry being liquid, with solids contents of 20% at the most and more typically 5%. The microbial population may be introduced with the organics as in sewage sludges or may be seeded into the digester when the substrate does not have a large microbial population of its own as in municipal solid waste. There are also uncontrolled anaerobic systems. For instance, the MSW organics in a landfill will naturally digest. Most landfills are deep enough to exclude oxygen from the bottom, have water available from infiltration, and are naturally seeded through pet and other wastes.

19.2.2 Basic Processes

The overall anaerobic process is not completely understood. Parts of the basic process are explained from the basic biochemistry of life processes; however, many of the bacteria that cause the conversion or produce the enzymes that catalyze the reactions have not been isolated, and the basic biochemistry of the isolated organisms are poorly understood. The first isolation of anaerobes was by Hungate in the 1940s. Subsequently, this first organism was determined to be two separate organisms that had a mandatory symbiosis to insure the survival of each. The overall lack of biochemistry information prevents genetic manipulation at this time.

In any event, the process has three steps. First, certain bacteria and their enzymes break down complex molecules, such as cellulose, to simple organics. Then, acidogenic bacteria convert those compounds to simple organic acids ending with acetic acid and hydrogen. The methanogens use the acetic acid and hydrogen to produce methane and carbon dioxide. The first two steps are completed by faculative (live in presence or absence of oxygen) bacteria. The methanogens are strict anaerobics and are killed by oxygen. For an anaerobic system producing digester gas, that fact is not important. For the future, it is possible that the system might be stopped at the simple organic acids that may then be converted to high-value liquid fuels such as octane and decane.

19.2.3 Feedstocks

The best feedstock is one which is a wet slurry with high volatile solids and low nonbiodegradables. Actually, a concentrated sugar or starch would be preferable but does not generally exist. The best current feedstock is sewage sludge or certain manures which are collected off the concrete floors of environmental feedlots. These are used without any further preparation.

Other feedstocks require preparation steps. Research is underway for means to pretreat wood and other organics to remove the lignin and to make the cellulose more readily accessible to bacteria. For further discussion, see Section 19.3.3, since wood and other plant matter are more likely to be feedstocks to alcohol processes.

The organics in MSW should be concentrated prior to digestion. It is necessary to remove the long stringy plastics and textiles. They are not biodegradable and, more importantly, wrap around moving parts. In two different tests, the stringy material wrapped around mixer shafts and built mats up to 6 ft (1.8 m) in diameter which destroyed mixing action. It required chain saws to remove these materials from the mixer shafts. In addition, as much inorganic material as feasible should be removed, since it uses digester volume and adds significantly to equipment wear.

Of course, the MSW in landfills requires no preparation, and all existing landfill gas projects have no modification to the buried waste. Research, however, indicates that shredding and the addition of buffering and nutrients seem to improve gas production. This latter research is in early stages and has not been applied at large scale as yet.

In any large system, the protein and undigested feed in manures are far too valuable to digest anaerobically. This material is washed out of the manure and refed to the animals. If the manure is gathered from a dirt feedlot, then processing to break up the dirt clods that tend to build around the manure is essential.

19.2.4 Products

The principal energy product is the digester gas which is saturated with moisture. It is typically 50 to 75% methane and 25 to 50% carbon dioxide. The contaminants may be 0 to 2% nitrogen and 0 to 1% oxygen from entrained air, 0 to 200 ppm hydrogen sulfide and 1 to 4% hydrogen. If a landfill gas extraction system is not properly controlled, the nitrogen levels may increase up to 20% with the methane reducing to 40% of the total. The oxygen in such instances will also increase, but only by 1 or 2 percentage points. Landfill gas may also contain a variety of other organic chemicals but at levels measured in parts per million or less.

Digester gas has many uses. With proper design to control condensate, it is used directly as boiler fuel and as an engine fuel in stationary spark-ignited or diesel engines. Initial full-scale tests are now underway to use raw gas in a turbine engine. Engine output has been used as mechanical power and to produce electricity. In addition, heat has been recovered from engine exhaust and cooling water for in-plant processes. No systems to date have been large enough to use combined cycle engines.

Digester gas has also been scrubbed of carbon dioxide and used as synthetic natural gas in existing pipeline systems. The same purified gas has been used compressed or liquified as a vehicle fuel. The carbon dioxide, particularly when recovered as dry ice, has some promise for use as a refrigerant.

Each anaerobic digestion system, except for a landfill gas system, has a waste product of its own that requires management. Sanitary sewage sludge digesters are used frequently as a means to stabilize and reduce the volume of sludge. However, the sludge from the digesters, although stable and basically odorless, became a disposal problem and the value of the methane prior to the major energy price increases in 1973 and 1978 was too low to make its use economic. Now, with the advent of new dewatering devices such as the belt press and plate and frame press for sewage and manure sludges and the belt and cone presses for high-fiber-content sludges such as those from MSW digesters, filter cake may be produced that is dry enough to be burned for site energy. The fibrous cake with its low nutrient level also has value as a soil conditioner. In the Far East where manure digestion is practiced extensively, the waste sludges have been used as soil amendments for centuries. Incidentally, the practices in those countries do not rely on the high levels of fertilizer and intensive mechanization of the American agricultural system.

The filtrate from the sludge dewatering systems is recycled to the head of the plant in sewage treatment, is recycled as digester makeup water in MSW systems, and is frequently used as irrigation in manure systems.

19.2.5 Reactor Types

Any air-tight tank or landfill with a surface seal may be an anaerobic digester. The principal differences concern the arrangements for introduction of feedstock and flow-through and removal of the sludge from the tank. Of course, the landfill is a simple batch-fed digester.

Perhaps the simplest of the digesters is the septic tank. It is fed very low solids content material, 0.5 to 1%. The tank acts as a settling tank so that the solids concentrate in the bottom and gradually digest. The liquid overflows out of the tank and percolates from a header system into the surrounding soil. The methane produced by these systems is very rarely captured and used. In the Orient, a slightly more sophisticated system (Gobar or cow manure converter) is fed human waste and manures (Figure 19.1). Fresh feed displaces digested material which flows out of the reactor. The gas is gathered under a floating cover and is subsequently used for cooling, lighting, and to operate irrigation pumps.

Another simple system is the plug flow digester that is emerging as a manure digester. Fresh

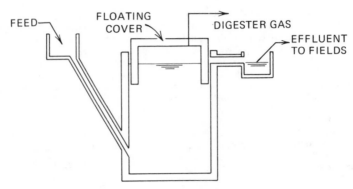

Fig. 19.1   Gobar converter.

manure from an environmental feedlot is about 18% solids and will barely flow. It is introduced into one end of a long tank. The weight of fresh feed serves to push the manure through the digester as a plug, and finally it overflows the far end of the tank. The manure undergoes the typical anaerobic digestion reactions and by the time it reaches the end of the tank is relatively stable and odor free. Also in the simple digester category is the multitank batch-fed system. Each tank is loaded with material and allowed to sit for the selected digestion period. Then it is opened up and the sludge removed prior to recharging.

Almost all large systems serving in wastewater treatment plants, at large feedlots, and in the test bed MSW system use continuously stirred tanks. The theory is that the stirring enhances the contact of organisms with the feed, provides uniformity in tank contents, and breaks up scum and other inhibiting conditions. One disadvantage is that fresh material may be expelled along with well-digested material. Figure 19.2 is a sketch of a continuously stirred reactor. The mixers are in many different forms. Most popular are mechanical mixers with horizontal paddle wheels and gas mixers which inject digester gas back into the reactor to cause a stirring action.

The next steps from the continuously stirred tank are the multitank or multiphase systems. The simplest of these empties the effluent from the first tank into an unheated, unstirred tank. Here the entire effluent has additional residence time for digestion. The solids tend to settle to the bottom of the tank and have an even greater time for digestion. This system is known as anaerobic contact stabilization. Another theory of multistage digestion is to separate the phases (that is, hydrolization, acidification, and methanation) into two or three tanks each optimized for the specific step. The desire is to increase digester gas production and methane content while decreasing overall tank volume. For instance, the acidogenic bacteria will double their population in less than a day while the methanogenic bacteria require 3 to 5 days retention time to double.

The packed, mixed, and expanded bed systems all introduce a media into the digester to which the bacteria may attach. Then high throughput rates do not wash the bacteria from the digester. The packed bed reactor uses coarse stone, plastic, or ceramic rings or other coarse media. The media

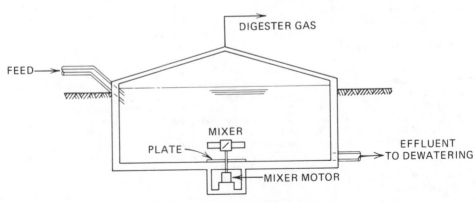

Fig. 19.2   Continuous stirred reactor.

also act as a filter so that the retention time of the liquids will be on the order of hours, while the retention time for solids will be years. Removing inert material from the reactor is difficult and the reactor has a tendency to channel. That is, the media becomes plugged and the effective reactor area is reduced. The mixed bed filter uses porous walls, netting, or strips of plastic as substrate for the bacteria. There is less filtering action. The expanded bed uses fine sand as a media and such high throughput that the bed expands and acts like quicksand or a fluid bed. The expanded bed has little tendency to channel, unlike the packed and to lesser extent the mixed-media bed. However, the feed solids have a greater tendency to wash out, and maintaining uniform fluidization and feed distribution across the section is difficult.

19.2.6 Design Parameters

With such a wide variety of feedstocks and reactor times, summarizing the actual design parameters is not feasible. Instead, this section will discuss the terms in general and then a large variety of design manuals, particularly for sewage sludge systems, are available. Actual parameters for the Department of Energy RefCoM™* test bed at Pompano Beach, Florida, have been published.

One of the early design parameters to be selected is the temperature of the reactor which may be heated or unheated. The temperature ranges of interest are psychophilic (under 68°F [20°C] or unheated), mesophilic (68 to 110°F [20 to 43°C]), and thermophilic (110 to 160°F [43 to 71°C]). Along with the temperature is the hydraulic retention time (HRT) or the time that it requires for the average particle of liquid to move through the receptor. This is equal to the reactor volume divided by the total volume fed per day. A classic curve of retention time versus temperature is available in many references. Using a standard for completeness of reaction, then, the digestion rate is about zero (over 100 days' retention time) at 40°F (4°C) and increases rapidly to 35 days' retention at 68°F (20°C). The increase is slower but peaks at 24 days' retention at 96°F (36°C). It then falls to about 26 days at 110°F (43°C) and then begins to increase again to a 13-day retention time at 140°F (60°C). From that point, activity decreases rapidly. At 160°F (71°C), digester activity ceases as the organisms are pasteurized.

A second time of interest is the solids retention time. In reactors that are not stirred and in reactors with beds (especially the packed bed), the solids may be retained in the reactor for a longer period of time than the liquid. This is known as solids retention time (SRT). It is a measure of the time the solids are exposed to the bacteria.

Typical design HRTs are on the order of 20 days at mesophilic temperature and 5 days at thermophilic temperatures. Thus, the trade-off is between the energy to heat the reactor to thermophilic temperature and the reduction in tankage volumes by 4 since retention time is one-fourth as long.

Another important parameter is the percentage of solids in the tank. As the solids content increases (doubles), the necessary tank volume decreases (halves). Most of the anaerobic digestion feedstocks are a viscous fluid at 10% solids and appear as solid at 15% solids by weight. If the reactor to be used is continuously stirred, the input feedstock to the tank is limited to about 10% solids. Since about 50% of the volatile solids are converted to digester gas, the tank contents will then be at about 5% solids. For sewage digesters achieving 10% solids for feed would require mechanical dewatering. For MSW and manures recycle filtrate is added to reduce the solids content.

Other design factors have been developed for sewage digesters. For instance, figures of 30 to 70 lb (13.6 to 31.8 kg) of volatile solids per cubic foot (0.028 m^3) of digester volume per day or 2 to 6 ft^3 (0.056 to 0.167 m^3) of digester volume per capita for heated tanks and double that for unheated tanks are cited in the literature. The amount of gas that may be produced varies widely. For an MSW feedstock with thermophilic temperatures and 8-day retention time, 6 ft^3/lb (.35 m^3/kg) of volatile solids fed may be expected. For sewage sludge, these figures increase to about 14 ft^3/lb (2.2 m^3/kg) of volatile solids destroyed.

19.3 FERMENTATION PROCESSES

19.3.1 Background

The fermentation processes are the oldest of the biochemical systems controlled by man, although the ethyl alcohol(ethanol) produced was used for consumption as liquor. The use of ethanol or other alcohols as a fuel is not new, although with the discovery of petroleum and the distillation of gasoline, the use of ethanol as a vehicle fuel ceased during the 20th century. The ethanol produced

---

*Trademark belongs to Waste Management, Inc.

for industrial purposes in the 20th century was virtually all produced by thermal processes from ethylene which was derived from petroleum or natural gas liquids. With the increase in oil prices and the availability of excess grain (particularly corn), the interest in the production of ethanol by fermentation processes as a vehicle fuel, particularly on a 10% substitute for gasoline (gasohol) basis, has rekindled.

19.3.2 Basic Processes

Unless starting with a sugar solution, all processes for the fermentation of anhydrous ethanol for use as a fuel have four steps. First, the feedstock is reduced to sugar solution. Second, the sugar solution is fermented to ethanol and carbon dioxide. Third, the ethanol is distilled from the fermented solution. This distillation is technically limited to achieving 95.6% ethanol and 4.4% water solutions. Finally, the remaining water is removed to produce anhydrous (dry) ethanol. After the production of the sugar syrup in the first step, the processes are the same. In comparison to anaerobic digestion, the various ethanol processes are much better understood.

19.3.3 Feedstocks

Many natural organic substrates can be used as feedstocks for the production of ethanol. The feedstocks with the largest potential are wood, agricultural residues, grasses, and the organic fraction of MSW. The first three are complex mixtures of cellulose, hemicellulose, and lignin. The hemicellulose is reasonably simple to remove and convert to 5-carbon sugar molecules such as xylose ($C_5H_{10}O_5$). Although removal of hemicellulose is done commercially today and it is converted to furfural, it is not commercially used to produce ethanol. Yeasts that will convert 5-carbon sugar to ethanol are not well developed and are in research at this time.

As noted under anaerobic digestion, the lignin and cellulose are closely linked. Lignin cannot be degraded to ethanol at this time and tends to shield the cellulose from microbial attack. Therefore, the cellulose must be freed from the lignin to undergo further attack. The means to free the cellulose may be either chemical or mechanical and the actual conversion of cellulose to sugars either chemical or biological. Some of the cellulose preparation steps being researched include grinding with various mechanical equipment, removal of lignin with concentrated acid or by heating in a water, sodium carbonate, and butanol mixture, and steam explosion. All of these are designed to prepare the cellulose for attack by enzymes of organisms that selectively convert the cellulose to glucose. Relatively high conversion efficiencies (90%) have been obtained at mild conditions of 160°F (71°C) in a few days' reaction time. The chemical pathway is through the use of an acid such as sulfuric to break down the cellulose. One group of researchers has reported high conversion of cellulose to sugar in a 10-sec reaction time at 400°F and 2 to 4% sulfuric acid. Unfortunately, all the acid hydrolysis processes are nonspecific as to end product and will convert sugar to furfural and other undesirable products.

The starches are the most common feedstock both for liquor and fuel-grade ethanol. Unfortunately, the starches from corn, wheat, rice, and other grains are both edible and grown on good land. In periods of excess grain crops, they are readily available but in years with crop failures and in face of expanding world population and hunger may not be reasonably available for the long haul. The process used is one of grinding to free the starch from the hull. The desired product is uniform particles with a minimum of flour and starch granules adhering to hulls. The starch granules are then slurried and heated with an amylase yeast. The mash is cooked at 160°F (71°C) to convert the starch to sugar. Finally, the temperature is raised to boiling to sterilize the sugar solution and mash in preparation for conversion to ethanol.

Finally, a sugar crop such as sugarcane, sugar beets, Jerusalem artichokes, and so on, may be raised and processed to a sugar syrup. Again, these sugars compete with food uses, although the lands used may be more marginal than those used for grains. In general, they require more energy to grow and refine to a useful sugar feedstock. Particularly, the concentration of the syrup is energy intensive.

19.3.4 Products

The principal products of today's fermentation processes are ethanol, carbon dioxide, and by-products such as distillers dry grain (DDG). Of course from an energy standpoint, the product of interest is ethanol. Its energy content is about 65% of that of gasoline. However, it has a higher octane rating and produces less pollutants because it burns cooler and has less contamination from trace compounds. When pure, it may be used as a fuel even with some remaining water. The water both cools and adds density to the fuel mixture. However, pure ethanol is harder to vaporize, requires carburetor modification and may deteriorate components of the fuel system. Therefore much of today's commercial use of ethanol is as gasoline/alcohol blends at about 90/10 proportions.

It must be anhydrous alcohol in the blend, since the presence of water can cause phase separation problems. However, additives that eliminate this problem are being developed and are in use. Other ethanol uses are as solvents and chemical intermediates in industry.

As with the carbon dioxide from an anaerobic digester, the gas from the fermenter can be used as dry ice, to carbonate beverages and to a small extent chemical processes. A growing need for carbon dioxide is in the petroleum industry where it is used to enhance oil recovery from older producing wells.

When grain or other starchy feedstocks are used, only the starch and sugar are converted to ethanol. The protein, the yeast cells used to convert starch and sugar to ethanol, and all the other components of the grain remain as a by-product. There are many ways to use these by-products but the principal use seems to be as cattle feed. If storage of more than 2 days is needed, the distillers grains (as the by-product is known) must be dried to DDG. Other options are to remove the protein before conversion steps or to potentially convert the DDG to vitamins, other alcohols, or other chemicals. Where corn is used as the feedstock, the tendency is to build integrated plants capable of producing corn oil, corn meal, ethanol, and other products. The actual product mix can then be tailored to the markets available.

19.3.5 Design Parameters

As in the anaerobic digestion biochemical processes, the wide variety of feedstocks and pretreatment steps makes a detailed discussion impossible. However, certain generic design figures are useful.

One pound (0.45 kg) of sugar theoretically produces 0.47 lb (0.21 kg) of ethanol. One gallon of ethanol weighs 6.6 lb (3 kg). The actual conversion of sugar to ethanol can be 85 to 95% efficient. The conversion of starch to sugar can be 80 to 90% efficient. Theoretically 1 lb of starch yields 1.11 lb (0.5 kg) of sugar.

A simpler set of conversion figures that 1 ton of corn stover or wheat straw will yield 47 gallons (178 L) of ethanol, 1 ton of sugarcane will yield 110 gallons (416 L), and one bushel of corn will yield 2.5 gallons (9.5 L).

There are various estimates of the energy required to produce ethanol from various feedstocks. One such estimate indicates that 42,000 BTU (4.2 MJ) of fuel and 1.3 kWh of electricity (13,000 BTU) (1.3 MJ) of coal at 10,000 BTU/kWh are necessary to produce 1 gallon (3.785 L) of ethanol with a lower heating value of 76,000 BTU (7.6 MJ). Thus, 29% of the energy is required for conversion of starch to sugar and 45% is required for distillation. The energy content of the feed, its transportation, and transportation of the ethanol to site of use are not included.

19.4 COMPOST

19.4.1 Background

Composting is a biochemical process that stabilizes the putrescible fraction of an organic material under controlled conditions. As with anaerobic digestion, it is an ancient natural process that has for millions of years broken down leaves and other organic material into humus.

Little scientific investigation was made of composting before 1950. It was largely through the efforts of a number of universities and federal agencies that research was begun. As a result, the basic scientific principles of composting are now well enough understood that the process can be utilized on a full-scale basis under favorable circumstances.

19.4.2 Basic Process

In the basic process, organisms break down the available biodegradable organics into simpler, more stable compounds and carbon dioxide. The organisms self-generate heat. The heat, if not controlled will inhibit the process. The available organisms that compost organics operate in the same mesophilic or thermophilic temperature ranges as anaerobic digestion. Unlike the anaerobic digestion process, composting may occur with either aerobic or anaerobic organisms. Since during their processing period the anaerobes generate offensive odors that are difficult to control in a composter, the normal practice is to use aerobic composting. Also the latter is more rapid and attains high temperatures (therefore tends to sterilize the end product better).

19.4.3 Process Description

Most composting operations consist of four basic steps: (1) preparation of the feedstock, (2) decomposition, (3) curing, and (4) finishing or product preparation. The preparation step consists of sorting the organic and inorganic fractions and adjusting the size of the feedstock if necessary, mechanical dewatering if necessary, addition of nutrients, principally nitrogen to feed the microorganisms to speed the decomposition, and addition of bulking agents particularly for wet feedstocks. The decomposition step is the critical step in which most of the microbiological action takes place; it includes the mesophilic, thermophilic, and cooling phases. To accomplish this step, several techniques have been used. In windrow composting, prepared solid wastes are placed in windrows in an open field. The windrows are turned by mechanical means to break up channeling, insure uniform reaction, and aid in aeration for a period of about 5 weeks. The third step, curing, accomplishes stabilization of the material. The time allowed for this step depends upon the proposed use, with a minimum time being 2 to 4 weeks, and an ideal time being about 2 to 3 months. Once the solid wastes have been converted to a humus material, they are ready for the fourth step of product finishing and preparation. This step may include screening to recover the bulking agent and fine grinding to remove oversize material, blending with various additives, granulation, bagging, storage, and shipping. As an alternative to windrow composting, mechanical equipment may be used for the digestion and/or curing steps to reduce operator costs and to insure more uniform reaction.

19.4.4 Feedstock

Any biodegradable organic is a suitable feedstock for composting, although the simpler and more readily accessible the organic, the faster and better the compost. For instance, leaves from a yard may require several seasons to stabilize while food wastes can be composted in a few weeks' time.

The two most prevalent feedstocks to composters today are sewage sludges and manures. The latter is likely to be a naturally composting pile in a farm yard whereas the former is likely to be a controlled process conducted in a designed and constructed site.

The composting process is basically one that operates on solid feedstocks; therefore sewage sludge must first be dewatered to 20 to 40% solids. Then to maintain aerobic conditions a bulking agent is added to help air to contact all parts of the sludge mass. The normal bulking agent is wood, 90% of which is recovered through screening and reused. Other potential bulking agents include shredded tires and prepared MSW. As noted, the microbes generate heat as part of their life process. This heat then assists in the evaporation of water. One test at a university demonstrated an initial water content of 76% being reduced to 22% through composting. The overall compost process then is one of drying sewage sludge at the expense of carbon converted to carbon dioxide by organisms.

When MSW is converted to compost, feedstock preparation is desirable. The inorganics have no use in the process and do not enhance the end product. Similarly the plastics, textiles, rubber products, and so forth, should also be removed. The principal research in MSW composting is being undertaken in Europe and Japan. The Japanese have been experimenting with a semiwet selective pulverizer which uses a rotating drum, injections of water and blades with different speeds, to recover a biodegradable compost feedstock.

19.4.5 Products

Except for the materials that may be recovered from the processing of MSW into feedstock, the sole product of the process is compost. It is a humus-like soil conditioner with very low levels of nutrients unless the compost is used as a carrier for chemical fertilizers. In addition, the principal sources of feedstock (sewage sludge and MSW) may contain disease-bearing vectors that survive the compost process and heavy metals and other contaminants of concern. This, plus the high price of the product when compared with chemical fertilizers, make marketing very difficult. Successful composting plants worldwide seem to develop a market and then to restrict their size carefully to match that market. For instance, a number of successful European composting plants supply soil conditioner to the vineyards. The high-fiber content probably acts as a blotter to both retain water for plant use and reduce runoff erosion on steep slopes.

19.4.6 Design Parameters

As in other biochemical processes, the design parameters are a function of the specific system. Land requirements vary widely depending upon the type of system, with mechanical systems requiring the least land. Windrow system size must be balanced between the need for natural insulation of the windrow and energy cost of forcing oxygen to all parts of the windrow. Oxygen supply also affects desired particle size. Maximum surface area for microbial attack is necessary, but the void volume must be large enough to permit maintenance of aerobic conditions. With certain feedstocks such as sewage sludges and manures, a bulking agent must be added to increase voids. In addition, the carbon/nitrogen ratio should be adjusted to about 30:1.

19.4.7 Reactor Types

Compost systems fall under three general categories: "open" or windrow/pile systems, "enclosed" or mechanical systems, and combination systems. The "open" or windrow systems are those in which the entire process is carried out in the open. The material is usually stacked in long windrows or elongated piles, preferably on a hard surface, such as asphalt. Ideally, but not necessarily, the surfaced area should be covered with an open-sided roof structure. Roofing is especially appropriate in areas subjected to heavy snowfalls, as melting snow can result in waterlogged windrows. A windrowed mass of composting material can be aerated by periodic turning and/or forced aeration. The efficacy of turning has been proven by long experience, whereas forced aeration remains to be demonstrated for many types of applications.

Mechanical systems place the prepared feedstock into closed containers. The contents are mechanically stirred and forced aeration is used to maintain aerobic conditions. When the compost is "cured" it is removed from the container.

The combined system retains feedstock for a period, say 3 days, in a mechanical device where temperature, moisture content, and oxygen/carbon dioxide levels can be carefully controlled. After this initial processing, the partially composted product is then allowed to cure in the open for 30 to 60 days.

Many are inclined to dismiss windrowing as inefficient because it requires extensive land usage and is quite likely to produce foul odors, thereby creating a public nuisance. However, mechanical systems are capital intensive, do not completely eliminate all odors, and require only a little less land than do the windrowing techniques. The "open" or windrow composting technique seems mainly to have a future in smaller rural communities, or in less developed countries with municipal solid waste high in putrescible materials, where capital is scarce and land is plentiful. Conversely, the advantages of the "closed" or mechanical composting technique lends itself to urban applications.

19.5 APPLICATIONS AND ECONOMICS

19.5.1 Anaerobic Digestion

The principal applications of anaerobic digestion in the United States has been in the sewage treatment process to stabilize and reduce the quantity of sludge. Generally the digester gas has been wasted. The true economies are not known since these systems are operated in the public domain as part of the service of sewage disposal. In addition, most of the more modern systems have received large federal grants for construction and the remaining capital has been raised using general obligation bonds, which are sold on the basis of the local tax base and not the soundness of the specific project. However, there are digesters in two-thirds (10,000) of American sewage plants today.

The projected economics of a plant anaerobically digesting MSW to produce methane were estimated in 1975. Those figures are presented in Table 19.1. The baseline was a 1000 TPD (892 t/day) plant constructed in the private sector. These data were updated in 1981 to reflect the results of the Department of Energy (DOE) test bed and are presented in Table 19.2.

Some explanation concerning the update is necessary. Capital amortization was increased based on the inflation rate and upon the conversion from private financing to a municipal revenue bond with equity. The operating costs were decreased based upon a simpler, more economic design. Filter cake and wastewater penalties were both deleted. The most logical plant will become part of a sewage treatment facility and, therefore, treated wastewater will provide any required makeup water and any excess filtrate will be recycled in the plant. The filter cake must be used productively for the plant to be economically sound. If one were to have to dispose of the filter cake, disposal would be most likely in a landfill on a volume basis and cost about $3 per input ton of waste (a penalty of $0.80/1000 ft^3). The cost of processing rejects has been increased to match the disposal cost (tipping fee) for waste delivered to the plant.

For credits, the disposal fee and sewage sludge disposal fees were increased although the increase is less than the rate of inflation over the period. These fees are believed to be the median real cost in the United States today (1983). The scrap iron sales were reduced to zero since there is a large scrap iron surplus in the United States and the steel industry is depressed.

These figures are believed to be conservative since the improved efficiencies at the Pompano Beach test bed are not included.

Landfill gas is now recovered, processed and sold at some 30 landfill sites around the United States. Many of these are privately developed projects by representatives of the gas utilities, oil companies, and waste collection and disposal companies. They are selling gas as a boiler fuel, as SNG and as electricity. In 1979, the expected selling price of landfill gas was $1.50 to 2.00/million

Table 19.1   Baseline Cost Estimate Anaerobic Digestion of MSW

		1975	
		$/km³	$/kcf[a]
Capital amortization		$ 77	$2.17
Operation costs		$ 68	$1.92
Penalties			
Filter cake	($30/dry ton)	$ 55	$1.55
Wastewater	($1/Mgal)	$ 1	$0.03
Processing rejects	($3.50/ton)	$ 9	$0.26
Credits			
Disposal fee	($10.65/ton)	($101)	($2.87)
Sewage sludge	($50/dry ton)	($ 18)	($0.51)
Scrap iron	(25/ton)	($ 16)	($0.26)
Total		$ 75	$2.09

Note: Based on 25% equity, 15% return on equity, and 9% interest on debt over 20 years.
[a] Thousand cubic feet (MCF).

Table 19.2   Estimated 1981 Baseline Cost

		1981	
		$/km³	$/kcf
Capital amortization		$177	$4.99
Operation costs		$ 74	$2.09
Penalties			
Filter cake	($30/dry ton)	$ 0	$ 0
Wastewater	($1/Mgal)	$ 0	$ 0
Processing rejects	($3.50/ton)	$ 39	$1.09
Credits			
Disposal fee	($10.65/ton)	($142)	($4.00)
Sewage sludge	($50/dry ton)	($ 36)	($1.01)
Scrap iron	($25/ton)	($ 0)	($0.0)
Total		$112	$3.17

BTU (25.2×10^4 k cal) for a medium BTU gas system and $3.00 to 4.00/million BTU (25.2×10^4 k cal) for pipeline quality SNG.

19.5.2   Fermentation

The use of grain and sugar as feedstocks for the conversion to fuel-grade ethanol is an expanding business opportunity, although the current stable costs of gasoline has slowed project implementation. Estimates in 1980 for a grain-based 50 million gallon/year fuel ethanol plant developed alcohol prices of $0.37 to 0.60 per gallon and for sugar based $0.55 to 0.94 per gallon (3.785 L).

There are no ethanol plants based upon cellulosic feedstocks; however, one set of cost calculations indicates a $0.93 to 1.29 per gallon (3.785 L) selling price. These prices should be multiplied by 1.5 to compare them directly to gasoline. However, there are tax advantages that make fuel alcohol production more attractive. For instance, $0.05 of the federal fuel tax is forgiven if the fuel contains 10% ethanol.

19.5.3   Composting

The composting of sewage sludge for subsequent sale as humus is expanding in the United States. A recent study conducted by the Agricultural Engineering Department of the University of Maryland and the Biological Waste Management and Organic Resources Laboratory of the U.S. Department of Agriculture indicates that in May 1983 there were 61 full-time sludge composting plants in operation in the United States. Within a year, an additional 29 plants are expected to begin full-

Table 19.3 Numbers of Composting Facilities of Different Types and Stages of Construction in the United States

Status of Facility	Aerated Pile	Types of Facilities		
		Enclosed Reactors	Windrow	Aerated Windrow
Operational	43	4	19	1
Under construction	2	0	0	0
Pilot	1	1	2	0
Design	14	7	0	0

Table 19.4 Capital, Operating, and Total Costs for Windrow Composting[a]

Plant Capacity (Mg)		Shifts (No.)	Capital Cost[b] ($/Mg/day)	Capital and Investment $/Mg	Operating[c] $/Mg	Total $/Mg
Day	Year					
47	12,260	1	20,484	7.59	14.78	14.36
45	11,790	1	16,974	6.75	15.05	21.79
90	23,580	2	9,404	3.47	11.58	15.05
90	23,580	1	11,047	4.06	10.91	15.07
180	47,164	2	6,019	2.22	9.59	11.81

[a] The data in the table are based on information from a U.S. EPA report.[18] Amounts based on actual costs of Johnson City composting plant with modifications with 7.5% financing over 20 years; straight-line depreciation of buildings and equipment over 20 years.
[b] Includes land costs estimated at $1960/ha (2.4 ac).
[c] Does not include costs for landfilling rejects.

scale operations. Table 19.3 summarizes the number, type, and operational status of composting facilities in the United States.

Very little cost information is available for composting plants. The most realistic values are contained in reports pertaining to the U.S. Environmental Protection Agency demonstration facility in Johnson City, Tennessee. The capital costs of the Johnson City facility and estimated capital costs for windrow plants of varying capacity are presented in Table 19.4. The projects are based on costs at Johnson City; actual costs would vary at other locations. Other cost variations would result from the use of different types of size reduction and other processing equipment. Costs are shown in 1969 dollars. The figures also indicate a pronounced effect of economy of scale on capital cost per thousand gallons (3.78×10^3 dm^3) of daily capacity.

19.6 CASE HISTORIES

19.6.1 Anaerobic Digestion

The only operating MSW anaerobic digester is the Department of Energy RefCOM test bed at Pompano Beach, Florida. Beginning in 1972, Dr. John Pfeffer of the University of Illinois at Urbana began laboratory scale research into the anaerobic digestion of MSW. His feedstock was refuse-derived fuel (RDF) collected from a pilot mechanical processing plant operating in St. Louis, Missouri. By 1974, Dr. Pfeffer's anaerobic digestion process had been scaled to 100-gallon (400 L) tanks that were producing 6 ft^3 gas/lb (0.37 m^3/kg) of volatile solids fed.

In 1975 Waste Management, Inc. was selected competitively to build a proof-of-concept facility. Waste Management already owned and operated a 15 TPH (tons per hour) vertical shredder which produced 3-in. (76 mm) minus material. Therefore, the proposed design included a magnet to remove ferrous metal, storage in an automatic reclaim silo, processing to include a trommel to remove undersize material, a secondary shredder to permit experiments testing the effects of different size particles, and an air classifier to refine further the organics. A pneumatic system delivered the processed material to a cyclone. The separated feedstock passed over a weigh feeder and was sent to a premix tank to be combined with nutrients and steam for heating. The digester had both underflow and overflow drains and the mixing system was designed to be gas operated. The underflow was to be the primary effluent with floating objects removed weekly by the overflow drain.

Several design changes were made prior to construction of the facility. The ferrous recovery system was deleted as being unnecessary to the experiment. However, the contractor upgraded his shredding facility to include a 60 TPH vertical shredder and a ferrous removal system for all the waste that was received at his landfill. The reclaim silo was deleted since it was not cost-effective, especially since shredded waste might be retained in it for extended periods, and it was replaced by a flat floor storage area. The overflow drain was deleted for unknown reasons. After a series of tests, a mechanical mixer was selected in lieu of gas mixing. Finally, a vacuum filter was selected for dewatering over microstrainers and a centrifuge.

Construction began in mid-1976 and was completed in mid-1978. There were no significant construction problems. The digesters were seeded with sewage sludge in July, 1978, and biogas first produced in August. The digesters have continuously produced gas since that time, although problems with the preparation system have prevented stressing the system.

There have been many lessons learned in this plant. The front-end system has been extensively modified and now consists of a primary shredder and a disk screen rejecting both an oversize component that is mostly rags, long stringy textiles and wire, large plastics, and a basically inorganic fine component.

The digestion process has worked very well. In the 3-month period from November, 1981, to February, 1982, the digesters were fed 18 TPD of municipal waste (with 80% volatiles) and about 50 lb (23 kg) of nutrients per day and produced 7.5 ft^3 (0.21 m^3) of biogas per pound (0.45 kg) of volatile solids with about 56% methane. The plant operated at 140°F (60°C) and an 18-day retention time. During the laboratory experiments at the same conditions, 7 ft^3 (0.2 m^3) of biogas per day was produced with about 53% methane.

The digesters were fed 5 days a week. Gas production fell dramatically during the weekend and recovered each Monday, clearly indicating that they were not stressed but also indicating that disruptions of two days in feed supply had no detectable adverse impacts on the digestion process.

During February and March, 1983, the digester was operated at a steady 10 TPD and an 8-day retention. It produced 6 ft^3 (0.17 m^3) of biogas/lb (0.45 kg) of volatile solids at about 55% methane concentration.

The following are case histories of two different landfill gas projects.

At the Azusa, California, landfill 3000 ft^3/min (1.42 m^3/sec) (4.3 million cubic feet/day, MMCFD) of landfill gas is extracted daily. The landfill gas is sold after condensate removal to Reichhold Chemical Company. Azusa land Reclamation Company is studying the possibility of generating electricity with the landfill gas at the Azusa landfill for sale to either Southern California Edison Co. or to the City of Azusa.

At Azusa the 17, 90- to 100-ft (27 to 30 m)-deep well collection system is designed to extract approximately 3000 ft^3/min (1.42 m^3/sec) of gas. The raw landfill gas is brought to the plant where free water is removed in an intake scrubber and then the gas is compressed. The compressor is operated by the landfill gas product. The compressed gas goes through coolers and the final water removal stage is a triethylene glycol process. The 500 BTU/scf gas is delivered in a 12-in. (30.5 cm) carbon steel pipe about 3000 ft (0.9 km) to the Reichhold Chemical Co. for fuel for their boilers.

The Monterrey Park, California, methane recovery facility began operations on August 15, 1979. It is designed to recover approximately 8 MMCFD (2.64 m^3/sec) of landfill gas and purify it to produce 4 MMCFD (1.32 m^3/sec) of pipeline quality (1000 BTU/scf) gas (8900 kcal/m^3). Product gas is delivered directly into the Southern California Gas Co. distribution system. The facility has been operating at full capacity since 1980.

The landfill gas collection system consists of 51 wells, half of which recover gas from the upper portion of the landfill with the other half pulling gas from the deeper sections. Gas is withdrawn from the wells through a network of nearly 3 miles of laterals and header pipe which route the gas to the plant inlet. The collection system is constructed to heavy-gauge polyethylene pipe. The collector system vacuum and process pressure is provided by two 1600 hp (1194 kW) gas-fueled, four-stage Ingersoll Rand reciprocating compressors. Gas purification is accomplished in two stages. The first stage is pretreatment, which removes moisture, heavy hydrocarbons, and trace contaminants by means of chilling and a proprietary pretreatment process step designed by Getty Synthetic Fuels. Following pretreatment the gas is compatible for processing by more conventional CO_2 removal processes. The Monterrey Park facility removes CO_2 by the selexol process licensed from Allied Chemical Corporation. This process was selected because of the higher gas volumes processed and the higher delivery pressures required. Following CO_2 removal, product gas is delivered to a Southern California Gas Co. metering station which measures volume, BTU content, and specific gravity. The Monterrey Park facility experienced a 90% on-stream factor during the first month of operation, and as high as 96% in some subsequent months.

19.6.2 Compost

Following are case studies of two operational sewage sludge composting facilities. Two different technologies are represented. The mechanical or "enclosed reactor" technology is represented by the Portland, Oregon, facility, whereas the "open" or aerated pile concept is used at Montgomery County Composting Facility in Silver Springs, Maryland.

For various reasons, Portland, Oregon, could not consider continuing sludge disposal at their landfill. The cost of disposal was expensive and increasing. Including the disposal fee, labor, material, and equipment cost, the city spent over $1.2 million in 1982 for sludge disposal. These costs were expected to increase to over $1.5 million in 1983. Second, Portland was required under Environmental Protection Agency (EPA) regulations to develop a long-term wastewater sludge disposal plan. Portland was also experiencing a problem with available landfill capacity. Third, the city's efforts to site a new landfill were hampered by legal and political problems; and fourth, and most important, the city council mandated in a policy directive that sewage sludge should be reused in a beneficial manner.

In 1981, the city selected the Taulman/Weiss Composting System. It uses a vertical circular tank with a closed top and a screw-type outlet device. This system is the first operational enclosed composting system in the United States.

It was constructed in the space that was originally intended for sludge incinerators. Odor control in the totally enclosed vessel was the most important consideration in the selection of this type of composting. The facility cost is $11.4 million (1982 dollars).

Taulman/Weiss will provide design and construction services as well as the marketing of the compost. Taulman has agreed to pay the city $10/ton of compost plus 50% of any gross income in excess of $16/ton for 20 years. The satisfactory operation of the completed facility is guaranteed by both Taulman and by Gebruder Weiss K.G., the holder of the proprietary rights to the system.

The city will lease land adjacent to the treatment plant to Taulman for storing, packaging, mixing, and distributing the compost. There is a set limit on the amount of compost to be stored on the site.

The Montgomery County, Maryland, Composting Facility opened in the spring of 1983 after many years of public opposition and court cases to prevent its construction. Design began about 1979. The facility currently receives 400 wet TPD (5 days a week) of undigested sludge from the Blue Plains Sewage Treatment Plant in Washington, D. C. Project design calls for the sludge received at the site to be 20% solids. This sludge is mixed with wood chips to attain a 60% moisture content. The design ratio of wood chips to sludge is 4.5 to 1.

The mixed wood chips and sludge are placed in 100×12.5-ft-high $\times$ 20-ft-wide piles ($30 \times 3.75 \times 6$ m). Oxygen and temperature sensors are implanted in each pile for control. Fifteen-horsepower blowers are used to force air through the pile. By careful control, the pile is kept at aerobic conditions throughout the process. During the first week, high (3500 to 4000 ft^3/hr/dry ton of sludge) (99 to 113 m^3/hr/dry ton) aeration rates are maintained to optimize decomposition and odor control. During the second week, aeration rates are decreased to increase pile temperatures to 120 to 140°F (50 to 60°C). During the third week, temperature is further raised to 140 to 160°F (60 to 70°C) to maximize drying. At the end of the third week, the design moisture content of the pile is 50%.

At this point, compost is screened in an enclosed building. Seventy percent of the wood chips are recovered and recycled. The compost is then returned to piles and cured for at least an additional 30 days. One-horsepower blowers keep the curing pile aerobic until the finished compost is sold. The largest users of the compost are landscapers, nurseries, and parks. It is approved for all uses except tobacco growing, although a state permit is required to use it on other agricultural crops.

The facility cost approximately $27 million to construct. Operating and maintenance costs are expected to be approximately $35/ton when the site is operating at design capacity. There are 34 employees at the plant when it is operating at 200 tons of wet sludge per day.

BIBLIOGRAPHY

1. *Energy Deskbook*, DOE/JR/05114-1, U.S. Department of Energy, Technical Information Center, Oak Ridge, Tennessee.
2. E. W. Steele, *Water Supply and Sewerage*, McGraw-Hill, New York, 1971.
3. L. C. Urguhart, *Civil Engineering Handbook*, McGraw-Hill, New York, 1967.
4. S. Bond and L. Straub, *Handbook of Environmental Control*, Vol. IV, *Wastewater Treatment and Disposal*, CRC Press, Boca Raton, Florida, 1966.
5. R. Lystax, *Environmental Engineers Handbook*, Vol. 1, *Water Pollution*, Chilton Book Co., New York, 1974.

6. *Energy from Biological Processes*, Vol. II, *Technical and Environmental Analysis*, Office of Technology Assessment, Congress of the United States.

7. *A Learning Guide for Alcohol Fuel Production*, Colby Community College, Colby, Kansas, 1976.

8. *Ethanol Fuel: Use, Production, and Economic*, SERI/SP-751-1018, Solar Energy Research Institute, Golden, Colorado, 1978.

9. *Full Scale Experimental Anaerobic Fermentation Facility*, DOE/ET/20009-T1, National Technical Information Service.

10. *Fuel Gas Production from Animal and Agricultural Residues and Biomass*, SERI/CP-231-1924, Solar Energy Research Institute, Tempe, Arizona.

11. *Biological Conversion of Biomass to Methane*, SERI/TR-98357-1, Solar Energy Research Institute, Tempe, Arizona.

12. *Fuel from Farms—A Guide to Small Scale Ethanol Production*, Technical Information Center, U.S. Department of Energy.

13. *Landfill Methane Utilization Technology Workbook CPE-7907*, The Johns Hopkins University, Applied Physics Laboratory, Baltimore, Maryland, 1979.

14. D. K. Walter, Anaerobic Digestion of Municipal Solid Waste to Produce Methane, VIII, Proceedings, Symposium of the Institute of Gas Technology, Washington, D.C., 1978.

15. L. F. Diaz, G. M. Savage, and C. G. Golueke, *Resource Recovery from Municipal Solid Waste*, Vol. II, *Final Processing*, CRC Press, Boca Raton, Florida, 1982.

16. Institute for Solid Wastes of American Public Works Association, *Municipal Refuse Disposal*, Public Administration Service, Chicago, Illinois, 1968.

17. D. G. Wilson, *Handbook of Solid Waste Management*, Van Nostrand Reinhold Co., New York, 1977.

18. P. H. McGaukey and C. G. Golueke, Reclamation of Municipal Refuse by Composting, Tech. Bull. No. 9, Sanitary Engineering Research Laboratory, University of California, Berkeley, June, 1953.

19. J. S. Wyley, Progress Report on High-Rate Composting Studies, Engineering Bulletin, Proceedings of the 12th Industrial Wastes Conference, Series No. 94, May, 1957.

20. K. K. Schulze, *Anaerobic Decomposition of Organic Waste Materials (Continuous Thermophilic Composting)*, Final Report, Michigan State University, Lansing, Michigan, April, 1961.

21. D. C. Wilson, *Waste Management, Planning, Evaluation, Technologies*, Clarendon Press, Oxford, 1981.

22. G. B. Willson and D. Dalmat, Sewage Sludge Composting in the U.S.A., *Biocycle* (September/October 1983).

23. U.S. Environmental Protection Agency, *Composting of Municipal Solid Wastes in the United States*, Waste Management Series, Publication SW-47r, Washington, D. C., 1971.

24. H. B. Gotaas, *Composting*, World Health Organization, Geneva, Switzerland, 1956.

25. C. G. Golueke and H. B. Gotaas, Public health aspects of waste disposal by composting, *Am. J. Health*, **44**, 339 (1954).

26. J. S. Wiley, Pathogen survival in composting municipal wastes, *J. Water Pollut. Control Fed.*, **34**, 80 (1962).

27. City of Industry Landfill Gas Recovery Operation, SCS Engineers, ANL/CNSV-TM-91 (November, 1981).

28. State of the Art Landfill Gas Recovery, EMCON Associates, ANL/CNSV-TM-85 (July, 1981).

29. The Environmental Impacts, Institutional Problems, and Research Needs of Sanitary Landfill Methane Recovery, ANL/CNSV-TM-86 (June, 1981).

30. Development of a Case Study for the Cinnaminson Landfill Methane Recovery Operation, ANL/CNSV-TM-73 (February, 1981).

31. The Review and Evaluation of a Bioconversion Process Demonstration Plant and Its Related Technologies, DOE/CS/20101-1 (February, 1983).

32. University of Iowa, Anaerobic Biological Treatment of Liquid Wastes from Pyrolysis Processes, NTIS #COO-4455-3 (September, 1980).

33. Pacific Gas & Electric Co. and Dynatech R&D Company, Development of the Utilization of Combustible Gas Produced in Existing Sanitary Landfills: Evaluation in the Use of Carbon Dioxide Produced in Sanitary Landfills, DOE/CS/20291-1 (October, 1982).

34. District of Columbia Department of Environmental Services, I-95 Landfill Gas Recovery and Utilization Feasibility Study, DOE/RA/50314-1 (February, 1983).

35. Idaho National Engineering Laboratory, Municipal Solid Waste Bioconversion Technologies, EGG-2193 (October, 1982).

APPENDIX 19.1 BIOMASS AS FUEL FOR ELECTRIC GENERATION: PLANNED AND EXISTING PROJECTS IN THE UNITED STATES*

James L. Easterly and Elizabeth C. Saris

A19.1.1 EXECUTIVE SUMMARY

This appendix summarizes current and planned use of biomass as a fuel to produce electricity in the United States. The specific biomass fuel types are woodwastes, agricultural residues, and animal manure. This appendix also summarizes the electric generating capacity represented by the use of municipal solid waste, landfill gas, and sewage methane as energy feedstocks. These municipal wastes can be classified under the general category of "biomass" feedstocks.

Information was obtained for utilities, municipalities, small power production, and cogeneration facilities which use biomass as a fuel to produce electricity. Findings indicate that total biomass-fueled generating capacity, from 196 existing and planned facilities, represents 3000 MW. This includes 18 electric utility projects representing 245 MW, or about 10% of this total capacity.

The study found that current utility use of biomass is typically on a relatively small scale—in a capacity range similar to small power production and cogeneration facilities, that is, 5 to 50 MW.

The following summary represents the total amount of kilowatts by biomass fuel type: wood, 1,385,077 KW (78 facilities); agricultural residues, 242,250 KW (15 facilities); animal manure, 935 KW (3 facilities); municipal solid waste, 1,244,360 KW (53 facilities); landfill gas, 89,245 KW (41 facilities); and sewage gas, 11,084 KW (6 facilities). In addition, data for each fuel type is summarized by geographic region. Further, the paper sets forth various analyses of the research data.

A19.1.2 BACKGROUND

The primary information sources used to identify projects using biomass to produce electricity included the Federal Energy Regulatory Commission (FERC), Electric Power Research Institute (EPRI), DOE's Energy Information Administration (EIA), and the National Conference of Mayors. Data were also collected by way of an extensive literature search of reports, newsletters, personal contracts, and phone interviews with plant managers at the facilities summarized in this paper. The information obtained from these contacts was then reviewed and tabulated, taking into account such factors as specific fuel type, facility type, electric generating capacity, geographic region, and operational status. Readers should note, however, that the data presented is not necessarily all-inclusive due to the everchanging (increasing) level of activity in this area nationwide. However, the authors believe the information contained herein does present a very thorough and comprehensive overview of the extent to which biomass is currently used as a fuel to produce electricity in the United States.

Utility projects were limited to those operating on biomass fuel on a sustained basis as opposed to experimental firing of various fuel types, but do include two demonstration projects. Those utilities engaged only in feasibility studies in the biomass area were excluded from this investigation.

This study summarizes the total electric generation capacity represented by biomass-fueled facilities (i.e., design capacity). Capacity factors associated with the biomass facilities (i.e., the ratio of actual biomass fuel usage to design capacity) are not addressed in this investigation. Since many of the facilities are still in the planning stages, any attempt to estimate their capacity factors would only be hypothetical.

The nonutility facilities summarized in this paper are primarily small power production and cogeneration systems that have been qualified by FERC for benefits under the Public Utility Regulatory Policies Act of 1978 (PURPA).

There are three primary benefits that PURPA provides to qualifying cogenerators and small power producers (QFs): (1) PURPA guarantees a market for the sale of electric power generated by

*Excerpted from a complete survey of the subject prepared for the U.S. Department of Energy, Biomass Energy Technology Division, under contract No. DE-AC01-83CE-30784.

QFs (utilities must pay a rate reflecting their full avoided costs); (2) utilities are not allowed to use discriminatory rates for back-up power to QFs; and (3) most QFs are exempt from the arcane intricacies of utility regulation (readers should refer to the Act for specific benefits and exclusions).

To qualify as a small power producer, a facility must not exceed 80 MW of power production and must receive 75% or more of its total energy input from renewable resources. To qualify as a cogeneration facility, PURPA delineates detailed criteria that must be met in regard to the relationship between electric power production and the extraction of useful thermal energy. There is no upper size limit to qualify as a cogenerator.

Municipal projects that have not filed for PURPA benefits were also identified during this investigation—25 facilities that burn refuse as a fuel to produce electricity and five facilities that burn landfill-methane to produce electricity. For the 25 facilities using refuse fuel (from municipal solid waste), the capacity in kilowatts was calculated on the basis of their refuse burning capacity in tons per day (TPD). Nine of the 25 are cogeneration facilities, hence it was necessary to estimate a reduced kilowatt capacity resulting from the split between electricity production and process heat production.

A19.1.3 SUMMARY OF FINDINGS

The total U.S. electric generating capacity fueled by biomass feedstocks for planned and existing facilities is shown in Table 19.5. For the six categories of biomass feedstocks listed, wood waste and municipal solid waste (MSW) represent by far the largest contributors to the total capacity. After wood and MSW, agricultural residues represent the next largest contribution to total electric generation capacity. The various agricultural residue projects use a wide range of feedstocks such as sunflower seed hulls, process waste from sugar cane and sugar sorghum, barley straw, cotton stalks, walnut shells, almond shells, and peanut shells.

Table 19.6 represents an overview of utilities using biomass in the United States. Utilities represent only about 10% of the total biomass-fueled generation capacity. Those utilities that use biomass as a fuel are generally in the same size range as the nonutility facilities—5 to 50 MW. This capacity range reflects the typical amount of assured biomass supplies that can be found at a favorable site. Assured feedstock availability is a primary concern of utilities with respect to their use of biomass as a fuel. Figure 19.3 depicts the regional boundaries of the sector distributions of utility biofuels used in the United States.

The large number of nonutilities that use biomass reflects the fact that biomass wastes are readily utilized on-site where they are initially generated or collected—at facilities such as lumber mills, paper manufacturers, and city sanitation departments. PURPA has also had a major impact in encouraging industries and cities to utilize their biomass wastes to produce electricity.

Table 19.5 Total Electric Generating Capacity using Biofuels for Planned and Existing Facilities in Kilowatts (Number of Facilities)

Fuel	Utilities	Small Power Production, Cogeneration, and Municipal Facilities	Total
Wood waste	180,725 KW (9)	1,204,352 KW (69)	1,385,077 KW (78)
Agricultural residues	22,500 (2)	219,750 (13)	242,250 (15)
Animal manure (to methane)	—	935 (3)	935 (3)
Subtotal	203,225 (11)	1,425,037 (85)	1,628,262 (96)
Municipal solid waste	33,800 (5)	1,210,560 (48)	1,244,360 (53)
Landfill methane	—	89,245 (41)	89,245 (41)
Sewage methane	7,970 (2)	3,114 (4)	11,084 (6)
Subtotal	41,770 (7)	1,302,919 (93)	1,344,689 (100)
Total	244,995 (18)	2,727,956 (178)	2,972,951 (196)

Table 19.6 Overview of Utilities using Biofuels

Utility	Fuel Description	Capacity	Status
New England			
Burlington Electric Dept. J. E. Moran Plant, Vermont	75% Wood waste, 25% oil or gas	20,000 KW	Operating
Burlington Electric Dept. Burlington, Vermont	100% Wood waste	50,000 KW	Planned
Mid-Atlantic			
Rochester Gas & Electric Corp. Rochester 7, New York	MSW and coal, MSW represents 2600 KW or 1% of total plant capacity	2,600 KW	Shut down temporarily
East North Central			
Columbus Sewage Treatment Columbus, Ohio	100% Sewage gas	4,970 KW	Operating
Northern States Power Co. Bayfront Plant, Wisconsin	Wood waste and coal, 1000 KW represents 56% of total boiler capacity	1,000 KW	Operating
Northern States Power Co. French Island Plant, Wisconsin	100% Wood waste burned in one boiler	1,300 KW	Operating
Madison Gas & Electric Blount Street Plant, Wisconsin	MSW and coal, MSW represents 6500 KW, 15% of the fuel for two boilers	6,500 KW	Operating
West North Central			
Northern States Power Co. Redwing Plant, Minnesota	100% MSW burned in one boiler	13,000 KW	Under construction
Bismarck Electric Co. Bismarck, North Dakota	Sunflower seed hulls, 10,000 KW represents 20% of total plant capacity	10,000 KW	Operating
City of Ames, Iowa Ames Plant, Iowa	MSW and coal, 6300 KW represents 5% of total plant capacity	6,300 KW	Operating
Mountain			
Montana Light & Power Co. Libby Plant, Montana	100% Wood waste	14,500 KW	Operating
Pacific			
Eugene Water & Electric Board Weyco Center Plant, Oregon	100% Wood waste burned in two boilers	50,000 KW	Operating
Washington Water Power Co. Kettle Falls, Washington	100% Wood waste	42,500 KW	Operating
Southern California Edison Co. Highgrove Station, California	100% Wood waste, wood gasifier demonstration unit	1,300 KW	Operating
Alaska Village Electric Cooperative, Anchorage, Alaska	100% Wood waste, wood gasifier demonstration unit	125 KW	Operating
Southern California Edison Co. Rosemead, California	MSW, direct combustion of MSW, sewage sludge, and sewage digester gas	5,400 KW	Under construction
City of Los Angeles Scattergood Plant, California	Sewage gas represents 7½% of total plant capacity	3,000 KW	Operating
Waialua Sugar Co. Mill Plant, Hawaii	100% Bagasse (sugar cane waste)	12,500 KW	Operating

Fig. 19.3 Regional boundaries.

Table 19.7 Biomass-Fueled Small Power Production, Cogeneration and Municipal Facilities: Status by Fuel Type

Fuel	Planned	Under Construction	Operating	Total
Wood waste	238,505 KW (17)[a]	115,630 KW (10)	850,217 KW (42)	1,204,352 KW (69)
Agricultural residues	102,700 KW (3)	56,000 KW (3)	61,050 KW (7)	219,750 KW (13)
Animal manure	—	10 KW (13)	925 KW (2)	935 KW (3)
Subtotal	341,205 KW (20)	171,640 KW (14)	896,192 KW (50)	1,409,037 KW (84)
Municipal solid waste	640,450 KW (24)	286,500 KW (12)	283,610 KW (12)	1,210,560 KW (48)
Landfill methane	39,550 KW (15)	29,650 KW (14)	20,045 KW (12)	89,245 KW (41)
Sewage methane	164 KW (2)	2,700 KW (1)	250 KW (1)	3,114 KW (4)
Subtotal	680,164 KW (41)	318,850 KW (27)	303,905 KW (25)	1,302,919 KW (93)
Total	1,021,369 KW (61)	490,490 KW (41)	1,216,097 KW (76)	2,727,956 KW (178)

[a] Number of facilities.

The number of biomass facilities that have filed for qualifying status under PURPA has approximately doubled each year since the act was implemented in 1978.

Table 19.7 summarizes the operating status, by fuel type, for the 177 biomass-fueled small power production, cogeneration, and municipal projects (that is, nonutilities). For the first three types of biomass fuels (wood waste, agricultural residues, and animal manure) 50 of the 83 facilities are already operating. For those that are not operating, two of the most common problems cited were delays and problems in obtaining adequate financing. For the last three fuel types (MSW, landfill methane, and sewage methane) only 25 of the 93 facilities are operating—a reflection of the fact that these municipal facilities generally involve a rather time-consuming approval process,

Table 19.8 Utilities using Biofuels: Status by Fuel Type

Fuel	Planned	Under Construction	Operating	Total
Wood waste	50,000 KW (1)[a]	—	130,725 KW (8)	180,725 KW (9)
Agricultural residues	—	—	22,500 KW (2)	22,500 KW (2)
Animal manure	—	—	—	—
Subtotal	50,000 KW (1)	—	153,225 KW (10)	203,225 KW (11)
Municipal solid waste	—	18,400 KW (2)	15,400 KW (3)	33,800 KW (5)
Landfill methane	—	—	—	—
Sewage methane	—	—	7,970 KW (2)	7,970 KW (2)
Subtotal	—	18,400 KW (2)	23,370 KW (5)	41,770 KW (7)
Total	50,000 KW (1)	18,400 KW (2)	176,595 KW (15)	244,995 KW (18)

[a] Number of facilities.

with public hearings and environmental assessments. In Table 19.8 it can be seen that relatively few new biomass-fueled systems are planned by utilities; 15 of the 19 utility systems are already operating.

The 19 utility plants include two that are wood-gasifier demonstration projects. It should be noted that of all the biomass small power production and cogeneration systems, there were only four that used thermal gasification, representing a total capacity of 3185 KW. Except for these four, all of the wood, agricultural residue, and MSW systems used direct combustion as the means for converting the biomass to energy.

Table 19.9 shows the geographic distribution of nonutility facilities. (The geographic regions used in Table 19.8 are defined in Figure 19.1.) The regional availability of biomass resources is reflected in the distribution of biomass-fueled facilities in Table 19.8. Wood resources and wood-fired generating systems are most abundant in the Northeast, Southeast, and Northwest. MSW and landfill resources are most abundant on the East Coast and West Coast—in dense population centers where land availability is becoming a problem with respect to waste disposal.

Figure 19.4 illustrates the frequency distribution of the electric generating capacity of the four largest categories of biomass-fueled facilities. Wood waste and agricultural residues are used most frequently in the smallest size range—less than 10,000 KW. For wood waste facilities, the medium size is 7500 KW and the mean size is 17,000 KW. For agricultural residues the median size is 9000 KW and the mean is 16,900 KW. Although the most frequent size range for facilities fueled by MSW is 10,000 KW or less, a large portion are of an intermediate size—20,000 to 50,000 KW. For MSW, the median facility size is 20,000 KW and the mean is 25,200 KW. The facilities fueled by landfill gas are nearly an order of magnitude smaller than the other system types. For landfill gas, the median facility size is 1500 KW and the mean is 2180 KW.

Appendices A and B* provide a detailed listing of each of the 196 existing and planned biomass facilities summarized in the report. Information such as facility type, location, status, capacity, and fuel description is included for each project. Points of contact are also provided for those seeking more information about particular installations. Appendix A includes projects fueled by wood waste, agricultural residues, or animal manure. Appendix B lists projects fueled by municipal solid waste, landfill gas, or sewage methane. In both appendices, the projects are grouped by geographic region. Within each region, projects are listed in subgroups based on fuel types.

---

*Appendices A and B are not included in this excerpted overall salient data but are included in the complete report prepared for the U.S. Department of Energy under contract No. D.E.-AC01-83CE-30784.

Table 19.9 Biomass-Fueled Small Power Production, Cogeneration, and Municipal Facilities: By Fuel Type Versus Geographic Region

Fuel	New England	Mid Atlantic	South Atlantic	East North Central	East South Central	West North Central	West South Central	Mountain	Pacific	Total
Wood waste	258,815[a] (14)[b]	33,675 (2)	242,700 (8)	41,900 (3)	211,590 (15)		171,905 (7)	62,500 (6)	181,267 (14)	1,204,352 (69)
Agricultural residues		20,000 (1)				9000 (1)	15,700 (4)		175,050 (7)	219,750 (13)
Animal manure (to methane)					25 (1)		910 (2)			935 (3)
Subtotal	258,815 (14)	33,675 (2)	262,700 (9)	41,900 (3)	211,615 (16)	9000 (1)	188,515 (13)	62,500 (6)	356,317 (21)	1,425,037 (85)
Municipal solid waste	201,100 (7)	275,000 (11)	343,000 (11)	153,100 (5)	550 (1)	82,460 (3)	17,700 (3)	3,500 (1)	134,150 (6)	1,210,560 (48)
Landfill methane	1,075 (2)	23,220 (13)		1,500 (1)					63,450 (25)	89,245 (41)
Sewage methane									3,114 (4)	3,114 (4)
Subtotal	202,175 (9)	298,220 (24)	343,000 (11)	154,600 (6)	550 (1)	82,460 (3)	17,700 (3)	3,500 (1)	200,714 (35)	1,302,919 (93)
Total	460,990 (23)	331,895 (26)	605,700 (20)	196,500 (9)	212,165 (17)	91,460 (4)	206,215 (16)	66,000 (7)	557,031 (56)	2,727,956 (178)

[a] Kilowatts
[b] Number of facilities.

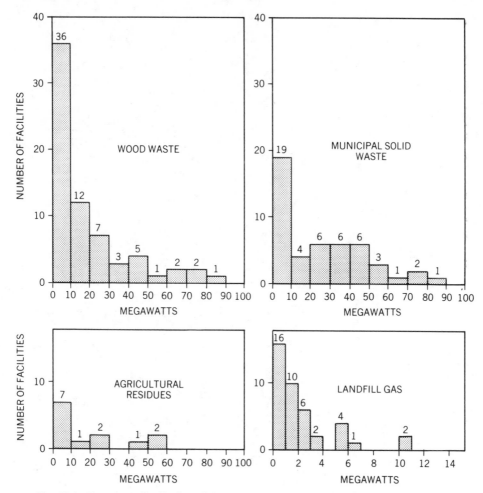

Fig. 19.4 Frequency distribution of the generating capacity of biomass-fueled facilities.

This report reflects the current status of facilities using biofuels to produce electric energy as of February, 1984. Due to the high level of activity in this area, 60 new biomass small power production and cogeneration projects have filed with FERC since that time. Updated versions of the report will be issued periodically to keep readers informed of the latest developments in the field.

PART 3
HAZARDOUS SOLID WASTE

CHAPTER 20

FEDERAL REGULATORY ISSUES

WILLIAM L. KOVACS

Jaeckle, Fleischmann & Mugel
Washington, D.C.

20.1 INTRODUCTION

The focus of this chapter is on the federal government as a regulator of hazardous waste from its point of generation, through its transportation, treatment, storage or disposal. In order to place the role of federal regulation in perspective, this chapter reviews hazardous waste management prior to the Resource Conservation and Recovery Act (RCRA), the reasons for federal involvement in hazardous waste management, the legislative history of RCRA, the structure and substance of RCRA as it regulates hazardous waste, the present status of hazardous waste regulations, the organization of the Environmental Protection Agency (EPA) for the management and enforcement of its hazardous waste regulations, the Comprehensive Environmental Response, Compensation, and Liability Act of 1980 (hereinafter referred to as "the Superfund Act"), the Superfund Act as it relates to RCRA, and current and future changes in the Federal regulation of hazardous waste.

20.2 HISTORY OF FEDERAL HAZARDOUS WASTE REGULATORY PROGRAM

The regulation of hazardous waste by the federal government literally begins with the enactment of the RCRA on October 21, 1976. Prior to RCRA, there was no federal regulation of hazardous waste and little information concerning the impact of improper disposal of hazardous waste on health and the environment. In fact, until the enactment of RCRA, the entire federal effort in the management of waste focused on research and development of resource recovery facilities and the awarding of state planning grants for the management of municipal waste. In 1975, however, the U.S. House of Representatives Subcommittee on Transportation and Commerce, the Congressional Subcommittee with substantive jurisdiction over the laws relating to the regulation of solid and hazardous waste, began hearings on the problems associated with solid waste management. During the course of these hearings, the Subcommittee began to receive unofficial damage assessment reports from scientists at EPA concerning the adverse health effects caused by the improper disposal of hazardous waste. Such reports caused the Subcommittee to inquire into the scope of the potential problems associated with the existing hazardous waste management system.

20.2.1 Past Practices

Although the hearings of the Subcommittee on Transportation and Commerce focused on the management of municipal solid waste and resource recovery, the Subcommittee used the informa-

The author was Chief Counsel of U.S. House of Representatives Subcommittee on Transportation and Commerce during the enactment of the Resource Conservation and Recovery Act. This Subcommittee has primary jurisdiction over solid and hazardous waste. The author presently practices law in Washington, D.C. with Jaeckle, Fleischmann & Mugel and is Chairman of the Virginia Hazardous Waste Facilities Siting Board and past Chairman of the American Bar Association Energy Resources Law Committee, Section Torts and Insurance Practice.

tion obtained from the EPA damage assessment reports to develop the legislative structure for its hazardous waste legislation. The hazardous waste provisions were then incorporated into RCRA as a separate subtitle of a comprehensive measure regulating solid and hazardous waste management, federal procurement of products made from recovered materials, and the development of standards for the manufacture of products containing recovered materials. Although the Subcommittee never held formal hearings on the hazardous waste portion of RCRA, its committee report detailed all of the damage assessments that formed the basis for moving forward with the hazardous waste portion of the legislation, and clearly indicated that the hazardous waste portion of the legislation was its overriding concern.

The companion Senate legislation proposing to regulate hazardous waste simply provided broad regulatory authority to EPA to regulate hazardous waste.

In particular, the Subcommittee found that other environmental laws required that hazardous wastes be taken out of the air and water, resulting in the pollutants being disposed of on land, and that such unplanned disposal of hazardous wastes could cause harm to humans and the environment through groundwater contamination via leaching; surface water contamination via runoff; air pollution via open burning; evaporation, sublimation, and wind erosion; poisoning via direct contact; poisoning via the food chain; and fire and explosion.

By 1975, the EPA estimated that approximately 201 million tons (wet) of industrial waste were generated per year in the United States and of that total, 37 million tons (wet) were hazardous. The growth rate of the generation of hazardous waste was estimated to be between 5 and 10% annually. Such hazardous waste falls into one or more categories of inorganic toxic metals, salts, acids or bases, synthetic organics, flammables, and explosives. Approximately 70% of the industrial hazardous wastes are generated in the mid-Atlantic, Great Lakes, and Gulf Coast areas of the United States. Most of the generated wastes are disposed of on land because such disposal has, for the most part, been unregulated, inexpensive, and its disposal in other environmental media has been prohibited.

Notwithstanding the large volume of hazardous waste generated, as well as the problems associated with the disposal of such wastes, few states regulated hazardous waste until the enactment of the RCRA. Although EPA could identify 25 states in 1975 with hazardous waste programs, there were only a total of 50 people employed in such programs, with one-third of such employees in the State of California. Further, the statutes of the states regulating hazardous waste illustrated almost total inconsistency. Each state had its own definition of "hazardous waste" and "hazardous waste management," thereby making it unnecessarily complicated to do business in more than one state.

20.2.2   Intent and Development of RCRA, Congressional Debate

Although the Subcommittee on Transportation and Commerce learned of the problems associated with hazardous waste after concluding its hearings on the municipal waste problem, the Subcommittee instructed its staff to develop legislation to regulate hazardous waste. Subsequently, the legislation regulating hazardous waste was included in the Subcommittee's draft legislation, notwithstanding the fact that hearings were not held on the subject. Instead, the draft of the hazardous waste provisions of RCRA were circulated by the Subcommittee staff to all interested parties and were negotiated prior to being made part of the draft bill being considered by the Subcommittee. Although individual employees of the EPA assisted Congress in drafting the hazardous waste provisions of RCRA, the EPA did not provide official agency comments on the House legislation. The Office of Management and Budget (OMB) recommended that EPA be permitted only to publish hazardous waste guidelines, with the states having the responsibility to implement such guidelines. Under OMB's recommendations, EPA could enforce its guidelines only against individuals and could not require the states to develop and enact hazardous waste management plans.

The Subcommittee bill, H.R. 14496, 94th Cong. 2d Sess. 1976 (hereinafter referred to as H.R. 14496) adopted the hazardous waste provisions of the draft legislation without amendment on August 30, 1976, and such provisions were passed by the full Committee on Interstate and Foreign Commerce by voice vote on September 9, 1976. However, because the House and Senate had entirely different versions of the legislation and there was only 1 week left in the 94th Congress, a conference between the House and Senate was impossible. Therefore, Congressman Fred Rooney (D–Pa.), Chairman of the House subcommittee that reported the bill, and Senator Jennings Randolph (D–W. Va.), Chairman of the Senate committee with jurisdiction over solid and hazardous waste management, requested that their staffs meet to determine whether a compromise bill could be agreed upon before the Congress adjourned. The staffs of the two committees were able to agree upon a compromise version of the House bill. In particular, the primary provisions of the compromise bill, that is, those relating to solid and hazardous waste management and the procurement of recovered materials, were almost identical to the provisions contained in H.R. 14496, the original Rooney bill.

With specific regard to the hazardous waste portions of the legislation, the Senate committee staff agreed to accept the entire House structure for regulating hazardous waste, provided that the word "reasonably" which preceded the phrase "protect human health and environment" was deleted. The House negotiators agreed to this request. The result of such change was the establishment of an absolute legislative standard to "protect human health and the environment" and the elimination of the judicial determination of the "reasonableness" of promulgated standards. Moreover, other factors such as technical or economic criteria were outside the legislative standard.

When the House of Representatives brought H.R. 14496 to the floor, Congressman Rooney moved to insert the agreed upon substitute in lieu of the provisions contained in H.R. 14496. By inserting and passing the agreed upon substitute in lieu of H.R. 14496, the House was able to send to the Senate legislation on which it could agree without a conference committee. The substitute was approved by both the House and Senate with only hours remaining in the 94th Congress and transmitted to President Gerald Ford for his signature. On October 21, 1976, President Ford signed RCRA into law.

Because RCRA was enacted without ever going to conference, its legislative history is incomplete and somewhat difficult to understand. If one is looking for the intent of Congress concerning a particular provision of RCRA, therefore, one must first determine if the provision in question originated in the House or the Senate and then look to the respective House and Senate bills and committee reports in order to determine its meaning. For some of the compromise provisions of RCRA, however, there is no legislative history.

20.3   THE ACT—ITS ORGANIZATION, SCOPE, AND CONTENTS

With the enactment of RCRA, Congress required the Administrator of EPA (hereinafter referred to as "the Administrator") to develop criteria for identifying hazardous wastes and to list the identified wastes within 18 months. The Administrator was also required to promulgate regulations for those who generate, transport, treat, store, or dispose of hazardous waste within the same 18-month period. The substance of such requirements established cradle-to-grave regulation of all hazardous waste.

20.3.1   Identification and Listing of Hazardous Wastes

Sections 1004(5) and 3001 of RCRA establish the legislative standard for the Administrator to develop the criteria for determining whether a waste is hazardous. The standard is whether a solid waste or combination of solid wastes which because of its quantity, concentration, or physical, chemical, or infectious characteristics and taking into account toxicity, persistence, and degradability in nature, potential for accumulation in tissue, and other hazardous characteristics, such as flammability and corrosiveness, may "(a) cause, or significantly contribute to an increase in mortality or an increase in serious irreversible or incapacitating reversible illness; or (b) pose a substantial present or potential hazard to human health or the environment when improperly treated, stored, transported, disposed of, or otherwise managed."

Based upon the above standard, the Administrator is required to develop criteria against which wastes are to be analyzed and a determination made as to whether to list a waste as "hazardous." Accordingly, the Administrator is also required to revise the list of hazardous wastes from time to time as may be appropriate.

20.3.2   Requirements Imposed on Generators of Hazardous Waste

Pursuant to Section 3002 of RCRA, the Administrator is required to promulgate regulations applicable to persons who generate identified or listed hazardous wastes. The regulations are to establish standards respecting: (1) recordkeeping practices that accurately identify those wastes or constituents thereof, which are identified as hazardous and the disposition of such wastes; (2) procedures to establish labeling practices for containers of hazardous wastes so that those who transport, treat, store, or dispose of such wastes are informed by the generator of the identity of the waste; (3) the use of appropriate container for hazardous waste by the generator; and (4) the initiation of a manifest, by the generator, that remains with the waste to its final place of storage, treatment, or disposal so as to insure that such waste is shipped to and arrives at a designated facility for which a permit has been issued or one operating under interim status; and the submission of reports setting out the quantities of hazardous waste generated and disposed of during a particular time period.

20.3.3   Requirements Imposed on Transporters of Hazardous Waste

Pursuant to Section 3003 of RCRA, the Administrator is required to promulgate regulations applicable to those who transport hazardous wastes. In promulgating such regulations, the Administrator is required to consult with the Secretary of Transportation, Department of Transportation

(DOT), and the governors of the states. The regulations are to include recordkeeping procedures for the transporter of such wastes, indicating the waste, its source and delivery points, and the establishment of a manifest which originates with the generator of the waste and accompanies the waste throughout transport to its final destination. This stature requires the generator to ship its waste to a facility holding a permit or operating under interim status and requires the transporter to transport the waste to such designated facility.

All regulations promulgated by the Administrator must be consistent with the regulations of DOT and published pursuant to the Hazardous Materials Transportation Act, 49 U.S.C. §1801 and following.

The regulations promulgated by Section 3003 of RCRA concerning the transportation of hazardous wastes are necessary for the development of a complete system for tracing the waste from point of generation to final disposal. Although violations of transportation safety regulations issued under the Hazardous Materials Transportation Act remain within the jurisdiction of DOT, the Administrator is empowered to promulgate and enforce regulations concerning the transportation of hazardous waste as are necessary to protect human health and environment. Further, the Administrator can recommend to the Secretary of DOT the addition of materials to be regulated under the Hazardous Materials Transportation Act.

20.3.4 Requirements Regulating Those Who Treat, Store, or Dispose of Hazardous Wastes

Pursuant to Section 3004 of RCRA, the Administrator is granted broad authority to issue regulations governing those who treat, store, or dispose of hazardous wastes. RCRA requires the Administrator to promulgate regulations establishing performance standards for the operation of treatment, storage, and disposal facilities. Such standards shall include, but need not be limited to requirements respecting: (1) maintaining records of all hazardous wastes treated, stored, or disposed of and the manner in which such wastes were treated, stored, or disposed of; (2) reporting, monitoring, and inspection and compliance with the manifest system; (3) operating methods, techniques, and practices; (4) location, design, and construction of the facility; (5) contingency plans to minimize unanticipated damage; and (6) qualifications as to ownership, continuity of operation, training for personnel, and financial responsibility.

The authority to issue regulations for the above specific activities combined with the legislative standard that the Administrator's authority "need not be limited" to these specific activities gives the Administrator considerable control over the actual treatment, storage, or disposal of hazardous wastes.

The Act is very clear that both owners and operators of a hazardous waste management facility are subject to its jurisdiction. In addition to subjecting owners and operators of hazardous waste management facilities to certain performance criteria, RCRA also requires that such persons provide assurances of financial responsibility and continuity of operation consistent with the degree and duration of risks associated with the treatment, storage or disposal of the waste treated. The Administrator can also deny authorization to operate a hazardous waste treatment, storage, or disposal facility unless he is satisfied that the owner or operator maintains sufficient insurance or is otherwise able to satisfy any liability resulting from the management of the operation. In addition, the Administrator can require estimates with respect to the composition, quantities, concentrations, time, frequency, or rate at which such hazardous wastes are proposed to be disposed of. All of the preceding requirements must be satisfied before the Administrator (or a state with an equivalent approved program) can issue a permit to operate the hazardous waste treatment, storage or disposal facility.

20.3.5 Permit Authority

Section 3005 of RCRA establishes two types of operating status—interim and permitted.

Interim status is available only to an owner or operator of a hazardous waste treatment, storage, or disposal facility which was in existence on November 19, 1980, who has complied with all the notification requirements of Section 3010 of RCRA, and has made an application for a permit under Section 3005 of RCRA. If the owner or operator meets the above requirements, the facility shall be treated as having been issued a permit (interim status) until such time as final administrative review of their permit application is made.

With regard to the requirements for obtaining a permit, RCRA requires the owner or operator of a hazardous waste treatment, storage, or disposal facility to apply for a permit by providing such information as may be required under regulations promulgated by the Administrator, including information respecting composition, quantities, and concentrations of hazardous waste proposed to be disposed of, treated, transported, or stored, and the time, frequency, and rate at which such waste is proposed to be disposed of, treated, transported, or stored and the location of the site.

Additionally, in order to receive a permit for a new landfill, new units of a landfill, or surface impoundment, as well as the expansion and replacement of an interim status landfill, impoundment, or pile, such facilities are required to install two or more liners and a leachate collection system (for landfills) above and between the liners.

Upon a determination by the Administrator of compliance by a facility for which the permit is applied under Sections 3004-3005 of RCRA, the Administrator shall issue the permit. A permit is limited to 10 years and the Administrator must review all land disposal permits no later than 5 years after issuance.

Section 3005 of RCRA also empowers the Administrator to modify or revoke said permit.

20.3.6 Authorized State Programs

Pursuant to Section 3006 of RCRA, a state that seeks to administer and enforce a hazardous waste program in lieu of federal implementation and enforcement may develop and submit to the Administrator an application for such authorization. Within 90 days after such application is submitted, the state is authorized to carry out such program in lieu of the federal program and to issue and enforce permits for the treatment, storage, and disposal of hazardous waste unless within such period the Administrator notifies the state that such program may not be authorized. Subsequent to such notice to the state, the Administrator must provide the state with an opportunity for public hearing. The state cannot administer its program in lieu of the federal program if the Administrator finds such program is not equivalent to the federal program, or that it is not consistent with the federal program or other state programs, or that such program does not provide adequate enforcement. Once the state's application is approved, the state implements and enforces its hazardous waste program in lieu of the federal program. However, the Hazardous and Solid Waste Amendments of 1984 impose many new requirements upon the states and the regulated community (ban on the disposal of certain waste in landfills, new technology requirements, and many others) which the states will be unable to immediately implement. Because of such new requirements, those states with fully authorized programs will still be able to implement their authorized program, but initially the state will be unable to implement the new RCRA requirements. Therefore, the requirements imposed by the 1984 amendments will be administered by EPA until the state's program is substantially equivalent to the new RCRA requirements and it receives EPA authorization to implement such requirements. Such a process means both the state and EPA will be jointly issuing permits until the respective states obtain final authorization for all provisions enacted by the 1984 amendments.

Interim authorization for the administration of state programs or portions of state programs in lieu of federal programs is also available under Section 3006(c). Under Section 3006(c) of RCRA, any state that has in existence a hazardous waste program pursuant to state law within 90 days after the date of promulgation of regulations under Sections 3002, 3003, 3004, and 3005 may submit to the Administrator evidence of such existing program and may request temporary authorization to administer such program.

If the state program is substantially the same as the federal program, the Administrator shall authorize the state to administer said program in lieu of the federal program for a period ending no later than January 31, 1986.

The state can apply for interim authorization to administer in lieu of the federal program part or all of the hazardous waste requirements promulgated by the Administrator under Sections 3002, 3003, 3004 and 3005 for the 24-month period.

20.3.7 Enforcement of RCRA

Enforcement of the hazardous waste regulations rests with the states, if they obtain EPA approval. EPA's enforcement authority in a state only occurs if the state does not seek such approval, fails to administer its program adequately, or does not have approval to administer all aspects of the federal program. In states with approved hazardous waste programs, those state requirements may be more stringent than the federal regulations.

As addressed in Section 3008 of RCRA, the primary federal enforcement tools are the compliance order, civil suit, and criminal prosecution. Civil penalties may be as high as $25,000/day for each day of noncompliance with an EPA order. Under the Act, any person who knowingly transports any hazardous waste to a facility that does not have a permit or disposes of any hazardous waste without having obtained a permit or is in violation of any material condition or requirement of a permit shall be subject, upon conviction, to a fine of $50,000 for each day of violation, imprisonment for not more than 5 years, or both. Further, if an individual or other entity makes any false statement or representation in any application, label, manifest, record, report, permit, or other document; violates interim status standards; fails to file required reports; or transports hazardous waste without a manifest, such individual shall, upon conviction, be subject to a fine of not more than $50,000 for each day of violation or imprisonment not to exceed 2 years, or both. For repeat offenders, the penalties double to $100,000 and 4 years imprisonment.

In situations where the violator is convicted of having knowingly violated RCRA provisions in circumstances that manifest an unjustified and inexcusable disregard for human life or an extreme indifference to human life, the punishment may be as high as a $250,000 fine and not more than 5 years in jail, or both. A defendant that is an organization can be fined up to $1,000,000.

In addition to federal or state enforcement of RCRA, Section 7002 of RCRA provides that any citizen may bring a lawsuit against any person or government agency for failure to comply with any effective regulation, standard, permit, condition, or requirement. A citizen may also bring suit against the Administrator of EPA for failure to perform any act or duty under the Act that is not discretionary. Further, the Hazardous and Solid Waste Amendments of 1984 also authorize citizen lawsuits when the past or present management of hazardous waste presents an imminent hazard to human health or the environment and it permits lawsuits against persons engaged in the open dumping of waste. Prior to initiating such lawsuit, the citizen must give 60 day's notice to EPA, the state in which the violation occurs, and the alleged violator. Citizen suits cannot be initiated when EPA or the state are diligently prosecuting the matter or have entered into a consent decree with the responsible parties. Finally, violations of standards concerning underground storage tanks (see Section 20.3.8) are subject only to civil penalties in the amount of $25,000 for noncompliance with an administrative order; $10,000 for a knowing violation of the notice requirements or noncompliance with any other standard.

20.3.8   The Hazardous and Solid Waste Amendments of 1984

In addition to the standards imposed by the initial enactment of RCRA in 1976, the Congress enacted the Hazardous and Solid Waste Amendments of 1984, Publ. L. 98–616 (November 8, 1984) which substantially strengthened EPA's ability to protect health and the environment.

Such amendments were enacted because of congressional concern that EPA's regulations, especially those standards regulating land disposal and groundwater contamination, were not sufficient to protect human health and the environment. Therefore, in the 1984 amendments Congress banned certain types of wastes from land disposal and required EPA to issue regulations concerning the safe disposal of other types of wastes, including specific levels or methods of treatment to substantially diminish the toxicity of the waste. If EPA fails to promulgate such regulations, such wastes are prohibited from being disposed of on land. This procedure has been characterized as a "hammer."

Although the Hazardous and Solid Waste Amendments of 1984 do not alter the regulatory structure of RCRA, they do substantially increase the scope of RCRA regulation and remove a substantial amount of regulatory discretion from EPA. Therefore, certain key provisions are addressed below so that the full scope of RCRA's impact is presented.

1. *Small Quantity Generators.* Section 3001 of RCRA is amended to require EPA to issue by March 31, 1986 standards for handling wastes from generators of between 100–1000 kg/month. Prior to this amendment, generators of less than 1000 kg/month were exempt from RCRA regulation. EPA, however, has the discretion to impose upon such small quantity generators standards that are less stringent than the standards imposed on those who generate over 1000 kgs of hazardous waste per month, provided that such standards protect human health and the environment. Prior to the effective date of the regulations implementing this provision, small quantity generators may only dispose of their waste at a permitted hazardous waste facility or a state approved municipal or industrial facility. If EPA fails to promulgate its small quantity generator regulations, all waste generated in excess of 100 kgs/month but less than 100 kgs/month shall be subject to compliance with the Uniform Hazardous Waste Manifest form; that disposal shall occur only at a permitted facility or a facility with interim status.

2. *Prohibitions on Land Disposal of Certain Wastes.* Section 3004 of RCRA is amended to prohibit within 6 months after the enactment of the Hazardous and Solid Waste Amendments of 1984 the placement of bulk of noncontainerized liquid hazardous waste in any landfill. In addition, Section 3004 of RCRA was also amended to prohibit within 12 months after enactment of the Hazardous and Solid Waste Amendments of 1984 the disposal of nonhazardous liquid waste in any hazardous waste landfill unless such landfill demonstrates it is the only reasonably available alternative for such liquids and the placement of such liquids in the landfill does not present a risk of contamination to any underground source of drinking water.

Further, within 15 months after the enactment of the Hazardous and Solid Waste Amendments of 1984 the Administrator is required to promulgate regulations: (1) which minimize the disposal of containerized liquid hazardous waste in landfills; and (2) minimize the presence of free liquids in containerized hazardous waste to be disposed of in landfills.

The new amendments to Section 3004 also ban the land disposal of certain enumerated wastes unless EPA, within specific time frames, issues regulations finding that prohibiting certain methods of land disposal is not required to protect human health and the environment.

EPA is also to determine whether to ban the land disposal of wastes according to a schedule which requires the review of one-third of all listed wastes within 45 months after the enactment of the 1984 RCRA amendments; two-thirds of all listed wastes within 55 months after the enactment of the 1984 amendments, and all listed and characteristic wastes within 66 months after the enactment of the 1984 amendments. If EPA does not issue its regulations within the required time frames, then the land disposal of such enumerated wastes is prohibited unless the generator certifies that there is no alternative capacity available, and disposal takes place at a facility in compliance with the minimum technological requirements concerning liners and monitoring.

Section 3004 was also amended to provide that each new landfill or surface impoundment or a new landfill or surface impoundment unit at an existing facility and each replacement, lateral expansion of an existing landfill, or surface impoundment is required to have at least two liners, a leachate collection system, and a groundwater monitoring system.

3. *Additional Land Disposal Restrictions.* The placement of bulk liquids in salt domes, salt beds, underground mines, or caves is prohibited.

Waste oil or other material contaminated with hazardous waste cannot be used as a dust suppressant or for road treatment.

4. *Hazardous Waste as a Fuel.* Section 3004 or RCRA was also amended to require the Administrator to regulate the owners and operators of facilities which produce a fuel from hazardous waste; facilities which burn hazardous waste as a fuel, and persons who market any fuel produced from hazardous waste. Further, the amendments require a warning label on such fuel indicating it contains hazardous waste.

5. *Length of Permits.* Section 3005 of RCRA was amended by the Hazardous and Solid Waste Amendments of 1984 to require the Administrator to process all land disposal permit applications within four years; all permit applications for incinerators within five years; and all other permit applications within eight years, Further, the length of a permit is limited to ten years and the Administrator must review all land disposal permits no later than five years after issuance.

With regard to land disposal facilities operating under interim status, their interim status terminates within 12 months after the enactment of the Hazardous and Solid Waste Amendments of 1984 unless the facility applies for a final permit within such time period and certifies that such facility is in compliance with all applicable groundwater monitoring and financial responsibility requirements.

Section 3005 of RCRA is also amended to prohibit within four years existing surface impoundments from receiving, storing, or treating any listed hazardous waste unless such impoundment meets the same technical standards (double lines, groundwater monitoring, and leak detection) as applied to new surface impoundments.

6. *Minimum Technology Standards.* A landfill unit or impoundment for which a final permit application has not been received by the date of enactment must have a double liner with a leachate collection system above and between the liners and a groundwater monitoring system. EPA must promulgate regulations concerning such technical requirements within two years of enactment of the 1984 RCRA amendments. Until such regulations are promulgated a synthetic/clay liner system is acceptable.

Interim status landfills, impoundments, and piles which are being expanded or replaced and which presently receive hazardous waste are subject to the same double liners and leachate collection systems as required of new landfills and impoundments.

Such interim states' facilities must notify EPA 60 days before receiving hazardous waste and file for a final permit within 60 days after notifying EPA that it intends to receive hazardous waste.

7. *Underground Storage Tanks.* The Administrator, by March 1985, is required to issue regulations concerning underground storage tanks. An underground storage tank is a tank with 10% or more of its volume underground. Further, the 1984 amendments require the establishment of a program to control underground storage tanks containing regulated wastes. The program requires the owners of each tank to provide notice of its existence to the identified state agency and to inform the agency of its use or of the fact that the tank was taken out of operation within the past ten years. The program also requires suppliers of regulated substances and tank sellers to inform tank owners and operators of their responsibility to notify the identified state agency within 30 days of bringing the tank into use. The Administrator must also develop tank standards that protect human health and the environment.

However, until the Administrator establishes the aforementioned program, a tank can be put into service only if the tank will prevent releases due to corrosion or structural failure for its operational life; is cathodically protected; constructed of noncorrosive material or designed in a manner to prevent the release of the stored substance; constructed of material compatible with the substance to be stored; or located in soil having a certain resistivity.

In addition, the 1984 RCRA amendments require that the regulations provide standards for leak detection or tank testing, recordkeeping, reporting, corrective action, closure, and financial responsibility.

As with all other parts of the hazardous waste program, states with standards no less stringent than the federal standards can apply to EPA for authority to implement the program in lieu of EPA.

8. *Miscellaneous Requirements.* Additional requirements imposed by the Hazardous Waste and Solid Waste Amendments of 1984 include the prohibition on the export of hazardous waste unless the exporter files a notice of what is to be exported and the receiving nation consents to accept such waste; the requirement that after September 1, 1985 all manifests must contain a generator certification that the volume and toxicity of the waste have been reduced to the maximum degree economically practicable and the management of the waste minimizes risk to the extent practicable; a ban on the underground injection of hazardous waste into or above any formation within a quarter mile of a well that is a source of drinking water; an authorization for the Administrator to issue permits for experimental facilities for one year without first issuing standards for permits for such experimental facility; and a requirement that the Administrator consider factors in addition to those for which the waste was listed as a hazardous waste when considering a delisting petition and to provide notice and comment before making the final decision.

20.4   HAZARDOUS WASTE MANAGEMENT REGULATIONS UNDER RCRA

The regulations imposing requirements upon those who generate and transport hazardous waste, as well as those who own or operate a hazardous waste storage, treatment, or disposal facility are found at 40 C.F.R. Parts 260–267. Permit requirements and standards concerning state plan approval are found at 40 C.F.R. Parts 270–271. Each part regulates a different facet of hazardous waste management. The regulations are initially published in the *Federal Register* and then compiled in Title 40 of the Code of Federal Regulations (C.F.R.) which is obtained from the Government Printing Office, located in Washington, D.C., or its regional offices.

20.4.1   40 C.F.R. Part 260, General Regulations for Hazardous Waste Management

Part 260 provides the definitions of the terms used throughout the Parts 260–267, 270–271 regulations, as well as:

1. The general standards EPA will use in making information public.
2. Information for those subject to the regulations who want to assert claims of business confidentiality with respect to information submitted to EPA pursuant to Parts 260–267, 270–271.
3. Procedures for petitioning EPA to amend, modify, or revoke any provisions of Parts 260–267, 270–271 and to establish procedures governing EPA's actions on such petitions.
4. Procedures for petitioning EPA to approve testing methods equivalent to those prescribed in Parts 261, 264, or 265 in terms of the sensitivity, accuracy, and precision of the test.
5. Procedures for petitioning EPA to exclude a waste at a particular generating facility from being listed as a hazardous waste.
6. Procedures to obtain variances from classification as a solid waste.

The most important aspects of the Part 260 regulations are those concerning confidentiality of business records and the delisting of a hazardous waste at a particular facility.

If a person seeks to establish the claim of business confidentiality covering all or part of the information submitted to EPA, all of the procedures set forth in 40 C.F.R. §2.203(b) must be followed. However, if a claim of confidentiality is not made, then the information received by EPA may be made available to the public without further notice to the person submitting such information.

Another important provision of Part 260 is the exclusion of wastes from a particular facility from being listed as hazardous wastes. In order to delist a waste successfully, the petitioner must demonstrate to the satisfaction of the Administrator that the waste produced by the particular generating facility does not meet any of the criteria under which waste is listed as a hazardous waste under Part 261.

20.4.2   40 C.F.R. Part 261, Regulations Identifying Hazardous Waste

Part 261 deals exclusively with the identification and listing of hazardous wastes and the criteria for identifying the characteristics of hazardous waste, as well as the exclusions from hazardous waste, including specific exclusions under Section 261.4, the small generators exclusion under Section 261.5, and the special requirements for reused, recycled, or reclaimed wastes contained in section 261.6.

Hazardous waste is only a portion of the solid waste stream and is regulated separately from the nonhazardous waste portion of RCRA (compare RCRA Subtitles C and D).

Section 261.3 of part 261 defines a waste to be hazardous if:

1. It exhibits any one of the following characteristics: ignitability, corrosivity, reactivity, toxicity.
2. It is listed in Part 261 C.F.R. Subpart D, the listing of specific hazardous wastes.
3. It is a mixture of solid and hazardous waste that is listed in Part 261, Subpart D, solely because it exhibits one or more of the following characteristics: ignitability, corrosivity, reactivity, or toxicity.
4. It is a mixture of solid and listed hazardous waste subject to Sections 307(b) and 402 of the Clean Water Act and certain listed wastes exceed certain parts per million standards.
5. It contains certain spent solvents.
6. It is one of the wastes listed in Section 261.32—heat exchanger bundle cleaning sludge from the petroleum refining industry.
7. It is discarded commercial chemical product listed in Section 261.32.
8. It is a solid waste that is not a hazardous waste, and not excluded from regulation, but becomes a hazardous waste when it first meets the criteria of those listed hazardous wastes or it is mixed with wastes listed as hazardous wastes.
9. It is not excluded from regulation as a hazardous waste under Section 61.4(b).

Sections 261.10 and 261.11 establish the criteria for identifying hazardous waste and for listing hazardous wastes, respectively. Specifically, the Administrator shall identify and define a particular characteristic (ignitable, corrosive, reactive, toxic) as "hazardous" only upon determining that each characteristic may:

1. Cause or significantly contribute to an increase in mortality or an increase in serious irreversible or incapacitating reversible illness.
2. Or pose a substantial present or potential hazard to human health or the environment when it is improperly treated, stored, transported, or disposed of.

Further, a waste may be listed as a "hazardous waste" when such waste possesses one of the characteristics of being ignitable, corrosive, reactive, or toxic; is found to be fatal to humans at low doses; or if it contains those toxic constituents listed in Appendix VII of Part 261.

Sections 260.20–261.24 describe the characteristics of ignitability, corrosivity, reactivity, and toxicity. Moreover, Part 261 in general contains seven appendices that discuss representative sampling methods, testing procedures, and the bases for determining and listing hazardous waste.

Notwithstanding the present broad characteristics of hazardous waste as well as the extensive listings of such wastes, there are numerous exclusions from the classification of what is a hazardous waste.

40 C.F.R. §261.4 provides specific exclusions and 40 C.F.R. §§260.20 and 260.22 provide a mechanism whereby petitions can be filed to exclude waste produced at a particular facility. Section 261.4 provides certain exclusions from the definition of solid waste for purposes of Part 261 regulation as well as exclusions from hazardous waste regulation. Those materials that are not deemed to be solid waste are:

Domestic Sewage.

Any mixture of domestic sewage and other wastes that pass through a sewer system to a publicly owned treatment works.

Point source industrial wastewater discharges subject to regulation under Section 402 of the Clean Water Act.

Irrigation return flows.

Nuclear waste.

Materials not removed from the ground as part of mining operations.

Spent sulfuric acid used to produce virgin sulfuric acid unless it is accumulated speculatively.

Those solid wastes which are not deemed to be hazardous waste for purposes of Part 261 are:

Household waste

Agricultural and animal waste returned to the soil as fertilizer.

Mining overburden returned to mine site.

Fly ash, bottom ash, slag waste, and flue gas emission control waste generated primarily from the combustion of coal or other fossil fuels.

Drilling fluids.

Solid waste from the extraction, benefication, and processing of ores and minerals.

Cement kiln dust waste.

Numerous other exclusions specifically related to certain manufacturing processes.

In addition, Section 261.5 provides that generators of less than 2205 lb (1000 kg) of hazardous waste per month are not subject to all the regulations, under Parts 262–265 concerning requirements upon those who transport, own, or operate a hazardous waste treatment, storage or disposal facility. However, if the generator generates over 1000 kg of hazardous waste per month or accumulates over 1000 kg of hazardous waste on-site, such wastes become subject to all EPA hazardous waste regulations. Notwithstanding any small generator exclusion, each small generator must still determine if his waste is hazardous as prescribed by Section 262.11 and either treat or dispose of his hazardous waste in an on-site facility or ensure delivery to a properly permitted off-site facility. Additionally, Section 261.6 provides that hazardous wastes being legitimately recycled or reclaimed or those being accumulated, stored, or treated prior to beneficial use or reuse, recycled, or reclaimed are exempt from the EPA regulations governing those who transport hazardous waste or own or operate a hazardous waste treatment, storage or disposal facility. This exemption, however, has been eliminated by Congress as of March 31, 1986. See Section 20.3.8.

20.4.3 40 C.F.R. Part 262, Requirements upon Generators of Hazardous Waste

40 C.F.R. Part 262 of the EPA regulations imposes standards upon the generators of hazardous waste. In particular, the regulations require that each generator of hazardous waste whose waste is transported, treated, stored, or disposed of off-site must:

1. Determine if his waste is hazardous as required by Part 261.
2. Obtain an EPA identification number as required by Part 261.
3. Comply with all manifest provisions of Part 262 which require the following information on the manifest: the generator's name and identification number, the EPA identification number for each transporter, the listing of the total quantity of waste by units of weight or volume, and the type and number of containers.
4. Obtain on the manifest the handwritten signature of the original transporter and the date of acceptance and give the remaining copies of the manifest to the transporter who is required to obtain the handwritten signature of the owner or operator of the facility designated to treat, store, or dispose of the waste and which is required to return to the generator a handwritten signature of the manifest evidencing receipt of the waste.
5. Comply with all DOT regulations concerning the packaging, labeling, marking, and placarding of hazardous waste.
6. Comply with all record-keeping provisions which require the generator to maintain all manifests for 3 years, and must file biennial reports with the Administrator.
7. Prepare and submit a biennial report to the Regional Administrator. Such report must identify each off-site treatment, storage, and disposal facility to which waste was shipped; the transporters used and the quantity and classification of the waste.

In addition, generators can accumulate hazardous waste on-site for 90 days without a permit or without having interim status, provided that the waste is placed in containers that comply with Subpart I of 40 C.F.R. Part 265 or the waste is placed in tanks which comply with Subpart J of 40 C.F.R. Part 265; the date upon which each period of accumulation is clearly marked; the container or tank is clearly marked "Hazardous Waste," and the generator complies with the emergency procedures and contingency plan requirements of 40 C.F.R. Part 265, Subparts C and D.

Generators who ship waste off-site and who do not receive a copy of the handwritten signature of the owner or operator of the designated facility within 35 days of the date the waste was accepted by the initial transporter must contact the transporter or designated facility to determine the status of the waste. If the handwritten signature of the owner or operator of the designated facility is not received by the generator within 45 days, the generator must sign an exceptions report with the administrator.

The manifest requirements and the DOT regulations concerning the labeling, marking, and packaging of hazardous materials do not apply to those generators who treat, store, or dispose of hazardous waste on-site.

20.4.4   40 C.F.R. Part 263, Requirements upon Transporters of Hazardous Waste

The 40 C.F.R. Part 263 regulations apply to transporters of hazardous waste other than the on-site transportation of hazardous waste. Each transporter of hazardous waste is required to obtain the EPA identification number; must not accept any hazardous waste unless it is accompanied by a manifest signed by the generator; and must sign and date the manifest acknowledging acceptance of the hazardous waste from the generator and return a signed copy to the generator before leaving the generator's property. Further, the manifest must accompany the hazardous waste to the designated disposal site where the transporter must obtain the signature of the owner or operator of the designated disposal facility. If for any reason the transporter cannot deliver the waste to the designated facility, the transporter must contact the generator for further instructions. However, all regulations continue to apply to the transportation of the hazardous waste until it reaches its final destination.

If during the transportation of the hazardous waste the transporter stores the hazardous waste at a storage facility for a period of more than 10 days, then such transporter is subject to the Part 264, 265 requirements regulating the storage, treatment, or disposal of hazardous waste.

Each transporter must keep copies of the signed manifest for a period of 3 years from the date the hazardous waste was accepted for transportation.

Finally, if there is a discharge of hazardous waste during transportation, the transporter must take appropriate immediate action to protect human health and the environment. In particular, the transporter must notify local authorities and clean up any discharge or take such other action as required or approved by appropriate government officials. Moreover, the transporter must report the discharge in writing to the Office of Hazardous Materials, Department of Transportation, Washington, D.C.

20.4.5   40 C.F.R. Part 264, Requirements upon Owners and Operators of Permitted Hazardous Waste Facilities

Part 264 establishes minimum national standards that define the acceptable management of hazardous waste for the owners and operators of permitted hazardous waste treatment, storage, and disposal facilities. Part 265 applies to owners and operators of hazardous waste treatment, storage, and disposal facilities that have not received final administrative disposition of their permit applications and operate under interim status. (see Section 20.4–6). The Part 264 regulations governing the owners and operators of hazardous waste treatment, storage, and disposal facilities provide standards concerning preparedness and prevention; contingency plans and emergency procedures; manifest and reporting requirements; ground-water protection; closure and postclosure of the facility; financial requirements; use of containers and tanks; surface impoundments, waste piles, land treatment, landfills, and incineration.

However, such standards are coordinated with numerous other environmental laws dealing with at-sea incineration, underground injections, and wastewater treatment facilities. The Part 264 standards, however, apply only to such activities to the extent required by the regulations governing at-sea incineration, underground injections, and waste treatment facilities. For example, with at-sea incineration, the Part 264 standards would apply to the treatment and storage of the hazardous waste before it is loaded onto the ocean vessel for incineration and the disposal permit would be obtained under the Marine Protection, Research, and Sanctuaries Act. The only other Part 264 standards (other than those relating to treatment and storage of hazardous waste before loading) would be those Part 264 standards incorporated into the permit by rule under 40 C.F.R. Part 270, relating to federally administered permit programs.

In addition to the limitations imposed upon the applicability of Part 264 by other environmental programs, the Part 264 standards do not apply to those persons who treat, store, or dispose of hazardous waste in a state with RCRA-authorized Phase I and/or Phase II hazardous waste programs to the extent of such state's authority to administer such parts of Phase I and/or Phase II. In such states, the authorized state program which is equivalent to the federal program applies to the extent of its authorization to those who treat, store, or dispose of hazardous waste. However, the Part 264 standards do apply within a state to those hazardous wastes activities for which the state does not have the authority to regulate. Part 264 standards also provide for certain exclusions for:

1. The owner or operator of a facility permitted by a state to manage municipal or industrial waste, if the only hazardous waste the facility treats, stores, or disposes of is excluded from regulation under Part 261.

2. The owner or operator of a facility managing recyclable materials described in Section 261.6(a)(2)(3).

3. Generators accumulating waste on-site for less than 90 days and in accordance with Section 262.34.

4. Farmers disposing of waste pesticides from their own use in compliance with Section 262.51.
5. A totally enclosed treatment facility as defined in Section 260.10.
6. The owner or operator of an elementary neutralization unit or a wastewater treatment unit as defined in Section 260.10.
7. A transporter storing waste at a transfer facility for 10 days or less.
8. Persons carrying out activities to immediately contain or treat a spill of hazardous waste or persons engaged in the treatment or containment activities in response to a discharge of hazardous waste.

General Facility Standards

Each owner or operator of a hazardous waste treatment storage or disposal facility covered by part 264 must;

Obtain an EPA identification number.

Inform the generator in writing that it has an appropriate permit for, and will accept, the waste the generator is shipping.

Obtain a detailed chemical and physical analysis of a representative sample of the waste before treating, storing, or disposing of any hazardous waste.

Repeat the sampling analyses as necessary to ensure that the samples are accurate and up to date.

Develop a security system to prevent unknown and unauthorized entry onto the portion of the facility that treats, stores, or disposes of hazardous waste.

Inspect the facility for malfunctions, deterioration, operator errors, and discharges that may be causing or may lead to release of hazardous waste constituents into the environment or cause a threat to human health.

Prepare a written schedule for inspecting monitoring, safety, and emergency equipment, and conduct such inspections often enough to prevent and correct problems before they occur.

Train facility personnel, either through classroom instruction or on-the-job training, to perform their duties in a way that ensures that the facility is in compliance with the requirements of Part 264.

Section 264.18 also prohibits new hazardous waste, storage, treatment, and disposal facilities from being located within 200 ft (60.9 m) of a fault or within a flood plain unless the facility is designed to prevent washout of any hazardous waste by a 100-year flood.

Preparedness and Prevention

Each owner or operator of a hazardous waste storage, treatment, or disposal facility subject to Part 264 must design, construct, maintain, and operate its facility to minimize the possibility of fire, explosion, or any unplanned sudden or nonsudden release of hazardous waste. As part of such planning process, each facility must:

Be equipped with an internal communication alarm system capable of providing immediate emergency instructions to facility personnel.

Have two-way communication with local police.

Have portable fire extinguishers and adequate water supply and pressure to supply water base streams.

In addition, safety and communication equipment arrangements must be established with the local police, fire, and emergency response teams for the provision of services in case of an emergency.

Contingency Plans and Emergency Procedures

Each owner or operator of a hazardous waste treatment, storage, or disposal facility must develop an emergency contingency plan designed to minimize hazards to human health or the environment from fires, explosions or any unplanned sudden or nonsudden release of hazardous waste. Each plan must;

Describe response actions of facility personnel.

Describe arrangements between the facility and local health and safety departments.

Identify the facility's emergency coordinators.

Describe all emergency equipment.

Be maintained at the facility and be submitted to all health and safety departments that may be called upon.

Each such contingency plan must be amended when the facility permit is revised, the plan fails in an emergency, or the facility changes its design, operation, or maintenance procedures.

Manifest System, Record Keeping, and Reporting

Subpart E of Part 264 applies to owners and operators of both on-site and off-site hazardous waste storage, treatment, and disposal facilities that receive hazardous waste. Facilities that receive wastes from off-site sources however, are required to sign the manifest that accompanies the waste and note any discrepancies between the waste received and the manifest, as well as to give the transporter and generator copies of such signed manifest. If there are any manifest discrepancies with the type and quantity of waste received, the owner or operator must attempt to reconcile such discrepancies with the transporter or generator within 15 days or submit a letter to the EPA Regional Administrator describing the discrepancies.

If a facility receives any waste for treatment, storage, or disposal from an off-site source without a manifest, and such waste is not excluded from the manifest requirement by Section 261.5, then the owner or operator must prepare a report of such incident for the EPA Regional Administrator within 15 days after receiving the waste.

Other paperwork required by Subpart E of Part 264 directs the owner or operator to keep a description of all hazardous waste received, its location, and all results of waste analyses performed on the waste and details of all incidents requiring implementation of an emergency contingency plan. Such records shall be maintained until the closure of the facility.

Further, biennial reports must be made by each owner or operator to the EPA Regional Administrator containing information regarding each generator sending its waste to the facility, and a description and quantity of waste received and the most recent closure estimate.

Groundwater Protection

Subpart F of Part 264 establishes certain groundwater protection standards that apply to all owners and operators of facilities that treat, store, or dispose of hazardous waste in surface impoundments, waste piles, land treatment units, or landfills. The only exemptions from the groundwater regulations are those not regulated by the hazardous waste regulations as specified in 40 C.F.R. §264.1 and those who design and operate surface impoundments in compliance with the requirement concerning "Double Lined Surface Impoundments," 40 C.F.R. §264.222; the "Waste Pile" requirements of 40 C.F.R. §264.250(c); the "Double Lined Piles" regulated by 40 C.F.R. §§262.252–262.253; or the "Double Lined Landfill" as regulated by 40 C.F.R. §264.302. Section 264.90(b)(3)(4) contains other technical exemptions from groundwater standards if it can be demonstrated that there is no potential for migration of liquid from a regulated unit to the uppermost aquifer during the life of the facility and its closure period.

The groundwater regulations apply to all hazardous waste treatment, storage, or disposal facilities regulated by Part 264 during the life of the facility, which includes its closure period. As part of the groundwater protection program, owners and operators must conduct a monitoring and response program which includes:

1. The initiation of a compliance monitoring program whenever hazardous constituents are detected at the facility's compliance point, which is determined by the EPA Regional Administrator pursuant to 40 C.F.R. §264.95.

2. The initiation of corrective action whenever the groundwater protection standards are exceeded, such standards being the standards established in the facility's permit that are designed to ensure that hazardous constituents entering the groundwater do not exceed the concentration limits specified by the EPA Regional Administrator under 40 C.F.R. §264.94.

3. The initiation of groundwater monitoring, detection monitoring, compliance monitoring, and corrective action programs required in the facility's permit by the EPA Regional Administrator.

The remaining portions of Subpart F of Part 264 define the "groundwater protection" standard; "hazardous constituents"; "concentration limits"; and give a detailed description of "groundwater monitoring, detection, and compliance monitoring" programs for groundwater protection.

Closure and Postclosure Requirements

Subpart G of Part 264 applies to all owners or operators of hazardous waste treatment, storage, or disposal facilities other than those excluded under 40 C.F.R. §264.1. In particular, the regulations require that each owner or operator close his facility in a manner that: (1) minimizes the need for further maintenance and controls; (2) minimizes or eliminates threats to human health and the environment; or (3) minimizes or eliminates the postclosure escape of hazardous waste, to include leachate land-contaminated rainfall. Each owner or operator must prepare a written "closure plan" which must be submitted with the permit application. A copy of the approved plan must be kept at the facility until closure is completed. The plan must give all necessary steps for completing or partially closing the facility at any point during its intended operating life. Additionally, the plan must include:

1. An estimate of the maximum inventory of waste in storage and treatment at any time during the life of the facility.
2. Steps needed to decontaminate facility equipment during closure.
3. An estimate of the expected year of closure.

All plans can be amended during the active life of the facility and must be amended if there are design or operating changes to the facility.

Within 90 days of receiving the final volume of waste, the owner or operator must treat or remove from the site, or dispose of on-site all hazardous waste in accordance with the approved closure plan. Longer periods for closure are within the discretion of the Administrator if certain technical reasons justify such longer periods, for example, if the closure activities will take longer than 90 days, or the facility has additional capacity to receive waste, or a person other than the owner or operator will recommence operation of the site.

When closure is completed, all equipment and structures must be decontaminated and an independent, registered professional engineer must certify that the facility is closed in accordance with the closure plan. Postclosure care must continue for 30 years and must consist of maintenance, monitoring, and reporting, as required by the regulations concerning groundwater protection, surface impoundments, waste piles, land treatment, and landfills, all of which are found within Part 264. The period of postclosure care can be reduced or extended if the Administrator finds that the reduced period is sufficient to protect human health and the environment. Each hazardous waste disposal facility must record as part of its deed to the property or in some other instrument filed with the deed a notice to all future purchasers that the site was used to manage hazardous waste. The land to which such instrument attaches is restricted in its use. For example, no person is permitted to remove the cap. A copy of the survey plat showing the location of the waste must also be attached to the instrument or other document giving notice that the site was used to manage hazardous waste.

Financial Requirements on Hazardous Waste Treatment, Storage, and Disposal Facilities

The financial requirements for closure apply to all facilities that treat, store, and dispose of hazardous waste not excluded by 40 C.F.R. §264.1. The financial requirements concerning the cost estimate for postclosure (40 C.F.R. §264.144) and the financial assurances for postclosure (40 C.F.R. §264.145) only apply to all disposal facilities; and piles, and surface impoundments from which the owner or operator intends to remove the wastes at closure. State and federal government are exempt from the financial requirements.

Under the financial requirements, each owner or operator must have a written estimate in current dollars of the cost of closing the facility in accordance with applicable closure requirements. The estimate must be of the point in time when the cost of closure would be most expensive. Further, the closure cost estimate must be adjusted annually for inflation.

In addition, each facility must establish financial assurance for the closure or, if required, the postclosure of the facility through one of the below-listed options for demonstrating such financial assurance:

Closure trust fund or postclosure trust fund.

Surety bond guaranteeing payment into a closure or postclosure trust fund.

Surety bond guaranteeing performance of closure or postclosure care.

Closure or postclosure letter of credit.

Closure or postclosure insurance.

Financial test and corporate guarantee for closure or postclosure.

Use of any combination of the above financial mechanisms.

Use of any combinations of the above financial mechanisms for multiple facilities.

In order to ensure compliance with the financial responsibility requirements, EPA developed specific wording for the financial instruments that demonstrate financial responsibility (see 40 C.F.R. Section 264.151).

In addition to establishing closure and postclosure financial responsibility, each facility must establish financial responsibility for bodily injury and property damage to persons for sudden and nonsudden accidental occurrences arising from the operations of the facility. The owner or operator must have and maintain liability coverage for sudden accidental occurrences in the amount of at least $1 million per occurrence with an annual aggregate of at least $2 million exclusive of legal defense costs. The above requirements can be satisfied by (1) liability insurance, (2) meeting the financial test for liability coverage in 40 C.F.R. §264.147(f), which requires analyses of the facility's net worth, available assets, and bond ratings, or (3) any combination of insurance and financial test permitted under 40 C.F.R. §264.147.

Section 264.147(c) provides for variances from the liability requirements from sudden and nonsudden occurrences if the owner or operator can establish that such financial responsibility requirements are not consistent with the degree and duration of risk associated with the treatment, storage, and disposal at their facility.

Finally, Sections 264.149–264.150 provide that an owner or operator of a facility can comply with State financial responsibility requirements in lieu of EPA's requirements if the Administrator determines that the state mechanisms are at least equivalent to EPA's requirement, or if the state assumes responsibility for the legal responsibility of the owner's or operator's compliance with the closure, postclosure, and liability requirements.

Use and Management of Containers

Sections 264.170–264.178 provide additional regulation for facilities that store containers of hazardous waste. These sections discuss the condition that the containers must be in to store hazardous waste, the compatibility of the waste with the container, and the management of the containers and the containment system that must be designed and operated in conjunction with the storage of waste in containers. These regulations also mandate that upon closure all remaining liners, bases, and contaminated soil must be decontaminated or removed.

Use of Tanks to Store Hazardous Waste

Sections 264.190–264.199 provide additional regulation for the storage of hazardous waste in tanks. Those sections discuss the design of the tank, general operating conditions when using tanks, and the inspections of the tanks. They also mandate upon closure that all hazardous waste residue must be removed from the tanks, discharge central equipment, and discharge confinement structures.

Surface Impoundments

Sections 264.220–264.230 apply to facilities that treat, store, or dispose of hazardous waste in surface impoundments. Section 264.221 requires that a surface impoundment (except the existing portion of a surface impoundment) must have a liner that is designed, constructed, and installed to prevent any migration of wastes out of the impoundment to adjacent subsurface soil or groundwater or surface water at any time during its active life. The remaining provision of Section 264.221 discusses construction materials and placement of the liner.

Double-lined surface impoundments meeting certain other requirements relating to location and monitoring, however, are exempt from the groundwater protection requirements contained in Sections 264.90–264.100. Sections 264.226–264.227 describe the requirements for monitoring, inspection, and emergency repairs to leaks in surface impoundments. Section 264.228 provides that at closure the owner or operator must: (1) remove or decontaminate all waste residues, subsoils, and liquids; and (2) cover the surface impoundment with a final cover designed to provide long-term minimization of migration of liquids through the closed impoundment. However, if some waste residues or contaminated materials are left in place at final closure, then the owner or operator must comply with all postclosure requirements including maintenance and monitoring during the postclosure period.

Waste Piles

Sections 264.250–264.259 apply to owners and operators that store or treat hazardous waste in piles. The regulations do not apply to owners and operators of waste piles that are closed with wastes left in place. Such piles are regulated as landfills. Nor do the regulations apply to waste piles inside or under a structure that provides protection from precipitation so that neither run-off nor leachate is generated, provided that the liquids are not placed in the pile and are protected from surface run-off.

Not unlike surface impoundments, a waste pile must have a liner designed, constructed, and installed to prevent any migration of waste out of the pile into adjacent subsurface soil or ground or surface water at any time during the active life of the waste pile. Moreover, like surface impoundments, the regulations establish criteria for the design, location, and monitoring of the waste pile, as well as exempting from the groundwater protection requirement those double-lined waste piles meeting certain conditions regarding location of the pile.

Land Treatment Units

Sections 264.270–264.283 regulate those facilities that treat or dispose of hazardous waste in land treatment units. Under such regulations, the owner or operator of the land treatment unit must design such unit to ensure that hazardous constituents placed in or on the treatment zone are degraded, transformed, or immobilized within the treatment zone. To achieve this goal, the EPA Regional Administrator will include in the facility's permit the elements of treatment, including the waste capable of being treated, design and operating practices, and monitoring provisions. Also, for each waste to be applied in the treatment zone, the owner or operator must demonstrate that the hazardous constituents can be completely degraded, transformed, or immobilized in the treatment zone. This demonstration can be made by the use of field tests, laboratory analyses, available data, or operating data.

During the closure period of a land treatment unit, specific operations concerning degradation, transformation, or immobilization of hazardous constituents in the treatment zone must be continuous, as well as those operations relating to run-off and run-off control systems. Further, the owner or operator during the postclosure period must continue all operations necessary to enhance degradation, transformation, and sustained immobilization of hazardous constituents in the treatment zone, as well as maintain the cover, a run-off and run-on control system, and continue monitoring the unit.

Landfills

Sections 264.300–264.317 regulate landfills (other than an existing portion of a landfill) and require that a landfill have a liner designed, constructed, and installed to prevent any migration of wastes out of the landfill to the adjacent subsurface soil or ground or surface water at any time during its active life. Section 264.301 also regulates the construction materials utilized, the location of the liner, and the location, use, and maintenance of the leachate removal and collection system. As in land treatment units and waste piles, landfills with double-liners are exempt from the ground-water protection requirements provided they comply with certain regulations regarding their location and the design, maintenance, and construction of a leak detection system.

During the construction and installation of liners and cover systems (coatings), the liner must be inspected for damage or imperfections. Throughout the operation of a landfill, the owner or operator must maintain a map containing the exact location of each cell with respect to permanently surveyed benchmarks, as well as the contents of each cell.

Liquid wastes, bulk or noncontainerized liquid wastes, or wastes containing free liquids are prohibited from being placed in landfills unless the landfill has a liner and leachate collection and removal system as required by Section 264.301 or before disposal the liquid waste is treated or stabilized so that the liquids are no longer present. Containers holding free liquids are also prohibited from being placed in landfills unless the liquid is removed or solidified or otherwise eliminated.

Upon closure of the landfill, the owner or operator must cover the landfill with a final cover constructed to provide long-term minimization of migration of liquids through the closed landfill, function with a minimum of maintenance, and, after closure, the owner or operator must monitor and provide maintenance throughout the postclosure period. Such maintenance includes maintaining the integrity of the final cover, maintaining a monitoring of the leak detection system, and maintaining operations of the leachate collection and removal system so as to prevent run-off and protect the surveyed bench marks used to locate the placement of the waste at the landfill. However, with the enactment of the Hazardous and Solid Waste Amendments of 1984, all landfill regulations will become more restrictive as EPA attempts to implement Congressional policy. Such regulation will include prohibiting the land disposal of many types of waste. See Section 20.3.8.

Incinerators

Sections 264.340–264.351 apply to owners or operators of facilities that incinerate hazardous waste. Section 264.340 identifies certain waste and waste characteristics that, if incinerated, may be exempted from the regulations governing incineration by the Administrator, except for the waste analysis and closure requirements.

Each incinerator must perform a waste analysis of feedstock. Such analyses must continue throughout normal operations to verify that the waste feed to the incinerator is within the chemical and physical composition limits specified in the facility's permit. Each incinerator must achieve a destruction and removal efficiency of 99.99% for each principal organic hazardous waste designated in its permit for waste feed.

Each incinerator can burn only those wastes specified in its permit and only under the operating conditions specified in the operating conditions of its permit that are established in Section 264.345. The operating conditions of the permit specify acceptable limits on the carbon monoxide level in stack exhaust gas, waste feed rate, combustion temperature, allowable variations in design and operating procedures, as well as the composition of the waste feedstock.

No other wastes may be incinerated other than those listed in the permit except for approval trial burns and those exempted by the Administrator under Section 264.340. In order to burn hazardous waste other than as permitted in the permit, a modification of the permit or a new permit is required.

Throughout the entire incineration process, the owner or operator must monitor combustion temperature, waste feed rate, the indicator of combustion gas velocity, and carbon monoxide at a point in the incinerator downstream of the combustion zone and prior to release into the atmosphere. The owner or operator must also inspect all associated equipment and test, at least weekly, the alarm systems.

20.4.6 40 C.F.R. Part 265, Interim Status Standards

The 40 C.F.R. Part 265 Interim Status Standards apply to all owners and operators of hazardous waste treatment, storage, and disposal facilities except those excluded under Part 261, that were in existence prior to November 19, 1980; those in existence on the effective date of statutory or regulatory changes that render the facility subject to the permit requirements; and those facilities that have applied for but have not yet received an EPA permit. The regulations are similar to the 40 C.F.R. Part 264 that apply to facilities whose permit application has been approved. However, the 40 C.F.R. 264 regulations concerning surface impoundments, land treatment, and landfills contain more stringent requirements than the interim status standards under 40 C.F.R. Part 265, especially concerning design and operating requirements. For example, Sections 264.300–364.316 require a liner as a basic requirement of a new landfill, whereas an existing landfill under Section 265.302 is only required to design, construct, and operate a system capable of preventing the flow of waste onto the active portion of the landfill during peak discharge for a 25-year storm. As EPA approves the permit applications of those on interim status, such facilities will become subject to Part 264 standards.

All facilities operating under interim status must apply for a final permit within 12 months after the enactment of the Hazardous and Solid Waste Amendments of 1984 and must certify that such facility is in compliance with all applicable groundwater monitoring and financial responsibility requirements.

20.4.7 40 C.F.R. Part 267, Interim Standards for Owners and Operators of New Hazardous Waste Land Disposal Facilities

Although now inoperative, the purpose of the Part 267 regulations, effective on August 13, 1981, was to establish standards regulating the permitting of new hazardous waste land disposal facilities until EPA had the opportunity to promulgate its final regulations concerning landfills, surface impoundments, land treatment facilities, and Class I underground injection wells. In addition to the Part 267 requirements when in effect, each owner or operator of a new hazardous waste land disposal facility was required to meet the preparedness and prevention requirements, contingency, and emergency procedures, and manifest, closure, postclosure, and financial requirements of Part 264.

The Part 267 regulations were to be in effect only until the Part 264 regulations for landfills, surface impoundments, land treatment, and underground injection facilities were promulgated or until February 13, 1983, whichever was earlier. The regulations concerning landfills, surface impoundments, land treatment, and underground injection facilities became effective January 26, 1983, thus making the Part 267 regulations inoperative. However, with the enactment of the Hazardous and Solid Waste Amendments of 1984 the existing landfill and land treatment standards will also be obsolete. See Section 20.3.8.

20.4.8 Interface of RCRA Regulations with State Programs (Part 271 Regulations)

Requirement for Final Authorization of State Program

40 C.F.R. Part 271 regulations provide a procedure whereby states can be authorized to implement and enforce hazardous waste management regulations in lieu of federal regulations and federal enforcement. Under the Part 271 regulations, a state's program can be approved by the Administrator for implementation and enforcement if its program is equivalent or more stringent than the federal requirements, including the requirements for permitting, compliance evaluation, public participation, sharing of information, as well as all the requirements of Parts 262–266 and 270. In addition to meeting or exceeding the substantive requirements, a state's application also must contain a complete program description as required by Section 271.6, an Attorney General's statement that the laws of the state provide adequate authority to carry out the described program, and a memorandum of agreement with the EPA Regional Administrator as to how the program will be executed. Sections 271.6–271.9 describe the contents required to be submitted by the state. In addition, Section 271.16 identifies the array of enforcement mechanisms needed by the state for approval of its program.

Prior to the state submitting its program to EPA for approval, the state shall issue a notice of its intent to seek public approval from EPA. The details of such public notice are contained in Section 271.20. Within 90 days after receipt of the state's complete program, the Administrator shall make a tentative determination as to whether he expects to grant such state authorization to implement its own program in lieu of the federal program. Subsequently, the Administrator shall give notice in the *Federal Register* of his tentative determination concerning the state's program which shall indicate that a public hearing will be held and that interested persons wishing to present testimony should file such request with the EPA Regional Administrator for such region in which the state is located. The notice should also afford the public 30 days in which to comment on the state's tentative submission and note the availability for inspection on the state's entire submission. Within 90 days after such notice in the *Federal Register*, the Administrator shall make a final determination whether or not to approve the state's program.

Section 271.21 establishes procedures for the revision of state authorized programs. Sections 271.22–271.23 provide procedures and criteria for the Administrator to withdraw approval of a state program and for a state voluntarily to transfer its responsibilities to implement its hazardous waste program to EPA.

Requirements for Interim Authorization of State Programs

Part B, Sections 271.121–271.137, provides a mechanism pursuant to which those states that do not meet the criteria (basically a program equivalent to the federal program) for final authorization of the entire state hazardous waste program can obtain authority to administer portions of the federal program. Interim authorization is divided into phase-I and phase-II authority.

The phase-I authority allows states to administer parts of a hazardous waste program in lieu of the corresponding portion of the federal program which covers the identification and listing of hazardous waste (40 C.F.R. Part 261), regulations relating to generators (40 C.F.R. Part 262) and transporters of hazardous waste (40 C.F.R. Part 263), and the interim standards for those hazardous waste treatment, storage, and disposal facilities (40 C.F.R. Part 265).

Phase II allows states to administer a permit program and the final standards for owners and operators of hazardous waste treatment, storage, and disposal facilities in lieu of the corresponding federal hazardous waste standards established at 40 C.F.R. Parts 264, 270.

The state has great latitude in determining how they will implement their programs in lieu of corresponding federal programs. A state need not receive phase I authorization before applying for final authorization. A state may apply for final (Phase II) authorization any time after such state promulgates the last component of its phase II regulations. However, with regard to the implementation of the phase-II program, qualified states have the option to either apply for interim authorization for each component of phase II as it is promulgated, wait until a group of components are promulgated, or wait until all components of phase II are promulgated before applying for approval to administer part or all of the phase-II authority. However, each time a state seeks authority to implement an additional component of the phase-II program it must amend the description of its program to reflect such amendment.

Further, each state seeking interim authorization under phase I and/or II shall submit from their respective Attorneys General a statement that the laws of the state provide adequate authority to carry out the described program and such state shall submit its Memorandum of Agreement to the EPA Regional Administrator.

EPA's orderly method of permitting the states to implement their respective hazardous waste programs in lieu of the federal program was somewhat dismantled by the Hazardous and Solid

Waste Amendments of 1984, which imposed numerous additional requirements upon the management of hazardous waste. Because it will take years for EPA to promulgate all the additional regulations the states will find it impossible to obtain final approval. Rather it will be EPA and the respective states jointly implementing the 1984 amendments for years to come. (See Section 20.3.8.)

20.5   EPA, ITS ORGANIZATION AND REGIONAL OFFICES

The federal hazardous waste program is administered by the Office of Solid Waste and Emergency Response of EPA which is headed by an assistant administrator and further divided into three offices—the Office of Solid Waste, the Office of Emergency and Remedial Response, and the Office of Waste Programs Enforcement. The three offices are further divided into 11 divisions each headed by a director. The 11 divisions are;

Office of Waste Programs Enforcement.
Office of Management Information and Analysis.
Hazardous and Industrial Waste.
Land Disposal.
State Programs and Resource Recovery.
Office of Emergency and Remedial Response.
Emergency Response.
Hazardous Site Control Division.
Hazardous Response Support Division.
Office of Policy and Program Management.

The Office of Solid Waste and Emergency Response administers all programs relating to solid and hazardous waste and Superfund, as well as emergency and remedial responses for hazardous waste clean up.

In addition to there being an Office of Waste Programs Enforcement under the Assistant Administrator for Solid Waste and Emergency Response, there is also an Assistant Administrator for Enforcement and Compliance Monitoring.

In addition to EPA's Washington, D.C. structure, which is primarily responsible for promulgation of the regulations, review of state plans, and general policy, the day-to-day work in the management of hazardous waste is undertaken by the regional offices headed by EPA Regional Administrators. It is these individuals that must enter into the Memorandum Agreement with the states concerning the scope and implementation of the hazardous waste programs of the respective states. There are 10 such regional offices geographically distributed throughout the United States which are located in Boston, Massachusetts; New York, New York; Philadelphia, Pennsylvania; Atlanta, Georgia; Chicago, Illinois; Dallas, Texas; Kansas City, Missouri; Denver, Colorado; San Francisco, California; and Seattle, Washington.

20.6   EPA'S PERMITTING PROCEDURES

40 C.F.R. Part 270 covers the basic EPA permitting procedures and standard permit conditions. The permitting requirements of Part 270 implement the substantive requirements of 40 C.F.R. Parts 260–267, 271.

Six months after the initial promulgation of the 40 C.F.R. Part 261 regulations, any person subject to such regulations must apply for or receive a permit in order to treat, store, or dispose of hazardous wastes. Any person subject to such regulations who has not applied for or received a RCRA permit is prohibited from conducting their activity until they obtain such permit.

20.6.1   The Permit Application

The permit application consists of Parts A and B. However, for Hazardous Waste Management facilities in existence prior to November 19, 1980; those in existence on the effective date of a statutory or regulatory change that renders the facility subject to the permit requirements; and those who have applied for but not yet received a final permit, the permit requirement is satisfied by submitting only Part A of the application. Part B must be submitted within the time frame established by the EPA Regional Administrator or State Director. This timely submission of the Part A application qualifies an existing Hazardous Waste Management facility to "interim status" which deems such facility to have an issued permit until EPA or a state with interim authorization for phase II or final authorization under Part 271 makes a final determination on the facility's permit application. However, because of the EPA delays in requiring the submission of Part B applications, the Congress enacted as part of the Hazardous and Solid Waste Amendments of 1984 a rigid schedule for the submission of the Part B permit applications. In fact, interior status terminates unless a land disposal facility submits its Part B application by October 1985; an incinerator must submit its Part B application by October 1986 and all other Part B applications must be submitted by October 1988.

Facility owners or operators with interim status must comply with the Part 265 requirements or equivalent state requirements when such state has received interim or final authority from the Administrator under Part 271.

Prior to requiring that Part B of the application be submitted by existing hazardous waste management facilities, the EPA Regional Administrator or the Director of a respective state must give at least 6 months, notice to such facilities. Permits are required for all who are subject to the provisions of the Act. Those excluded, as discussed previously, are not required to obtain a RCRA permit.

Compliance with a RCRA permit during its term constitutes compliance for the purposes of enforcement with Subtitle C of RCRA. However, a permit may be modified, revoked, or terminated during its term.

The contents of Part A are basically informational. Part A requires name, address, telephone number, indication of whether the facility is new or existing, a description of the process to be used for storing, treating, or disposing of hazardous waste, and the identification of the waste being treated, stored, or disposed of, and a list of all other permits received or applied for.

Part B of the permit application consists of additional general information and specific information necessary for the EPA Regional Administrator or the State Director to determine compliance with the Part 264 standards.

The general information required consists of a description of the facility, a chemical and physical analyses of the hazardous waste to be handled at the facility, a description of the security procedures, the inspection schedule, a copy of the contingency plan, the procedures and equipment used to prevent hazards in unloading, run-off, contamination, and undue exposure of personnel to hazardous waste, as well as other safety precautions to prevent harm to health and the environment. In addition, each Part B application must contain specific information depending upon how the facility treats, stores, or disposes of hazardous waste. In particular, additional information is required by Section 270.14 for containers, Section 270.16 for tanks, Section 270.17 for surface impoundments, Section 270.18 for waste piles, Section 270.19 for incinerators, and Sections 270.20–270.21 for land treatment facilities.

20.7 EPA'S INSPECTION AUTHORITY, REPORTING REQUIREMENTS, AND ENFORCEMENT

20.7.1 Inspections

Section 270.30(i) requires the permittee to allow the EPA Regional Administrator or the State Director, upon presentation of credentials, to enter at reasonable times upon the permittee's premises where the regulated facility or activity is located or conducted or where records are kept, to have access to and copy any records required to be kept, to inspect the facility, and to sample or monitor for purposes of assuring permit compliance.

20.7.2 Reporting Requirements

In addition to inspections and monitoring by the EPA Regional Administrator or the State Director, the permittee shall retain all monitoring information and records of all data used to complete the application for the permit for a period of at least 3 years from the date of the sample, measurement, report, or application. Further, the permittee shall maintain records from all groundwater monitoring wells and associated groundwater surface elevations for the active life of the facility and for the postclosure care period in the case of disposal facilities.

In addition to records retention, all permittees must give notice to the EPA Regional Administrator or State Director of all (1) planned changes to the facility; (2) any anticipated noncompliance that may occur as a result of the facility's activities; (3) transfers of ownership; (4) monitoring reports; (5) compliance schedules; (6) any noncompliance, within 24 hr, which may endanger health or the environment; (7) discrepancies in the manifest reports within 15 days after an attempt to resolve the discrepancies; (8) unmanifested waste reports within 15 days after receipt of the unmanifested waste; (9) biennial reports covering all facility activities; and (10) any other relevant facts that permittee failed to discuss in its application.

Further, each permit shall specify monitoring requirements, including the proper use and maintenance of monitoring equipment and the frequency of such monitoring, as well as additional reporting requirements.

20.7.3 Enforcement

Whether it be EPA or the state operating under final authorization of its approved hazardous waste program or phase-I and/or phase-II interim authorization of its hazardous waste program, both governmental entities have a wide array of powers in which to enforce violations of permit requirements. Section 3008 of RCRA provides civil, criminal, and administrative penalties for violation of any RCRA provisions, as well as injunctive relief for the Administrator against violators of RCRA or regulations promulgated to implement RCRA.

40 C.F.R. §271.16 requires the state implementing hazardous waste programs in lieu of federal programs to possess similar authorities.

20.8 THE SUPERFUND PROGRAM

The Comprehensive Environmental Response, Compensation, and Liability (CERCLA) Act of 1980, Pub. L. 96-510, is known as the "Superfund" because it created a fund whose revenues were derived from a tax on crude oil and petroleum products and used to remedy the environmental problems caused by hazardous waste dump sites. Superfund was enacted into law on December 11, 1980. The primary difference between Superfund and RCRA is that RCRA is a prospective cradle-to-grave regulatory scheme to ensure the proper transportation, treatment, storage, and disposal of hazardous waste. RCRA, however, does not provide any funding for the immediate clean up of abandoned sites or releases of hazardous substances into the environment. Further, RCRA applies only to hazardous waste whereas Superfund applies to any hazardous substance, which includes all those wastes identified or listed under RCRA, as well as those contaminants identified under all other federal acts regulating the environment. (Compare Section 1004(5) of RCRA, the definition of hazardous waste with Section 1001(14) of Superfund, the definition of hazardous substance.) However, like RCRA, Superfund was enacted at the end of a legislative session and there is no conference report between the two houses of Congress. Also like RCRA, the two houses of Congress informally worked out their differences so that when the Senate amended their version of the legislation because of time constraints, S.1480, 94th Cong., 2nd Sess. (1980), by inserting a substitute bill in lieu of the original version, S.2631, 94th Cong. 2nd Sess., the House Committee on Interstate and Foreign Commerce had already agreed to accept the Senate substitute legislation in order to avoid a conference committee. On November 24, 1980, the Superfund legislation passed the Senate. On December 3, 1980, the House concurred with the Senate amendments. Superfund became law on December 11, 1980. The primary differences between the House and Senate versions of Superfund were the scope of the legislation. Such differences were primarily the result of the fact that the Senate Committee on Environment and Public Works had broader jurisdiction than the House Committee on Interstate and Foreign Commerce. Therefore, the Senate expanded the scope of the House legislation which dealt only with abandoned sites and imposed strict liability upon all releases of hazardous substances into the environment that cause damage or injury to natural resources, including the reasonable costs of assessing such injury, loss, or destruction arising from a release of a hazardous substance into the environment; as well as liability for all governmental costs of removal and remedial action and other costs of response necessarily incurred and not inconsistent with the national contingency plan.

20.8.1 Key Superfund Provisions and the Agencies that Implement It

The three primary facets of Superfund are the fund itself and how it works, the liability provisions of the Act, and the National Contingency Plan.

The Superfund

The Superfund Act established a $1.6 billion fund for 5 years to be used for the payment of response costs, remedial actions, and damages to natural resources caused by the release of hazardous substances into the environment. Taxes on crude oil and the petrochemical industry would amount to 87.5% of the total fund with 12.5% of the total fund being derived from general revenues. The fund is to provide the vital resources necessary for a proper response to a release of hazardous substances into the environment. The monies in Superfund are available for use as payment of governmental response costs whenever: (1) a hazardous substance is released into the environment or where there is a release or substantial release into the environment of any pollutant or contaminant that may present an imminent or substantial danger to public health or welfare; (2) payment of necessary response costs incurred as a result of carrying out the national contingency plan; or (3) for the payment of any claim against the fund that is not paid by any person liable under Section 107 of the Superfund Act and deemed a valid claim under Section 112 of the Act.

The Liability Provisions

The liability provisions of Superfund provide that (1) the owner and operator of a vessel or facility; or (2) any person at the time of disposal of any hazardous substance who owned or operated any facility to which such hazardous substance were disposed of; or (3) any person who by contract, agreement, or otherwise arranged for disposal or treatment or arranged with a transporter for transport for disposal or treatment of hazardous substances owned or possessed by such person, by any other party, at any facility owned or operated by another party, at any facility owned or operated by another party and containing such hazardous substances; or (4) any person who accepts or accepted any hazardous substances for transport to disposal or treatment facilities selected by such person, from which there is a release of a hazardous substance, shall be liable for *all* costs of removal or remedial action as well as necessary costs of response incurred by any person consistent with the national contingency plan and damages for injury or loss of natural resources. Superfund does not provide a cause of action for individuals injured by the release of hazardous substances into the environment.

The only defenses to the liability imposed by Superfund are an act of God; an act of war; or the act or omission of a third party other than an employee of the defendant or one whose act or omission occurs in connection with a contractual relationship existing directly or indirectly with the defendant.

Simply, the degree of liability for those potentially responsible parties listed above is strict liability. Although the legislative history does not resolve whether the liability is also joint and several, the authors were clear that such determination shall be controlled by the evolving principles of common law. Recently, however, several Federal District Courts have imposed joint and several liability upon the responsible parties.

The National Contingency Plan

Section 105 of Superfund requires the President to develop a national contingency plan for the removal of oil and hazardous substances released into the environment. The plan shall include discovering and investigating facilities at which hazardous substances have been disposed of; methods of analyzing the relative cost of remedying any releases or threats of releases from facilities that pose substantial danger to health or the environment; and criteria for determining the appropriate extent of removal or remedy of such released hazardous substances and the appropriate roles for federal, state, and local government in responding and remedying releases of hazardous substances into the environment. Finally, the National Contingency Plan must establish criteria for determining priorities among releases or threatened releases throughout the United States for the purpose of taking remedial or removal action. Each year, pursuant to Section 105 of Superfund, EPA publishes its list of Superfund sites, in order of priority, that need remedial or removal action because of the release or threat of release of hazardous substances into the environment.

Agencies with Responsibilities under Superfund

Although EPA is the primary agency involved in implementing Superfund, numerous other agencies have duties under the Superfund Act. In particular, notice of the release of a hazardous substance must be reported to the National Response Center. Also, the Superfund Act establishes within the Public Health Service an Agency for Toxic Substances and Disease Registry that is to effectuate and implement the health-related authorities of the Superfund Act.

The Secretary of Transportation and Secretary of Treasury are required to deny entry to any vessel that does not have proper certification of financial responsibility. Also, the DOT is required, within 90 days after date of enactment of Superfund or at the time a substance is listed as hazardous, to list such substance as a hazardous material under the Hazardous Materials Transportation Act.

The Department of the Treasury has the responsibility for all aspects of Title II of Superfund concerning the imposition of taxes on petroleum and certain chemicals, and the establishment of the Hazardous Substance Response Trust Fund and the Post-Closure Tax and Trust Fund.

Additionally, the Secretary of Interior is responsible for developing guidelines concerning the determination of damage of loss of natural resources.

20.8.2   The Relation of Superfund to RCRA

RCRA and Superfund, as statutes, fit together to provide the comprehensive regulation of hazardous waste from point of generation to its final treatment or disposal while applying strict liability to not only a breach of a permit requirement, but also to the release of any hazardous substance that causes harm to the environment. Simply, Superfund picks up where RCRA leaves off. RCRA is prospective in that it establishes the standards by which industry must operate. Further, a breach of a RCRA standard subjects one to loss of a permit as well as an administrative compliance order, civil and/or criminal penalties. The civil penalties can be as high as $25,000 for each violation and the criminal penalties can be as high as $250,000 and 5 years' imprisonment if the conduct manifests an extreme indifference to life. However, nothing in RCRA provides penalties or other financial resources for the removal of the released hazardous waste, whether from an active or abandoned site, a spill, or any other type of release of a hazardous substance into the environment. Although RCRA regulations require the discharger of the hazardous waste to perform clean up, and subjects the discharger to administrative, civil, and criminal sanctions for failure to do so, it does not provide funds for the government to undertake the clean up and seek reimbursement from the discharger for the cost of the clean up and any damages for injury to natural resources. Superfund, however, provides such statutory authority.

RCRA sets the requirement by which the industry must act and provides the sanctions for failure to act in conformity with such standards. Superfund provides the reserve of money to do what is needed, when needed, to protect the environment, as well as the statutory authority to recoup damages for harm to the environment and for clean up costs.

Further, Superfund's coverage relates to hazardous substances as defined by Section 101(4) of the Act, which includes; (1) any substance designated under Section 311(b)(2)(A) of the Federal Water Pollution Control Act; (2) any substance designated under Section 102 of the Superfund Act; (3) those wastes listed as hazardous pursuant to Section 3001 of RCRA; (4) any toxic pollutant listed under Section 307(a) of the Federal Water Pollution Control Act; (5) any hazardous air pollutant listed under Section 112 of the Clean Air Act; and (6) any imminently hazardous chemical substance or mixture with respect to which the Administrator has taken action pursuant to Section 7 of the Toxic Substances Control Act. Because of Superfund's broad applicability, it acts as a backup to all environmental laws to ensure that there is a remedy for any release of a hazardous substance into the environment.

RCRA and Superfund also complement each other in the regulation of post-closure liability. RCRA's requirements, discussed previously, detail the responsibilities of the owner and operator of the hazardous waste disposal facility after closure. Superfund, on the other hand, provides that if a facility is closed in accordance with RCRA regulations and permit conditions and the surrounding area has been monitored for a period of 5 years after closure to demonstrate that there is no substantial likelihood of any migration off-site or release from confinement of hazardous waste, the liability for such facility shall be transferred and assumed by the Post-Closure Liability Fund established by Superfund. Such transfer shall be effective within 90 days after the Administrator makes such determination. During such 90-day period, the Administrator or the state in which the facility is located are able to review such determination and impose additional conditions upon the owner or operator of the facility during the post-closure period, if necessary.

Therefore, RCRA and Superfund work together to ensure compliance with all standards concerning hazardous waste management and remedies for their breach or the release of any hazardous substance into the environment. However, neither provides a federal cause of action to individuals injured by releases of hazardous substances into the environment.

20.9 CURRENT CHANGES AND FUTURE FEDERAL ROLE

20.9.1 Changes by the Reagan Administration

Although the EPA was rocked by scandal during the Reagan administration, progress continued on environmental matters throughout the United States during that time period. At the federal level, EPA announced its status report on the identification of hazardous waste sites and the progress made on its Superfund sites in November, 1983. The statistics indicate considerable activity. In particular, EPA has:

Identified 16,300 hazardous waste sites in the United States.

Estimated a total of 22,000 hazardous waste sites in the United States.

Conducted preliminary assessments of 6,800 of the 16,300 identified hazardous waste sites.

Rated 56 of the identified sites for inclusion on the National Priorities List because they pose relative danger of air, groundwater, and surface-water contamination.

Reached settlements with the states totaling $175.6 million worth of clean up at Superfund sites since December, 1981.

Issued 67 administrative orders for the clean up of uncontrolled or abandoned sites.

Approved 220 emergency-type actions at Superfund sites since December, 1980.

By December, 1983, moreover, EPA has granted 43 states interim authorization to implement Phase-I standards and 19 states were granted interim authorization to implement parts of the Phase-II standards.

As the states begin administering their own hazardous waste programs and initiating lawsuits under state law and Superfund against parties responsible for the release of hazardous waste and hazardous substances into the environment, the political maneuvering in Washington will be less newsworthy and will have less of an effect on the implementation of the laws regulating hazardous waste. Further, the citizen suit provisions of RCRA, as well as the ability of citizens to put pressure on state government to utilize the provisions of Superfund to clean up releases of hazardous substances, and to enact state Superfund statutes, will give the remedial provisions of environmental law a life of their own irrespective of the attitude of any particular administration toward environmental protection.

20.9.2 Future RCRA Regulatory Program

The Hazardous and Solid Waste Amendments of 1984 focused on the most serious problems with the improper disposal of waste on land. Not only did Congress strengthen EPA's authority to regulate land disposal, it required the ban on many types of land disposal unless EPA's regulations provide for the protection of health and the environment. The future challenge will be whether the EPA has the will to implement its new authority concerning land disposal, and if EPA does not have the will to implement such authority it will be worth watching whether Congress will let its prohibitions on certain wastes being disposed on land go into effect or whether it will exert its deadlines. As the nation faces serious water contamination problems in the next decade there will be considerable citizen pressure to maintain congressional prohibitions and limitations on the disposal of wastes on land. Also, implementation of such prohibitions and limitations will for the first time make new technology for the treatment and disposal of hazardous waste competitive and necessary in order to manage our hazardous waste. Simply, if EPA and the Congress remain steadfast on their prohibitions and limitations on land disposal for certain wastes, the new technologies for treating and disposing of hazardous wastes will flourish. If Congress and EPA do not remain steadfast, Congress will be under constant pressure to permit the continued land disposal of all types of hazardous waste and the problems associated with land disposal will only by made worse.

20.9.3 Future Superfund Program

Representative James Florio (D–N.J.), Chairman of the Transportation, Commerce, and Tourism Subcommittee of the U.S. House of Representatives, has introduced legislation in the 98th Congress which would extend Superfund's authorization from 1985 to 1990 at the rate of $1.8 billion/ year. Large increases in the program are likely to be enacted. However, because of the strict liability and joint and several liability provisions of Superfund, and because of the consistent findings of various district courts upholding such legal principles, EPA and the states will be able to initiate negotiations with potentially responsible parties and obtain through negotiation the clean up of a great number of the hazardous waste sites without tremendous increases in available funds. The additional funds, however, will allow EPA and the states to clean up those sites upon which agreement with the potentially responsible parties cannot be reached and then bring a lawsuit for reimbursement under Superfund.

Another possible change to Superfund would be to provide compensation to victims who suffer personal injury from the improper disposal of hazardous waste. Such a proposal has already been introduced in the 98th Congress. Victims' compensation legislation is the last piece of a comprehensive environmental policy that imposes strict liability upon those who generate, transport, treat, store, or dispose of hazardous waste for harm caused by the release of their waste or substances into the environment.

20.10 SUMMARY

The costs of hazardous waste disposal will dramatically increase not only because of increased costs of disposal, but also increased liabilities that responsible parties will face when damage is caused to human health or the environment. Such increased disposal costs and liabilities will make alternative technology for the treatment or disposal of hazardous waste more economically able to compete with landfill. When this occurs the nation will begin manageing its hazardous waste in a manner that protects human health and the environment.

REFERENCES

1. Solid Waste Disposal Act of 1965 §202(b)(2), 42 U.S.C. §3251(b)(2) (1970).
2. U.S.E.P.A., *Disposal of Hazardous Wastes*, 1974, pp. ix, 3, 41–46.

3. Lazar, Damage incidents from improper land disposal. *J. Hazardous Materials* **1** (1975–76).
4. U.S.E.P.A., Hazardous Materials Damage Reports (SW-151) (SW-151.3) (1976).
5. W. Kovacs, Federal Control on the Disposal of Hazardous Wastes on Land, in *Resource Conservation and Recovery Act: A Compliance Analysis*, 1979, pp. 16–24, 31 fn. 3.
6. H. Rept. No. 1491, 94th Cong., 2nd Sess. 17–23 (1976).
7. S.2150, 94th Cong., 2nd Sess. (1976).
8. Staff Report 20, U.S. House of Representatives, Committee on Interstate and Foreign Commerce, Materials Relating to The Resource Conservation and Recovery Act of 1976, 94th Cong., 2nd Sess. 22 (1976).
9. Kovacs and W. Klucsik, The new federal role in solid waste management: The Resource Conservation and Recovery Act of 1976. *Colum. L.R.* **2**, 205, 216–220 (1977).
10. 42 U.S.C. §§6903(5).
11. 42 U.S.C. §§6921-6931.
12. 42 U.S.C. §6972.
13. 40 Code of Federal Regulations §§270.70–270.73.
14. H. Rept. No. 1491, 94th Cong. 2d Sess., pp. 29–30.
15. 40 Code of Federal Regulations §260.2.
16. 40 Code of Federal Regulations §260.10.
17. 40 Code of Federal Regulations §260.20.
18. 40 Code of Federal Regulations §260.21.
19. 40 Code of Federal Regulations §260.22.
20. 40 Code of Federal Regulations §261.2.
21. 44 *Fed. Reg.* 14472 (April 4, 1983).
22. 48 *Fed. Reg.* 47894 (Oct. 17, 1983).
23. 40 Code of Federal Regulations Part 261, Appendices I–VIII.
24. 40 Code of Federal Regulations §261.6.
25. 40 Code of Federal Regulations §261.11.
26. 40 Code of Federal Regulations §261.12.
27. 40 Code of Federal Regulations §§262.10–262.11.
28. 40 Code of Federal Regulations §§262.20–262.21.
29. 40 Code of Federal Regulations §§262.30–262.34.
30. 40 Code of Federal Regulations §§262.40–262.43.
31. 40 Code of Federal Regulations §§263.10–263.12.
32. 40 Code of Federal Regulations §263.12.
33. 40 Code of Federal Regulations §263.21.
34. 40 Code of Federal Regulations §263.22.
35. 40 Code of Federal Regulations §263.30.
36. 40 Code of Federal Regulations §264.1.
37. 40 Code of Federal Regulations §264.3.
38. 40 Code of Federal Regulations §264.12.
39. 40 Code of Federal Regulations §264.13.
40. 40 Code of Federal Regulations §264.14.
41. 40 Code of Federal Regulations §264.15.
42. 40 Code of Federal Regulations §264.16.
43. 40 Code of Federal Regulations §§264.31–264.32.
44. 40 Code of Federal Regulations §264.51.
45. 40 Code of Federal Regulations §264.51.
46. 40 Code of Federal Regulations §§264.52–264.54.
47. 40 Code of Federal Regulations §§264.71–264.73.
48. 40 Code of Federal Regulations §264.75.
49. 40 Code of Federal Regulations §§264.91–264.94.
50. 40 Code of Federal Regulations §§264.97–264.99.
51. 40 Code of Federal Regulations §§264.111–264.117.
52. 40 Code of Federal Regulations §264.120.

53. 40 Code of Federal Regulations §264.140.
54. 40 Code of Federal Regulations §§264.142–264.147.
55. 40 Code of Federal Regulations §264.151.
56. 40 Code of Federal Regulations §§264.250–264.254.
57. 40 Code of Federal Regulations §§264.271–264.273.
58. 40 Code of Federal Regulations §264.280.
59. 40 Code of Federal Regulations §264.301.
60. 40 Code of Federal Regulations §264.303.
61. 40 Code of Federal Regulations §264.309.
62. 40 Code of Federal Regulations §264.310.
63. 40 Code of Federal Regulations §264.314.
64. 40 Code of Federal Regulations §§264.341–264.347.
65. 40 Code of Federal Regulations §271.1.
66. 40 Code of Federal Regulations §§271.5–271.6.
67. 40 Code of Federal Regulations §§271.60–271.16.
68. 40 Code of Federal Regulations §§271.121–271.129.
69. BNA Environmental Reporter, Federal Laws, pp. 61-1651–61-1652.
70. 40 Code of Federal Regulations §270.1.
71. 40 Code of Federal Regulations §270.4.
72. 40 Code of Federal Regulations §270.10.
73. 40 Code of Federal Regulations §270.13.
74. 40 Code of Federal Regulations §270.14.
75. 40 Code of Federal Regulations §270.43.
76. 40 Code of Federal Regulations §270.30(h).
77. 40 Code of Federal Regulations §270.30(j).
78. 40 Code of Federal Regulations §270.31.
79. 126 Cong. Rec. S. 14948-S. 14986 (Nov. 24, 1980).
80. 126 Cong. H11773-H11803 (Dec. 3, 1980).
81. Weekly Compilation of Presidential Documents, Vol. 16, No. 50: Dec. 11, Presidential Statement.
82. Pub. L. 96-510, 96th Cong. 2d Sess. §§104; 42 U.S.C. 9604.
83. Pub. L. 96-510, 96th Cong. 2d Sess. §§105; 42 U.S.C. 9605.
84. Pub. L. 96-510, 96th Cong. 2d Sess. §107(a); 42 U.S.C. §9607(a).
85. Pub. L. 96-510, 96th Cong. §107(k) (1980); 42 U.S.C. §9607(k).
86. Pub. L. 96-510, 96th Cong. 2d Sess. §111; 42 U.S.C. §9611.
87. Pub. L. 96-510, 96th Cong. §§221–223 (1980).
88. 42 U.S.C. §§9631-9633.
89. Pub. L. 96-510, 96th Cong. §306 (1980), 42 U.S.C. §9656.
90. Pub. L. 96-510, 96th Cong. §§201-232 (1980); 42 U.S.C.
91. The National Contingency Plan was published at 47 *Fed. Reg.* 10972-10995 (1982) and the EPA priority list of sites needing remedial or removal action was published by EPA Dec. 20, 1982.
92. 40 Code of Federal Regulations §263.31; 40 Code of Federal Regulations §§264.50–264.56.
93. *U.S.* v. *Wade*, 20 ERC 1277 (Dec. 20, 1983).
94. EPA New Superfund Rept. (Nov. 1–27, 1983).
95. EPA, State Programs Branch, Office of Solid Waste.
96. H.R. 2867, 98th Cong. 1st Sess. §§3, 5, 6, 8, 24 (1983); H. Rept. 98-198, 98th Cong. 1st Sess. 15 (1983).
97. H.R. 4813, 98th Cong. 1st Sess. (1983).
98. H.R. 4813, 98th Cong. 2d Sess. (1983).
99. The Hazardous and Solid Waste Amendments of 1984, 98th Cong., 2d Sess. (Pub. L. 98-616), November 8, 1984.
100. H. Rept. No. 98-1133, 98th Cong. 2d Sess. (1984).

CHAPTER 21
STATE AND LOCAL REGULATORY ISSUES

JAMES REYNOLDS

HDR TECHSERV, Inc.

H. LANIER HICKMAN, JR.

Governmental Refuse Collection and Disposal Association

21.1 INTRODUCTION

This chapter addresses issues that involve state and local hazardous waste management practices. The first section briefly describes some general trends in changing state regulatory programs and identifies some programmatic issues that should give the hazardous waste generator, transporter, and owner/operator of hazardous waste treatment, storage, and disposal facilities some insight into the major changes that have been made in state hazardous waste management programs. The second section identifies, for the local governmental official, some areas of hazardous waste management that directly affect him/her and in which he/she should be involved.

21.2 STATE PROGRAM DEVELOPMENT

21.2.1 Life Before the Resource Conservation and Recovery Act

With the advent of the massive federal program to standardize the identification and regulation of hazardous wastes, the work of the state regulator is defined. Each state had its own ideas for addressing hazardous waste management prior to the Resource Conservation and Recovery Act (RCRA). Some states believed they had no hazardous waste generated within their boundaries (there were nearly as many definitions of hazardous waste as there were states), others chose to regulate the wastes on a case-by-case basis as problems arose; some required filing notices and requests for disposal approval but such paperwork was usually handled differently in each state and, in some cases, within different agencies in the same state. These differences from state to state not only made it complicated for generators and disposers of hazardous waste to do business in more than one state, but also made it difficult for state agencies to assist each other in regulating the management practices of interstate businesses.

The result was that most businesses were left to manage hazardous waste as they thought best. Occasional inspections by state agencies sometimes helped keep public health and environmental protection in perspective. However, little requirement was placed upon generators and disposers to notify regulatory agencies prior to their actions. Of course, different businesses managed hazardous waste in various ways which sometimes changed as the needs of the businesses changed. Because an organized, comprehensive hazardous waste regulatory program was either not in place or not consistent, such wastes were sometimes disposed of without proper consideration for public health or the environment. As discussed in Chapter 20, when Congress began to receive reports of environmental harm it became evident that a nationally standardized program to manage hazardous wastes was needed. This need was addressed in Subtitle C of RCRA (1976) and required the development of a sophisticated, comprehensive program to identify and control the handling of hazardous wastes.

21.2.2 Standardization

One of RCRA's intents was to change these inconsistencies into a "standardized," nationally approved program. Definitions of hazardous waste, operating requirements for owners/operators of treatment, storage, and disposal (TSD) facilities reporting requirements, all assisted states in standardizing their programs. In addition, before states could be funded for "approved" hazardous waste management programs, the federal government required state regulatory agencies to provide various program features such as manpower levels, methods of tracking hazardous waste activity and enforcing regulations, permitting processes, and coordination and sharing of information with other state and federal agencies. Without such features, state agencies were threatened with nonapproval of their state hazardous waste programs and, therefore, the regulation of the hazardous waste activity within their states by the federal government. Of course, state laws would still be enforceable under a "nonapproved" state program. The intent of "standardization" would quickly develop into major confusion and chaos and, in some cases, it was envisioned that the resulting regulatory programs could become worse than before RCRA.

In effect, then, the federal program required states to develop strict hazardous waste management programs and to scale up their enforcement of violations of law and the prevention of potential harmful effects on public health and the environment. Many states have satisfied these requirements by adopting federal regulations by reference or by reprinting them in their appropriate sections of state law. Some states, however, have taken a different approach. While the federal program offers much in the way of a standardized, national approach to hazardous waste management, each state has the authority to promulgate laws and regulations that are more strict than the federal requirements. It may seem trite to say that without such authority our governmental system of sovereign states would certainly cease to function. However, with such authority we can only approach a truly standardized national program. This means, of course, that each state program is somewhat different, and any generator or transporter of hazardous waste, or owner/operator of a TSD facility must recognize these differences and must keep abreast of state as well as federal developments in hazardous waste management requirements.

Although differences do occur from state to state, the federal regulations have given an important standard to which generators, transporters, and owners/operators of TSD facilities can turn. RCRA regulations are probably the most complex and comprehensive environmental regulations developed to date. With a reasonable understanding of the federal hazardous waste management regulations, a generator, transporter, or owner/operator of a TSD facility should have little trouble in knowing the scope of responsibility placed upon him for proper management of hazardous wastes.

21.2.3 Effects of RCRA

The effect of RCRA on each state program is unique. However, one can generally say that states have increased the number and specialization of employees addressing hazardous waste. Many states have made significant changes in their sampling and analysis capabilities, personnel training, and facility monitoring procedures. Where inspections of a generator's businesses and TSD facilities used to include a semiannual "walk around," some inspections now require sophisticated portable analyzing equipment, several-page inspection forms, and more than 1 day to complete. Follow-up inspections are often conducted monthly or weekly if the situation warrants. It is not uncommon to have inspections made by teams of personnel, each of whom may have an expertise in a certain area of public health or environmental protection.

Some states utilize such state-of-the-art procedures as aerial photography (including satellite and infrared), nationally linked computer data management systems, interagency assistance agreements, and 24-hr emergency response teams. Other states tend to continue their program on a more simplified level of compliance monitoring and enforcement. Some states have siting boards for finding and approving hazardous waste TSD facility sites and promoting the establishment of such facilities in areas of the state where there is a large quantity of hazardous waste generated. Other states assist in offering facility operator training seminars and waste management courses. Still others have sophisticated public education programs for environmental and public health. Some states are aggressively fining violators through administrative action such as violation notices and litigating cases where alleged violators may protest the fine or where the violation is "too severe" for administrative action alone.

This type of change in state programs, both in complexity and, in many cases, in the quickness of such change, does not come without some problems for the state program administrator and for businesses in each state affected by the new program. It is important to keep in mind that states are in the process of developing their programs—even after several years of such development. Federal regulations have taken more than 6 years to "gel." They are still being altered. Changes to state programs did not start immediately after RCRA was enacted and will require several more years to catch up. Also, each state is in a different phase of development.

The result of these differences of program maturity and administrative philosophy is that any single source of information is too limited to provide an up-to-date and comprehensive understanding of hazardous waste management in the United States on a statewide level. The generator, transporter, and owner/operator of TSD facilities should be disciplined to search continuously and systematically for and evaluate several sources of information, and should spend some time getting to know the state system—the organization, administrative and regulatory philosophy, and the types and locations of information on file and available. These items should be updated from time to time. Appendix 21.1 lists the agencies and contacts that administer the solid and hazardous waste programs in each state.

Information also should be gathered from a national perspective. Treatment and disposal techniques and personnel training assistance are examples of areas where technology and services are rapidly changing. A national base of information offers more chance for getting information on state-of-the-art technology and practices that have been tried elsewhere and have failed or proven less than satisfactory.

21.3 POLICY ISSUES OF CONCERN TO LOCAL GOVERNMENT

21.3.1 Introduction

Although local government is not burdened by the formal regulation of hazardous wastes, local government does have a role to play through participating in the review of regulations development, impacts on generators, emergency response responsibilities, abandoned waste shipments, and orphaned hazardous waste sites. The following discussion briefly outlines that role and the issues associated with that role.

21.3.2 Facility Siting

Siting of hazardous waste management facilities continues to be the primary responsibility and challenge of the generating and service segments of hazardous waste management. However, even though the Environmental Protection Agency (EPA) and state government have, for the most part, the authority to permit hazardous waste management facilities, local government is in most cases the ultimate authority to determine zoning matters and other local land use issues concerning siting and operating privately owned hazardous waste management facilities. Public opposition to the siting of such facilities is severe to say the least. The opposition often comes from the public fears of toxic materials leaching into the environment, causing pollution that goes unnoticed until it is "too late," or catastrophic effects on public health caused by major operational problem or "unavoidable human error." These fears are somewhat inconsistent with the intent of the regulations under RCRA since hazardous waste facilities that are properly designed and operated would generally represent a much lower risk to the community than many activities involving the manufacture or use of hazardous substances, or facilities that accept hazardous wastes from the community, but do not recognize the need for proper management of such wastes. A concern of communities that is more difficult to address is the disposing of hazardous wastes from other areas of the state, or even the country, in one community's "back yard." Issues of eminent domain and monetary and health remunerations are complex and have quickly become legal issues which are easier to argue than to resolve. In some states, the ultimate determination of a permit for a hazardous waste management facility rests with local government. Local government, therefore, does have a role, a role which in most cases it is not technically, financially, or politically prepared to fulfill.

21.3.3 Economic Impact on Industry

The hazardous waste regulatory program requires that hazardous wastes be managed in a prescribed fashion. All of the players in the game are required to provide for the proper "cradle-to-grave" control of their hazardous wastes. Since existing facilities are limited and since it is becoming increasingly difficult to site new facilities anywhere in the nation, one can expect treatment and/or disposal costs to continue to rise at a rapid rate.

This is especially true for those industries whose location, waste volume, and economic resources will not permit on-site treatment or disposal. Local industries falling into this category may eventually find themselves in an inferior competitive position with industries having easier and less costly access to permitted disposal sites. Local government should be concerned about the costs of hazardous waste management for their industries.

21.3.4 The Exempted (Small) Generator

The current RCRA hazardous waste regulations exempt small-quantity generators from most of the hazardous waste regulations if they generate less than 2205 lb (1000 kg) of hazardous wastes within one calendar month. The hazardous waste regulations, however, require that these wastes be disposed in a permitted, hazardous or nonhazardous waste disposal facility. However, disposal of a hazardous waste in a nonhazardous waste disposal facility should not normally occur without the explicit approval of the state and owner/operator (government or private company) of the site. Further, no such approval would be given by the state unless it can be proved that the facility can properly accommodate such wastes.

It should be noted that acutely hazardous wastes are not subject to the 2205 lb (1000 kg) exemption. Rather, a 2.2 lb (1 kg) limit has been established for these wastes. These acutely hazardous wastes are listed in 40 C.F.R. Part 260 and are included as part of state regulations where states have an approved program.

The fate of hazardous wastes from small-quantity generators is essentially unknown. Although it can be assumed that some percentage is being managed in approved hazardous waste management facilities, it is far more likely that these "phantom" hazardous wastes are winding up in municipal solid waste collection, transfer, recovery, and disposal facilities. This means that facilities that may not be designed for hazardous wastes are becoming hazardous waste management facilities. Communities need to address this issue seriously to determine if a problem exists or if the amount of those wastes going into their municipal solid waste streams is insignificant.

Consistent with this determination, local government will have to decide what course of action to pursue relative to the small generator. This suggests that local government needs to develop a hazardous waste management plan to address all hazardous waste issues within their jurisdiction.

21.3.5 Closed and Abandoned Hazardous Waste Disposal Sites and Orphaned Hazardous Wastes

Perhaps one of the most perplexing and frustrating hazardous waste management issues facing local government is the one associated with: (1) orphaned hazardous wastes and (2) closed or abandoned hazardous waste disposal sites. In the former, it is quite common for a community to be faced with an amount of unknown waste products which suddenly appear in a vacant lot or along the roadside. Most communities are poorly equipped to conduct the appropriate analytical efforts to determine the nature of the material.

Frequently, state government is not available or able to assist in this determination. Attempting to sample and analyze an unknown container offers many threats to personnel safety, yet something must be done to protect public health. Eliminating the presence of the orphaned hazardous wastes will require technical and financial resources. Assistance from the Superfund program (CERCLA, see Chapter 20) may or may not be available. Regardless, local government will be faced with the costs for talent and corrective measures. Procedures need to be developed to face this issue.

Sites that appear, such as Love Canal and the Valley of the Drums, perhaps many years after such a site is closed or abandoned have been major contributors to the emergence of a national hazardous waste management effort. These sites now cause, or could cause, major economic, health, and environmental impacts on a community. The solutions to the problems that may be presented by these sites are complex, time-consuming, and costly. Often, ownership cannot be assigned nor responsibility established. In these instances, it becomes the problem for the local community to find ways and means to see to the correction of any problems existing. State and federal government assistance may or may not be available. It is clearly evident that the amount of resources available from Superfund, as it is currently structured and financed, will not even start to provide resources necessary to fix the many sites that may need correction. Some states have established their own funds to assist in these matters; however, most of these funds are also inadequate to meet the need.

Where does the money and talent come from to deal with this problem? The answers are neither available nor apparent at this time. However, it is obvious that local government will wind up paying for a big portion of the bill.

21.3.6 Emergency Response and Contingency Plans

The first line of defense in the management of emergency responses is almost always local government. Unless the emergency response mechanisms of the state and federal governments are called upon to respond to an emergency occurrence, local government will be responsible for protecting public health and environmental quality. Within a state master plan for emergency response, local government has a defined role. This role assumes a certain ability on the part of local government that must be a part of its overall emergency response plan.

By regulation, emergency occurrences that arise from hazardous waste management activities are to be handled by the owner/operator with local involvement as specified in the required contingency plan for hazardous waste management facilities. (Transporters do not have contingency plans due to the wide range of potential occurrences and mobile nature of the activity. Transporters are, however, responsible for cleanup of spills.) All facilities that store, treat, or dispose of hazardous wastes must have contingency plans. This includes many generators because many operate hazardous waste management facilities.

By regulation, local government and its emergency service agencies must have been given an opportunity to participate in the development of these contingency plans. Moreover, copies of contingency plans were to have been filed with the involved local government and agencies by November 19, 1980. These plans are also on file at each hazardous waste management facility. A list of facilities within a state should be available from the state hazardous waste management agency. Under the interim status provisions of federal regulations, hazardous waste management facility contingency plans are not required to be filed with the state, but are reviewed by the state during the normal course of inspection of these facilities. Where no plans have been developed, the state guides the facility owner/operator to develop a plan. The state also works with the owner/operator to assure that local government is involved in developing the plan. The determination of the acceptability of a contingency plan will ultimately rest with the state along with all other requirements associated with final permitting of a facility. Until then, however, local government must assure itself, through discussions with the owner/operator, of the acceptability of what is included in a contingency plan. These plans may very well call for local government and agency involvement in coping with an emergency at a permitted hazardous waste facility. It is also possible that a plan may call upon local resources that may not be adequate or available. Consequently, it is in the best interest of local government to determine the nature and content of these plans.

Local government will have an immediate and vital role to play in emergencies arising from the transport of hazardous material. (Again, hazardous materials are composed of hazardous wastes and hazardous substances.) The role will be almost identical whether there is a hazardous waste or a hazardous substance involved. Until the threat of injury or death is over, it makes no difference whether the hazard comes from substance or a waste.

Fire service units, emergency response coordinators, emergency health service forces, solid waste management programs, and many others may be called upon to respond to emergency incidences that can occur from a spill, at a hazardous material storage terminal, or at a hazardous waste management facility. Consequently, an understanding of hazardous material response procedures and what may be present at one of these facilities is an essential local government action.

Except to protect workers, the gainful use of hazardous substances is not regulated. As with other incidences, however, emergencies arising from these activities remain the responsibility of local government. If, as a result of an emergency incident, these substances become wastes and the resulting waste products become hazardous as defined by regulations, the state government has a defined responsibility and local government cannot act without their involvement.

21.3.7 Summary

In summary, therefore, local government is faced with certain issues related to hazardous waste management:

1. The role of local government in participating in the development of hazardous waste management facility contingency plans as well as the responsibilities of local government in the implementation of a contingency plan should the need arise.

2. Emergency response roles associated with permitted hazardous waste management facilities.

3. Emergency response roles associated with spills, fires, or other sudden releases from non-permitted hazardous waste sources such as exempt generators; from the transport of hazardous wastes or substances; and from generators and users of hazardous substances.

4. The relationship between hazardous waste management and municipal solid waste management systems.

5. The potential impact on selected industries associated with compliance with the federally mandated hazardous waste management system as prescribed by the EPA hazardous waste regulations under RCRA.

6. Dealing with orphaned hazardous wastes, and closed and abandoned hazardous waste disposal sites.

7. The presence of the exempted (small) hazardous waste generator; the waste management practices of those generators; and the impact of those practices on a municipal solid waste management system.

8. The siting and permitting of hazardous waste management facilities within the jurisdiction of a community.

All of these issues are exacerbated by the involvement of several entities of local government. A clear understanding of the respective roles is one of the best ways of promoting effective and harmonious relationships between the state and local government and industry. Open communications in all directions are extremely important.

INDEX

Acid rain, 608, 609, 610, 617, 618
Air classification, 428, 497
Air pollution, 613
  acid gases, 618
  control systems, 614–618
    comparison, 617
    dry Scrubber–electrostatic precipitator
      tandem, 616
    electrostatic precipitator, 614
    fabric filters (Baghouse), 615
    wet Venturi scrubber, high energy,
      615
  particulate matter, 613
Air pollution emissions, 595
  carbon monoxide (CO), 607, 617
  combustible particulate, 605, 618
  dioxins, 611, 612, 613
  heavy metals, 610, 619
  hydrocarbons – PNH, PCB, 611, 617
  hydrochloric acid (HCL), 609, 610
  inorganic particulate, 597
    ash content, 597
    burning rate, 605
    firing methods, effects, 597, 604
    grate type, 605
    incinerator size, 605
    underfire air, 604
    volatization, metallic salts, 605
  micro pollutants, 609
  nitrogen oxides (NOx), 608, 617
  sulfur oxides (SOx), 608
  total particulate, 606
Air pollution regulations, 595
  BACT, 596, 598–602
  federal, 595
  state, local, 597–602
Ash residue, 665, 689
ASTM classification, guidelines, 484–496
Auto shred residue, 743, 745, 747

Baling, 327, 338
Biofuels, electric generation, 763
  definitions, 763, 764
  scope, 763, 764
  surveys, 764–768
Biological processes, 749
  anaerobic digestion, 750
    basic process, 750
    design parameters, 753
    feedstocks, 750
    products, 750

    reactor types, 751
  applications/economics, 757
    anaerobic digestion, 758
    composting, 759
    fermentation, 758
  case histories, 759, 760, 761
  composting, 758
    basic process, 755
    feedstock, 756
    products, 756
    reactor types, 756, 757
  definitions, 750
  fermentation process, 753
    basic process, 754
    design parameters, 755
    feedstocks, 754
    products, 754
Boiler types, dedicated units, 557
Bulky waste disposal, 697
  aborted projects, 734
  case histories, 703–734
  process experience, 700
    bale, 702, 719
    maul, 702
    shear, 702, 721, 725
    shred, 702, 703, 706, 714, 719, 725, 734,
      735, 736
  status, 698
  waste types, 697

Capital cost, 93
  construction costs, 99
    control of, 87, 105, 106, 107
    labor, 87, 102
    schedules, 100, 101, 103, 104
  consulting engineer, 78, 96
  control of, 97
  Europe, 94
  foreign financing, 96
  North America, 93, 115, 116, 117, 118,
    119
  operating costs, 107
    control of, 115, 120
    debt service, 61, 114
    government owned, operated, 109
    labor, 113
    materials, 113
    operating organization, 110
    overhead, indirect expense, 114
    privately owned, operated, 109
    private operation, government owned, 109

Capital cost, operating costs (*Continued*)
  spares, 113
  working capital, 114
  plant financing, 95, 121
  plant types, 94
  preproject engineering, 58, 59, 63, 96
  preproject planning, 95
  purchasing procedures, 99
  redundancy, reliability, 97
  revenues, 109
    energy sales, steam, electricity, 52, 53,
      62, 111
    RDF sales, 111, 112
    recovered materials, 52, 53, 55, 112
    tipping fee, 109
Carbon monoxide (CO), 607, 617
Cement industry, refuse fuels, 411, 737
  experience, 737
    in England, 737
    in North America, 746
    in West Germany, 743
Codisposal, 567, 568, 569
Cofiring, 383, 567, 569
Cogeneration:
  industrial plants, 668
    costs, operation and maintenance, 673
    economics, 672
    regulatory factors, 672
  municipal waste, 383, 384, 565, 668, 672,
    673
Collection/disposal:
  franchise administration, 156, 157, 163
    administrative organization, 160
    collector/hauler delivery tickets, 160
    collector/hauler registration, 160
    delivery tickets, purchase, return, 161
    disposal site owner, ticket redemption,
      162
    disposal site transaction, 161
    franchise petition, 157
    rate setting, 158
      the formula, 159
      rate computation, 158
      uniform rate advantages, 158, 159
    reevaluation, uniform rate, 159
  intermunicipal agreement, 163–171
    accounting, weight scales, records operation,
      168
    charges, payments, 167, 168
    county deliveries, activity, 166
    definitions, 165
    further assurances, 168
    miscellaneous, 169, 170
    term of agreement, 170
    transfer of facilities, 168
  landfill environmental trust fund, 172
  landfill indemnity bond sample, 171
Collection of solid waste, 178–192
  contracts, residential service, 183
    specifications, 183, 184, 185
    type, level, 183
  cost accounting, 179
  maintenance, 185
    equipment, 186
    facilities, 185

managing change, 178, 182
  optimization of service, 187–193
    productivity measurement, efficiency,
      187–191
  unions, 180
    collective bargaining, 180
    labor relations, 181
    strikes, 180
Compost, 755–759
Conveyers, 444–449

Densified refuse derived fuel (dRDF), 461–
  473
  characteristics, 464–468
  combustion, 471
  costs, 469, 470
  densification equipment, 462
    feedstock quality, 464
    performance, 462, 463
    power, maintenance, 467
    problems, 462, 467
  installations, 466, 472
  storage, handling, 469
Dioxin, 611, 612, 613
DOE federal statutes, solid waste, 16, 17
  DOE Act, 1978, Civilian Applications, Publ.
    L 95–238, 18
  DOE Organization Act, Publ. L 95–91, 17
  Energy Security Act, Publ. L 96–294, DOE,
    16
  FERC (Federal Energy Regulatory Commis-
    sion), 18
  National Energy Conservation Policy Act,
    Publ. L 95–619, 18
  PURPA (Public Utilities Regulatory Policies
    Act of 1978), 18

Energy marketing law, federal, 622, 625
  purchase rates, 627
  qualifying status, 626
  right to contract, 627
  state agency implementation, 628
  system emergencies, 628
  utility cost data, 627
  utility purchase obligation, 627
Energy marketing principles, 621
  energy sales contracts, negotiations,
    625
  legal orientation, 622
  political orientation, 622
  potential electricity customers, 623
    identification, 624
    investigation, 624
  potential steam customers, 623
    identification, 623
    investigation, 623
Energy recovery:
  mass burn, 557
    furnace design, 575, 581
      cold furnace syndrome, 586
      combustion volume, intensity, 581,
        582
      convection heat transfer, 586, 587
      emissivity (absortivity), furnace walls,
        586

flame emissivity, 586
open pass design, 587
overfire (secondary) air, 586, 587
radiant heat transfer, 586, 587
rear arch design, 587
rotary kiln air distribution, 587
stefan-boltzman constants, 583, 584
view factor, 586
methods/products, 562-569
coburning, 567, 569
codisposal/sludge, 567
cogeneration, 565
constraints, 556, 561
dedicated units, boiler types, 557
electricity, 564
modular units, small scale, 590, 677-680
refractory furnace, convection boiler, 558
retrofits, existing units, 560
steam, energy balance, 562
waterwall furnace, 558
Stoker design, 578, 579
air flow, 577, 578
combustion efficiency factors, 559
combustion process, 575
fuel bed depth, 578
fuel characteristics factor, 578
heat release, 577
rotary combustor, refractory lined, 577, 583
rotary combustor, water cooled, 587
systems and sizing, 571, 572
crane design, 573
crane feed cycle, 574
field erected boilers, 571
fire, ventilation, 57
oversized material, 572, 697
receiving, storage (pit/bunker), 572
processed fuels, 377
bioconversion, 412, 749, 763
case histories, 388
cement kilns, 411, 737
codisposal, 386, 567
cofiring, 383, 567, 569
cyclone furnace, 409
dedicated units, 380
economics, 388
fluidized bed, 405
processed fuel types, 398
pyrolysis, 410
retrofit, existing units, 380
spreader stoker (partial suspension firing), 401
suspension firing, 404
small scale, 561, 677, 678, 679, 680
application constraints, 593, 677-680
combustion concepts, 590
combustion systems, 591
emissions control, 591
raw material, 590, 670-680
feed, 590
receiving, 590
retrieval, 590
storage, 590

Energy sales contract negotiations, 636
basic rules, 638
contract elements, long term sales, 638
sources, utilities information, 636
tactical dealing, utilities, 637
Energy values, 627, 628
electricity to end users, 635
electricity to utilities, 630
avoided cost calculations, 631
escalators, 633
other valuation issues, 634, 649
energy and capacity factor, 628
steam, 629
EPA, 15, 19, 20, 791
Explosions, 532, 533, 534, 535, 542, 546, 550

Feasibility study, 58-63
collection, transport analysis, 60, 178, 179, 195
costs modeling, 58, 59, 63
alternatives, 61
capital cost, debt service, 61, 93
economic analysis, pro-forma, 63-67, 114
life cycle cost analysis, 66, 93, 99, 107
microcomputer, sensitivity analysis, 66, 68
net system, 58
operations, maintenance, 61
primary haul, 60, 178, 179
revenues, 62, 109
secondary haul, 60, 195
demographic projections, 48, 50
employment, 48, 49, 50
population, 48, 49, 50
disposal facility alternatives, 56
landfill, 56, 57, 58, 267
resource recovery, 56, 57, 58
facilities siting, 51
landfill, 52, 267
resource recovery, 52
transfer station, 52
landfill study, 45, 155, 267
existing operation, 45, 49, 50
expansion potential, 57
site life projections, 45, 50, 57
sizing, 57
resource recovery study, 52-56
energy types, methods, markets, 52, 53, 111, 621
facility sizing, 27
materials types, quantities, markets, 52, 53, 55, 482, 683
residue landfill requirements, 57
solid waste survey, 44-50, 507, 515
combustion characteristics, 46
composition, 44, 47, 515
data gathering, 44
densities, 45
generation, per capita, 48
quantity, 45-48, 507, 649
sampling, 46, 47
seasonal variation, 46
source separation, 44, 215
survey scope, 45, 46

Filters, air pollution, 614–618
Financing, resource recovery/disposal
      projects:
   accounting, 126
      builder/operator ownership, 126
      joint venture leasing, 126
      legal relationships, project participants,
         126
      leveraged lease structure, 125
   bond ratings, 139–150
   builder/operator ownership, 125
   case analyses/financing alternatives, 127
      landfill only, 127, 128, 131, 132
      resource recovery, 127, 128, 129, 130
   leveraged lease, 124
      revenue stabilization fund, 125
      structure, 125
      tax perspective, 124
      tax questions, 124
   private ownership, 123
      capital cost recovery, 123
      equity marketing impact, 124
      service agreements, 124
   public ownership, 121
      general obligation bonds, 121
      project revenue bonds, 122
      security, 122
   risk assessment, 133
      managing risks, 136, 137, 138
      project risks, 134
      risk exposure, 133
Flail mill, 441, 443
Furnaces, 558, 575, 577, 581, 583, 587,
      590

Hazardous waste:
   federal program, 773
      EPA, 15, 19, 20, 791
         inspection, reporting, enforcement, 792
         organizations, regions, 791
         permitting, 791
      federal regulatory history, 773
         congressional intent, RCRA, 774
         past practice, 773
      RCRA, interim standards, 789
      RCRA, Amendments 1984, 778
      RCRA regulations, 780
         disposal owners, operators, 40 CFR,
            Part 264, 783
         general, 40 CFR Part 260, 780
         Generator 40 CFR Part 262, 782
         Identification 40 CFR Part 261, 780
         Transporter 40 CFR Part 263, 783
      RCRA scope, contents, 775
         authorized state programs, 777
         disposal, storage, treatment, 776
         enforcement, 777
         future regulatory program, 796
         generator, 775
         identification listing, 775, 780
         permit authority, 776
         transporter, 775
      RCRA and state standards, 790
      requirements, facilities operators/owners,
         780

         closure, postclosure financing, 786
         container management, 787
         emergency, contingency procedures,
            784
         general standards, 784
         groundwater protection, 785
         incinerators, 788
         landfills, 788
         prevention, preparedness, 784
         records, manifesting, reporting, 785
         surface impoundments, 787
         tanks 20.8, 787
         waste piles, 787
      superfund, 793
         agencies, 794
         future program, 796
         liabilities, 794
         national contingency plan, 794
         RCRA relationship, 794
      Reagan administration changes, 795
   state/local, 799
      intent, scope, 800
      local government issues, 801
         closed, abandoned disposal sites,
            802
         emergency response, contingency
            plans, 802
         exempt (small) generator, 802
         facility siting, 801
         industry economic impact, 801
         ophaned unknown material, 802
      before RCRA 21.1, 799
      RCRA effects, 800
Health factors, 530
   accidents, injury, illness – conclusions,
      539
   dusts, 536
      contaminents, organic, metals, 537
      OSHA, 537
   explosions, 532
      preventions, 533
      protection, 533
      shredders/venting, 533, 534
      types, 534, 535
   microbiological aerosols, 537, 538
   noise, 538, 539
   pathogens/infectious disease, 537, 538
   records, health and safety, 539
   sampling, 538

Industrial wastes, energy recovery, 653
   ash removal, 665, 689
   as boiler fuel, 653
   boiler types, 663
   case histories, 680
   energy types, 662
   feedwater, 664, 666
   incinerators:
      controlled air, 655
         dual chamber, excess air, 655, 677–680
         rotary kiln, 587, 657
         starved air, 655, 677–680
      early designs, 654
   operation, maintenance, 664
   stoker types, 665

Industrial wastes incineration, 666
concept choices, 666
mass burn, 666
processed fuel, 667
Interstate commerce, 153
market participant, 154
market regulator, 154
state's natural resources, 154
IRS (Internal Revenue Service), 19
Industrial Development Bonds, Code
Section 102, 19
Recycling Credit, Energy Tax Act, Publ
L. 95-618, 19
solid waste, definition, 19
statutory mandate, 19

Landfill:
bales, 338
balefill, 370
baling process, balers, 338, 341
building foundation potential, 373
comparison – balefill, shredfill, 347, 370
costs – capital, operating, 345
transport of bales, 343
environmental control, 266
guidelines, federal, state, 266, 267
finished site liabilities, 156, 329-334
insurance, 156
surety bonds, 156
trust funds, 156
leachate control, 294-308
collection systems, 308
geophysical methods, 312
groundwater monitoring, 310, 311, 312
groundwater pollution, remedial action,
312, 313
hydrogeologic investigation, 298-302
liner systems – soil, membrane, 302-
308
natural attenuation, 297, 298
treatment disposal, 308-311
leachate formation, 286
leachate composition, 286-289
leachate quantity, 289-294
limiting generation, surface techniques,
294-297
methane gas, 313
composition, 313, 314
energy recovery, 320
gas safety guidelines, 318
movement through waste, soil, 315
production rates, 315
venting, collection, 316
operator safety, 327
principles of operation, 260
biological/chemical processes, 263
methods, 260, 261
area, 262
combination, 262
operating costs, 263, 264
trench, 261
site closure, 329
checklist, 330
closure factors, 329
continuing liabilities, maintenance,

environment, 155, 156, 332, 333, 334
end uses, 334
financial liability, 155, 156, 334
Landfill compaction, 306
equipment and methods, 307, 308
Landfill development, 267
comparing sites, 273, 274
land requirements, service area, quantities,
268, 269
quantities, 268, 269
service area, 268
plan preparation/regulatory approval, 280-
286
public involvement, 274-280
siting procedures, 269-274
Landfill operations, 321
equipment selection, utilization, 323
on-site processing, 326
baling, 327, 338, 370
material recovery/recycling, 326
shredding, 326
Leachate, 266, 286-313

Magnets, 408, 501, 502, 503
Mass burn, 557, 653, 659, 665, 666
Materials recovery, municipal solid waste,
452, 482, 497
equipment, operation, 425, 426, 452, 455
air separators, 428, 497
disc screens, 458
magnets, 498, 501, 502, 503
shear shredder (rotary shear), 425, 442,
443, 452, 702
trommel screens, 456
vibrating screens, 455
specifications, markets, 482, 643
aluminum, 486, 503
ash, 506
ferrous metals, 483, 498
glass, 484, 505
nonferrous metal, 503
paper, corrugated cardboard, 487,
504
plastics, 505
textiles, 491, 492
systems, operation, 497
ash, 506
ferrous metal, 498
glass, 484, 505
nonferrous metal, 503
paper, corrugated cardboard, 487,
504
plastics, 505
RDF, air classification, 428, 497
Materials recovery (scrap industry viewpoint),
643
disincentives, 648
engineering unpredictable material, 649
flow control aspects, 649
quality, saleability, 648
raw material supply, technologies, 645
recovered materials, the hard facts, 650
risks, 648
scrapmarket criteria, 482, 644
energy recovery process effects, 643, 645

Materials recovery, scrapmarket criteria
    (*Continued*)
    ferrous metal, 644, 645
    nonferrous metal, 645, 648
    strategic planning, models, 645, 646, 647
Methane, 263, 264, 313–321
Modular units, 561, 590, 591, 593, 677–680

Nitrogen oxide (NOx), 608, 617

Permits, 280–286, 595, 597, 801, 802, 803
Precipitators, 614, 616
Processed fuels, 398, 416–422
Processed fuel systems, 377, 398, 416–428
Procurement, solid waste facilities, 68–77
    A&E, 68, 69, 70
    competitive negotiation, 74, 75, 76
    full service, 70, 73
    life cycle cost analysis, 69
    methods, 73
    risks, 72
    sealed bids, 73, 74, 75
    sensitivity analysis, 68
    sole source negotiation, 74–77
    turnkey, 70, 71
    two-step formal advertising, 74, 75, 76
Public perceptions/community relations, 31
    acceptance, resource recovery, 31
        Europe, 31
        United States, 31
    case analyses, 34
        abandoned projects, 34, 35
        abandoned sites, 36, 37
        projects in doubt, 39, 40
        successful projects, 38, 39
    case studies, implications, 33
        approval, resistance, economics, 33, 34
    community/education, 34, 114
    facilities investigated, 31
    survey questions, 32
    survey techniques, 32
PURPA, 18, 21, 622, 623, 624, 763

RCRA:
    future issues, 23, 153, 154
        citizen suits, 23
        liabilities, 23
    guidelines and planning, 13
        compliance, 13
        financial assistance, 13
        goals, 13
        minimum standards, 14
        technical assistance, 13
    hazardous waste management, 10, 11
        financial assistance, 13
        generators, 11
        identification, hazardous waste, 10, 11, 775
        RCRA amendments Pub. L 98–616 (November 8, 1984), 11, 12
        record keeping, 11
        transporters, 11
        treatment, disposal, storage, 11

implementation, 19
    DOE program, 21
    EPA participation, 19, 20
    EPA hazardous waste program, 20
    new federalism, impact, 21, 22, 23
    procurement, recovered material, 15, 16, 21, 26, 153, 154
legislative history, 9
recovered material procurement, 15
    DOC (Department of Commerce) role, 16
    EPA role, 15
    market development, 16
    OPP (Office of Procurement Policy) role, 15
    specification, DOC standards, 16
    technology transfer, methods, 16
standards/enforcement, 12
RDF, dry process equipment, 423–428
    air classifications/shredding anomalies, 428
    design/operating improvement needs, 437, 449
    flail mill, 441, 443
    performance, 424, 425
    raw material receiving systems, 444
    receiving conveyers, 446, 449
    shear shredder (rotary shear), 442, 443, 702
    swing hammermills, 430, 433–436
        operating characteristics, 432
        power requirements, consumption, 431
        sizing and selection, 431, 432
    systems, 424–428
    transfer conveyers, 444–449
RDF distribution/feed, 479, 481
RDF process systems, 424–428
    combined dry/wet process, 417, 419, 421
    dry process, 416, 417, 424–428
    energy output comparisons, 419, 422
    materials by-product recovery, 419, 423
    operating characteristics, 419
    RDF market resistance, 419
    RDF storage/retrieval, 422, 473
    status, 415, 416
    wet process, 417, 421
RDF storage retrieval, 473
    Atlas storage bin, 475
    Bunker storage, 477
    Hooper live bottom bin, 480
    Miller Hofft storage bin, 417
    Miller Hofft surge bin, 479
    return burner distribution/storage, 422, 479
    Sprout Waldron surge bin, 479
    surge storage, burner front, 479
    tip floor storage, 499
RDF transport, 474
    remote steam plant, 474
    same site, process and steam plants, 474
Recycling, 215, 216
    definitions, 216
    New Jersey statewide program, 238
        Appendix 10.5, program structures, 256
        citizen education, 248
        drop-off centers, suburban, 258
        goals, 238, 240, 246, 247, 248

implementation, 246, 247
legislation, 240, 246, 247
markets, 247
municipal curbside collection, 257
municipal curbside drop-off centers, 256
scope, 240, 242, 244, 248
Regulations:
air pollution, 12, 14, 595, 596, 597–602
federal, 9–23, 773–791
land pollution, 9–23, 266, 267
state, 11, 12, 3, 266, 267, 596–602, 800–
802

Safety factors, implementation, 541
explosion containment, 532, 550
explosion remedial measures, 552
hazards potential, raw material, 542, 546,
551
personnel safety, 542
post-explosion procedures, 554
presort, hazardous material, 546
safety rules, practices, 542
surveillance, raw material, 548
Screening, 455–460
Scrubbers, 615, 616
Shear shredders (rotary shear), 425, 442,
443, 446, 452, 702
Shredding, 326, 430–436, 702
Site selection, landfill, resource recovery, 51,
52, 269
Solid waste composition survey, 515
composition evaluation, 527, 528
equipment, personnel, facilities, 523
sampling procedures, 523
sorting program, 522
Solid waste quantification survey, 507
costs, 512
current practice, 507
haulers, 509
personnel, equipment, 512
planning, 508, 509
scope, 509
site, scales, 511, 512
totals, type and season, 514
vehicle traffic, 515
weigh program, 513
Source separation:
benefits, 219, 220, 223
case studies, 229–238
collection/processing, 237
cost analysis, 238–245

materials:
percent, 235
ton/month, 234
public awareness, 236
collection, buy back programs, 228
comparison, centralized resource recovery
facilities, 221
contracts, 226
definitions, 216
industry relationship, 216
materials, markets, 223, 224, 225
perceptions, analysis, status, 215
Stokers, 577, 578
Sulfur dioxide (SOx), 608
Superfund, 793, 794, 795, 796
Surveys, solid waste, 44–50, 507, 515

Transfer station, 195
advantages, 195
ancillary facilities, 202, 203
cost, case studies, 211–214
cost analysis, 208–211
construction, 209
labor, 210
transfer haul, 208, 211
design, 197
concepts, 200, 201
process options – shred, bale, 203, 204
sizing, 197, 198, 199, 202
types, 199
location and users, 196
maintenance, 207, 208
materials handling, 206, 207
site development, 202, 203
Transfer vehicles, 204
maintenance, 207, 208
number required, 205
travel time, trips/day, 206
types, 204

Waste flow control, 22–26, 51
Akron, Ohio, 22
Boulder, Colorado, 22
competitive tipping fees, 151
legislative controls, 152
Municipal Action Exemption Publ. L 98–544,
24, 25
private agreements, contracts, 151
recyclers, 24, 26
size, waste/energy facilities, 24, 25
state action doctrine, 152, 153